Institute of Cast Metals Engineers

AA002387

67th World Foundry Congress

wfc06

"Casting the Future"

June 5 – 7, 2006
Harrogate, UK

Volume 1 of 2

Printed from e-media with permission by:

Curran Associates, Inc.
57 Morehouse Lane
Red Hook, NY 12571
www.proceedings.com

ISBN: 978-1-60423-676-7

Some format issues inherent in the e-media version may also appear in this print version.

Institute of Cast Metals Engineers

The Institute of Cast Metals Engineers, ICME, has been proud to organise the 2006 World Foundry Congress and present to you this CD-Rom of the Congress Proceedings.

ICME is the hub for the casting industry professionals in the UK and worldwide. It offers membership and internationally recognised qualifications as well as a range of technical publications for the castings industry.

ICME, National Metalforming Centre
47 Birmingham Road
West Bromwich
West Midlands
B70 6PY
United Kingdom

Tel: +44 (0) 121 601 6979
Fax: +44 (0) 121 601 6981
Email: info@icme.org.uk

www.icme.org.uk

Institute of Cast Metals Engineers

Membership of ICME

ICME is the professional members institute for individuals in the castings industry. ICME's members are now part of a global industry, involved in a wide variety of sectors within the castings industry. Our members *are* traditional foundrymen, but they are also design engineers, metallurgists, moulders, patternmakers, CAD technicians, methods engineers, researchers and suppliers to the industry.

ICME is able to offer its members a host of tangible and less tangible benefits. This includes guidance with professional development, training and education, technical support (through our network of members and our library; our staff will always aim to help with technical enquiries). Members are able to pursue registration with the Engineering Council UK, as professional engineers at Chartered, Incorporated Engineers or Engineering Technician level. All members also receive the monthly journal, Foundry Trade Journal. The branch network ensures that members are able to meet with like-minded individuals from the industry in both technical and social events.

www.icme.org.uk

Institute of Cast Metals Engineers

ICME Publications

The Institute has several major industry publications in its portfolio, including the **Foundry Trade Journal** of which ten English language issues are published annually and two Chinese language issues, widely distributed in China and the Far East. This journal has a very strong features list and contains industry news, features and technical papers presented in an accessible way. It is also the main voice for a number of other organisations, including PMMMA, the pattern, mould and model manufacturers association and EICF, The European Investment Casters Federation.

Diecasting World, also published by ICME, is received by over 3000 die-casters in all parts of the world.

The Foundry Yearbook and Casting Buyers Directory, produced annually, is a major reference book for all those who work in the castings industry, including, as it does, detailed listings for all the foundries in the UK, contact details for trade associations from around the world as well as international suppliers to the industry categorised in a variety of useful ways.

ICME also publishes the **Castings Buyer**, the only publication whose sole aim is to promote castings and the castings industry to engineers, buyers, component designers and specifiers. The Castings Buyer features case studies showing where fabrications have successfully been replaced by castings and articles on properties and applications of castings alloys.

ICME sees this as an important service to the industry and will be ensuring that our next issue is available as widely as possible.

www.foundrytradejournal.com

www.diecastingworld.com

www.castingsbuyer.com

www.icme.org.uk

67th World Foundry Congress

5-7 June 2006
Harrogate International Centre, Harrogate, UK

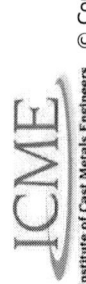
INDUCTOTHERM® GROUP

- Welcome
- Organising Committee
- Programme
- Proceedings Sponsor
- About ICME
- Disclaimer

Disclaimer

The opinions expressed in this publication are those of the authors, and do not represent the views of the Institute of Cast Metal Engineers (ICME), its Council or its Officers, the Organising Committee of the World Foundry Congress 2006 (WFC) or the World Foundrymens Organization, except where explicitly identified as such.

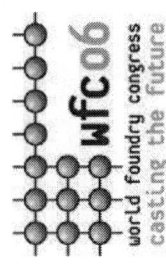

© Copyright Institute of Cast Metals Engineers

ICME
Institute of Cast Metals Engineers

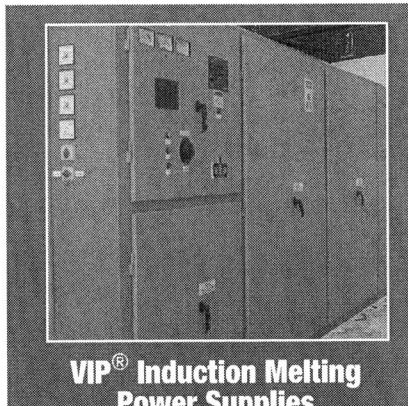

VIP® Induction Melting Power Supplies

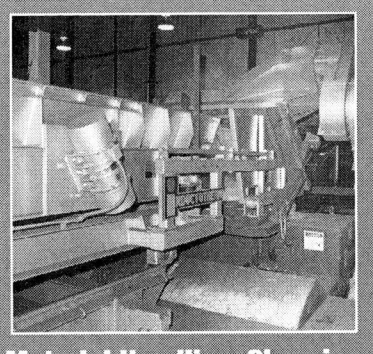

Material Handling, Charging and Preheating Systems

Coreless Melting Furnaces for All Metals

Aluminum Melting Systems

Holding and Duplexing Systems

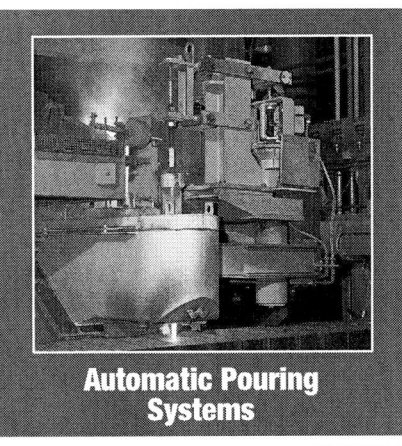

Automatic Pouring Systems

Computer Control Systems

Cooling Systems

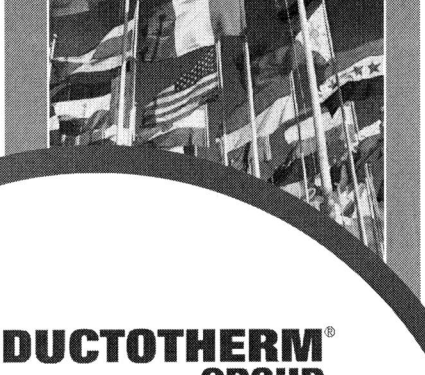

i INDUCTOTHERM® GROUP

Contact Inductotherm Today!

Inductotherm builds induction melting, holding, heating and pouring systems for virtually all metals, including, gray and ductile iron, steel, copper and copper-based alloys, aluminum, zinc, reactive metals and precious metals.

As the world's largest manufacturer of the induction equipment shown here, only Inductotherm can offer **proven** efficient, reliable and effective systems for all your melt shop needs locally with our global presence.

For more information, visit www.Inductothermgroup.com

INDUCTOTHERM® GROUP Leading Manufacturers of Melting, Thermal Processing & Production Systems for the Metals & Materials Industry Worldwide.

Important: Personal Protective Equipment (PPE) must be worn by anyone in proximity to molten metal.

67th World Foundry Congress

5-7 June 2006
Harrogate International Centre, Harrogate, UK

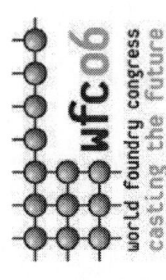

wfc06
world foundry congress
casting the future

INDUCTOTHERM® GROUP

Organising Committee

- Welcome
- Organising Committee
- Programme
- Proceedings Sponsor
- About ICME
- Disclaimer

World Foundry Congress Organising Committees

Chairman	Dr David Rowlands CEng, FICME, MIMMM
Chair of Technical Committee	Dr Pam Murrell FICME
Chair of Funding Group	Mr Peter Nix FICME
Congress Administrator	Mr Matthew Poole

WFC 2006 Steering Committee

Dr David Rowlands CEng, FICME, MIMMM
Mr Bob Brown CEng, FICME, BSc, MIRefE
Mr. Mike Clifford CEng, Hon FICME, BSc, MIM, MCIM
Mr Bob Jordan IEng, FICME, CBIM, MIM, FIDir, FRSA
Dr Pam Murrell FICME
Mr Peter Nix FICME
Dr Tom Paterson FICME, FIMMM, CSci
Dr Philip Ramsell CEng, FICME, FIMechE
Mr Colin Steed IEng, FICME
Mr Barrie Williams CEng, MICME

ICME
Institute of Cast Metals Engineers

© Copyright Institute of Cast Metals Engineers

67th World Foundry Congress

5–7 June 2006
Harrogate International Centre, Harrogate, UK

INDUCTOTHERM GROUP

Organising Committee

- Welcome
- Organising Committee
- Programme
- Proceedings Sponsor
- About ICME
- Disclaimer

WFC 2006 Technical Committee

Dr Pam Murrell FICME (Secretary General and Chair)
Dr Bill (W D) Griffiths CEng FICME, MIMMM (Vice Chair)
Eur Ing Malcolm Bird OBE, CEng, MIMMM
Mr Roger Davies IEng, AMICME, AIMMM
Mr Martin Fallon CEng, FICME
Mr Roger Kendrick IEng, MICME MIMMM
Mr Malcolm Macnaughtan CEng, FICME, MIMMM
Mr Simon Olive MICME
Dr Philip Ramsell CEng, FICME, FIMechE
Prof Rachel Thomson, MA, PhD Cambridge

Institute of Cast Metals Engineers

President	Mr Barrie Williams CEng, MICME
Senior Vice President	Mr Peter Nix FICME
Junior Vice President	Mr Willie Howson IEng, FICME
Hon Treasurer	Mr David Fletcher FREng, CEng, FICME, FIMMM
Institute Manager	Dr Pam Murrell FICME

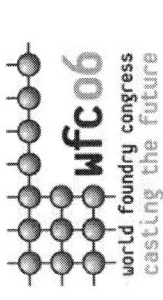

ICME
Institute of Cast Metals Engineers

© Copyright Institute of Cast Metals Engineers

67th World Foundry Congress

5-7 June 2006
Harrogate International Centre, Harrogate, UK

Introduction

The papers can be read on screen or printed using any PDF file reader software. The copyright of the CD contents remains with the Institute of Cast Metals Engineers, from whom permission should be obtained before any paper, or part of a paper is published or reproduced.

I would like to thank all of the authors for their hard work and for the time in preparing and presenting these papers. Special thanks are due to the Technical & Scientific Papers Group of ICME for their long, patient and successful production of such a programme of papers. In particular I wish to acknowledge the support of Inductotherm in the preparation of this CD of Congress Proceedings.

David Rowlands
Organising Chairman
WFC June 2006

- Welcome
- Organising Committee
- Programme
- Proceedings Sponsor
- About ICME
- Disclaimer

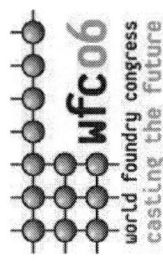

© Copyright Institute of Cast Metals Engineers

Institute of Cast Metals Engineers
67th World Foundry Congress
wfc06

TABLE OF CONTENTS

Volume 1

Using Stress Simulation to Tackle Distortion and Cracking in Castings 1
Dr. Ing, A. Egner-Walter, S. Olive

The Use of Different Computer Simulation Software Packages to Predict Casting
Filling and Solidification 11
S. Oxley, P.M. Haigh

Prediction of the Infuluence of Microstructure, Porosity and Residual Stresses on
Strength Properties of aluminium Casting 21
R. Baehr, M. Todte, H. Stroppe

Intelligent Riser/Chill/Gating Design System Using Simulations and Discrete
Optimisation Algorithm 31
C. Lim, J. Nam, I. Cho, S. Yoo, J. Choi

Research on Mould Filling Process of Melt in Vertical Centrifugal Casting 41
G. Jingjie, L. Changyun, W. Shiping, F. Hengzhi

Nonlinear Modeling with Hydrodynamics and Flow Control Using Inverse Pouring
Dynamics of Tilting-Ladle-Type Automatic Pouring Process 51
Y. Noda

Three-dimensional Modelling and Simulation of Die-Casting Processes for Al-Si
Alloys 61
H.Y. Hwang, J.K. Choi, E.I. Marukovich, A.M. Branovitsky, I.L. Zakharov

Empirical Model for Tensile Property Prediction in Cast and Heat Treated Al-Si-
Cu-Mg Alloys 70
J. Fang, H.D. Brody, J.E. Morral

Simulation of Solute Redistribution during Casting and Solutionizing of Multi-
phase, Multi-component aluminium Alloys 80
F. Yi, H.D. Brody, J.E. Morral

Computer Simulation Study Upon the Influence of Geometry on the Critical
Velocity for Molten Aluminium 90
R. Cuesta, A. Delgado, J.A. Maroto, D. Mozo

Porosity Criteria Functions Revisited 100
J.T. Berry, R. Luck

Simulation of Casting Solidification Using Different Boundary Conditions 108
J.S. Suchy, A. Gradowski, J. Lelito

Mathematical Modelling of Compacting Process of Greensand Molding 118
L. Wenzhen, Z. Kelin, W. Junjiao

Improving Casting Performance Through Customized Insulation Shapes and
Advanced Simulation Techniques 126
J. Prat, A. Meléndez, A. Seoane, E. Anglada, A. Beeson, M. Arrieta, J. Galaz, A. Jorge, T.
Vicario

Prediction of Shrinkage Defect in Steel Casting for Marine Engine Cylinder Cover by Numerical Analysis 136
K.H. Kim, J.H. Hwang, J.S. Oh, D.H. Lee, I.H. Kim, Y.C. Yoon

Production of Ductile Iron Castings in Green Sand Molds Without Feeders 145
R. Sillen

Developments in the Design of Steel Castings 155
M. Blair, R.W. Monroe

Multiloop Approach for Automatic Mold Filling in Ferrous Foundry 165
C. Debray, M. Dussud, M. Biardeau, P. Canon

Cast Iron Material Standards for a New Millennium 174
M.P. Macnaughtan, C. Eng

Safety Cast Components for the Automotive Industry. the Metallic Charge, the Presence of Micro Elements and Their Most Relevant Effects. 184
R.S. Creo, J.I. Maguregi, J.A. Goñi Güemes

New Engineering and Standards Developments in Austempered Ductile Iron (ADI) 194
K.L. Hayrynen, J.R. Keough, A. Rimmer

How to Link OSHA Requirements to Kyoto's Demands for Protection of Climate 204
R. Kurtsiefer

Occupational Exposure to Chemical Agents in the Portuguese Foundry Industry 212
J.C. Costa

Improving Environmental Performance and Satisfying Regulatory Requirements Through the Continuous Monitoring of Particulate Emissions from Foundry Processes 219
W. Averdieck, J. Harshorne, S. Werrell

Novel Approaches in Reducing Pouring Emissions 229
J.H. Helber

The Use of Tilt Filling to Improve the Quality and Reliability of Castings 239
R.A. Harding

The Influence of Heat Treatment on the Structure and Tensile Properties of Cast Titanium Alloy Ti-511 249
A.C. Robinson, E.J. Czyryca, D.A. Koss

Magnesium Alloy Castings - Past and Present 259
P.J. Thompson

Magnesium Alloy R&D Challenges - Aerospace Spinoff 269
S. Sundarrajan

Alpha-Case Controlled Titanium Casting 278
S.Y. Sung, Y.J. Kim

Surface Tension and Viscosity of Mg Alloys 288
S. Park

Mechanical Properties and Microstructural Evolution of Rheo-Diecast AZ91D Magnesium Alloy During Heat Treatment 295
Y. Wang, G. Liu, Z. Fan

Effect of Grain Refinement on the Mechanical Properties of Magnesium Alloys and Its Alloys 305
Y. Han

Green Manufacturing for Magnesium Alloys 315
S.K. Kim, J.K. Lee, Y.O. Yoon, H.H. Jo

Role of Carbon as Solute Element in Carbon Grain Refinement of Mg-Al Alloy 324
S.Y. Shim, Q. Jin, S.G. Lim

Wicking of Liquid Polystyrene Degradation Products into the Pattern Coating in the Lost Foam Casting Process 331
P.J. Davies, W.D. Griffiths

Mould Filling in the Lost Foam Casting of aluminium Alloys 341
M.J. Ainsworth, W.D. Griffiths

Rapid Shell Build for Investment Casting: Revolutionizing an Ancient Process 351
S. Jones, C. Yuan, S. Blackburn

Fabrication of Ni-Al Intermetallic Compounds on the Al Casting Alloy by SHS Process 361
G.S. Cho, K.R. Lee, K.H. Choe, K.W. Lee, A. Ikengaga

Influence of Viscosity-Increasing Processes on Metal Foam Stability 371
K. Kadoi, M. Tayama, H. Nakae

Study on Vacuum Die Casting Process of aluminium Alloys 380
B. Hu, S. Xiong, M. Murakami, Y. Matsumoto, S. Ikeda

Investigation on the Flow Pattern in the Shot Sleeve of the Cold Chamber HPDC Process 389
J.H. Hong, Y.S. Choi, H.Y. Hwang, J.K. Choi

Definition and Development of an Innovative Coating for Optimised Tooling, Used in Aluminium Die-Casting 396
H. Delorme, C. Héau, E. Neto, K. Metzgar

A New Future for Gravity Molding 405
M. Bakrim, J. Vervier, C. Vandenhaute, F. Ngirabacu, D. Vervier

Titanium Matrix (TiB + TiC) Composites Shot Sleeve for Al Alloys Die Casting 412
S.Y. Sung, Y.J. Kim, B.J. Choi

AWB - An Environment-Friendly Core Production Technology 422
T. Steinhäuser

Advances in Thin-Wall Sand Casting 430
R.E. Showman, R.C. Aufderheide, N.P. Yeomans

The Significance of Total Carbon in Greensand Systems 440
A. Brown

A New Generation of Advanced Polyurethane Coldbox Binders for Aluminium Castings 450
A. Schrey

A Technological Advantage for the Environmental Age 460
L.R. Horvath

Dispersive Mixing of Natural Molding Sand - An Optimised Preparation Process 467
M. Mueller

New Innovative Solutions for Foundries by Inorganic Concepts 473
J. Müller, R. Stotzel

Phenolic Urethane Cold-Box Binders - A Study of Global Properties, Variables, Causes and Effects 483
M. Stancliffe

Modeling and Identification of Pouring Flow Process with Tilting-Type Ladle for an Innovative Press Casting Method Using Greensand Mold 493
Y. Matsuo, Y. Noda, K. Terashima, K. Hashimoto, Y. Suzuki

Inorganic Binders: Properties and Experience 503
K. Löchte, R. Boehm

Inclusive, Innovative and Sustainable Environmental Management System 510
V. Narasimhan

Foundry Management by Internet 517
T.D. Law

South African Aluminium Foundry Industry: An International Perspective 527
T. Paterson

A Novel Aluminium Matrix Composite Synthesized by Magnetochemical Melt Reactio in the System Al-Zr-O-B 528
Y. Zhao, Y.M. Yousseff, R.W. Hamilton, P.D. Lee

Compo-Casting Method for Alumina Ceramics Inserted in Cast Iron to Reduce Thermal Stress 538
Y. Tomita, H. Sumimoto, K. Nakamura, S. Kiguchi

Production of Hybrid Metal Matrix Composites and its Wear Behavior 548
K.V. Mahendra, Dr. K. Radhakrishna

Effect of Volume Fraction and Particle Size of Reinforcement on Thermal Analysis and Heat Transfer Parameters of Gravity Die Cast Hypereutectic Al-22% SI Alloy Matrix Composites 558
P.K. Subramanya, S. Hedge, K.N. Prabhu

Processing of Al-Mg/Al$_2$O$_3$ Interpenetrating Composites by Pressureless Infiltration 568
H. Chang, J.G.P. Binner, R.L. Higginson, R. Sambrook

Spontaneous Infiltration Mechanism of Al-Si Melts into SiCp Preform 576
H. Nakae, Y. Araoka, Y. Sugiyama

Fabrication of Short Alumina Fibre and In-Situ Mg$_2$Si Particle-Reinforced Magnesium Alloy Hybrid Composite and Its Strength Properties 585
K. Asano, H. Yoneda

Production, Evaluation and Comparison of Mechanical Wear and Corrosion Properties of Al-Flyash1 and Al-Flyash2 Metal Matrix Composite 593
M. Ramachandra, K. Radhakrishna

Evaluation of Mechanical Wear Properties of Al-Si(12.2%)-Graphite Metal Matrix Composite Synthesized Using Stir Casting Method 603
M. Ramachandra, K. Radhakrishna

Volume 2

Energy Saving Potential of Melting Medium-Frequency Coreless Induction Furnaces 613
F. Donsbach, D. Trauzeddel

Advances in the Melting of High Quality Grey Iron Automotive Castings at Precision Disc Castings 623
M.P. Macnaughtan, D. Eggleston, N. Richardson

Oxygen Technologies: Reduce Melting Cost and Emissions 633
  T. Niehoff, P. Keena

Development and Use of a New Optical Sensor System for Induction Furnace
Crucible Monitoring 643
  W. Schmitz, F. Donsbach, H. Hoff

Development of Coke Alternate Material Using Woody Biomass 653
  Yasufumi Yamaguchi, Shoji Kiguchi, Hirotoshi Murata

Accurate and Complex NET-SHAPE Castings for Challenging Markets 662
  M. Horacek, J. Cilecek

Innovations in Machine Learning and Defect Diagnostics 672
  R.S. Ransing, M.R. Ransing

Effect of Ultrasonic Vibration on Structure Refinement of Metals 681
  Q.M. Liu, Y. Zhang, Y.L. Song, F.P. Qi, Q.J. Zhai

HIPing - A Potent Post Casting Treatment for High Integrity Aluminium Castings 689
  R.M. Pillai

Metal Castings' Secret Ingredient 699
  W.W. Sorenson

People and Skills for Today's Industry - An Indian Experience 705
  K. Gnanamurthy

How Metals Employers Engage in the UK Skills Agenda 715
  E. Bonfield

Latest Trends in Industrial Skills Development Techniques 716
  Dr. C. Ashley, Eur Ing C.J.C. Bale, N. Millan, T.M. Williams, R.J. Hendley

OVOTRAIN On-line Virtual Vocational Training System 736
  K. Bako, K. Lengyel

Modeling Microstructural Evolution in Cast Alloys 741
  Rachel Thomson

Maximising Supply Chain Competitiveness and Market Opportunities by
Exploitation of Technology 760
  Dr. M.C. Ashton

The Auto Sector in a World of Ultra Competition: Nowhere for the Inefficient to
Hide 767
  Garel Rhys

Fluidised Bed for Stripping Sand Casting Process 768
  G. Belforte, M. Carello, V. Viktorov

Robot Based Oxy-Fuel Cutting and Stub-Grinding for Castings in Low Series 778
  Dr. B. Lauwers, H. De Baerdemaeker, Dr. P. Haigh, R. Wallis, K.U. Leuven

Automatic Visual 3-D Inspection of Castings 788
  D. vom Stein

Thermomechanical Treatment of Austempered Ductile Iron 798
  A.A. Nofal, H. Nasr El-Din, M.M. Ibrahim

New Adaptive Machining Methods for the Foundry Industry 808
  P. Dicken

Development of a Low Alloy Cast Steel for Automobile Blanking, Drawing and Trimming Die 816
J.S. Shin, C.B. Song, B.H. Kim, S.M. Lee, B.M. Moon

Squeeze Casting of Aluminium Alloy Safety Critical Components for Automotive Applications 824
R. DasGupta, C. Barnes, P. Radcliffe, P. Dodd

Effects of Pouring Temperature and Squeeze Pressure on the Properties of Al-8%Si Alloy Squeeze Cast Components 834
A. Raji, R.H. Khan

On the Effect of Cooling Conditions and Variation of Alloying Elements on the Microstructural and Mechanical Properties of Al-7%Si Cast Alloys 844
S. Seifeddine, I.L. Svensson

A Study on the Hot Tearing in Al-1% Sn Cast Alloys 854
G.L. Datta, D. Benny Karunakar

A Study of Double Oxide Film Defect Behaviour in a Quiescent Aluninium Melt 864
R. Raiszadeh, W.D. Griffiths

Effect of Melting and Casting Conditions on Aluminium Metal Quality 874
D. Dispinar, J. Campbell

Strontium Effect on the Solidification Path of a 319-Type Aluminium Alloy 884
J.J. Montes-Rodríguez, M. Castro-Román, M. Herrera-Trjo

Effectiveness of Zn-Ti Based Refiner of Al and Zn Foundry Alloys 894
W.K. Krajewski, A.L. Greer, J. Zych, J. Buraś

Effects of Process Parameters on the Morphology of TiAl₃ Particle During the Production of Al-Ti-B Master Alloy by Flux Reaction 904
M. Ryou, S.H. Choi, M.H. Kim

The New Method and Device for Fast Evaluating Inoculation Result of Eutectic Al-Si Alloys 914
L. Dayong, S. Dequan, W. Lihua, Z. Yutong

Advances in the Determination of Hydrogen Concentrations in Aluminium Alloys 921
A. Froescher

The Effects of HIP on Bifilms in Aluminium Castings 929
J. Staley

The Effect of Using Bubbling and AlCuP for Refining Primary Silicon at Al-18%Si Alloy 939
S.M. Kim, J.P. Choi, T.W. Nam, E.P. Yoon

A Study of the Structural Controlling of Al-Si Alloy by Using Electomagnetic Vibration 946
J.P. Choi, T.W. Nam, E.P. Yoon

Investigation of Crack Generation Under the Influence of Thermal Stress During Cast Process 955
S.Y. Kwak, N.U. Ho, S.W. Lee, J.T. Kim, J.K. Choi

The Prefil Technique for Molten Metal Cleanliness Measurement 965
J. Pickering, P. Enright

Fabrication of Silicon Ingot by CCCC Method 975
B.M. Moon, D.S. Lee, B.H. Kim, J.S. Shin, S.M. Lee

Microstructure Observations and Refining Performances of Al-Ti-C Master Alloys Prepared by Salt Route ... 983
A. Sharma

Effects of Inoculation and Solidification Rate on the Thermal Conductivity of Grey Cast Iron ... 993
D. Holmgren, A. Diószegi, I.L. Svensson

Thermochemistry and Kinetics of Iron Melt Treatment ... 1003
S.N. Lekakh, D.G.C. Robertson, C.R. Loper Jr.

The Filtration of Large Grey and Ductile Iron Castings ... 1013
E. Wiese

Study of the Occurrence and Suppression of Metal Reoxidation in Ferrous Castings ... 1023
T. Elbel, J. Senberger, A. Zadera, L. Kocian

New Austenitic Flake Cast Iron with Manganese ... 1033
P. Sriram

Carbon Recovery and Inoculation Effect of Carbonic Materials in Cast Iron Processing ... 1041
M. Chismera

Streamlined Gating Systems and Improved Yield - Dimensioning and Experimental Validation ... 1051
N. Tiedje

The Advantages of SiC-Quenching Plates (Chills) When Compared with Grey Iron Quenching Plates (Chills) ... 1061
T. Reuther

Chunky Graphite in Ductile Iron Castings ... 1071
R. Kallbom

A Comparison of Thixocasting and Rheocasting ... 1081
S.P. Midson, A. Jackson

Near Net-Shaping Aerospace Alloys by Thixoforming ... 1091
P. Kapranos, M. Farnsworth

Solidification Structure and Mechanical Properties of Semi-Solid Processed Cast Iron ... 1099
M. Ramadan, H. Nomura, M. Takita

Microstructure and Mechanical Properties of Rheo-Diecast (RDC) Aluminium Alloys ... 1109
X. Fang, J. Patel, Z. Fan

Industrial Application of Thixoforming at SAG ... 1119
J. Wöhrer, A. Kraly

Direct Chill Rheocasting (DCRC) and Extrusion of AZ31 Mg-Alloy ... 1128
S.M. Zhang, Z. Zhen, Z. Fan

Microstructure Evolution in Al-7Si-0.3Mg Alloy During Partial Melting and Solidification from Melt: A Comparison ... 1138
S. Nyamannavar, M. Ravi, K. Narayan Prabhu

Net Shape Forming of Iron and Steel for Clean Production ... 1148
J. Youn, Y.J. Kim

A Study on Semi Solid Squeeze Forging of High Strength Brass 1157
K.H. Choe, G.S. Cho, K.W. Lee, Y.J. Choi, K.Y. Kim, M.H. Kim

The Aerospace Industry and Its Work with Global Suppliers 1165
D. Jakstis

Castings Under Viewpoint 1166
M. Kennedy

New Solutions in Ductile Cast Iron for the Retrievable Storage of Radioactive Waste 1167
C. Macke-Bart, G. Regheere, D. Linxe, A. Beziat

The Intelligent Casting Production: How to Use Automation to Improve the Economics of Near Net Shape Castings 1176
K. Holmen

Double Cavity Casting of Transmission Case: A Techinical solution to a Foundry Capacity Problem 1185
M. Ahmed

Development of Tilt Casting Technology for High Performance Sport Wheels 1195
K.K. Tong, M.S. Yong, Y.W. Tham, R. Chang, K.W. Wee

Process Development for Highly Stressed Aluminium Castings Under Consideration of the Increase in Performance of the Diesel Engines 1205
F. Mnich, H.C. Saewert, A. Tamez, R. Bähr, E. Krebs

Author Index

**Using Stress Simulation to tackle
Distortion and Cracking in Castings**

Dr.-Ing. A. Egner-Walter[*], S. Olive[**]
[*] MAGMA GmbH, Germany, [**] Maxima Engineering, UK

Abstract

The use of stress simulation in casting design and manufacturer has found wide spread use in not only the automotive industry but also more general engineering castings. The development of light-weight body structures in the automotive industry has lead to a new generation of large thin walled complex aluminium and magnesium castings with demanding dimensional and stiffness tolerances, which require particular attention to casting design and production techniques. To achieve economic production long die tool life is required and thermo mechanical fatigue cracking 'heat checking' is often the limiting factor. Minimum distortion and maximised die life can be achieved by using simulation tools and optimisation techniques during casting design and production. A group of companies with competences in casting simulation, material testing, design and high pressure die casting has worked together on the overall design and manufacturing process and examples will show the success of this approach. The paper will give a review of the current capabilities of the application of a numerical simulation to stress related problems in casting manufacture.

Keywords

Casting Simulation, Distortion, Die Constraints, Visco-Plastic Simulation, Die Life

Introduction

With the production of thin walled structural components the foundry industry has been able to substitute classic pressed parts with castings. The generally large surface area complex castings with demanding dimensional and stiffness requirements place high demands on design and production by high pressure die casting. Along side the formation of hot tears, the distortion places great demands on the process. Certain parts that are particularly prone to hot tearing which has lead to the replacement of non heat treatable alloys to ones requiring heat treatment. This increases the productions cost due to the addition of heat treatment, which can also lead to distortion of component. Tackling the distortion requires an accurate understanding of the distortion mechanism of each individual production stage as well as how they interact. Due to the numerous parameters, simulation of the complete production process is a powerful tool. With this in mind a group of companies with expertise in simulation, materials testing, design and production of die casting dies formed a consortium as part of a BMBF funded research project. Some of the results of this project are presented in this paper.

Formation of Residual Stresses and Distortion in Castings

The formation of residual stresses and distortion can be easily understood by the simple example given in figure 1. The solidification and cooling to room temperature of all castings occurs with an inherent inhomogeneous temperature distribution. When areas which are cooling faster, such as the thinner side arms of the lattice, thermally contract and are constrained, tensile stresses are formed. At high temperatures where yield stress is low thermal stresses are relieved to some degree by plastic deformation. The highest stresses and plastic deformation occur when the temperature difference between faster and slower cooling rate areas is at a maximum (figure 1 $t=t_1$). As cooling proceeds, the temperature difference between the two areas decreases. This initially leads to stress relief (reduction in tensile stress level) and then to a change in the stress state from tensile to compressive in the thinner, faster cooling outer arms. In the slower cooling thick arm the initial compressive stress changes to tensile. This stress transition is due by plastic deformation at high temperatures. After complete cooling to room temperature, figure 1 $t=t_2$, compressive residual stresses arise in the thin side arms areas, whereas tensile residual stresses arise in the thick central arm /1/. The residual casting stresses can be affected by subsequent processes such as clipping, straightening, heat treatment, machining and shot blasting. The residual stresses can be redistributed during clipping and machining but additional tensile and compressive stresses can be introduced into the casting by machining and clamping forces or heat treatment. In the worst case cracks can form in the casting during machining /2/. The calculation of residual stresses caused by the casting process is carried out in a decoupled 2 stage process. Firstly, the temperatures in casting and die casting tool are calculated by a filling and solidification simulation. Secondly, these temperatures are used for the stress calculation. A knowledge of temperature and microstructure dependent thermal and mechanical properties of the casting material as

well as the consideration of the shrinkage constraints ca
are crucial parameters for the accurate calculation of residual stresses.

Heat Treatment Residual Stresses Formation and Distortion
Most structural castings do not meet the required mechanical properties after casting and subsequent heat treatment is used, which usually consists of a solution anneal, quenching and artificial ageing or precipitation treatment. During the solution anneal process, the casting is heated up to a temperature below solidus temperature. The yield stress of material reduces with increasing temperature and residual stresses are relieved by plastic deformation. At these temperatures the casting can deform under its own weight. In the subsequent quenching inhomogeneous cooling again leads to the formation of new residual stresses /3/ - /5/. These stresses can lead to further distortion. In the subsequent ageing process, the residual stresses can be reduced by 20-30% due to creep and plastic deformation, causing further distortion. The level of stress relief and distortion is dependent on the residual stress after quenching, the ageing temperature and the ageing duration.

Simulation of the Distortion of a Structural Component
Starting point for any simulation is a three-dimensional geometry model of the casting, the runner and overflows, the die casting tool with the die heating and cooling circuits. Figure 2 shows a die cast rear door lock panel for a passenger car. The whole geometry model is subsequently enmeshed for the calculation. The simulation considers all relevant process conditions, including the shot profile, the temperature of melt, inhomogeneous die temperature and die cooling, die spraying as well as the casting cycle time. The simulation should reflect the steady state condition of the process after sufficient 'preheating' cycles as is carried out in die casting production. Once steady state is reached, the filling, the solidification, and the cooling of the casting are simulated. The calculated temperatures over time are applied to casting and runners as external loads in the subsequent stress calculation. The areas, that cool down fastest, solidify and start to contract rapidly. The contraction is at least partly constrained by the die. Consequently tensile stresses form. Immediately before ejection from the die, tensile stresses occur in nearly the whole component, Figure 3. After ejection, the shrinkage constraint caused by the die is removed and the casting and runner can now contract freely. The runners and gates influence the contraction of the casting, with the casting being bent towards the shot slug. The temperature field at ejection, figure 4, is the basis for the formation of stresses and deformation during cooling down to room temperature. As the runner is generally hotter, it contracts more in the subsequent cooling phase and pulls the casting inwards towards it, Figure 5. The swan neck moves in an upwards direction, as the upper side of the casting, which faces the shot slug, has a higher temperature than the lower side after ejection. Figure 6 shows the deviation from the required dimension for a contour of 23 measurement points after casting. The diagram displays the average

value of all measurements (15 castings, blue line), the
limit of the measurements (dashed lines), and the results of the simulation
(red line). The datum points are also shown in the diagram. It can easily
be seen that the calculated distortion corresponds well to the measured.
After the removal of the runners and gates during clipping, the component
opens outwards towards its starting point slightly, Figure 7. However, a
certain bending towards the gate can still be seen. The correlation
between the measured and simulated distortion values is still very high
after clipping.

Simulation of Distortion during Heat Treatment of a Cast Bulk Head
The production process for automotive high pressure die cast components
often includes a heat treatment stage. The reduction in as cast residual
stress levels occurs as the heating of the casting during annealing reduces
the necessary stress level to initiate creep deformation. Using a new
simulation model, visco-plastic strain has been taken into account for the
first time. As an example the annealing of a 'bulk head' casting is shown
here. During heat treatment the casting is mounted on a horizontal bar on
both sides, as it can be seen in Figure 8. The simulation considered the
symmetry of the geometry and only half the model is shown and was
simulated. After analyzing measurement data, it was found that the
'solution anneal' process has the largest and most significant influence on
the overall distortion. This process step was simulated using the visco-
plastic model which can also simulate creep in castings at elevated
temperatures, as occurs in annealing. Figure 9 shows the stresses and
deformation after solution annealing. The maximum stresses in Figure 9
are only ~0.9 MPa. This is can be compared to typical stress after casting
of 20 to 50MPa. The distortion which occurred during the heat treat anneal
is also shown with the maximum displacements of approximately 4.6 mm.
The simulated dimensional deviations from the nominal, as well as the
average, the lower and upper measured values are displayed in Figure 10.
It can be seen that the simulated distortion value of the casting correlates
well with the measured average values. Without the inclusion of the visco-
plastic material law, an accurate simulation would have not been possible.

Fatigue Life of High Pressure Die Casting Tools
Die casting tools usually have a limited life due to the thermal fatigue
cracking or 'heat checking' of the die surface, which is caused by high
cyclical thermal loads. The main features of the temperature and stress
cycling are shown in figure 11. The surface of a die is exposed to a rapid
increase in temperature as the metal fills the die and the casting solidifies.
After ejection the die surface cools as the source of heat, the casting, is
removed. If as is normally the case, the die surface is sprayed and
experiences very rapid cooling as the water base of the die spray
evaporates. As the die heats up during filling and solidification
compressive stresses develop in the surface of the die as it tries to expand
but is constrained by the cooler subsurface areas. After ejection the
compressive stress level reduces and as spraying occurs the stress

becomes tensile, peaking just after the end of spraying, as the surface tries to contract but is constrained by the now hotter subsurface areas. This process repeated over a few thousand cycles causes thermal fatigue cracking in the surface of the die and the appearance of heat checking. The cracks that form are effectively stress relieving the die surface. Not only the thermal load plays an important role in tool life, but also the manufacturing stages of die machining, eroding, finishing, hardening, the die steel used, the casting alloy, as well as the design of the die /6/-/8/. The calculation of the anticipated tool life of high-pressure die casting dies consists of three steps. Firstly, the temperature distribution in the die during filling, solidification, cooling, and spraying is calculated. Secondly, these temperatures are applied as external loads for the calculation of thermal stresses in the die tool. Thirdly, an analysis is performed of how the calculated cyclical stress loading is expected to influence the onset of heat checking. The thermal stresses of the die casting tool of the earlier described rear door lock panel (figures 3 to 5) were calculated for one casting cycle. Based on the results, the expected life time of the tool is established. The calculated life times were compared with the crack patterns of the actual die after 50,000 cycles, Figure 12. Critical areas are shown in the evaluation in blue, less critical areas in yellow to white. As can be seen in Figure 12, the area around the ejector pins on the runner shows radial cracks, with a calculated life time of about 2,000 cycles. A critical area is where the runner splits and has it largest cross section. The areas on the right of the die photo and simulation image in the figure 12 also shows a critical area where the runner splits into 3 ingates, light blue. Figure 12 also shows a critical area in the swan neck of the casting. The geometry has 3 notch type features. The notch effect will lead to higher stress concentrations and lower tool life time which is seen in the resulting die photo and in the lower calculate life cycles.

Conclusion
The control of the distortion of large thin walled structural components places great demands on the producers of such castings. For a successfully approach, a prerequisite is an exact understanding of the mechanism by which the distortion occurs as well as its design dependant effects. With the help of the latest methods of casting simulation, it is now possible to accurately predict the deformation due to the casting, clipping and heat treatment processes. Based on this understanding and capability it is also possible to determine preventative actions to distortion such as a pre-emptive deformation of the designed geometry, *'bend the die to straighten the casting'* and pre-emptive development of heat treatment orientations and jigging. As well as prediction of residual stress and distortion, today it is possible to predict areas prone to hot tearing and thermal fatigue stress 'heat checking' in die tools. With these developments simulation can support the high pressure die casting in securing an increased field of application in thin wall cast structural components.

References

/1/ *Egner-Walter, A.*: „Berechnung der Entstehung von Spannungen beim Gießen" Hoppenstedt, Gussprodukte `99 (1999).

/2/ *Wieckowicz, P., Olive,* S.: „Residual Stress and Casting Cracking during Machining. A Case Study", published in MAGMAtimes 03/2003, Vol. 8 Publication of MAGMA GmbH, Sept. 2003

/3/ *Maaßen, F.; Loeprecht, M.; Egner-Walter, A.:*"Rechnerische Optimierung von hochbelasteten Zylinderköpfen aus Aluminium – Dauerlaufsimulation". 9. Aachener Kolloquium Fahrzeug und Motorentechnik, Herausgeber, Ort, (2000), S. 805-841)

/4/ *Auburtin, P., Morin, N.*: "Simulation thermomécanique du traitement thermique de culasses en aluminium", Mécanique & Industries 4 (2003), S. 319-325.

/5/ *Fent, A.:*"Einfluss der Wärmebehandlung auf den Eigenspannungszustand von Aluminiumgussteilen", Dissertation Tech. Universität München (2002)

/6/ *Nogowizin, B., Friebe, G.*: „Standzeiten der Druckgießformen", Druckguss-Praxis 4/2002, S. 101-108

/7/ *Pries, H., Liluashvili, S., Dilger, K.*: „Brandrissentstehung an Druckgießformen", Druckguss-Praxis 3/2003, S.135-140

/8/ „Verschleißmechanismen an Druckgießformen", Abschlussbericht zum AiF-Forschungsvorhaben 12685, TU Braunschweig (2003)

Acknowledgements

The authors thank the German Federal Ministry of Education and Research for funding this project as part of the framework concept "Research for tomorrow's production" under FOGL (No. 02PD2141).

Figures

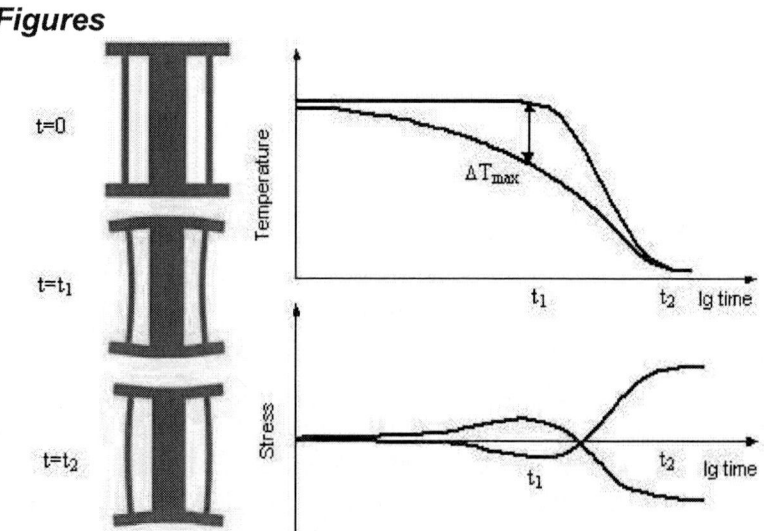

Figure 1:   Formation of residual stresses and distortion in a stress lattice

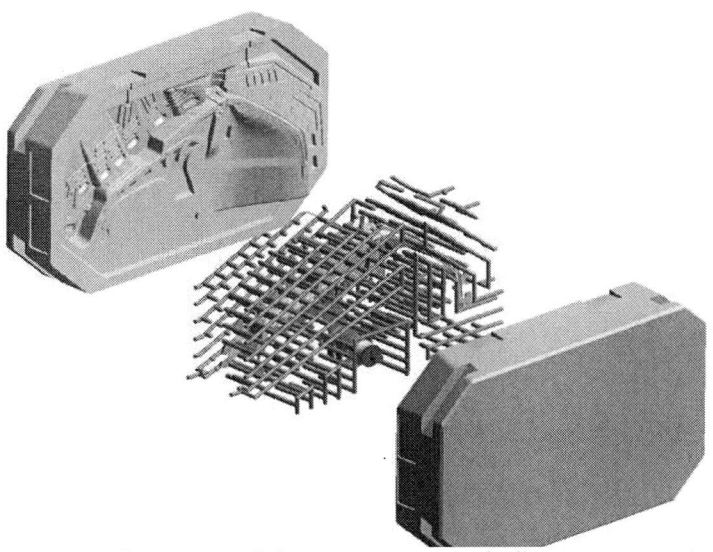

Figure 2: Geometry of the component including runner system, water cooling and oil tempering control channels, and the two die halves

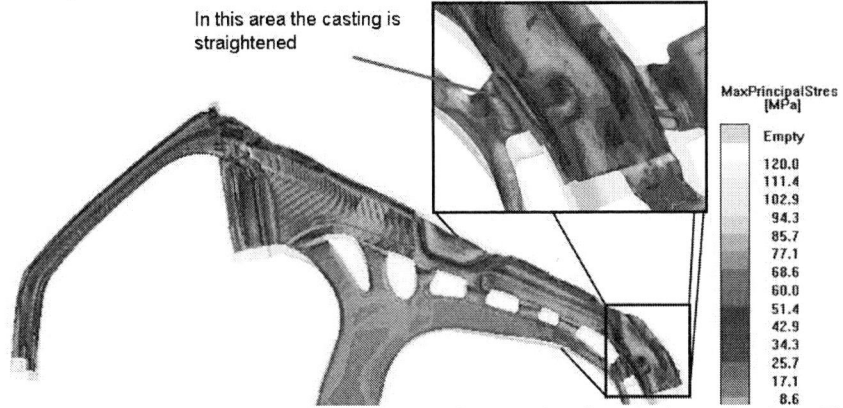

Figure 3: Maximum Principle Stress in the casting just before ejection. The original geometry and the deformed geometry showing the relative shape change and areas of contact and constraint.

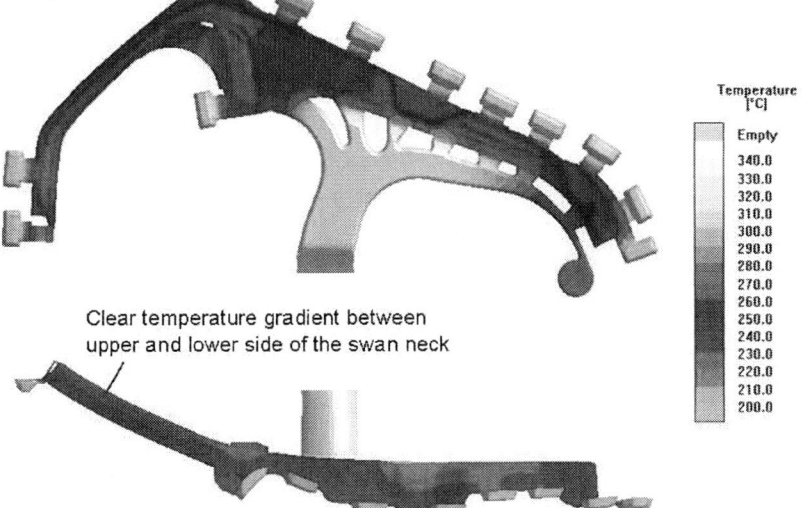

Figure 4: Temperature distribution in the casting and runner at ejection.

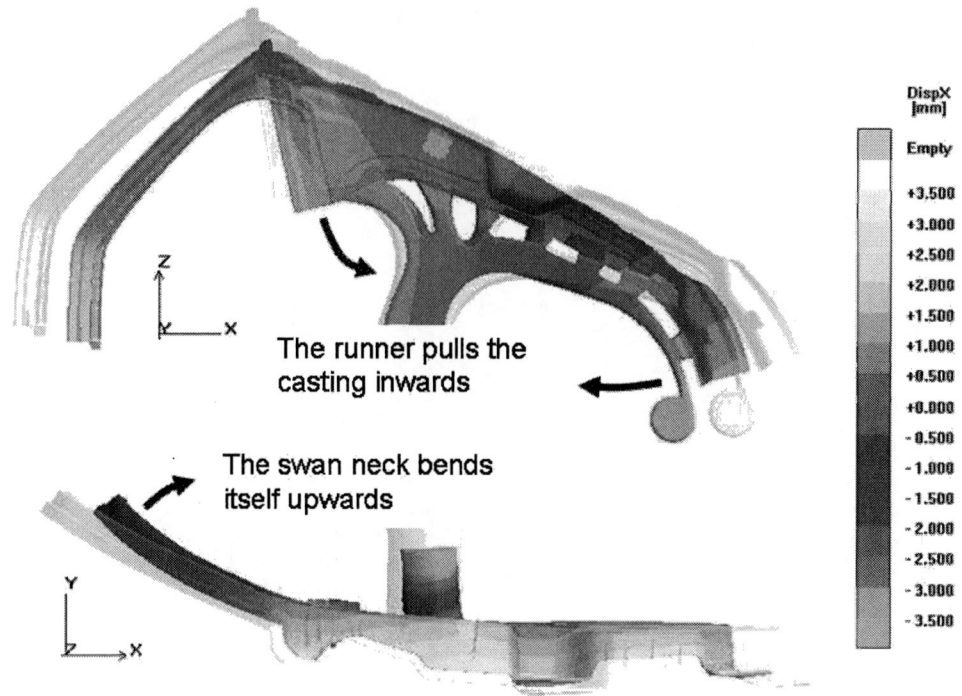

Figure 5: Deformation of the casting after cooling to room temperature shown as displacement from the starting point in the x direction. The original geometry is shows as semi transparent grey with the deformed geometry coloured.

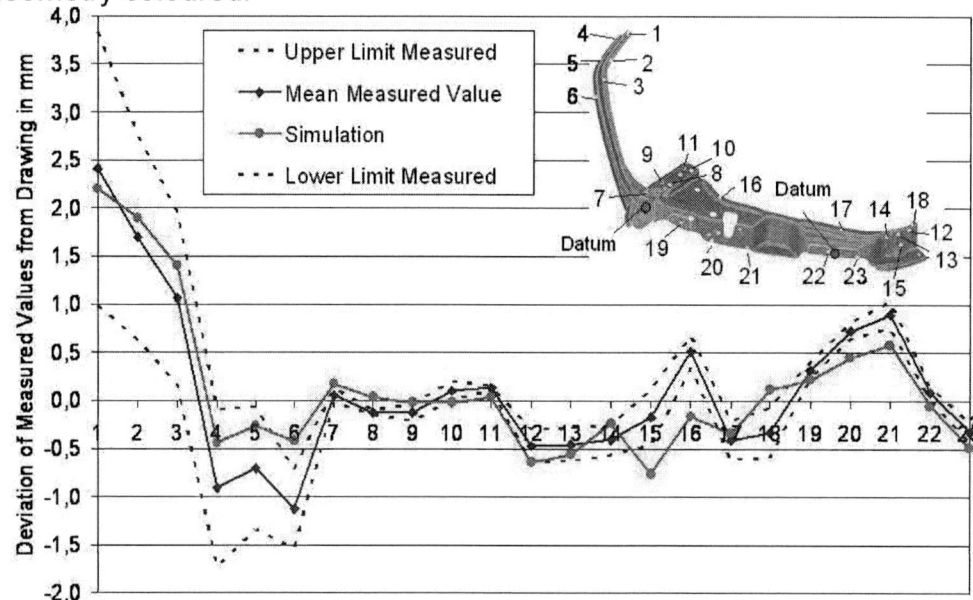

Figure 6: Comparison of the measured and calculated deviation from the required design geometry for 23 measuring points of the rear body lock panel.

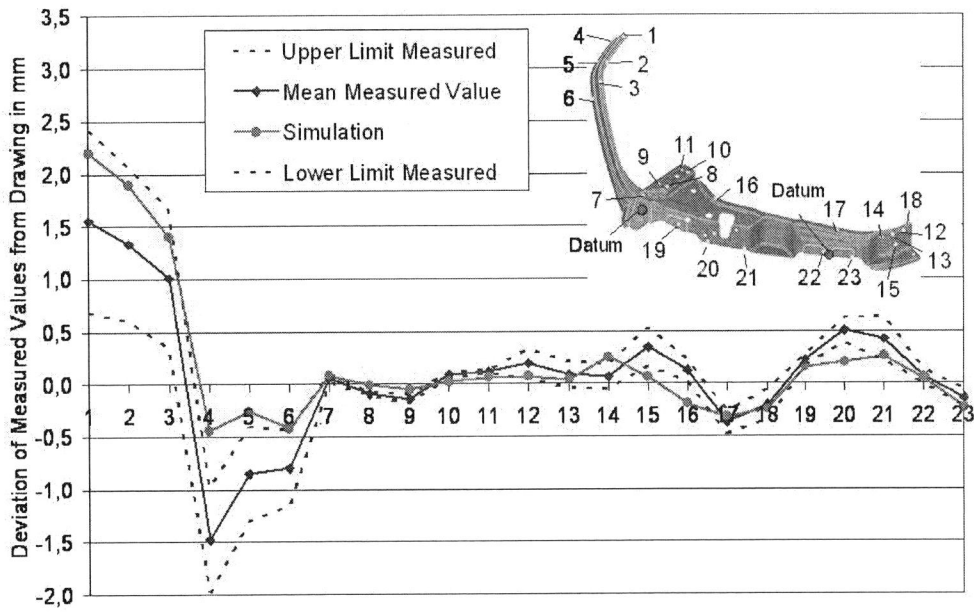

Figure 7: Comparison of calculated and measured distortion after clipping.

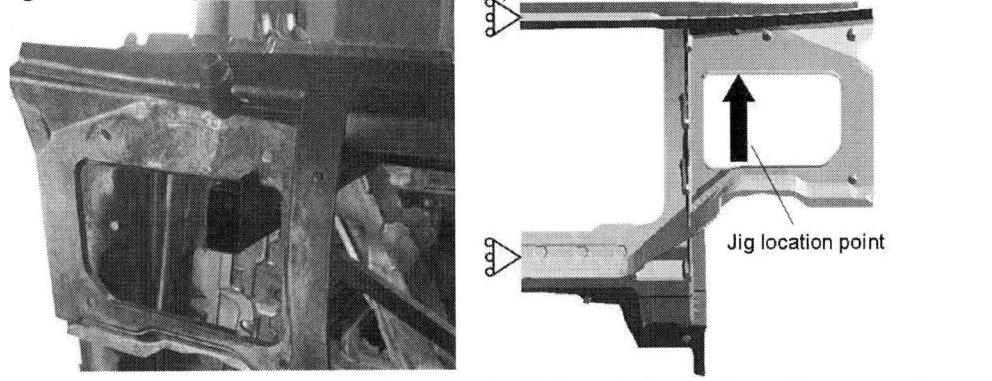

Figure 8: Jig location points the bulk head during heat treatment.

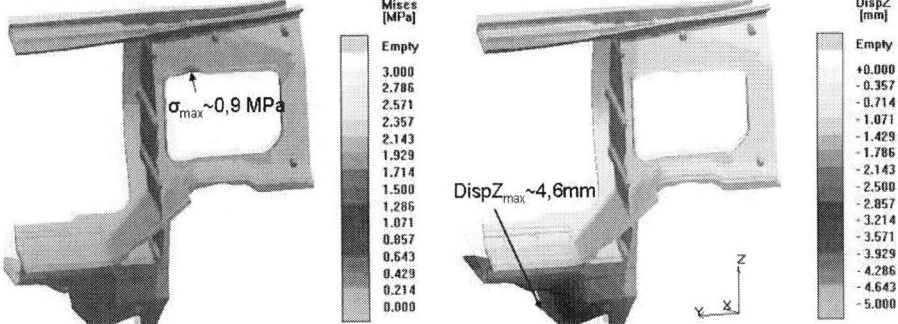

Figure 9: Stresses and deformation of the bulk head after solution heat treatment. The original geometry of the bulk head is shown in semi transparent grey.

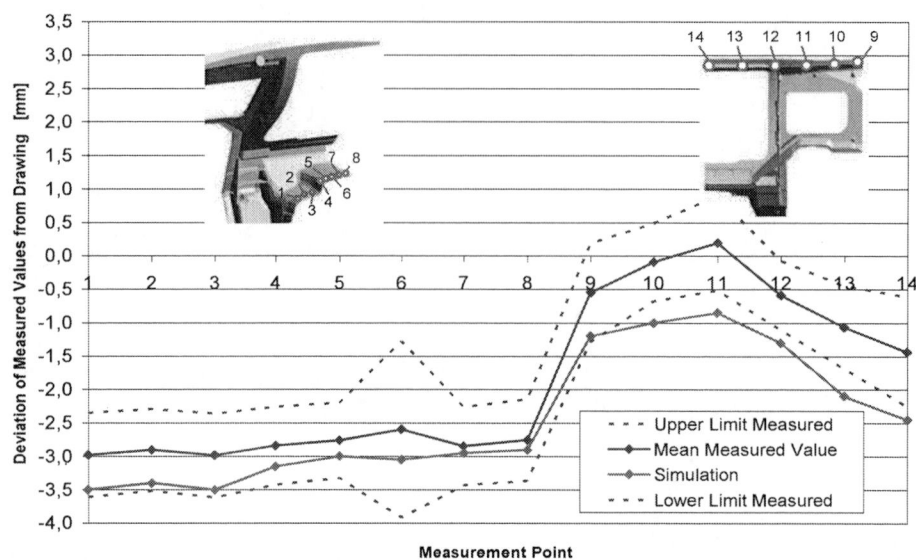

Figure 10: Calculated and measured distortions after heat treatment.

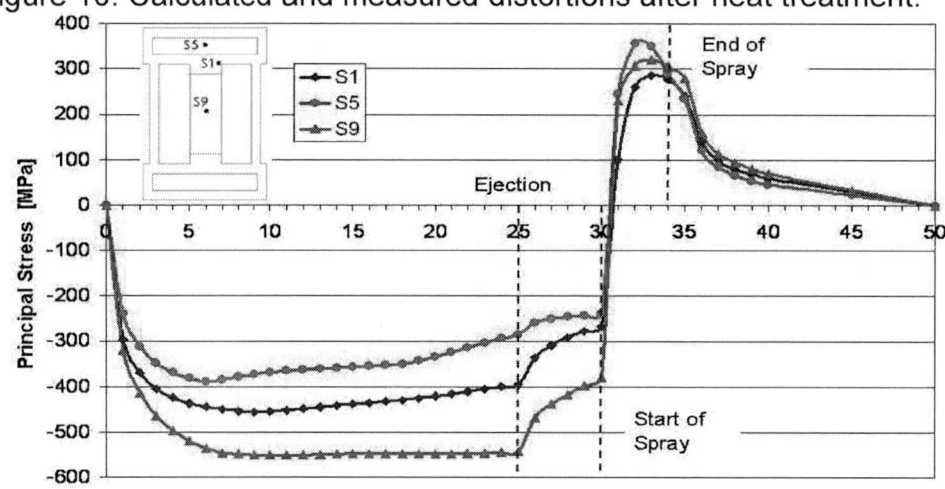

Figure 11: Typical stress changes in the die surface during one cycle.

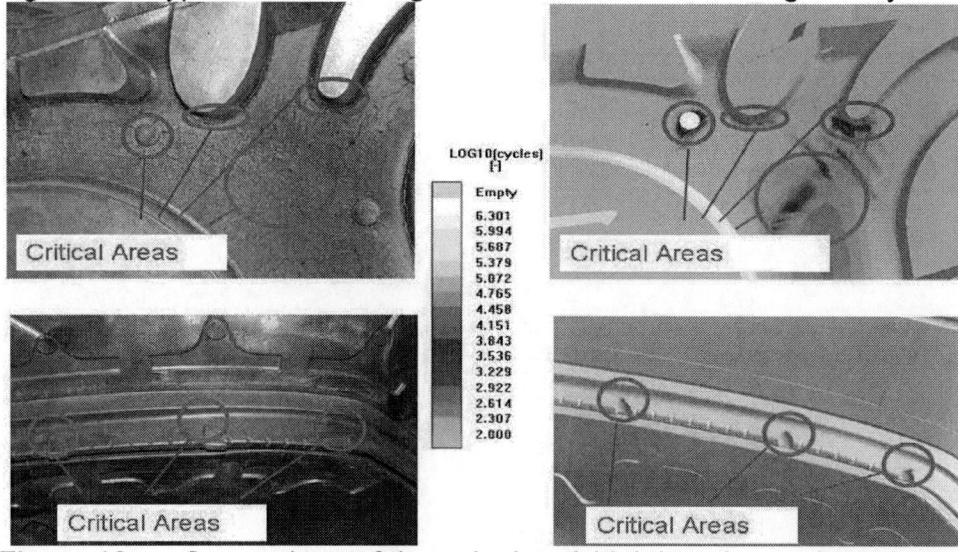

Figure 12:   Comparison of the calculated (right) and actual crack pattern in the area of the gate and the swan neck (left). The most critical areas are marked with red circles.

The use of Different Computer Simulation Software Packages to Predict Casting Filling and Solidification

S Oxley and P M Haigh,

Castings Technology International

Abstract

A series of filling and solidification simulations have been carried out using different simulation programs to evaluate the effectiveness of simulation for precision castings.

A four casting geometry was designed to be a problematical system to create a very poor filling pattern and with a feeding method which will produce areas containing shrinkage defects. Results were determined using six different commercially available software programs:

Thermophysical data sets for the aluminium LM25 alloy were used from data obtained through the National Physical Laboratory in London. These datasets have full traceability to measurement techniques and data calculation methods.

The results are presented as snapshots at specific positions throughout the simulations. Castings were also produced and inspected to assist with the verification of the simulations carried out.
The paper reviews the results obtained.

Key words (5 maximum) Simulation, Solidification, Castings, Aluminium.

Introduction

A series of filling and solidification simulations have been carried out using different simulation programs to produce information to evaluate how simulation usage may be better adopted in the European precision casting industry. This activity was carried out within the European Thematic Network "SARE".

A four casting geometry was designed to be representative of a component that will provide information related to the effective use of the simulation programs. The component and production method was deliberately designed to be a problematical system using a gating system designed to create a very poor filling pattern and a feeding method which would produce areas containing shrinkage defects. [1] The geometry was supplied to all of the SARE project partners interested in conducting the simulation. The simulations run using the following software programs:

- MAGMAsoft - Cti
- ProCAST – Inasmet, Calcom, VTT
- Flow3D - HUT
- PAM-CAST - CTIF
- MAVIS – Swansea University
- View Cast – WTCM

The list order above was not intended to represent any preference or ranking order of the simulations carried out. Each simulation result was assessed against a particular the stage of the simulation, e.g. bottom of the downsprue, the first metal entering cavity, the last part of the cavity to fill, etc. It was never the intention to directly compare each simulation programs results against each other, only to collect information to allow better use of simulation programs.

Experimental

The models to be used were supplied to the project partners as a casting and gating system component and a ceramic shell model. The property dataset for the LM25 alloy were supplied from data obtained through the National Physical Laboratory in London. The dataset has a full traceability to measurement techniques. There is currently little measured property data available for ceramic shell materials. The property data for the ceramic shell mould was therefore supplied from data developed and used by Cti. Cti also supplied the material boundary conditions as being representative of those that should be used to undertake a simulation of this nature. [2]

The simulation results were obtained as snapshots at specific positions throughout the simulations. A screen image from each simulation program was takenat the most representative image nearest to the time of the event being monitored. Due to the simulations being carried out by several partners it was not possible to display images at exactly the same time in each case. The results are therefore the closest available fit between each simulation, as the examples shown in figures 1 and 2.

The resuts of the filling simulation were evaluated and at the "bottom of the downsprue" position:

- All simulations filled the downsprue, and begin to fill the top runner bar.
- All simulations begin to top fill the casting cavity
- There were significant differences in the amount of the pouring cup that has been filled

At the position of the first metal to enter the cavity:

- There were significant differences in the amount of the pouring cup that was filled
- There were significant differences between the simulations when comparing the amount of metal which entered the casting cavity related to the amount of metal in the lower runner bars.
- There was a general agreement between the simulations that top filling of the cavity was occurring
- Differing amounts of metal were predicted to have entered the casting cavity

At the last part of the casting cavity to fill

- There were significant differences in the amount of the pouring cup which was filled
- There were significant differences between the simulations when comparing the amount of metal which entered the casting cavity related to the amount of metal in the lower runner bars.
- There were significant differences in the amount of turbulent metal which was predicted during the filling cycle
- There were differences between the simulations related to the filling pattern
- There were disagreements between the simulation related to whether top filling, or top & bottom filling of the cavity was occurring
- Differing amounts of metal were predicted to have entered the casting cavity

At the end of the filling cycle:

- There were significant differences in the amount of the pouring cup which was filled
- The predictions at fully filled position show differences in the actual temperatures predicted at the end of the filing cycle
- There was a general agreement of the overall temperature distribution within the mould cavity

The results obtained for the solidification simulation were evaluated in a similar manner.

For the first part to solidify:

- There was a general, but not unanimous, agreement that solidification begins in the thin side walls of the casting
- Some results were not available for two of the simulation programs

The isolated sections within the casting revealed:a general agreement between the simulation programs of the solidification pattern during the initial stages of solidification [3]

The results for the prediction of shrinkage defects showed:

- There was a general agreement that the majority of shrinkage defects were predicted in the middle of the casting as shown in figure 3
- There were differences between the number of shrinkage defects predicted in the middle of the casting
- There were significant differences between the shrinkage defects predicted in the top section of the casting
- There were significant differences between the shrinkage defects predicted in the lower shaft guide boss area

In order to verify the results of the computer simulations, a set of verification castings were produced using a ceramic shell process to make test castings, figure 4. These castings were subjected to radiographic and metalllurgical examinations. The radiographic examination did not exhibit any severe gross shrinkage cavities, however microporisity defects were distributed throught a number of regions of the casting, figure 5.

The metallurgical examination was carried out to further investigate the shrinkage defects predicted in the simulations carried out. Micro porosity defects were clearly seen in the sections taken through the casting, figure 6. It should be noted that the defects predicted in the simulation programs should be referenced as micro shrinkage, and not macro shrinkage

There was a general agreement between the simulation programs relating to the distribution of shrinkage defects predicted across the central section of the casting.These were in agreement with the micro porosity found in the casting. There were also predictions of defects in the top and bottom sections of the casting, for which there is little common agreement between the simulations. These defects shown to be in the actual castings were present as distributed micro porosity only.

Conclusion

The results have show that whilst there is a general agreement between the overall predictive capabilities of the simulations there are a number of potentially significant differences between the simulations.

The rate at which the pouring cup is filled differs in all of the simulations. This suggests that the program is treating the pressure buildup at the bottom of the downsprue in different ways. Some are allowing metal to pass through this 'choke' and others are using this are as a flow restrictor.

The filling cycle behaves differently in many programs relating to the severity of the turbulence predicted during the filling cycle. [4,5] The test geometry was deliberately designed to produce a very turbulent fill, and has produced significant oxide defects in the test castings. Some of the programs predict a very turbulent fill, whilst others do not show significant metal stream breakup. It is doubtful that all of the simulations would have predicted the severity of the surface defects achieved. The last areas to fill in the mould cavity also differ across the simulations.

The solidification cycle predicts differences in the areas that become isolated during solidification, which therefore influence the prediction of defects. Shrinkage in all the simulations was shown to be generally concentrated around the middle of the casting but the programs all differ in the severity of defects predicted in the area.
The simulations also predicted of defects in the top and bottom sections of the casting for which there was little common agreement between the different software rprogrmmes.

References

1   R.W. Wlodawer. (1966) *Directional Solidification of Steel Castings.* Pergammon press.

2   D.G.R. Sharma and M. Krishnan (1991) *Simulation of Heat Transfer at the Casting Metal-Mould Interface.* AFS Transactions.

3   P.N. Hansen and P.R. Sahm, (1998) *How to Model the Feeding Behaviour in Casting to Predict Shrinkage and Porosity.* Modelling of Casting and Welding Process IV. The Minerals and Materials Society.

4   B.D. Nichols, C.W. Hirt, R.S. Hotchkiss. (1980) *SOLA-VOF: A Solution Algorithm for Transient Fluid Flow with Multiple Free Boundaries.* Technical Report LA8355, Los Alamos Scientific laboratory.

5   J.Wang, S.F. Hansen, P.N. Hansen. (1988) *3D Modelling and Simulation of Mould Filling using PC Computers.* Modelling of Casting and Welding Processes IV. The Minerals and Metals Society.

s.oxley@castingstechnology.com; pm.haigh@castingstechnology.com

Acknowledgements

The computer simulation and casting manufacture were carried out within the frame of the EU Thematic Network SARE (GTC2-2000-33028) as part of the European Community funding project " Competitive and Sustainable Growth"(1998-2002) The project was coordinated by INASMET and included a large number of partners across the EU.

Figures

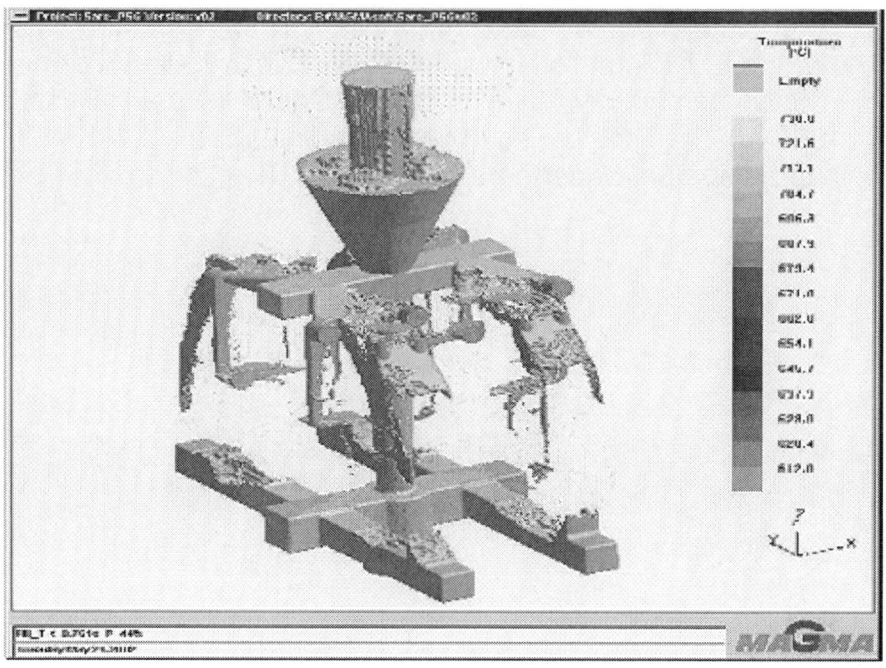

Figure 1   Filling simulation at the bottom of the downsprue

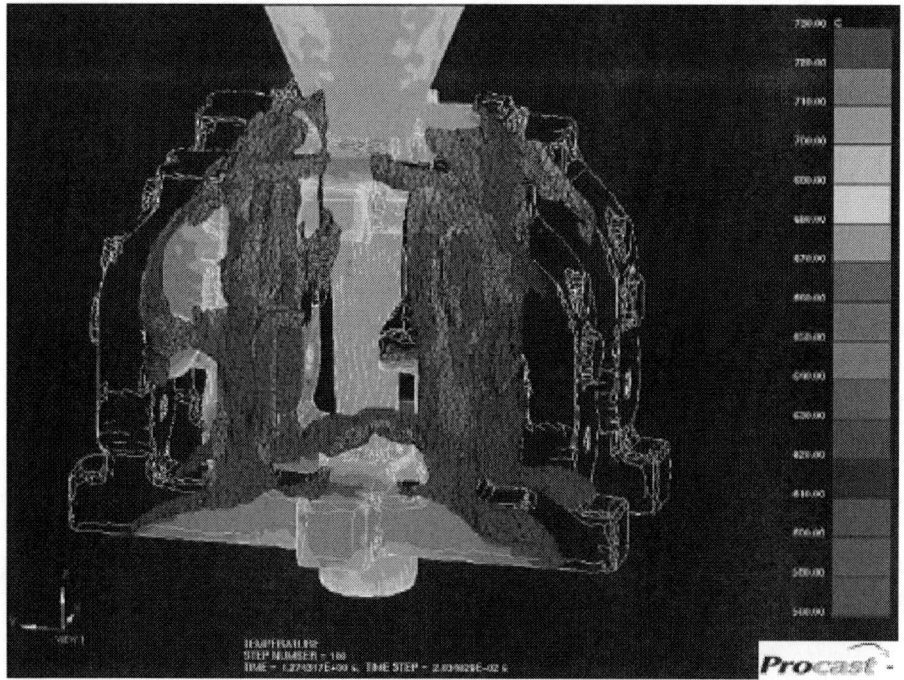

Figure 2   Filling simulation at the botom of the downspr

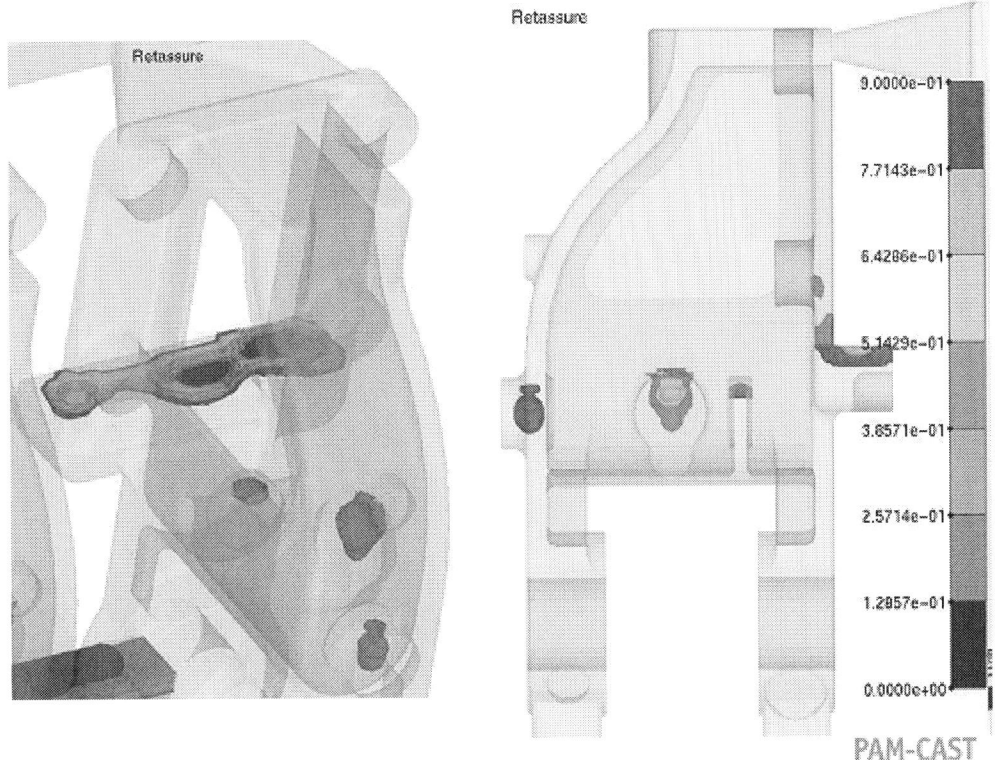

Figure 3   Prediction of shrinkage defects

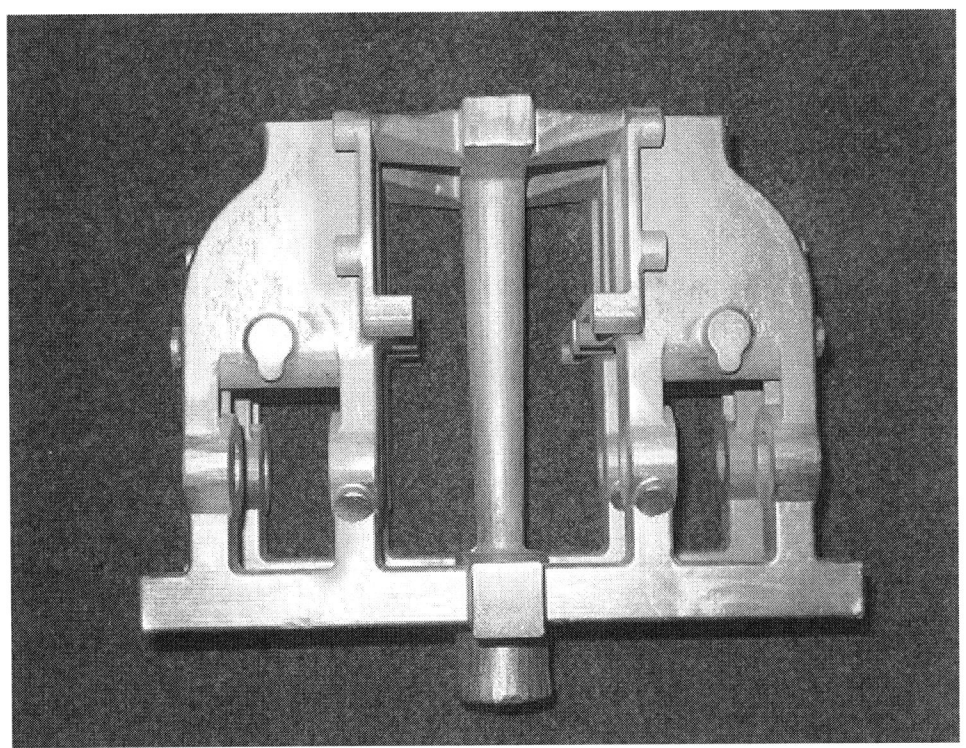

Figure 4   LM25 verification test castings

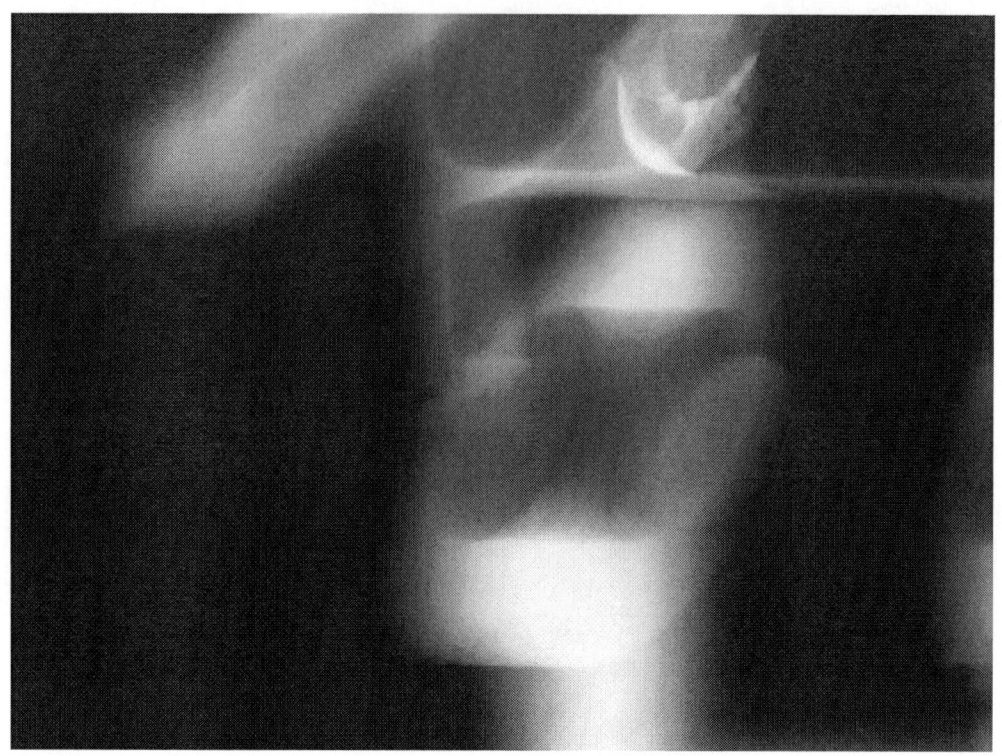

Figure 5   X-ray image of test casting showing porosity defects

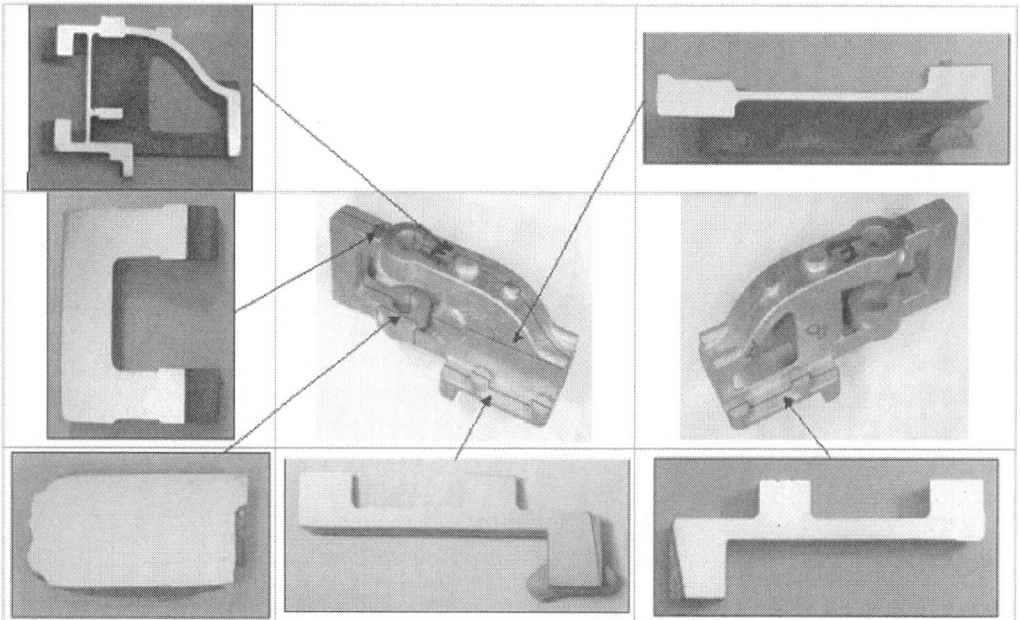

Figure 6   Metallurgical examination of LM25 test castings

Prediction of the influence of microstructure, porosity and residual stresses on strength properties of aluminum castings

R Baehr *, M Todte * and H Stroppe **.

* University of Magdeburg, Casting Technologies, Germany,
** Institute of High Technologies and Education e.V. Magdeburg, Germany

Abstract

It is necessary for a design engineer of castings to precisely know the mechanical properties (yield strength, tensile strength, elongation to fracture, strain hardening exponent and so on) of every part of the casting. These are very closely connected to the locally very different cooling- and solidification preconditions of the melt. Estimating strength properties through all the varying characteristics of the cast structure (dendrite structure, morphology of pores, eutectic phases, residual stresses and so forth) via numeric simulation is yet an unsolved task of great theoretical and practical importance.

The introduced research results have created a possibility to already predict achievable properties of the casting during the construction phase. By taking the primary simulation results and calculating their mechanical properties and quality index, several production techniques, casting alloys and geometrical variants of castings will virtually be analyzed on a computer. Results of own analyses will be referred to in this article.

Key words

Simulation, mechanical properties, microstructure, residual stresses, fatigue life.

Introduction

Simulation of casting processes has become an important tool in foundry technology worldwide. They make it possible to numerically calculate complex activities of flow and numerous phenomena of solidification and cooling of castings. In order to achieve a practical realization of the results, adequate criteria functions have to be used. By evaluating the calculated primary temperature- and flow fields including their mathematical-physical regularities, certain casting properties can be estimated.

Regarding casting products, especially in the automotive industry, there's a trend recognizable towards the production of highly specialized parts that reach the limit of their strength. This requires the development and use of qualitative novel simulation tools. Therefore, the faculty of foundry technology of the Otto-von-Guericke University Magdeburg, in close collaboration with industrial partners (like NEMAK, Porsche), is intensively engaged in the estimation of mechanical properties that appear during the solidification of the melt (yield strength, tensile strength, elongation to fracture, strain hardening exponent and so on), whereas, so far, we were only able to simulate the residual stress of the cast part, as well as certain structure properties such as dendrite arm spacing, porosity and so on.

One cannot assume that the properties of castings are homogeneous throughout the entire part. This holds true especially for geometrically complicated cast parts, due to the different local cooling- and solidification conditions of the melt (**Figure 1**). Furthermore, inner tensions, which can take pretty high values, develop within the casting and may influence the casting's strength behavior.

Since not only specific structure properties (dendrite arm spacing, porosity), but also residual stress of the casting can be determined via simulation, we have the opportunity to predict the impact of the casting process on the static and cyclic strength properties of a casting via virtual reality.

Correlations between mechanical properties, structure properties and solidification parameters

The above mentioned mechanical properties of a casting are significantly dependant on its structural properties. These are, in turn, influenced by the chemical composition, the metallurgic treatment of the melt (decrease of grain size, refinement and so on), the different preconditions and parameters for the process during the casting- and solidification activities, as well as on the after-treatments (heat treatment, HIP) of the casting.

The structure of AlSi casting alloys is basically characterized by the primary dendritic growth of the α-phase in aluminum, as well as by the formation of an interdendritic eutectic. The secondary dendrite spacing (DAS), as well as the formation structure and distribution of hard eutectic Silicon particles and intermetallic phases respectively are the substantial structural properties, which have an effect on ductility- and strength properties of the solidified casting alloy. Additionally, there are the feeding- and solidification properties.

All the mentioned structure properties are dependant on each other; they correlate with each other via local solidification time. As the solidification time increases, both dendrite arm spacing (**Figure 2**), and the eutectic or intermetallic particles and pore volume increase as well. These structure properties in turn have a quantifiable impact on the solidification properties (**Figure 3**). Thus, the solidification time turns out to be a basic decisive parameter for the quality of the casting structure, with which a prediction of mechanical properties in a simulation of the casting- and solidification process can be made; provided that the correlations between these properties and the structure parameters are known (**Figure 4**).

Influence of porosity on mechanical properties
The physical behavior of cellular structures is, next to the structure building material, characterized by its porosity, which is characterized by portion, arrangement, size and shape of each pore. Porosity in casting parts is, next to dimensional deviation and surface irregularities, one of the most frequent reasons for rejection. A degradation of the mechanical properties of the casting is, as known, adherent to this porosity. Therefore, this phenomenon was included in the simulation model. Research results from former works [1] formed the basis. If one applies the derived mathematic-physical relations between mechanical properties and porosity to the research results, you receive a high concordance of experimental and the actual, calculated values (**Figure 5**).

Implementation of criteria-functions into the simulation system
Appropriate correlation functions between solidification parameters, structural properties and mechanical casting properties were deduced by Todte (with porosity involved) and integrated to the known simulation program SimTec/WinCast® of the RWP GmbH Germany [2]. That way we are able to predict the mechanical properties, as well as the quality index in dependence on the alloying additions and the solidification process [3 to 5]

Aiming at the verification of the newly developed correlations, locally calculated characteristic values have been compared to values estimated by simulations. **Figure 6** shows an example: The tensile strength of two chosen measuring points of a cylinder head. Despite possible errors (such as subjectively conditioned tolerances, systematic errors), all results show high and even very high correlations between experiment and theory [2].

Estimation of fatigue life of engine components
As the power output of modern combustion engines increases, the material-appropriate design and the dimensioning of cylinder heads made of aluminum casting alloys grow more and more important. On the one hand, the installation- and operation preconditions (like pretensioning force, cylinder pressure and oscillating mass-forces) have a higher-frequented,

largely isotherm fatigue of the material. On the other hand, temporally and locally inhomogeneous fields of temperature and stress, which develop in, for example, start-stop- and load changes, result in a low cycle fatigue, as well as in creep resistance. Those aren't only dependent on temperature and level of stress, but also to a great extent on the structural properties of the material.

Calculating the fatigue life of castings, especially of cylinder heads, during cyclic stress in dependence on material properties and the amount of load, is yet a problem that has to be solved. Most of the theories that have been developed so far base on macroscopic-phenomenological points of view, in which the material's characteristics, such as residual stress, porosity and structure formation, plus their impact on locally static and dynamic strength properties, are disregarded. [6, 7]

Highly stressed cylinder heads usually consist of a hypoeutectic AlSi-alloy. The structure's fineness, which significantly influences the fatigue strength, is primarily conditioned by the secondary dendrite arm spacing. Since the dendrite growth correlates with the local solidification time of the melt, it is controllable through the casting process. When considering the influence the structure has on the analysis of the operational stability, there has to be used a material model in dependence on the dendrite arm spacing for every FE-structure element. In order to realize this, complex tests to calculate the material-S-N curve on test items are necessary for every alloy in different states of solidification.

Based on an analytic model of structure-dependent material fatigue developed by Stroppe [8] for aluminum alloys, calculation methods are being developed that allow to calculate S-N curves of hypoeutectic AlSi-casting alloys not only within the time strength range, but also in the area of the low cycle fatigue for any tension depending on the dendrite arm spacing (**Figure 7**). The new theoretical model also allows us to take the influence of porosity and residual stresses on material fatigue into consideration.

So far, no analytic approach in casting simulations is known that makes it possible to calculate S-N curves for different kinds of stress from structure parameters (like dendrite arm spacing) via a model based on material-physics. An attempt to estimate dynamic characteristic values for a cylinder head is shown in **Figure 8**. By introducing specific model theories into the simulation system SIMTEC/WinCast®, it is possible, for instance, to estimate the permanent stress cycles in dependence on dendrite arm spacing (and therefore on casting- and solidification processes).

Summary
Based on new mathematic-physical models to describe the developing structure- and cast part properties during the solidification of the melt, it is possible to already predict mechanical properties of cast parts during the development phase via numeric simulation. In order to get these predic-

tions, one used to be dependent on long-winded and cost-intensive analysis of the material. Furthermore, not all areas of the casting have the geometrical preconditions for taking samples. Therefore, "trial and error-loops" still have to be run trough until new components can be produced. With the presented research results, the numeric simulation of casting processes has been enhanced by the creation of a physically based calculation of mechanical properties and the quality index.

Due to the increasing competitive pressure on the one hand and the present development status of hard- and software on the other hand, basic innovation in the process of product design is to be expected within the coming years. In the nearer future it's going to be possible to make realistic statements concerning essential product characteristics already during an early state of development.

References

1. Stroppe H, *Einfluss der Porosität auf die mechanischen Eigenschaften von Gusslegierungen*, Gießereiforschung 52 (2000) 2, pp 58 – 60.

2. Todte M, *Prognose der mechanischen Eigenschaften von Aluminium-Gussteilen durch numerische Simulation des Erstarrungsprozesses*, Dissertation (2003), Otto-von-Guericke University Magdeburg.

3. Bähr R; Scheel B; Mnich F; Stroppe H; Ambos E and Todte M, *Optimization of the mechanical properties of highly stressed castings through direct control of the casting process*, Proceedings of the 65th World Foundry Congress, Gyeongju, Korea 2002.

4. Stroppe H; Todte M; Ambos E and Bähr R, *Einfluss von Erstarrung und Gefüge auf den Qualitätsindex von Aluminiumguss*, Gießerei Forschung, 55 (2003) 4, pp 151 – 155.

5. Todte M; Stroppe H and Honsel C, *Prognose der mechanischen Eigenschaften von Aluminium-Gussteilen durch numerische Simulation des Erstarrungsprozesses*, Gießerei-Praxis, (2003) 6, pp 263 – 269.

6. Nefischer P; Steinparzer F and Kratachwill H, *Neue Ansätze bei der Lebensdauerberechnung von Zylinderköpfen* ,12. Aachener Kolloquium für Fahrzeug- und Motorentechnik 2003.

7. Minichmayr R and Eichlseder W, *Lebensdauerberechnung von Gussbauteilen unter Berücksichtigung des lokalen Dendritenarmabstandes und der Porosität*, Giesserei 90 (2003) 5, pp 70 – 75.

8. Stroppe H, Lebensdauerberechnung von Gussbauteilen unter Berücksichtigung des lokalen Dendritenarmabstandes und der Porosität, Magdeburger Forschungsinstitut für Fertigungsfragen MFF e.V., (2004) previously unreleased.

Figures

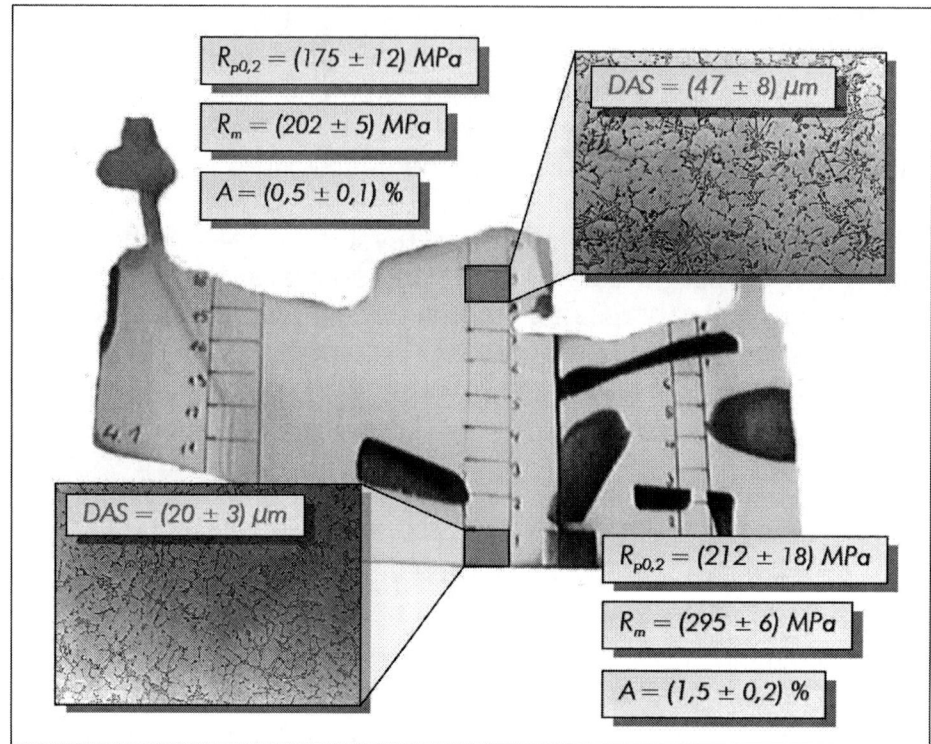

Figure 1: Experimentally determined, locally different mechanical properties within a cylinder head made of an Al-alloy
(DAS = secondary dendrite arm spacing)

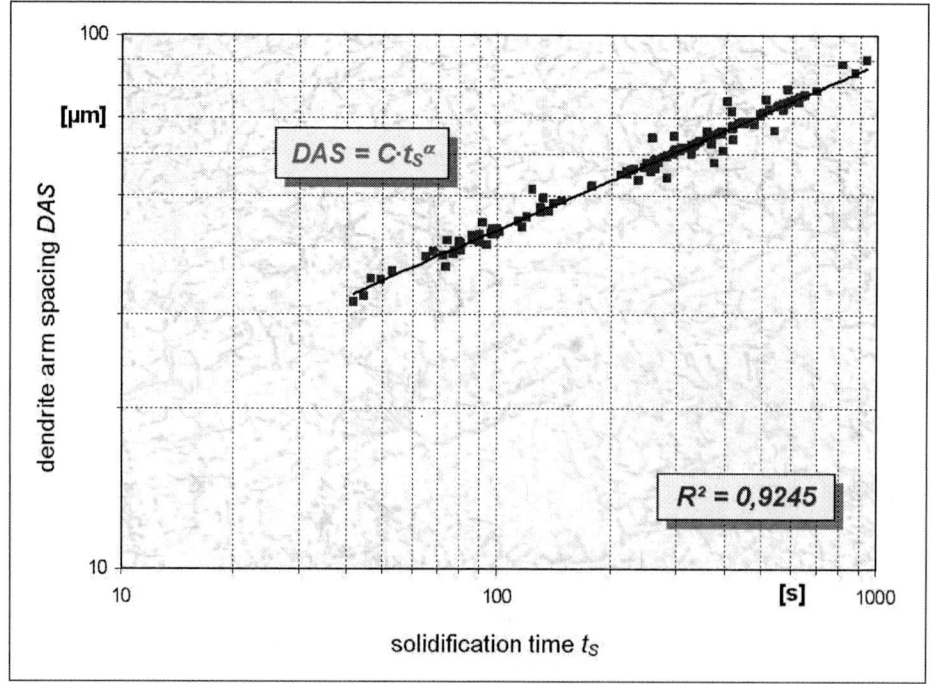

Figure 2: Experimentally determined correlation between dendrite arm spacing DAS and the local solidification time t_S in an Al-alloy

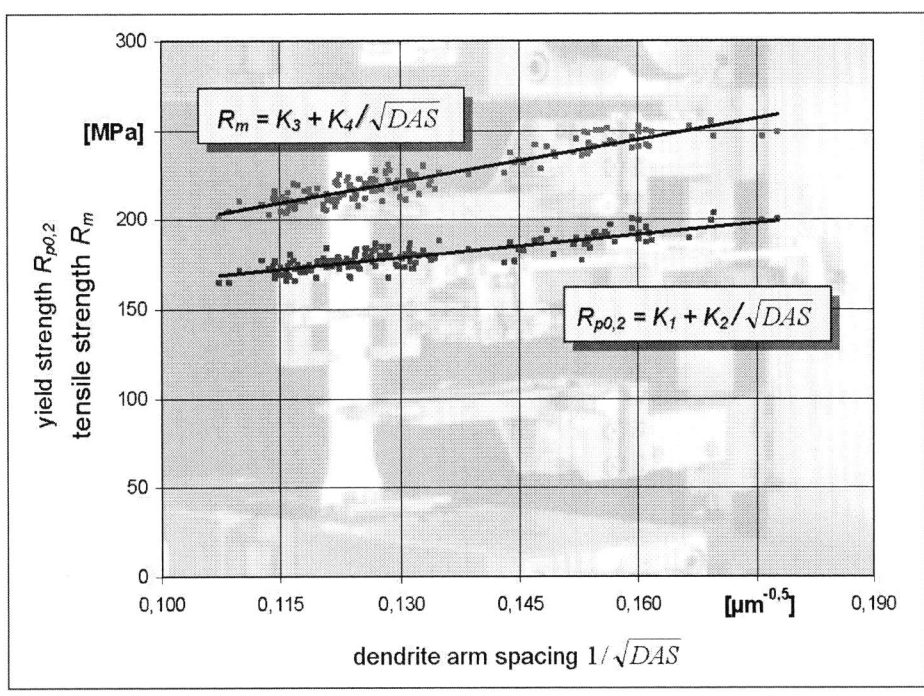

Figure 3: Experimentally determined correlation between yield strength, tensile strength and dendrite arm spacing (DAS)

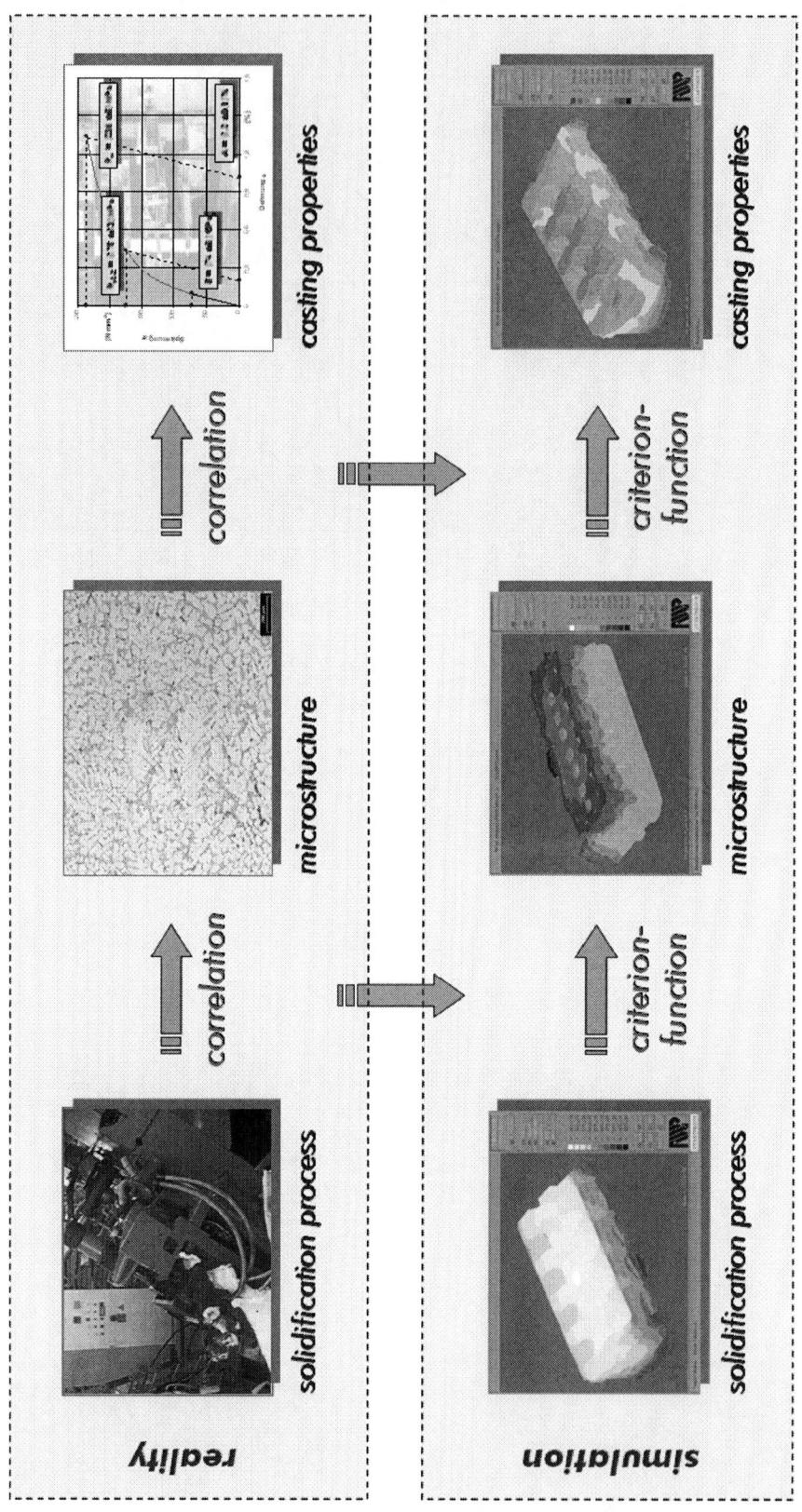

Figure 4: Prediction of mechanical properties of a casting through results of numeric simulation during the solidification process

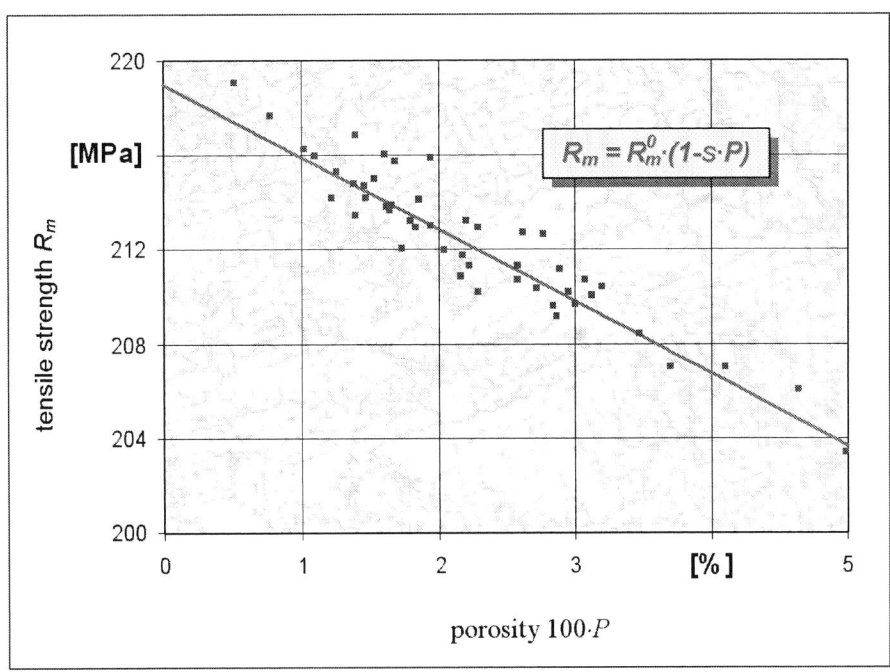

$$R_m = R_m^0 \cdot (1 - s \cdot P)$$

Figure 5: Experimentally determined and theoretically calculated dependency of the tensile strength R_m on porosity P (s = shape factor)

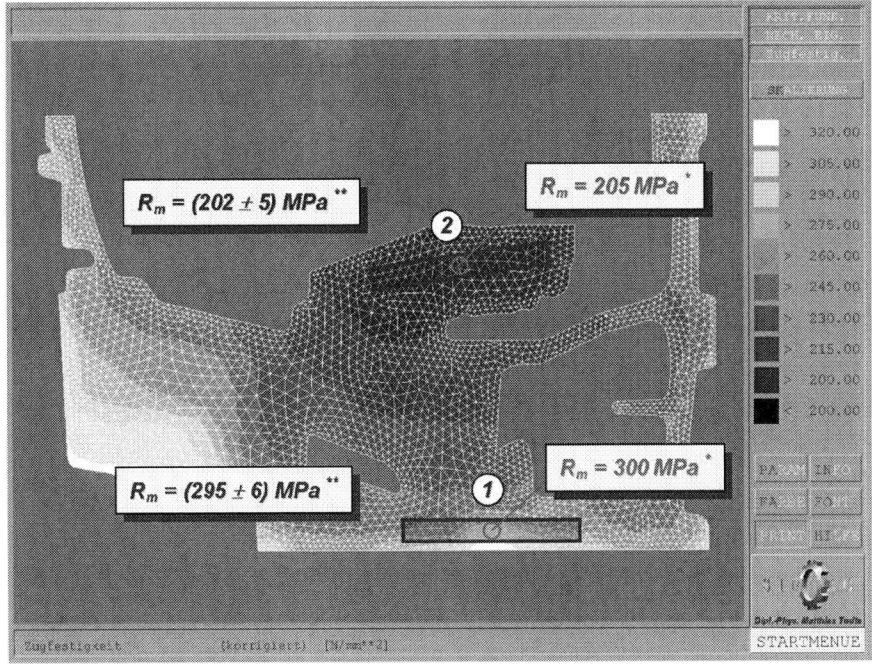

Figure 6: Comparison of the calculated* and experimentally determined** values of the yield strength

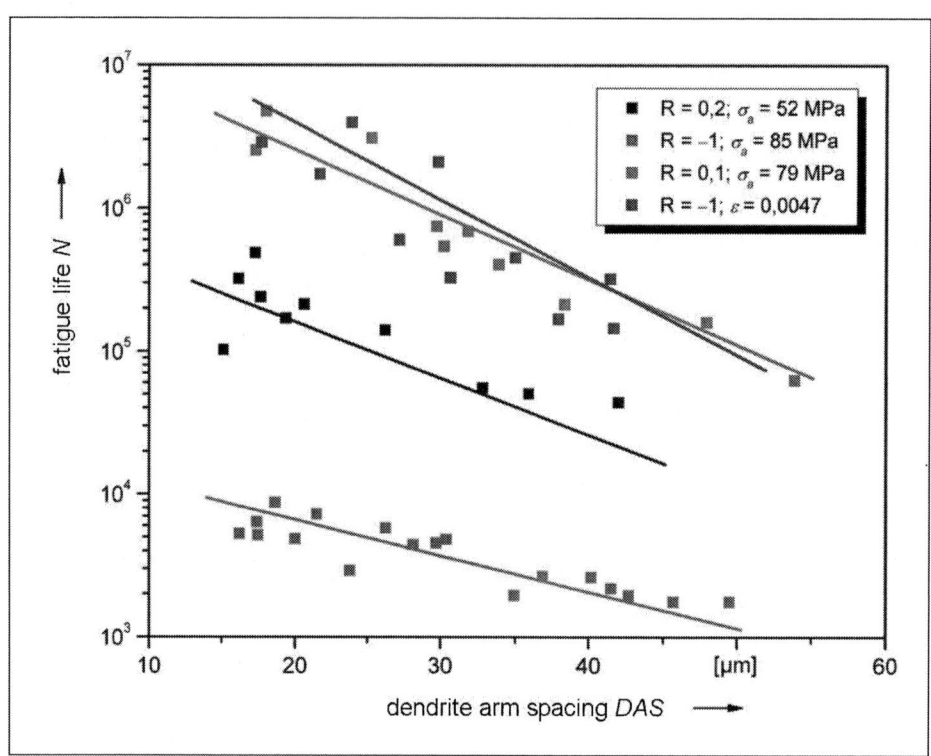

Figure 7: Dependency of the cycles of fracture on the dendrite arm spacing DAS for the alloy GK-AlSi7Mg0,3 wa (measuring point taken from [7], curve progression calculated according to Stroppe [8])

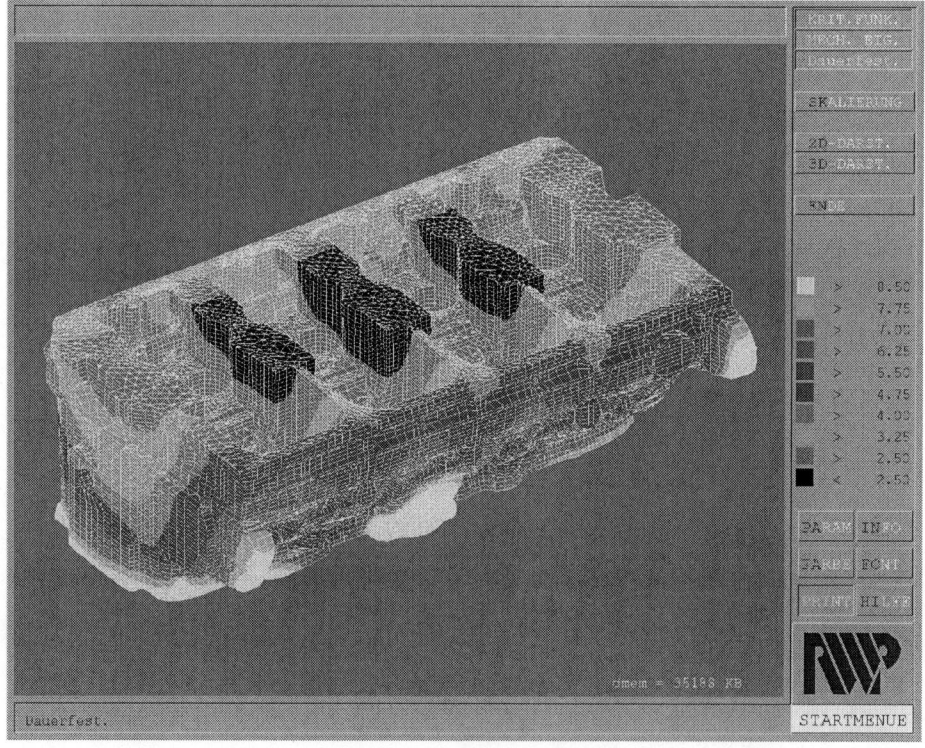

Figure 8: Fatigue life (cycles of fracture) of a cylinder head, calculated with the program SIMTEC/WinCast®

Intelligent Riser/Chill/Gating Design System Using Simulations and Discrete Optimization Algorithm.

Chaeho Lim, Jeongho Nam, Insung Cho, Seungmok Yoo, Jeongkil Choi

*Korea Institute of Industrial Technology, 994-32, Dongchun-Dong, Yeonsu-Gu, Incheon, Korea

Abstract

In the casting industry field, a lot of researches have been carried out to maximize soundness and productivity and improve quality stability of mass produced products. The researches on the optimum design by using computer simulation have achieved a lot of accomplishments especially. Their results are utilized to mass-produce sound products. There are some difficulties to use the simulation in order to make the design of product. In general, repetitive calculations of computer simulation require much analyzing time in order to make sound design of product. Another difficulty is that only expertise technicians can analyze results. To overcome these difficulties, a lot of researchers carry out studies to make design by connecting simulation with optimization algorithm. We have performed such studies to compose the system which connects solidification simulation with the optimization algorithm called SOA(Sequential algorithm with Orthogonal Array). We also have standardized various cast design rules of documents and experts in permanent mold casting. By using them, we have composed the IES(Intelligent Expert System) where a computer makes an initial design of a riser, a chill and a gate with the minimum intervention of a user. After that we achieve the optimum design via simulation and optimization. This study will introduce what is mentioned above, the general idea of the system and some examples to apply it to real products.

Key words : riser, chill, gate, optimization, SOA

Introduction

Compared to the other hi-technology industries, the casting industry today seems to get less important. However, various effective casting manufacturing technologies are more needed along with rapid development of other industries.

Casting production has some difficulties because lack of understanding of the casting industry makes it difficult to obtain the demand for manpower in many industrial fields. Also advanced casting design technology is needed and the competitive circumstances demand improvement in productivity. Today many industries use the simulation via a computer in an attempt to overcome these difficulties. It is not enough to perform only simulation in order to get the solutions of design. It becomes more necessary to develop the IES system to graft vast database on to an intelligent algorithm.

Many studies are in full swing to use simulation programs and optimization algorithms. This study is about installing a riser and a chill automatically by the established casting design rules and making a sound design by using it. It also introduces how to decide the location and size of a riser by using the solidification simulation. [1] [2] Also, it shows how to accomplish the design goal when just the riser installation cannot obtain a sound product. SOA, an optimization algorithm is applied and developed to reduce repetitive analysis time, which will be introduced in this study.

We also have been building the vast product design database in order to obtain an effective initial design. We previously studied how to utilize the database via the pattern recognition method so that we can easily obtain initial designs of a riser, a chill and a gate of a product. [1] [4]
Examples to use the intelligent expert system built as the study result are examined.

Composition of IES

Fig.1 is the composition of the IES system we are developing now. The IES system is composed of three layers: analysis, design and artificial intelligent. First, analysis layer is the well-known simulation analysis composed of pre and post processor and analysis layers. Second, the design layer decides several designs of a riser, a chill, a padding and a gate based on the suggestion of the artificial intelligent layer which can be deduced from the first analyzed results and the database.

This design layer is largely divided into the modeler function to produce various designs and variable decision layer to accomplish the optimization. Finally the artificial intelligent layer is the deduction layer composed of rule-based and case-based deductions. It is the core part in the whole IES system. Regarding the case-based deduction, the information of a similar product that we want to design is retrieved by the automatic shape recognition program. Then the information of the retrieved similar product is used to design the product what we want to make. [4] The design rules related automatic generation of a riser and a chill, some part of rule-based deduction will be examined in the next chapter.

Automatic installation of a riser

At this section, solidification simulation, riser location calculation, and interference confirmation are studied. Through solidification simulation, the generated shrinkage cavity cells are categorized into several groups. The volume and the location of each group are calculated. The results of the calculation are used to determine the location and the size of a riser. (Since complex riser images have numerous related variables, the system uses only the riser of some basic shape.) When the location and the size of a riser are determined, the interference between risers and betweena riser and a cast are confirmed.   Then, the possible points to set a riser can be shown to an engineer.

The proper size of an initial riser can be decided by calculating the module of a product. It has been studied for a long time and its empirical formulas have been well-known. In order to use a modulus values, we have to solve the big problem. The problem is that a product should be divided. The effective method to divide a 3D product has not known yet. But the method to divide the 3D product have been developed by using the "shell wrapping method". This method is opposite of thining product.
When a module of each divided part is calculated, the size of a riser can be decided.

Fig.2 is an example of an initial riser installation in an actual product via the expert system. First, STL is formed with voxels. Then, the solidification simulation is performed and can be found the shrinkage defects as shown in Fig.2 (a). The shrinkage defects are grouped to cover them with voxels. When a product is divided as shown in Fig.2 (b), the module of each divided part is calculated to decide the size of a riser.

When the location of a riser is decided in an above manner, a riser can be installed as shown in Fig.2 (c). However, All of the generated risers are not used. If some riser collide them that are automatically removed. Also, one riser is fixed, rest risers are automatically removed. Or a user can remain or remove the generated risers manually. Then, an initial riser can be automatically determined as shown in Fig.2 (d).

Automatic installation of a chill and SOA

After an initial riser is generated via the above method, the solidification simulation is performed. If the result of the solidification simulation does not expect the shrinkage defect, the number and the size of a riser are optimized by using SOA(SOA is an optimum algorithm that will be introduced below). If a defect is expected, a riser is reinstalled around the defect. If a riser cannot be installed in the location of a defect (Due to the interference between a riser and a product), a chill is installed. Fig.3 shows the steps to install a chill automatically. First, a cell that is closest to the surface of a product is found from the center point of a shrinkage defect. Then a cell where a product in contact with a mold is found among cells (It is indicated as '"1" in Fig.3 (d)). If this method is performed based on a cell that is newly founded, a chill can be installed automatically around a product.

This process is repeated until the surface area of a chill becomes 50 % of a shrinkage defect surface area. There are two methods to decide the chill installation location. Both of them are explained about the first found cell of chill location. The first method is to find the cell closest to the product surface from the center point of shrinkage defect. The second one is to install a chill at the bottom of a product located in the shrinkage defect. A user chooses one of them in consideration of a riser location and a product shape. Fig. 4 is the example to automatically install a chill to "gear box cover" product by these two methods.

After a riser and a chill are installed, the solidification simulation is performed. If the result does not expect a shrinkage defect, the optimization via SOA is performed. SOA is the optimization method that is newly developed as a discrete optimization algorithm. One of the most well known discrete optimization algorithms is a genetic algorithm. While it is easy to make for system but it needs much time for optimization.

To overcome this problem, a sequential algorithm with orthogonal array is used. It organizes an orthogonal array table to find an optimum. This algorithm improves the discrete algorithm's drawback to demands much time to find an optimum.

Fig.5 shows the efficiency of the SOA algorithm. Compared with both ASA (Advanced Simulated Annealing) and MIGA (Multi-island Genetic Algorithm) are the discrete optimization algorithm used a lot in various areas. Although ASA and MIGA are efficient in some special cases. Fig.6 proves that SOA is the most efficient algorithm when it is related to general analysis problem. We apply this algorithm to IES so that we solve the problem of the discrete optimization algorithm that it took much time to find the optimum solution. [3], [5]

Study on gate installation by using IES
As mentioned briefly in the introduction, we want to make the database the important part of the IES system in order to determine the design of an initial riser and gate of a product.

First, the characteristics and keywords of collected product patters are abstracted to store the pattern characteristic keyword database in advance. Based on it, the pattern of a product which a engineer wants to produce through a user interface is made and inputted. Then 3D and 2D pattern data is abstracted among input pattern characteristics. After that the characteristic vector of the abstracted input pattern is compared with the 3D modelling data. Then the 3D modeling data of the casting design with the optimum correspondence can be abstracted as shown in Ref [1],[4].

The 3D modelling data of this casting design has a lot of information related to a riser and a gate, which is very useful to determine the initial location and size of each riser and gate. Moreover, a product-related keyword is inputted into the pattern data constituting the database so that the text keyword retrieval makes it possible to retrieve data on the effective design through linkage with the vast text database.

IES system provides two types of modeller to link the database results with casting design modelling of a product. (The modeller which we are developing now can be used only to sand and metal mold method. Die-casting method with the complicated gate pattern is being studied.)
The first modeller is used for database construction. In order words, this modeller quantifies casting design information among the collected product modelling data so that it is easy to be stored in the database. The quantitative information is obtained according to the determined rules by using dots and lines from the modelling database of a product. When a product is in the form of the modelling data including STL, it is possible to obtain the quantitative information immediately. Although a products is not formatted as a CAD data, the quantitative information can be obtained by performing the reverse engineering through Industrial CT in our research center to obtain the modelling data.

The second modeller has the function of CAD system to utilize the design data in the database making the modelling in our system possible. This modeller make the use of the pre-determined information (regarding a riser : patterns and figures of truncated corns and track corms, and regarding a runner and a gate : patterns and figures of each runner and a gate) to perform the initial modelling for the purpose of looking for the optimum design via the simulation. Fig.6 (a) shows marks of dots and lines to obtain the quantitative information on the given product's design. Fig.6 (b) shows the quantitative information on the product's design.

Application and results
Fig.7 is the examples of installing a riser as well as a chill and of an optimum design by using optimization. The riser installation method above decides a riser as shown in Fig.7 (a). When the solidification simulation is performed, a defect on the lower part is expected as shown in Fig.7 (b). However, a riser cannot be installed in this part due to the interference between a riser and a product. Therefore, a chill is installed in the above way as shown in Fig.7 (c). Fig.7 (d) and (e) show the results that the number and the size of a riser are optimized by SOA.

The following example is the real application of IES system. The product in Fig.8 is the frame of a electro-optical measuring system. We were asked to produce a prototype for product development, we selected the initial riser design through IES system as shown in Fig.8 (a) and (b), and obtained the optimum riser design in (c) through the optimization process. We decided the final casting design in Fig.8 (d) by using this riser design. Fig.9 (a) shows the process to produce a real prototype and Fig.9 (b) is the final result. As shown in Fig.9 (c), the prototype is proved to be sound by a CT defect test.

Conclusions

By using the result of the solidification results of a product, we determine an initial a riser and a chill. It also briefly examines the process to analyze an expected defect after the solidification simulation to install a riser, chill and then to optimize the location. Also the size of the riser and chill can be decided which are chosen at an initial process.

We do not know the performance of IES thoroughly because we haven't experimented and applied various products and our whole system including the modeller system for effective gate design is not completed. However, we make an effort to complete the system and it will be completed soon.

Also, we will be performed to apply and complement more various rules to utilize it. We confirm that this IES will actually help to non-experts in the casting engineering field.

References

1. C.H.Lim, Y.C.Lee and J.K.Choi, *Intelligent expert system for casting design using simulation and optimization* : Istanbul, Turkey, 6-9 Sept, 2004, pub Toksad: The Foundrymen's Association of Turkey, 2004.
2. C.H.Lim, Y.C.Lee and J.K.Choi, *Basic construction of an IES for riser design using database system and optimization tools* : IJCMR, pp195-201, Volume 18, No. 4, 2005
3. K.H.Lee, B.S.Kang, J.S.Park and G.J.Park, *An optimization algorithm using orthogonal arrays in discrete design space for structures*: Finite element in analysis and design, Vol.40 2003
4. C.H.Lim, Y.C.Lee and J.K.Choi, *Database construction and retrieval method for optimal design of casting*: IJCMR, (is accepted)
5. Jasbir S. Arora, *Introduction to optimum design* : McGRAW-HILL

Figures

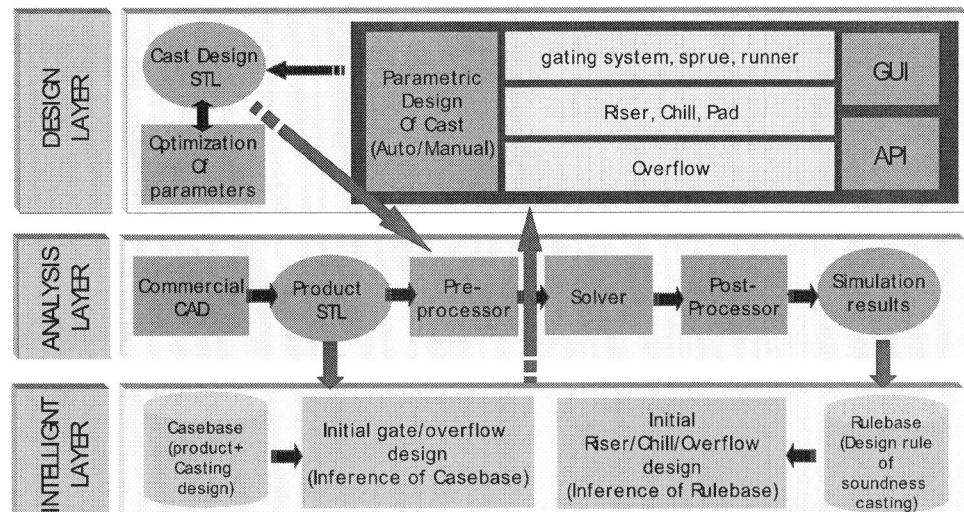

Fig.1. The composition and flow chart of the IES

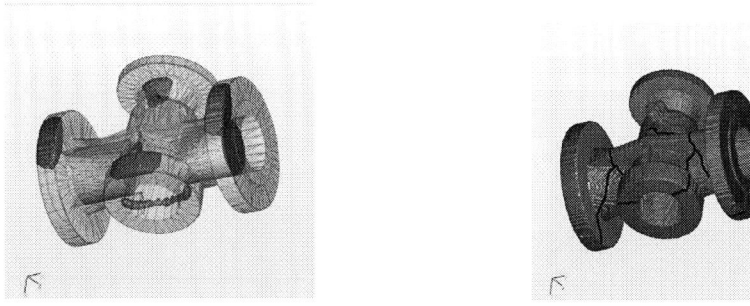

(a) Retrieval of shrinkage cavity result

(b) product division by shell wrapping method

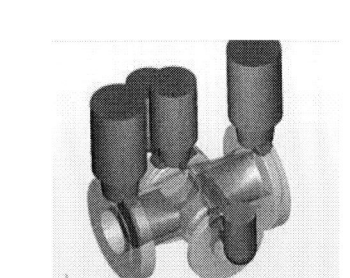

(c) Automatic riser generation

(d) initial riser generation

Fig.2. The riser installation process using product geometry

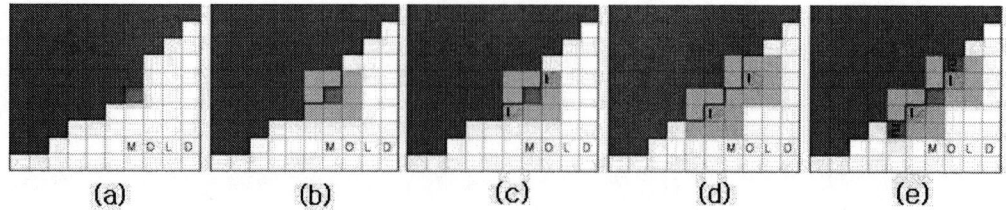

Fig. 3. The automatic chill installation process

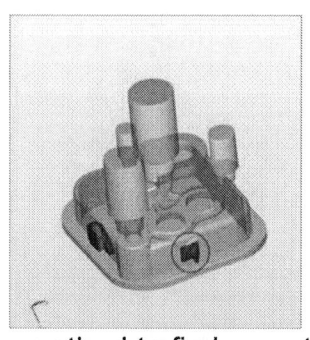

(a) The method to find a center cell closest to a product surface.

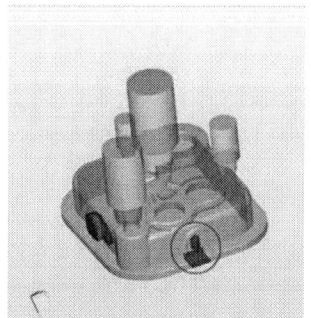

(b) The method to find a center cell at the bottom of a product.

Fig. 4. Examples of the automatic chill installing

Find $\quad x_1, x_2, x_3, x_4$

to minimize $\quad f(x) = 100(x_2 - x_1)^2 + (1 - x_1)^2 + 90(x_4 - x_3^2)^2$
$$+ (1 - x_3)^2 + 10.1((x_2 - 1)^2 + (x_4 - 1)^2) + 19.8(x_2 - 1)(x_4 - 1)$$

subject to $\quad -10 \le x_i \le 10 \qquad i = 1, 2, 3, 4$

method	Number of call function	Optimum variables				Fuction value
		x_1	x_2	x_3	x_4	
SOA	28	0	1	2	-1	-44
ASA	358	0	1	2	-1	-44
MIGA	1000	-1	1	2	-1	-24

ASA : Advanced Simulated annealing / MIGA : Multi-Island Generic Algorithm

Fig.5. Efficiency of the SOA algorithm

	No	Shape	Scale
RISER	1	cylinder	r1=49.5mm , r2=49.5mm
	2	cylinder	r1=49.5mm , r2=49.5mm
	3	cylinder	r1=49.5mm , r2=49.5mm
	4	Conical+ sphere	r1=41.25mm , r2=60mm
Ingate	1	conical	r1=50mm , r2=42.5mm
Sprue base	1	cylinder	r1= 55mm
Runner	1	trapezoid	a=31mm , b=55mm , h=65mm
	3	trapezoid	a=31mm , b=55mm , h=65mm

(a) Marks of dots and lines to obtain the quantitative information	(b) The quantitative information on the product's design

Fig.6 Modeller to link the database results with casting design modelling of a product

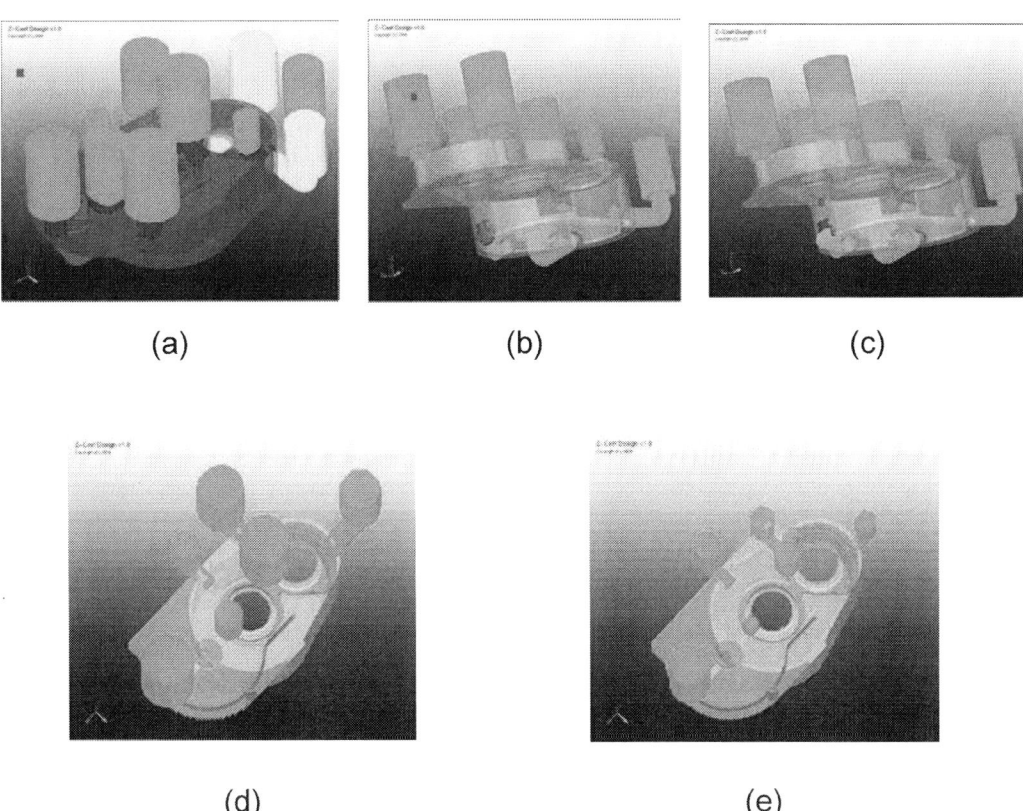

(a) (b) (c)

(d) (e)

Fig.7. The optimization process after setting an riser and chill

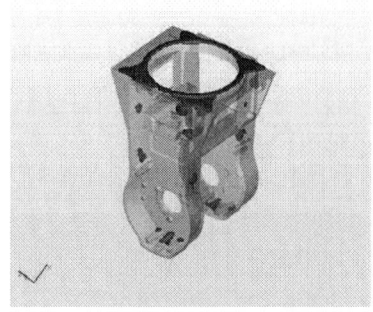 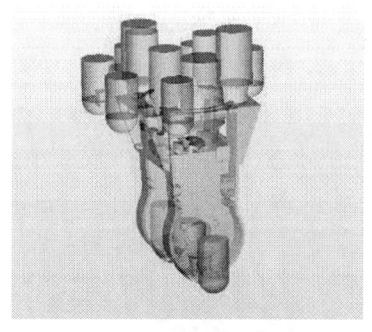

(a) Retrieval of shrinkage cavity result   (b) Automatic riser generation

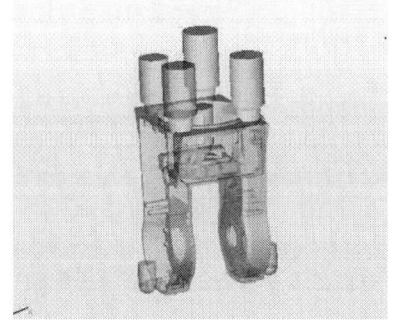 

(c) Optimum riser design   (d) Final casting design

Fig.8 Obtained the optimum riser design by using the IES

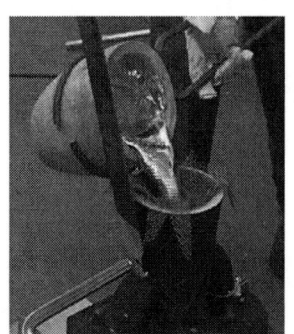  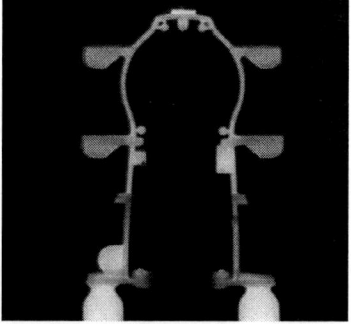

(a) The process to produce a real prototype   (b) Real product   (c) Investigate the defect of product

Fig.9 Make the real prototype and Investigate the product

Research on mould filling process of melt in vertical centrifugal casting

Guo Jingjie, Li Changyun, Wu Shiping, Fu Hengzhi

School of Materials Science and Engineering, Harbin Institute of Technology, Harbin, 150001, P.R.China

Abstract

Mould filling of melt in vertical centrifugal casting is a complex process, which is governed by a number of processing variables. In this paper, a complete analysis of mold filling process of melt in vertical centrifugal casting is accomplished by combining theoretical analysis, numerical simulation and experiments. Results show that the melt fills the cavity with varying cross-sectional area and inclined angle. The filling process is divided into two steps in vertical centrifugal casting: forward filling and back filling. The cross-sectional area decreases and the inclined angle increases with increasing of the rotational velocity during forward filling. Moreover, the filling process of liquid has been simulated based on SOLA-VOF technique. The comparison between the experimental and the numerical results shows good agreement.

Key words Mold filling, vertical centrifugal casting, theoretical analysis, numerical simulation, experimental result

Introduction

High-strength and lightweight titanium alloy castings are widely applied in aviation, astronavigation and civil fields. [1,2] Lighter weight components can improve performance and reduce fuel consumption as well as protect the environment. Titanium alloy is not only structural materials, but also functional materials. [3~5] Now more and more researchers and engineers are studying the titanium alloy, and many researchers have devoted themselves to study its forming characteristics.[6,7] However, few researchers study the melt flow mechanism during titanium alloy vertical centrifugal casting, which is important for analyses of solidification and defects. Titanium alloys are inherently difficult materials to cast, [8~12] mainly high activity in molten state and poor fluidity. Centrifugal casting can improve alloy melt's flow rate, save raw material, reduce production costs and can also increase casting's precision, simultaneously, the fluidity of titanium alloy melt is enhanced with increasing rotational velocity of mold. [13~16] In order to produce high quality, thin-walled castings, it is necessary to study the filling mechanism of titanium alloy and its effects on formation of defects.

In this paper, based on the force analysis of melt mold filling process in vertical centrifugal casting, the Navier-Stokes (N-S) equation and the continuity equation as well as on the SOLA-VOF method, a mathematical model for numerical simulation of melt mold filling in vertical centrifugal casting was developed and the experiments using an adjustable-velocity centrifugal machine, high-speed camera, some liquid, and two kinds of gating systems have been carried out. The comparison between the experimental and the numerical results shows good agreement.

Force analysis

The flowing of liquid metal can be considered as group movement of mass particles. Three kinds of force act on the liquid metal in centrifugal field, including gravity, centrifugal force and Coriolis force. In a rotating system with an angular velocity ϖ and the distance between the moving particle and rotating center is r. The Coriolis acceleration is $2 \times \omega \times v_r$, which only depends on the velocity; and the centrifugal acceleration is $r \times \omega \times \omega$, which only depends on the position and its direction pointing to rotating center; the gravity acceleration is g. It can be seen from Fig.1 that the mold filling process of melt during vertical centrifugal casting is a composite motion process of fluid under three forces. According to force analysis and the principle of force composition, the composite force of liquid particle is $\vec{F}_n$.

For simplification, it is assumed that: ①Melt fills the cavity continuously; ②Melt is in a fully developed steady state flow; ③Dynamical viscosity coefficient μ and density ρ of melt are both constants; ④ Melt is regarded as an incompressible fluid; ⑤Do not consider the friction loss during the filling process. In the cross section, the center of gravity should be the maximum velocity point, which is similar to that at cavity entrance. $m_i(x, y, z)$ is a point on the maximum velocity stream line (contains all cross-sectional centers of gravity) and locates in cross section dx_i. $\vec{F}_{gc}, \vec{F}_{ge}, \vec{P}$ and $\vec{F}_n$ terms express the forces acting on it and following relation should be satisfied: [17]

$$\vec{F}_{gc} + \vec{F}_{ge} + \vec{P} + \vec{F}_n = 0 \qquad (1)$$

Where: $\vec{F}_{gc}$ - Coriolis force; $\vec{F}_{ge}$ - Centrifugal force; $\vec{P}$ - Gravity; $\vec{F}_n$ - resultant force of above three forces acting on m_i, whose direction is the same as that of cross-sectional free surface normal, so $\vec{F}_n \times \vec{n}_s = 0$ ($\vec{n}_s$ is unit normal vector on free surface). Therefore, the expression of $\vec{F}_n$ is obtained:

$$\vec{F}_n = M_i \left(\vec{\omega} \times (\vec{\omega} \times \vec{r}) - 2\vec{\omega} \times \vec{v}_r - \vec{g} \right) \qquad (2)$$

Mathematical model for numerical simulation

Consider the stationary volume element within a fluid moving with a velocity having the components v_x, v_y, v_z and the density of fluid in the engineering is often constant, then the continuity equation reduces to

$$\frac{\partial v_x}{\partial x} + \frac{\partial v_y}{\partial y} + \frac{\partial v_z}{\partial z} = 0 \qquad (3)$$

Momentum conservation is often referred to as the Navier-Stokes's equation. In the form of Equation (4), we can recognize it in the form *mass* (ρ) × *acceleration* ($\frac{Dv}{Dt}$) equals the *sum of forces*, namely, the pressure force ($-\nabla P$), the viscous force ($\eta \nabla^2 v$), and the gravity or body force ρg

$$\rho \frac{Dv}{Dt} = -\nabla P + \eta \nabla^2 v + \rho g \qquad (4)$$

The mold filling process of vertical centrifugal casting is a transient flow process of fluid with free surface. VOF method was used to calculate free surface in this paper. The equation of volume of fluid-function is:

$$\frac{\partial F}{\partial t} + v_x \frac{\partial F}{\partial x} + v_y \frac{\partial F}{\partial y} + v_z \frac{\partial F}{\partial z} = 0 \qquad (5)$$

Where: F is the volume function. For full cell $F = 1$, for empty cell $F = 0$, and for unfilled cell $0 < F < 1$

Experimental process and simulation results

The rotational velocity range of centrifugal machine that used in the experiment is 0-1200 rpm, so different rotational velocities can be selected in the experiment and find which one is the best for filling. The high-speed camera used in the study can take 1000 (320×320) pictures per second, which can meet the need of the experiment. The experimental schematic diagram is shown in Fig.2. Two different filling methods were used in the experiment so as to study the melt's filling state, including top filling method and bottom filling one. Results show that bottom filling is better than top filling, which can achieve stable filling. In order to find the regularity of filling in vertical centrifugal casting, three kinds of rotational velocities were used in bottom filling method. And the rotational direction was anti-clockwise. Fig.3-Fig.4 show states of bottom filling at 163rpm and 245rpm respectively, which indicate that the volume of melt mould filling increases with increasing rotational velocity of mold.

Based on the mathematical model for numerical simulation, 3-D simulation software was developed to simulate the mold filling process in vertical centrifugal casting. The mold applied in simulation is that of the experimental one which can be used to verify the accuracy and reliability of the software. Firstly, using Proe or AutoCAD to build mold and output STL file, then using self-developed software to split the mesh and the mesh spacing is 2mm, the total number of mesh is $170 \times 47 \times 32 = 270720$. The simulation results are shown in Fig.5 and Fig.6.It can be found from the simulation results that the filling volume increases with increasing of filling time and rotational velocities, which match with the experimental results. Fig.7 gives the fact that the cross-sectional area of melt is large at inlet of cavity and decreases with the increase of filling length.

Discussion

Two quantities named cross-sectional area of melt and inclined angle of free melt surface respectively are defined and their variations must be found out first to describe the vertical centrifugal filling process. Actually, the free surface of any cross section whose geometry is dependent on its distance to the rotational axis should be a curve due to the centrifugal filling characteristics in the cavity. Now the free surface is defined as a line and the cross section can be regarded as an arc whose area is equal to that of an actual one and is defined as the cross-sectional area. Fig.7 and Fig.8 show that the whole filling process in centrifugal field can be divided into two steps, one is forward filling and the other is back filling. The experimental and simulation results show that during forward filling, the cross-sectional area is larger at first, and then with the increase of the filling length, the cross-sectional area of melt becomes smaller gradually, and at last the liquid sticks to the back-wall of cavity or runner which is opposite to the rotational direction to finish forward filling process. Consider the inlet of runner is overflow, during the forward filling the cross-sectional area at optional position [17] is

$$s_i(t) = \frac{2v_0 Lh}{(v_0 + \omega L_0)e^{\omega t} + (v_0 - \omega L_0)e^{-\omega t}} \tag{6}$$

Where $s_i(t)$ is cross-sectional area, L is the runner width, h is the runner thickness. Equation (6) shows that the cross-sectional area decreases with increasing filling length, and the decreased value is reduced gradually. It also can be found that with the increase of rotational velocity of mold, the cross-sectional area at the same position reduces correspondingly. The relationship between the cross-sectional area and filling length is shown in Fig.9. It can be seen from curves that the cross-sectional area of melt decreases gradually with increasing filling length and rotational velocity, and the experimental results show good agreement with the calculation results. During the back filling the cross-sectional area is unchangeable, and the back filling velocity is a constant.

The cross-sectional inclined angle $\theta_i(t)$ is defined as the angle between the arc line normal direction and the rotating axis (k axis). The inclined angle $\theta_i(t)$ can be expressed: [17]

$$tg\theta_i(t) = \frac{F_{gc}}{M_i g} = \frac{\omega(v_0 + \omega L_0)}{g}e^{\omega t} + \frac{\omega(v_0 - \omega L_0)}{g}e^{-\omega t} \tag{7}$$

Where: v_0 is initial velocity and L_0 is initial displacement which can be seen from Fig.1. $\vec{F}_{gc}$ produced by the centrifugal field is critically dependent on velocity vector $\vec{v}_r$ and increases with the distance from cross section to the rotating axis that makes the inclined angle increase gradually. $\omega \to \infty$ then $\theta_i(t) \to \pi/2$, which indicates that the cross-sectional free surface tends to be vertical at the cavity end if the rotating velocity is high enough or the cavity length is long enough. The relationship between the inclined angle and filling length is shown in Fig.10. The curves show that the inclined angle increases with increasing filling length and rotational velocity during the forward filling.

Bottom filling is better than top filling because forward filling and back filling were both occurred in the cavity in top filling, which lead to the liquid crash each other and it will cause defects forming during the titanium alloy production. However, the liquid crash were occurred in runner in bottom filling, and the back filling in the cavity is a uniform velocity filling process, which assure the stable filling and reduce defects forming during the production of titanium alloy. The data picked in the bottom gating method are shown in Fig.11. It can be found from curves that filling volume increases with increasing rotational velocity. But the rotational velocity should not too high, and the effect of over-high rotational velocity on filling volume is the further work to do.

Conclusions

Hydraulic simulation experiments were carried out in this paper. Results show that the whole filling process is divided into forward filling and back filling in vertical centrifugal casting. During the forward filling, liquid cross-sectional area decreases with the increase of filling length, and the filling velocity becomes faster and faster, as well as the back filling is a uniform velocity filling process in that the cross-sectional area of free surface is a constant value. Moreover, the filling velocity increases with the increase of rotational speed of mold. In addition, with increasing rotational velocity, the inclined angle increases too. Based on the forces analysis in vertical centrifugal casting, the mathematical model of melt mould filling process in vertical centrifugal casting is established and the simulation results show good agreement with the experiment results.

Acknowledgements

Authors would like to thank Project 973 (No. 51319) for financial support.

References

1. Zhou Yanbang. *Titanium Alloy*. Publishing House of Aviation Industry. 2000.
2. NanHai, Xie Chengmu. *The Foundry Titanium Alloy and Its Application and Development Overseas*. China Foundry Equipment and Technology. 2003(6): 1-3
3. W.Huisman, T.Graule&L.J.Gauckler. *Alumina of High Reliability by Centrifugal casting.*Journal of European Ceramic Society.1995(15):811-821
4. Zou Jianxin. *The Present State, Perspectives and Suggestion of Titanium and Titanium Alloy material at Domestic and Overseas*. Technology of Aerospace Material. 2004(1): 23-25
5. Zhang Yansheng. *The Application of Titanium and Titanium alloy in Automotive Industry*. Progress of Titanium Industry. 2004.21(1): 16-17
6. Pinggen Raoa , Mikio Iwasa, Takahiro Tanaka, Isao Kondoh. *Centrifugal casting of Al2O3 - 15 wt.%ZrO2 ceramic composites*. Ceramics International. 2003(29): 209-212
7. P.E.Jones, W.J.Porter III, and D.Eylon etc. *Development of a Low Cost Permanent Mold Casting Process for TiAl Automotive Valves*. Gamma Titanium Aluminides. 1995:53-62
8. W.R.Chinnis and R.C.Eschenbach. *Plasma Heart Melting of Seeded Titanium*. Proceeding of the Nith International Vacuum Metallurgy Conference, San Diego, CA,1988, April:11~15
9. C.H.Entriken. *The Removed of High Density Inclusions by Hearth Refining Process*. in Electron Beam Melting and Refining-State of the Art 1985, Ed. by R.Bakish:40~54
10. Sheng Wenbin, Guo Jingjie, Su Yanqing etc, *Influences of Pouring*

Power （ISM）on the Quality of TiAl-Based Alloys Casting Surface. Special Casting & Nonferrous alloys. 1998(6):4~7

11. Guo jingjie, Sheng Wenbin，Su Yanqing etc. *Analysis of Overflow Critical Value for TiAl Based Alloy During the Process of Centrifugal Casting.* Trans. Nonferrous Met. Soc. China. 1999(6) Vol.9 No.2:207~212

12. C.M.Austin and T.J.Kelly. *Progress in Implementation of Cast Gamma Titanium Aluminde.* Gamma Titanium Aluminides.1995:21-32

13. K. Watanabe, O. Miyakawa, Y. Takada, O. Okuno, T. Okabe. *Casting Behavior of Titanium Alloys in a Centrifugal Casting Machine.* Biomaterials. 2003(24): 1737-1743

14. A. Halvaee, A. Talebi. *Effect of Process Variables on Microstructure and Degregation in Centrifugal Casting of C92200 Alloy.* Journal of Materials Processing Technology. 2001(117): 123–127

15. Li Xinian. *The Technology and Application of Special Shaped Casting in Vertical Centrifugal Casing.* Special Casting and Nonferrous Alloys. 2000(5): 30-32

16. Li Baoquan. *Research on "Free Flow Direction" of Metal Fluid in High-Frequency Centrifugal Casting.* Journal of Bai Qiuen Medical University. 1996.22(2): 142-153

17. Sheng Wenbin. *Study on Centrifugal Casting Process of Permanent Mold for TiAl Based Alloy Exhaust Valves.* Doctoral Dissertation .2000, 33-64.

Figures

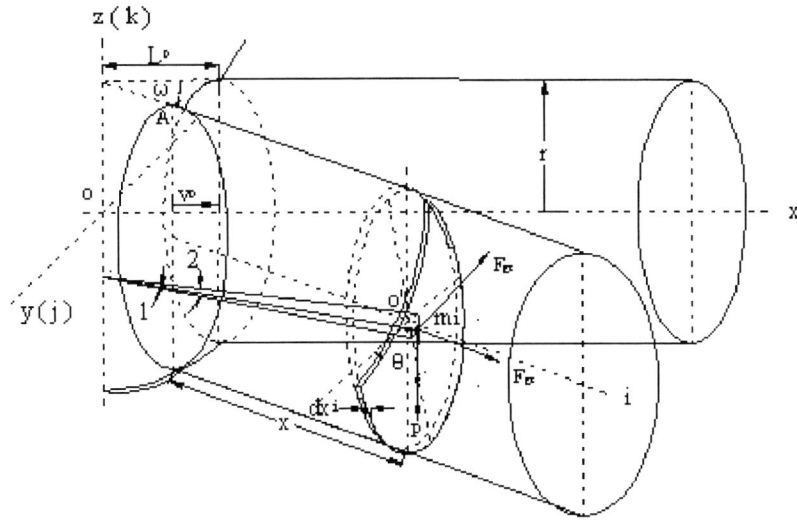

Fig.1 Schematic diagram for flow analysis

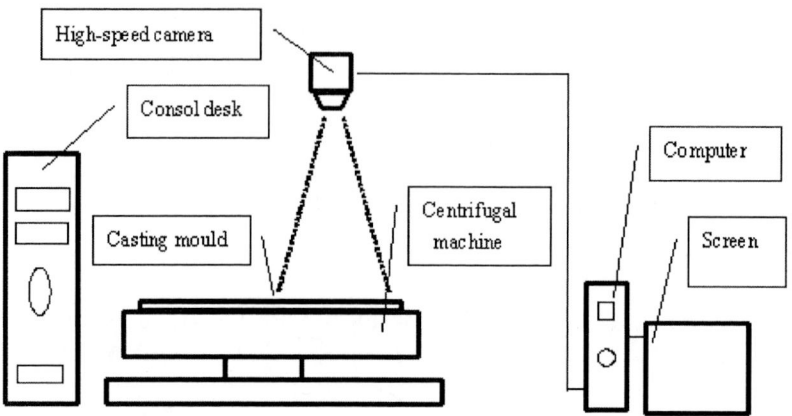

Fig .2 Schematic diagram of hydraulic simulation in centrifugal casting

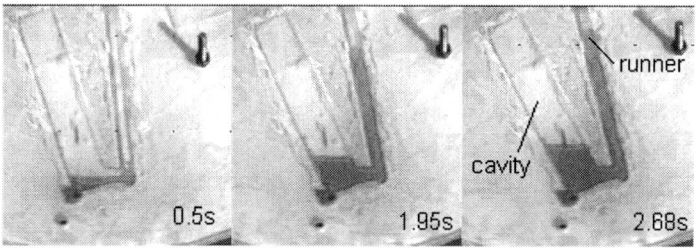

Fig.3 Photographs of mould filling of liquid at different time at 163rpm

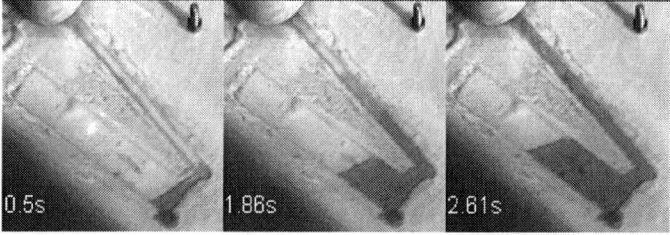

Fig.4 Photographs of mould filling of liquid at different time at 245rpm

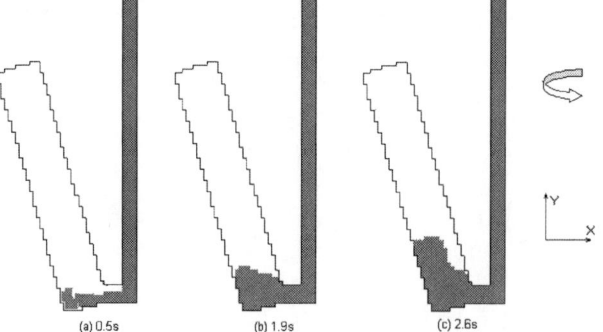

Fig.5 Simulation results of liquid filling at the x-y planes for different times at 163rpm

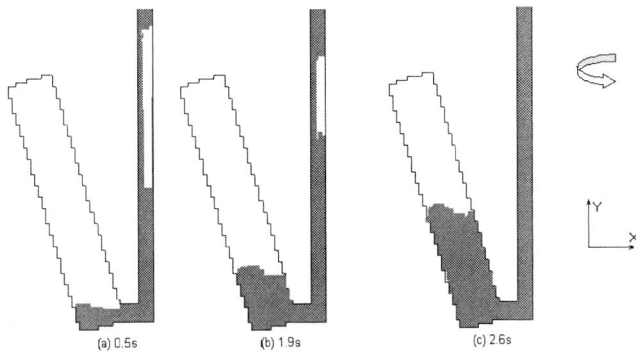

Fig.6 Simulation results of liquid filling at the x-y planes for different times at 245rpm

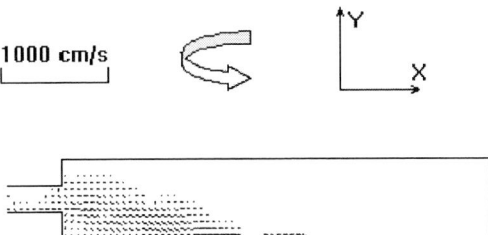

Fig.7 Changeable melts cross-sectional area in the cavity (w=163rpm)

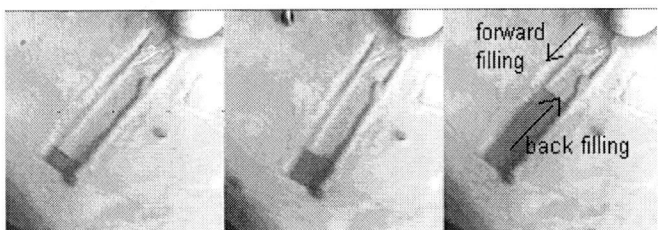

Fig.8 Forward filling and back filling in vertical centrifugal casting (rotational direction is anti-clockwise)

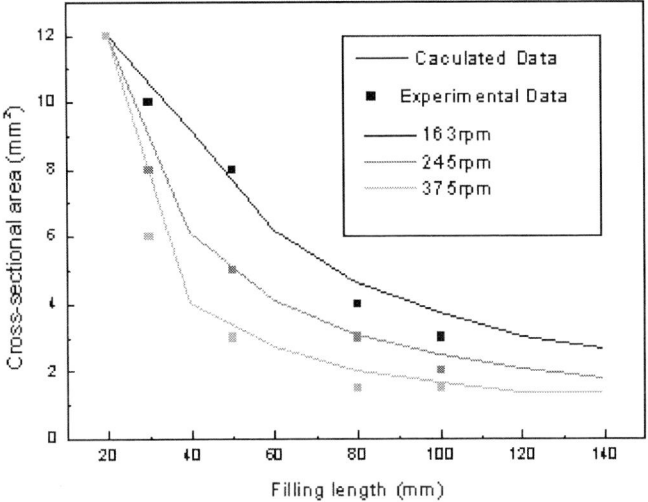

Fig.9 The variation of cross-sectional area with filling length

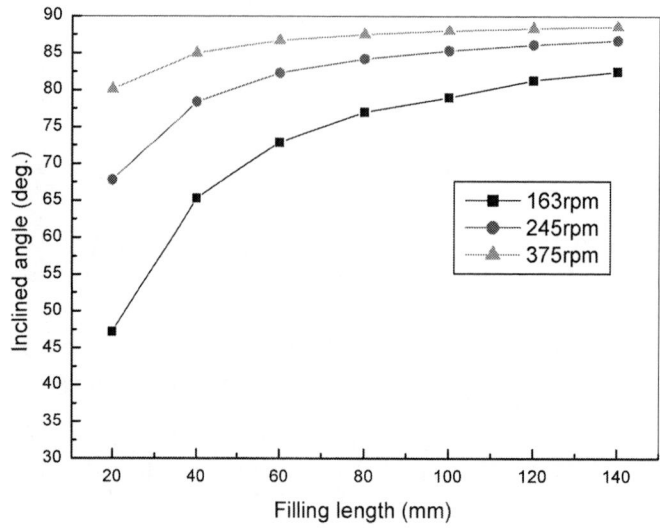

Fig.10 The variation of inclined angle with filling length

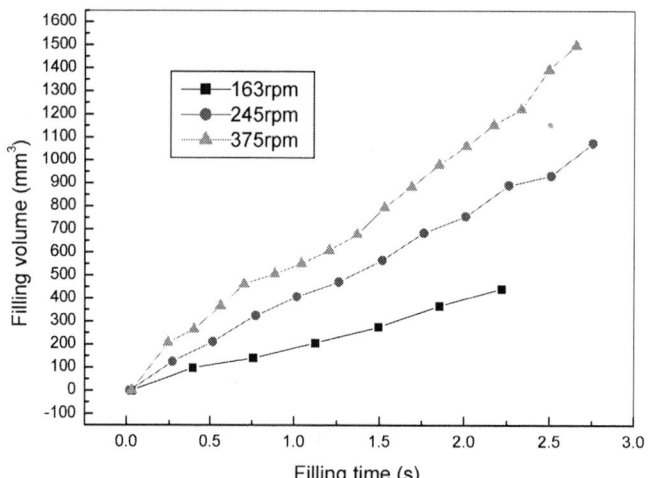

Fig.11 Filling volume *vs* filling time in bottom gating

Nonlinear Modeling with Hydrodynamics and Flow Control Using Inverse Pouring Dynamics of Tilting-Ladle-Type Automatic Pouring Process

Yoshiyuki Noda*, and Kazuhiko Terashima*,
* Deptertment of Production Systems Engineering,
Toyohashi University of Technology

Abstract

This paper is concerned with the pouring control of tilting-ladle-type automatic pouring systems in casting industries. A new approach to the modeling and control system of the pouring process is proposed. In the approach, the flow rate model is described by a nonlinear system with hydrodynamics. The flow rate control system is built by a feedfoward control using the inverse dynamics of the flow rate model. In this control system, control parameters are obtained from the shape of the ladle. Therefore, the present approach does not reqpuire many experiments to obtain the control parameters. In addition, this system enables application for ladle with a complex shape. The proposed method is applied to a tilting-type automatic pouring system with a cylindrical ladle, and the effectiveness of the present system is then demonstrated through experiments.

Key words: Automatic Pouring System, Batch Type Pouring Process, Flow Rate Control, Modeling

Introduction

Batch-type automatic pouring systems have been used to produce a disk drum, a cylinder block of engine, and other things. There are various pouring methods in the batch-type automatic pouring system, such as the stopper type, the pressure type, and the tilting ladle type [1]-[3]. The advantages of the tilting ladle type pouring system is that the cost of building the pouring system is low, and it is easy to alternate the present molten metal in the ladle to the different molten metal with another property [1]. However, pouring control of this system is difficult, since the molten metal has to be accurately poured into a sprue cup in a mould.

Some studies on the pouring control of the tilting-ladle-type automatic pouring system have been published. An automatic pouring system called the Mel-pore system was developed [4]. In this system, the pouring pattern was constructed by a teaching play-back method. In another study [5], in order to make the initial flow rate of outflow from a ladle, a device that lets a tilting center move up and down was fitted to the pouring mouth of the ladle. A method of achieving liquid level control in the sprue cup has been developed by authors [6][7]. However, the above pouring control systems have been constructed for a pouring system with a fan-type ladle. it is easy to achieve pouring control using a pouring system with a fan-type ladle, because the relation between the tilting angle of the ladle and the outflow volume from the ladle is linear. However, because the fan-type ladle has large area for the same volume, a larger machine must be built and the temperature of the molten metal quickly decreases. Therefore, a pouring system with cylindrical ladle such as a simple shape has been developed. However, it is difficult to achieve the pouring control, because the relation between the tilting angle of the ladle and outflow volume is non-linear. Some researches on a pouring system with a cylindrical ladle have been published. Fuzzy control is applied to the pouring control of a cylindrical ladle [8]. However, numerous experiments are needed to establish fuzzy rules. In the literature [9], a flow rate model of the pouring process has been constructed using a LPV (Linear Parameter Varying) model, and a feedforward control by using an inverse dynamics of the LPV model has been applied to the pouring system with a cylindrical ladle. In this study, numerous experiments for model identification were required, and since the LPV model for the flow rate is an approximated model, the system cannot be controlled precisely. Therefore, it is necessary to construct the control system by performing several experiments, and to precisely control the flow rate of the outflow from the ladle.

In this paper, nonlinear modeling with hydrodynamics to the outflow rate in the pouring process is proposed, and then a flow rate feedforward control system using inverse dynamics of the nonlinear flow rate model is constructed. The nonlinear model for the flow rate is derived from the general shape of a ladle. Therefore, several model parameters are calculated from the shape of the ladle, and only once experiment for identification of the flow rate coefficient is performed. Once experiment is safe for workers, and it saves energy for the pouring process treating molten metal. In addition, it is possible to apply a pouring system using ladles of different shapes. Furthermore, it is not necessary to use many sensors in the proposed feedforward control system.

To verify the effects on the proposed flow rate control system, the control system is applied to a tilting-type automatic pouring system with a cylindrical ladle, and several experiments are performed.

Tilting-Ladle-Type Automatic Pouring System

The tilting-ladle-type automatic pouring system in this paper is shown in Fig.1. The automatic pouring system can transfer the ladle on two dimensions (Y and Z-axes) and rotate the ladle (Θ-axis). The rotary direction of the Θ-axis is driven by an AC servomotor. The drive force of the AC servomotor can be amplified by reducing the gear ratio. The center of the ladle's rotation shaft is placed near the ladle's center of gravity. When the ladle is rotated around the center of gravity, the tip of the pouring mouth moves in a circular trajectory. It is then difficult to pour the molten metal into the mould, as the pouring mouth is moved by tilting. Then, the position of the tip of the pouring mouth is controlled invariable during pouring by means of a synchronous control of the Y- and Z- axes for rotational motion around the Θ-axis of a ladle [9]. The rotation angle is measured by an encoder installed in the AC servomotor. The Y- and Z-axes in the automatic pouring system are also driven by AC servomotors. However, the driving force of each of these motors is amplified through the ball and screw mechanism. Each axis can be independently moved.

The object liquid in the present experiments is water. The kinematical viscosity of the water (293[K]) and molten metal (1673[K]) are 1.004×10^{-6}[m^2/s] and 0.970×10^{-6}[m^2/s], respectively. Therefore, the fluid behavior of the water is nearly equal to that of the molten metal. In order to measure the weight of the liquid in the ladle, a load cell is fitted to the arm of the automatic pouring system. The weight of liquid in the ladle is obtained by taking off the weight of the ladle, motor, and arm from the whole weight measured by the load cell.

The block diagram of the pouring process in the automatic pouring system is shown in Fig. 2. In Fig. 2. P_m is the motor for the tilting ladle as follows.

$$\frac{d\omega(t)}{dt} = -\frac{1}{T_m}\omega(t) + \frac{K_m}{T_m}u(t) \tag{1}$$

, where $\omega(t)$[deg/s] is the angular velocity for the tilting ladle and $u(t)$[V] is the input voltage applied to the motor. T_m[s] is the time constant, and K_m[deg/sV] is the gain. In the experimental apparatus in this paper, the time constant T_m is 0.0006[s] and the gain K_m is 24.58[deg/sV]. P_f is the flow rate model in the pouring process. In the model P_f, the flow rate $q(t)$[m^3/s] is obtained from the tilting angular velocity $\omega(t)$[deg/s] and tilting angle $\theta(t)$[deg]. The flow rate model is discribed in more detail in the next section. In order to show the weight of the outflow liquid from the ladle, the flow rate $q(t)[m^3/s]$ is integrated and then increased $K=1.0\text{X}10^3[Kg/m^3]$ times as the water's density. P_L is the dynamics of the load cell fitted to the arm of the pouring system as follows.

$$\frac{dw_L}{dt} = -\frac{1}{T_L}w_L(t) + \frac{1}{T_L}w(t) \tag{2}$$

, where $w_L(t)[Kg]$ is the weight of the outflow liquid measured by the load cell, and $T_L[s]$ is the time constant as the response of the load cell. In this paper, T_L is 0.10[s].

Flow Rate Model in Pouring Processes

A cross-section diagram of the pouring process is shown in Fig. 3. In Fig. 3, θ[deg] is the tilting angle of the ladle. $V_s(\theta)[m^3]$ is the volume under the pouring mouth, and $A(\theta)[m^2]$ is the horizontal area to the pouring mouth. The volume $V_s(\theta)[m^3]$ and the area $A(\theta)[m^2]$ are changed depending on the tilting angle θ [deg]. $V_r[m^3]$ is the volume over the area $A(\theta)[m^2]$ and h[m] is the height from the area $A(\theta)[m^2]$ to the surface of the liquid in the ladle. When there is liquid over the pouring mouth, the liquid outflows from the ladle. $q[m^3/s]$ is the flow rate of the liquid poured from the ladle.

In Fig. 3, a mass balance of liquid in the ladle from time t[s] to $t+\Delta t$[s] is shown in Eq. (3).

$$V_r(t) + V_s(\theta(t)) = V_r(t + \Delta t) + V_s(\theta(t + \Delta t)) + q(t)\Delta t \tag{3}$$

Therefore, on $\Delta t \to 0$, it follows that

$$\lim_{\Delta t \to 0} \frac{V_r(t + \Delta t) - V_r(t)}{\Delta t} = \frac{dV_r(t)}{dt} = -q(t) - \frac{dV_s(\theta(t))}{dt} = -q(t) - \frac{\partial V_s(\theta(t))}{\partial \theta(t)}\frac{d\theta(t)}{dt} \tag{4}$$

The tilting angular velocity $\omega(t)$[deg/s] of the ladle is shown in Eq. (5).

$$\omega(t) = \frac{d\theta(t)}{dt} \tag{5}$$

By substituting Eq. (5) into Eq. (4), the following equation is obtained.

$$\frac{dV_r(t)}{dt} = -q(t) - \frac{\partial V_s(\theta(t))}{\partial \theta(t)}\omega(t) \tag{6}$$

Moreover, the volume $V_r[m^3]$ can also be shown as Eq. (7).

$$V_r(t) = \int_0^{h(t)} A_s(\theta(t), h_s)dh_s \tag{7}$$

The area $A_s[m^2]$ is the horizontal area of liquid on the height h_s[m] from the area $A[m^2]$ in Fig. 4. The area $A_s[m^2]$ is separeted into the area $A[m^2]$ and the changing area $\Delta A_s[m^2]$ from the area $A[m^2]$ as shown in Eq. (8).

$$V_r(t) = \int_0^{h(t)} (A(\theta(t)) + \Delta A_s(\theta(t), h_s))dh_s = A(\theta(t))h(t) + \int_0^{h(t)} \Delta A_s(\theta(t), h_s)dh_s \tag{8}$$

In general ladles such as a fan type, the cylindrical ladle, and other similar types, the changing area $\Delta A_s[m^2]$ is extremely small compered with the area $A[m^2]$. Therefore, it follows that

$$A(\theta(t))h(t) \gg \int_0^{h(t)} \Delta A_s(\theta(t), h_s) dh_s \qquad (9)$$

Then, Eq. (8) can be shown as follows.

$$V_r(t) \approx A(\theta(t))h(t) \qquad (10)$$

Therefore, the height h[m] is shown as Eq. (11).

$$h(t) \approx \frac{V_r(t)}{A(\theta(t))} \qquad (11)$$

By using Bernoulli's theorem, the flow rate q[m^3/s] at the height h[m] is shown as Eq. (12).

$$q(t) = c \int_0^{h(t)} (L_f \sqrt{2gh_b}) dh_b = \frac{2}{3} cL_f \sqrt{2g} h(t)^{3/2}, \qquad (0 < c < 1) \qquad (12)$$

In Eq. (12), h_b[m] is the depth from the surface of the liquid in the ladle and L_f[m] is the width of the pouring mouth. In addition, c is the flow rate coefficient and g[m/s^2] is the gravity acceleration.

From Eqs. (6), (11), and (12), the flow rate model for the pouring process can be derived as follows.

$$\frac{dV_r(t)}{dt} = -\frac{2cL_f \sqrt{2g}}{3A(\theta(t))^{3/2}} V_r(t)^{3/2} - \frac{\partial V_s(\theta(t))}{\partial \theta(t)} \omega(t) \qquad (13)$$

$$q(t) = \frac{2cL_f \sqrt{2g}}{3A(\theta(t))^{3/2}} V_r(t)^{3/2}, \qquad (0 < c < 1) \qquad (14)$$

In Eq. (13) and (14), the parameters $V_s(\theta)$[m^3], $A(\theta)$[m^2], and L_f[m] are obtained from the shape of the ladle. Therefore, only the flow rate coefficient c is an unknown parameter and can be identified by doing the experiment only one time.

Identification of Flow Rate Model in Cylindrical Ladle

In order to identify the parameters of the cylindrical ladle as shown in Fig. 5, the volume $V_s(\theta)$[m^3], the area $A(\theta)$[m^2], and the width L_f[m] of the pouring mouth can be calculated. The volume $V_s(\theta)$[m^3] and the area $A(\theta)$[m^2] at the tilting angle θ[deg] obtained from the shape of the ladle are shown in Fig. 6. In Figs. 6(a) and (b), $A(\theta)$[m^2] shows the horizontal area in the pouring mouth and $V_s(\theta)$[m^3] the volume under the area $A(\theta)$[m^2], respectively.

The flow rate coefficient c can be identified once an experiment is performed. In the experimental condition, the liquid is poured at a constant angular velocity $\omega(t)$=1.75[deg/s] and an initial tilting angle of θ=39.0[deg]. The experimental results are shown in Fig. 7. In Fig. 7, the tilting angular velocity is denoted by $\omega(t)$[deg/s], the tilting angle by θ[deg], and the flow rate by q[m^3/s], which is solved by simulation using Eqs. (13) and (14). (d) is the simulation result and experimental result with respect to the weight of the outflow liquid. In (d), the dotted line is the simulation result and the solid line the experimental result. By fitting the simulation result to the experimental result, the flow rate coefficient could be obtained as c=0.70.

Feedforward Control by using the Inverse Dynamics of the Pouring Process

By using the flow rate model and the motor model, the feedforward control system for the pouring process is constructed. The block diagram of the feedforward control is shown in Fig. 8. The control system works for generating the input voltage $u(t)$[V] into the motor to tilt the ladle, where the actual flow rate $q(t)$[m³/s] realizes the desired flow rate $q_{ref}(t)$[m³/s]. In Fig. 8, the inverse dynamics P_m^{-1} of the motor is shown as Eq. (15).

$$u(t) = \frac{T_m}{K_m}\frac{d\omega_{ref}(t)}{dt} + \frac{1}{K_m}\omega_{ref}(t) \tag{15}$$

The inverse dynamics P_f^{-1} of the flow rate model in the pouring process is derived by Eq. (13) and (14). The volume $V_{rref}(t)$[m³] to generate the desired flow rate $q_{ref}(t)$[m³/s] is obtained from Eq. (14). The volume V_{rref}[m³] is derived by Eq. (16).

$$V_{rref}(t) = \frac{A(\theta(t))}{(\frac{3}{2}cL_f\sqrt{2g})^{2/3}}q_{ref}(t)^{2/3} \tag{16}$$

By using the desired flow rate $q_{ref}(t)$[m³/s] and the volume $V_{rref}(t)$[m³], the angular velocity $\omega_{ref}(t)$[deg/s] is obtained from Eq. (13), and is shown in Eq. (17).

$$\omega_{ref}(t) = -\frac{\dfrac{dV_{ref}(t)}{dt} + q_{ref}(t)}{\dfrac{\partial V_s(\theta(t))}{\partial\theta(t)}} \tag{17}$$

Experimental Results of Flow Rate Feedforward Control

In order to verify the effectiveness of the control system proposed in the previous section, the present flow rate feedforward control system is applied to the automatic pouring system, as shown in Fig. 1. At first, the desired flow rate $q_{ref}(t)$[m³/s] is determined as follows.

$$q_{ref}(t) = \begin{cases} \dfrac{Q_r}{2}(1-\cos(\dfrac{\pi t}{T_r})) & (0 \le t < T_r) \\ Q_{st} + \dfrac{Q_r - Q_{st}}{2}(1+\cos(\dfrac{\pi t}{T_{st}-T_r})) & (T_r \le t < T_{st}) \\ Q_{st} & (t \ge T_{st}) \end{cases} \tag{18}$$

, where, T_r[s] is the time of the rising flow rate and Q_r[m³/s] is the flow rate at time T_r[s]. T_{st}[s] is the time of the stable flow rate and Q_{st}[m³/s] is the stable flow rate. This desired flow rate is obtained from [6]. In this paper, T_r=2[s], Q_r=3.5[m3/s], T_{st} =3[s], Q_{st} =1.5x10^{-4}[m^3/s].

As the comparison, the feedforward control system of the flow rate by using the LPV model[9] is applied to the automatic pouring system. The model parameters in the LPV model are the same as those in the case of the literature[9].

The experimental results are shown in Fig. 9. The initial tilting angle of the cylindrical ladle as shown in Fig. 5 is 39.0[deg]. In Fig. 9, the input voltage is described by $u(t)$[V] applied to the motor to tilt the ladle, the angular velocity to tilt the ladle by $\omega(t)$[deg/s], the angle to tilt the ladle by $\theta(t)$[deg], and the weight of the liquid poured from the ladle by $w_L(t)$[Kg], which is measured by the load cell. The chain lines in Fig. 9 are the experimental results of the control system using the previous LPV model, and the solid lines are the experimental results of the proposed control system. The dotted line in Fig. 9(d) is the desired weight which is derived from the desired flow rate $q_{ref}(t)$[m^3/s] in Eq. (18) through the integrator and dynamics of the load cell in Eq. (2). In the experimental results, the weight of the outflow liquid from the ladle obtained by the proposed control system tracks precisely to the desired weight, while LPV does not agree with the reference.

Conclusions

In this paper, the nonlinear model for the flow rate in the tilting-ladle-type automatic pouring system was built and the feedforward control system for the flow rate proposed. It was shown in the experiments that the flow rate of the proposed control system tracks precisely to the desired flow rate.

The proposed control system has some advantages, in that it is possible to both apply the general shape of the ladle and build the control system by doing an experiment only one time for identification of the control parameter. Based on these results, it was shown that this control system could be useful in industrial applications.

References

1.   W. Lindsay: Automatic Pouring and Metal Distribution Systems, Foundry Trade Journal, February, 1983, pp.151 - 165.
2.   Elmar Neumann and Dietmar Trauzeddel: Pouring Systems for Ferrous Applications, Foundry Trade Jounal, July 2002, pp. 23-24.
3.   Yury S. Lenrner and Cedar Falls: Ironing Out the Pouring Options, Modern Casting, Nobember, 2003, pp.44 - 47.
4.   Jiro Watanabe and Kenichi Yoshida: Automatic Pouring Equipment for Casting "Mel Pore System", Industrial Heating (in Japanese), Vol. 29, No. 4, 1992, pp. 19 - 27.
5.   Masao Matsuda and Sadatoshi Koroyashu: Approach for an Increase in Flow Rate at Start of Pouring from Auto Pouring Equipment by 2-Stage Up-Down Mechanism of Tilt-Center,

Journal of Japan Foundry Engineering Society, Vol. 71, No.7, 1999, pp.443 - 448.

6. Yu Sugimoto, Ken'ichi Yano, and Kazuhiko Terashima: Liquid Level Control of Automatic Pouring Robot by Two-Degrees-of-Freedom Cotrol, Proceedings of the 15th IFAC World Congress, 2002-7.

7. Kazuhiko Terashima and Ken'ichi Yano: Supervisory Control of Automatic Pouring Processes, Presented at 66th World Foundry Congress, 2004-9.

8. Kazuhiro Shinohara and Hiroyuki Morimoto: Development of Automatic Pouring Equipment, Journal of the Society of Automotive Engineerings of Japan, Vol. 46, No. 11, 1992, pp.79 - 85.

9. Yoshiyuki Noda, Ken'ichi Yano, and Kazuhiko Terashima: Control of Self-Transfer-Type Automatic Pouring Robot with Cylindrical Ladle, Preprints of the 16th IFAC World Congress, 2005-7.

10. Kazuhiko Terashima, Ken'ichi Yano, Yu Sugimoto, and Mitsuaki Watanabe: Position Control of Ladle Tip and Sloshing Suppression During Tilting Motion in Automatic Pouring Machine, Preprints of 10th IFAC Symposium on Automation in Mining, Mineral, and Metal Processing, 2001, pp.182 - 187

Figures

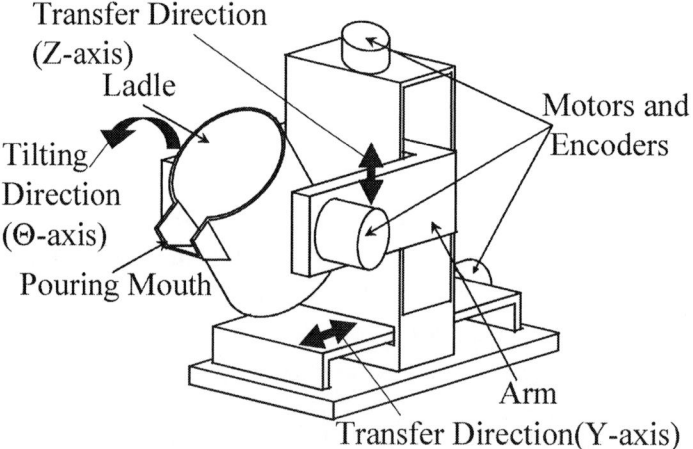

Fig. 1 Automatic Pouring System

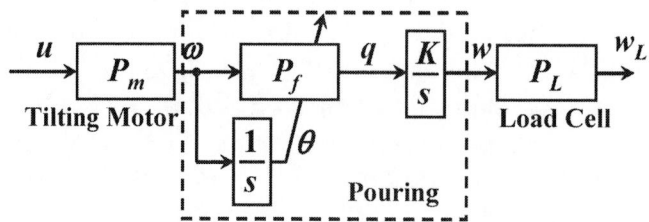

Fig. 2 Block Diagram of Pouring Process

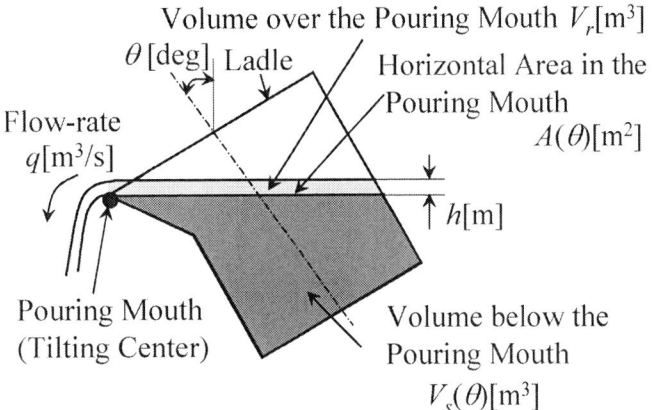

Fig. 3 Cross-Section Diagram of a Ladle in the Pouring Processes

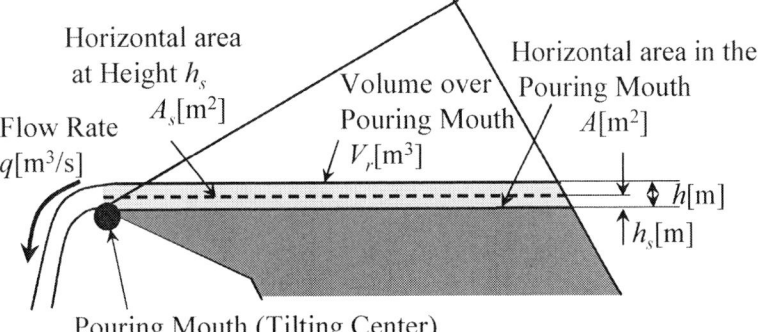

Fig. 4 Detail of Liquid over Pouring Mouth

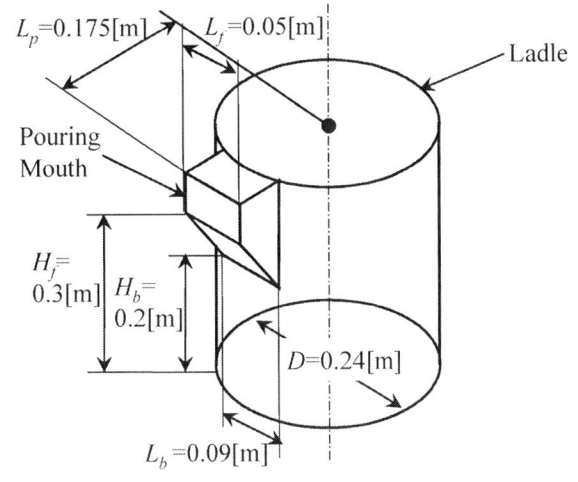

Fig. 5 Cylindrical Ladle

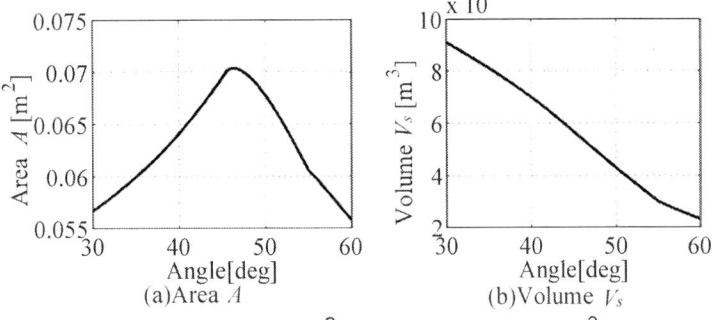

Fig. 6 Area A[m^2] and Volume V_s[m^3]

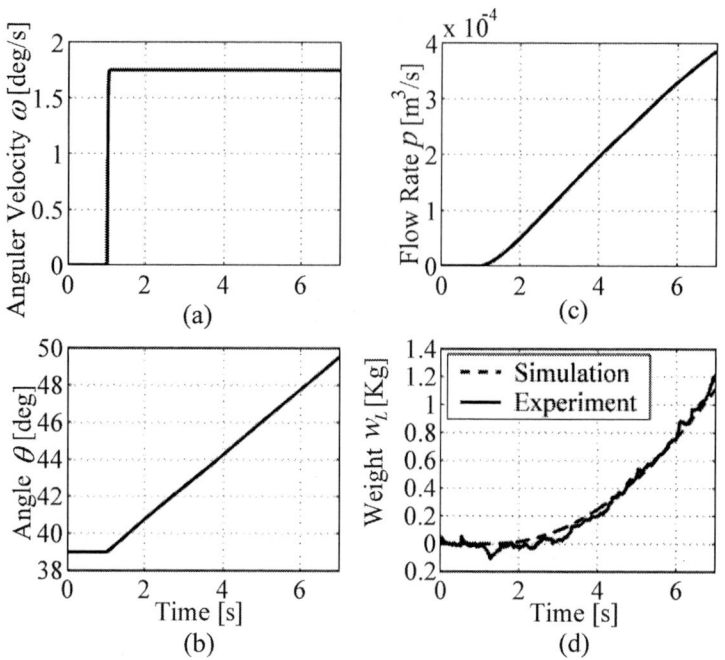

Fig. 7 Simulations and Experimental Results for Identification
(In the Case of Initial Angle 39[deg])

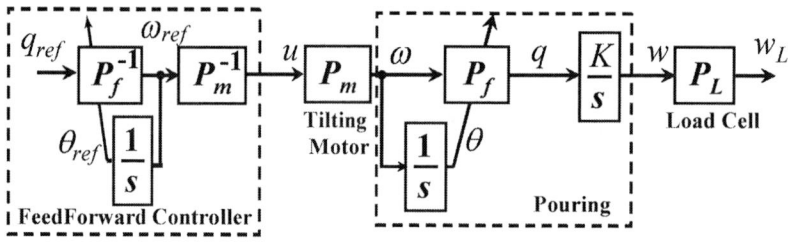

Fig. 8 Feedforward Control for Flow-Rate

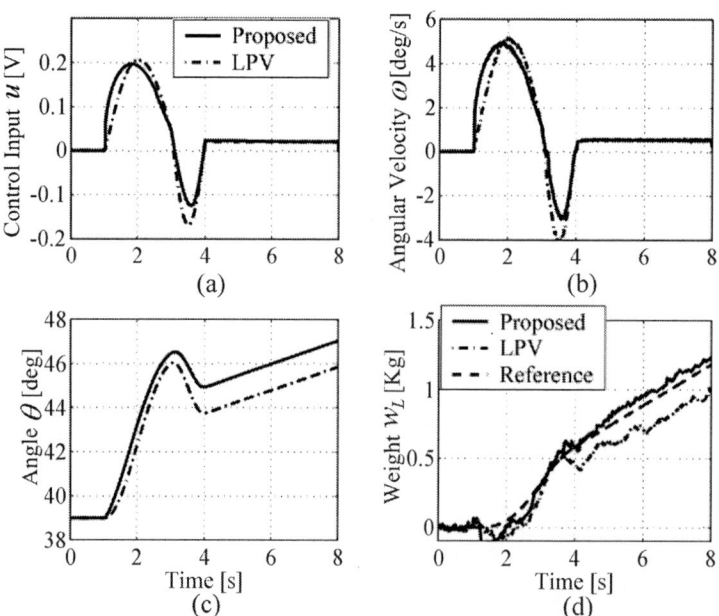

Fig. 9 Experimental Resuts of Flow Rate Control
(In the Case of Initial Angle 39.0[deg])

Three-dimensional modelling and simulation of die-casting processes for AlSi alloys

H Y Hwang*, J-K Choi*,
E I Marukovich**, A M Branovitsky** and I L Zakharov**.

* Korean Institute of Industrial Technologies (KITECH), Republic of Korea,
** Institute of Technology of Metals of National Academy of Sciences of Belarus (ITM NAS of Belarus), Belarus.

Abstract

We propose a novel method for simulation die-casting process of AlSi alloys with preset microstructure. This approach based on two separated steps, including microstructure modelling required for production alloys with possibility of quenching heredity solidification, then hydrodynamic simulation of mold filling with parameters of modelled alloy. Developed method can be used for the process optimization of die-casting. Advantage of proposed procedure is the possibility to predict as microstructure as porosities formation in billet. Water analogue modelling was used for software identification in simulation of die-casting process. Three-dimensional visualization of water analogue modelling results of die-casting shows good agreement with simulations. Developed method can be used for die-casting industrial process optimization.

Key words Die-casting, Water Analogue Modelling, Simulation.

Introduction

Properties of aluminum alloys billet essentially depends on the following factors: 1 - porosity, 2 - presence of impurity, 3 - microstructure of the received material. Therefore, each of factors demands the careful control over casting process. The given work is devoted to the simulation and analysis of two factors - a microstructure and porosity formation. The microstructure is one of the key factors determining mechanical properties. For designing process, the minimization of defects formation and automation production, the mold filling control is needed. For physical simulation of fluid flow in casting processes water analogues are widely used. The water analogue gives satisfactory results for the cases of isothermal mold filling in die casting process [1]. Water modelling of molten aluminum flow is correct for large-scale flow features such as surface waves and when changes of parameters of liquid are not significant in the case of real molten aluminum. The process of modelling should be strongly three-dimensional as the three-dimensional casting mold leads to three-dimensional (3-D) behavior of liquid [2]. Image processing techniques were applied for 3-D reconstruction from several 2-D images of water flow in the tasks of metal casting modelling. The non-linear recursive method which takes into account *a priori* information was applied for 3-D reconstruction [3]. Experiments have shown a possibility of reconstruction of surface of water and heterogeneities in it [3]. Analogue modelling can be successfully used in the next tasks:

- to optimize gate and runner geometry for flow separation in the problem of porosity and cavities formation;
- in designing process for die-casting to control the undesirable phenomena over filling as splashing;
- for validation of numerical models for die-casting, the results can be used for industrial optimization or mold design;
- for comparison of various models on modelling by water.

Our procedure of simulation the die-casting process includes two main steps. Given work is organized as follows: at first we consider microstructure modelling and after short describe numerical simulation of die-casting. Discussion after experiments with water modelling makes it possible to validate simulation results. In Conclusions we summarize our results and indicate perspectives.

Modelling of microstructure of AlSi alloy

There are many of special ways of using modifiers for improving a microstructure of aluminum alloys. It is known, that magnification of a cooling rate results in occurrence of finer microstructure of silumins [4, 5], including eutectic growth. As a rule, sizes of structure of eutecticum can be characterized by inter-lamellar distance. We used method of producing of finer structure of casting by increasing of speed of cooling [4]. Thus, there are greater numbers of nuclei of crystals and their growth rate increases. Simultaneously the increasing speed of cooling results in reduction of the sizes of zone in which there is a crystallization, that complicates diffusion of components, and provides reduction of the sizes of crystals [5]. Such reduction is observed both for dendritic crystal formations, and for eutectic. In many aluminum alloys with near eutectic composition of silicon breaks occur in a view of break eutectic silicon plates [6].

As it has been shown in [4] this alloy has possibility of heredity for quenching solidification method. To predict microstructure we model of crystal growth of AlSi (about 12% of Si). Samples of a microstructure have been obtained by grinding fragments of bar (Fig.1). Grinding was carried out with the help of an emery paper and then with the help of diamond paste. Samples are recorded by CCD, maintained on microscope with magnification in 300 times. The area of plates of primary silicium in investigated structure (dark continuous sites) is much less than area of eutectic formations of silicium. On this base it is possible to consider, that practically all silicium is crystallized in eutectic form [7, 8].

For modelling process of producing silumin with required structure as in Fig. 1, we used software (developed in ITM NAS of Belarus), which makes it possible to calculate amount specified crystals. In accordance to model of Johnson-Mehl, the speed of crystals growth W depends on current T and eutectic T_e temperatures as

$$W = b\,[\,T_e - T\,]^2 \tag{1}$$

with coefficient b for certain alloy. At the same time, for diminution of cooling rate not only crystals of silicium tend to grouping, but also sizes of the continuous sites occupied in aluminum, as dendritic. It enables generalizations of concept inter-lamellar distances in our case, considering instead of inter-lamellar the average distance describing size of aluminum zones (white zones in figures). Such introduction of a quantitative estimation of geometry of a microstructure can be used for describing of our registered images (Fig.1).

Simulation of die-casting process
For numerical simulation of water filling in mold, we used software "Z-Cast", version 2.5. Parameters of water and mold have been taken identical to real die-casting process. Viscosity of water can be changed by pouring glycerin in required proportion. For melt of AlSi12, we have taken viscosity which can be received by solution about 8% of glycerin.

A numerical model based on the SOLA-VOF technique was adopted to simulate mold filling in casting process. Continuity equation for fluid flow is as follows:

$$\nabla \cdot V = 0\,, \tag{2}$$

where V is velocity vector. Navier-Stokes equation is given by

$$\frac{\partial V}{\partial t} + (V \cdot \nabla)V = G - \frac{1}{\rho} \cdot gradP + \nabla^2 \cdot V, \tag{3}$$

where G, ρ and P are vector of gravitational acceleration, density of molten metal and pressure. Energy Equation is

$$\rho c \frac{\partial T}{\partial t} - k\nabla^2 T + \rho c V \cdot \rho L \frac{\partial f_s}{\partial t} = 0, \tag{4}$$

where T, c and L are temperature, specific heat of fluid and latent heat. Fraction of solid f_s and v is kinematics viscosity and k thermal conductivity. Volume of fluid (VOF) equation is written by

$$\frac{\partial F}{\partial t} + \nabla \cdot VF = 0, \tag{5}$$

where t and F are time and fraction of fluid volume. The VOF method use the fractional volume function $F(x, y, z, t)$ for tracking free boundaries. When averaged over the cells of a computational mesh, the average value of F in a cell is equal to the fractional volume of the cell occupied by fluid. A unit value of F corresponds to a cell full of fluid, whereas a zero value indicates the cell contains no fluid. The cells with F value between zero and one contain a free surface.

$$F = \frac{Volume\,of\,fluid\,in\,a\,cell}{Volume\,of\,a\,cell} \tag{6}$$

The VOF method considers the amount of F to be fluxed through the face of a cell during a time step of duration t. The amount of F in one time step is δF times the face cross section area, where

$$\delta F = MIN\left[F_{AD}|V_x| + CF, F_D \delta x_D\right] \tag{7}$$

and

$$CF = MAX\left[(1.0 - F_{AD})|V_x| - (1.0 - F_D)\delta x_D, 0.0\right] \tag{8}$$

The choice of the time increment necessary for stability is governed by several restrictions. The calculation is converged when the following conditions are fulfilled.

$$\Delta t < MIN\left(\frac{\Delta x_i}{|u|}, \frac{\Delta y_j}{|v|}, \frac{\Delta x_k}{|w|}\right) \tag{8}$$

and

$$\Delta t < \frac{3\gamma}{4} \frac{\Delta x_i^2 \cdot \Delta y_j^2 \cdot \Delta z_k^2}{\Delta x_i^2 \cdot \Delta y_j^2 + \Delta y_j^2 \cdot \Delta z_k^2 + \Delta z_k^2 \cdot \Delta x_i^2}. \tag{9}$$

Experimental results of water analogue modelling

For analogue water modelling of die-casting processes the experimental setup was developed (Fig. 2). It consists of water pump, controlled by PC, acrylic model of mold, four synchronized cameras with frame grabbers and four PC for storing pictures [3].

Simultaneously registration of water stream by four cameras makes it possible to record several images of water surface and heterogeneities from different views (Fig. 3(c)). For camera calibration, it is important to find a transformation from the 3-D world coordinates to the real image coordinates [9]. After calibration procedure we can reconstruct water distribution in volume by our algorithms [3, 10]. For comparison of modelling and simulation results we selected three levels of filling (Fig.4).

Discussion

When volume is reconstructed we can visualize water surface, cross-sections, heterogeneities (bubbles and splashing) *etc*. Also it is possible to see sections of volume for detection of bubbles [3]. In Fig. 4 the comparison of results water analogue modelling and numeric simulations is shown. Numerical simulations give satisfactory results for calculation of level of fluid. However volume reconstruction of water filling of mold shows a presence of bubbles and splashes. In the beginning of filling (Fig.4 (a)) splashing of water is biggest than at the end filling (Fig.4 (c)). Since middle of filling level (Fig.4 (b)) we can see bubbles in water. Therefore, for modelling of porosity formation problem the water analogue method has better possibility than application of numerical simulation.

Conclusions

In the given work the results of modelling and simulation for die-casting of AlSi alloys are brought. The main advantage of received results is the possibility of prediction of alloys microstructure for numerical simulation. The method of simulation is based on combination of technique of numeric simulation of filling and solidification followed by micro-macro modelling of crystals growth. We consider the possibility of heredity quenching for AlSi solidification. Microstructure prediction where verified by casting received from real process. The numerical simulation of 3-D mold filling was validated by compison with water analogue modelling results. For this purpose 3-D water distribution in the filling process was reconstructed from four images simultaneously captured by synchronized cameras. Received results are correct for large-scale flow features such as surface waves, when changes of parameters of liquid are significant in the case of AlSi molten alloy.

The results of work can be applied for industrial design of gating system for complex shaped molds and for billet quality optimization in die-casting processes.

Future directions of work are realization of auto calibration procedure and software for industrial optimization of developed approach.

References

1. Schmid M and Klein F, *Fluid Flow in Die Cavities Experimental and Numerical Simulation*, 18th NADCA International Die Casting Congress and Exposition, Indianapolis, IN, USA, 1995, pp 93-99.
2. Jong S and Hwang W, *Three dimensional mold filling simulation for casting and its experimental verification, AFS Transactions*, pp 117–124. 1991.
3. Choi J-K, Choi K-Y, Hwang H-Y, Marukovich E I, Branovitsky A M, Dovnar D V and Zakharov I L, *Three-dimensional image reconstruction for water modelling of metal casting processes*, International Conference On Modelling And Simulation MS'2004, 27-29 April 2004, Minsk, Belarus, pp 216-219, pub BSU, Minsk 2004, ISBN 985-6107-33-4.
4. Marukovich E and Stetsenko V, *Casting of aluminum-silicon alloys with nanostructure eutectic silicon,* 66th World Foundry Congress, , Istanbul, Turkey, 6-9 Sept, 2004, pp 1349-1354, pub Toksad: The Foundrymen's Association of Turkey, Proceedings of 66th World Foundry Congress, 2004.
5. Boettinger W, Warren J, Beckermann C and Karma A, *Phase-Field Simulation of Solidification*, Annu. Rev. Mater. Res, 2002, vol 32, pp 164-194.
6. Hague M and Ismail A, *Effects of cooling rate on structure and properties of Aluminium-Silicon Alloys, 65th* World Foundry Congress, Proceedings of 65th world foundry congress, 2002, pp 379-386.
7. Griffiths W D and Lai N W, *Mould filling: the critical ingate velocity of aluminium and magnesium alloy castings*, Foundry Trade Journal, 178, No 3618, Oct 2004, pp344-348.
8. Ceceres C and Wang QG, *Solidification Conditions, Heat Treatment and Tensile Ductility of Al-7Si-0.4Mg Casting Alloys*, AFS Transactions, vol. 104, 1996, pp 1039.
9. Choi J-K, Hwang H-Y, Choi K-Y, Zakharov I and Branovitsky A, *The technique of calibration multi-camera imaging system for fast water flow registration and reconstruction,* 8th International Conference on Pattern recognition and Information Processng, PRIP'2005, Minsk, 18-20 May, 2005, pp 177-180, ISBN985-6329-55-8.
10. Zakharov I, Dovnar D, and Lebedinsky Yu, *Super-resolution image restoration from several blurred images formed in various conditions*, ICIP 2003, Barcelona, Spain, September 14-17, 2003, Vol. II, pp 315-318, pub IEEE International Conference on Image Processing, 2003.

Acknowledgements

Authors bring up acknowledgements to Contract No 5-3 between ITM of NAS of Belarus and KITECH (Republic of Korea).

Figures

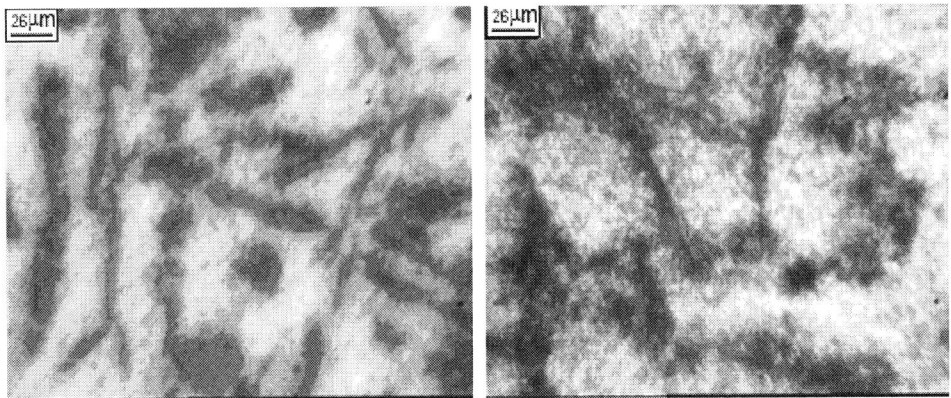

Figure 1. Samples of a microstructure (magnification X300): at edge of casting (on the left), in the central zone (on the right).

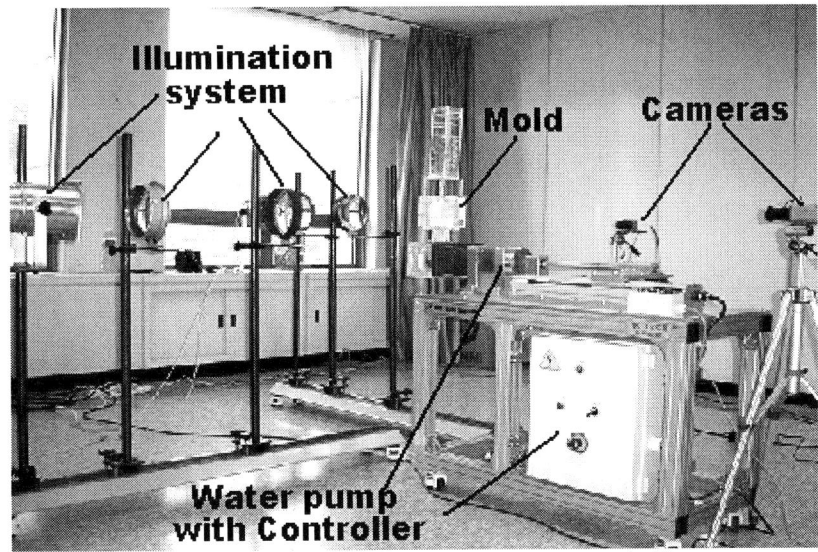

Figure 2. Experimental setup for water analogue modelling.

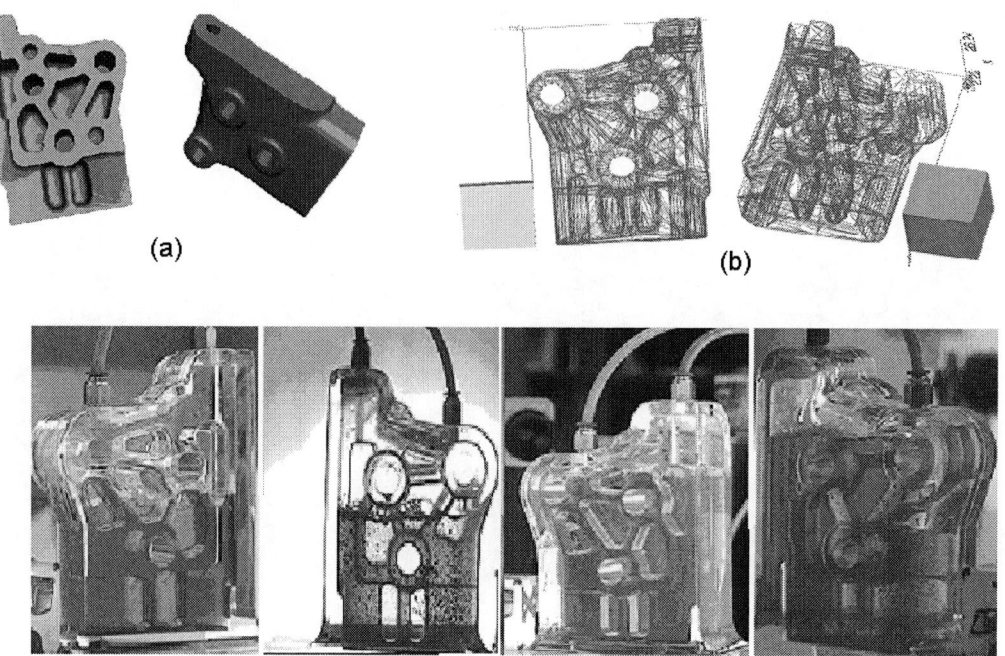

(c)

Figure 3. Images of water analogue modelling: (a) – two views of AutoCAD draw of developed mold; (b) – converted and calibrated draw with cube (test pattern) by software for 3-D reconstruction; (c) – registered images of acrylic mold with water.

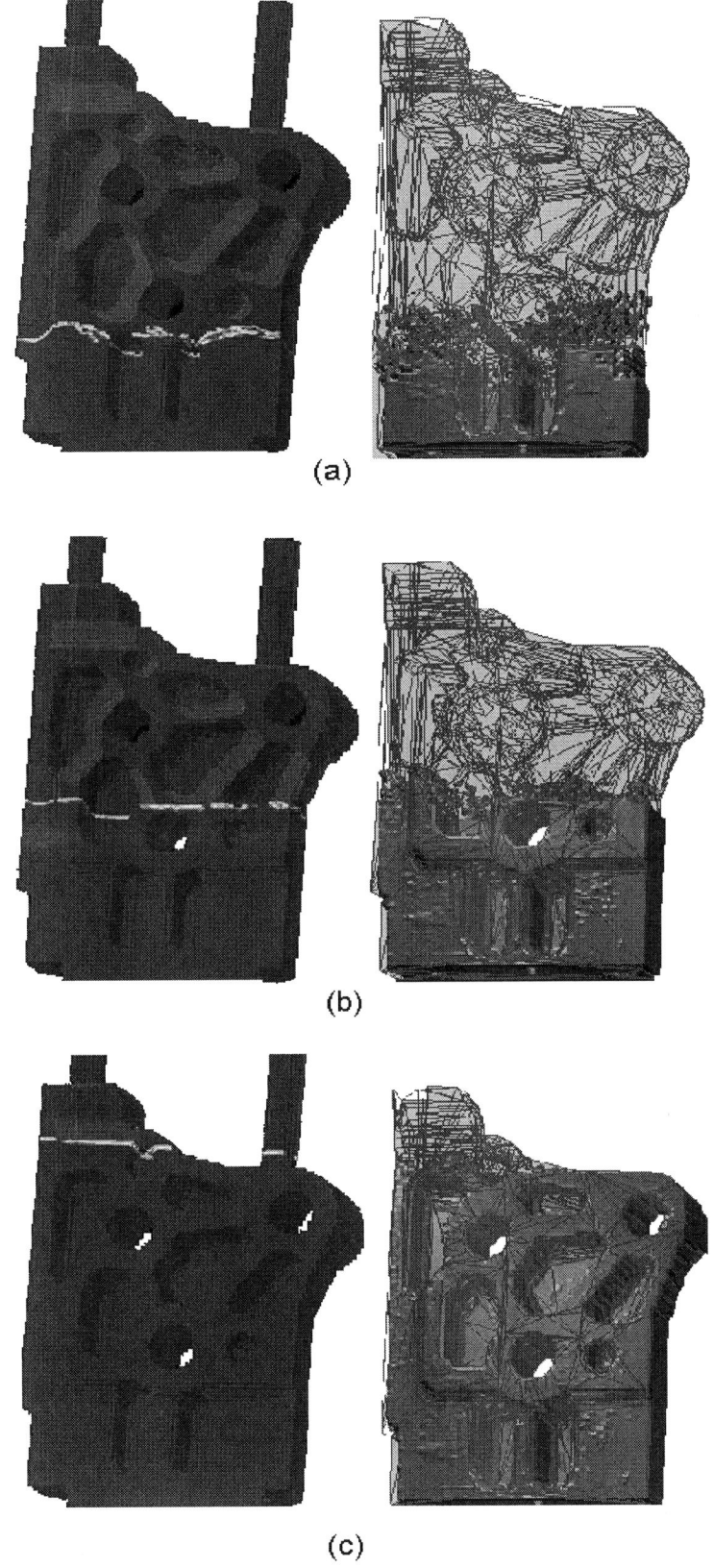

(a)

(b)

(c)

Figure 4. Comparison numerical simulation with water analogue modelling results.

Empirical Model for Tensile Property Prediction in Cast and Heat Treated Al-Si-Cu-Mg Alloys

J. Fang*, H. D. Brody* and J. E. Morral**
* University of Connecticut, Storrs, CT 06269-3136 USA
** The Ohio State University, Columbus, OH 43210-1178 USA

Abstract

An empirical relation has been developed for Al-Si-Cu-Mg alloys to facilitate the computer aided design of casting and heat treatment processes that can be used to achieve specified tensile properties in critical locations of cast components. The empirical relation is one element of a collaborative program to develop, verify and market an integrated system of software, databases, and design rules to enable quantitative prediction and optimization of the heat treatment of aluminum castings to increase quality, increase productivity, reduce heat treatment cycle times and reduce energy consumption.

End-chilled cast plates were used to produce as-cast microstructures with secondary dendrite arm spacing (das) ranging from 20 micrometers to 70 micrometers and volumetric porosity (%P) ranging up to 1.2%. The nominal composition studied was Al-7%Si-3.5%Cu-0.33%Mg-0.5%Fe. Coupons cut from the cast plates were given solutionizing times (θ_S) from 0 to 32 hours, quenched, aged, machined into cylindrical test bars, and tested in tension. Ultimate tensile strength, yield strength, and % plastic elongation were recorded. Three parameters were selected to represent the casting and heat treating process conditions and the tensile strength (TS) was fit to an expression of the form

$$TS = a \cdot das + b \cdot \ln(1 + \%P) + c \cdot \theta_S + d$$

where a, b, c, and d are the adjustable parameters.

The measured tensile properties of newly cast plates were compared to the tensile properties predicted by the empirical model derived from the original set of cast plates. The slopes of the correlation curves are one with $R^2 > 0.95$ for the compositions for which duplicate plates were cast.

Key words Tensile properties, aluminum castings, solution treatment, computer-aided design, aluminum alloys.

Introduction

The Al-Si-Cu-Mg family of cast aluminum alloys, including aluminum alloy-319, are increasingly used in automotive applications.[1,2] Components cast from this family of alloys often are used in the as-cast condition or the castings can be heat treated to improve mechanical properties. Cast structural components typically are designed to have a complex geometry and stringent mechanical properties are specified in critical locations. Computer-aided process design routines can be used to select the combined casting and heat treating parameters to obtain specified properties in critical locations. The ability to successfully predict properties rests upon application of validated relations between process parameters and properties.

The scope of a collaborative university-industry-government program is to develop, verify and market an integrated system of software, databases, and design rules to enable quantitative prediction and optimization of the heat treatment of aluminum.[2,3] The software predicts the thermal cycle in critical locations of individual components in a furnace, the evolution of microstructure, and the attainment of properties in heat treatable aluminum alloy castings. The system is built upon a quantitative understanding of the kinetics of microstructure evolution in complex multicomponent alloys, on a quantitative understanding of the interdependence of microstructure and properties, on validated kinetic and thermodynamic databases, and validated quantitative models. Validated process-property maps need to be tailored to the alloys and practices of individual foundries.

The specific objective of the work described herein is to determine and validate a design rule that can be used to predict the tensile strength of cast and precipitation hardened Al-Si-Cu-Mg alloys. The design rule is based on the minimum set of casting and heat treatment parameters needed to adequately predict tensile strength. The procedure used to determine the design rule for this family of aluminum alloys is general and can be followed to develop design rules for other families of cast aluminum alloys and for other cast alloy systems.

In the typical microstructure of cast and heat treated 319 the major constituents are the terminal solid solution α-aluminum that solidifies dendritically, coarse silicon-phase particles, iron-rich and copper rich particles that are found in the interdendritic regions.[1-5] Interdendritic microporosity is a common feature of the microstructure, the extent and size of the porosity depending on the local solidification time and the amount of hydrogen dissolved in the melt. Coarse particles that form on solidification and porosity reduce ductility and strength. Submicroscopic intermetallic precipitates that form on aging raise the strength of the heat treated alloy. The solidification rate of the casting determines the scale of the dendrite structure, the silicon-phase particles, the interdendritic intermetallic particles and the interdendritic porosity. The solidification rate can be predicted by solidification software which, in turn, can be used to predict the scale of the as-cast microstructure. Alternatively, the dendrite arm spacing can be measured to represent the solidification rate and other aspects of the as-cast microstructure. That fraction of the interdendritic particles that are the result of nonequilibrium solidification can be dissolved by solution heat treatment.[3-6] The increased alloy content of the aluminum-rich matrix that results from dissolution of intermetallic-phases during solution treatment can contribute to strengthening of the alloy during aging and can increase the ductility by removing or rounding of intermetallic particles.

The empirical approach to determine property-process relations for Al-Si-Cu-Mg alloys follows the lead of Passmore and Flemings.[4-6] End-chilled plate castings are used to provide a range of solidification conditions. Coupons cut from the cast plates are solution treated for times ranging from 0 to 32 hours, quenched, and given the same aging treatment. Microstructural parameters, including dendrite arm spacing and volume percent porosity, are measured. Tensile properties are measured for the full range of solidification conditions and solutionizing times. A multi-dimensional least square analysis is used to develop an empirical relation, i.e. a design rule relating tensile properties and a minimum set of microstructure and processing parameters. The empirical relation is validated by casting, heat treating, and testing additional sets of test plates and comparing the second set of measured properties to the properties predicted by the relation determined from the first set of measurements.

Experimental

An end-chilled plate (230 x 180 x 25 mm) is used to prepare test coupons with a wide variety of as-cast microstructures and mechanical properties. A copper block (230 x 50 x 50 mm) is placed in the mold against one end (180 x 25 mm) of the pattern. A tapered cylindrical riser (90 mm in dia.) is placed over the other end of the plate. Note Figure 1 for a sketch of the plate casting. Solidification rates and dendrite arm spacings are comparable to the thermal conditions, microstructures, and properties found in permanent mold and sand cast aluminum alloys for automotive applications.[2]

An unpressurized gating system is used to bring the molten alloy into the mold. The ceramic filter, runner and ingate are in the drag. The plate, riser, tapered rectangular sprue, and copper chill are in the cope. Carbon dioxide is used to set the sodium silicate bonded silica sand (AFS 80). The flaskless molds are set on bottom boards with a ten-degree tilt. The alloy runs uphill from the riser end toward the chilled end.

The alloy is melted in a 3000 Hz induction furnace in an alumina crucible. All charge materials are preheated to at least 200°C before they are added to the crucible. Melts are degassed with an argon purge from a preheated graphite lance inserted to the bottom of the melt. In a few of the initial melts Ar-10%Cl was used as the purging gas. When used, Al-10%Sr modifier is added after degassing. The molten alloy is tilt poured (730°C, approximately) into the mold through a preheated alumina funnel placed directly over the pouring basin in the sprue. Generally, seven end-chilled plate molds are poured at a time so that each plate has the same composition, including the same dissolved hydrogen content.

Plate castings are allowed to cool overnight before they are removed from the mold. Plates are cut into two sets of six coupons each. One set (180 x 25 x 25 mm) is used for tensile testing. The second set (180 x 13 x 13 mm) is used for microstructural analyses. The sectioning scheme for the plate castings is shown schematically in Figure 1.

The nominal alloy composition selected for this study is Al-7%Si-3.5%Cu-0.33%Mg with 0.5%Fe. Microstructures and properties for alloys richer and leaner in Cu and Si and with no or very low Mg and Fe also have been examined. The nominal alloy is not modified and not grain-refined. Either modified, grain-refined, or both modified and grain-refined alloys also have been examined.

Most coupons are solution treated at 505°C for solutionizing times ranging from ½ hr to 32 hrs prior to a warm water quench. Some samples are aged in the as-cast condition (0 hrs. solution treatment time). Solutionizing and aging are carried out in a forced air resistance heated furnace. Coupons are tied in groups by wire with a separation of at least 25 mm between coupons. At the end of the solutionizing period a bundle of samples are pulled from the furnace and quenched within a few seconds in warm water. All coupons are given the same aging heat treatment, held at 230 °C for 3 ½ hrs.

Tensile coupons are machined after heat treatment into standard cylindrical test specimens (0.505" or 12.8 mm dia with 2" or 50.8 mm gauge length) and pulled with an extensometer attached to the gauge length at Westmoreland Testing. If the reported elastic modulus is not within 10% of the handbook value, the tensile test data for the coupon is not used. The extensometer is used to measure total strain to fracture. Elastic strain is subtracted from total strain to compute plastic strain.

Dendrite arm spacing and dendrite cell spacing in the as-cast alloys are measured on the coupons prepared for microstructural analysis. Dendrite cell spacing is measured by inscribing random lines over the microstructure and measuring the average distance between intersections of the lines with interdendritic regions. Alternatively, dendrite arm spacing is measured by identifying primary arms with several perpendicular secondary arms in the plane of view and measuring the average perpendicular distance between secondary arms. Volume percent porosity is measured by comparing the density of sections cut from the metallography coupons of the end-chilled cast plate to the density of a control sample of the same composition (barring macrosegregation along the length of the plates). The control sample is made by remelting and then chill casting a small piece of the cast plate.

Results

Typical results for dendrite arm spacing are plotted in Figure 2. Dendrite arm spacing increases continuously with distance from the chilled end of the plate. Sections close to the chill end have a dendrite spacing and, thus, a local solidification time comparable to permanent mold castings. Sections of the plate further from the end-chill have dendrite arm spacings and local solidification times comparable to sand castings.

Typical results for volume percent porosity are plotted as a function of distance from the chill in Figure 2. Both the size and volume fraction of pores increase continuously with distance from the chill, as does dendrite arm spacing. Dendrite arm spacing and volume percent porosity do not change independently in the end-chilled plate castings.

The tensile properties of the nominal alloy are plotted simultaneously against dendrite arm spacing and solution treatment time at 505°C in Figure 3. The contour plots illustrate a significant increase in tensile strength for fine dendrite arm spacing and extended solution treatment time. Qualitatively similar results are found for other compositions and conditions studied. The lowest values in each contour plot represent samples that were aged as-cast (without prior solution treatment). The remainder of the contour surface demonstrates that tensile strength increases continuously as dendrite arm spacing and volume percent porosity decrease and solution treatment time at 505°C increases.

The yield strength increases significantly for the first increment of solution treatment and then reaches a plateau. The decrease of yield strength with increasing dendrite arm spacing and volume percent porosity is not as dramatic as the variation of tensile strength. Significant increase in yield strength occurs between samples that are aged without solutionizing and those that are given a two-hour solution treatment.

Elongation properties peak at fine dendrite arm spacing and long solution treatment times. Increasing solution treatment time at 505°C has a significant effect on ductility (% plastic eleongation at fracture) for locations with fine dendrite arm spacing. Samples from locations with coarse microstructures and long local solidification times have very low ductility. Ductility is not improved significantly by extended solution heat treatment.

Discussion

Our starting hypothesis is that for a fixed alloy composition, pouring practice, quenching practice and aging treatment, the primary factors that influence the local tensile properties of the cast and heat treated alloy are the local thermal parameters during solidification (local solidification time and temperature gradient), the amount of dissolved hydrogen in the melt as the alloy enters the mold, and the solution treatment thermal cycle. The local solidification time determines the scale of the as-cast microstructure: dendrite arm spacing, silicon-phase particle size, copper-rich and iron-rich intermetallic particle size, volume percent porosity, and size of porosity. Dendrite arm spacing is selected here to represent local solidification time and the size scale of the as-caast microstructure.

Volume percent porosity in the as-cast microstructure is a direct result of dissolved hydrogen, local solidification time, and temperature gradient and is a predominant factor in limiting alloy ductility. Volume percent porosity in the as-cast microstructure can be determined by quantitative metallography or by density measurement.

The solution treatment thermal cycle influences the effectiveness of the dissolution of nonequilibrium phases, the formation of equilibrium phases, and the distribution of strengthening elements through the terminal solid solution matrix (α-aluminum dendrite cores). Time at the solution treatment temperature is the single parameter used to represent the solution treatment process.

A multidimensional least square procedure is used to express the tensile strength as a linear function of dendrite arm spacing (DAS), volume fraction porosity (%P), and solution treatment time (θ_S):

$$UTS = a(DAS) + b(1 + \%P) + c(\theta_S) + d$$

where **a**, **b**, **c**, and **d** are parameters to be determined by least square analysis. The least square equation determined for the nominal alloy is shown in Figures 4 and the result is used to plot curves of tensile strength versus solutionizing time for different dendrite arm spacings. The points plotted in Figure 4 are the measured tensile data. The computed tensile strength is compared to measured tensile strength in Figure 5 and the 45-degree line shows excellent agreement, with an R^2 value of 0.97.

The comparisons in Figures 4 and 5 are between the predictions of the least square relation and the data that were used to develop it. To validate the empirical relation tensile properties were measured for another set of plates of the nominal composition that were cast and heat treated according to the experimental procedures described above. Tensile properties measured on the second set of plate castings and the predicted properties based on the least square fit to the original data set are compared in Figure 6. Again the comparison fits a line of slope equal to one (45-degrees) and now the R^2 value is 0.95, not quite as strong a correlation as when comparison is made to the original data set.

The parameters used to represent as-cast microstructure, dendrite arm spacing and volume percent porosity, do not vary independently for our standard practice, note Figure 2. During the course of the program plates have been cast where the dissolved hydrogen exceeded our standard practice. In these plates, different values for volume percent porosity are obtained for the same dendrite arm spacing. In Figure 7 the least square fit is used to predict iso-strength lines on a plot of volume percent porosity versus dendrite arm spacing. Then the measured tensile strength is marked over the points that indicate the combination of porosity and dendrite arm spacing for the location tested. The correlation of predicted values and measured values is surprisingly good.

Conclusions

For a family of alloys around Al-7%Si-3.5%Cu-0.33%Mg-0.5Fe tensile strength increases for extended solution treatment and by achieving fine as-cast microstructures; yield strength increases rapidly for short solution treatment times and plateaus; and that ductility increases significantly for extended solution treatment for locations with fine as-cast microstructure and not at all for locations with coarse dendrite arm spacing.

An equation developed to fit the measured ultimate tensile strength to just three parameters, dendrite arm spacing, volume percent porosity, and solution treatment time, represents the measured data very well. When the empirical relation fit to the data is used to predict the tensile strength of plates cast with the same composition but a wider variety of as-cast microstructures, the predictive capability is very good.

Similarly good fits and predictive capability are found for other compositions close to the nominal alloy composition. It is suggested this approach to developing design rules for predicting tensile properties based on a limited set of process or microstructural parameters will be effective for other aluminum casting alloys and for the casting practices of other foundries, once calibrated for local conditions. A series of measurements, such as the ones described herein, can be used to calibrate the design rule for local practices.

References

1. Gauthier J; Louchez P R; Samuel F H, *Heat Treatment of 319.2 Aluminum Automotive Alloy. I. Solution Heat Treatment,* International Journal of Cast Metals Research, 8 (1995) 91-106.

2. Cheng T T, Morral J E, and Brody H D, *Effect of Microstructure on Properties of the 319-Family of Aluminum Casting Alloys: A Survey of Recent Literature,* in Final Report, University of Connecticut to Center for Heat Treating Excellence (2004).

3. Yao Z et al, *Integrated Numerical Simulation and Process Optimization for Aluminum Alloy Solutionizing,* ASM Heat Treating Society Conference Proceedings (2005).

4. Flemings MC, Solidification Processing, McGraw-Hill, New York,1974.

5. Flemings M C et al, *Rigging Design of a Typical High Strength, High Ductility Aluminum Casting,* Trans. AFS, 65 (1957) 550-555.

6. Passmore E M et al, Fundamental Studies on Effect of Solution Treatment, Iron Content, and Chilling of Sand Cast Aluminum-Copper, Trans. AFS, 66 (1958) 96-104.

Acknowledgements

The authors are grateful to our sponsors: the Department of Energy under contract DOE DE-FC36-01ID14197 and the Center for Heat Treating Excellence. We are grateful to the members of the industry focus groups for these programs, chaired by Dr. Scott MacKenzie, Houghton, for the DOE project and Dr. Paul Crepeau, GM Powertrain, for the CHTE program and to the program teams at WPI, UConn, and OSU. Important contributions to this project have been made by UConn team members: Dr. Dingfei Zhang, Sudhir Adibhatla, and Yong Ma.

Figures

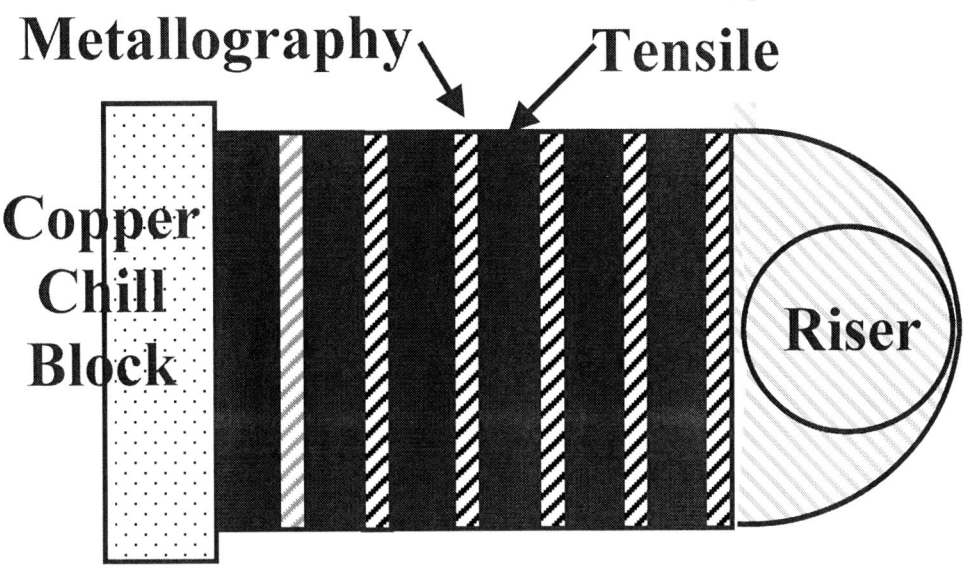

Figure 1: Design and sectioning scheme for the test plate castings.

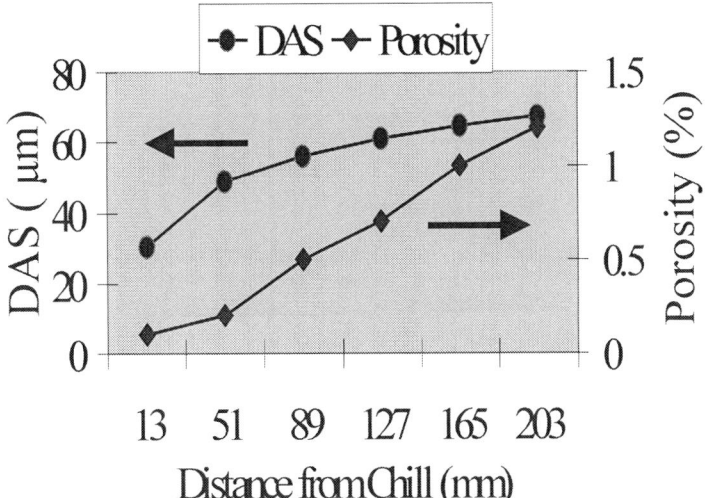

Figure 2: Variation in dendrite arm spacing and volume percent porosity with distance from the chill in end-chilled plate castings.

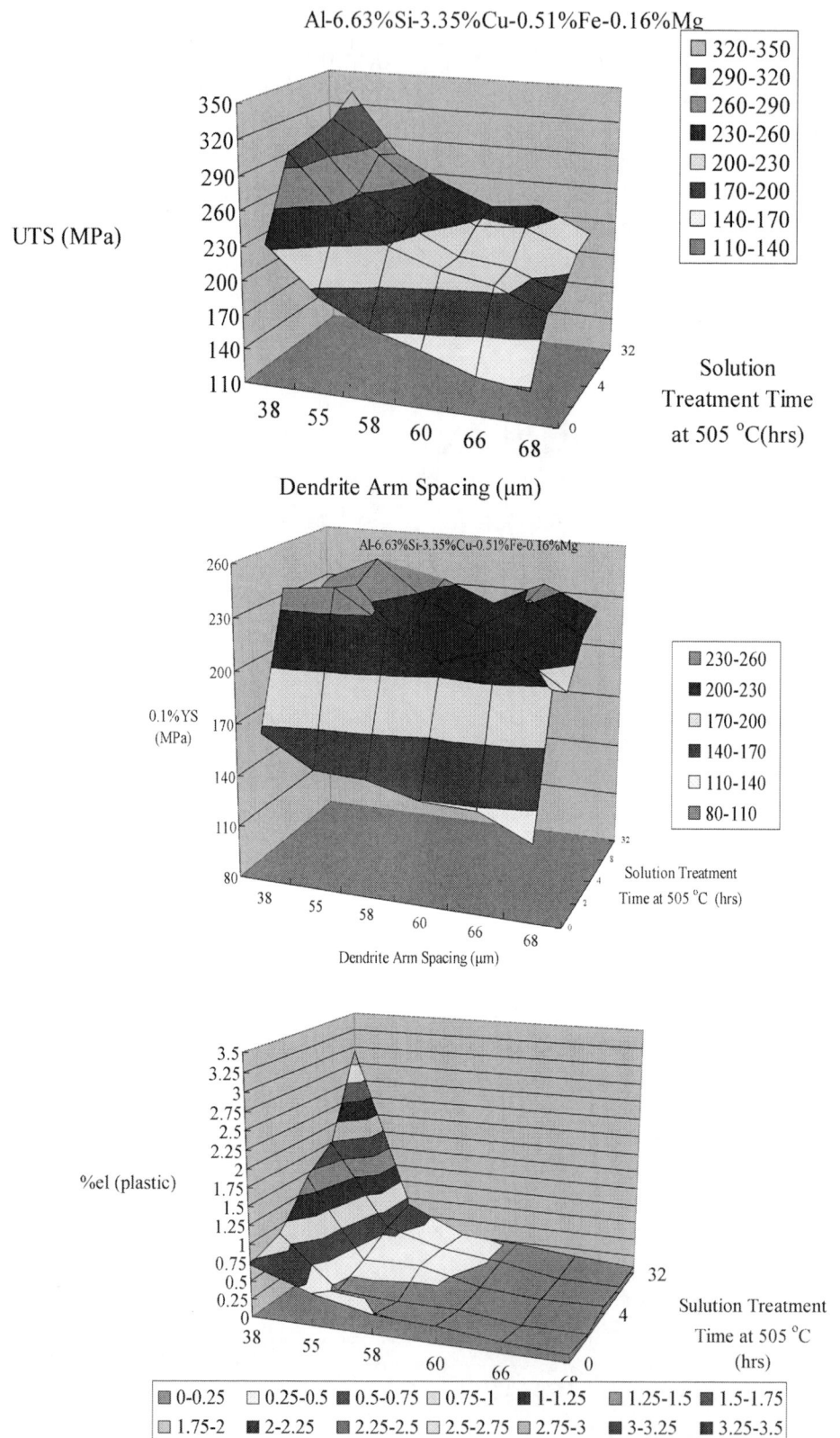

Figure 3: Tensile strength, yield strength (0.1% offset), and plastic strain (%) vs. solution treatment time and dendrite arm spacing.

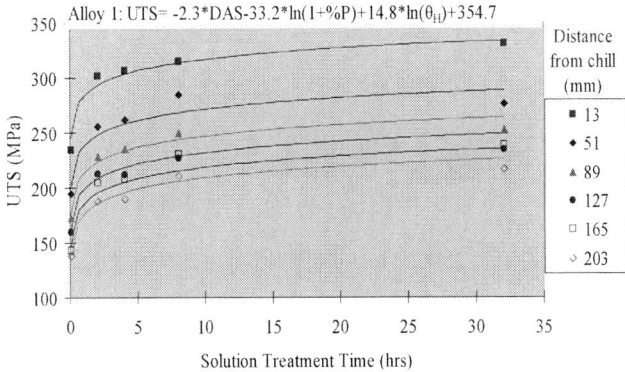

Figure 4: Comparison of measured temsile strength (points) and computed tensile strength (curve) vs. solution treatment

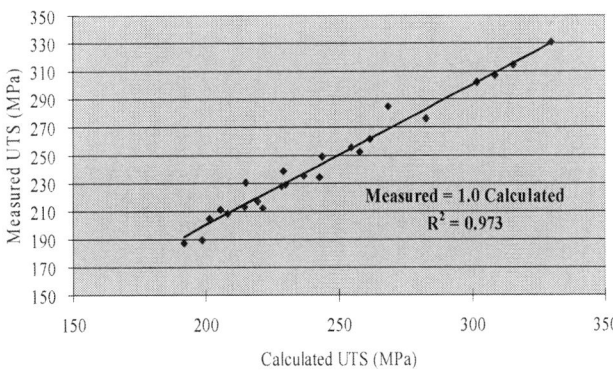

Figure 5: Comparison of measured temsile strength and computed tensile strength.

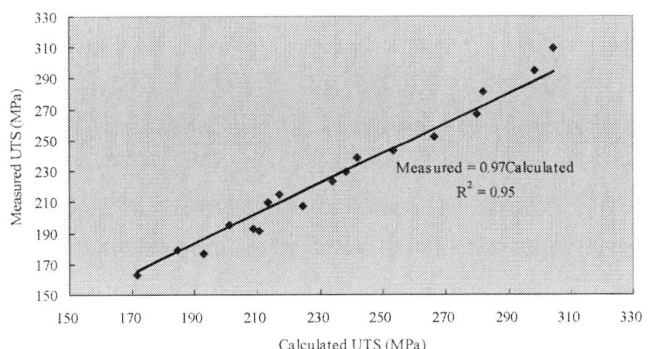

Figure 6: Validation: Comparison of measured temsile strength and predicted tensile strength for additional cast plates

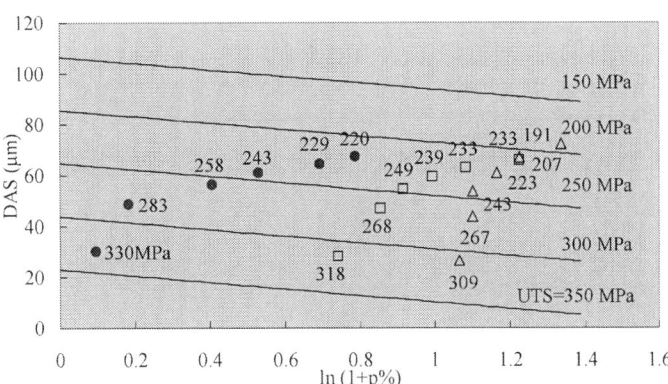

Figure 7: Comparison of measured tensile strength and iso-strength lines computed with the empirical relation for as-cast dendrite arm spacing and volume percent porosity.

Simulation of Solute Redistribution during Casting and Solutionizing of Multi-phase, Multi-component Aluminum Alloys

F. Yi,* H. D. Brody* and J. E. Morral**
* University of Connecticut, Storrs, CT 06269-3136 USA
** The Ohio State University, Columbus, OH 43210-1178 USA

Abstract

Numerical simulations of solute redistribution and microstructure evolution during solidification, post-solidification cooling, and solution heat treatment have been extended to multi-phase, multi-component Al-Si-Cu-Mg alloys. The basic models that have been applied to binary and ternary systems have been modified to handle the extra degrees of freedom in quaternary and higher order systems, which are more representative of commercial casting alloys.

During solidification Si and Mg diffusion is extensive and microsegregation of Si and Mg in the dendrite cores at the end of solidification is negligible. Diffusion of Cu through the dendrite cores controls the extent of microsegregation of Cu in the as-solidified alloy. The presence of small amounts of Mg enhances diffusion of Cu, and the diffusivity of Cu in multi-phase interdendritic regions is several multiples of the diffusivity through the single-phase dendrite cores.

Substantial solute redistribution occurs during post-solidification cooling in sand or permanent molds. Si, in particular, diffuses from dendrite cores to the interdendritic regions as the solubility of Si in the interdendritic α-Al phase decreases sharply with decreasing temperature.

During solution heat treatment the redistribution of solute that occurs during post-solidification cooling is reversed quickly. The nonequilibrium amounts of θ-phase and Q-phase that are distributed in the interdendritic regions between secondary dendrite arms at the end of solidification are reduced to their equilibrium amounts within a few hours. Nonequilibrium phases distributed in the grain boundaries and between primary dendrite branches require ten or more hours to reduce to their equilibrium amounts.

Key words Aluminum castings, solution treatment, solidification simulation, aluminum alloys, diffusion.

Introduction

As part of a collaborative program to develop, verify and market an integrated system of software, databases, and design rules to enable quantitative prediction and optimization of the heat treatment of cast aluminum alloys, we are simulating solute redistribution during the solidification, post-solidification cooling, and solution heat treatment of Al-Si-Cu-Mg alloys.[1,2] The ultimate goal is to simulate the casting and heat treatment processes in multicomponent multiphase alloys with sufficient accuracy to enable computer-aided-design and control of process parameters to achieve specified properties in critical locations. Currently, the simulation routines are used iteratively with experiment to check the influence of process and material parameters on microstructure evolution.

In developing and applying models for solute redistribution to solidification and heat treatment processes a key step is assuming a geometry and critical dimension(s) for a characteristic volume element to represent the overall process. Typically a one-dimensional model is chosen and the characteristic dimension (d) of the volume element is more important than its shape (plate-like, cylindrical, or spherical) in determining the predictions of the model. Diffusivity of the solute in the primary dendritic phase is the key material parameter influencing the degree of as-cast microsegregation and the effectiveness of solution treatment in reducing microsegregation for typical commercial casting and solutionizing conditions.[1,3]

Models appropriate to solutionizing of castings use the as-cast solute distribution and phase fraction as the initial condition. As in solidification models, diffusion of solute through the dendrite cores and the dimension of the characteristic volume element control the rate of dissolution of nonequilibrium phases and homogenization of the matrix phase.

Comprehensive solidification and heat treatment models have been developed for binary alloys.[4] Extension of the models to multi component alloys presents challenges to account for the extra degrees of freedom and the interactions between alloying elements to modify equilibrium phase relations and diffusivity.

Experimental

Models are developed iteratively with experimental studies to increase understanding of basic processes and their interaction with materials parameters; to build databases of phase relations, diffusivity, and process parameters; and to validate assumptions and predictions against measurements on test castings. As example, phase relations --- temperatures and order of phase appearance --- are determined by interrupted solidification. Diffusion coefficients are determined from electron microprobe measurements on diffusion couples. Quantitative metallography, scanning electron microscopy, and electron microprobe analyses of sections taken from end-chilled test plate castings are used to validate predictions of the model.[5]

Model

The characteristic volume element used for most of the results presented here is the 1-d plate-like geometry with characteristic dimension d taken as one-half the secondary dendrite arm spacing. The characteristic volume element is divided into N equal diiferential elements of width $\Delta\lambda = {d}/{N}$. Considering the rate of solidification to be linear and the final eutectic reaction to be isothermal, the rate of solidification is $\dfrac{df_S}{dt} = \dfrac{(1 - f_E)}{\theta_f}$ where θ_f is the local solidification time and f_E is the fraction eutectic microconstituent. Above the eutectic temperature the time for each differential volume element to solidify is $\Delta t_S = \dfrac{\theta_f}{N(1 - f_E)}$. The eutectic microconstituent freezes instantaneously. The contribution of dendrite coarsening to solute redistribution during solidification is not considered here. The local thermal cycle $T = T(t)$ during post-solidification cooling and during solution heat treatment is input from measured cooling and heating curves herein (but could be input from heat transfer.

Initially, the characteristic volume element is considered to be 100% liquid of composition C_{Oi} where the subscript, i, refers to the solute elements Si, Cu, Mg and/or Fe. The liquidus temperature for the alloy is determined by interpolation for the initial alloy and each subsequently computed liquid composition by accessing the stored phase relations database (temperature and solute concentrations) used to represent the liquidus surface. Undercooling of the primary aluminum phase due to nucleation or curvature is assumed negligible. The solid phase or phases in equilibrium with the liquid are determined from the stored phase data and that composition is assigned to the differential volume element at the solid/liquid interface. Secondary phases solidify without undercooling at the point of saturation of the liquid. Two or more solid phases form within the solidifying volume element in the proportions dictated by the tie lines for the average composition of the solid phases. When solubility decreases in a volume element that is multi-phase there is no barrier to growth or dissolution of existing phases. On-the-other-hand, when a solid volume element becomes saturated with respect to a new secondary phase, the barrier to nucleation is assumed infinite (unless otherwise stated). When desired a solid state precipitation model can be coupled to the simulation.

After each differential volume element solidifies, diffusion through the primary solid phase is computed, iteratively, for the time interval, Δt_S.

Then solute balances are computed that simultaneously satisfy the equilibrium phase relations. The concentration of the solute in the α-phase is used to compute the solute flux in Fick's Laws and the average composition is used in solute balance computations.

Based on diffusion couple data [1] obtained for ternary and quaternary alloys the diffusivities (m²/s) used for the α-phase, the terminal FCC solid solution around aluminum, are $D_{CuCu} = 1.45x10^{-6} \exp\left(-112,300\middle/RT\right)$;

$D_{SiSi} = 6.3x10^{-6} \exp\left(-113,000\middle/RT\right)$; $D_{MgMg} = 8.57x10^{-6} \exp\left(-119,400\middle/RT\right)$.

The cross coefficients D_{CuSi}, D_{SiCu}, D_{CuMg}, D_{MgCu}, ... are negligible. The variation of D_{CuCu} with Cu content is negligible over the limited solubility range. Small additions of Mg (<0.5%) increase the diffusivity of Cu in the α-phase by a factor of 2 to 4. Also, diffusivity of Cu through the two-phase α + silicon region is increased by a factor of 5 to 20.

As sketched in Figure 1, an alternative 2-D volume element used here is rectangular in cross-section. The characteristic dimensions, d_1 and d_2, are taken as one-half the primary dendrite arm spacing and one-half the secondary dendrite arm spacing, respectively. The differential volume element dimensions for solidification, $\Delta\lambda_1$ and $\Delta\lambda_2$, are taken in the same ratio, which is 3/1 for the results reported here. The differential volume elements used for diffusion computations are $\Delta\lambda_1$ x $\Delta\lambda_1$.

To account for divorced solidification, the sequence of solidification of the differential volume elements is modified, as illustrated in Figure 2. When the composition of the liquid phase becomes saturated with respect to silicon, the composition of the α-phase and the equilibrium ratio of α+silicon phases are computed to satisfy, simultaneously, the solute balances and the stored phase equilibria data. Predetermined differential volume elements ahead of the α/L interface are considered to transform to an α+silicon colony in a ratio determined by quantitative metallography. The observed fraction of α-phase in the binary eutectic colonies is below the equilibrium value, the remainder of the α-phase continues to form at the original α/L interface.

Results
The 1-D model is applied first to simulation of solute redistribution during solidification for the ternary alloy Al-6.5%Si-3.5%Cu. The solidification path for the alloy is shown on the projections of the □-liquidus and □-solidus surfaces in Figure 3. Also shown in Figure 3 is the solidification curve, i.e. phase fraction versus temperature through the solidification range. Freezing begins at the liquidus with solidification of □-phase, which is more dilute in Si and Cu than the liquid. As solidification proceeds through the □+L region the concentrations of silicon and copper increase in the solid and liquid phases. The line of two fold saturation (the eutectic valley) is reached when 42wt% □-phase has solidified, As temperature decreases along the eutectic valley,□-phase and essentially pure silicon solidify as a binary eutectic, which is about 90wt% □-phase. The Cu concentrations continue to increase and the Si *concentrations decrease slightly in the□ and liquid phases. At the ternary eutectic temperature the*

remaining liquid transforms isothermally to ☐-phase, ☐-phase (CuAl₂), and silicon, in the ratio by weight 39/56/5.

The predicted distribution of Cu and Si in the α-phase at the end of the ternary eutectic solidification in an Al-7%Si-3.5%Cu alloy is shown in Figure 4. The Cu concentration in the α-phase increases continuously through the primary α-phase and two-phase α+silicon regions and reaches a plateau at the ternary eutectic concentration. The Cu distribution at the end of solidification is little changed from that predicted by the Gulliver-Scheil model.[3] The Si concentration is essentially uniform at the equilibrium ternary eutectic composition for the α-phase, which is consistent with the data that the diffusivity of Si in the α-phase at the liquidus temperature is four times that of Cu.[1] Because the solubility for Si in the α-phase decreases during solidification of the binary eutectic, the silicon particles in the α+silicon region grow larger and fine silicon precipitates can nucleate within the primary α.

The predicted distributions of Cu in the α-phase after an Al-7%Si-3.5%Cu alloy has solidified and then cooled to 580K in the mold are shown in Figure 5 for two positions in an end-chilled plate casting. By measurement position 1 has a dendrite arm spacing of 38μm, a local solidification time of 514s, and cools time from the eutectic to 580K in 1848s. The comparable measurements for position 6 are 71μm, 769s, and 1194s, respectively. Comparison of the Cu distributions in Figures 4 and 5 show post-solidification diffusion of copper into the primary phase to reduce the concentration gradient and diffusion of copper from the α+silicon region to the α+θ+silicon region as the solubility of Cu in the α-phase decreases with decreasing temperature.

Comparisons of the predictions of the 1-D model in Figures 3-5 with the results of interrupted solidification studies and quantitative microscopy on sections from end-chilled plate castings indicate a major discrepancy.[1,5] The apparent fraction of primary α-phase in as-cast plates is over 70% as compared to the predicted 42% and in the interrupted solidification studies the two-phase α + silicon colonies are observed to grow in the liquid ahead of the α/L interface with a silicon fraction much greater than the predicted 10%. Using quantitative metallography data as a guide, the divorced eutectic model, Figure 2, is being used to simulate solidification and solute redistribution. Figure 6 presents a result for Al-7%Si-3.5%Cu.

The 1-D plate geometry and the 2-D plate geometry (Figure 1) are being used to simulate phase formation and dissolution and solute redistribution during silidification and post-solidification solution treatment. Both the 1-D and 2-D simulations consider divorced binary eutectic solidification and enhanced Cu diffusion in the binary eutectic region (due to the silicon particles). Predicted rates of dissolution of θ-phase during solutionizing (at 505°C) for a region close to the chill in an Al-7%Si-3.5%Cu test plate casting (38μm DAS), Figure 7, are consistent with observations of the dissolution of copper-rich θ-phase during solution heat treatment of Al-Si-

Cu and Al-Si-Cu-Mg alloys. The θ–particles distributed in the interdendritic regions between secondary arms dissolve within two hours and θ-particles within clusters of eutectic microconstituent distributed at primary dendrite subboundaries and at grain boundary junctions require eight or more hours to dissolve.

Similar results are found for the quaternary alloy Al-7%Si-3.5%Cu-0.5%Mg. Solidification of primary α-phase is followed by the binary eutectic α+silicon, the ternary eutectic α+silicon+Q ($Al_5Cu_2Mg_8Si_6$), and finally the quaternary eutectic α+silicon+Q+θ. The predicted distributions of Cu in the α-phase for the 1-D model at the end of quaternary eutectic solidification are shown in Figure 8 for two cases, (i) using the expression D_{CuCu} given above (LSD) and (ii) using 2 x D_{CuCu} (ESD) to account for the enhanced Cu diffusivity with small additions of Mg.

The predictions for Cu distribution in the quaternary alloy for a location near the chill in a test plate casting (38μm DAS) using (i) the divorced eutectic model (LSD), (ii) enhanced diffusivity for copper in the α-phase due to the Mg addition (ESD1), and (iii) enhanced diffusivity of Cu in the α-phase within the two-phase α+silicon region (ESD2) are shown for the simulation of solidification and post-solidification cooling (to 300°C) in Figure 9. Also shown in Figure 9 are a series of electron microprobe measurements across a typical dendrite arm.[6] The agreement is surprisingly good.

Conclusions
Experiment and predictions of simulations of solute redistribution and phase appearance for solidification and post-solidification cooling of multi-component and multi-phase Al-Si-Cu-Mg alloys using (i) a 1-D model, (ii) a characteristic dimension equal to one-half the secondary arm spacing, (III) measured solute diffusivities and phase relations, (iv) enhanced diffusion of copper in the FCC α-phase due to small additions of Mg and/or the presence of silicon particles, and (v) a model for divorced solidification of α+silicon colonies. A 2-D model with two characteristic dimensions for solution treatment of as-cast microstructre is consistent with observations.

References
1. Brody H D and Morral J E, *Solution Heat Treatment of Aluminum Alloys: Effect on Microstructure and Properties*, Final Report, Univ. of Connecticut to Center for Heat Treating Excellence, 2004.
2. Yao Z et al, *Integrated Numerical Simulation and Process Optimization for Aluminum Alloy Solutionizing,* ASM Heat Treating Society Conference Proceedings, 2005.
3. Flemings MC, Solidification Processing, McGraw-Hill, New York,1974.

4. Gandin CA et al, *Modelling of solidification and heat treatment for the prediction of yield stress of cast alloys*, Acta Materialia 50(5): 2002, 901-927.
5. Fang J, Brody HD, *Morral JE, Empirical Model for Tensile Property Prediction in Cast and Heat Treated Al-Si-Cu-Mg Alloys*, This Volume, WFC#35.
6. Ma Y., University of Connecticut, Personal Communication, 2006.

Acknowledgements

The authors are grateful to our sponsors: the Department of Energy under contract DOE DE-FC36-01ID14197 and the Center for Heat Treating Excellence. We are grateful to the members of the industry focus groups for these programs, chaired by Dr. Scott MacKenzie, Houghton, for the DOE project and Dr. Paul Crepeau, GM Powertrain, for the CHTE program and to the program teams at WPI, UConn, and OSU. Important contributions to this project have been made by the UConn and OSU team: Dr. X. Pan, Dr. D. Zhang, M. Qian, S. Adibhatla, C. Lin, and Y, Ma.

Figures

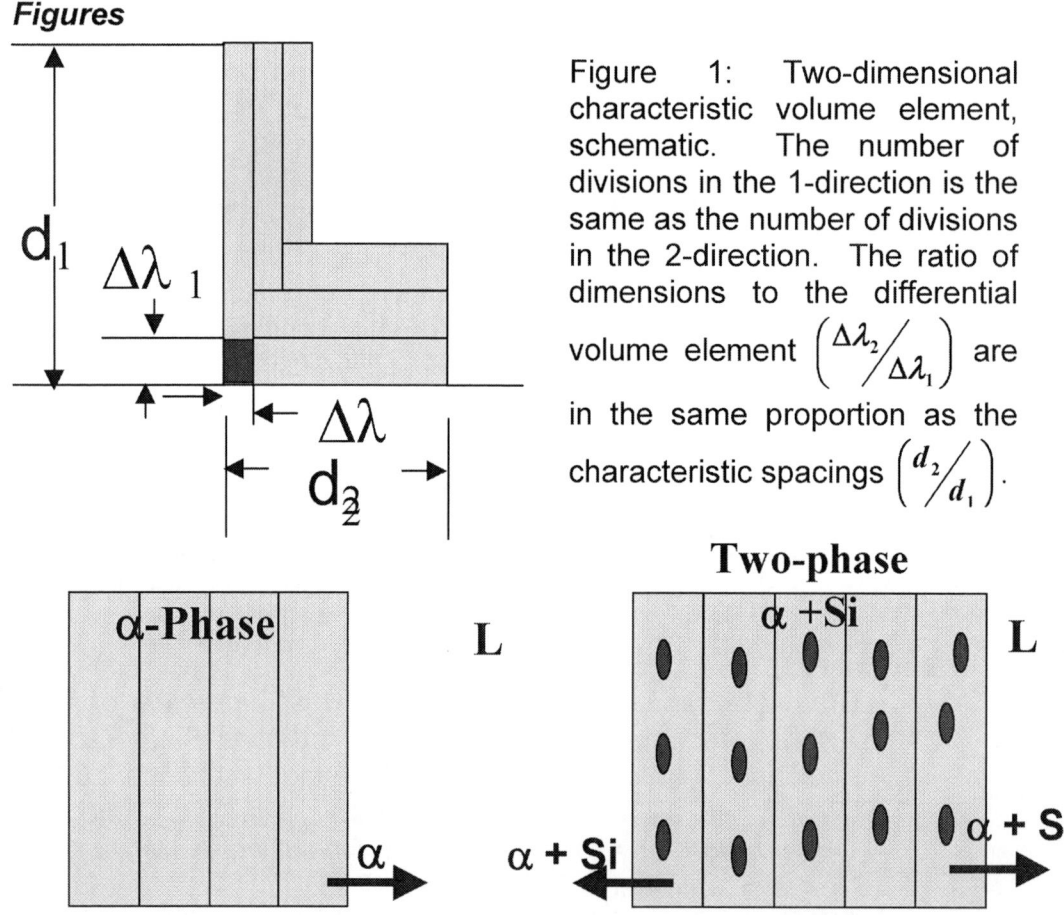

Figure 1: Two-dimensional characteristic volume element, schematic. The number of divisions in the 1-direction is the same as the number of divisions in the 2-direction. The ratio of dimensions to the differential volume element $\left(\Delta\lambda_2 \big/ \Delta\lambda_1 \right)$ are in the same proportion as the characteristic spacings $\left(d_2 \big/ d_1 \right)$.

Figure 2: Schematic representation of 1-D characteristic volume element for divorced solidification of α + silicon binary eutectic microconstituent.

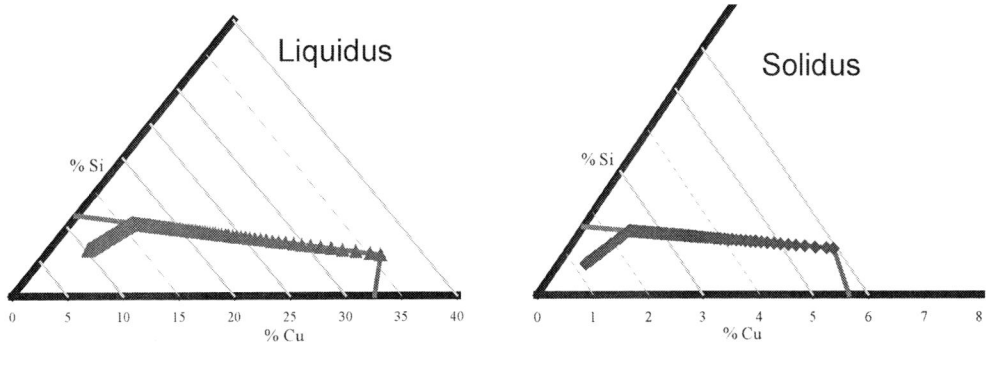

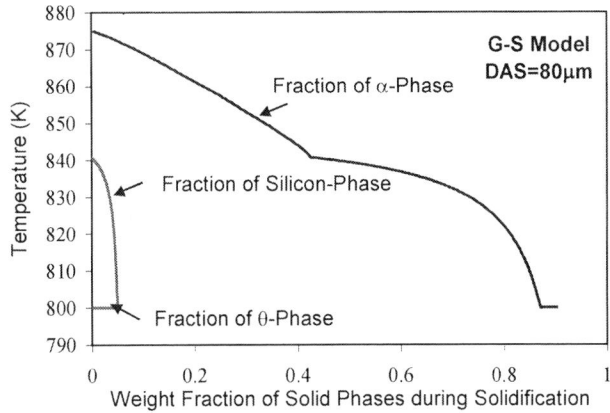

Figure 3: Solidification path along solidus and liquidus surfaces and solidification curve for ternary alloy Al-6.5%Si-3.5%Cu, dendrite arm spacing equal to 80μm.

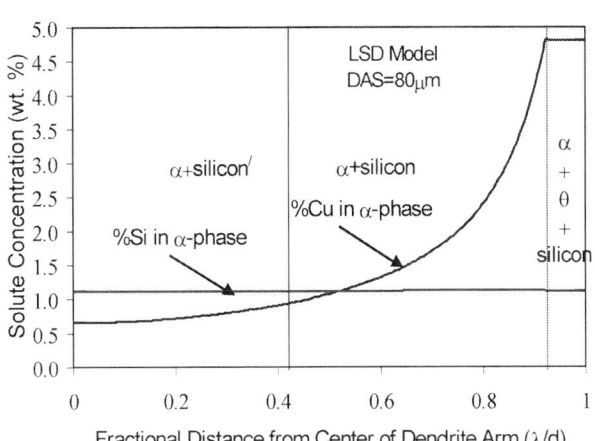

Figure 4: Distribution of Cu and Si in the α-phase after solidification to the ternary eutectic temperature for an Al-6.5%Si-3.5%Cu alloy.

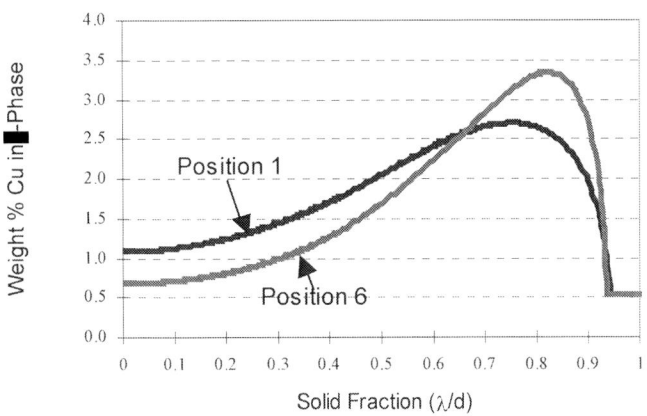

Figure 5: Copper concentration in the α-phase across the characteristic volume element after post-solidification cooling to 580K for two positions in an end-chilled Al-7%Si-3.5% Cu casting.

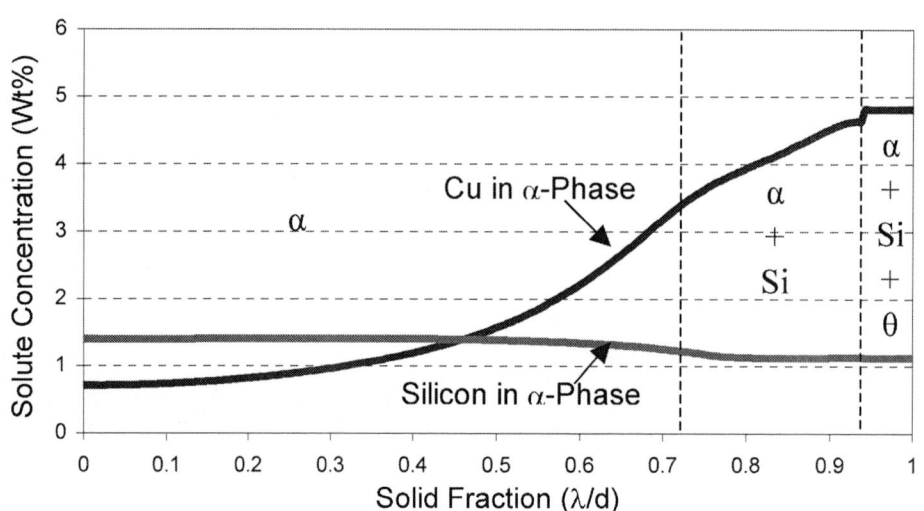

Figure 6: Cu and Si distribution after solidification to ternary eutectic temperature for Al-7%Si-3.5%Cu alloy using divorced eutectic model.

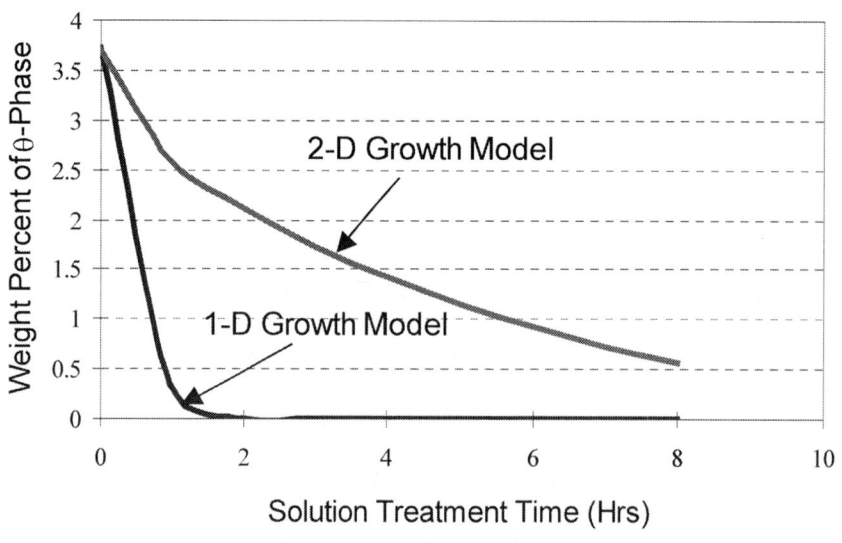

Figure 7: Dissolution of θ-phase during solution treatment of Al-7%Si-3.5%Cu at 505°C: Comparison of 1-D and 2-D models.

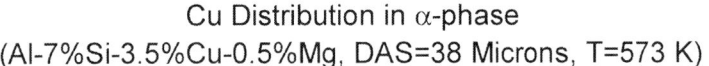

Figure 8: Cu distribution in α-phase for Al-7%Si-3'5%Cu-0.5%Mg alloy after solidification to the quaternary eutectic temperature: Comparison of use of D_{CuCu} (LSD) measured for ternary alloy without Mg and use of enhanced diffusivity ($2 \times D_{CuCu}$) for Mg addition (ESD).

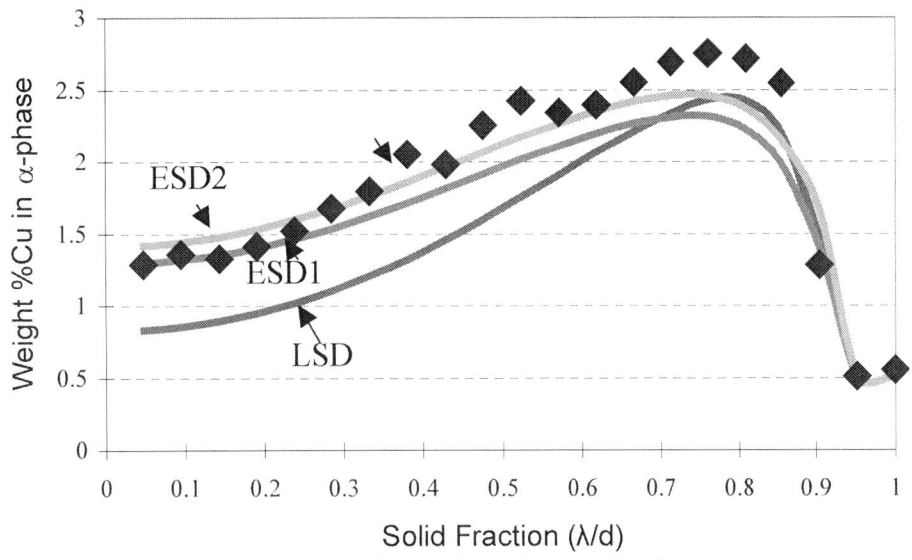

Figure 9: Comparison of predicted Cu distribution in α-phase for as-cast Al-7%Si-3.5%Cu-0.5%Mg with microprobe data (points). Model ESD2 includes divorced solidification. enhanced diffusivity due to Mg addition and enhanced diffusion in α+silicon region.

Computer Simulation Study upon the Influence of Geometry on the Critical Velocity for Molten Aluminium

R Cuesta, A Delgado, J A Maroto, D Mozo

Foundation for the Research and Development in Transport and Energy, CIDAUT, Spain.

Abstract

In aluminium casting processes, the entrainment of surface oxides into the bulk of liquid metal during mould filling determines a severe reduction of the strength of the cast part. In order to avoid this, melt velocity must be kept below a certain value, namely critical velocity, which is widely assumed to be equal to 0.5 m/s. In this paper we investigate, by means of computer simulation, the dependence of the critical velocity for pure aluminium with geometrical features of running channels, such as size and shape. The critical velocity is found to decrease with size for flat channels. Round channels seem to have larger values of critical velocity than flat ones. The variations in critical velocity are explained because of variations in the Webber number of the liquid metal flow.

Key words

Critical velocity, aluminium, entrainment, oxides

Introduction

The filling stage of casting processes has a big influence in the mechanical properties of cast parts, especially in those alloys with a special propensity to get oxidized in its contact with air, such as aluminium and ductile iron alloys [1]. Whenever the molten metal gets in contact with air, its surface gets instantly covered by a layer of oxide of high melting point. If the free surface enfolds over itself during filling, a piece of oxide will be irremediably entrained into the bulk liquid. Whenever that happens, the mechanical reliability of the cast parts is expected to decrease sharply.

As far as oxide entrainment is concerned, one of the most critical types of flow consists in a sudden widening in the molten metal path. The behaviour of molten metal flowing in this kind of geometry has been researched in [2], where molten aluminium emerging from ingates at different speeds into a much large vessel was filmed. Once the aluminium penetrates into the vessel, it doesn't count with the restraining action of the side walls of the mould, therefore, under certain conditions, the liquid might flow side wards and form an air pocket, which is dragged afterwards by the liquid stream entraining the oxides on its surface. This kind of turbulence has been named as surface turbulence to distinguish it from the turbulence present in the bulk liquid [1]. Increasing the melt velocity makes more likely the surface turbulence to occur. In [1], the aluminium velocity above which entrainment occurs has been settled in 0.5 m/s. This value, named critical velocity, is widely assumed in the foundry industry for the design of the casting process, the geometry of the mould and pouring conditions alike.

In this study we research the dependence of the critical velocity with the geometry of the running channel. We analyze, by means of 2D computer simulations, the flow of pure aluminium in a sudden widening of the liquid metal path for two different types of symmetry, plane and cylindrical, and two different sizes of channels. We explain the results obtained making use of the non dimensional fluid dynamic Webber number.

Simulation

The commercial software FLUENT was used to perform fluid dynamic simulations of the filling of simple, bidimensional (2D) cavities with pure liquid aluminium. The cavities feature a vertical channel, bottom filled, connected to a much large vessel in its upper end. Two types of symmetries were considered: cylindrical and plane. For the case of plane symmetry, two channels of 10 and 25 mm width were studied. Likewise, channels with 10 and 25 mm diameter were considered for the cylindrical symmetry. For each channel, several inlet velocities were imposed. In each case, the velocity was taken constant along the cross section of the bottom of the channels, which were designed sufficiently long (150 mm) to ensure a more realistic velocity profile at the end, just in the transition to the vessel. The dimensions of the vessel were kept constant for all the simulations and equal to 130 (horizontal) x 65 (vertical). The whole list of cases studied appears in table 1.

The K-ε model was used for the momentum transfer equations. The movement of the free surface between air and aluminium was calculated using the VOF method. Molten aluminium was considered to be at 700ºC. Thermo physical variables taken for the melt were: 0.91 N/m for surface tension, 2380 kg/m3 for density and 5.57e-7 m^2/s for kinematic viscosity; all these properties were taken constant throughout the filling. The enmeshment of all the cavities was made using square elements of 0.25 mm size.

Results

Figure 1 represents the position of the free surface in the area of the sudden widening at the selected times for three of the simulations run, all featuring a channel of 10 mm size. The color code for each cell of the enmeshment means the volume fraction of aluminium. Dark grey means the cell is totally full of liquid, black means it's totally full of air. The cells only partially filled with liquid define the position of the free surface. The flows found in the simulations can be classified according to three different types, which are described next:

Type 1: at low velocities the aluminium floods smoothly over the horizontal wall of the vessel and no air entrainment occurs. This type of flow can be seen in figure 1a, which corresponds to the simulation of the 10 mm flat channel with an inlet velocity of 0.3 m/s. A very similar flow pattern was found for the simulation of the 10 mm round channel.

Type 2: at higher velocities, the fluid particles along the free surface of the melt move essentially vertically at the start of the filling of the vessel, with very little or no contact with the horizontal wall. The melt flow can therefore be classified as a jet. Later on, the particles at both sides of the jet start to move horizontally and finally they have a movement which is mixed downwards and horizontal. This causes the formation of an air pocket between the aluminium and the horizontal wall, which is assumed to be entrained into the bulk liquid at a later stage of filling. In the real casting this situation would determine a structural damage for the cast part. The figure 1b illustrates this type of flow, till the instant when an air pocket is formed. The pictures correspond to the simulation of the 10 mm flat channel with an inlet velocity equal to 0.7 m/s.

Type 3: in this type of flow, the start of the filling is jet-type, but in this case once the melt has gained certain height, the particles on the sides of the jet move essentially downwards. In this third type of flow no air pocket is predicted by the simulation, nevertheless we can anticipate a situation of risk of entrainment in the real casting, provided that any source of asymmetry in the real flow occurs, such as bubbles traveling in the liquid or simply a non symmetrical distribution of velocities in the channel, which could provoke folds along the surface. Therefore, this flow type could be qualified as non-safe, as far as entrainment is concerned. Flow type 3 is illustrated in figure 1c, corresponding to the simulation with a 10 mm round channel and inlet velocity equal to 0.8 m/s.

One very useful parameter to quantify the damage caused to the casting due to oxide entrainment is the length of the free surface between the melt and air in the region of the air pocket. The higher this length, the less reliable the casting will be. This length can be calculated directly from the simulation results by simply counting the number of cells partially or totally full of aluminium that have at least one side bordering the air pocket region, and multiplying this number by the cell size, 0,25 mm. The calculated values for the length of the free surface between the melt and air in the region of the air pocket, variable to be named as L from now on, are presented in table 1 for all the simulations run. Obviously, only the simulations with a flow of type 2 have L values different from cero.

In the case of flat channels, L represents the surface of oxide entrained per unit of length. Therefore, this value would enable us to estimate the total surface of oxide entrained for a 3D flat channel, provided it's *sufficiently* long, just multiplying L by the length of the channel. In the case of cylindrical channels, the calculation of the total surface of oxide entrained must follow a not so obvious calculation.

Discussion

As table 1 shows, out of the four channels considered, in three of them, -both flat channels (10 and 25 mm) and the 25 mm round channel-, a flow pattern of type 2 was found for some simulations. Flow type 3 was only encountered in round channels.

Figure 2 plots the length of the free surface in the region of the air pocket, L, versus inlet velocity, for the flat channels. As the figure shows, L grows steadily with velocity, being this length several times larger for the 25 mm channel. Due to linearity between the L values different from cero and velocity, if a linear regression is made taking exclusively the points with $L \neq 0$ (flow type # 2), the cross of the regression line with the velocity axis gives the critical velocity, V_{crit}. These regression lines for each channel size are represented in fig. 2. The reader can prove that for the 10 mm flat channel, the critical velocity calculated by this method yields a value of:

$$V_{crit} = 0.47 \text{ m/s} \tag{1}$$

Which is very close the previously established critical value of 0.5m/s.

Proceeding the same way for the 25 mm flat channel we obtain a critical velocity considerably lower for that channel:

$$V_{crit} = 0.30 \text{ m/s} \qquad (2)$$

Which is 40% lower than 0.5 m/s.

Going through the results found for round channels, in the one of 10 mm diameter, the flow type changes from #1 to #3 for a velocity somewhere between 0.4 and 0.5 m/s. For the 25 mm diameter channel, the flow type changes in the same range of velocities, but in this case the change is from type 1 (V=0.4 m/s) to type 2 (V=0.5 m/s). Therefore, the critical velocity should be between 0.4 and 0.5 m/s for the two sizes of round channels studied. The simulation with 0.6 m/s yields an increase of oxide entrainment respect the 0.5 m/s simulation. For higher velocities, the flow pattern found is type 3.

In the round channels, once the critical velocity has been exceeded, the amount of oxide entrained in the casting is larger for the 25 mm diameter; actually the simulations with the 10 mm channel show no entrainment in all the range of velocities considered. In any case, the increase of oxide entrainment with channel size found for round channels is much smaller than in flat channels.

The results presented above show an influence of the size and geometry of the channel in the flow structure and therefore in the critical velocity. It has been proposed [1] that the movement of the free surface, and therefore any eventual entrainment phenomena, is governed by two kinds of fluid dynamic forces: the inertia forces and the surface tension forces. The non dimensional number that picks the ratio of these forces is the Webber number. The expression for the inertia forces is [1]:

$$F_i = \frac{\rho V^2}{2} \qquad (3)$$

Where ρ is density and V velocity of the liquid respectively. The surface tension forces have the expression:

$$F_s = \sigma(\frac{1}{R_1} + \frac{1}{R_2}) \qquad (4)$$

Where σ is the surface tension of the liquid and R_1 and R_2 are the main radii of curvature of the free surface. In the case of flat channels we can assume one of the radii of curvature to be infinite. If the channel is cylindrical, we can assume the shape of the free surface to be nearly spherical. Figure 3 schematically represents the shape of the free surface when metal arrives to the point of transition between the channel and the vessel. Following figure 3b, we can choose R_1 to be equal to half of the width of the channel and R_2 to be infinite for flat channels. For round channels, both radii can be considered the same and equal to the radius of the channel. Thus, we easily can combine equations 3 and 4 to obtain an expression for the Webber number, We:

$$We = \frac{\rho V^2 T}{n \sigma}$$

(5)

Where, for a flat geometry T is the width of the channel and n gets the value of 4. For cylindrical channels, T is the diameter and n is equal to 8. Table 1 shows the calculated We numbers following equation 5 for all the simulations performed. For a given molten metal, We combines the three degrees of freedom of the flow: size and geometry of the channel and liquid metal velocity. The higher the We, the higher the inertia forces versus the surface tension forces and, following [1], the more propensity to oxide entrapment. It has been reasoned [1] how $We = 1$ seems to define the transition to surface turbulence. Thus, as long as We depends on geometry, thus critical velocity should depend on geometry as well.

The discussion that follows next analyses in a quantitative manner the influence of We on the existence and eventual intensity of surface turbulence.

Figure 4 plots the dependency of L with Webber number for the channels where transition to surface turbulence came defined by a change in flow type from #1 to #2. The figure clearly points the existence of a transitional value of We below which no air pocket forms. This value is around 1.4.

The reader can check that the We corresponding to the critical velocities found for the 10 and 25 mm channels (equations 1 and 2) are 1.44 and 1.47, respectively.

In order to find the transitional We for the 10 mm diameter we must follow a separate treatment, since no air pocket was encountered. Anyway, as it has been discussed, the critical velocity has been estimated in the range 0.4-0.5 m/s for that channel, which corresponds to a range of We number between 0.52 and 0.82 (see table 1).

Considering the minimum and maximum transitional We presented above, we can propose an interval of We between 0.5 and 1.5 to define the transition to surface turbulence: staying below 0.5 would mean to be safe from a filling stand point. On the other side, We numbers above 1.5 would imply the occurrence of surface entrapment. We numbers between 0.5 and 1.5 could determine oxide entrainment depending on kinematical and/or geometrical features of the liquid metal flow.

In order to illustrate how the occurrence and intensity of entrainment is more related with We rather than to melt velocity, we can examine the simulations with velocity equal to 0.4 m/s for the flat channels and find a different type of flow for each size: in the 10 mm channel no entrainment is found, whilst in the 25 mm channel, for the same velocity, the simulation yields an entrainment of 7 mm of surface of oxide per unit length. *We* for the 10 mm channel is equal to 1.05, which falls within the transitional range of 0.5-1.5 described above; however, in the 25 mm channel we have *We* = 2.62. This increase in *We* clear above the transitional range of 0.5-1.5 can explain the apparition of entrainment upon enlargement of the channel size, in spite of keeping velocity constant. Thus, this can also explain the existence of a different critical velocity for each flat channel (0.47 m/s for 10 mm and 0.3 m/s for 25 mm).

If, for a given type of channel (round or flat), we assume that *We* above which surface turbulence occurs is approximately constant regardless the size of the channel, then following equation 5, the critical velocity should fall with T, and more in particular can be assumed to be proportional to $T^{-1/2}$. This relationship is accurately satisfied by the figures of critical velocities and channel sizes for the flat geometry, since the ratio of the channel widths elevated to exponent -0.5 almost equals to the ratio of critical velocities found in equations (1) and (2):

$$(\frac{10}{25})^{-0.5} = 1.58 \approx (\frac{0.47}{0.3}) = 1.57 \tag{6}$$

Comparing the critical velocities between flat and round channels, the minimum critical velocity found, 0.3 m/s, corresponding to the 25 mm flat channel, lies below the range of critical velocities for round channels: 0.4 – 0.5 m/s. This seems to indicate that, for the same channel size, the critical velocity is lower for flat channels vs. round. In fact, the much larger values of *L* found for flat channels (see table 1) points the bigger propensity of this kind of channel vs. the round one towards surface turbulence, reassuring the concept that for a flat geometry the critical velocity should be lower. This can be explained because of the higher *We* for flat channels, which doubles versus round ones for the same channel size and melt velocity (see equation 5), due to exponent n (8 for round, 4 for flat).

Once the critical velocity has been exceeded, we can have an estimation of the deleterious effect caused to the cast part by studying the values of *L*, in the cases where the flow undergoes from type 1 to type 2. Figure 4 shows that whenever *We* number grows above that 1.4, *L* grows steadily. Interestingly the growing tendency is approximately the same regardless the size or the type of symmetry of the channel, which reinforces the strong influence of *We* number over the structure of the flow.

Conclusions

- It has been researched, by means of computer simulation, the critical velocities for round and flat channels of two different sizes: 10 mm and 25 mm. The minimum critical velocity found in the study was 0.3 m/s, corresponding to the 25 mm flat channel. The critical velocity for the 10 mm flat channel was calculated as 0.47 m/s. The critical velocities for round channels lie in the range 0.4 – 0.5 m/s.

- There seems to be an interval of Webber numbers above which surface turbulence occurs. This interval has been proposed to be 0.5 – 1.5. In the simulations performed, the increase of *We* above that range implied an increase of oxide entrainment for flat channels.

- Based on Webber number, it has been proposed a quantitative relationship that establishes a diminution of the critical velocity with channel size. This relationship has been accurately confirmed by the simulations for the case of flat channels.

- The minimum value for critical velocities has been found to be lower in the case of the flat channels. This has been explained because of the higher *We* of flat channels vs. round ones.

- Further research can be accomplished to reassure the range of proposed transitional *We*, by extending the study to a broader interval of channel sizes and different cast metals.

References

1. Campbell J, Castings, 2nd Ed, Butterworth-Heinemann Ltd, Oxford 2003, ISBN 0 7506 4790 6.
2. Ruynoro J and Campbell J, *The running and gating of light alloys.*
3. Crespo A, Mecanica de Fluidos, E.T.S. de Ingenieros Industriales de Madrid, 1994, ISBN 84-7484-061-9.
4. Lide D., Handbook of Chemistry and Physics, 78th ed, David R. Lide (editor in chief); H.P.R. Frederikse (associate ed.) Boca Ratón, Florida (1997). ISBN: 0849304784.

e-mail of the main author: rafcue@cidaut.es

Tables

Table 1: results of the simulations

Type of channel	Size (mm)	Velocity (m/s)	Type of flow pattern	L (mm)	We
Flat	10	0,3	1	0	0,59
Flat	10	0,4	1	0	1,05
Flat	10	0,5	2	1,25	1,63
Flat	10	0,6	2	4	2,35
Flat	10	0,7	2	7,5	3,20
Flat	10	0,8	2	13	4,18

Flat	10	0,9	2	14	5,30
Cylindrical	10	0,3	1	0	0,29
Cylindrical	10	0,4	1	0	0,52
Cylindrical	10	0,5	3	0	0,82
Cylindrical	10	0,6	3	0	1,18
Cylindrical	10	0,7	3	0	1,60
Cylindrical	10	0,8	3	0	2,09
Cylindrical	10	0,9	3	0	2,65
Flat	25	0,4	2	7	2,62
Flat	25	0,5	2	11	4,09
Flat	25	0,8	2	31,5	10,46
Cylindrical	25	0,4	1	0	1,31
Cylindrical	25	0,5	2	4,75	2,04
Cylindrical	25	0,6	2	6,75	2,94
Cylindrical	25	0,7	3	0	4,00
Cylindrical	25	0,8	3	0	5,23

Figures

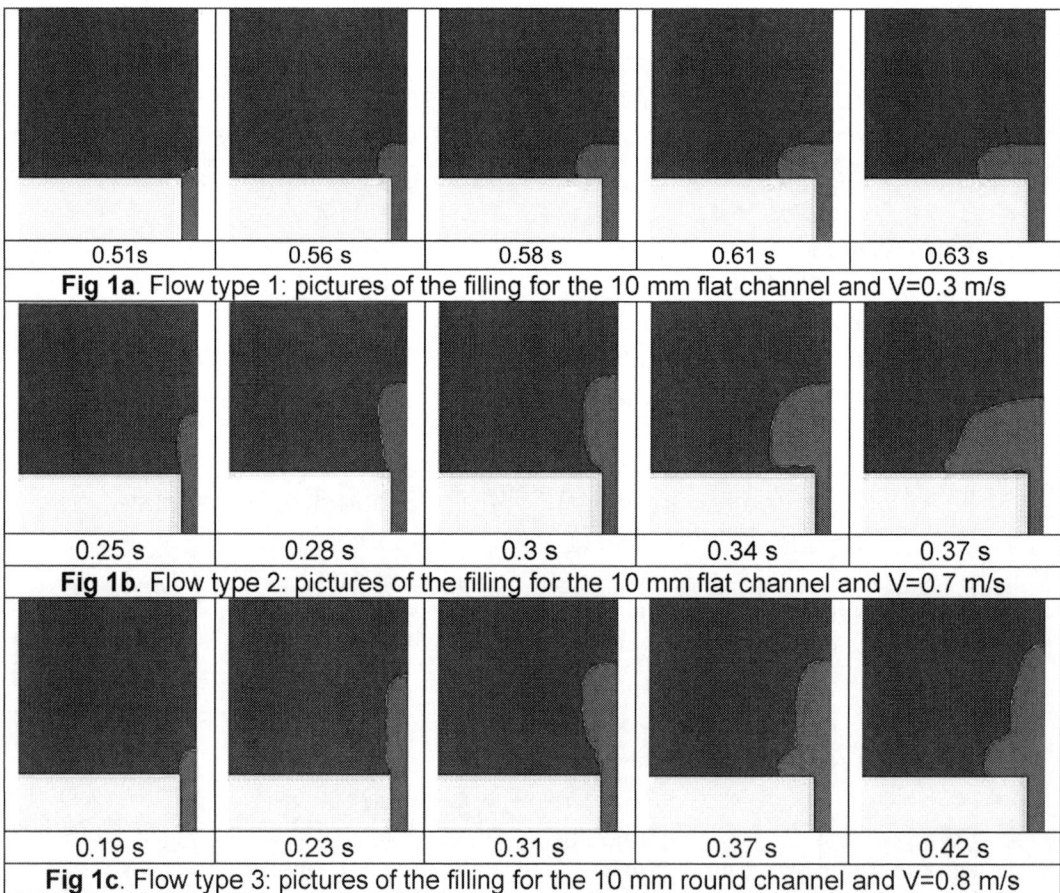

| 0.51s | 0.56 s | 0.58 s | 0.61 s | 0.63 s |

Fig 1a. Flow type 1: pictures of the filling for the 10 mm flat channel and V=0.3 m/s

| 0.25 s | 0.28 s | 0.3 s | 0.34 s | 0.37 s |

Fig 1b. Flow type 2: pictures of the filling for the 10 mm flat channel and V=0.7 m/s

| 0.19 s | 0.23 s | 0.31 s | 0.37 s | 0.42 s |

Fig 1c. Flow type 3: pictures of the filling for the 10 mm round channel and V=0.8 m/s

Fig 1: Filling pictures illustrating the three types of flow.

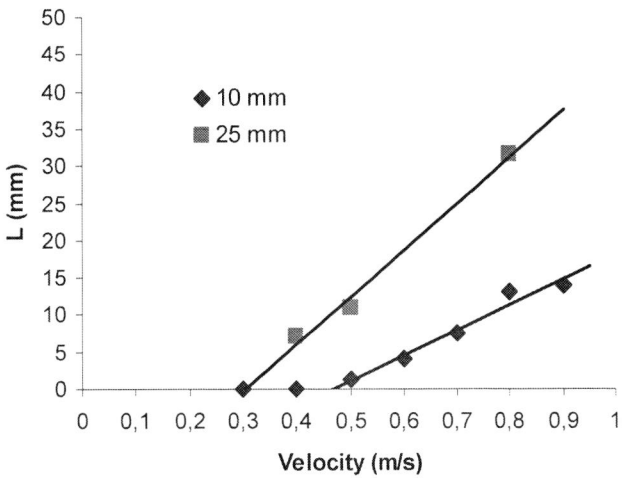

Fig. 2: Dependence of L with melt velocity for 10 mm and 25 mm flat channels. Regression lines for the L values different from cero for each channel are shown.

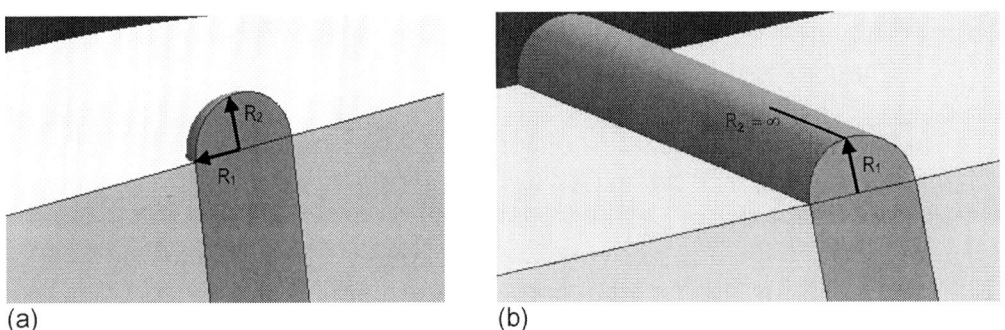

(a)   (b)

Fig 3: Schematic representation, in cross section, of molten metal surface emerging form a channel into a much large vessel. (a) Cylindrical. (b) Flat. The radii of curvature of the surface are shown.

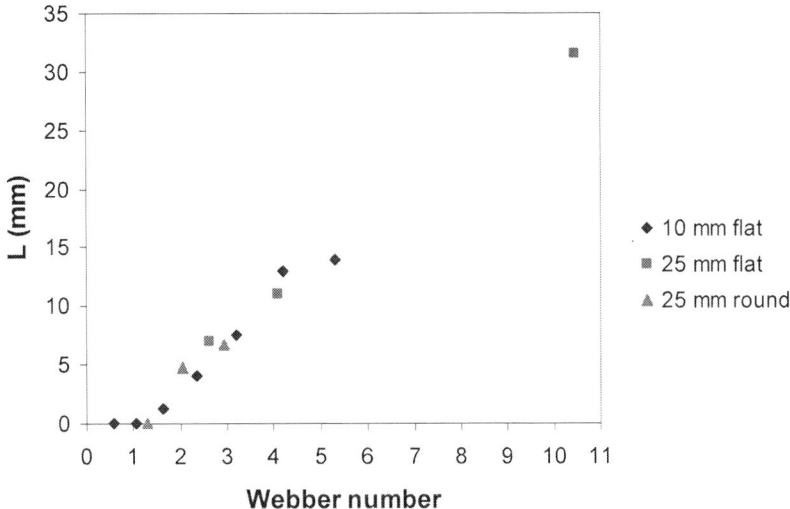

Fig 4: Dependence of L vs. We number for the simulations where transition to surface turbulence was defined by a change in flow type from #1 to #2.

Porosity Criteria Functions Revisited

J. T. Berry and R. Luck

Department of Mechanical Engineering, Mississippi State University, USA

Abstract

Casting modeling software for porosity minimization often results in critical parameter values, for example local values of temperature gradient, solidus velocity, solidification time, gate velocity etc. Such parameter combinations are termed criteria functions (CFs).

Work by Campbell on oxide bifilms and liquid metal damage motivated the authors to re-examine experimental data accrued over fifteen years on porosity in aluminum castings.

Results of tests on plate castings melted using distinctly differing methods were re-examined. Melting methods included:
- (a) induction melting
- (b) reverbatory furnace melting
- (c) resistance furnace melting
- (d) gas fired melting
- (e) heated ladle melting

Results indicate that use of an electrical resistance heated ladle melting, where minimal disturbance occurs in melting plus a reduced number of transfer operations, leads to significantly lower porosity in castings poured and more reliable CFs.

Key words Porosity, Criteria Functions, Melting Methods.

Introduction

The term 'criteria function' as applied to the modeling of the pouring and solidification of castings appears to have come into use in the nineteen eighties. It was mentioned in 1980 in a paper by one of the present authors and Dr. P.N. Hansen (1). Essentially and in the solidification modeling context, it refers to combinations of parameters or possibly a single parameter, such as a minimum temperature gradient (G) or rate of movement of liquidus (Vs) Cooling Rate (R) which are required to be maintained at some location within a casting to ensure that it is free from porosity.

Niyama and his co-workers (2) developed such a criterion for avoiding center-line shrinkage in low-carbon steel castings (G/√R). This criterion improved upon a criterion of a minimum temperature gradient ($G_{min.}$) which was proposed by Pellini and co-workers at the US Naval Research Laboratory in the nineteen fifties. (3)

Early attempts to obtain criteria for longer freezing range alloys than low carbon steels, for example, Aluminum-10% Mg sand casting alloy(4), also specified minimum temperature gradients during freezing.

Since that time many workers have suggested and have attempted to validate a number of other criteria. Spittle, Berry and their respective co-workers have reviewed certain of the criteria proposed (5,6). Like many others they pointed out that no one parameter or combination of parameters appeared to be capable of providing anything other than rough guidance to the foundry methods engineer involved in designing rigging systems for aluminum alloy castings. Many of the objectives raised by other workers hinged upon the insensitivity to dissolved hydrogen content, casting geometry and melting practice.

Experimental

In an analysis of experiments involving various modeling derived criteria functions (CFs) and their application, which included their re-derivation, Taylor at alia (7) mention that the best overall correlation was seen with results of a small scale melt with especially low hydrogen content (8). Figure 1 shows the results of this specific melt, while Table I presents correlative data (7).

Consequently, and in view of the great disparity of results obtained in the course of an extensive investigation, the writer and his colleagues suggested an upper-bound approach. In this approach an upper boundary of all results is drawn where porosity is plotted as a function of the criteria function.

Figure 2 shows such a boundary drawn above results collected for several aluminum base casting alloys poured under both industrial and laboratory scale conditions. The upper bound represents a level of porosity which is not liable to be exceeded for a particular level of the criterion chosen (Viz. the Niyama criterion). The castings concerned were principally end-gated and fed plates of thickness ranging between 15 and 50mm. The alloys concerned were A356, 319, and 206. Various melt treatments were involved, as were different melting and handling routines. Certain castings were chilled, while most heats were monitored closely for dissolved hydrogen content. Details of the experiments concerned may be found in the reports to the US Department of Energy (DOE), which supported this investigation (8, 9).

Results

Drawn in figure 3 is the range of CFs for which an approximately linear upper bound could be drawn for all heats involved in the subject investigation. It includes all revised data for alloy 319.

> (a) induction melting
> (b) reverberatory furnace melting
> (c) resistance furnace melting
> (d) crucible/gas melting
> (e) heated ladle melting

Of particular interest to the writers have been the results for plate castings poured directly from a heated ladle (i.e. the heat was not reladled, lifted and transported to the pouring station). These results fall below those for the other melting methods.

Following the period during which the many castings in the DOE supported investigation were poured. Professor Campbell and his colleagues developed a completely new approach to the problem of producing high quality aluminum castings which has been based on a critical gating velocity of 0.5m/s. The seminal investigations of Runyoru and Campbell (10) established that if this critical speed were exceeded, the bend strengths of samples excised from castings experiencing these velocities were significantly lower than those excised from castings experiencing velocities below 0.5m/s. The underlying mechanism proposed by Campbell and co-workers behind these results was the damaging of the molten metal through the inadvertent occlusion of oxide bifilms during pouring. The bifilms concerned provide sites for porosity formation. The term 'nucleation', usually contained in most theories of how dispersed porosity form initially, is deliberately not used in this context, as the air space between the folded oxide films (termed 'bifilms') provides a ready-made site for further gases expelled during cooling and solidification (such a hydrogen from molten aluminum) to precipitate. According to Campbell these partially occupied sites will then grow by the folded bifilms inflating and/or unfurling into finite pores. Bifilms which do not inflate and/or unfurl will still constitute surfaces of weakness in the subject casting. Consequently, surface turbulence must be avoided at all costs. Several papers in a recent TMS symposium honoring Professor Campbell speak to this topic (11).

Elapsed time and local pressure effects, together with those of dissolved hydrogen are explained in the term of the unfurling the suppression and/or inflation of the oxide bi-film packets during freezing. Clearly, the ingestion or inadvertent occlusion of oxide bi-films and their associated entrapped air will not be confined to metal flow within the mold. All stages of melting, laundering, ladling and other aspects of molten metal transfer will be involved.

In the light of these aspects the results of the DOE funded investigation (8, 9), together with those kindly supplied by Dr. Spittle of the University of Wales, Swansea (12) have been carefully re-examined. Spittle's data involves small scale heats where resistance heating was employed in melting the charge.

Discussion

In re-examining the above results the entire sequence of melting, ladling and transporting was considered.

Four basic sequences to have been exampled:
- (a) Induction hearing melting, lifting/transporting and pouring
- (b) Crucible/gas melting, lifting, transporting, and pouring
- (c) Reverberating furnace melting, ladle dipping, transporting and pouring
- (d) Heated ladle melting and pouring

Careful examination of the heat histories concerned suggests that the plate castings associated with multi-stage movement of molten metal possesses dispersed porosity at higher levels than those associated with minimal movement. Furthermore, the additional circulation of molten metal that occurs within the crucible through electromagnetically induced action in induction melting provides a serious initial contribution to the above effects.

Significantly, the lowest of the upper bound results referred to earlier is associated with metal that was both melted in and poured from one vessel, a heated ladle. The results of Spittle and co-workers (12) which were associated with laboratory scale experiments involving resistance furnace melting for castings of similar thickness (15mm) and composition (LM25) to those associated with the heated ladle melt heat above (A356) are contained by the bounds drawn for the latter heat. (fig 4) The scatter band, which is significantly narrower than those observed for other melt/pour sequences, will still reflect the variation in porosity level which arises from the varying oxide bifilm population within individual castings. The experiments concerned (12) would appear to be associated with minimum liquid metal damage. (The dissolved hydrogen contents of both heats involved were maintained between 0.1 and 0.2 mls/100g. It is gratifying to observe that porosity levels of castings poured and model predictions of criteria made several thousands of miles apart correspond so closely.

Conclusions

Although the effects of handling, that is 'disturbing' or even 'damaging' the liquid metal meniscus during and immediately after melting need to be confirmed by independent experiments, the results of the above re-examination appear to point out that an in-situ approach to melting and pouring would be preferred to one involving unnecessary movement, such as reladling or large scale laundering.

References

1. Hansen P N, and Berry J T, *The Simulation of Heat Transfer in Castings and Weldments-Some Thoughts on Needed Research*, Modeling of Casting and Welding Processes, 1981, pp 497-501.

2. Niyama E, Uchida T, Morikawa M, Saito S, *A Method of Shrinkage Prediction and Its Application to Steel Casting Practice*, AFS Cast Metals Research Journal, 7, No. 3, 1982, pp 52-63.

3. Pellini W S, *Factors which Determine Riser Adequacy and Feeding Range*, Trans. Am. Foundry Soc., 61, 1953, pp 61-80.

4. Johnson W H, Kura J G, *Some Principles for Producing Sound Al-7Mg Alloy Castings*, Trans. Am. Foundry Soc. 67, 1959, pp 535-552.

5. Spittle J A, Brown S G R, Sullivan J G, Applications of Criteria Functions to the Prediction of Microporosity Levels in Castings in Solidification Processing 1997 (J. Beech, H. Jones, eds) 1997, pp 251-255.

6. Berry J T, Huang H, *Evaluation of Criteria Functions to Minimize Microporosity Formation in Long Freezing range Alloys*, Trans. Am. Foundry Soc. 101, 1993, pp 669-675.

7. Taylor R P, Berry J T, Overfelt R A, *Parallel Derivation and Comparison of Feeding Resistance Porosity Criteria Functions for Casings*, Proc. ASME 31st National Heat Transfer Conference, 1996, Vol. 1 HTD-vol 323, pp 69-77.

8. Berry J T, (editor), *Final Report on Determination of Correlation Factors for the Prediction of Shrinkage in Castings*, July 1991 Metal Casting Technology Center, Univ. of Alabama.

9. Berry J T, *Criteria Functions for Defect Prediction,* in Final Report Metal Casting Competitive Research, August 1994 (T.S. Piwonka, Ed.) DOE Report DOE/ID/13163-1 (DE 95016652)

10. Runyoru J, Boutorabi S M A, Campbell J, *Critical Gate Velocities for Film Forming Casting Alloys*, Trans. Am. Foundry Soc. 100, 1992, pp 225-234.

11. Tiryakioglu M, Crepeau P N (Eds.) *Shape Casting-The John Campbell Symposium*, TMS, Warrendale, Pa, 2005. ISBN 0 87339 583 2

12. Spittle J, Private Communication, November 2005.

Acknowledgements

The support of the USDOE and MSU's Center for Advanced Vehicular Systems is gratefully acknowledged, as is the help of Ms. D. Youngblood and Messrs. A.R. Williams and B. Dawsey in preparing material for the manuscript. A special thank you must be added for the additional data supplied by Dr. John Spittle of the University of Wales, Swansea.

Tables

Table 1 Criteria Functions in Original and Rederived Simplified Forms

Criterion	Original Form	Simplified Form	R^2 Value
Niyama et al. (NUMS)	$\dfrac{G}{R^{1/2}}$	$\dfrac{V_s}{G}$	0.86
Lee et al. (LCC)	$\dfrac{G \cdot t_s^{\frac{2}{3}}}{V_s}$	$\dfrac{V_s^5}{G}$	0.87
Suri (Dimensional) (SD)	$\dfrac{V_s^{1.63}}{G^{0.052}}$	$\dfrac{V_s^{1.63}}{G^{0.052}}$	0.85
Suri (Non-dimensional) (SND)	$V_s^{-0.37} \cdot G^{-0.052}$	$V_s^{-0.37} \cdot G^{-0.052}$	0.78
G	G	G	0.78
V_s	V_s	V_s	0.84

Figures

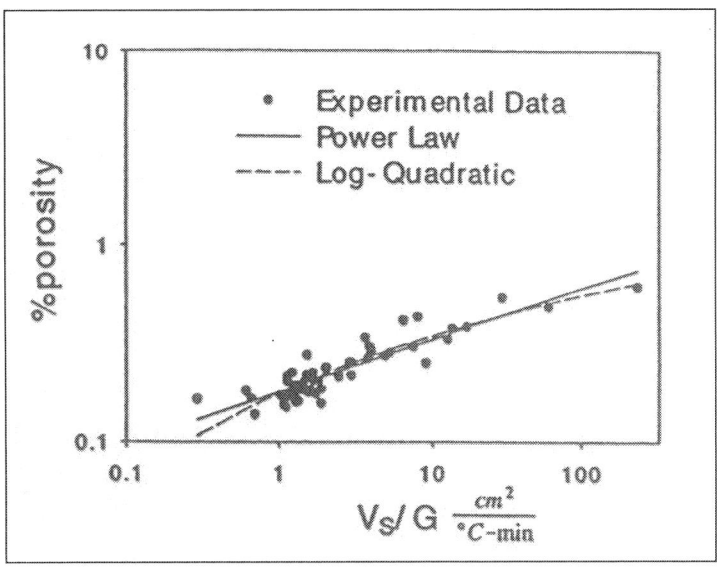

Fig. 1 Porosity v. modified Niyama Criterion (Vs/G) for data of Lee for heat of A356 equivalent with exceptionally low hydrogen content (0.01ml/100gm). See ref. 7.

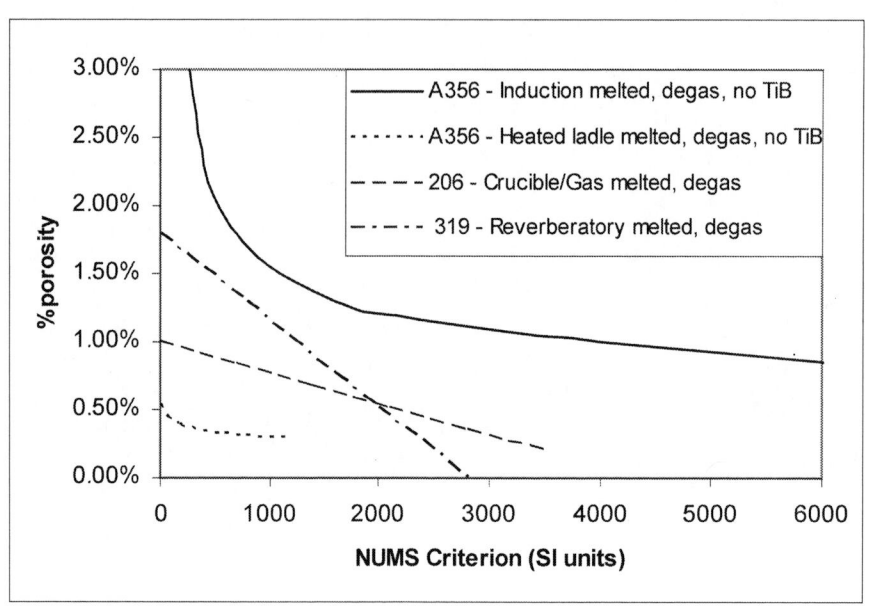

Fig. 2 Upper bounds for a variety of aluminum cast alloy heats

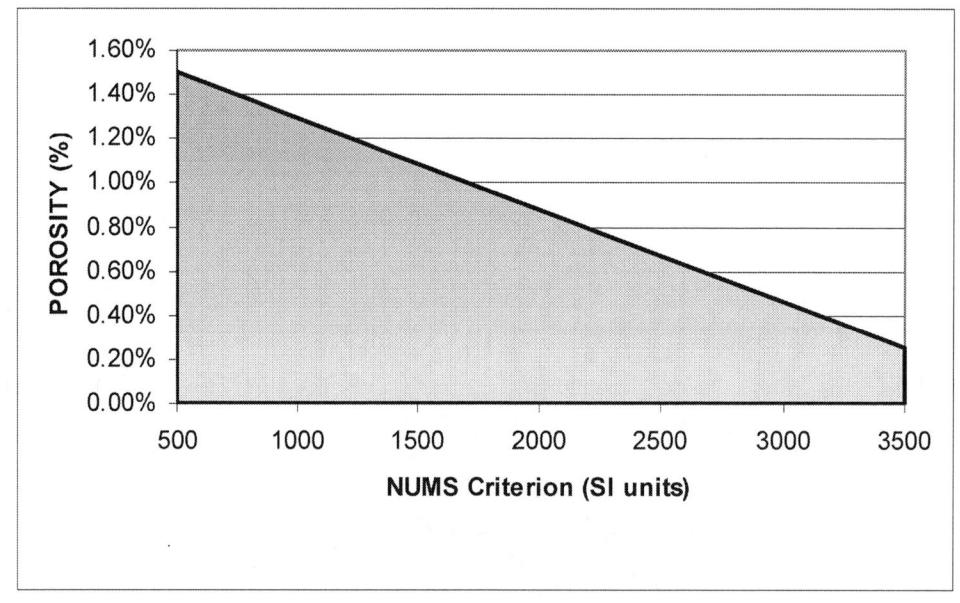

Fig. 3 Upper bound for all heats in figure 2

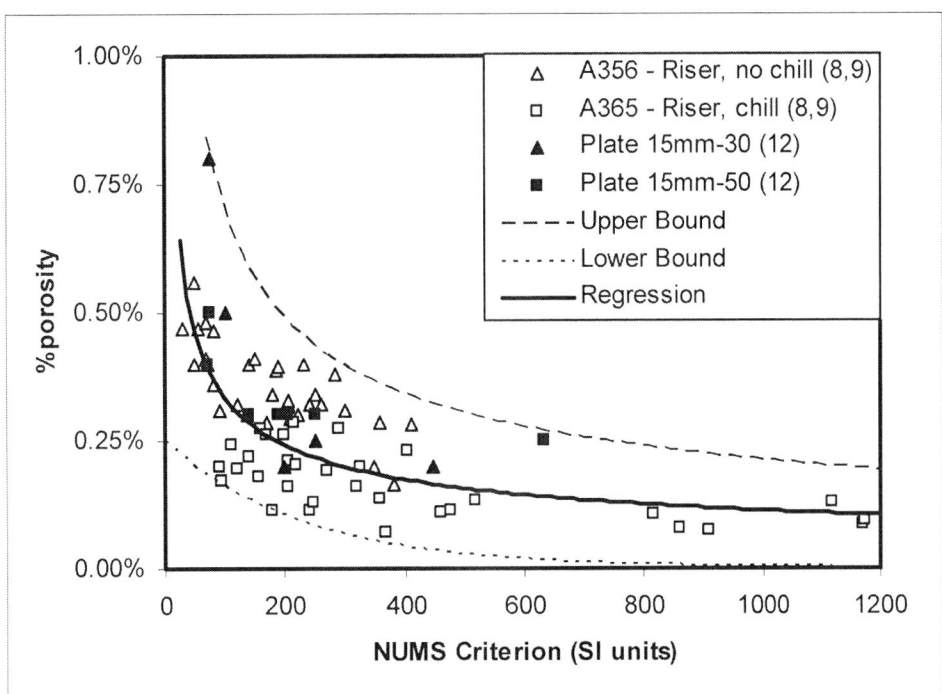

Fig. 4 Bounds for A356 heat melted and poured from heated ladle (8, 9). Also included are data points from laboratory scale experiments of Spittle (12) for LM25 castings. Both sets of castings were degassed and TiB treated. Hydrogen contents were maintained between 0.1 and 0.2 ml/100gm.

Simulation of Casting Solidification Using Different Boundary Conditions

J S Suchy *, A Gradowski * and J Lelito*.

* AGH, University of Science and Technology, Department of Modelling of Foundry Process, PL.

Abstract

Correctness and credibility of results of computer simulation of cooling and solidification process of casting, first of all depends on the appropriate selection and correct formulation of model's conditions, among which initial – boundary conditions are of great importance, because they reflect kinetics of the heat flow in cast-mould system. Defining geometrical and physical conditions does not create serious problems. However, the procedure of selection of boundary conditions has not been convincingly defined.

In this paper we have attempted to define the procedure of selecting parameters which describe boundary conditions for modelling heat processes, through the analysis and comparison of simulation results gained while using modifications of boundary conditions.

Keywords simulation, solidification, boundary conditions, ductile iron, hot spot.

Introduction

Results of numerical simulation of casting cooling and solidification depend on assumed parameters of a mathematical model: the so called geometrical, physical, initial and boundary conditions. In the paper, an influence of physical and boundary conditions on heat flow in solidifying casting is discussed.

The aim of the paper is an attempt to define parameters describing above mentioned conditions of mathematical modelling of heat flow by analyse and comparison of results of simulation using different physical and boundary conditions. The results of simulation will be compared with measurements of temperature in a casting of diffuser made of ductile cast iron as well as its mould. Measurements have been performed in a company WSK Rzeszow, manufacturer of the diffusers.

The following mathematical models of casting-mould heat exchange have been assumed:

a) a 2D model for a characteristic casting cross - section of the casting using the 4th type boundary conditions (an ideal thermal contact between casting and mould),

b) a 2D model for the same characteristic casting cross - section using the 3rd type boundary conditions between casting and environment ("virtual mould"),

c) an analytical solution of the simplified model described in the point b),

d) 3D model of the complete casting and mould geometry using MAGMASOFT® software.

Different mathematical models may lead to different results. Basing on results of temperature measurements, parameters of the models will be changed in such a way, to obtain similar values of solidification time as well as coordinates of last-solidified area. Thermo-physical parameters of moulding sand have been calculated basing on measurement of solidification time of the casting.

Experiment

To fulfil the assumed aims of the paper, the cooling curve of the selected hot spot of the cast was measured. To do so, we conducted the melting process in industrial conditions in the WSK Foundry in Rzeszow. It was done in an induction furnace of 1 tonne capacity, using pig-iron, ferrous scrap and technical pure silicon. Spheroidization and modification were conducted with the Sinter – Cast technique, using 2 elastic wires. One of the wires contained magnesium core, whereas the other one - a modifier. The chemical composition of such ductile iron is presented in Table 1.
Green sand with 4.5% water content was used. The casting of the diffuser including gating system and feeders shows the Figure 1.

Obtained casting had the initial pearlite – ferrite microstructure of the matrix with typical nodular graphite (Figure 2).

To measure the temperature, thermocouples of type K with wire diameter of ϕ 0.3 mm were used. The thermocouple registering the change in temperature in the solidifying alloy was secured from the liquid metal with a quartz tube encapsulated on one side. The thermocouple was placed 308 mm away from the cast's axis on the depth shown in Figure 3 (point A). In the case of the thermocouple registering the change in temperature within the mould, its hot junction was in direct contact with the moulding sand. The thermocouple was placed 418 mm away from the cast's axis, on the same level as the measuring point in the cast.

A registering AGILENT 34970A device was used to register the thermocouples voltage, its conversion to temperature values and also to store data for later work.

The curves of temperature changes versus time for casting – mould system are presented in Figure 4. The temperature interval (T_e = 1140 °C) connected with eutectic reaction can be seen on the cooling cast curve. In the case of the temperature changes curve in the mould we can also see the temperature interval occurring at 100 °C. It corresponds with the process of water evaporating from the moulding sand. These curves are the basis for evaluating the heat accumulation coefficient in the material of the mould.

2D numerical model of casting cooling and solidification

Mathematical modelling will be applied to a cross-section of a working part of diffuser, i.e. the blade rim. The point A showed on the Figure 3 has been selected as a check point to analyze the process of solidification according to two kinds of boundary conditions assumed. In the point A a thermocouple has been positioned as well.

The simulation of solidification for 2D geometry and the 4th type boundary conditions

A computer programme has been written to simulate field of temperature as well as solidification within the area of selected hot spot for 2 dimensional mathematical model of heat exchange using the 4th type of boundary condition. The following assumptions have been assumed:

1. The liquid metal has uniform initial temperature.
2. Thermo-physical parameters of casting and mould are independent of temperature.
3. Shrinking gap arising during casting solidification is very small, giving negligible heat resistance.

A finite difference method has been used to simulate the heat transfer process. A general differential equation based on control volume method has been used [1].

Physical and initial conditions are following:
a) casting is made of ductile cast iron with eutectic temperature equal to 1140 ºC,
b) initial temperature of liquid metal is 1330 ºC,
c) green sand has constant density equal to 1600 kg/m^3,
d) initial temperature of mould is 20 ºC.

The calculations have been repeated for assumed range of heat accumulation coefficient $b_m = \sqrt{\lambda_m c_m \rho_m}$ varying from 800 to 1400 $W \cdot s^{1/2}/(m^2 \cdot K)$, where λ_m is heat conduction coefficient, c_m is specific heat and ρ_m is mass density. A constant value of specific heat of green sand has been used.

The aim of numerical simulation was to establish the following parameters:
  a. solidification time at the check point A (Fig. 3),
  b. total solidification time of the examined hot spot,
  c. relative solidification time of the check point A,
  d. solidification modulus of the hot spot $M_{hs} = K_o \cdot \sqrt{\tau_{sol}}$,
  e. coordinates of last solidified point (x_{3w}, y_{3w}), compare Fig. 3,
  f. solidification constant K_o, according to Chvorinov rule.

Space steps equal to 2 mm, both in casting as well as in mould, while time step equal to 0.19 s have been assumed. Results of numerical simulation have been shown in the Table 2.

Varying the heat accumulation coefficient of the green sand from 800 to 1400 $W \cdot s^{1/2}/(m^2 \cdot K)$, the solidification parameters will change as follows:
  a) solidification time of the check point A changes from 1512 to 545 s,
  b) relative solidification time of the check point A, i.e. $S_1 = \tau_{3A}/\tau_{3w}$ changes in a small range from 0.631 to 0.594,
  c) solidification modulus also changes a little, from 1.71 to 1.85 cm, what means that it depends not only of geometry,
  d) to obtain the same calculated solidification time of the check point A as in experiment, the value b_m = 1310 $W \cdot s^{1/2}/(m^2 \cdot K)$ should to used,
  e) coordinates of the last solidified point K are the same (10, -3 mm), independent of a value of the coefficient b_m, i.e. independent of casting cooling intensity,
  f) solidification constant is varying proportional to value of heat accumulation coefficient, according to its definition.

The calculated values of thermal solidification module vary along to variation of heat accumulation coefficient, i.e. along to heat transfer intensity. Since the modulus slightly increase with the value of the coefficient, we can assume that influence of non-uniform cooling of the casting (area of mould limited by the three interfaces, see the Fig. 1) depends on intensity of casting cooling. It can be concluded, that known in the literature methods of calculation of thermal modulus [2] should be extended taking into account this effect.

Simulation of solidification for 2D geometry and the 3rd type boundary conditions

The second mathematical model of casting cooling and solidification also concerns 2D geometry but the 3rd type of boundary condition casting-environment has been assumed for the case. It means that the real mould has been substituted by a "virtual mould" one, what is a concept already known in the literature. In such a case, thermal behaviour of the mould is equivalent to an effective (average) heat transfer coefficient α, ensuring the same solidification time. Calculations have been performed for the values of the coefficient α varying in the range $30 \div 70 \ W/(m^2 \cdot K)$.

Results of numerical modelling of the process are presented in the Table 3.

As follows from the Table 3, the measured value of the solidification time of the check point A equal to 610 s can be obtained in simulation for the case of the "virtual mould", assuming the value $\alpha = 41.5 \ W/(m^2 \cdot K)$. It was not obvious that the coordinates of the last solidified point are quite different in the case (x = 11 mm and y = -11 mm), regardless of values of heat transfer intensity. For the case of the "virtual mould", calculation of solidification modulus is impossible.

Comparison of results of calculation for the two types of applied boundary conditions confirms that proper choice of the conditions has key influence on the results of calculation.

Analytical calculation of an average value of heat transfer coefficient for the casting – shrinkage gap – mould system

We are seeking an average value of heat transfer coefficient in the casting-shrinkage gap-mould system to obtain the same solidification time for both the 3rd as well as 1st type of boundary conditions. To define it, we will write the heat flux using two equations:

a) for the mould surface according to the 1st type boundary condition:

$$q_{face2} = \frac{b_m \left(T_{face2} - T_{2p} \right)}{\sqrt{\pi \tau}} \tag{1}$$

where:

T_{face2} – temperature on the inner surface of mould,

T_{2p} – initial temperature of the mould.

b) for the shrinkage gap:

$$q_{gap} = \frac{\lambda_{gap}}{X_{gap}} \left(T_{face1} - T_{face2} \right) = R_{gap}^{-1} \left(T_{face1} - T_{face2} \right) \tag{2}$$

where:

λ_{gap} – effective coefficient of heat conduction of shrinkage gap including heat transfer through radiation,

X_{gap} – gap thickness, changing according to the square root principle,

R_{gap} – heat resistance of the shrinkage gap,

T_{face1} – temperature of the casting surface.

Both the heat fluxes (1) and (2) are the same. According to Newton's law (boundary condition of the 3rd type) the same heat flux can be described by the equation:

$$q_{3r} = \alpha\left(T_{face1} - T_{2p}\right)$$ (3)

From the equations (1), (2) and (3) it follows:

$$\alpha = \frac{b_m}{\sqrt{\pi\tau} + b_m R_{gap}}$$ (4)

This equation (4) allows to obtain an average value of the heat transfer coefficient:

$$\alpha_{average} = \frac{1}{\tau_{sol}} \int_0^{\tau_{sol}} \alpha(\tau)d\tau$$ (5)

Assuming that the thickness of the shrinkage gap X_{gap} is a linear function of a solidified layer, therefore it can be written in the form:

$$X_{gap}(\tau) = \varepsilon M \sqrt{\tau/\tau_{sol}}$$ (6)

where:

M – geometrical modulus of solidification,

ε – dimensionless value of the linear shrinkage of the casting (cast iron),

τ_{sol} – solidification time of the casting.

Substituting the equations (4) and (6) into the equation (5) we obtain the average value of the heat transfer coefficient:

$$\alpha_{average} = \frac{1}{\tau_{sol}} \int_0^{\tau_{sol}} \frac{1}{\dfrac{\dfrac{\sqrt{\pi}}{b_m} + \dfrac{b_m \varepsilon M}{\lambda_{gap}\sqrt{\tau_{sol}}}}{\sqrt{\tau}}} d\tau$$ (7)

If we assume: b_m = 1310 W s$^{\frac{1}{2}}$/(m^2 K), ε = 0.01, λ_{gap} = 0.074 $W/(m\ K)$, M = 10 mm and τ_{sol} = 1022 s, we will obtain the average value of the coefficient $\alpha_{average}$ = 44.8 W/(m^2 K).

If the shrinkage gap was not considered, then the approximate value of this coefficient would be equal to 46.2 $W/(m^2\ K)$. The analysis allows to select the boundary condition for the mathematical model of casting cooling and solidifying corresponding to the idea of "virtual mould".

Simulation of solidification for 3D geometry

MAGMASOFT® software was used to perform 3D numerical calculations. The geometry of the real casting in the STL format was input to the programme. After positioning virtual thermocouples in the casting – mould system in locations corresponding to real ones, the differential mesh was generated. The total number of control volumes was 609280, where 81276 control volumes were in metal. The generated differential mesh is shown on the Fig. 5.

The filling of the mould's cavity with liquid alloy was controlled by the percent filled (Fig. 6).

Next, the process parameters have been selected, such as: chemical composition of cast iron (Table 1), the pouring temperature (T_{pour} = 1330 °C) and thermo-physical properties of the casting as well as mould from the MAGMASOFT® database [4].

As a result of simulation, a cooling curve in the check point A has been obtained, that is compared with the measured cooling curve for the same point (Fig. 7). Both the curves have the same slope in the initial stage of casting cooling. The solidification times, i.e. calculated and measured are similar, having the value about 610 s.

For such a case of simulation, a 3-dimensional field of solidification time can be shown (Fig. 8). The check point A is out of the last solidified area.

Conclusions

1. Three variants of numerical simulation of the process of solidification of the casting have been compared:

   a) a 2D model of casting-mould heat transfer using boundary conditions of the 4th type for a selected, characteristic cross-section of the casting,

   b) a 2D model of casting-environment heat transfer using boundary conditions of the 3rd type for the same cross-section of the casting, (a "virtual mould" has been applied),

   c) a 3D model of heat transfer using variable boundary conditions of the 4th type for the whole volume of the casting.

When appropriate physical as well as boundary conditions are applied, similar solidification times can be achieved, despite different mathematical models used.

2. Although both the 2D mathematical model used allow to obtain the same time of solidification of the selected checkpoint on the casting cross-section, location of the last solidified point depends on boundary condition applied.

3. In our opinion, for the 2D geometry, the 4th type boundary condition gives values of solidification process parameters (solidification time as well as position of the last solidified point) closer to reality, than the 3rd type boundary condition (or "virtual mould") gives.

4. 3D modelling using MAGMASOFT® gives solidification time equal to 610 s. To obtain the same value using 2D models, it is necessary:

   a) set value of heat transfer coefficient α = 41.5 $W/\left(m^2 \cdot K\right)$ for the 3rd type boundary condition (casting - virtual mould),

   b) set value of the heat accumulation coefficient of the sand mould b_m = 1310 W·s$^{1/2}$/(m^2·K) for the 4th type boundary condition.

5. To calculate the value of heat transfer coefficient α we have defined the equation (7). Then, instead of the casting – shrinkage gap – sand mould system, an equivalent casting – environment system can be applied (virtual mould).

6. Influence of boundary conditions on process of solidification will be further investigated using a simpler shaped casting. It would minimize errors arising from numerical representation of geometrical conditions of the examined casting – mould system.

References

1. Gradowski A, *Numerical Optimization of Casting Feeding by Application of Risers Solidifying in an Insulating Layer and Risers Coated with Insulation Mould Powder*, Przeglad Odlewnictwa, No 1, 2003, pp 7-12.

2. Longa W, *Solidification of Castings*, Katowice, 1985, pp 357-379.

3. Mochnacki B, Suchy J, *Numerical Methods In Computations of Foundry Processes*, Krakow, 1995.

4. Manual of MAGMASOFT® 4.2, Aachen, 2002.

Tables

Table 1. Chemical composition of ductile iron

Results of analyses [wt. %]													
C	Mn	Si	P	S	Cr	Ni	Cu	Mg	Mo	Ti	Sn	Pb	V
3.48	0.63	2.58	0.029	0.017	0.06	0.04	0.32	0.054	0.07	0.015	0.006	0.001	0.009

W	Zn
0.038	0.003

Table 2. The parameters of the casting solidification process according to boundary conditions of the 4th type (b_m and λ_m are input values)

b_m, W·s$^{1/2}$/(m^2 K)	800	1000	1200	1400	1310
λ_m, W/(m·K)	0.364	0.568	0.818	1.11	0.975
τ_{3A}, s	1512	980	702	545	610
τ_{3W}, s	2397	1615	1183	917	1022
$S_1 = \tau_{3A}/\tau_{3W}$	0.631	0.607	0.593	0.594	0.597
M_{hs}, mm	17.1	17.5	18.0	18.5	18.3
x_{3w}, mm	-10	-10	-10	-10	-10
y_{3w}, mm	-3	-3	-3	-3	-3
K_o 10^4, m/ s$^{1/2}$	3,49	4.37	5.24	6.11	5.72

Table 3. The parameters of the casting solidification process according to the boundary conditions of 3rd type (α is input value)

α, W/(m$^2 \cdot$ K)	30	40	50	60	70	41.5
τ_{3A}, s	828	632	512	433	377	610
τ_{3W}, s	1022	784	641	545	475	758
$S_1 = \tau_{3A}/\tau_{3W}$	0.810	0.806	0.800	0.794	0.794	0.805
x_{3w}, mm	11	11	11	11	11	11
y_{3w}, mm	-11	-11	-11	-11	-11	-11

Figures

Fig. 1. Diffuser with gating system and feeders

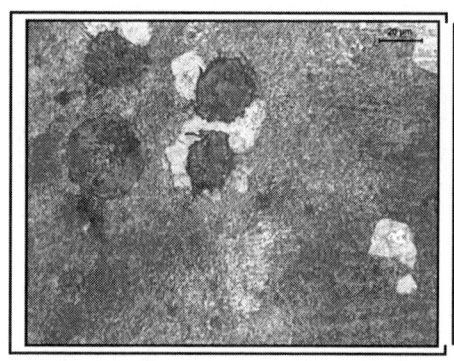

Fig. 2. Microstructure of ductile iron with pearlite - ferrite matrix, etchant: 4% nital

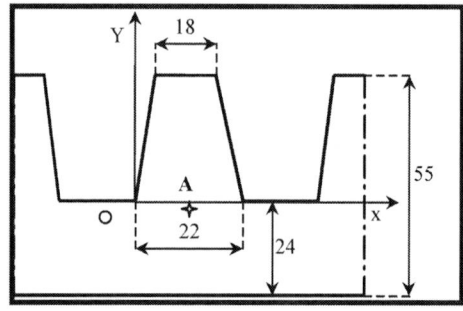

Fig. 3. Vertical cross-section of the examined hot spot of the casting

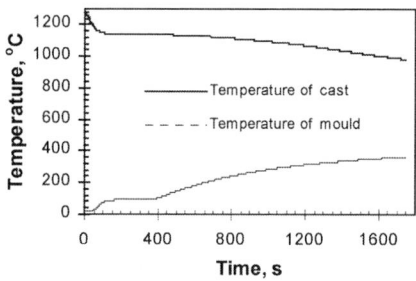

Fig. 4. Temperature plots in the casting and mould

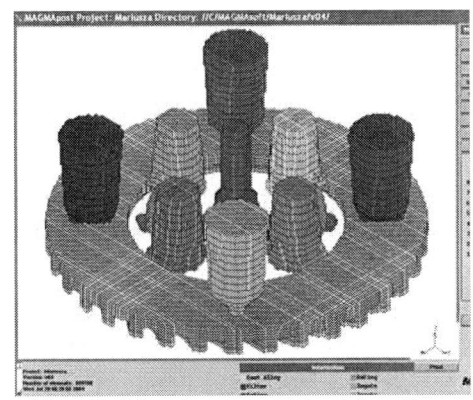

Fig. 5. Mesh generated for the casting

Fig. 6. Percent filled of alloy during the mould filling

Fig. 7. Measured and calculated cooling curves at the check point A.

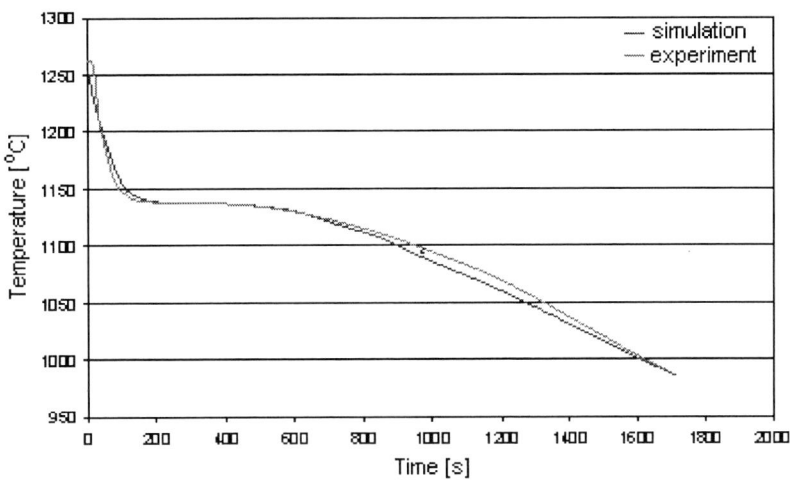

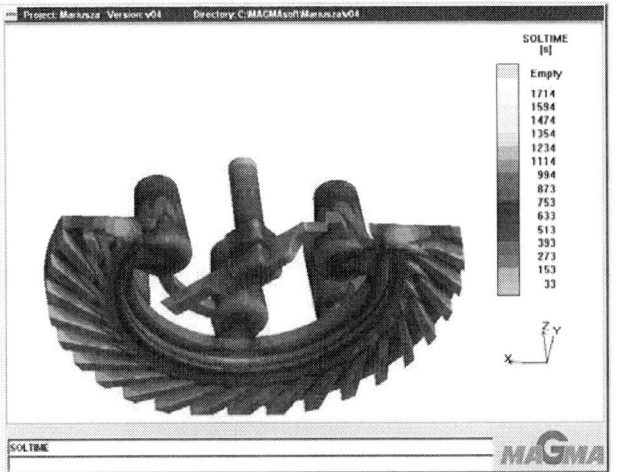

Fig. 8. Solidification time on a cross-section of the casting.

Mathematical Modeling of Compacting Process of Green Sand Molding

Li Wenzhen, Zhou Kelin, Wu Junjiao.

Department of Mechanical Engineering, Tsinghua University, Beijing 100084, P.R. China.

Abstract

Green sand molding has ever been playing an important role in modern foundry, especially in the casting production of automobile parts. The aim for a modern green sand molding requires the even distribution of mold hardness with expected mold strength. For loose green sand, several nonlinear problems must be met, including molding materials, molding process and mold geometry. In order to improve the efficiency of calculation, especially the accuracy of predicted results, some mathematical models should be set up. In this paper, the mathematical models of compacting process of green sand molding were developed by experiment. Based on these models, finite element method was used to simulate the compacting process and predict the hardness distribution of green sand mold. The predicted results have a good agreement with the measured ones.

Key words Green sand molding, compacting process, mathematical modeling, finite element method.

Introduction

Green sand molding and casting is an important processing method for the production of machine parts, especially for the casting production of automobile parts. The quality castings depend to a large extent on the quality mold. The aim for a modern green sand molding requires the even distribution of mold hardness with expected mold strength. Some new molding machines have been proposed to satisfy the above requirement quite well. However for a higher productivity of molding and for a higher mold quality of complicated castings, there is a lot of work to be done to improve the molding machine design. Numerical simulation can provide some good support to it. During last decades, some researches have been done on this field, with focus to be put on the numerical methods such as finite element method (FEM), distinct element method (DEM), and so on [1-4]. For loose green sand, several nonlinear problems must be met, including molding materials, molding process and mold geometry. In order to improve the efficiency of calculation, especially the accuracy of predicted results, some mathematical models should be set up. In this paper, the mathematical models of compacting process of green sand molding were developed by the squeezing test of molding sand. Based on these models, FEM simulation of the compacting process have been used to predict the hardness distribution of green sand mold.

Experiment

The constitutive relationship of green sand is the basis of FEM analysis of green sand compaction process [2]. In this paper, the constitutive relationship has been tested by the Institute of Materials Processing, Tsinghua University. The test appratus is shown in Figure 1.

Firstly, the loosed green sand is put into the sand canister. Then load is applied slowly from the top of the canister with the hydraulic pressure machine. During this process, the load sensor will record the load P and the displacement sensor will record the displacement U. The curves between U and P will be outputted by the X-Y recorder. The curves of U and P can be converted into the relationship between stress σ and strain ε through the following equations:

$$\sigma = P/S \tag{1}$$

$$\varepsilon = U/h \tag{2}$$

in which S stands for the area of board, h for the height of the sand canister.

The main properties of green sand are as follows: compactablity, 40% and bentonite content, 8%. The initial density is 0.9g/cm^3, 1.0 g/cm^3, 1.1 g/cm^3, and 1.2 g/cm^3 respectively. The experiment results are shown in Figure 2. One of the processed results is shown in Figure 3 (the initial density is 0.9g/cm^3). The empirical formula of constitutive relationship is as follows:

$$\sigma = e^{a(\varepsilon - b)} \tag{3}$$

$$a = 28.18 - 77.02\rho_S \tag{4}$$

$$b = \rho_W \tag{5}$$

in which σ is stress in MPa, ρ_s is bentonite content, ρ_w is compatability.

Mathematical model for numerical simulation

The squeezing and compacting process of green sand during molding is a typical dynamics problem. In this process, inertia force plays a very important role. Because the deformation of green sand in the compacting process is quite large, the finite element model considering finite deformation theory has been established.

Up to now, the finite deformation theory has been used in many engineering fields, such as materials engineering, structure engineering, mechanical engineering and geological engineering. It is well known that a structure can not endure large deformation when it is being used. The nonlinear analysis used in structure engineering is material nonlinear. But there is geometrical nonlinear besides the material nonlinear. The finite deformation problem is a geometrical nonlinear problem. The Langrange method and/or the Euler method can be used in the finite deformation theory.

In classical finite deformation theory, the Almansi Strain is:

$$e_{ij} = \left[\frac{\partial U_j}{\partial x_i} + \frac{\partial U_i}{\partial x_j} - \frac{\partial U_k}{\partial x_i} \frac{\partial U_k}{\partial x_j} \right] / 2 \tag{6}$$

The Green Strain is:

$$E_{ij} = \left[\frac{\partial U_j}{\partial x_i} + \frac{\partial U_i}{\partial x_j} + \frac{\partial U_k}{\partial x_i} \frac{\partial U_k}{\partial x_j} \right] / 2 \tag{7}$$

in which U is displacement, x is coordinate.

If the Euler Stress τ_{ij} is the actual stress related to the deformation. Then the Langrange Stress Σ_{mj} is :

$$\Sigma_{mj} = J \frac{\partial x_m}{\partial x_i} \tau_{ij} \tag{8}$$

in which J is Jacobi determinant as follows:

$$J = e_{ijk} \frac{\partial U_i}{\partial X_1} \frac{\partial U_j}{\partial X_2} \frac{\partial U_k}{\partial X_3} \tag{9}$$

The Kirchhoff Stress S_{ij} is :

$$S_{lm} = \frac{\partial X_l}{\partial x_i} \Sigma_{ij} = J \frac{\partial X_l}{\partial x_i} \frac{\partial X_m}{\partial x_j} \tau_{ij} \tag{10}$$

In large deformation dynamical FEM analysis, the incremental method is used to simulate the compacting process of loose green sand. The dynamical differential equation of Incremental method is:

$$M \ddot{U}_{t+\Delta t} + C \dot{U}_{t+\Delta t} + K\Delta U = R_{t+\Delta t} - Rs_{t+\Delta t} \tag{11}$$

$$\Delta U = U_{t+\Delta t} - U_t \tag{12}$$

in whcih M is mass matrix, C is hysteresis damp matrix, K is stiffness matrix

Simulation results
In this paper, the compacting process of the green sand mold as shown in Figure 4 was simulated In order to obtain an evenly compacted sand mold, the relationship of A/B=a/b has been assumed.

During compacting process, there exists friction force between sand particles and flask wall which cannot be neglected in numerical simulation. The friction coefficient μ is assumed to be 0.5.

As the initial conditions used in the simulation, the compactablity and the bentonite content are 40% and 8% respectively. The initial density of green sand is assumed to be evenly distributed in the flask and the filling frame. In this paper, the initial density of green sand is set to 1.0 g/cm^3.

As there is friction force between sand particles and flask wall, the density of the mold usually increases from the bottom to the top when the sand is compacted from the top side of the flask. The green sand which is located at the bottom of the flask is usually less compacted. In order to improve the quality of the mold, molding sand can be compacted both from the top side of the flask and from the pattern side of the flask. Therefore in this paper, two loading were simulated, which are loading only from the top of the flask and loading from both the top side of the flask and the pattern side of the flask. It is assumed that the ration of the displacement of the squeeze plate to the displacement of the pattern plate is 3 over 1.

The constitutive relationship curve can be seen as several straight lines as shown in Figure 3, the load related to the displacement can be calculated accordingly. After the calculation of the preceding step is finished, the result will be used to change the position of the nodes, the model will be remeshed, and the elastic module will be changed. Before the next step of calculation, the stress field of the preceding step is applied to the new model as initial stress. This procedure is repeated until the end of the calculation.

The calculated results of the two loading ways are shown in Figure 5 and Figure 6, in which the unit of stress is in Pa. In order to compare the results of the two loading ways, the values of the stress along the red line in Figure 4 are picked out, as shown in Table 1. From these results it can be seen that the stress in the softline area (the orange and red area, especially in the corner of the pattern) is much lower than that in the other part of the mold, which is also known as the " bridge phenomenon ", one of the common defects in green sand molding. The calculated results is in good agreement with the measured results of Zhang's [2].

Conclusions

1. The mathematical models used for simulation the squeezing and compacting process of green sand during molding were built through expriments. The constitutive relationship were obtained.

2. Based on the mathematical models, FEM was used to simulate the practical compacting process of green sand. The calculated results of two loading ways have been compared and it can be seen that to squeeze molding sand both from the top of the flask and from the pattern side can improve the quality of the mold obviously.

3. The simulated results showed that the stress in softline area is much lower than in the other part of the mold. The calculated results is in good agreement with the measured ones.

References

1.   Wu J, Jiang J and Yang G, *Application of FEM to predict hardness distribution of air-impact, compacted green sand molds*, AFS Transactions, 104, 1996, pp491-495.

2.   Zhang Y and Wu J, *Numerical simulation of squeeze molding and air-impact molding with the finite element method*, AFS Transactions, 111, 2003, pp11-14.

3.   Leone M and Lewis R L, *Numerical modeling of green sand during compaction*, AFS Transactions, 96, 1988, pp763-774.

4.   Makino H, Maeda Y and Nomura H, *Numerical analysis of blow molding using distinct element method*, 63rd World Foundry Congress, Budapest, Hungary, 12-18 Sept, 1998, pp5-6.

Tables

Table1. The value of the stress in MPa along the testing line in Figure 4

Height from bottom (m)	0	0.0371	0.0743	0.0111	0.148	0.185	0.222	0.247
Top side Squeezed	0.308	0.371	0.491	0.682	1.044	1.298	1.529	1.732
Both side squeezed	0.469	0.468	0.599	0.728	0.889	1.238	1.495	1.707

Figures

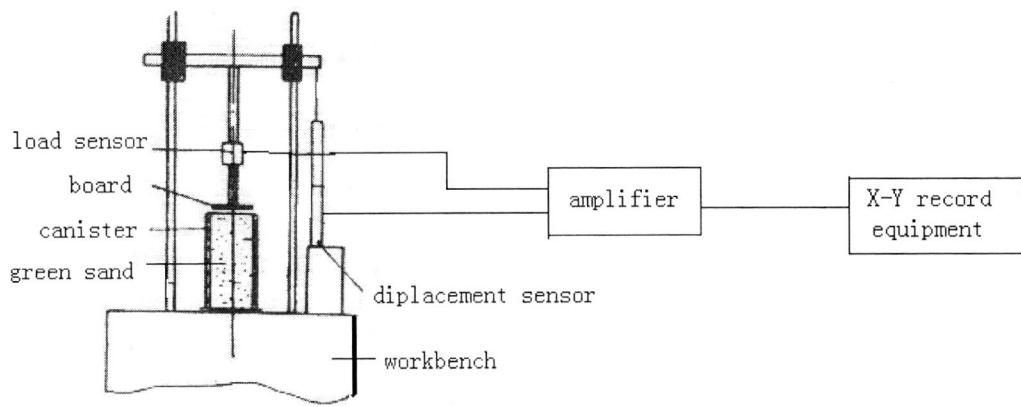

Figure 1. The equipment for testing the constitutive relationship of green sand

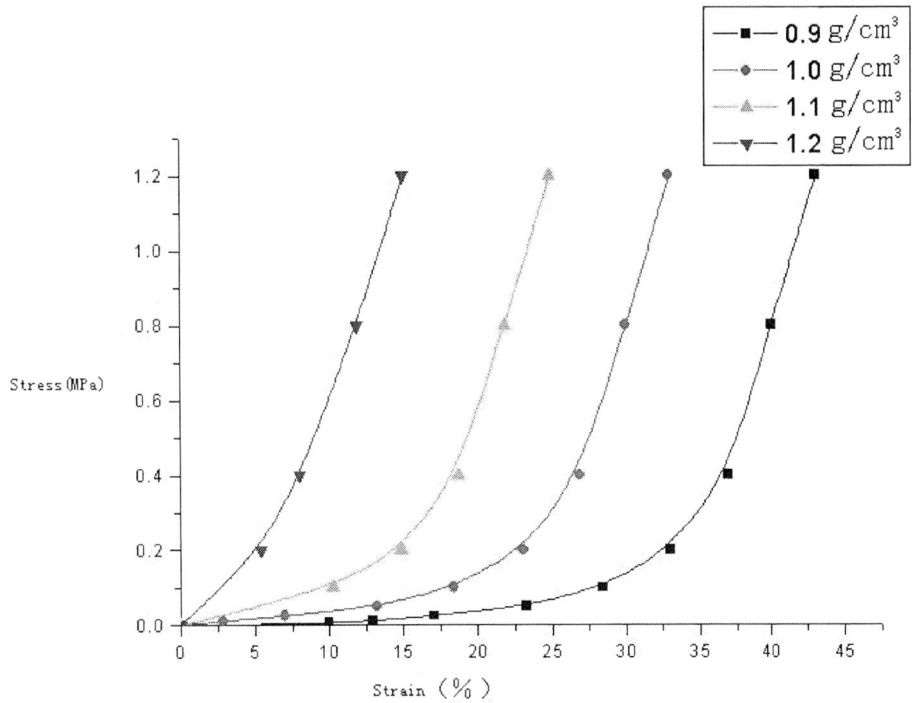

Figure 2. The constitutive relationship of green sand

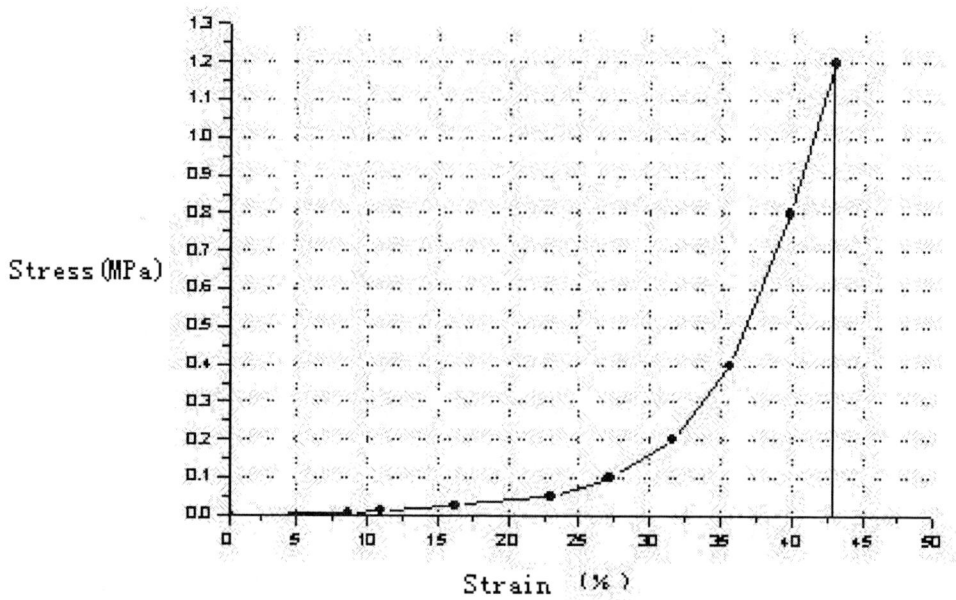

Figure 3. The processed constitutive relationship of green sand

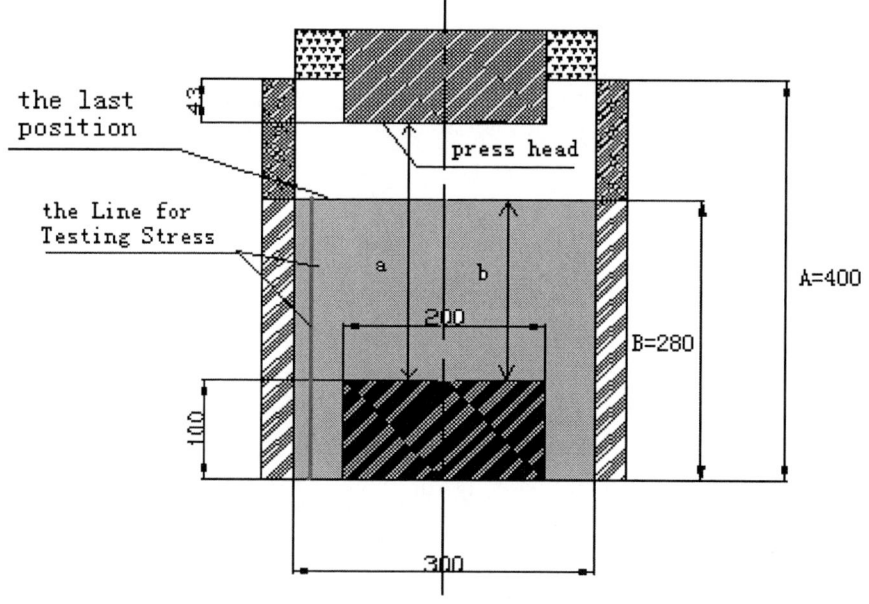

Figure 4. The geometrical model, with dimension of 300mm(in Width) by 400mm (in Height)

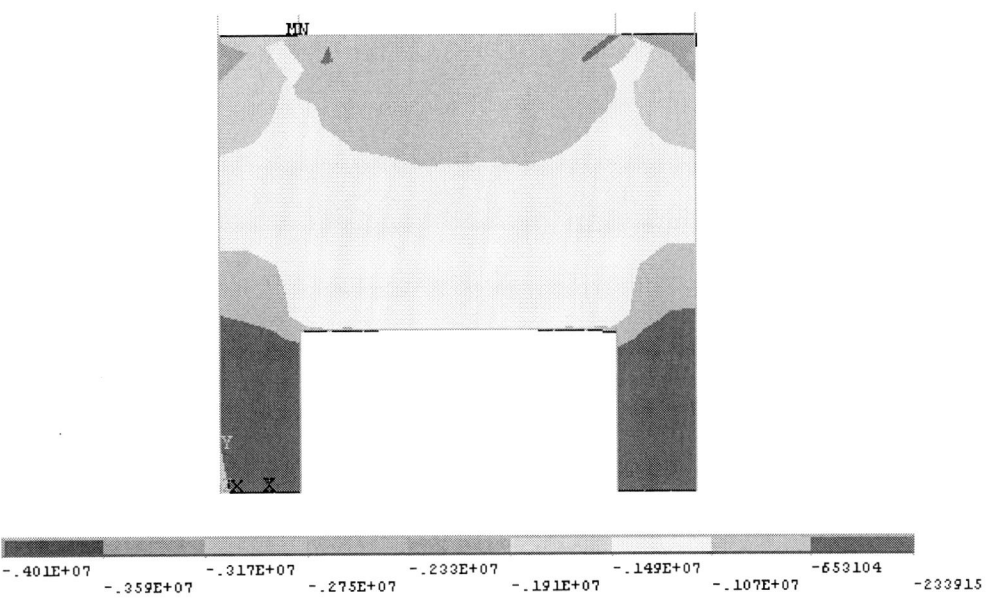

Figure 5. The stress field of the green sand where the mold is squeezed from the top side

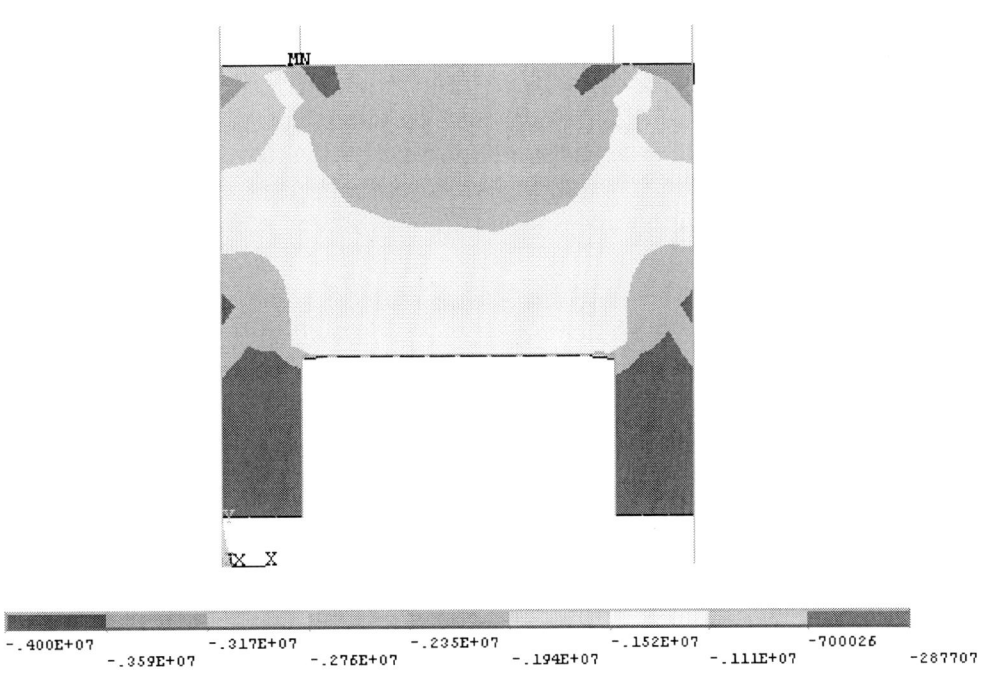

Figure 6. The stress field of the green sand where the model is squeezed from both sides

The email address of the author is zqqlwz@tsinghua.edu.cn.

Improving Casting Performance Through Customized Insulating Shapes And Advanced Simulation Techniques

Jaime Prat*, Antton Meléndez*** & Alberto Seoane**, Eva Anglada***, Andrew Beeson*, Mitxel Arrieta*, Jon Galaz *, Asier Jorge** & Teresa Vicario***

*ASHLAND CASTING SOLUTIONS, Spain, **BETSAIDE S.A.L., Spain, ***INASMET-TECNALIA, Spain

Abstract

Over the years foundries have been looking for ways to improve their process. IBERIA ASHLAND and INASMET TECNALIA, together with BETSAIDE S.A.L. have implemented an innovative solution for methoding castings to ensure the integrity of the finished parts by using their latest development of customized insulating shapes combined with advanced simulation techniques. This study will show how improvements can be achieved in the mechanical characteristics of a casting by eliminating graphite deformation and shrinkage defects through the use of customized sleeves and insulating cores made of advanced moulding materials. Additionally, by utilizing the latest in simulation techniques, various compositions of the insulating materials can be "tested" effectively, allowing us to compare the final casting, without the need and related cost to produce multiple iterations of the part.

Key words: (Core-sleeves, insulating sleeves, inverse calculation, simulation, and shrinkage)

Introduction

ASHLAND *Casting Solutions*, is already offering a solution to enhance savings and optimise the feeding systems of castings: *EXACTCAST*® precision riser sleeves.[1] This solution can be extended to other parts of the core and mould providing dual functions in that it can be used not only to make riser sleeves but it can also be used to make parts of the mould and/or core package. This solution has special advantages in those areas of the mould cavity where the melt arrives having already lost some of its initial pouring temperature and where specific areas within the casting are difficult to access using conventional risers or sleeves. Normally feeding of the casting is accomplished from the outside of the mould.

EXACTHERM® sand additive, a new, technically viable alternative was developed from low-density alumina silicate ceramic (LDASC) materials used in Ashland's patented riser sleeve technology.[2,3] This material can be used to not only address and resolve design problems which arise from the need to feed areas of the part that are inaccessible from the outside of the mould, but it also presents a solution that aims at the elimination of contamination products in the green-sand system stemming from the use of conventional exothermic riser sleeve materials. Contaminants such as fluorine and aluminum can produce undesirable mould-metal reactions that may lead to unintentional localized alterations of the microstructure of the casting.[4-7]

These solutions are easy to apply in practice, and can be simulated on a computer before a casting is ever poured. In support of the modelling process Ashland, in collaboration with INASMET-TECNALIA, has developed an Electronic Catalog which provides the thermophysical characterisation of these products, so as to ensure more precise use in later simulations.

Experimentation Details
Goals of the Project

The initial series of parts was chosen with the following factors in mind:

> ➢ Complex geometry.
> ➢ The existence of a central core from which the different areas of the casting could be fed simultaneously.
> ➢ The need for multiple risers.
> ➢ The possibility of using a non-exothermic sleeve.
> ➢ The selection of the pilot part (Fig. 1).

Uniqueness of the chosen part

Aside from the structural complexity of the casting, additional difficulty resulted from the transfer of the manufacturing of the casting from a horizontal moulding system (considered appropriate for this kind of part) to a DISAMATIC vertical moulding line. This change in the orientation of the casting increased the problems of feeding the critical areas.

Methodology Used in the Project
The following steps were taken:

1. Simulation and analysis of the problem areas of the part in its initial configuration. Initially, the part had five conventional risers, as shown in Fig. 2. An analysis of this configuration using a combination of ProCAST (ESI Group) by INASMET-TECNALIA R&D Centre and MAGMA (MAGMA GmbH) by FUNDICIONES BETSAIDE S.A.L. confirmed the advantages of using the Ashland electronic catalogue and helped to validate its use for any numerical calculation system. The use of these two tools simultaneously meant, on the one hand, that the R&D centre had the security of using the data available and, on the other, that the foundry had the confidence of being able to use its usual simulation tool. Comparing these calculated results to the actual results of previous casting production helped to calibrate the above-mentioned tools conservatively. With the confirmation of the reliability of these tools, future simulations could be run on the proposed improvements

2. Calculations and study of possible solutions, on the basis of the above-mentioned goals. During this second stage, all the data obtained by INASMET-TECNALIA was used for the characterisation of the insulating sleeves and incorporated into the simulation analysis using the ProCAST Inverse Calculation Module.[8] Of this data we should highlight the "Thermal Module vs. Geometric Module" equivalence, calculated and applied to optimise the weight of the new riser inserted in the core and the size of the feeder necks of each of the areas to be fed. The solution that was finally chosen was to replace four of the five risers with a single central riser incorporated within the central core thus creating a core-sleeve. This drastically reduced the total weight of feed metal to feeding. The fifth riser was outside the area of influence of the core-sleeve, meaning that it could not be replaced and remained in its initial location. See Fig. 3 with the project prototype and the chosen solution.

3. New simulation of the proposals and adjustments made to obtain a scalable prototype.[9] See Fig. 4 for the solution and the results of the simulation.

4. Construction of the tools for the prototype (Fig. 5). Provisional tools were generated that allowed the core-sleeve to be manufactured quickly, economically and geometrically identical to the existing one.

5. Manufacture of the core sleeves (Figs. 6 and 7) was achieved using a blend of sand and LDASC. Fifteen units were manufactured, which were enough to resolve all the problems that arose during the fine-tuning of both the tools used and the material. From these core-sleeves, eight were chosen for the in-plant trials. It should be noted that the gas evolution of the new core sleeve was similar to the original sand core (See Tables 1 and 2 – Gas Evolution and Density).

6. In-plant trials took place at FUNDICIONES BETSAIDE S.A.L. using the same facilities and materials as the normal standard parts, with six parts being poured using the new sand/LDASC blend core-sleeves. See Fig. 8 (detailed view of the core-sleeve in the DISAMATIC core setter), Fig. 9 (the core-sleeve in one of the casting moulds) and Fig. 10 (the cast part).

7. Checking the results. Sections of the parts (4) and the initial part (1) were cut using the metallographic cut-off machine (Fig. 11). All of the areas that were identified as problematic by the computer simulations were examined by the technical staff from the team carrying out the project. They were compared against a standard part manufactured using the original method and tools. The remaining castings (2) manufactured for the project were X-rayed and did not show any defects.

Results

The project was a success. Precise cuts were made in the areas identified as areas of risk by the simulation programmes, so as to allow the analysis of both the initial part that incorporated the five risers and of the prototype parts manufactured with the new core sleeves. Fig. 12 shows the prediction by the BETSAIDE S.A.L. simulator of one possible defect, Fig. 13 shows photographs of the cuts obtained from the initial and prototype parts, left and right respectively, and Fig. 14 demonstrates the elimination of shrinkage in the prototype.

The predictions of the simulations were conservative, as in the two cases, both with the initial and the project castings, shrinkage defects were less numerous than predicted. However, one of them, which appears in one of the areas of greatest risk, was indeed visible (see to the left of Fig. 13 circled in red) and has been used to compare the best results obtained (see Fig. 14).

In addition, it should be noted that the risers in the project design are half of the weight of the initial part. This reduction in weight can have a positive impact in the handling and setting of core sleeves in this and other future designs.

Discussion

The results of this project open a new and exciting method for designing and making castings and for providing solutions to casting defect problems as shown in Fig. 14.

It is also very important to add that the foundry can now evaluate other additional benefits, such as:

- ➤ A reduction in the weight of a part, due to the use of LDASC materials and its lower density, namely 4.1 kg for the original sand core versus 1.7 kg for the sand/LDASC blend core-sleeve. This compensates for the industrial costs of the solution adopted as well.
- ➤ The new core sleeve design allows for reducing the core size and the number and size of risers which in this example also made room for two castings per mould instead of one, with consequential cost decrease. See Fig. 9.
- ➤ There is no mould-metal reaction that adversely affects the quality of the casting, as the core sleeve material is a mixture of silica foundry sand and LDASC additive.
- ➤ The binder system in both cases is a phenolic urethane cold box binder.
- ➤ In both cases dimensional accuracy is achieved.
- ➤ Shake-out debris stemming from cores containing LDASC material has no negative effect in the green-sand system.
- ➤ Unlike the original cores used to make this casting, the cores produced with LDASC did not require a refractory coating ("core wash") in order to achieve the same casting surface finish quality (See left side of Fig. 15, and Fig. 16, casting finish without core wash).

Conclusions

To sum up, the following are the real and potential savings derived from the use of LDASC materials in core and or mould components as was demonstrated in the solution proposed by the above project:

- ➤ Better casting quality where shrinkage was drastically reduced.
- ➤ Improved yield with 50% reduction of feeding needs.
- ➤ Reduced core weight with subsequent handling advantages.
- ➤ Reduction of surface area needed on the pattern thus allowing one more casting per mould.
- ➤ In this particular case, due to beneficial effect of the sand/LDASC blend, no core-wash was needed in order to achieve a good casting finish and no veining.
- ➤ Elimination of the core-wash application, drying process and subsequent handling for the new cores.

The joint work on projects such as this, between Ashland Casting Solutions, the INASMET-TECNALIA Casting Unit and foundries, such as BETSAIDE S.A.L., opens a revolutionary method to producing sound castings, with optimised costs using simulation techniques combined with new materials such as LDASC .

References

1. Patent: EP 9601607 (ASHLAND)
2. Patent: EP 97948913.5 (ASHLAND)
3. Prat J, *Moulding sand for making casting moulds and cores*, Patent RU2202437. 2003-04-20.
4. Aufderheide R, Showman R, Close J, Zins J, *Eliminating Fish-Eye Defects In Ductile Castings*, AFS Transactions, Paper 02-047 (2002)
5. Aufderheide R, Showman R, & Twardowska, *New Developments in Riser Sleeve Technology*, AFS Transactions, 395-400 (1998)
6. Showman R., Lute C, Aufderheide R, *Exothermic Riser Sleeves Can Cause Flake Graphite In Ductile Iron*, AFS Transactions, Paper 01-086 (2001)
7. Aufderheide R, Showman R, *A Process for Thin-Wall Sand Castings*, AFS Transactions, Paper 03-145 (2003).
8. INASMET Foundation, article on Inverse Calculation methodology (PAM-TALK N° 24, ESI Group, 2003, http://www.esi-group.com/Corporate/pam-talk/Pamtalk_24.pdf)
9. Prat J, Meléndez A, *New Casting Solutions: Numerically simulated EXACTCAST© core-sleeves eliminate critical problems with automobile high-security components (Patented)*, 66th World Foundry Congress, Istanbul, Turkey, 6-9 Sept, 2004, pp 45-57, pub Toksad: The Foundrymen's Association of Turkey, 2004

Acknowledgements

We would like to express our sincere thanks to the following people for the help received during this Project: the staff and management of AUXMAK and MODELOS LEBAITI. Thanks are also due to Iker SAGARDUY and Pedro GACETABEITIA (FUNDICIONES BETSAIDE S.A.L.), as well as Ion HIDALGO (INASMET-TECNALIA) for their valuable collaboration.

Tables

Table 1

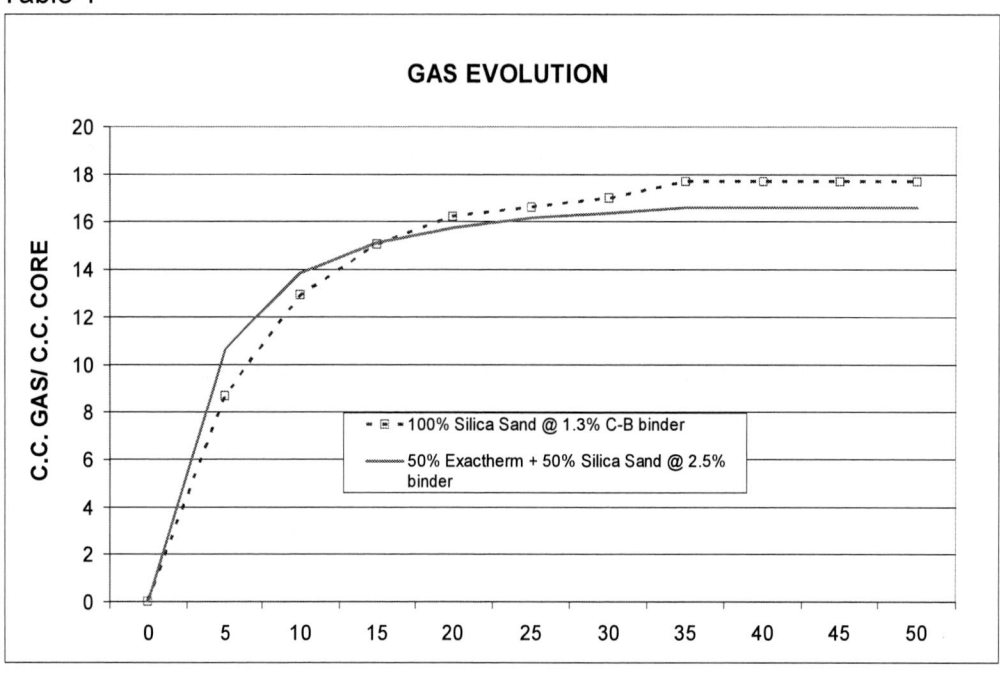

Table 2

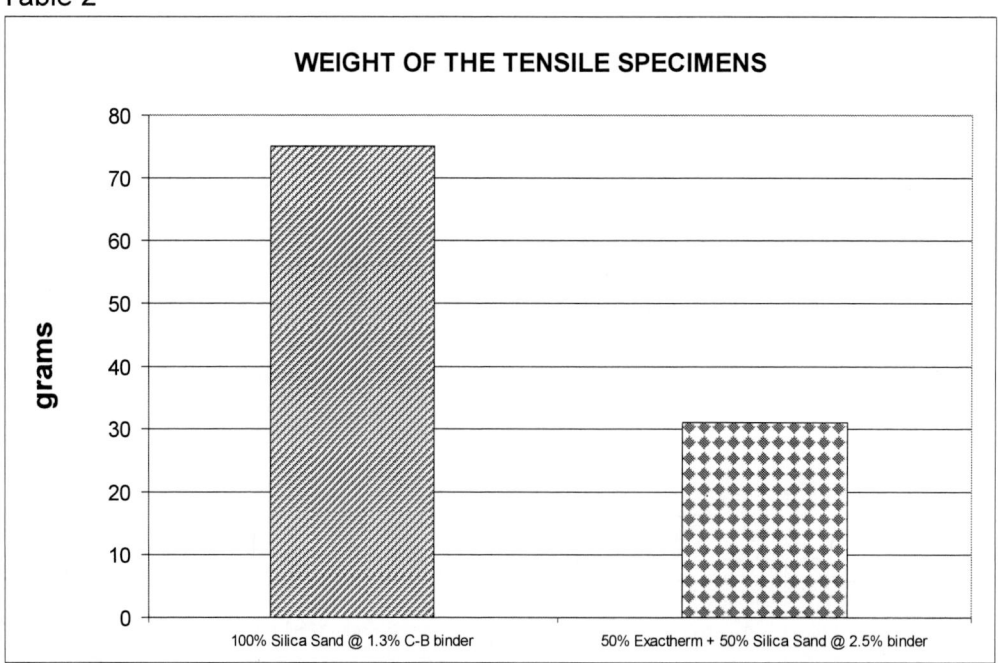

Figures

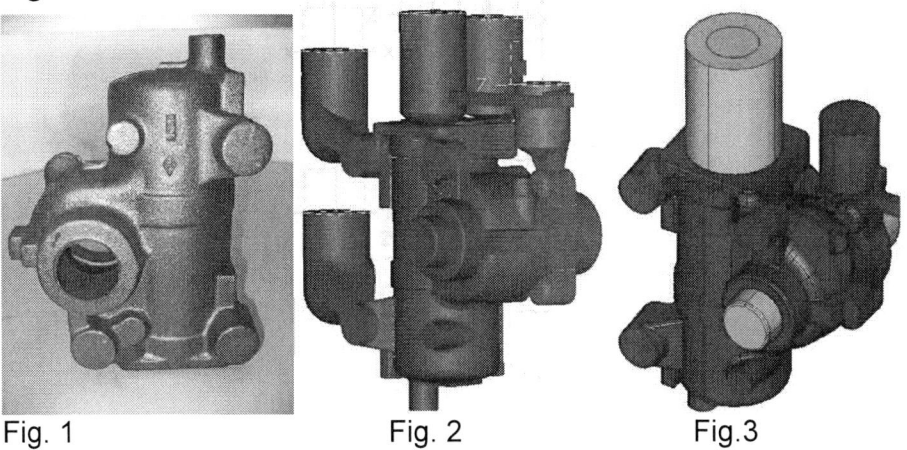

Fig. 1　　　　　Fig. 2　　　　　Fig.3

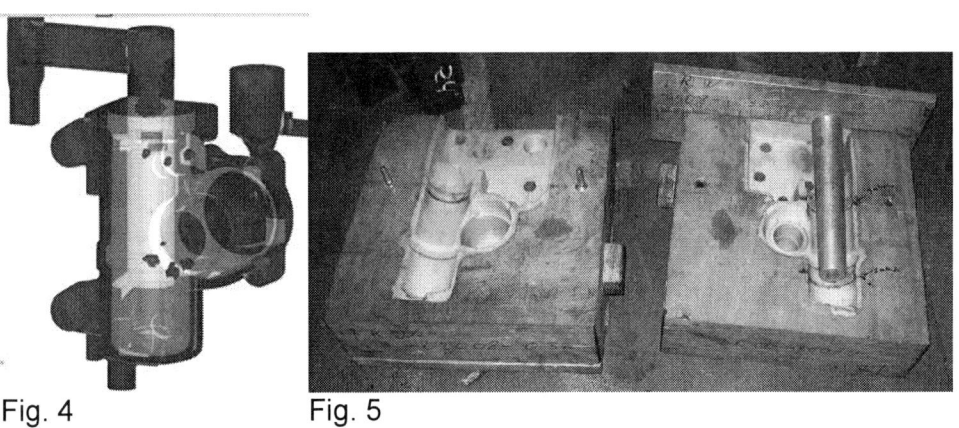

Fig. 4　　　　Fig. 5

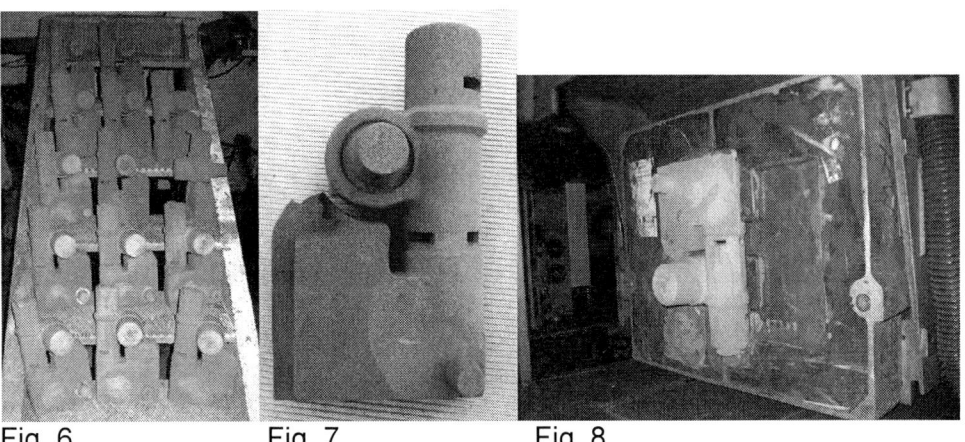

Fig. 6　　　　Fig. 7　　　　Fig. 8

Fig. 9   Fig. 10

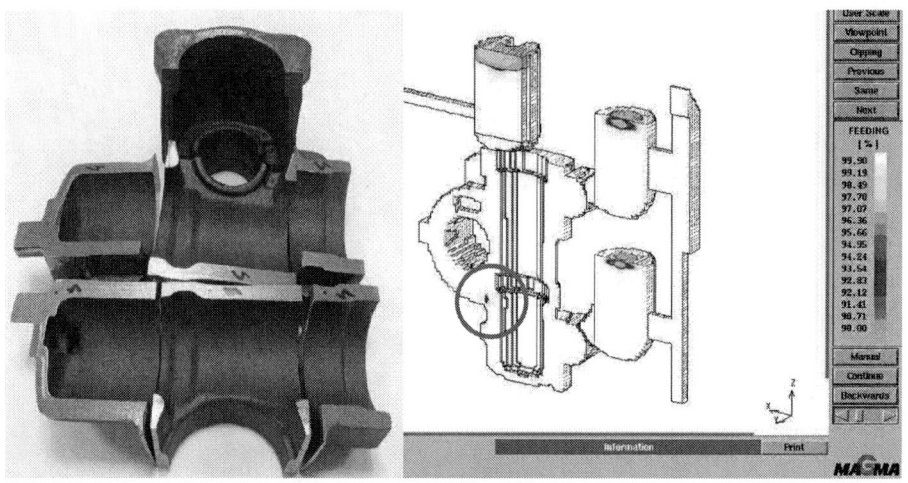

Fig. 11   Fig. 12

Fig. 13

Fig. 14   Fig. 15

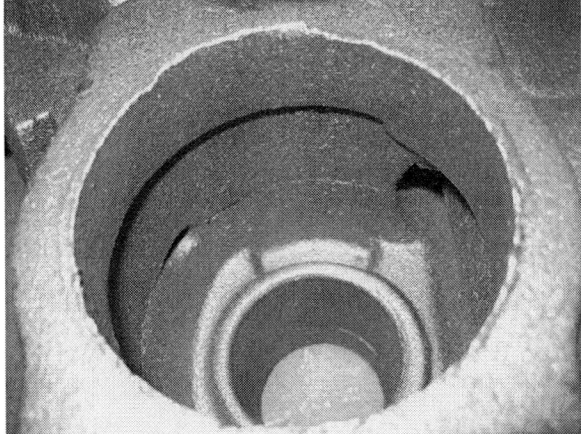

Fig. 16
jprat@ashland.com

Prediction of Shrinkage Defect in Steel Casting for Marine Engine Cylinder Cover by Numerical Analysis

K H Kim*, J H Hwang*, J S Oh**, D H Lee***, I H Kim* and Y C Yoon*

* Hyundai Industrial Research Institute, Hyundai Heavy Industries Co., Ltd., Korea
** Engine & Machinery Division, Hyundai Heavy Industries Co., Ltd., Korea
*** Hankuk Heavy Machinery Ind., Korea

Abstract

Shrinkage defect in an alloy steel casting for marine engine cylinder cover was estimated by numerical analysis with reference to casting variables such as riser and chiller. Retained melt modulus was employed as a prediction parameter in this study. Shrinkage defects were developed at the final solidification region. This is associated with hot spots, which could be controlled more effectively by adjusting the height of riser and by attaching chiller to the casting outside. Based on the numerical analysis, sound castings were produced successfully and economically.

Key words

Cylinder cover, Steel casting, Numerical analysis, Retained melt modulus, Defect potential

Introduction

Forging is commonly adopted as a manufacturing method of cylinder cover for large marine diesel engines although casting can be an economic manufacturing method. This is mainly attributed to high mechanical properties of forged products due to the continuous metal flow line. Actually, a forged product shows higher mechanical properties than a cast one for the same composition. By the modification of chemical composition and/or control of casting defects, however, it is possible to produce the cylinder cover by casting.

Advanced computational simulation technique brings a lot of changes in casting industries. Numerical analysis helps to enhance mechanical properties by reducing greatly casting defects through the precise prediction. This makes it possible to produce near-net shape product using a casting process and finally contributes to increase of its productivity.

There are various prediction parameters being applied in forecasting shrinkage defects of steel castings [1]. Niyama criterion is one of them but it is affected by the size and shape of casting [1-3]. Therefore in this paper a retained melt modulus that can improve the accuracy of the prediction is applied to establish the optimum casting condition from the results of analysis [4].

The aim of this study is to establish a suitable casting plan for the cylinder cover and then to produce the sound products.

Experimental

Figure 1 shows dimensions of the cast cylinder cover for a large marine diesel engine simulated in this study. The total casting weight is about 5 tons. The material for the casting is a 1Cr-0.25Mo steel.

Table 1 is the casting variables studied for analyzing the effect of chiller and riser on the casting quality. The numerical models with a variation of the chiller's shape and position are as shown in Fig. 2.

Table 2 is physical properties used for the calculation. Three-dimensional numerical model of the casting consists of 5 million meshes, as shown in Fig 3.

To verify the numerical analysis results, two full size castings were produced under the casting condition as summarized in Table 3.

Results & Discussion

Filling patterns

The in-gate of the casting was designed circumferentially at the bottom position of the casting in order to keep the continuous filling of the melt casting as shown in Fig. 3. Figure 4 shows an example of filling pattern at the 15% filling condition. The melt flowing velocity was about 50 cm/s. The melt well rotates along the bottom side of the casting and no turbulence is developed during the filling. Turbulence easily makes defects such as slag inclusion, pores due to the entrapped gases. Therefore, smooth filling is necessary for producing sound castings [5].

Prediction parameter for shrinkage defect

Equation 1 is the expression of the retained melt modulus [4, 6]. In equation 1, M_R is retained melt modulus parameter, V_R is retained melt volume and S_R is retained melt surface area. The lower the value of M_R is, the higher the possibility of defect formation is.

$$M_R = \frac{V_R}{S_R} \quad (1)$$

Defect potential criterion was also used in order to get more realistic predictive results. Defect potential is a probabilistic defect parameter affected by the rate of material's shrinkage. It is defined as the multiplication of shrinkage rate during solidification by casting constant, which is changed by material and casting conditions. It reveals highly probable defects by searching all the elements having lower M_R [6, 7].

Prediction of the defects probability

Figure 5 shows the defect distribution predicted by the retained melt modulus and defect potential with respect to chiller location. Without chiller, the shrinkage defect is developed within the product part as shown in Fig. 5(a). But for all the cases with chiller the shrinkage defects go up toward the riser part and then the product parts are free from the defects. Among them, the case 3 attaching the chiller circumferentially at the outside surface of the casting gives the most desirable result.

From Fig. 6 showing the defect distribution with a variation of riser height for bottom chiller, it is found that the location of possible defects moves toward more upside and inside surface of the riser part. This implies that the soundness of the product part increases as the riser height increases.

Based on these results, a desirable casting plan for the cylinder cover was finally established based on both the soundness of the product quality and its productivity.

In order to verify the simulation results, two castings were made and followed by nondestructive evaluations (NDE) such as UT, MT and PT. Figure 7 shows the NDE result. The cast with 400mm of the riser height without chiller (case 1) possesses a lot of defects at the boundary region between the product and the riser part. However no defect at the same position is found for the cast with bottom chiller and 500mm of the riser height (case 5). These results well coincide with the simulation results as shown in Fig. 5(a) and Fig. 6(b), respectively.

Conclusions

A cylinder cover casting was successfully produced based on the numerical analysis and the following conclusions are made.

(1) Retained melt modulus as a defect prediction criterion is a suitable parameter for effectively predicting casting defects and establishing the casting plan for the cylinder cover.
(2) Safety margin of the product part from the defects is the highest when the chiller is applied at the outside surface of the casting.
(3) The soundness of the product part increases as the riser height increases.

References

1. J K Lee, J K Choi and C P Hong, *Thermal parameter-based quality criteria for the prediction of shrinkage defects in steel castings*, J. of the Korean Foundrymen's Society, Vol. 18, No. 1, Feb 1998, pp77-84.
2. K D Carson, S Ou, R A Hardin and C Beckermann, *Development of new feeding-distance rules using casting simulation: Part 1. Methodology*, Metallurgical and materials transactions B, Vol. 33B, Oct. 2002, pp731-740.
3. Z Ignaszak and P Mikolajczak, *Chosen aspects of gradient criteria correlation with shrinkage defects in post processing procedure of simulation code.*
4. S M Lee, K H Kim, T D Park, W J Lee, S C Park and D S Park, *Defect prediction of Al-bronze sand casting by numerical analysis*, The 66th world foundry congress, Vol. 1, Sep. 2004, pp259-267.
5. H F Taylor, M C Flemings and J Wulff, *Foundry* engineering, 3rd edition, 1966, pp229-258.
6. AnyCasting Inc., *AnyCasting user's manual & technical reference*, 2003.
7. Y C Lee, H J Lee and C P Hong, *Solidification analysis and prediction of shrinkage cavity in aluminium alloy castings*, J. of the Korean Inst. Of Metals, Vol. 26, No. 8, 1988.

Tables

Table 1 Simulation cases with a variation of chiller and riser height

Simulation	Chiller		Riser height (mm)
	Location	Number	
Case 1	None		400
Case 2	Bottom	8	
Case 3	Outside	12	
Case 4	Inside	1	
Case 5	Bottom	8	500
Case 6			600

Table 2 Physical properties used for the calculation

Material	Casting (1Cr-0.25Mo steel)	Mold (silica sand)
Density (g/cm^3)	7.87	1.48-1.52[1]
Specific heat (cal/g·°C)	1.16-1.43[1]	1.62-2.94[1]
Thermal conductivity (cal/s·cm·°C)	0.12-0.13[1]	1.4×10^{-3}-1.8×10^{-3} [1]
Liquidus/Solidus(°C)	1513/1473	-
Latent heat(cal/g)	49.47	-

[1] Variable with temperature

Table 3 Casting conditions for test casting

Material	1Cr-0.25Mo alloy steel
Filling weight (ton)	5.4
Pouring temperature (°C)	1560±10
Filling time (sec)	70

Figures

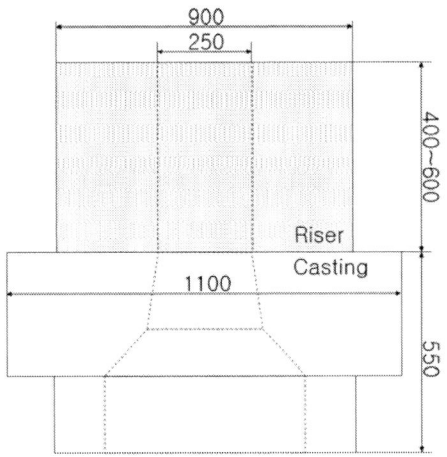

Fig. 1 Schematic of the cylinder cover used for this study (unit ; mm)

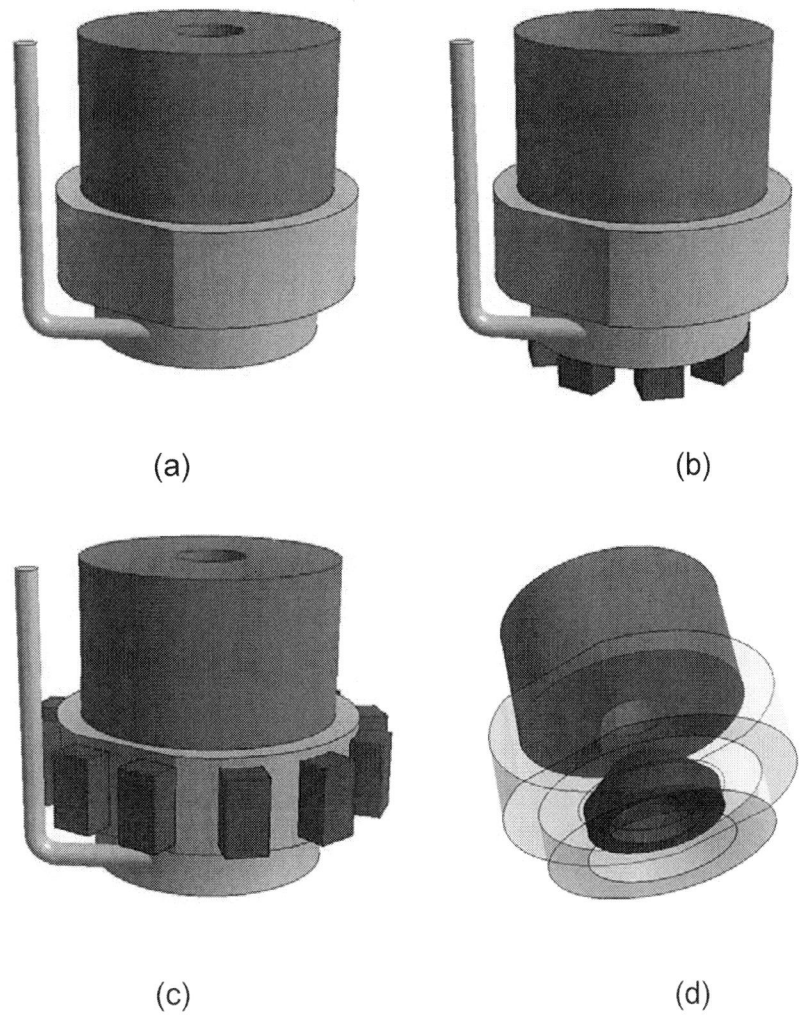

(a)   (b)

(c)   (d)

Fig. 2 The position and shape of the chiller ; (a) without chiller, (b) bottom chiller, (c) outside chiller, (d) inside chiller

Fig. 3 Numerical model of the casting

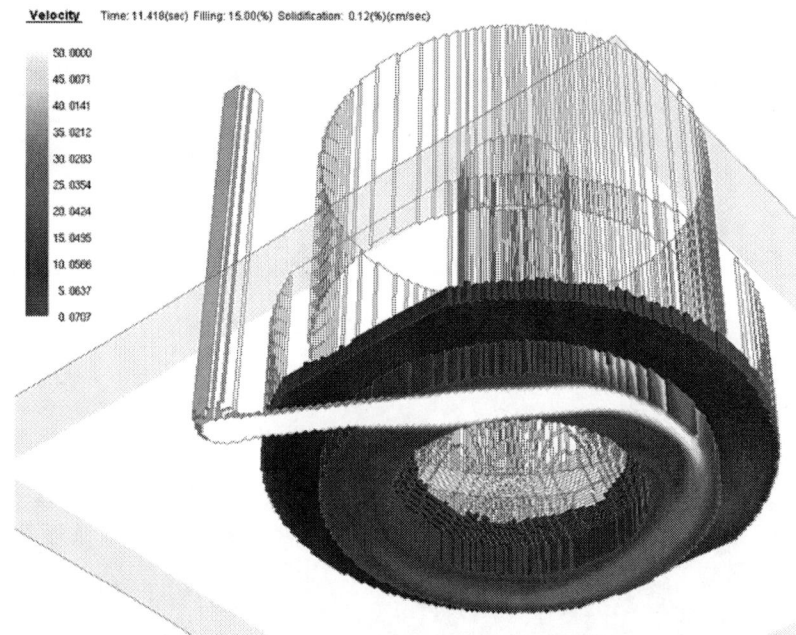

Fig. 4 Velocity profile at the 15% of filling condition

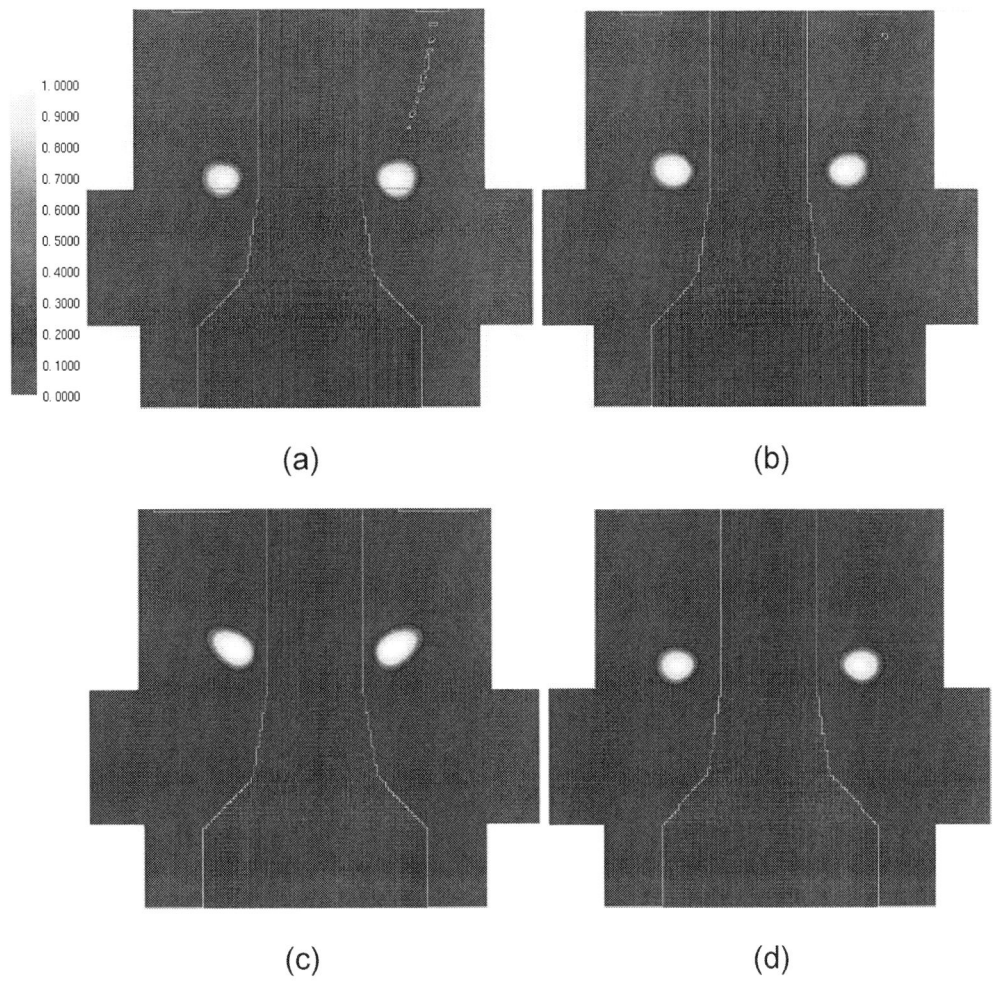

Fig. 5 Defect predicted using retained melt modulus and defect potential for 400mm of riser height ; (a) Case 1 ; without chiller, (b) Case 2 ; bottom chillers, (c) Case 3 ; outside chillers, (d) Case 4 ; inside chiller.

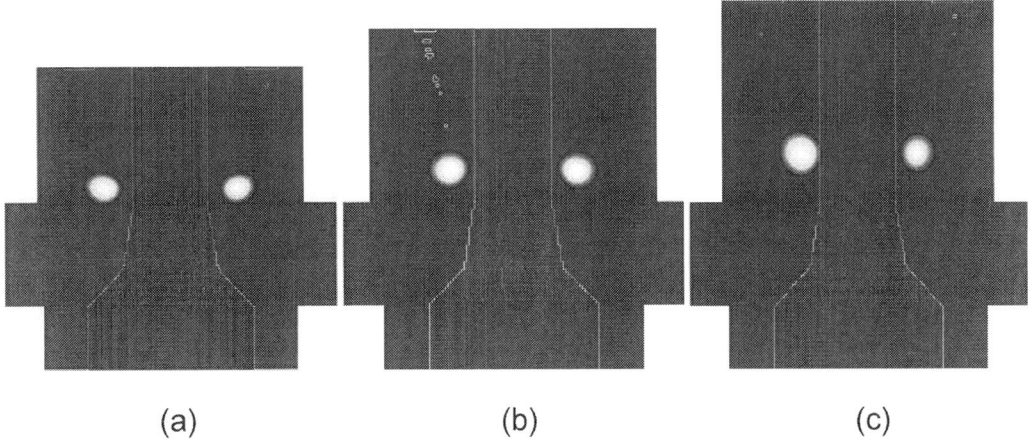

Fig. 6 Defect predicted with reference to riser height for bottom chiller ; (a) 400mm, (b) 500mm, and (c) 600mm

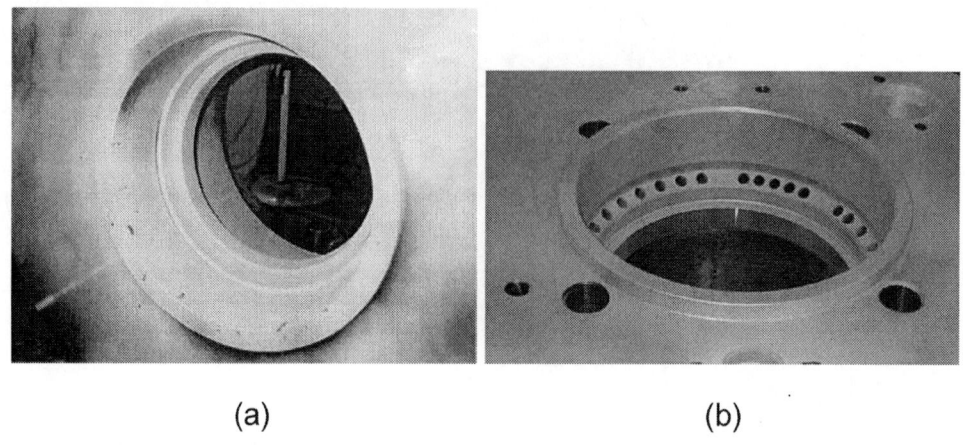

(a)   (b)

Fig 7 Nondestructive results of the test casting ; (a) Case 1 ; a lot of defects at the product surface, (b) Case 5 ; no defects after machining

Production of ductile iron castings without feeders

Possibilities of producing ductile iron castings in green sand mould without feeders by means of "ingate" feeding and control of the graphite precipitation pattern by means of advanced thermal analys is.

Rudolf Val. Sillén*

* NovaCast Technologies AB, Sweden

Abstract

The paper focuses on the mechanisms behind shrinkages in cast iron and how they can be avoided. Grey and ductile iron are unique casting alloys due to their solidification behaviour. During solidification the alloys initially contract when the temperature drops from the pouring temperature to the liquidus temperature. This contraction must always be compensated for by supply of feed metal from the gating system and often also from a feeder (riser). The paper will demonstrate that this type of feeding can only be functional until the amount of solid phase reaches a certain level. This usually happens slightly after reaching the low grey eutectic temperature. When this stage has been re ached a balanced precipitation rate and amount of graphite will make it possible to compensate for contraction of remaining liquid and the eutectic austenite. The proper balance is crucial as the precipitation of graphite is associated with a volume expans ion, which if too high will cause mould wall movement and increase shrinkage tendency. Too low precipitation will lead to micro-shrinkages. The paper describes how optimization of the gating system in many cases makes it possible to eliminate feeders for s upply of the initial feed metal. After the low eutectic temperature is reached the contraction can be eliminated by careful control of graphite precipitation using a combination of chemical and adaptive thermal analysis. Through proper control of the feedi ng sequences it is possible to produce certain ductile iron castings in green sand moulds without feeders!

Key words
Feederless ductile iron, green sand

Introduction

Shrinkages are one of the most common casting defects. The main cause of shrinkage cavities is that all commercial alloys contract when a casting cools from the pouring temperature to solidus. The contraction is usually between 1-5 % depending on the type of alloy and the pouring temperature. The main contraction occurs between the pouring temperature and the liquidus temperature. If the contraction is not compensated for by feed metal, either by supplying feed metal or by forming a depression on the outer surface, a shrinkage cavity will occur. Cast iron alloys, which solidify with a precipitation of graphite, represent a more complex behaviour, the reason being that the dissolved carbon partly precipitates as graphite with a lower density than the base iron. The precipitation is therefore associated with an increase in volume, which in some cases partly might offset the contraction of the liquid and the austenite. By careful control of the mould filling and the precipitation of graphite it should be possible to produce ductile iron castings in green sand moulds without the use of feeders. In this paper we will study what happens during solidification, as well as the mechanisms behind shrinkages and how to avoid them.

What happens during solidification of cast iron?

When liquid metal cools off, the temperature is reduced, causing energy to be released. The temperature represents the total amount of thermal energy in a melt and is related to the kinetic energy of the molecules. When a melt cools down, the thermal agitation of its molecules is reduced. This energy is expressed as specific heat e.g. as KJ/ Kg and degree C. When the liquid reaches a temperature called liquidus, the bond between the atoms becomes more rigid on a macro scale level. More energy is then released until the liquid is transformed into solid state. The energy released at this stage is called latent heat of fusion, fusion enthalpy or just latent heat for short. Theoretically, the temperature stays constant until the transformation is completed. Latent heat is measured as KJ/Kg. The temperature where the metal or a precipitated phase is completely solid is called solidus. The precipitation of various phases in grey and ductile iron is to a large extent dependent on factors such as nucleation that can not be estimated using the chemical composition. In this investigation we have therefore used advanced thermal analysis as a tool to understand and control the progression of the solidification. Thermal analysis is based on recording temperatures at certain time intervals during the solidification process. Cooling curves can thereby be constructed and used to analyse and classify an alloy. A cooling curve is a plot of the temperature as a function of time for a sample of an alloy poured into a standardized mould with a thermocouple, usually positioned in the center. Arrest temperatures such as liquidus and solidus in a cooling curve, as well as cooling rates during various phases of the solidification can be used as metallurgical attributes to classify a melt and to correlate it to the behaviour when poured in a mould.

When a casting cavity has been filled with liquid cast iron the temperature is reduced until the liquidus temperature (TL) is reached, then austenite crystals start to form if the alloy is hypoeutectic. If the alloy is eutectic both austenite and graphite are precipitated from the melt. If the alloy is hypereutectic then the initial phase is graphite. The latent heat for austenite is fairly low, about 200 KJ/kg. Therefore almost no recalescence (R) occurs at TL. The latent heat for graphite is very high, about 3600 KJ/kg. Thus when graphite precipitates, heat is released, which cause s the temperature to increase and cause s recalescence. Precipitation of graphite is also associated with a volume expansion as the density of graphite is about 2.2 g/cm^3 versus about 6.9 for the liquid melt.

Let us study the progression of solidification using a hypereutectic alloy as an example. The cooling curve shows what happens at each moment in the centre of the sample cup. A typical cooling curve and its first derivative looks as follows:

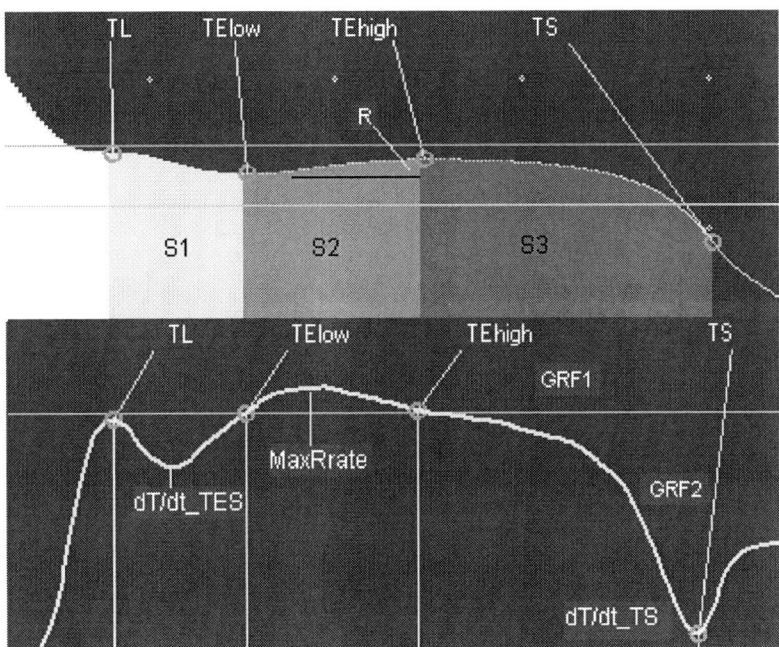

When the liquidus temperature is reached the cooling curve shows a horizontal plateau. The length of the horizontal plateau is a function of the time it takes for austenite to grow from the walls of the cup to the centre where the thermocouple is located. The melt contracts first in the liquid state and then during the crystallisation of primary austenite, which continues until the low eutectic point (TElow) is reached. At that time the eutectic reaction where simultaneously austenite and graphite a re precipitated has just started. The temperature increases due to release of latent heat until the high eutectic temperature (TEhigh) is reached. The increase in temperature is called recalescence (R). The eutectic solidification then continues until no more liquid remains and the solidus temperature (TS) is reached. The GRF1 factor measures the eutectic behaviour during the second phase of eutectic. GRF2 is a marker that indicates the effect of the eutectic at the very end of freezing. During the eutectic phases (S2 and S3) the liquid can expand its volume provided a sufficient amount of eutectic graphite is precipitated. The available amount of eutectic graphite can be estimated using the formula:

$$\text{Eutectic Graphite \%} = \text{Carbon\%} - (2.1 - 0.11 \cdot \text{Si\%}) \quad (1)$$

Thus if Si is 2% and C is 3.8% then the maximum amount of eutectic carbon is about 2%. Assuming that the density for the liquid is 6.9 g/cm^3 and 2.2 for graphite, the volume expansion from the graphite is about 6.2%. This expansion can, if carefully controlled, compensate for the contraction of remaining liquid and austenite in a casting.

The solidification progression can be illustrated as follows:

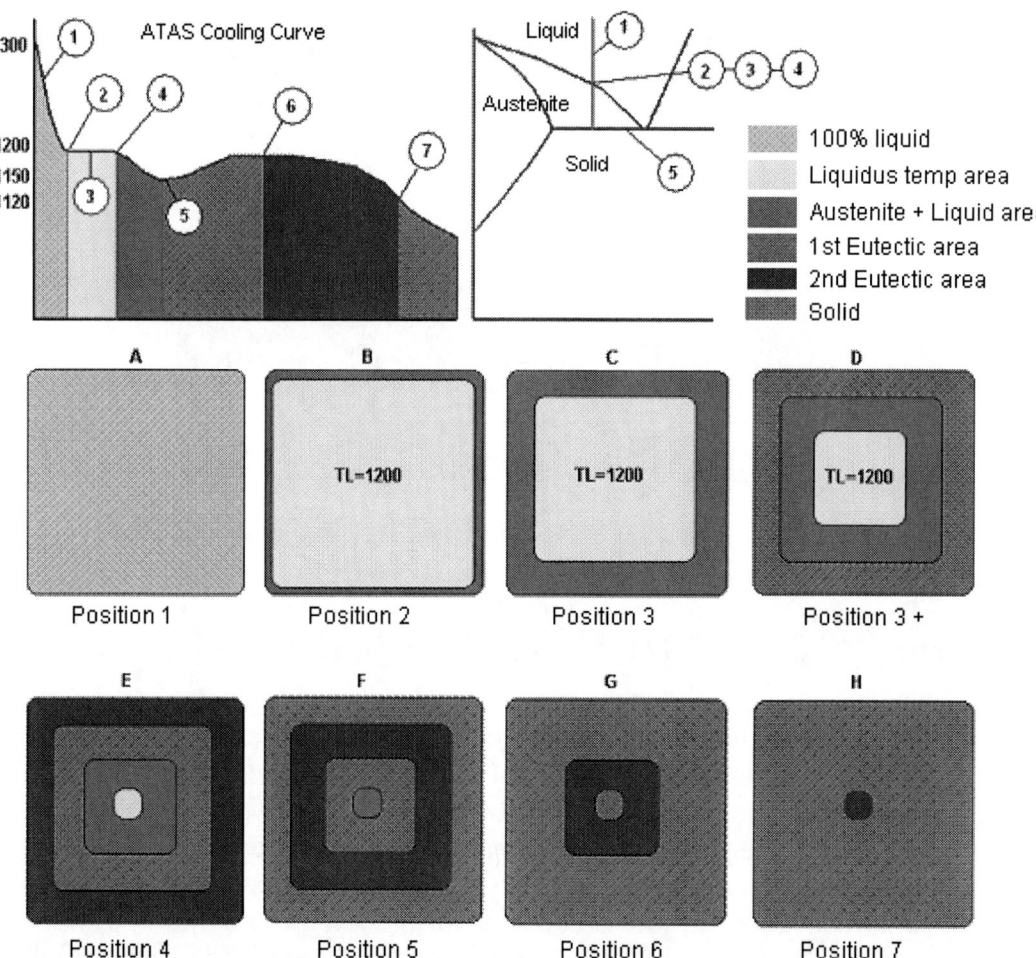

The illustrations show the situation in a section through a cup used for thermal analysis. Figure A shows the melt at position 1 when the cup has just been filled. At position 2 the temperature has reached liquidus. The temperature gradient is then zero due to thermal currents. The zone with metal at the liquidus temperature is gradually reduced, which is shown in positions 2 to 4. Note that the temperature in the liquidus zone is constant. Thus there is no contraction in this zone! In position 3 austenite has started to grow inwards. In position 3+ the low eutectic temperature has been reached at the walls and a zone which is expanding has been created. In position 4 also secondary eutectic is formed and expanding. In positions 5 - 7 a solid phase also appears. In a casting several of these zones appear at the same time. If an initial amount of feed metal ca n be supplied until position 3+ is reached then it is likely that the casting can be made without any feeders.

The challenge is to be able to supply the initial need for feed metal from the gating system and to control the nucleation and thereby the precipitation rate of graphite so that it can balance the contraction of austenite and remaining liquid. Accodring to the author this is possible by means of simulation systems for optimizing gating systems and by using a combination of chemical and advanced the rmal analysis for controlling the solidification process.

Which variables can be monitored by thermal analysis?

Thermal analysis can be used to monitor the metallurgical variables that influence the development shrinkages. The ATAS thermal analysis system (developed by NovaCast in cooperation with the Swedish Foundry Association) is specially designed to monitor and interpret cooling curve data and predict the potential risk for shrinkages, as well as

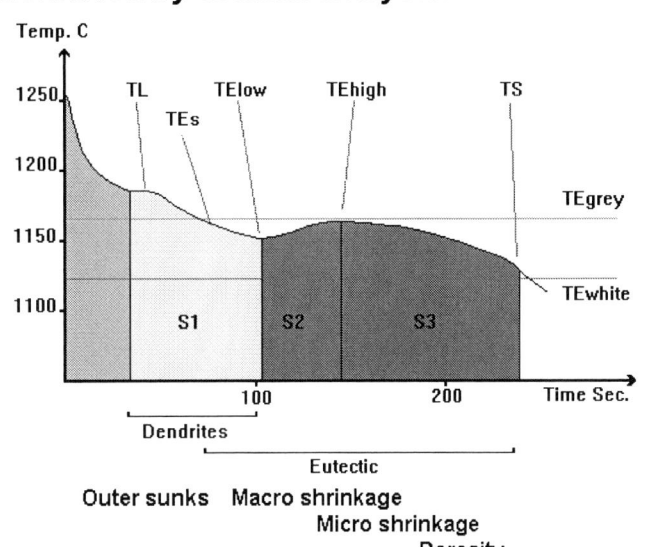

other problems. The system is based on analysis of "gr ey" cooling curves. The illustration shows a typical cooling curve for a hypoeutectic iron and the "time-position" for various types of shrinkages.

Shrinkage mechanisms

In the following we have classified shrinkages in four basic types. We will discuss how they occur and how they can be avoided. The four basic types are Outer sunks (pull downs), Macro shrinkages, Micro shrinkages and Porosities.

A. Outer Sunks

A1 – Definition and location

 Outer sunks (pull downs, sinks) can be seen on the outside of the casting, usually as a smooth depression in the casting surface. They are normally located on thick sections of the casting and on surfaces located on the top of the casting during pouring. Outer sunks are also referred to as "pull downs" and they can occur not only on horizontal top faces of the casting but sometimes also on vertical surfaces.

A2 – The basic mechanism behind outer sunks

The metal starts to solidify at the surface of the mould cavity and a thin skin is formed. The temperature drops further, which causes the liquid and the semi-liquid metal inside the casting cavity to contract further. If no feed metal is available either from the gating system or from a feeder then contraction will cause a negative pressure inside the cavity. In order to equalise the pressure difference between the atmosphere and the interior the solid outer skin will be "pulled" inwards. The effect is that the contraction is compensated for by a reduction of the volume of the casting. Thus outer sunks develop at an early stage of the solidification process before the massive eutectic freezing has commenced.

A3 – Variables that influence the creation of outer sunks

The major variables that influence the likelihood for development of outer sunks and their effect are as follows:
1. Too high pouring temperature. The volumetric contraction is about 1.4% per 100 C for cast iron.
2. Insufficient amount of feed metal available at early stages of the solidification. <u>In order to be able to produce ductile iron castings in green sand mould without feeders it is essential that feed metal can be supplied from the pouring cup through the gating system during the initial contraction in liquid state and until major parts of the casting have reached a temperature close to the eutectic temperature</u> .

B. Macro shrinkages

B1 – Definition and location

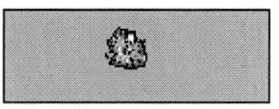 Macro shrinkages are usually found inside the casting and close to heat centres. They appear as larger holes, usually with rough surfaces, often dendritic and are often larger than 5 mm. If the iron is hypereutectic then the shrinkage might be rounded with smooth walls. Macro shrinkages are usually not revealed unless the casting is machined or deliberately cut through sections of heat centres. They can also appear in or close to ingates.

B2 – The basic mechanism behind macro shrinkages

Macro shrinkages develop after the initial solidification on the surface. A shell, that can not be deformed by the pressure difference to the atmosphere, has then been established. If no more feed metal is available at that point in time then the contraction of the liquid and semi liquid metal as well as contraction of the already solidified parts result in a cavity. Gas dissolved in the metal might also diffuse into the shrinkage cavity.

B3 – Variables that influence the creation of macro shrinkages
Shrinkages are influenced by the behaviour both of the alloy and the mould. The dominating variables for macro shrinkages are:

1. Insufficient supply of feed metal – too small feeder modulus or feeder neck or feeder does not pipe due to too large ingate modulus or height position.
2. Feeding path closed too early – wrong position of feeder. More dendrites than usual at an early stage. Too low active carbon equivalent (ACEL= the true carbon equivalent measured with ATAS). For ductile iron, ACEL should be eutectic or slightly hypoeutectic and carbon minimum 3.6%.
3. Mould hardness during solidification. Soft moulds that favour mould wall movement. A green sand mould, hard at room temp can be soft during solidification due to the high water level in the condensation zone.
4. Mould weighting or clamping insufficient to stand the pressure during solidification. Eutectic pressures up 50 kg/cm^2 have been claimed!
5. Higher liquidus temperature (TL) than normal. In hypoeutectic irons this means too much primary austenite and more difficult feeding. In hypereutectic it means more primary graphite, which reduces the amount of eutectic graphite.
6. Higher amount of primary austenite (S1) than normal. Too low ACEL.
7. Too high recalescence and recalescence rate causes expansion from graphite to occur too early. (In ductile - too much late inoculation with low ACEL).
8. Too low eutectic temperature might cause some primary carbides to form, which reduces the amount of eutectic graphite.

B4 - How can macro shrinkages be avoided?
The remedies are basically the same as for outer sunks. However, to avoid macro shrinkages the feed metal must be available longer than for Outer sunks because macro defects occur at a later stage during solidification. A solidification simulation e.g. using NovaSolid is highly recommended to ensure that feed metal can reach critical areas and that feed metal is available during the initial contraction period. In order to produce ductile iron castings without feeders the active carbon equivalent should be adjusted to the castings modulus so that the solidification is eutectic. A low gas content in the metal is also essential.

C. Micro shrinkages

C1 – Definition and location

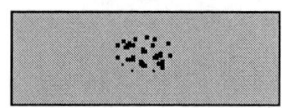

Micro shrinkages are smaller cavities with irregular surfaces, often with signs of dendrites. Sizes are often less than 3 mm. The defects are usually located close to heat centres in the casting. Micro shrinkages are often referred to as porosity or "leakage" as castings with this type of defect often leak during a pressure test. Micro shrinkages are usually not revealed unless the casting is machined or deliberately broken.

C2 – The basic mechanisms behind micro shrinkages

The defects occur at the latest stages of solidification. It is therefore more difficult to solve micro shrinkage problems by changing the gating or feeding system. <u>Micro shrinkages are more of a metallurgical problem</u>. There are three basic mechanisms behind micro shrinkages:

The first and most common mechanism behind micro shrinkages is that the contraction of the austenite (primary and eutectic) at the end of freezing can not be fully compensated for by the precipitation of carbon into graphite, which is associated with an increase in volume. The precipitation pattern of graphite from the start of eutectic freezing until the end of freezing is therefore very important.

C3 – Variables that influence the creation of micro shrinkages

1. Low mould stability which might cause mould wall movement. Too soft moulds due to high moisture content or low compression during moulding,.
2. Too low levels of sea coal addition in the green sand.
3. Bentonite (Calcium) with low wet compression strength in green sand. Sodium bentonite are optimal for reducing shrinkages.
4. Insufficient weighting or clamping of moulds.
5. Too little eutectic graphite especially at the end of freezing; its precipitation can not compensate for the shrinkage of the austenite. A true eutectic composition is. (Note that the eutectic point is a function of the thermal modulus!). Often a high C/Si ratio can reduce the risk for shrinkages.
6. Too high hypereutectic composition in ductile iron. If too high then some of the dissolved carbon will be precipitated as primary graphite and the amount of eutectic graphite might be insufficient. Evidenced by too many large size nodules which have been growing early in the liquid.
7. Too much magnesium in ductile iron. Levels of Mg and RE must be consistent with the thermal modulus of the casting. Low levels of Ce and high levels of La (0.005 – 0.010) can be very effective in reducing shrinkages in ductile iron.

8. Too high amount of phases that exhibit solidus temperatures below 1100 C. e.g. Fe_3P.
9. Too high recalescence and recalescence rate causes expansion too early and consumes a high amount of the carbon so that the expansion at the end of freezing is unsufficient.
10. Too low eutectic temperature might cause some primary carbides to form. A low eutectic temperature also means that the contraction in liquid state increases.
11. Too small Graphite Factor 1 (GRF1) indicating too low amounts of eutectic graphite during the second part of the eutectic.
12. Too high Graphite Factor 2 (GRF2) indicating too little eutectic graphite precipitation at the end of freezing.
13. Too low solidus (TS) which may induce carbides at the last portions to freeze. Too high levels of Mg, Nb, V or similar elements tend to segregate to the grain boundaries and form carbides that contract during their solidification. For ductile iron TS should be above 1100 C.
14. Too high silicon will increase segregation of carbide forming elem ents, which increases the tendency for micro shrinkage.

C4 - How can micro shrinkages be avoided?

The first condition is to have a hard mould. A green sand mould is compressible, however hard at room temperature, because of the formation of a condensatio n zone. In the condensation zone, which travels from the surface of the cavity and inwards, the moisture content can be up to 3 times higher than the initial value. This means that the bentonite layer becomes semi-fluid and can easily be compressed. It is recommended to use a sodium bentonite, sea coal as additive and a low moisture

Metallurgically, the most important factor is to ensure that a sufficient amount of carbon is precipitated as graphite during solidification. It is important that the initial growth rate is not too high. The metallurgical factors can be influenced by selection of charge materials, charging sequence, the melting cycle (temp/time steps) as well as type and amount of alloying materials, inoculants and FeSiMg. It is also important to avoid phases that are liquid below the main solidus temperature.

D. Porosities

D1 – Definition and location

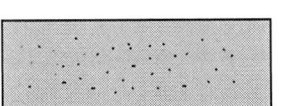
Porosities are small, dispersed cavities with irregular or rather smooth surfaces, often less than 1 mm in size. The defects are usually located close to heat centres and in grain boundaries. They are more dispersed than micro shrinkages. Typically, the defect is not discovered until the casting is subjected to a leakage test with water or air.

D2 – The basic mechanism behind porosities

The defects occur at the very latest stages of solidification. Therefore it is not possible to solve porosity problems with changes in the gating or feeding system (unless a very steep temperature gradient is maintained). Porosities are a metallurgical problem mainly depending on the chemical composition. The main mechanism is that due to the composition of the iron, one or more phases solidify at a lower temperature than the austenite-graphite eutectic. In ductile iron too high magnesium levels can cause similar problems as magnesium segregates to the rest melt and can induce formation of carbides which contracts and creates porosity.

Production of ductile iron castings without feeders in green sand.

By applying the principles outlined in this paper the author b elieves that ductile iron castings even with high modulus can be produced in green sand mould without feeders. The first condition is that the mould is sufficiently compressed and that that the sand properties are optimized.
The second condition is that the modulus of all parts of the gating system must be higher than about 40% of the dominating modulus of the casting. This ensures that most of the feed metal needed to compensate for contraction of the metal between the pouring temperatures down to liquidus can be supplied from the pouring cup. The third condition is that the metal expands sufficiently during solidification in order to match the contraction of austenite and remaining liquid phase in all parts of the casting. The expansion that comes from pr ecipitated graphite must not only be sufficient in volume, the precipitation must also be balanced in order to avoid mould wall movement and exhibit an expansion pattern until the end of freezing. The control of the solidification progression can be achieved by using a combination of chemical and thermal analysis of grey samples. The essential metallurgical factors to consider are:

- ?? The active carbon equivalent must be selected as a function of the modulus of the casting. Basically the active carbon e quivalent should be eutectic. Truly hypereutectic solidification must be avoided.
- ?? The carbon/silicon ratio should be high.
- ?? The nucleation level in the base iron must be sufficient so that the low eutectic temperature is high, preferably higher than 1140 C.
- ?? The recalescence must be less than 5 C.
- ?? The magnesium level must be as low as possible.
- ?? The graphite precipitation pattern in the final iron must be controlled so that a sufficient amount of eutectic graphite is precipitated after reaching the high eutectic temperature (GRF1 should be high and GRF2 should be low). These factors usually are at their peak about 3 minutes after Mg-treatment.

rudolf.sillen@novacast.se

Developments in the design of steel castings
M. Blair, R.W. Monroe

The Steel Founders' Society of America, Crystal Lake, Illinois, USA

Abstract

The development of designs regardless of product form has essentially been an evolutionary rather than revolutionary process. Designers have usually pushed the envelope in small increments building on their current designs. As designs have developed and inadequate performance has been expereinced, designs have become more and more conservative. This usually manifests itself in the specification of more restrictive non-destructive inspection requirements. The assumption in requiring more and more restrictive non-destructive testing requirements is that the different quality levels have some predictable effect on performance. This paper provides a summary of current research work on steel castings, that is directed at quantifying the effects of indications and illustrates ways in which casting designs may overcome some common problems such as hot tearing.

Key words (Design, steel, radiography, indications).

Introduction

Regardless of the product form designers make a number of assumptions in their design criteria, these might include;

1. Components are homogenous,
2. Any variance from homogeneity can be compensated for by applying factors of safety,
3. Non-destructive testing standards will ensure that the manufacturer complies with the requirements,
4. The use of specifications and standards will ensure that the component will be endowed with some guarantee of performance.

The work described here has been part of the Steel Founders' Society of America's research program and has employed both the University of Iowa (UI) under the leadership of Professor Christoph Beckermann and the University of Alabama-Birmingham (UAB) under the leadership of Professor Charles Bates.

Definitions

It is appropriate to clarify one point regarding the terms that are used for describing either internal or surface indications. Often the term "defect" is used to describe an indication. Within ASTM the definition of a defect is - an indication that exceeds the specified maximum dimension. When a component fails and there is a feature on the fracture surface, such as shrinkage or an inclusion, these are not defects and were not the cause of failure. A part will always fail through the weakest point, the reason for the failure is that the part was overstressed.

Shrinkage indications

Perhaps one of the most common non-destructive testing requirements invoked by designers is the requirement to meet certain radiographic quality levels, Most frequently a foundry will be asked to supply a casting to a shrinkage quality level of 3. On occasions the designer will ask that the quality level should be 2, the lower the number the less indications are allowed. The foundry may then respond that their experience in making similar parts for many years has shown that a shrinkage quality level of 3 has been satisfactory. At this point the designer may try to establish what is the difference between the different quality levels and how do they affect the intended performance of the part. In the Steel Founders' Society of America (SFSA) research program one of the projects is aimed at improving yield without reducing quality. The initial part of this work consisted of the casting of plates of various lengths at member foundries, using amongst other methods radiography to verify the quality. A sub objective was to determine if it was possible to determine the radiographic quality levels by simulation. In order to normalize the results of all the interpretations of the radiographs of the test plates, the radiographs were sent to different film readers who were asked to interpret the film and assign shrinkage type and levels to the films.

The results of the normalizing process are shown in Figure 1 which clearly indicates that a trained film reader can only judge to ± 1.4 levels which level that will be assigned to a film[1]. From Figure 1 it will be noticed that a new level was added, i.e. "0". This level is to include films that did not have any indications at all. If the data is analyzed using only levels 2, 3 and 4 then the variability of interpretation increases to >± 2 levels. Additional effort was brought to bear on this problem by subjecting the ASTM standards to computed image analysis and other manual techniques, it was not found to be possible to characterize the different quality levels.

The answer to the designer, regarding the difference in performance when moving from one quality level to another is, it is not possible to know whether there has been a movement from one quality level to another. So the question regarding a performance change is moot.

Regarding the possibility of being able to predict the quality level by simulation, this also becomes moot, however it is possible to indicate that a shrinkage quality level of 1 is achievable if the Niyama value is ≥ 0.1. This statement may only be true for Niyama levels determined by Magmasoft as the different software developers may calculate the value at different points during solidification.

The design of castings is not only concerned with strength levels it also concerned with issues such as leakage in components such as valve bodies. Case history studies[2] show that when a valve maker experiences leakage in a valve it is often not detectable by current NDE methods. Recently leaks were found in valve bodies that had clear radiographs. The foundry examined the leaks and found microshrinkage. The remedy was to re-engineer the risering of the casting which on simulation showed that to eliminate the microshrinkage a Niyama of 1.0 was required, Figures 2 and 3.

Effect of indications and Niyama on Mechanical Properties
The Niyama value is the relationship between the cooling rate and the thermal gradient:

$$N_y = G/ \sqrt{T} \quad \text{(where T is the thermal gradient)}$$

Two paths were defined for the future work:

1. Determination of the effect of the Niyama value on tensile properties. The test samples would be taken from material where the casting was radiographically sound, i.e. zero indications. The results of this work are shown in Figures 4 and 5[3]. The ultimate strength levels are not affected until values of 0.5 are reached whereas ductility values fall away much

earlier at values of 1.8 or less. Microstructural examination revealed the presence of microshrinkage in the fracture surfaces.

2. Determination of the effect of shrinkage which is visible on a radiographic on the fatigue properties of steels. The results of some of this work are shown in Figure 6[4]. The figures indicate that the properties follow the laws of mixing.

Given that shrinkage has an effect on the mechanical properties the issue is still not straight forward, until recently solidification simulation tools were not able to predict the size and location of shrinkage, they merely suggested that there is the likelihood of shrinkage in a particular location, compounding the problem is that finite element analysis (FEA) tools have not been able to deal with parts that have holes in them. Development of a model to predict shrinkage size and location has been completed and is available and is being incorporated in commercial software. The shrinkage model not only predicts shrinkage due to solidification but also includes the effect of dissolved gasses on the size of the pores.

Cracks and tears
To date a study to characterize the nature of surface indications has not been carried out. Because of this it is not possible to say with any degree of certainty what the nature of the indications are and to what degree do they extend. At this time the University of Alabama -Birmingham (UAB) is gathering samples to determine their root cause. When considering the surface of the casting, reference is often made to various inspection standards. The effect of surface indications is important, but the magnitude of their effect on the performance of a part has not been quantified. The standards that are used to set quality levels for surface are also workmanship standards and do not indicate the relative differences in their effect on performance. When agreeing on quality levels the negotiation that takes place between the purchaser and supplier simply reflects what they foundry is capable of and what the purchaser can live with. As shown with the radiographic standards the Gage R&R of the standards is not good. Determination of the Gage R&R of the surface inspection techniques, i.e. magnetic particle and dye penetrant techniques is important. Not only is it important to understand the limitations of current standards but it should also be used to determine the quality levels in any new standards that might be developed.

Hot tears are often classified as cracks but are essentially solidification issues. There are many rules of thumb that are used when trying to deal with the hot tearing problem. Such rules of thumb are; keep sulfur levels low, reduce the pouring temperature, increase the pouring temperature, use cracking brackets, use chills. All of these may have some effect but in many cases they are neither predictable nor effective. Hot tears are a function of design. Cracking brackets can be effective and are often built into patterns, the cracking bracket is a change in the casting design.

Software developers have made attempts to predict the likelihood and location of hot tears. Their success has been limited as they have usually just considered the strain that may occur during solidification. To determine where a hot tear will occur it is necessary to consider not only strain but the moment in time when feeding cut off. If strain produces a void in the casting it may be filled if the feeding path is still open. The practical results of this wok at UI[5,6] have provided a model which accurately predicts hot tears, this is shown clearly in Figures 7 and 8 which show CA15 castings where the new model predicted the tears experienced by the foundry and also suggested ways of eliminating them.

Inclusions

Inclusions are ubiquitous, they are one of the major problems experienced by purchasers of castings. The elimination of inclusions has been the subject of extensive work and is not addressed in this paper. However it is worth noting that air entrainment is one of the major contributors to the formation of inclusions. Water-modeling results and analysis of these results show that velocity is the key contributor to air entrainment, Figure 9[7].

$$\text{Air Entrainment is proportional to } th^{2.5}$$

$$\text{Where t = time and h = head height}$$

This equation requires that castings should be poured in the shortest time at low velocity. From other investigations it was determined that if runner systems were filled as quickly as possible then less inclusions would be formed. With respect to gating ratios, little if any effect could be seen by changing gating ratios to predict the occurrence or location of inclusions. As to the effect of inclusions on performance, it can be argued that the effect is likely to be similar to that of shrinkage porosity - they are voids that are filled with oxides. Therefore it is important to determine their size and location. Work has been continuing at UI[8] to develop a model for inclusion formation and their distribution in the castings, Figure 10.

Discussion

Clearly one of the most difficult issues faced in design is how to specify non-destructive examination requirements. It is not only problematic in assessing the effect of different quality levels but with the ability to reliably dicriminate between levels. In fact the inability to disriminate makes the first problem impossible to answer. The question is how can the effect of indications be assessed and how can useful specifications be written.

The effect of surface indications and how they can be dealt with is being investigated. One suggestion has been that it maybe that if these indications are found to extend into a casting to some predetermined depth then the surface of the part may be discsrded to this depth in the design calculation. This may be too pessimistic but might be a way.

For internal indications it is evident from the work at UI that if the size of shrinkage and inclusions can be determined then the laws of mixing may be applied to the modulus of elasticity, E. Where there are concerns regarding ductility this may be related to the Niyama values.

Conclusions
Most non-destructive examination stanadrds do not provide the information required by designers of castings.

Clearly the work on surface indications is in progress and the approach to their treatment will need to be developed.

For internal indications the approach is clearer. It can be hypothesised that the way of dealing with internal indications is to use simulation with radiography as a quality assurance check. The use of radiography for quality assurance will also require the developmnt of a new standard with the ability to reliably discriminate between different quality levels.

References
1. K. Carlson et al., *"Analysis of ASTM X-ray Shrinkage Rating for Steel Castings"*, Int. J. Cast Metals Res., 14 (3) (2001), 169-183.
2. K. Carlson et al., *"Developemnt of a Methodology to Predict and Prevent Leaks Caused by Microporosity in Steel Castings"* Paper presented at the 55th Steel Founders' Society of America National Technical and Opeating Conference, Chicago, Illinois, November 2001.
3. J.A. Griffin, R.D. Griffin, and C.D. Pears, *"Correlation of Tensile Properties to Niyama and Fracture Surface Properties"* Paper presented at the 54th Steel Founders' Society of America National Technical & Operating Conference, Chicago, Illinois, November 2000.
4. Blair et al., *"Designing Reliable Castings"* The John Campbell Symposium, TMS (The Minerals, Metals and Materials Society, San Francisco, 2005).
5. C. Monroe, C. Beckermann, *"Developemnt of a Hot Tear Indicator for Use in Casting Simulation"* Paper presented at the 58th Steel Founders' Society of America National Technical and Operating Conference, Chicago, Illinois, November 2004.
6. G. Catherwood, *"Hot Tears in CA15 Casting"*, Paper presented at the 58th Steel Founders' Society of America National Technical and Operating Conference, Chicago, Illinois, November 2004.
7. Monroe, Blair, Bates, Beckermann, *"Dross and Inclusions in Iron and Steel Castings"* BCIRA International Conference, England, April 1994.

Figures

Figure 1 Average one-sided confidence level of ASTM X-ray level ratings, grouped by average x-ray level. Average for 128 films is 1.42 rating levels.

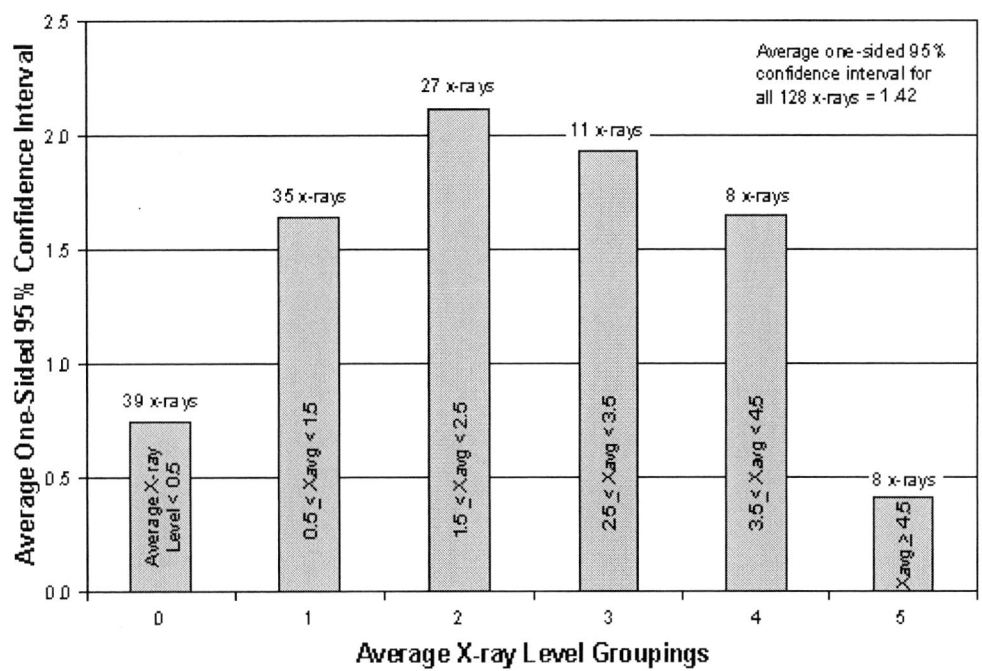

Figure 2 Leakage – original riser system

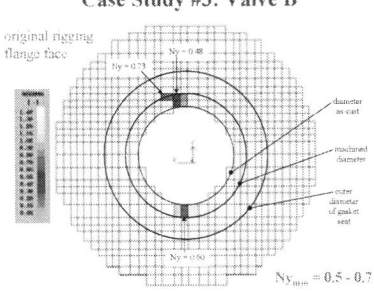

Figure 3 Leakage – revised riser system

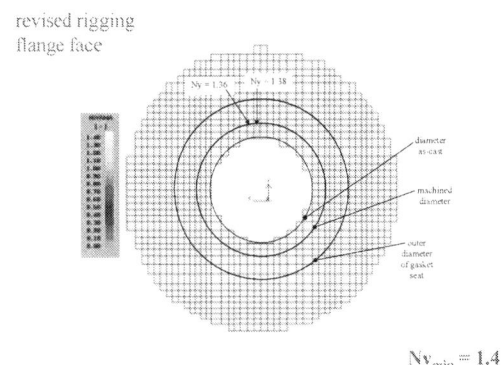

Figure 4 Effect of Niyama on Tensile Properties

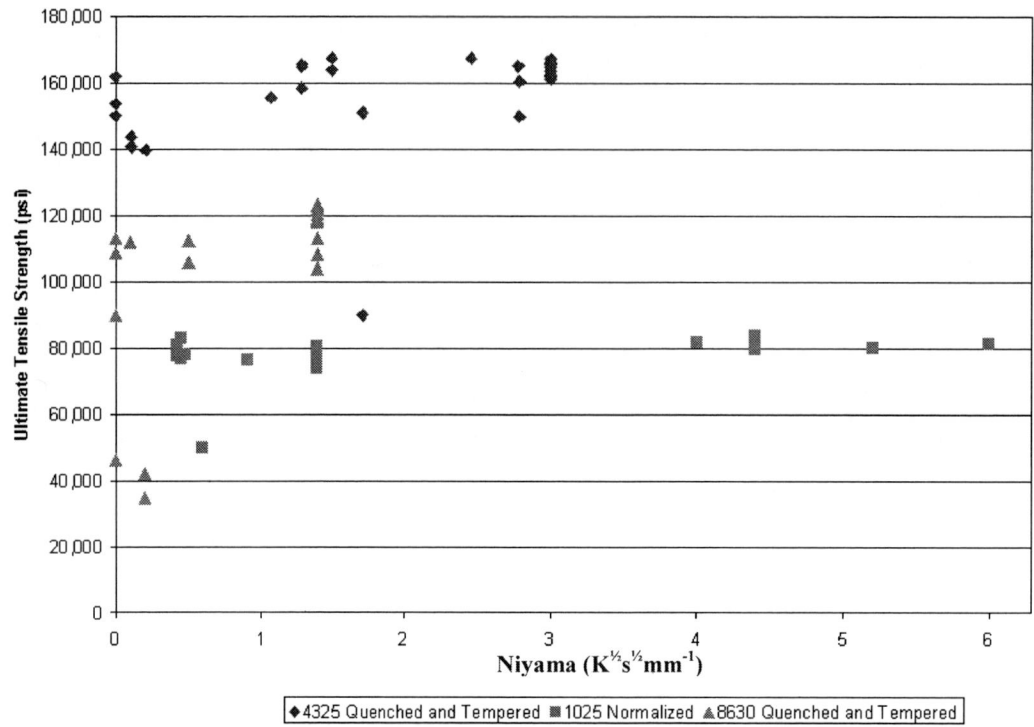

Figure 5 Effect of Niyama on Elongation

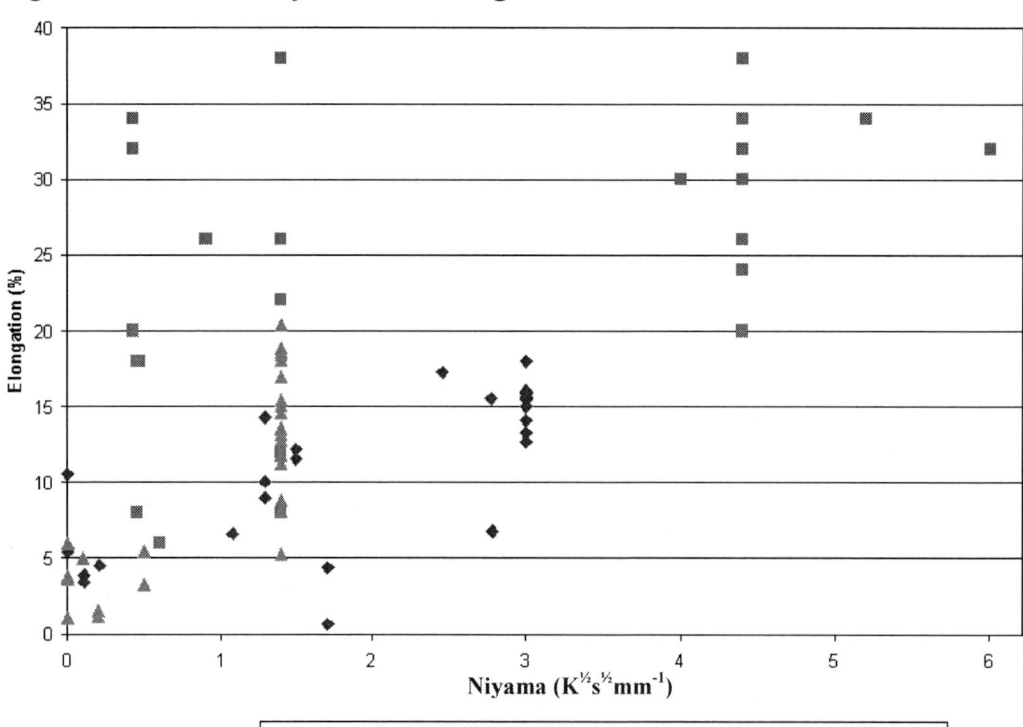

Figure 6 Effect of Percentage Porosity on Elastic Modulus and Fatigue

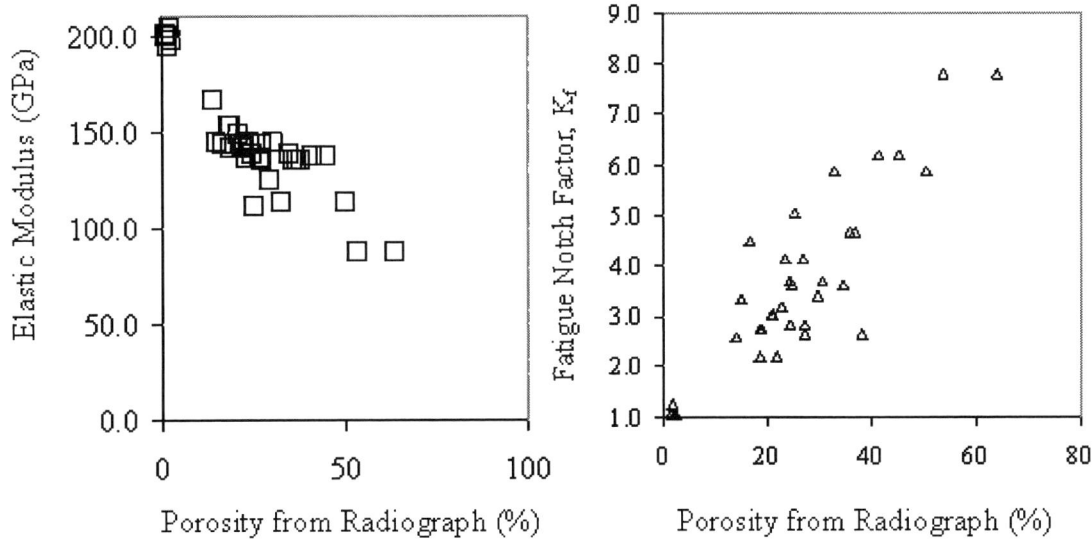

Figure 7 Casting Radiographically Sound

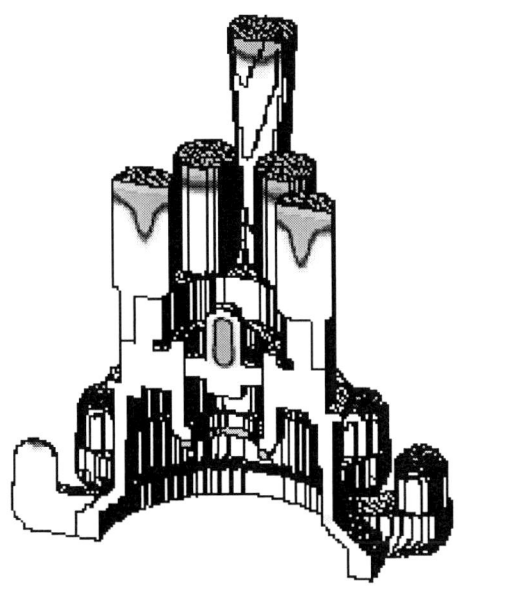

Figure 8 New Rigging – No Hot Tears

Figure 9 Relationship of Air Entrainment Results from Water Modelling Trials to the Pouring Time and Head Height for all Ladle Types

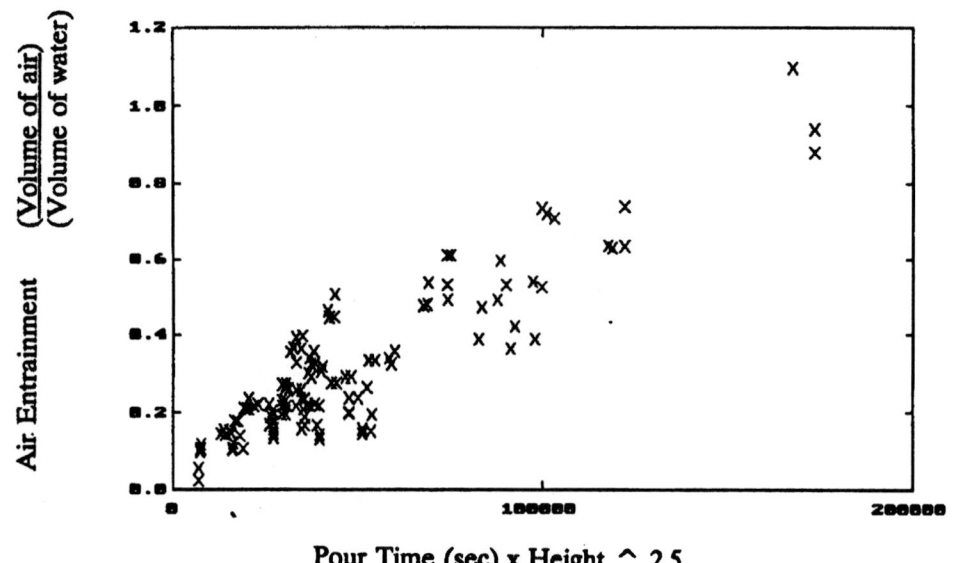

Figure 10 Distribution of Reoxidation Inclusions – Cast Steel Plates and Simulation

Multiloop approach for automatic mould filling in ferrous foundry.

C. Debray*, M. Dussud*, M. Biardeau*, and P. Canon**

* SERT, France, ** ACI Le Mans, France

Abstract

Foundries expect such performance from automatic pouring that their requisites are going beyond the reach of a classic approach to automatic control. A particular shortcoming that has been evidenced is that an automatic system based on a single level sensor is not apt to cope with the phenomena related to mould filling speed, such as refeeding. Multiloop strategy answers those needs: it implies a second camera, called post-pouring sensor, installed a little downwards after the pouring point. The results obtained with that solution at ThyssenKrupp Waupaca (USA) and ACI Le Mans (France) have shown a drastic reduction in pouring-related downgrading, reduced human involvement and optimised yield.

Key words

Automatic pouring, multiloop control, camera.

Introduction

Based on constant technical progress and better understanding of the process of gravity pouring with stopper, automatic mould pouring is tending to anticipate the foundries' ever-growing demands in terms of productivity: reduced downgrading due to pouring, optimised yield and pouring time, less human involvement…

Most automatic pouring systems are based on the same trio: sensor, controller, and actuator. As a general rule, the sensor gives a real-time value for the metal height in the cup, during pouring [1], [2]. Such a system, said single-loop, does not always answer the foundries' quiz, i.e.: *Minimised pouring time + optimised yield + minimised human involvement = maximum productivity.*

This paper first presents the context which has led to build a new approach to automatic pouring, more complete than the classic single-loop vision. Afterwards, it describes the new approach, based on 'multiloop' control. Then, it focuses on how a second sensor is implied and implemented in this solution. Finally, based on 2 application examples, practical results will illustrate how meaningful this approach is.

The limits of classic control

On vertical moulding lines, like Disamatic or Loramendi, it often happens that the sequence time of the moulding line is considered the reference sequence time. In one word, pouring is not supposed whatsoever to slow down the moulding machine. With that aim in view, the objective of the automatic pouring device is to lose as little time as possible: it is out of the question to do with a sort of compromise pouring time / machine cleanness. The point is to pour a filled cup, in the shortest possible time, and the final level in the cup after pouring must meet the minimum required by the model pattern, without metal waste : nor too much, nor too little. Now, the hydrodynamic behaviour of the iron within the mould frequently induces iron 'refeeding' after pouring. In some cases that provokes a level drop in the pouring cup, the importance of which depends on the pouring speed, but also on variable parameters, independent from pouring, such as sand porosity, or temperature, and thus iron viscosity. If the pouring cup does not have the necessary volume, or if the pieces and the feeding are very high in the mould, that excessive refeeding implies defects linked to short-poured moulds (misrun, blow holes…).

Even worse, if it is not detected, the controller may endlessly repeat the same mistakes. It may thus generate whole downgraded series. Therefore, the operator must keep his attention on pouring and rectify the settings in case a drift is observed. This means, to the eyes of an automation specialist, that to some extent those classic systems work in open loop.

Multiloop control

Since 1996, French company SERT-METAL has marketed an automatic pouring solution called UCERAM, about 50 units of which have been sold in Europe and in the United States. It is a turn-key automation solution for mould pouring in ferrous foundry, comprising a camera sensor, an electric stopper actuator, an advanced controller and all the instrumentation necessary to obtain a quality pouring, respecting pouring criteria set up with the foundrymen. The sensor is based on a multimeasure image analysis device, able to deliver to the controller some real-time information about the pouring spot (iron level in the cup, stream width, nozzle leakage, etc).

Nothing revolutionary about this architecture; it is pretty much the same as a classic system. However, what makes the difference is that the controller combines two controllers, each having its proper scope: the first one is the standard real-time controller, able to apply a pre-defined pouring strategy, but also able to cope with inopportune drifts as compared with that strategy. The second controller already constitutes in itself a first multiloop approach, as it is able to use the previous pouring, and the corrections that were made, to bring further fineness to the strategy followed by the first controller. This is the self-adaptation of the system to slow disturbance of the process [3].

From there on, it thus appears appropriate to close the loop identified in the previous paragraph, which gives even more autonomy to the system, optimizes the quantity of metal necessary for good mould filling, all the while gaining pouring time. To do so, a second image analysis sensor is installed, aiming at the cup that has just been poured, and measuring the iron level once it is stabilized, free from hydrodynamic fluctuation and the choking is over. That measure is taken in blind time, while the following cup is being poured, without delaying the sequence.
The final approach is represented in figure 1.

Implementing a post-pouring sensor

The equipment used for that post-pouring function is the same as the pouring sensor. It is an intelligent camera, within a protection sleeve. The detection software is loaded in both sensors, which make them interchangeable, and thus makes it easier and cost-effective when dealing with spare parts.

As shown in figure 2, the mechanic implantation of the second sensor does not pose any particular problem. The pouring camera is installed perpendicular to the moulding line, facing the nozzle. The post-pouring camera is placed a little downwards, by a distance corresponding to the average of the minimum and maximum widths of the moulds used on the machine. Since the camera does not move as compared with the nozzle, its lens must be chosen so that the visual field covers all the possible positions of the pouring cup used with the different pattern plates.

When the moulding line pushes forward, the just-poured cup gets into the visual field of the post-pouring sensor. It is then necessary to wait for the end of the refeeding phenomena before launching the level measure procedure. That waiting time can vary from 1 to 4s from the end of the mould push.

There begins the detection process, based on advanced techniques of image analysis, on the one hand to isolate the pouring cup in the image, on the other hand to determine the level in that cup, knowing that the scene may occasionally be 'polluted' by possible iron spillage on the mould, slags, or inoculant excess on the metal surface.

The measure of the final iron level in the pouring cup is sent to the pouring controller, which uses it to optimise its end of pouring rules. The possible variations as compared with a given setpoint are thus taken into account in real time for the following mould which is being poured.

Practical Results
Results at ACI le Mans
French plant ACI Le Mans operates a vertical moulding line (Disa 2130), producing brake parts. As everywhere else, pouring time is a productivity criterion to which they must pay particular attention. The solution ACI opted for, to optimise the pouring time, has consisted in using oversized pouring cups, creating an iron reserve available in the end of pouring. Those moulds can thus be poured with some overflow, so as to obtain a full cup when the stopper closes. At the end of pouring, the reserve in the cup goes down into the mould, thus completing the filling off-line.

Controlling that process manually can be tricky: the operator must provide for that overflow during pouring and close the stopper just in time. Given the high pouring flows, the slightest approximation provokes either excessive emptying of the mould, thus making it a misrun, or a waste of iron with negative impact on the yield rate. That problem cannot be solved with a Teach-In mode. Indeed, given the process variations (sand hygrometry, iron temperature, metallostatic height ...), Teach-In requires frequent corrections, and thus all-instants unfailing attention.

The advantage brought by the post-pouring sensor now clearly appears. It acts so that the metal quantity poured into the mould gives quality pieces, all the while providing an optimised yield. Further to the implementation of the multiloop method at ACI, measures taken after the shake-out have shown an iron gain over 2 kg per mould poured, while dividing by 1.8 the misrun rate (see figures 3 & 4).

Results at Waupaca Plant 3

Waupaca Plant 3 is located in Wisconsin (USA). It is part of ThyssenKrupp Waupaca group, that has 6 foundry plants in the United States, and mainly works in the automotive sector.

Plant 3 has 7 production lines. In 2001, one of them was equipped with the automatic pouring system UCERAM, without post-pouring sensor as that technique was under evaluation at the time. The productivity gains given by that first system led not only the plant management to duplicate the system on their other six moulding machines, but also the group management to equip their other plants. Today, some twenty Uceram units have been purchased by Waupaca group.

Within their continuous improvement programme, plant 3 wished to make a further step, 3 years after the first installation, with the purpose to reduce even more drastically their rate of downgrading directly linked to pouring. Besides, just like any other western industry, Waupaca is bound to reduce human involvement as much as possible, and thus needed to find a way to limit the moulding machine adjustments, implied by slow and uncontrollable evolutions of the process.
The post-pouring sensor was there to meet both those expectations.

First trials were carried out on one strand and the results were very conclusive: while they used to have 2000PPM short-poured moulds, the addition of the post-pouring sensor brought that result down to about 670PPM, which corresponds, given the production rhythms of the plant, to less than 1 short-poured per 8 hours shift (see figure 5).

A computerized tracking also showed that human intervention for adjustments were reduced by 50% after that modification (see figure 6). And most of them correspond to start of campaign adjustments, after which, unless exceptional disturbance, the pouring machine is completely autonomous.

Further to those trials, Plant 3 management decided to adopt that new equipment on all of their 7 production machines.

Conclusion

Based on those application examples, it can be observed that the multiloop approach brings considerable benefits, and there is now a system available which can correct the pouring:
1.   in real time
2.   against slow drifts of the process
3.   taking into account the iron flow phenomena occurring after pouring.

Since the first trials of that innovative technique, some twenty installations were equipped with it, with an outstanding return on investment for such industrial equipment.

It is particularly adapted in situations where pouring times determine the cycle time of the machine, or when the user's expectations in terms of performance are to the limits of the process.

The examples of ACI le Mans and Waupaca3 clearly show how that multiloop approach gives even more meaning to automatic pouring. The interest of that technique has been proved for foundries aiming at different targeted priorities.
In a few words, the UCERAM multiloop system offers repeated production of quality pieces, all the while minimising sequence time, metal quantity, and human involvement.

References
1. Goran Lowback, *Automatic Tilt- or Bottom Pouring?*
    Foundry Management&Technology, January 2005, pp36-37.
2. William Pflug, Emad Tabataba, *The Challenge for Automatic Pouring Systems,* Foundry Management & Technology, January 2005, pp33-34.
3. C.Debray, P.Simonnin, *Updated generation of automatic mould filling,* Foundry Trade Journal, July 2002.

Acknowledgements
We wish to warmly thank Mr. Jannick BEAUGE & Mr. Patrick CANON from ACI Le Mans, and Mr. Jeff WALTERS & Charlie KERNEEN from ThyssenKrupp Waupaca, for their support to prepare this paper.

Tables & figures

Figure 1 multiloop approach of the automatic pouring controller

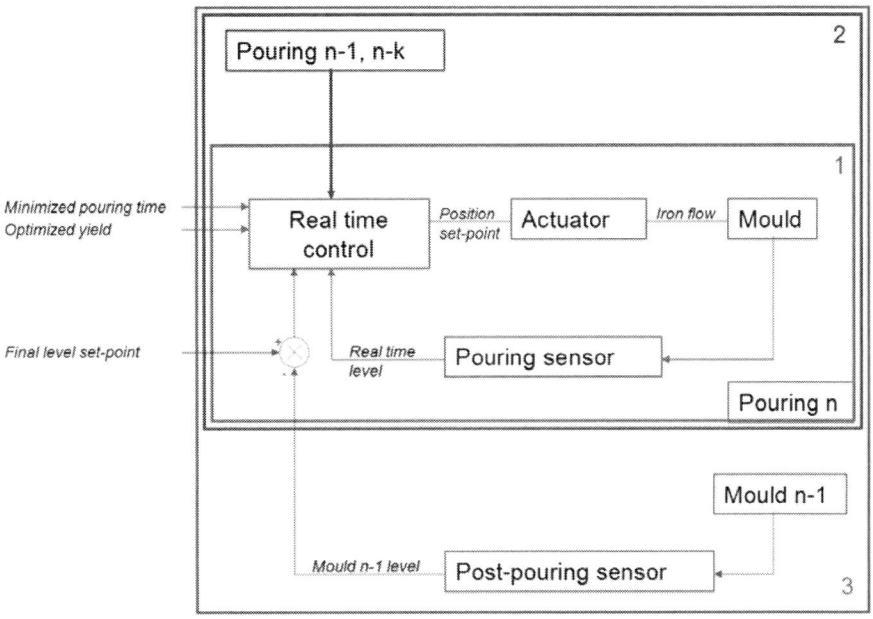

Figure 2 Lay-out of the pouring and post pouring sensors

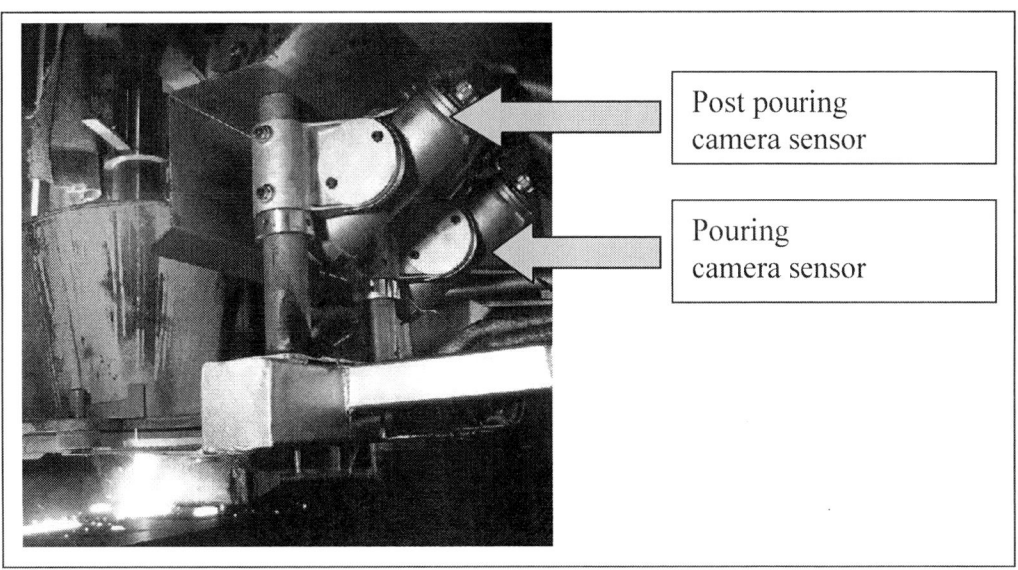

Figure 3 Evolution of the yield at ACI Le Mans (France).

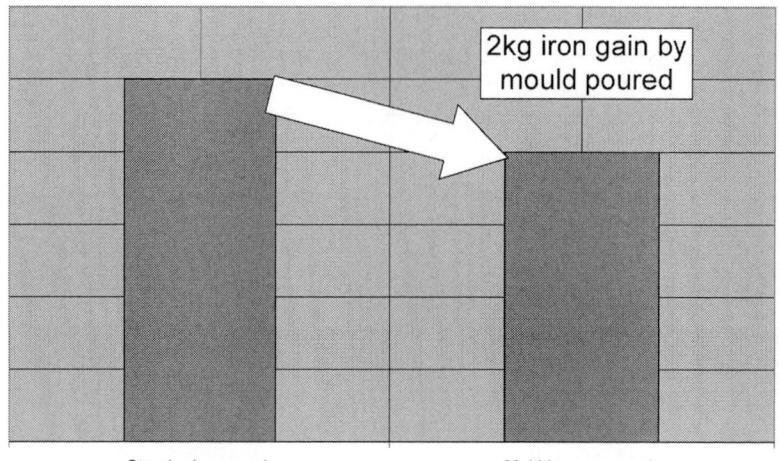

Figure 4 Evolution of misrun rate at ACI Le Mans (France).

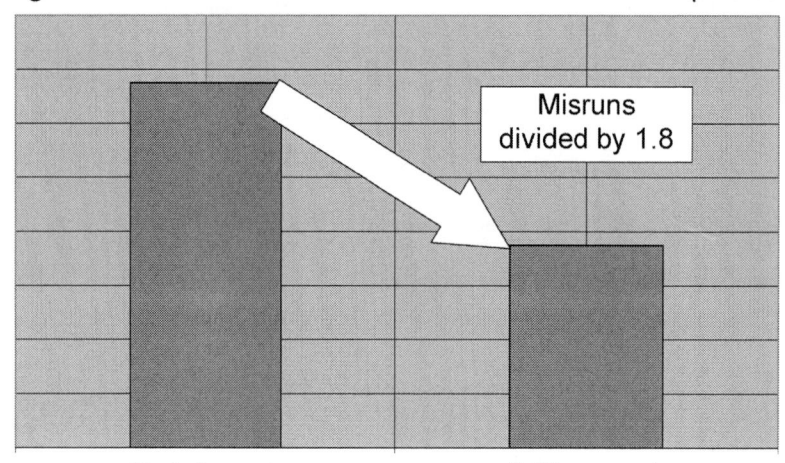

Figure 5 Evolution of the short-poured rate at Waupaca Plant 3 (WI, USA).

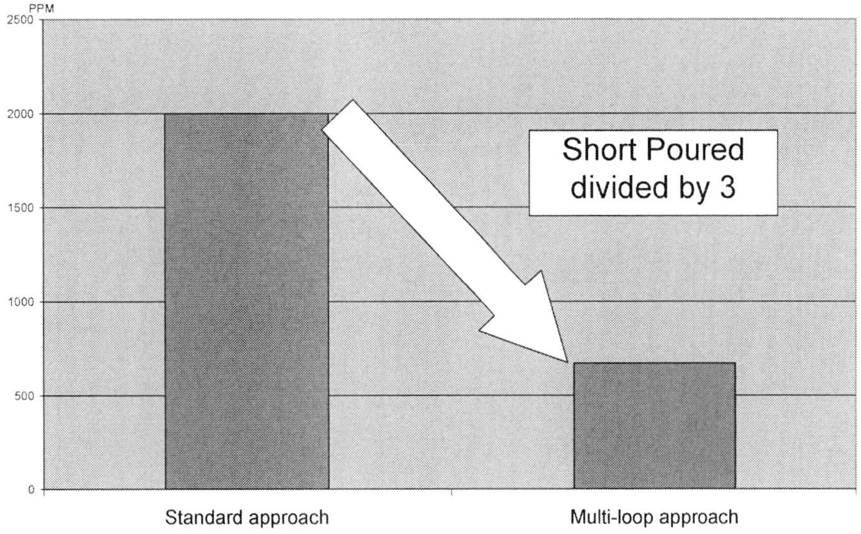

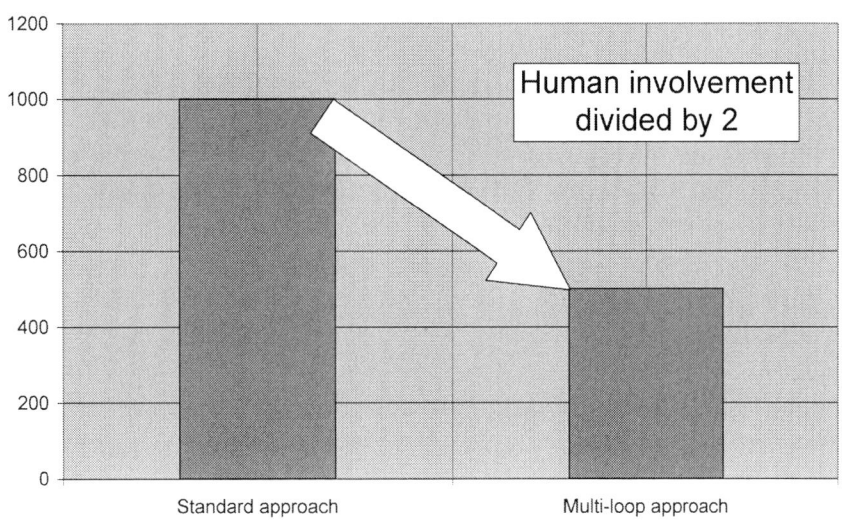

Figure 6 Evolution of human involvement at Waupaca Plant 3 (WI, USA).

Authors' email addresses:
C.DEBRAY – c.debray@sert-metal.com
M.DUSSUD – m.dussud@sert-metal.com
M.BIARDEAU – m.biardeau@sert-metal.com
P.CANON - patrick.canon@auto-chassis.com

Cast iron material standards for a new millennium

M P Macnaughtan C.Eng MIMMM FICME

Chairman of ISO/TC25 – Cast irons and Pig irons
Group Technical & Research Manager - EURAC Ltd

Abstract

A shift in the production of cast iron components away from the western industrialised nations, and the emergence of Asian countries as producers of significant volumes of cast iron, resulted in the reactivation of the International Organization for Standardisation committee ISO TC 25 under the Secretariat of the United Kingdom.

The first plenary of ISO/TC 25 held in 1998 agreed that a review and revision of the published ISO standards for cast iron materials would take place, standards would also be written for materials not covered. ISO/TC 25 has 50 member countries that account for 94% of the global production of cast iron. This paper will review the work of TC25 whose aim has been to introduce cast iron material standards that are both globally acceptable and technically competent.

Key words.
Material standards for cast iron

Introduction

Cast iron, a material that has been with us for several thousand years, was once described as "a simple material having in its nature many mysteries". Fortunately, the passage of time, and the involvement of many skilled Foundrymen and scientists, has solved the majority of these mysteries. Today's industrial engineers have inherited an extremely versatile and cost effective casting material with a wide range of benefits that are unequalled among casting materials.

Cast irons are important, strategic, versatile materials that cover the widest possible spectrum, from safety critical components for the automotive industry, to specialised pipe and valve fittings for the energy industry. A whole host of general engineering components too numerous to mention are also easily and competitively produced in a wide range of both alloyed and unalloyed iron materials. Clearly, the industrialised world would be a much poorer place without the existence of the wide range of cast iron materials available today.

The 38th Census of world casting production performed in 2003 listed thirty-five countries who produced between them some 56.6 millions tonnes of iron castings [1]. ISO/TC 25 members accounted for 94% of this production. The work of TC 25 can therefore be seen as globally relevant.

Why do we need international standards?

An expansion of trade, and in particular the demand for cast iron components to satisfy the requirements of the global market place, suggests that the need for a common terminology, measurement techniques and material descriptors increases as globalization and international trade expands. This expansion also implies that there is a requirement for International Standards of the type produced by ISO, that is, standards produced by co-operation between both countries, and experts and stakeholders within those countries. The need for such standards within the global cast iron market has been recognized and is encompassed in the mission statement of ISO/TC 25 which states that TC 25 will 'Raise the awareness of industry to the image and importance of cast irons and assist in this awareness, by the preparation of International standards that are technically, and commercially acceptable to the global market'. The need for truly global standards has involved experts from many countries who have adopted the slogan "do it once, do it right, do it internationally".

International Organisation for Standardisation

The International Organization for Standardisation (ISO) was founded in 1946 by delegates from 25 countries, and began operating in 1947. ISO develops international standards for materials, products, processes, systems and a wide range of organisational and quality practices. These standards help ensure the vital characteristics of components, facilitate trade and disseminate technology.

ISO, which is based in Geneva, Switzerland, operates a three-tier membership structure comprising of member bodies, corresponding members and subscriber members.

- Member bodies – a member body of ISO is the national standardisation organisation most representative of standardisation in a particular country, for example in the UK the member body is BSi. Member bodies have full voting rights and are entitled to actively participate in the standards and policymaking process. Once ISO standards are approved and issued, member bodies can adopt the documents as the basis for their own national standards.

- Correspondent member – Correspondent members do not take an active part in the technical or policy development process, they are however entitled to be kept fully informed of work in progress and can use any ISO standards as a basis for their national standards.

- Subscriber members – This is the basic level of membership and allows countries with small economies that cannot support a fully developed standards making activity to maintain contact with international standardisation.

The technical work of ISO is overseen by the Technical Management Board, which is responsible for setting up technical committees, appointing committee chairmen and monitoring the progress of work programmes within the technical committees. A technical committee can in turn establish subcommittees to focus on specific areas its work programme. Further subgroups known as working groups can then be established to focus on specific items within the work programme.

All ISO member bodies and correspondent members are entitled to appoint representatives to ISO committees. There are, in practical terms, two types of committee membership:

- Participating members (P-members) - members who play an active roll in the work of a technical committee or sub-committee. P-members nominate experts to sit on technical committees, these experts help develop a standard and provide feedback to their national standards body.

- Observer members (O-members) – members who follow the work of a particular technical committee without actively participating in the work programme. This type of membership is open to both ISO member bodies and correspondent members.

ISO/TC 25

The International Organisation for Standardisation (ISO) committee TC 25 – Cast irons and Pig irons – was formed in 1947, and as such was one of the first committees formed after the formation of ISO. The main work of TC 25 was completed in the mid-nineteen eighties, and the committee then lay dormant until reactivation in 1998 under the Secretariat of the United Kingdom. The first plenary of the reactivated committee was held in London on the 24th September 1998, when a mission statement and business plan were formally adopted. Since then, the plenary has convened annually and pursued an aggressive revision of the cast iron material standards in its portfolio.

Currently, ISO/TC 25 has a membership of 50 countries, 29 Participating (P) members and 21 Observer (O) members, a full list of the membership will be found in Tables 1 and 2. The aim of TC25 has been to produce a comprehensive range of globally acceptable cast iron material standards that are technically competent and informative.

Why reactivate ISO/TC25?

Discussions with industrialists in Europe revealed several areas of concern that potentially presented a threat to the use of cast iron materials in certain market sectors. These concerns are summarised as follows:

- The increasing use of plastic materials for the construction of underground pipe work systems replacing iron components
- The increased use of aluminum and other light weight alloys in the automotive industry
- The designer in the supply chain is frequently not well versed in the use and potential of cast iron materials and therefore either relies unnecessarily on fabrications or designs castings with a high safety factor resulting in a product that is unnecessarily heavy and therefore uncompetitive.
- The worldwide expansion of the iron casting industry has resulted in the need for global standards acceptable on a global basis.

The committee structure of TC25

There are currently seven subcommittees and one work group in the structure of TC25, these committees and the international standards they are responsible for are:

- SC1 – Malleable cast iron ISO 5922
- SC2 – Spheroidal graphite cast irons ISO 1083
         Ausferritic Graphite cast irons ISO 17804
- SC3 – Grey cast irons ISO 185
- SC4 – Pig irons ISO 9147
- SC5 – Technical conditions of delivery
- SC6 – Austenitic cast irons ISO 2892

Abrasion resistant cast irons ISO 21988
- SC7 – Compacted graphite cast irons ISO 16112
- TC25/WG1 – Designation of microstructure of graphite ISO 945
- Ad-hoc work groups – formed as required

TC 25 meets annually at a plenary session held in one of the member countries. The various subcommittees and work groups meet more frequently in order to ensure that their respective work programmes meet the targets defined in the ISO Directives and the TC25 business plan.

The work of the TC25 subcommittees will now be described in greater detail.

SC1. Malleable Cast iron
The secretariat of SC1 is currently held by Japan (JIS). The standard for malleable cast iron, ISO 5922, was updated in 2005 and specifies five grades of white heart malleable and ten grades of blackheart malleable cast iron. The standard covers the production of castings produced in sand moulds or a material of similar diffusivity. An informative annex includes information about the impact resistance of both the malleable types covered by the ISO 5922.

SC2. Spheroidal graphite cast iron and Ausferritic Spheroidal graphite cast iron
The remit of SC2 was to implement a full revision of the standard for spheroidal graphite cast iron ISO 1083 and write a new standard to cover ausferritic spheroidal graphite cast irons. The French member body AFNOR currently holds the secretariat of SC2, which has two work groups WG1 and WG2.

A detailed revision of ISO 1083 was performed by WG1, and resulted in the second edition, which was published in 2005. The revised document covers fourteen grades of spheroidal graphite cast iron defined by mechanical properties, minimum impact resistance values are also specified for four of these grades. ISO 1083 also includes three normative and four informative annexes. The normative annexes cover spheroidal graphite irons with high silicon content, the relationship between elongation values obtained from test pieces where the gauge length is both four and five times the diameter of the test piece and the classification of ten grades specified by hardness. The four informative annexes cover material toughness, guidance values for 0.2% proof stress for test pieces cut from castings, nodularity and finally additional information providing mechanical and physical property data. A bibliography listing is also provided at the end of the standard. The French member body of TC25 currently holds the secretariat of WG1.

At the plenary meeting of TC25 held in London on the 4th / 5th October 1998 it was proposed that a standard should be written for ausferritic spheroidal graphite cast irons. The work item was subsequently approved after a ballot of the membership, and a second work group WG2 with a Dutch secretariat and Italian chairman was formed to perform the task. Ausferritic spheroidal graphite cast iron, also known as austempered ductile iron (ADI) is basically a cast alloy, iron and carbon based, with the carbon mainly present in the form of spheroids. The material has the benefits of higher strength and toughness properties, which are the result of an austempering heat treatment process. Ausferritic spheroidal graphite cast iron has been used in the manufacture of gear components for many years, and has recently found increased use in the manufacture of modern high-pressure diesel engines for the automotive industry. The new ausferritic graphite cast iron standard ISO 17804 was approved for publication towards the end of 2005.

ISO 17804 covers six grades of ausferritic spheroidal graphite cast iron with one of these grades also being qualified by room temperature impact resistance value. There are two normative and eight informative annexes. The normative annexes specify abrasion resistant grades of ausferritic spheroidal graphite cast iron and minimum elongation values for test pieces with an original gauge length of four times the diameter. The informative annexes cover subjects such as guidance values for Brinell hardness, a procedure for determination of the hardness range, guidance values for tensile strength and elongation for test pieces cut from a casting, unnotched impact testing, nodularity, machinability, additional information on mechanical and physical properties, and a cross reference between the grades specified in ISO 17804 and current EN, ASTM, JIS and SAE specifications. A bibliography listing is also included.

SC3. Grey cast irons
The standard for grey cast iron ISO 185 has been completely revised by SC3, under the watchful eye of the secretariat, held by the German member body DIN. This second edition of ISO 185, which was published in 2005, embodies a major revision of the initial document first published in 1988.

The number of grades classified by tensile strength has been increased from six to eight with the introduction of the intermediate grades for 225 N/mm² and 275 N/mm². In addition to the eight grades classified by tensile strength six grades classified by hardness have been introduced to the normative section of the standard. These grades were previously only classified in the informative annex of the first edition of ISO 185. Additional information useful to both designers and users of grey cast iron components such as mechanical and physical property data has also been included in the informative annexes, as has additional information on the relationships between hardness and tensile strength, and hardness tensile strength and the wall thickness of castings.

SC4. Pig irons.

This sub-committee has been inactive since TC25 started work in 1998. The standard for pig irons ISO 9147:1987 was reconfirmed by the members in 2004 for a further five year `period.

SC5. Technical conditions of delivery

This sub-committee has also been inactive since the re-convening of TC25. The members of TC25 were balloted in 2003 with a view to writing an ISO standard to cover technical conditions of delivery but following a vote of the membership the majority required to allow a work item to proceed was not obtained.

SC6. Austenitic cast irons and Abrasion resistant cast irons.

ISO 2892, the austenitic cast iron standard was first published in 1973. Austenitic cast irons are a range of highly alloyed cast irons containing nickel and, according to the grade, manganese, copper, chromium, niobium and an elevated silicon content. Carbon is present in the form of either lamellar flakes or spheroidal graphite nodules. In some grades carbon is present in the form of iron carbides. Under the watchful eye of the secretariat held by the United Kingdom and the chairmanship of the German member body, ISO 2892 is now in the final stages of the revision process.

The proposed second edition of ISO 2892 specifies two lamellar and ten spheroidal grades of austenitic cast iron. Five informative annexes cover subjects such as the properties and applications of the austenitic grades of cast iron, heat treatment schedules, additional mechanical and physical property data useful for both design engineers and castings users, a cross reference table relating ISO 2892 to other austenitic iron standards and information about the preparation of samples for testing austenitic cast iron materials.

Abrasion resistant cast irons are widely used in the mining, earth moving, milling and manufacturing industries where resistance to highly abrasive materials is required. The abrasion resistance of these materials is dependant on the structure which is normally a white iron having a specific chemical composition and hardness range. The abrasion resistant cast iron material standard, which is at the final draft international stage (FDIS), has been allocated the number ISO 21988.

The document classifies thirteen grades of abrasion resistant white cast irons specified in terms of chemical composition and hardness. Six of the grades covered by the standard are high chrome irons, five are nickel-chrome irons and two are unalloyed or low alloy abrasion resistant irons. The standard contains five informative annexes detailing topics such as heat treatment schedules for tempering, soft annealing and annealing, a conversion table for comparing Brinell, Vickers and Rockwell hardness values, the relationship between casting thickness and chemical composition for nickel-chromium cast irons, typical microstructures of abrasive resistant cast irons and cross reference between ISO 21988 and other standards.

SC7. Compacted (vermicular) graphite cast iron

SC7, the secretariat of which is currently held by the United States member body, first convened in March 2002 in London where a work item to write a standard for compacted (vermicular) graphite cast iron (CGI) was proposed and approved by the membership. The draft document is currently in the final stages of the standards making process and will be published in 2006 as ISO 16112.

ISO 16112 will be first international standard to be written for compacted (vermicular) graphite cast irons (CGI) and classifies fives grades of the material in accordance with their mechanical properties obtained from either separately cast samples, cast-on samples or samples cut from the body of the casting. Important information such as procedures to evaluate graphite nodularity, the influence of metallurgical variables on the machinability of CGI, properties and typical applications for CGI materials and a cross reference of ISO 16112 grade designations to other standard grades of compacted graphite (vermicular) graphite cast iron are also included in a series of five informative annexes. There is also a comprehensive bibliography cataloging thirteen useful reference works.

TC25/WG1. Designation of microstructure of graphite ISO 945

Following acceptance by the membership of TC25 of the work item to write a standard for compacted (vermicular) graphite cast iron it quickly became apparent that the ISO standard for the designation of microstructure of graphite ISO 945 would require revision. A resolution to this effect was presented to the plenary of TC 25 and duly accepted. Work is currently underway to completely revise ISO 945 in order to take account to compacted irons. The opportunity has also been taken to consider a fundamental revision of ISO 945 by splitting the standard into two parts. Part 1 will cover the classification of graphite by visual analysis and Part 2 the classification of graphite by image analysis. Advantage will also be taken of digital technology during this revision, and photographic images of graphite forms and structures will be incorporated into the revised standards.

Ad-hoc work groups

Two ad-hoc work groups have been formed within the structure of ISO/TC 25. The first group was formed to consider the matter of designations. It was recognized by the plenary of TC25 at an early stage during its allotted programme of work that it would be an advantage to have a consistent designation system for both the new and revised ISO cast iron material standards. With this in mind the ISO Technical report ISO/TR 15931, Founding – Designation system for cast irons and pig irons was written. The ISO Central Secretariat formally approved the document for publication in 2004.

A second ad-hoc group has been formed to consider the publication of a comprehensive document containing basic metallurgical knowledge, manufacturing guidance, technical information and data that will highlight the advantages and disadvantages of the whole range of cast iron materials covered by the current ISO standards. The document will be targeted at designers and engineers, with the specific aim of making them more aware of the benefits of each material. It is hoped that this strategy will result in more innovative and imaginative uses of cast iron materials.

Conclusion.
The work programme of ISO/TC 25 – Cast irons and pig irons is now well advanced. It is hoped that by modernising existing standards and introducing new standards for materials not previously covered, the committee will have moved closer to achieving its main aim of developing quality, market relevant standards suitable for international acceptance by the global casting industry.

References

1.   *38th Census of World Casting Production – 2003.* Modern Casting, December 2004, pp25 – 27.

Bibliography

1.   ISO 5922, Malleable cast iron
2.   ISO 1083, Spheroidal graphite cast iron – Classification
3.   ISO 17804, Founding – Ausferritic spheroidal graphite cast irons – Classification
4.   ISO 185, Grey cast irons – Classification
5.   ISO 21988, Abrasion resistant cast irons – Classification
6.   ISO 2892, Austenitic cast irons – Classification
7.   ISO 16112, Compacted (vermicular) graphite cast iron – Classification.
8.   ISO/TR 15931, Founding – Designation system for cast irons and pig irons

Acknowledgements
The author would like to thank all the experts from the many countries involved in ISO/TC 25 for their input into the work of the committee. Their contributions have been made in a friendly and professional manner and as such have been invaluable in helping to make the work of TC25 a truly global co-operation. The support of the Cast Metals Federation (CMF) in

providing funding and resources for the secretariat of ISO/TC25 is gratefully acknowledged as is the support of EURAC Ltd.

malcolm@eurac-group.com

More information about the work of ISO and TC25 can be found on the ISO web page at www.iso.org

Tables

Table 1. Participating (P) members of ISO/TC 25 and member bodies

Australia	SA	Japan	JIS
Austria	ON	Korea, Democratic People's Rep.	CSK
Belgium	IBN	Korea, Republic of	KATS
China	SAC	Norway	SN
Czech Republic	CNI	Poland	PKN
Egypt	EOS	Russian Federation	ASRO
Finland	SFS	Serbia & Montenegro	ISSM
France	AFNOR	South Africa	SABS
Germany	DIN	Spain	AENOR
Greece	ELOT	Sweden	SIS
Hungary	MSZT	Turkey	TSE
India	BIS	United Kingdom	BSI
Indonesia	BSN	USA	ANSI
Iran, Islamic Republic of	KATS	Venezuela	TCVN
Italy	UNI		

Table 2. Observer (O) members of ISO/TC 25 and member bodies

Argentina	IRAM	Portugal	IPQ
Bulgaria	BDS	Romania	ASRO
Croatia	HZN	Saudi Arabia	SASO
Cuba	NC	Slovenia	SISI
Denmark	DS	Sri Lanka	SLAI
Malaysia	DSM	Switzerland	SNV
Mexico	DGN	Tanzania, United Republic of	TBS
Mongolia	MASM	Thailand	TISI
Netherlands	NEN	Tunisia	INORPI
Pakistan	PSQCA	Viet Nam	TCVN
Philipines	BPS		

Safety cast components for the automotive industry. The metallic charge, the presence of micro elements and their most relevant effects.

Ramón Suárez Creo, Julián Izaga Maguregi, José A. Goñi Güemes.

AZTERLAN Metallurgical Research and Foundry Centre

Abstract

The Automotive Industry keeps on demanding increasing requirements to the cast components. The critical situation generated by the extraordinary price increase of the raw materials has not meant any flexible position, on the contrary, new stronger demands reappear according to the constant development of the market.

Economical reasons, market demands and weight reduction policies introduce new concepts in the sheet steels used in the car bodies. Many of these new sheet steels incorporate macro and micro alloys as a common source. We are talking about "new materials" in which elements as Al, Ti, Zn, Nb, and Mn are present in a very significant way.

This work establishes the relationship between some of the mentioned trace elements, incorporated to the iron through the metallic charge, and the solidification model and metallurgical characteristics. An special attention is paid in certain particular influences and on the development of a wide range of iron casting defects related to these trace elements.

The final conclusions of this paper refer to the metallurgical influence from the residual elements:

- The amount of Zn dissolved in the iron depends not only on the metallic charge, but also on the melting operative and on the metal conditioning treatments.
- The micro-segregation mechanisms associated to the presence of Nb, V and Ti have as well an special influence on machinability and on the formation of reticular carbides.
- The trace elements affect the metallurgical quality of the iron and its solidification model.

State of the art

The Automotive Industry is constantly aiming to reduce costs and this situation affects directly the Foundry Industry as it has to endure the aggressive purchasing policies used. This situation is very well known within all the industrial sectors but, lately, some other factors such as the emerging and developing countries, delocalisation, raw material's crisis and the production cost increase are distorting this scenario even more.

In order to face this complex and, up to a point, "new" situation, it has been necessary to introduce new concepts that take into consideration this latest reality, and introduce process modifications that affect all the foundry conceptions conventionally accepted. Productivity improvements, design changes, rejection rate reduction, or elimination of non added value operations could be mentioned within these improvements (Figure 1).

One of the main working lines from this technical work is directly related to the production of the liquid metal; and more precisely, to the metallic components of the charge, the usage of iron conditioners, and the role of recarburation. Metallurgical criteria have been traditionally considered when choosing the most appropriate metallic charge. In this context, pig iron has been widely used since it adds features to the liquid iron such as easy melting properties, homogeneity, high cleanliness levels, good nucleation capacity and the minimum presence of disturbing elements.

Economical considerations, the actual pressure from the market, offer limitations and in some cases, difficulties in the material supply, are forcing a big change in the composition of the metallic charge. The steel consumption and more precisely the so called "automobile steel scrap" are exponentially increasing. Nowadays it is relatively frequent to work with sinter charges, that is, melt iron produced only with returns and steel as metallic charge.

The steel used in the automobile chassis has been also developing according to market demands. The steel producers have had to take special actions in order to meet the corrosion resistance and the increase of mechanical strength demands. In order to achieve these requirements, some relevant changes had to be introduced as far as residual elements are concerned (Figure 2).

The raw materials crisis from 2004 and 2005, along with the unbearable strong price increases have forced the foundry technicians to revise many metallurgical concepts, being forced to introduce among several changes a partial or a complete substitution of pig iron for steel scrap.

This presentation is intended to cater for this new situation. The conclusions reached on this work, allow to relate the quantity of steel scrap added in the metallic charge, with the presence of trace elements and their main metallurgical effects.

Laboratory scale experiments with variable quantities of steel in the melt have been carried out and later validation on industrial praxis has been performed. The conclusions achieved are directly related to the metallurgical effects of the following trace elements: Al, Ti, Nb, V and Zn.

Experimental plan
The scope of the present study is limited to the ferritic and ferritic-pearlitic ductile irons. Gray and compact graphite irons are expressly not taken into consideration.

Concerning the evaluation of the obtained results, besides the usual tools and techniques commonly used in foundry, new elements as thermal analysis, the chemical composition of the trace elements, and the establishment of the shrinkage tendency by means of analytic tools such as X-ray and scanning electron microscopy have been used.

Initial situation
In order to introduce modifications in the process parameters and to evaluate their outcome, it is essential to define properly the initial stage. For that purpose, 4 different laboratory scale experimental melts have been carried out; and from each one of them 3 Y2 samples in Cold Box-moulding sand, as well as 3 cross test specimens have been prepared.

A medium frequency melting equipment with a 250 kg capacity has been used to conduct the tests. In the 4 melts from the reference batch (Batch A) the following metallic charge has been used: 50 % returns, 20 % forming steel and 30 % highly purity pig iron (Table 1). The magnesium treatment is done by the Sandwich method and the monitoring is carried out by means of thermal analysis techniques, paying special attention to the $T_{(E\ min)}$, $T_{(S)}$ and the graphitising factor k.

The results obtained are used as reference when comparing the different experimental heats.

Experimental melts
The working methodology and the control plans applied in all the melts are similar to the ones used for the definition of the initial state.

A total of 4 x 4 melts (Batches B, C, D and E) are prepared on laboratory scale. The most significant difference in all of them is the metallic charge (Table 1).

With the aim of evaluating the role played by the microelements, the steel used in the charge is High Strength Steel, whose distinctive feature is the variable content of Ti, Nb and V. In all the cases, the steel sheet has a superficial zinc coating against corrosion. The carbon and silicon adjustments are conducted in the own furnace, although preconditioning is done by controlled additions of Desulco 9012S.

The nodulization treatment is performed by the Sandwich method; where a FeSiMg 5-6 % alloy is used. The inoculation is carried out during the transfer to the pouring ladle, using 0,20 % of a Zr base commercial inoculant.

Three test samples of the Y2 type and three cross test specimens are poured out of each melt. The following Control Plan (Table 2) is applied to the casting process and outcome materials from the experimental phase.

Industrial trials

After modeling the process on laboratory scale, the same working methodology is carried out to an industrial level. Two limitations are established regarding the metallic charge; from one side the obligation of consuming all the returns generated in the foundry (50%) and from the other side the need to mix soft forming steel with high strength steel (HSS).

The control plan is mainly focused on the results obtained in the own castings, not paying so much attention to the values obtained in the Y2 test samples and the cross specimens, which were very useful for the characterization of the experimental melts. The inspections done within the castings are orientated basically to quantify micro-shrinkage and the possible presence of carbides.

Results

The use of sinter metallic charges bring about economic advantages that cannot be rejected. During the experimental stage, it has been noticed that significant quantities of steel can be used in the metallic charge, keeping always under control the metallurgical quality of the resulting iron.

In this direction, monitoring the melts by means of thermal analysis has been a determining factor, since the adjustments of the iron preconditioning are carried out bearing in mind the $T_{(E\ min)}$, R_{ecal}, $T_{(E\ max)}$ and the graphitising factor k.

Chemical composition

The use of sinter charges gives more importance to the recarburisation and to the silicon adjustment. Both operations, apart from allowing the chemical composition to meet the specification requirements, must be considered as an opportunity for the iron pre-conditioning, in such a way that the different solidification models adapt to the initial reference batch (Batch A - metal with pig iron).

As the steel quantity increases so does the presence of trace elements characteristic in the new materials. Therefore, in the chemical composition of these castings, significant amounts of Mn, Ti, Al, Nb and V appear.

Pre-conditioning of the iron

When ensuring that the solidification model of the iron produced with important quantities of steel is adjusted to the reference model, different pre-conditioning materials have been required. The Desulco 9012S graphite has been selected due to its easy handling and its proven efficiency. Different quantities have been added, both in the furnace adjustments as well as in the ladle itself. Related to the effects of this graphite, both the recarburising efficiency and the pre-conditioning effect must be highlighted.

The modelling of the nuclei effects of this graphite has been carried out by means of thermal analysis. The nucleation capacity related to this pre-conditioning operation has been specially effective since the solidification model changes significantly; the presence of graphite increases in the same measure as the micro segregation decreases.

Metallurgical quality

There is a direct correspondence between the metallic charge and the metallurgical quality of the resulting iron, in such a way that graphitisation and the k factor are directly conditioned by them. From this point of view, the thermal analysis becomes a powerful tool for the foundryman as it allows to establish metallurgical differences among irons that have a similar chemical composition.

The synthetic metallic charges, that is, those with important steel quantities; apart from the effects caused by the trace elements, force the use of pre-conditioning materials as it is not possible to obtain the desired characteristics with the simple adjustment of the chemical composition.

The effects of two industrial recarburisers have been evaluated: a petroleum coke and a Desulco 9012S. Although it is true that as far as the efficiency of the recarburation is concerned, different behaviours are observed mainly in the final adjustment operations, the most significant features are related to the iron pre-conditioning and its solidification.

Those irons in which the carbon adjustment and the pre-conditioning has been carried out with Desulco 9012S (0,20 % + 0,10 %) show a higher k graphitising factor. The solidification models show as well that the tendency to generate carbides or shrinkage diminishes.

Graphitising factor

Shrinkage if the most frequent defect occurring in ductile cast irons. Its development is conditioned by the casting's design, the mould characteristics, the metallic charge and the iron metallurgical quality itself.

From the metallurgical point of view, the defects associated to the secondary contraction are directly related to the eutectic arrest and therefore, the degree of nucleation of the iron is conclusive.

The balance between the expansion due to the graphite formation and the eutectic contraction is quantified by means of the k graphitising factor. As the steel quantity is increased in the metallic charge, the k factor diminishes and consequently the shrinkage extension is increased. This tendency is so certain that in some industrial trials it has been necessary to stop the production as the number of faulty parts was unbearable (Figures 3 and 4).

Secondary contraction
The image analysis technique (image J1.33u) adapted to the inspections conducted by means of fluoroscopy allows to evaluate the micro-shrinkage evolution in relation to the steel increase in the metallic charge (Figures 5 and 6).

It is outstanding that in cast irons with similar chemical compositions, when reducing the ingot and logically, increasing the steel quantity, the area taken up by the secondary porosity increases significantly.

Carbides formation
Bearing in mind the high segregation factors from V, Nb and Ti, the effects of these elements on the carbide eutectic temperature have been studied.

The above mentioned elements segregate in the residual liquid and thus, they increase the corresponding carbide eutectic temperature. This behaviour justifies the development of intercellular carbides in the thermal centres (hot spots) of castings with a high thermal modulus. With the help of the SEM, V, Nb and Ti have been identified as the elements that systematically appear around the carbide areas (Figure 7).

Residual elements
The addition of certain residual elements to the liquid iron introduces important and, at the same time unknown, metallurgical changes. Results corresponding to Zn, Al, V and Ti are presented as they are the most relevant ones.

Residual zinc
Most of the metal sheets used in cars' bodyworks are zinc coated; the thickness or quantity of zinc by m^2 is variable, reaching values up to 350 gr/m2 (coated in both sides). The type of platform, the car maker own characteristics, or the market demands are some of the factors that determine the final coating characteristics.

During the cast iron production process, most of the zinc sublimes (Ts=906 °C) and later on it oxidizes; anyhow some zinc is incorporated to the liquid bath and a minimum quantity of this element tends to provoke different disturbances.

Metallurgical effects

The quantity of zinc dissolved in the liquid iron depends on several factors, although the melting procedure, the iron temperature and the time that the iron stays in liquid state, are determining elements (Figure 8).

In the case of ductile iron, the Mg treatment exerts complementary effects. The corresponding agitation during the treatment always results in a reduction of the contents of Zn. Some of the metallurgical effects provoked by the zinc dissolved in the liquid iron have been identified in this study; the following ones must be pointed out:

- **Graphite shape**. The effects of zinc are unnoticeable since other factors determine its typical spheroidal form.
- **Effects on the structure**. Zinc acts as a moderate ferretizing agent; it makes the pearlite become coarser.
- **Mechanical properties**. According to the conventional mechanical properties, the zinc presence becomes obvious in relation to hardness. Indeed, as the steel quantity is increased in the metallic charge, so is the zinc quantity increased in the iron, and in that same measure the hardness values are reduced.
- **Defects**. In the case of ductile iron, there is a synergic effect between Al and Zn. Some pinhole problems have occurred in ductile iron castings in which the Al presence was inferior to 0,009%. The presence of water vapour inside the mould is quite common and therefore, the following reaction can take place:

$$Zn + H_2O \text{ (vapour phase)} ---\rightarrow ZnO + H_2$$

Aluminium in the metallic charge

The extensive use of steel containing aluminium as a deoxidiser element, along with the use of ductile iron returns that have not been shot peened produce, in some circumstances, the build up of the acid refractory in those areas in which wear problems usually occur. The root cause can be found in the reaction taking place between the silica of the castings returns and the Al of the steel according to the following reaction:

$$SiO_2 + (4/3) Al \rightarrow Si + 2/3 Al_2O_3.$$

The reaction drifts right until the equilibrium point is reached. If the alumina adheres to the silica refractory, it will protect it from subsequent attacks and thus start a build up process, not observed regularly so far, in certain areas of the furnace.

Effects from the presence of Vanadium and Titanium

These elements which quite often segregate during the solidification time (particularly in the castings with a high thermal module), promote the generation of inter cellular carbides. The Vanadium as well as the Titanium segregate on the border of the cells from the latest solidification zones, rising considerably the formation temperature of the eutectic carbide from this regions and balancing the influence from the Silicon and from the graphitising elements on the temperature of the eutectic carbide formation.

The use of the "color attack" technique allows to visualize the last solidification areas, and therefore, the location of these carbides. For the correct identification of the carbides, micro-analysis techniques must be always used.

Conclusions
✓ From the trace elements point of view, it is more and more frequent to find charge materials that add disturbing elements to the melt, since the use of micro-alloy steels is increasing progressively.
✓ The foundry metallurgy must be able to respond to the usage of new synthetic charges in a way that the pre-conditioning of the iron becomes a basic support tool when modifying the iron solidification pattern.
✓ The high tendency of elements such as Nb, Ti and V to micro-segregate forces a change in the iron solidification pattern. In order to minimize this tendency, it is essential to improve the graphitic density.
✓ The k graphitising factor diminishes as the quantity of steel used in the metallic charge increases. It has been checked that the risk of micro-shrinkage occurring increases in the same proportion.
✓ The steels treated against corrosion with Zn add varying quantities of this element to the liquid iron, it is quite frequent to have figures higher than the 0,200 %.Insert text here.

References
1. C.R. Loper, Jr. University of Wisconsin - Madison. *Graphite Pretreatment of Ductile Irons.* AFS Transactions.
2. Andreas Jentsch. Superior Graphite Europe Ltd. *The influence of carburisers on the microestructure, quality and overall production cost cast iron parts.* AFC conference Bangokok Thailand (Sept. 2003).
3. Nastac L., D.M. Stefanescu. 1993 Trans. AFS 101: 329,933.
4. Mampey F. 55th International Foundry Congres. Moscow.
5. Gerval V. and Lacaze J. Solidificacion processing 1997.
6. Skaland T., F. Grong and T. Grong 1993 Metall. Trans 2321 y 2347.
7. Y. Yang, A. Louvo, T. Rantala. *The effects of alloying and cooling rate on the microstructure of low alloy gray iron.* 57th World Foundry Congress. Osaka.
8. R.W. Heine. *Carbon, silicon, carbon equivalent, and thermal analisys relationships in gray and ductile iron.* AFS transactions 462-470.
9. Ramón Suárez y Julián Izaga. *Piezas de seguridad para el sector de automoción. Mejora de propiedades mecánicas y vida a fatiga.* II International Foundry Technical Forum-Bilbao 2005.

Tables

Table 1. Metallich charge in the reference and experimental melts.

Melts	Metallic charge			Chemical composition (in the casting)	
	Pig iron	Steel	Returns	Carbon	Silicon
Batch A	30	20	50	3,72	2,48
Batch B	50	---	50	3,77	2,39
Batch C	30	20	50	3,70	2,52
Batch D	10	40	50	3,68	2,43
Batch E	---	60	40	3,65	2,53

Table 2. Summary of the control plan applied to the reference and experimental melt.

	Type of control	Remarks
Process	Chemical composition. Recarburation performance. Mg performance. Thermal analysis. Zn evolution.	The residual elements are included. Relationship between steel and recarburiser. Carbon yield. Variations in $T_{(E\,min)}$, $T_{S,}$ and K graphitising factor. Relationship between Zn and the melting praxis.
Y2	Metallographic inspection. Structure from the last solidification zones (colour attack). Segregations and intermetallic compounds. Mechanical properties.	Graphite shape. Perlite / ferrite contents. Presence of carbides. Precipitates identification. Ductility and resilience.
Cross specimens	Secondary shrinkage. Micro-shrinkage volume (image analysis).	Relationship between the metallic charge and the micro-shrinkage area.

Figures

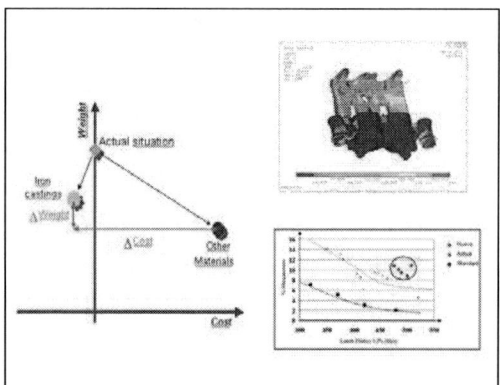

Figure 1. Weight reduction policies can be an opportunity for the iron foundries.

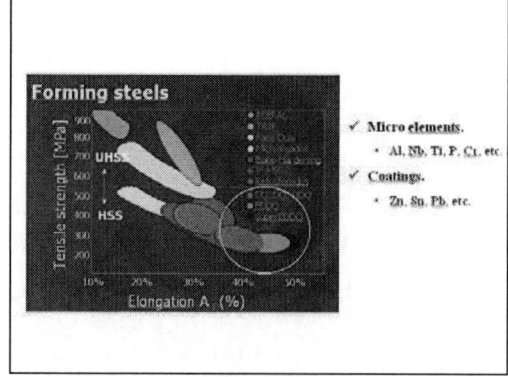

Figure 2. Forming steels development for the automotive industry.

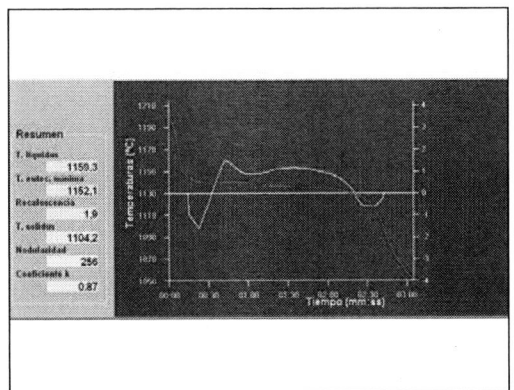 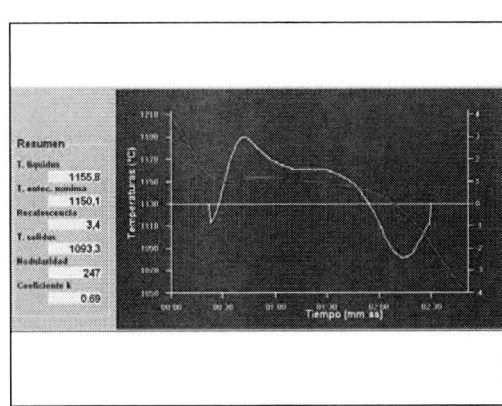

Figure 3. Solidification model from a DI with low micro-shrinkage tendency (k factor = 0,87).

Figure 4. Solidification model of a DI with a high micro-shrinkage tendency (k factor = 0,69).

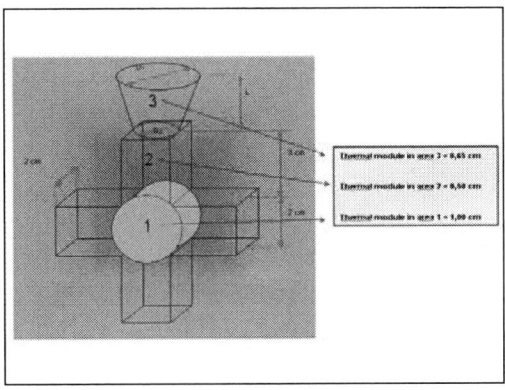 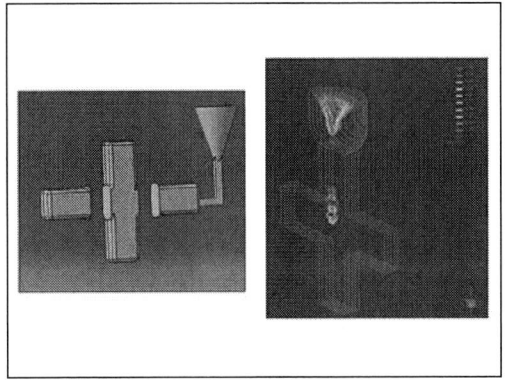

Figure 5. Cross test specimen used to quantify the micro-shrinkage area.

Figure 6. X-ray image analysis techniques are used to evaluate the micro-shrinkage volume.

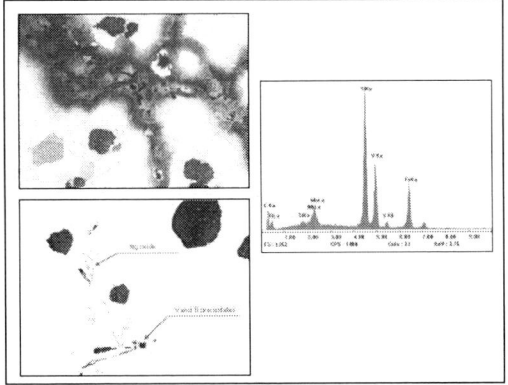 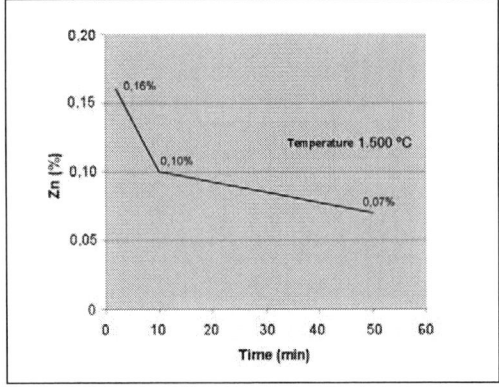

Figure 7. On castings with a thermal module higher than 1,5 cm, V and Ti precipitates appear on the border of the cells.

Figure 8. Medium frequency induction furnace. Zn evolution at a 1.550 °C temperature.

New Engineering and Standards Developments in Austempered Ductile Iron (ADI)

Kathy L. Hayrynen, PhD*, John R. Keough, PEng*, Arron Rimmer, PhD**.

* Applied Process Inc.-Technologies Division, Livonia, Michigan, USA; ** ADI Treatments Ltd., Birmingham, West Midlands, UK.

Abstract
Austempered Ductile Iron (ADI) has only a 30 year commercial history, making it one of the younger engineering materials currently deployed. New theoretical and applied research has revealed important property and material behavior information that is useful to the metallurgist, the machinist and the design engineer. Work on wear resistance, machinability, Carbidic and ferritic hybrids, environmental effects, gear design and other investigations will be summarized. Current ASTM, SAE and ISO standards will also be reviewed with comments relevant to the new engineering research.

Key words
ADI, Ausferrite, Ductile Iron, Austemper

Introduction

ADI is tough, has a high strength-to-weight ratio and is significantly lower in cost than steel or aluminum. It is also 100% recyclable and every grade of ADI can be made from 100% recycled materials. It is, however, an unknown commodity to many design and metallurgical engineers. Investigators are continuing to uncover the many unique properties of the material.

ADI and its Metallurgical Hybrids

The metal matrix of Austempered Ductile Iron consists of acicular ferrite and high carbon austenite. This matrix structure is called ausferrite. It is this unique microstructure that provides the strength, toughness and wear properties of ADI.

Carbidic ADI (CADI) was developed to improve the wear performance of conventional ADI. It is produced by austempering ductile iron that contains an engineered volume of carbide. The wear performance increases with increasing carbide volume, however impact properties will correspondingly decrease with increasing carbide volumes.

Grade 750 ADI was developed to improve the machinability of conventional ADI. It is produced by austenitizing in an intercritical region, which results in a final microstructure of proeutectoid ferrite and ausferrite.

Microstructures of conventional ADI, CADI and Grade 750 ADI are shown in **Figure 1**.

Machining of ADI

While the austenite in ADI's Ausferrite matrix is thermally stable to near absolute zero, the austenite can transform to Martensite when acted upon by a high normal force. This must be understood when machining ADI because a very thin chip can be very hard and tough, resulting in excessive tool wear and high turning forces.

ADI has a thermal conductivity that is somewhat lower than that of ductile iron or steel. Therefore, the workpiece/tool interface will run hotter than that of ductile iron, gray iron or aluminum. Utilizing high volumes of coolant can mitigate this effect. If dry cutting is utilized, then tools capable of high interface temperatures must be employed.

ADI has a coefficient of thermal expansion that is higher than that of steel or ductile iron, so an increase in workpiece temperature will result in growth during machining. This needs to be accounted for in operations such as deep-hole drilling.

ADI's yield (proof) strength is higher than most steels, but its Young's Modulus is 20% lower, making it prone to vibrations in machining. Therefore, ADI requires extremely rigid holding and clamping devices and similarly rigid tool holders with short tool bending moments. Failure to understand, and account for, reduced workpiece stiffness will result in undesirable vibrations during metal removal which accelerates tool wear and results in poor surface finish and increased dimensional variation.

Machining of ADI results in a compact, discontinuous chip that is magnetic and can be 100% recycled. The machinist must be aware that ADI chips can have a density of 2.5 tonnes/m^3 which can be up to ten times more than the stringy, continuous chip produced by the machining of steel or aluminum.

Tool materials for drilling, turning, milling and tapping ADI are selected based on the ADI hardness grade. Conventional tool steels can be used for ADI up to 300 HBW. Hard coated steel tools can be successfully utilized in machining ADI with hardnesses of up to 400 HBW. Ceramic tooling can be used for cutting ADI at hardnesses of up to 500 HBW. Ceramic tooling can experience damage when used for interrupted cuts; however, collaborative research in the US, Italy, Germany and the UK has shown that aluminum oxide tooling with silicon carbide whiskers yields acceptable results in both rough and finished machined cutting over the entire range of ADI hardnesses. [1]

Generalizations with respect to deep drilling and broaching are more complex. Users are, in fact, deep drilling and broaching using coated metal tooling, but the techniques are proprietary and application specific.

The machine speed algorithm displayed graphically in **Figure 2** provides the machinist with a starting point for setting up to machine ADI. It assumes that the user has experience of machining ductile iron. The modified set-up for ADI will be based on that ductile iron experience.

Assumptions:
- Hard coated or ceramic tooling with coolant
- S_o = Original speed for the successful ductile iron application
- F_o = Original feed for the successful ductile iron application
- M_o = Machining speed coefficient for a successful cast iron job with a known Brinell hardness.

Procedure:
- $S_{ADI} = S_o (M_{ADI} / M_o)$
- $F_{NEW} = F_o$

If the combination of S_{ADI} and F_o produces adequate tool life and surface finish, increase the feed speed to $F_{NEW} = F_o \times 1.05$. Repeat the trial, increasing the feed by 5% each time until the tool life or surface finish becomes unacceptable. Procedurize the speed and feed combination based on S_{ADI} and the highest feed rate that produced acceptable results.

Environmental Effects and ADI

Ductile irons, particularly ADI, have been reported to be prone to environmentally assisted embrittlement. In order for this phenomenon to occur, three conditions must be met: (1) a liquid is in contact with the material; (2) a stress is applied that approaches the yield strength of the material and (3) the stress is applied at a low strain rate. Several researchers have reported reductions in tensile ductility of 70 – 80% for Grade 1 ADI when the aforementioned conditions were met. [3]

Ferritic ductile iron does not exhibit this embrittlement phenomenon. As such, one might anticipate that Grade 750 ADI, which contains some proeutectoid ferrite, might offer some resistance to embrittlement. Both Druschitz [4] and Hayrynen [5] have published results for Grade 750 ADI with Druschitz reporting a smaller relative reduction in elongation than Hayrynen. The likely explanation for this difference is the amount and network of ferrite within the specimens tested. Druschitz reported the presence of a network of proeutectoid ferrite in the specimens tested, while Hayrynen's specimens had proeutectoid ferrite present that was fragmented and not as continuous.

Although the embrittlement phenomenon has been documented to exist for various grades of ductile iron, including ADI, it should not result in reluctance to use ADI. Design engineers typically use design safety factors to ensure that components will not be loaded near the yield strength of the material. Extra precautions should be taken if an ADI component will be used in a liquid-containing environment to ensure that the three conditions mentioned above do not occur simultaneously.

ADI Gears and Powertrain

ADI has been used in gearing applications since the 1970's, but the gear design community is largely ignorant of the properties of ADI. ISO and American Gear Manufacturing Association (AGMA) committees are working hard to develop practical standards for ADI gearing so that gear designers can exploit the advantages of ADI to reduce cost and/or improve the performance of their products.

The AGMA 2701-04 mill gearing standard includes life data for ADI gears in an annex. [6] ISO 17804-05, the standard on "Austempered Ausferritic Spheroidal Graphite Iron" includes references to allowable gear properties in its annex. [7] The AGMA Helical Gear Rating Committee is working to publish an information sheet that will allow gear designers to develop workable design models for ADI gearing. The document is due to be balloted and published in 2006.

Figures 3 and 4 show the graphic summary of the 99% probability, allowable stress levels for both tooth root bending and contact fatigue in ADI over a range of hardness. These curves, derived from work done at the ASME Gear Research Institute [8], show that ADI is a competitive material in bending and contact fatigue. Anecdotal data indicates that ADI's fatigue performance in bending and contact is better than that of through hardened steels and similar to that of nitrided steels. [9]

Wear Properties

Wear properties of ADI and competitive materials have been tested in both low stress and high stress environments. Low stress testing was completed per ASTM G65-00 for Dry Sand/Rubber Wheel (DSRW) testing. High stress abrasion testing (Pin Abrasion method) was completed per ASTM G132-96(2001).

Results for abrasion testing are presented in **Figures 5** and **6**. **Figure 5** shows low stress results (DSRW). **Figure 6** shows results from pin abrasion testing in a high stress environment.

In the low stress abrasion test mode (DSRW in **Fig. 5**), the curves for the austempered irons have a steeper slope than those for the pin abrasion testing (**Fig. 6**). This occurs because the normal forces applied during testing are not sufficient to initiate the austenite to martensite transformation. As a result, the wear performance of ADI and its competitive materials is dependent solely on the bulk hardness of the material in low stress environments. On the other hand, the applied forces during pin abrasion testing are sufficiently high to cause the formation of martensite on the surface. This results in a relatively flat curve or the wear performance of ADI being relatively independent of bulk hardness.

ADI Standards Developments

ISO, ASTM [10] and Society of Automotive Engineers [11] standards for ADI have been developed and updated to reflect industry practice. Although the various standards differ, somewhat, with subsequent releases and revisions they are becoming more similar. **Table 1** compares the various grades in the ISO, SAE and ASTM standards.

The ADI standards are meant to codify the properties of the various grades of ADI so that the engineer may practically specify the use of the material. The ultimate strength, proof strength and elongation listed in the standards are the specified minimums. Methods for confirming those properties in production are specified within the standards. These standards also define the range of hardness for each grade as either typical or specified, (depending on the standard).

Work has also been done to develop useable fatigue design data for ADI. In the US, the Ductile Iron Society [12] and the American Foundry Society [13] have collaborated to develop the fatigue design coefficients and exponents to be used in finite element analysis models.

When we combine the monotonic property standards, the aforementioned gear properties data and formulae that define the fatigue behavior of ADI we begin to have a robust property portfolio for product design. Investigators and producers on six continents have contributed to the ADI body of work referred to here. These data and standards will allow for the continued growth of ADI applications worldwide.

Conclusions

Basic and applied research in the field of austempered ductile iron is ongoing. The full potential of the material has yet to be exploited commercially due, largely, to the slow dissemination and codification of the already developed information on ADI. This paper has attempted to familiarize the reader with ADI material and some of that ongoing work. It is hoped that this information can be used to expand the knowledge base so that the market may benefit from the cost and energy savings that properly produced ADI components can deliver in specific applications.

References

1. Keough J, "ADI- Oh No, Not Another Hard-to-Machine, High Performance Material". Industrial Tooling 2003, Southampton, UK, June 2003.
2. "Machining of Austempered Ductile Iron (ADI) – A Tutorial" produced by Applied Process, June 2005.
3. Gagne M and Hayrynen K, "Environmental Embrittlement of Ductile Irons – A Review of Available Data", DIS News, Issue 2, August, 2005.
4. Druschitz A P, tenPas, D J, "Effect of Liquids on the Tensile Properties of Ductile Iron", SAE Paper # 2004-01-0793.
5. Hayrynen K, and Boeri R, unpublished results, May 2005.
6. AGMA 2701-04 "Gear Power Rating for Cylindrical Shell and Trunion Supported Equipment", Annex J, 2004.
7. International Standards Organization; ISO 17804 "Ausferritic Spheroidal Graphite Iron Castings", issued 2005.
8. "Data Base of Critical Technical Information for Austempered Ductile Iron", Breen et al; ASME Gear Research Institute Project A4001, June 1989.
9. "Ductile Iron Data for Design Engineers"; Rio Tinto Iron and Titanium, revised and copyrighted 1990.
10. American Society for Testing Materials; ASTM A897/A897M "Standard Specification for Austempered Ductile Iron Castings", Revised 2003.

11. Society of Automotive Engineers; SAE J2477 "Automotive Austempered Ductile (Nodular) Iron Castings (ADI), revised May 2004.

12. Ductile Iron Society; "Monotonic and Cyclic Design Data for Ductile Iron Castings", Tartaglia J., Ritter P. and Gundlach R., February 2000.

13. American Foundry Society; "Strain-Life Fatigue Properties Database for Cast Irons", DeLao J., ISBN 0-87433-267-2, published 2003.

14. "Abrasion Resistant Cast Iron Handbook", American Foundry Society, ISBN 0-87433-224-9, 2000.

Acknowledgements

The authors acknowledge the employees of the Applied Process Companies and ADI Treatments Ltd. for their assistance with this paper.

Contacts

jkeough@appliedprocess.com
khayrynen@appliedprocess.com
arron.rimmer@aditreatments.com

Tables

Table 1- Comparison of ISO, SAE and ASTM ADI Standards
[The convention in this table is: Ultimate Strength (MPa)-Proof Strength (MPa)-Elongation (%)]

ISO 17804 (Issued 2004)	SAE J2477 (Revised 2004)	ASTM-A897/A897M (Revised 2004)
800-500-10	750-500-11	
900-600-08	900-650-09	900-650-09
1050-700-06	1050-750-07	1050-750-07
1200-850-03	1200-850-04	1200-850-04
1400-1100-01	1400-1100-02	1400-1100-02
	1600-1300-01	1600-1300-01

Figures

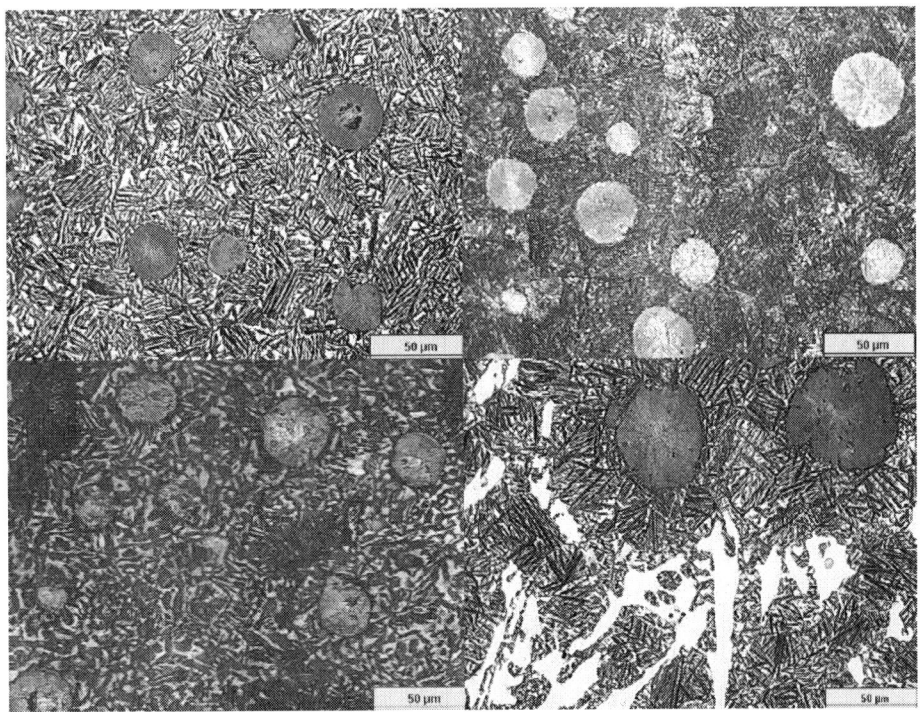

Figure 1 : Photomicrographs (Top left – clockwise) of Grade 1 ADI, Grade 5 ADI, Carbidic ADI and Grade 750 ADI. (Etched with 7% Nital.)

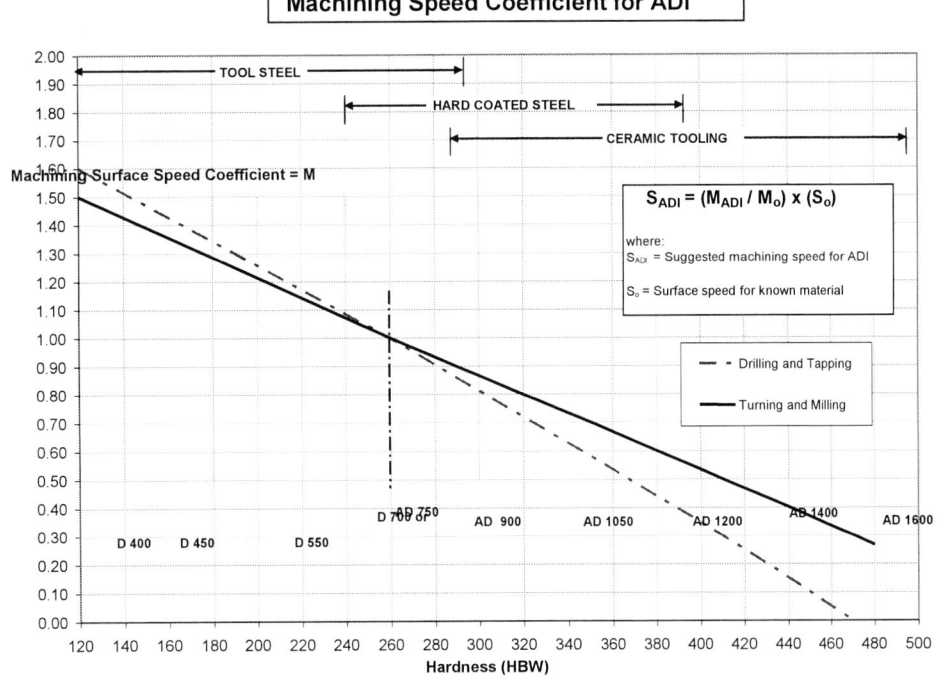

Figure 2 graphically displays the method for calculating surface speed for machining ADI by comparing it to the machining characteristics of as-cast ductile iron. (The SAE Grades of ductile iron and ADI are referenced in the figure).

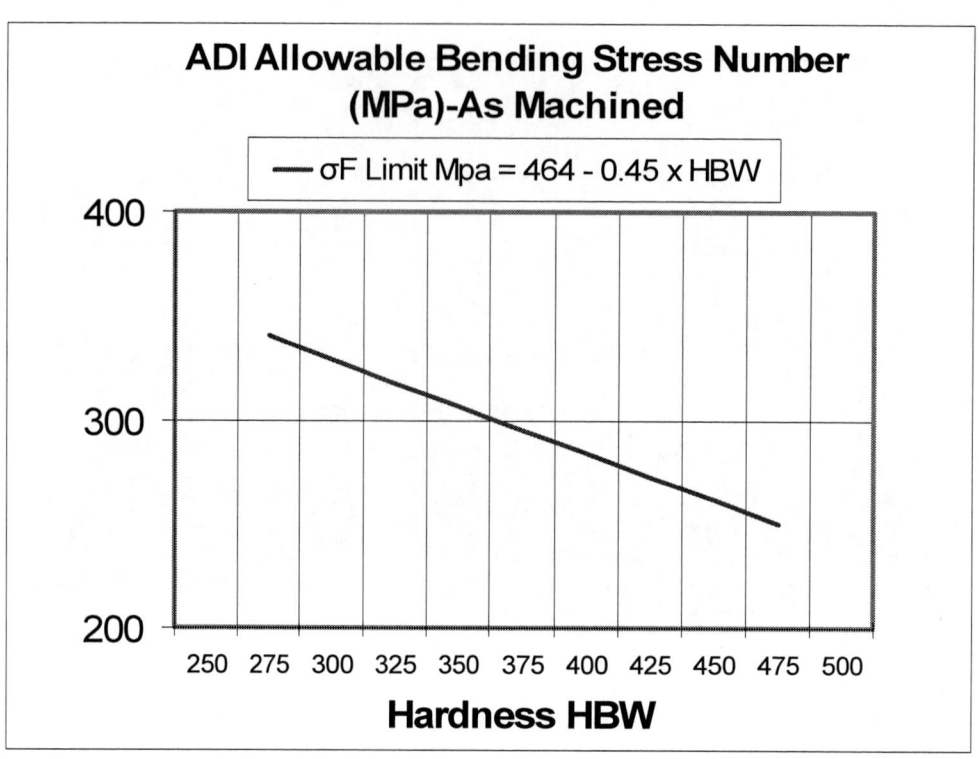

Figure 3 graphically summarizes the allowable bending stress for ADI over a range of hardnesses. (99% probability)

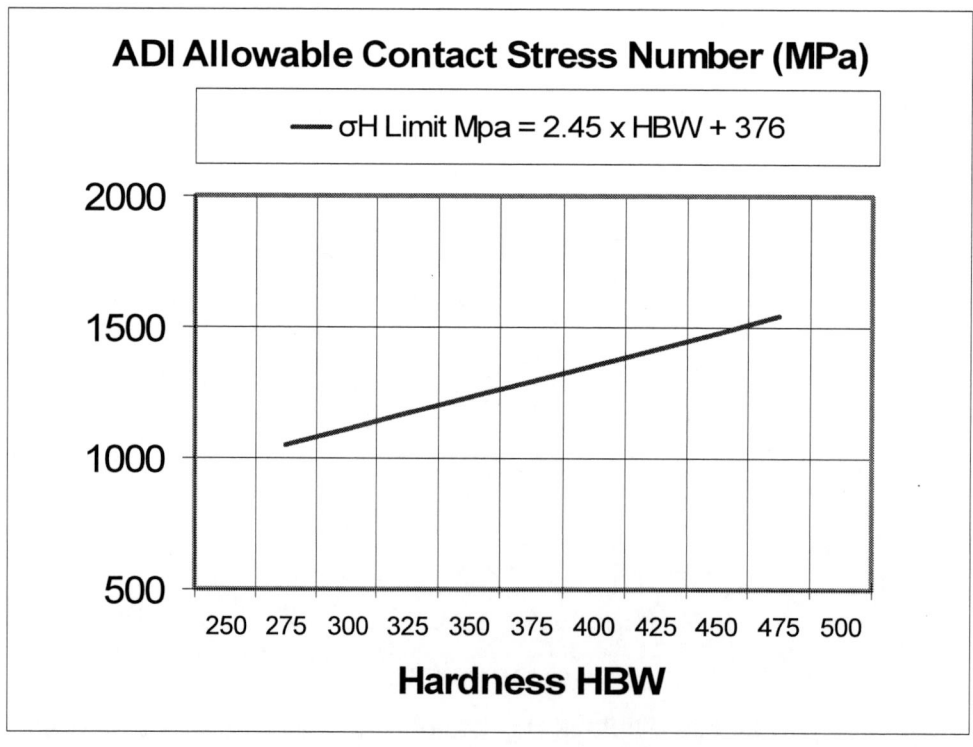

Figure 4 graphically summarizes the allowable contact stress for ADI over a range of hardnesses. (99% probability)

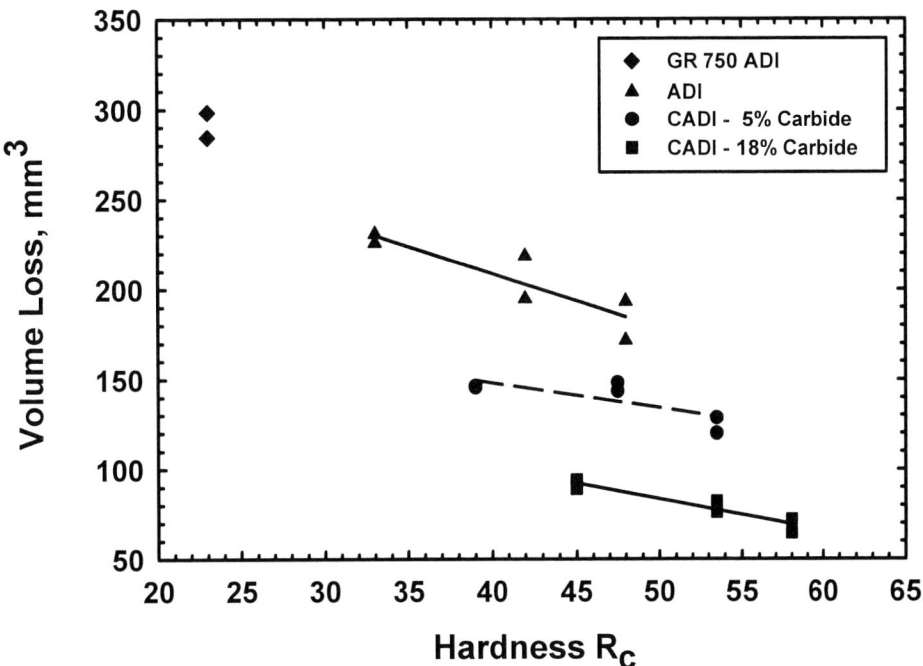

Figure 5 : Dry Sand/Rubber Wheel Abrasion Test Results for various austempered ductile irons.

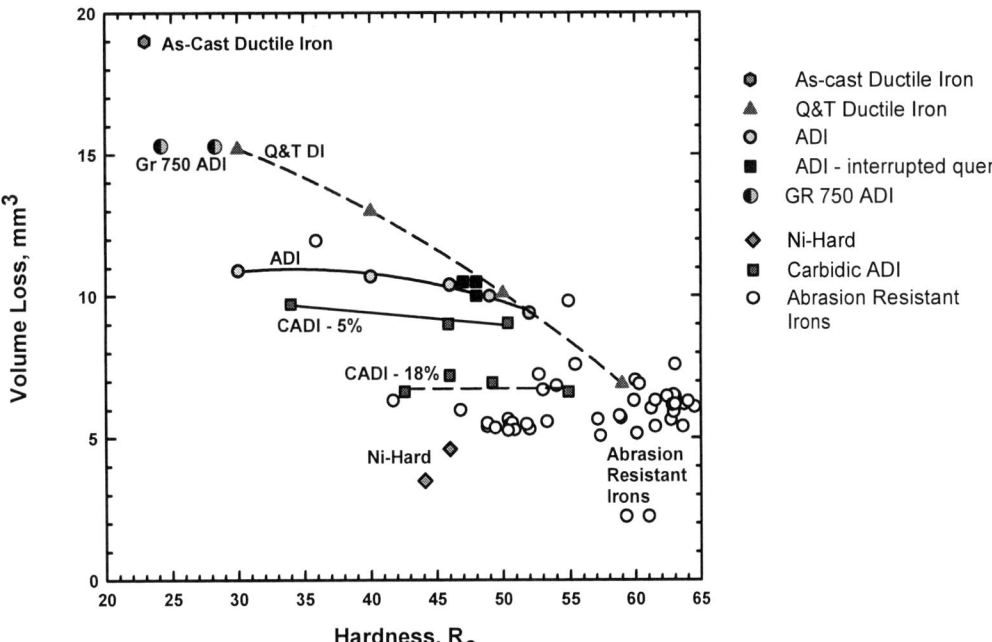

Figure 6 : Pin Abrasion Data for ductile iron, austempered ductile irons and abrasion resistant irons. Abrasion resistant iron data from the Abrasion – Resistant Cast Iron Handbook. [14]

How to link OSHA requirements to Kyoto`s demands for protection of climate?

Rolf Kurtsiefer.

KMA GmbH, Koenigswinter, Germany.

Abstract

Since the idea of the "white foundry" was born, managers focussed a lot of their environmental activities on meeting the OSHA-requirements for occupational health: A look into a modern foundry shows impressive improvements of air quality inside the shop.

Nevertheless the foundries activities for fume removal often could not yet solve the problem of working in accordance with the aims of sustainability: Increasing exhaust air rates in modern foundries cause an equivalent raise of energy-wasting makeup air and hence thwart Kyoto´s aims for climate protection.

This paper shall outline new ways how the foundries can solve this conflict. It will demonstrate, that improved methods of waste air treatment allow foundries giving a substantial contribution to Kyoto and at the same time offer impressive cost savings in the shop.

Key words
Exhaust-ventilation, OSHA, waste air treatment

Introduction

At first sight exhaust air treatment for occupational protection and engagement into environmental protection seem to be two expressions for the same matter. On closer inspection however it becomes obvious that activities for improvement of the air quality at the workplace will not compellingly lead into positive contributions to environmental protection.

Astonishingly, they sometimes even may generate negative impacts on the objectives of environmental protection. This statement becomes clearer when considering that a foundry running in compliance with the regulations for occupational safety often will use big exhaust ventilation systems in order to remove fumes and dust to keep the air inside the shop clean.

Looking at the workplace, the foundry has solved its problem. Looking at the outside, it has generated two new problems:
- the exhausted waste air pollutes the atmosphere with fumes and oil aerosols
- exhausting waste air into the open is always accompanied with losses of energy; in winter because the warm air inside the shop becomes replaced by cold makeup air from outside, in hot areas because the makeup air requires air condition treatment. In both cases replacement of waste air by makeup air from outside requires huge quantities of energy and hence go against the efforts of climate protection.

Experimental

Let us have a look on a typical die casting foundry in order to illustrate the above statements. We make the following assumptions for our model:
- The foundry is operating 10 die casting machines, each with 1200 tons closing force
- The foundry shop shall have a size of 1200 m² (12.900 square feet) and 8.5 m height (28 ft), hence the volume is approximately 100.000 m³
- Each die casting machine shall be equipped with a fume extraction hood, extracting 10.000 m³/h per machine in order to maintain the inside air quality well below the OSHA-limits.
- The average waste air pollution of the fumes extracted from the die casting machines is 12 mg of oil aerosols/m³
- The foundry has a yearly operating time of 6480 hours (3-shift day, 6 days per week, 45 weeks per year)
- The average outside temperature in the foundries location (which shall be Munich) during the heating period (mid October – mid April) is 1.6°C. The temperature inside the shop is 17°C

As the foundry is furnished with waste air ventilation and fume extraction hoods above all die casting machines, the shops air quality will be well in compliance with the OSHA-standards (residual concentration of aerosols <2mg/m³ in workers respiration area).

Let us now look, how the foundry is working with the environment:
- every day the foundry is emitting 28.8 kg oil aerosols into the environment, making a total pollution of approximately 7.7 tons of oil per year
- the average loss of heat energy due to the replacement of waste air by makeup air is 470.000 kcal per hour. The loss of heat generates an additional consumption of heating oil of approximately 220.000 kg per winter period

We arrive at the conclusion, that our model foundry may be exemplary regarding occupational protection, but poor regarding environmental protection.

For that reason let us modify our model: We extend the foundries ventilation system with a demister for waste air filtration. The demister shall absorb in our model approximately 65% of all particulate emissions (this is a realistic assumption for 2-stage stainless steel demisters with integrated cip-cleaner or scrubbers).

Let us look on the result:
- the air quality inside the shop is still equal to the state before we modified the model
- But, looking on the environment, there is an improvement: the emission of oil fume aerosols to the area around the foundry went down from 7.7 tons to 2.7 tons per year.
- Nevertheless, the integration of the waste air filter did not reduce the negative secondary effects on the environment, produced by emission of waste heat: our foundry still requires an additional 220.000 kg heating oil per winter in order to condition the makeup air.

Therefore we modify our foundry again: Instead of using a demister or scrubber for waste air treatment, we decide to invest into a filter system of higher efficiency and choose an electrostatic precipitator (a HEPA-filter would do the same job, but we prefer the electro filter due to the fact that automatic cleaning of the filter is an absolute need and only electro filters are available with integrated cip-cleaners). As a result the absorption efficiency of the filter now is much higher than 90% (clean air residual concentration <1mg/m³) and the process air may recircle into the shop after being treated in the filter. In order to prevent a long-term increase of concentration inside the shop we decide nevertheless to exchange the shops air volume twice per hour. This can be managed either by ventilating 80% of the clean air back to the shop (mixed with 20% makeup air) or returning 100 % of the clean air to the shop and using an independent exhaust air system with only 20% capacity of the waste air treatment filters.

Results of our modification:
- The air quality inside the shop is still in compliance with OSHA-standards
- The emission of oil aerosols into the open now is drastically reduced to less than 0.02 kg per hour (less than 130 kg/year). That is a reduction of pollutants of 98% compared to the status of simple waste air ventilation!
- At the same time the energy balance of our foundry became improved decisively: as the volume of exhausted air was reduced from 100.000 to 20.000 m³/h, the loss of waste heat produced by makeup air dropped down by 80%. On top of it the remaining exhaust air of 20.000 m³/h now is sufficiently clean to pass a heat-exchanger which in addition will return 50% of the remaining loss of heat (this procedure is not applicable when using a simple ventilation system because of dirt problems in the heat exchanger)

Results

The experiment proves that the choice between different proceedings of waste air treatment will have a strong impact on the environmental compatibility of the foundry.

We should notice that all proceeding described above take care for a clean atmosphere inside the foundry shop and hence work in compliance with OSHA.

In addition we should notice that at least the later two proceedings(use of a demister/scrubber or use of an electrostatic precipitator) are in accordance with the today regulations for environmental protection in the EU, USA and other countries. But only when appliyng a waste air treatment system which is able to recircle the clean air back to the shop the foundry will give a substantial contribution to the aims of the Kyoto-protocol: Here the combination of high efficient removal of pollutants and at the same time prevention of excessive air exchange offers a strong reduction of primary and secondary emissions.

Discussion

Beside of all positive aspects, in times of limited budgets the foundry management will take into contribution particularly the cost combined with this investment. It will investigate how much more expansive our favored investment is in comparison to the more simple solutions one and two.

As the first alternative – simply ventilate the waste air into the open – is no solution that points the way ahead, because it would not comply on the long run with the environmental regulations of most countries, the management will focus on the two proceedings of waste air treatment systems. But comparing the two solutions we are astonished to find out that the investment cost differences between a "simple" waste air filter and a "high end" solution are much smaller than expected. The reason becomes clear if we take into consideration that both solutions require equally

- a big fume extraction hood for each die casting machine. Due to reasons of service or die change, the hood must allow easy access to the mould area of the machine and hence needs to be retractable on rails.
- A ventilation system for exhausting the waste air
- Both solutions require a waste air filter system. Here, in fact is a substantial difference between the solutions: The "high end" filter needs a higher investment, particularly because the investment will incorporate a cip cleaning system in order to hold maintenance cost down.
- Both solutions necessitate more or less equal cost for assembly on site
- Finally we must take into consideration that the demister or scrubber solution will require additional investment into the makeup air equipment (Fan, air ducts, heating system), because the volume of makeup air is five times higher in comparison to the solution with the electrostatic precipitator.

As a result we have examined in a lot of comparisons that the additional costs of including a heavy-duty electrostatic precipitator into the waste air ventilation is hardly ever more than 15% of the total investment.

This amount, however, opens on the other hand the chance of long-term savings of the foundries operation costs: In our foundry described above, the savings of heating energy add up to more than 100.000 € per year (based on the current cost for heating oil of 0,45€/Kg)!

In this way the extra investment of 15% pays back in just one heating period.

Conclusions

This paper shall set us thinking and demonstrate that environmental precautions will not lead inevitably to excessive cost.

On the contrary, it proves in our model that the foundry which realized in the best way both occupational protection and environmental protection, realized at the same time the best relation of TOC (total cost of investment, which means initial investment plus operating cost of the first ten years of use).

Certainly the more simple solutions without air circulation still present the today standard in a lot of foundries, but they are obviously inferior both regarding their environmental compatibility as well as regarding their costs.

Tables

our experimental foundry:	
location:	Munich/Germany
foundry shop size	1200 m³
foundry shop volume	10.000 m³
number of dcm	10
size of dcm	1200 tons
exhaust volume per dcm	10.000 m³/h
daily operation time	3 shift/24 h
weekly time	6 days
yearly operation time	6480 h (45 weeks)
desired air temperature inside shop	17°C
average outside winter temp.(oct-apr)	1,6°C

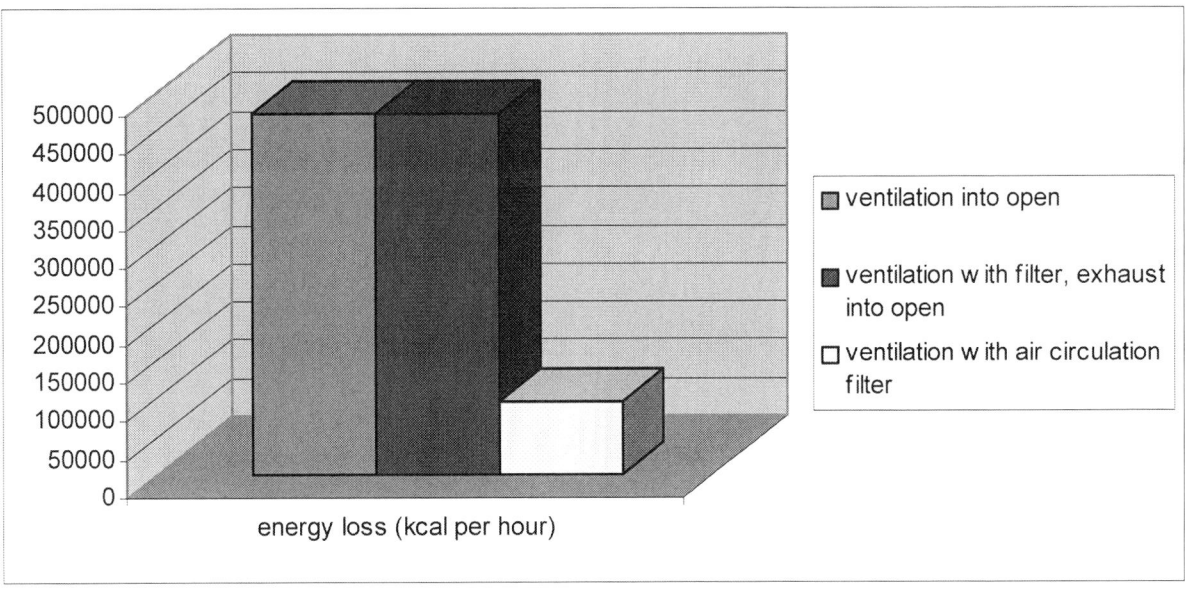

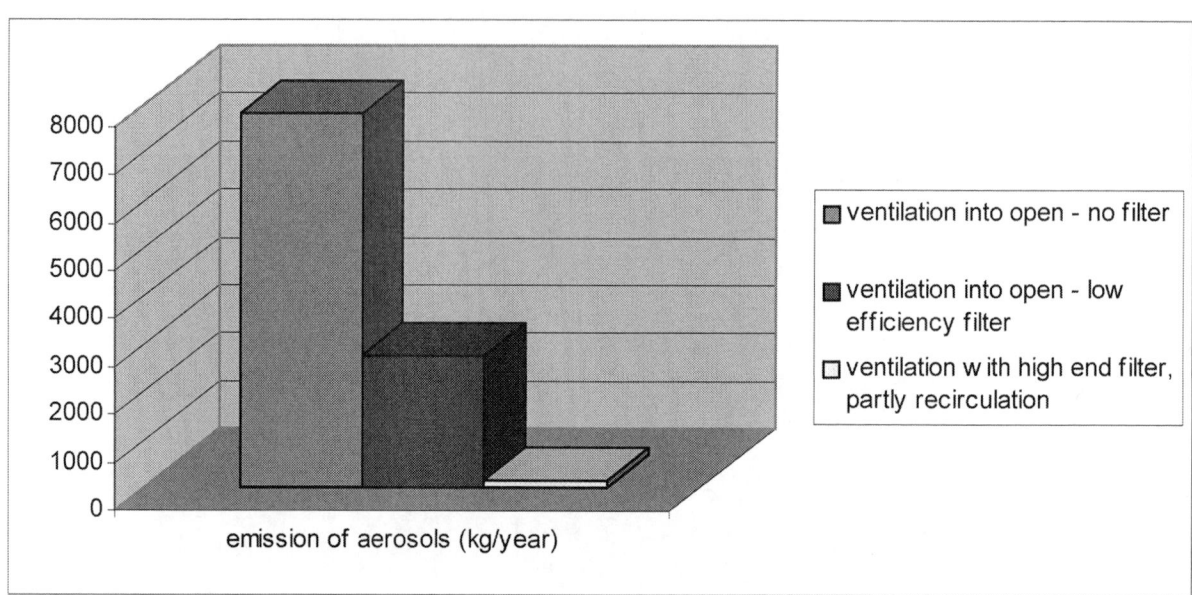

emission of aerosols (kg/year)

- ☐ ventilation into open - no filter
- ■ ventilation into open - low efficiency filter
- ☐ ventilation with high end filter, partly recirculation

Figures

pict.1: die casting machine with fume extraction hood, ventilation into open without waste air treatment

pict.2: die casting machine with fume extraction hood, waste air filter prepared for exhaust into the open

pict.3: die casting machines with fume extraction hoods, waste air filters made for air recirculation

Occupational Exposure to Chemical Agents in the Portuguese Foundry Industry

J C Costa*

* Instituto de Soldadura e Qualidade, Pt.

Abstract

This paper reports on a project on occupational exposure to chemical agents developed in the Portuguese foundry industry on 2000-2002, involving a sample of 15 foundries. A total of 148 exposure profiles to crystalline silica, metallic dust and fumes, total dust and mineral oil mist were determined through monitoring within the 234 similar exposure groups identified. The profiles are described in terms of statistical distributions whenever possible. The strategy followed allows for an adequate risk assessment fully in accordance to applicable legal requirements and to the risk assessment process. Actual practice reviews together with risk assessments provide the chance for improvements on companies practice and for public administration management aspects of compliance checking.

Occupational exposure, chemical agent, risk, risk assessment, foundry.

Introduction

The Portuguese foundry industry comprised in 2000 130 companies and employed 4100 direct workers. These companies delivered a total production of 125000 tons of casts, representing almost 4% of UE total production, for a total turnover of 300×10^6 €. This production showed a steady increase in the period 1998-2000 [1].

The production of metal castings is a complex process that is long associated with worker injuries and illnesses related to exposure to chemical agents [2]. In Portugal a study was carried out on the ferrous sub-sector during 1980-1981, showing several exposures that would be considered as non acceptable [3].

In Portugal legal requirements on occupational exposure are based upon the framework directive on occupational safety and health [4] complemented for chemical agents by the related individual directive [5]. The transposition of these documents into national law [6], [7] did keep their technical approaches even if they coexist with former legislation whose technical approach is no longer adequate [8]. This was partially a motivation to develop the project "Professional Exposure to Chemical Agents in the Portuguese Foundry Industry" [9].

Methodology

The project "Professional Exposure to Chemical Agents in the Portuguese Foundry Industry" was composed by four interrelated phases, as described:

- Phase 1 - Sector characterization, describes the Foundry economic and social aspects, presents the technologies involved and related substances used in order to identify the hazards and risks within this context;
- Phase 2 - Occupational safety and health situation, describes the actual practices on this issue for the selected sample of foundries, addresses similar studies identified as bibliographic references and the incidence of occupational diseases in Portugal;
- Phase 3 - Professional exposure profile, determines the exposure profiles for similar exposure groups identified in the sample companies;
- Phase 4 - Strategy to legal requirements conformity, describes the suggested actions to attain conformity to legal requirements in a way that can be either compatible to a global approach of risk management or evidenced to any interested party.

Actual risk management should be considered as a set of four processes where risk assessment is the first building block [10]. Risk assessment in the project was based on the traditional monitoring of the agent concentration in worker breathing area [11], [12]. However a coherent and well defined exposure assessment strategy was adhered to in order to obtain data expressing in fact the health risk associated to the agent exposure [13]. This strategy comprehends two very important and complementary phases: qualitative risk characterization and quantitative risk determination. Qualitative risk characterization encompasses a basic characterization, the identification of similar exposure groups (SEG) and priority setting, while the quantitative risk determination encompasses the actual monitoring and data treatment to get a reliable exposure measurement, then the comparison to the occupational exposure levels. This allows for a risk assessment that is in full compliance with the requirements laid down in the legislative references.

Experimental

The basis to experimental work started with information gathering for each foundry related to:

- Workplace - technology, processes, equipment and products used and work activities carried out;
- potentially exposed workers;
- agents identification and characterization, including definition of occupational exposure levels;
- exposure controls in place;
- exposure history, if any.

The information gathering was based upon relevant documentation and validated through actual visits to the foundries to locally witness activities, practices in use.

The analysis of the information gathered and observation of actual conditions in the foundries led to:

- SEG identification;
- priority setting as to the SEG exposure determination.

Table 1 provides an example of the set of SEG identified (eleven for this case) in a ferrous sub-sector foundry. Table 2 provides an example of the chemical agents identified for a specific SEG of a ferrous sub-sector foundry with the information on what exposure profiles were determined.

The experimental work itself consisted on monitoring through atmosphere sampling in the workers breathing area of a selected SEG and for a specific agent. The samples obtained were subjected to laboratory analysis as specified in order to allow for a further determination of time-weighted averages. The sampling strategy used a sample dimension of 6 to allow the definition of the statistical distribution (whenever possible).
The sampling parameters and laboratory analysis carried out followed the specifications of the National Institute of Occupational Safety and Health analytical methods [14].
Occupational exposure limits play a very important role in risk assessment of exposure to chemical agents as they are the risk criteria to which the

profiles are essentially compared to. In this project the Threshold Limit Values from the American Conference of Governmental Industrial Hygienists, edition 2001 [15] were retained as they have been the reference most widely used in Portugal and is deemed to be a very consistent system on this subject.

Results

The methodology used allowed the identification of 234 SEG in the 15 foundries studied. From these a total of 148 exposure profiles (88 in the FSS, 60 in the NFSS) were determined. The exposure profiles were spread over a total of 47 profiles on metals, 20 on silica, 20 on dust, total and 1 mineral oil mist for ferrous foundries and 37 on metals, 7 on silica, 11 on dust, total and 5 on mineral oil mist for non-ferrous foundries.

For each profile a descriptive statistics was performed followed by a Wilk-Shapiro fitting test to a lognormal distribution. In the case of adequate fit the distribution parameters were calculated (geometric mean and standard deviation) and a statistical inference carried out to obtain the values for analysis. This consisted on the upper limit of the estimated arithmetic mean as derived from the Land method. In the case of a non fit situation the profiles were considered represented through the range of time weighted averages and their maximum value.

Table 3 gives examples of these two situations for silica exposure in two different foundries. Foundry 1 has two SEG exposed (Abatement and Grinding) fitting to lognormal distribution which upper control limit show that exposures are not controlled (0,10 and 0,17). Foundry 2 has one SEG exposed that does not fit a lognormal so the description of the profile is no longer obtained trough the lognormal parameters. In this case the profile is not controlled as well.

Discussion

The project made clear that risk assessment of occupational exposure to chemical agents calls for a large set of activities involving different sectors of the companies and the consideration of very different but concurring aspects. The strategy used, by considering hazard identification, establishing risk estimation, evaluation thus allows for a further risk assessment based upon a very important set of data, including numerical data, statistically meaningful. In consequence it represents an important tool to exercise an effective risk management by supporting the activities related to the specific legal requirements.

In the ferrous sub-sector 49 (56%) of the profiles determined represent controlled exposures. Uncontrolled exposures occur either to metals or to silica. The finishing operations allow for important levels of exposure. In the non-ferrous sub-sector 46 (77%) of the profiles determined represent controlled exposures.

The actual practices deserve a technical analysis in a way to allow for improvements in the overall approach to risk assessment of occupational exposure. This improvement should concentrate on the basic methodology to follow and the corresponding evidences that should be produced.

Conclusions

The Portuguese Foundry industry developed a project aimed at achieving a global knowledge of its situation in the field of occupational exposure to chemical agents. The results obtained suggest that this subject is a part of the overall risk management activities but the actual practices have an important space for improvements.

These improvements should be centered on the adoption of a strategy that is in conformity with actual legal, technical requirements and can be made evident to interested parties such as the public administration. The quantitative analysis allowing for a robust and a dynamic risk assessment is of utmost importance.

The industry is willing to disseminate the knowledge acquired with this project and widen its use namely to other agents such as the organic type ones.

References

1. European Foundry Association – CAEF. *Data on Foundry Industry*. 2001.
2. National Institute for Occupational Safety and Health (NIOSH): *Recommendations for Control of Occupational Safety and Health Hazards – Foundries*. DHHS (NIOSH) publ. no. 85-116. U. S. Department of Health and Human Services. Centers for Disease Control, 1985.
3. Ataíde, António, Silva, Eduardo D., Ribeiro, Fernando P., Rodrigues, Manuel V., Macedo, Ricardo, *Riscos de Doenças Profissionais nas Fundições Portuguesas de Ferro e Aço (Avaliação e Prevenção)*: 2.ª Ed. Caixa Nacional de Seguros de Doenças Profissionais, Lisboa, 1988.
4. Council Directive 89/391/EEC of 12 June 1989 *on the introduction of measures to encourage improvements in the safety and health of workers at work*, OJ EC L 183 29.6.89.
5. Council Directive 98/24/EC of 7 April 1998 *on the protection of the health and safety of workers from the risks related to chemical agents at work* (fourteenth individual Directive within the meaning of Article 16(1) of Directive 89/391/EEC) OJ EC L 131 5.5.98.
6. Decree Law nº 441/1991 of 14 November - Transposition to internal legal framework of Council Directive 89/391/EEC.
7. Decree Law nº 290/2001 of 16 November -Transposition to internal legal framework of Council Directive 98/24/EC, and Commission Directives 91/322/EEC and 2000/39/EC.
8. Regulamento de Segurança e Higiene no Trabalho dos Estabelecimentos Industriais (Portarias nº 53/71 and 702/80) INCM 1980.
9. Costa J C, Dias A M, Peixoto A R, Chaves M B, Ribeiro, C S, Malheiros, L F, Costa, H M: *Exposição Profissional a Agentes Químicos na Indústria da Fundição Portuguesa*. ISHST. 2005.

10. International Organization for Standardization (ISO): *Risk management – Vocabulary. Guidelines for use in standards*. ISO DGUIDE 73, 2002.
11. European Committee for Standardization (CEN): EN 1540:1998. *Workplace atmospheres – Terminology* 1998.
12. European Committee for Standardization (CEN): EN 689:1995. Workplace atmospheres – *Guidance for the assessment of exposure by inhalation to chemical agents for comparison with limit values and measurement strategy*, 1995.
13. Mulhausen, J R and Damiano, J: *A Strategy for Assessing and Managing Occupational Exposures*, Second Ed., 1998.
14. U. S. Department of Health and Human Services: *NIOSH Manual of Analytical Methods* – 4th Ed. DHHS (NIOSH) Publ. N° 94. 1994.
15. American Conference of Governmental Industrial Hygienists (ACGIH): *2001 – TLVs and BEIs, Threshold Limit Values for Chemical Substances and Physical Agents, Biological Exposure Indices*. 2001.

Acknowledgements

The author wants to thank all support provided for the project and this paper by Manuel Botelho Chaves and Carlos Silva Ribeiro (Associação Portuguesa da Fundição) e Ana Maria Dias (Instituto de Soldadura e Qualidade). A specific thanks goes to Instituto para a Segurança e Higiene do Trabalho whose financial support made the project possible altogether.

Tables

Table 1 – Identified similar exposure groups – ferrous foundry

Similar exposure group (SEG)	Composition (Operators)
1	Fusion
2	Pouring
3	Furnace and accessories maintenance
4	Green sand preparation
5	Green sand casts abatement
6	Other sand types preparation
7	Other sand types casts abatement
8	Core making
9	Grinding
10	Welding and cutting
11	Wood preparation

Table 2 – SEG chemical agents exposure

Operation / workplace	Chemical agent	SEG	Notes
Fusion	Dust, total Dust, respirable Iron oxide (fumes) Manganese Mg oxide (fumes) Cr (III) Cr (VI) Ni, metal Ni, insoluble compounds Mo Carbon dioxide Carbon monoxide Silica Ba Al (dust and fumes)	1	Professional Exposure to: • Iron oxide (fumes) • Manganese • Cr (VI) • Dust, total

Table3 – Silica exposure - profile characteristics

Foundry	SEG	TWA $(mg.m^{-3})$	DF	Distribution		TWA$_{max}$ $(mg.m^{-3})$	Profile control
				GSD	UCL		
F1	5/ Abat.	0,01-0,090	Y	2,2	0,10	-	N
	9/ Grind.	0,035-0,11	Y	1,6	0,17	-	N
F2	2/ Fusion	0,01-0,034	N	-	-	0,034	N

Improving environmental performance and satisfying regulatory requirements through the continuous monitoring of particulate emissions from foundry processes.

Author: William Averdieck,
Speakers: John Hartshorne and Steve Werrell
PCME Ltd, UK

Abstract

Operators of many ferrous and non –ferrous foundry processes have made significant environmental improvements in recent years fitting high performance fabric filter dust collectors (bagfilters) to furnace, cupula, shotblasting, sand cleaning and finishing processes. Emissions from such processes are highly abated only when bagfilters operate to their design condition, and therefore the current issue is how to effectively monitor the condition of the bagfilter to ensure it is operating to design conditions and demonstrating, on an ongoing basis, that particulate emissions are below legislative limits.

This paper focuses on how Electrodynamic type particulate monitors are used by foundries in the UK, Germany, Japan, and US to continuously monitor the performance of bagfilters as well as help maintenance personnel diagnose the location of leaking or faulty bag rows hence reducing re-bagging costs. It also covers how such monitoring satisfies environmental legislative requirements resulting from PPC (UK), IPC (Europe) and MACT (US) as well as improves environmental performance.

Emission sources in Foundry processes
Particulate emission applications

A foundry process has multiple sources of particulate. The major sources include furnaces, poured metal, shot-blasting, machining and fettling operations, core production and sand plant recovery. In many plant, these processes are spread across an industrial site resulting in each emission source being individually controlled, although in newer larger plant the emissions sources may be ducted to common arrestment plant. In both cases arrestment plant is used to control emissions of particulate from the plant. Arrestment plant type includes Electrostatic precipitators (now less often used to problems of meeting lower emission limits), wet scrubbers (now becoming less used due to resulting water pollution problems) and fabric filters or baghouses (the most common type of arrestment plant and one installed in most new applications).

The foundry manufacturing process and
particulate emission sources

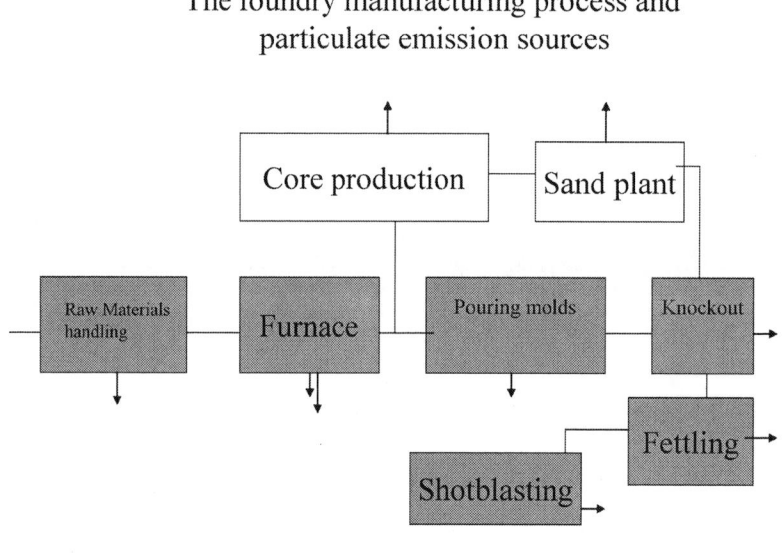

Fig 1

Continuous particulate monitors are fitted on the clean air side of the bagfilter or other type of arrestment plant to monitor the levels of particulate emissions from the arrestment plant, provide records of emissions for emission reporting and trigger alarms on changes in emissions (associated with arrestment plant failure).
Continuous monitoring provides these same benefits whether it done on a large bagfilter used to control a sintering plant in a steel plant or a small bagfilter used to control emissions from a shotblasting process on a ferrous foundry.

Figure 2 : Photographs of bagfilters on Sinter plant and shotblasting applications with Electrodynamic instrument superimposed.

Oil mist applications

The emissions of Oil mists can be an added concern on large foundries with machining operation (eg automotive engine plant). This is as a result of machining oils used when machining castings, for lubrication and cooling purposes. Air in the vicinity of this machining is extracted to oil mist collectors so as to maintain a healthy working environment in the factory and these emissions are cleaned before being emitted to atmosphere.

Continuous monitors are installed after the oil mist collectors to monitor changes in emissions from the plant and indicate any changes in arrestment plant condition to avoid unwanted emission incidents.

Continuous particle and/ oil mist monitoring instruments
Requirements

Instruments for continuous monitoring of particulate (and oil mist) exist using a variety of measurement principles. However in all cases, the instrument does not measure the mass of particles but measures a parameter which can be correlated to particle emissions. The absolute calibration in mg/m^3 is application specific. The generic accuracy of the instrument is defined in type approval schemes such as UK MCERTS and German TUV

Of key importance to operators, is rugged, reliable operation since instruments are required to measure with large periods of unattended operation and require minimal maintenance.

While Opacity type instruments were historically known, over the past 15 years these types of instruments have played a less significant role in foundry processes since emissions have fallen below the resolution limits of the instruments and opacity instruments generally require high maintenance due to the optical surfaces. Meanwhile Electrodynamic instruments have been used in a growing proportion of industrial applications, since the instruments have sufficient resolution to monitor the low emissions after bagfilters, are tolerant to contamination and may be fitted with a single stack connection to an existing duct. They provide an effective method of monitoring particulate and oil mist from foundry processes.

Electrodynamic instruments
Principle of Operation:

In an Electrodynamic system a grounded metallic sensing probe is installed across part of the stack of interest and this rod is connected to signal processing electronics capable of amplifying and measuring an AC current of RMS magnitude in the order of 10pA. Particles in the stack to be monitored carry charge as a result of upstream activity and these particles induce an AC signal as they pass the rod. The magnitude of the AC signal is a function of the average charge per particle and the variation in the spatial distribution of the particles. This AC signal is proportional to total mass concentration in conditions where the charge per particle remains constant (a function of particle type, particle and size and the process conditions) and and stack conditions where the particle number concentration is small (since the particle distribution follows a Gaussian distribution in steady flow conditions). The proportional relationship between particle concentration and instrument response has been validated in regulatory approvals in Germany by TUV and UK by MCERTS [2]. Since the AC signal is primarily derived from charge induction from particles passing the rod (unlike Triboelectric instruments which measure the direct current caused by particles colliding with the rod), the related problems of rod contamination and velocity dependence are minimised.

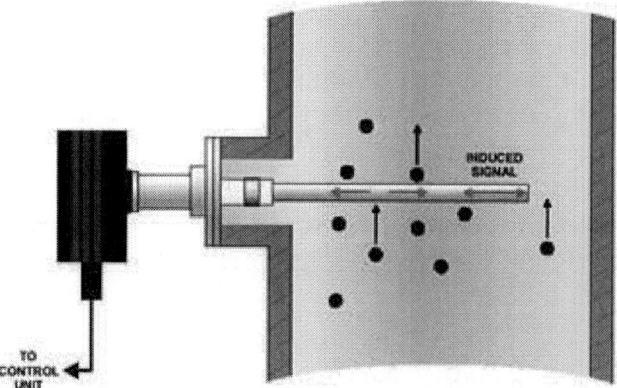

Figure 3: Schematic of charge induction in Electrodynamic sensor caused by charged particles passing the rod

In many foundry processes especially those controlled by ceramic and fabric filters where the filter element surface acts to pre-condition the particle charge, the charge per particle in the final emission stack is sufficiently constant to permit a reliable calibration in mg/m^3 by comparison to the results of an isokinetic sample (gravimetric sample under matched velocity conditions).

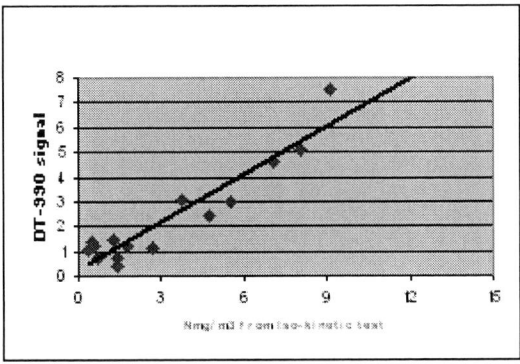

Figure 4: Calibration of DT-990 Electrodynamic instrument with isokinetic test

Practical Considerations for foundry applications:

- The sensor rod can be made of an FeCrAl alloy which can tolerate temperatures of up to 1200degC.
- The sensor rod can tolerate contamination without reduction in performance since the measurement signal derives from induction rather than collision. This means that as the rod becomes coated with contamination there is no reduction in measurement capability.
- In oil mist applications the rod is coated with a layer of Teflon to minimise the effects of sensor shorting due to the moisture. The induced signal still is created by this type of sensor and therefore the instrument is capable of detecting the change in oil and particulate emissions associated with the failure of oil mist eliminators.

Use and limitations of Technology:

Electrodynamic instruments are used to satisfy qualitative and measurement requirements on Bagfilters and ceramic filters in the metals as well as mineral and chemical industries. Their adoption in UK is extensive and their use in Europe, Japan and Australia is widespread. Regulatory approvals exist for instruments in the UK and Germany according to MCERTS and TUV (BlmSchV 17) respectively. Technical limitations are as follows:

- The use of Electrodynamic technology for particulate measurement requires applications with predictable particle type and pre-charge, non-condensing conditions and a minimum velocity of 5m/s. There are only minor effects of changing velocity if the velocity is greater than 8m/s

- The instrument cannot be used for measurement with the presence of water droplets, however this is rarely an issue in high temperature applications. The instrument can discriminate between solid particles and water vapour.
- The process limits for the technique are that it measures all particles from 0.1micron and larger (response is inversely proportional to particle size), measures particle concentration from below $0.1mg/m^3$ to over $1000mg/m^3$ and should be used in applications where there is a minimum velocity of 3m/s. These conditions are met in bagfilter applications.

Construction of Electrodynamic instruments

An Electrodynamic instrument comprises of a sensor which is mounted in the stack via a coupling or flange connection. The version of the sensor used at temperatures to 1200 C includes a ceramic insulation (to isolate the rod from the stack wall), a FeCrAl sensor rod and a heat shield to protect the sensor electronics from the stack temperature. Other variants of the sensor are available for operation up to 250degC, 400degC and 800degC.

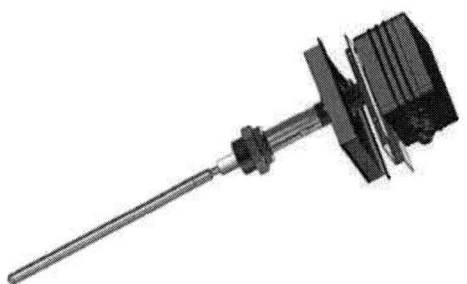

Fig 5   High temperature Electodynamic particulate sensor capable of monitoring in stack temperatures to 1200degC

The sensor is connected to a control unit via a single cable which provides all sensor power and communication with the sensor. In versions using modbus for communication the cable length can be up to 2000m in length. The control unit is used for all user interface, instrument set up and provides 4-20mA, RS232/485 and Ethernet outputs for connection to external plant control systems, PLCs and Plant LANs. Multiple sensors can be connected to the same control unit which provides a cost effective solution for foundry processes with multiple emission points.

Fig 6 Multi sensor Electrodynamic particle emission system

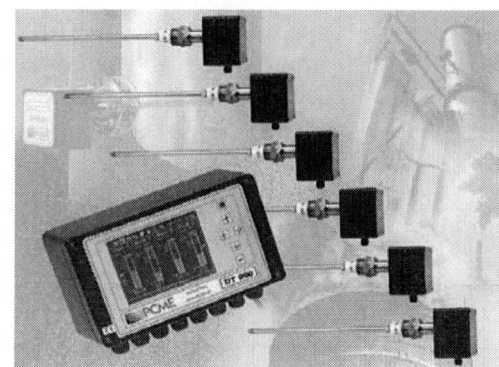

Reducing industrial emissions through predictive continous monitoring

In addition to satisfying emissions legislation, installation of particulate emission monitors can significantly reduce the cost of operating large bagfilter systems by reducing the number of replacement filter bags. Instead of replacing all bags periodically the plant can be operated so that bags are only replaced when there is a need to. The instrument assists maintenance personnel locate the leaking element in a large filter system. This is done by synchronising the output from the particulate monitor to the cleaning sequence of the filter and using the dynamics of the dust signal to pin point which row of elements when cleaned is causing high dust levels and hence is beginning to fail.

High peaks in dust are associated with the cleaning of a bag row with a failing filter bag and therefore maintenance personnel can easily determine which bag row requires maintenance before a major incidence occurs.

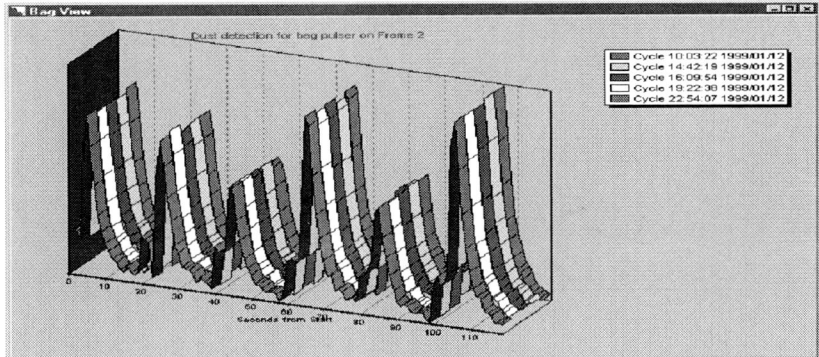

In addition in large multi-compartment bagfilters, multi-probe dust monitoring systems can be used to detect and isolate the compartment causing elevated emissions which again allows emissions to be controlled.

Plant operators use the graphics screens on Electrodynamic instruments to view emissions trends and hence diagnose filter condition. Once the row has been identified and the suspect fabric filters changed, the same method can now be used to confirm;

- Correct problem area identified and repaired.
- New fabric filters are not / have not been damaged.
- New fabric filters are fitted correctly ie Seal integrity is good.

PC software is also available to help this analysis. It is designed to clearly show the status of the baghouse at a glance and enables records to be generated.

After running this analysis for some time, a picture of how the baghouse is working can be built. The time for deterioration of filters can be assessed such that maintenance can be scheduled for natural shutdowns or when the process is least affected.

Legislative requirements for continuous particulate emission monitoring
Types of continuous monitoring
Legislative requirements for continuous monitoring of particulate from foundry processes are converging in most parts of the industrial world. With the growing body of evidence linking metal and carbon particles to reduction in human health the regulatory pressures are very likely to become more stronger. Legislation is both focussed on ensuring emissions are below prescribed mg/m^3 limits, but also more fundamentally on ensuring that arrestment plant is functioning correctly (in which case the emissions will be designed to be well below defined emission limits). There are therefore two types of monitoring:

- Quantitative (instruments are calibrated in mg/m^3 by comparison to an isokinetic sample)

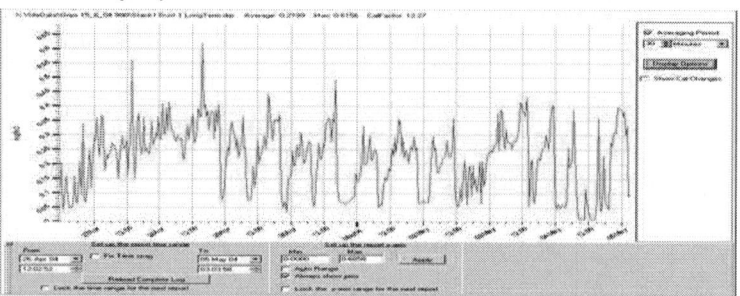

- Indicative in which the trend in emissions from the bagfilter is monitored and alarms set based on changes in emissions). This type of monitoring is referred to as Qualitative monitoring in Germany and Bag failure monitoring in USA.

The choice in monitoring can depend on the local national legislation and the size of the emission point.

An overview of legislation for Foundries in UK, Germany and US is as follows. These requirements mirror are similar to the legislative approach in most other industrial countries.

Legislation in United Kingdom
Continuous monitoring of particulate is widely implemented in the UK. With the implementation of the Environmental Protection Act in 1990, continuous monitoring of particulate was required in the majority of industrial stacks since it was considered BATNEEC (Best Available Technique Not Encurring Excessive Cost). The regulatory focus in the UK on foundry processes is 3 levels:

1. Larger foundry processes are regulated under PPC regulations (UK implementation of the EU directive on Integrated Pollution and Control- IPPC) by the Environment Agency. Typically, continuous measurement is required but for smaller emission points broken bag detection is sufficient. Very few industrial stacks have no continuous monitor.

   The UK Environment Agency provides incentives (in the form of less regulatory attention) to industrial operators that use continuous monitoring instrumentation which are MCERTS approved. MCERTS, is a certification scheme which has operated in the UK since 1999. This Scheme defines standards to which continuous monitors must perform. Instruments obtain a certificate for specific processes and measurement ranges based on a laboratory and three month field test overseen by an independent test body (SIRA). ISO-10155 is used as a basis for the test standards against which particulate monitors are tested. Measurement and qualitative instruments are covered by this scheme.

2. Smaller foundries are regulated for air emissions by local authorities and regulations require;
   - Continuous measurement in stacks with air flow greater than $300m^3$/min
   - Qualitative monitoring and broken bag detection in stacks with air flow greater than $50m^3$/min

3. Certain foundry processes which burn waste as a fuel source for furnaces also fall under the European Waste Incineration Directive (WID) These plants are regulated by the Environment Agency and must meet emission limits and monitoring protocols defined in the EU Directives. Continuous monitoring is to be done with the measurement uncertainty assessed. CEM systems are being upgraded in the period 2005- 2008 to meet these new requirements which are detailed in the European standard EN-14181. This standard sets procedures for instrument certification (QAL1/MCERTS), calibration (QAL2) on going quality assurance (QAL 3) and an annual audit (AST)

Legislation in Germany
Like in the UK, regulatory limits are based on particulate concentration limits although the need for continuous measurement is based on local air pollution issues (ie stack is close to residential area) or on stacks when the total mass emissions of particulate is likely to exceed defined limits. These limits depend on the toxicity of the particulate. As in the UK and other European countries Incineration and Power plant falling under the WID and LCPD directive must fit continuous monitors according to EN-14181

Specific national regulations impacting the use of continuous particulate monitors are:

- BlmSchV 17: Incineration Plant and emissions below 10mg/m^3
- BlmSchV 27: Qualitative monitoring of particulate after filter plant

A type approval scheme exists in Germany. This scheme is widely respected, due to the importance placed on field testing and quality assurance issues (such as instrument checks). Particulate monitors are tested by independent test authorities (eg TUV) against standards and for measurement ranges defined by each of the above regulations. Test certificates note any restrictions on the use of an instrument.

Legislation in United States

Historically the UK and Germany have led the US in terms of experience with particulate monitors, since US emission limits and monitoring methods have been specified in terms of Opacity (colour) for large power plant. However continuous particulate monitoring is now becoming a regulatory issue on many metals processes (including foundries) due to new regulatory rules which focus on continuous monitoring of particles:

- Specified industries (including Steel and foundry processes) must apply new MACT (Maximum Achievable Control Technologies) rules. Many of these rules especially in the metals industry require that baghouses be fitted with appropriate filter failure monitors. Qualitative Particulate monitors are used to satisfy these requirements
- Title V plants (the major metals, chemical, mineral and combustion processes) are required under the new CAM (Compliance Assurance Monitoring) regulations to develop a method to ensure the continuous compliance of their particulate arrestment plant (eg Baghouses and Electrostatic Precipitators). It is likely **qualitative particulate monitoring** will be chosen by many sites as a pragmatic solution to this new requirement when Title V permits are renewed.

Conclusion

The emissions of particulate from foundry processes can be effectively continuously monitored using Electrodynamic type instruments. These instruments provide robust, and accurate continuous monitoring of particulate emissions and arrestment plant condition (satisfying MCERTS and TUV approvals requirements). Instruments are used to both monitor particulate from bagfilters and also oil mist from mist eliminators. Continuous monitoring instruments are used to satisfy specific legislative monitoring requirements in UK, Germany, US, and most other industrial nations. Plant operators may use the predictive capability of continuous monitors to reduce emissions from arrestment plant by appropriate preventative maintenance and reduce the costs for replacement filter bags.

William Averdieck: williama@pcme.co.uk, John Hartshorne: johnh@pcme.co.uk, Steve Werrell: stevew@pcme.co.uk

Novel approaches in reducing pouring emissions

J H Helber
IfG – Institute of Foundry Technology, Germany, Duesseldorf

Abstract

The paper presents a basic understanding of some thermal degradation processes within the mould. I. e. the mechanism of the formation of lustrous carbon was investigated as well as the formation of aromatics. The suppression of those reactions and the state of the art in modelling the gas phase will be addressed respectively the problems which still have to be overcome.

The scope of the presented brand new investigational results is regarded to be fundamental knowledge within the fields of gas defect analysis, emission reduction, and mould material design.

Key words

Mould - gas reactions – modelling – emission reduction

Foreword

The formation of the gas phase in the mould during and after mould filling and its reactions with core gases and hard surfaces are still very much a "black-box". The reactions leading to the generation of pouring gas and its constituents are not well understood in detail. Predictions of pouring gas production rates, of gas induced casting defects and, finally, of atmospheric emissions have remained empirically based. These empirical findings ought to be substituted, or at least expanded upon, with reliable knowledge. But to model effects like those mentioned above, it is necessary to possess data that can only be gained via a clear understanding of the thermochemical reactions involved.

It can hardly be expected that these very complex processes will be fully understood anytime soon. However, significant advances may be achieved within the next few years. This presentation is intended to describe a small area of the investigations currently conducted to clarify the physico-chemical processes in the sand mould. A presentation of further experimental results - partly derived from comparative testing of binder and moulding auxiliaries - is to be reserved for another occasion.

Also outlined in this paper are a number of thermal degradation processes taking place within the mould. Besides, the mechanisms underlying the formation of lustrous carbon and aromatics are investigated. The suppression of these reactions and the state of the art in gas phase modelling will likewise be addressed to some extent.

Introduction

The investigations relate to sand moulds with resin-bonded cores and bentonite-bonded moulds for casting iron. The basic thermophysical and thermochemical processes in the gas phase were examined, *inter alia*, as part of two publicly subsidized projects [1, 2]. The aim was to determine the fundamentals for creating a numeric model whereby the formation of gas as well as the emission of pouring gases could be simulated.

Key questions relate to the in-mould transformation and flow of materials. While the convective gas transport is driven by pressure differences, diffusion processes are based on Braun's molecular movement and gravitation effects. Gas transport is naturally directed from the inside toward the outside, in line with the heat flow (but see notes below). The transformation of materials is driven by the heat flux from the metal into the moulding material.

Since gases and vapours are produced in the mould through the evaporation of water and the thermal decomposition of organic and instable inorganic compounds, transport mechanisms and chemical processes are directly linked and mutually interdependent. To arrive at a physical description, it is necessary to know the sources and sinks for each material, or in other words, to know which materials will form and disappear where and when. Let us first take a look at the gas phase and the role of gas condensation and adsorption phenomena.

Results

The integral increase in gas volume *in the mould* which can be observed after the pouring cycle as a result of the casting-to-mould heat transfer is defined by three basic mechanisms, i.e.

1. (Permanent) gas expansion
2. Formation of water vapour
2. Coal decomposition

Given the many influencing factors, this volume increase will vary on a case-by-case basis. On the other hand, components 1 and 2 will become zero once again as soon as a thermal balance with the ambient environment has been achieved (i.e., when room temperature prevails both on the outside and on the inside).

Calculation example:

Composition of the moulding material

Active binder concentrations (binder composition) in the mould:

•	Bentonite(B)	8 %		
•	Water (W)	3 %		
•	Lustrous carbon former (LCF)	1.7 %	Σ =	12.7 %
Support medium:	Quartz sand			87.3 %
			Σ =	100.0 %

Pouring conditions for modelling calculations:

Pouring temperature	1420 °C
Casting wall thickness	10 mm

Mould characteristics:

Initial mould temperature	20 °C	(293 K)
Mould pore volume	35 %	
Mould density	1.5 g/cm³.	

Simplified physical and chemical effects assumed for the simulation:

Release / decomposition of:

•	Water	3.0 % m/m at 100 °C
•	Volatiles (FB) from LCF	0.6 % m/m at 400 °C
•	Coke (residue) from LCF	1.1 % m/m at 400 °C
•	Water from bentonite	0.8 % m/m at 400 °C
•	Residue from bentonite (residue)	7.2 % m/m

Total volatiles from LCF and bentonite at 400°C   1.4 %

It follows from the above that at 400 °C or 673 K, respectively - i.e., during the passage of the (already much diminished) heat front through the moulding material - the gas volumes formed (per 1 ml of moulding material) can be approximately quantitated thus:

0.45 ml due to thermal expansion of gas in the pore space
160 ml due to water vapour
29 ml of gases and vapours attributable to coal dust (lustrous carbon former)

This increase in gas volume *inside* the mould is accompanied by a gas volume increase *outside* the mould that is commonly known (and can be measured) as pouring gas emissions. Various instruments exist for determining these emissions; the results obtained vary accordingly. For the purposes of this paper, the equipment shown in **Picture 1** was used.

Below we shall see how these assumptions tally with the experimental findings (**Fig. 2**). The diagram shows the measured gas emissions produced by various combinations of moulding material constituents, i.e., pure quartz sand (yellow), moulding sand + bentonite / water (blue), and moulding sand + bentonite / water + lustrous carbon former (magenta), under identical conditions.

Evidently, there can be no gas release from pre-annealed quartz sand alone. Rather, this volume increase is due to the thermal expansion of the gas (yellow line). The blue line describes the gas expansion plus the permanent gas formation resulting from water (vapour) reactions with the metal, whereas the violet curve additionally plots the emission of gas from the lustrous carbon former and the results of their reactions.

In the trials, about 50 ml of gas was obtained after 6 minutes of pyrolysis. The calculated gas phase volume is equal to more than four times the actual measured volume. In the presence of coal, air and (subsequently) water vapour are displaced by the coal pyrolysis gas. As a result, they can no longer react with the metal but only with the remaining gases and with the moulding material. Otherwise the gas volumes measured would have to be much greater since the coal alone cannot be expected, in theory, to produce approx. 20 - 50 ml of gas per gram of coal. However, it is not this effect alone which contributes to the reduced measurements compared with the primary gas formation in the heat zone. Condensation and adsorption effects also function as gas-phase sinks, although the extent to which each of these contributes to the result cannot be determined externally. A clarification of this issue will remain reserved to future studies.

The above raises the question which gases are involved in the formation of gas volumes. The following chart **(Fig. 3)** compares the contribution of the individual moulding material constituents to the gas emission. This time, however, the comparison is related not to the total volume but to the composition of the gas phase.

Bentonite in dry sand releases hydrogen and CO_2 by splitting key constituents, i.e., water of crystallization and (sodium) carbonate. The coal releases further amounts of hydrogen and CO_2 in addition to methane, ethyne, benzene and other aromatics.

Due to the initial uniformity of the gas flow, the concentrations measured simultaneously by mass spectrometry essentially correspond to the amounts of material or gas volumes released at the start of the test.

The addition of water, common in practice, will not change these patterns fundamentally but reduces the levels of unsaturated hydrocarbons and CO_2. Thus, moist bentonite as a necessary constituent of green sand results in clearly diminished benzene and ethyne outgassing, whereas the amounts of hydrogen and methane remain unchanged.

At this point let us consider a new generalized gas phase model. **Fig. 4** gives a graphic explanation of the gas-phase processes we are currently discussing with regard to green sand.

Coal decomposition - lustrous carbon formation
At low temperatures, coal pyrolysis is a very slow process. The composition doesn't become complete until about 1900°C. Depending on the decomposition temperature, different breakdown products are obtained. As the thermal load increases, only the most stable hydrocarbon species "survive" these harsh conditions **(Fig. 5)**.

Fig. 6 illustrates a simplified carburizing mechanism on a steel surface having a temperature of 900°C. The carburizing medium used is propane [3]. We find that the general reaction mechanism described by other authors in the context of flame investigations, diamond synthesis or case-carburizing is fully confirmed with regard to green sand moulds. The synthesis of benzene is only one of several (potential) "reaction paths" toward the formation of lustrous carbon. Interrupting it would be tantamount to suppressing the formation of lustrous carbon at least partially.

Interruption of formation mechanisms
Generally speaking, the options available for preventing or reducing benzene emissions include the following:

- burn-off or collection of pouring gas outside the mould
- elimination of the use of lustrous carbon formers
- retention if substances in the moulding material by absorption or condensation mechanisms

- destroying substances within the mould before they can be released
- preventing their formation.

The "Advanced Oxidation" method developed in the U.S., as far as it is understood today, relies on the formation of active charcoal from the coal [4]. However, it involves the introduction of an oxydant into the sand near the end of the sand reconditioning cycle. This may result in the destruction of compounds adsorbed by the active charcoal. The exact mechanism is unknown.

An interruption of the formation mechanism may be achieved with the aid of carbon dioxide **(Figs. 7 a) + b))**. However, this must be done in such a manner that the deposition of lustrous carbon is not disrupted, i.e., in the central mould section (in the outer mould areas, temperatures are too low for reactions to take place). A patent has been granted on this process [5].

The following diagrams illustrate the instant interruption of the benzene formation mechanism upon introduction of CO_2. The formation of benzene resumes as soon as the CO_2 supply is discontinued.

Conclusion

The results aim to provide a differentiated view of the thermally induced processes taking place in green sand after pouring. On the way to establishing a physico-chemical simulation model, some key milestones have been reached:

- The benzene formation process during cooling is now understood in principle. There are different reaction channels at different temperatures $T1 < 900\,°C$; $T2 > 900 - 1300\,°C$.
- Benzene development can be suppressed only gradually.
- Scientific advances have been relied upon to formulate new LCA with reduced benzene emissions.
- The chemical principles can be transferred to polymer binders to some extent.
- A further conclusion has been to interrupt the reaction channels by means of additives.

References

1. J. Helber, C. Rogers, E. Bruemmer, "Reduzierung von Emissionen aus bentonitgebundenen Formstoffsystemen durch Design optimaler Glanzkohlenstoffbildner", Final report (P-302); published by IFG, Duesseldorf, Germany II/2005.
2. "Entwicklung emissionsreduzierter PU-Coldboxbinder", Final report (P-300); published by IFG, Duesseldorf, Germany IX/2003
3. F. Graf; S. Bajohr; R. Reimert, "Pyrolyse des Aufkohlungsgases Propan bei der Vakuumaufkohlung von Stahl", HTM 58 (2003), p. 20-23

4. S. Lewallen, J. Furness, "Incorporating Advanced Oxidation into Your Operation for Management and Engineers", Presentation XI/2004; Purdue Symposium V7.01 (internet access: http://www.google.de/search?hl=de&q=furness+Lewallen&btnG=Google-Suche&meta=)

5. J. Helber, J. Winterhalter, W. Hunck, Offenlegungsschrift DE 103 20 067 A1 "Unterdrückung der Aromaten- und Polyzyklenbildung bei Hochtemperaturprozessen mittels CO_2"

Figure 1: Induction furnace, shown with a gas collecting bag. The bag can be easily replaced with a burette for measurement of gas volumes

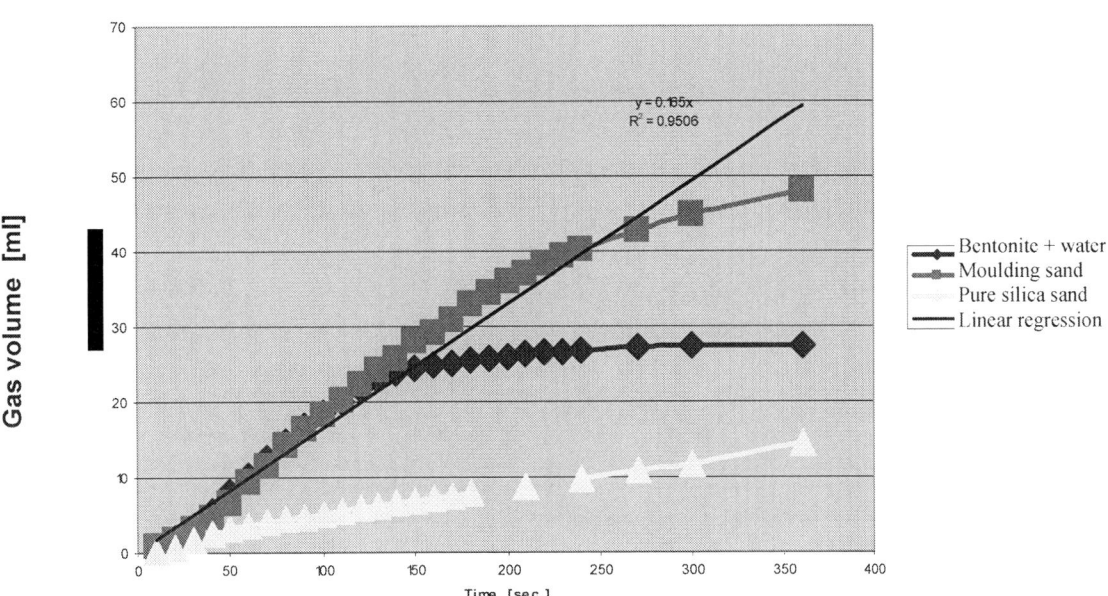

Figure 2: Comparative study of gas release from moulding sand over 6 min with rapid heating and a temperature gradient in the moulding material (Quartz crucible 14 x 100 mm, 1.4 g Fe surrounded by 6 g moulding sand, sand / bentonite / water and pure quartz sand; induction-heated at full power input; gas volume measured with a water-filled burette)

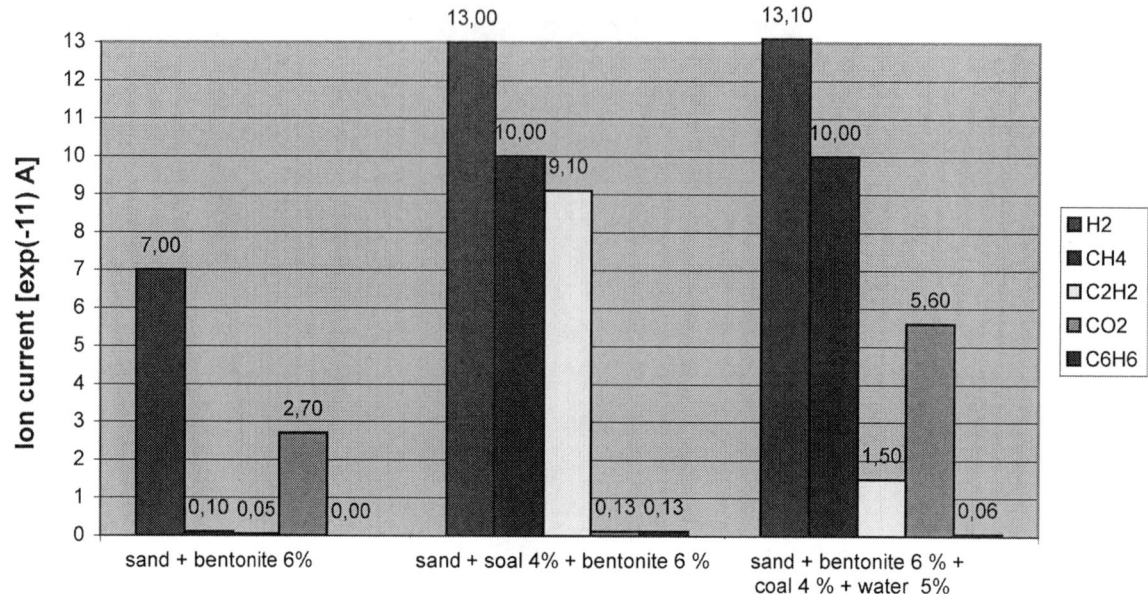

Figure 3: Coal produces methane, hydrogen, C2 alkanes and aromatics; moist bentonite produces CO_2 and hydrogen

(Silica crucible 14 x 100 mm, 1.7-2 g Fe surrounded by 6 g sand with additives, induction-heated)

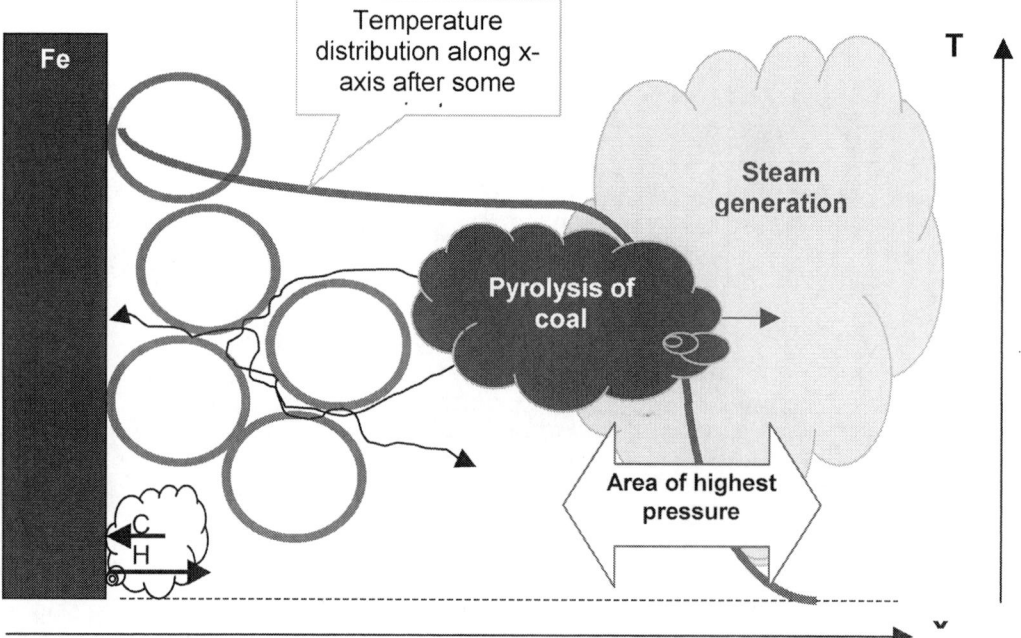

Figure 4: Simplified heat and gas flow model

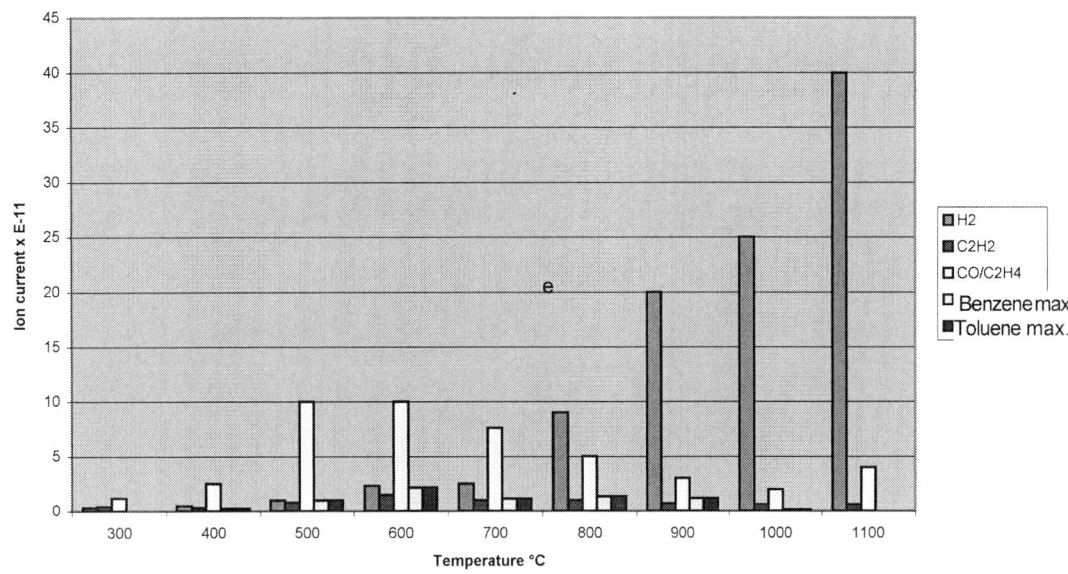

Figure 5: Coal pyrolysis at different temperatures

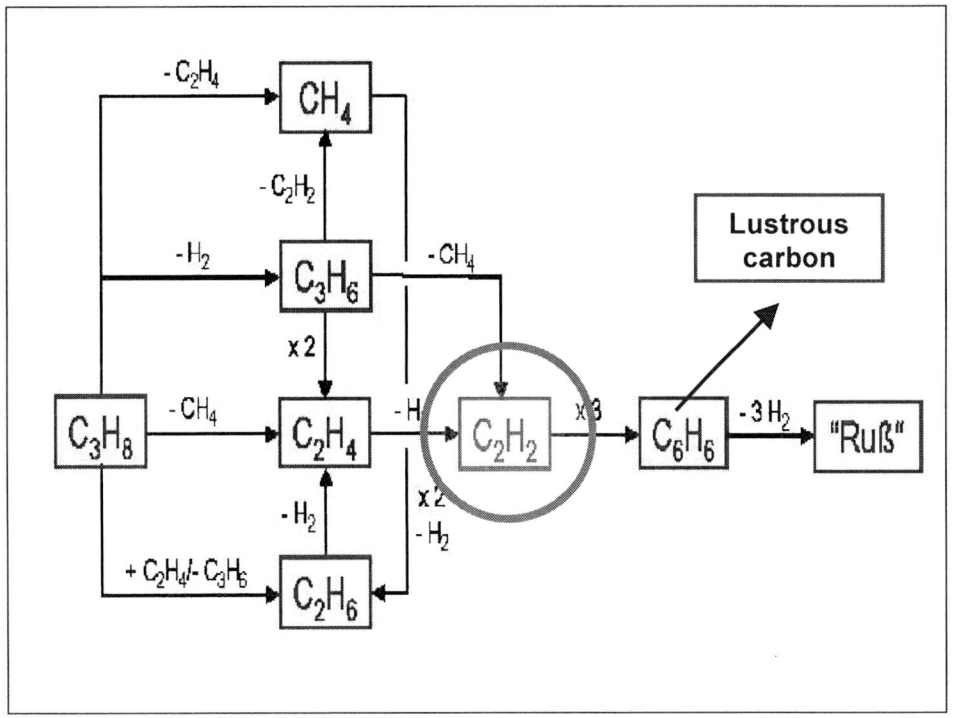

Figure 6: According to steel carburizing trials with propane conducted at EBI Karlsruhe, the carbon deposition process is based on build-up (synthesis) reactions and the formation of benzene [3]

Methane pyrolysis +

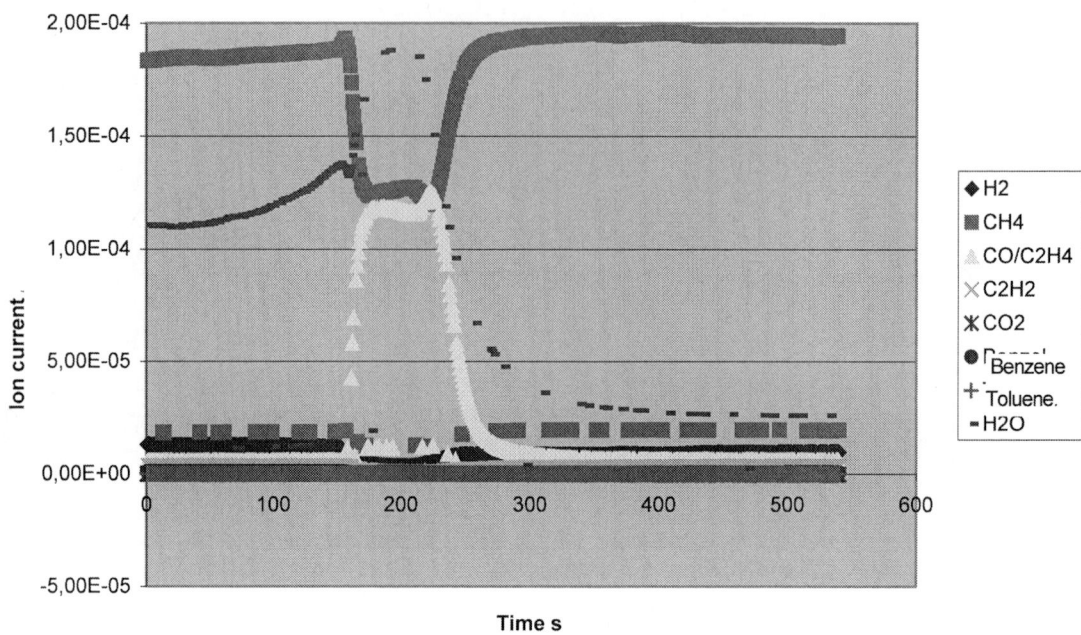

Methane pyrolysis + CO2

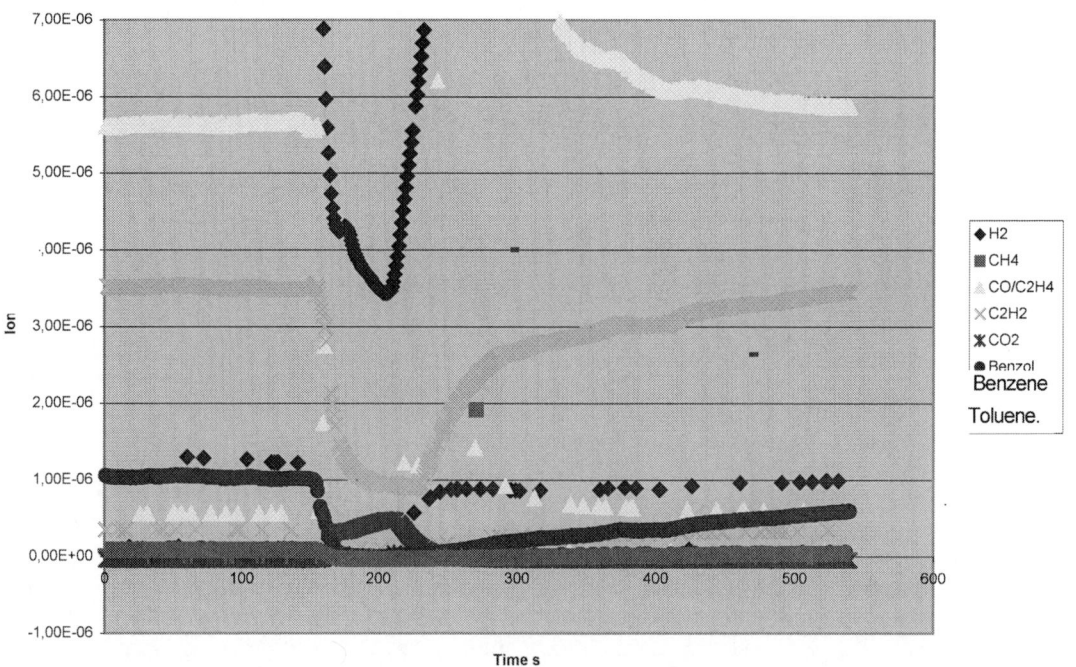

Figure 7 a/b: Reaction scheme in a laboratory flow reactor system which simulates basic reactions in the mould with respect to benzene suppression

The Use of Tilt Filling to Improve the Quality and Reliability of Castings

R.A. Harding

IRC in Materials Processing, The University of Birmingham, UK.

Abstract

There is currently an interest in using quiescent mould filling to overcome problems inherent in the casting of Ti alloys, particularly bubble entrainment during the rapid mould filling necessitated by limited superheat. As a first step, the Durville tilt casting process has been re-evaluated for casting Al alloys in sand and ceramic shell moulds and has enhanced the understanding of the process parameters. Computer simulations of surface turbulence during mould filling have been validated by real-time X-ray radiography of practical casting trials using a computer-controlled rotating wheel. Weibull's statistical technique has been used to characterise the scatter in the tensile properties of large numbers of test bars. This shows that careful selection of the tilt filling parameters can markedly improve the reliability of castings compared with conventional gravity castings.

Key words Tilt casting; rollover casting; reliability; aluminium alloys; titanium aluminide.

Introduction

Casting titanium alloys

The research described in this paper was originally motivated by the difficulties of casting Ti alloys, particularly titanium aluminides [1]. Since all Ti alloys are very reactive in the molten condition, they are usually melted in a water-cooled copper crucible in processes such as Induction Skull Melting (ISM) to obtain low gas and inclusion contents. However, this results in low superheats (typically <50°C) [2] and so the metal is usually 'dump' poured into the mould. The resulting surface turbulence can entrain bubbles of the furnace atmosphere which are then trapped in the casting by rapid solidification. The defects can be removed by Hot Isostatic Pressing but this creates unacceptable surface depressions ('HIP sinks').

Preliminary casting trials [3] with a prototype Vacuum Induction Rollover (VIR) furnace resulted in bubble-free castings, Fig. 1, in contrast to similar castings made using an ISM furnace. This initiated further research to clarify whether this was because the VIR furnace gave a higher superheat, or because the rollover mould filling was less turbulent. This led to a more general assessment of tilt filling as a means of improving casting reliability.

Effect of surface turbulence on the reliability of castings

It is well established [4] that there is a critical velocity of ~0.5 m s^{-1} for most molten metals and alloys. Once this is exceeded, the surface of a flowing molten metal becomes increasingly unstable, resulting first in the formation of surface undulations, then breaking waves and droplets. Since most molten metals are covered in an oxide film, surface turbulence leads to the film being folded onto itself, resulting in double oxide film defects ("bifilms"). These persist in the solidified castings and not only nucleate other defects such as hot tears and gas porosity, but also cause premature failure in service. Bifilms occur randomly and it is now clear that they are a major factor controlling the reliability of castings.

The effect of surface turbulence on the reliability of castings was powerfully demonstrated [5] by using the Weibull statistical technique [6] to analyse the tensile strength data of large numbers of castings. The two parameter form of the Weibull distribution is:

$$F_W = 1 - exp\left[-\left(\frac{x}{\sigma}\right)^{\lambda}\right]$$

(1)

where F_W is the fraction of specimens that fail at or below a given value of x (e.g. a measured tensile strength), σ is a characteristic value of x at which 62.8% of the population of specimens have failed and λ is the Weibull modulus. By plotting $\ln\{\ln[1/(1- F_W)]\}$ against $\ln(x)$ it is normally possible to obtain a straight line with a gradient of λ. High λ values represent a narrow spread of properties, i.e. more reliable castings. Traditional casting methods, in which the metal is poured from a ladle into the top of the mould via a conical pouring basin, Fig. 2a, inevitably create surface turbulence which produces unreliable castings, with a typical λ of ~10 in sand castings [5] and ~25 in investment castings [7]. In contrast, bottom gated running systems, Fig. 2b, provide more tranquil filling and a corresponding increase in λ to ~50 which matches the bottom end of the range of 40–100 for forgings. Thus, well-made castings can be as reliable as forgings.

The key features of bottom gated moulds include an offset weir pouring basin and stopper, a tapered sprue, a thin runner bar, a filter and large ingates to reduce the velocity to <0.5 m s^{-1}. These are relatively easy to use in sand moulds but more difficult to use for investment casting because the slender wax running systems often fail during mould manufacture. Nevertheless, bottom gated investment castings have been made and resulted in more reliable Al, steel and Ni alloy castings [8].

The search for more reliable castings - of both TiAl alloys and more common materials - has led to a re-evaluation of the use of tilt filling to reduce entrainment defects.

Tilt casting
Tilt casting is not new. Durville [9] patented a method in 1919 for tilt filling ingot moulds to overcome the susceptibility of Cu-10%Al coinage to oxide inclusions. His process has subsequently been used for sand castings [10], aluminium die castings [11] and for casting billets of air-melted Ni alloys [12]. Tilt casting has been used by some investment casting foundries, originally with rocking indirect arc furnaces [13] and, more recently, with induction rollover melting furnaces [14]. Tilt casting is particularly useful for thin section castings, possibly because the absence of a transfer ladle improves control over the pouring temperature, although the reduced surface turbulence may also be beneficial.

This paper summarises two major research projects which were undertaken to gain a better understanding of tilt casting, with the ultimate object of using this process to cast TiAl.

Experimental
In one project [15-17] an Al-4.5%Cu alloy was tilt cast in sand moulds containing ten 10 mm diameter test bars, Fig. 3. In the second project [18], an Al-7%Si-0.4%Mg (2L99) aluminium alloy was tilt cast in hot ceramic shell moulds containing eight bend test samples, Fig. 4. For comparison, sand and investment castings were also made using gravity filled moulds of the designs shown in Fig. 5. A 1 m diameter tilt casting wheel driven by a computer-controlled servo-motor was used for tilt casting and gave highly reproducible filling conditions, Fig. 6. It had fixtures to hold a crucible of molten metal and either the sand mould or pre-heated ceramic shell mould clamped in contact with it. The pair could then be rotated about their plane of contact to transfer the metal into the mould using either single or multi-stage mould filling cycles. Constant velocity tilting

cycles tended to cause 'sloshing' of the metal within the mould and therefore more complex velocity-time profiles were developed to minimise this. In the case of the sand moulds, it was possible to start with the mould in different initial orientations, Fig. 7.

A Computational Fluid Dynamics software package (Flow 3D) was used to simulate and to optimise mould filling. Real-time X-ray radiography was used to visualise metal flow during mould filling and the filling sequences were compared with those predicted by computer simulation.

Large numbers of test pieces were cast, solution treated and aged. The sand cast Al-4.5%Cu bars were machined into 6.75 mm diameter bars and tensile tested. The investment cast 2L99 plates were tested in 4 point bending in the unmachined condition since investment castings are often used without being machined. The measured properties were analysed using the Weibull technique (eq. 1).

Results
Sand casting
Excellent agreement was found between the flow patterns observed by X-ray radiography and those predicted by Flow3D, as illustrated by Figs. 8 and 9 for turbulent filling. This meant that computer simulation could be used with confidence for analysing the flow and for optimising tilt casting.

In the sand casting trials, X-ray radiography showed that when the mould was initially angled above the horizontal (Fig. 7a), much of the runner was filled quiescently before the mould passed through the horizontal, Fig. 10. This horizontal metal transfer is a unique feature of tilt casting. Less control was achieved when the mould was initially aligned with the horizontal, Fig 7b, and surface turbulence developed in the metal flow, Figs. 8 and 9. When the mould was initially angled below the horizontal, Fig 7c, the flow became very unstable. The Weibull moduli of the tensile strengths of test bars cast using different combinations of initial orientation and rotation speed are plotted against the metal velocity at the end of the sprue in Fig. 11 which shows a dramatic decrease in reliability (a drop in λ from ~36 to ~15) once the velocity exceeded ~0.5 m s^{-1}.

The results of the X-ray radiographic observations and the Weibull analysis were combined to define an operational map for tilt pouring, Fig. 12, which is specific to the alloy and the pouring temperature used (730°C). Curve A defines the combination of parameters below which incomplete filling occurred. Reliable castings can be produced by ensuring that the tilt casting parameters fall between curve A and line B.

Investment casting
The different mould geometry used for investment casting precluded changing the starting mould orientation to minimise surface turbulence during filling so more effort was placed on developing multi-stage filling

cycles. Again, the Flow3D modelling results correlated well with the real-time X-ray observations. It was found that a poor choice of the tilt casting variables created considerable turbulence, thus stressing that the process must be carefully implemented to obtain the best results. For example, Fig. 13 shows frames from a simulation of tilt pouring at a constant, relatively high speed (28°/s). It can be seen that the metal level was not horizontal as would be the case if flow were tranquil. Metal flowed downhill into the runner bar and a back-wave reflected off the far end of the runner travelling back over the incoming metal stream, potentially leading to entrainment defects. In contrast, quiescent filling was achieved by multi-stage cycles comprising uniform velocity stages separated by pauses to stabilise the metal level. Fig. 14 shows the simulation of a filling cycle which took 40 s for the full 0-180° rotation. The mould was rotated quickly to 10° above the horizontal and more than half the runner was filled without the metal falling under gravity. A 5 s pause dissipated the swell set up by the momentum of the first stage. The mould was then rotated slowly to 2° below the horizontal in 15 s, Fig. 14*d*, in order to minimise reflected back-waves. After a 2 s pause, faster rotation was used to complete the cycle in a further 15 s.

The results of the Weibull analysis of the bend strengths are given in Fig. 15 and the Weibull moduli (λ) are summarised in Table 1. The values for the gravity filled moulds produced at the same time show that λ was significantly improved (from 23 to 49) when the top gated design was replaced by a bottom gated design. All tilt filling cycles gave more reliable castings than top gated gravity filled castings. Furthermore, there was a significant effect of the tilt filling cycle on the reliability of castings. Fast single stage filling gave a λ value of 31 which was improved to 45 when the rotation speed was reduced to increase the filling time to 5 s. When a multi-stage cycle with a similar filling time (5 s) was used, λ dropped slightly to 42, but it is questionable whether this is significant. However, when a slower multi-stage cycle was used (filling time of 15 s), λ was improved to 55.

Discussion
Most foundries currently use gravity filled, top gated moulds but it is clear from the present and previous research that this produces the least reliable castings. The reliability of castings can be significantly improved by using carefully designed bottom gated running systems, although they may not be very easy to implement in practice, particularly in investment casting foundries due to the fragility of the wax assemblies. Tilt filling offers a significant improvement in casting reliability compared with top gated, gravity filled moulds and a relatively small increase in reliability over well designed filtered bottom-gated, gravity filled moulds.

Tilt filling offers a number of practical advantages, particularly for investment casters. Compared with bottom gated gravity filled designs, the mould designs are much simpler and easier to fabricate, the less fragile wax assemblies are more likely to survive ceramic shelling, and the casting yield is higher. Tilt filling has considerable potential for further development and for producing castings having a more complex geometry than the simple test bars cast in this project. These might require different flow rates in different parts of the mould at various stages of filling cycle which should be possible to achieve using computer-controlled equipment similar to that used in the present work.

In the next phase of this research, the process is being adapted for use inside a vacuum chamber in conjunction with an Induction Skull Melting furnace for casting Ti-base alloys.

Conclusions
- A re-evaluation of tilt filling has showed that it is capable of producing castings that are always more reliable than those produced by traditional top gated gravity filled methods.
- Careful attention to the mould filling cycle also enables tilt filling to produce castings that are more reliable than those produced by optimised bottom gated gravity casting methods.

References
1. Harding R A, Wickins M and Li Y G, *Progress towards the production of high quality γ-TiAl castings*, Structural Intermetallics 2001, Ed. K.J. Hemker *et al.*, TMS, 2001, 181-189.

2. Harding R A and Wickins M, *Temperature measurements during the induction skull melting of titanium aluminide*, Mat. Sci. Tech., 2003, **19** (9), 1235-1246.

3. Kuang J P, Harding R A and Campbell J, *Examination of defects in gamma titanium aluminide investment castings*, Int. J. Cast Metals Research, 2000, **13**, 125-134.

4. Campbell J, Castings, 2nd Ed., 2003, Butterworth-Heinemann, Oxford. ISBN 0 7506 4790 6.

5. Green N R and Campbell J, *Statistical distributions of fracture strength of cast Al-7Si-Mg alloy*, Mat. Sci. Eng., 1993, **A173**, 261-266.

6. Weibull W, *A statistical distribution function of wide applicability*, J. Applied Mechanics, 1951, **18**, 293-297.

7. Cox M, Harding R A and Campbell J, *Optimised running system design for bottom filled aluminium alloy 2L99 investment castings*, Mat. Sci. Tech., 2003, **19**, 613-625.

8. Cox M, Wickins M, Kuang J P, Harding R A and Campbell J, *Effect of top and bottom filling on reliability of investment castings in Al, Fe, and Ni based alloys*, Mat. Sci. Tech., 2000, **16**, 1445-1452.

9. Durville P M G, British Patent 23,719 (1913)

10. Cox M and Townsend D W, *Design and manufacture of nickel aluminium bronze sand castings*, Ship Department Publication 18 MoD, 1975, updated as Ministry of Defence Standard 02-747(NES 747) pt 5, 1 April 2000, 252 Pages.

11. Stahl G W, *Twenty-five years tilt pouring aluminum*, AFS Trans., 1986, **94**, 793-796.

12. Siddall R J, Private communication, October 2005.

13. Bidwell H T, Investment Casting, The Machinery Publishing Co. Ltd., London, 1969. SBN 85333 196 0.

14. Beeley P R and Smart R F, Investment Casting, The Institute of Materials, London, 1995. IBSN 0 901716 66 9

15. Mi J, Harding R A and Campbell J, *The tilt casting process*, Int. J. Cast Metals Research, 2002, **14** (6), 325-334.

16. Mi J, Harding R A, Wickins M and Campbell J, *Entrained oxide films in TiAl castings*, Intermetallics, 2003, **11**, 377-385.

17. Mi J, Harding R A and Campbell J, *Effects of the entrained surface film on the reliability of castings*, Met. Mat. Trans. A, 2004, **35A**, 2893-2902.

18. Cox M and Harding R A, *The influence of tilt filling on the reliability of 2L99 aluminium investment castings*. FOCAST 3rd Mini-conference, University of Birmingham, 5-6 November 2002. [to be published].

Acknowledgements

The author would like to acknowledge the contributions of his colleagues (particularly Prof. J. Campbell, Mr. M. Cox and Mr. M. Wickins) and former students (Dr. J.P. Kuang and Dr. J. Mi). The financial support of EPSRC (GR/M60101/01), the Rolls-Royce UTC and the School of Metallurgy and Materials, University of Birmingham is also gratefully acknowledged. Thanks are also due to Aeromet International plc for undertaking some of the casting trials.

E-mail: r.a.harding@bham.ac.uk

Table 1: Weibull moduli (λ) and average bend-test breaking stresses in 2L99 investment castings produced using gravity and tilt casting.

Filling method	Rotation time from 85 to 180°, s	λ	Average breaking stress, MPa
Gravity - top gated	-	23	59.8
Gravity - bottom gated	-	49	59.3
Tilt - fast single stage	1.6	31	57.7
Tilt - slow single stage	5	45	59.6
Tilt - fast multi-stage	5	42	60.0
Tilt - slow multi-stage	15	55	59.1

Fig. 1: Reduced level of defects in TiAl castings when cast in VIR furnace instead of ISM furnace.

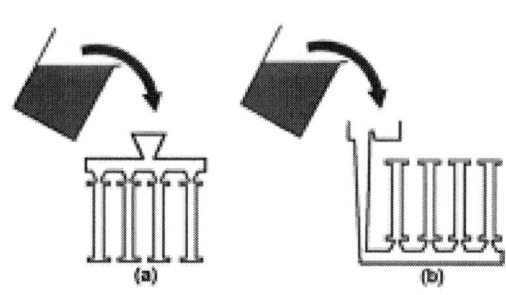

Fig. 2: Comparison of gravity casting (a) top gated (b) bottom gated.

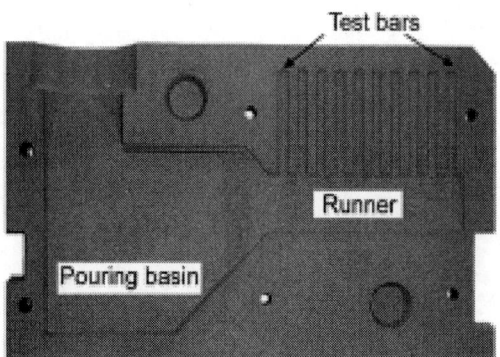

Fig. 3: Sand mould used in tilt casting trials.

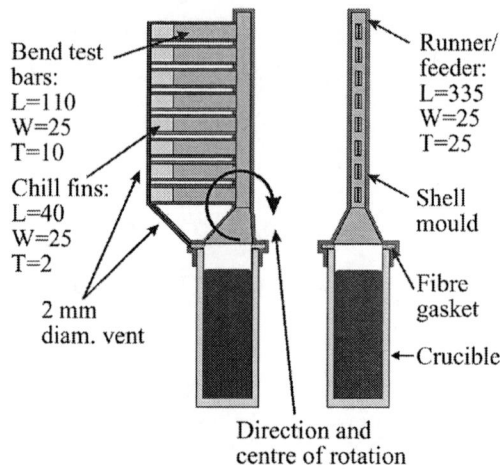

Fig. 4: Investment tilt casting mould design and method of mounting on crucible.

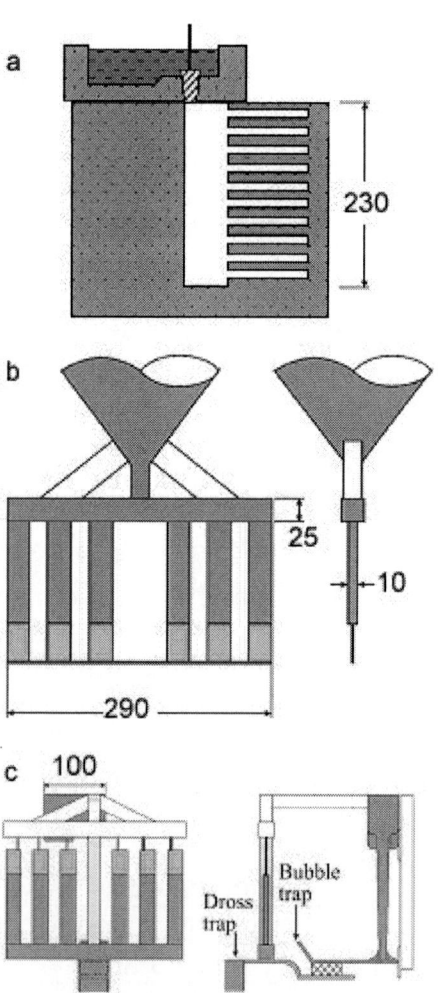

Fig. 5: Gravity filled moulds (a) top gated sand (b) top gated investment (c) bottom gated investment.

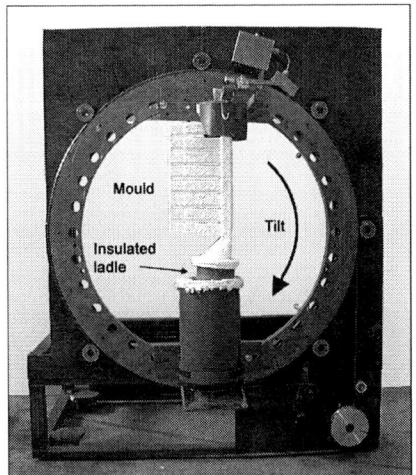

Fig. 6: Computer-controlled casting wheel for tilt casting.

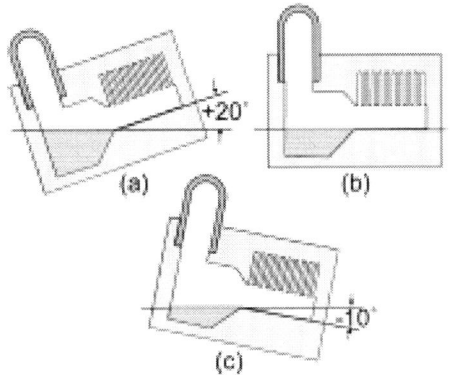

Fig. 7: Initial mould orientations used for tilt casting sand moulds.

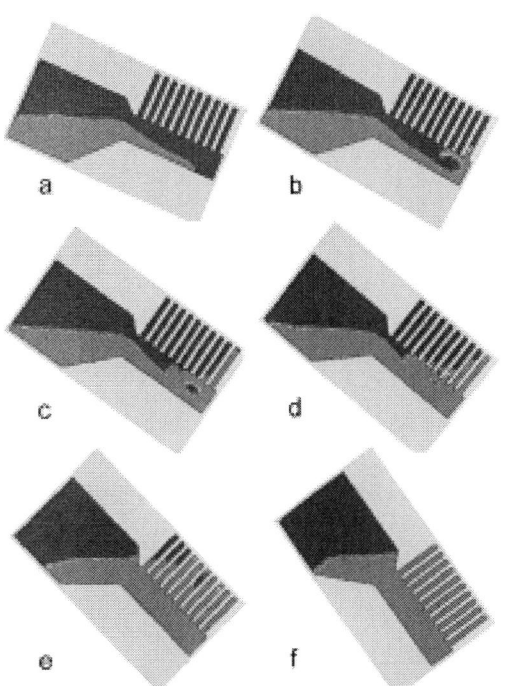

Fig. 9: Flow-3D simulation of the turbulent metal flow during tilt pouring; initial mould orientation: 0°; speed: 43°/s.

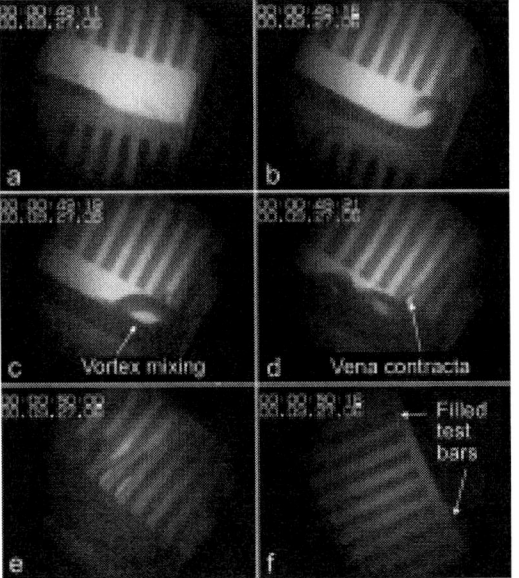

Fig. 8: Frames from an X-ray video showing turbulent mould filling when tilt pouring Al-4.5%Cu; initial mould orientation: 0°; speed: 43°/s.

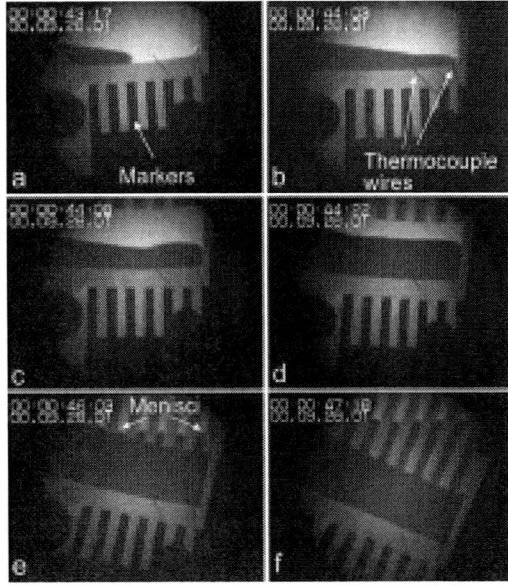

Fig.10: Frames from an X-ray video showing tranquil mould filling when tilt pouring Al-4.5%Cu; initial mould orientation: 20°; speed: 7.1 °/s.

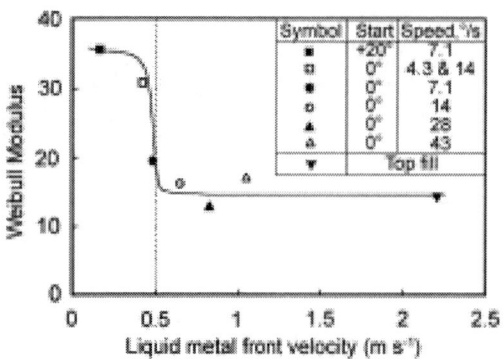

Fig. 11: Effect of metal velocity at end of runner bar on Weibull modulus of Al-4.5%Cu castings.

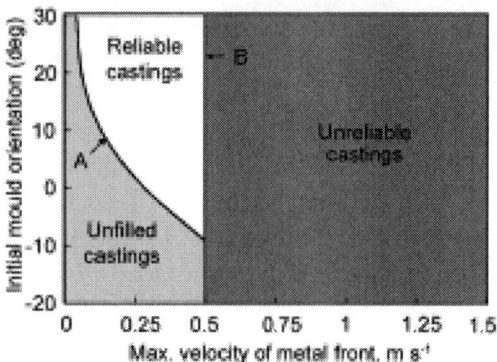

Fig. 12: Map showing effect of tilt filling parameters on casting quality.

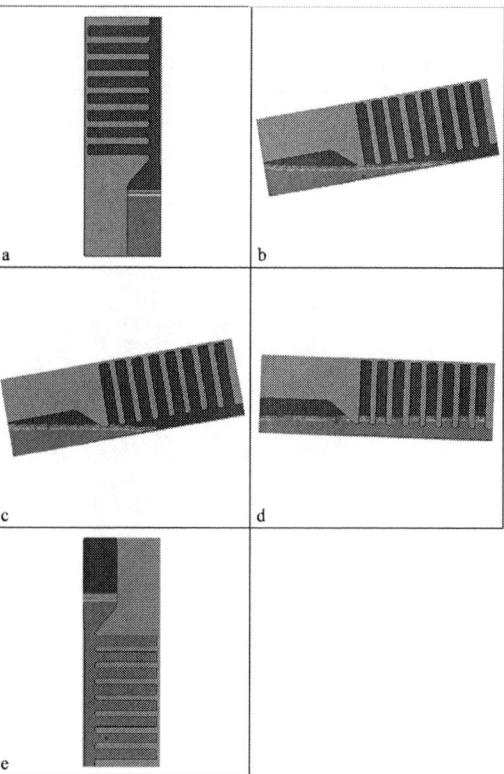

Fig. 14: Flow3D simulation of filling during a multi-stage uniform velocity rotation cycle: (a) 0°; (b) and (c) start and end of 5 s pause at 80°; (d) 92°; (e) 180° rotation.

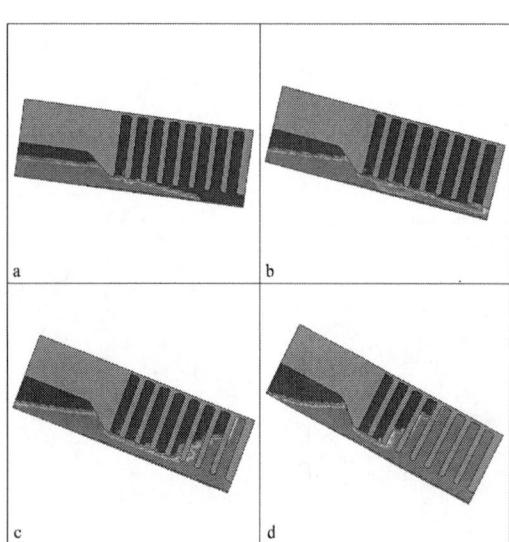

Fig. 13: Flow 3D simulation of a single-stage cycle rotating 180° in 6.35 s (28°/s).

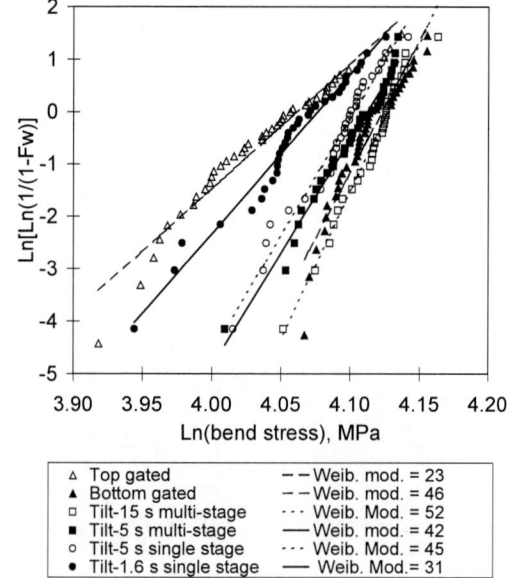

Fig. 15: Effect of gravity and tilt filling on Weibull modulus of 2L99 investment castings.

The Influence of Heat Treatment on the Structure and Tensile Properties of Cast Titanium Alloy Ti-5111

A.C. Robinson[1], E.J. Czyryca[1], and D.A. Koss[2]

[1]Naval Surface Warfare Center, Carderock Division, USA
[2]The Pennsylvania State University, USA

Abstract

The influence of heat treatment on the microstructure and tensile properties of the cast Ti-5111 alloy has been examined. The effects of annealing temperature and cooling rate on the microstructural evolution in this near-alpha alloy are assessed, and the relationships between microstructure and the corresponding tensile properties and fracture behavior are explored. The results indicate good strength in both the as-cast condition and after six different heat treatments. The study also indicates heat treatments promoting a large prior beta grain size also result in an occasional low value of tensile ductility in which failure occurs due to crack initiation and growth.

Keywords

Cast titanium, microstructure, fracture, Ti-5111, tensile ductility

Introduction

The titanium alloy Ti-5111 (Ti-5Al-1V-1Sn-1Zr-0.8Mo) has relatively high specific strength, is non-magnetic, weldable, and corrosion resistant, and is lower in cost than other Ti alloys having these attributes. As such, Ti-5111 is attractive for high performance marine machinery and structural components. Ti-5111 is currently being used in naval service using wrought and fabricated products in several applications. More extensive use of Ti-5111 is limited in large part by cost, and one way of realizing both cost and schedule savings is to utilize the Ti-5111 alloy in the *cast* form, where near-net-shape, complex components can be produced. Although the cast version of the Ti-5111 alloy was initially used for prototype unmanned undersea vehicle hulls, increased use of cost-effective cast Ti-5111 components is desirable. Therefore, a more thorough understanding of the influence of processing on the microstructure and properties of Ti-5111 is highly desirable, and that is the purpose of this study.

Based on their experience with cast Ti-6Al-4V and due to foundry reluctance to risk heat treatments outside their standard practice, the producers of investment cast Ti-5111 use heat treatments developed for the Ti-6Al-4V castings. Because Ti-5111 and Ti-6Al-4V alloys are chemically and structurally different, the optimum heat treatment for Ti-6Al-4V may not be applicable to Ti-5111. This paper focuses on understanding the effects of controlling heat treatments (i.e. annealing temperatures and cooling rates) on cast Ti-5111 plate material. The microstructure of a cast plate is characterized for the as-cast condition and six different thermal treatments, and the effect of critical microstructure features (prior beta grain size, grain boundary alpha thickness, alpha lath thickness, and volume fraction of each phase) on the fracture behavior of tensile specimens is assessed.

Background

The Ti-5111 alloy is considered a near-alpha alloy due to its high concentration of alpha stabilizers (5 wt% Al) and relatively low level of beta-phase stabilizers, such as V or Mo. Wrought Ti-5111 products receive a β anneal at 1010°C followed by α/β anneal at 954°C. As a result, the room temperature cast microstructure typically consists of some version of a Widmanstatten α structure in which the hexagonal close packed (hcp) α phase (usually in the form of laths or plates) is the dominant phase. However, important characteristics of that microstructure will be sensitive to both cooling rates after casting and subsequent heat treatments [3,4,5,6]. For example, the alpha plates can develop in well-defined colonies (Widmanstatten α) of parallel plates/laths nucleating from the prior beta grain boundaries, or they can develop with numerous random orientations (basketweave α) independent of the grain boundaries. These laths can also range in thickness from fine to coarse and in their length scale. The α laths are typically surrounded by thin layers of retained β, which is enriched in β-stabilizing elements [7]. These layers form as a result of diffusion of the β stabilizing elements ahead of the migrating α interface. In general, critical microstructure features such as prior beta grain size, alpha lath thickness, grain boundary alpha thickness, and volume fraction of beta phase are sensitive (to some degree) to the time and temperature of the heat treatment as well as the subsequent cooling rate. The resulting microstructure changes can obviously affect the mechanical properties and fracture behavior of the alloy.

Materials Investigated

For this study, Wah Chang cast a 3.2-cm thick Ti-5111 plate (Heat R286) using a graphite mold (GM). The plate was subsequently HIPed at 899°C for 2 hours at a pressure of 103 MPa. The chemistry of this plate, along with the ASTM Standard B 265 Grade 32 for Ti-5111 plate is shown in Table 1. The plate was cut into 16.5 x 10.1 x 3.2 cm plate pieces and heat treated by PCC Structurals, Inc. as outlined in Table 2. Additionally, an as-cast plate piece was included as a benchmark in determining how each of these microstructure properties change as a function of heat-treatment temperature and cooling rate. The heat-treatments in Table 2 were intended to vary the critical microstructure features by altering the solution-treating temperature and cooling rate. Table 2 also provides resulting prior beta grain size, volume fraction of alpha phase, and Rockwell hardness, discussed later in this paper.

Experimental Procedure

Microstructure Characterisation

Clemex® imaging software was used in coordination with optical microscopy to determine the volume fraction of the alpha and beta phases in the microstructures. One hundred regions of a metallographic specimen were measured and averaged to determine the volume fraction of alpha phase with the remainder being the retained beta phase. The prior beta grain size was measured in accordance with ASTM E 112 using the Planimetric Procedure method. A region approximately 25 mm by 20 mm in size was averaged to obtain a representative grain size.

Mechanical Properties

Mechanical property evaluation of each casting included hardness measurements and tensile tests. The hardness measurements were conducted using a Rockwell-C indenter with a load of 150 kg. A minimum of three measurements on each specimen was taken to obtain representative hardness values. For each cast plate, three tensile specimens were tested in accordance with ASTM E 8 using 6.4 mm diameter specimens. The tests were conducted at a strain rate of 10^{-3} /s and ambient temperature.

Results and Discussion
Microstructure

Figure 1 shows microstructures resulting after the heat treatments detailed in Table 2. All specimens exhibited a Widmanstatten colony structure with grain boundary alpha phase decorating the prior beta grain boundaries. The as-cast specimens (GM-1) exhibit a fine colony structure with thin alpha laths and a thin grain boundary alpha phase. Some of the alpha colonies are large and cover a significant portion of the prior beta grain, Fig. 1(a). The largest difference in microstructure of the heat- treated specimens was observed as a function of cooling rate. The specimens cooled slowly at a rate of 1°C/min exhibited significantly coarser alpha laths and thicker grain boundary alpha phase. For example, specimens GM-2, Fig. 1(b), and GM-3, Fig. 1(c), were both β annealed at 1010°C and cooled at two different rates. Specimen GM-2 (14°C/min) shows a very fine lath structure with thin grain boundary alpha, while the specimen cooled at 1°C/min (GM-3) exhibits a rather coarse structure with thicker grain boundary alpha phase. Similar cooling rate differences can be noted between the specimens given a single α/β anneal or duplex anneal. The micrograph of specimen GM-7, Fig. 1(g), shows minimal grain boundary alpha and a rather fine lath structure, consistent with these features being minimally affected by an α/β heat treatment alone.

Comparisons of the prior beta grain size, volume fraction of alpha phase, and the macrohardness of each heat-treated specimen that are reported in Table 2 show the following:

- The as-cast material has the highest hardness and with a relatively large prior beta grain size compared to the other plates. It also has a comparatively low volume fraction of fine-scale α phase and thin grain boundary alpha.
- The β-annealed structures have beta grain sizes smaller than the as-cast material. The β anneal decreased the hardness even though the volume fractions of each phase was similar to the as-cast condition.
- Increasing the cooling rate typically decreased the volume fraction of α phase that formed during cooling to room temperature, as expected. Decreased cooling rates from the β-phase field increased the scale of the alpha laths and grain boundary α phase.
- The α/β anneal increased the volume fraction of α phase and resulted in a small increase in hardness above that of the β anneal condition. Omitting the β anneal heat treatment and relying solely on an α/β anneal after casting resulted in similar hardness levels to those specimens that had been β annealed, but with larger prior beta grains.

Mechanical Properties

Consistent with the hardness results (Table 2), the yield and tensile strengths of the as-cast condition exceeded the strength of heat-treated castings, as shown in Table 3. Note in Table 3 the tensile properties for all castings consistently exceeded the minimum yield and tensile strength for Ti-5111 wrought plate by ASTM B 265, Grade 32.

Table 3 shows the large variation of ductility values (% elongation and % reduction of area after fracture) that resulted both within a single casting set and between castings. In some cases, both elongation to failure and reduction of area values varied by a factor of two within the three specimen test set. Several specimens were below the 10% minimum elongation benchmark for adequate ductility in shock applications. As in previous studies of the Ti-6Al-4V alloy [4,9], low ductility resulted mainly from tests where microstructures exhibited large prior beta grain sizes.

In to understand the large variation in ductility, the fracture surfaces of tensile specimens from casting GM-3 showing extreme differences in % elongation were analyzed. Figure 2 and Figure 3 contrast the beta-annealed specimen (subject to a slow 1°C/min cooling rate) exhibiting 13% elongation with a companion specimen that failed after only 7% elongation. Figure 3 shows the presence of a large, relatively flat facet on the fracture surface of the low ductility test. The facet, which is outlined in Figure 3a and extends ~1.8 mm along the specimen surface, suggests a microstructural feature on the scale of 1-2 mm may deform and crack in a coordinated manner such to create a large, relatively planar crack that subsequently propagates and limits ductility.

The comparatively small difference between percent elongation (7%) and reduction of area (11%) in this case supports the hypothesis that this tensile specimen fractured by crack initiation and crack growth. Thus, we believe the large facet fractured first and initiated a large crack, whose crack-tip stress field tended to confine the crack plane promoting a smoother overall fracture surface. In contrast, the fracture surface of the higher ductility specimen shows a rough, tortuous fracture path with no large, smooth fracture facets evident. Therefore, it can be concluded that the high ductility specimen deformed without a single, dominating crack.

To relate the large, flat facet on the fracture surface to its corresponding microstructure, the specimen was cross-sectioned perpendicular to the fracture surface, and subsequently ground, polished, and etched, as shown in Figure 3c & d. This figure shows the large facet on the fracture surface directly corresponds to a large prior beta grain with a linear intercept of 1.5 mm at the fracture surface. This grain was oriented in such that a crack nucleated and propagated through the entire grain with minimal resistance.

The fracture surfaces, such as those shown in Figure 2 and Figure 3, suggest that the presence of a large microstructural feature on the scale of 1-2 mm can fracture and initiate a large crack that then significantly limits ductility by subsequent crack growth. As an example, Figure 4 shows two adjacent Widmanstatten colonies that share an orientation relationship allowing neighboring, planar slip bands to extend 2 mm across the boundary between two large colonies with no apparent deviation. Such long slip band distances are known to promote crack initiation [8], and in this case, the result is a planar fracture surface and comparatively low tensile ductility (percent elongation of this specimen was 8%). It is likely the flat fracture facet outlined in Figure 3a is the result of such a planar slip deformation/fracture process that extends across an entire beta grain. Thus, decreasing the prior beta grain size should improve ductility, and Tables 2 and 3 suggest such a relationship where castings given a β anneal at 1010°C followed by α/β anneal showed smallest prior beta grain size and highest ductility.

Conclusions

A graphite mold cast Ti-5111 plate was examined in the as-cast condition as well as after six different heat treatments. Specimens given a β anneal (either solely or in conjunction with an α/β anneal) had prior beta grain sizes smaller then either the as-cast plate or those given solely an α/β anneal after casting. Specimens given only an α/β anneal performed the lowest in ductility, while the as-cast specimens had some of the best properties. The ductility of cast Ti-5111 is controlled by the size and orientation of the prior beta grains. Large grains oriented unfavorably to the tensile axis can nucleate a slip-initiated crack of sufficient size that limits ductility by crack growth. Therefore, reducing the prior beta grain size improves ductility and reduces scatter within groups of specimens given the same heat treatment.

References

1   Stauffer, A.C., Czyryca, E.J., and Koss, D.A., "The Influence of Processing on the Microstructure and Properties of the Titanium Alloy Ti-5111," *3rd Inter. Conf. for Adv Mat and Proc*, (2005) 547-551.

2   Gaies, J.G., Stauffer, A.C., and Czyryca, E.J., "Fracture Toughness of Cast Titanium Alloy Ti-5111 and Cast Commercially Pure (CP-2)," NSWCCD-61-TR-2004/25 (2004).

3   Weinem, D., Kumpfert, J., Perters, M., and Kaysser, W.A., "Processing Window of the Near-α Titanium Alloy TIMETAL-1100 to Preclude a Fine-grained β-structure," *Mat. Sci. and Eng.* **A206** (1996) 55-62.

4   Lütjering, G., "Influence of Proc. on Microstructure and Mechanical Properties of (α+β) Ti Alloys", *Mat. Sci. and Eng.*, **A243** (1998) 32-45.

5   Greenfield, M.A., Pierce, C.M., and Hall, J.A., "The Effect of Microstructure on the Control of Mechanical Properties in Alpha-Beta Titanium Alloys," *Tit Sci and Tech,* **3** (1973) 1731-1743.

6   Rogers, D.H., "The Effects of Microstructure and Composition on the Fracture Toughness of Titanium Alloys," *Ti Sci and Tech*, Ed. by R.I. Jaffee and H.M. Burte, **3** (1973) 1719-1730.

7   Hammond, D. and Nutting, J., "The Physical Metallurgy of Superalloys and Titanium Alloys," *Metal Science*, October (1977) 490.

8   Chestnutt, J.C., "Relationship Between Mechanical Properties, Microstructure, and Fracture Topography in $\alpha+\beta$ Titanium Alloy," *Fractography ASTM STP600* (1976) 99.

9   Cotton, J.D. and Johanson, B.M., unpublished research, The Boeing Company, 2002.

Acknowledgements

The work discussed in this paper was conducted in support of the Office of Naval Research through the Seaborne Materials Technology Program under Dr. Julie Christodoulou and in support of the In-house Laboratory Independent Research (ILIR) program at the Carderock Division, Naval Surface Warfare Center under John Barkyoumb. The authors also acknowledge Mr. David Lee and Mr. Boyd Mueller from the Howmet Research Center for helpful discussions.

Tables

Table 1: Chemical Composition (wt. %) of the Graphite Mold Cast Ti-5111

Element	Al	Sn	V	Zr	Mo	Fe	Si	O	C	N	H
ASTM B 265 Grade 32	4.5-5.5	0.6-1.4	0.6-1.4	0.6-1.4	0.6-1.2	0.25*	0.06-0.14	0.11*	0.08*	0.03*	0.015*
3.2 cm-thick castings	4.98	1.03	0.99	0.97	0.78	0.06	**	0.09	0.01	<0.01	0.002

*max   **not determined

Table 2: Heat Treatments for As-Cast Ti-5111 Plates, along with Prior Beta Grain Size, Volume Fraction of Alpha Phase, and Rockwell Hardness

Plate ID	Phase(s) at Temp.	Temp. (°C)	Cooling Rate (°C/min)	Prior Beta Grain Size (μm)	Vol. Frac. α Phase	Hardness (HRC)
GM-1	AS-CAST + HIP	AS-CAST + HIP	AS-CAST + HIP	1200	0.69	32.0
GM-2	β	1010	14	1120	0.62	26.7
GM-3	β	1010	1	1130	0.74	26.0
GM-4	β / α/β	1010 / 954	14 / 14	920	0.78	29.3
GM-5	β / α/β	1010 / 954	1 / 1	1090	0.80	29.0
GM-6	α/β	954	14	1210	0.76	27.0
GM-7	α/β	954	1	1370	0.73	28.0

*All times at temperature are 1 hour

Table 3: Tensile Properties of Ti-5111 Castings

Plate ID	Ultimate Tensile Strength (MPa)	0.2% Yield Strength (MPa)	Elongation (%)	Reduction of Area (%)
GM-1	814	807	11	21
	834	807	10	20
	834	800	8	x
GM-2	779	710	11	18
	793	731	11	14
GM-3	765	717	7	11
	779	731	13	26
	772	717	12	20
GM-4	772	703	11	23
	786	724	12	14
	807	738	11	17
GM-5	765	717	13	24
	779	724	10	26
	772	724	12	22
GM-6	793	717	5	11
	779	724	10	19
GM-7	772	727	8	19
	793	738	9	21
	779	724	11	18
ASTM B 265, Grade 32	689 minimum	586 minimum	10 minimum	-
Ti-5111 wrought plate (typical range)	825 to 890	715 to 810	12 to 18	20 to 40

Figures

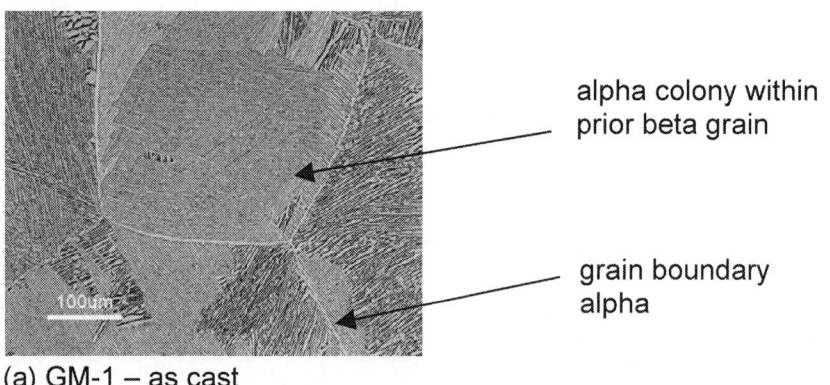

alpha colony within prior beta grain

grain boundary alpha

(a) GM-1 – as cast

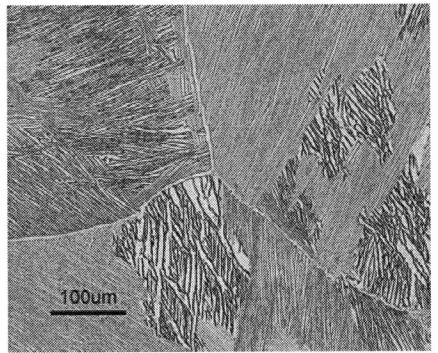 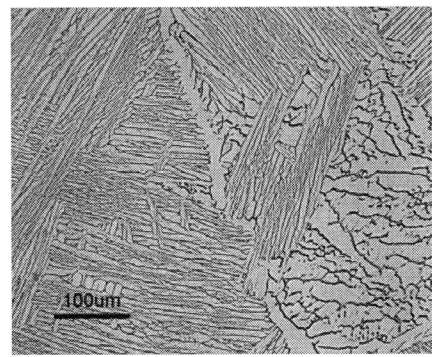

(b) GM-2 – β anneal, cool 13.6°C/min   (c) GM-3– β anneal, cool 1°C/min

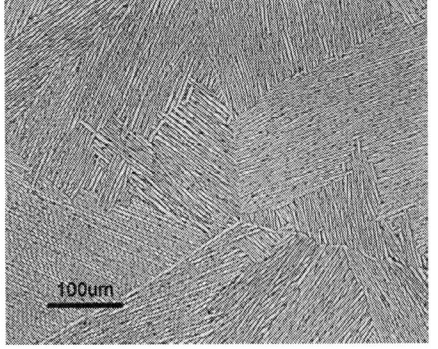 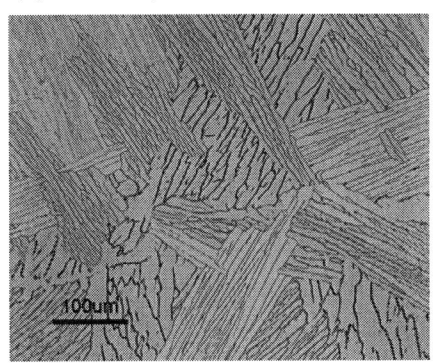

(d) GM-4 – β, α/β anneal, cool 13.6°C/min (e) GM-5 – β, α/β anneal, cool
1°C/min

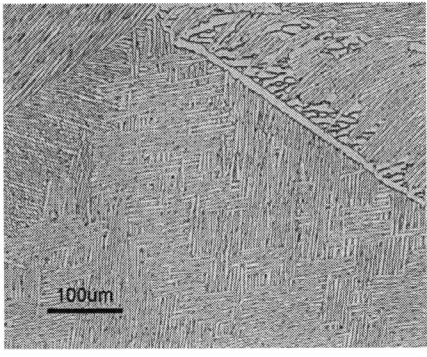 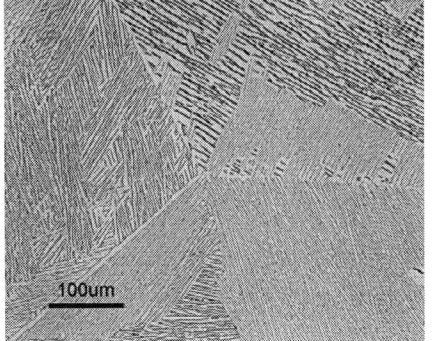

(f) GM-6 – α/β anneal, cool 13.6°C/min   (g) GM-7 – α/β anneal, cool 1°C/min
Figure 1: Micrographs of each heat-treat condition of Ti-5111 casting

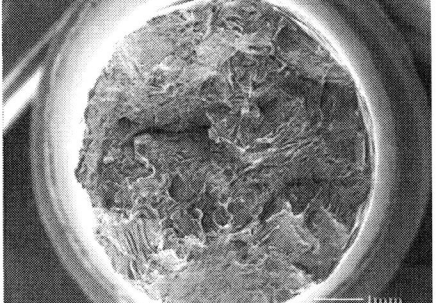

Figure 2: Fracture surface of specimen GM-3-4 exhibiting 13% elongation

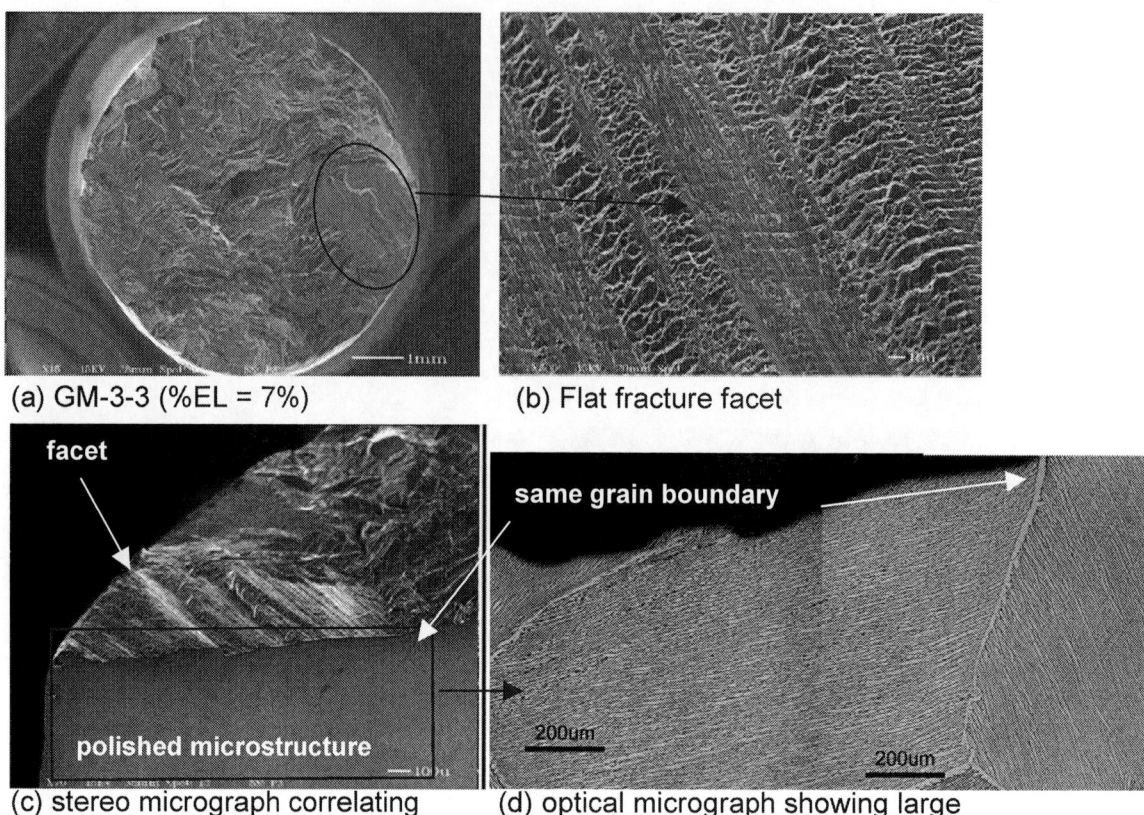

(a) GM-3-3 (%EL = 7%)

(b) Flat fracture facet

(c) stereo micrograph correlating facet on fracture surface with the underlying microstructure

(d) optical micrograph showing large prior β grain correlating to large facet

Figure 3: The fracture surface and corresponding microstructure of tensile specimen GM-3-3 that exhibited low ductility

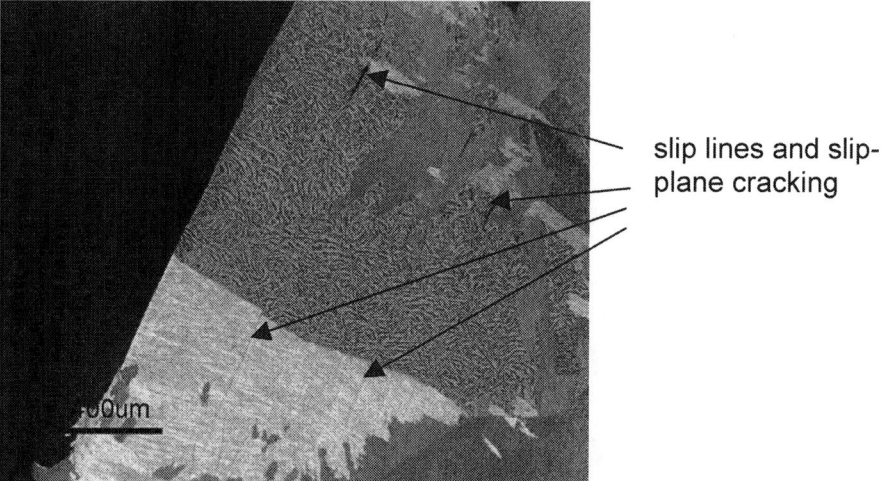

Figure 4: The cross-section of as-cast tensile specimen (GM-1-5) with low 8% elongation showing slip bands parallel to planar fracture surface that extend across two large colonies

Magnesium alloy castings – past and present

Peter John Thompson*

* Castings Technology International

Abstract

Magnesium is the lightest of the structural metals. Sand cast and HPDC Magnesium is used in a wide range of markets and applications, from automotive to aerospace. In all cases, weight saving is the primary reason for using Magnesium alloys.

Precision cast Magnesium is not in widespread use. Main reason for this lies in the potential reaction of molten Magnesium with Silica found in most shell moulds. Recent work has shown that by modification of mould preparation procedures and use of low reactivity Alloy Elektron 21, metal / mould reaction can be completely avoided.

Elektron 21 is corrosion resistant, produces uniform properties and can operate up to 200°C. This combination of alloy properties, light weight and precision casting technology will be attractive to many market sectors where weight is a key factor.

Introduction

Magnesium has useful potential for weight saving in many areas of industry. The greatest increase in the use of magnesium is in the automobile industry. The need here is to comply with the stringent weight requirements laid down in legislation concerning emissions and energy consumption.

Magnesium is distinguished by its comparitively low density, which is approximately 25% that of steel and aproximately 60% that of aluminium. This property makes magnesium have great potential for lightweight construction.

BMW's new inline six-cylinder spark-ignition engine sets high standards in specific power, weight and fuel consumption. The new design is based on the latest technologies with the major innovation being the composite magnesium-aluminium crankcase. The engine will be installed in virtually all BMW vehicles. The requirements for the new inline six-cylinder spark-ignition engine were 'stronger, lighter and more economical'. The world's first composite Mg/Al crankcase is a fundamentally new technology and the basis for a significant reduction in weight. This amounts to 24% weight saving compared to the equivalent aluminium crankcase.

The primary reason for the rapid growth over the past 10-15 years for High Pressure Die Cast automotive applications, apart from its lightweight properties, is the development of alloys having excellent corrosion resistance and for some alloys, elevated temperature performance to 250° C, which is superior to most aluminium alloys. At the same time magnesium enjoys sustained use in the sand cast form for aerospace and speciality applications. Aircraft costs can be considered in terms of price per unit of weight, as a result there is a relationship between weight saving and cost. The value of 1 lb in weight saving has been estimated to range in savings of $300-3000, depending on the type of commercial aircraft [1] and significantly more for Military and space applications. It is therefore no surprise that Magnesium alloys are used as major components of jet engines(intermediate casings, gearboxes, covers) and for helicopter components (structural and ancillary gearboxes) – refer to figs 1,2,3 &4.

The applications so far described are produced either by High Pressure Die Casting (HPDC) or Sand casting. By comparison, precision castings (i.e. using investment shell and plaster block moulds) in magnesium are currently produced on a limited basis. Aluminium on the other hand is seeing a growth in the use of this technology. [2]

Part of the reason why magnesium is not extensively used for precision casting applications lies in a lack of general education on the attributes and existing use of magnesium alloys. More specific reason is the potential reaction of magnesium with the commonly used silica containing constituents of plaster/shell mould materials. Silica is present in most slurry mixes as either a constituent of the binder or refractory - silica reacts with magnesium to produce magnesium oxide and magnesium silicide.

Several foundries have overcome this reactivity issue by proprietary techniques, successfully producing magnesium from plaster block moulds. Ceramic shell moulds have also been produced in more recent times by careful process control and/or modifications of primary shell compositions.

Alloy Types
Two groups of Magnesium alloys are commonly used; these are Magnesium-Aluminium and Magnesium-Zirconium.

Magnesium-Aluminium.
This group of alloys are the oldest and rely upon the strengthening effect of Aluminium addition to Magnesium. Tensile strength reaches a peak at approximately 10% Aluminium addition, after which tensile strength and ductility fall dramatically, thus setting the upper limit for alloying. The most commonly used alloy is AZ91 (9% Aluminium, 1% Zinc), whilst lower Aluminium containing alloys such as AM50 and AM60 (5% and 6% Aluminium respectively) are preferred, where improved ductility is required.

AZ91 can be used for HPDC, sand casting and has been used for precision casting. High purity versions (low in cathodic impurities, such as iron and nickel) can exhibit good corrosion performance.

Limitations of this group of alloys include poor elevated performance (maximum 120 °C) and difficulty in achieving good grain refinement. The consequence of coarse grain size is a reduction in mechanical properties. For HPDC components, grain refinement is less of an issue because the rapid cooling rate in the metal die is generally enough to generate a good grain size. For sand casting, however, comparatively slow cooling rates can result in coarse grain size, which varies through thick (slow cool) and thin sections.

Magnesium-Zirconium
The basis of this group of alloys is the potent grain refining effect which zirconium has on magnesium. Addition of less than 0.6% is satisfactory to achieve grain sizes less than 80 microns in sand cast parts. This grain refining effect is not strongly affected by the cooling rate of the alloy, so castings have a fairly consistent grain size irrespective of cooling rate. Similar properties might therefore be expected in sand castings and slower cooled precision castings.

A range of alloys exists from this alloy group. Alloy development over the past 40 years has focussed upon achievement of ever increasing strength and suitability for applications subjected to elevated temperatures. Over the last twenty years, corrosion performance of new alloys has been improved to a similar level to that of aluminium based alloys. Some examples are given below:-

- Elektron RZ5 (4% Zinc, 1.3% Rare Earth, 0.6% Zirconium) –Pressure tight, good castability, useful operating temperature up to 130 °C-150 °C.
- Elektron WE43 (4.3% Yttrium, 2.3% Neodymium. 1% Heavy Rare Earth,0.6% Zirconium) – pressure tight, good corrosion performance, useful operating temperature to 250 °C.
- Elektron 21 (2.8% Neodymium, 1.4% Gadolinium, 0.3% Zinc, 0.5% Zirconium) – pressure tight, good castability, good corrosion performace, useful operating temperature to 200 °C [3].

A comparison of mechanical properties and corrosion performance of the above alloys is listed in Figure 5 and Table 1.

Alloys Suitable for Investment Casting

For components operating at less than 120 °C where mechanical properties are not of great concern, AZ91 is a potential choice and is indeed used. The alloy can be specified in High Purity form (AZ91E), which with good melt control and processing, will offer good corrosion performance. The liquidus temperature of AZ91 is low due to high alloy content. This allows pouring temperatures to be as low as 700°C , reducing the opportunity for potential metal mould reaction.

Limitations of this alloy type have already been highlighted in terms of grain size and mechanical properties. Fantetti et al [5] have further illustrated the poor properties achieved in AZ91 castings made from plaster moulds.

For components requiring improved mechanical properties, whether at room temperature or elevated temperature, the Magnesium- Zirconium alloys are favoured.

For any new application, corrosion resistant alloys are a pre requisite, which eliminates RZ5 alloy. Elektron WE43 offers the best property package. Elektron 43 is, however, one of the more difficult alloys to handle in the foundry due to the presence of Yttrium, which has a tendency to oxidise easily. This means that special metal handling techniques are required for both molten metal handling and mould preparation.

Elektron 21 offers most of the property benefits of Elektron 43 including good corrosion resistance and elevated temperature performance. An important additional aspect for precision casting is that Elektron 21 has low oxidation characteristics [3], which may reduce reaction with mould material and certainly facilitates easy molten metal handling for the foundry. For these reasons, Elektron 21 was chosen for this precision casting evaluation.

Mould Materials
Problems occur when molten magnesium is in contact with moisture (water), iron oxide, or silica.

The constituents of a typical light alloy shell mould can contain several of these components in the particulate and binders used. Similarly, plaster block moulds are often produced from aqueous solutions and contain silicate binders.

Means of avoiding reaction by modifying the mould make-up is one approach, which is used.

Techniques include:-

Inhibitors – Silica sand moulds are effectively inhited by addition of 1 –2 % Sodium Fluorosilicate or Potassium Borofluoride. The use of Fluoride containing inhibitors in plaster block moulds is also possible.Options include the aforementioned, plus Ammonium Bifluoride and Aluminium Fluoride. Knowledge and control of inhibitor content is important [6], because they can affect setting times. For shell moulds effective inhibitors are more problematic, since they tend to dissociate at <500 $^{\circ}$C. Typical firing temperatures of 1000°C will therefore remove the inhibitors.

Drying – Removal of free water is a key stage.

Avoidance of Silica – Removal of Silica containing materials from the shell mould is existing/developing technology for the production of Nickel-Aluminium alloys. Rosefort & Kort have reported good results with magnesium using the same type of technology [7].

Experimental – Precision Casting Trials
1. Introduction
Evaluation was carried out with Elektron 21 alloy in two stages:

Stage 1 was to define whether improvements in alloy type (Elektron 21) made this alloy suitable for precision casting.

Stage 2 was to evaluate the effect of best practice/ improved mould technology on the produceabiliity of precision cast Elektron 21 components.

2. Alloys
Casting trials were carried out with several precision casting foundries, which had experience of casting magnesium.

Mould preparation (including inhibition and drying cycles) was proprietary to each facility. The objective was to compare Elektron 21 in terms of metal handling, castability and reactivity with AZ91 type alloys currently used.

Feedback was favourable. Elektron 21 was considered easy for the foundries to handle. Mould reaction occurred in one foundry associated with procedural issues, all the other foundries reported good results with no metal mould reaction.

The author's comments on handling Elektron 21 were ' during melt down and holding period, Elektron 21 appears to be cleaner and less prone to oxidation (than other alloys). Furthermore the standard practice of feeding/ gating as applied to the simple older RZ5 alloy works very well with Elektron 21 – castings of various wall sections and complexity have been produced and have met radiographically the AMS-STD 2175 Grade A & B standards.

3. Mould Materials
Casting Technology International (Cti) has the facilities to produce a wide range of investment shell types and the ability to carry out development / pre-production casting of magnesium alloys. Stone Foundries Limited [8], a large producer of Magnesium castings, has supported the development programme by supply of complete and part assemblies for the various trials.

The following investment casting factors were considered and evaluated using investment shell moulds: -

Moulds were produced using two shell mixes. Primary layers were free of silica to prevent potential reaction. Initial trials gave successful results. No reaction occurred, however, some localized shell delamination / damage was observed on some castings.

Whilst these results were promising, limited shelf life of the slurry mixes and cost of the materials are limiting factors.

4. Mould Inhibition
Oxidation of Magnesium can be avoided by distribution of a protective gas onto the melt surface, pouring stream and/or inside the mould. This is typically a mixture of $1 - 2\%$ Sulphur Hexafluoride (SF_6) in a carrier gas of Carbon Dioxide (CO_2). Higher levels of SF6 have been used to purge plaster moulds to successfully prevent reaction [9]. For the current evaluation, some moulds were cast without passing protective gas into the mould, some with CO_2/SF_6, others employed SF_6. In each case, gas was introduced into the mould immediately prior to pouring metal.
Moulds were produced using conventional shell systems by Cti and proprietary shell system provided by Stone Foundries Limited.

5. Temperature Variables
Mould temperatures were evaluated in the range $50^{\circ}C - 500^{\circ}C$. Metal temperature varied between $710^{\circ}C$ and $780^{\circ}C$.

6. Section Thickness

Metal / Mould reaction can increase when larger castings are produced because there is more time at temperature for reaction to occur. Using the technique developed during this work, thick section castings were produced in Elektron 21 –refer to Figure 8.

7, Property Comparison

Both separately cast test bars and samples cut from castings were tested to determine the properties achievable from the alloy cast into precision moulds. A selection of results from Elektron 21 castings is summarized in Tables 2,3 and 4.

Results

Results initially showed some variability, with mould reaction sometimes occurring in moulds when no protective gas was used in the mould – refer to Figure 6. Subsequently, tests showed a trend whereby no reaction occurred with moulds into which SF6 had been introduced immediately prior to pouring – refer to Figure 7.

No reaction was observed in the thick section castings (Figure 8).
In summary, Elektron 21 did not react easily with moulds under the wide range of parameters evaluated. In terms of mould / processing technology, proprietary silica – containing and silica – free shell moulds could yield reaction free castings. Castings made from moulds purged with SF_6 consistently gave good results.

The mechanical properties summarized in Table 3 are similar to sand cast components when subjected to the recommended solution and ageing treatment. All section thickness easily surpassing AMS 4429 specification. Alternative heat treatments were evaluated specifically because the use of a hot water quench could lead to distortion in thin walled investment cast parts. This showed that air-cooling from the solution treatment temperature yielded properties in section thickness of 7 mm which exceeded the AMS 4429 minimum requirements (Table 4).

Material in the as cast condition could not achieve AMS 4429 minimum properties (Table 2), however, can still exceed those of AZ91 investment cast parts in the fully heat treated condition. This could be beneficial for prototyping of HPDC AZ91 in investment Elektron 21.

Discussion

Mould technology is already available to produce magnesium castings without reaction issues. The trials carried out at Cti have improved the knowledge of shell mould technology required for magnesium and shown that by careful control of process parameters and use of effective purging techniques (with SF_6), high quality, reaction free castings to meet aerospace quality standards can be produced.

Alloy development has complimented these activities. Elektron 21 offers low oxidation characteristics, which appear of benefit in the precision casting process. This alloy also opens up the following opportunities, which were not readily available to the precision casting industry.

Prototyping of HPDC AZ91 parts. Elektron 21 offers the properties achieved by the AZ91 HPDC component from a precision part. This was not previously achievable using AZ91.

New Applications where the end user requires a lightweight alloy, which can give consistent properties at room temperatures, operate up to 200°C and exhibit general corrosion resistance similar to aluminium alloys.

Replacement of aluminium and other materials where a lightweight solution is required. The Western world Aluminium investment casting market has been estimated to be worth approximately $400 million [2]. Even a small segment of this market is attractive proposition for producers of competitive magnesium alloy components.

Acknowledgement
Paul Lyon of Magnesium Elektron for his help in preparing the paper.

References
1. K.Weiss et al. Magnesium Castings in Aeronautics Applications – Special Requirements. 12th Automotive Seminar. Germany – September 2004.
2. Steven Kennerknecht. Structural Aerospace Aluminium Investment castings 10-years supply & demand evolution – AeroMat 2001.
3. Paul Lyon. Elektron 21 – New alloy for Aerospace and speciality applications. TMS Charlottesville USA.2004.
4. Ken Clark. AZ91E Magnesium Sand casting alloy, the standard for corrosion performance. 43rd World Magnesium Conference, LA, USA 1986.
5. Nick Fantetti , Properties of Magnesium Plaster Castings. Technical paper 910413.SAE Detroit USA 1991.
6. John King. Investment/Plaster Casting of Magnesium Alloys. Magnesium Elektron internal MR10/Data report. 1997.
7. M.Rosefort etal. An Innovative Technology for investment casting of Magnesium – annual Investment Casting Conference, Edinburgh, UK 26 May 2005.
8. Mike Randal. Investment Manager at stone Foundries Ltd.
9. Neelameggham US patent US4579166. 1986.
10. INCAST Magazine. P12. October 2002.

Table 1 – Corrosion Comparison of Magnesium & Aluminium Alloys

Corrosion Rate [4]		
14-Day Salt Fog		
Alloy	MPY	MCD
AZ91C	1300	16.3
ZE41	480	6
AZ91E	<20	<0.25
WE43	<20	<0.25
Elektron 21	<30	<0.38
C355	<10	<0.19
A357	<10	<0.19

Table 2 - Tensile properties relating to as cast specimens

Specimen No.	Section thickness mm	0.2% PS MPa	TS MPa	Elongation %	Comments fracture
1	14	113	192	5	Clean
2	14	110	181	4	Clean
14	7	107	197	9	Clean
15	7	110	204	9	Clean
16	7	112	208	9	Clean
AMS 4429 Sand Cast Minimum		145	248	2	

Table 3 - Tensile properties relating to T6 (HWQ) specimens

Specimen No.	Section thickness mm	0.2% PS MPa	TS MPa	Elongation %	Comments fracture
3	14	171	272	3	Clean
4	14	164	262	3	Clean
5	12 on 75	163	268	3	Clean
6	12 on 75	159	243	2	Oxide
7	12 on 75	167	277	3	Oxide
17	7	174	306	5	Clean
18	7	175	323	6	Clean
19	7	171	312	6	Clean
AMS 4429 Sand Cast		145	248	2	

Table 4 - Tensile properties relating to T6 (Air Cool) specimens

Specimen No.	Section thickness mm	0.2% PS MPa	TS MPa	Elongation %	Comments fracture
8	12 on 75	143	270	8	Clean
9	12 on 75	135	246	5	Oxide
10	12 on 75	143	254	5	Oxide
11	7	153	294	8	Oxide
12	7	155	265	5	Clean
13	7	150	270	5	Clean
AMS 4429 Sand Cast Minimum		145	248	2	Clean

Fig.1 - Lynx Helicopter Gearbox Cover & Case (RZ5)

Fig.2 - Sikorsky S92 Main Gearbox ElektronWE43

Fig.3 - F22 Fighter – Engine mounted Auxiliary drive (EMAD) ElektronWE43

Fig.4 - F22

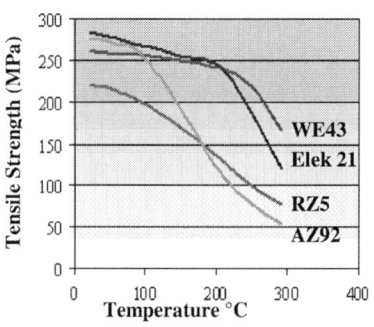

Fig.5 - Tensile strengths of Sand Cast Magnesium Alloys

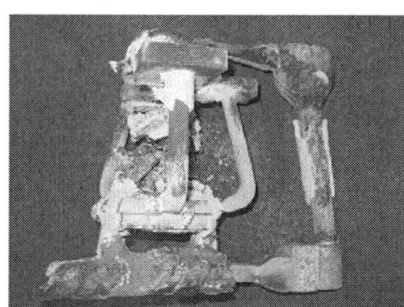

Fig.6 - Cti Shell. No mould purge. Shell temp 500°C. Pouring temp 760°C.

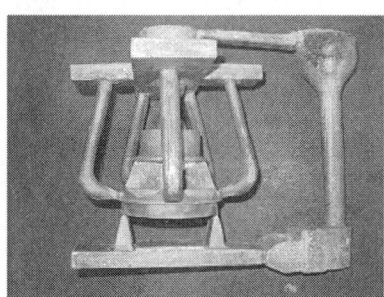

Fig.7 - Cti shell. Mould purged with SF6. Shell temp 500°C. Pouring temp 730°C

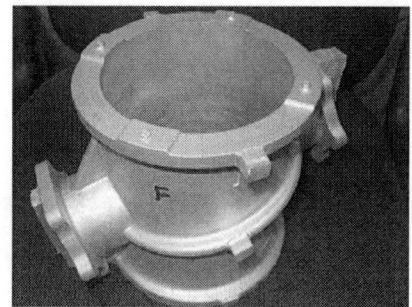

Fig. 8 - Thick section Casting in Elektron 21

Magnesium Alloy R&D Challenges – Aerospace Spinoff

S. Sundarrajan
Defence Research and Development Laboratory (DRDL)
Hyderabad, India

Abstract

Magnesium alloy castings are emerging as major players in global business by replacing aluminium alloys in aerospace and automobile applications. International trend indicates upward demand for Magnesium alloy die castings. Indian aerospace programme has gone in for designing Magnesium components directly and establishment of production technology through networking of industries and research institutions. The paper deals with an outline of applied experimental research carried out on casting characteristics like fluidity, mould filling ability and shrinkage behaviour and implementation of recommendations by adopting Taguchi techniques of design of experiments. Experimental findings have very good practical significance for reduction of rejection levels. The Indian scenario on Magnesium Research provides an outline of R&D capabilities established in the country. Spin off recommendations as strategies for future have been provided for value added and commercial products.

Key Words Magnesium alloy, casting characteristics, spinoff

Introduction

Magnesium alloy castings are emerging as major players in global business by replacing Aluminium alloys in aerospace and automobile applications. Many components in transportation equipments, consumer products, machine tools, electronic products, construction and material handling sectors are currently produced out of Magnesium alloys and there are many more identified as potential users of Magnesium castings [1]. The growing market is due to the lower weight, good damping capacity and excellent thermal diffusivity characteristics. Indian missile programme provided the nucleus for the growth of R&D in this field with demanding requirements from the component designers. While the global attention is on substituting Aluminium alloys with Magnesium alloys, Indian approach has been towards the design of components with Magnesium alloys, thus exploiting its full technical potential for aerospace programme. Apart from carrying out applied R&D in the areas of product development and productionisation through technology transfer, the programme has also established a network of academic institutions, R&D laboratories and industries, thus laying down a strong technical foundation for techno commercial exploitation of spin off potential. This paper provides a review of R&D work carried out in this area towards product development for the missile programme and also outlines the Indian R&D capabilities. With global networking for commercial application, India has strategic advantages of serving the international community in this area.

Application Oriented Experimental Research Products Outline & Design Approach

The products for aerospace applications fall under the following categories.

- Cylindrical outer shells of aerospace vehicles mainly from weight reduction point of view
- Contoured plate structures for wings from vibration / flutter damping point of view
- Electronic package components for heat dissipation / thermal diffusivity
- Vibration fixtures and tables for elimination of unwanted modal vibration during environmental testing of aerospace components

Photographs of some of these components are given in Fig. 1.

The design for all these components follows AVP 32 aerospace standards, satisfying the requirements of operations at hot [+ 60°C] and cold [-40° C] temperatures, vibration, rain and electromagnetic protection requirements. The 'fail-safe' design calls for specification of defect levels as per radiographic standard ASTM E-155. While designing for operation life of 10 years, corrosion protection factors are taken into consideration. The electronic package housings contain printed circuit boards and other items which emit heat during operation. The heat, if not diffused out affects the overall performance of the system. In addition, high precision components like gyroscopes and accelerometers perform well only in particular thermal environment. Transient iteration optimisation analysis of thermal elements is carried out during thermal analysis. Conduction, convection and radiation parameters are defined by boundary conditions and the temperature acquired by the chips and heaters are calculated through experimental data.

Experimental Research

Based on the component requirements, a major research programme has been executed with studies on casting characteristics [fluidity, mould filling ability and shrinkage] as the focus. Process parameters like pouring temperature, moulding method, sand fineness, mould coats and alloy additions have been optimised using Taguchi method of design of experiments. Issues arising out of corrosion have been tackled by launching a time bound R&D on corrosion schemes and sequence of corrosion protection methods. Mg-Al systems form the basis for all investigations.

Salient research findings are outlined below:

Flow and Filling Characteristics : Large size thin walled castings (where length to thickness ratio is greater than 100) require adequate flow of liquid metal. Reproduction of contours is important for the electronic package castings having heat dissipation fins. Spiral fluidity test method (Fig. 2) and pin test piece (Fig. 3) have been used to evaluate the fluidity and mould filling characteristics. Aluminium addition up to 5% decreases fluidity and thereafter it increases whereas the mould filling ability values increase with increase in Aluminium addition [2]. The decrease in fluidity values is attributed to change in solidification morphology from exogenous to endogenous shell up to 5% and to endogenous pasty subsequently. Investigations also revealed that the fluidity and mould filling ability are improved by using air setting sand made with fine sand and alumina based mould coats and with increase in pouring time.

Shrinkage Characteristics: Volume deficit and its distribution in to macrocavities, surface sinks, volumetric contraction and shrinkage porosity have been determined experimentally for cylindrical, cubical and rectangular shapes for Mg-Al alloys (Fig. 4) with increase in Aluminium content, the total volume deficit and macrocavities decrease and surface sink increases with no appreciable variation in internal porosity and volumetric contraction. With increase in pouring temperature, macrocavities decrease whereas the surface sinks and volumetric contraction increases with no appreciable variation in internal porosity. Macrocavities decrease with chilling power of the mould whereas the surface sinks and volumetric contraction increase with no appreciable variation on internal porosity [3]. Table 4 gives summary of recommendations favourable for reducing the defects.

Process Parameter optimisation through Taguchi Technique

Alloy additions, degree of superheat, fineness of sand used for moulding, moulding method and mould coats are the major process parameters which affect the casting characteristics. In order to optimise the process parameters to suit the shop floor conditions, Taguchi method of design of experiments is an effective procedure. Table 1 provides the details of major factors and their levels. Table 2 provides the optimum process parameters arrived at after carrying out analysis of variance and signal-noise ratio analysis. As far as shrinkage is concerned, the objective is to increase the macrocavity and decrease the shrinkage porosity. Taking this aspect alone into consideration, the desirable combination of factors are listed at Table 3 while increased Aluminium content upto 10 % is found to be advantageous from fluidity and mould filling ability points of view, Mg-2% Al was found to be ideal from shrinkage aspect. Green sand moulds and Zircon based coating are suitable for tackling shrinkage defects while air set mould and alumina coat are desirable from fluidity and mould filling ability points of view.

Practical Significance of Experimental Findings

During technology transfer phase of productionisation of shell castings, spongy shrinkage in the middle portion led to 80% rejection. This level was reduced to 20% by incorporating the following recommendations based on experimental study.

1. Operate at the lowest level of Aluminium content for AZ91C alloy.
2. Use of Zircon coating of green sand mould at localised areas.
3. Improved methoding practice to reduce pouring time.

Magnesium Research – Indian Scenario

CECRI and NML, leading national laboratories have developed production technology for Magnesium metal through electrochemical and silicothermic technology respectively. Production units are facing problem of high cost of metal due to low volume of production. DMRL has developed and characterised Mg-Li alloys and ZM21 products and productionised through MIDHANI. Rapid prototype activities have been initiated at GTRE for development of magnesium alloy cast products. HAL has developed fluxless melting technique and Regional Research Laboratory is working on Magnesium matrix components. Extensive work has been carried out at ISRO on machining and protective treatments. Studies on process maps for rolled plates cut of cast ZM21 blocks have been carried out at IISc. IIT (Chennai), IIT (Mumbai) and RRL (Tiruvananthapuram) have developed methoding software for processing of Magnesium cast components. CECRI has carried out extensive studies on corrosion characteristics of Magnesium alloy cast products in various environments and have come out with recommendations for sequence of corrosion prevention procedures for cast products. HAL is the leading industry manufacturing magnesium alloy components in its divisions at Bangalore and at Koraput. Network of the industries, R&D labs and academic institutions has given a strategic advantage to India for entering global market.

Spin off- Strategies for future

Valued added products: Increase in usage of Magnesium alloy components will be as value added products for aerospace applications. This trend will be there for all components, requiring weight reduction and damping capacity. Spin off in the areas of medical for stretchers, light weight cots and wheel chairs and for light weight reflecting head gears for surgeons performing brain operations may be expected during the next 2 years. Machine tool plates and vibration tables are major business areas. The limited application as on now may be due to limited international players in this business. India is in a better position to cater to the needs of Asia and South east.

Commercial Products: For automobile applications, Magnesium alloys have to compete with Aluminium alloys on cost aspects too. For heavy volume production like die castings, Magnesium alloys have an edge over Aluminium alloys due to the following reasons:

- Less energy for melting with fluxless melting energy efficiency is still increased due to reduction in melt losses.
- Tool life of dies for Magnesium alloy die castings is 3.3 times more than Aluminium alloy die castings resulting in one third reduction of cost per shot when die cost is amortised over the life of the tool.

The international trend shows an upward increase in demand pattern for magnesium die castings.

Conclusions

- Adapting Magnesium alloys at the design stage itself has better advantages than substituting aluminium alloys with magnesium alloys at the product stage.
- Application oriented experimental research has been carried out on Mg-Al alloys on fluidity, mould filling ability and volume deficit and its distribution as macrocavities, surface sinks, volumetric contraction and internal porosity.
- Process parameters have been optimised through Taguchi techniques. Using the results it was possible to reduce rejection level from 80% to 20% by selective incorporations of recommendation in problem areas.
- Indian scenario of Magnesium research indicates that there are institutions with core knowledge on Magnesium components development and production. Network of industries, R&D labs and academic institutions provide strategic advantage to India to enter global market.

- Spin off strategies for future are in two areas. The first is on development of value added products for aerospace and medical applications and the second is for cost effective alternative to Aluminium alloys for automobile and commercial applications.

References

1. Sundarrajan S, Ganapathy RS and Krishnadas Nair CG, *Strategies for Magnesium.* A report by the National Materials Policy Project, Source Book on Magnesium Alloy Technology, Dec 1995, pp 1 – 14.
2. Sundarrajan S, Rangarao NTV and Roshan H Md and Ramachandran EG *Evaluation of fluidity of alloys – a statistical approach,* Proc of Institute of Indian Foundrymen, 33rd Annual Convention Vol-2, 1984 pp 109 – 114.
3. Sundarrajan S, Roshan H Md and Ramachandran EG *Studies on shrinkage characteristics of Magnesium-Aluminium alloys,* Transactions of the Indian Institute of Metals, Vol. 37, No. 4, August 1984, pp. 373 – 381.

Acknowledgements

The author sincerely acknowledges the support and encouragement provided by many individuals and institutions. He gratefully remembers Late Prof. E.F. Emley, considered to be the father of Magnesium Technology, who technically guided him to enter the field and dedicates demanding requirements of research had come from Dr. V.K. Saraswat and Mr Prahlada, Programme Directors and currently Distinguished Scientists and Chief Controllers of DRDO. He records his thanks to Director, DRDL for his kind permission to present the paper.

E-mail address: s_sundarrajan@hotmail.com

Table 1 Major Factors and their levels

Factors	Level I	Level II	Level III
Alloy	Mg-Al4	Mg-Al7	Mg-Al 10
Degree of superheat	80°C	100°C	150°C
Sand fineness No	25AFS	57 AFS	75 AFS
Moulding Method	CO_2	Green	Air set
Mould coats	Alumina	Sulphur	Graphite

Table 2 Optimum Process Parameters
(Fluidity and Mould filling ability)

Alloy	Mg – 10% Al
Degree of superheat	150°C
Sand fineness	75 AFS
Moulding method	Air set
Mould coat	Alumina base

Table 3 Optimum Process Parameters
(Increased macrocavity and reduced shrinkage porosity)

Alloy	Mg – 2% Al
Degree of superheat	100°C
Sand fineness	75 AFS
Moulding method	Green sand
Mould coat	Zircon base

Table 4 Summary of Recommendations favourable for reducing defects

	FLUIDITY	MOULD FILLING ABILITY	SHRINKAGE
Aluminium addition	Beyond 5 %	Up to 10 %	Up to 2 %
Moulding method	Air setting	Air setting	Green sand [with localised zircon sand in problem areas]
Mould coats	Alumina based	Alumina based	Graphite based
Pouring temperature [degree of superheat]	High (150°c)	High (150°c)	Low (50°c)

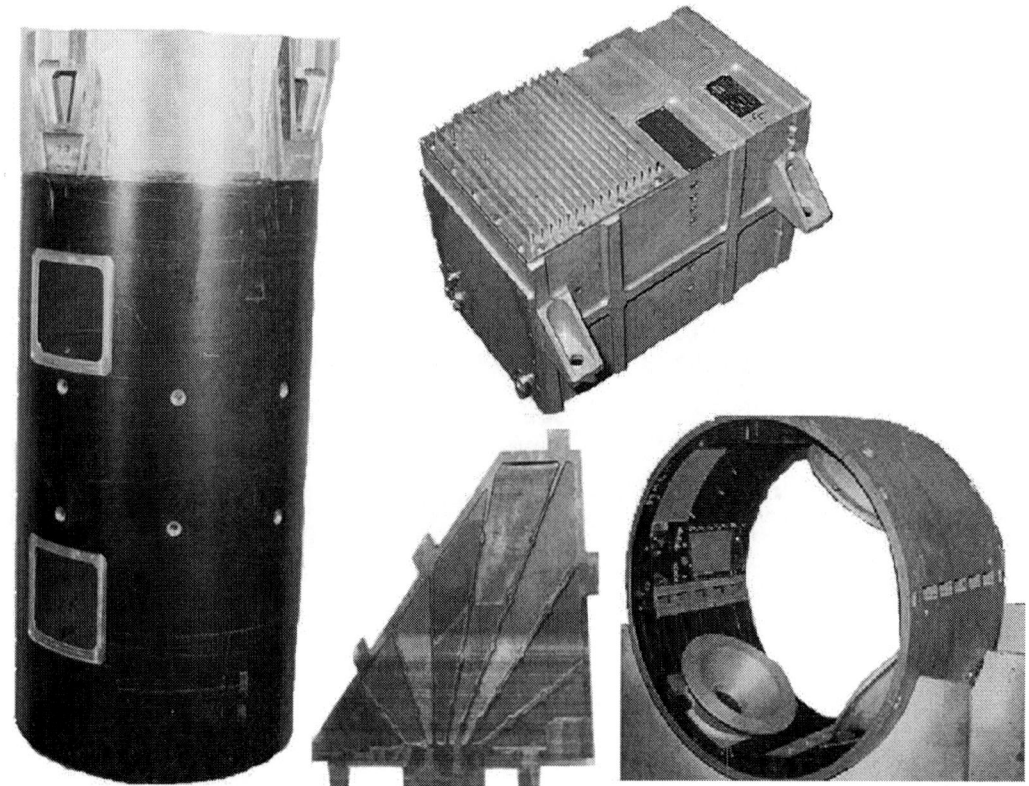

Fig. 1 Magnesium Alloy Components

Fig. 2 Fluidity Test Setup

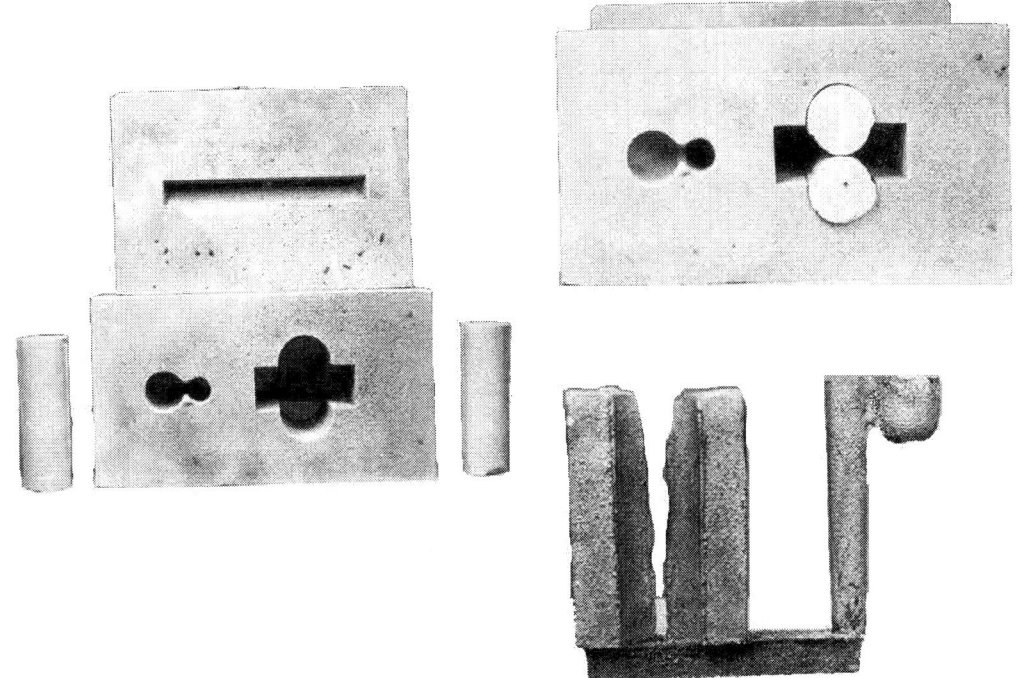

Fig. 3 Mould Fillers Ability Test Set Up

Fig. 4 Shrinkage Test Set Up

Alpha-case Controlled Titanium Casting

Si-Young Sung and Young-Jig Kim

Department of Advanced Materials Engineering, Sungkyunkwan University, Suwon, Korea

Abstract

The alpha-case formation mechanism was elucidated for the titanium casting. The α-case formation reaction between Ti and Al_2O_3 mold was examined in a plasma arc melting furnace. The reaction products were characterized by electron probe micro-analyzer and transmission electron microscopy. The α-case generation between Ti and Al_2O_3 mold was not able to be explained by the conventional α-case formation mechanism, which is known to be formed by the interstitials, especially oxygen dissolved from mold materials. However, from our experimental results and thermodynamic calculations, it was confirmed that the α-case is formed not only by an interstitial element but also by substitutional metallic elements dissolved from mold materials. Our newly established α-case formation mechanism will surely lead to a variety of significant applications of the α-case controlled Ti casting.

Key words : titanium, investment casting, alpha-case, interfacial reaction

Introduction

The major thrust in the area of titanium development has been aimed at achieving cost reduction rather than developing alloys with enhanced properties. The cost reduction can come from either innovative techniques for the fabrication or from the reduction in the cost post-treatment such as HIP, chemical milling, welding repair and machining [1]. The cost of techniques for the shape forming such as casting process itself, is only a small fraction of the total cost of a final component completion with the post-treatment after casting. This phenomenon is caused by the absence of casting technique which can omit the post-treatment. In the case of permanent mold, there are the heat checking, thermal deterioration, soldering and low productivity problems. The investment casting also has a drawback, called 'α-case' which needs to be removed by chemical milling and machining. On these reasons, the casting of titanium not a net shape forming but only a near net shape forming process. However, the cost problem cannot be overlooked any more for the widespread use of titanium, especially in the field of non-aerospace engineering.

In a viewpoint of cost efficiency, the investment casting of titanium could be the most economic net shape technology rather than permanent mold and vacuum die-casting, since the investment casting allows complexity, prototype, rapid tooling and high reliability. In order to develop the economic net shape forming technique of titanium by casting process, it is necessary to take a much closer look into the α-case formation mechanism. Until now, to avoid the α-case problem, the expensive ceramics, such as CaO, ZrO_2 and Y_2O_3 have been adopted as mold materials since the standard free energy changes of the formation of their oxides are more negative than TiO_2. Regardless of thermodynamic approaches, the α-case formation reaction still remains to be eliminated in the chemical milling processes. Therefore, the exact α-case formation mechanism must be examined and the development of α-case controlled mold materials is required for practical applications of titanium.

In this study, in an effort to optimize the investment casting of Ti alloys, the α-case formation mechanism was investigated through microstructures, hardness profiles, electron probe micro-analyzer (EPMA) mapping and transmission electron microscopy (TEM) analysis.

Experimental

The wax patterns for the examination of α-case reactions were made by pouring molten wax into a simple cylindrical silicon rubber mold (Ø15 × 70 mm). Subsequently, the patterns were inspected and dressed to eliminate any imperfection or contamination, and coated with Al_2O_3 slurry. The Al_2O_3 shell molds were dried at a controlled temperature (298 K ± 1 K) and a relative humidity (40% ± 1%) for 4 hrs. Dipping, stuccoing and drying procedures were repeated three times. After the primary layer coating, the patterns were coated with the back-up layers by the chamotte. To prevent the shell cracks, the dewaxing process of the shell mold was carried out at around 423 K and 0.5 MPa in a steam autoclave. Finally, the shell molds were fired at 1223 K for 2 hrs.

In order to prevent any contamination from refractory crucibles, the evaluation of α-case formation was examined in a plasma arc melting (PAM) furnace with drop casting procedure. The atmosphere in the PAM was controlled in the range 1.33×10^{-1} Pa by a rotary pump and then in order to minimize the effect of oxygen contamination, high purity argon was backfilled to the pressure of 4.9×10^{3} Pa. The schematic diagram of the apparatus for the evaluation of metal-mold reaction is shown in Figure 1. The microstructure in the reaction region of the castings was observed by Olympus PME3 microscopy, and its hardness was measured using a Mitutoyo MVK-H2 microvickers hardness tester with the condition of 100 g load and 50 μm intervals. For a closer examination of the alpha-case formation, the distribution state of the elements analysis at the alpha-case region was performed by SHIMADZU EPMA 1600 and the phase structures of reaction products were identified by JEOL JEM-3011 TEM.

Results and Discussion

1. Evaluation of α-case reaction

In this study, Al_2O_3 was selected for the mold material, because the standard free energy changes of the formation of its oxides are more negative than TiO_2. In addition, Al_2O_3 features suitable strength, permeability and collapse-ability, which can ensure dimensional accuracy of castings. Fig. 2 shows the microstructure and hardness profile on the surface of pure titanium investment castings poured into Al_2O_3 mold. The microstructure of a distinct reaction layer is about 200 μm thick on the interface between Ti and Al_2O_3 mold. In addiction to the reaction layer, there shows a hardened layer about 300 μm thick. The reaction layer and the hardened layer together are called the α-case of titanium castings. However, when the molten titanium (around 2,000 K) is poured into the Al_2O_3 investment mold, and if the α-case results between Ti and the interstitial oxygen, the reaction could be described as follows:

$$Ti(l)+O_2(g)=TiO_2(s) \qquad \Delta G^{o}_{TiO2}= -585.830 \text{ kJ/mol (1)}$$

$$4/3Al(l)+O_2(g)=2/3Al_2O_3(s) \qquad \Delta G^{o}_{Al2O3}= -1028.367 \text{ kJ/mol (2)}$$

$$Ti(l)+2/3Al_2O_3(s)=TiO_2(s)+4/3Al(l) \qquad \Delta G^{o}_{F}= +99.748 \text{ kJ/mol (3)}$$

The above calculations utilized the joint of army-navy-air force (JANAF) thermochemical tables [2]. According to the equation (3), the α-case formation by interstitial oxygen cannot occur spontaneously. Therefore, the reason why the α-case reaction is generated cannot be explained by the conventional α-case formation mechanism.

2. Alpha-case formation mechanism

For the clear examination of the α-case formation mechanism, the distribution of the elements of the α-case and its chemical composition were investigated by EPMA elemental mapping as shown in Fig. 3. On the α-case region, the oxygen element was uniformly distributed. Also, the Si element originated from the colloidal silica binder was scarcely detected on the surface. However, the concentrated Al contamination layer about 30 μm thick was detected on the interface. The EPMA mapping result shows that not only the interstitial oxygen elements but also the substitutional Al elements dissolved from the mold material affect the metal-mold reactions. Until recently, the effect of substitutional metallic element dissolved from mold materials have been ignored as negligibly small [3]. However, the Fig. 2 and 3 indicate that the effect of metallic elements cannot be overlooked any more.

The phase identification of the detected Al element will be the very core of α-case formation mechanism. The phases of the detected Al on the interface were examined by transmission electron microscopy (TEM). In order to observe α-case region precisely, the titanium castings specimen was grinded from inside to the surface until about 30 μm and final thinning was carried out using ion milling. Fig. 4 is the cross-sectional bright field TEM image of the α-case region. To examine what kind of phases are composed in the α-case region, C2 aperture was temporarily removed. Fig. 5 (a) shows ring and spot patterns on the TEM image without C2 aperture. The definite contrast of continuous ring pattern could not be found on the TEM image since the ring pattern was extremely small size polycrystalline TiO_2 phase. In the case of spot patterns on Fig. 5 (b), the contrast could be distinguished on the TEM image. And the indexed pattern was hexagonal close-packed (HCP) phase in the [2$\bar{1}\bar{1}$0] beam direction as shown Fig. 5 (b). The convergent beam electron diffraction (CBED) analysis was carried out for the verification of the phase of spot patterns with primitive cell volume method [4]. Primitive cell volume is characteristic material constant. So, the phase of interest can be easily identified by comparing the measured primitive cell volume of an unknown phase using CBED pattern with the known values of the possible phases. This method is accurate (error range of <10%) for the phase identification. In the CBED pattern where zero order Laue zone (ZOLZ) disk and high order Laue zone (HOLZ) ring are present, by measuring the distance of ZOLZ disk from the transmitted beam to the diffracted beam (D_1, D_2), an internal angle (ANG) between D1 and D2, and HOLZ ring's radius (CRAD), the primitive cell volume of unknown phase can be easily determined using the equation (4). Camera Length (CL) was calibrated with Au standard sample at 200 keV.

Fig. 6 (a) shows the CBED pattern of the HCP phase. There are two possible HCP phases of Ti and Ti3Al. The measured primitive cell volume (61.08 ang3, CL=521.4 mm, D_1=D_2=6.5 mm, ANG=60°, CRAD=23 mm) from CBED pattern corresponded to the theoreticvalue of Ti_3Al (66.60 ang^3). This phase identification was in accordance with EDS spot analysis as shown in Fig. 6 (b). The phase of the detected Al from EPMA mapping was Ti_3Al Phase. Thus, considering the microstructure, hardness profile, EPMA mapping and TEM, the α-case is not wholly TiO_2, but TiO_2 and Ti_3Al between Ti and Al_2O_3 mold.

In this study, from the synthesis of the microstructure, hardness profile, EPMA mapping and TEM, it could be confirmed that the α-case is formed by not only interstitial oxygen element but also substitutional metallic elements dissolved from mold materials as shown in Fig 7.

The α-case formation mechanism was applied to the development of α-case controlled SKK mold and verified by titanium castings with the α-case controlled SKK mold materials. Fig. 8(a) indicates that in the case of CaO stabilized ZrO_2, the expensive and thermally stable mold, the externals of Ti castings lost metallic luster as the result of the α-case formation reactions. However, in the α-case controlled mold, its characteristic luster of Ti was well preserved as shown in Fig. 8 (b). This external examination result of the α-case controlled mold is in a good accordance with the microstructure and hardness profile as shown in Fig 9.

Therefore, by the α-case controlled SKK mold which contained the α-case reaction products, the α-case formation mechanism can be verified for α-case controlled titanium casting.

Conclusions

Regardless of thermodynamic approach, the reason why the α-case reaction is generated cannot be explained by the conventional α-case formation mechanism. However, from the gathering up the threads of the microstructure, hardness profile, EPMA mapping and TEM, we suggest that the α-case is formed by not only interstitial oxygen element but also substitutional metallic elements dissolved from mold materials. The development of α-case controlled mold materials and clear and economic Ti casting can be possible on the basis of the mechanism. Our newly established α-case formation mechanism will surely lead to a variety of significant applications of the α-case controlled Ti casting.

References

1. Christoph L, Manfred P, Titanium and titanium alloys, Wiley-VCH, WeinHeim, 2003, ISBN 3 527 30534 3.
2. Chase MW, Davis CA, Downey JR, Frurip DJ, McDonald RA, Syverud AN, JANAF Thermochemical Tables, American Chemical Society and American Institute of Physics, New York 1985, ISBN 0 88318 473 7.
3. Suzuki K, *An introduction to extraction, melting and casting technologies of titanium alloys*, Metal. & Mater. Int., 7 No 6, Nov 2001, pp587-604.
4. Williams DB, Carter CB, Transmission Electron Microscopy, Plenum Press, New York, 1996, ISBN 0 306 45247 2.

Figures

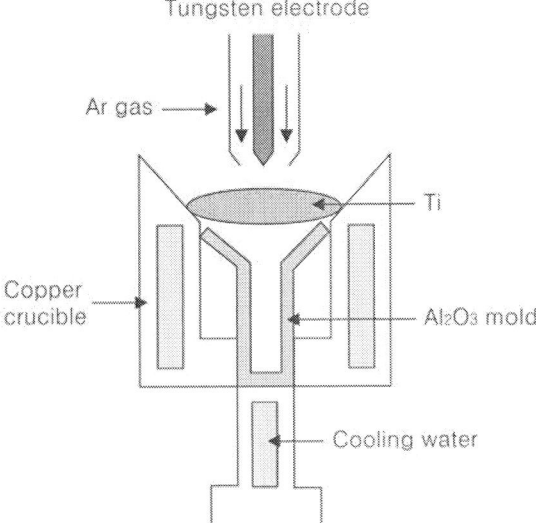

Figure 1. Schematic diagram of drop casting procedure of titanium with a plasma arc melting furnace.

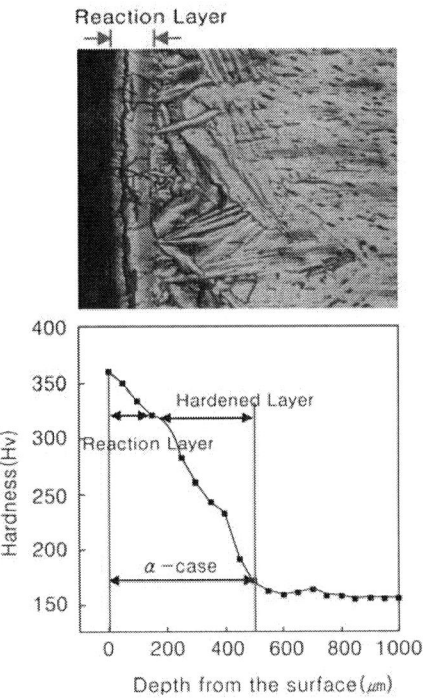

Figure 2. Microstructure of the interface between titanium and Al_2O_3 mold, and hardness profile.

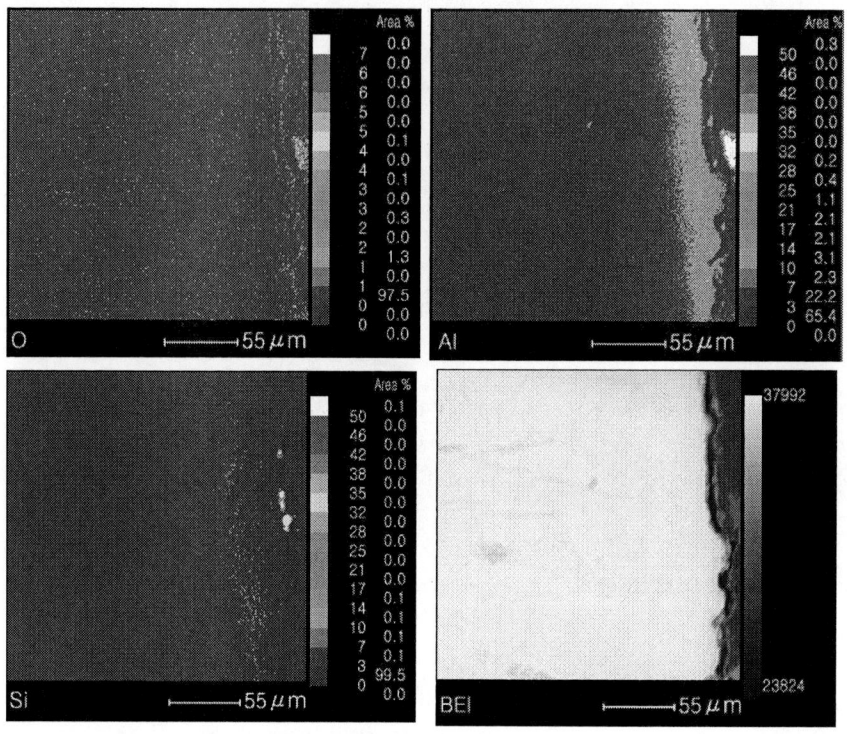

Figure 3. Comparison of elemental mapping images of O, Al and Si in Ti castings into Al_2O_3 mold and BEI image by EPMA.

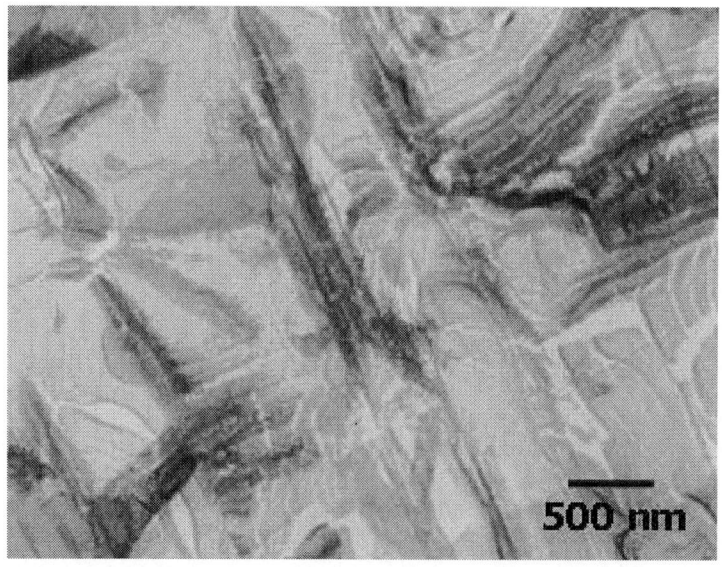

Figure 4. Bright field TEM image of the α-case region.

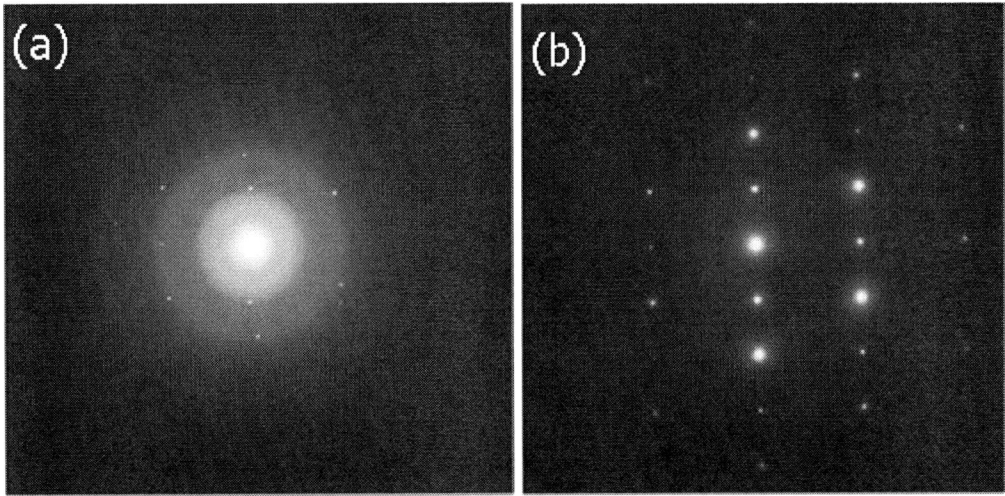

Figure 5. Results of analytical transmission electron microscopy of (a) ring and spot patterns on the TEM image without C2 aperture on the bright field image, and (b) spot pattern on the TEM image was HCP phase in the [$2\bar{1}\bar{1}0$] beam direction.

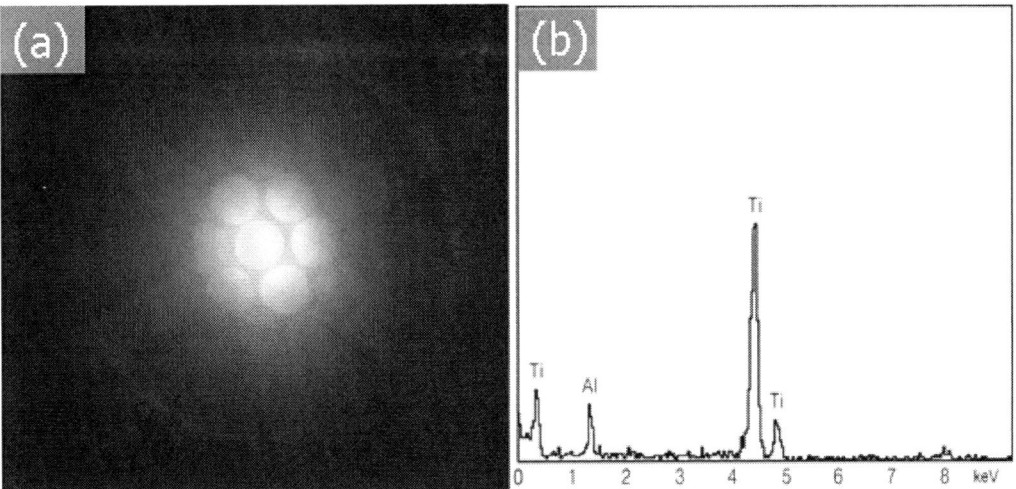

Figure 6. Convergent beam electron diffraction analysis with primitive unit cell volume method (a) The measured primitive unit cell volume, 61.08 ang^3 (CL=521.4 mm, D_1=D_2=6.5 mm, ANG=60°, CRAD=23 mm, λ=0.0251 Å), and (b) EDS spot analysis result.

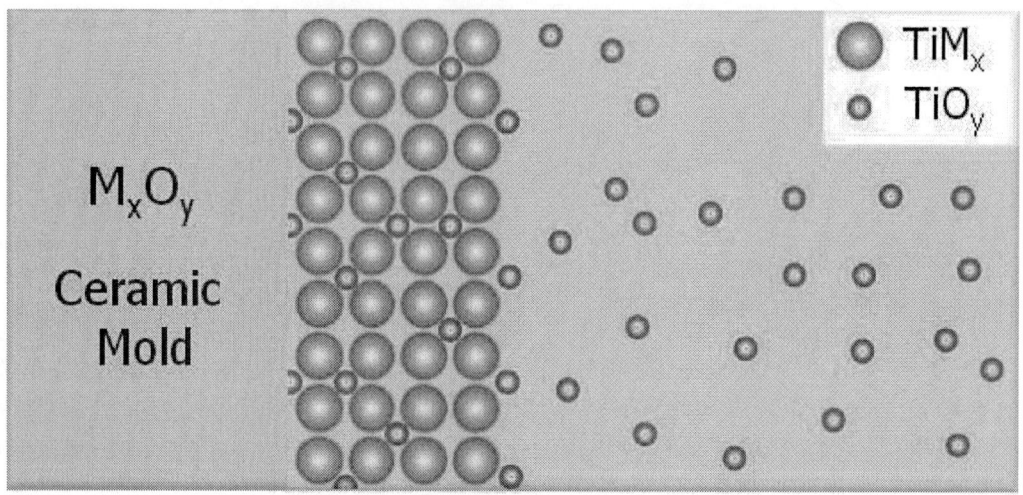

Figure 7. Schematic diagram of the interstitial and substitutional α-case formation mechanism of titanium castings.

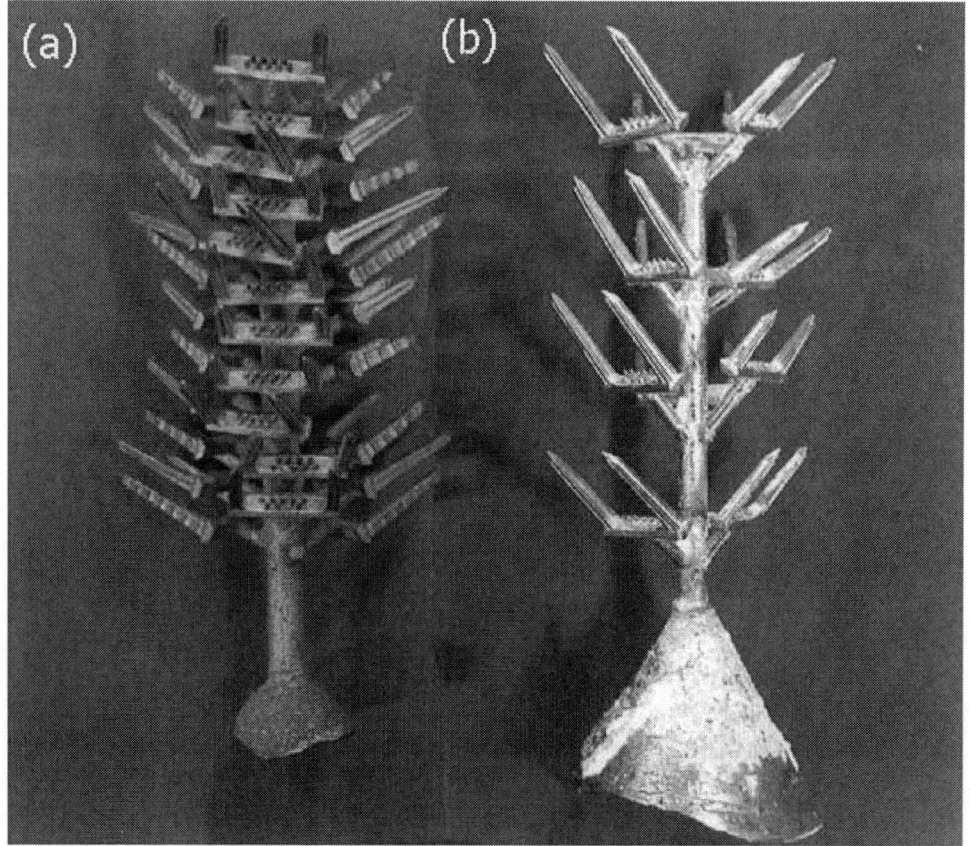

Figure 8. Comparison of externals of titanium castings (a) CaO stabilized ZrO_2 mold, and (b) the α-case controlled SKK mold.

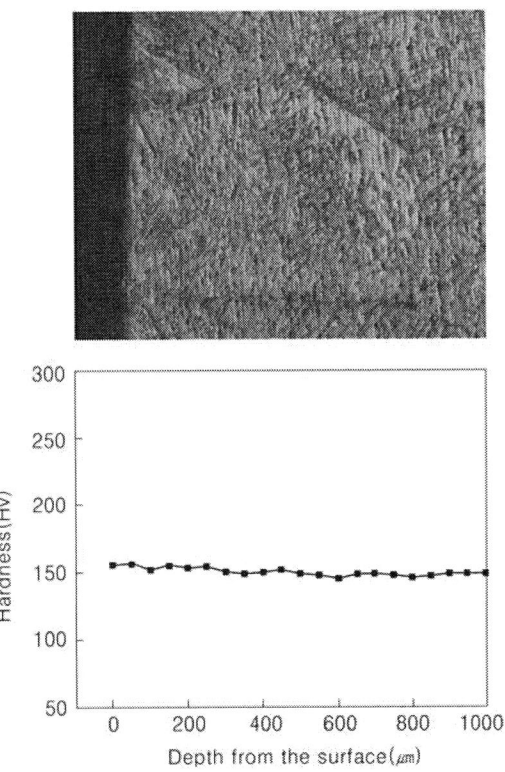

Figure 9 Microstructure and hardness profile between pure titanium and α-case controlled SKK mold.

Surface tension and Viscosity of Mg alloys

Soo-han Park and Bo-young Hur

K-MEM R&D Cluster-GSNU, AMRC, Division of Advanced Materials Engineering, Gyeongsang National University, Jinju, 660-701, Korea

Abstract

The rheological characteristics are the most important factors in casting process and metallic foam manufacturing especially. The surface tension (by the ring method) and the viscosity (by the rotation method) of molten Mg alloys (AZ91 and AM60) have been measured under pure Ar or SF_6 +CO_2 atmosphere. The surface tension and the viscosity of Mg alloys were investigated in the temperature range of 600-850°C, and the effects of the additional elements were investigated at the 660~680℃. The result show that the surface tension and viscosity of these alloys decrease with increasing temperature together. The viscosity of the AZ91 is about 6.12 ± 0.5 cp and AM60 is 4.91 ± 0.4 cp. The effect of additional elements has the tendency that is the surface tension decreased and the viscosity increased.

Keywords: *metallic foam, surface tension, viscosity, Mg alloys.*

Introduction

Thermopysical properties are very important factors in casting process. Such as molten temperature, pouring temperature, density, viscosity and surface tension in molten metals. Many metallic foam produced by Directly Melt Foaming method have coarse and irregular cell structures. A primary current aim is to fabricate foams with more uniform structure and cell size. It is important to understand the mechanisms and factors controlling. For the control of the bubble in molten metal, such as birth (foaming agent, TiH_2), life, death, shape and size in molten metal, the physical properties of liquid metal which have great influence on fabricating properties of metal foam must be given adequate attention[1]. Thus this paper will investigate the bubble behavior of the molten metal and it's the most important two parameters: surface tension and liquid viscosity.

These two factors are considered with two liquid mechanisms operating in foam. The first is gravity-driven melt flow from the top to base of foam column. The second is capillarity-driven melt flow from cell face to plateau borders. This leads to cell face thinning and often to cell face rupture [2,3]. Therefore, we investigated relation between driving force for a melt flow and two parameters to make metal foam with fine cell structure.

In this experiment, the drop weight method and the rotational method were used because of simple measurement method. They are also directly applicable to the fabrication process. Viscosity and surface tension of Mg alloy, which is used as foaming material, were investigated. Influences of temperature to viscosity and surface tension were studied.

Therefore in this paper, the rheological characteristics, namely viscosity and surface tension of Mg-Al alloy (AZ91 and AM60) were investigated in function of temperatures. Of course in the actual foaming process the appearance of the rheological characteristics are very much influenced from the existing of particles [4]. However the fundamental rheological data do not lose the importance.

Experimental

Surface tension was measured by the drop weight method, which applies the capillary phenomenon to measure the maximum force and wetting angle when the ring is pulled out from the melt surface [5,6]. Fig. 1 shows a schematic illustration of the experimental apparatus for the surface tension measurement.

Viscosity was measured by the rotational method, which measures the viscosity by calculating the torque acting on the rotation rod by the melt [5,7]. In this experiment, the high purity argon or SF_6+CO_2 gas sealing was used to prevent surface oxidation. The flowing rate was set to 25(ℓ/min). The temperature range of measurement was set from 600°C to 850°C for Mg alloys. The temperature was measured by Pt-Rh thermocouples, which were put them into the bottom and the side of the crucible.

The maximum force (F_{max}) that is measured by the ring method can be recalculated to a surface tension (σ_{st}) by equation (1).

$$\sigma = \frac{F_{max}}{4\pi R \cdot \cos\theta} \cdot f\left(\frac{R^3}{V}, \frac{R}{r}\right) \qquad (1)$$

Where $4\pi R$ is wetted length, F_{max} is the total maximum force, f is the Harkins Jordan factor [8] and θ is the wetted angle. Viscosity (η_{visco}) can be calculated from the measured torque (T) on the rotation rotor.

$$\eta_{visco} = \frac{15T}{\pi^2 r_1^3 N \left(\dfrac{r_1}{ab} + \dfrac{h}{a}\right)} \qquad (2)$$

Where T is the measured torque and N is the revolutions per min of the rotor. r_1 is the radius of rotor : 26mm, a is side gap : 4mm, b is bottom gap : 5mm and h is wetted height : 100mm.

In addition, to compare the foamability among Mg alloys, foaming tests were carried out. The testing conditions were taken as table 1.

Results and discussion

Fig. 2 and 3 shows the measured surface tension and viscosity of AZ91 and AM60 Mg alloy with the change of temperature. The comparison between the measured data of AZ91 and AM60 magnesium alloys is shown as well. As shown in Fig. 2, the surface tension of AZ91 is about 473mNm^{-1} near the melting point, and with increasing temperature, it decreased lineally following the relationship of σ_{st} = 473 - 0.545(T-T_m), and the surface tension of the AM60 follows the relationship σ_{st} = 557 - 0.468(T-T_m). Compared with reference [6], the surface tension of pure Mg corresponds to the data measured under slight surface monolayer existence, not ultra height vacuum. As the result, error range of data is comparatively wide. The surface tension of AZ91 alloy was lower than that of AM60 alloy. It may be caused by formation of impurity elements and aluminum atom cluster having higher melting point than magnesium. The decreasing tendency of surface tension for increasing temperature is similar.

Fig. 3 shows the variation of the viscosity of AZ91 and AM60 Mg alloys with the change of temperature. As shown in Fig. 3, the viscosity of AZ91 near the melting point is about 6.12 ± 0.5 cp and the viscosity of the AM60 Mg alloy is 4.91 ± 0.4 cp. The viscosity value of AZ91 was higher than that of the AM60, which may be caused by adsorption behavior of solute (aluminum and zinc) in liquid AZ91 [5]. In the case of Mg alloys, it rapidly changes at near the melting point.

Mg alloy is satisfied with required value for metal foam fabrication. Viscosity value of optimal conditions for the metal foam manufacturing is about 10~14[cp]. In the case of Al and alloys, it is possible that the optimal conditions of the viscosity can be obtained through controlling the amount of adding elements (Ca or Mg) [9-12]. But the interrelation of the surface tension and viscosity shows that Mg alloy needs proper selection of addition elements, as a result of present research

The foaming test results are shown Fig. 4. AZ91 foam is produced by adding Ca to increase the melt viscosity and blowing agent (TiH_2) to generate gas at 650°C. Because hydrogen can react with Mg at high temperature, more amount of hydrogen generated by TiH_2 is needed, so 3wt.% TiH_2, which is twice as much as that used for Al foaming process, is added.

The foaming test results are shown Fig. 4. AZ91 and AM60 foam is produced by adding Ca to increase the melt viscosity and blowing agent (TiH_2) to generate gas at 650°C. Because hydrogen can react with Mg at high temperature, more amount of hydrogen generated by TiH_2 is needed, so 3wt.% TiH_2, which is twice as much as that used for Al foaming process, is added.

Fig. 4(c),(d) shows the photographs of foamed AZ91 and AM60 of a horizontal section. Compared with foamed Al, Mg alloys foamed by optimal condition have very bad pore structures. Fig. 4(a),(b) shows photographs of the foamed pure Al and alloy column. According to the previous study, optimal conditions of pure Al and alloys have viscosity of 10~14 cp and surface tension of 550~650mNm^{-1} [9-12].

The specimen heights increased to more than eight times for Al and alloys and three times for AZ91 with Ca and two times for AM60 that of the original column height. The porosity including skins of the foamed columns was evaluated by Archimedes' principle to be (a) 0.92, (b) 0.90, (c) 0.51 and (d) 0.63, respectively. A few larger pores of about 4mm exist at the center of Al and alloy though the remaining small pores of less than about 3mm are dispersed uniformly around them.

The AZ91 and AM60 alloy foams have coarse and shows big pore and more irregular cell structures than foamed Al and alloys. Foamed Mg alloys satisfied viscosity level, but surface tension is too low. We can see that foamed Mg alloys has ruptured and coalescence pores

Conclusion

The purpose of this study was to measure the physical properties of molten Mg alloys. The results are:

The surface tension of AZ91 and AM60 followed the relationship $\sigma=473-0.545(T-T_m)$ and $\sigma=557-0.468 (T-T_m)$.

The viscosity of the AZ91 and AM60 is about 6.12±0.5cp and 4.91±0.4 cp respectively.

The porosity including skins of the foamed columns was evaluated by Archimedes principle to 0.51 and 0.63, respectively. A few larger pores of about 4mm exist at the center of Al and alloy though the remaining small pores of less than about 3mm are dispersed uniformly around them.

References

1   J. Banhart, M. Ashby and N. Fleck: *Cellular Metals and Metal Foaming Technology* (Verlag MIT Publishing 2001).
2   H. Fusheng and Z. Zhengang, J. Mater. Sci. 34 (1999) 291.
3   M. Meier, D. Hille and G. Wallot, Cellular Metals: Manufacture, Properties, Applications, MIT-Verlag Publication, 2003, p. 65.
4   J. Banhart, Journal of Metals 52 (2000) 22-27.
5   T. Iida and P. I. L. Guthrie: *The Physical Properties of Liquid Metals* (Clarendon press. Oxford 1988).
6   J. P. Anson, R. A. L. Drew and J. E. Gruzleski: Met. & Mater., Trans. B, 30 (1998), pp. 1999
7   Y. Shiraishi: J.of Kor. Inst. Met & Mater., Vol. 25 (1987), pp. 11
8   W. D. Harkins and H. F. Jordan: J. of Am . Chem. Soc., Vol. 52 (1930), pp. 1751
9   B. Y. Hur, H. J. Ahn, D. C. Choi and S. Y. Kim: Limat (2001), pp. 671
10   B. Y. Hur, S. H. Cho and K. B. Kim: Proceedings of Fall Conference. Vol. 1 (2001), pp. 246
11   B. Y. Hur, H. J. Ahn, D. C. Choi, S. H. Cho, K. D. Park, Y. J. Kim and S. H. Jun: Proceedings of the Symposium on Solidification Process of Metals (2000), pp. 87
12   S. Y. Kim, B. Y. Hur, C. K. Kwon, D. K. Ahn, S, H. Park and A. Hiroshi: Proceedings of the 65th World Foundry Congress (2002), pp. 499
13   J. W. Gibbs: *Thermodynamics* (The Collected Works of J. W. Gibbs, vol. I, Yale Univer. Press, New Haven, CT 1948).
14   P. Kozakevitch: J. of Met. (1969), pp. 57
15   D. Skupien and D. R. Gaskell: Met & Mater., Trans. B, Vol. 31 (2000), pp. 921
16   H. C. Birkman: J. Chem. Phys., Vol. 20 (1992), pp. 571
17   Y. J. Lee and S. H. Yi: J. of Kor. Inst. Met & Mater., Vol. 35 (1997), pp. 8

Acknowledgements

This work was supported by grant No. RTI04-01-03 from the Regional Technology Innovation Program of the Ministry of Commerce, Industry and Energy (MOCIE)

Tables

Table 1. The conditions of foaming test.

Thickening	Agent	Ca (1.5wt% add.)
	Stirring	15minutes
Blowing	Agent	TiH_2 (3wt% add. to Mg alloy)
	Stirring	20 Seconds
Temperature	Thickening & Blowing	650°C for Mg alloys
	Curing	10min. at 500°C for Mg alloys

Figures

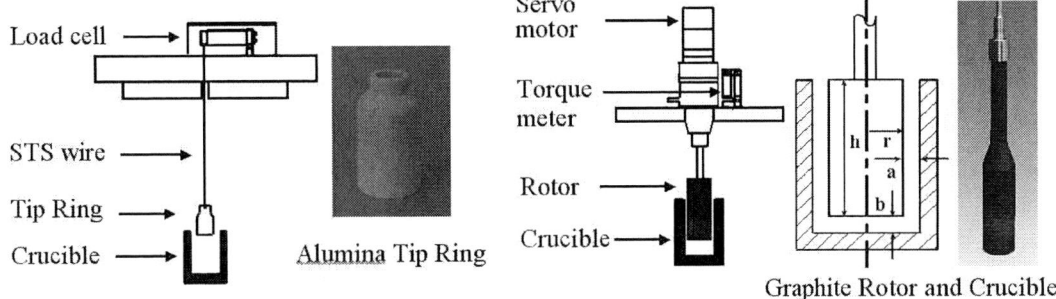

(a) Surface tension measurement apparatus.   (b) Viscosity measurement apparatus.

Fig. 1. Schematic diagram of the apparatus used for the surface tension and viscosity measurement.

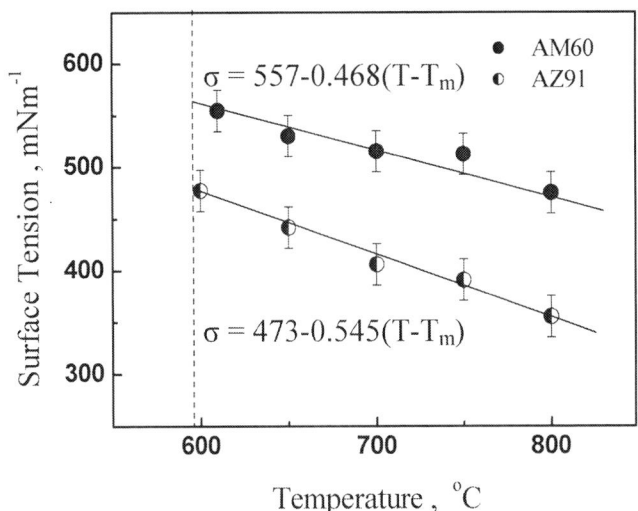

Fig. 2. Temperature dependencies of surface tension of Mg alloy.

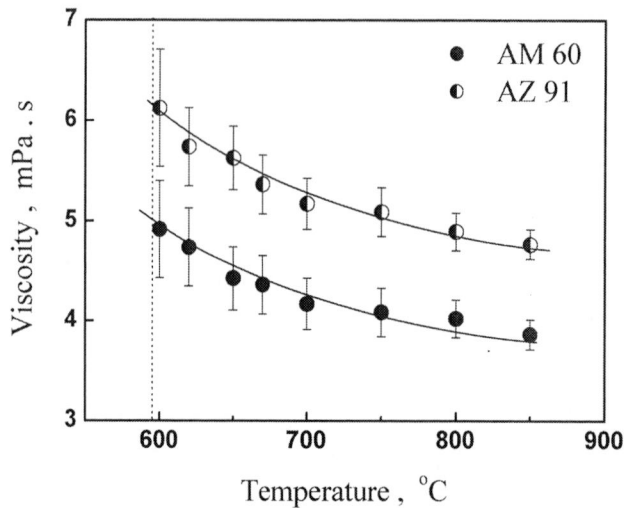

Fig. 3. Temperature dependencies of viscosity of Mg alloy.

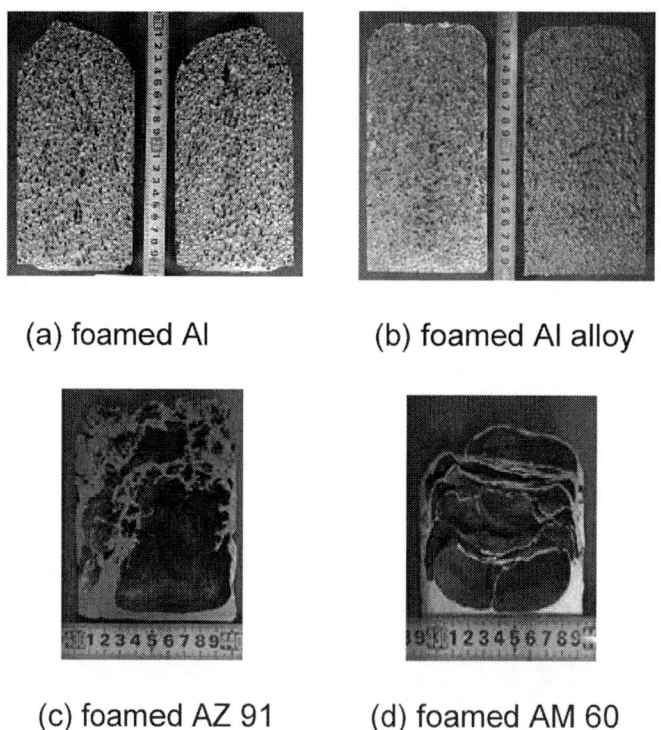

(a) foamed Al   (b) foamed Al alloy

(c) foamed AZ 91   (d) foamed AM 60

Fig. 4. Photographs of foamed Al and Mg alloys column.

Mechanical Properties and Microstructural Evolution of Rheo-diecast AZ91D Magnesium Alloy During Heat Treatment

Y Wang, G Liu and Z Fan

Brunel Centre for Advanced Solidification Technology (BCAST),
Brunel University, Uxbridge, Middlesex, UB8 3PH, UK

Abstract

Mechanical properties and microstructural evolution of rheo-diecast (RDC) AZ91D alloy during heat treatment (HT) have been studied. The semi-solid processed alloy exhibits a unique solidification microstructure featuring the uniformly distributed primary α-Mg globules, the fine secondary α-Mg grains and the β-$Mg_{17}Al_{12}$ network. Compared with high-pressure die-cast AZ91D alloy, the RDC alloy displays a faster solution response and accelerated precipitation during HT. A new HT schedule Tx, has been developed, which was carried out at a temperature close to the solvus to break up the β-phase network. HT offers improvement to mechanical properties and the Tx treatment gives a superior combination of strength and ductility. The accelerated HT kinetics for the RDC alloy and the lower temperature and shorter time of the Tx indicate a considerable reduction in HT cost.

Key words: Semi-solid processing; Rheo-diecasting; Magnesium alloys, Heat treatment

Introduction

AZ91D offers a good combination of castability, mechanical strength and ductility [1]. Currently, the applications of the Mg alloys are mainly achieved by high-pressure diecasting (HPDC), which is characterised by high efficiency, high production volume and low production cost. However, the HPDC components contain a substantial amount of porosity due to gas entrapment during die filling and hot tearing during the solidification in the die cavity [2]. Such porosity not only deteriorates mechanical properties, but also denies the opportunity for property enhancement by subsequent heat treatment. A new semisolid processing technology, rheo-diecasting (RDC), has been recently developed by BCAST at Brunel University [3-5] and offers advantages, such as extremely low porosity, fine and uniform microstructure and enhanced mechanical properties [6]. More importantly, the resulting RDC products have close-to-zero porosity and therefore can be heat treated [6,7]. This opens up new opportunities for die-cast Mg alloys in high performance applications. Previous studies have been extensively carried out on kinetics and mechanisms of the β-$Mg_{17}Al_{12}$ precipitates for Mg-Al based alloys, including AZ91 [8-13]. During aging, the β-phase precipitates out from the super-saturated α-Mg solid solution in two forms, discontinuous precipitation (DP) and continuous precipitation (CP) with the two types of morphology competing with each other depending on Al content and aging temperature [11]. DP occurs preferentially along the grain boundaries at the early stage. The cellular growth of alternating layers of the β phase and near-equilibrium magnesium matrix in the DP regions stops relatively early in the precipitation process, with CP forming in the remaining regions of the matrix grains [13]. It is known that different casting processes, especially semisolid processes, will result in a variation of the as-cast microstructure, which in turn influences dissolution and precipitation kinetics during subsequent heat treatment. An understanding of the as-cast microstructure and its response to heat treatment, for the RDC AZ91D alloy, is required. The aims of the present study are to characterise the kinetics of dissolution and precipitation of the β-$Mg_{17}Al_{12}$ phase during solution and aging treatments, and to assess the currently used heat treatment schedules. The optimised heat treatment procedure and a new heat treatment schedule can therefore be established for the RDC AZ91D magnesium alloy.

Experimental

The composition of the AZ91D alloy used in the present study was Mg-8.8wt%Al-0.67wt%Zn-0.22wt%Mn-0.03wt%Si-0.002wt%Fe. The details of the RDC process have been reported elsewhere [4-6]. In the RDC process, AZ91D ingot was melted at 675°C under the protection of N_2+0.5vol%SF_6 gas mixture and fed into a twin-screw slurry maker at 630°C. The slurry maker was operated at 589°C, corresponding to a solid volume fraction of 23% for the AZ91D alloy, and at a rotation speed of

500rpm, producing high shear rate and high intensity of turbulence. The semisolid slurry was sheared for 35 seconds and then transferred into a 280-ton cold chamber HPDC machine, which was used for casting standard tensile test samples. The alloy samples were then solution treated at 413°C (T4) and aged at 216°C (T6). The age-hardening response was determined by Vickers hardness measurements. Based on the experimental findings from the conventional heat treatment, a new procedure, denoted as Tx, was also carried out. This utilised a temperature close to the solvus for a short time, typically at 365°C for 2 hours. The mechanical properties of the alloy were characterised with tensile tests carried out with a Lloyd Instrument EZ50 machine, using the standard ϕ6mm samples with 50mm gauge length. The microstructural evolution was examined by optical microscopy (OM), scanning electron microscopy (SEM), transmission electron microscopy (TEM) and X-ray diffractometry (XRD). A Zeiss optical microscope was utilised for the OM observations and the quantitative metallography, while a Jeol JXA-840A scanning electron microscope was used to perform the SEM examinations. For TEM, 3mm diameter discs were hand polished and finally ion beam thinned. TEM was carried out on a JEOL FX2000 microscope operated with an accelerating voltage of 200kV.

Results

3.1 As-cast microstructure of RDC AZ91D alloy

Fig.1a shows the typical microstructure of the as-cast AZ91D alloy produced by the RDC process. The RDC alloy exhibits a homogeneous microstructure with the primary α-Mg globules being uniformly dispersed in the matrix. The solidification occurring in the RDC process can be considered in two stages. The first is the nucleation and growth of the primary α-Mg globules in the slurry maker, and the second is the solidification of the liquid portion of the slurry inside the die cavity, which involves the final eutectic reaction. Inside the slurry maker, the turbulence of the shearing imposed on the slurry resulted in not only a uniform dispersion of the solid α-Mg globules, but also a homogeneous distribution of the solute elements in the liquid portion of the slurry. This in turn gave a homogeneous nucleation and growth for the secondary α grains and the final eutectic. SEM revealed the fine secondary α grains and the β-phase network formed inside the die cavity, Fig.1b. TEM revealed that the β-$Mg_{17}Al_{12}$ phase was located along the secondary α grain boundaries and was virtually free of dislocations. A relatively high dislocation density was observed inside some of the α grains close to the β-phase.

3.2 Response to solution treatment

Fig. 2 shows the variation of Vickers hardness against solution time. It is seen that the hardness decreases faster for the RDC alloy than the HPDC alloy, indicating an accelerated dissolution of the β-phase for the RDC alloy. XRD revealed that the β-$Mg_{17}Al_{12}$ phase has been completely dissolved after 24 hours at 413°C, resulting in a single phase material. Microstructural evolution of the RDC AZ91D alloy during solution treatment involves four processes: (a) dissolution of the β phase, (b) formation of new α grains as a result of the β phase dissolution, (c) morphological change of the primary α-Mg globules and (d) grain growth. The dissolution of the β phase left sharp grain boundaries behind and the structure of the secondary solidification region eventually evolved into the new fine α grains. The α grains coarsened with increasing solution time. Quantitative metallography revealed that the mean grain size increased from 7.3 mm to 34.1 mm as the solution time increased from 0.5 hours to 24 hours. The main phenomenon concerning the primary α globules was diffusion of Al from the secondary solidification area, so that Al-rich halos were seen in OM and SEM micrographs. The halo eventually disappeared when the solution time was longer than 5 hours, with their initial globular morphology being replaced by traditionally well-defined α grains.

3.3 Response to aging

Fig.3 is the age-hardening curve of the alloy, solution treated at 413°C for 24 hours and aged at 216°C. The RDC alloy reaches its peak hardness at 5.5 hours, compared to 11 hours for the same alloy produced by HPDC. This indicates an accelerated age-hardening response for the RDC alloy, which corresponds to the faster solution treatment response. Table 1 lists the important features of the age-hardening curves. Fig.4 shows the microstructure of the RDC AZ91D alloy at under-aged and peak-aged states. The discontinuous precipitation (DP) of the β phase was seen to initiate preferentially at grain boundaries, with lamellar growth into the grains, Fig.4a. With the increase of aging time, the DP ceased and continuous precipitates (CP) were found to form in the remaining regions of the grains. At the peak aging time, there are lamellar DP cells spreading from the grain boundaries, and the fine CP needles inside the grains. In the over-aged condition, the microstructure was essentially the same as the peak aged state. In Fig.5, TEM bright field images clearly show the two types of the β precipitation in the peak-aged RDC alloy.

3.4 Microstructural evolution during Tx treatment

The precipitation of the β phase usually gives a relatively poor strengthening effect due to the direct transformation of equilibrium β precipitates without G.P. zone or intermediate phases occurring during aging. Solution treatment followed by aging (T6) usually improves strengths slightly with a considerable loss of ductility for the AZ91D alloy. In consideration of modifying the β phase morphology, a new heat treatment, Tx, was developed in the present study. Fig. 6 shows the breaking up of the β network, with the majority of the β particles being situated at the grain boundaries after Tx treatment at 365°C. A typical composite structure produced is presented. Due to the shorter treatment time and lower temperature, severe grain growth was prevented. This fine composite structure is expected to offer superior combination of strength and ductility.

3.5 Mechanical properties of RDC AZ91D alloy

Table 2 gives the mechanical properties of the RDC samples under as-cast and different heat treatment conditions. Compared with the as-cast condition (RDC), T4 heat treatment (solution) improves ductility substantially but decreases strength. In contrast, T6 (solution plus aging) improves the ultimate tensile strength (UTS) but with a reduction in yield strength and elongation. Furthermore, T5 (artificial aging) slightly reduces both strength and ductility of the as cast alloy under the experimental conditions of the present study. Tx treatment, however, gives the as cast alloy a further substantial improvement in ductility at no sacrifice of ultimate tensile strength (UTS) and with slightly reduced yield strength, offering an improved combination of yield strength, UTS and elongation [14]. Due to the low temperature and short time, an additional advantage of the Tx treatment is the reduction in heat treatment cost.

Discussion

The fluid flow inside the slurry maker is characterised by high shear rate and high intensity of turbulence. Consequently, both temperature and composition are uniform throughout the entire volume of the melt. Therefore, heterogeneous nucleation occurs throughout the entire volume of the liquid, and every nucleus will survive and contribute to the final microstructure. Under intensive forced convection, such nuclei will grow rapidly in a spherical manner until a certain solid volume fraction, corresponding to the semisolid processing temperature, is reached [15, 16]. Therefore, primary solidification in the RDC process offers semisolid slurry, in which fine and spherical solid particles are uniformly dispersed in a liquid matrix. Secondary solidification occurs inside the die cavity under high cooling rate ($\sim 10^3$K/sec), but without shearing. However, the remaining liquid was intensively sheared previously, and its temperature and composition should be uniform throughout the entire semisolid slurry. Once nucleation is trigged by the mould wall, heterogeneous nucleation occurs throughout the entire liquid with a much higher nucleation rate. The high nucleation rate and the high nuclei survival rate ensure the each nucleus can only grow to a limited size before the eutectic reaction cuts in, without reaching the morphological instability. Consequently, fine and equiaxial α-Mg grains (secondary) are produced.

The faster dissolution rate of the β phase for the RDC samples can be explained in terms of the size and the distribution homogeneity of the β phase. Some coarse lumps of the β-phase were found in the HPDC

AZ91D alloy, which take a longer time to dissolve completely. However, the uniform distribution and a finer β phase for the RDC alloy reduce the diffusion distance and increase the diffusion area, so that a shorter time for dissolution of β phase is needed for the RDC sample. During the solution treatment, dissolution of the β phase is accompanied with the redistribution of Al solute from the secondary α grains into the primary α-Mg globules. The dissolution of the β phase leaves sharp grain boundaries behind, and the initial secondary α-Mg morphology is transformed to take the shape of classic grains bound by sharp grain boundaries. The final grain size is dependent upon the solution temperature and time. Age-hardening curves, Fig.4, indicate an accelerated age-hardening kinetics for the RDC alloy compared to the HPDC alloy. Similar accelerated aging kinetics has been observed by Cerri et al [17] in their thixocast AZ91D alloy. They suggested that the faster solution kinetics was attributed to the higher degree of super-saturation for the semisolid processed alloy. However, such accelerated aging kinetics is not attributed to the higher degree of super-saturation due to much longer solution treatment time (24 hours compared to 2 hours) at a similar temperature used in the present study. Instead, the accelerated aging kinetics would be attributed to the finer and more uniform grain structure in the RDC alloy. Despite the grain growth after solution treatment at 413°C for 24 hours, the RDC alloy exhibited an average grain size as fine as 34.1μm in diameter, which was finer than that of the HPDC alloy, being measured to be 57.6μm after the same solution treatment. The finer grain structure provides more grain boundaries along which the DP of the β phase takes place, and gives more opportunities for CP to occur inside the grains, resulting in faster age-hardening kinetics. In contrast, there are some very coarse α grains in the solution treated HPDC AZ91D alloy, resulting from the inhomogeneity of its as-cast microstructure. It takes a longer time for the precipitation reaction of the β phase, either DP or CP, to complete in the coarse grains, and consequently results in an extended peak aging time.

The β network has been effectively broken up after Tx treatment and the remaining β grains were fine and uniform locating at the α grain boundaries, mostly at the triple grain boundary junctions, Fig.6b, forming a composite structure. It is obvious that the fine (average 2.8μm in size) and homogeneously dispersed β grains contribute to strengthening, although β phase is not an ideal reinforcement [18]. In addition, the fine grain structure of the secondary solidification area has been retained. The substantially improved ductility, together with the high level of strength, provided by the RDC process plus the subsequent Tx heat treatment, can promote wider applications of Mg alloys. More importantly, Tx is a single step process carried out at low temperature for a shorter time, which not only introduces less thermal stresses and increases the dimensional precision of the heat treated components, but also reduces the heat treatment cost. In practice, temperature and time for Tx can be selected to suit specific cast components, offering technological flexibility.

Conclusions

The RDC process produces a unique microstructure for the AZ91D Mg alloy, with the primary α-Mg globules, formed inside the slurry maker at primary solidification, being uniformly distributed in the matrix of the fine secondary α-Mg grains and the eutectic β-$Mg_{17}Al_{12}$, the result of secondary solidification.

RDC AZ91D alloy exhibits a faster dissolution rate of the β phase, which is attributed to the fine and uniform as-cast microstructure as well as the increased homogeneity in chemistry offered by the RDC process. Compared with HPDC AZ91D alloy, the aging kinetics of the β-$Mg_{17}Al_{12}$ precipitate for the RDC alloy is accelerated, as a result of the finer and more uniform α grain structure after solution treatment. Both discontinuous and continuous precipitates of the β phase were observed at the aging temperature of 216°C.

Mechanical properties of the RDC AZ91D alloy can be further improved by subsequent heat treatment. T4 improves ductility substantially but decreases strength, and T6 increases ultimate tensile strength but reduces yield strength and ductility. A new heat treatment Tx, developed in the present study, improve the mechanical properties of the RDC alloy. Ductility is increased and ultimate tensile strength is maintained, thus offering an improved combination of strength and elongation in comparison to T4, T5 or T6. More importantly, Tx is carried out at low temperature for a short time, therefore it not only decreases thermal stresses in alloy components but also reduces heat treatment cost.

References

1. Polmear IJ, Mater. Sci. Tech. 10 (1994) 1.
2. Friedrich H, Schumann S, J. Mater. Processing Tech. 117 (2001) 276.
3. Balasundaram A, Gokhale AM, In: Hryn J, (ed.), Magnesium Technology 2001, TMS, 2001. p 155.
4. Fan Z, Bevis SJ, Ji S, PCT Patent, WO 01/21343 A1, 1999.
5. Fan Z, In: Proc. 9th Inter Conf Al Alloys and their Appl, Brisbane, Australia, 2004. p. 1092.
6. Fan Z, Ji S, Liu G, Mater. Sci. Forum, 488-489 (2005) 405.
7. Fan Z, Liu G, Wang Y, J. Mater. Sci. 2005. In press.
8. Clark JB, Acta Metall. 16 (1968) 141.
9. Crawley AF, Lagowski B, Metall. Trans. 5 (1974) 949.
10. Crawley AF, Milliken KS, Acta Metall. 22 (1974) 557.
11. Duly D, Simon JP, Brechet Y, Acta Metall. Mater. 43 (1995) 101.
12. Bettles CJ, Mater. Sci Eng. A348 (2003) 280.
13. Celotto S, Acta Mater. 48 (2000) 1775.
14. Wang Y,Liu G, Fan Z, Scripta Mater. 2005. In press.

15. Fan Z, Liu G, Acta Mater. 53 (2005) 4345.
16. Das A, Ji S, Fan Z, Acta Mater. 50 (2002) 4571.
17. Cerri E, Barbagallo S, Cabibbo M, Evangelista E, In: Kaplan HI (Ed.), Magnesium Technology 2002, The Minerals, Metals and Materials Society, Warrendale, PA, USA, 2002, p.221.
18. Lu YZ, Wang QD, Ding WJ, Zeng XQ, Zhu YP, Mater. Lett. 44 (2000) 265.

Acknowledgement

The financial support from EPSRC (UK), Ford Motor Co and Magnesium Elektron Ltd (UK) is acknowledged with gratitude.

Table 1. Important features of the aging behaviour obtained for the AZ91D alloy*

Alloy Produced by	Vickers Hardness after T4	Vickers Hardness at Peak	Incubation Time (hour)	Time to Peak Hardness (hour)
RDC	57.3±2.5	81.7±3.2	<0.5	5.5
HPDC	57.1±2.4	79.5±3.0	~1.0	11

* Solution treated at 413°C for 24 hours followed by aging at 216°C for up to 144 hours

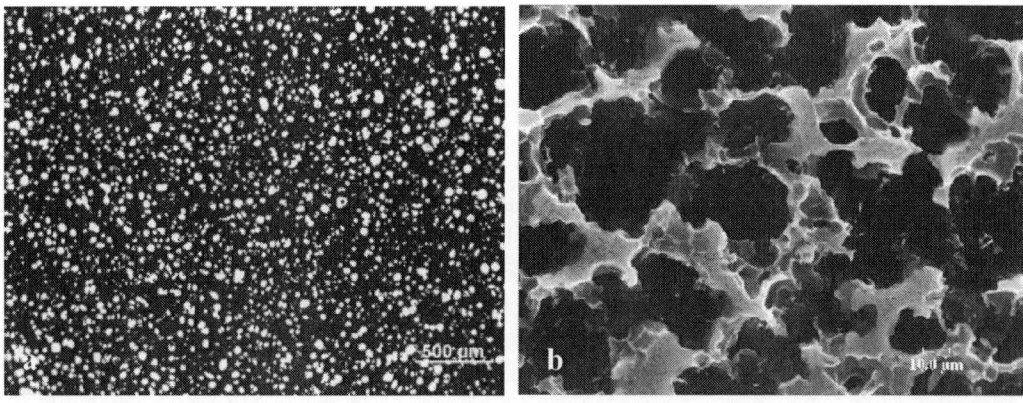

Fig.1 (a) Optical and (b) SEM micrographs showing the general solidification microstructure and the β-$Mg_{17}Al_{12}$ network, respectively, of the as-cast RDC AZ91D alloy.

Table 2. Mechanical properties of RDC AZ91D alloy heat-treated under different conditions.

Samples	Conditions	YS (MPa)	UTS (MPa)	Elongation (%)
RDC	As-cast	145	248	7.4
RDC+T4	413°C, 5 hrs	91	230	11.2
RDC+T5	216°C, 5 hrs	133	236	6.5
RDC+T6	413°C, 5 hrs + 216°C, 5.5hrs	134	255	6.7
RDC+Tx	365°C, 2hrs	132	249	9.1

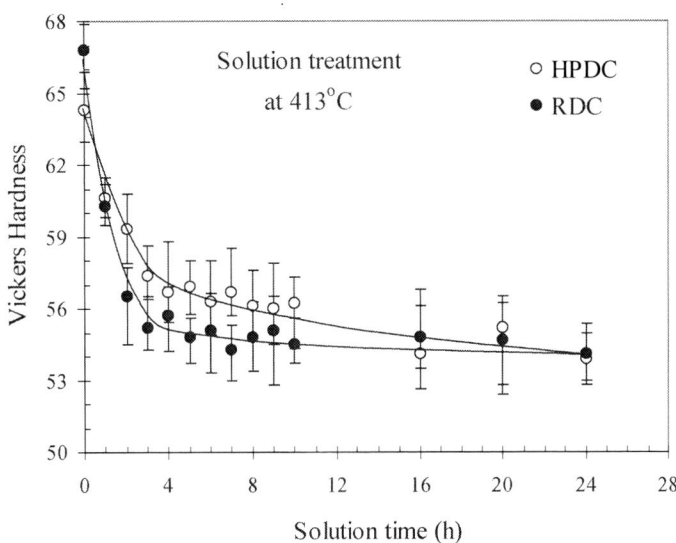

Fig. 2. Variation of the Vickers hardness of the AZ91D alloys against time during solution treatment at 413°C.

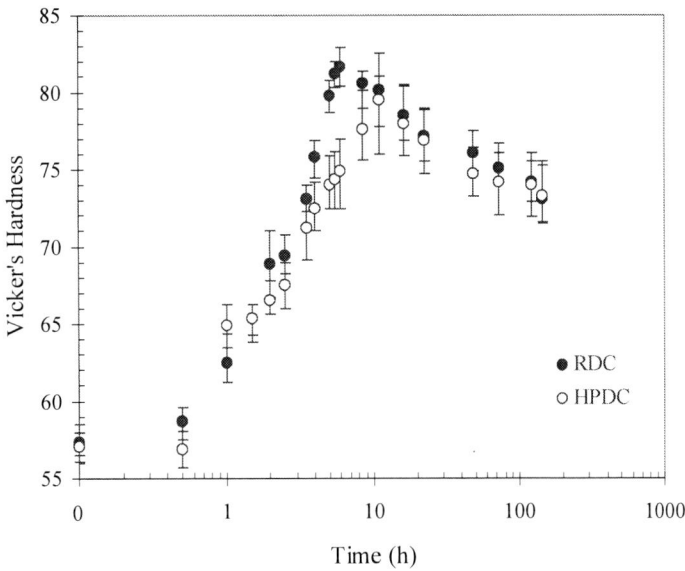

Fig. 3. Age-hardening curves for the AZ91D alloy aged at 216°C after solution treatment for 24 hours at 413°C.

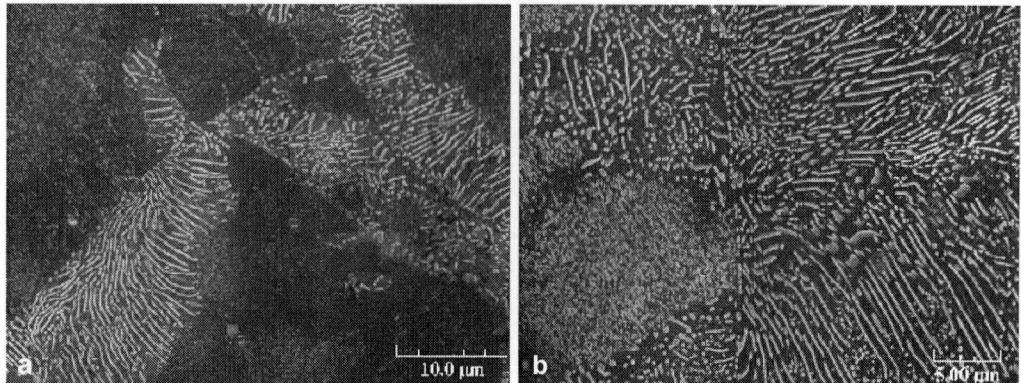

Fig. 4. SEM micrographs showing the microstructure of the RDC AZ91D alloy solution treated at 413°C for 24 hours followed by aging at 216°C for (a) 0.5 hours (under aged); (b) 5.5 hours (peak aged).

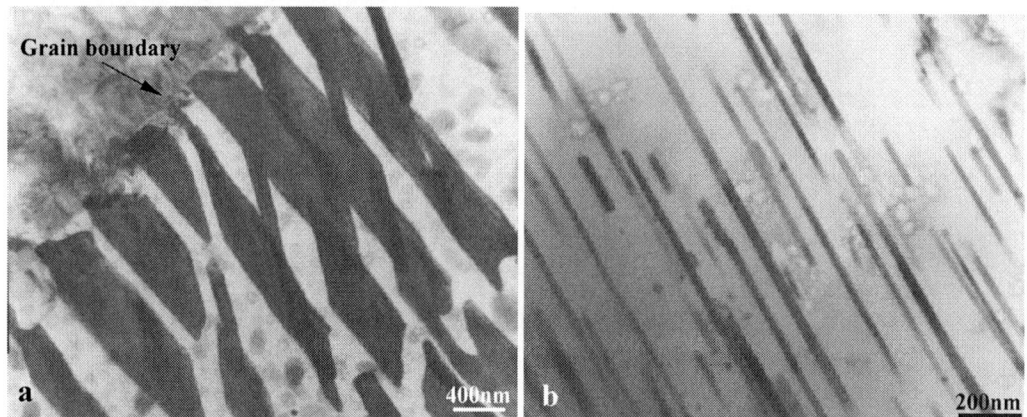

Fig. 5. TEM bright field images of the RDC alloy, solution treated at 413°C for 24 hours and aged at 216°C for 5.5 hours, showing (a) discontinuous precipitation and (b) continuous precipitation.

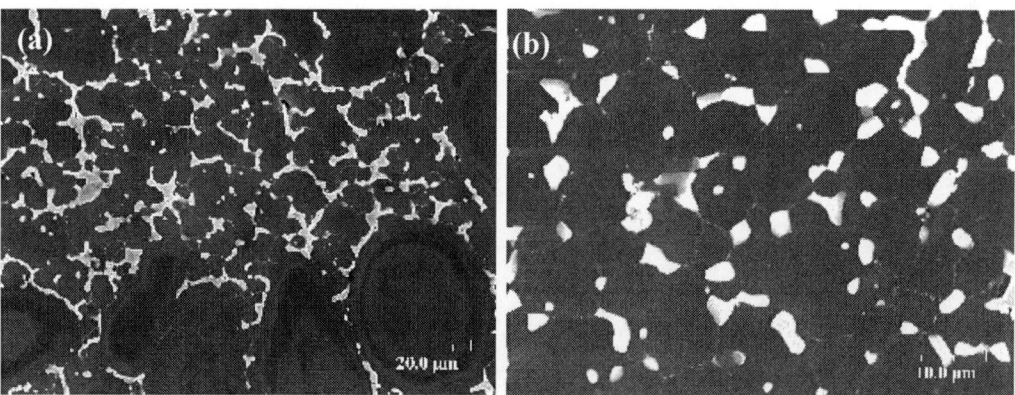

Fig. 6. SEM micrographs showing microstructure of the RDC AZ91D alloy after Tx treated at 365°C for (a) 1 hour and (b) 2 hours.

Effect of grain refinement on the mechanical properties of magnesium and its alloys

Yun Sung Han

Ulsan Innovation Industrial Agency, Ulsan Industry Promotion Technopark, Small & Medium Business Center, 758-2, Yeonamdong, Buggu, Ulsan, 683-804, Korea

Abstract

The majority of magnesium components is generally used in the automotive and electronic industries by conventional casting process. However, there is a strong need to develop high strength casting alloy for wide-spread application of Mg alloys. The present study was carried out to investigate the influence of Al-3%Ti-0.15%C and Al-5%Ti-1%B master alloys on the grain structure and mechanical properties of high purity magnesium, Mg-9Al-1Zn and Mg-6Al-0.35Mn alloy. It has been found that grain refinement was successful when using small addition of Al-3%Ti-0.15%C and Al-5%Ti-1%B master alloy on the AZ91D and AM60B alloy. The tensile strength and yield strength of AZ91D and AM60B alloys were increased with Al-3%Ti-0.15%C and Al-5%Ti-1%B master additions. However, mechanical properties of both alloys were decreased with 0.3w.%Ca addition.

Key words: Mg alloys, Grain refinement, Al-Ti-C, Al-Ti-B, AM60B

1. Introduction

Magnesium and aluminium alloys found applications in automotive industries for lowering the weight to reduce the fuel consumption and emissions[1]. Magnesium alloys have the longer mold life, faster cycle time than aluminium alloys. However, it is very difficult to produce sound casting due to its high oxidation during the melting. Recently, these problems were overcome using SF_6+CO_2 gas for protection the high melting oxidation and developing various new processes[2-4]. As the results, magnesium is presently used in a wide range of automotive applications such as steering wheel core, seat frame, transmission housings[5-8].

The most commonly used magnesium alloys are AZ91D and AM50/60. These alloys exhibit good castability and have reasonable room temperature mechanical properties and corrosion resistance in many environments. Thus, these alloys are mainly used to produce such parts as engine crankcase, instrument panel, airbag housing, which do not have as same requirements on relatively high strength and high toughness a automobile wheels in mechanical property[9,10]. Improving the mechanical properties of magnesium alloys is one of important research fields for us in order to enlarge their application scope in the automobile industry.

The grain size of a casing alloy has an important effect on its mechanical properties[11], thus grain refinement is an important proves for magnesium and its alloys. Accordingly, many studies on the grain refinement of magnesium alloys have been carried out[12-14]. Must of grain refining test was performed by some elements additions such as Ca, Si or Sr. But, the grain refining effect is still unclear[15]. Thus, a new and superior process of grain refinement is required. During the development of magnesium alloy, we pound that the grain of AZ91D and AM60B alloy casting was refined by Al-Ti-C and Al-Ti-B master alloy.

The present work is focused on the effect of grain refinement on the mechanical properties of AM60B and AZ91D alloy with Al-Ti-C, Al-Ti-B and Ca addition.

2. Experimental Procedure

The materials used in this study were high purity magnesium(HPMg), AM60B, AZ91D alloy and Al-Ti-C master alloy used for grain refinement. The chemical compositions of the alloys are given in Table 1. The Ca addition in AM60B and AZ91D alloys was supplied as an Mg-3%Ca master alloy.

The high purity magnesium, AZ91D and AM60B melts were prepared in a steel crucible using an electric resistance furnace under SF_6 gas. 10 kg of each sample was heated to 720 °C. After the samples were completely molten, the melt was poured into a preheated cylindrical shaped steel mould as shown in Figure 1. The experiments were repeated with the same procedure for various compositions obtained by adding an Al-3%Ti-0.13%C and Al-3%Ti-1%B master alloy. The melt was held for 10 minutes and stirred before pouring. The castings were produced by a directional solidification technique. The directional solidification apparatus consisted of a steel sleeve with inside dimensions of 100 mm (high) × ϕ 35 mm × 1.5 mm wall thickness. The sleeve was wrapped in ceramic wool to ensure good thermal insulation and minimise radial heat loss. A water-cooled copper chill plate formed the bottom part of the sleeve which was attached beneath the steel frame.

The grain structure was revealed by mechanical polishing and subsequently etched with prepared etchant(12 ml acetic acid, 4.2g picric acid, 10 ml H_2O, 70 ml ethanol). Mechanical test specimens for Mg alloys were melted in an electric resistance furnace under SF_6 gas at 720°C and cast into Y-block mold which were preheated to 250 °C. Tensile test was carried out at an initial strain rate of 1 (10^{-3}/s using Instron type test machine. Specimens for testing were prepared by turning to a cylindrical shape with 6mm diameter and 32 mm gauge length.

3. Results and Discussion

Grain refinement of magnesium and its alloys with Al-Ti-B and Al-TiC master alloys

The grain refinement of HP(high purity) magnesium was investigated by adding 0.15~0.3wt.% Al-Ti-C master alloys. As shown in Figure 2 (a), the microphotographs of alloys with 0.15 wt.% addition consist of two regions, namely a region of columnar grains grown from the chilled end and a region of coarse equiaxed grains below the columnar grains in the bottom section. In the case of the alloys to which 0.3wt.% master alloy added, the coarse equixed grain structure exists from the bottom to the top surface, as shown in Figure 2 (b). Figure 3 shows the general macrostructure of grain refined or non-grain refined AM60B alloys containing Al-Ti-C or Al-Ti-B master alloy. As shown in Figure 3 (a) non-grain refined AM60B alloy has large equiaxed grains. The grain refined AM60B alloys have fine equiaxed grain structures as shown in Figure 3 (B) and (c). Figure 4 shows the grain structure of AZ91D alloys at different Al-Ti-C master alloy addition.

As shown in Figure 4 (a), non-grain refined AZ01D alloy has large equiaxed grains and columnar grains. The grain refined AZ91D alloys have fully equixed grain structures as shown in Figure 4 (b). On increasing to more than 0.15wt.%Al-T-C in AZ91D alloy, grain size was slightly reduced.

For the investigation of influence of Ca element on the grain structure of AM60B alloys, 0.3wt.% Mg-3%Ca master alloy was employed. It is found that grain refinement is very sensitive to the Ca element in AM60B alloy. Figure 5 (c) shows that the equiaxed grains were coarser and columnar grains were formed in the middle of sample with 0.3 wt.% Ca. Therefore, 0.3wt.% Ca addition causes the detrimental effect of grain refining efficiency in AM60B alloy.

Mechanical properties of magnesium and its alloys with grain refiner
Figure 6 shows the mechanical properties of HPMg compared to the grain refined HPMg. HPMg containing 0.15 and 0.3 wt.%Al-Ti-C master alloy show a increased tensile strength and yield strength. However, Elongation is slightly decreased with increasing Al-Ti-C master alloy addition.   Figure 7 shows the effect of grain refiner addition on the mechanical properties of AM60B alloys. The addition of 0.15wt.% Al-Ti-C and Al-Ti-B master alloy increases the strength and ductility of the AM60B alloys. However, 0.3 wt.%Ca addition is found to decrease the mechanical properties in AM60B alloys. The result shows that the Al-Ti-B master alloy was more efficient than Al-Ti-C master alloy in AM60B alloy. Figure 8 shows the effect of grain refiner addition on the mechanical properties of AZ91D alloy with different Al-Ti-C master alloy addition rates. The tensile strength and yield strength were increased with increasing Al-Ti-C master alloy. However, there is no significant difference in the elongation between unrefined and refined AZ91D alloys. From the present experimental results, the addition of Al-Ti-B and Al-Ti-C master alloy results in improvement in mechanical properties of AM60B and AZ91D alloy.

In a future work it is planned to study the effect of heat treatment on the mechanical properties of AM60B and AZ91D alloys.

4. Conclusions

The performance of Al-3Ti-0.13C and Al-5Ti-1B refiner has been examined in high purity magnesium, AZ91D and AM60B alloys. Based on the present study, both Al-3Ti-0.13C and Al-5Ti-1B master alloy are an effective grain refiner of AZ91D and AM60B alloys. The results obtained and the conclusions drawn are summarized as follows :

1. The results reveal that grain size decreased with increasing grain refiner addition levels in both AM60B and AZ91D alloy.

2. Grain refined AM60B and AZ91D alloys have substantially better mechanical properties than non-grain refined alloys.

3. 0.15 wt.% addition level of Al-Ti-B master alloy in AM60B alloy results in better mechanical properties than 0.15wt.% Al-Ti-C master alloy.

4. A small amount of Ca (0.3 wt.%) has a detrimental effect on grain refining performance in AM60B alloys. Thus, the mechanical properties was decreased with 0.3 wt.% Ca addition.

References

1. Aghion E, Bronfin B and Eliezer D *The role of the magnesium industry in protecting the environment*, Journal of materials Processing Technology, 117, Nov 2001, pp381-385.

2. Bolstad J, Cashion S, Kettler C and Dunlop G *Cost Effective Solutions for Handling Magnesium alloys in Diecasting Plants*, 62th Annual world Conference Proceedings, Berlin, Germany, 22-24 May 2005, pp85-89.

3. Braun A H *Premium Quality Magnesium Alloy Castings*, Recent Advances in Magnesium Technology, Proceedings of the AFS/CMI Conference, California, Jun 1985, pp43-58.

4. Bartos S C Building a Bridge for Climate Protection U.S. EPA and Magnesium Industry, Proceding of 59th International Magnesium Association, Montreal, 2002, pp22-24.

5. Friedrich H and Schumann S *Reserch for a new age of magnesium in the automotive industry*, Journal of Materials Processing Technology, 117, Nov 2001, pp276-281.

6. Esdaile R J *Magnesium Casting Applications in the Automotive Industry*, Light Metal Applications for the Automotive Industry: Aluminum and magnesium, 2001, pp1-5.

7. Schumann S and Friedrich H *Current and Future of magnesium in the automobile Industry,* Materials Science Forum, 419-422, 2003, pp51-56.

8. Luo A A *Magnesium:Current and Potential Automotive Applications,* JOM, Feb 2002, pp42-48.

9. Jacques R P and DasGupta *Magnesium in Automotive Components,* SAE technical paper, No.960416.

10. Kaneko T and Suzuki *Proceeding of the Second Osaka International Conference on Platform,* Science and Technology for Advanced Magnesium Alloys, 2003, pp67-72.

11. Peete K and Winkler L *Magnesium alloy AM50 die cast experiment shows improved mechanical characteristics over previous finding,* Magnesium Properties and Applications for Automobiles, SAE, SP-962, pp71-77.

12. Bamberger M *Structural refinement of cast magnesium alloys,* Materials Science and Technology, 17 Jan 2001, pp15-23.

13. Xue F, Du W and Sun Y *Microstructure Refinement of Magnesium Based Alloy,* Proceedings of the International Conference on Magnesium, Science, Technology and Applications, Beijing, China, Sep 2004, pp143-146.

14. Nishino N, Kawahara H, Shimizu Y and Iwahori H *Grain Refinement of Magnesium Casting Alloys,* Magnesium Alloys and their Applications, Munich, Germany, Sep 2000, pp59-64.

15. Zhong Y, Ozturk K, Liu Z K and Luo A A, Magnesium Technology 2002, TMS, 2002, pp69-73.

E-mail: emxysh@ultra-net.org

Table 1.   Alloy composition as determined by chemical analysis

Alloys	Ti	B	C	Fe	Si	Mn	Zn	Al	Mg
HP Mg	0.05	0.03	0.001	0.005	-	-	-	-	bal.
Al-Ti-C	3.09	0.01	0.13	0.97	0.05	0.01	0.01	95.73	-
Al-Ti-B	5.1	1.1	-	0.15	0.19	-	-	93.46	-
AZ91D	-	-	-	0.006	0.01	0.12	0.85	8.7	bal.
AM60B	-	-	-	0.0013	0.03	0.35	0.118	5.83	bal.

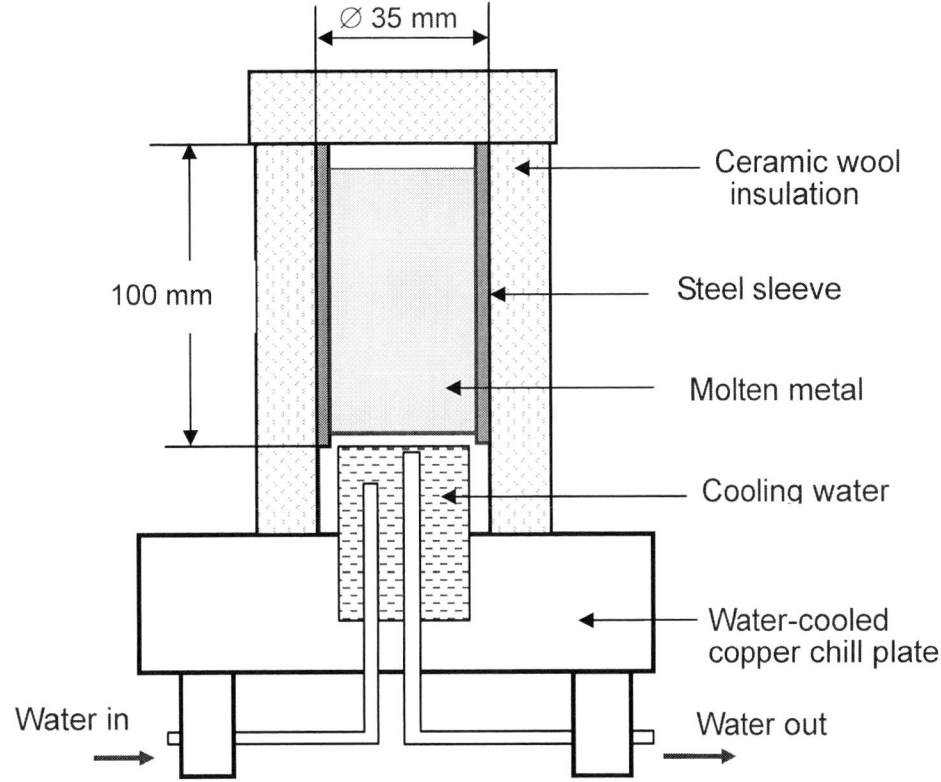

Figure 1.   Schematic of experimental set-up for grain refining test casting.

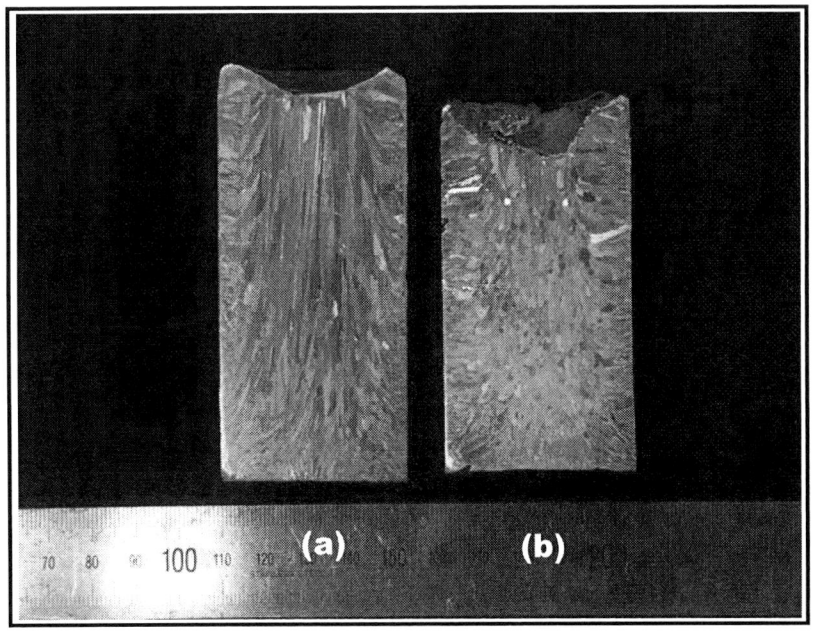

Figure 2.   Macrophotography of High Purity Magnesium(HPMg) with Al-Ti-C master alloy (a) HPMg+0.15wt.%Al-Ti-C, (b) HPMg+0.3wt.%Al-Ti-C.

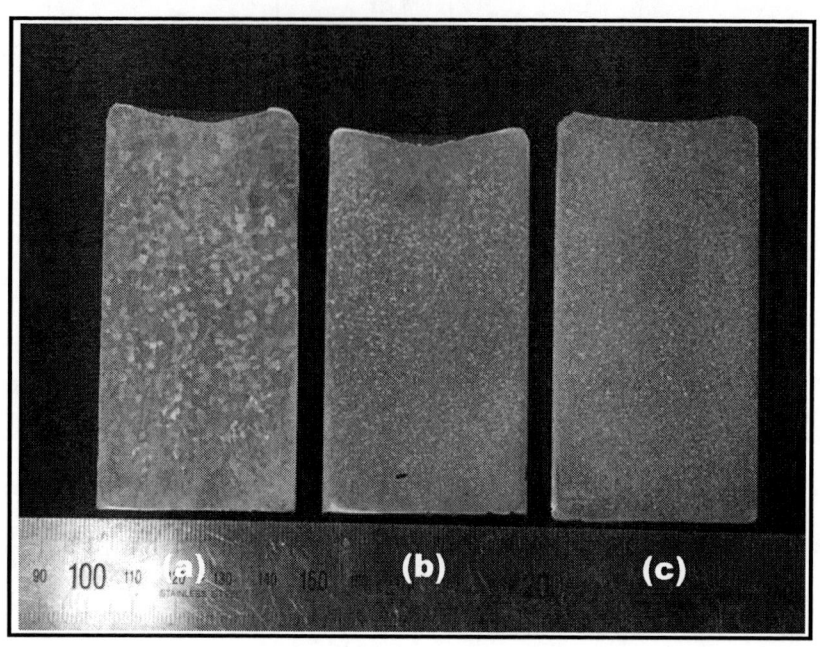

Figure 3. Macrophotography of AM60B alloy without or with
Al-Ti-C and Al-Ti-B master alloy (a) AM60B, (b) AM60B
+0.15wt.% Al-Ti-B and (c) AM60B+0.15wt.%Al-Ti-C

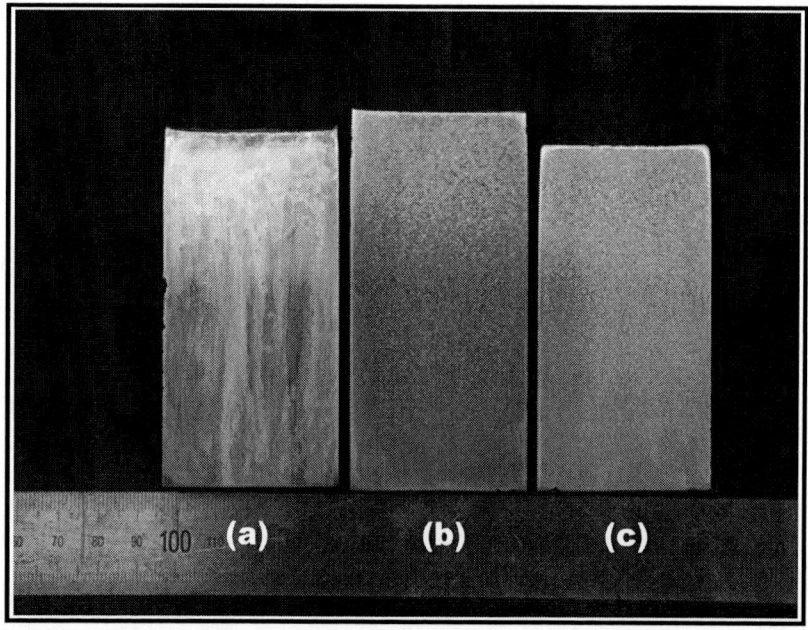

Figure 4. Macrophotography of AZ91D alloy without or with
Al-Ti-C master alloy (a) AZ91D, (b) AZ91D+0.15wt.%
Al-Ti-C and (c) AZ91D+0.3wt.%Al-Ti-C

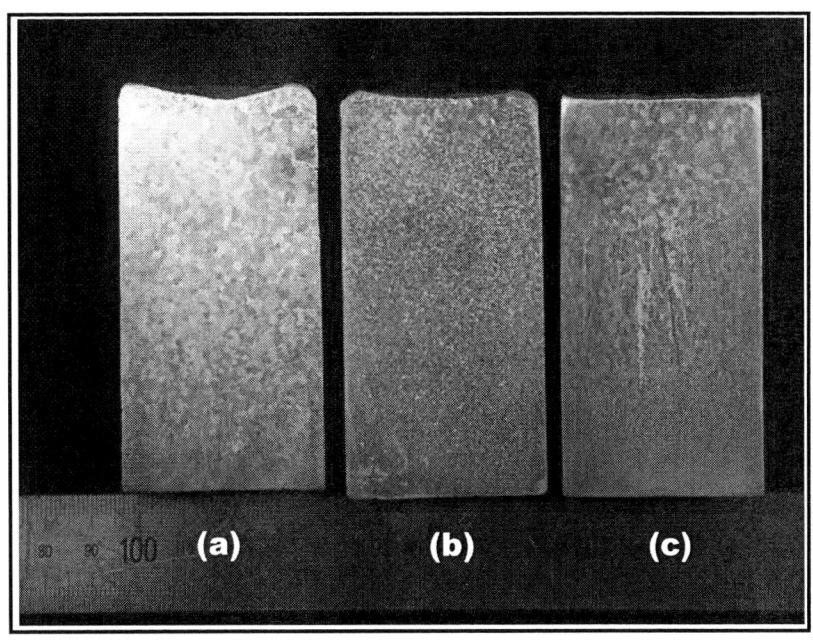

Figure 5. Macrophotography of AM60B alloy with Al-Ti-C master alloy and Ca addition (a) AM60B, (b) AM60B+0.15wt.% Al-Ti-C and (c) AM60B+0.15wt.%Al-Ti-C+0.3wt.%Ca

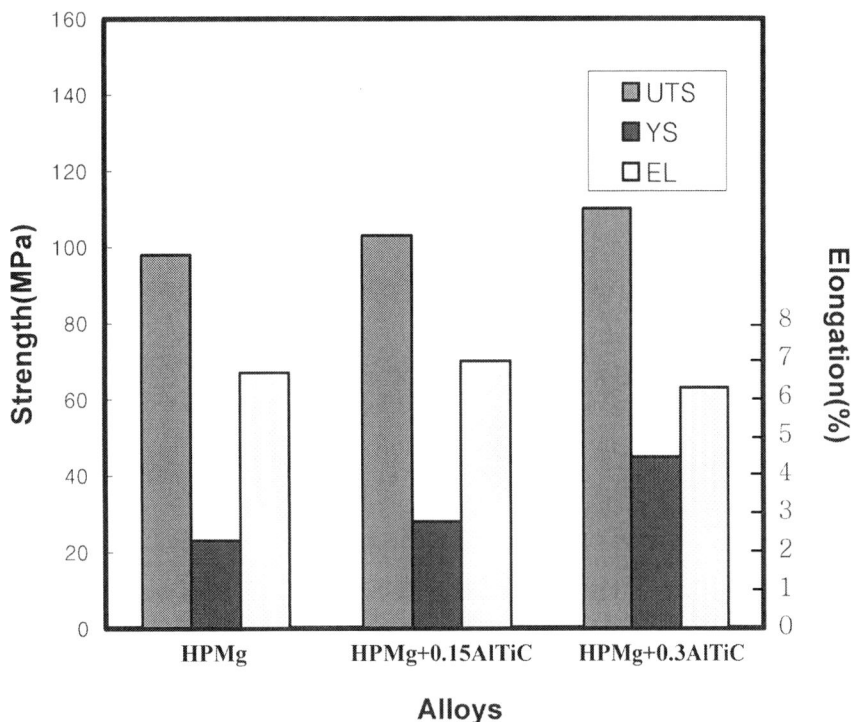

Figure 6. Effect of grain refiner addition on the mechanical properties of high purity magnesium

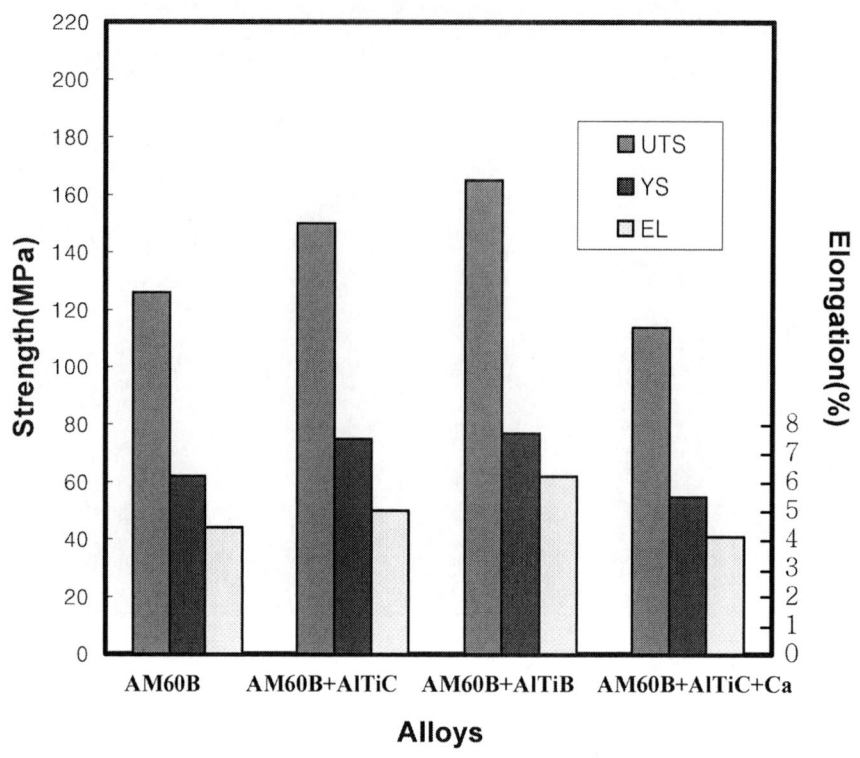

Figure 7. Effect of grain refiner addition on the mechanical
properties of AM60B alloys

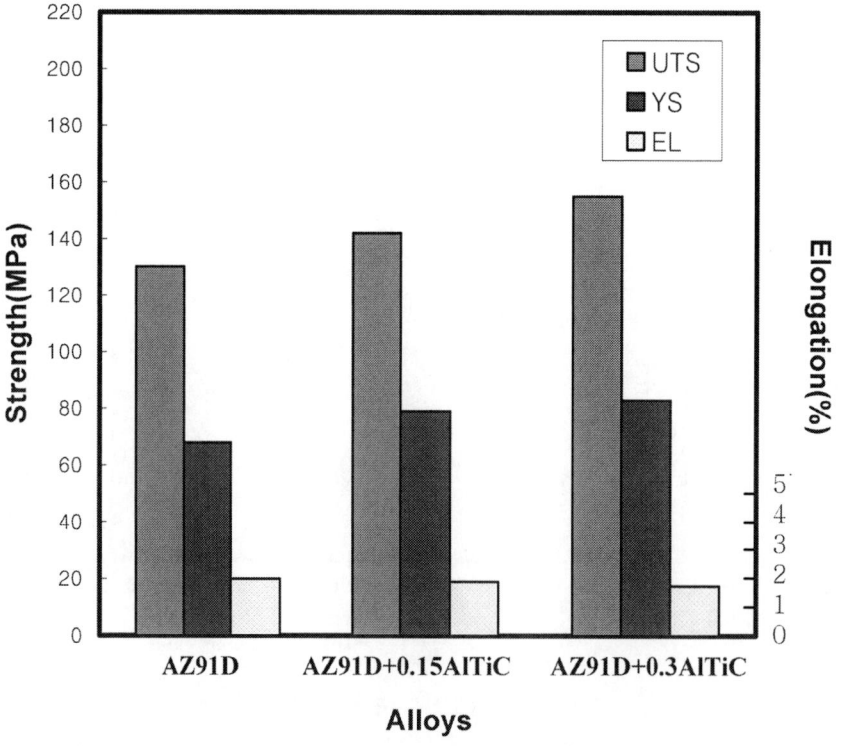

Figure 8. Effect of grain refiner addition on the mechanical
properties of AZ91D alloys

Green Manufacturing for Magnesium Alloys

Shae K. Kim, Jin-Kyu Lee, Young-Ok Yoon and Hyung-Ho Jo

Korea Institute of Industrial Technology, Korea.

Abstract

Mg alloys present a number of interesting properties, such as low density, high specific strength and good castability, etc. Despite these properties, Mg alloys are used with many precautions of high chemical reactivity, especially during melting and casting procedures. Research has been directed to solve the problem by replacing commonly used protection gas SF6, which is very effective in protection but has high global warming potential and long life time.

The aim of this study is to manufacture CaO added Mg alloys. CaO added Mg alloys could be well manufactured by conventional melting and casting procedures and the effects of CaO on green manufacturing of burning behavior, oxidation resistance, protection gas usage and even fluidity were investigated.

Key words Mg alloys, Calcium oxide, Green manufacturing, Ignition temperature, Protection gas

Introduction

Magnesium alloys present a number of interesting properties, such as high specific strength, good castability, low density, etc. With theses properties, magnesium alloys are gaining increased importance for uses in the transport industry and in electronics applications where low inertia is required for rapidly moving parts [1-5]. Despite these properties, magnesium alloys are used with precautions of high chemical reactivity, limited high strength and creep resistance at elevated temperatures.

Research has been directed to improve these properties by alloying and/or developing protective gas of low environmental burden. It is well known that calcium and strontium, though their high cost, are effective for oxidation resistance and elevated temperature properties [6-8]. However, the relatively high cost of these materials might prevent widespread industrial applications. Generally, there are two factors accounting for the high cost; the raw material cost of magnesium and calcium or strontium and the processing cost. Furthermore, the handling of calcium or strontium is not easy.

The aim of this study is to manufacture CaO added magnesium alloys and to investigate the behavior of CaO in the alloys. In terms of properties, it is intended to increase burning temperature (to improve oxidation resistance) and to reduce protection gas through processes. In terms of manufacturing, it is intended to use cheap CaO instead of calcium and to manufacture CaO added magnesium alloys via conventional foundry processes to minimize the capital costs associated with processing. The effects of CaO on green manufacturing of burning behavior (oxidation resistance), protection gas usage and even fluidity were investigated.

Experimental

Molten AZ91D and AZ31 magnesium alloys were prepared in a steel crucible in an electric resistance furnace at 680 deg C under HFC-134a protective atmosphere (3,000ppm). The desired fraction of CaO particles were added with a feed rate of 1 g/min. CaO particles wrapped in aluminum foils, instead of CaO particles themselves, were added. By this way, the amount of CaO particles that were fluttered in the furnace and piled-up in the top edge side of the melt surface was greatly reduced.

CaO particles, which were dehydrated, pulverized and stored in vacuum with average diameters of under 100 μm, were used. In order to avoid a big temperature change during the addition of the particles, they were heated to 250 deg C prior to addition.

During melting procedure, the effect of CaO on the minimum amount of protective gas was examined for CaO added AZ91D magnesium alloy. As shown in Fig. 1, the minimum amount of protective gas not to make burning in the melt was evaluated in the case of using generally used oxidation-inhibiting sulfur hexafluoride at 720 deg C, at which general melting and casting procedures are made. This experiment was done by adjusting the amount of SF_6 gas with synthetic air both in the sealed and unsealed conditions.

The melts of AZ91D magnesium alloy prepared were cast into a spiral fluidity mold at room temperature. The fluidity results are based on the average of 3 tests. Fig. 2 shows the schematic illustrations of the horizontal channel and vertical cross section of the spiral fluidity mold and the enlarged cross sectional spiral channel. The mold is designed to remove entrapped oxides and slags insofar as possible and to introduce the composite melt into the mold as smoothly as possible, such that it always just fill the cross section at all times and does not react excessively with the atmosphere and the mold.

The melts of AZ91D and AZ31 magnesium alloys prepared were cast into a 32mm diameter cylindrical mold at room temperature. The alloy chips were prepared from the as-cast ingots by drilling without cutting oil for ignition test. Chips on a copper boat were inserted into an electric resistance furnace in an ambient atmosphere. Temperature change was monitored by a thermocouple equipped with a copper boat and ignition was detected by abrupt temperature rise. The obtained ignition temperatures were based on the average of 3 tests. DTA tests were also done to generalize burning behaviors of CaO added magnesium alloys under nitrogen gas atmosphere.

Results & Discussion

Fig. 3 shows the result about the effect of CaO on the minimum amount of protective gas, which is necessary not to make burning in the melt. As mentioned before, this experiment was done in the case of SF_6 gas with synthetic air in sealed and unsealed conditions. Fig. 3 clearly demonstrates how much the minimum required amount of SF_6 gas is lowered by adding CaO of just 0.182wt% to AZ91D magnesium alloy both in the sealed and unsealed conditions. As expected, reduction of SF_6 gas, which is regarded as having high global warming potential with long atmospheric life time, can be obtained for cleaner or environmental-friendly manufacturing melting and casting procedures.

The mass fraction of CaO particles was deduced based the amount of calcium detected by ICP-mass. The resultant spiral fluidity lengths are 246mm for AZ91D magnesium alloy and 145mm for 0.056wt% CaO, 177mm for 0.154wt% CaO and 145mm for 0.356wt% CaO added AZ91D magnesium alloys in Fig. 4. The change in the viscosity is the major factor to determine the spiral fluidity in the present investigation. It is generally accepted that the addition of CaO particles results in a decrease in the spiral fluidity due to the increase in the viscosity. Therefore the spiral fluidity decreases with increasing CaO particle mass fraction. In all cases, however, there is no one to one correspondence between the mass fraction of CaO particles and the spiral fluidity. Another factor, which is necessary for defining the spiral fluidity in particle added melts, is distribution of the particles in presumably in the melts. Although the result is not yet clearly explained, the proceeding results of the spiral fluidity seem to suggest that CaO particle added magnesium alloys could be cast at 680 degC into variety of shapes and sizes. A better understanding of the experimental observations can be made when careful melting/casting environments and detailed examinations for various mass fractions of CaO particles are taken into account.

The dependence of ignition temperature on the mass fraction of CaO particles in AZ91D magnesium alloy is given in Fig. 5. It could be seen that the ignition temperatures are significantly improved over AZ91D magnesium alloy itself. Regardless of the test methods, there is one to one correspondence between the ignition temperature and the mass fraction of CaO particles. It was reported that calcium addition to magnesium alloys retarded the oxidation rate during melting process by the formation of thin and dense CaO film on the melt surface. You et al. used a thermogravimetric analyzer (TGA) to study the oxidation of the Mg-Ca alloys in the Ca range from 0 to 3 wt% at 440-500 deg C for up to 7 hours in air. They confirmed that a dense and compact protective MgO/CaO layer formed at elevated temperature, although the oxidation behavior of calcium bearing magnesium alloys could not be explained by Pilling-Bedworth theory. In this study, however, CaO is present in the form of particles and, more than that, CaO particles are located not on the surface but inside. Also there might be no calcium which could react with oxygen to form CaO. Much work should be carried out to understand the oxidation behaviors of Ca and CaO bearing magnesium alloys.

Fig. 6 shows the result of DTA test to verify the burning behavior of CaO added AZ31 alloy under N_2 atmosphere. It could be seen that the ignition temperatures are significantly improved over the alloy. Fig. 6 show clearly that in these samples the linear relationship holds between ignition temperature and CaO content on DTA results. It should be noted that ignition occurs in the liquid state. Therefore CaO added AZ31 melt could be safely handled at operation temperature over 650 deg C under N_2 atmosphere without any protective gases of high global warming potential and high cost.

Conclusions

The feasibility study to manufacture CaO added magnesium alloys and to investigate the behavior of CaO in magnesium alloys was carried out. CaO added magnesium alloys could be well manufactured by conventional melting and casting procedures. Although further study is indispensable, the effects of CaO on minimum protective gas amount, fluidity and ignition behavior of magnesium alloys could be investigated. In terms of property and process, it could be possible to obtain:

increasing burning temperatures of alloys for ensuring safety during manufacturing and application;

reducing amount of protective gas during manufacturing (melting and casting);

eliminating the use of protective gas during subsequent forming processes of extrusion and rolling, etc.;

maintaining or improving elevated temperature properties with lower amount of Ca or Sr or rare earth metal elements, compared with conventionally developed high-temperature magnesium alloys;

easy alloying of inexpensive CaO instead of expensive Ca with high oxidation tendency even during alloying.

References

1. I. J. Polmear: *Magnesium alloys and applications*, Materials Science and Technology, Vol. 10, 1994, pp1-16.

2. I. J. Polmear: *Recent Developments in Light Alloys*, Materials Transactions, JIM, Vol. 37, 1996, pp12-31.

3. B. L. Mordike and T. Ebert: *Magnesium Properties-application-potential*, Materials Science and Engineering, Vol. A302, 2001, pp37-45.

4. H. Friedrich and S. Schumann: *Research for a new age of magnesium in the automotive industry*, Journal of Materials Processing Technology, Vol. 117, 2001, pp276-281.

5. E. Aghion, B. Bronfin and D. Eliezer: *The role of the magnesium industry in protecting the environment*, Journal of Materials Processing Technology, Vol. 117, 2001, pp381-385.

6. B. L. Mordike: *Creep-resistant magnesium alloys*, Materials Science and Engineering, Vol. A324, 2002, pp103-112.

7. R. Schmid-Fetzer and J. Grobner: *Focused Development of Magnesium Alloys Using the Calphad Approach*, Advanced Engineering Materials, Vol. 3, 2001, pp947-961.

8. B. S. You, W. W. Park and I. S. Chung: *The effect of calcium additions on the oxidation behavior in magnesium alloys*, Sctripta Materiallia, Vol. 42, 2000, pp1089-1094.

Figures

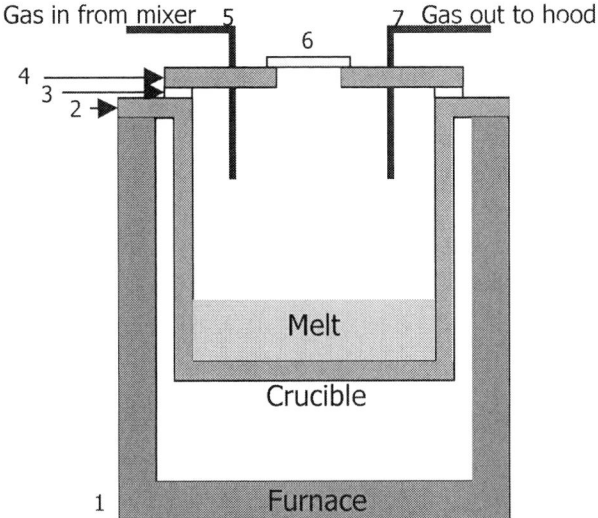

Figure 1: Schematic illustration of optimum protective gas experiment apparatus; 1 furnace, 2 crucible, 3 ceramic wool, 4 steel cover, 5 gas-in tube, 6 quartz plate, 7 gas-out tube.

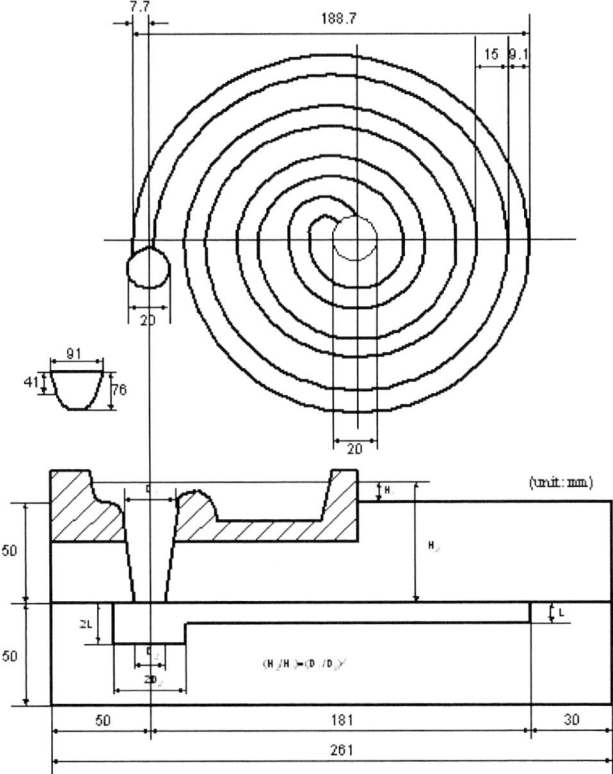

Figure 2: Schematic illustration of spiral fluidity mold: dimensions in millimeters.

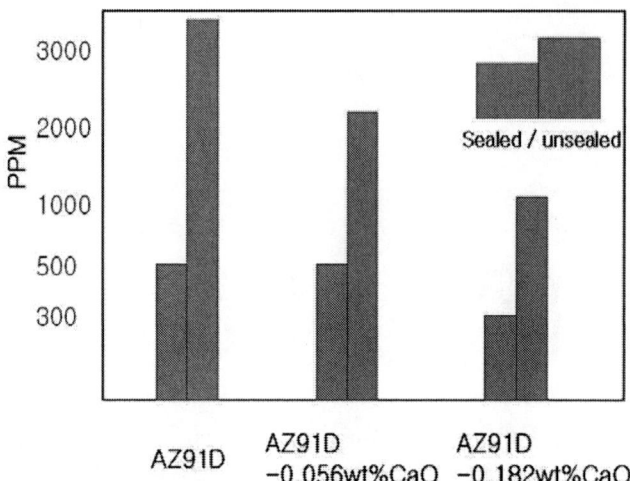

Figure 3: Results of optimum protective gas experiments at the melt temperatures of 720 in CaO added AZ91D magnesium alloys.

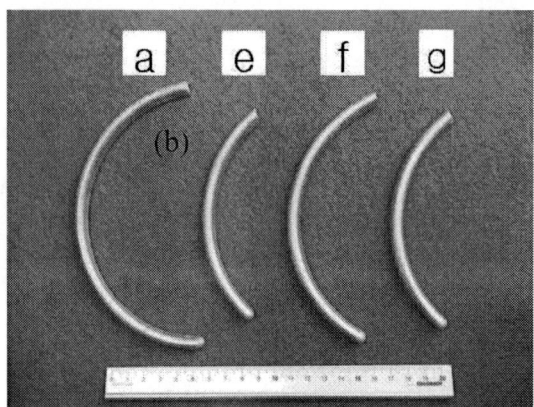

Figure 4: Photographs of as-cast spirals of the alloys; (a) 0wt% CaO, (e) 0.056wt% CaO, (f) 0.154wt% CaO and (g) 0.356wt% CaO added AZ91D magnesium alloys.

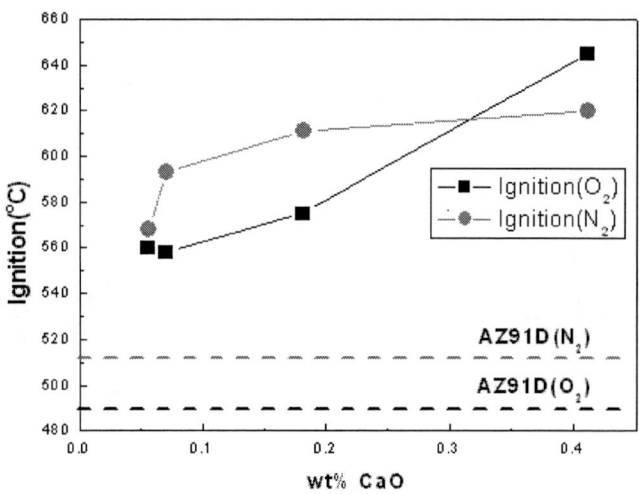

Figure 5: Results of chip ignition test under an ambient atmosphere and DTA test under a nitrogen atmosphere in CaO added AZ91D magnesium alloys.

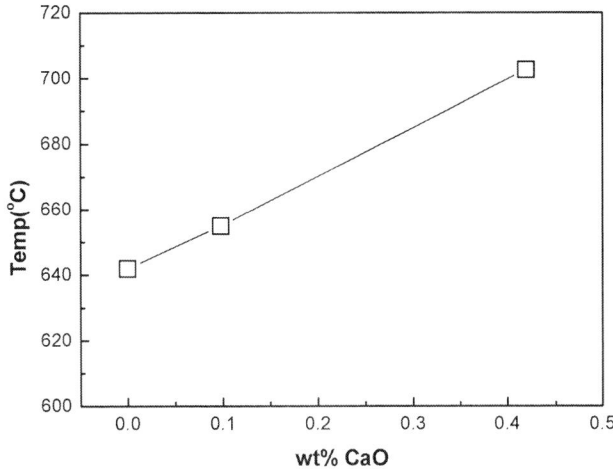

Figure 6: Result of DTA test under a nitrogen atmosphere in CaO added AZ31 magnesium alloys.

Role of Carbon as Solute Element in Carbon Grain Refinement of Mg-Al Alloy

Seong-Yong Shim, Qinglin Jin, Su-Gun Lim*

Department of Materials Science and Engineering, Research Center for Aircraft Parts Technology, Gyeongsang National University, Jinju, Gyeongnam, Korea

* email: suglim@nongae.gsnu.ac.kr

Abstract

Grain refining effects of carbon (C_2Cl_6) addition were investigated using AZ31 magnesium alloy. The addition of C_2Cl_6 to AZ31 magnesium alloy significantly reduced the grain size from 400µm to about 120µm. The tensile properties were increased due to grain refinement. EPMA analysis showed that apparent segregation of carbon element was occurred in AZ31castings treated with carbon. It was considered that the carbon may have strong segregating power that restrict the grain growth and affect the constitutional undercooling which, in turn, results in grain refining effects.

Key words: AZ31 alloy; Carbon addition; Grain refinement, Segregation.

Introduction

Grain refinement is an important practice to improve the properties of magnesium alloys. Many grain refining methods including super-heating, violent agitation, Elfinal or ferric chloride inoculation, carbon inoculation and so on, have been developed for the Mg-Al alloy system. Among them, the addition of carbon containing agents to the melt, the so-called carbon addition method, offers many practical advantages because of the lower operating temperature and less fading. Various carbon-containing agents such as C_2Cl_6, CCl_4, CaC_2 and granular graphite, have been reported to produce successful grain refinement in Mg-Al alloys [1, 2,] and many different types of carbon-based grain refiners have been continuously developed until now.

A number of hypotheses have been proposed to explain the mechanism by which carbon addition methods cause grain refinement. The results of the work at Battelle Memorial Institute led to the hypothesis that aluminum carbide, Al_4C_3, was the compound responsible for the refining effects largely on the basis that both aluminum and carbon had to be present in order to make such treatments effective [3]. So far it has been widely accepted, but no conclusive experimental data has been found to substantiate this supposition due to lack of understanding on the fundamental factors involved and the difficulties in identifying the active nucleants.

Recent work on the grain refinement of magnesium and aluminum alloys show that grain refinement can be facilitated by two factors. The first is the number and potency of the nucleant particles and the other is the solute elements present in the melt [4]. The effect of solute elements on grain size is believed to be as important as that of nucleant particles, since the addition of solute elements generates constitutional undercooling in diffusion layer ahead of advancing solid/liquid interface. The constitutional undercooling may restrict grain growth because slow diffusion of the solute limits the rate of grain growth. In addition, further nucleation may occur in front of the interface in the diffusion layer because nucleants in the melt are more likely to survive and be activated in the constitutionally under-cooled zone. Some alloying elements, such as aluminum, Calcium and silicon have been reported to produce grain refinement in Mg alloys due to there

At present, all the hypotheses about carbon addition methods have focused on the nucleant particles, like Al_4C_3, which may act as potent nucleation sites and none have thought that the carbon, which may also have a strong tendency for segregation, could greatly affect the constitutional undercooling and consequently restrict grain growth during the solidification process. In this study, the mechanism of the carbon addition method was investigated paying specific attention to the role of carbon as a solute element in grain refinement of AZ31 magnesium alloy.

Experimental

In the experiments, a commercial AZ31B magnesium alloy was melted in an electrical furnace using mild steel crucible and mold under an argon (Ar) gas atmosphere. The most commonly used carbon refiner, C_2Cl_6, were added into the melts and held at 780°C for 30 minutes then poured into the mold. An ingot with a size of 30 × 30 × 90 mm was fabricated. Tensile specimens with a gage section of 12.7×5×2mm were cut directly from the casting.

Optical microscopy samples were etched with acetic-picral etchant [5]. The mean linear intercept method was used to measure the average grain size. However, for the unrefined casting, the very large grain sizes meant that only a few grains could be measured. For these samples, the grain size was estimated directly from the casings and so no indication of the error in the measurement is given. Characterization of microstructure and qualitative analysis were conducted on selected specimen using optical microscope and JXA-8100 Electron Probe Micro Analyzer (EPMA).

Results and discussion

Fig. 1 shows the macrograph of as-cast AZ31 magnesium alloy with and without the addition of C_2Cl_6. It is evident that the carbon addition is an effective way to refine the grain size of AZ31 alloy. Fig. 2 shows the microstructures of as-cast AZ31 magnesium alloy after homogenization treatment. The grain size significantly decreased from approximately 400 µm (Fig. 2 a) to 120µm (Fig. 2. b) when 0.6 wt.% C_2Cl_6 is added.

There are seldom studies have been made of grain size strengthening of Mg and its alloys. It may be due to the range of grain size has not always been very wide (reflecting the lack of an effective grain refiner for Mg-Al alloys). Fig. 3 shows that addition of C_2Cl_6 has beneficial effects on the mechanical properties of AZ31 magnesium alloy. At condition of the 0.6 wt.% C_2Cl_6 addition, the ultimate strength of 243 MPa, yield strength of 80 MPa were achieved. Grain refinement has great influence on elongation, which was significantly improved from 10 % to 21 %.

EPMA area analysis was performed to trace the aluminum, zinc, magnesium, oxygen and carbon distribution in as-cast AZ31 after the addition of 0.6 wt.% C_2Cl_6 and the result is shown in Fig. 4. Fig. 4 a is a scanning electron image of as-cast AZ31 after the addition of 0.6 wt.% C_2Cl_6. As shown in the figure, the fully divorced eutectic phase ($Mg_{17}Al_{12}$) exhibits a wide range of morphologies. Al-Mn particles (white particles) are located in the α Mg. Apparent segregation of aluminum can be clearly observed in Fig. 4 c. Zn rich phase usually locates in the center of the eutectic phase (Fig. 4 e). The oxygen concentration is lowest in eutectic phase (Fig. 4 f).

In the carbon mapping results, as shown in Fig. 4 b, it is found the carbon concentration is the highest in the Al-Mn phase (red particles). The high carbon concentration of the Al-Mn particles suggests that carbon may easily form stable carbides with Al and Mn. These Al-Mn particles are generally considered to have no relationship with grain refinement or even to deteriorate the grain refining effects of carbon addition [4].

The EPMA results of this study show that there was no evidence of so-called Al_4C_3 particles being present in the casting after carbon treatment.

In Fig. 4 b, it is notable that the carbon concentration is also high in the eutectic phase (yellow-green) and this indicates that segregation of carbon occurs during solidification. The carbon concentration increased progressively from the central dendritic arm (dark area) to the edge of the dendritic arm (blue area) and to the eutectic region (yellow-green area). This observation is important because it provides evidence that carbon has a limited and temperature dependent solubility in magnesium. It can, therefore, be considered that when a nucleated crystal forms during solidification, the carbon is rejected at the liquid-solid interface. The rejected carbon will greatly affect the constitutional undercooling at the solid-liquid interface and restrict grain growth. The liquid around the crystal is therefore further undercooled, thus allows other crystals to nucleate more easily in this region. This process will continue until the carbon element combines with the eutectic $Mg_{17}Al_{12}$ phase which is formed by the eutectic reaction at about 426□, or the Al-Mn phase.

Regarding the common belief associated with the Al_4C_3 hypothesis, and that, many observations can be explained with it, no clear evidence of Al_4C_3 acting as nucleation site has been found [5]. This hypothesis also ignores the possible role of carbon as a solute, which may significantly affect the amount of constitutional undercooling further restricting grain growth during the solidification. Thus this aspect of carbon's role in the grain refinement of Mg-Al magnesium alloys should not be neglected and needs further research.

Conclusion

Carbon addition is an effective way to refine the grain size of AZ31 alloy. The tensile properties of AZ31 alloy was increased due to the grain refinement. The present study shows that segregation of carbon may occur in grain refined AZ31castings. It is considered that the carbon may have a strong segregating power that increases the amount of constitutional undercooling and restricts grain growth, which in turn results in grain refinement.

References

1. Jeong-Pil Eom, Seng-Kyu Jeong. Su-Gun Lim, Hee-Tack Shin:*Effect of carbon inoluction in C_2Cl_6 addtion on the grian refinement of AM60 Mg alloys*. Journal of the Korean Foundrymen's Society. 19, 1999, pp.263-268.

2. Jeong-Pil Eom, Seng-kyu Jeong, Su-Gun Lim, Hee-Tack Shin, Deuk-Soo Jeong*: Grain refinement and mechanical properties of AM60 Mg alloy by $CaCN_2$ addition*. Journal of the Korean Foundrymen's Society. 14, 1999, pp383-388.

3. C. E. Nelson: *Grain size behavior of Mg alloy*: Transactions of the American Foundrymen's Society. 56, 1948 pp1-23.

4. Arne K. Dahle, Young C. Lee, Mark D. Nave, Paul L. Schaffer and David H. StJohn: *Development of the as-cast microstructure in magnesium–aluminium alloys* 1, 2001, pp 61-72.

5. ASM Specialty Handbook, Magnesium and Magnesium alloys, p 37, Materials Park,OH, USA, 1999.

Acknowledgements

This work was supported by grant No. RT104-01-03 from the Regional Technology Innovation Program of the Ministry of Commerce, Industry and Energy (MOCIE).

Figures

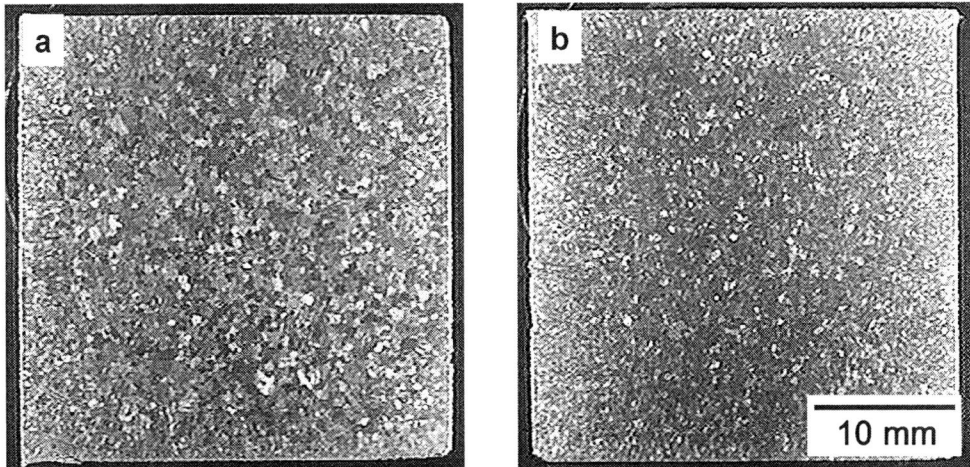

Fig. 1. Optical macrograph of as-cast AZ31 Magnesium alloy.
(a) without C_2Cl_6 addition   (b) with 0.6 wt.% C_2Cl_6 addition

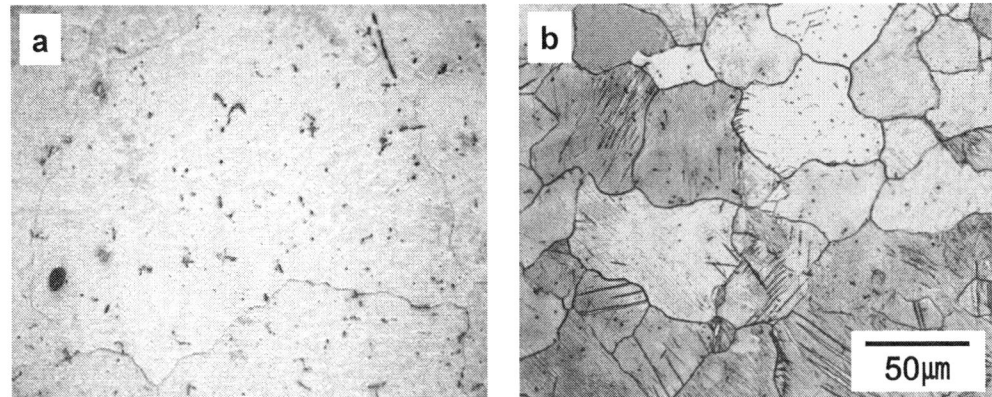

Fig. 2. Optical micrograph of as-cast AZ31 Magnesium alloy after homogenization treatment at 380°C for 8 hours.
(a) without C_2Cl_6 addition   (b) with 0.6 wt.% C_2Cl_6 addition

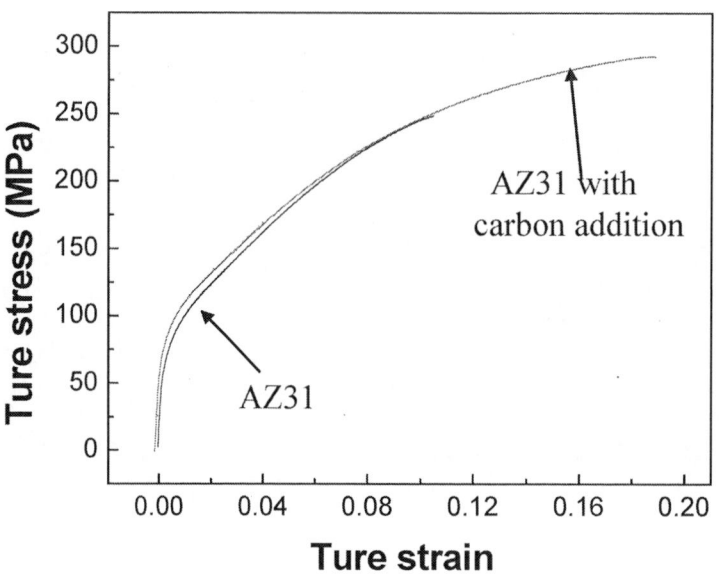

Fig. 3. True stress true strain curve of AZ31 and AZ31 with 0.6wt% carbon addition.

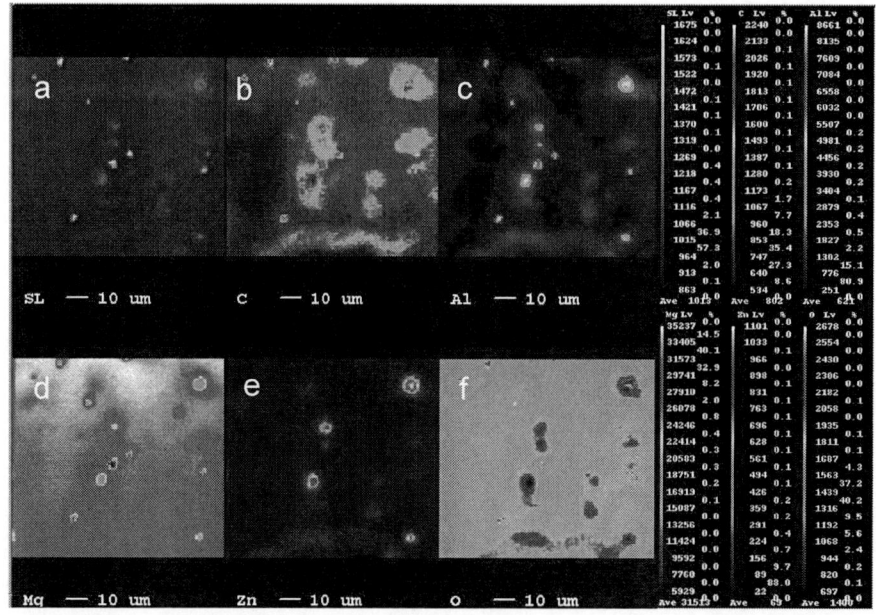

Fig. 4. Result of EPMA area analysis.
(a) SEM image   (b) C   (c) Al   (d) Mg   (e) Zn   (d) O

Wicking of Liquid Polystyrene Degradation Products into the Pattern Coating in the Lost Foam Casting Process

P J DAVIES and W D GRIFFITHS

Department of Metallurgy and Materials, School of Engineering,
University of Birmingham, Edgbaston, Birmingham, B15 2TT, UK.

Abstract
Entrapment of liquid polystyrene degradation products during filling is detrimental to the quality of Aluminium Lost Foam castings. In this study, the degradation of expanded polystyrene under different heating conditions, and the ability of two pattern coatings of different permeability to absorb the partially degraded liquid polymer, was determined using gel permeation chromatography and scanning electron microscopy. It is suggested that, with the degradation of the polystyrene to a sufficiently low molecular weight liquid, significant absorption of the liquid polystyrene can occur with each coating. A number of simple plate patterns were cast in a real-time x-ray facility to determine the effect of each coating type on filling behaviour and casting quality. Similar tensile properties were found in each case, although the high permeability coating was associated with a filling velocity about twice that associated with the low permeability coating.

Key words Lost Foam, Aluminium, Polystyrene degradation, Wicking, Pattern coatings

Introduction

Lost Foam Casting offers numerous advantages over conventional casting processes, in particular the ability to consolidate several components into a single, highly complex, cast part [1]. Lost Foam Casting uses an expendable foam pattern (made from Expanded Polystyrene and coated with a ceramic wash), which is placed in a mould box, and surrounded with unbonded sand, compacted by vibration. Liquid metal is poured directly onto the pattern, and heat causes it to decompose to both liquid and gaseous degradation products which are thought to be transported through the thin porous pattern coating (typically only 200 μm thick) with the pattern eventually being completely replaced by the liquid metal.

Unfortunately, the porous pattern degradation products often get trapped in the liquid metal during mould filling, and this can lead to the formation of a variety of casting defects including folds, porosity, surface dimples and misrun, (due to premature solidification of the liquid metal front) [2,3]. Whilst several casting variables can affect the severity of these defects, it has been recognised that the pattern coating plays a key role in controlling mould filling and defect formation, primarily by controlling how the pattern degradation products are removed from the mould [4].

Several researchers have shown that, when pouring aluminium alloys, the major expanded polystyrene degradation product is a viscous liquid residue [5,6], (rather than gaseous products such as are generated when pouring iron at higher temperatures). However, the mechanisms of elimination of the viscous liquid residue from the mould cavity are unclear.

Sun et al. [7] proposed that at a critical combination of polymer residue temperature and coating temperature, the residue is able to wet the coating and, with a further increase in temperature, is absorbed or "wicked" into the coating. This wetting and wicking model is shown in Figure 1A. In contrast to this, Zhao et al. [8] proposed that the liquid degradation product is unable to penetrate the coating, but instead forms small globules of liquid residue that are trapped against the coating surface as the metal flows through the pattern. As these become heated, the surface of the globules vaporise, and the gases produced are able to pass through the small area of exposed coating into the sand. As this process continues, the globules reduce in size, until all residue is removed from the mould cavity, as shown in Figure 1B.

The objective of the current work was to clarify how the liquid polymer residue is removed by the pattern coating, and to determine how the ability of the coating to remove the liquid residue affects casting quality.

Experimental

10 mm x 10 mm expanded polystyrene samples were cut from a 10 mm thick plate having a nominal density of 23.4 kgm^{-3}. Three layers of a low-permeability (9×10^{-14} m^2) mica-based coating were applied to a single face of each sample, giving a total coating thickness of approximately 500 μm (three coating layers were necessary since a single coating layer was found to break up when the sample began to shrink). The coated samples were heated in argon at atmospheric pressure, in a resistance-heated furnace equipped with a quartz window, at a rate of 10 K per minute between room temperature and 700 °C. Changes in the structure of the sample were recorded using a digital camera. These experiments were also repeated for samples coated with a high-permeability (5×10^{-13} m^2) mica-based coating.

Samples of coated expanded polystyrene were also placed in a resistance-heated tube furnace and heated at different rates to temperatures of between 300 and 500 °C. The rate of heating was controlled by varying the temperature of the pre-heated furnace to between 400 and 1000 °C. At the required sample temperature a water-cooled chill was used to rapidly quench the sample and minimise any further degradation of the polymer, thus preserving the characteristics of the polymer at that temperature and time.

For these experiments 20 mm diameter x 10 mm thick samples were prepared as described for the shrinkage experiments. A type K chromel-alumel thermocouple with a bead diameter of 80 μm, constructed from 50 μm diameter wire, was embedded in the second coating layer while still wet and was used to monitor the temperature of the coating during its heating and cooling. The coated polystyrene samples were inserted into the apparatus, as shown schematically in Figure 2. After the experiments, samples were fractured through their centres, mounted, and then coated with gold using standard techniques, for scanning electron microscopic examination.

Gel Permeation Chromatography was used to determine the average molecular weight of the starting pattern materials and the polymer remaining in each coating after being heated. All chromatography work was undertaken by Rapra Technology Ltd using tetrahydrofuran as a solvent, with a column set suitable for the analysis of medium/high molecular weight polymers. In cases of non-wetting of the liquid polymer on the coating, the polymer residue was removed from the surface of the coating prior to dissolving in the solvent. In all other cases, the polymer was extracted from the coating by placing the whole coating sample in the solvent and filtering the solution. The Gel Permeation Chromatography system used for the work was calibrated with polystyrene.

Finally, a number of simple expanded polystyrene plates (440 mm x 180 mm x 10 mm) coated with ~500 μm thickness high- and low-permeability coatings, were poured in a flat, horizontal orientation at 850 °C using untreated 319 aluminium alloy (Al-6wt.%Si-3.5wt.%Cu). A hollow downsprue was attached to one end of each plate using hot-melt adhesive. Mould temperatures experienced during filling were measured by the same 80 μm diameter type K thermocouples, which were embedded in the pattern and the top surface of the coating, along the centre line of the plate, at distances of 90, 190, 290 and 390 mm from the downsprue/plate joint, (see Figure 3). Filling behaviour was recorded using a real-time x-ray apparatus and a high-speed digital camera capable of recording video at a rate in excess of 1000 frames per second. The cast plates were examined for any surface defects, and were then sectioned along their length and machined into standard test bars for tensile testing.

Results
Degradation of Expanded Polystyrene Patterns
The effect of heating a coated expanded polystyrene sample can be seen in Figure 4. The foam structure initially collapsed, by approximately 90%, forming a highly viscous residue which did not immediately wet the coating. Several fragments of coating also became detached from the main coating layer as the expanded polystyrene collapsed (see Figure 4B). With further heating, a significant amount of the polymer residue was observed to boil on the coating surface (see Figure 4D).

The original molecular weight of the expanded polystyrene pattern material was determined by the Gel Permeation Chromatography technique to be 280,000 – 290,000. Tables 1 and 2 show the average molecular weights of the polystyrene degradation products remaining in the samples after the heating experiments. It should be noted that these values are an average of the molecular weight of the polymer both on the surface of the coating and, where wicking of the polymer had taken place, within the coating pores. The shaded areas in the table indicate which samples showed evidence of wetting only, and in which cases wetting and wicking occurred. In a few cases, no molecular weight could be determined, due to insufficient polymer remaining in the sample after heating.

Tables 1 and 2 show that the extent of degradation of the polymer residue was dependant upon the furnace temperature (i.e., heating rate), and the temperature of the sample. Decreasing the furnace temperature increased the extent of polymer degradation for a given sample temperature, probably the result of the additional time taken to reach the target sample temperature, which allowed for greater degradation of the polymer molecules. Increasing the sample temperature was also found to cause further degradation of the polymer.

Tables 1 and 2 show that there was a critical molecular weight at which wicking of the polymer residue occurred. For the low-permeability coating (Table 1) this was estimated to be between 10,000 and 15,000. Wicking was always observed below this molecular weight, for example, see Figure 5, where polymer residue was clearly seen in the pores of the coating to a depth of approximately 150 μm. However, above a critical molecular weight of 10,000 to 15,000, only wetting occurred, resulting in a smooth layer of polystyrene on the surface of the coating with no penetration of the liquid residue into the coating pores. The critical molecular weight at which wicking was able to occur in the high-permeability coating was found to be approximately 7 times greater (between 70,000 and 75,000).

Mould Filling Behaviour

Figure 6 shows a sequence of enhanced real-time x-ray images showing the filling behaviour as the liquid aluminium alloy filled an expanded polystyrene plate pattern coated with the low-permeability coating. In this example, and for the pattern coated with a high-permeability coating, there were no obvious signs of entrapment of pattern degradation products by the liquid metal. It was observed that the average metal velocity was much higher in the pattern coated with the high-permeability coating, about 18.0 mm s^{-1}, compared to 7.5 mm s^{-1} in the pattern coated with the low-permeability coating.

Figure 7 shows heating curves for a mould coated with the low-permeability coating. As the metal reached a given position, the measured temperature within the mould increased very rapidly (in less than 0.5 seconds), whereas the temperature of the coating rose much more slowly, requiring more than 30 seconds to reach a maximum temperature, which was about 400 °C.

Casting Quality

Examples of the surface quality of cast plates produced using both high- and low-permeability coatings are shown in Figure 8. The plates produced using the low-permeability coating were generally free of surface defects, whereas the plates produced using the high-permeability coating had markings indicative of liquid polystyrene trapped at the metal-coating interface at some time during mould filling.

The Ultimate Tensile Strengths of a series of tensile test bars produced from each cast plate were determined and used as a measure of the internal quality of the plates. Figure 9 shows that, despite the fact that the filling velocity was different for each pattern/coating combination, the mechanical properties of the cast plates were generally very similar, and were consistent along the length of the plates, suggesting very few internal defects. Isolated reductions in Ultimate Tensile Strength could be attributed to fold defects in the cast plates, as highlighted in Figure 8.

Discussion

The results from the current work indicate that the thermal degradation of the expanded polystyrene pattern is a complex process, with the rate of degradation being dependant on both the temperature of the polymer and also the amount of time that the polymer has been heated; i.e. the amount of heat the polymer has experienced. The partially degraded polystyrene residue was absorbed by each mica-based coating, as proposed by Sun et al. [7] and this contradicted the mechanism proposed by Zhao et al. [8]. However, wicking only occurred after the polymer had degraded sufficiently such that the molecular weight, and hence the viscosity of the polymer residue, was low enough to allow penetration of the pores of the coating. In this investigation, a critical molecular weight of between 10,000 and 15,000 was needed for wicking into the low-permeability coating to occur, while a molecular weight of between 70,000 and 75,000 was needed for the high-permeability coating.

The slow rates of temperature increase in each pattern coating during mould filling suggested that very little polymer was absorbed into the coating near the metal front, and that the polymer would be, instead, trapped at the metal/coating interface for a period of time after the metal front had passed, leading to surface defects on the casting, where absorption of the liquid polymer residue into the coating was not able to take place before complete solidification of the metal. Since higher metal velocities were associated with the high permeability coatings, less time was available for the degradation and wicking of the liquid residue into the coating at the metal/foam interface and, consequently, a larger number of surface defects were produced using this coating type.

Metal velocities in the pattern coated with the high-permeability coating were approximately double those where the low-permeability coating was used. This was most likely due to the fact that the insulating gases generated at the metal/foam interface could more easily escape from the high-permeability coating, thereby degrading the pattern more quickly than when the low-permeability coating was used. However, differences in metal velocity did not affect the stability of the metal/foam interface, which would explain why there were no significant differences in the internal quality of the castings produced.

Conclusions
1. The liquid polymer residue generated during the degradation of the expanded polystyrene pattern was observed to wet and wick two mica-based coatings.
2. Wetting and wicking occurred only after the polymer degraded sufficiently that a critical molecular weight of between 10,000 and 15,000 for the low-permeability coating, and between 70,000 and 75,000 for the high-permeability coating, had been obtained.
3. High metal velocities were detrimental to surface quality due to insufficient degradation of the polymer for absorption into the coating.

References

1. Rogers R C and Heine J, Foundry Management and Technology, October 1990, pp20-27.
2. Hill M, Vrieze A E, Moody T L, Ramsay C W and Askeland D R, AFS Transactions, 114, 1998, pp365-374.
3. Warner M W, Miller B A and Littleton H E, AFS Transactions, 161, 1998, pp777-785.
4. Liu J, Ramsay C W and Askeland D R, AFS Transactions, 137, 1997, pp419-425.
5. Shivkumar S, Yao X and Makhlouf M, Scripta Metallurgica et Materialia, 33, 1995, pp39-46.
6. Yang J, Huang T and Fu J, AFS Transactions, 128, 1998, pp21-26.
7. Sun Y, Tsai H-L and Askeland D R, AFS Transactions, 167, 1992, pp297-308.
8. Zhao Q, Burke J T and Gustafson T W, AFS Transactions, 83, 2002, pp1399-1414.

Acknowledgements

The authors are grateful to Rapra Technology Ltd, EPSRC, Foseco Morval, and Mr Adrian Caden of the University of Birmingham

Tables

Table 1. Weight average molecular weights of residues from expanded polystyrene samples coated with a low-permeability coating. Light grey shading = Wetting only, Dark grey shading = Wetting and Wicking.

Molecular Weight						
		Sample Temperature / °C				
		300	350	400	450	500
Furnace Temperature / °C	400	156000				
	450	138000	93500			
	500	167000	128000	31700		
	600	179000	170000	117000	11900	No signal
	800	221000	182000	168000	96400	13100
	1000	211000	191000	193000	169000	52000

Table 2. Weight average molecular weights of residues from expanded polystyrene samples coated with a high-permeability coating. Light grey shading = Wetting only, Dark grey shading = Wetting and Wicking.

Molecular Weight						
		Sample Temperature / °C				
		300	350	400	450	500
Furnace Temperature / °C	400	154000				
	450	153000	120000			
	500	163000	150000	39400		
	600	177000	172000	110000	20000	No signal
	800	189000	180000	177000	70500	13900
	1000	208000	215000	207000	152000	19900

Figures

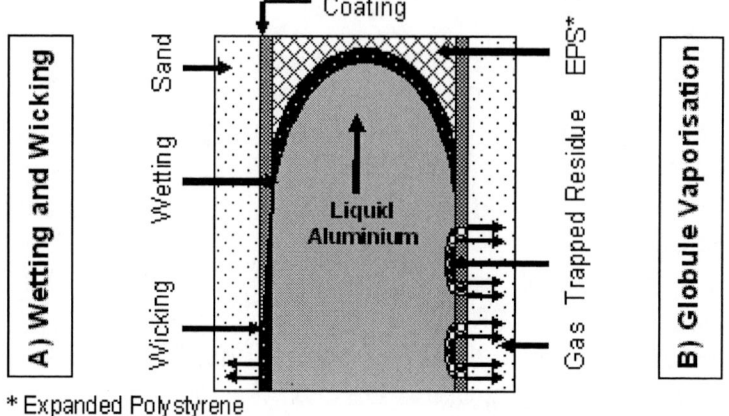

Figure 1. The two theories proposed for the removal of liquid degradation products by the pattern coating. A) Wetting and Wicking, B) Vaporisation model. Arrows indicate the flow of hot gases.

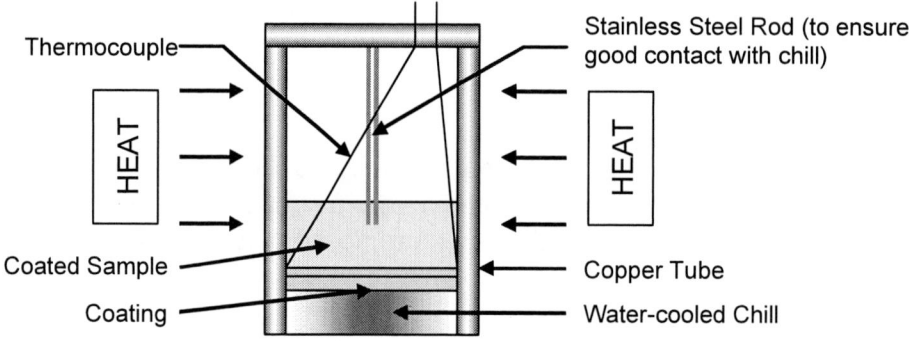

Figure 2. Schematic of the apparatus used in the polymer degradation experiments.

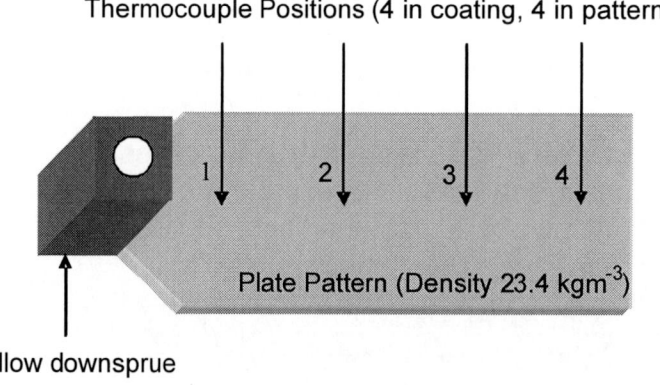

Figure 3. Schematic of the plate pattern showing thermocouple positions.

338

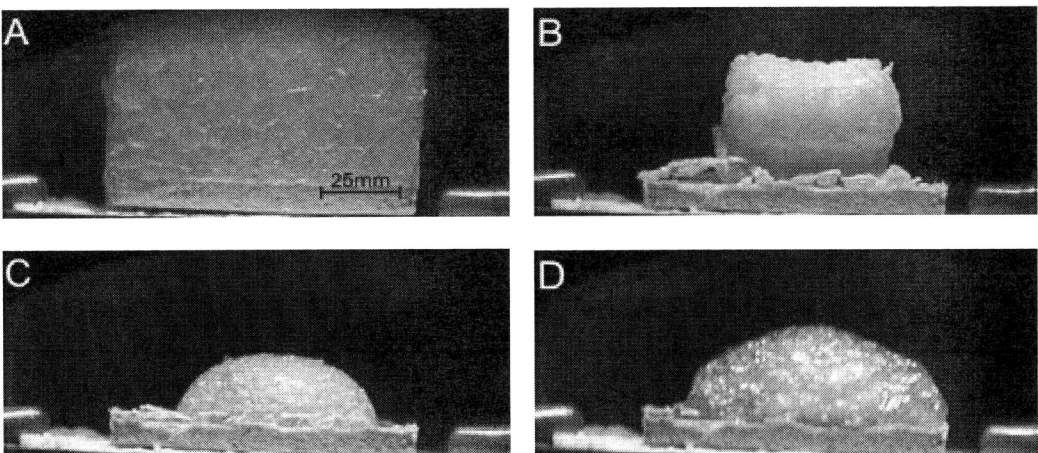

Figure 4. The appearance of expanded polystyrene coated with a low-permeability coating at various temperatures during heating. A) 25 °C, B) 160 °C, C) 350 °C, D) 390 °C

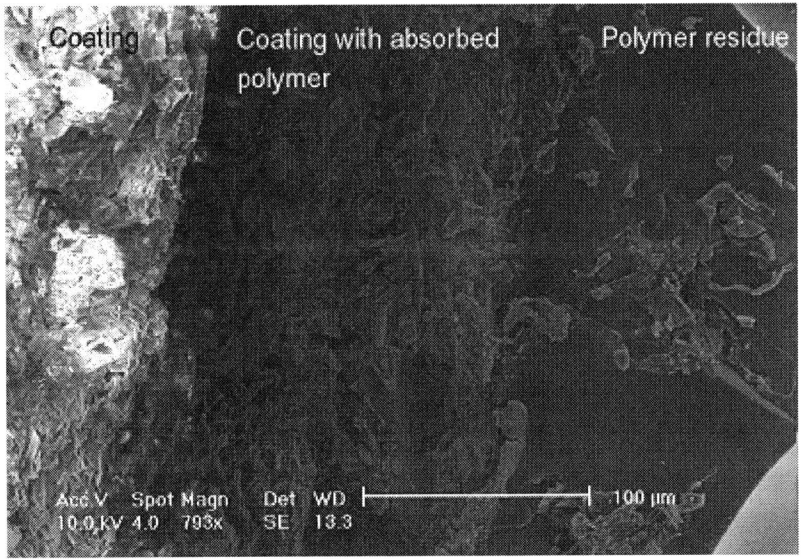

Figure 5. Coated polystyrene sample after heating, showing absorption of polystyrene residue to a depth of approximately 150 μm. (Some polystyrene residue remained on the coating surface).

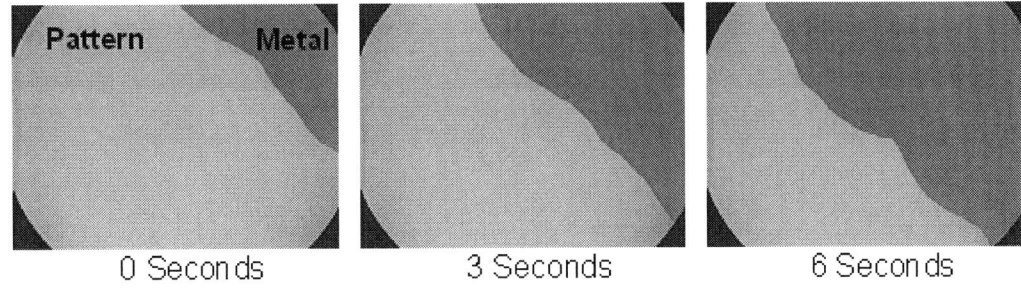

Figure 6. Observed mould filling behaviour with the low-permeability coating. Average metal velocity = 7.5 mm s^{-1}

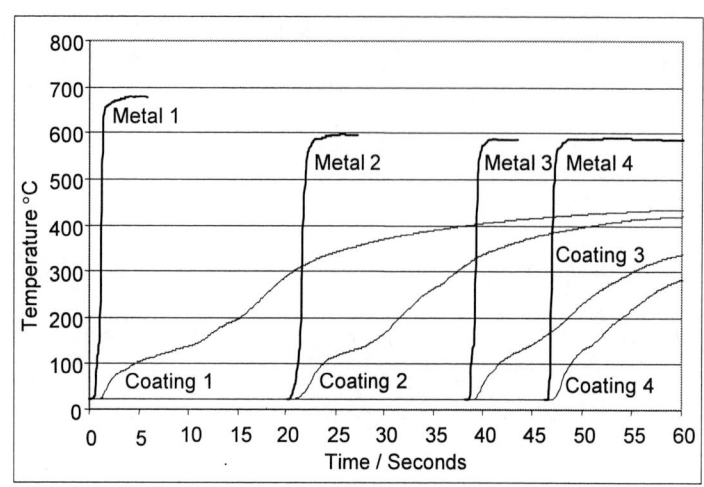

Figure 7. Temperature measurements made during mould filling in a pattern coated with the low-permeability coating.

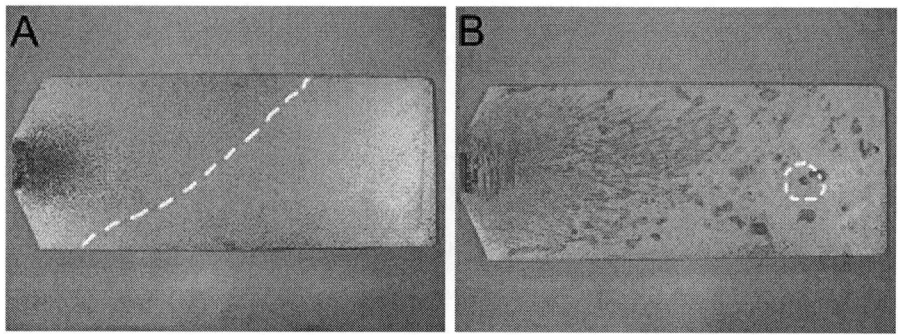

Figure 8. A) Cast plate produced with the low-permeability coating. B) Cast plate produced with the high-permeability coating. Dashed lines highlight fold defects.

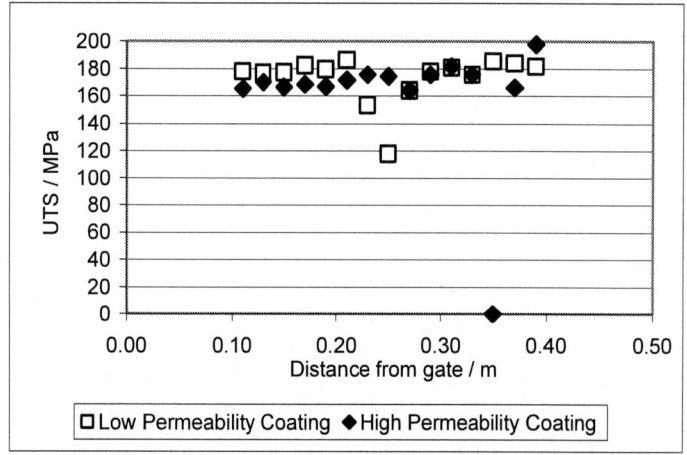

Figure 9. Variations in Ultimate Tensile Strength along the length of cast plates produced using different pattern coatings.

PJD862@bham.ac.uk, W.D.Griffiths@bham.ac.uk

Mould Filling in the Lost Foam Casting of Al Alloys

M. J. Ainsworth[1] and W. D. Griffiths[2]

1. Gemco Cast Metal Technology, Science Park Eindhoven 5053, Son, The Netherlands.

2. Department of Metallurgy and Materials Science, School of Engineering, University of Birmingham, United Kingdom. B15 2TT.

Abstract

In the Lost Foam casting process the filling of vertically oriented plates was viewed using a real-time X-ray technique to examine the interaction of the cast Al with the polystyrene foam pattern. The plates were cast by means of a counter-gravity technique and bottom-gated running system.

Real-time X-ray images showed that the advancing metal front was associated with the development of finger-like protrusions as mould filling velocity was increased. This inevitably led to the entrapment of the degradation products of the polystyrene pattern. Tensile properties of the plates showed a significant reduction when an unstable metal front was present during filling.

A similar type of instability was encountered during the displacement of liquid glucose by mercury in a Hele-Shaw cell. This analogue model of the process can be used to discover the conditions under which interface instability can occur in the Lost Foam casting process.

Key words Lost Foam casting, Aluminium, Real-time X-ray, Mechanical Properties.

Introduction

In Lost Foam casting the pattern is made from a expanded polystyrene shape, coated with a thin porous refractory coating, and placed in a moulding box and surrounded by loose dry sand. The liquid metal is then poured directly onto the pattern, vapourising it as it fills the mould. Lost Foam casting can therefore be used to create complex castings, that could not be made by a conventional casting route that requires the pattern to be removed from the mould.

Casting of aluminium into Lost Foam moulds, which is performed at significantly lower temperatures than with ferrous alloys such as cast irons and steels, has the effect of heating and melting the foam with little associated gasification and creates a viscous two-phase (vapour and liquid) residue ahead of the advancing metal front [1]. These polymer products are often entrapped by the advancing liquid metal.

In conventional open mould castings ingate number and position control the filling of the mould cavity, and greatly influence the quality of the final casting. In Lost Foam casting metal front velocity is an important factor in determining the quality of the final product, but the size, number and position of the ingates are not significant in controlling the metal fill velocity [2]. The "process choke" has been determined to be either the metal-foam interface or the metal-foam-coating interface, depending on whether the decomposition of the foam or removal of the decomposition products was the most influential [3]. The main focus of control of metal velocity has been on pattern composition, coating composition and coating thickness [4].

This aspect has been studied by a number of authors [2,4,5] who reported that critical metal velocity windows between 12 and 23 mms^{-1} have produced fully filled castings with a low incidence of defects. Misrun and sand collapse was observed below the minimum figure, and inclusions encountered above the upper limit.

Thus the velocity of filling of Lost Foam moulds has a significant effect on the level of pattern degradation products found in the final casting. Almost all metal front morphology studies have made use of a series of thermocouple or timing probes that have been placed in a pre-determined array in the pattern prior to casting. This method of profile morphology examination has a number of constraints that means that results can only be approximations even in the best cases, and lack definition, meaning that a precise relationship between metal front morphology and entrapment of degradation products has not yet been obtained.

Real-time X-ray imaging can overcome many of these limitations and, when applied to Lost Foam casting, has revealed that the metal front profile was planar at very low velocities, but "finger-like" as velocities increased[6]. The incidence of entrapment defects in castings was also higher when unstable metal fronts were observed.

The main focus of this work has been to investigate (i) the effects of various metal front morphologies on the quality of the castings produced, as indicated by their mechanical properties, and, (ii) the mechanism of metal front morphology variation during the filling of the mould. To achieve this counter-gravity filling techniques were applied to the filling of simple plate patterns oriented in the vertical plane with the metal front profile recorded by means of a real time X-ray instrument. Furthermore, a physical model was constructed to simulate, at room temperature, the effects of variation in velocity on the interface morphology of an advancing metal front.

Experimental
A simple plate pattern was employed to evaluate the metal front profile behaviour and the mechanical properties of castings filled by counter-gravity techniques. The patterns were made from pre-expanded, T-grade polystyrene beads on an industrial pattern moulding machine that was equipped with horizontally split aluminium tooling. All patterns were produced in one batch, and had a bead size of approximately 1 mm and a nominal pattern density of 29.9 kg/m^3.

After ageing for 1 day each pattern was attached to its own running system, made up of a short expanded polystyrene upsprue with a 10 mm diameter hollow center, and a single expanded polystyrene ingate of 10.4 cm^2 cross section. Pattern and running system components were manually assembled using a proprietary hot-melt glue. An alumino-silicate refractory, suspended in water, was applied by manually dipping each pattern until all but the bottom face of the upsprue was immersed, and resulted in coating thicknesses ranging from 200 – 275 µm. The pattern cluster was positioned in a flask and loose, dry, silica sand was used to fill the flask while vibration was applied to compact the sand.

Counter-gravity filling was achieved by using a pressure vessel containing a silicon carbide crucible (see Figure 1), that contained a pre-modified Al-7wt.%Si alloy into which one end of a refractory coated steel tube was inserted. The other end of the tube was fitted to the bottom inlet of the mould. Compressed air was used to pressurize the chamber to drive the liquid metal up the connecting steel tube into the mould at a controlled and reproducible velocity.

All castings were filled at a temperature of 785°C, and the filling process monitored using a 160 kV real-time X-ray with the image captured using a S-VHS video recorder at a frame rate of 50 Hz.

Specimens for tensile strength determination were taken from two plate castings, with one of the plate castings being filled with an unstable advancing liquid metal front, (obtained using a chamber pressure of 27.6 kPa), and the other being filled more slowly to obtain a planar advancing metal front, (using a chamber pressure of 24.1 kPa). The plate that filled with a planar metal front did not fill completely and exhibited a misrun defect along its upper edge. In this case samples were only obtained from the lower part of the plate. The tensile values obtained from the test bars were analysed using a Weibull distribution[7], an approach that more accurately describes the distribution of the tensile strengths of castings when compared with the Gaussian distribution.

A physical model was constructed to simulate the interaction of molten metal and liquid polystyrene observed during the counter-gravity casting of the plates, (shown in Figure 3). The model consisted of two parallel plane glass sheets, (known as a Hele-Shaw cell), with the same dimensions as the cast plates, and contained a layer of mercury overlaid with a layer of glucose. The Hg was chosen to represent the liquid Al because its viscosity of 1.55 mPas at 20°C [8] was relatively close to the viscosity as liquid Al (1.22 mPas at 20°C) [9], and because its density was below that of the glucose. The glucose, (with a viscosity of 95 Pas), therefore represented, with respect to the Hg, the higher viscosity, but lower density, polystyrene degradation products produced in the Lost Foam casting of Al.

Results
Countergravity Mould Filling

The application of pressures of 34.5 kPa and above within the casting unit gave rise to metal fronts in the polystyrene patterns with long finger-like protrusions that were present throughout the filling of the plate, as shown in Figure 4. In this case the average filling speed, captured from the real time X-ray image of a plate filled using a chamber pressure of 69 kPa, was 24 mms^{-1}, and the rate of fill was such that the liquid metal reached the top of the plate in 21 s. However, the liquid and gaseous pattern degradation products preceding the front could not escape through the mould coating at the same rate and became temporarily trapped in this upper region. It took a further 7 s for this trapped material to leave the mould cavity and pass into the surrounding sand.

The average filling velocity of plates cast with chamber pressures of between 27.6 and 31.0 kPa was 13 mms^{-1}. In these cases the metal front profile assumed a slightly convex shape, across the complete width of the plate, with the leading region being situated towards the centre of the plate (see Figure 5) and it was characterized by a series of small protrusions, approximately 5 mm in height, that gave it a "cellular" appearance.

A reduction of the chamber pressure by only 3.4 kPa to 24.1 kPa reduced the average filling velocity to approximately 5 mms^{-1}. This produced an almost horizontal metal front that had a profile devoid of any protrusions (shown in Figure 6) and was classified as a "planar" type front. However, the filling velocity was so slow that it resulted in the arrest of the metal front approximately 30 mm from the top of the plate and an incomplete casting.

Mechanical Properties

The Ultimate Tensile Strength values obtained from two cast plates are shown in Table 1, which shows that the properties associated with filling with a "cellular" metal front were significantly lower than those where a "planar" metal front was present during filling. The Weibull distribution was plotted for each of the two sets of UTS results and has been shown in Figure 7. The Weibull modulus associated with a "cellular" advancing metal front was only 10.6, whereas the value related to a planar metal front was higher, at 23.2, indicating a narrower spread of mechanical properties and hence fewer internal defects.

Metal – Foam Interface Modelling

The analogue casting model with the liquid mercury advancing into the glucose syrup produced a planar interface up to a vertical velocity of 20 mms^{-1}. As the speed of the metal front increased above this value, the interface became unstable. This was apparent by the formation of a series of fingers (see Figure 8) that were observed across the width of the interface. The displacement speed had a direct effect on the number and length of these fingers. At 22 mms^{-1}, one finger was seen emerging from the central area of the metal front. Its length grew to a maximum of 15 mm ahead of the main front during filling. In contrast, a velocity of 39 mms^{-1} caused the initially planar metal front to break up completely into 5 distinct fingers. These fingers were relatively long in comparison to their width and also increased in length during the displacement cycle. The largest finger was situated centrally along the metal front, directly above the ingate, and protruded at its maximum length to about 70 mm ahead of the tip of the nearest trailing finger.

Discussion

The main aim of these experiments was to determine the level of benefit, in terms of casting quality improvement that could be achieved, by application of counter-gravity filling to Lost Foam moulds, and to simulate the interaction of the metal front and liquid polystyrene during the filling process in a room temperature model.

The casting experiments showed that, as the velocity of the liquid Al front increased above a critical speed, it became unstable and began to break up into a number of fingers that preceded the main body of flow. The recombination of these individual metal streams is likely to have caused entrapment of degradation products, as well as leading to the incorporation of the oxidised surface of the liquid metal.

These problems were avoided when a filling velocity of 5 mms^{-1} was applied, but the metal front was arrested approximately 30 mm short of the top of the mould cavity, resulting in a short run defect in the final casting as a result of lack of metal fluidity in the final stages of filling caused by heat loss to the pattern and the mould. It is interesting to note that where the metal front was relatively stable, ie,. in planar or cellular form, its profile was convex, meaning that the pattern degradation products would be displaced to the sides of the pattern where they could be removed via the coating. The location of the ingate and the relatively controlled and constant flow achieved through counter-gravity filling were considered to be primary factors in achieving this.

Published literature has shown that castings manufactured in open cavity moulds, under controlled conditions and with well-designed running systems, have been associated with Weibull moduli of up to about 38. This contrasted strongly with the results obtained in this set of experiments, where a Weibull modulus of 10.6 was obtained from a plate associated with cellular filling and a value of 23.2 achieved by planar filling. These results indicated that any type of non-planar metal fill profile has a serious, deleterious effect on the reproducibility of the casting properties.

The room temperature model used to simulate the interaction of liquid aluminium and liquid polystyrene clearly demonstrated that displacement of a viscous fluid, (i.e., the polystyrene degradation products), by another liquid of a lower viscosity, (liquid Al), caused an instability to occur, once a particular displacement velocity was reached. This phenomena has been researched before[10], and is known as Saffman-Taylor Instability.

Despite the initial attractiveness of filling a Lost Foam casting from below, and the expectation that this would produce better castings, the casting experiments and the room temperature analogue experiments suggest that entrapment of the polystyrene degradation products by the advancing metal flow would be inevitable, if the filling velocity was too high, owing to this instability effect. However, by recognising the existence of this instability in the Lost Foam casting process, the results do suggest some criteria with which a Lost Foam casting can be designed and its filling controlled, to produce good quality castings.

Conclusions

1. Filling of Lost Foam plates from the bottom, with controlled counter-gravity conditions, showed that only very slow filling conditions were associated with reproducible mechanical properties, as characterized by the Weibull modulus approach. Counter-gravity filling with a velocity of about 5 mms^{-1} was associated with a Weibull modulus of 23, still much less than can be obtained with an open cavity shape casting.

2. Examination of the filling process using real-time X-ray equipment showed that liquid Al front velocities of around 13 mms^{-1} were associated with a break-up of the advancing liquid metal front into

short fingers, which would lead inevitably to the entrapment of pattern degradation products.

3. The real-time X-ray images showed that the uniform tensile strengths and higher Weibull modulus associated with the slower metal front velocity of 5 mms^{-1} were the result of a planar advancing liquid metal front, which would be associated with a reduction in the trapped degradation products.

4. Modelling the effect of velocity on the metal front profile with a Hg-glucose analogue of the Lost Foam casting process revealed similarities to the filling of Lost Foam moulds with aluminium. It is suggested that the irregular advancing liquid metal fronts in Lost Foam casting are a case of Taylor-Saffman Instability.

References

1. Zhao Q, Burke J T and Gustafson T W, *Foam Removal Mechanism in Aluminum Lost Foam Casting,* AFS Trans., vol. 110, p. 1399 – 1414, (2002).

2. Hill M W, Lawrence M, Ramsay C W and Askeland D R, *Influence of Gating and Other Processing Parameters on Mold Filling in the LFC Process,* AFS Trans., vol. 105, p. 443 – 450, (1997).

3. Lawrence M, Ramsay C W and Askeland D R, *Some Observations and Principles for Gating of Lost Foam Castings,* AFS Casting Conference, May 1998, Pre-print No. 98-112.

4. Wang C, Ramsay C W and Askeland D R, *Process Variable Significance on Filling Thin Plates in the LFC Process – The Staggered, Nested Factorial Experiment,* AFS Trans., vol. 105, p. 427 – 434, (1997).

5. Ramsay C W, Askeland D R and Tschopp M A, *Mechanisms of Formation of Pyrolysis Defects in Aluminum Lost Foam Castings,* AFS Casting Conference, April 2000, Pre-print 00-131.

6. Sun W L, Littleton H E and Bates C E, *Real-Time X-ray Investigations on Lost Foam Mold Filling,* AFS Trans., vol. 110, p. 1347 – 1356, (2002).

7. Weibull W, *A Statistical Distribution Function of Wide Applicability,* Journal of Applied Mechanics, vol. 18, p. 293 – 297, (1951).

8. Elert G, *The Physics Hypertextbook*, 1998.

9. Dinsdale A T and Quested P N, *The Viscosity of Aluminium and Its Alloys – A Review of data and Models,* J. Mat. Sci., vol. 39, p. 7221 – 7228, (2004).
   Saffman P G and Taylor G, *The Penetration of a Fluid into a Porous Medium or Hele-Shaw Cell Containing a More Viscous Liquid,* Proc. Royal Society A, A245, p. 312 – 329, (1958).

Tables

Ultimate Tensile Strength (Mpa)												
Fill Pressure	Test Piece Position											
	A	B	C	D	E	F	G	H	I	J	K	L
24.1 kPa	227	196	205	197	203	212	199	198	206	209	205	207
27.6 kPa	106	133	126	126	109	108	117	119	126	121	123	135
	M	N	O	P	Q	R	S	T	U	V	W	
24.1 kPa	215	222	211	190	220	202	217	212	n/a	n/a	n/a	-
27.6 kPa	130	140	128	94	128	134	131	137	140	144	124	-

Table 1 Ultimate Tensile Strength achieved at various locations
in plates cast in a counter-gravity fashion at two different filling pressures.

Figures

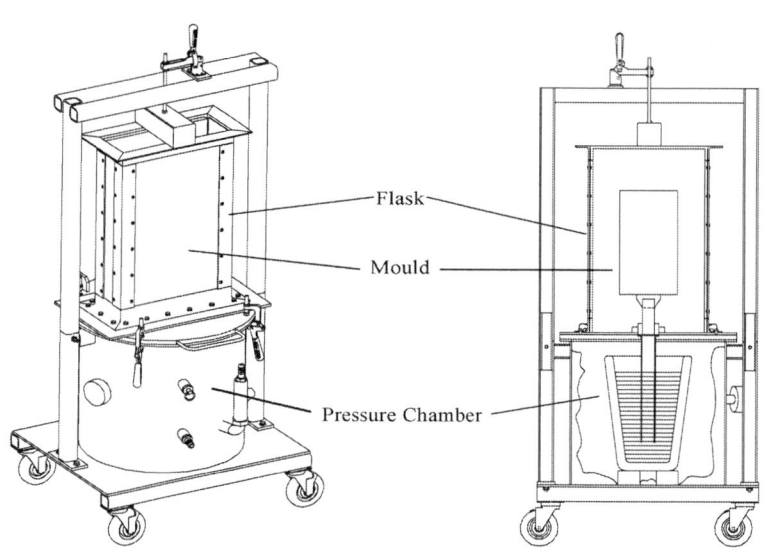

Figure 1(a). Isometric view of the counter-gravity casting machine with mould in position.

Figure 2. Section view of the counter-gravity casting machine with mould in position.

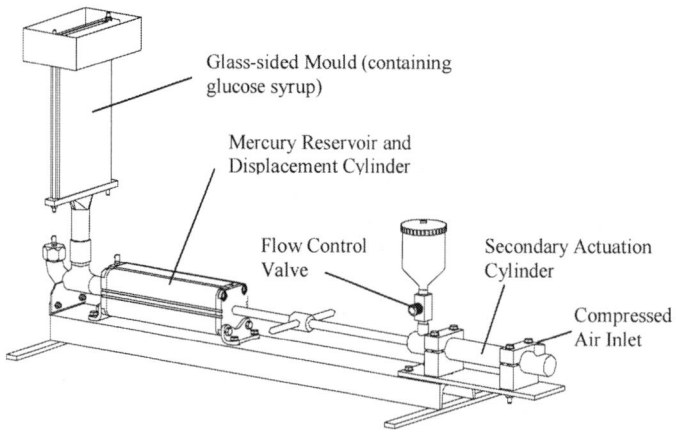

Figure 3. Isometric view of the physical model used to observe fluid interface morphologies at varying displacement velocities.

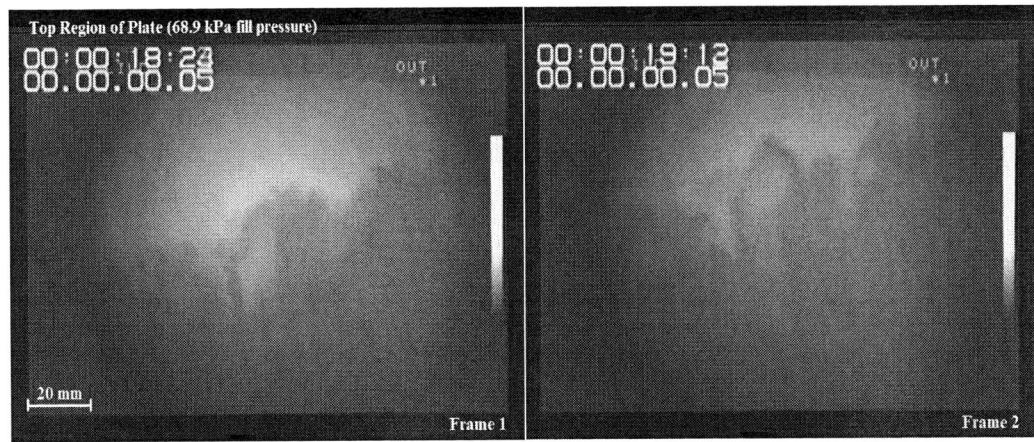

Figure 4. Finger-type front in the upper region of a plate; filling pressure = 69 kPa.

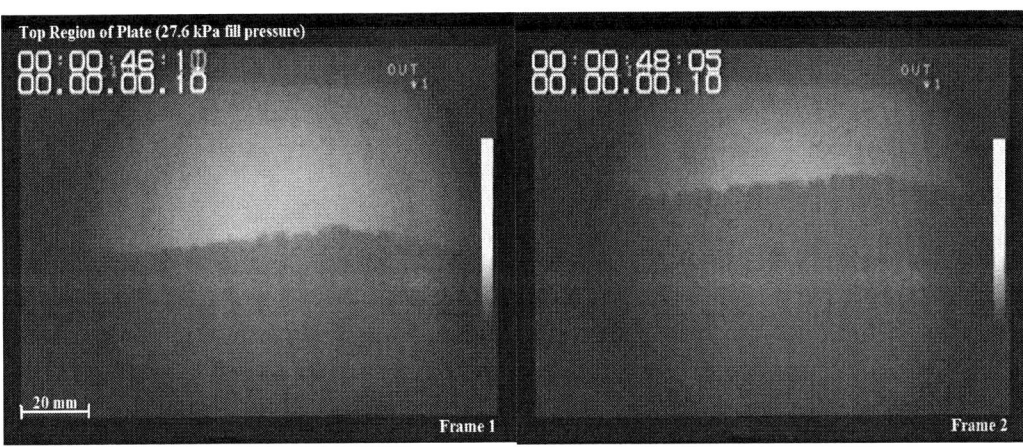

Figure 5. Irregular metal front in the upper region of a plate; filling pressure = 27.6 kPa.

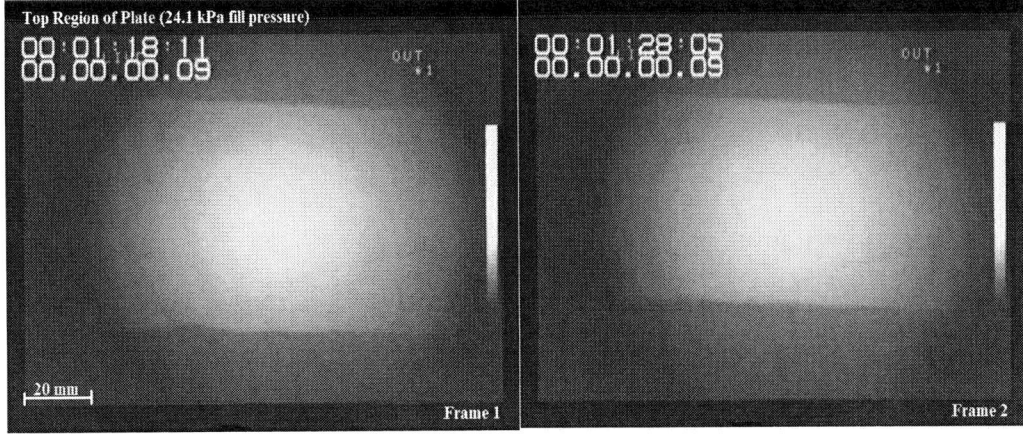

Figure 6. "Planar" metal front in the upper region of a plate; filling pressure = 24.1 kPa.

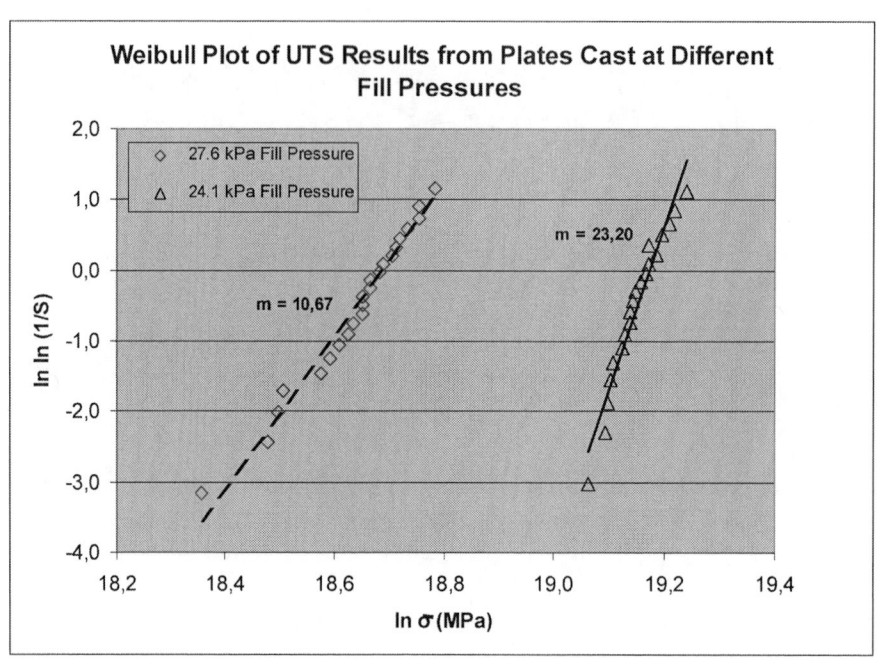

Figure 7. Weibull plots of the Ultimate Tensile Strength of samples taken from plates filled in a countergravity fashion at two different filling pressures.

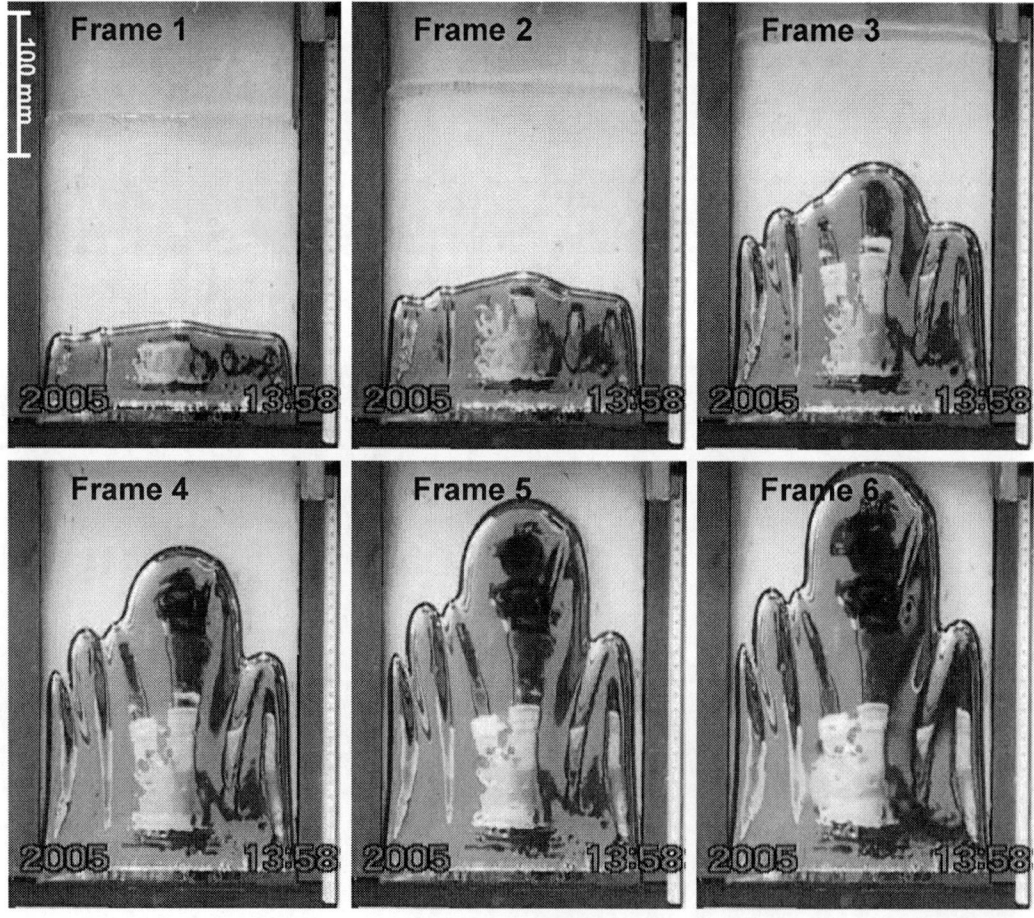

Figure 8. Saffman-Taylor Instability observed during the displacement of glucose syrup (viscosity 95 Pas) by liquid mercury (viscosity 1.55 mPas) in a Hele-Shaw Cell.

Rapid Shell Build For Investment Casting: Revolutionizing an Ancient Process

S Jones, C Yuan and S Blackburn

IRC in Materials Processing, University of Birmingham, UK.

Abstract

Investment casting is a time-consuming, labour intensive process, which can produce complex, high value-added components for a variety of specialised industries. Drying and strength-development of each coat during shell mould production is the most significant rate-limiting factor in the reduction of lead times and production costs for industry. As such, improvements which reduce cost and cycle times, open up new opportunities for product development.

A new shell build technique using a super absorbent polymer additive has been developed for mould production, such that moisture removal is not required to cause binder gellation for each individual mould coating. Therefore the drying time before a subsequent coat can be applied is dramatically reduced. All coats can be applied using the same novel additive thus leading to a large cumulative saving in overall mould processing time. This technology may allow the industry to successfully compete for the first time with other high volume forming methods as the traditional barrier of long processing times will be reduced. This paper introduces the technique, details shell development and results and follows the process through various iterations to the final proven concept.

Key words

Investment, superabsorbent, gellation, colloidal, binder

Introduction

Investment casting is a time-consuming, labour intensive process, which can produce complex, high value-added components for a variety of specialised industries. The principles can be traced back to 5000BC[1] when Early Man employed the method to produce rudimentary tools. This was followed by centuries of use for jewellery and artistic components[2] before the military requirements of the 2nd World War saw the process developed for aerospace and subsequently engineering components[3]. The term 'investment' casting derives from the characteristic use of mobile ceramic slurry, or 'investments', to form a mould with an extremely smooth surface[4], replicated from precise component patterns and transmitted in turn to the final metal casting itself. The process allows dimensionally accurate components to be produced and is a cheaper alternative than forging or machining since waste material is kept to a minimum[5]. The technique itself has tremendous advantages in the production of quality components and key benefits of accuracy, versatility and casting structural integrity. As a result, the process is one of the most economic methods of forming a wide range of metal components. Environmental[6] and economic[7] pressures have, however, resulted in a need for the industry to improve current casting quality, reduce manufacturing costs and explore new markets for the process. Optimisation of the mechanical and physical properties of the ceramic shell, whilst reducing process/material costs, will be instrumental to achieving these aims.

Following the introduction of the Environmental Protection Act in 1992, the degree of waste gases allowed from industrial processes was drastically reduced and resulted in the increased use of water-based (colloidal) binders within the UK industry to reduce volatile organic compound (VOC) emissions. Unlike previous alcohol based binders, water-based colloids are not chemically set and require sufficient moisture removal between coat applications to achieve gellation. Drying conditions, such as temperature, humidity and air speed need to be carefully controlled to avoid shell cracking and as a consequence the inter-coat dry times vary between 2 and 24 hours dependant upon which layer is being applied. Increased automation, worldwide foundry competition, rapidly increasing energy costs and the need to recycle constituent materials, has placed increasing pressures upon the industry. The incentive to reduce inter-coat dry times is tremendous and a variety of new processes aimed at reducing production times have been suggested and developed. However, the need to remove sufficient moisture to gel the colloidal binder and develop sufficient ceramic strength for re-dip has remained a fundamental time consuming and rate limiting feature. Here work carried out to explore the possibility of rapid gellation of the binder by alternative means, which at the same time produces shells which have sufficient strength to be re-dipped almost immediately removing the need for costly and time consuming moisture removal at this stage of the process is reported. The aim was wax to de-wax in less than 60 minutes, significantly reducing production costs by an estimated 75%.

As the basic technique itself is inherently expensive, improvements that reduce cycle times open up new product opportunities for foundries through a reduction in processing costs, lead times and an increased ability to be involved in rapid component prototyping. A new technique using super absorbent polymer additives[8] has been developed for mould production such that conventional moisture removal is not required to cause binder gellation and the drying stage is dramatically reduced. All investment coats can be carried out using the same novel additive thus leading to a large cumulative saving in processing time.

Conceptual Background

Gelcasting[9] was developed as a shaping method for advanced ceramics to overcome long production times associated with more traditional methods. A concentrated slurry of ceramic powder in a solution of organic monomer is poured into a mould and polymerised in situ to form a green body in the shape of the mould cavity. Extending this concept of gel solidification to include colloidal sol binders is achieved by utilising the effects of polyacrylamide and related absorbent polymers[10] upon water containing bodies. Colloidal silica can be rapidly flocculated[11] using polyacrylamide, with the added advantage that the resultant gel absorbs moisture and "locks" it within the structure whilst maintaining a degree of rigidity. Consequently, if the water absorbing properties of these materials could be used produce a ceramic body with sufficient strength to be autoclaved immediately then an advantageous process would be forthcoming. Moisture could then be removed during the standard firing schedule of the moulds.

Superabsorbant polymers are essentially synthetic hydrophilic non-toxic cross linked polymers that can absorb up to 800 times their own weight in water, but cannot dissolve because of their three-dimensional polymeric network structure. All superabsorbers are based upon the acrylate monomer[12] (Figure 1), which is an ethyl chain with a carboxylic acid side chain. The polymerisation of the acrylate monomers produces a linear molecule that has a very high molecular weight usually greater than one million monomer units, with high quantities of carboxylic side chains forming the polymeric network and giving the superabsorbant properties. In the dry powdered state the chains of the polymer are coiled and lined with carboxyl groups or (-COOH). When hydrated with water, the carboxyl groups dissociate into negatively charged carboxylate ions (COO^{-1}) which form bonds with the hydrogen atoms in water, 'locking' the water and preventing the polymer being taken into solution. The carboxylate ions repel one another along the polymer chain, thereby widening the polymer coils, allowing more water to move into contact with more of the carboxyl groups. As the polymer continues to uncoil the swelling forms a suspension with a gel-like consistency.

The different polymers available are modifications of polyacrylate achieved by changing of the chain length or by modifications of the carboxylic side group. Two common examples are sodium polyacrylate and polyacrylamide. The difference between sodium polyacrylate (Figure 2) and the polyacrylamide polymer (Figure 3) is the ion used to replace the hydrogen bonded onto the oxygen of the carboxylic acid, sodium polyacrylate has a sodium ion (Na^+) while the polyacrylamide has an amide (NH_2). When these polymers are hydrated with water, the amide group /sodium ion dissociates to form positive ions which form bonds with the OH^- ions in the water, 'locking' the moisture from within the water and preventing the polymer being taken into solution. Again the dissociated ions repel one another along the polymer chain, thereby widening the polymer coils. The sodium ion works more efficiently due to the greater ionic charge and allows the polymer to absorb approximately 800 times its weight in distilled water. The amide ion is considerably less efficient and such polymers will only absorb around 40-200 times their weight in water. This lack of ionic strength in the polyacrylamide polymer greatly reduces its water absorbing characteristics.

In the Rapid Shell Build technique[13], a superabsorbant polymer is incorporated into the stucco material before application. This material then rapidly uptakes moisture from the previous slurry coat upon contact to immediately gel the colloid, thus eliminating the time needed for controlled drying between each individual coating to achieve the same effect. Instead of removal, in essence the binder/slurry moisture is retained within the stucco until 300°C[14,15] and above when the polymer will release the bound water during the mould firing cycle. The use of these polymers for this technique is novel and innovative, even though the materials themselves were developed for widespread use in disposable nappies in the early 1980's[16]. Currently, the major uses for superabsorbant polymers are in medical science, production of soft contact lenses, disposable nappies, moisture retentive materials in agriculture and as filtration media for paper manufacture[17]. This represents a significant redevelopment of the basic principles of a casting technique that has remained largely unmodified for centuries. This paper introduces the technique, details shell development and follows the process through various iterations to the complete patented concept.

Experimental Shell Development
Colloidal silica can be rapidly flocculated using polyacrylamide, with the added advantage that the resultant gel absorbs moisture and "locks" it within the structure whilst maintaining a degree of rigidity. As such, the water absorbing properties can be tailored to rapidly gel individual layers of ceramic moulds. Any retained moisture can then be removed during controlled firing after complete mould production giving much reduced process times overall. To test this hypothesis, standard aluminium ceramic slurries were prepared and ceramic mould samples built by repeated dipping and stucco coating according to standard processing conditions. This gave a total preparation time of 2750 minutes (46 hours). At the same time a 10wt% polyacrylamide addition, in the form of a coarse particulate

material (0.3 to 1 mm), was made to the secondary stucco material (30/80 mesh alumino-silicate refractory) and comparable samples were prepared. The primary coat was produced conventionally (1240 minutes), however, to prevent excessive shell cracking and reduction of overall strength. Here the total preparation time was 1280 minutes (21 hours) and shells with reasonable structural integrity were formed (Figure 4), but there appeared to be high levels of re-absorption of moisture both during manufacture and de-wax. De-lamination and particle swelling occurred in this initial system, which led to a loss in strength and failure of the primary coat during de-wax and firing (Figure 5). Firing also appeared to induce extensive breakdown of the ceramic structure, possibly due to pyrolysis products emanating from the polymer particles weakening the ceramic through void formation.

The green shell was absorbing large quantities of excess water from the slurry, leading to unacceptable swelling and structural damage during shell build. In order to reduce this an alternative polymer was sought. Polyacrylamide was originally developed to grow plants in arid environments and has been refined to last longer and absorb water at a high rate than other classes of absorbant. Polyacrylate is a closely related member of the same polymer group but is capable of absorbing greater amounts of liquid and yet breaks down easily on heating. The greater absorption allows the particle size of the additive to be reduced significantly, whilst at the same time allowing a more controllable polymer removal stage. Further, because of the superior absorbency less material was required (2.5% by weight) to achieve the desired result. Polyacrylate addition modified ceramic mould samples were prepared. The polyacrylate had a average particle size of <300 mm and the primary coat was also polymer modified to further reduce shell preparation time by removing the need for the slow, controlled drying of this individual coat. This gave a total preparation time of 22.5 minutes (Figure 6). Though the shell build time was rapid, delamination of the primary coat still occurred during the shell firing schedule, which would be unacceptable for an investment mould.

The de-lamination and 'stripping' (removal of previous gelled coats) during shell manufacture and de-waxing was considered to be due to the volume expansion of the individual polymer particles as water is absorbed and the particles 'swell'. The stripping effect may also have been due to the polymer being introduced as a 'discrete' particle, preventing all the moisture from the slurry layer being removed from the colloidal binder before the next coat was dipped. There is a limit to the rate of moisture transport through a capillary network and the use of a small number of large polymer particles would have increased the average depth through which water molecules would have to travel before being 'locked' into the polymer structure. As the next layer is dipped, there would be an excess of unlocked moisture within the colloidal network, preventing binder gellation and promoting 'breakdown' of the already gellated structure. The delamination and cracking of the shell structure during firing was possibly due to a thermal mis-match between ceramic/colloid/polymer addition or the promotion of excessive expansion due to volatilisation of the polymer particles. Discrete particle additions would have a high concentration of polymer in one particular location leaving holes as this is removed, leading to weakening of the ceramic structure.

Coating the individual stucco particles with the polymeric material, rather than having discrete particles within the shell structure can achieve a more even distribution of the polymer. Such a distribution would reduce expansion cracking and thermal mis-match during production and firing. This would also distribute a larger percentage of polymer throughout the entire coating structure, reducing the average transport distance for water molecules, preventing re-wetting and stripping of previous coats during rapid dipping. A mixture of polymer, stucco and deionised water was produced in a high shear mixer and force dried to remove excess moisture. The acrylate content (0.25 wt%) was an order of magnitude less than that used in previous trials where discrete particles were added. The coating was accomplished by using the polymers ability to form a liquid gel by hydration and dehydration. Once in gel form the stucco is added and the resultant mixture agitated. Dehydration of the gel produces a solid mix of stucco in polymer. This material is re-graded into the original particle distribution (by light grinding) to produce a stucco material, which is now *partially* covered by superabsorbant material (Figure 7). Ceramic mould samples were dipped in 40 minutes with another 1080 minutes (18 hours) of a final dry before de-wax to stabilise the shell system. With the much reduced polymer content in the stucco phase, the modified samples did not crack at all during de-wax (Figure 8). The entire shell, both primary and secondary layers were intact as a result of the new polymer addition process. After firing the shell structure (including the primary coat) was intact, producing a structurally sound mould for investment casting.

Conclusions
Drying and strength-development of each coat in the manufacture of investment shell moulds is one of the most significant rate-limiting factors in the reduction of lead times and production costs. As such, improvements which reduce cost and cycle times, open up opportunities for product development, cost savings and the environmentally sound practice of reduced energy consumption. An alternative method of individual slurry coat colloidal binder gelation, using a super absorbent polymer additive to rapidly remove binder moisture has been developed for investment mould production, such that time consuming removal by individual coat drying in controlled atmospheres is no longer required. The system has been proven as an industrial alternative, requiring little capital cost or equipment replacement, as current systems can easily be adapted. There is potential for decreased labour requirements and material costs and the current lead times from wax/casting can be greatly decreased allowing current components to be produced faster. Improvements that reduce cycle times open up new product opportunities for foundries

through a reduction in processing costs, reduced energy consumption and an increased ability to be involved in rapid component prototyping. This represents a significant redevelopment to the basic principles of a casting technique that has remained largely unmodified for centuries.

References

1. Taylor P R, An Illustrated History of Lost Wax Casting, Proc. 17th Annual B. I. C. T. A. Conference, September, 1983.
2. Kotzin E L, Metalcasting and Molding Processes, American Foundrymens Society Inc., Illinois, USA.
3. BARNETT S O, Investment Casting - The Multi Process Technology, Foundry Trade Journal International, **11** (3), 1988, pp33-37.
4. Beeley P R and Smart R F, Investment Casting, Institute of Materials, 1st Edition, 1995, ISBN 0901716669.
5. Jones S and Marquis P M, The Role of Silica Binders in Investment Casting, British Ceramic Transactions, **94**, No 2, 1995.
6. Environmental Protection Act, Crown Copyright 1990
7. Rosskill Information Services, 2004, http://www.roskill.com/news
8. Jones, S., Patent Filing: Improved Investment Casting Process, O2004014580 published 19th February 2004.
9. Omatete O O, Janney M A and Nunn S D, Gelcasting: From Laboratory Development Toward Industrial Production, J. Europ. Ceram. Soc., **17** (1997), pp407-413.
10. Dalian Guanghui Chemical Company Ltd, Literature, http://www. ghhx.com.cn/e_noname2.htm
11. Baltar C A M and Oliveira J F, Flocculation of colloidal silica with polyacrylamide and the effect of dodecylamine and aluminium chloride pre-conditioning, Minerals Engineering, 11, no. 5, 1988, pp255-267.
12. Mukerjee M , 'Superabsorbers' article, http://www.Sciam.com
13. Jones, S, Patent Application: GB0402516.9, filed 5th February 2004.
14. Klug, F, Method for removing volatile components from a gel-cast ceramic article, US Patent US2002/0109249A1 August 15th 2002.
15. J Boisvert J, Persello J, Castaing J and Cabane B, Dispersion of alumina-coated TiO_2 particles by adsorption of sodiumpolyacrylate Colloids and Surfaces A: Physicochemical and Engineering Aspects,**178**, (1-3), March 2001, pp 187-198
16. Butterworth G A M and Elias R T, Disposable Absorbent Pad, US Patent 3,967,623 July 1976.
17. Macro Galleria. The University of Southern Mississippi. http://www.psrc.usm.edu.

Acknowledgements

The authors wish to gratefully acknowledge EPSRC (GR/R81480/01 ROPA Realising Our Potential Awards) for funding this project.

Figures

$$--[CH_2--CH]_n--$$
$$|$$
$$C=O$$
$$|$$
$$O$$
$$|$$
$$H$$

Figure 1: The simplest acrylate – Poly (acrylic) acid

$$--CH_2--CH(CO_2Na)$$

Figure 2: The monomer for sodium polyacrylate

Figure 3: The monomer for sodium polyacrylate

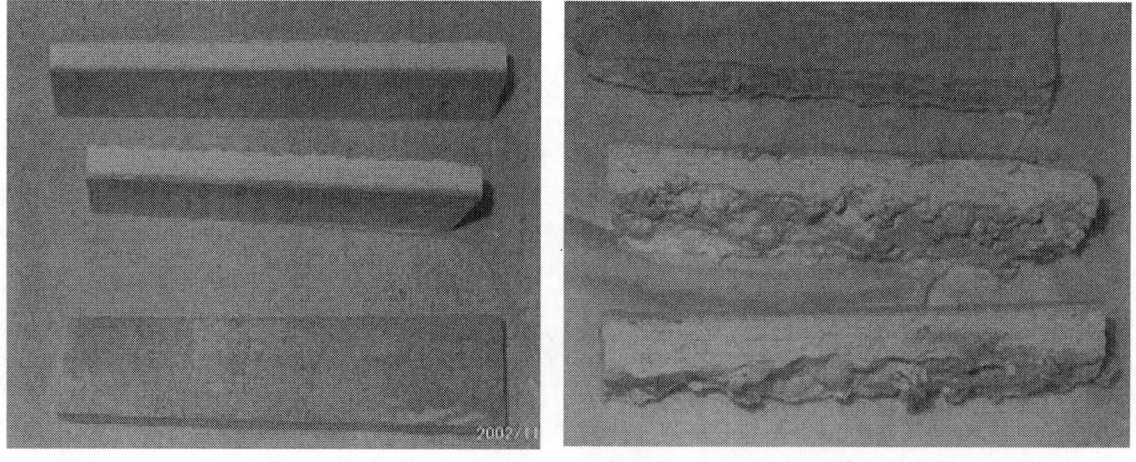

(a)   (b)

Figure 4: Comparison of initial aluminium shell build (a) Standard ceramic shell (b) Initial development shell (10 wt% acrylamide particles)

Figure 5: Initial development shell (10 wt% acrylamide particles) showing extensive delamination after de-wax

(a)   (b)

Figure 6: Comparison of initial aluminium shell build (a) 2.5wt% shell made in 22.5 minutes and de-waxed immediately (b) 2.5wt% shell made in 22.5 minutes and de-waxed after 12 hours

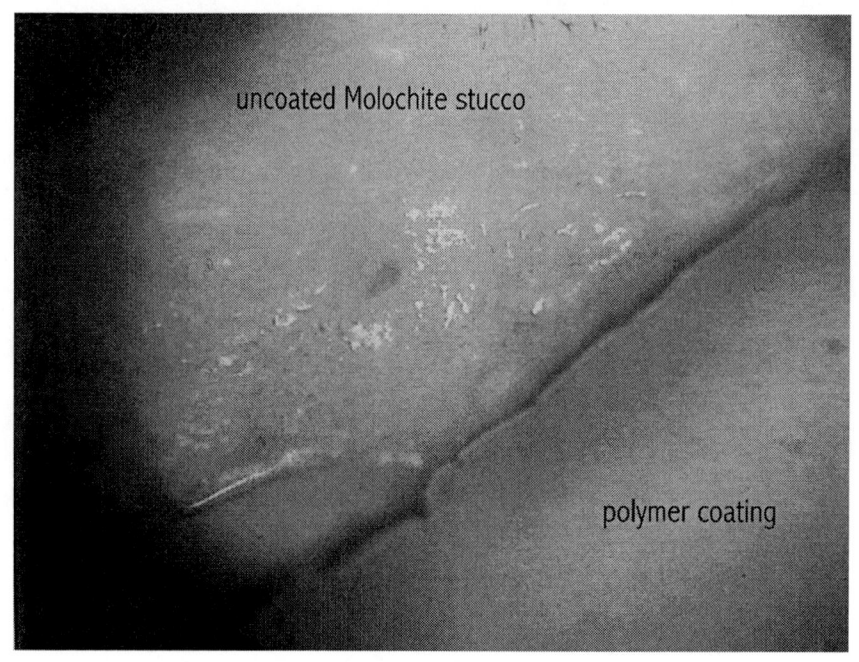

Figure 7:   Optical image of partially coated stucco particle

(a)   (b)

Figure 8:   (a) green 0.25% modified shell (b) fired 0.25% modified shell

Fabrication of Ni-Al Intermetallic Compounds on the Al Casting alloy by SHS Process

G.S. Cho[*], K.R. Lee[*], K.H. Choe[*], K.W. Lee[*] and A. Ikenaga[**]

*Advanced Material R/D Center, KITECH, 994-32 Dongchun-dong, Yeonsu-gu, Incheon, 406-130, Korea

* Osaka Prefecture University, 1-1 Gakuen-cho, Sakai, Osaka 599-8531, Japan

Abstract

Combustion synthesis with the advantages of time and energy savings has been recognized as an attractive alternative to the conventional methods. In order to improve the surface properties of Al casting components, Ni-Al intermetallic compounds are fabricated by the SHS(Self-propagating High temperature Synthesis) process using the reaction heat of the elemental powders. Three kinds of nickel aluminides, Ni_3Al, NiAl and $NiAl_3$ were produced by the emission heat from the Al molten metal and a coating layer of intermetallic phase was simultaneously formed on Al casting alloy surface. Microstructure and phase formation behavior of Ni-Al based intermetallic compounds synthesized by combustion reaction were investigated in terms of thermal and phase analysis using scanning electron microscope(SEM), energy dispersive x-ray spectrometer (EDS), Electron Probe Micro Analyzer (EPMA) and X-ray diffractometer(XRD). The microstructures of nickel aluminides were varied by casting temperature, mixing condition and mixing ratio of elemental powders. The reaction layer formed between Al alloy and the powder mixtures of nickel aluminide was observed to be different depending on process conditions such as green density of elemental powders, molten metal temperature and chemical composition. The differences of junction behavior with pouring casting alloys were compared and the schematic model of bonding reaction was defined.

Key words

Ni-Al intermetallic compounds, combustion reaction, casting process, Al alloy, reaction layer.

Introduction

Nickel aluminide intermetallic compounds are regarded as promising candidates for the development of the next generation of high-performance high-temperature structural materials[1,2]. In the Ni-Al systems, there are three typical intermetallic compounds, $NiAl_3$, NiAl and Ni_3Al. Lately, considerable interest has been focused on the reactive synthesis of these materials. In reactive processing, two intimately mixed metallic reactants A and B react to form an intermetallic product A_xB_y. If the reaction is highly exothermic, the process is called self-propagating high temperature synthesis(SHS)[3,4]. By utilizing the SHS reaction, it is expected that near net shaped compound can be obtained from the elemental powder. If Ni-Al intermetallic compound is formed on the surface of casting alloy, it is anticipated that its casting surface will be improved and show good wear resistance. In the previous work[5,6], Ni-Al intermetallic compounds were formed onto the solid substrate by hot pressing method for joining intermetallic compound and for surface modification via solid-state process. The solid-state joining process needs large equipment and high cost to apply it on the commercial components.

In the present work, by setting the mixture of elemental Ni and Al powders in a casting mold, the powder mixture reacted to form Ni-Al intermetallic compound by SHS reaction ignited by the heat of molten casting metal and simultaneously bonded with the Al casting alloy. By the presence of hard intermetallic compound on Al casting surface, wear property will be remarkably improved. By applying the combustion synthesis to the dissimilar bonding, it is expected that joining can be easily achieved under easier bonding condition, i.e., lower bonding temperature or shorter bonding time. We attempted to investigate in-situ joining method for surface modification via the formation of Ni-Al intermetallic compound on the Al casting alloy.

Experimental

Table 1 shows the chemical composition of Al casting alloy. The sizes of Ni, Al and Si elemental powders with 99.9% purity were 3, 3 and 10μm respectively, which were mixed in the composition of Ni - 75at.%(Al-0, 4, 8, 12at.% Si), Ni-50at.%(Al - 0, 4, 8, 12at.% Si), and Ni - 25at.%(Al-0, 4, 8, 12at.% Si). Composition of these mixtures is shown in Table 2. These mixtures were then cold-pressed with a load of 700MPa for 60s into disc-shaped compact of 20mm in diameter and 3-5mm in thickness. The relative density of the compact was in the range of 82-92%. These green compacts were set in a sand mold and molten Al alloy was poured at 700, 800 and 900 . The schematic drawing of sand mold and green compact for in-situ joining is shown in Fig.1. The molten pure Al was poured at 750 to compare the differences of junction behavior with various casting alloys. The Al casting alloy and pure Al bonded with the Ni-Al intermetallic compound was sectioned and observed by optical microscopy and scanning electron microscopy (SEM). The composition of the intermetallics in the coated layer and reaction layer formed in the interface were identified by energy dispersive spectroscopy (EDS), Electron Probe Micro Analyzer (EPMA) and X-ray diffraction (XRD) analysis. The effect of Si addition on the reactivity of the powder compact was examined by differential thermal analysis(DTA) at the heating rate of 20K/min under Ar flow atmosphere.

Results and Discussion

Fig.2 shows the as-cast cross sections of Al alloy bonded with different composition of Ni-Al based compacts poured at 800 . Bonding interface of Ni-75at.%Al and Ni-50at.%Al compact is not clearly found between the reacted Ni-Al compact and Al casting alloy. The formed Ni-Al intermetallic compounds are mixed with Al casting alloy and swelled off during the SHS reaction. It is considered that the liquid Al casting alloy is penetrated into and mixed with the Ni-Al intermetallic compound during the combustion synthesis. But Ni-25at.%Al compact corresponding with the equilibrium phase of Ni_3Al showed a curved linear interface and well bonded with the Al casting alloy.

Fig.3 shows the as-cast macrostructures of Ni-25at.%(Al-0at.%Si) compact with different pouring temperatures. When the pouring temperature increased to 900 , the bonded coating layer was fractured at the inside of Al casting alloy. Also the bonded interfaces formed at lower pouring temperatures were curved. It was probably due to the difference of thermal expansion between Ni-Al intermetallic compound and Al alloy. The two kinds of intermetallic compounds form in Ni-75at%(Al-0at%Si) and Ni-50at%(Al-0at%Si) compacts[7]. The $NiAl_3$ and Ni_2Al_3 intermetallic compounds were formed by the SHS combustion synthesis ignited from the heat of Al casting alloy. Ni_2Al_3 phase was surrounded by the $NiAl_3$ intermetallic compound

Fig. 4 shows typical XRD patterns of intermetallic compounds formed in Ni-75at.%(Al-0at.%Si), Ni-50.at%(Al-0at.%Si) and Ni-25at.%(Al-0at.%Si) compacts. For the sample with a composition of Ni-75at.%Al, diffraction peaks of Al and $NiAl_3$ were mainly detected but Si peak from the Al casting alloy was also recognized due to the mixed compacts as shown in Fig.2. The phase of Ni_2Al_3 was detected at the composition of Ni-50at.%Al compact. There was no peak of unreacted Ni in the XRD patterns, which means that the Ni particle is fully reacted with Al and consumed by Ni-Al intermetallic compound formation. Only the Ni_3Al phase was formed at the composition of Ni-25at.%Al synthesized compact. It was considered that the Al peak of Ni-50at.%Al and Ni-75at.%Al specimens came from the melted Al powders and mixed Al casting alloy.

Fig.5 shows the optical microstructures and SEM image showing the bonded interface between Ni-25at.%Al intermetallic layer and Al casting alloy. Pores in the upper part of Ni-25at.% coating layer were observed near the bonded interface. The reaction synthesized materials often contain significant amounts of porosity. This porosity can arise from various sources such as intrinsic pores, solidification shrinkage and volume change[1]. It is considered that pores remained near the bonded interface are under a much lower compression stress state than the lower part of the reacted compact due to the residual thermal stress.

Fig.6 shows the results of EPMA mapping around the bonded interface of Ni-25at.%Al compact. Si element is enriched near the bonded interface, which means that liquid Al alloy near the interface solidified in the end as Al-Si eutectic structure. It is considered that the exothermic heat of Ni-Al intermetallic formation and pore distribution near the interface decreases the solidification rate of liquid Al alloy near the Ni-25at.%Al compact. Also the thermal residual stress will be applied to the bonded interface during cooling, the weak Al-Si eutectic structure near the bonded interface will be cracked and fractured.

Fig.7 shows microstructure of the reaction layer formed between the Ni-25at.%Al compact and Al casting alloy. $NiAl_3$ intermetallic reaction layer of 25μm thickness was formed between Ni-25at%Al compact and Al alloy. The $NiAl_3$ phase will be immediately formed through interdiffusion of the liquid Al and solid Ni elemental powder as the Al casting alloy is poured into the sand mold. Also the crack line was observed in the Al-Si eutectic layer near the bonded interface.

Fig. 8 shows the cross section of the interface between the Ni-25at%Al compact and pure Al. The Ni particles diffused into the pure Al during the reaction and formed a diffusion layer about 700μm in thickness near the interface, which means that liquid pure Al near the interface solidified in the end as $NiAl_3$-Al eutectic structure. In this $NiAl_3$-Al eutectic layer, some cracks and fractures were observed near the interface by the thermal residual stress.

Fig.9 shows the bonded interface between the Ni-25at%Al compact and pure Al. The $NiAl_3$ and Ni_2Al_3 reaction layers were sequentially formed at the interface with 10μm in thickness respectively. Ni_3Al intermetallic compound was observed in the compact because the molten Al was reacted with the Ni particle in the compact. But the ratio of Ni particle was higher at inner part of the compact than surface, so $NiAl_3$, Ni_2Al_3 and Ni_3Al intermetallic compounds were formed sequentially.

Fig.10 shows the schematic modeling of the junction behavior at the interface between Ni-25at.%Al compact and casting alloys when the combustion synthesis started at the surface of the compact and the reaction was completed. The ignition is started when the Al powder melted in the compact enough to accumulate the heat from the molten Al alloy. This reaction heat was propagating into the inner part of the compact and also remelting the Al alloy were cooling after pouring into the sand mold near the interface. So, in this remelting zone, Al-Si and NiAl$_3$-Al eutectic layer was formed due to the liquid Ni-Al intermetallic compounds of the compact interdiffused with the casting alloy near the interface. Moreover, these weak eutectic layers near the bonded interface will be cracked and fractured.

Conclusions

1. Initial disc-shape of Ni-75at%Al and Ni-50at%Al compacts was lost. However, initial disc-shape of the Ni-25at%Al compact was hold.

2. NiAl$_3$ and Ni$_2$Al$_3$ intermetallic compounds were observed in the Ni-75at%Al and Ni-50at%Al compacts, respectively. And Ni$_3$Al intermetallic compound was formed in the Ni-25at.%Al compact and NiAl$_3$ intermetallic reaction layer with about 20 μm in thickness was observed at the bonded interface.

3. Al-Si eutectic layer was formed after completed solidification in Al alloy near the interface and some cracks were observed at this Al-Si eutectic zone.

4. Ni is diffused into the pure Al and formed NiAl$_3$ intermetallic layer. The NiAl$_3$ and Ni$_2$Al$_3$ reaction layers are sequentially formed at the interface.

References

1. K. Morsi: Review reaction synthesis processing of Ni-Al intermetallic materials, , Materials Sci. and Eng., A299, 2004, pp1-15.
2. N.S.Stoloff, C.T.Liu and S.C.Deevi: Emerging applications of intermetallics, Intermetallics, Vol.8, Issues 9-11, 2000, pp.1313-1320
3. Z. A. Munir: Ceram.Bull., 67, 1988, pp.342
4. A. G. Merzhanov: Int.Chem.Eng., 20, 1980, pp.150
5. T.Kimata, K.Uenishi, A.Ikenaga and KF.Kobayashi: Mater.Trans., 44, 2003, pp.407
6. T.Kimata, K.Uenishi, A.Ikenaga and KF.Kobayashi: Intermetallics, 11 (2003), p.947-952
7. G.S. Cho: *in press*

Tables

Table 1 Chemical composition of Al casting alloy (mass%)

Si	Mg	Fe	Cu	Cr	Al
7.22	0.428	0.168	0.013	0.035	Bal.

Table 2 Compositions of Ni-Al-Si mixtures

Sample name	Chemical composition (at.%)		
	Ni	Al	Si
Ni-75at.(Al-12at.%Si)	25	66	9
Ni-50at.(Al-12at.%Si)	50	44	6
Ni-25at.(Al-12at.%Si)	75	22	3

Figures

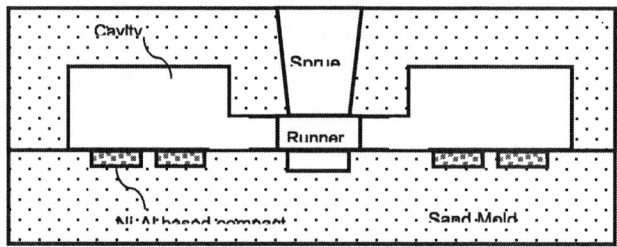

Fig. 1. Schematic illustration of sand mold and green compacts.

Fig. 2 As-cast cross sections of the synthesized products with different composition of (a) Ni-75at.%(Al-4at.%Si), (b) Ni-50at.%(Al-8at.%Si) and (c) Ni-25at.%(Al-8at.%Si).

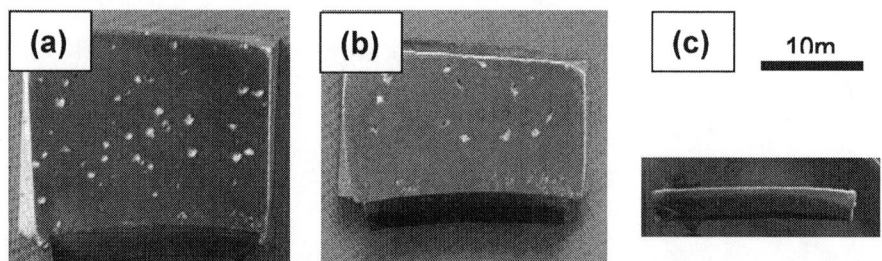

Fig. 3 As-cast cross sections of Ni-25at.%(Al-0at.%Si) compacts with different pouring temperatures of (a) 700 , (b) 800 , (c) 900 .

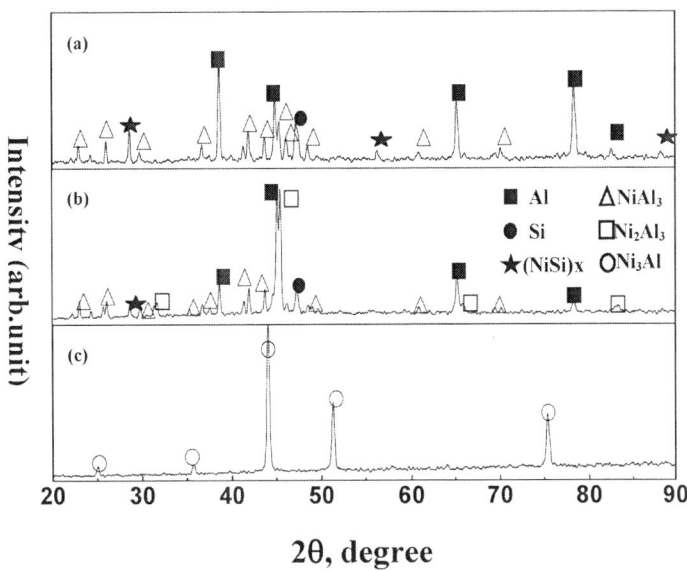

Fig. 4. X-ray diffraction patterns of combustion synthesized compacts with different composition of (a) Ni-75at.%Al, (b) Ni-50at.%Al, (c) Ni-25at.%Al

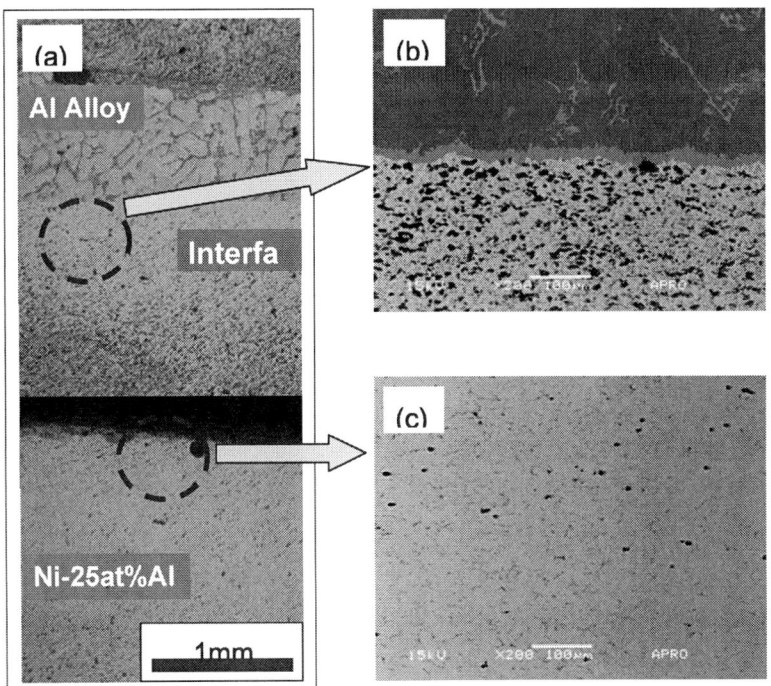

Fig. 5. Bonded interface of Ni-25at.%Al compact showing (a) optical microstructure, (b) SEM image of interface and (c) SEM image of inner part of the reacted Ni-25at%Al compact.

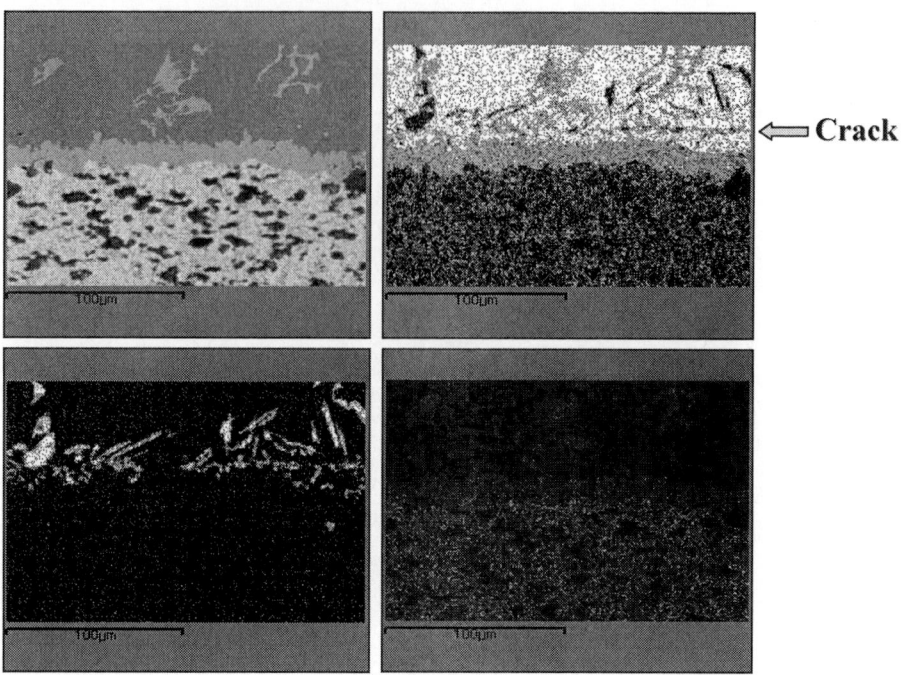

Fig. 6 EPMA mapping showing the bonded interface on Ni-25at.%Al compact

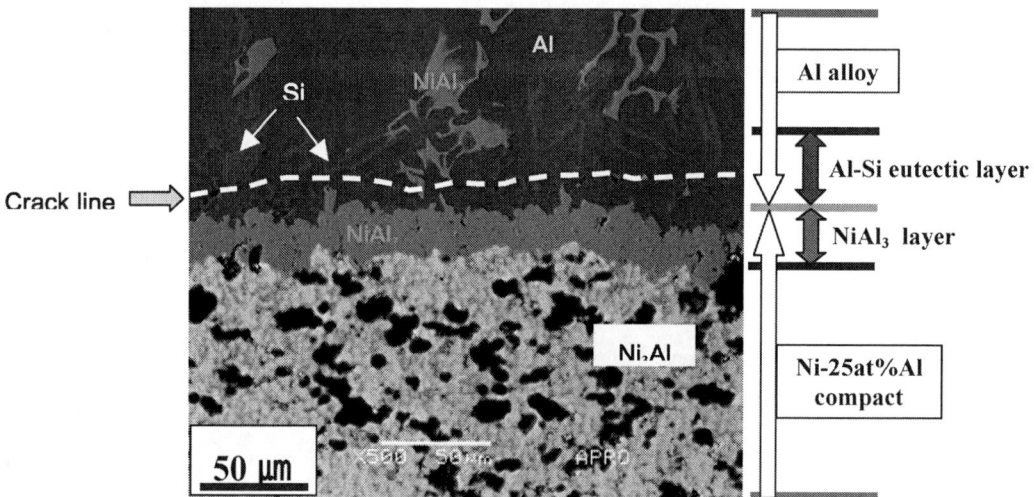

Fig. 7 Reaction layer formed between the Ni-25at.%Al compact and Al casting alloy.

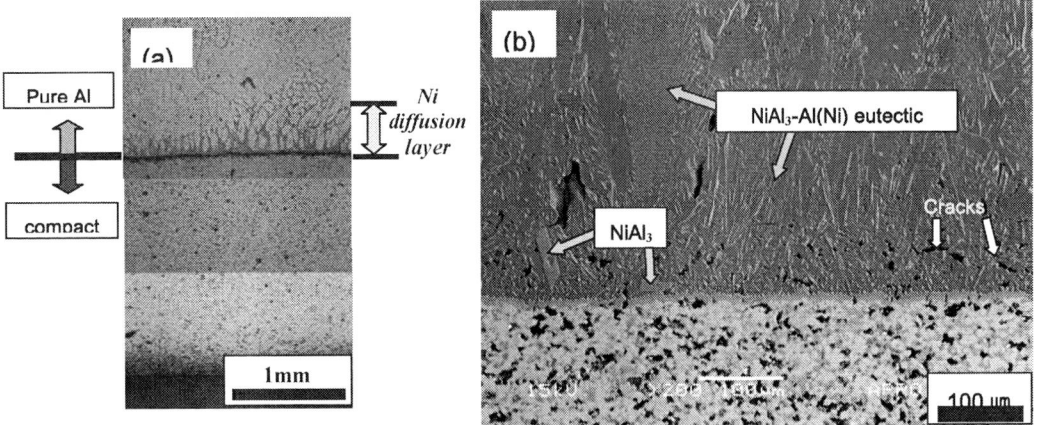

Fig. 8 The cross section of the interface between the Ni-25at.%Al compact and pure Al
(a) optical microstructure and (b) SEM image of Ni diffusion layer

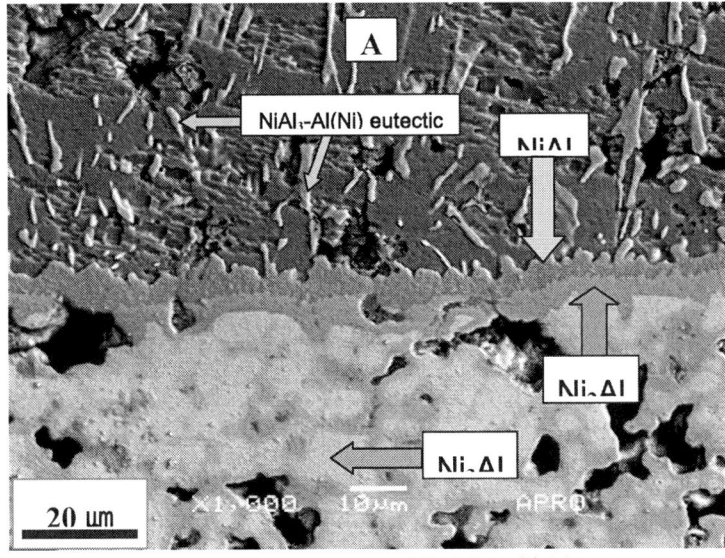

Fig.9. The bonded interface between the Ni-25at.%Al compact and pure Al.

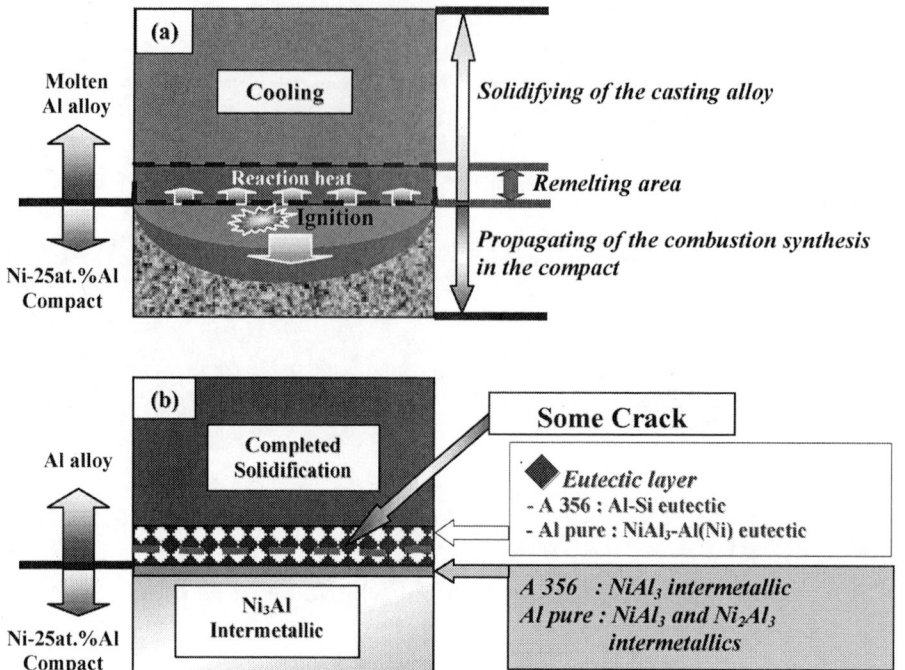

Fig.10. Schematic Modeling of the junction behavior at the interface between Ni-25at.%Al compact and casting alloys when (a) the combustion synthesis started at the surface of the compact and (b) the reaction was completed

Influence of viscosity-increasing processes on metal foam stability

Kota Kadoi*, Mitsukaze Tayama*, Hideo Nakae**
* Graduate Student, Dep. of Material Science and Engineering, Waseda University, Japan, ** Dep. of Material Science and Engineering, Waseda University, Japan

Abstract

The significant features of closed cell aluminum foams are their low density, high energy absorption, etc. These characteristics are affected by the size of the foams, namely, the smaller the better. Generally, the low surface tension and the high viscosity of liquid metals can be appropriate for the production of fine metal foams. Therefore, it is necessary to study the effect of the viscosity on the stability of the foams. In this paper, the influence of the viscosity-increasing processes on the stability of the film was investigated using aqueous solutions. The viscosity increased by the addition of polyvinyl alcohol and ceramic powders. As a result, the longer lifetime of the film is found with the addition of good wettable particles than that of poor particles or polyvinyl alcohol.

Keywords: Metal foams, Stability, Viscosity, Particles

Introduction

The significant features of closed cell aluminum foams, which are fabricated by various methods, are their low density, high specific strength, high energy absorption, etc[1, 2, 3]. Due to these advantages, aluminum foams are very attractive materials for automobile and aerospace applications, where weight reduction and improvement in safety are requested. However, in order to enhance the practical use, reduced production costs and a higher energy absorption are necessary. The properties of the foam such as energy absorption, etc., are affected by the size of the foam, namely, the smaller the better[4, 5].

A fine foam will be formed by using a low surface tension and high viscosity liquid[6-8]. In the case of aluminum foams, it is well known that thickening is indispensable for producing a fine foam. The lifetime of the foam film depends on the viscosity, namely a high viscosity liquid prevents drainage, and a low surface tension.

Goumiri reported that if an Al_2O_3 monolayer exists on the surface of aluminum melt, the surface tension decrease from 1050 mN/m to 865 mN/m[9]. Moreover, Lang investigated the influence of alloy elements on the surface tension as shown in Fig.1 and 2[10, 11]. Nevertheless, the surface tension of the aluminum melt including a surfactant, such as Bi, is not enough to stabilize the foams.

Iida reported the viscosity of liquid metals (Table 1)[12]. Ueno et al. who invented the particle decomposition in the melt process, reported that the addition of 1.4% Ca increased the viscosity up to 2.5 times the original value[13-18]. Yang et al. reported that the addition of aluminum powders led to fine Al_2O_3 particles[19, 20], and that 4.6 % Al_2O_3 particles increased the viscosity up to 3 times the original value. From this experimental evidence, it seems that all of the metal foams are produced by solid particle additions, but the mechanism or the behavior of the particles in the film is not clear. Moreover, the role of the viscosity or the surface tension on the stability is under dispute[21].

The influence of powder addition on the thickness of the aluminum foam films was discussed by Babcsán et al. using 20 vol% SiC and Al_2O_3 powders with a diameter from 10 to 20 μm[21, 22]. The membrane became thicker with the Al_2O_3 addition vs. that of the SiC due to the formation of clusters on the surface of the film. S.W.Ip et al. investigated the influence of the SiO_2 powder addition on the stability of air bubbles using an aqueous solution model[23]. In that research, the stability was measured by the change in the bubble layer height due to the collapse of the foams. They reported that good wettable particles could stabilize the liquid foams, however, the details are not clear.

Therefore, a systematic study is necessary to discuss the influence of the additional particles on the lifetime of the foam film. In this paper, in order to investigate the factors which affect the foam stability, the lifetime of the foam film was measured under visible conditions using the aqueous solution. As for the viscosity increasing agent, polyvinyl alcohol and the ceramic powders, which have various diameters, shapes and wettabilities, were used. As the first step, the influences of the volume fraction and the kinds of particles on the viscosity were examined. Furthermore, the lifetimes for the particles added to aqueous solutions were measured.

Experimental procedure

The particles of SiO_2, Al_2O_3 and MgO with diameters from about 0.3 μm to 10 μm were used. Their characteristics are shown in Table 2. The contact angles were measured by the sessile drop method using a solid ceramic plate. For the SiO_2 and Al_2O_3 particles, two types of shapes were used, namely spherical SiO_2 (amorphous) and granular SiO_2 (cristobalite) as shown in Fig.3. It was impossible to produce a foam film without the addition of a surfactant into the pure water, therefore, the aqueous solution contained 0.2 mass% sodium oleate. The surface tension of this solution is 26.4 mN/m which is one-third that of pure water, 73 mN/m.

The stability of the foam film was based on the time to collapse, namely its lifetime at room temperature. For making the foam film of a solution, we put steel rings into the aqueous solution which increased the viscosity and then pulled them out. The diameter of rings was 20 mm. The aqueous solution was stirred using a magnet stirrer to mix the ceramic powders. The viscosity value was measured within 5 sec after the stopping the stirrer using a turning fork-type viscometer; the measurement range was from 0.3 to 10,000 mPa·s and its accuracy was ±1%. The viscometer was calibrated using standard viscosity liquids before the measurements. The average lifetime and the two times standard deviation band range were used for the discussion.

As a preliminary step, the influence of the humidity on the lifetime was investigated before the main experiments. The humidities were 5, 50 and 100%. The humidity was controlled using silica gel, Ar gas, a NaBr saturated aqueous solution and/or a humidifier. In order to control the experimental atmosphere, the lifetime was measured within a glove box, and the different humidities were monitored by an electric capacity type humidity meter whose accuracy was ±3%. Moreover, in order to examine the influence of the drainage on the lifetime, the rings were kept vertically or horizontally. For the main experiments, the influence of the particles and volume fraction was examined. Furthermore, the influences of the kinds of ceramics, namely the size, shape or contact angle, on the lifetimes vs. the addition of particles were compared.

Results and discussion
Preliminary experiments
The PVA contents in the solutions were changed to 1, 2 and 3 mass%, and their viscosities were 3.42, 15.0 and 58.5 mPa•s, respectively. The lifetimes of the films were measured under the atmospheres of 5, 50 and 100% humidity. The lifetimes increased with an increase in the humidity. Moreover, the influence of the draining on the lifetime was investigated by holding the sample vertically or horizontally. The lifetime of the vertical film is only 10 to 20% of the horizontal one. This result shows that the effect of the holding conditions, namely vertically or horizontally, on drainage is very significant, so that the actual foam should collapse in the vertical position. In the case of an aluminum melt, evaporation never occurs due to its low vapor pressure. Therefore, in order to neglect the evaporation, the measurement of the lifetime should be measured at a 100% relative humidity using aqueous films. Furthermore, the lifetimes were measured when held vertically at a 100% relative humidity.

Influence of particle on lifetime for foam film
The influence of the particle volume fraction on the viscosity was examined, and the results are shown in Fig.4. For both the MgO and Al_2O_3, the addition of smaller particles is an effective way to increase the viscosity. This tendency is also found for the addition of spherical SiO_2 particles, 0.9 SiO_2 and 8.0 SiO_2. However, the addition of 1.8 SiO_2 is more effective than that of the 0.9 SiO_2. This is caused by the difference in the specific surface area volume of the particles. As shown by these results, the smaller and larger specific surface areas of the particles are preferable for increasing the viscosity. This is explained by the viscosity formula suggested by Zettlemoyer and Lower[24], as shown by equation 1).

$$\eta_c = \eta_0 [1 + K_1 (1 + K_2 A) V_c] \tag{1}$$

where, η_c: relative viscosity, η_0: specific viscosity, K_1 and K_2: constant, A: surface area of particle, V_c: volume fraction.

This formula is in good agreement with the fact that the larger surface area and the higher volume fraction is effective for increasing the viscosity.

The influence of the viscosity on the lifetime is shown in Fig.5. In the case of Al_2O_3 and spherical SiO_2, a longer lifetime occurs with the addition of smaller particles vs. that of larger ones. In spite of the smaller size of the1.8 SiO_2 particles vs. that of the 8.0 SiO_2 particles, the lifetime of the 8.0 SiO_2 is greater than that of the 1.8 SiO_2. An identical tendency was found for the addition of Al_2O_3 particles, the 1.1 Al_2O_3 (spherical) vs. the 1.0 Al_2O_3 (granular). By using these results, we can discuss not only the influence of viscosity, but also the kinds of ceramic powders and the surface areas. When comparing 0.9 SiO_2 to 1.1 Al_2O_3, the lifetime with the addition of 0.9 SiO_2 particles is remarkably long. This shows that the wettability of the particles, namely the contact angle, affects the lifetime. Therefore, this means that not only the increase in the viscosity, but also the wettability of the particles affect the stability of the foam. On the surface of the film, the particles should act as the notch for the fracture, as shown in Fig.6. Furthermore, the addition of good wettable, small particles is effective for stabilizing the foams.

Conclusions

The lifetime of the foam film was measured under visible conditions using an aqueous solution. The obtained results obtained are as follows:
1. The addition of large surface area and small particles is effective to increase viscosity.
2. The good wettable and small particles stabilize the foam film.
3. The shape of additional particles affects the lifetime.
4. The increase in the viscosity should not necessary effective to increase stability of the foam.

References

1. M.F.Ashby, A.Evans, N.A.Fleck, et al.: Metal Foams, a design guide, Butterworth-Heinemann Ltd, 2000
2. T.W.Cline, F.Simancik: Metal Matrix Composites and Metallic Foams, Wiley-Vch, 2000
3. J.Banhart: Progress in Material Science, No.46, 2001, pp.559-632.
4. E.Andrews, W.Sanders, L.J.Gbson: Materials Sci. and Eng., A270, 1999, pp .113-124
5. P.R.Onck, E.W.Andrews, L.J.Gibson: Int. J. Mech. Sci., No.43, 2001, pp.681-699
6. C.V.Boys: Soap Bubbles, Soc. for Promoting Christian Knowledge, 1920
7. J.J.Bikerman: Surface Chemistry, 2nd, Academic Press Inc., 1958
8. S.Perkowitz: Universal Foams, William Morris Agency, 2000
9. L.Goumiri, J.C.Joud: Acta Metall, Vol.39, 1982, pp.1397-1405
10. G.Lang: ALUMINIUM, Vol.50, 1974, pp.731-734
11. G.Lang: ALUMINIUM, Vol.49, 1973, pp.231-238
12. T.Iida, R.L.Guthrie: The Physical Properties of Liquid Metals, Oxford, Sci. Pub., 1988
13. H.Ueno, S.Akiyama, S.Osada, A.Kitahara, K.Imagawa: Report of Kyusyu Industrial Res. Inst., No.35, 1985, pp.19-25
14. H.Ueno, S.Akiyama: Report of Kyusyu Industrial Res. Inst., No.37, 1986, pp.2355-2361
15. H.Ueno, S.Akiyama: J.Light Met., Vol.37, No.1, 1987, pp.42-47
16. M.Itoh, T.Nishizawa, K.Morimoto, S.Akiyama, H.Ueno: Bulletin of Japan. Inst. Met., Vol.26, 1987, pp.311-313
17. S.Akiyama, K.Imagawa, A.Kitahara, S.Nagata, K.Morimoto, T.Nishizawa, M.Itoh: European Patent Application, 0.210.803 A1, 1987
18. S.Akiyama, K.Imagawa, A.Kitahara, S.Nagata, K.Morimoto, T.Nishizawa, M.Itoh: US Patent Application, 4.713.277, 1987
19. C.C.Yang, H.Nakae: J. Mat. Pro. Tech., Vol.141, 2003, pp.202-206
20. H.Nakae, C.C.Yang: J.JFS, Vol.74, 2002, pp.782-788

21. N.Babcsan, D.Leitlmeier, J.Banhart: Colloids and Surfaces A, Vol.261, 2005, pp.123-130

22. N.Babcsan, D.Leitlmeier, H.P.degischer: Mat.-wiss. u. Werkstofftech, Vol.34, 2003, pp.22-29

23. S.W.Ip, Y.Wang, J.M.Toguti: Canadian Metallugical Quasrterly, Vol.38, No.1, 1999, pp.81-92

24. A.C.Zettlemoyer, G.W.Lower: J. Colloid Sci., Vol.10, 1955, pp.29-45

Tables

Table 1 Viscosity of liquid metal at melting point

Metal	Viscosity (mPa·s)		
	$\mu_{cal}{}^{a}$	$\mu_{cal}{}^{b}$	μ_{obs}
Na	0.68	0.62	0.70
Mg	1.22	1.39	1.25
Al	1.90	1.79	1.2-4.2
K	0.50	0.50	0.54
Fe	6.37	4.55	6.92
Co	5.93	4.76	4.1-5.3
Ni	5.64	4.76	4.5-6.4
Cu	4.07	4.20	4.34
Zn	2.65	2.63	3.50
Ga	2.00	1.63	1.94
Ag	3.53	4.07	4.28
In	1.97	1.97	1.80
Sn	2.04	2.11	1.81
Sb	2.23	2.68	1.43
Au	5.50	5.80	5.38
Hg	2.31	2.06	2.04
Tl	2.55	2.85	2.64
Pb	2.52	2.78	2.61
Bi	2.13	2.54	1.63

Table 2 Characteristics of viscosity increasing agents

	notation	density g/cm³	diameter d, μm	shape	contact angle	
					water	SLS add
SiO₂	0.9 SiO₂	2.20	0.9	spherical	15°	4°
	1.8 SiO₂	2.32	1.8	granular		
	8.0 SiO₂	2.20	8.0	spherical		
MgO	0.7 MgO	3.65	0.7	granular	58°	31°
	5.6 MgO	"	5.6	"		
Al₂O₃	0.3 Al₂O₃	3.99	0.3	"	45°	30°
	1.0 Al₂O₃	"	1.0	"		
	1.1 Al₂O₃	3,30	1.1	spherical		
	3.0 Al₂O₃	3.99	3.0	granular		

Figures

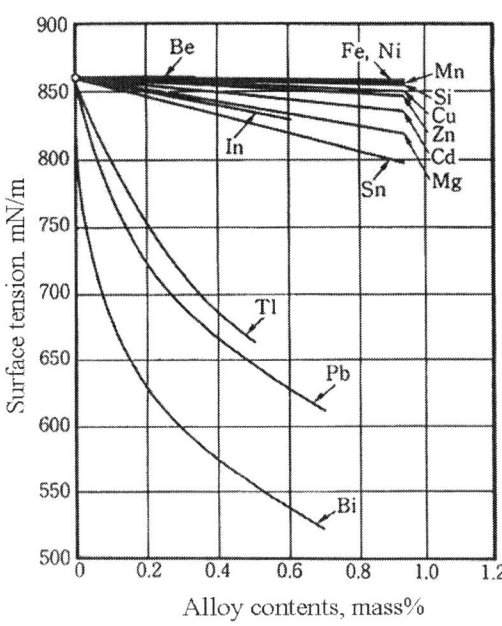

Fig.1 Influence of alloying elements on surface tension for aluminum

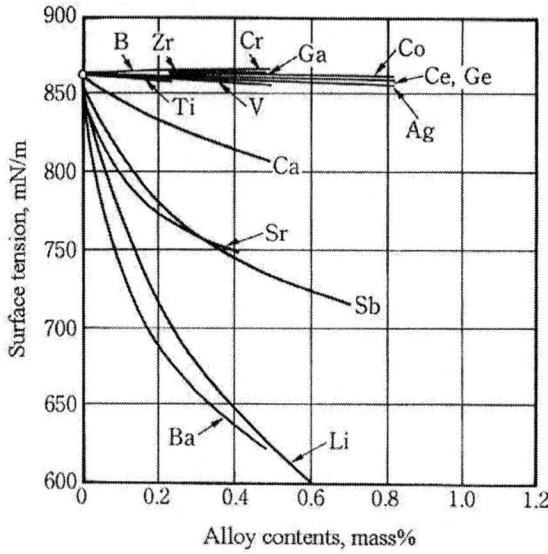

Fig.2 Influence of alloying elements on surface tension for aluminum

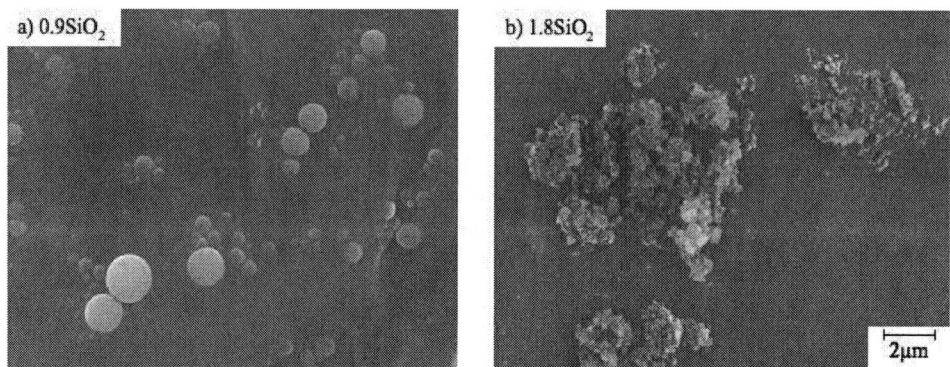

Fig.3 Shapes of SiO$_2$ particles

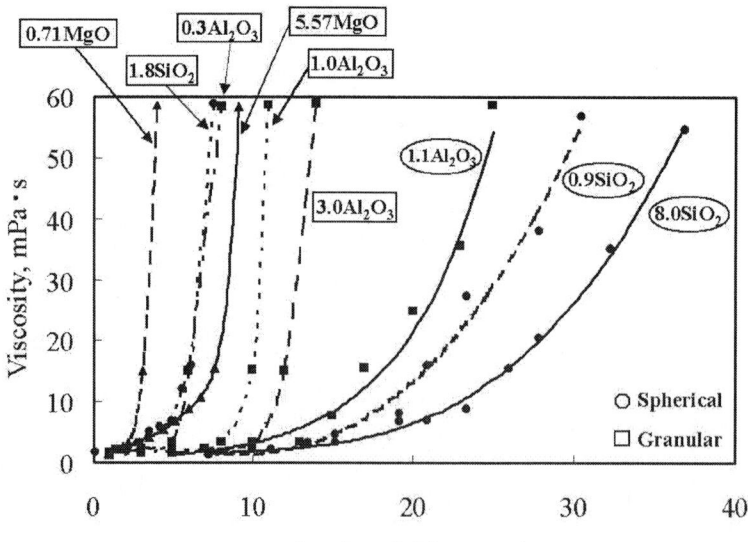

Fig.4 Influence of particles and volume fraction on viscosity

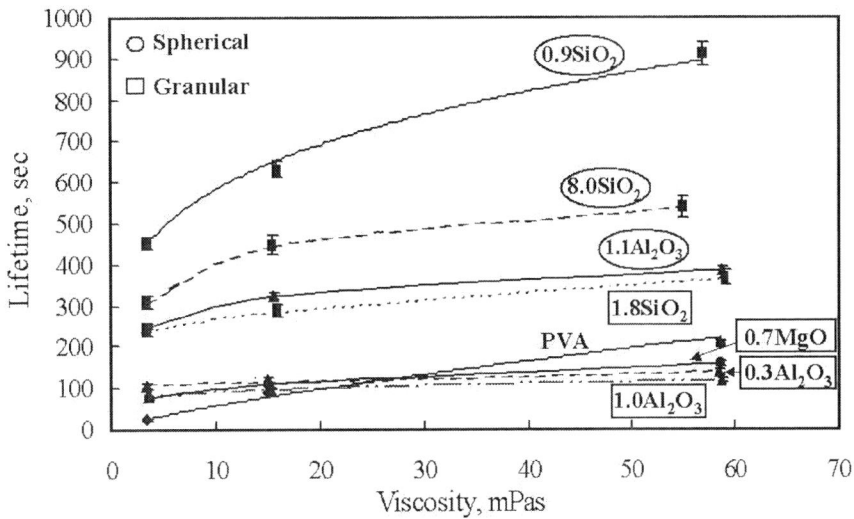

Fig.5 Influence of additive on lifetime of foam film

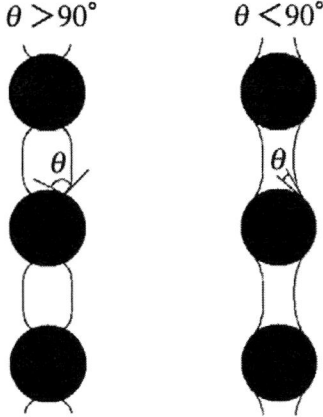

Fig.6 Effect of wettability on solid/particle at liquid/vapor interface

Study on vacuum die casting process of aluminum alloys

HU Bo*, XIONG Shoumei[1]*,
Masayuki Murakami**, Yoshihide Matsumoto**, and Shingo Ikeda**.

* Key laboratory for advanced manufacturing by materials processing technology, Department of Mechanical Engineering, Tsinghua University, Beijing 100084, P.R. China
** TOYO Machinery & Metal Co. Ltd, Japan

Abstract

The influences of process parameters and vacuum level in vacuum die casting technology on the density and mechanical properties of ADC12 alloy die castings were studied. Standard die casting tensile testing specimens were produced on a 650t cold chamber die casting machine with different vacuum pressures from 5kPa to 100kPa and different casting pressures from 66.7MPa to 22.7MPa. Other operating conditions were kept constant. Density, ultimate tensile strength, yield strength and elongation were measured. The relationship between operating conditions and casting properties was discussed and the conclusion was indicated that vacuum die casting is able to improve the property of castings and reduce the equipment tonnage while maintaining the property.

Key words

Vacuum Die Casting, Aluminum Alloy, Porosity, Mechanical Property.

Introduction

High pressure die casting (HPDC) is a high efficiency, near net-shaped process, which has been widely used in modern industry. Aluminum alloys are widely used in die casting process because of their good combination of strength and light weight. Porosity is the main obstacle for the application of die castings in welding, heat treatment and pressure test. Without heat treatment, the strength of aluminum die castings is not good enough for the manufacturing of components with high importance. Studies on die casting process and methods to improve die casting properties have been carried on for many years[1,2]. Besides, particular attention was paid on the process of air venting during die casting to investigate the mechanism of porosity formation and reduce gas content[3-6]. However, gas entrapping is almost impossible to be eliminated in conventional die casting. With gas pores inside, casting strength is weakened and further machining or heat treating is impossible.

Vacuum die casting is a kind of high integrity die casting process which is able to significantly reduce gas content of die castings. In vacuum die casting process, cavity gas is exhausted through a vacuum valve with a vacuum pump. A typical work cycle of vacuum die casting is shown in Figure 1[7]. Along with the vast development of die casting application, attempts to improve and extend vacuum die casting never stopped. NISSAN Motor manufactured aluminum structural parts with vacuum die casting[8]. RYOBI LIMITED reported the MIG welding characteristics of aluminum vacuum die castings[9]. VAW Aluminum studied heat treatment properties of aluminum vacuum die castings[10].

In spit of the extensive application foreground, the process of vacuum die casting still requires improving. In this paper, the relationship between operating conditions and mechanical properties of vacuum die castings of ADC12 alloy is systematically investigated, and the influence of vacuum level on selecting traditional process parameters is discussed.

Experimental
Experimental Condition

The experiment was carried on a TOYO 650t cold chamber die casting machine with a VCS vacuum system which has controlling linkage to a die casting machine. Vacuum pressure in the cavity was measured by a KEYENCE vacuum sensor, which has a minim precision of 0.1kPa. With a sealing plate on the back of the ejector die, a stable vacuum pressure lower than 5kPa was maintained during work cycle. Figure 2 is a layout of the test mold with the position of vacuum sensor. The test mold was made according to ASTM standard B 577-02a with H13 steel[11].

In order to realize the vacuum level inside the cavity during work cycle, two channels of vacuum system were connected to the cavity. The vacuum sensor was located at the terminal of the right channel as shown in Figure 3. A vacuum pump was connected to the left channel. In this way, the vacuum sensor was able to show the vacuum pressure at the end of the cavity. Black arrows show the direction of air flow inside the cavity when the vacuum pump is working.

Figure 4 gives a picture of the casting. The second bar from the left is round-sectioned and all of the mechanical properties were tested on this bar. Figure 5 gives detailed sizes of the specimen.

Density measurements were performed according to Archimedes principle. Mechanical properties were measured on a CC-5510 electronic experimental machine at 1mm/min drawing speed. For each operating condition, five specimens were tested to get an average result.

Experimental Procedure
Standard process parameters were kept constant in this experiment as shown in Table I. Slow phase injection speed was much less than ordinary value because vacuum pump requires a minim slow phase injection time to exhaust cavity gas. If the slow phase injection speed is not low enough, melt metal will flow into the cavity before a satisfying vacuum level is reached. Vacuum pressures and casting pressures were set as shown in Table II.

Results and Discussion
Density
Density of vacuum die castings increases with the increase of vacuum level as shown in Figure 6. As vacuum level increases from 100kPa to 5kPa, casting densities at all five casting pressures increase monotonously. At each vacuum level, casting density increases with the increase of casting pressure from 22.7MPa to 66.7MPa, which is similar with conventional die casting.

It is know that porosity in die castings includes two parts: gas porosity and shrinkage porosity. Under vacuum conditions, the physically entrapped gas can be minimized with the increase of vacuum level. Therefore, the density of castings will increase with the increase of vacuum level at each of the five casting pressures.

Shrinkage porosity could be reduced with the increase of casting pressure. Higher casting pressure influences the process of porosity formation, raises metal-mold interfacial heat transfer coefficient (IHTC), affects solidification phenomenon and reduces grain size. In this experiment, increasing casting pressure from 22.7MPa to 66.7MPa always contributes to the increase of casting density.

Ultimate Tensile Strength
Ultimate tensile strength of vacuum die castings increases with the increase of vacuum level at high casting pressure. At the casting pressure of 66.7MPa, ultimate tensile strength increases with vacuum level

monotonously as shown in Figure 7. However, the relationship is not clear at lower casting pressures.

Casting pressure has great influences on the ultimate tensile strength of aluminum die castings. With a lower casting pressure, gas porosities and shrinkage porosities increase in the castings, therefore ultimate tensile strength of vacuum die castings does not show clear relationship with vacuum level at casting pressures of 55.7MPa to 22.7MPa.

Yield Strength
Yield strength of the vacuum die castings increases monotonously with the increase of both vacuum level and casting pressure as shown in Figure 8. At each casting pressure, yield strength increases obviously with the increase of vacuum level. At a particular vacuum level, yield strength increases with the increase of the casting pressure. The influence of vacuum level on yield strength is greater than that of the casting pressure.

Elongation
Elongation of vacuum die castings increases monotonously with the increase of vacuum level as shown in Figure 9. At each vacuum level, elongation increases with the increase of the casting pressure. The influence of vacuum level on elongation is also greater than that of the casting pressure.

Machine Tonnage Requirement
Since vacuum level is proved to be efficient in improving mechanical properties of die castings, it is reasonable to produce eligible components at a lower casting pressure in vacuum die casting process. Figure 10 shows the relationship between vacuum pressure and casting pressure under different yield strengths.

Curves in Figure 10 show the same tendency from the bottom-left to the top-right with increasing yield strength. This indicates that increasing of either vacuum level or casting pressure has the same effect of improving casting yield strength. In conventional die casting process, if casting pressure is limited by machine tonnage, the application of vacuum condition will be capable to produce eligible castings at available casting pressure and to meet the requirements for the casting properties.

Conclusions
1. Density of aluminum vacuum die castings increases with the increase of vacuum level at each of the five fixed casting pressures. Under a particular vacuum level, density increases with the increase of the casting pressure from 22.7MPa to 66.7MPa.
2. Ultimate tensile strength of aluminum vacuum die castings increases monotonously with the increase of vacuum level at the casting pressure of 66.7MPa. Yield strength and elongation increase with both vacuum level and casting pressure. The

influence of vacuum level on yield strength and elongation is greater than that of the casting pressure.
3. The application of vacuum condition on aluminum die casting process is able to reduce machine tonnage requirement for castings with the yield strength maintained.

References

1. V D Tsoukalas, The effect of die casting machine parameters on porosity of aluminum die castings, Int J Cast Metals Res, 15, 2004, pp581-588.
2. Haavard T Gjestland et al, Effects of casting temperature, section thickness and die filling sequence on microstructure and mechanical properties of high pressure die castings, Die Casting Engineer, 7, 2004, pp56-65.
3. G Bar-Meir, E R G Eckert and R J Goldstein, Air venting in pressure die casting, J Fluids Eng, 1997, pp473-476.
4. G Bar-Meir, E R G Eckert and R J Goldstein, A model of vacuum pumping, ASME J Manuf Sci Eng, 1996, pp259-265.
5. A Nouri-Borujerdi and J A Goldak, Modeling of air venting in pressure die casting process, ASME J Manuf Sci Eng, 2004, pp577-581
6. J Hernandez, J Lopez and F Faura, Influence of unsteady effects on air venting in pressure die casting, ASME Trans 2001, pp884-892
7. X P NIU et al, Vacuum assisted high pressure die casting of aluminum alloys, J Materials Processing Technology, 105, 2000, pp119-127
8. Tetsuo Sakamoto, Keiji Kira, and Hiroshi Kambe, Development of automotive suspension part by high vacuum die casting, Journal of Japan Foundry Engineering Society, 76, 2004, pp283-288
9. Toru Komazaki, MIG welding characteristics of aluminum vacuum die castings, Journal of Japan Foundry Engineering Society, 76, 2004, pp289-295
10. Wolfgang Schneider and Franz Josef Feikus, Heat treatment of aluminum casting alloys for vacuum die casting, Light Metal Age, 1998, pp12-37
11. ASTM Standard, Standard test methods of tension testing wrought and cast aluminum- and magnesium-alloy products [Metric], ASTM B 557M-02a

Acknowledgements

The research was financially supported by Tsinghua-TOYO R&D center of magnesium and aluminum alloys processing technology jointly established between Tsinghua University and TOYO Machinery & Metal Co. Ltd.

The study was also supported by National Natural Science Foundation of China (50475016) and Program for New Century Excellent Talents in University.

Tables

Table I: Die casting process parameters in the experiment

Parameter	Melting Temp. ( )	Mold Temp. ( )	Slow phase plunger velocity(m/s)	Fast phase plunger velocity(m/s)
Value	680	80	0.02	1.0

Table II: Vacuum levels and casting pressures in the experiment

Casting Pressure (MPa) / Vacuum Pressure (kPa)	66.7	55.7	44.7	26.8	22.7
5	①	②	③	④	⑤
20	⑥	⑦	⑧	⑨	⑩
40	⑪	⑫	⑬	⑭	⑮
60	⑯	⑰	⑱	⑲	⑳
80	㉑	㉒	㉓	㉔	㉕
100	㉖	㉗	㉘	㉙	㉚

Figures

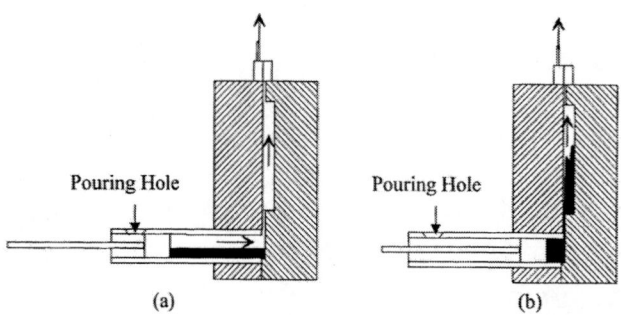

Figure 1. Plunger position at vacuum pump on (a) and off (b) state in vacuum die casting process

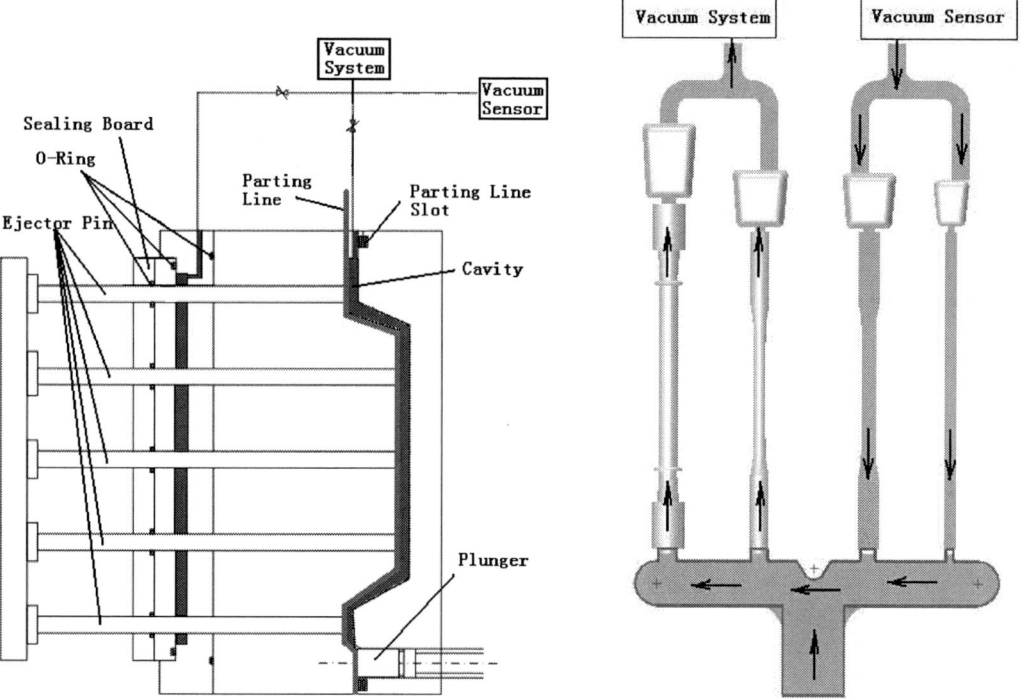

Figure 2. Layout of the vacuum system

Figure 3. Vacuum sensor position and air flow direction in the cavity

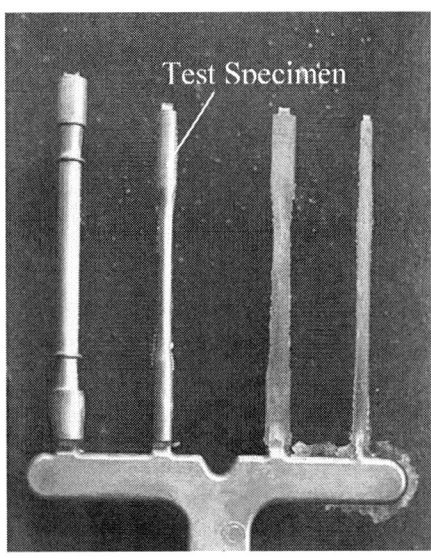

Figure 4. Picture of a specimen casting

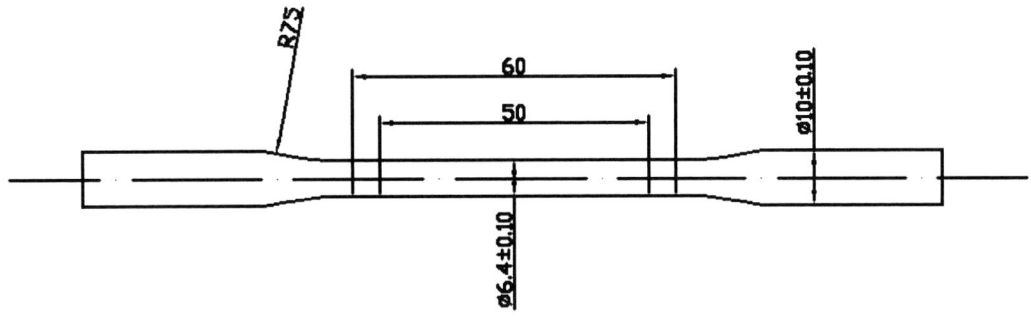

Figure 5. Dimensions of the ASTM standard tensile specimen

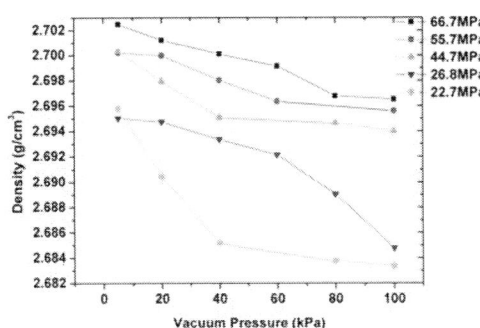

Figure 6. Density result of castings at different vacuum levels and casting pressures

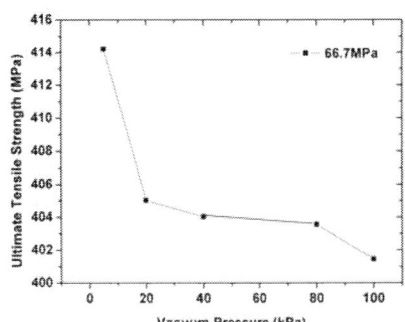

Figure 7. Ultimate tensile strength result of castings at different vacuum levels and casting pressure 66.7MPa

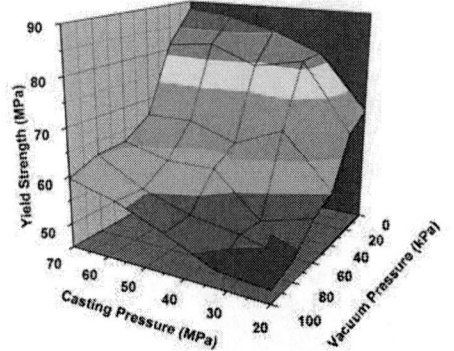

Figure 8. Three dimensional model of yield strength result at different vacuum levels and casting pressures

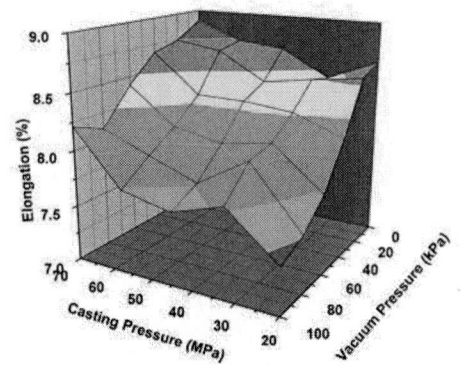

Figure 9. Three dimensional model of elongation result at different vacuum levels and casting pressures

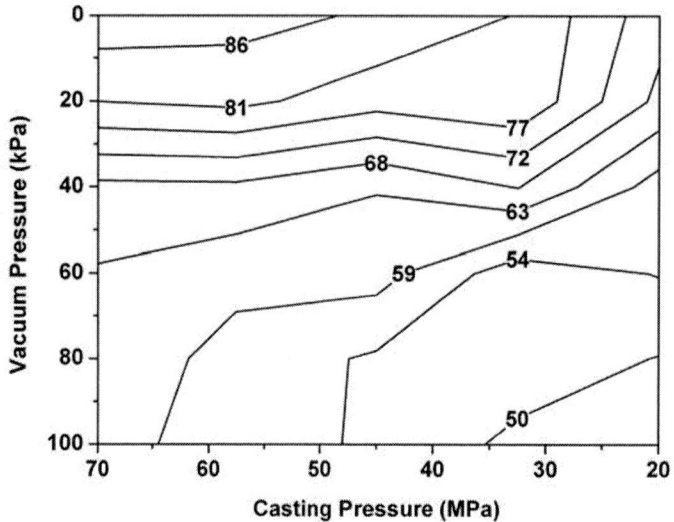

Figure 10. Relationship between vacuum level and casting pressure of particular yield strength

Investigation on the flow pattern in the shot sleeve of the cold chamber HPDC process

Jun-Ho Hong, Young-Sim Choi, Ho-Young Hwang, Jeong-Kil Choi

Center for e-Design, KITECH (Korea Institute of Industrial Technology), 994-32, Dongchun-Dong. Yeonsu-Gu, Incheon, 406-800, South Korea.

Abstract

The cold chamber high pressure die casting process is one of the most general processes to make a lightweight material alloy casting. This process usually guarantees the short cycle times and good surface quality. However, the gas porosity defect appears easily in this process due to the air entrapment during the injection stage. The flow pattern of molten metal in the shot sleeve is closely related to the air entrapment. Generally, the flow patterns in the shot sleeve are concerned with the various plunger speeds and fill rate of molten metal. To investigate the mechanism of the gas porosity generations in the shot sleeve, the numerical simulation is introduced in this study. The numerical simulation results were shown good agreement with the experimental results using water model. From the comparison results, we conclude that the developed mathematical model for the flow pattern simulation in the shot sleeve of the cold chamber high pressure die casting process is useful to determine the proper condition of the plunger speed in some rage of the molten metal fill rate.

Key words shot sleeve, gas porosity, various plunger speed, numerical simulation, water model

Introduction

Many recent research works in material engineering have been carried out to find out more effective casting methods for the lightweight materials such as aluminum alloys[1]. The cold chamber high pressure die casting process is one of the most general processes to produce the lightweight material cast. It injects the liquid molten metal into a mold at high speeds. From this process, we can usually get the short cycle times and good surface quality of casting products. This process, however, has a problem that the gas porosity defect appears easily[2]. The gas porosity is occurred mainly due to the entrapment of air in the molten metal as a consequence of the high speed injection. The gas porosity usually causes the pin-hole defects that are harmful to the mechanical properties and surface quality etc.

The flow pattern of molten metal in the shot sleeve is closely related to the air entrapment. The flow pattern of molten metal in the shot sleeve depends on the die casting operation conditions; filling rate and plunger speed. The main aims of this study are the development of a numerical simulation system for the flow pattern analysis in the shot sleeve and the accuracy verification by comparing to the experimental results from a water model.

Numerical model

A numerical model based on the SOLA-VOF technique[3][4] was used to simulate the molten metal flow in the shot sleeve. Continuity equation for the mathematical modeling is the equation (1).

$$\nabla \cdot V = 0, \tag{1}$$

where V is the vector of velocity. Volume of fluid equation is written by

$$\frac{\partial F}{\partial t} + \nabla \cdot VF = 0, \tag{2}$$

where t and F are time and fraction of fluid volume, respectively. Navier-Stokes equation is given by

$$\frac{\partial V}{\partial t} + (V \cdot \nabla)V = G - \frac{1}{\rho} \cdot gradP + \nabla^2 \cdot V, \tag{3}$$

where G, ρ and P are vector of gravitational acceleration, density of molten metal and pressure.

Energy equation is

$$\rho c \frac{\partial T}{\partial t} = \frac{\partial}{\partial x}\left(K\frac{\partial T}{\partial x}\right) + \frac{\partial}{\partial y}\left(K\frac{\partial T}{\partial y}\right) + \frac{\partial}{\partial z}\left(K\frac{\partial T}{\partial z}\right) + \rho L \frac{\partial f_s}{\partial t} = 0, \tag{4}$$

where T, c and L are temperature, specific heat of fluid and latent heat. f_s is the fraction of solid, and k is the thermal conductivity.

The VOF method is using the fractional volume function F(x, y, z, t) for tracking free surface boundaries. When averaged over the cells of a computational mesh, the average value of F in a cell is equal to the fractional volume of the cell occupied by fluid. A unit value of F corresponds to a cell full of fluid, whereas a zero value indicates the cell contains no fluid. The cells with F value between zero and one contain a free surface.

$$F = \frac{Volume\ of\ fluid\ a\ cell}{Volume\ of\ a\ cell}. \tag{5}$$

The VOF method considers the amount of F to be fluxed through the face of a cell during a time step of duration Δt. The amount of F in one time step is δF times the face cross section area, where

$$\delta F = MIN\left[F_{AD}|V_x| + CF, F_D \delta x_D\right] \tag{6}$$

and

$$CF = MAX\left[(1.0 - F_{AD})|V_x| - (1.0 - F_D)\delta x_D, 0.0\right]. \tag{7}$$

To deal with the problem where the domain keeps decreasing, a very simple moving wall boundary condition is employed. It is emphasized that in the description of plunger moving the grid system in the domain remains unchanged. We chose a method of maximum preserving the calculating system before. In this method, the moving wall is treated such as the second fluid in the multi-phase flow calculations. Fig. 1 shows the flow chart of the calculating code.

Experimental model

The water modeling instrument for the visualization experiments was made to prove the accuracy of the developed shot sleeve flow pattern simulation module as shown in Fig. 2. It includes a transparent plastic acrylic shot sleeve, a plunger, and the 1D robot system which can control the plunger speed of shot sleeve. The space between the plunger and the sleeve end is 60mm in diameter and 500mm in length.

Plunger speeds were set in 30, 50, and 70 cm/sec. Then the fill rate of shot sleeve was 30% unique. The setting of the injection conditions in the shot sleeve for the experiments and numerical simulations is shown in Table 1.

Results and Discussion

Fig. 3 shows the 3D shot sleeve model and the 30% fill rate state of molten metal. In the first, we have done the verification test through the comparison with the theoretical plunger displacement and the numerical simulation result. The 3D grid system (210 x 40 x 40) was used in this calculation.

The molten metal is poured initially into the shot sleeve through the left side gate system in the Fig. 4. The right side one is a vent system. The molten metal pushed out by plunger is exhausted through this vent system. The plunger shape is not drawn in this figure while the molten metal flow pattern in the shot sleeve is expressed. When the plunger speed is 50 cm/sec, the plunger displacement has to be 20.5 cm after 0.41 seconds theoretically. As we can see in the Fig. 4, the numerical simulation result shows the plunger displacement exactly.

Fig. 5 shows the flow patterns in the shot sleeve for the plunger speed of 30 cm/sec. The molten metal in the shot sleeve is strongly suppressed by the plunger as the plunger start to move. Then the molten metal is pulled up to the upper wall of sleeve. In this plunger speed, the pushing pressure of the plunger is not so high yet, so that the front of molten metal pulled up by the plunger does not touch the upper part wall of the sleeve. When the plunger moves further, the wave moves faster than the plunger and hits the end of the shot sleeve. After that, the end part of the shot sleeve is full up firstly with the molten metal remaining the empty part in front of the plunger. The void region in the shot sleeve is undesirable in the casting operation. And, as you can see in the Fig. 5, the numerical simulation results and the experimental results are showing a good agreement.

Fig. 6 shows the flow patterns in the shot sleeve for the plunger speed of 70cm/sec. As the plunger starts to move, the molten metal in front of the plunger is suppressed high and the top surface of the molten metal touches the upper wall of the shot sleeve. After that, the molten metal pushes the air in front of the melt wave and exhausts the air from the shot sleeve. In the case of fill rate of 30%, it has been reported from several previous studies[5][6] that the air in the shot sleeve is exhausted desirable by the wave propagation of molten metal in critical plunger speed 65cm/sec. The critical plunger speed is very close from this case plunger speed condition.

And so, in this case, the vortex shedding phenomenon from which the molten metal flow is separated from the top wall of the shot sleeve is seen. It is not too serious yet, however, compared with the vortex shedding phenomenon in more higher plunger speed[5].

From the comparison with the two cases, the results from the numerical simulations and experiments are corresponding very well.

Conclusions

In this study, numerical simulation system for the flow pattern analysis in the shot sleeve is developed and the accuracy of the numerical simulation is compared with the experimental observations from a water model. In the numerical simulation system, the moving wall is treated as the second fluid in the multi-phase flow calculations. From this study, we have the following conclusions;

1) Air void region in the shot sleeve is appeared when the plunger speed is 30 cm/sec with fill rate of 30%. This plunger speed is less than the critical plunger speed, 65 cm/sec.

2) In the case of plunger speed 70 cm/sec, the vortex shedding phenomenon, which means that the molten metal flow is separated from the top wall of the shot sleeve, is seen.

3) The developed numerical simulation system for the flow pattern analysis in the shot sleeve is very useful to determine the proper condition of the plunger speed in some range of the molten metal fill rate.

References

1. K. Fukizawa and H. Shiina, J. Soc. Auto. Eng. Jpn. 46(5), 1992, pp. 66.
2. A. Kaye and A. Street, Die Casting Metallurgy, Butterworths, London, 1982, pp. 231-235.

3. B. D. Nicholas and C. W. Hirts et. al, Tech. Report LA-8355, Los Alamos Scientific Lab., 1980.
4. R. A. Stoehr and C. Wang, MCWASP, 1991, pp. 725-732.
5. J. H. Kuo, S. M. Pan and W. S. Hwang, MCWASP, San Dego, California, 1998, pp. 149-156.
6. J. R. Brevick, M. Duran and Y. Karni, Transaction of 16th Int. Die Casting Cong. and Expo., 1991, pp. 399-404.

Table

Table 1. The setting of the plunger speed for the water model.

	Fill Rate of Shot Sleeve	Plunger speed (cm/sec)
1	30 %	30
2	30 %	50
3	30 %	70

Figures

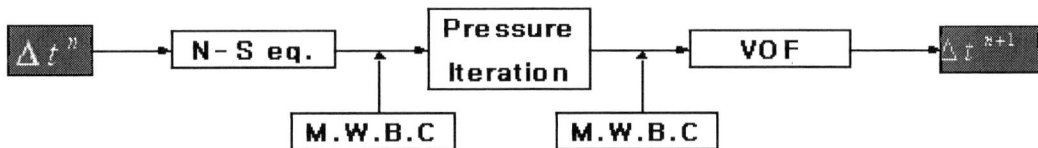

Figure 1. The flow chart of calculating code.

Figure 2. Water model apparatus.

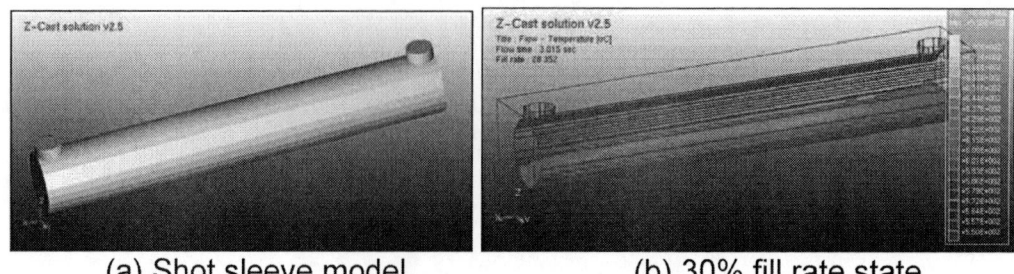

(a) Shot sleeve model   (b) 30% fill rate state

Figure 3. Shot sleeve model and 30% fill rate state of molten metal.

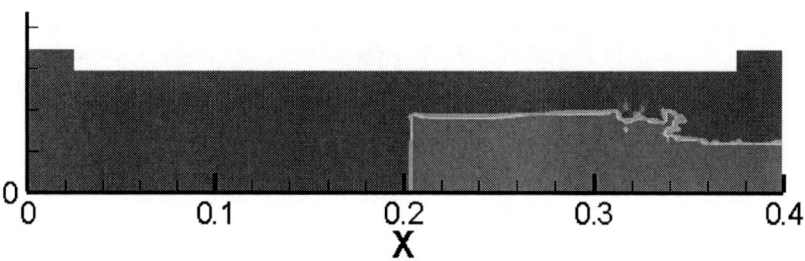

Figure 4. Displacement verification of numerical simulation result: plunger speed is 50 cm/sec.

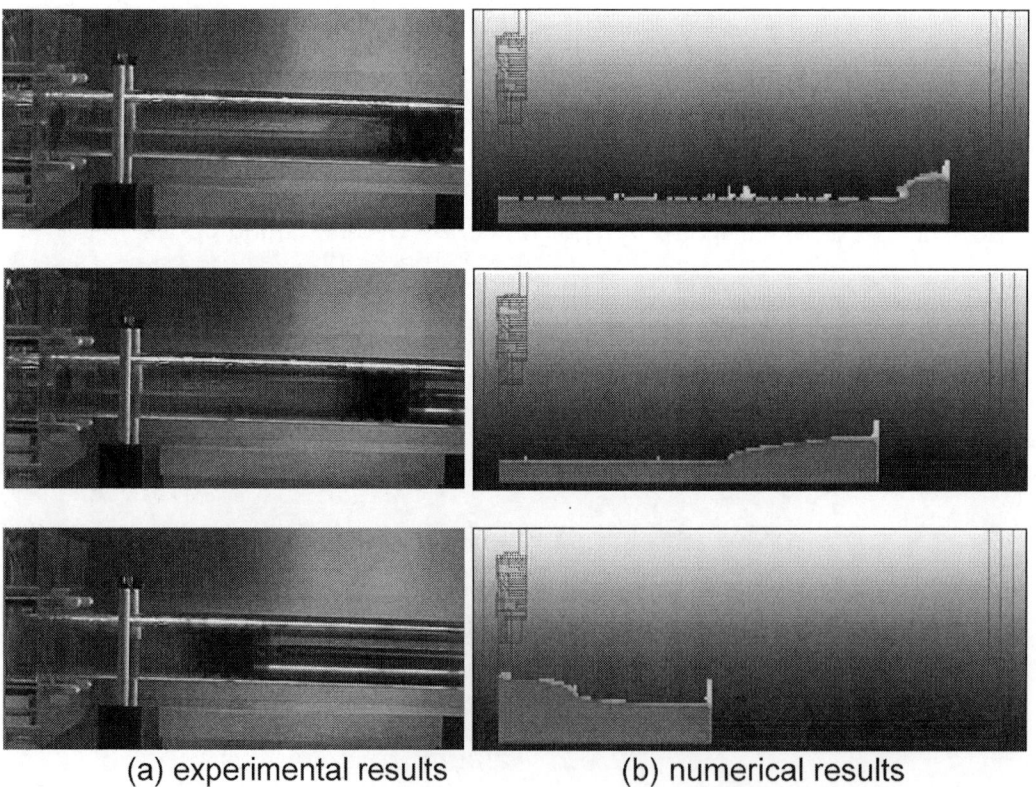

(a) experimental results   (b) numerical results

Figure 5. Time history of flow pattern: plunger speed is 30 cm/sec and the fill rate is 30%.

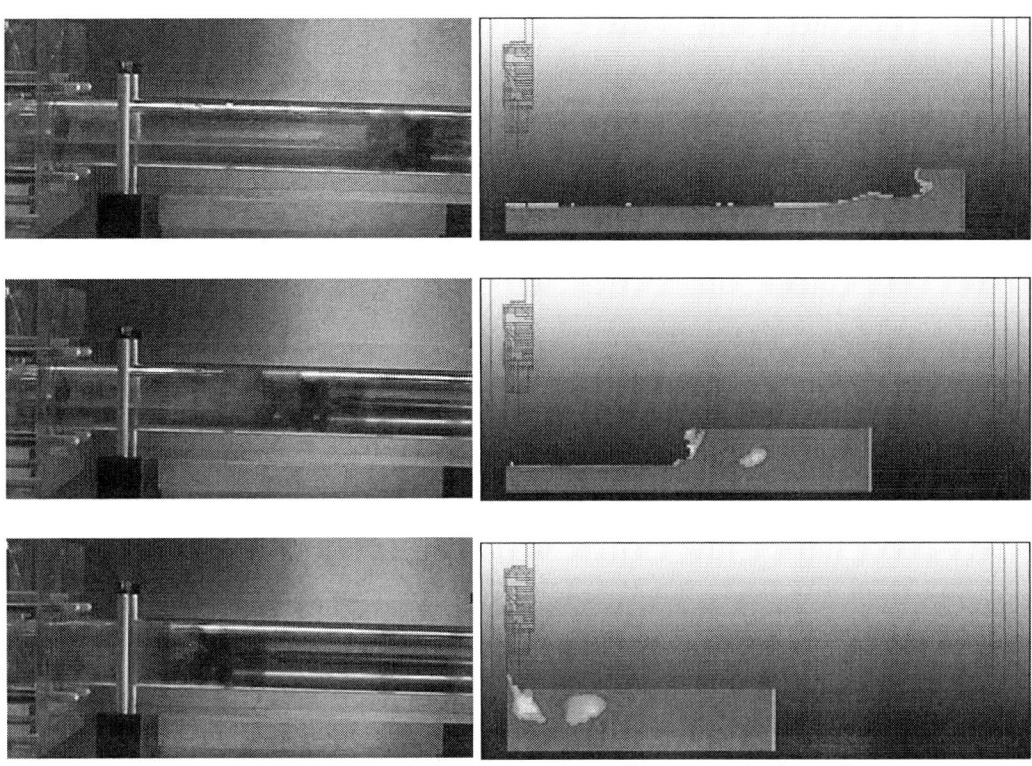

(a) experimental results   (b) numerical results

Figure 6. Time history of flow pattern: plunger speed is 70 cm/sec and the fill rate is 30%.

Definition and development of an innovative coating for optimised tooling, used in aluminium die-casting.

H.Delorme *, C.Héau ** ,E.Neto*** and K.Metzgar****.

* ExproHEF, France, ** HEF R&D, France, *** HEF do Brasil,****HEF USA

Abstract

The present paper is dealing with the enhancement of aluminium die casting tools, thanks to a patented coating, carried out by the PVD process.

The choice of the coating constituents is based on a very low chemical affinity with aluminium and alloys. The composition is chosen so as to provide the highest hardness, and by consequence, the best abrasion resistance. The thermodynamic stability is also controlled up to 800 °C (1472 °F). The combination of those properties makes this nano-crystalline coating the ideal solution to increase the performances of the tooling.

After definition of the coating parameters, the coating development is conducted using the P.E.M.S.® (Plasma Enhanced Magnetron Sputtering) technology; state of the art solution for industrial deposition.

After laboratory experimentation, fields tests have been done and demonstrate the high performance of the solution:

- The protected moulds show a significantly increased resistance to soldering.
- The thermodynamic stability eliminates (to a large extent) the risk of thermal fatigue damages (cracks)
- The extremely high hardness of the coating makes it a very good protection against abrasion, in particular close to the gates.

Key words : Soldering, Abrasion, Coating, Protection, Productivity.

Introduction

The processing of aluminum and its alloys presents several issues. Die Casting at high and medium pressure reveals two major problems: soldering phenomenon and abrasive wear:

- Soldering phenomenon is initiated when iron and aluminum interfere, iron as part of the mould's steel and aluminum as the injected material.

- Abrasive wear generally occurs when molten aluminum enters the mould, and is drastically increased when die casting high silicon content alloys, especially at high pressure.

Most of the time, the damage caused to the tooling is due to these two combined phenomena and occurs in the severely stressed areas very close to the gates. These problems are in many cases initiated by checkings or cracks due to oxydation.

Numerous tests and studies have been carried-out in order to understand those phenomena and to solve the problems they create. Although the damage mechanisms are now well understood [1] [2], no satisfactory solution has yet been proposed.

For this reason we set out to achieve a realistic and industrially compatible (reliable) solution.

Experimental
Soldering:

When the molten alloy enters the mould, the temperature of the extreme surface of the cores and cavity increases enough to allow an interaction between iron and aluminum, then an aluminum-iron phase is created.

Monitoring the difference between the die casting temperature parameter and the mould temperature in order never to reach the solidus of the Iron-Aluminum or Iron-Aluminum-Silicon alloy is a very interesting idea [2]. The most difficult part of the problem is to make it fit in with the practical reality.

The approach that seems to be the most rational and realistic to avoid the Aluminum and Iron interaction, is to make sure Iron and Aluminum will never meet by creating a barrier in between the mould and the product. This protection is ideally made out of materials with absolutely no chemical affinity with aluminum and are to be absolutely free of Iron, Copper and of course first of all, free of Aluminum, as those materials are readily reactive with Aluminium and create alloys with the injected Aluminium which then would initiate soldering.

Abrasive wear:

To be a durable solution, the choosen protective barrier (coating) is to be resistant to the mechanical stresses and strains of abrasion and thermal fatigue.

For this reason, the coating must have the highest possible hardness and be thermo-dynamically stable at the die casting processing temperature. This way, it becomes a suitable solution to avoid oxidation and cracks (checking) which are the initial damage that promotes abrasive wear.

Choice of the composition of the coating:

Some compounds such as Titanium di-Boride (TiB2) and Boron Carbide (B4C) are very well known to demonstrate good behaviour when used in molten aluminum processing [3]. However, these materials have a bad reputation due to their difficulty to be processed as a coating.

HEF group developed a material the composition of which is Boron, Titanium and Nitrogen based, conforming to the resistance against soldering and abrasive wear.

This coating is implemented by physical vapor deposition at a temperature not exceeding 350°C (662°F), a temperature allowing this coating to be applied on moulds without any need of modifying the material or the heat treatment. Current B4C coatings are implemented by thermal Chemical Vapor Deposition techniques at a temperature incompatible with mould steels (about 1000°C, 1800°F).

Hardness:

There are many different availiable coatings currently on the market. The hard coating that is of main interest for our subject is the one offering the highest hardness in combination with a thermo-dynamic stability when submitted to the usual extremes of temperature of die casting.

In other words, the grain size and the mechanical properties of the layer have to remain the same after reaching the temperatures associated with the die casting process cycles, and the layer is to remain inert to molten aluminum.

The implementation of the coating has to be possible directly on a finished mould, ie. existing tooling, and has not to require a changing of the thermal state or of the dimensions and tolerances of the moulds.

Thermo-dynamic stability :

The substitution of Nitrogen atoms by Boron atoms in a face-centered cubic (fcc) structure as Titanium Nitride (TiN) is of major interest to fulfil the thermo-dynamic stability and inertia requirements for aluminum die casting. (see thermal stability page 5).

In addition, Boron atoms are bigger than nitrogen and contribute to the increased hardness of the coating.

Note : excess of Boron or Nitrogen can lead to a different structures made of fccTiN and amorphous Boron Nitride (BN). Those structures are far less stable than the single phased structure .

More particularly when maintaining or during storage, coatings may be submitted to ambient humidity which leads to decay of the amorphous phases as BN, turning them to Boric Acid. Such an amorphous phase is undesirable and represents a serious risk for the layer.

The chosen coating technology is the one developed by HEF R&D and used as an industrial method worldwide: Plasma Enhanced Magnetron Sputtering (PEMS ®).

This technology is characterized by generating the metallic vapor independently from the plasma generation, the plasma is assisting the growth of the coating.

The so-deposited coatings are nano-structured (this means the crystal size is in general in between 5 and 10 nanometers). Having a nano-structured layer allows very high hardnesses and in addition a highly elastic behaviour.

Those properties lead to a particularly efficient abrasive wear resistance. As an example, a simple and classical TiN implemented by PEMS reaches 3200 Vickers instead of 2500 if implemented by arc evaporation. PEMS deposits nanostructured TiN layers with 8 to 10 nanometer grains and arc evaporation leads in general to 20 nanometer grains.

The composition of the coating HEF developped is a f.f.c. TiN structure with Boron atoms in substitution of Nitrogen ones, in a proportion making the hardness optimum and keeping a single phased material (no BN amorphous phase).
Graph 1 shows the influence of the nitrogen flux on the grain size

We can notice that an increase in nitrogen content leads to a smaller grain size, which appears at first surprising, but is explained by the appearance of the BN phase around the TiN nano-crystals, so crystals turn from Ti-B-N to TiN.

Graph 2 shows the hardness and the wear rate of the coatings related to the Nitrogen content.

For the patented single phased composition, hardness is already at the maximum and wear rate is the smallest, this is directly related to the friction ability of this material in ambient conditions.

As soon as the Nitrogen content increases, wear resistance decreases even if the hardness is the same: appearance of a double-phased material puts together Ti-B-N and amorphous BN.

This material is very hard but unstable, decaying when submitted to air humidity, and has a lower wear resistance.

In the third part of the the graph, nitrogen content is increased again, and now TiN appears plus amorphous BN, no more TiBN. The wear resistance becomes increasingly weak.

This phenomenon can be described as follows:
Adding nitrogen in the fcc TiN structure already including boron on some Nitrogen sites is only possible for a certain amount of atoms. Past this proportion, adding Nitrogen ousts Boron atom out of the face-centered cubic structure. Boron reacts with nitrogen and creates a BN type amorphous phase around grains.

This reaction is undesirable but unavoidable for every Ti-B-N compound if out of the single phased material coating. The trade name of the coating is "CERTESS SD", the characteristics of which are given hereafter:

> Hardness : 4500 HV
> Young modulus : 390

Thermal stability :
Graph 3 shows the comparison of thermal stability of the usual layers as Chromium base Nitrides (CrN_x and Cr_2N_x) and Titanium Nitride (TiN) versus the developped solution.
It is important to notice that grain parameters of this coating (by way of consequence its structure and also hardness) are not affected when submitted to die casting processing temperature and even higher: 1100°kelvin = 1520 °F = 820°C

Results
After the coating was developed in the Materials and Processes laboratory, it was extensively laboratory tested before field tests were launched, in an industrial application.

The coating is applied on existing moulds, core pins and slides, with no particular specification regarding surface roughness and tolerances.

The following characteristics are listed as possibilities of productivity gain:
a) *Soldering protection :*
   Graph 4 gives the results of comparative tests on a core pin, located close to the gate. Result is given in terms of number of

shots before stopping the mould for cleaning maintenance due to soldering and out of specification tolerances.

b) *Abrasion resistance* :
Graph 5 shows the results of tests on core slide, in front of the gate, with 12% silicon content aluminum alloy.

c) *Checking resistance* :
Graph 6 gives the amount of shots before the cracks quantity and depth becomes prohibitive, test is conducted on a sealing area and stops when a machining operation is needed to keep watertightness in the specifications.

The feedback from those industrial tests also highlighted other non expected and surprising results:

- Cleaning release agent burning becomes less frequent.
- Possible reduction (sometimes significant) of the release agent.
- Better filling in of the cold areas.

A dedicated and additionnal study will be necessary to evaluate those possibilities.

Conclusions
Aluminum Die Casting has always experienced problems with soldering and wear.

The existing range of solutions did not solve the problems, mainly due to the difficulty of application in industrial fields and conditions.

The performance of Boron based plasma nano-structured coating has exceeded any other industrial solution since it's launch.

Choosing a nano-structured material leads to a high hardness and choosing the right composition avoids the creation of alloys with the injected aluminum and the iron from the steel mould.
The coated moulds behavior is very promising and suggests high possibilities of productivity gains.

After this important step, some additional studies are to be done for friction and rheology for a better understanding of the possible additional gains.

References

1. Y.L.Chu, P.S.Cheng,R.Shivpuri, *Soldering phenomenon in aluminum die casting: possible causes and cures*, Transactions Rosemont,Illinois: NADCA 1993, pp361-371.
2. Q.Han S.Viswanathan, *Analysis of the mechanisme of die soldering in aluminum die casting,* Transactions A, Metallurgical and Materials, Volume 34 A, p 139, January 2003.
3. G.Mackiewicz Ludtka, V.K.Sikka, Aluminum soldering performance testing of H13 steel as boron coated by cathodic arc technique, SVC 2004.

Figures

Graph 1: Influence of the nitrogen flux on the grain size.

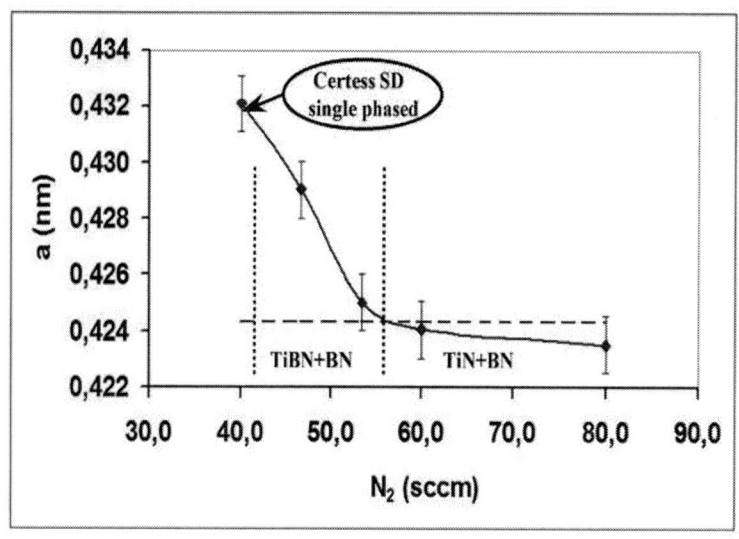

Graph 2:
Hardness and wear rate of the coatings related to the nitrogen content.

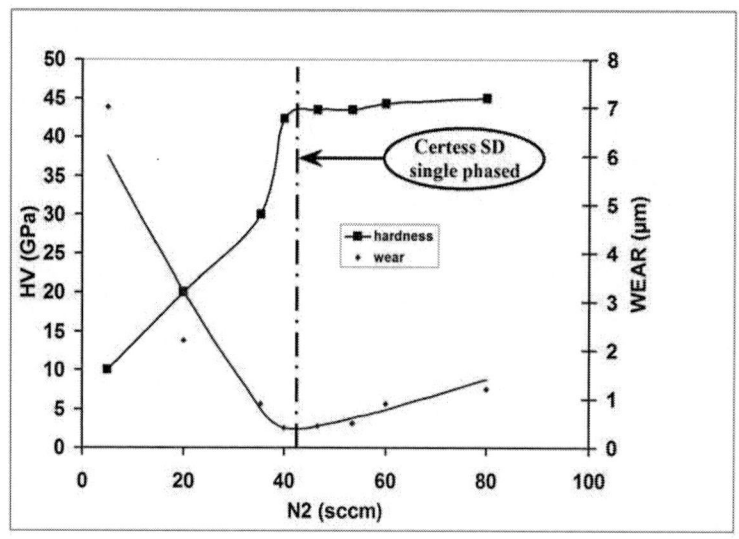

Graph 3: Comparison of thermal stability of usual layers versus CERTESS SD.

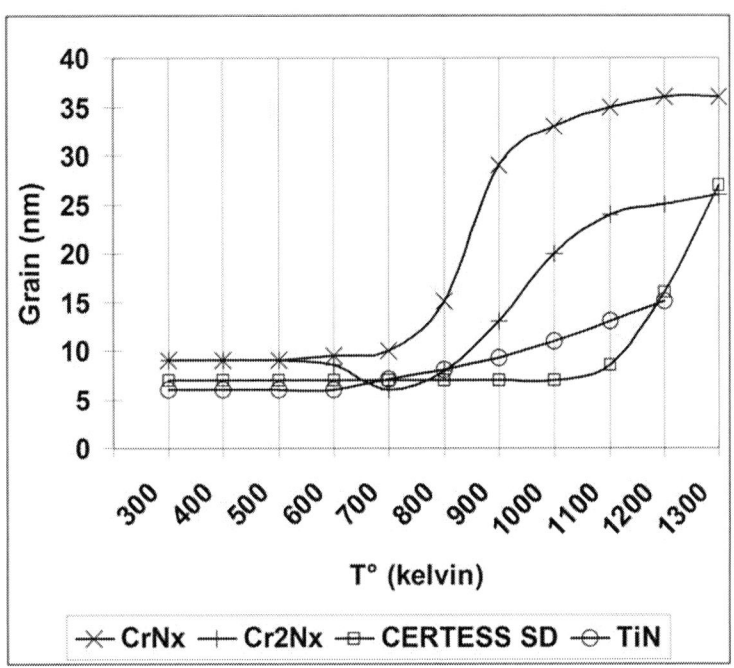

Graph 4 :
Results of comparative tests on a core pin, located close to the injector (gate). Result is given in terms of number of shots before stopping the mould for cleaning maintenance due to soldering and out of specifications tolerances.

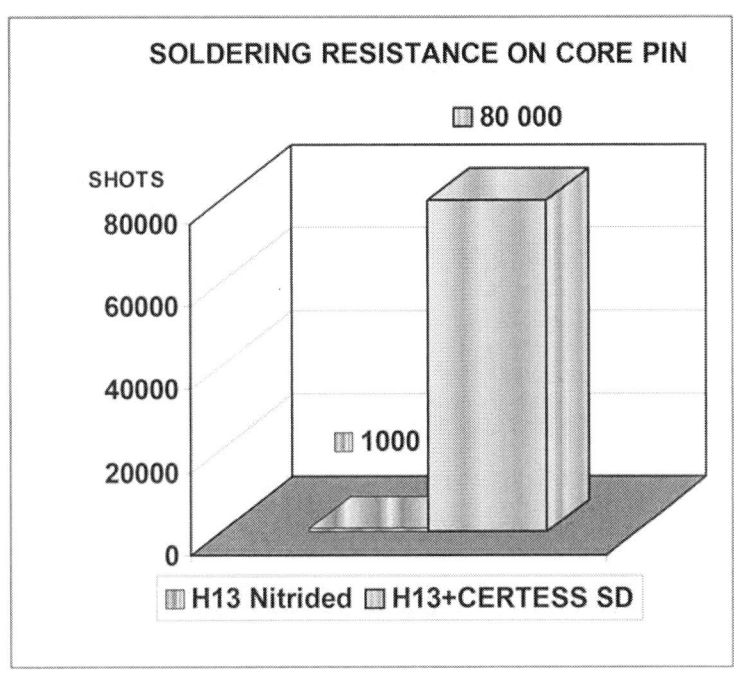

Graph 5 :

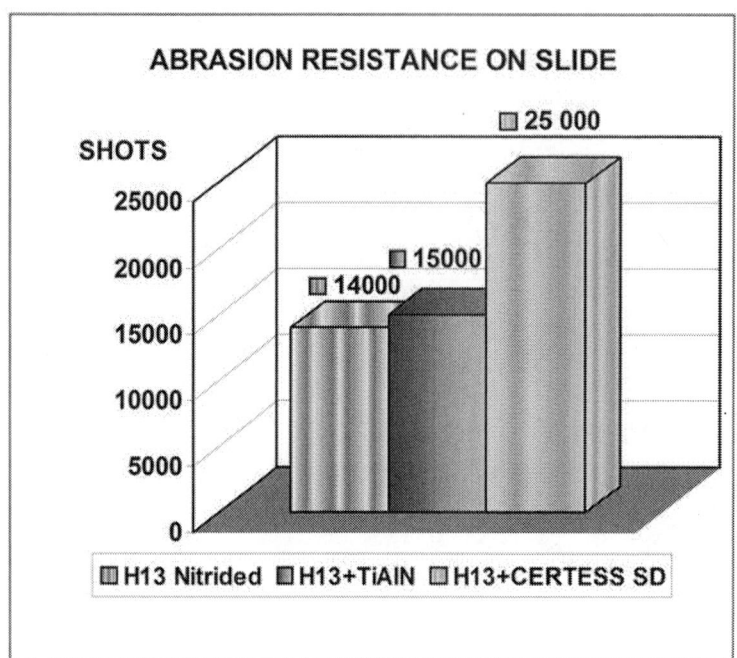

Results of test on core slide, in front of the injector (gate), with 12% silicon content aluminum alloy.

Graph 6 : Amount of shots before the cracks quantity and depth becomes prohibitive, test is conducted on a sealing area and stops when a machining operation is needed to keep watertightness in the specifications

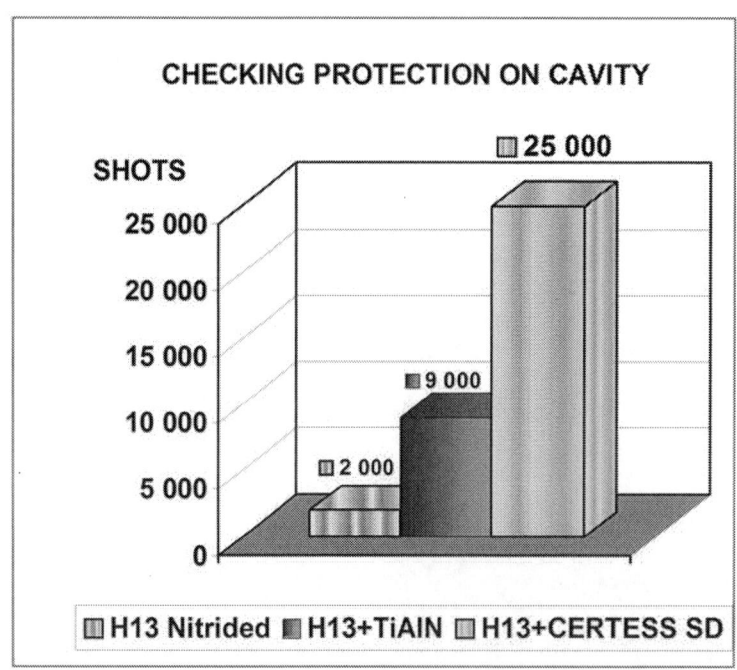

hdelorme.exprohef@hef.fr

A new future for gravity moulding

M Bakrim, J Vervier, C Vandenhaute, F Ngirabacu and D Vervier.

Atelier de la Mécanique (ALM SA), BE

Abstract

Gravity moulding process should be automated for 2 main reasons.

Firstly the large series of injected moulded parts are leaving Europe for the Far East countries. The European foundries have to concentrate on smaller series AND keep their production prices as competitive as possible.

Secondly manual moulding faces a manpower problem. It is difficult to convince people to have a physical work in such a harsh environment.

Original solutions have been developed by company ALM to make this automation as productive as possible.

1. A variable volume dosing system (patented by ALM)
2. Integration in a single enclosure of: The melting oven, a holding zone under controlled atmosphere and, the dosing system.
3. Development of the proprietary AluShieldTM surface treatment resistant to aluminum molten metal
4. Integration of all needed techniques (PLC, Touchscreen MMI, Thermal metrology and controls)

Key words (Automated gravity, moulding; melting oven; holding; dosing).

Introduction

The pieces production of foundry is made considering to expending results. This is why the process choices of moulding is difficult because of the combination of technical, economical and human context.

The process of moulding by gravity in chill consists in tipping out, with the help of ladle, a melting metal in duct of mould cavity.

In order to answer to criterion of quality, price, delay and working conditions face to Far East Countries, the European foundries have to re – examine their production system and provide the technical solutions to economical and human problems.

This is point of view that ALM, a limited company at Fleurus in Belgium developed the automatic gravity moulding machine dedicated to alloys at low melting point.
Based on patented "VERVIER" process, this machine consists of a melting oven with the holding area and high precision variable volume proportioning system (figure1). This would use in metallic mould feeding and/or sands; injection mould feeding in chills (it is then used upstream from pressurized die casting).

So that, the process allows improving the cost production feeding until four different moulds on carrousel and avoiding with certain parts to use pressure injection whose hourly cost is unsuited to small and medium size series.

Characteristics of process

The process allows the moulding of any type of metal and alloys which melting point is lower than 800°: aluminum, zinc, magnesium, lead, tin, indium, cadmium,

The machine can operate in differed starting what allows a starting up of burners before the arrival of operators. The liquid metal is set in the metal crucibles covered by a specific surface coating. This function saves important time as well as saving energy. The crucibles present one double lifespan compared to the graphite crucibles.

The innovating machine includes the most modern techniques of control (figure 4) and offers a great flexibility: it acts of a process coming close to uninterrupted casting system which offers:
- control casting parameters
- record of "recipes" including all parameters relate to each order
- operation: semi or full automatic cycles which saves the human services.

The machine allows synchronization with peripheral equipment like ingots, conveyor, carrousel … All controls are accessed through a screen type mobile operator mould (figure 3).

The crucibles operate in double room, what allows a better control of maintenance and melting points, a weak fall of temperature at during the introduction of ingots or solid returns and a better homogeneity of liquid metal.

The tank of maintenance is kept under inert atmosphere and is separated physically from the tank of fusion where ingots and gates are discharged, what ensure the system of proportioning to take clean metal and free from oxides. The purification of the liquid metal in the tank of fusion is done by gas bubbling and solid introducing of flow.

This proportioning system patented of ALM, extracts the liquid metal from the crucible of maintenance to send it in a mould or launders with exactitude and an unequalled repetitivity of the flow of metal.

The system allows on the one hand, maintaining an optimal rate of continuous control of rate of filling and that in a constant way (figure 5) The proportioning precision is lower than 1%. Such regularity can not be assured when the operation is made with the hand.

Tests in foundry
Comparison of two situations (A: manual and B: automation) on a site of moulding of alloys aluminum parts AS7G03.

Situation A

The site before automation (figure 6) consists of six operators turning in two teams. While two operators carry out of parts of 1 kg out of 2 manual post distributed around a central crucible, the third separates the jet from cast and the gates from the cast parts and assure the feeding of ingots in crucible.

Data of production:
Type of production: medium size series: 45000 pieces per month (achievement of two pieces over 45 second).
Daily production: +/- 2000 kg of aluminium
Energy used: fuel - oil
The bath of aluminium is cleaned regularly by manual additional of flow.

Every operator carries out the moulding according following sequences:
1) To close the mould and set up the cores
2) To take the ladle, to fill it of metal, to swivel and fill the mould and finally to deposit the ladle on the edge of crucible
3) T wait for the end of natural cooling of casting
4) To open cores and mould, to leave the parts. To close the mould and set up the cores.

Situation B

The preceding site was automated by the installation of an automatic machine for moulding by gravity making possible to supply four metal moulds on a carrousel
The device allows to release two operators and the sequences of 1, 2, 3 and 4 described above, do not require any more a human intervention.

This is to say a reduction of 33% of the cycle time on the whole moulding site (achievement of three parts every 45 seconds)

The only operator available carries out the separation of the jets of gates
The first post is that of the filing and the post n°2 and 3° allow the cooling of the casting before release from the mould

An alarm engages if the level of metal is low in the melting crucible. The operator ensures the filing by introducing a new ingot there.
This function can be also replaced by the introduction of an automatic feeding system.

Results
We broke up the expenditure of means of production and the expenditures of operators for both situations: situation A before automation and situation B after automation

2 teams : 4000 h / year cost :	Situation A Crucible oven 50.000 €	Situation B Vervier Process 70.000 €
Power	300 KW	300 KW
Starting up / day	3 hours	1 hour
Labour Production	2100 kg	3600 kg
Damping over 5 years	16.120,00 €	22.560,00 €
Maintenance (crucibles, collar, …)	4.000,00 €	6.500,00 €
Energy	73.920,00 €	84.480,00 €
Peripheral dumping over 5 years	4.830,00 €	8.052,00 €
Hiring surface	32 m2 1.920,00 €	16 m2 960,00 €

| Workforce | 216.000,00 € | 72.000,00 € |

| Consumable Ladle, flow, die coating, mold cleaning, small tools, ... | 3.800,00 € | 1.900,00 € |

| Reject | 5% of production 16.030,00 € | 2% of production 3.929,00 € |

| Total | 336.620,00 € | 200.381,00 € |

Conclusions

Mechanization and automation are justified economically from data of productivity: unit cost price (expect raw material overhead costs) is 0, 53 € with the manual moulding machine (situation A) and 0, 21 € with the automatic machine (situation B), is a ratio of 2, 5 (table 1).

The suppression of the repetitive and forced movement, the intense radiation, the air pollution, projections of level of noise made it possible to improve the working conditions and the well being of operators. These are released for other less hard tasks with the reduction in the risks of personal injuries by suppression of projection of metal and intoxication.

Automation also allowed an improvement of quality by a better control of technical parameters as well as a reduction of the costs of energy which are considerable.

The calculation of return on investment of the automatic machine of moulding by gravity is of six months and half.

References

1. Pechiney Rhenalu, l'aluminium dans l'automobile, Paris, 1998.
2. STEFANESCU D. M. coord., casting, Volume 15 in ASM Handbook 9th Edition, Americain Society for Metals, USA, 1998.

Tables

Process \ ASPECTS	"Ladle"	"Injection"	"VERVIER"
Investment	1	10	1.4
Series	Small size	Large size	Small, Middle, larges size
Cost of value added (*)	1	0.6	0.4

Table1: Performances of various current processes of moulding

(*) Considering pieces of 2 kg, operation of 16 hours per day and damping of investment over 5 years

Figures

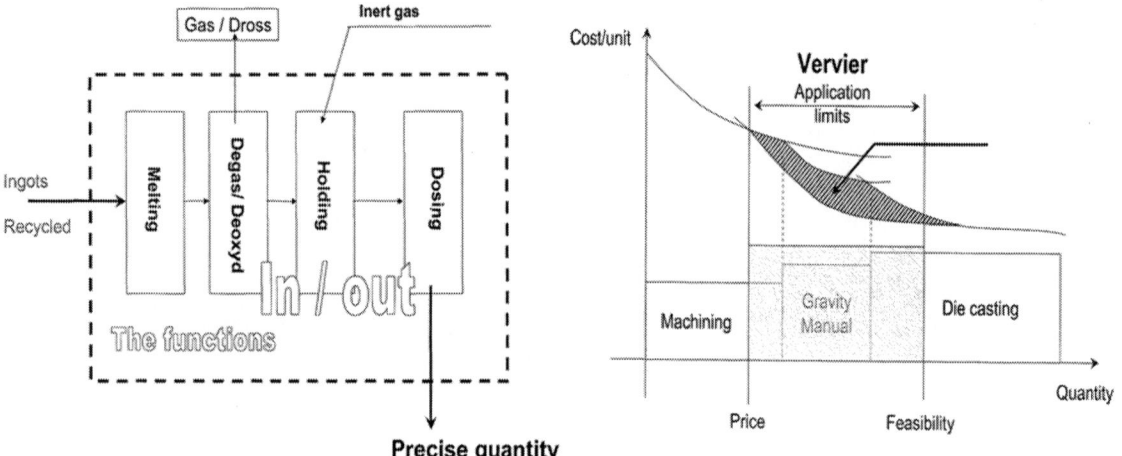

Figure 1 : Shematic drawing of aluminium Vervier- machine

Figure 2: Distribution of production costs

Figure 3 : General sight of moulding site

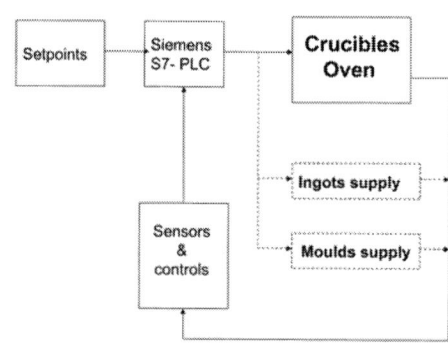

Figure 4 : System of process control

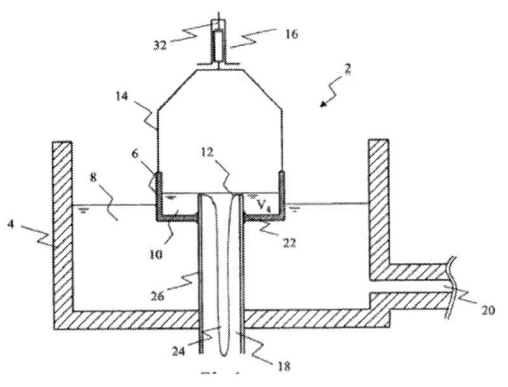

Figure 5: VERVIER proportioning system

Figure 6: Situation A

Figure 7: Situation B

Titanium Matrix (TiB+TiC) Composites Shot Sleeve for Al Alloys Die-casting

Si-Young Sung, Bong-Jae Choi and Young-Jig Kim

Department of Advanced Materials Engineering, Sungkyunkwan University, Suwon, Korea

Abstract

The applicability of titanium matrix composites (TMCs) sleeve for Al alloys die-casting was investigated by interfacial reaction, and wear and mechanical test. (TiC +TiB) reinforced TMCs were in-situ synthesized and net-shape formed using a horizontal vacuum induction melting system. The resistance-ability of (TiC+TiB) reinforced TMCs to molten Al alloys attack was also examined. Their reactions were carried out in a furnace at 993 K for times varying from 0 to 1200 s. In the case of conventional sleeve material, H13 steel, there were severe interfacial reactions and erosion after 60 s. On the other hand, the resistance of TMCs to interfacial reactions and erosion by molten A380 alloy was significantly increased. Moreover, the results of wear and mechanical properties of TMCs met all requirements for Al alloys die-casting shot sleeves.

Key words : titanium matix composites, shot sleeve, investment casting, interfacial reaction

Introduction

One of the most major issues in the shot sleeve of aluminum alloys die-casting process is that the molten aluminum alloy adheres to the surface of H13 tool steel, eventually causing the failure of the shot sleeve, which is known as "aluminum soldering" [1]. In particular, aluminum soldering phenomenon has a negative effect not only on the shot sleeve service life but even on the casting quality. Besides, at high process temperature, and rapid heating and cooling cyclic environments, H13 tool steel has poor mechanical properties, which causes the inferior resistance of the shot sleeve to heat checking, thermal fatigue, erosion and distortion. The poor performance of H13 tool steel results in frequent shot sleeve replacements [2]. Until now, in the case of die-casting dies, many technological solutions are provided by surface coating engineering process to improve the resistance to erosion by molten aluminum alloy, and the mechanical and wear resistance of H13 tool steel [3, 4]. However, the shot sleeves are subject to the rapid heating and cooling cycle and dynamic load by the plunger. Thus, the coating process makes it impossible to ensure long-term protection as to real tool, because they do not meet the requirements of adequate thickness, excellent bonding, suitable mechanical strength and toughness, and thermal shock resistance.

TMCs shot sleeve in the aluminum alloys die-casting machine are expected to reduce energy consumption during the die-casting process, while increasing the productivity and the service lives of shot sleeves since TMCs meet all the requirements listed above [5]. Unfortunately, the wide application of TMCs has been limited due to their expensive manufacturing cost.

In this study, the in-situ synthesis and the alpha-case controlled investment casting system of TMCs was examined for the economic net-shape forming of TMCs shot sleeves. In the present study, the immersion test of TMCs and H13 tool steel in a molten aluminum alloy, wear and mechanical test were carried out for the applicability of TMCs shot sleeve for aluminum alloys die-casting. After immersion test, the specimens underwent the interfacial reaction analysis by an optical microscopy (OM) and an electron probe micro-analyzer (EPMA) to make a comparative evaluation.

Experimental

Titanium and 1.88 wt.% B_4C were prepared for the synthesis of 10 vol% TiC and TiB hybrid TMCs. In-situ synthesis and investment casting of TMCs were carried out in a VIM furnace. The immersion test of TMCs and H13 in a molten A380 alloy were carried out in a furnace at 993 K for times varying from 0 to 1200 s. OM and EPMA were employed to examine the interfacial reaction products between A380 alloys, and TMCs and H13 tool steel. The upper specimen is fixed by a cantilever type of load-cell, and the TMCs and H13 disk are rotated by a constant velocity at 0.7536 m/s. The constant load of 10 N and 20 N were applied to ball-on-disk and cone-on-disk type, respectively. The wear volume was estimated by measuring the size of specimens before and after the test using a micrometer. The tensile strength and elongation were evaluated by the Hyundai HD 200 universal testing machine with ASTM-E8M standard [6].

Results and Discussion
1. In-situ synthesis and net shape forming

Figure 1 shows the SEM image of the synthesized TMCs from titanium and B_4C by induction melting and casting. The spherical and needle-like reinforcements were distributed uniformly in the titanium matrix. In addition, no interfacial reaction could be observed between the matrix and reinforcements.

In order to identify the phases of synthesized reinforcements, EPMA elemental mapping was carried out as shown in Fig. 2. The results of EPMA mapping show that the carbon element is detected at the spherical reinforcements, while the boron at the needle-like and large, many-angled reinforcements. For clear identification of the synthesized reinforcements, they were analyzed by TEM. Fig. 3(a) is a bright field TEM image of the carbon-detected spherical reinforcements. In the right region of Fig. 3(a), titanium and carbon are detected by EDS as shown in Fig 3(b). Also, the spot diffraction pattern is TiC phase in the [$0\bar{1}1$] beam direction. Fig. 4(a) shows a bright field TEM image of the boron-detected needle-like and large, many-angled reinforcements region. Titanium and boron are detected in the facet morphology as shown in Fig 4(b). The diffraction pattern at the facet region is TiB phase in the [$00\bar{1}$] beam direction. The EPMA and TEM results prove that the spherical reinforcements are TiC and the needle-like and large, many-angled reinforcements are TiB phase. These experimental results were evaluated through thermodynamic calculations utilizing the joint of army-navy-air force (JANAF) thermochemical tables [7]. If B_4C is added into molten titanium at 2000 K, three reactions are possible, which can be described as follows:

$$3Ti(l)+B_4C(s)=2TiB_2(s)+ TiC(s) \qquad \Delta G_f = - 585 \ kJ/mol \qquad (1)$$

$$5Ti(l)+B_4C(s)=4TiB(s)+TiC(s) \qquad \Delta G_f = - 689 \ kJ/mol \qquad (2)$$

$$TiB_2(s)+Ti(l)=2TiB(s) \qquad \Delta G_f = - 46 \ kJ/mol \qquad (3)$$

These thermodynamic calculations show that (TiC+TiB) synthesis reaction is more favorable than (TiC+TiB_2) synthesis reaction, and the dissolution of the synthesized TiB_2 reinforcement to TiB occurs spontaneously.

In this study, the α-case controlled mold was designed in consideration of the interstitial and substitutional α-case formation mechanism. The α-case controlled mold contained the interstitial reaction products, TiO_2 and the substitutional reaction products, TiAl and Ti_3Al, on Al_2O_3 base [8]. In the Al_2O_3-based α-case controlled mold, Fig. 5 shows clear the interfacial microstructure between TMCs and mold.

There are no α-case problems between TMCs and the α-case controlled mold. However, the conventional static casting in the VIM furnace causes the defects such as misrun and cold-shut due to the low fluidity of TMCs melts as shown Fig 6(a). The mold preheating and superheating control of TMCs melts are not a practical solution since the procedures lead α-case formation problems. Among various methods, centrifugal force control can be a practical solution. Thus in-situ synthesis and investment casting of TMCs were carried out in a horizontal centrifugal VIM furnace. 10 G of centrifugal force was applied in order to overcome the low fluidity of TMCs. The TMCs casting with centrifugal force shows no misrun and cold-shuts. Moreover, its characteristic luster of metals was preserved as shown in Fig 6 (b). The results of the in-situ synthesis and the investment casting of TMCs show that melting and casting route can be an effective approach for the economic net-shape forming of TMCs shot sleeves.

2. Interfacial reaction

The useful life of H13 tool steel is limited because molten aluminum adheres to the surface of the steel, eventually causing the failure of the sleeve. This adhesion of aluminum is termed "aluminum soldering." In the case of H13 tool steel, there were severe interfacial reactions after 60 s at 993 K as shown in Fig. 7(a). However, (TiC+TiB) TMCs show clear interface with molten A380 alloy. Only about 5 μm thick layer is observed at the TMCs substrate and molten A380 alloy interface while under the same conditions the depth of reaction layer of H13 tool steel is about 80 μm thick as shown Fig. 8. In the case of H13 tool steel, there are three stages of the interfacial reactions. In the first stage, the interfacial reaction is under the control of diffusion by H13 tool steel elements into molten A380 alloy. After 600 s holding, the reactions occurred by not only diffusion of H13 tool steel elements but also dissolution of H13 tool steel by molten A380 alloy. That is, after the second stage, H13 tool steel shot sleeve is eroded by molten A380 alloy. In the last stage, the erosion rate of shot sleeve is accelerated with holding time and the diffusion jumping occurs as shown in Fig. 9. EPMA elemental mapping result shows that the diffusion jumping products are the intermetallic compounds between molten A380 and H13 tool steel. Eventually, the intermetallic compounds cause the deterioration of castings quality. The order of interfacial reaction process between H13 tool steel and A380 alloy can be summarized as shown in Fig. 10.

The reason why no interfacial reactions were generated in TMCs can be explained by the difference of solubility in molten aluminum and the formation route of intermetallic compound. Firstly, the solubility of iron in liquid aluminum is 3.2 wt.% at 993K. However, that of titanium in liquid aluminum is only 0.2 wt.% [9]. The low solubility of Ti makes the liquid layer adjacent to the TMCs more readily saturate with Ti solute. Its saturation inhibits the further dissolution of TMCs. Moreover, the low solubility suppresses the growth of the intermediate layer between TMCs and molten A380. Secondly, If the H13 tool steel is submerged in molten A380 alloy, the eutectic reaction products such as θ-FeAl$_3$, η-Fe$_2$Al$_5$ and ζ-FeAl$_2$ intermetallic compounds are formed by the reduction of free energy of the system with the dissolution of Fe element [10]. On the contrary, the formation of peritectic products such as TiAl and Ti$_3$Al, implies an increment in the free energy with dissolution.

From the viewpoint of endurance against molten aluminum alloys, the TMCs shot sleeve has definite advantages over the conventional technology, especially in its resistance to aluminum soldering and erosion.

3. Mechanical properties

Fig. 11 shows the variation of the friction coefficient during a sliding test to about 900 m. The wear tests were carried out in no lubricant environment such oil and water. The friction coefficient and wear volume are not absolute values but correlated values with hardness, load, lubricant and friction material and test type. The friction coefficient of H13 tool steel and TMCs was the same value of 0.3 in the ball-on-disk type. However, in the case of cone-on-disk type, the friction coefficient of TMCs was higher than that of H13 tool steel. There are no larger differences of the friction coefficient both ball-on-disk and cone-on-disk type. Also, wear volume of H13 tool steel and TMCs has the same value in the ball-on-disk type. However, the wear volume of H13 tool steel was 8 times higher than TMCs in the cone-on-disk type at the load of 20 N as shown Fig 12. In particular, the diamond cone was repeatedly fractured after 900 m sliding test on TMCs disk.

Titanium and its alloys are notorious for possessing poor tribological properties. In sliding wear test, titanium badly deforms and readily transfers materials to the counterface in unlubricated tribosystems. Furthermore, the great affinity of titanium for oxygen results in the formation of oxide contamination surface layer, which also readily transfers and adheres to metallic and non-metallic surfaces resulting in severe adhesive wear. Moreover, titanium and Ti6Al4V alloy wore at a rate 15 times higher than AISI D2 tool steel in the abrasion resistance [10]. This poor tribological behavior has been attributed to the low hardness and absolute values of tensile and shear strengths of titanium and its alloys. However, in this study, it could be confirmed that the wear property of TMCs is superior to H13 tool steel. In TiC plus TiB reinforced TMCs, TiC reinforcement plays an important part in lubricant with spherical morphology. Moreover, faceted needle-like TiB reinforcement is an extremely hard boride; hence, the resultant composite is expected to possess the excellent mechanical properties and good wear resistance. Especially, lubricant behavior of TiC reinforcement can be confirmed in Fig. 11. As to both ball-on-disk and cone-on-disk, the friction coefficients of H13 tool steel show the steep rise since there are no lubricant agents. However, the friction coefficients of TMCs show the constant value like in lubricant environments.

Fig. 13 presents tensile and elongation results of titanium and TMCs. The increase in the tensile strength compared with the pure titanium is a result of much higher work hardening rate at lower strains resulting from the constraint exerted by the elastic reinforcing particles on the plastic flow of the matrix. The elongation of TMCs is much higher than conventional power route TMCs (below 0.5%). These superior mechanical properties of

TMCs can be explained by the difference of fabrication route between powder route and melting route. In the case of powder route, the distribution of reinforcements is not homogenous due to their agglomeration. Also, there are fatal drawbacks such as the porosity and the interfacial reaction problem between matrix and reinforcements. The presence of the agglomeration of reinforcements is crucial for the low elongation behavior since it represents higher local reinforcement volume fraction areas than the mean value. The important features of the composites fracture behavior are local matrix void coalescence and interfacial reaction between matrix and reinforcement. Consequently, the correlation of agglomeration of reinforcement, porosity and interfacial reaction cause poor mechanical properties of powder route TMCs. However, in the case of in-situ synthesis of TMCs with melting route, there are no agglomeration of reinforcement, porosity and interfacial reaction. Therefore, in-situ synthesized TMCs by melting route can apply for the structural materials.

From the evaluation of wear and mechanical properties of TMCs, we can affirm that the H13 tool steel shot sleeve can be substituted with TMCs for aluminum alloys die-casting.

Conclusions

Until now, the powder metallurgy monopolizes the manufacturing of TMCs with ex-situ and in-situ synthesis processes. However, the spherical TiC plus needle-like and large, many-angled facet TiB reinforced TMCs can be synthesized with Ti and B_4C by a melting route. Moreover, the economic net-shape forming of TMCs can be possible by a casting route. We evaluated the possibility of the economically in-situ synthesized (TiC+TiB) reinforced TMCs for shot sleeve materials of aluminum alloys die-casting. In-situ synthesized TMCs presents significant advantages over the conventional H13 tool steel shot sleeve, especially in its resistance to aluminum soldering and erosion. Moreover, the wear and mechanical properties of TMCs meet all the requirements for shot sleeve such as the resistance to heat checking, thermal fatigue and distortion. Through the evaluation of the endurance against molten aluminum alloys, and wear and mechanical properties, it is confirmed that the conventional H13 tool steel shot sleeve can be substituted with in-situ synthesized TMCs shot sleeve fabricated by melting and casting route.

References

1. Shankar S, Apelian D, *Mechanism and preventive measures for die soldering during Al casting in ferrous mold,* Journal of Metal, 54, No 8, Aug 2002, pp47-54.
2. Herman EA, Die Casting Dies, Society of Die Casting Engineering, Michigan 1985.
3. Salas O, Kearns K, Carrera S, Moore JJ, *Tribological behavior of candidate coatings for Al die casting dies,* Surface & coating technology, No 172, July 2003, pp117-127.

4. Batista JCA, Joseph MC, Godoy C, Matthews A, *Micro-abrasion wear testing of PVD TiN coatings on untreated and plasma nitrided AISI H13 steel*, Wear, No 249, Nov 2001, pp971-979.

5. Alman DE, Hawk JA, *The abrasive wear of sintered titanium matrix–ceramic particle reinforced composites*, Wear, No 225-229. pp629-639.

6. Bailey SJ et al, editor. Annual book of ASTM standards Vol. 03.01, ASM International, Baltimore 2002, ISBN 0 8031 3166 6.

7. Chase MW, Davis CA, Downey JR, Frurip DJ, McDonald RA, Syverud AN, JANAF Thermochemical Tables, American Chemical Society and American Institute of Physics, New York 1985, ISBN 0 88318 473 7.

8. Kim YJ, Sung SY, Kor Patent No. 10-2004-0069535, 2004.

9. Yan M, Fan Z, Review Durability of materials in molten aluminum alloys, Journal of Materials Science, 36, No 2, Jan 2001, pp285-295.

10. Lumeda KC, Bayer RG, Wear of materials, Am Soc Mech Eng, New York, 1991.

Figures

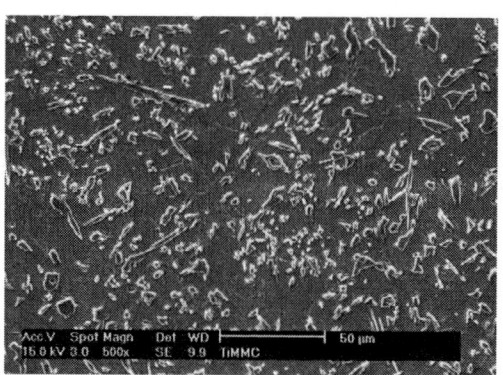

Figure 1. SEM image of microstructure of in-situ synthesized (TiC+TiB) reinforced TMCs between titanium and B_4C.

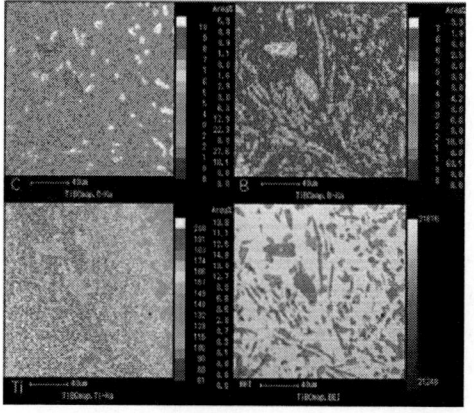

Figure 2. Comparison of elemental mapping images of C, B and Ti by EPMA.

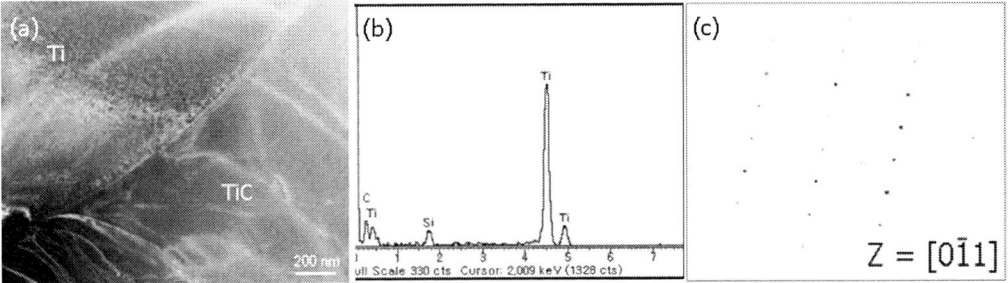

Figure 3. The results of TEM analysis: (a) bright field image of Ti and TiC, b) EDS peak and (c) spot diffraction pattern of synthesized TiC.

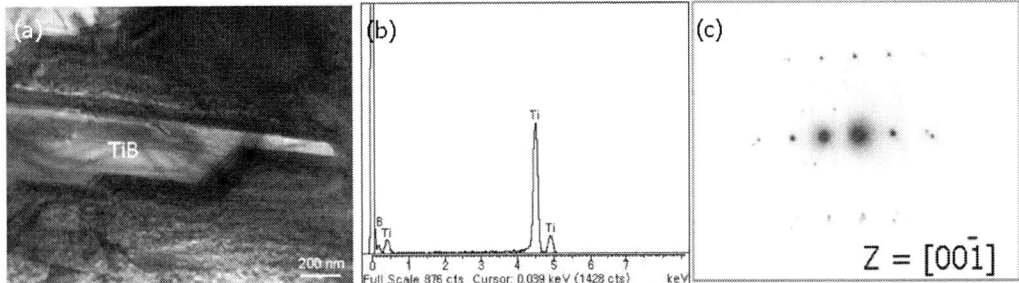

Figure 4. The results of TEM analysis: (a) bright field image of Ti and TiB, b) EDS peak and (c) spot diffraction pattern of synthesized TiB.

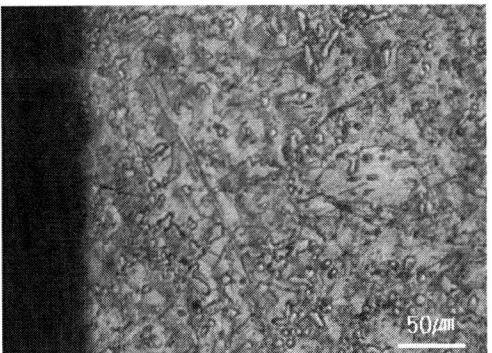

Figure 5. Microstructure of the interface between (TiC+TiB) reinforced TMCs and α-case controlled mold.

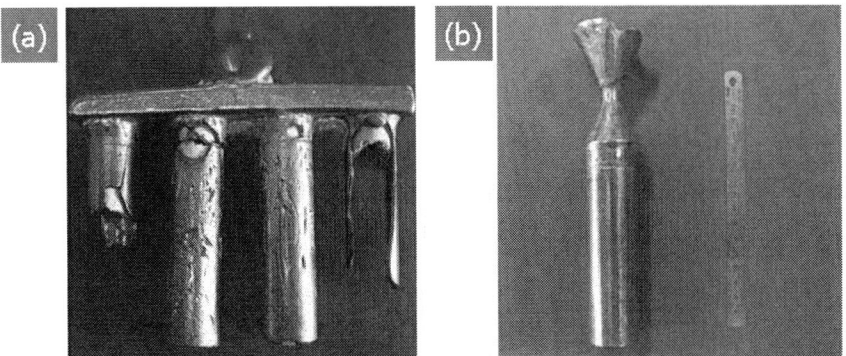

Figure 6. Photographs of (TiC+TiB) reinforced TMCs shot sleeve castings poured into α-case controlled mold by (a) static casting and (b) centrifugal casting.

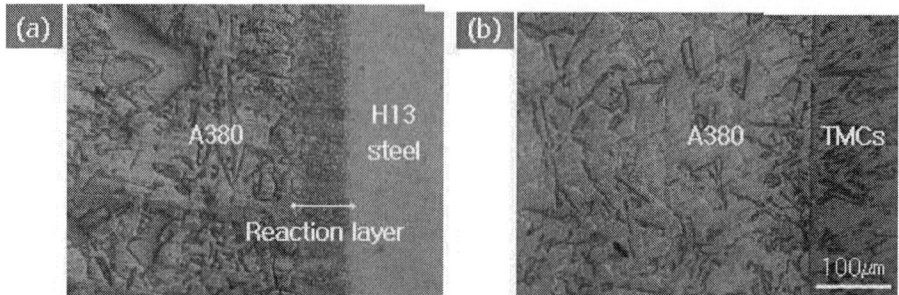

Figure 7. Microstructures of the interfacial morphology of (a) H13 tool steel and (b) (TiC+TiB) reinforced TMCs in molten A380 alloy for 1200 s at 993 K.

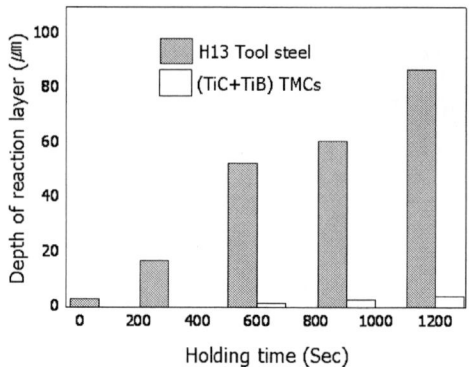

Figure 8. Depth of interfacial reaction layer of H13 tool steel and (TiC+TiB) reinforced TMCs in molten A 380 at 993 K for times varying from 0 to 1200 s.

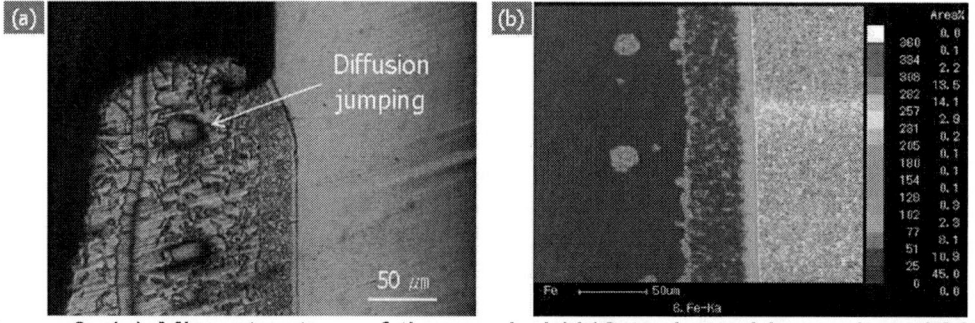

Figure 9. (a) Microstructure of the eroded H13 tool steel by molten A380, and diffusion jumping of the interfacial reaction products and (b) EPMA elemental mapping images of Fe on the interfacial reaction for 900 s at 993 K.

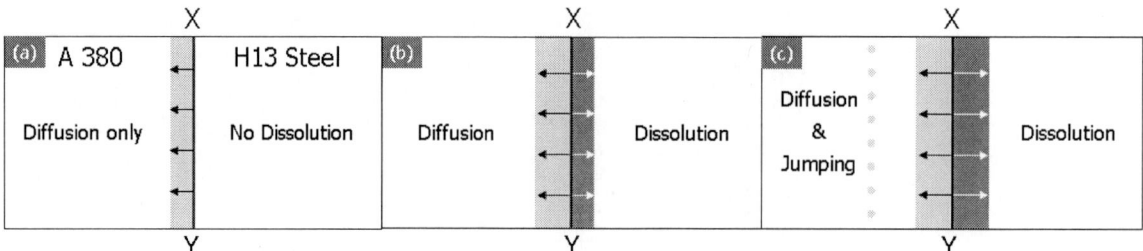

Figure 10. Schematic diagram of the order of interfacial reaction process between H13 tool steel and A380 alloy (XY: initial interface).

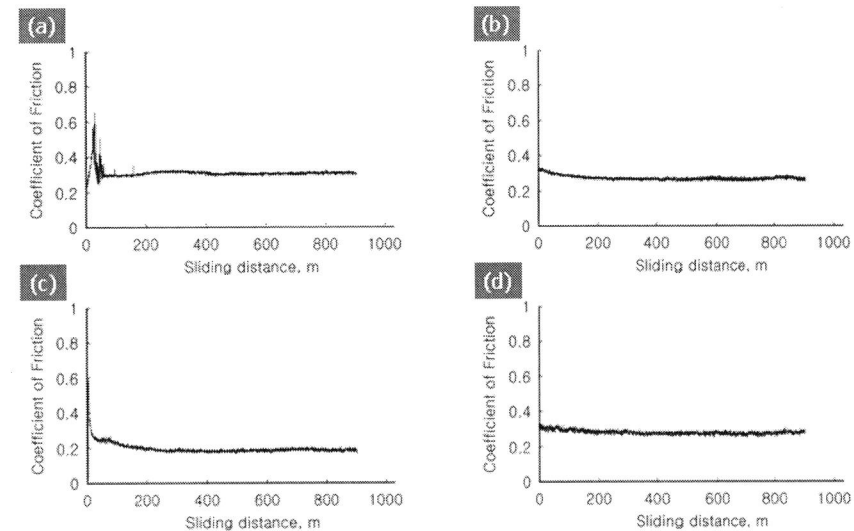

Figure 11. Variation of friction coefficient of H13 tool steel and (TiC+TiB) reinforced TMCs against AISI 52100 steel ball diamond cone: (a) H13 with steel ball at the load of 10 N, (b) TMCs with steel ball at the load of 10 N, (c) H13 with diamond cone at the load of 20 N and (d) TMCs with diamond cone at the load of 20 N.

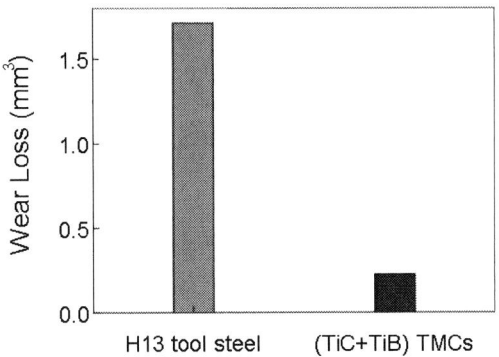

Figure 12. Wear volume of H13 tool steel and (TiC+TiB) reinforced TMCs in the cone-on-disk type at the load of 20 N.

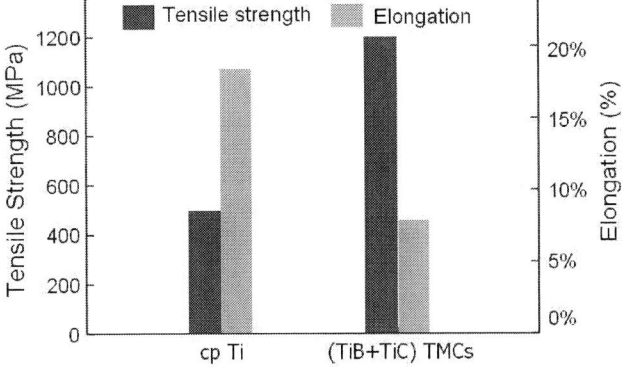

Figure 13. Tensile strength and elongation of cp Ti and (TiC+TiB) reinforced TMCs in the room temperature.

AWB – an environment friendly core production technology

Th. Steinhäuser *, A. Wolff *

* Institut für angewandte Materialtechnik, Universität Duisburg - Essen, Germany.

Abstract

The AWB core binder system is based on sodium silicate binders. The hardening of the cores is done by dehydration, there is no chemical reaction. This makes the hardening process reversible and reclamation simple. Dehydration is done by a combination of heated core box, vacuum and microwave treatment. The binder system has been widely tested at the University Duisburg – Essen and at the labs of Hydro Aluminium. A great number of cylinder heads in aluminium gravity die-casting have been made to prove suitability for serial production.

Key words

AWB, inorganic core binder,

Introduction

Because of rising interest in environmental issues within the foundry industry during the last decade, there has been a lot of development especially in core binder systems. One direction was the reduction of emissions on organic binders. The other way was the development of inorganic core binders with competitive performance [1]. Patented already in 1997, the AWB process is one of the first new inorganic binder systems [2]. Due to now eight years of development it has become an efficient, practical, suitable system that can be used in serial coremaking with only minor modifications to the existing equiment.

Experimental

The basis of the technology is a modified sodium silicate.

The AWB-binders are of a distinctly lower viscosity than conventional sodium silicate core binders and are therefore much easier to mix and homogenise with sand. This also serves to improve core shooting characteristics of the sand-binder mixtures which are in the range of cold-box mixtures.

With AWB even thin-walled and complicated cores such as water jackets for cylinder heads can be produced.

Sodium silicate binders are based on the chemical formula $Na_2O_xxSiO_2xH_2O$. During conventional hardening by gassing with CO_2, the strength is generated through precipitation of Na as Na_2CO_3 by changing the "module", i.e. the Na_2O/SiO_2 ratio. This procedure has decisive disadvantages:

- The chemical reaction is irreversible.
- The "solvent" water largely remains within the core.
- Only low strengths can be attained.
- Na_2CO_3 and SiO_2 can form a glassy phase at higher temperatures which makes decoring difficult
- Reclamation of such core sands is difficult or impossible.

In the AWB-process, hardening is done only by the removal of water (Fig.1). This has the following advantages:

- Physical drying reaction is reversible.
- AWB-cores are nearly free of water and thus can be stored for a longer period.
- Due to their dry state, cores emit much less core gas compared with organic binders.
- Distinctly high core strengths are attainable.
- Easy decoring possible due to lack of formation of sodium carbonate.
- Reclamation easy, remaining binder can be reactivated by addition of water.

Hardening starts in the heated core box at temperatures of approx. 140 – 200 °C with a connected vacuum to absorb water vapour. After approx. 10 – 60 seconds – depending of core weight and geometry – a solid shell has been formed so that the core has sufficient handling strength and can be removed from the core box. Final dehydration takes place in a microwave at low power. This step is very important because any water remaining in the core may solve binderbridges and reduce storability.

After cooling to room temperature, the core has reached its final high strength and can be used for casting.

Results

Characteristics of cores produced with the AWB-process as compared to cores produced with other binders have been evaluated in several extensive test series performed both at University Duisburg – Essen and Hydro Aluminium's R&D in Bonn. Cold-box cores with a binder content of 0.4 / 0.4 % (resin and activator) and 0.8 / 0.8 % were used as benchmarks. Standard bending test pieces size 180 x 22.5 x 22.5 mm were used at a constant shooting pressure, silica sand H32 was used.

The tests showed that in order to reach the same strength as with the above reference mixture, 1.5 and 2.5 % respectively of AWB-binder is required [3].

To produce strong cores in short cycle times it is important to keep the hardening time in the heated core box as short as possible. Fig. 2 shows the influence of the hardening time on the bending strength without microwave treatment. In practice, usually 100 N/cm^2 is good enough to handle the core without problems. After microwave treatment, the core will have reached its full strength and will keep it over storage time. In Fig. 3 you can see, that the shorter the hardening time in the core box, the higher the strength after the microwave treatment.

Particularly when casting aluminium alloys, the gas surge of vaporised binder material can lead to massive problems. The gas evolution was measured during casting in accordance with the COGAS®-system. Fig. 4 shows the gas emission during casting compared to those of a Hot-Box and a Cold-Box core. The results show that AWB drastically minimises the risk of gas-induced casting defects. In fact, it can be assumed that the measurable gas evolution is attributable to the expansion of the pore volume within the core sample.

In order to document the suitability for serial production at Hydro Aluminium Mandl & Berger GmbH in Austria, slightly modified Hot-Box core tools were used to produce all cores for an automobile cylinder head, e.g. waterjackets, oil-galleries, inlet, outlet and cover cores.

Cycle times of the core production corresponded to those of the current serial production. The connected vacuum helped to reduce the core shooting pressure by approximately 0.5 – 1 bar. Hardening time in the microwave (1.5 kW) took 1 – 3 minutes, depending upon the core weight. This microwave time is independent from the cycle time of the core blower and can be performed either in batches or by a run-through microwave. Fig. 5 shows the process flow chart for the cylinder head production. It was proven that an uninterrupted core production with AWB on a standard core shooting machine is possible with the quality of the cores being as good as those of the current serial production with organic binders.

The results of extensive tests of aluminium gravity die casting at Hydro Aluminium Mandl & Berger with AWB-cores proved that the process:

- Does not produce emissions other than water vapour.
- Does not cause any condensate deposits within the die.
- Allows production of defect-free castings.
- Decoring can be done with standard equipment.

Recently there have been made successful tests with core package castings (CPS) for engine blocks at the Hydro Aluminium foundry in Dillingen [4]. AWB has also shown very good results after several trials of University Duisburg-Essen in iron, steel, stainless steel and other metal alloy castings. Fig. 6 + 7 show different castings produces with AWB. First tests to determine the compatibility of AWB bonded core sand with bentonite bonded systems have shown no negative effects on the compactibility and the wet tensile strength when adding up to 25% of AWB core sand. These tests have not yet been finalised, a definite report on their outcome will be published at a later stage.

Conclusions
The AWB-process is a new inorganic process, suitable for serial production. Based on the purely physical bonding, the sand can be regenerated without any problems. Core strength and cycle times are comparable to those of organically bonded cores. The production of filigree and complicated cores creates no problems. Serial production of aluminium cylinder heads and test production of cast iron, stainless steel, steel and non-ferrous metals have been very successful. The inorganic AWB-binder does not cause health-injurious emissions, neither during core production nor during casting.

References
1.   *Anorganische Binder – Durchbruch oder Hoffnung*, VDG-Fachtagung, Wuppertal/Germany 14.11. 2002.
2.   *EP 0 917 499 Optimierte Kernherstellung*, Prof. Dr.-Ing. Thomas Steinhäuser, 26.07.1997
3.   Wolff A., Steinhäuser T., *AWB – ein umweltverträgliches Kernherstellungsverfahren*, Giesserei 91 (2004) Heft 6, Seite 80 - 84

4.   Gosch R., *Inorganic core manufacturing technology for engine blocks in core package process CPS*, WFO/VDG Conference, Hannover/Germany, 23.-24.11.2005

5.
Figures

Fig. 1

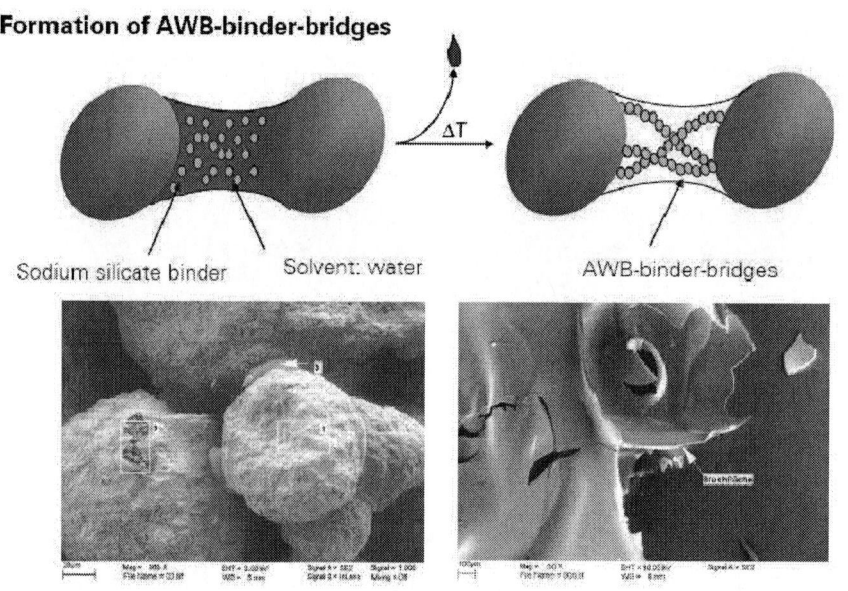

Fig.2

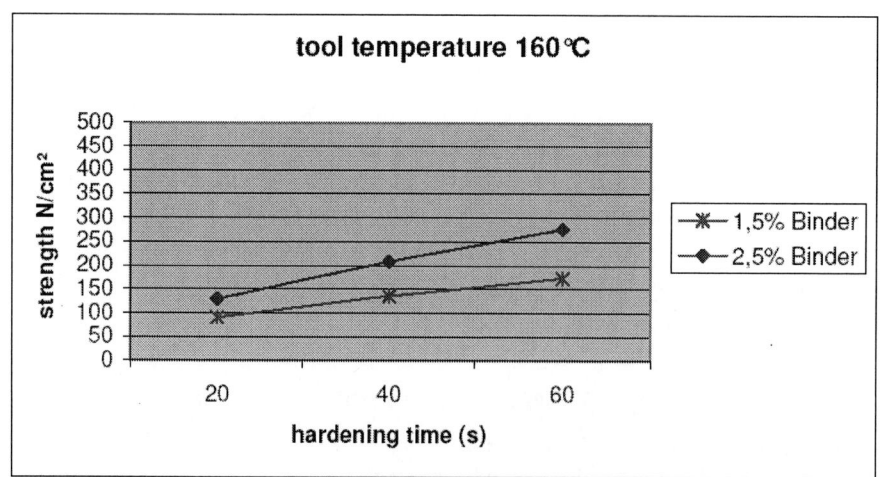

Bending strength before microwave treatment versus hardening time.

Fig.3

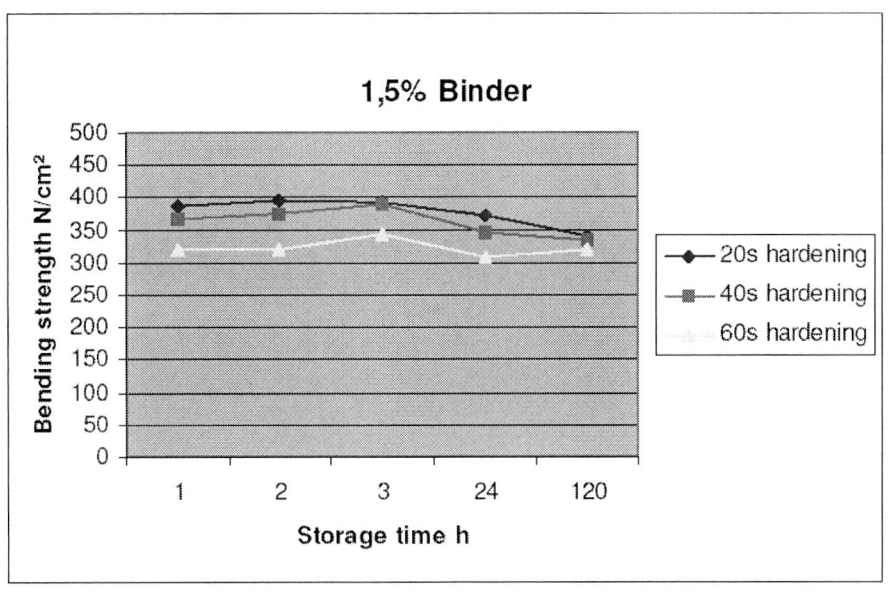

Bending strength after microwave treatment versus storage time.

Fig.4

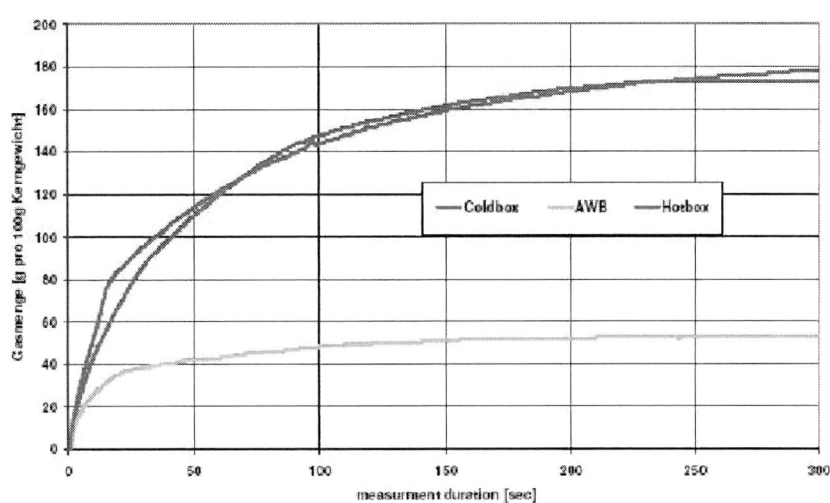

Gasvolumes

Fig. 5

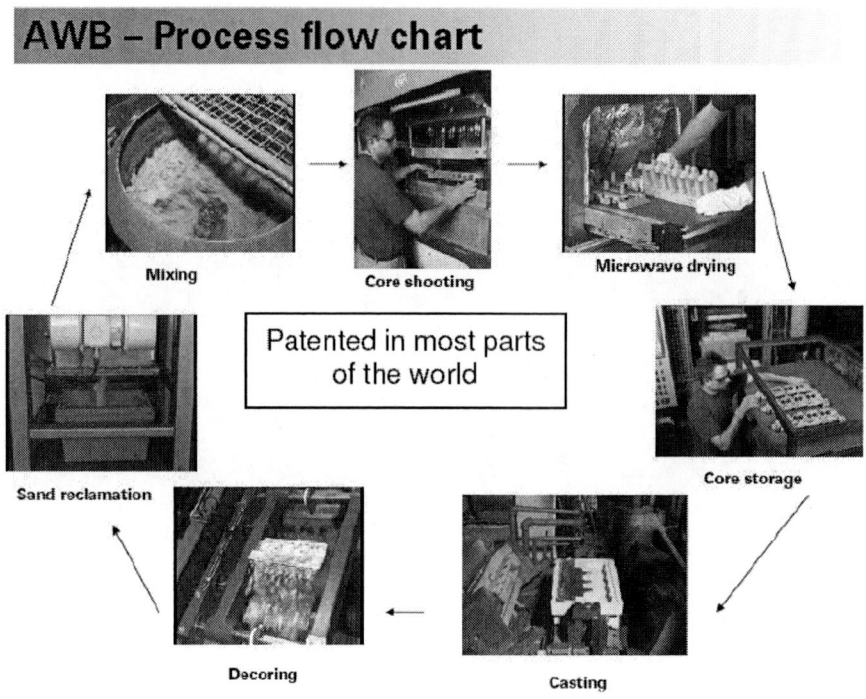

Fig.6

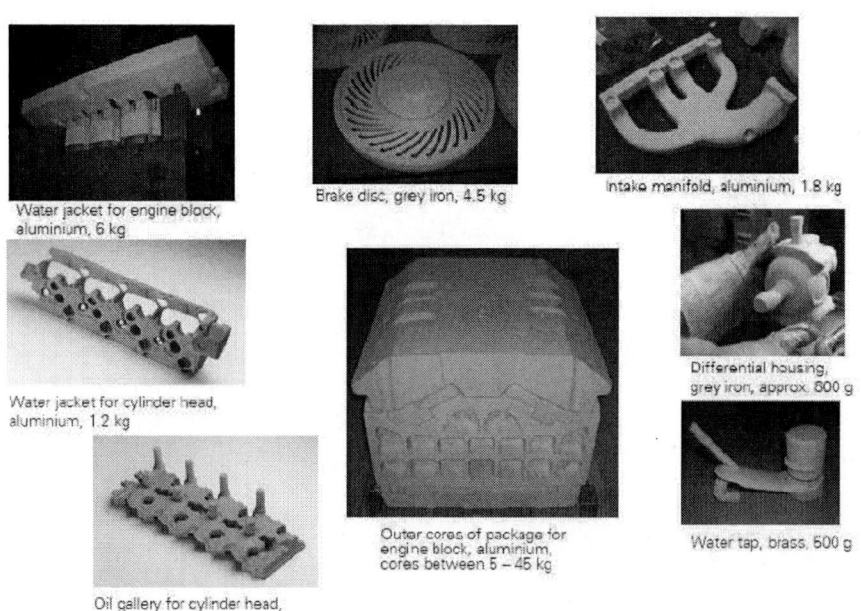

Fig.7

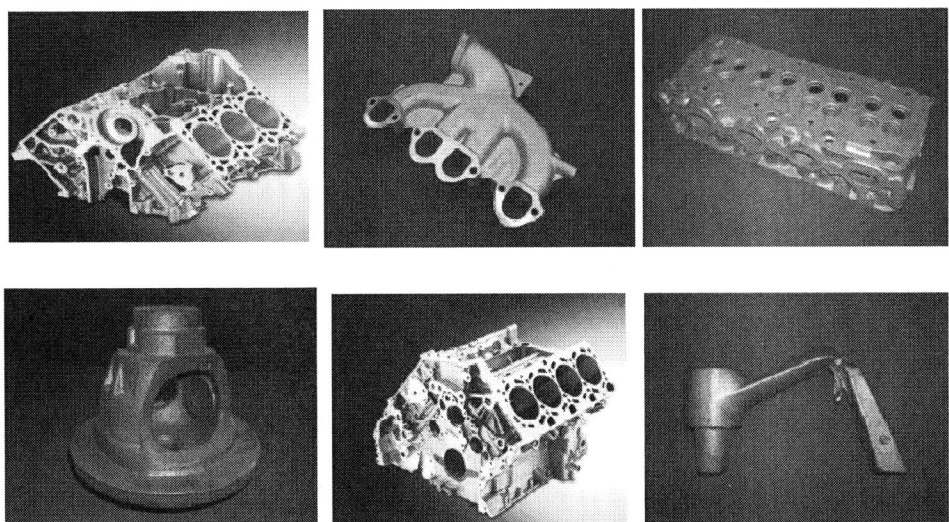

Advances in Thin-Wall Sand Casting

R.E. Showman, R.C. Aufderheide, N.P. Yeomans

Ashland Casting Solutions, Dublin, Ohio, USA

Abstract

Economic and environmental pressures continue to force metalcasters to search for ways to produce castings with thinner walls and lighter weight, but with equal or improved mechanical properties. Previous studies have shown that the use of low-density alumina-silicate ceramic (LDASC) as a sand additive/replacement can dramatically modify the thermal properties of sand molds and cores and can enable the production of thin (<3mm) sections without sacrificing metallurgical or mechanical properties.

This paper reviews laboratory testing conducted with LDASC in the sand mold or core and summarizes effective thermal properties, resulting cooling rates and casting properties. Additional new data is provided detailing the use of LDASC to effectively enhance gating and feeding systems through the development of thermal channels within the mold cavity.

Key words – thin-wall, LDASC, thermal-channels, sand-castings

Introduction

The demand for light-weight, thin-wall castings continues to grow. Section thickness reductions for automotive and other castings are driven by the need to reduce weight, tighten dimensional tolerances, enhance performance and reduce overall cost. This has resulted in a continuing shift toward more aluminum and magnesium castings and toward gravity and high-pressure diecasting. However, the strength-to-weight ratios for ductile and compacted graphite irons make them attractive if thin-wall (<3mm) parts can be produced with good dimensional accuracy and controlled physical and microstructural properties and precision sand casting offers a number of advantages over other casting methods[1].

Efforts by a number of researchers [2 - 5] have shown that thin-wall iron castings can be reliably produced in sand molds by using the proper metal chemistry, inoculation practice, gating design, and mold and core materials. These controls may tighten the processing "window" for iron foundries, but do not pose insurmountable problems. The procedures simply reflect an extension of existing practices rather than totally new manufacturing processes. However, foundries have been slow to adopt these practices and move toward thinner castings without more proven examples.

Review of Previous Work with LDASC

Earlier work by the authors [6] has shown that LDASC can be used as a sand additive/replacement to modify and control the thermal properties of molds and cores, and to make the production of thin-wall iron castings a reality. LDASC has a density about $\frac{1}{4}$ of silica sand and a corresponding reduced specific heat and thermal conductivity. When used at levels of 20% – 100% (volume %) with sand, the heat extraction characteristics are dramatically changed, making thin-wall iron castings with good metallurgical and physical properties possible. The change in thermal properties slows the cooling and solidification rates as shown with ductile iron. (Figure 1 [7])

The ability to produce thin-wall iron castings with LDASC has been demonstrated in several ways. First, standard fluidity spiral castings were produced with various blends of LDASC and silica sand. These blends reduced the cooling and solidification rates in these castings, enabling much increased metal flow. Where a standard sand mold produced a flow distance of approximately 71 cm, a 40% sand, 60% LDASC (volume %) blend resulted in flow all the way to the end of the spiral, or about 147 cm. (Figure 2)

Thin-wall plate castings were also produced to show the effectiveness of LDASC additions. The test casting was approximately 200mm x 200mm x 1.5mm with two ingates. The mold was produced in phenolic urethane no-bake, first with sand and then faced with 100% LDASC on both the cope and drag surfaces. The sand mold produced a misrun when poured with ductile iron at 1400° C. The mold faced with LDASC filled completely, with no indications of misruns or coldshuts. (Figure 3)

Somewhat more complex thin-wall castings were produced to show that the LDASC material could be used either on the mold or core surfaces. A manifold test casting was used that was approximately 380mm long by 55mm in diameter with a 2mm wall (Figure 4). By gating into all three heavy sections, the casting could be 90% – 95% filled with ductile iron. However, by using at least 25% LDASC in either the core or mold facing, the casting filled completely using a single gate, nearest the sprue.

Even in thin-wall castings, the use of LDASC controlled the solidification rates and resulting microstructures of iron castings. This was first demonstrated with gray iron "chill wedges." These castings are typically used to show the chilling tendencies of gray iron with different chemistries or inoculation. However, they can also be used to show the effects of slower cooling with LDASC additions on casting microstructure, particularly on carbide formation. (Figure 5)

The effects of LDASC additions on ductile iron microstructure were most dramatically shown in the thin-wall manifold casting in the 2mm section. When both a 100% sand core and mold were used, the 2mm section contained massive carbides in fine pearlite. When a 50/50 (by volume) mixture of sand and LDASC was used in both the core and mold, the 2mm section showed small graphite nodules in ferrite and pearlite, with some small retained carbides. However, when 100% LDASC was used in both the core and mold, the 2mm section contained large graphite nodules in a typical "bulls-eye" structure that might be expected in section of 5mm and above. (Figure 6)

The physical properties of thin-wall casting are dependent on the cooling rates and resulting microstructures. Hardness is often used as an indicator of tensile strength in castings where it is impractical to cut and test actual sections from the casting. For the 2mm manifold test casting, hardness values were recorded from the 2 mm section and plotted against the volume percentage of LDASC in contact with the casting. The resulting graph shows a dramatic reduction in hardness, which would indicate a corresponding increase in ductility and yield. (Figure 7)

Labrecque, Gagne & Javaid [8] also performed tensile tests on samples cut from thin-wall ductile iron castings produced with varying amounts of LDASC. They found a significant decrease in ultimate tensile strength (UTS) and yield strength (YS) and a corresponding increase in percent elongation (%E) with increasing amounts of LDASC. (Figure 8)

Thermal Channels
When tests were conducted on the thin-wall manifold casting, there was an improvement between castings made with LDASC in the core versus the castings with LDASC in the mold facing, also covering the gating

system. The LDASC appeared to improve the performance of the gating system. The use of LDASC in the mold or core reduced heat extraction from the metal and thus reduced cooling and solidification rates. In effect, the area with LDASC "acted" like a heavier section. This concept generated the idea to use LDASC inserts as "thermal channels." The inserts could be placed on the mold surface leading from the ingates, across thin sections of the casting. The inserts would cause the area to "act" like a heavier section, thus creating a thermal channel for the metal from the gate to flow through. Thermally, the insert would create an effective heavier section, but the actual casting section would remain thin and uniform.

An extreme test casting was chosen for pouring trials. The 200mm x 200mm x 1.5mm casting could not be poured with a sand mold in either gray or ductile iron at normal pouring temperatures without misruns. It was an "impossible" casting without the use of special techniques. Pouring trials in normal no-bake sand molds produced castings that showed the expected flow pattern spreading from the ingates, but the metal cooled and solidified before the entire mold could fill. (Figure 9)

Thermal channels were created by placing 25mm x 12mm bars of 100% LDASC as inserts on the casting surface. Four configurations were developed: extension from one gate, extensions from both gates, one cross piece between extensions, and two cross pieces between extensions. Initially molds were made with the extensions only in the drag, but a second set of molds were made with the extension in both cope and drag. (Figure 10) The molds were poured in Class 30 gray iron at approximately 1370° C. The castings with the inserts in the drag only showed improved metal flow and fill distance and the castings with the inserts in both the cope and drag showed even greater improvement. It was apparent that gate extensions had "acted" as thicker sections and helped fill the "impossible" casting. (Figure 11)

Thermal channels can also be used to control feeding in thin-wall castings. Typical thin-wall castings are prone to centerline shrinkage [9] because of the very low modulus (volume to surface area ratio) and the inability to maintain feeding pathways through the thin sections. The use of LDASC in the mold or core can change the effective modulus of the section and thus improve feeding through the section. This is similar to the traditional use of "padding" to improve feeding characteristics. By selectively using LDASC inserts to create thermal channels through the thin sections, feed pathways can be maintained and directional solidification will occur toward the appropriate gates, risers or heavy sections. (Figure 12)

Modeling Thermal Channels
The thin plate casting was modeled using Novaflow® to show the flow and cooling patterns. The initial model of the plain no-bake mold showed flow

extending from the ingates and cooling to create a likely misrun. This was very similar to the fill patterns seen in the actual test castings. (Figure 13)

LDASC inserts were then added to the model in the same configurations as the four-test casting. The density and thermal properties of the inserts were adjusted for the LDASC material to show its effects on the fill pattern. (Figure 14)

The resulting casting models showed fill patterns very similar to the actual castings (Figure 15). This confirmed the ability of the model to show the effects of the gate extensions and indicates that the modeling of gate extensions can be applied to complex castings.

Conclusions
- Previous work has shown the effectiveness of LDASC to modify the thermal properties of sand molds and to reliably produce thin-wall iron castings with good metallurgical and physical properties.

- LDASC can be used to create thermal channels that improve the flow and feeding through thin-wall sections. This provides an additional design and process option for thin-wall castings including a reduction in runner size, reduction in the number of ingates and improved feeding through thin sections.

- Models can be used to show the effects of the thermal channels and to predict flow and feeding for design purposes.

References
1. Busby, A.D., Archibald, J.J., "Expanded Opportunities for Precision Sand Casting Utilizing Coldbox Binders", AFS Transactions, 05-045, 2005
2. Charoenvilaisiri, S., Stefanescu, D., Rusanda, R., Piwonka, T., "Thin-wall Compacted Graphite Castings", ASF Transactions, 02-176, 2002
3. Javaid, A., Thomson,J., Davis, K., "Critical Conditions for Obtaining Carbide-Free Microstructures in Thin-Wall Ductile Irons", AFS Transactions, 02-028, 2002
4. Katz, S., "Thin-wall Iron Castings – Planning the Future", Foundry Management & Technology, June 1997, pp34-36
5. Ruxanda, R., Stefanescu, D., Piwonka, T., "Microstructure Characterization of Ductile Thin-wall IronCastings", AFS Transactions, 02-177, 2002
6. Showman, R.E., Aufderheide, R.C., "A Process for Thin-Wall Castings", AFS Transactions 03-145, 2003
7. Aufderheide, R.C., Showman, R.E., "Controlling the Skin Effect on Thin-Wall Ductile Iron Castings", AFS Transactions 05-043, 2005
8. Labrecque, C., Gagne, M., Javaid, A., "Optimizing the Mechanical Properties of Thin-Wall Ductile Iron Castings", AFS Transactions 05-116, 2005

9. Woolley, J.W., Stefanescu, D.M., "Microshrinkage Propensity in Thin Wall Iron Castings", AFS Transactions 05-094, 2005

Figures

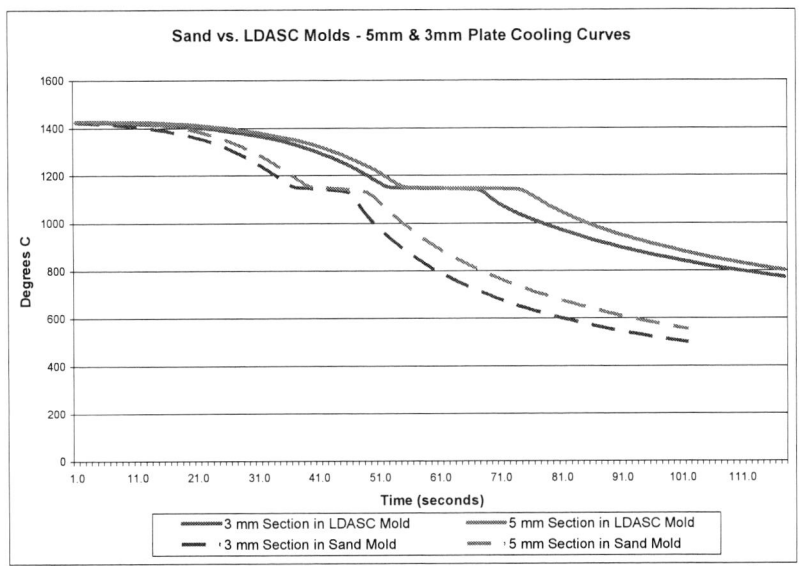

Figure 1. Cooling curves for 3 & 5 mm ductile iron plates[1]

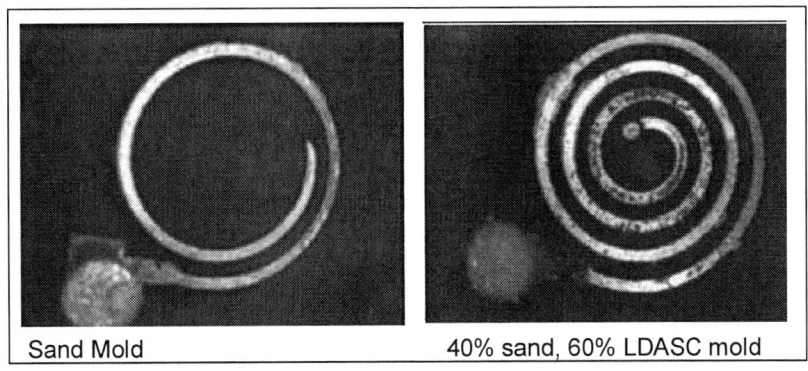

Figure 2. Fluidity spiral castings, gray iron

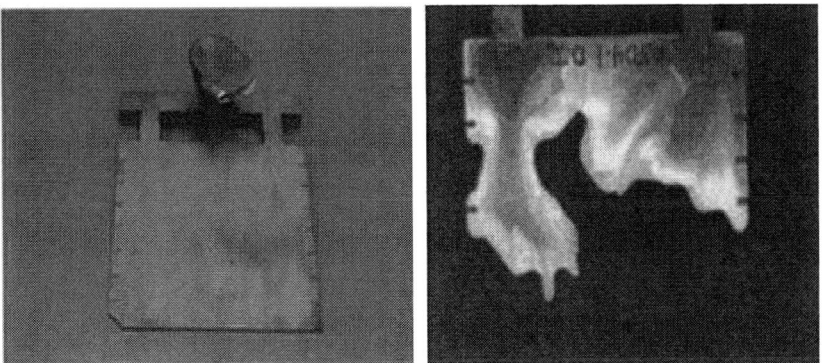

Figure 3. Thin-wall plate castings: LDASC mold left, sand mold right

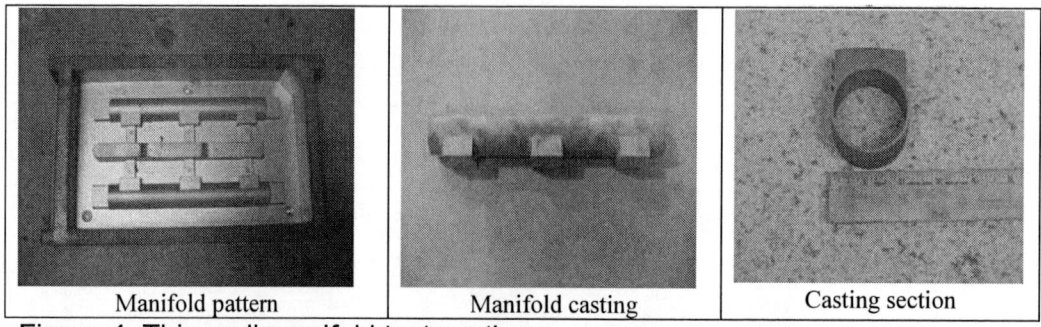

| Manifold pattern | Manifold casting | Casting section |

Figure 4. Thin-wall manifold test casting

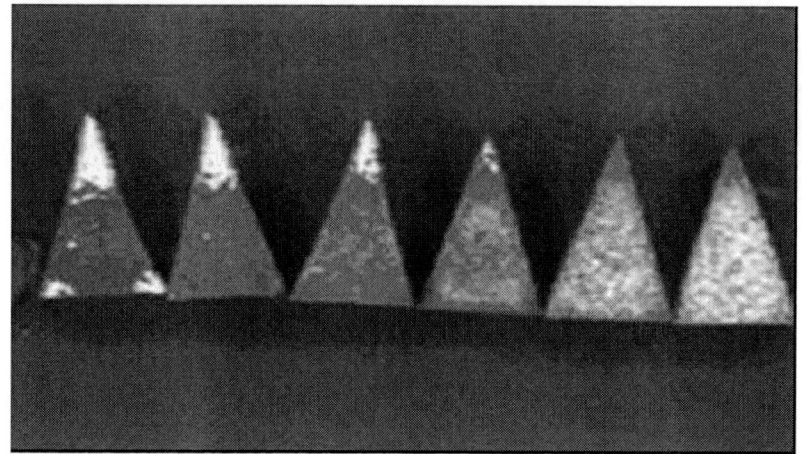

Figure 5. Gray iron "chill wedge" castings from 100% sand to 100%

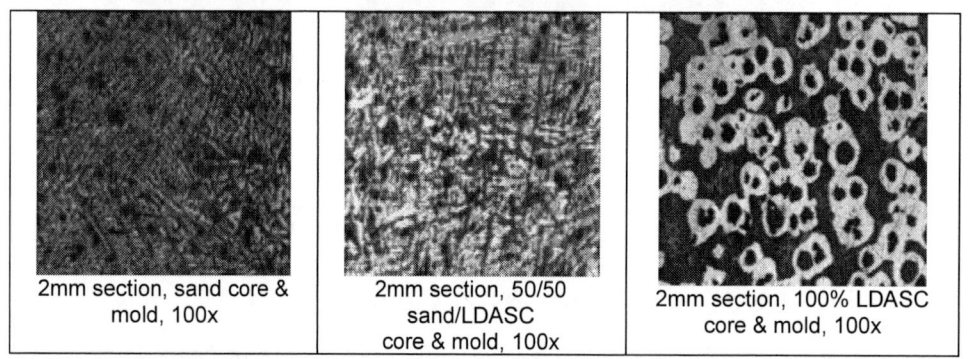

| 2mm section, sand core & mold, 100x | 2mm section, 50/50 sand/LDASC core & mold, 100x | 2mm section, 100% LDASC core & mold, 100x |

LDASC in 20% increments

Figure 6. Ductile iron microstructures from 2mm manifold test casting

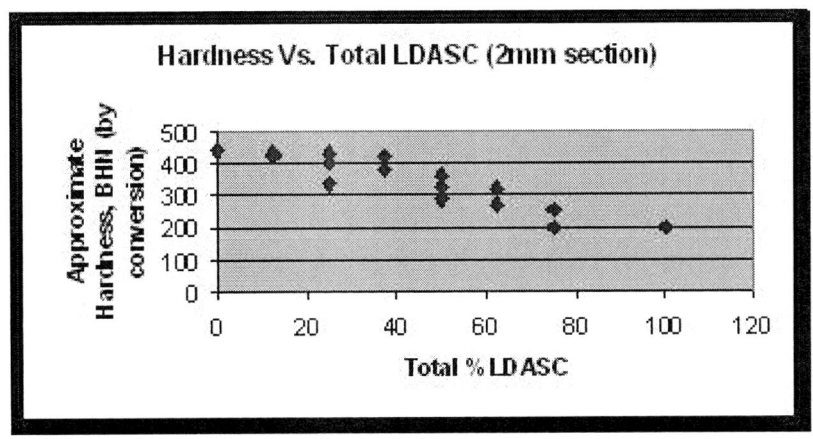

Figure 7. Graph of casting hardness vs. LDASC %

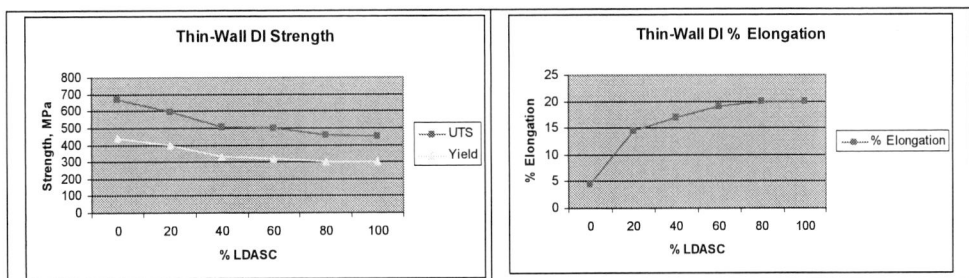

Figure 8. Thin-Wall DI casting properties as a function of % LDASC in the mold sand

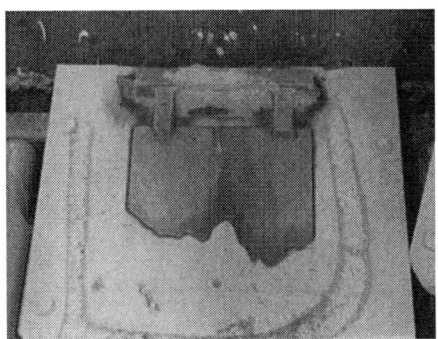

Figure 9. Thin plate casting poured in no-bake sand mold

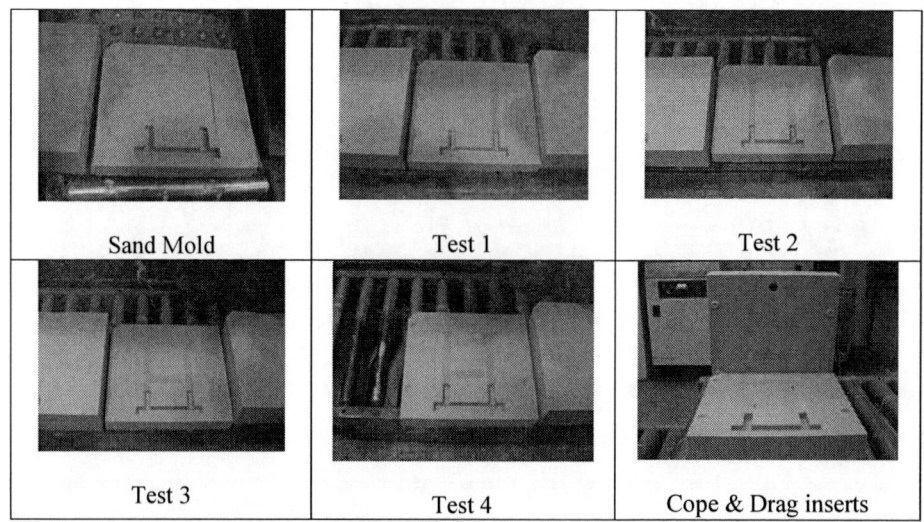

Figure 10. Test molds with different gate extension configurations

Figure 11. Thin plate castings poured with gate extensions; top row - drag only, bottom row – cope & drag

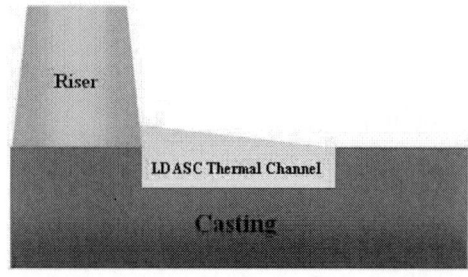

Figure 12. Sketch of thermal channel to promote feeding through thin section

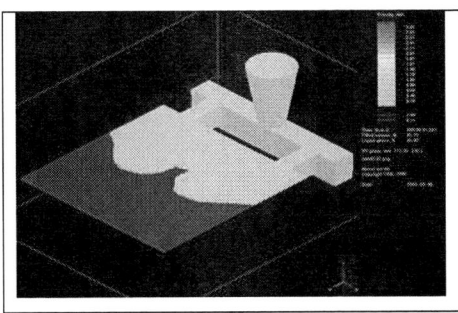

Figure 13. Modeled fill pattern in thin plate casting with sand mold

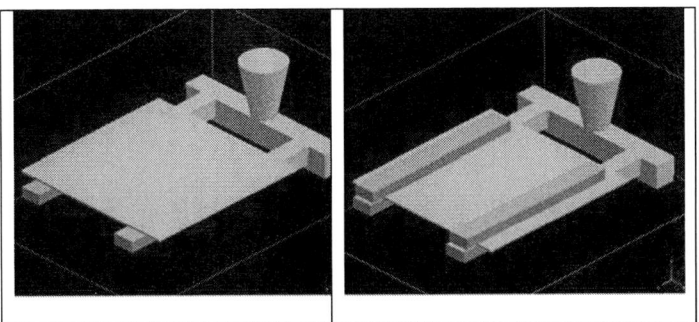

Figure 14. Examples of the model with gate extension in the drag and both cope and drag

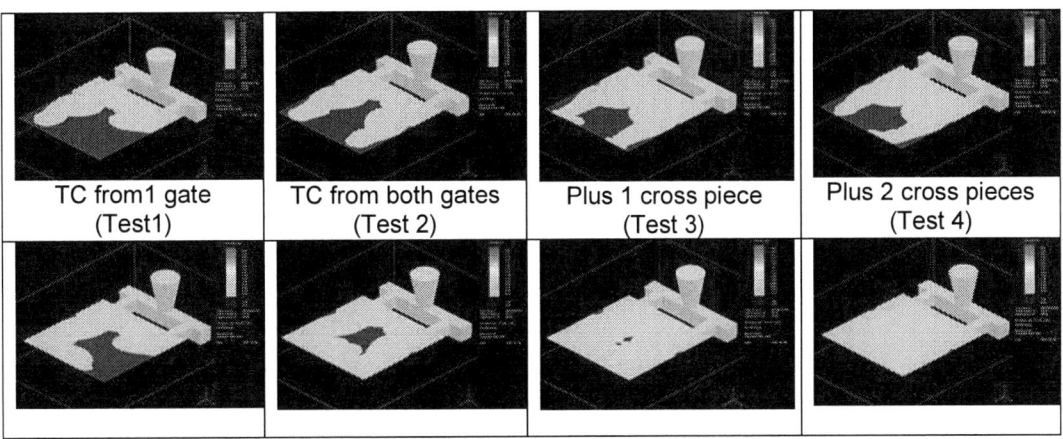

| TC from1 gate (Test1) | TC from both gates (Test 2) | Plus 1 cross piece (Test 3) | Plus 2 cross pieces (Test 4) |

Figure 15. Models of fill patterns with thermal channels in the drag (top row) and both cope and drag (bottom row)

The significance of total carbon in greensand systems
by Alexander Brown FICME
James Durrans and Sons Limited, UK

Abstract

The need for control of raw materials is critical to the success of iron casting production from greensand systems. The base silica sand is often overlooked with the main focus on bentonite additions. Carbonaceous additives can be considered a "necessary evil" to ensure a good surface finish and reduction in sand related surface defects. Other additives are used when systems get out of balance and these in turn add further to the complex nature of greensand systems.

For castings requiring cores this becomes a bigger issue, as many differing resin systems are employed for core production and these must be taken into consideration when controlling both the carbonaceous levels and the overall grading of the sand system. Possible changes on the loss on ignition properties and the effect on the carbon level needs careful understanding and control.

Various test methods are discussed including volatile and loss on ignition (LoI), and control methods such as total carbon are reviewed.

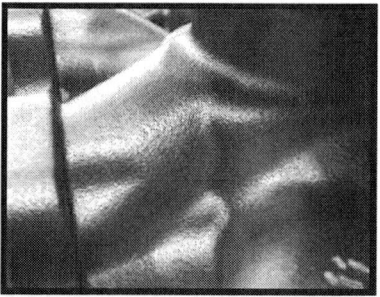

The quality of additives and their role and more importantly their interaction is highlighted, as this is an area often neglected as foundrymen battle for success to produce consistent quality castings. The use of automatic mill control has simplified the production of usable moulding sand but this cannot replace the correct balance of additives and having sand at the mill, prior to additions, in the optimum condition

Also reviewed is the interpretation of results and the action required to ensure control for consistent quality castings from greensand systems, with the focus on the use and understanding of carbonaceous additive on casting performance.

Key words

Graded Coal, Bentonite, Greensand, Total Carbon, Volatile

Introduction

Control of raw materials

Much has been documented on additives to greensand systems. Typical areas frequently discussed are listed

- **Poor sampling methods / frequency and timing of samples**
- **Lack of calibrated testing**
- **Focus on the wrong control areas**
- **Lack of understanding of primary and secondary sand tests**
- **Little focus on silica sand or sand grading**
- **Bentonite focus detriment to the carbonaceous additive**
- **Poor in-coming test procedures and over reliance on suppliers**

All of the above are important but equally good castings can be produced from systems with little or no control. By control, means a basic understanding of all the raw materials used allied to in house testing and/or approved certification, coupled with good consistent casting performance. This is monitored by general scrap rates and the costs associated with knockout and shot blasting, costs so often not taken into consideration when selecting the raw materials to be used in the system.

System mass balance

So much information can be gathered from doing a complete mass balance on a Greensand system. This exercise picks up so much useful data that it should be a regular exercise in all foundries, especially if the casting weights or size of castings alters over time. Consider the simple question "What is the total sand weight in the greensand system"? This normally gets either a wild estimate or a complete blank response.

Mass balance check list

From the data gathered a much clearer picture can be seen and importantly those controlling systems can actually monitor the burn out rates of bentonite and carbonaceous additives. Most observers agree that additions at the mill require time to be effective and by understanding how the sand reacts to varying sand to metal ratios, steps can be taken to be proactive. Various foundries have used predictive software or even a simple traffic light system to monitor heat load, with good results. The author believes it is not the actual system that matters but importantly understanding what is happening and being one step ahead.

- **Return sand storage silo/bunker capacity**
- **Return sand temperature/moisture (at various points)**

- **Weight of sand on line/in boxes etc**
- **Core weight input (if applicable)**
- **Fines extraction/Sand losses**
- **Additive control/weigh calibration/stock against usage**
- **Sand carryover at knockout**
- **Shot blasting times/consumption**
- **Scrap levels/sand related defects**
- **Sand to metal ratio/casting weight data**
- **Volatile/Loss on ignition data at mill and in return sand**

Base Silica sand

Those foundries without cores are obviously in a much better position than those with core input. Cored systems simply have to accept coarser sand as a dilution but with careful selection this does not have to be a problem. More important is knowing the AFS (Average Fineness Number) and AGS (Average Grain Size) and determining what produces the best casting surface at the most economical cost. The AFS clay grade washing and sieving of the washed sand needs regular review and these tests along with optimum additive rates are the keys to success in a greensand system.

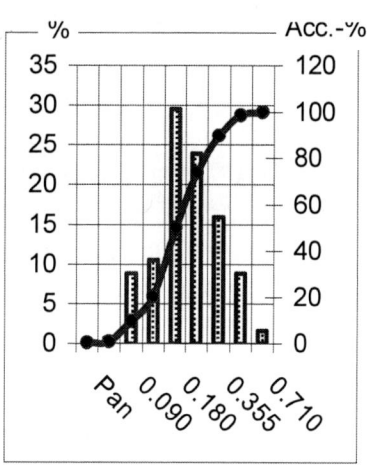

Figure 1 Typical washed system sand sieve graph (AFS = 65.91   AGS = 0.268 mm)

Bentonite types and addition rates

The suppliers will claim many advantages for their products but many differing bentonites are used with success throughout the iron casting world. That said, consistency is the main aspect to control and that is entirely another matter. Few foundries have the specialist equipment for bentonite analysis, so testing should be limited to meaningful values and working with your supplier will establish a working specification.

Carbonaceous types and addition rates

Many products are available with many claims. Coal, either as high quality bituminous or lower grade material is the main additive, and grading selection should also be tailored to the castings being produced. Coal is by far the cheapest and safest carbonaceous additive to greensand systems. Its unique all round properties make this easy to use and its use is so under-rated that in most cases it is under used and systems do not benefit fully from its various beneficial properties.

Other carbon additives are mostly high volatile products normally added to poorer quality coals to enhance levels of volatile content. These can cause additional problems in high concentrations due to hydrogen/nitrogen gases and are quickly eliminated from a system with out contributing to the coke build up (total carbon). This can be an expensive route for a carbonaceous additive and too often the foundrymen has no idea what products he/she is using as the blends are often "trade secrets". To understand a system and its interaction with additives you simply have to know what you are adding and why. The classic system of one single source bentonite and coal is by far the simplest and most cost effective route to producing quality castings.

Mixer controls and testing areas

The need for control in this area is obvious as changes here are reflected throughout the system. Regular mixer maintenance coupled with calibration of additives and usage checks against actual stock purchases can often spot problems before they give concern. This should be part of the mass balance procedure that should be a routine feature.

Primary test	Secondary test
Sieve analysis	Compactability
Active Clay	Moisture
AFS Clay Grade	Green Compression Strength
Volatile	Green Shear Strength
Loss On Ignition	Wet Tensile Strength
	Shatter Index
	Permeability

Table 1   Primary and Secondary sand tests

Getting the primary and secondary testing balance right is difficult for foundries. Too much and the testers see no positive advantages other than routine test control. Too little and when things go wrong there is no starting point for investigation and correction.

Concentrate on the primary tests to ensure consistency. Especially during good casting performance periods use the time to monitor what makes the system work. Look to the secondary tests to reinforce the casting quality and only test if you are prepared to react on the results. Meaningless testing is a waste of resources. Ensure proper reporting and graphing of data. All sand testing is ultimately showing a "trend"

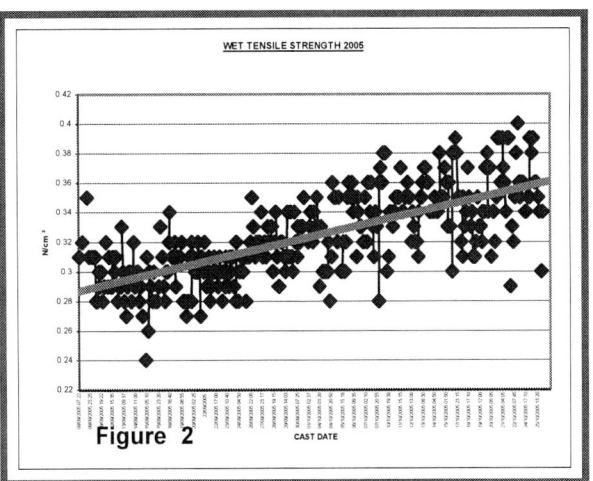

(example **Figure 2**). Even if the testing methods are not perfect if conducted on a consistent basis it will show the trend line which you are able to react on.

Focus on additive testing - Bentonite

Bentonite testing has been well documented. One area for control is in the raw material acceptance testing at the foundry. This can be established in conjunction with the supplier and tests such as swelling, sieve grading and specific gravity are useful in association with active clay calibrations.

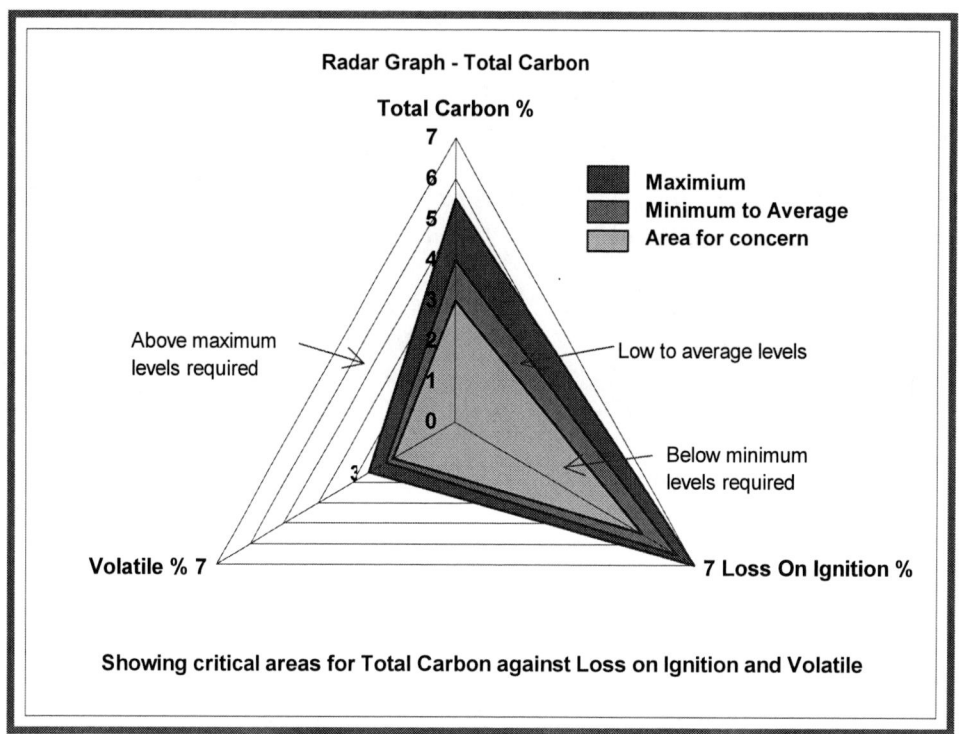

Figure 3 Radar graph for total carbon

Focus on additive testing - Carbonaceous additive

One of the most difficult tests to get consistent results is the volatile test. So many foundries still use this as the guide for carbonaceous additions irrespective of the casting quality. With a 4 decimal place measurement and a weight loss at a specified temperate (910°C) and time (7 minutes) on a small sample weight (1 or 2 grams) it can be a difficult test, especially between different laboratories. This volatile test is normally used in conjunction with the loss on ignition test and investigations into an alternative method to supplement these tests will ensure focus on this critical area for consistent casting performance.

Discussion - Total Carbon method

This method was part of the George Fischer (Lincoln) Ltd control on their LM1 and LM2 moulding lines starting around 2001. This was introduced as a routine test for their straight coal/clay system. Over the following year it became obvious that the interaction of total carbon, sulphur, volatile and loss on ignition were a key combination to the surface quality of the castings.

The radar graph shown in **Figure 3** was used to monitor the relationship and it was concluded that if the total carbon, loss on ignition and volatile were kept in the desired ranges, the casting performance was acceptable. If any results fall into the grey area then positive action had to be taken.

Testing units

The Leco SC144 unit was selected because it determines carbon/sulphur in various organic matrices from low concentrations eg bentonite to high levels such as coal and coke, as well as higher carbon products such as graphite. This unit offers a simple solution that determines carbon and sulphur simultaneously using direct combustion and infrared detection.

No hazardous chemicals are used and accurate results are provided in 3 minutes from a small sample weight of 0.3 grams, using reusable refractory sample boats, without accelerating elements. Coupled to the unit is a weighing unit with a dedicated software package, allowing for data storage, statistical analysis and customised operating parameters.

SC 144   Sample

Samples (such as bentonite, coal, silica sand, as well as system and return sand and extraction fines) can be tested on a daily basis to ensure control, coupled with the routine sand testing. Analysis begins with a nominal weight of 0.3 grams being weighed in a combustion boat. The sample is placed in a pure oxygen environment regulated at 1350°C. The combination of high temperature and oxygen flow cause the sample to combust and it goes through an oxidation and reduction process that causes carbon and sulphur bearing compounds to break down and free the carbon and sulphur. The carbon then oxidises to form CO_2 and the sulphur forms SO_2.

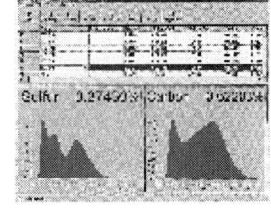

The SC144 can be classed as high end, multi-element detection instrument and investigating alternative units various laboratories worldwide were checked. Many have older type carbon and carbon/sulphur units, mostly doing metal analysis. It was decided to check samples on both an SC144 and a CS244 to determine the difference.

The CS 244 and other units in the CS series are ideal for smaller lower

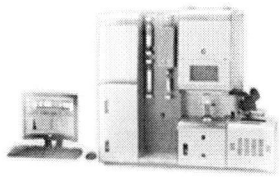

volume laboratory looking for a cost effective solution, without sacrificing precision, reliability and accuracy. These CS units use induction heating element and a larger sample weight of 1 gram. Also used are accelerating

elements such as high purity iron, tungsten or copper with a temperature in excess of 2000°C for 15 seconds, to ensure complete combustion. Any

CO/SO_3 is converted and only CO_2/SO_2 are measured. The graph of readouts shows the difference in both machine types.

Sand testing

As a control measure foundries have relied on loss on ignition and volatile for measuring carbonaceous addition. Coupled with AFS clay grade and active clay measurement, this was considered adequate testing, along with the usual series of permeability, strength and moisture. Experience has shown that active clay levels in most systems are between 1 and 2% higher than required and these levels are regarded as a safety feature and as we know, foundrymen make excellent mouldable sand. That of course does not always translate into top quality castings.

Volatile testing can be a difficult procedure and total carbon determination removes the need for total reliance on this method. A common problem with volatile testing is the use of a wide necked crucible with a poor fitting lid. These always give a larger false reading and the use of the standard parallel sided crucible is urged, with a tight fitting lid. Loss on ignition is of course an easier test and this can be backed up with further testing on washed and unwashed sand samples. Conduct both volatile and loss on ignition before and after washing the clay grade out to determine the contribution from the carbonaceous additive. Approximately 1% of volatile in a sand system actually comes from the bentonite addition.

Interpretation of total carbon results

One foundry does not have all the answers so it was important that sand samples be collected from as wide a group as possible and by selecting countries such as Malaysia, Thailand, Czech Republic, Denmark, South Africa, India as well as the UK it was possible to include all available core binders were applicable and also differing silica sand, bentonites and carbonaceous additives, including those containing lustrous carbon.

All participating foundries were assessed for optimum use of raw materials and control of sand testing and moulding properties to ensure a level playing field. The key measurables were casting scrap related to sand condition, surface finish quality, sand carryover at knockout and shot blast times. Of course not all of this can be attributed to carbonaceous additive but in the end I settled on a good, average and poor system.

Category	Scrap % (Sand)	Surface Finish	Sand carryover	Blast times
Good	< 1%	Excellent	None	Low
Average	3%	Acceptable	Some	Middle
Poor	>3%	Rough	Lots	High

Table 2 Foundry performance categories

This may at first look seem too simple a table to be useful. The results and conclusions justify the selection, but further work now needs to be done on putting numbers against the various selected categories.

Having defined the foundry, samples were collected over the year and washed and unwashed samples were tested for volatile and loss on ignition as well as using the SC 144 unit to determine the total carbon and sulphur levels. The author in order to monitor casting performance and to cross check our own laboratory results personally visited all foundries. It became apparent that the categories matched the level of total carbon.

Sample	Description	SC Carbon	CS Carbon	SC Sulphur
Blank	Blank	0.000	0.000	0.000
Blank	Blank	0.000	0.000	0.000
Standard	CaCO$_3$ Standard	12.027	12.000	0.011
Standard	CaCO$_3$ Standard	12.027	12.000	0.011
A	System sand 1	4.699	4.800	0.071
B	System sand 2	2.382	2.300	0.013
C	Natural Graphite A	94.948	93.200	0.290
D	Natural Graphite B	89.498	88.200	0.522
E	System sand 3	3.850	4.100	0.076
F	System sand 4	2.806	2.950	0.532
G	System sand 5	1.848	2.075	0.045
H	Raw Coal import	70.614	67.300	0.859
I	Coal/Bentonite Blend	50.108	50.700	0.705
Standard	CaCO$_3$ Standard	11.796	12.000	0.011
Coal STD	Coal Std 74%C 0.98%S	74.131	72.560	0.980
Blank	Blank	0.000	0.000	0.000
Range of samples from both SC and CS units at one session				

Table 3   Various samples for one session of testing on SC/CS units

Results - *Measured Total Carbon*

No database of information was available so over 20 foundries were tested with total carbon ranging from a low 1.84% to a high of 5.20%. Out of interest and because the SC 144 also does sulphur at the same time we recorded the numbers and they have ranged from 0.052 to 0.113%. Against each sample we did a volatile and loss on ignition test to see if there was a relationship and to check the foundries own results, which in itself was a useful exercise, and quite differing results were obtained. Three independent laboratories were used, including foundries and suppliers to verify results over the year.

Table 3 show the results from one session using various samples from low to high carbon content on the SC 144 unit. The CS results are to show the difference in the detection systems but the accuracy shows the possibilities and the versatility of the units for rapid testing and control. The main list in **Table 4** shows the results from the average of samples taken from each of the foundries system sand. Both volatile and loss on ignition tests were also carried out to supplement the total carbon results. All these results come from the SC 144 unit, so we can be confident we have the total carbon.

Performance	Foundry	Foundry Type	Total Carbon	Sulphur	Volatile	Lol
Acceptable	A	H SG GI H	3.85	0.075	1.85	5.85
Poor	B	H GI M	2.80	0.054	1.72	4.03
Poor	C	H SG M	1.84	0.052	1.55	3.45
Acceptable	D	V SG GI L	3.56	0.076	2.03	6.04
Good	E	V SG GI L	4.65	0.078	2.25	6.50
Good	F	H SG GI H	4.90	0.065	2.25	6.10
Acceptable	G	H GI L	4.70	0.085	2.40	7.20
Good	H	V GI L	3.67	0.075	2.15	6.34
Good	I	H SG H	4.54	0.083	2.35	5.95
Good	J	V GI H	3.50	0.113	2.15	7.40
Good	K	V SG H	3.67	0.854	2.34	6.43
Acceptable	L	H GI M	3.60	0.064	1.95	5.95
Good	M	H SG H	5.20	0.073	2.35	6.40
Poor	N	H GI H	3.34	0.076	1.65	4.56
Acceptable	P	H SG GI M	3.54	0.075	2.10	6.34
Acceptable	Q	H SG GI M	3.15	0.084	1.83	7.20
Poor	R	H GI H	2.95	0.082	1.87	5.20
Poor	S	V GI M	2.87	0.075	1.76	5.60

H=Horizontal V=Vertical SG=Ductile GI=Grey
H=Heavy +50 kgs M=Medium +25 kgs L=<25 kgs
Performance= scrap, surface finish, sand carryover and shot blast times

Table 4   List of data from selected foundries

Evaluation of Volatile and Lol properties
The whole question of the use of carbonaceous additives and their purpose has been well documented. This work simply adds to the conclusion that there is an interaction between the total carbon in a sand system, allied to a level of active effective volatile and a loss on ignition maximum for successful casting production. Often overlooked in problem solving is the amount of volatile, which must come from the bentonite, and typically this is classed as 10% of the AFS clay grade. To investigate this further washed and unwashed tests reveal the true active volatile content from the carbonaceous additive. Many

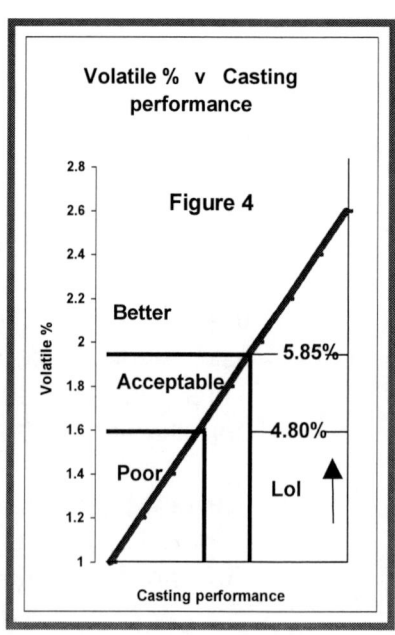

foundries are simply unwilling to move their LoI numbers above 5% which limits the active volatile, which in turn limits casting performance.

The graph in **Figure 4** shows the approximate levels of volatile for good casting performance from results obtained in one foundry. Each foundry could construct its own graph based on the parameters selected. Obviously other factors come into play but it is clear that quality active volatile release is one of the major contributors to casting excellence. The "classic" loss on ignition (LoI) to volatile ratio of 3:1 still holds true in most cases and when linked to total carbon levels the picture becomes clearer.

Conclusions

Greensand systems can be both simple and complex but they are still the most cost effective method for volume iron casting production. Simple in terms of additives and control, complex in the amount of variables possible and the need for constant vigilance.

The research shows that **total carbon** can be a very useful measuring and control tool for greensand systems, coupled with accurate volatile and loss on ignition tests. Foundries categorised as acceptable had total carbon figures in excess of 3%. In foundries categorised as good it was no coincidence that the volatile and loss on ignition figures were also in what the author would describe as "a good area."

Foundries would do better armed with meaningful data as far as carbonaceous additives are concerned. Total carbon determination could include all aspects of a sand system, including raw materials, and this would make decision making easier. Areas such as return sand and fines extraction could be easily monitored with better control.

All sand systems have differing moisture sensitivity. Using bentonite from known approved sources coupled with correctly graded coal, low in ash content (less than 3%), ensures the coal does not complete against the bentonite for available water. The key for foundries is to understand the raw materials used and to control the consistency. Once this has been established, the minimum amount of additives to achieve the goal of consistent quality castings will be possible. Ensuring adequate carbonaceous levels is not difficult but for many foundries it still remains only a secondary focus. World class performance is being demanded by casting buyers and those surviving need to ensure they operate "in the correct zone" with effective known source additives.

Alexander Brown FICME   abrown@durrans.co.uk

A new generation of advanced Polyurethane Coldbox binders for Aluminium castings

A. Schrey, Foseco GmbH, Borken (Germany)

Abstract

Introduction

When casting Aluminium there are specific problems of core breakdown and removal, that are not found with ferrous castings. The lower pouring temperatures of aluminium result in lower core sand temperatures and reduced resin binder breakdown from thermal decomposition [1]. With less breakdown, the cores retain higher strength after casting and can be difficult to remove with mechanical vibration at shake-out **(Figure 1)**. Additional time and/or effort may be needed to completely remove cores from narrow passages, increasing casting costs [2]. Thin-walled castings prone to damage during shake-out and high sand-to-metal ratios can be particularly troublesome.

A new class of binders have now developed to meet the particular demands of aluminium foundries, addressing the need for exceptional breakdown after pouring as well as reduced emissions of hazardous compounds. See Figure 2.

Health and environmental aspects have been taken into consideration and extensive studies of emission levels have been carried out during the work. Owing to a new solvent system it was possible to reduce the odour as well as lower the emissions of benzene, toluene and xylene, during the development.

Along with the development of the resins new methods to determine differences in terms of breakdown were established. Thus a reliable test, called the "dip test" was developed to determine the breakdown properties.

Experimental

The development work for a new generation of Urethane-Cold-Box-Binders focused on two major objectives: Initially the new binders had to meet the specific requirements of the Non-ferrous foundries to achieve better shakeout characteristics and secondly to reduce the environmental impact by the addition of less harmful materials.

To provide information for improved core removal a new test method had to be established which was both practical and reliable. The so-called "dip test" proved useful and is described below.

The whole procedure is split into 6 steps the first three steps being as follows:

Step 1 (**Figure 3a**,) shows two dip test cores that are equal and will be tested simultaneously. The cores were produced by means of a small lab core shooter and are cylindrical but slightly conical. The density must be homogeneous and noted at the start of the test.

Step 2 (**Figure 3b**) shows both test cores clamped in the dipping device. The cores are fixed and cannot be moved, additionally the outlet for core gases is placed in the upper part of the clamping device.

Step 3 (**Figure 3c**) shows the actual casting process. The clamping device with both cores has to be immersed into the aluminium manually or by means of an automatic device. The immersion length has to be constant and the dipping device must not be moved once it has been dipped into the melt. The melt temperature and the dipping time have to be recorded.

Additionally the evolution of gas and condensates is measured by means of the COGAS®-device **(Figure 4)**. The COGAS® equipment [3] simulates the casting process, where a test core is fully enveloped by molten metal.

The gas evolved by decomposition of the organic binder is collected in a measuring tube **(Figure 5a and 5b)** and the quantity of gas produced during the various stages of decomposition is measured and recorded. A cooling trap precipitates condensable material, this can be weighed and analysed.

Breakdown:

Step 4 Figure 6a, shows two dipped cores after a given dipping time. The cores are covered with a thin layer of solidified aluminium during cooling.

Step 5 (**Figure 6b**) has to be conducted very carefully: The dipped cores are un-wrapped, the thin aluminium foil being removed manually. It is important to remove the foil completely without damaging the core surface.

Step 6 (**Figure 6c**) serves to assess the breakdown properties of the binder. Every core has to be brushed by means of a soft brush until every loose sand grain has been removed. This has to be done very carefully without use of any force. Afterwards every core has to be re-weighed.

The figure to assess the breakdown is the ratio of core weight after casting divided by the core weight before treatment as a percentage. The statistical error is about ± 1,5 % related to the absolute value.

Figure 7 shows the appearance of cores made by means of a resin that has slightly better thermal decomposition properties than normal   versus a core bonded with a resin from the latest development. Both cores were originally immeresed for 120 s into molten Al at 720° C.

Results and discussion

Figure 8 shows the results of a dip test for four different recipes. Each column shows the residual core percentage for four different dipping times from 60 to 150 seconds at intervals of 30 s.

The y-axis indicates the residual core in percent, and as described previously low figures indicate improved core breakdown.

As expected the core breakdown improves by prolonging the dipping time. In Figure 8 the columns on the very left-hand side show the initial recipe with aromatic solvents, followed by the present low aromatic system and ending at the stage-2-low-aromatic binder with exceptional breakdown properties.

 The replacement of solvents and the addition of special additives resulted in a significant improvement in breakdown performance. The reduction from 79 % residual core weight for the initial recipe after 150 s at 696° C down to 75 % for the current system is in accordance with foundry practice where the theoretical figures were substantiated in reduced cycle time for shakeout. As is seen in Fig 8 the stage-2-resin provides a complete decomposition of the chemical bondings after 150 s.

Clearly all of the previously achieved benefits (sufficient strength, excellent core release out of the core box, and long bench life) were maintained while the breakdown properties were improved significantly and the emissions reduced dramatically.

COGAS® and environmental aspects

The binder residues collected in the cooling trap (**Figure 9a**) were weighed and afterwards analysed by means of the GC-MS-device shown in **Figure 9b**.

Figure 10 shows the results of the quantitative determination of gas and condensates by means of the COGAS® equipment for four cores with four different binders after the immersion into molten Aluminium at 720° C for a 3 minutes period.

Interestingly the amount of gas evolved during heating seems not to be affected by the modifications to the various binders that lead to improved breakdown, whereas the amount of condensates increased from the initial recipe with poor breakdown to the recipe with exceptional breakdown.

The loss of bonding power due to the thermal impact on the polyurethane molecule reduces the amount of binder left in the core (loss-on-ignition) at the expense of binder that has been converted either in gas or condensates.

Thus improved breakdown will always cause a difference in binder remaining in the core to an increased amount of released binder in the form of gas or condensates.

The graph (**Figure 11)** shows significant improvement in the avoidance of harmful substances by comparing the initial recipe with aromatic solvents and the latest recipe with low aromatic solvents.

The orange (toluene) and green (m/p-Xylene) columns of the low aromatic based binder clearly indicate the successful results of the development work in order to achieve a more environmental friendly binder system, which can out-perform competitive Coldbox binders.

Conclusion

This paper summarises the development of a new class of cold box binders for the non – ferrous foundry industry to improve both breakdown of cores after casting and improve environmental conditions.

The new binders have already proved successful at a number of foundries passing quite demanding quality control tests. Cores based on the new recipes are providing better breakdown and emmitting lower levels of harmful substances.

A "dip test", using COGAS equipment to measure emitted gas and condensates, has been established for quick and reliable quality control and to develop improved resins in the future.

References

1. Boenisch, D.: The Coldbox-Plus-Process: Higher Quality Cores with Lower Binder Levels , Research Development Reports 1986, RWTH Aachen
2. Simpson, B.: Recent developments in the application of the Polyurethane Cold Box process, Foundry Practice, Issue 243, p. 1-7
3. N.N.: COGAS-System, mk Industrievertretungen, Stahlhofen, 2004.

Decoring of aluminium castings

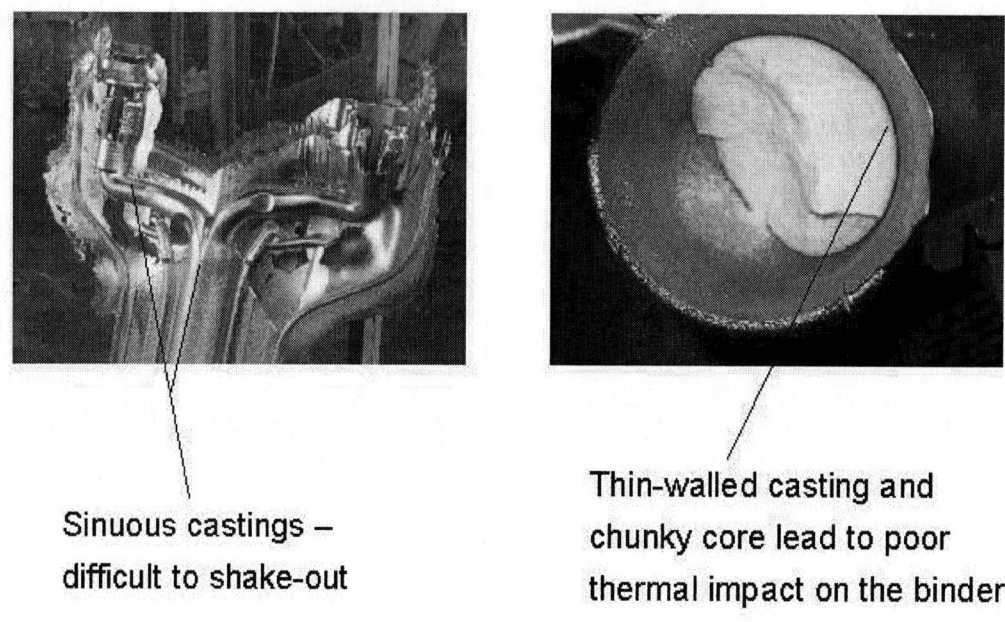

Sinuous castings –
difficult to shake-out

Thin-walled casting and
chunky core lead to poor
thermal impact on the binder

Figure 1. Difficult shake-out in castings with bends and thin walls

Demands

Low odour / less emissions

Easy core breakdown

No resin wipe off

Good core release

High as gassed strength

Sufficient surface quality

True to dimensions

Low gas evolution

Figure 2: Demands upon cores

Dip test for determination of breakdown – method

Step 1:   Step 2:   Step 3:

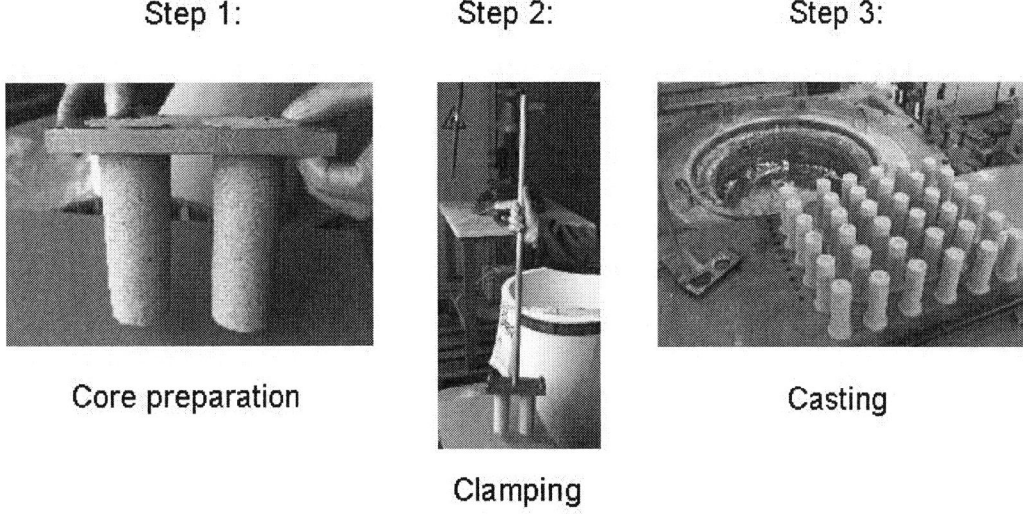

Core preparation   Casting

Clamping

Figures 3a, 3b, 3c. Preparation of cores for the so-called dip test

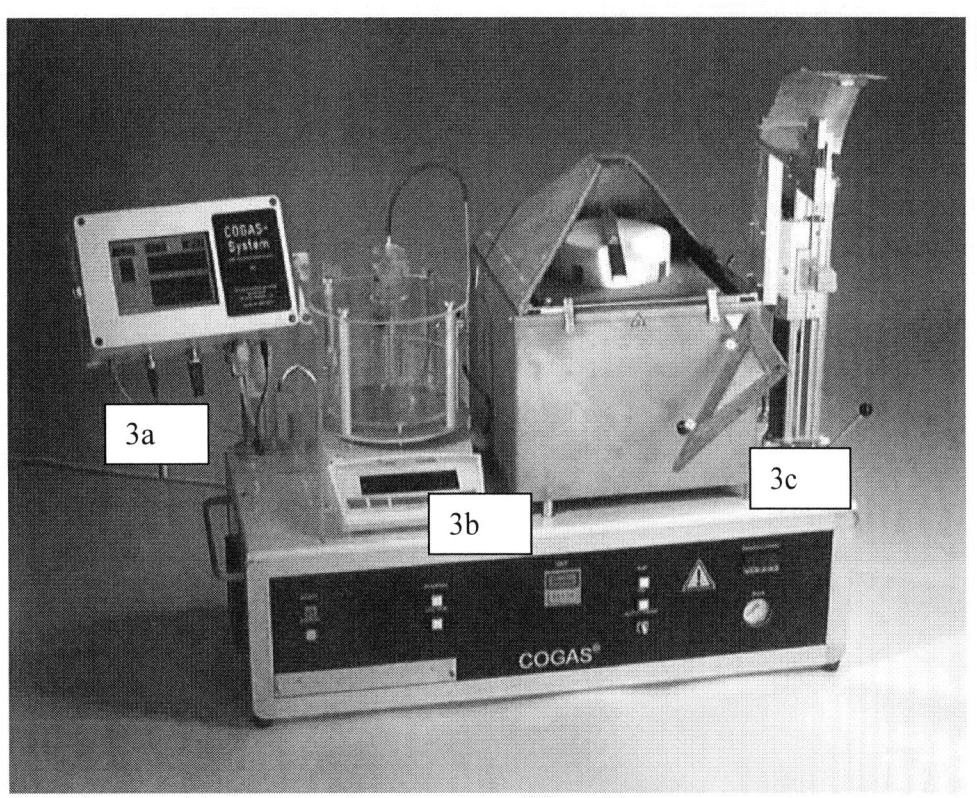

Figure 4. COGAS-device for determination of gas and condensates

Figure 5a. Clamped test core and trap for condensates (Figure 5b)

Dip test for determination of breakdown (2) – method

Step 4:   Step 5:   Step 6:

After casting - cooling

6a

Un-wrapping

6b

Brushing and re-weighing

6c

Figure 6. Dipped breakdown test cores

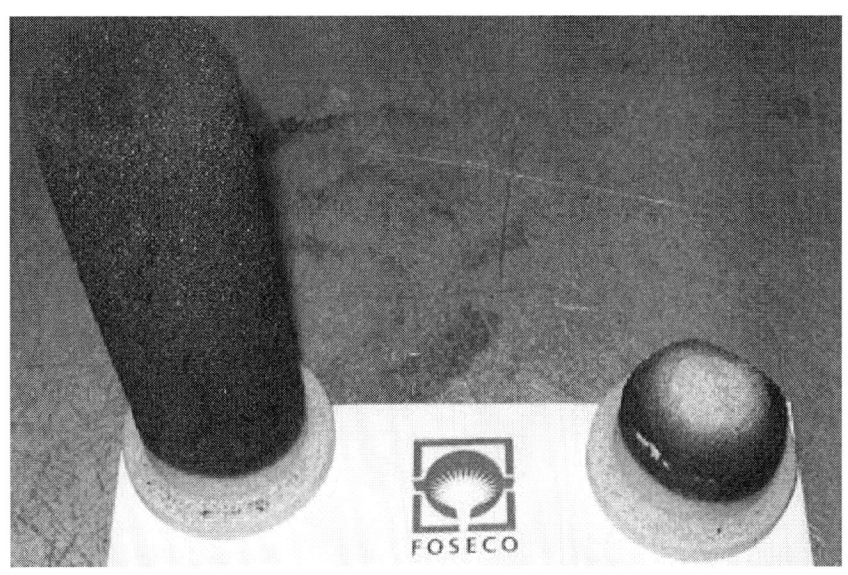

Figure 7. Stage-1-resin bonded core (left) vs. stage-2-resin bonded core (right)

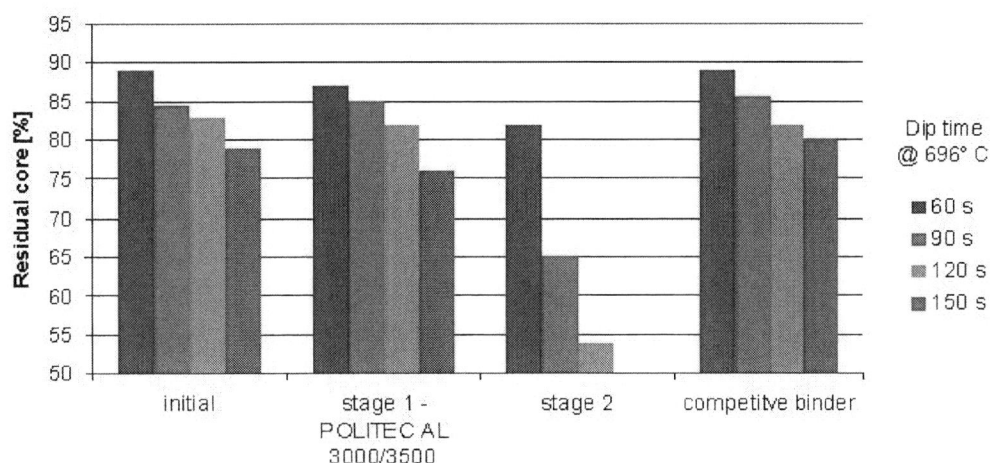

Figure 8. Breakdown properties for different binders at different dipping periods

Figure 9a. COGAS-test: core after determination of gas and condensates

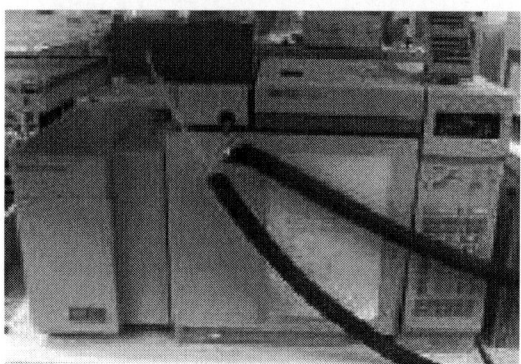

Figure 9b. Condensates / tar collected by means of COGAS and determined by GC/MS

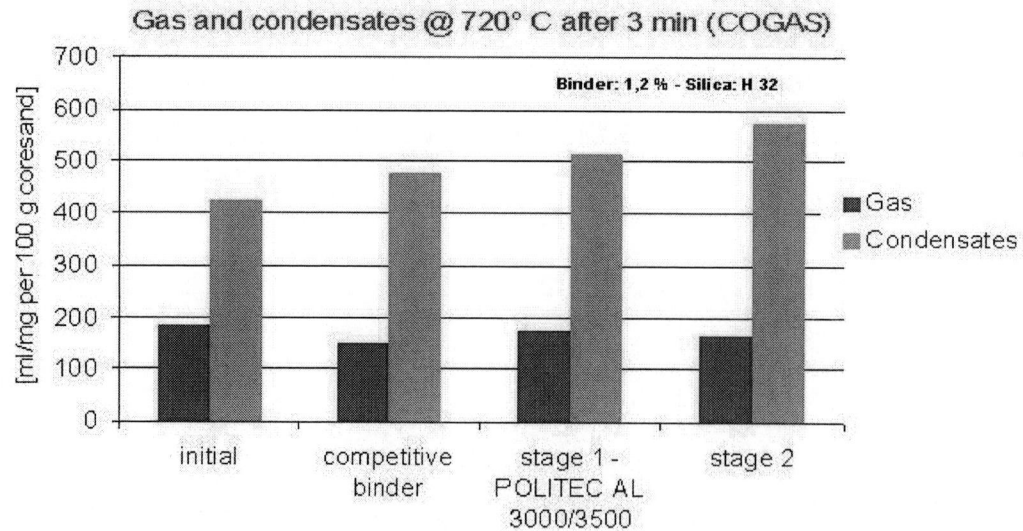

Figure 10. Slightly increased condensates due to better breakdown

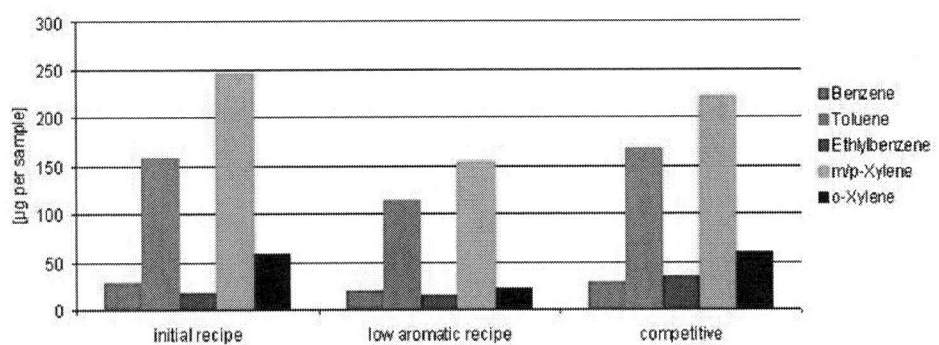

Figure 11. Determination of Benzene, Toluene and Xylene in the lab by means of gas chromatography and mass selective detection

Email:
Alexander.Schrey@foseco.com

A Technological Advantage for the Environmental Age

L.R. Horvath

Ashland Casting Solutions, Columbus, Ohio, USA.

Abstract

In today's world of the environmentally conscious foundryman, technological progress is required to reduce waste, energy use and scrap rates. As foundries are forced to turn away from solvent-based products due to volatile organic compounds (VOC) and exposure limits, water-based products continue to grow as the consumables of choice. Significant progress has recently been made to address the current and future needs of foundries struggling to compete in a global casting market.

Refractory coatings are currently a necessity for making saleable castings in most metals. They can have several benefits: eliminating metal penetration, veining defects and improving casting surface finish.

All refractory coatings must be thoroughly dried after application. No-bake and jobbing foundries have traditionally used solvent-based refractory coatings because they are typically air-dried, ignited or torch-dried. There was no need for expensive ovens or the high cost of running and maintaining this equipment. Currently, there is a shift from solvent-based coatings to water-based coatings for both environmental and safety reasons.

The main problem foundrymen face when using a water-based coating is complete drying and how to determine when the coating is thoroughly dried. A new coating technology that shows a complete color change as it transitions from wet to dry phases has recently been developed. A diagnostic indicator that visually shows exactly when the foundrymen can close and pour the mold is invaluable. Other advantages of this technology are reduced cycle times, lower oven temperatures (energy savings) and lower scrap rates caused by moistures related defects (namely blows and porosity). An additional feature of the coating is recognition of mold and core density gradients.

As foundries are continuing to move towards water-based coatings to satisfy environmental regulations and safety requirements, technological advances are imperative in order to meet these challenges while, at the same time, improving process efficiencies.

Key words: color-change, refractory coatings, water-based, solvent-based, drying, environmental.

Introduction

The use of engineered refractory coatings (or "core wash") is a necessity in today's modern foundries. These consumables play an important role in the prevention of many different types of defects, including metal penetration, veining and gas porosity. Approximately 225,000 metric tons of foundry coatings are consumed globally each year.

In many of today's foundries alcohol-based refractory coatings are still being used mainly because of the convenience in their ease of drying. These alcohol carriers consist mainly of isopropanol, ethanol and methanol. With the continuing push of environmental, health and safety regulations to lower VOC emissions and exposure limits for workers, the transition from solvent-based to water-based products intensifies. This being said, the ratio of refractory coatings is still around 70% water-based to 30% solvent-based, with the use of solvent-based coatings possibly being higher in less regulated countries.

Several years ago, some foundries had even used 1-1-1 trichloroethane as a solvent to decrease the flammability inherent in alcohol-based coatings. These products air-dried quickly, but the toxicity of this solvent lead to it's elimination as a carrier option. That left foundries with the option of using either flammable alcohols or water as a solvent. Those foundries with no means to properly dehydrate a water-based product were forced to use an alcohol-based coating.

Today, we are seeing new water-based coating technologies created that can make the needed transition from alcohol-based consumables to water-based ones a little easier. This paper will first discuss some of the pros and cons of using an alcohol-based coating versus a water-based one. Then, it will discuss new technologies in engineered water-based refractory coatings that will help transition foundries into a more highly regulated era of metal casting.

Experimental

When weighing the pros and cons of using alcohol or water as main carriers for a foundry coating there are some obvious differences, and some that are not so easily discovered. We will begin with a discussion of alcohol-based coatings.

The most obvious benefit of using alcohol as a carrier is the fast drying capability. Most foundries either air-dry, torch or light-off the coating to burn off the solvent. Due to the nature of alcohols, they also provide adequate penetration of the coating into the core or mold without the need for expensive surfactants used to reduce surface energy. Both of these benefits also lead directly to some of the more important disadvantages of alcohol-based coatings. The fast-drying capability comes at the cost of releasing VOC into the workplace if proper ventilation is not installed. The excess penetration seen when using an alcohol carrier can lead to the observation that all the solvent has burned off after light-off, when in fact there is still residual solvent left within the core [1]. When this occurs, there is the potential to produce a "spalling" defect where the coating actually pops off the core as the residual solvent is flash-evaporated during pouring.

As with most solvents, flammability can be a problem with alcohol-based coatings. Depending on the country, some regulations impose limits on the amount of flammable material that can be stored in one single area. This can cause inventory problems depending on the consumers' usage levels and the storage space available.

Other potential problems associated with alcohol as a coating carrier is the chance of over-spray and inhalation of the coating "mist" formed during spray applications. In some cases, regulations are in place that can require capital expenditures for ventilation hoods to extract the over-spray mist.

Lastly, the price volatility of the chemical market can cause a significant strain on the cost to produce alcohol-based coatings since most solvents used in these coatings are derived from oil. This has never been as evident as in recent years when skyrocketing solvent costs have, at times, made the production and usage of some alcohol-based coating formulations economically unattractive.

As we turn our attention to water-based coatings, there are several obvious advantages that stand out. The most beneficial being the environmental benefit of using water as the main carrier instead of an alcohol. Water is the lowest cost carrier possible for refractory coatings, and provides a much safer product for workers to handle and apply.

Another great advantage of a water-based coating is the flexibility in formulating. With some of the advances in suspension agents, leveling agents, and anti-settling products a coating can be formulated to do almost anything for any application. Significant improvements were made in these areas in recent years.

There are also several drawbacks to using a water-based product. One of them is the negative effect that water can have on certain binder systems. The most notable of these is the weakening of the sand-to-binder bond and subsequent lowering of tensile strengths of cores using phenolic urethane cold-box binder systems. It is critical to remove the water as quickly as possible upon application of the coating, as longer exposure of the bonded sand to the water increases the degradation of the cold-box core [2].

Additionally, care must be taken when using water carriers in environments with very low temperatures. Storage of water-based coating in areas of temperatures below freezing can lead not only to the water carrier being frozen, but in some cases irreversible damage to the refractory coating. Some formulations fare better than others upon freeze/thaw cycling, but it is best to avoid this situation all together by storing the water-based coating in a temperature-controlled area, ideally somewhere between 10°C - 30°C.

The main drawback in using water-based products is the need to drive off all the moisture before pouring any metal against the core or mold. Any drying operation that consumes energy, whether it is a standard forced-air convection oven, microwave oven or torch, will add to overall cost and slow productivity. Today, not many castings producers can afford the time to air-dry a refractory coating. The question of how to dry a coating leads to another critical issue: when do we know the coating has been completely dehydrated and the core or mold is void of any moisture?

In the past, the answer to this question has been debatable. In many cases a foundry could only tell that moisture was truly present inside a mold or core when they saw the porosity, gas or penetration defects show up in the cleaning room (See Figure 1). New ways to answer this question have been developed in order to help advance the change from alcohol-based coatings to more environmentally-friendly water-based consumables.

Results
The new coating technology described within uses specific moisture indicators in the refractory coating that allow a highly visible color change to occur when progressing from the wet stage to the dry stage (See Figure 2-4). This technology provides the foundry operator the peace of mind that the protective refractory coating just applied to the core or mold is completely dry before sending it down the production line to the pouring area. If properly used, the color-changing coating can in most cases eliminate moisture-related defects like gas porosity (blows), scabbing, and metal penetration.

There are many benefits that have been discovered on the foundry floor when using this type of technologically advanced refractory coating. Not only can the operator be certain that all the moisture has been removed form the core or mold, but he may see areas of poor sand compaction as indicated by certain areas of the mold/core that do not change color, or change color more slowly than other, denser areas. This is due to the fact that the poorly compacted areas are more porous or "spongy", causing them to absorb more water and making them more difficult to dehydrate.

Another benefit of this coating technology is relevant to foundries which must apply more than one coat of the refractory barrier. This is seen mostly in heavy section iron and steel castings where high metal-to-sand ratios exist and more protection is needed. Knowing when the initial coat of the refractory material is completely dry and the second can be applied is critical to eliminating the possibility of trapping moisture under the second or third layer of coating. Knowing exactly when you can close a mold or set a core will eliminate any "buffer drying time" included in your process to ensure complete drying. This can ultimately lead to productivity gains that more than offset the cost of having to dry the water-based coating. The color change phenomenon can also tell you if your drying equipment is working properly. Noticing that the color change is not occurring as it normal does when the core/mold exits the oven or heating area can be an indication that the oven is not working properly, energy is being wasted, and poor drying is the result.

Discussion

There is no question that both water-based and solvent-based refractory coatings each have their pros and cons from a usage and production standpoint. From an economical standpoint there is no doubt that the rising costs of energy and natural gas can lead to more cost involved in drying a water-based coating through the use of many different forms of ovens or heaters that are designed to drive off moisture. This cost can be offset to some extent by the low price of water-based coatings, which can be up to 50% less than the cost of solvent-based formulations. At this point in time most foundries have the option to choose a solvent or water-based product based on the overall cost-to-use within their process. For those who desire to make the transition to water-based coating, new technologies like the water-based refractory coating that changes color when it dries are a stepping stone that foundries can use to help them achieve the mandate of a more environmentally friendly workplace.

Conclusions

Environmental laws and regulations are becoming stricter each and every year. Many foundries are choosing to switch to more water-based products to protect not only the environment, but also their workplace and their employees. Depending on the environmental health and safety regulations in your country, this choice may not be optional for much longer. To become more prepared for that day, we as foundrymen must search for new technologies that help us transition into the "environmental age" of metal casting, without sacrificing productivity and quality. This new refractory coating technology is just one way to help accomplish this task. The faster we pursue other foundry consumables with unique process diagnostic features like the color-change coating, the more likely we are to progress to the next level of casting production.

References

1.   Penko T, *Measurement of Emissions Associated With Application of Flammable Solvent-Based Core and Mold Coating*, AFS Transactions, Vol. 106 (1998), pp 173-179

2.   Guyer O B and Adamson B and Cieplewski J and Rebholz K W and Willkomm, *Exploring the Effects of Water Solutions on*

Coldbox Core Strength, Modern Casting, Oct 1999, pp 30-32

Figures

Figure 1: Wet Coating Defects

Figure 1: 100% Wet Coating

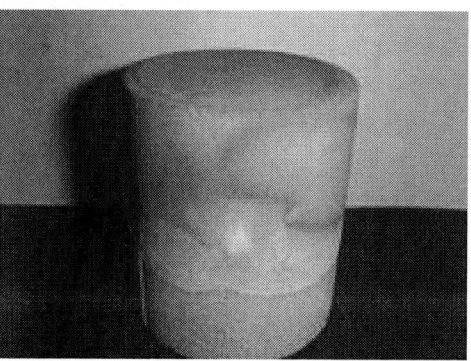

Figure 2: 50% Dry Coating

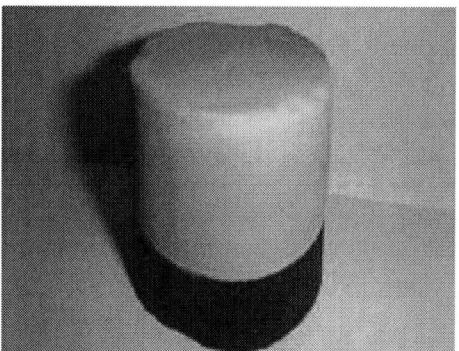

Figure 3: 100% Dry Coating

Author's E-mail Address*: lhorvath@ashland.com*

Dispersive mixing of natural molding sand – an optimized preparation process

M. Mueller

Maschinenfabrik Gustav Eirich GmbH & Co KG, Hardheim, Germany

Abstract

Natural molding sand with a grain spectrum of 0.5 to 2.0 mm is used for aluminium casting, particularly if the surface of the castings is required to meet high standards. But the processes of molding and aluminium casting cause the formation of hardened agglomerates in the natural sand.

This is particularly challenging for the solids mixer, because the agglomerates need to be disintegrated first, before all solid components can be homogenized by distributive mixing and get wet with water.

The paper presents a new preparation concept developed for a producing aluminum foundry. By means of tests on laboratory scale a concept is developed for a system that yields a higher throughput rate with optimized mixer operation.

An up-to-date control system for the mixer optimizes the preparation process, while keeping the energy input constant under changing material and machinery parameters.

Key words

New preparation concept for aluminium industry
Higher throughput rate with optimized mixer operation
Constant energy input under changing material and machinery parameters

Introduction

Natural molding sand with a grain spectrum of 0.5 to 2.0 mm is used for aluminum casting, particularly if the surface of the castings is required to meet high standards. But the processes of molding and aluminum casting cause the formation of hardened agglomerates in the natural sand. In the sand preparation mixer the return sand needs to be mixed with fresh natural sand from the pit and additional water until the optimum moisture content (6 %) for the molding line is adjusted.
(Figure 1)

Experimental

This is particularly challenging for the solids mixer, because the agglomerates need to be disintegrated first, before all solid components can be homogenized by distributive mixing and get wet with water.
(Figure 2)

The preparation quality, however, is only achieved when the silica grains are evenly coated with clay. So coating is another task the mixer must accomplish in order to establish isotropy in the sand mold.

The preparation is demanding, because varying material conditions must be taken into account when introducing energy in the mixer. The agglomerates are hard and embedded in fine-grained return sand. The application of energy via the mixing tools is buffered by the high percentage of fines, a fact that makes dispersive mixing difficult.
(Figure 3)

At the same time, new sand has to be mixed in. The new sand is too moist and the percentage of clay contained is much too high, compared to prepared sand. Unlike the agglomerates, the new sand is very soft and plastic because of its high clay content and tends to agglomerate when getting into contact with the fine, dry return sand. At the same time, the soft, plastic clay needs to be dispersively mixed to obtain a homogeneous distribution.
(Figure 4 / Figure 5)

Conclusions

A new preparation concept is developed for a producing aluminum foundry. By means of tests on laboratory scale a concept is developed for a system that yields a higher throughput rate with less power consumption of the agitator motor.
(Figure 6)

An up-to-date control system for the mixer optimizes the preparation process while keeping the energy input constant under changing material and machinery parameters.

Down-breaking of agglomerates when increasing the spec. mixing work from 1,5 kWh / 1 Mg to 5,6 kWh / 1 Mg and the spec. mixing energy from 9 kW /100 kg to 13,5 kW/100 kg.

Decreasing the agitator velocity from 40 m/s to 20 m/s and decreasing the velocity of the aerator from 1500 min^{-1} to 900 min^{-1}.

No additional improvement by increasing the spec. mixing work from 5,6 kWh / 1 Mg to 11,5 kWh / 1 Mg and the spec. mixing energy from 13,5 kW/100 kg to 27,5 kW/100 kg.

New plant design:
External down-breaking of agglomerates after sieving (0,5 mm) with heated screens.
Reduction of the spec. mixing work is 50 %
The muller preparation of 4 kW/100 kg and 3,3 kWh/ 1 Mg is not strong enough for down breaking of the agglomerates.
(Figure 7)

References

1. Grefhorst C. and Mueller M., *Comparison between a conventional and vacuum molding sand preparation*, cp+t, volume 16 no. 1 2000, March, page pp4-11
2. Mueller M., *Vakuumaufbereitung bentonitgebundener Formstoffe*, Gießerei-Erfahrungsaustausch, 6/2005, pp2-7

Figures

Return sand:
moisture: ca. 1,0 –1,5 %
clay content: 12 – 13 %
bulk density: 1120 g/l

New sand:
total content: 1 – 2 %
moisture: ca. 18 – 20 %
clay content: 16 – 17 %
bulk density: 800 g/l

Figure 1

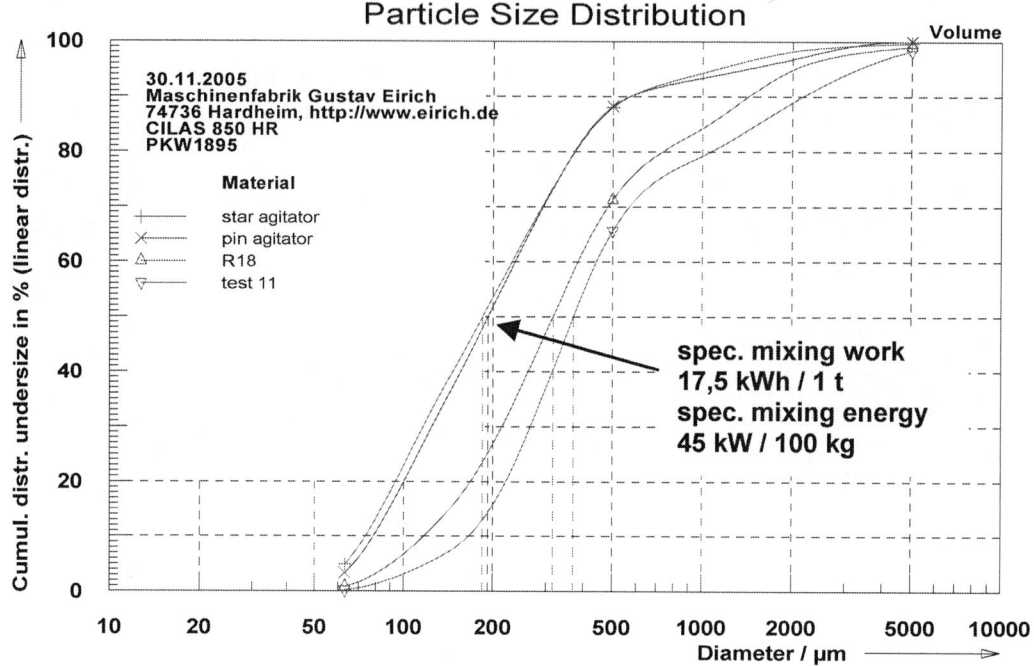

Figure 2

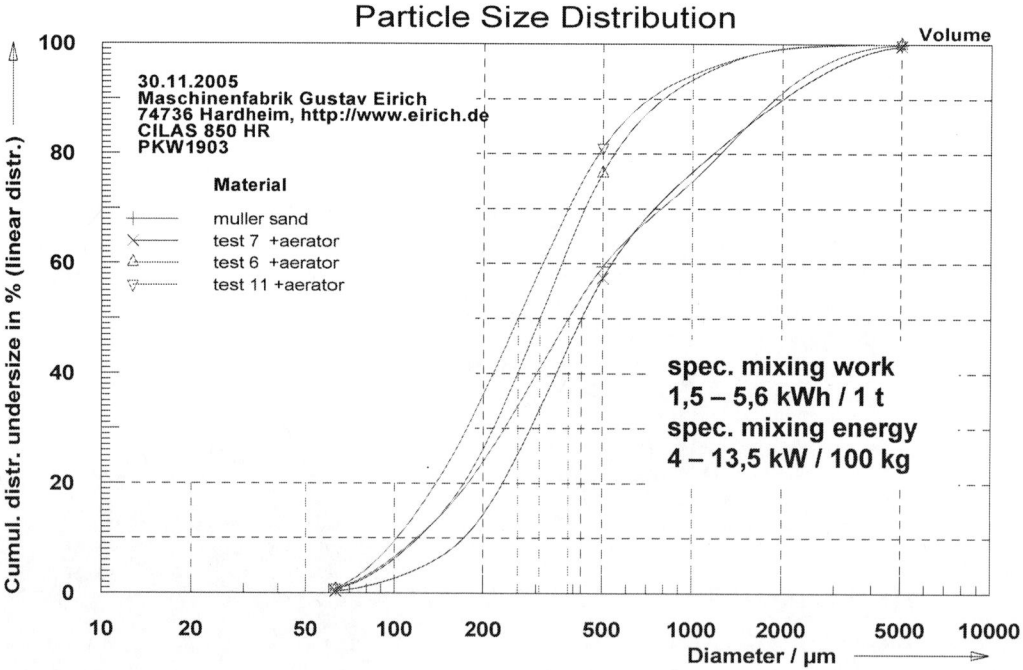

Figure 3

Test parameter:

spec. mixing energy / spec. mixing work:

Production:	40 m/s	1050 min^{-1}	1,5 kWh / 1 t
			90 kW / 1 t
Muller-Mixer:	1,4 m/s	26 min^{-1}	3,3 kWh / 1 t
			40 kW / 1 t
R 11-Mixer:	9 m/s	300 min^{-1}	3,0 kWh / 1 t
			70 kW / 1 t
6 beaters	20 m/s	660 min^{-1}	5,6 kWh / 1 t
			135 kW / 1 t
8 beaters	20 m/s	660 min^{-1}	11,5 kWh / 1 t
			275 kW / 1 t

Figure 4

Prepared sand:
moisture: ca. 5,5 – 5,8 %
clay content: 12 – 13 %
bulk density: 1120 g/l

Figure 5

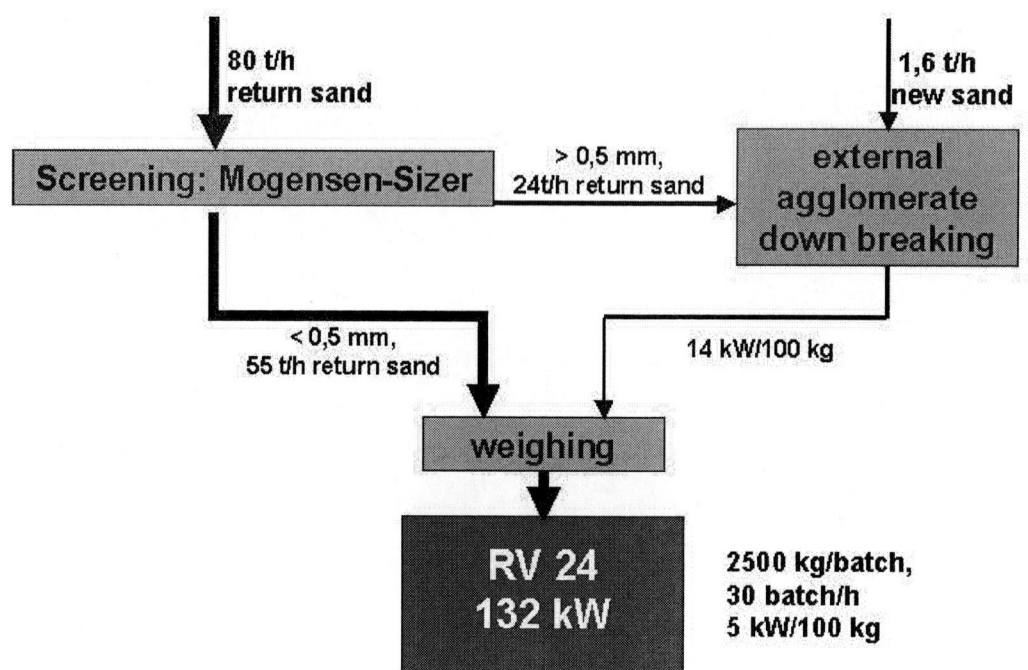

Figure 6

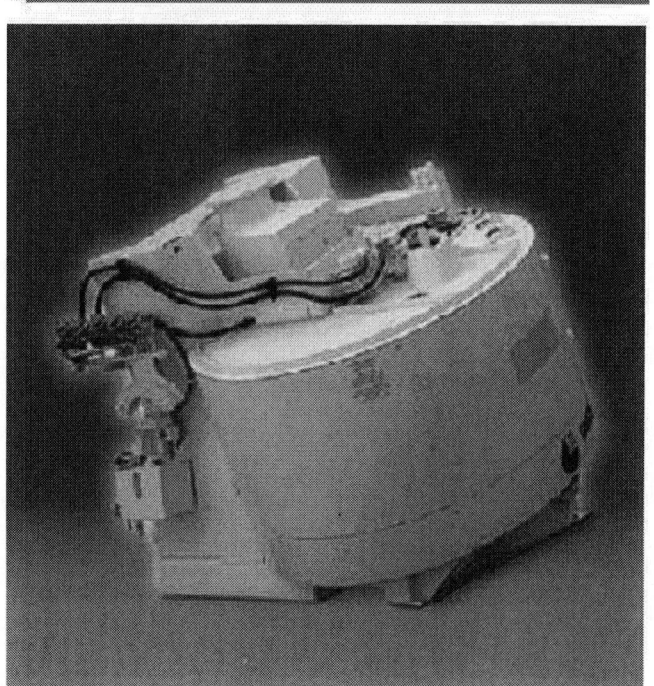

Figure 7

Email address: marcus.mueller@eirich.de

New innovative solutions for foundries by inorganic concepts

Jens Müller* and Reinhard Stötzel*

*ASK GmbH, Hilden

Abstract

In today's core manufacturing, the high-performance cold box systems are very well established. However, they are confronted with increasing demands regarding lower smell and reduced emissions, especially in aluminium die casting. ASK develops a solution to these problems with INOTEC, a new inorganic binder system which is based on consequent and detailed investigations of inorganic principles. Even in demanding brass casting, a serial production has been successfully started in 2005.

Also based on the broad inorganic know-how, the application of newly developed ceramic coatings, and zirconium silicate-free high refractory coatings as an integral part of the entire process have improved the performance of the foundries. A new alcohol-diluted water-based, but flammable coating has been a result of these developments.

Keywords: cold box, inorganic binders, emission, refractory coating, casting defect.

Inorganic Technology

Due to environmental aspects and the proximity to residential areas, an expansion of core manufacturing using conventional organic binder systems is associated with higher demands regarding emission control and waste disposal. In recent years, ASK accelerated the development of ecological binder systems that meet the requirements of a high-tech coremaking. One result of these efforts is INOTEC®, an inorganic binder system which is now in the stage of product launch.

This inorganic binder basically consists of three components: a silicate mixture (quantities: 1,8-2,5%), different Promotors (quantities: 0,1-1,0%) and, if necessary mineral molding materials, that are added in quantities of 1-30 percent by weight to the silica sand (figure 1).

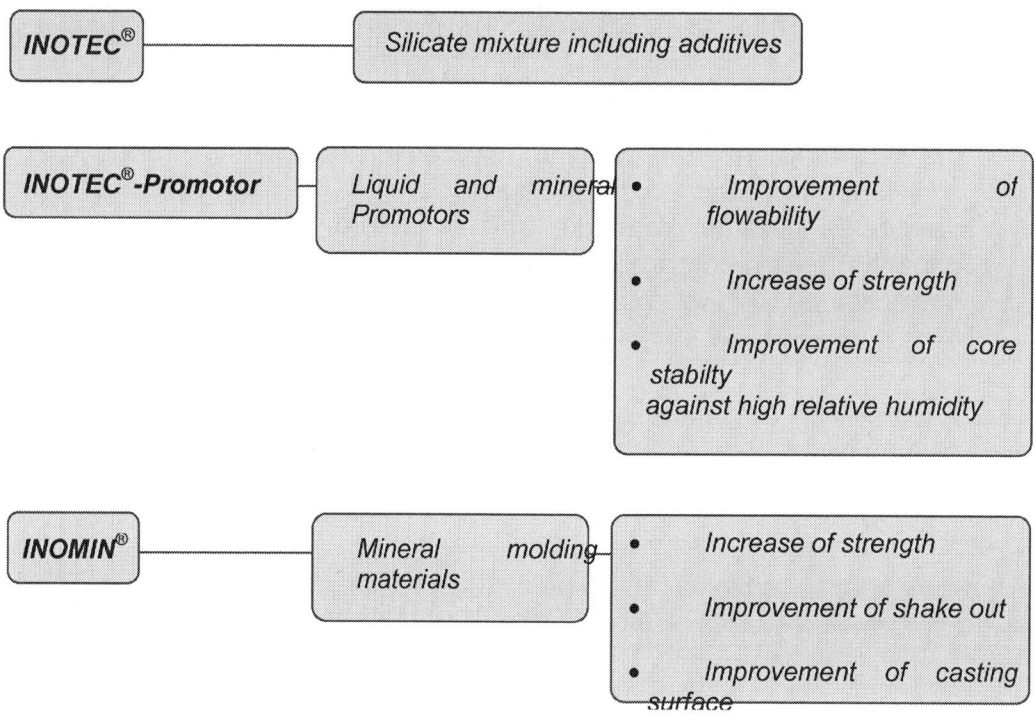

Figure 1: INOTEC®: 3-Component-System consisting of Binder, Promotor und the optional component INOMIN®.

These two or three components can be varied similar to a construction kit among each other, and can be individually adjusted to the technical demands of the respective core production and casting. Especially the immediate strength and the stability against high relative humidity can be influenced significantly by the right choice of binder and Promotor.

The curing takes place in a dehydration process, contrary to the established silicate/CO_2 or the ester-curing methods. Removing water by means of a combination of heatable tools and hot air (180-200°C) has been proven of value. High immediate strength, short cycle times and the resulting high productivity show the advantages of this technique. Due to this special curing process, a homogenous hardening of the whole core can be achieved. Cores made by this method are nearly water-free and emit low gas amounts and no condensate during the casting.

For core manufacturing with this inorganic binder, mixing and dosing equipments of established designs can be used. The core box should be (electrically) heatable and connected to a hot air purging unit. This means that hot box machines can be used by installing a hot air purging unit, and modern cold box machines can be fitted subsequently with a heatable core box. In conclusion, the binder can be described as a low-maintenance system that runs with current machine technology.

Shake-out is primarily done mechanically. The shake out can be regarded as excellent compared to the established organic systems like cold box or epoxy-SO_2,. If required, e.g. for metallurgical reasons, castings made with this binder can also be de-cored with the aid of water.

In brass casting, the binder has already been launched successfully in a serial production at IDEAL Standard (Wittlich, Germany), a leading fittings manufacturer (figure 3).

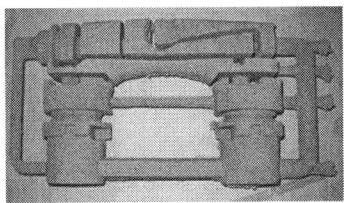

Figure 3: Core making and casting with the inorganic binder at IDEAL STANDARD (Wittlich, Germany).

Only minor adjustments on existing core shooters have been necessary to realize the change-over from the organic hot box process to the inorganic core production with this binder. Further to the evident reduction of odour and emissions during core making and casting, more advantages can be observed. Hot Box cores had to be coated in some cases to give satisfying casting results but not the cores made with the in organic binder under consideration. Figures 3 and 4 show the high quality of the casting surface.

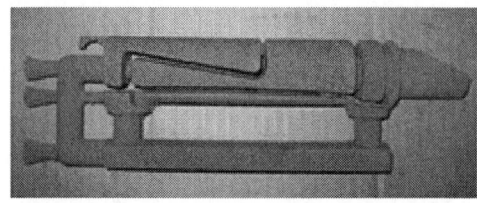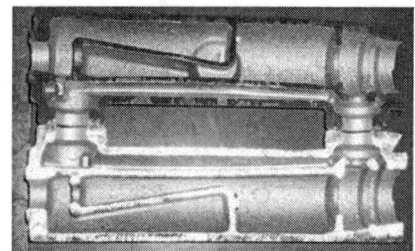

Figure 4: Core made with inorganic binder and respective casting (IDEAL STANDARD, Wittlich).

Coating was necessary with the former Hot Box system. With the inorganic binder the coating can be omitted. Even on very complicated parts of the casting, no sand adhesions could be found. Omitting coating saves time, work and material expenditures.

On the other hand, shutdown of die casts caused by cleaning and maintenance work could be decreased significantly. Problems vanished with the formation of condensate which in summary leads to an increase of productivity. Existing cycle times in the core making process could easily be kept, and in spite of the very complicated geometries the (mechanical) de-coring proceeds excellently.

The series production with the binder in light metal casting is planned for 2006. Large-scale tests are already running on site at several leading automotive manufacturers. BMW, for example modified a cold box series system to fit the demands of INOTEC®. The benefits this change-over to the new binder system had in productivity and lower emissions are clearly visible: Odorless and emission-free core making, emission and smoke-reduced pouring and the simplified cleaning of tools. By the use of special Promotors, immediate/hot strengths can be achieved (>200 N/cm^2) which guarantee an optimal handling of the cores and a process-reliable production. Existing cycle times of the cold box system can be kept, and the de-coring properties are similar to the experiences in brass casting, and very good. Die casts which formerly had to be laboriously cleaned regularly with dry ice under high pressure, can now run with significant less interruptions.

An important question is recycling of the used core sand. A project together with foundries is planned and will show reliable results until end of 2006.

Summary

Due to environmental requirements and increased public pressure, many core manufacturers and foundries are forced to look for alternatives to organic binder systems. In this context, inorganic binders could be in the course of development an alternative to organic binder systems. Just to name a few advantages: emission-free core making, reduced smoke, emissions and gas release during pouring, simplified cleaning of tools (core box, die casts) and, as showed in the pictures, a good surface quality of the castings. Studies concerning the recycling of used sand are currently running.

The product launch of the new inorganic binder is focused on gravity die casting, both in brass and in light metal casting. In the case of sand casting, the influence of the binder on the bentonite has yet to be tested. But also the application of the binder in cast iron is quite conceivable as cores can be water-coated in consideration of specific parameters. A precise statement about this issue can be made in the foreseeable future.

Ceramic Coatings

To enable foundries to produce effective castings (beside the necessary use of binders) coatings have to be applied to reduce casting defects. Especially in coatings inorganic and ceramic concepts have been shown as a powerful development. First step was the application of so called impregnation coating to avoid penetration. Here a correspondingly fine-grained coating is applied at thermally loaded places. The coating remains not at the surface of the sand, but penetrates the cavities and fills them up. Figure 5 shows impressively the difference between a conventional coating being able to penetrate the sand by only 1 - 2 mm, and the impregnation coating KERAFILL AZ which penetrates several millimeters and thus prevents metal burn-in defects. These coatings are used as pre-treatment coating at exposed areas in order to reduce so the fettling expenditure enormously.

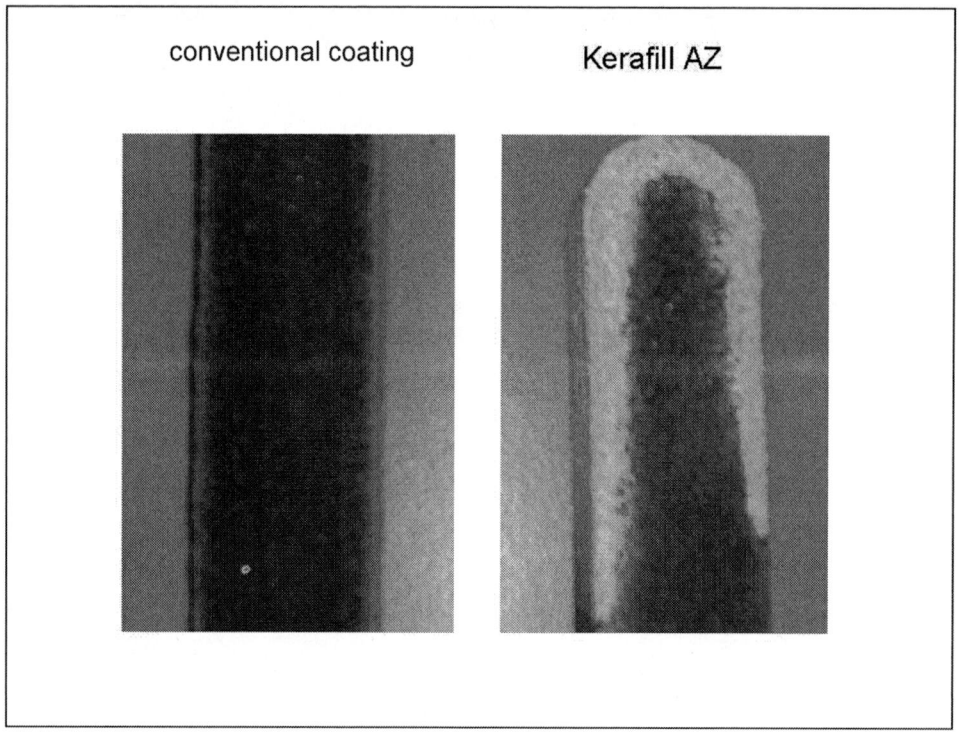

Figure 5: Comparison of serial and impregnation coating.

The consistent development of ceramic principles led to the so-called ceramic coatings. These are characterized by a ceramic bond. One of such coating formerly developed is Kerntop V 107. This usually has a density of about 2 g/cm³ as for zircon silicate coatings. Through the consistent utilization of ceramic principles the density could be increased on noticeable 3 g/cm³, what corresponds to an enormous packing density of the refractory components. It could be achieved, that even with a layer thickness of about two to three millimeters per application no or almost none of the normally usual drying cracks arise (figure 6).

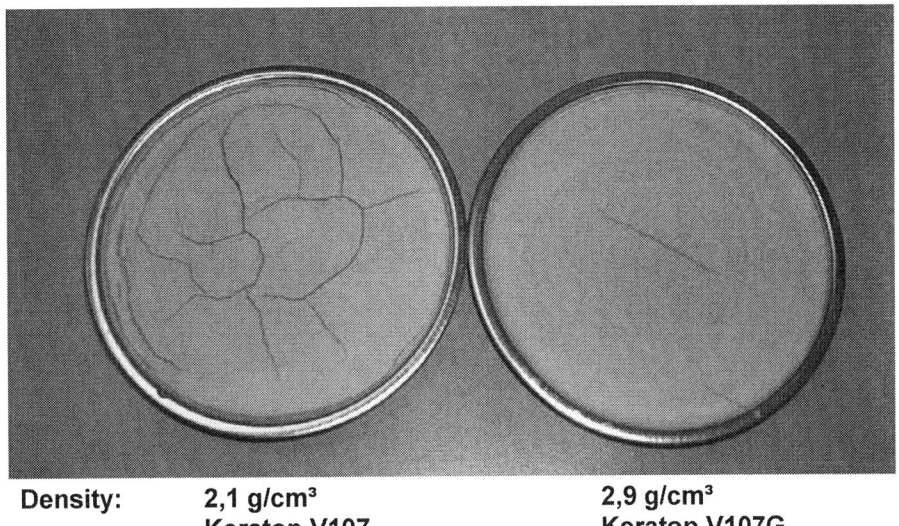

Density:	2,1 g/cm³	2,9 g/cm³
	Keratop V107	Keratop V107G

Figure 6: Re-developed ceramic coating Keratop V 107 G.

Application finds these Kerntop V 107 G coating for the deposition of cores for roll pegs. These roll pegs are loaded by an extremely high metal static pressure and tend in particular to extreme metallization at the edges that lead to a very high wear on the ceramic cutting tools during the processing. Through utilization of the Kerntop V 107 G and its dense ceramic structure the castings could be produced without defects.

It kept on turning out that this new coating performs outstanding services also in the melting shop. Through surface coating of the furnace snouts and of rammed ladles the endurance could be prolonged around the factor three to five.

Zirconium silicate -free high refractory coatings
In the course of the year 2005 the raw material prices went up for zirconium silicate as well as of isopropanol as solvents extremely. Hence the availability of zirconium silicate was subjected to increasingly stronger variations also in the past, a development project was carried out to find alternative high-refractory filler materials.

To that pouring-experiments with the so-called hexagon test were carried out in cooperation with a foundry. During this test coated compression cores are stuck into a hexagon-shaped mould and are over-poured with about 500 kg of GJL within 1.5 minutes (figure 7).

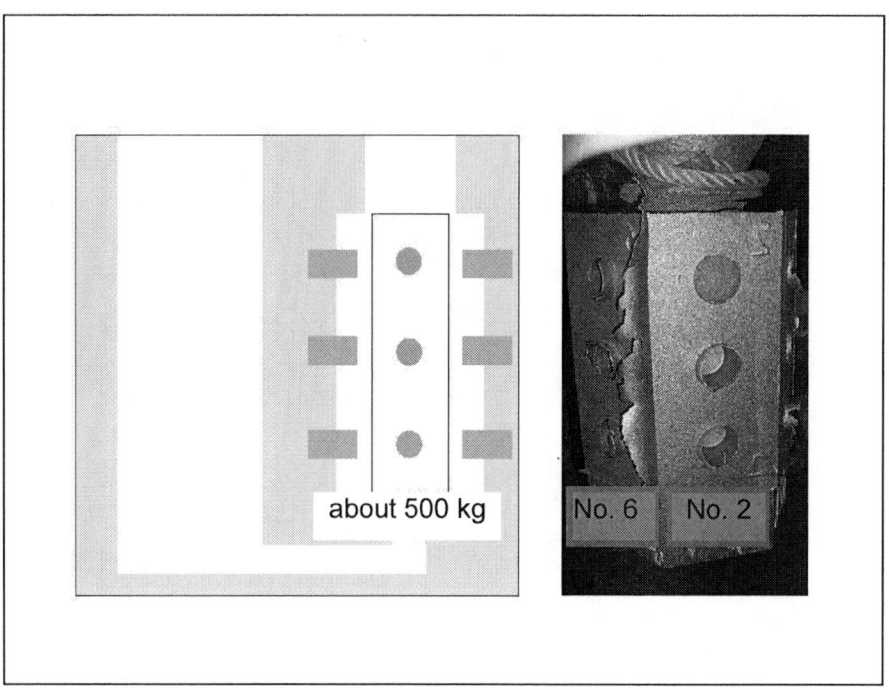

Figure 7: Test assembly to determine the penetration.

Under these conditions extreme loads with regard to penetration, metallization, erosion and finning occur. During the development a refractory combination could be found, which provide under these extreme conditions as good or in part even better pouring results than a serial coating of conventional kind (figure 8).

Figure 8: Development steps of the coating composition optimization to avoid penetration.

This combination is a patented specific mixture of high-refractory aluminium silicates.

The corresponding coating - family is available among following denotation:

Trioflex WK-FF   Water coating
Silico HP L   Alcohol coating
Trioflex HP   Alcohol diluted inflammable water coating

With that very good casting results could be achieved under practice conditions.

An essential advantage of these coatings is the lower density in comparison with the normally usually used zirconium silicate coatings.
In table 1 a calculation was carried out exemplary, that the tremendous saving shows in the consumption and with that in the costs for the coating.

Table 1: Exemplary cost calculation of alcohol based coatings.

	Zirkon-coating	Silico HP L	Savings
Price (€/kg)	1,10	1,10	
Amount VOC (%)	25	40	
Dilution (%)	12	20	
Price of Dilution (€/kg)	1,00	1,00	
Cost of Dilution (€/kg)	0,12	0,20	
Total Costs (€/kg)	1,22	1,30	
Density (kg/l)	1,90	1,20	
Costs / l (€/l)	2,32	1,56	0,76
Annual Consumption (t)	100	59	41
Annual Cost for Coating (€)	122	77	45
Annual Amount VOC (t)	37	35	2

The calculated savings in the order of 25 to 40 percent confirmed themselves in different foundries fully in the case of introduction of the coating, which is hardly to be realized in another way.

A much redesigned concept could be realized in this coating family, which is the principle of an alcohol diluted, however nevertheless inflammable water based coating. This setting of a task appears simple, is, however, a real innovation since the anti-settling, binding and rheological additives are not compatible for water and/or alcohol coatings. ASK succeeded in developing such coating under the name Trioflex HP. This coating can be diluted for brushing or over-pouring with water and/or if required for brushing or over-pouring or dipping with isopropanol.
Also at this coating the refractory components are based on of the high-refractory aluminium silicate combination with low density.
Through implementation of this Trioflex HP a further step to convert alcohol based coatings by water based has been succeeded in No-bake foundries. Also the transportation of hazardous good and the stock

keeping of dangerous material are gone. These advantages mount up to an immense potential of savings.

Summary
The capability of enterprises is the key for the success in the contest of the chances and threats of a global market.
To increase newly developed coatings as Kerafill AZ, Keratop V 107 G, Trioflex WK FF, Silico HP L and Trioflex HP the capability of the foundry according to the leverage principle (small expenditure - great effect) is an integral component of the entire process.

Not some, but *the one*, carefully selected, best suitable coating leads to the success in the mould and core shop and with that for the entire foundry.

The implementation of inorganic concepts, whether by the use of new binder systems or by the use of newly developed coatings, can be a powerful tool for foundries in the future.

**Phenolic Urethane Cold-box Binders –
A Study of Global Properties, Variables, Causes and Effects**

M. Stancliffe

Ashland Casting Solutions, Kidderminster, England

Abstract

The complexity of phenolic urethane cold-box binder materials, properties, processes and applications around the world make the development of binders to cover the expanding range of global industry requirements a challenge. The recent shifts in geographic casting production are continually raising new combinations of property requirements or more difficult combinations of conditions under which phenolic urethane cold-box binders must deliver these customer requirements. A model study was undertaken to link customer requirements, from a more general or global perspective, to verifiable phenolic urethane cold-box core making product and process design characteristics. Design for Six Sigma (a registered trademark of Motorola, Inc.) techniques and tools were applied and an experimental design was developed to provide the basis for the systematic evaluation of the factors binder composition, binder level, global sand source, and global location of test piece manufacture and testing. Testing included typical core-making properties and test castings. The results were analyzed using statistical methods and effects, and the contribution of the design factors to the variation of the test results is discussed.

Key words: Cold-box, sand binders, sand testing

Introduction

The history and development of phenolic urethane cold-box (PUCB) binders has seen many twists and turns as new technology advancements come and in some cases go. Successful phenolic urethane cold-box binders are those that deliver the optimal performance in the maximum number of properties while not having any properties that are clear "show stoppers." It is extremely rare that any adjustment in phenolic urethane cold-box technology will only result in property improvements. There are almost always one or more property compromises. A thorough knowledge of all foundry applications and processes as well as a well defined ingredient/property relationship matrix is required. Also required is a knowledge of geographic-specific factors such as local sand sources and qualities and environmental conditions and their modified requirements from the binder. Phenolic urethane cold-box binders have evolved into their current form and have been the result of many trial-and-error-type experiments and trials. In addition the requirements of foundries on phenolic urethane cold-box binders have changed and become much stricter in their expectations of productivity and scrap levels. Finally, the ability to investigate and test core-making and casting properties have greatly improved over recent years. In addition, experimental design and statistical analysis techniques, such as those employed within the Design for Six Sigma (a registered trademark of Motorola, Inc.) concept, have also improved the quality and reliability of the data analysis and interpretation.

The authors have undertaken to map out the various differences and commonalities resulting from binder composition and different sands through equipment settings and test methodology to differences in ambient environmental conditions around the world. [1]

Consideration Of Experimental Factors

Sand binders are often underestimated as to their influence throughout the foundry process and the quality and efficiencies in producing metal castings. Focusing on making savings in one isolated area may well have a bigger cost penalty in another area.

The phenolic urethane cold-box binder system is a two-part binder system. Part 1 typically comprises a phenolic resin, solvents and performance additives. Part 2 contains isocyanate, solvents and performance additives. The binder, when cured, is best described as a phenolic urethane polymer which has profound impact on core mechanical properties, resilience of the core to atmospheric conditions and, of course, the performance of the core when it interacts with the liquid metal. Depending on the type of phenolic resin and the isocyanate, a large spectrum of different phenolic urethane polymers can be created. Solvent technology is also being viewed as critically important for the performance of the binder, both at core making and during subsequent stages of storage and at casting. Solvents make up 20% - 40% of the total binder composition and provide the basis for many property variations.

Performance additives generally account for only a small portion of the entire binder composition, but result in "big" effects. As these additives are also very costly it was viewed that the approach to this re-assessment of the technology should not include the effects of additives or at least keeps their effects to a low and standardized level within any comparison.

The fact that a typical phenolic-urethane bonded core may consist of 99% by weight of sand and only 1% by weight of binder suggests that the sand source and quality might play a major part in the resultant properties of the resulting core and possibly has an impact on the final casting as well.

In terms of core making process controls and variations, resin addition (i.e., the "binder level") is known to have an important effect, as does the physical location in the world of the core machine due to extreme variations in ambient conditions. Based on the above, five variables of immediate interest and global impact were identified:

-Sand   -Resin addition
-Binder solvents   -Location of manufacture and test
-Binder polymer

Designed Experiments

Design of experiment techniques provide an approach to efficiently designing complex, large-scale experiments which will improve the understanding of the relationship between product and process parameters and the desired performance characteristics.

For the performance comparison of different phenolic urethane cold-box binders on different global sands the following factors and levels were applied in the study:

1. Sand source
    Four subangular silica sands with AFS GFN 50-60 were chosen from various locations around the world. The analysis data for each sand is shown in Figures 1 and 2 and Table 1. Sands were picked that were fairly similar as the effects of big differences in sand quality are well documented.[2,3] The object of this study was to look for more subtle differences that can still have a significant effect on phenolic urethane cold-box binder performance.

2. Binder solvent package
    Phenolic urethane cold-box binders require an appropriately designed solvent package that solvates the phenolic resin, the polyisocyanate and the curing phenolic urethane polymer. Whilst this can be a complex blend of solvents each with varying properties, in this case, two specific and simplified packages were chosen to

provide a comparison of the two main solvent package families currently employed in phenolic urethane cold-box binder technology. Solvent systems based mainly on aromatic hydrocarbon solvents and ester solvents were considered, denoted S1 and S2, respectively.

3. Phenolic urethane polymer
A significant proportion of phenolic urethane cold-box technology relates to the design of the phenolic base resin and the subsequent phenolic urethane polymer when reacted with a typical grade of isocyanate. For the purpose of this experimental design, two globally successful and yet different phenolic base resins were utilized, of specific interest to the authors, denoted in this case as B1 and B2.

Combined with the two solvent package types, this resulted in four different binders. Each binder was optimized in formulation ratios to give a maximum immediate strength within a specified viscosity range. Viscosity was limited because it controls the property of mixed sand flowability which could influence or mask performance due to variations in equipment design. Each formulation was also limited by having to achieve a practical level of cold temperature stability without separation.

4. Resin addition
2 levels: 1% and 1.5%, measured as weight percent of binder based on sand, of Part I and Part II applied at a ratio of 1:1. The adjustment of binder additions is known to have significant effects, both advantageous and disadvantageous on several properties.

5. Location of test piece manufacture and test
Three test laboratories in the U.S., Europe and Asia. By the inclusion of three testing facilities in different locations, the comparative level of contribution to variation in certain properties could be established in proportion to the other factors examined. By allowing these differences into the experimental design we can at least pinpoint worst cases to their location and start to investigate and hopefully eliminate their cause.

The experimental design was a L16 mixed 2-4 level Taguchi design based on a fractional factorial experiment which allows an experiment to be conducted with only a fraction of all possible experimental combinations of parameter values.

The core-making and casting tests employed to comparatively evaluate the performance of the various factors are described below.

Core-making tests: Casting tests:

Immediate tensile strength [4]   Gray iron erosion wedge casting [5 7]
Immediate transverse strength [4]   Gray iron penetration casting [5]
Tensile strength at 24 h [4]   Gray iron warpage block casting [8]
3-hr mixed sand bench life [4]   Aluminum shake-out casting [9]
Tensile strength @ 90%RH [4]   Hot distortion [8]
Release and resin wipe-off [6]
Amine consumption [6]
Curing efficiency [6]

Veining, penetration, erosion and surface finish were visually rated using a scale of 1 to 5 with 1 as the "best" and 5 as the "worst." Warpage was measured as deflection of the core cavity in the resulting casting.

Analysis of Variation
The raw data gained from the experiments were statistically analyzed using analysis of variation (ANOVA) techniques. A design factor with a large difference, in contribution to the total variation, from one level to another indicates that the factor is a significant contributor to the performance characteristic in question. Contributions derived from the results of the core- making and casting tests are shown in Figure 3.

All factors had a significant effect on at least one and usually more core-making properties. The binder level had the largest effect on the greatest number of properties. The variation from the different sands had a surprisingly smaller effect than expected on many of these properties. Sand variation had the greatest effect on benchlife performance on immediate tensile strength.

The variation between test locations was fairly consistent in most properties. Any testing that includes exposure to humidity features an inherently larger error than testing under normal conditions. Humidity resistance was also clearly a property greatly affected by binder formulation variations. The results of resin wipe-off testing indicate that prediction of this property is least reliable. The larger variation associated with the two release and transversal test locations was immediately investigated and verified and found to be due to a design / part specification difference in the test piece manufacturing equipment resulting in an immediate benefit from this experiment.

The contribution to variation arising from the casting tests and evaluation of casting properties resulted in a very different picture. Most notably the error contribution is considerably larger. Some casting properties were mostly sand specific, while others were affected by all factors. The contribution of binder technology and binder level was significant in properties such as erosion resistance and core warpage on casting and the binder level clearly dominated shake-out. Veining proved to be completely independent of phenolic urethane cold-box binder choice and addition, and largely dependent upon the sand.

Effects
The significance of each factor and level can be visualized in an Effects plot which shows the effect of each level per factor per test measurement discussed in Figures 4-8.

Phenolic base resin B1 gave rise to better erosion performance than B2. Dalin sand had a very beneficial effect on veining. The lower level of purity in the Dalin sand is a likely reason for this. Phenolic base resin B2 had a lower level of core warpage on casting than resin B1, which is opposite to the comparison of erosion resistance. Nugent sand had an unusually low level of core warpage on casting. Solvent package S1 and Arena sand gave rise to improved levels of shake-out performance.

Additional Properties and Tests
Once the general structure of this experiment had been designed, additional tests were run and their results introduced providing further depth to the effects of the various factors. For example:

Hot distortion tests were performed on the sand, resin and % factors. The hot distortion, or hot deformation test, involves the measurement of deflection of one end of a sand core test piece when a gas flame is applied to the centre of a bar shaped sand core test piece. This test has long been the discussion topic of foundries and binder manufacturers alike as to the relationship of different parts of the graphs with particular casting properties. An example of the hot distortion curve typical for a PU bonded core is shown in Figure 9. The curve contains 3 distinct measures: maximum distortion height (upward deflection of the core); time to zero distortion (return of the core end to its original position); time to failure (excessive sagging or breakage of the core).

The effect plots for each of the graph measurements are shown in Figure 9. The plots for each measurement on the graph clearly responded very differently to the different levels for each factor. The plots can be used almost like finger prints which can be compared to the casting effect plots in Figures 7 and 8 to find a reasonable match. Out of all the casting plots only core warpage on casting, not surprisingly, resulted in a good match and only in the measurement of time to zero deformation.

Conclusion
The enormous amount of data within this study and the subsequent statistical analysis has already generated a wealth of information as well as several directions for future study. The data and analysis presented is only a sample of what was obtained and further mining of this data along with additional experimentation will be required to extract full value.

The knowledge accrued forms a secure foundation on which further experiments in sand, binder and process variables can be based and added to. Future binder development can only benefit from studies such as this, providing more secure knowledge of phenolic urethane cold-box binder performance and the factors which can be adjusted to more accurately deliver customer requirements, regardless of global location and sand used.

References

1. Copyright 2006 American Foundry Society (www.afsinc.org): Stancliffe M; Kroker J; Wang X; "Phenolic Urethane Cold-box Binders-Using Design for Six Sigma Methodology to Evaluate Core Making Needs for the Global Market," AFS Transactions (2006)
2. Carey P, Swartzlander, M W, "Sand Binder Systems Part II – Resin/Sand Interactions", *Foundry M & T* (1988).
3. Garner T E, "Mineralogy of Foundry Sands and its Effects on Performance and Properties", *AFS Transactions* (1977).
4. "Mold and Core Test Handbook," 3rd Ed, American Foundrymen's Society, Des Plaines, IL.
5. Tordoff W L, Tenaglia R D, 25th Anniversary Paper: "Testing Casting Evaluation of Chemical Binder Systems", *AFS Transactions* (2006).
6. Blackburn P A, Henry C M, "A More Productive Phenolic Urethane Coldbox Process", *AFS Transactions* (1996).
7. Henry C M, Showman R E, Kahles D, Nikolai M., "Process Variables Affecting the Erosion Resistance of Phenolic-Urethane Cold Box Cores", *AFS Transactions* (2003).
8. Rigel J A, Sturtz G P, "Developing New PUCB Binders for Aluminum Casting Applications", *AFS Transactions* (2005).
9. Henry C M, Showman R E, Wandke G, "Process Variables Affecting Al Casting Shakeout of Coldbox Cores", *AFS Transactions* (1999).

Tables

Table 1. Compositional analysis by X.R.F.

XRF TEST RESULTS				
	Dalin Industrial	Arena C-55	Congleton HST 50	Nugent 480
	%	%	%	%
SiO2	91.70	95.38	97.67	95.41
Al2O3	4.46	2.37	1.16	2.46
K2O	2.24	1.45	0.69	1.25
Na2O	0.86	0.32	0.11	0.45
Fe2O3	0.28	0.06	0.22	0.25
CaO	0.20	0.12	<0.05	0.21
MgO	0.08	0.05	0.07	0.10
TiO2	0.07	<0.05	<0.05	<0.05
Cr2O3	<0.05	<0.05	<0.05	<0.05
Mn3O4	<0.05	<0.05	<0.05	<0.05
BaO	0.05	<0.05	<0.05	<0.05
ZnO	<0.05	<0.05	<0.05	<0.05
V2O5	<0.05	<0.05	<0.05	<0.05
P2O5	<0.05	<0.05	<0.05	<0.05
ZrO2	<0.05	<0.05	<0.05	<0.05
SrO	<0.05	<0.05	<0.05	<0.05

Figures

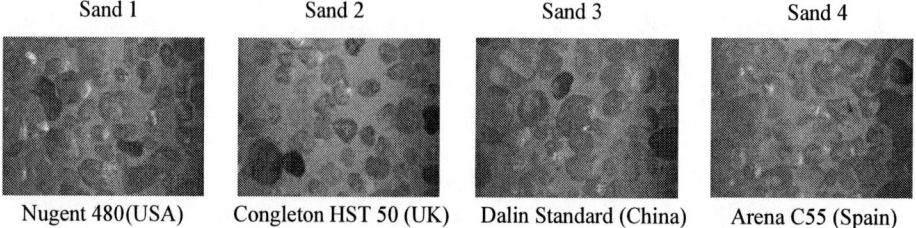

Figure 1: Micrographs of the four sands used in the experiment.

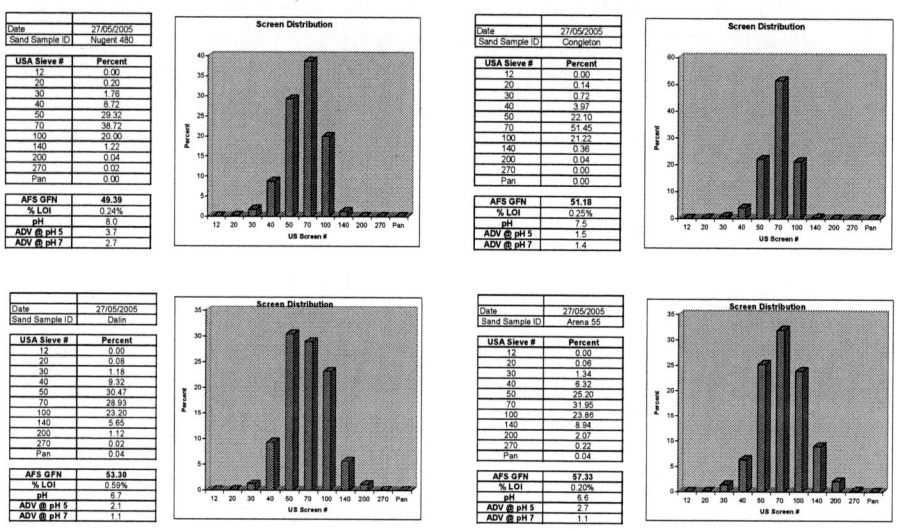

Figure 2: Sand granulometry, physical and chemical analysis data.

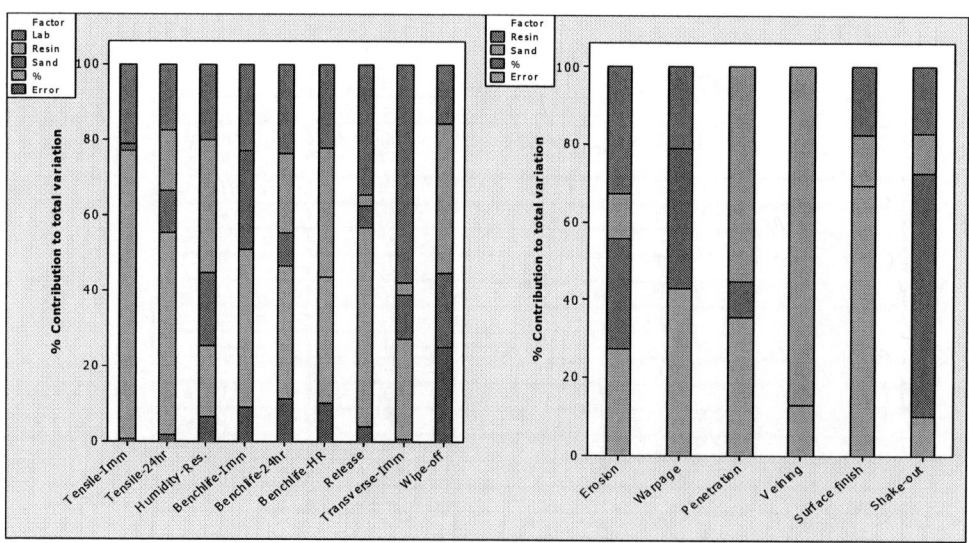

Figure 3: Contribution to variation vs. factor per casting property.[1]

All tensile strengths measured at the Euopean location were unusually high. Reasons for this may include equipment design as the processing parameters were generally considered quite similar.

Resin B1 gave higher 24hr tensiles strengths than B2. Solvent Package S2 gave superior humidity resistance than S1 and the interaction of B1 and S2 gave the highest performance. The same trend was observed in humidity resistance testing.

Dalin sand gave noticeably poorer 24hr and 24hr@90% RH tensile strengths.

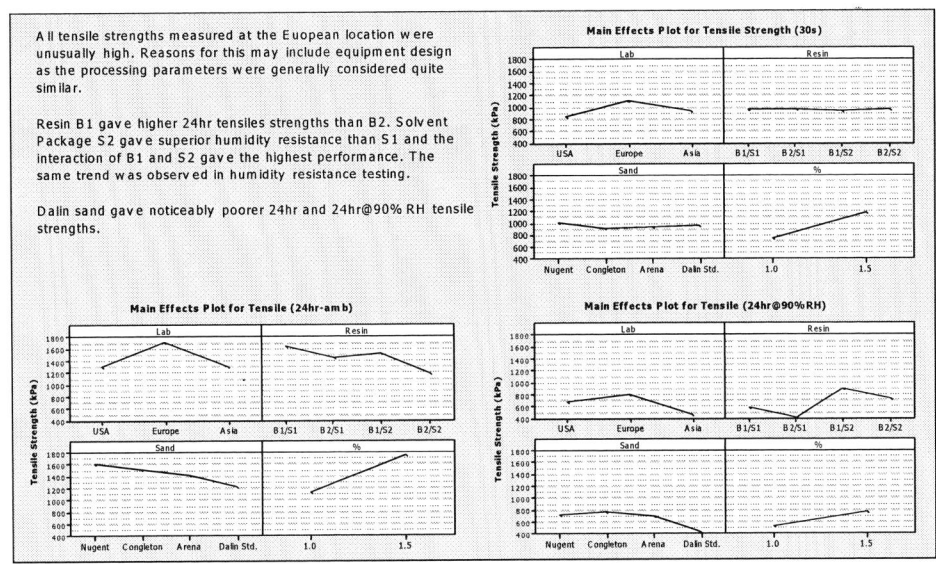

Figure 4: Effect plots for core strength properties.

Benchlife properties were much lower in the Asian testing location. Ambient environmental conditions showed no obvious correlation as to why. Free amine levels in the vicinity of the stored sand could be responsible.

Solvent package S2 gave rise to higher benchlife performance in terms of 24hr and 24hr@90% RH conditions. Nugent and Congleton sands were generally better in benchlife performance even though both sands pH's measured higher than the Arena and Dalins sands. No correlation with ADV was observed.

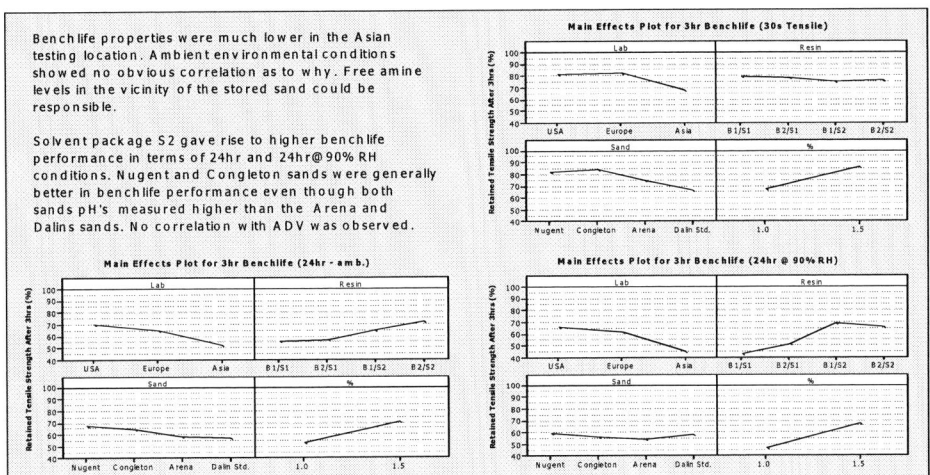

Figure 5: Effect plots for benchlife properties.

The difference between USA and European test equipment has been identified as a difference in part specification which resulted in the difference in test data obtained.

Solvent package S1 gave higher transverse strengths than S2.

Nugent and Congleton sand gave higher transverse strengths than Arena and Dalin sands. This difference was not observed in the tensile data.

Resin B1/S1 and Arena sand both gave unusually high levels of resin wipe-off.

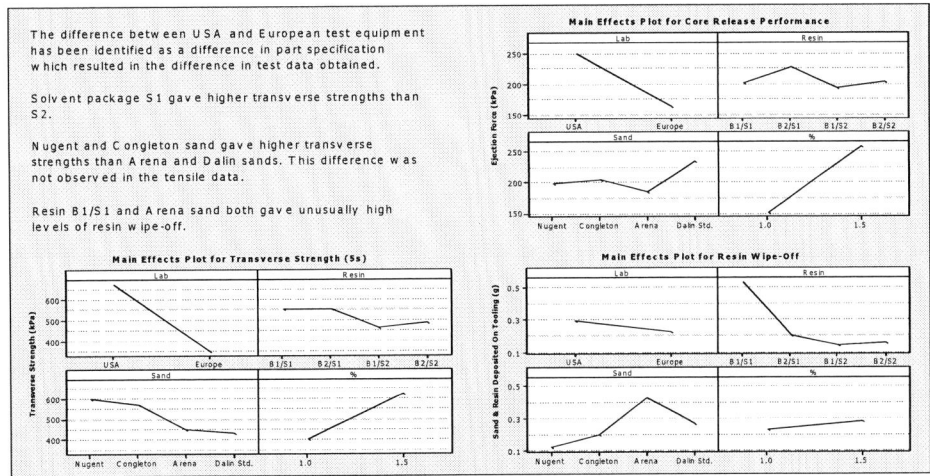

Figure 6: Effect plots for tooling properties.

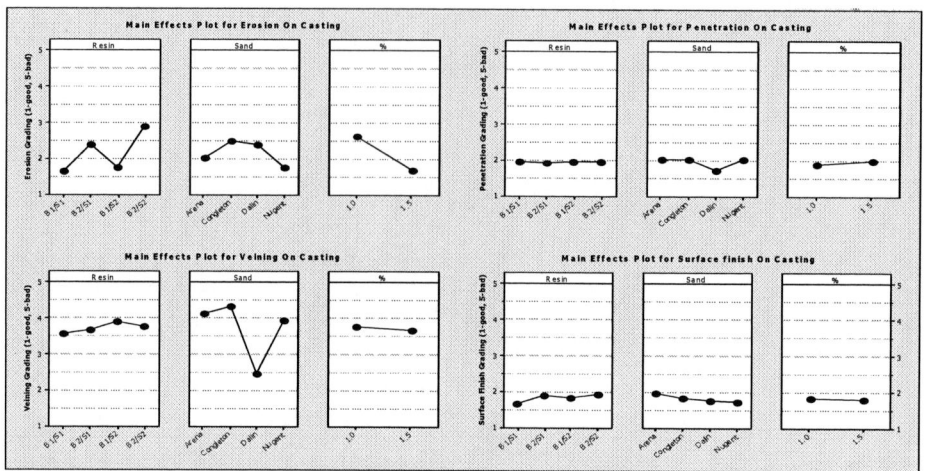

Figure 7: Effect plots for casting quality properties.

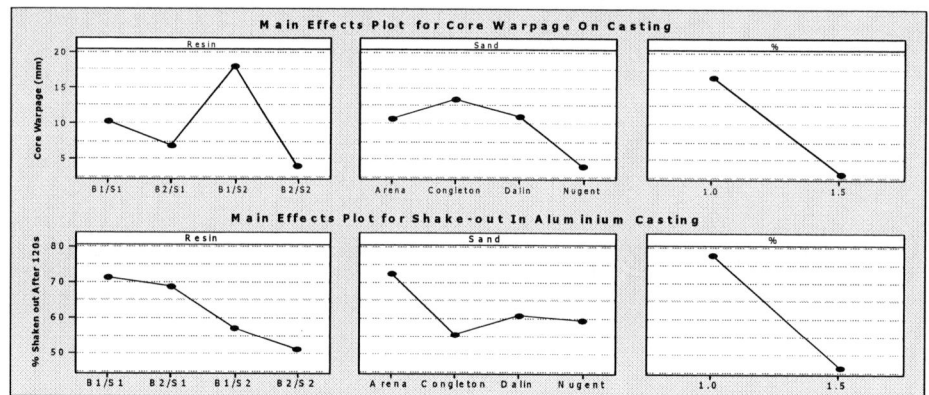

Figure 8: Core warpage on casting and shake-out performance.

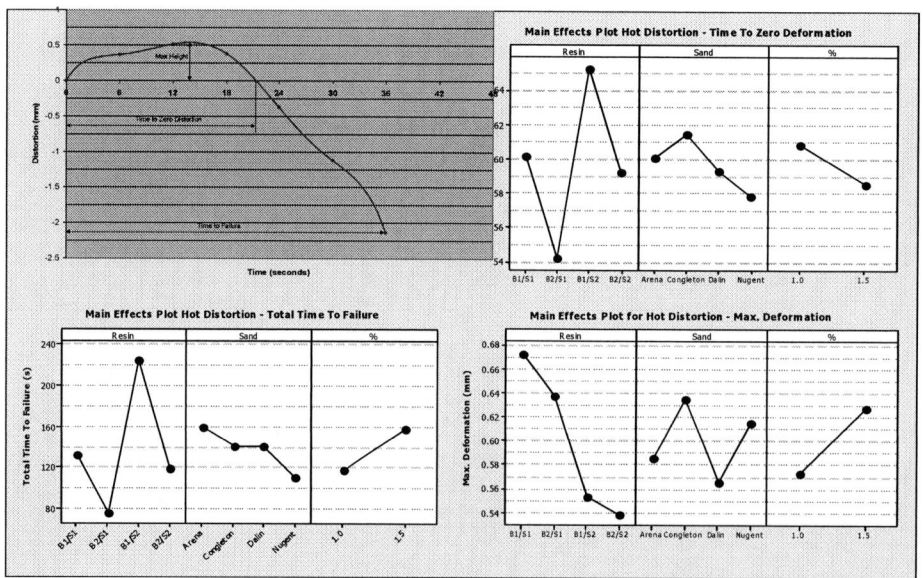

Figure 9: Typical hot distortion behavior & effects plots / measurement

Author's E-mail Address: *mstancliffe@ashland.com*

Modeling and Identification of Pouring Flow Process
With Tilting-Type Ladle
for an Innovative Press Casting Method Using Greensand Mold

Yusuke Matsuo*, Yoshiyuki Noda*, Kazuhiko Terashima*,
Kunihiro Hashimoto**, Yuji Suzuki***
* Dept. of Production Systems Engineering, Toyohashi Univ. of
Technology
** Sintokogio, Ltd.
*** Aisin Takaoka Co., Ltd.

Abstract

Recently, a new method of a press casting using greensand mold was proposed by our group as a means of ferrous casting. This method has advantages in terms of energy savings and cost because a sprue cup and runner channel are not needed. Still, it is important to pour accurately and quickly in this method. Therefore, a feedforward outflow control is presented by using a mathematical model and inverse dynamics. The outflow is measured by a loadcell under the ladle. However, the loadcell data has a large uncertainty that is caused by the tilting motion. Therefore, a compensation for the loadcell is proposed. The determination of the back-tilting time is optimized by a dichotomy method. Finally, the effectiveness of the proposed system is shown through simulation and experiment.

Key words: press casting, automatic pouring, outflow volume control, batch-type pouring, modeling

Introduction

Recently, a new innovative casting method involving a press casting using greensand mold has been proposed and studied as a ferrous casting method by our research group. This method involves a process in which molten metal in a ladle is first poured into a lower mold, and then the upper mold is pressed down on the lower mold which fills the molten metal. This method has the advantage of low cost, because a sprue cup and runner channel are not needed. The cast is produced in this method without a runner channel, which is the source of much waste in the traditional approach. As a result, the pouring temperature can be reduced. Therefore, this new casting method costs less than the traditional approach and saves energy. However, it is important when using this method to accurately and quickly pour the molten metal into the mold. Therefore, outflow control of the liquid in the automatic pouring process plays a more important role in this new casting process.

On tilting-type automatic pouring machines used in industrial plants, a teaching and playback method is used. However, much time is spent on the teaching in this process, and It is difficult to achieve target outflow with accuracy because the human error.

Outflow control has been largely studied by using feedback control[1],[2]. In the literature [2], the velocity curve of the ladle's tilting is switched several times during pouring by using feedback information of the molten metal. However, the appropriate velocity is determined by the trial-and-error method in the experiment. On the other hand, our group proposed a feedforward control using a fun-type ladle. A fun-type ladle is a simple shape that seems to be easy to control. A better method is required to control ladles of general shapes. In addition, our previous paper provides a method for predicting the pouring quantity that flows out after a ladle is back-tilting to cut the pouring. However, because the previous method requires real-time measuring of molten metal weight and model calculation, applying it to real Industry seems difficult.

Therefore, in this paper, outflow control of the liquid in an automatic pouring process is proposed. Modeling by using the system identification method is presented, and then feedforward control using the inverse dynamics of the pouring process is achieved. The automatic pouring machine has a servo-motor to drive a ladle and a loadcell for measuring outflow.

However, the outflow measured directly by the loadcell has a large level of uncertainty, because the data includes the reaction force caused by the tilting. Therefore, a compensation method[3] incorporating a mathematic model of the loadcell is being newly proposed here.

Finally, the effectiveness of the proposed system is shown through simulation and experiment. It is shown that the method proposed in this paper can be applied to pouring systems used in industry, and the cost of molten metal can be reduced by the proposed optimum outflow control.

Automatic Pouring System

The laboratory water experimental apparatus and the industry molten metal apparatus used in this paper are shown in Figs.1(a) and (b) respectively. The rotary direction of the T-axis is driven by an AC servomotor in both apparatuses. The driving force of the AC servomotor can be amplified by reducing the gear ratio. The center of the ladle's rotation shaft is placed near the ladle's center of gravity. When the ladle is rotated around the center of gravity, the tip of the ladle nozzle (or mouth) moves in a circular trajectory. It is then difficult to pour the molten metal into a mold if the pouring mouth is moved by tilting. Then, the position of the tip of the ladle nozzle is maintained during pouring by means of a synchronous control of the Y- and Z-axes for rotational motion around the T-axis of the ladle[4]. The rotation angle is measured by an encoder installed in the AC servomotor. Y-, Z- and T-axes are also driven by AC servomotors, but the driving force of each of these motors is amplified through the ball screw mechanism. Each axis can be independently moved.

The weight of fluid in a ladle is measured by a loadcell located under the pouring machine. The measured values from the encoder of each axis and the loadcell are input into a computer with an A/D converter. A control input is sent to a motor driver via a D/A converter, driving the AC servomotor. In the automatic pouring machine with water used as the substance being pouring, a control instruction is carried out in the DSP through the AD/DA converter and an up/down counter. In the apparatus used industrially in plants, a control instruction is carried out in the PLC through the AD/DA converter and an up/down counter. And the apparatus used in industry is controlled by position control. So, the tilting angle obtained by the water experiment is directly applied to the industry apparatus.

A Series of Models of the Total Pouring Process

In this section, modeling of the pouring process is carried out using water. The kinematical viscosity of the water (293[K]) and the molten metal (1673[K]) are $1.004 \times 10^{-6} [m^2/s]$ and $0.970 \times 10^{-6} [m^2/s]$, respectively. Therefore, the fluid behavior of the water is nearly identical to that of the molten metal.

Here, two models, from the input voltage $u(t)$[V] for the motor to the flow rate $q(t)$ are divided into two parts.

1. A model showing the relation between the input voltage of the AC servo motor $u(t)$ and the tilting angular velocity of the ladle $\omega(t)$: G_M
2. A model showing the relation between the tilting angular velocity of the ladle $\omega(t)$ and the flow rate into a under mold $q(t)$: G_q

$G_M(s)$ and $G_q(s)$ are respectively referred to here as the motor model and flow rate model.

MOTOR MODEL :
The relationship between the tilting angular velocity of the ladle and the input voltage to the motor is described in the following equation:

$$G_{MT}(s) = \frac{\Omega_T(s)}{U_T(s)} = \frac{K_{MT}}{1+T_{MT}s} \tag{1}$$

, where $\omega(t)$[rad/s] is the tilting angular velocity, $u(t)$[V] is the input voltage for the T-axis, K_{MT}[rad/(sV)] is the gain of the motor, and T_{MT}[s] is the time constant of the motor. Further, the motor model of the X-, Y-, and Z-axis is described in the same form as in Eq. (1).

FLOW RATE MODEL :
In this paper, the ladle shown in Fig.2 is used for the pouring. Here, with regard to the fun-type ladle, controlling the tip of the ladle nozzle invariably, the surface area of the liquid in the ladle is constant while the ladle is tilted. Therefore, the pouring flow rate can be considered to be constant when a ladle is rotated at a constant angular velocity. Hence, the flow rate model can be described by a first-order transfer function as follows.

$$G_{qf}(s) = \frac{Q(s)}{\Omega_T(s)} = \frac{K_{qf}}{1+T_{qf}s} \tag{2}$$

, where K_{qf}[m³/rad] and T_{qf}[s] are the gain and the time constant of the fun-type ladle's flow rate model, respectively.

However, the surface area of the liquid in a practical ladle is not constant while the ladle is tilted. Therefore, the ladle's flow rate model can be expressed by varying the gain and the time constant of the fun-type ladle's flow model in correspondence with the tilting angle. The ladle's flow rate model can be described by the Linear Parameter Varying function (LPV model) as follows.

$$G_q(s) = \frac{Q(s)}{\Omega(s)} = \frac{K_q(\theta)}{1+T_q(\theta)s} \tag{3}$$

, where $K_q(\theta)$[m³/rad] and $T_q(\theta)$ are the gain and the time constant of the flow rate model, respectively.

Derivation System of Feedforward Input for Outflow Control

In this paper, a feedforward system is proposed as shown in Fig.3. The inverse model is used for deriving the control input. The inverse model is shown in Eqs.(4) and (5).

$$u(t) = \frac{1}{K_M}\omega_{ref}(t) + \frac{T_M}{K_M}\frac{d}{dt}\omega_{ref}(t) \tag{4}$$

$$\omega_{ref}(t) = \frac{1}{K_{qf}(\theta)}q_{ref}(t) + \frac{T_{qf}(\theta)}{K_{qf}(\theta)}\frac{d}{dt}q_{ref}(t) \tag{5}$$

In the outflow control, the flow cuts the pouring flow through ladle back-tilting in order to obtain the target outflow. However, during the back-tilting, the pouring flow continues initially. Therefore, we must predict the flow quantity during back-tilting. We call this flow quantity the "after-flow quantity". Thus, we must consider the after-flow quantity as we begin the back-tilting. In this paper, the optimum back-tilting timing is detected by a dichotomy method[5] that is a general numerical solution method for algebraic equations. A block diagram of the outflow control system is shown in Fig.4. In the simulation, the control input is obtained by a dichotomy method, and outflow control is realized by giving the control input to the apparatus by means of feedforward control. Flow cutting is implemented to switch from forward-tilting to back-tilting at a given time t_s[s]. The outflow when the flow is cut at t_s is denoted by $f(t_s)$ [m³]. Here, Eq.(6) is solved by a dichotomy method, where f_{ref} [m³] is the target outflow.

$$f_{ref} - f(t_s) = 0 \tag{6}$$

Compensation System of Load Cell

The gain $K_q(\theta)$ and time constant $T_q(\theta)$ of the flow rate model as shown in Eq.(3) are obtained by experiment. The parameters are identified using the loadcell data. However, there is a problem in the loadcell data, as they are influenced by the acceleration in the ladle's up and down movement under synchronous control as shown in Fig.5. Therefore, the exact weight of the ladle cannot be measured. Therefore, in this paper a system of compensation for the loadcell system is proposed, as shown in Fig.6. The exact weight of the molten metal in a ladle is obtained by subtracting the output of the empty ladle model from that of the original loadcell.

Here, the relationship between the Z-axis acceleration and the load cell output is described by the second order transfer function in the following equation.

$$G_l(s) = \frac{F_l(s)}{A_z(s)} = \frac{K_l\omega_{nl}^2}{s^2 + 2\zeta_l\omega_{nl}s + \omega_{nl}^2} \tag{7}$$

, where K_l[kg·s^2/m], ω_{nl}[rad/s] and ζ_l[-] are the gain, angular velocity and dumping rate of load cell model, respectively. The parameters are identified using a simplex method[6]. The evaluation function is described in Eq.(8).

$$J = \int_0^t (f_{l0}(\tau) - f_l(\tau))^2 d\tau \qquad (8)$$

, where $f_{l0}(t)$[m^3] represents the experimental data of the loadcell when the empty ladle is moved. On the other hand, where $f_l(t)$[m^3] represents the output if the empty ladle mold calculated in Eq.(7), $F_l(s)=L[f_l(t)]$ and $L[-]$ denotes the operator of the Laplace transformation. Good compensation results were obtained, and they are shown in Fig.7.

Flow Model Parameter Identification
The flow model parameter is identified by using the proposed loadcell compensation system. The parameter identification sequence is as follows;

[Procedure of parameter identification]:
1. The ladle is tilted from the given angle of the ladle to 3[deg].
2. The outflow is measured by using a loadcell when Step 1 is effectuated.
3. The obtained data in Step 2 are approximated in the first order system.
4. The gain and time constant of the first order system are same as the gain and time constant of the LPV model in the initial tilting angle.
5. The steps from Step 1 to Step 4 are repeated at every 1[deg] interval for the operable angle (from 20 to 42 [deg] in this paper).

The identification results obtained by the above sequence are shown in Fig.8.

Experimental Results and Discussion
In this section, the experimental results of the outflow control are shown. The simulation and experimental results of the outflow control with water, using the water experimental apparatus, are shown in Fig.9. The designed flow rate reference is shown in Fig.9. Here, as the experimental condition, the initial angle, target outflow and back-tilting motion are 26[deg], 0.783[kg] and 3[deg] for 1.5 [s], respectively. As a result, the actual outflow is 0.758[kg] for reference 0.783[kg]. The pouring error is 3.2[%].

The control input used in the water experiment is straightforwardly given to the molten metal apparatus in industry. The experimental results obtained by using the molten metal are shown in Fig.10. As the experimental condition, the initial angle, target outflow and back tilting motion are 26[deg], 5.48[kg] and 3[deg] for 1.5[s], respectively. As a result, the actual outflow is 5.74[kg] for reference 5.48[kg]. The pouring error is 4.7[%].

Both the results in laboratory and industrial results are considered to be good, because the pouring error is 5[%]. This means that scrap loss is less than 5[%], and then the yield rate is 95[%]. In the traditional casting process using greensand molding, the yield rate is below 50[%]. Hence, the proposed innovative process is very effective.

Conclusion

This study presented a series of outflow control systems for automatic pouring machine used in an industrial plant setting. The obtained results are summarized as follows.

(1) The untrue loadcell data for Z-axis motion could be compensated for by the proposed loadcell compensation system.

(2) The pouring machine could pour the target outflow with an error rate of about 5[%].

(3) The system proposed in this research involves sensorless feedforward control, which can be easily applied to the existing automatic pouring machine currently used in industry.

References

1. Kazuhiko Terashima and Ken'ichi Yano: Supervisory Control of Pouring Proscess, 66th World Foundry Congress, Istanbul, Turkey, 6-9 Sept, 2004

2. Jiro Satoh, Kenichi Yoshida: Automatic pouring equipment for casting "Mel Pore system", Industrial Heating, 1992, vol.29, No.4, pp19-27 in Japanese

3. Janxin Sun, Yoshihiro Fujioka, Toshiro Ono, Takeyoshi Nagao and Toru Kohashi: On the Fast and High Accurate Mass Measurement under the Conditions of Floor Vibration, Journal of The Japan Society for Precision Engineering, 1998, vol. 64, No.4, pp567-572 in Japanese

4. Kazuhiko Terashima, Ken'ichi Yano, Yu Sugimoto and Mitsuaki Watanabe: Position Control of Ladle Tip And Sloshing Suppression During Tilting Motion in Automatic Pouring Machine, 10th IFAC Symposium Automation in Mining Mineral and Metal Processing (MMM2001), Japan, Tokyo, 4-6 Sept, 2001, pp182-187

5. Brice Carnahan, H. A. Luther, James O. Wilkes: Applied Numerical Methods, John Wiley & Sons, Inc., 1969

6. L. Collatz, W. Wetterling: Optimization Problem, Spring-Verlag New York, 1975

Table

Table1 Model parameter

Parameter	Symbol	Value
Motor gain (T-axis)	K_{MT}	0.4290 [rad/(sV)]
Motor gain (Y-axis)	K_{MY}	0.0830 [m/(sV)]
Motor gain (Z-axis)	K_{MZ}	0.0828 [m/(sV)]
Time constant (T-axis)	T_{MT}	0.006 [s]
Time constant (Y-axis)	T_{MY}	0.006 [s]
Time constant (Z-axis)	T_{MZ}	0.007 [s]
Loadcell gain	K_l	26.42 [kg·s^2/m]
Loadcell angular velocity	ω_{nl}	66.80 [rad/s]
Loadcell dumping rate	ζ_l	0.28

Figures

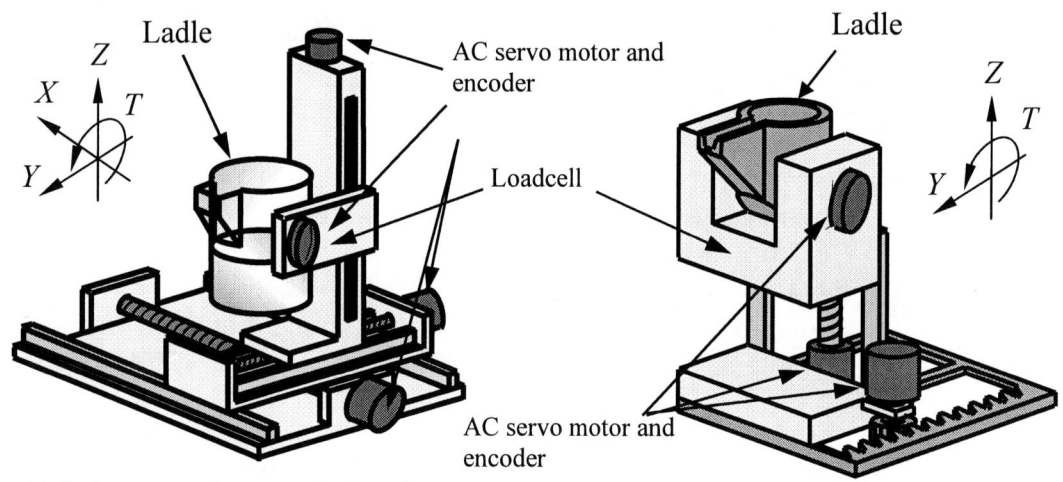

(a) Laboratory Automatic Pouring
Machine by Using Water

(b) Industry Automatic Pouring
Machine by Using Molten Metal

Fig.1 Automatic Pouring Machine Outline

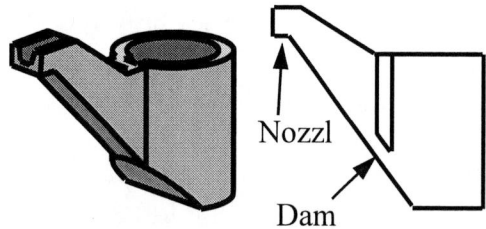

Fig.2 Ladle Outline

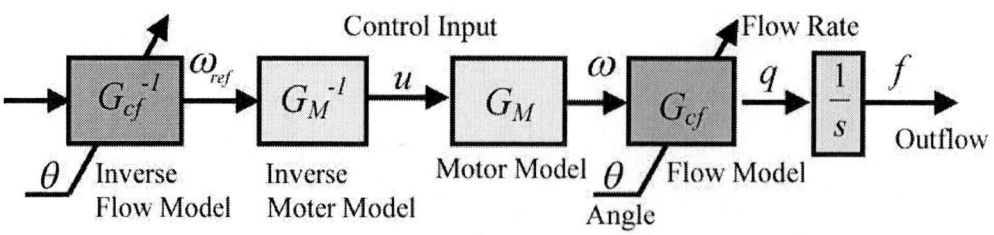

Fig.3 Structure of Pouring System and Inverse System

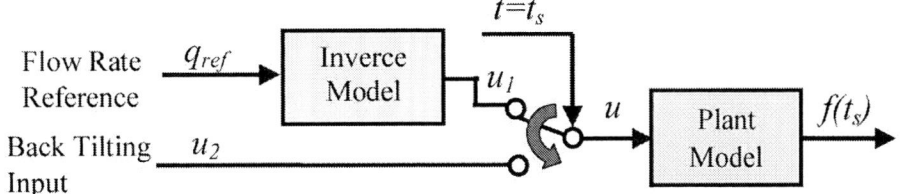

Fig.4 Schematic Diagram of Outflow Control

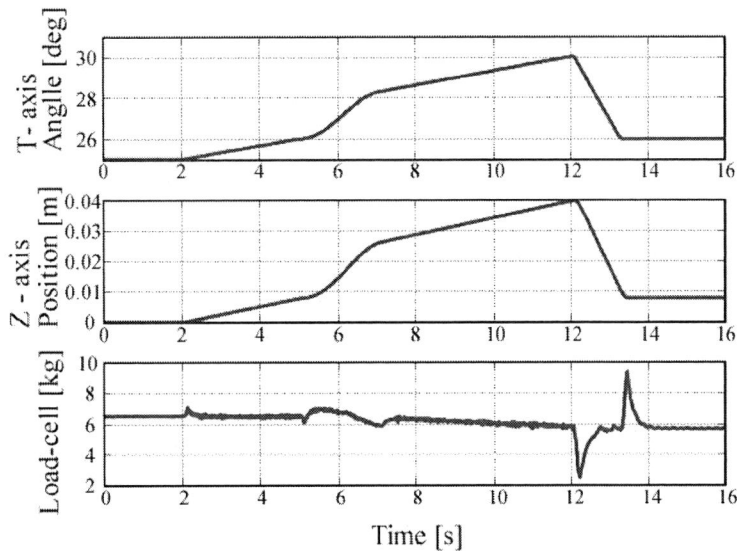

Fig.5 Effect of Synchronous Control of Ladle on the Loadcell Sensor

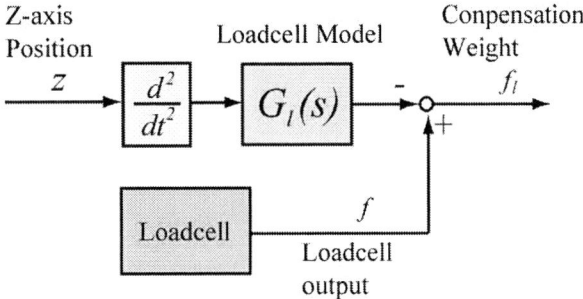

Fig.6 Compensation System of Loadcell

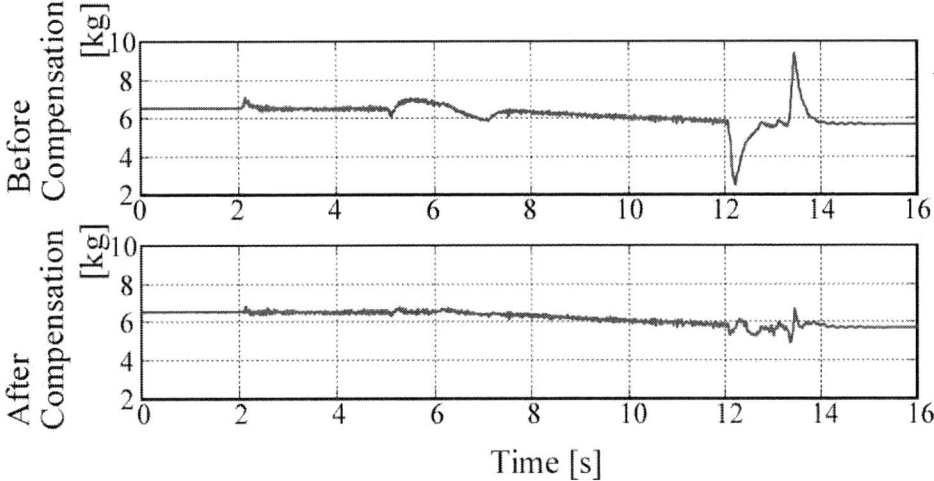

Fig.7 Experimental Results by Compensation System of Loadcell

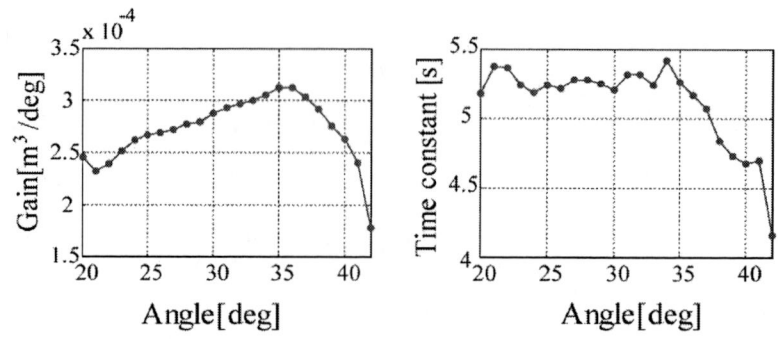

Fig.8 Flow Model Parameter

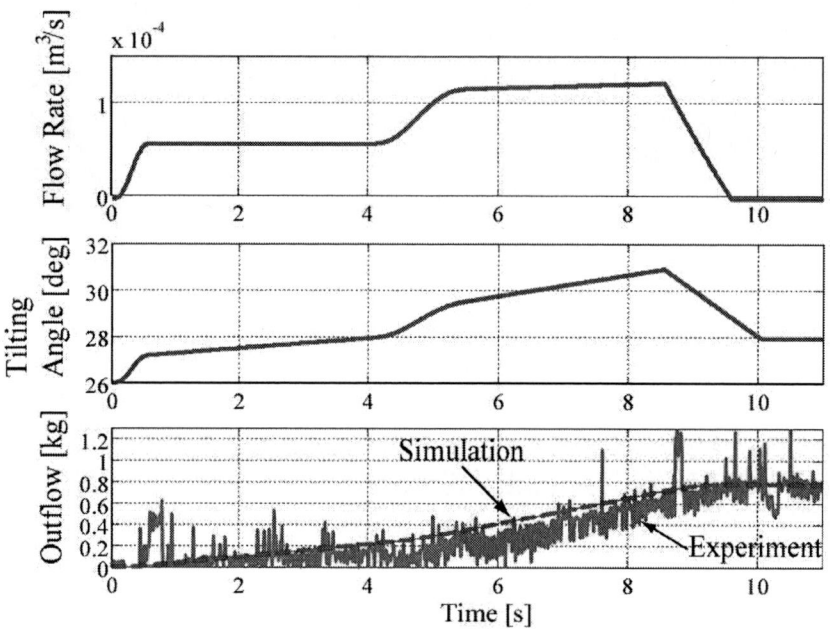

Fig.9 Experimental Results by Using Water

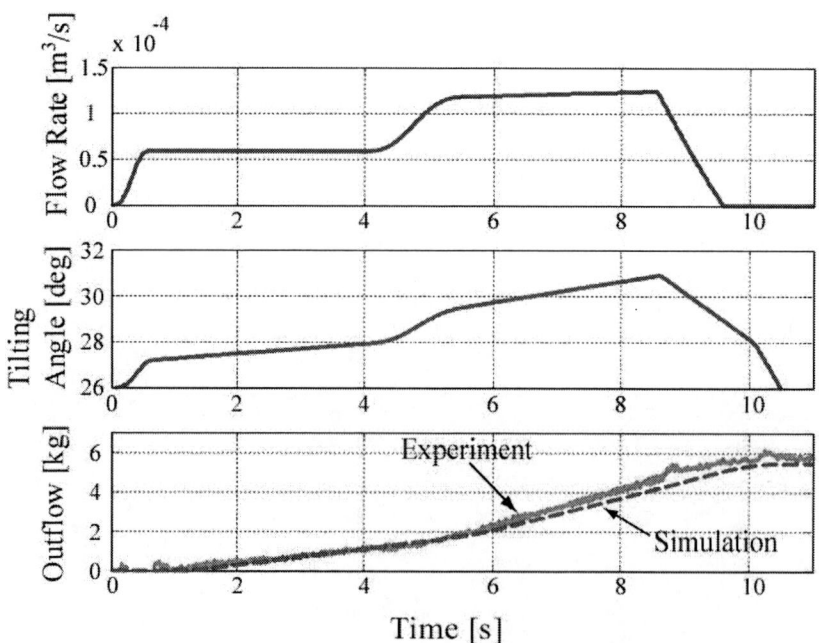

Fig.10 Experimental Results by Using Molten Metal

Inorganic binders: Properties and experience

K Löchte *, R Boehm *

* Hüttenes-Albertus Chemische Werke GmbH, Germany

Abstract

Foundries are constantly presented with new challenges in terms of design requirements and in connection with the increasingly stringent environmental and safety regulations.

Particularly problematic are evaporation, low-temperature carbonization and cracking products released during casting and subsequent cooling from the sand cores, which today are mainly produced from resin-bonded silica sands.

Emissions and the related odours are not only released from the sand cores during casting, but also during the coremaking process. Although major efforts have been undertaken and considerable improvements in the overall process achieved during the last few years, there are still a number of problems to be solved. As a result the demand for alternative binder systems has increased during the last few years, with especially inorganic binders moving into the centre of attention.

Their use holds out the expectation of lower emissions and less odours arising during coremaking and, above all, during casting, especially of aluminium.

Key words

inorganic binder – Cordis – productivity – environmental-friendly - condensate

Introduction

In November 2002 a much recognized VDG conference, which was to guide inorganic core binders out of their niche position, discussed the use of inorganic binders in foundries under the fateful question "Inorganic binders - breakthrough or everlasting hope?" [1].

Suppliers of inorganic binders presented their binder systems and the application potential of the new products. The great interest in and the positive response to this topic also became obvious at GIFA 2003, where it was a central issue [2].

At both the VDG conference and at GIFA 2003 the "Cordis" inorganic binder system was introduced to experts of the industry.

To promote the market breakthrough of inorganic binders, the development of this system was aimed at using existing and established techniques and equipment in the manufacture and handling of cores made with inorganic binders.

The idea is to use the binders in conventional core shooting machines and decore the castings on equipment existing in the foundries, the preferred process here being that of dry decoring.

The fundamental precondition for the establishment of an inorganic binder in the market is, however, that it produces casting qualities comparable to those achieved with the organic systems currently available on the market. With these aspects in mind, further development of the binder has been intensively pursued especially during the last few years.

Experimental

The basic development of this binder took already place in the 1990ies.

The fundamental idea was to realize a binder system with water as the only solvent and a completely inorganic binder matrix. This idea has completely materialized in the most recent generation.

Depending on the binder type, the matrix consists of a combination of phosphate, silicate and borate groups. The properties of the individual binder types can be set by objectively combining and varying the structure of the binder matrix. Further inorganic additives, either integrated in the binder or directly added during the coremaking process, are used to further enhance and sophisticate the properties. Additive combinations enable the objective setting of important properties such as flowability of the moulding material mixture or the storability of the cores (Figure 1 a+b).

Especially the use of inorganic binders calls for excellent homogeneity of the moulding material mixture. Therefore it is recommended that the mixture be prepared in a suitable mixing unit.

The cores can be made in commercial core shooting machines (provided that they include a heated core box).

The moulding material is blown into a heated core box. Depending on the core geometry, the core box temperature is between 120 and 160 °C.

A homogeneous heat distribution in the core box is to be aimed at. When the moulding material mixture, the sand and aqueous binder are in the box, a hardened case forms along the outer contour of the core (Figure 2). This first curing step is the effect of a drying process resulting from the extraction of moisture - the solvent water - from the core. This is a purely physical phenomenon.

In addition to this, depending on the binder type used, a chemical curing process may be initiated. Then the core's handling strength becomes higher than in the case of a purely physical strengthening process (drying).

Depending on the used sand and type of binder, with binder additions between 1.5 and 3.0 % (volume fraction) cold bending strengths between 350 and 550 N/cm² are achieved.

The curing process can be strongly accelerated when the core is gassed with hot air.

The gassing improves the heating of the moulding material and at the same time efficiently drains off the water released during the curing of the core. The curing time of inorganic cores very strongly depends on their geometry and the used core boxes. The larger and more compact a core is, the more difficult the setting process becomes. Due to its good flowability, this inorganic binder can be used without problem for cores with filigree contours, such as water jacket cores.

In addition to intake manifolds (Figure 3), which are already produced in series, the binders have been used for a wide range of other core types. They range from highly filigree cores, such as retarders and channel cores, via cores for chassis parts through to complete core assemblies for cylinder heads.

In aluminium alloy casting uncoated Cordis cores can be used.

Should it still be necessary to use coatings, both alcohol and water-based coating are suitable.

Similar to organic cores, inorganic cores loose a little bit of their strength when being coated.

With water-based coatings the loss of strength is higher than with alcohol-based coatings.

Special attention must be paid to the storage of the cores made with inorganic binders. Due to their hydrophilic property (the binder solvent is water) they are very susceptible to high atmospheric humidities.

This is why inorganic cores should not be exposed to the open air or extreme atmospheric humidities over longer periods.

The development of inorganic binder systems therefore includes a storability test by means of test bars held in simple climate or humidity chambers, in which they are stored in atmospheres of defined humidities. After this, the properties of the test bars are checked. In comparison with other inorganic binders, the Cordis binder has proved resistant to elevated air humidities in laboratory tests (Figure 4).

The cores stored in an atmosphere of high relative humidity for 24 h loose only about one third of their strength.

The environmental advantages of inorganic binders in cores become even more salient during the casting than during the coremaking process. A cylinder head was gravity die cast in aluminium with a complete core package. No fumes or odours arose during the casting process.

Moreover, even after several castings there were no condensates in the die.

Discussion

It is a wide-spread opinion and a controversially discussed issue that, in contrast to organic cores, inorganic cores can be cast without a gas shock. Investigations have, however, demonstrated that a gas shock can also be observed with inorganic cores.

A very impressive observation here is the dependence of the gas shock on the grain size of the used sand. With coarse-grained sand (AFS 43) the gas shock is much less intensive than with fine-grained sand (AFS 74) (Figure 5).

The fact that both specimens contain identical amounts of binders makes clear that the grain size of the sand has a dominant effect on the intensity of the gas pressure.

The cores exhibit different gas shock behaviour than organic cores.

Hot box cores, for example, start to gas as soon as they get into contact with the melt - and then continue to gas. Inorganic cores, on the other hand, cause the gas pressure to rise only after approx. 30-40 s. The pressure-falls again as soon as the pressure peak is reached (Figure 6).

Results

Besides the already described positive experience with these cores used in aluminium gravity die casting, also in aluminium low-pressure die casting and aluminium green sand casting very good results have been achieved. The casting surfaces are comparable with or better than those achieved with organic cores.

Decoring does not pose any problems, as the available decoring equipment can be continued to be used. The castings can be decored without any prior heat treatment.

It is, however, recommended that foundries using reclaimed sand should handle organic and inorganic core sands separately due to their different binder matrixes.

Conclusions

The use of inorganic binders in coremaking has already reached a high level of sophistication, but calls for a new approach to the process chain. At the same time it must be admitted that the performance potential of organic binders (making and storage of the cores) has not been entirely explored to date.

The breakthrough of inorganic binders is nearing. However, the related cost and the requirements on the equipment are not to be underestimated.

References

1. *Umweltverträgliche anorganische Bindemittel zur Form – und Kernherstellung.* Wuppertal, 14.11.2002.: Giesserei 90 (2003) Nr. 10, S. 42 – 46.
2. Franken M, Giesserei 90 (2003) Nr.6, S. 182 – 184

Acknowledgements

We would like to express our special thanks to the companies supporting our development activities, such as Volkswagen Gießerei Hannover, Rautenbach-Guss Wernigerode GmbH, Dipl.-Ing. Roman Kohlisch, Cold Box Coremaking, Eurokern Gießereitechnik GmbH and Hähnel & Leon GmbH Modell- und Werkzeugbau.

Tables and Figures

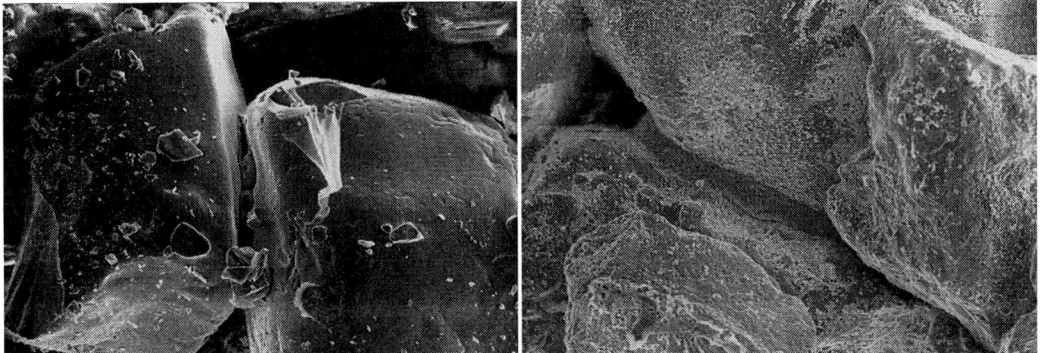

Figure 1 a+b: SEM images showing the binder bridges of two different Cordis binder types

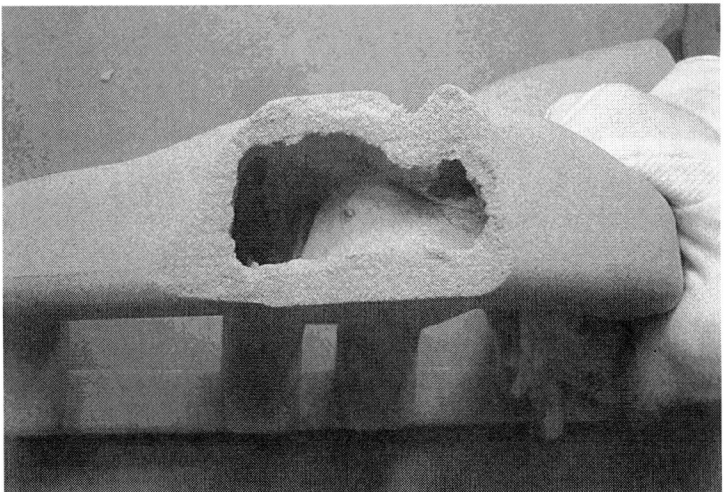

Figure 2: A Cordis intake manifold core forms a hardened case immediately after being taken out of the machine

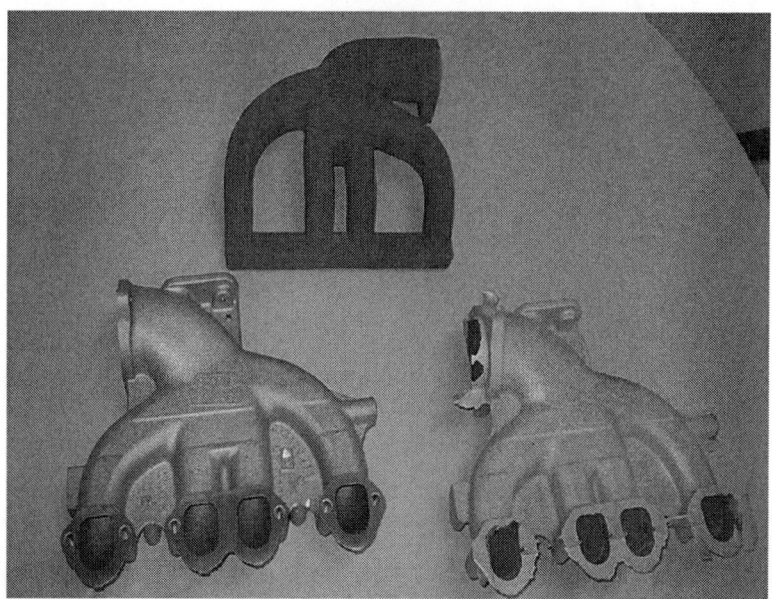

Figure 3: Intake manifold casting using the Cordis binder system: Cordis core (top), Cordis core after casting (right), casting produced using a Cordis core (left)

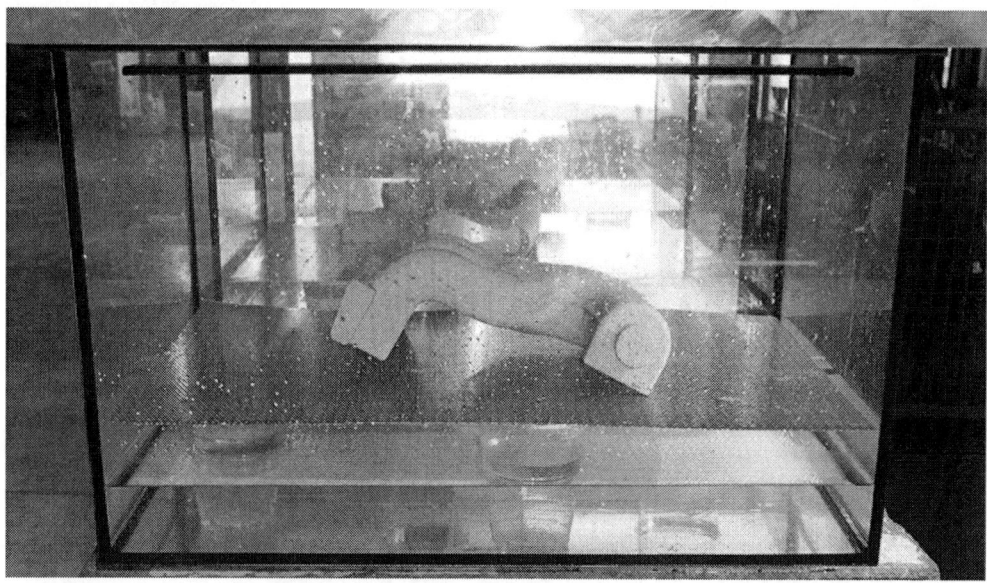

Figure 4: Storability test of a Cordis core in a high-humidity climate chamber

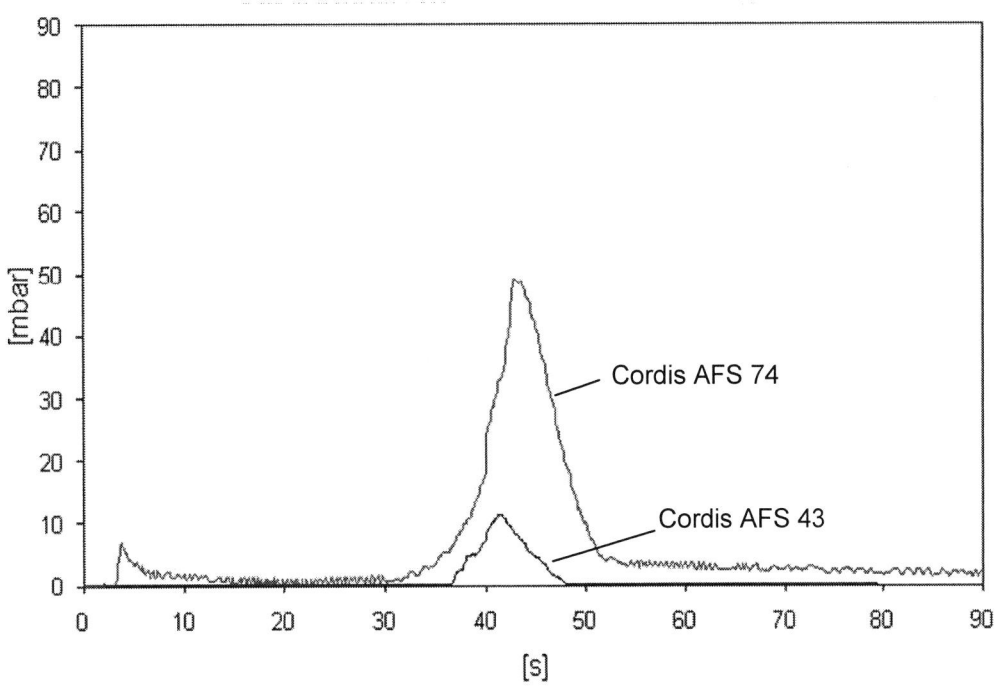

Figure 5: Gas pressure curves of two Cordis cores with different grain sizes: AFS 74 (fine-grained sand); AFS 43 (coarse-grained sand)

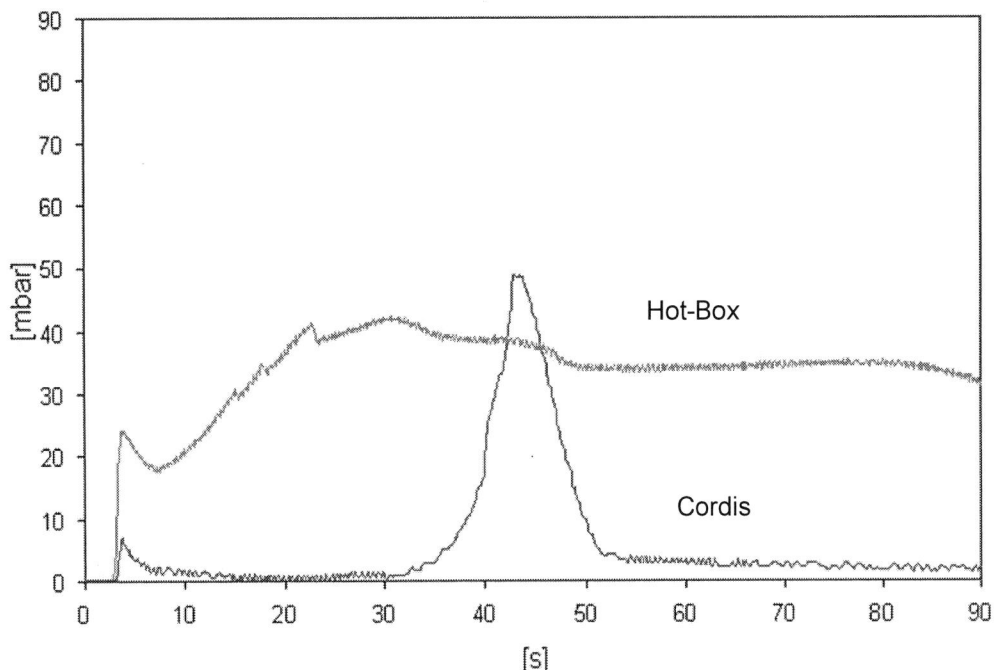

Figure 6: Gas shock with Hot-Box and Cordis cores

kloechte@huettenes-albertus.com
rboehm@huettenes-albertus.com

Inclusive, Innovative and Sustainable Environmental Management System

V Narasimhan
Brakes India Limited, Foundry Division, Sholinghur, India.

Abstract

Brakes India Foundry (BIF) has a magnetic vision, which is a synthesis of the visions of all the stakeholders. BIF is proud of its role as a responsible corporate citizen. BIF believes that profit should not be the only motive and the company should have noble goals beyond the immediate financial results. BIF is fully committed to conservation of non-renewable resources, reduction of pollution, continually improving on the Environmental performance and the total involvement of everyone in this effort. BIF has become a zero discharge foundry of pollutants and solid wastes and taking advantage of the recyclability, is working towards creating a new value proposition. BIF have succeeded in developing stabilized mud blocks and green belt from solid waste. The accent has been on innovation, involvement of the villagers and unique Environmental Management System.

Key words:

Responsible Corporate Citizen
Conservation of non-renewable resources
Total involvement
Zero discharge foundry
Recyclability

Introduction
BIF is involved in the manufacture of Permanent mould grey iron castings and High pressure moulded grey iron and ductile iron castings. The wastes generated inside the premises and their applications are as follows

Source of generation		Application
Rejected sand from sand plant And Slag from Cupola	⇒	Brick manufacturing, Concrete manufacturing, Construction of roads, Development of green belt.
Dismantled lining material from furnaces, ladles, HMR and Tundish	⇒	Construction
Withered leaves and discharge from S.T.P.	⇒	Manure for Green Belt Development.
Packing wood	⇒	For Making Stillages
Stationary waste from office	⇒	Recycling

Experiments
Idea :
Gainful use of the waste generated inside the foundry such as return sand and slag for construction.

Experiment:
An experiment was conducted with return sand & slag from foundry, soil and cement by mixing it in a proprietary proportion to make bricks. The brick was analysed at Indian Institute of Science, Bangalore with various mixing proportions by varying the curing time and size to obtain optimal properties of brick such as compression strength, wet & dry density, percentage of water content.

Designed Experiment with different levels of ingredients.

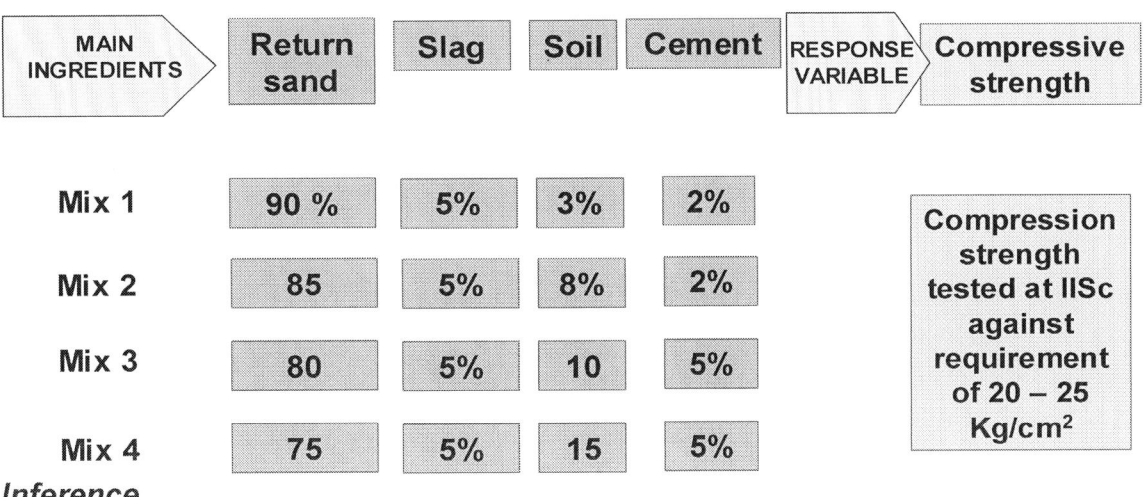

	Return sand	Slag	Soil	Cement		Compressive strength
MAIN INGREDIENTS					RESPONSE VARIABLE	
Mix 1	90 %	5%	3%	2%		Compression strength tested at IISc against requirement of 20 – 25 Kg/cm²
Mix 2	85	5%	8%	2%		
Mix 3	80	5%	10	5%		
Mix 4	75	5%	15	5%		

Inference

MAIN INGREDIENTS ▷	Return sand	Slag	Soil	Cement	RESULT ▷	Compressive strength
Mix 1	90 %	5%	3%	2%		<18 Kg/cm^2
Mix 2	85	5%	8%	2%		20-25 Kg/cm^2
Mix 3	80	5%	10	5%		22-27 Kg/cm^2
Mix 4	75	5%	15	5%		30-35 Kg/cm^2

Mix 4 with return sand 75%, slag 5%, soil 15% and cement 5% has yielded very high compressive strength, hence the "Stabilised Mud Block Technology"was devised to manufacture the high strength bricks.

The desired properties like wet strength and dry density were obtained by varying the curing time.

Table 1

S No	Curing time	Size (cm)	Density (gm/cc)		Water content (%)	Wet strength (Kg/cm^2)
			Wet	Dry		
1	7 days	7.6 X 7.6 X 7.6	2.11	1.86	13.65	18.21
2	21 days	7.6 X 7.6 X 7.6	2.12	1.86	14.04	23.53

Fig 1

Brick – Manufacturing flow

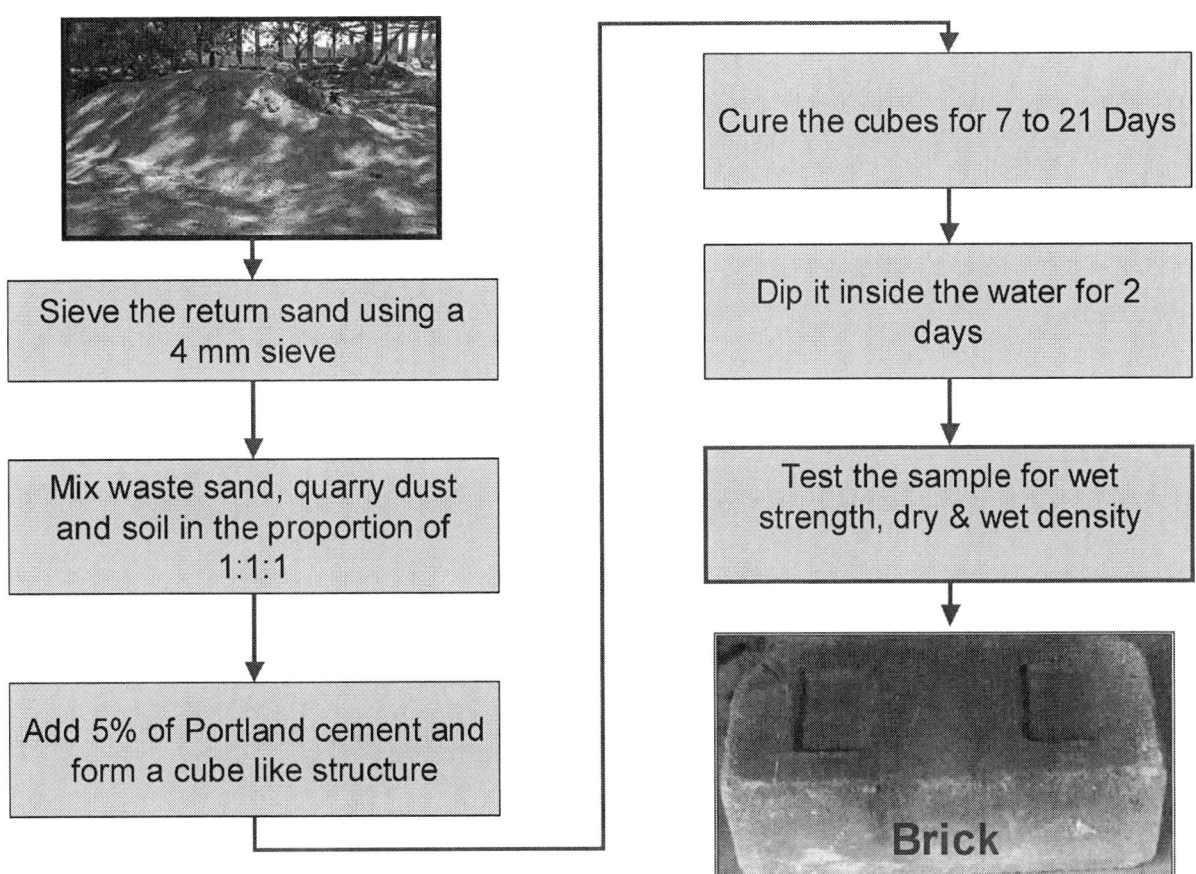

Sieve the return sand using a 4 mm sieve

Mix waste sand, quarry dust and soil in the proportion of 1:1:1

Add 5% of Portland cement and form a cube like structure

Cure the cubes for 7 to 21 Days

Dip it inside the water for 2 days

Test the sample for wet strength, dry & wet density

Brick

Results

Table 2

Property	Burnt brick	Brick made of return sand
Composition	Red earth	Soil ~20% +80% return sand+ 5% cement
Specific gravity	2.0 g/cc	2.1g/cc
Water absorption ratio	10 to 15%	2 to 3%
Compressive strength	20 to 30 kg/ cm^2	30 to 35 kg/cm^2 with 5% cement

Advantages:

- ➤ No air pollution during the manufacturing process
- ➤ No fuel is required for manufacturing. Burning firewood to dry the bricks was avoided.
- ➤ This can be manufactured in the site itself.
- ➤ Curved profiles like cornice blocks, corbelled blocks can be made
- ➤ Variation in the size is minimum (< 0.5 mm in length and breath and <1.5 mm in depth)

Applications:

Used for the construction of buildings:

Fig. 2 Dining Hall

Fig. 3 Boiler Room

Fig. 4. Canteen

Fig. 5 Training Centre

Conclusions

From the experimental analysis one can conclude that, with Standard Operating Procedures, the solid wastes generated in the foundry such as return sand and slag can be utilized to manufacture bricks at the lowest cost.

BIF is on the process of training the villagers in this sustainable technology so that it can be commercialised to provide gainful employment. By implementing such techniques, we have become a ZERO discharge company

Discussion

1. Stabilized mud block technology

➤ The above results reveal that there is no significant deviation in the properties of brick as follows

➤ Only 4% variation in the specific gravity of the brick

➤ Water absorption ratio was also good when compared to conventional burnt brick;

To differentiate between light / medium load bearing walls different mixing propositions as illustrated below were identified.

i. Walls bearing no load (Fig: 6)

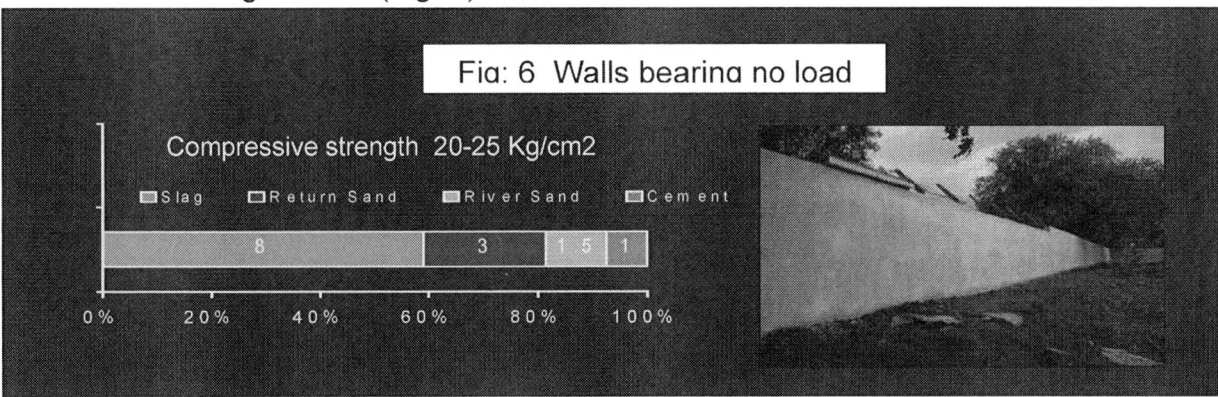

Fig: 6 Walls bearing no load

ii. Walls bearing light load

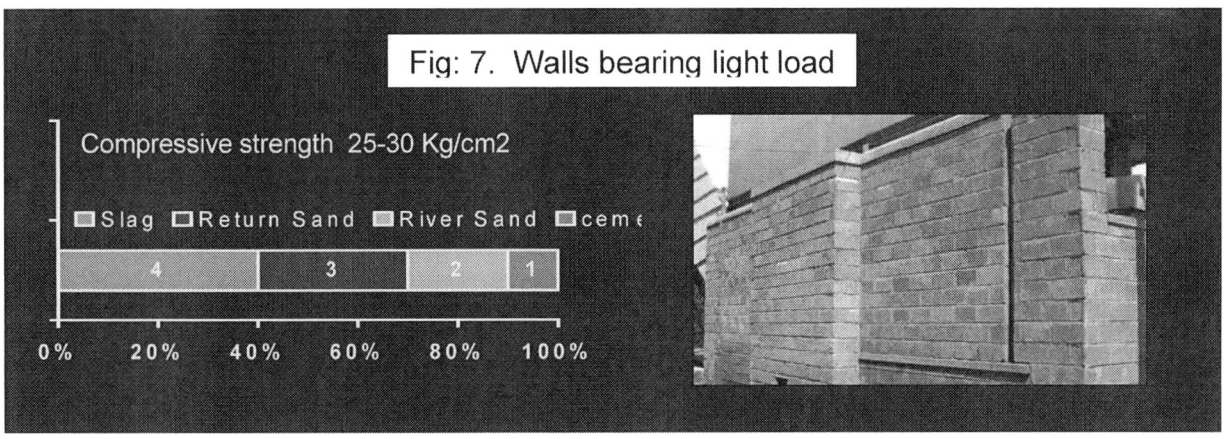

Fig: 7. Walls bearing light load

iii. Walls bearing medium load.

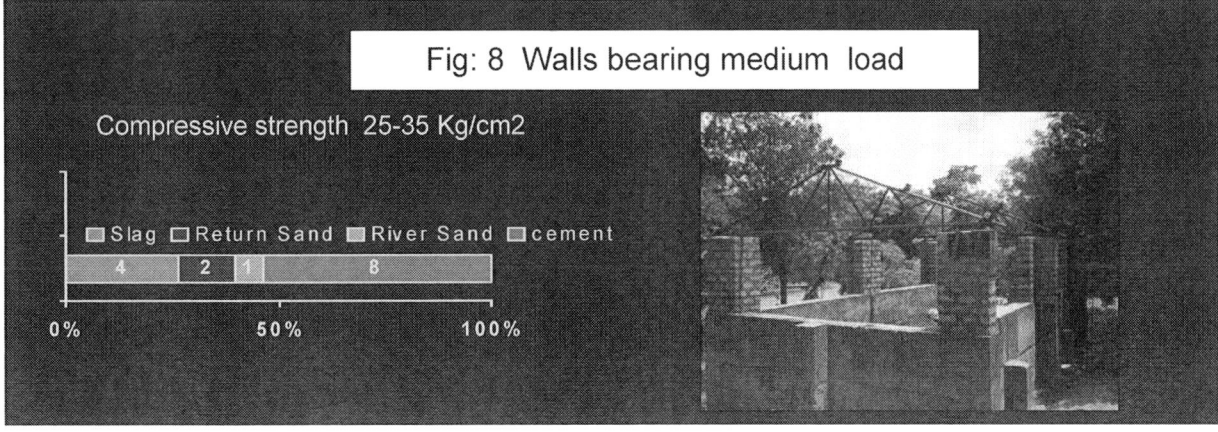

Fig: 8 Walls bearing medium load

Normally the return sand fines can only used for landfills. But since the fines contains bentonite and no harmful substance, an experiment was made to develop a lawn with return sand. Bentonite retains moisture and the amount of water required to tend the lawn reduced considerably the experiment was repeated for greens, kitchen garden and flowering trees.

In India we are wasting power for pumping water from 100m depth and wasting it on non-cultivable land. Is this innovation the answer to Indian scene? BIF is developing a green belt with return sand over an extent of 10 Hectares.

Fig 9: Promotion of green belt from solid wastes

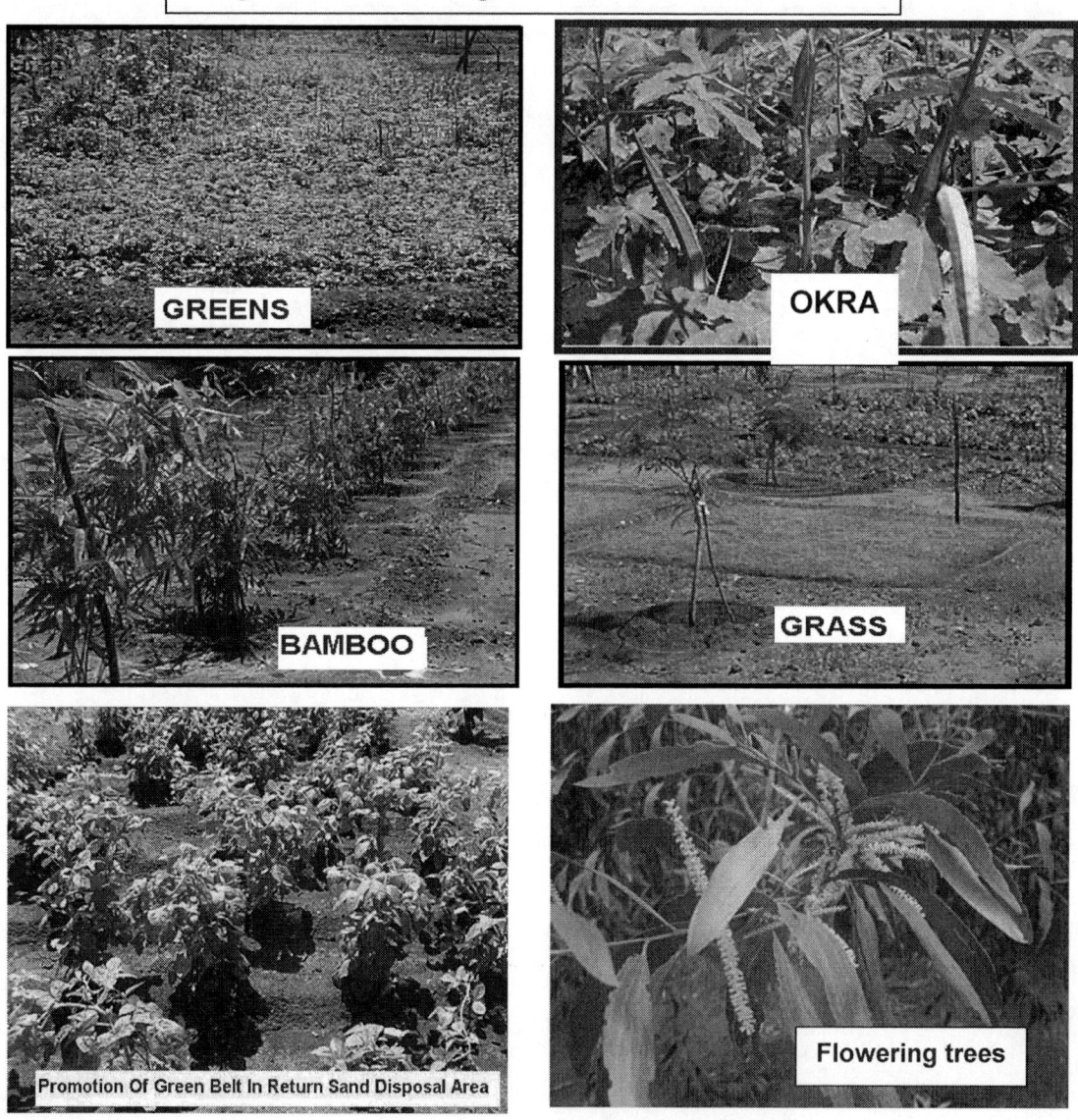

References
1.   IS 1077:1992 common burnt clay building bricks (Fifth revision)
2.   IS 3495:1992 Methods of testing for burnt clay building bricks
3.   Guidance from the professors of IISc Bangalore and
4.   Technical input from Auroville Pondicherry.

Foundry Management by Internet

T D Law
AFSoftware Limited

Abstract:

This paper argues that computer software, including programs designed specifically for foundries, has 'grown up' over the past few years. It is now quite straightforward to supply and support foundry management software, 'at a distance' – even when the distances involved are world-spanning. This is true of course, only if the software supplied is a **true package** which is easily supportable, in the way that an accounting package is a true package where all users operate with an identical software product...

The suppliers of a major foundry management package (Synchro 32 for Windows), now support users of their software on **four continents** via Internet technology. For the past five years this has been achieved with remarkable success. Customer responses show the high satisfaction levels achieved to date in Europe, Australia and the Americas – while Asian experience is as yet too recent to have been assessed. Previous attempts at supporting software via conventional methods had been very unsatisfactory when great distances were involved!

The main mechanism, which has revolutionised software support, is the **Internet** – involving the use of websites, software downloads and email data transfer.

The paper uses case-studies at three very different foundries in countries on three continents – in Western Australia, in the American Mid-West, and back home in the UK, to show the problems that were encountered and how they have been overcome.

The case-studies encompass several major areas of management control including costing, production scheduling, quality assurance and commercial management in foundries involving sand casting, investment casting and die casting.

Keywords: Foundry, Management, Software, Worldwide, Internet

Foundry Computing:

Foundries use software of three distinct types only one of which is both foundry specific <u>and</u> requires support worldwide

- Commercial software (accounts packages, payroll systems, etc) are usually limited to use in one country for legislative reasons. They also operate no differently in foundries than in any other industry. As a result locally produced and supported systems are the norm and there is little call for support of individual packages on a worldwide basis.

- Foundry Technical Software (methoding, castings design and solidification simulation systems) are highly specific to the demands of the foundry industry and are straightforward to support regardless of the location of the user and the support team.

- Foundry Management Software (order processing, production scheduling, production monitoring, and invoice creation) are both industry specific and critical to the wellbeing of the company. As a result any failure of the system or non-availability of software support when needed is crucial to the foundry's existence.

Major Characteristics of Foundry Management Software:

Foundry management software is less proscribed in its design than either accounting software or technical design software. Accounts software must obey both legal requirements and accounting custom and practice. Technical design software must follow the immutable laws of physics and thermodynamics. While foundry management software has certain general requirements regardless of the type of foundry (the need to track orders for instance) and common management reporting requirements (such as a list of jobs still to be made), there is no requirement for conformance to a given system design.

The following are critical to a successful foundry management system:

1. The creation of temporary tooling (moulds) must be allowed for – this is a unique characteristic of castings manufacture.

2. Casting designs are usually exclusive to a customer. Products therefore are normally made against customer orders rather than for 'stock' as is the case in most other industries.

3. Foundry operation rarely allows accurate data recording. In fact accurate 'shop floor' feedback is often a major problem.

518

4. MRP software widely used in many industries is all-but useless for foundries because of it's poor 'fit' to their needs. One foundry manager recently lamented (within the hearing of the author) the behaviour of his foundry's MRP system... "Our (expletive deleted) system demands that we instruct it to **make some moulding sand** before it allows us to schedule a batch of castings"

5. Failure of a production control system means shipments do not occur, invoices are unprocessed and people and machines stand idle. Company health is rapidly and seriously affected.

The ways in which foundry management systems normally satisfy these needs are dealt with in the following section:

The Basic Functions of Foundry Operations Management:

Leaving aside the associated functions of computerised costing and estimating, Computerised Foundry Management software normally creates a plan for manufacture that satisfies the demands of a Quality Assurance (QA) System, while processing jobs through production in an efficient manner. The major component parts of a foundry manufacturing system are outlined in Fig 1 and described below:

Stage 1: Customer orders are 'translated' by an order entry program that checks against a database to ensure details of products, prices, materials etc are correct. This satisfies the requirement for 'Contract Review' under the QA system, by checking out the demands of the new order with data held on the software system database (Fig 2). Finally the computer system can undertake preparation and printing of a 'Customer Order Acknowledgement' document.

Stage 2: New orders are integrated into the overall production plan by a 'scheduling' facility in the computer program. This normally requires a dialogue between human planner and computer to achieve the best result. Graphical planning tools (see Fig 3) help an efficient dialogue. A printed 'Make to List' will then be issued.

Stage 3: Any materials required for manufacture will be identified and listed for purchasing etc. If required, the computer program can undertake purchase ordering and stock control.

Stage 4: Traveller documents, often required to ensure correct processing of components, can now be printed – with digital photos if needed. The computer system may also hold video clips of critical operations to ensure no one forgets how to perform critical activities.

Stage 5: The job now goes to the 'shop floor' and W.I.P data is entered to the computer system so the order processing status can be updated. Castings made, rejected and ready for shipment are recorded.

Stage 6: The computer analyses the data and prepares management reports, detailing production efficiency, problems and necessary actions.

Stage 7: The computer prepares shipping documents: advice notes, invoices, test certificates, etc. The customer details, addresses, product weights, material types etc all come from the database.

Additional Functions of a Foundry Operations Management System:

Increasingly foundries require linkage of production to an accounting system, integrating with an estimating and quoting program, or providing connectivity to their major customers, via a website presence.

Some foundry computer systems offer their own accounting functions, others (including the globally applied subject of this paper), provide linkage to any accounting program of choice. This is important when installing in different countries. Few foundries would wish to use an overseas accounting software package, so linkage to a system of choice is critical

Estimating calculations in foundries are often unique and repetitive. This mitigates towards a computerized estimating system, but specifically demands that different arithmetic can be set up easily for each user foundry. The system discussed here uses a spreadsheet integrated with the package. Fig 4 shows a sample spreadsheet used in an iron foundry.

Purchasing and inventory control systems integrated with production scheduling are now seen as important in foundries, as is the need to audit supplier performance – one of the key requirements of ISO 9000 and associated quality standards. Timely ordering of materials needed for recently added orders can be ensured routinely by the system.

Creation of Test Certificates is often important to foundries. A computer driven management system should have the ability to link with any spectrographic equipment and produce suitable certification to accompany shipments of castings.

Internet-enabled foundry programs allow website display of current orders, production and delivery status etc, where they can be accessed by (empowered) customers. As yet fairly uncommon in foundries, such facilities will become important under pressure from the more demanding customers. The rapid growth in numbers of broadband connections will certainly accelerate use of this feature in future.

Development of Global Foundry Software:

There are several reasons why UK developed foundry software has become a leading player in foundry industries around the world.

Language: The dominant status of English worldwide meant that only a program based on English would have a sufficiently widespread uptake to ensure an initial spread around the globe. Since 1998 the software described here has been used in more than a hundred foundries in USA, Canada, UK, Australia, New Zealand, Mexico, West Indies, and is now beginning to spread into non-English speaking areas with newly developed 'Language Versions' of the package. .

Geography and Time-Zone: For many years geography was a major stumbling block to providing successful ongoing support for foundry management software operating around the world (Figure 5). Expensive telephony and the use of airmail for update programs meant that in the 1980s and early 1990s support of far-flung users was virtually impossible.

In the last few years the availability of Internet and Email services, plus cheap telephony (VoIP etc) have meant that support of software users half a world away is a practical proposition.

The Effect of System Design: The choice of programming language and associated tools for functions such as management reporting and document generation is of major importance. Also critical is the choice of operating system, although currently Windows is a default choice. A powerful and easy to use report generator (Crystal Reports) allows customers to customize their own system.

The Influence of Internet, Email and Computer to Computer Linking:

Support Website: The Internet has revolutionised software support. Websites devoted to support of far-flung customers are becoming commonplace. Updates and corrections can be provided regularly and instantaneously to all customers. This is vastly different to the early '80s when an airmailed program update could take upwards of a week to reach a recipient in Australia.

Email Facilities: Email systems impact powerfully on long distance support. Rather than a program support team waiting for telephone calls, now all support calls are either via emails or digitally recorded phone messages from customers sent to the support team as email attachments.

Computer to Computer Links: Recently released software simplifies linking together computers in different parts of the world. Training can be conducted at distance - both users sharing a particular screen and software but having *joint control* via dual keyboards and dual mice. A voice connection – often via a free-to-use *VoIP (Voice over Internet Protocol)* service completes the linkage.

User Support Case Studies:

Currently there are more than one hundred examples of the Synchro 32 system operating around the world. All users have stories to tell about their installation and implementation experience, including those below...

USA – T&L Foundries, Inc:

This aluminium sand-caster from Oklahoma attended a software seminar in Chicago during late 1999; following which they decided to install a system in mid 2000. They had previously used Excel based systems. All of their spreadsheet data was transferred to the new system database, cutting down dramatically on data entry. Two on-site installation sessions, each of three days were required to go live.

Problems experienced were limited to hardware a malfunction due to increased traffic on the original network – a consequence of more enthusiastic user activity. In the first twelve months, T & L linked from their new system to an existing accounts package with help from the supplier.

UK – Lestercast Limited:

Lestercast is an investment foundry, based in Leicester, UK. They demonstrate the rapid progress possible when everyone is committed to successful implementation.

Lestercast was purchased in mid-2001 by new owners, who had previous experience of working with the author's company at other foundries.

An early decision that a production control system was essential was followed by rental of the software package. Database creation took eight weeks. By December 2001 the system was ready to go live and this coincided with the installation of a new computer network at the company.

Australia – Wundowie Foundry Pty Ltd:

Wundowie Foundry Pty, an iron and steel foundry based in Western Australia had used an earlier DOS-based system from the author's company since 1995. They had frequently complained about support difficulties due to time-zone differences and program update response.

Since 2001 the company has been supported by Internet and email - initially direct from the UK, but since 2003 with assistance from a Brisbane (Queensland) based local agent for some operational questions.

They have reported dramatic improvements in support performance, credited equally to the effect of Internet/Email – based support, and the availability of a local agent.

User Support Feedback from Customers:

While individual case studies are interesting, a far better view can be obtained by systematic collection of customer opinions on a periodic basis, using a structured approach. With this in mind surveys of customer opinion have been conducted, (Fig 6) using the support website to ask on each occasion:

"How well are we doing on new development and support?" Give us a 'mark out of 5' where 1 = Awful and 5 = Excellent. (See Fig 7)

In practice results have averaged between 4 and 4.5 out of 5, for all respondents. Almost all users have responded to the questionnaire.

"Which functional areas would you like to see tackled next?" – Users are asked about several potential areas of improvement or additional functionality and provided clear guidance. Obviously not all users will agree with the priorities given by other foundries – but there is often a fair measure of agreement (see Fig 7). The majority of subsequent development has been based upon such customer feedback.

Conclusions:

The author believes that valuable conclusions can be drawn from the experience related in this paper. While it is tempting to attribute benefits to technological improvements alone, it would not be accurate... Substantial benefits have also arisen from the following...

Standardisation and a refusal to entertain **'one-off modifications'** - however beguiling the request – has paid great dividends. Who would now consider making custom changes to an accounts package?
If a standardized computing approach is necessary with accounts, it is also necessary for production software.

Systems Analysts must accept responsibility for controlling software development. Programmers should never be allowed the freedom to **'make improvements'** because they often fail to see the bigger picture.

Regular updates of the software package are the corollary of standardization. So far it has been found expedient to issue monthly updates to every user foundry simultaneously via the Internet website.

Routine Feedback from customers should be sought regularly. If you don't know you are going wrong, you can't fix it, so asking "how well are we doing" and "what else can we do" should help considerably.
If these questions are asked and honest attempts are made to respond to the findings, we believe the product and its support service will not go too far wrong!

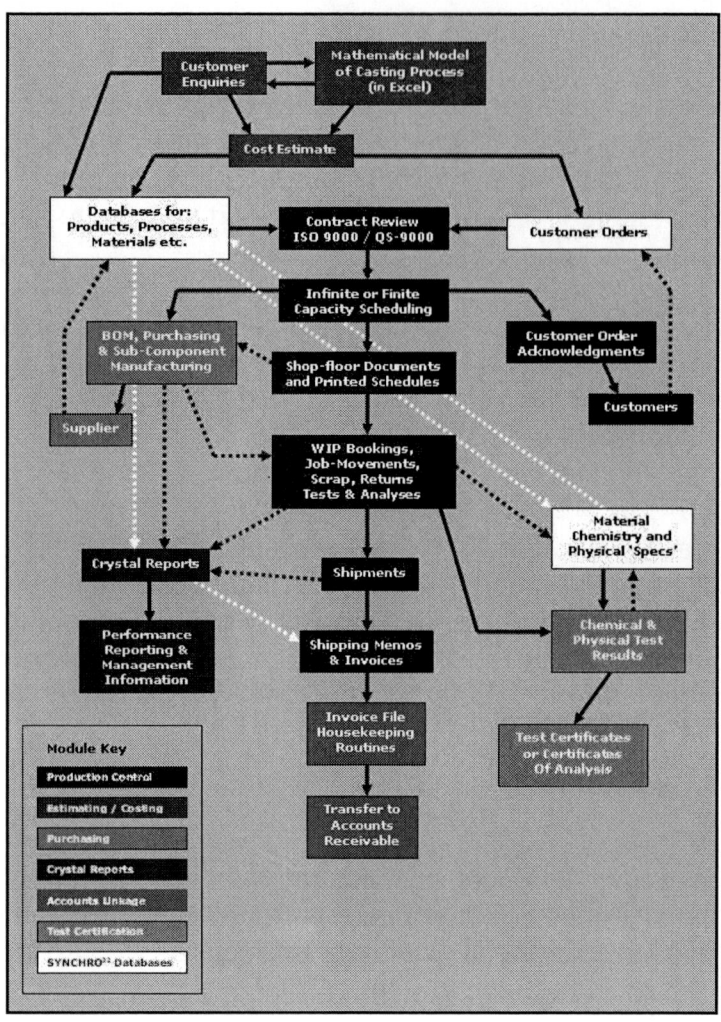

Fig 1. Major components of a foundry management system

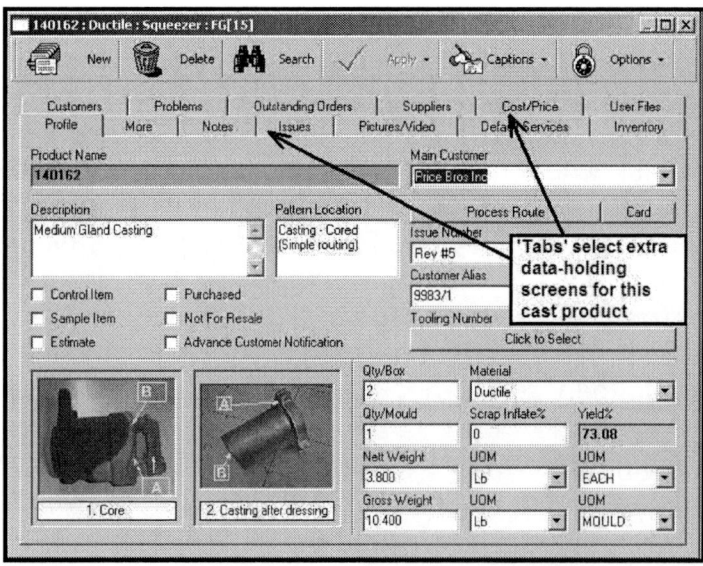

Fig 2. A typical product record in a computerized foundry database

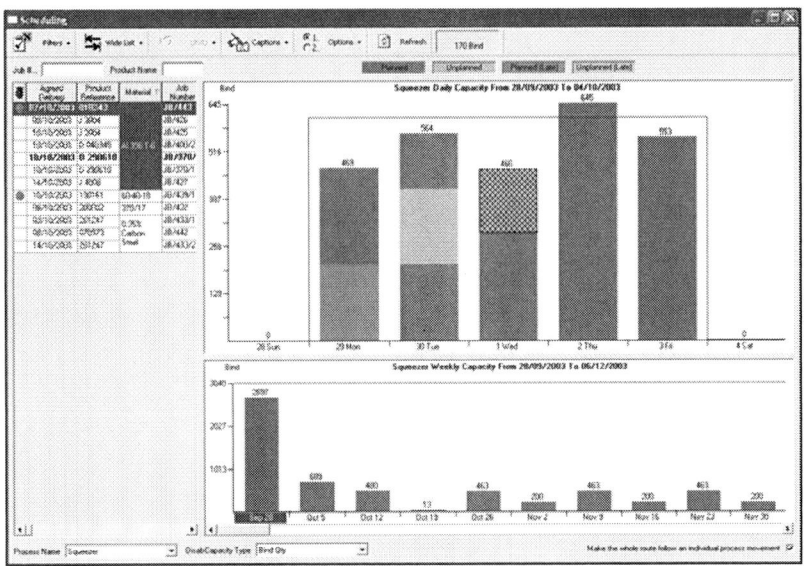

Fig 3. A graphical planning tool for foundry scheduling

	A	B	C	D	E
1		SAND DEMO ESTIMATE			
2	Product Name	140162		Customer Ref:	ST ESTIMATE G78225
3	Description	Medium Gland Casting		Our Ref:	CE/7
4	Metal Type	Ductile		Enquiry Date	3-Jun-03
5	Net Weight	3.80		Box Weight	10.4
6	No per Box	2		Yield %	73.08
7	Material Description	Material Weight		Material Cost	Value
8	Ductile	10.40		1.70	17.68
9	Metal Returns	2.80		0.64	-1.79
10	0	0.00		0.00	0.00
11	0	0.00		0.00	0.00
12	0	0.00		0.00	0.00
13				Material Cost	15.89
14					
15	Process	Time		Rate	Value
16	Tooling Setup	0.00		0.55	0.00
17	CO2 Core	4.00		0.55	2.20
18	Shell Core	9.00		0.55	4.95
19	Core Dressing	2.00		0.55	1.10
20	Pin Lift	16.00		0.55	8.80
21	Dressing / Cleaning	12.00		0.55	6.60
22	Assembly	0.00		0.55	0.00
23	Inspection	3.00		0.55	1.65
24	Shipping	1.50		0.55	0.83
25	0	0.00		0.55	0.00
26				Process Cost	26.13
27					
28	Additional Charge	Charge Rate		Based on	Value
29	Scrap %	0.00		39.58	0.00
30	Admin Charge	1.50		42.01	0.63
31				Extras Total	0.63
32					
33				Total Cost	42.64
34					
35	Rep Commission	2.50			
36	Profit Margin (%)	Selling Price		Profit Margin	Selling Price
37	10	48.60		25	58.32
38	15	51.45		30	62.48
39	20	54.67		40	72.89

Fig 4. A typical cost estimating spreadsheet for an iron foundry

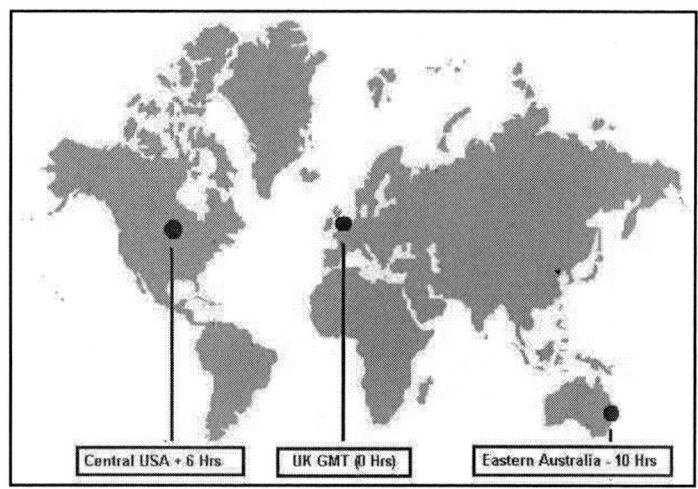

Fig 5. Time-Zones for software user support

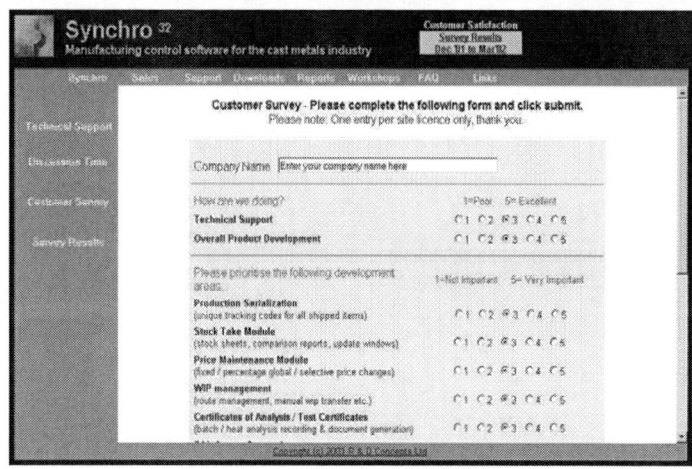

Fig 6. Questions posed to customers in support performance survey

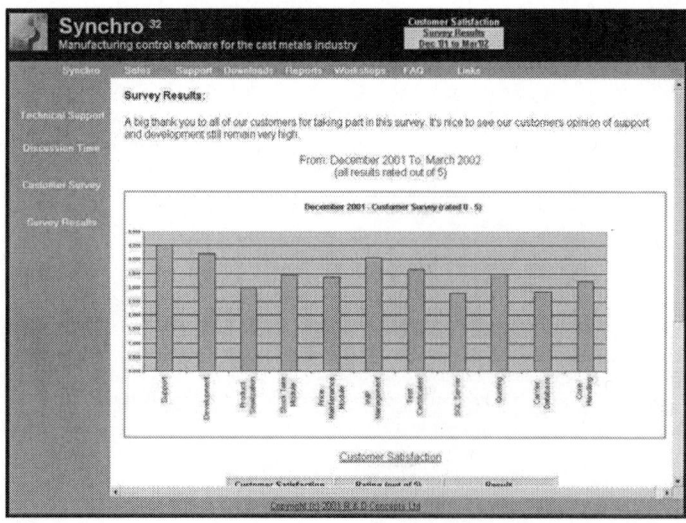

Fig 7. Summary of answers in support performance survey

The South African Aluminium Foundry Industry - An International Perspective"

Dr Tom Paterson, FIMMM, FICME, C.Sci
Striko Westoven GmbH.

This paper will consider the question 'can South Africa make it as a global supplier of aluminium castings'. Points shall be raised concerning how important castings are to designers and purchasers of components and a consideration of the foundries of the future.

There are a number of key challenges to be overcome by the South African Foundry Industry and the author will discuss how South African foundries investing in new technology can become global guppliers of castings. Although primarily focusing on South Africa, the questions raised are also relevant to foundries in the world over.

A novel aluminium matrix composites synthesized by magnetochemical melt reaction in the system Al-Zr-O-B

Y T Zhao *, Y M Youssef **, R W Hamilton ** and P D Lee **[1].

* School of Materials Science and Engineering, Jiangsu University, Zhenjiang 212013, China, ** Department of Materials, Imperial College London, SW7 2AZ, UK.

Abstract

A novel aluminium matrix composite reinforced with a mixture of *in-situ* formed ZrB_2, Al_2O_3 and Al_3Zr particulates was synthesized using a magneto-chemical melt reaction. Using an electro-magnetic assisted melt reaction resulted in a uniform distribution of *in-situ* particles with no indication of clustering and an equivalent diameter in the range of 60 nm to 2 μm. Using this inexpensive *in-situ* manufacturing route, a composite with significantly enhanced mechanical properties was obtained. Increases of up to 50% in σ_{uts} and 70% in σ_y were obtained for an Al-4Cu based composite containing 15 vol% particulate in comparison to the unreinforced alloy. A negligible reduction in elongation was observed (from 5.4 to 4.9%). After a T6 heat treatment, σ_{uts} and σ_y showed further increases of 11% and 23%, respectively over the as-cast properties.

Key words *in-situ* synthesized composites; magneto-chemical melt reaction; aluminium alloys; metal matrix composite; mechanical property.

---

[1] Corresponding author. Tel: +44 20 7594-6801; Fax: +44 20 7594-6758
Email address: p.d.lee@imperial.ac.uk

Introduction

Particulate-reinforced metal matrix composites (PMMCs) possess high elastic modulus, high strength, good wear resistance and excellent properties at elevated temperature compared with conventional monolithic alloys [1]. Aluminium alloy based PMMCs are considered to be light-weight, high-performance advanced materials for aerospace, electronics and transportation applications. Potential uses include aircraft components and more recently key parts in automobiles, such as brake rotors and drums. The casting process is used in the manufacture of most PMMCs components because of its excellent near-net-shape production capability. Near-net-shape casting processes can be classified as *ex-situ*, where the reinforcing particles are added to the melt as a powder (e.g. pressure infiltration and melt stirring) or *in-situ*, where particles are synthesized within the melt. The *in-situ* process has the advantage of forming thermodynamically stable reinforcements with superior mechanical properties from inexpensive raw materials [2]. Because of the great potential that *in-situ* PMMCs offer for a variety of applications, many techniques have been developed to produce them. Examples include the exothermic dispersion (XD^{TM}), the directed melt/metal oxidation (DIMOXTM), the vapour-liquid solid (VLS) reaction process, the London Scandinavian Metals Ltd (LSM) process, the self-propagating high-temperature synthesis (SHS) and the direct melt reaction (DMR) process [3-5].

The DMR process is considered one of the most promising *in-situ* synthesis techniques for commercial applications due to its simplicity, low cost and near net-shape forming capability. However, the main limitation of the DMR process is that it requires long stirring time at high temperature in order to obtain the full incorporation and thorough reaction of the added reactants in the molten metal [6]. A new *in-situ* synthesis method based on a magneto-chemical melt reaction technique was developed to eliminate the requirement for long mixing time at high temperature.

Recently, researchers have highlighted the dramatic influence that magnetic fields can have on chemical reactions in gases, liquids and solids [7]. In the gaseous state, magnetic fields improve the combustion process in various engines. An external field can affect the shape and distribution of carbon nanotubes synthesized by arc-discharge. In the liquid state a magnetic field can influence the population balance of opposing-chirality organic molecules. It can also affect the phase and morphology of electrodeposited phases. In the solid state, and in particular in solid flame combustion reactions, magnetic fields can modify reaction pathways and product compositions, as well as a lot of the structural and magnetic properties. However, there is as yet no report on the influence of magnetic field on *in-situ* synthesized metal matrix composites. Moreover, until now the *in-situ* reactions considered are mainly from Al-Ti-X systems, for example, Al-Ti-O, Al-Ti-B and Al-Ti-C. Thus the focus is on a few *in-situ* formed reinforcement particulates such as Al_3Ti, Al_2O_3, TiB_2 and TiC [8,9].

In the present work, novel Al and Al-4Cu based composites reinforced with a mixture of $Al_2O_3+Al_3Zr+ZrB_2$ particles were fabricated by a magneto-chemical melt reaction in the new reactive system of $Al-Zr(CO_3)_2-B_2O_3$. The distribution and morphology of *in-situ* formed reinforcements were analyzed. The mechanical properties of the composites were characterised and the influence of electromagnetic field on the microstructure was investigated and discussed.

Experimental Technique

Raw materials are aluminium ingot (purity 99.85%), commercial purity Al-4wt%Cu alloy, zirconium carbonate, $Zr(CO_3)_2$ powder (purity 99.97%, average particle size 50μm), and boric oxide, B_2O_3 powder (purity 99.95%, average particle size 45μm). The $Zr(CO_3)_2$ and B_2O_3 powders were dehydrated by pre-heating at 250°C for 3 hours, and then the calculated stoichiometric powder mixture of $Zr(CO_3)_2$ and B_2O_3 was thoroughly mixed for 8 hours in a planetary ball mill. At the same time, the aluminium ingot was melted in an RL100-60H graphite crucible using an electric furnace, and held at 850°C for 20 min. Then the powder mixture of $Zr(CO_3)_2+B_2O_3$ was added to the Al melt. Subsequently, liquid aluminium reacted with the $Zr(CO_3)_2+B_2O_3$ powder to produce *in-situ* synthesized particles at a temperature of 850°C for an in-melt reaction time of 25 min. The melt-reaction process was conducted under electromagnetic field with a magnetic frequency of 10 Hz, and a magnetic current of 300 A. Finally the composite melt at 720°C was poured into a permanent mould, and directly cast into 12 mm diameter bars. A schematic diagram of the *in-situ* magneto-chemical reaction apparatus is shown in Fig. 1.

The composites were characterized by X-ray diffraction (XRD, D/Max2500PC) and scanning electron microscopy (SEM, JEOL-JXA-840A) equipped with an energy dispersive X-ray (EDX) analyzer. The particle size analysis was carried out by an image analyzer using the linear intercept method. Tensile tests were conducted at room temperature in a computer-controlled electronic tensile testing machine (DWD-200). The specimens with a gauge diameter of 6.35 mm (0.25 in.) and length of 25.4 mm (1.0 in.) were prepared according to the ASTM standard E8 specifications. The tensile properties reported are the average of three tests at each condition.

Results and Discussion
Microstructure and X-ray Diffraction

A typical secondary electron (SE) SEM micrograph showing the even distribution of the *in-situ* formed particles in the Al matrix is illustrated in Fig. 2a. Fig. 2b shows the size and morphology of the different particles during the *in-situ* reaction. The chemical composition of the formed particles was confirmed using bulk X-ray diffraction analysis (see Fig. 3). The analysis confirmed that the particles were a mixture of α-Al_2O_3, Al_3Zr and ZrB_2.

The particulate concentration in the *in-situ* synthesized composite was approximately 14.9 vol% with an average equivalent diameter in the range of 60 nm to 2 μm.

The suggested reactions taking place in molten Al to yield the reinforcement particles are the following:

$$Zr(CO_3)_2 = ZrO_2 + 2CO_2\uparrow \qquad (1)$$

$$3ZrO_2 + 4Al_{(l)} = 3[Zr] + 2Al_2O_3 \qquad (2)$$

$$B_2O_3 + 2Al_{(l)} = Al_2O_3 + 2[B] \qquad (3)$$

$$[Zr] + 2[B] = ZrB_2 \qquad (4)$$

$$[Zr] + 3Al_{(l)} = Al_3Zr \qquad (5)$$

The overall reaction can be expressed by:

$$3Zr(CO_3)_2 + 2B_2O_3 + 11Al_{(l)} = 6CO_2\uparrow + Al_3Zr + 4Al_2O_3 + 2ZrB_2 \qquad (6)$$

Influence of Electromagnetic Field

The influence of electromagnetic field on the microstructure of the *in-situ* synthesized composite is illustrated in the SEM examination shown in Fig. 4. The SEM micrographs show a clear beneficial effect, in terms of the even distribution of particles in the matrix and a reduction in the particle size to a sub-micron level when the electromagnetic field is initially applied.

However, comparing Fig. 4c with Fig. 4d it can be seen that the incremental effect on particle size reduces at higher electromagnetic fields. No significant particle size reduction can be observed beyond an electromagnetic field of frequency 10 Hz, and a magnetic current of 300 A. Fig. 4c also shows that the *in-situ* particles synthesized under electromagnetic field are more evenly distributed than those synthesized under zero field (Fig. 4a). In addition, the agglomeration of particles in the matrix is significantly reduced by the applied electromagnetic field. This is considered a major benefit as agglomeration can be a significant issue in cast composites. The electromagnetic field results in a strong shearing effect. This shearing effect is expected to help in preventing clustering and breaking up existing clusters, unlike other *in-situ* synthesis methods, such as the LSM method, which suffer from a significant clustering tendency problem [10]. Furthermore, the larger electromagnetic force also gives rise to considerable agitation and turbulence flow in the melt. This in turn improves significantly the diffusion between reactants leading to fine and evenly distributed particles [5].

Mechanical properties of the composites

The mechanical properties of *in-situ* composites in the as-cast and T6-treated conditions are shown in Figs. 5, 6 and 7. The results indicate that the tensile and yield strengths of the Al and Al-4Cu based composites were significantly enhanced with increasing particulate concentration. At a particulate volume fraction of 15 %, the tensile properties of the Al base composite were σ_{uts}=152 MPa and σ_y=112 MPa, which is markedly higher (95% and 167% respectively) than that of the pure aluminium matrix as shown in Fig. 5. Similarly, the tensile properties of the Al-4Cu base composite were σ_{uts}=394 MPa and σ_y=317 MPa, which is higher by 72% and 53% than that of the matrix alloy. After T6 heat treatment, σ_{uts} and σ_y showed further enhancement. For the Al-4Cu base composite σ_{uts} and σ_y increased by a further 11% and 23%, respectively, as shown in Fig. 6.

Fig. 7 shows the elongations of *in-situ* synthesized Al and Al-4Cu based composites. For the case of the CP-Al matrix, the elongation reduces for particle volume fractions above 5%. However, the reduction in elongation for the Al-4Cu composite is negligible up to 15 vol% particulate. For the as-cast condition the elongation was reduced from 5.4 to 4.9% at 15 vol% particulate (i.e. reduced by approximately 9%) which is considered insignificant.

The microstructure of the Al-4Cu composite containing 15 vol% particles for the as-cast condition is shown in Fig. 8. Fig. 8a shows that the *in-situ* formed particles are uniformly distributed in the Al-4Cu matrix. The white coloured phase observed in back-scattered (BS) mode around the grain boundaries, shown in Fig. 8b, was confirmed by EDX analysis to be the $CuAl_2$ intermetallic phase. The *in-situ* formed particles on the same micrograph are the dark phase observed on the lighter grey background and are shown to be well distributed in the matrix.

Conclusions

A novel route for the manufacture of Al and Al alloy-based composites reinforced with a mixture of ZrB_2, Al_2O_3 and Al_3Zr was developed. The *in-situ* synthesis method uses a magneto-chemical reaction in molten Al. The *in-situ* particles formed via this route are evenly distributed with no evidence of clustering, with an equivalent diameter ranging from 60 nm to 2 μm.

Using this inexpensive *in-situ* manufacturing route a composite with enhanced mechanical properties could be obtained. Increases of up to 50% in σ_{uts} and 70% in σ_y were obtained for an Al-4Cu composite containing 15 vol% particulate. The reduction in elongation as a result of the reinforcing hard phase was negligible. After a T6 heat treatment, σ_{uts} and σ_y showed further increase of 11% and 23%, respectively.

References

1. Tjong SC and Wang GS, High-cycle fatigue properties of Al-based composites reinforced with in situ TiB2 and Al2O3 particulates, Mater. Sci. Eng., A386, 2004, pp48-53.
2. Wang X, Jha A and Brydson R, In situ fabrication of Al3Ti particle reinforced aluminium alloy metal-matrix composites, Mater. Sci. Eng., A364, 2004, pp339-345.
3. Tong X, Al-TiC composites in situ-processed by ingot metallurgy and rapid solidification technology. II. Mechanical behavior, Metall. Mater. Trans., A29, 1998, pp893-902.

4. Tjong SC and Ma ZY, Microstructural and mechanical characteristics of in situ metal matrix composites, Mater. Sci. Eng., R29, 2000, pp49-113.
5. Zhao YT, Cheng XN, Dai QX, Cai L and Sun GX, Crystal morphology and growth mechanism of reinforcements synthesized by direct melt reaction in the system Al-Zr-O, Mater. Sci. Eng., A360, 2003, pp315-318.
6. Tsunekawa Y, Suzuki H and Genma Y, Application of ultrasonic vibration to in situ MMC process by electromagnetic melt stirring, Mater. Des., 22, 2001, pp467-472.
7. Pankhurst QA and Parkin IP, Chemical Reactions in Applied Magnetic Fields, in 'Magnetism: Molecules to Materials IV', Miller JS, Drillon M, Eds, Wiley-VCH Verlag Gmbh & Co., 2002, pp467-479.
8. Yu P, Mei Z and Tjong SC, Structure, thermal and mechanical properties of in situ Al-based metal matrix composite reinforced with Al2O3 and TiC submicron particles, Mater. Chem. Phys., 93, 2005, pp109-116.
9. Kennedy AR, Weston DP and Jones MI, Reaction in Al-TiC metal matrix composites, Mater. Sci. Eng., A316, 2001, pp32-38.
10. Youssef YM, Dashwood RJ and Lee PD, Effect of clustering on particle pushing and solidification behaviour in TiB2 reinforced aluminium PMMCs, Composites Part A, 36, 2005, pp747-763.

Acknowledgements

This research was sponsored by the National Natural Science Foundation of China (No.50471050), the Foundation of the Ministry of Education of China (No.00170), Jiangsu Provincial Foundations of Industry Key Project and Science and Technology (No.BE2002039 and JH02-039), and the Foundation of China Scholarship Council (No.2003832124). The authors would also like to thank Dr. M.G. Ardakani of the Department of Materials at Imperial College London for his assistance.

Figures

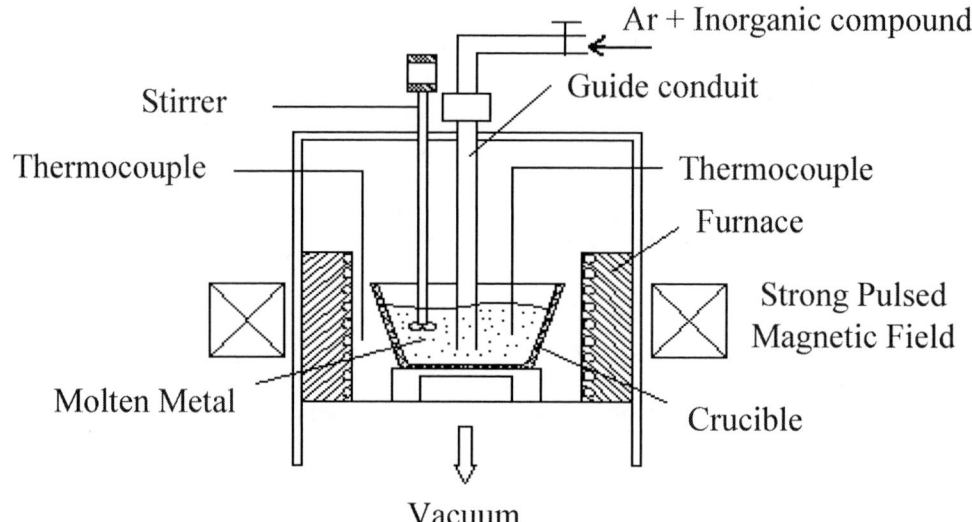

Fig. 1. A Schematic diagram of the *in-situ* magneto-chemical reaction apparatus.

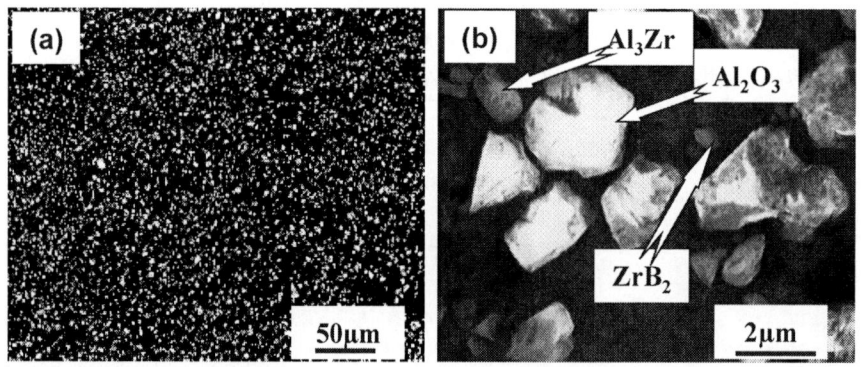

Fig. 2 SE-SEM micrographs of the as-cast microstructure showing: (a) particle distribution in the matrix and (b) the different particles formed.

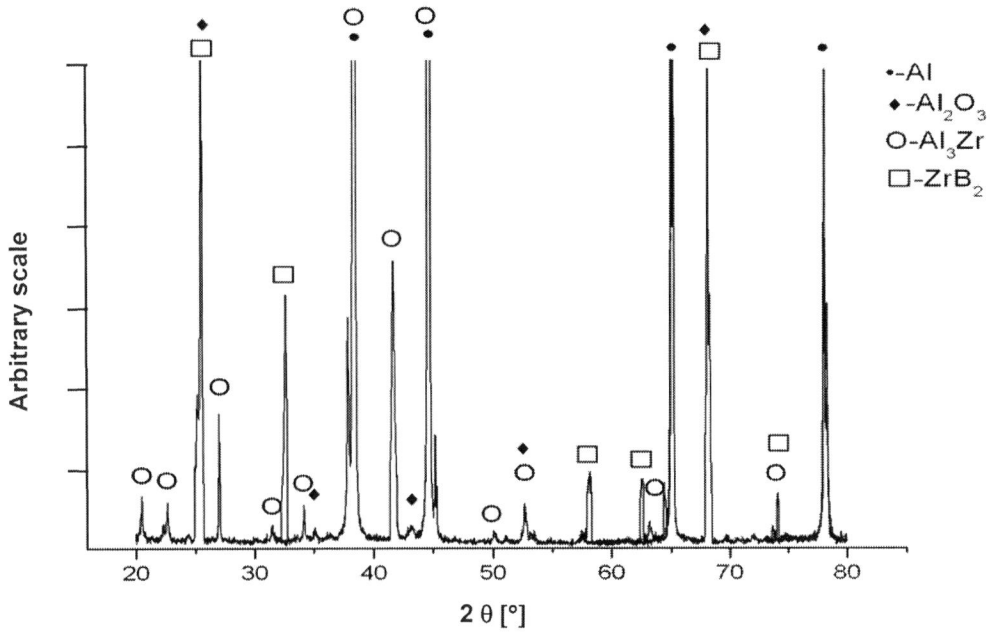

Fig. 3. X-ray diffraction analysis showing the particles synthesized in the Al-Zr(CO$_3$)$_2$-B$_2$O$_3$ system.

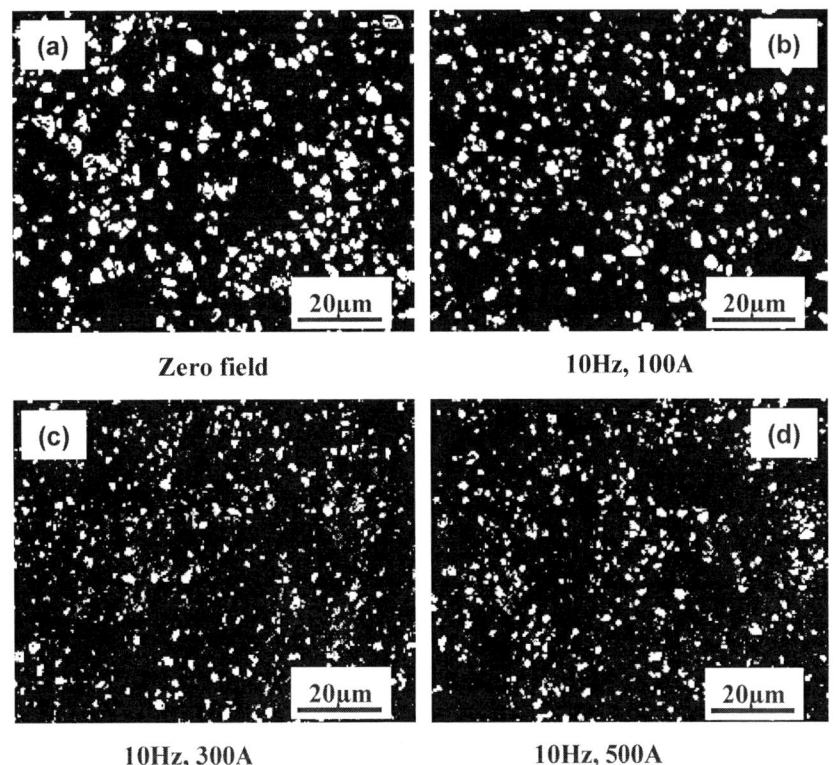

Fig. 4. SE-SEM micrographs showing the Influence of electromagnetic field on the microstructure of a 15 vol% (Al$_2$O$_3$+Al$_3$Zr+ZrB$_2$)p/Al composite.

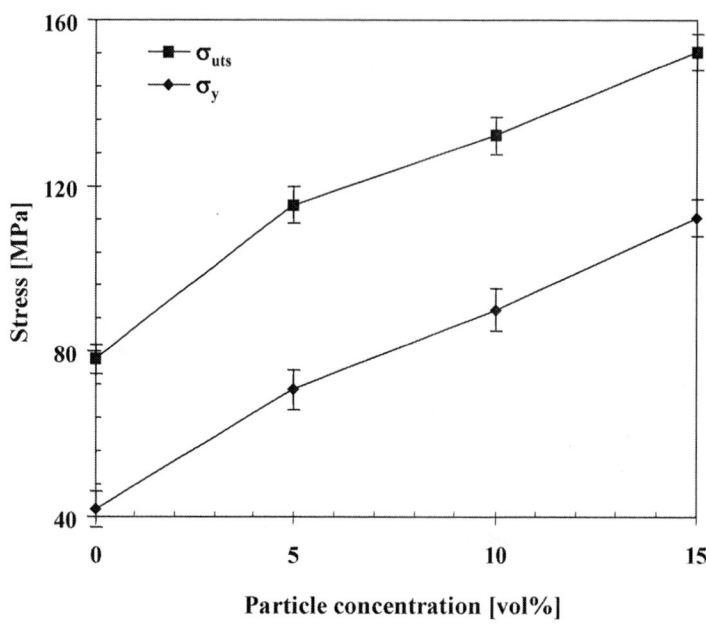

Fig. 5. Tensile and yield strengths of $(Al_2O_3+Al_3Zr+ZrB_2)p/Al$ composites.

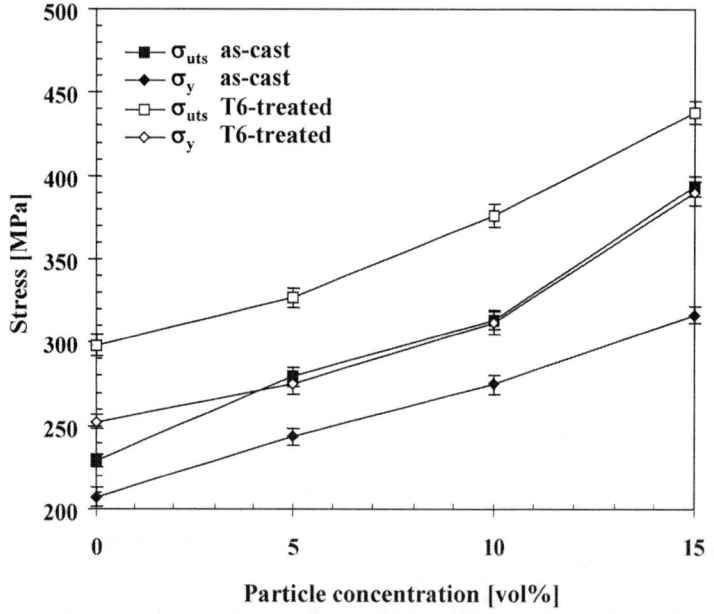

Fig. 6. Tensile and yield strengths of $(Al_2O_3+Al_3Zr+ZrB_2)p/Al$-4Cu composites for the as-cast and T6-treated conditions.

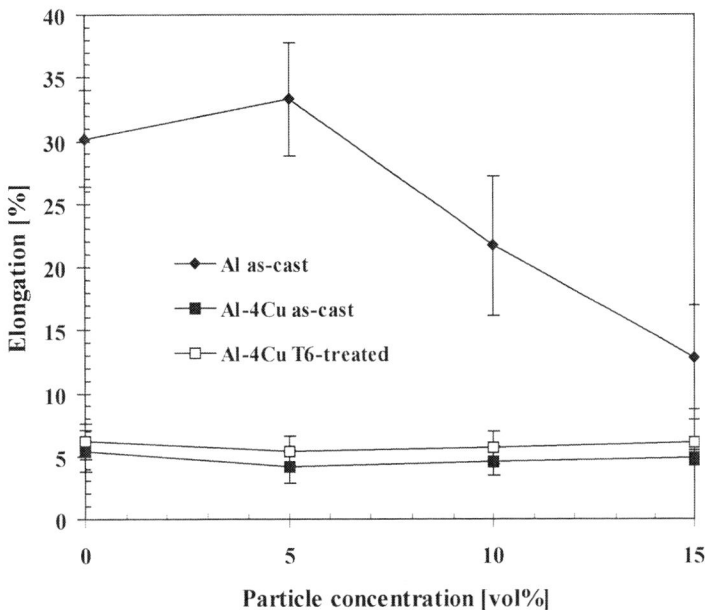

Fig. 7. Elongations of *in-situ* composites synthesized in the system Al-$Zr(CO_3)_2$-B_2O_3 for the as-cast and T6- treated conditions.

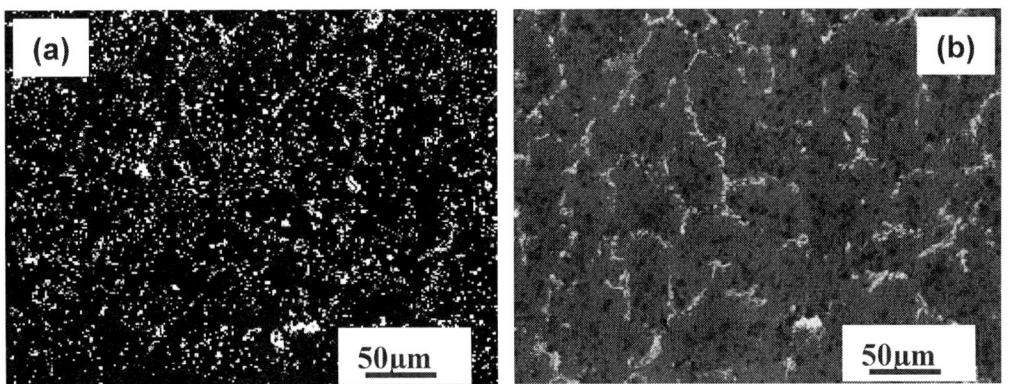

Fig. 8. SEM examination of the as-cast microstructure for Al-4Cu composite with (a) a SE-SEM micrograph showing the particulate distribution and (b) a BS-SEM micrograph showing the CuAl2 intermetallic phase at the grain boundaries.

Compo-casting Method for Alumina Ceramics Inserted in Cast Iron to Reduce Thermal Stress

Y Tomita *, H Sumimoto * , K Nakamura ** and S kiguchi *.

* Kinki University, JAPAN, ** Honorary Professor of Kinki University, JAPAN.

Abstract

The largest problem in compo-casting method is cracking caused by thermal shock. Although this cracking can be prevented by reducing the thermal stress by preheating the ceramics, the necessary preheating temperature is high and its precise control is difficult at practical foundries. In this study, we tried to estimate numerically the critical preheating temperature of ceramics using a thermal stress analysis during transient thermal conduction and Newman's diagram. We also found that preheating of the ceramics to reduce thermal stress can be replaced by placing appropriate cast iron covers around the ceramics.

We describe a new compo-casting method that needs no preheating of the ceramics and has been prove useful experimentally. Also, a strength evaluation of the compo-casting material was done.

Key words

Cast Iron, Alumina Ceramics, Compo-casting, Heat Stress analysis, Hybridization

Introduction

Recently, the demand for mechanical property improvement of cast iron has arisen. However, there is a limit to the strength development of cast iron. Ceramics have been introduced to the compo-casting research using the old cast iron as a method of adding the function of ceramics to the cast iron. However, it is necessary to prevent the cracking that is generated in ceramics in the compo-casting situation with ceramics. In addition, practical use examples are not readily available because evaluation of the integrated casting is difficult.

The biggest problem in the compo-casting of ceramics and cast iron is the generation of cracks in the ceramics caused by significant thermal shock. We have already experimentally clarified that the generation of cracks can be prevented by preheating the ceramics[1]. We have also recognized that the cracks are generated by a large temperature difference between the surface and the center of the ceramics, and we also clarified that the critical temperature difference for the generation of cracks can be numerically estimated by a thermal stress analysis of transient thermal conduction[1].

In this study, we numerically estimated the critical preheating temperature for the non-generation of cracks using Newman's diagram[2] which was obtained from transient thermal conduction equations and the analysis mentioned above. Also, we examined the results by measuring the temperature difference between the surface and the center of the ceramics. Moreover, we also applied the estimation results for the development of a new compo-casting method which requires no preheating of the ceramics.

2.Examination of Preheating Temperature of Ceramics

The cracking is caused by a large thermal stress due to a large temperature difference between the surface and the center of the ceramics. Our previous experimental results showed that alumina ceramics can be compo-cast without cracks when the preheating temperature is higher than 1273 K. This thermal stress can be obtained from a thermal stress analysis during transient thermal conduction, and the critical temperature difference (ΔT_{max}) for the non-generation of cracks can then be estimated from the equation (1).

$$\Delta T_{max} = \frac{\sigma_B (1-\nu)(2.0+4.3k/rh)}{E} \qquad (1)$$

The estimated values for the alumina ceramics is 227K[1]. This means that preheating of the ceramics is necessary to make the temperature difference less than these values. We then tried to find a way to estimate the appropriate preheating temperature to make the temperature difference less than ΔT_{max}.

The ceramic is heated by the cast metals and its temperature changes with time. If the ceramic body is assumed to be a cylindrical bar with an infinite length, we can estimate the heat transfer phenomenon by a transient thermal conduction equation for a one-dimensional ordinate system. The solution for the particular case of a cylindrical bar with an infinite length is expressed as a Newman diagram, and the solution can be obtained from the diagram.

Fig.1 shows the relation between the maximum temperature difference and the preheating temperature, which also includes T_{cr}. The figure indicates that the critical preheating temperature appropriate for compo-casting of the ceramic is 1002 K for alumina ceramics. This estimated value is some 250 K lower than 1273 K which is the critical preheating temperature obtained by the experiments. This lower estimation should result from the temperature drop in the ceramic body during the pouring operation, which means that the temperature of the ceramics was assumed to be the same as the preheating temperature in the experiment, but it might have dropped some 200 K until the molten metal touched the ceramic body. Although the estimation results were somewhat lower than those from the experiments, the estimation method that combines the thermal stress analysis during transient thermal conduction and Newman's diagram, which are obtained from the transient thermal conduction equation, is a useful way to estimate the critical preheating temperature in the compo-casting of a ceramics body.

3. Compo-casting of Ceramics Using a Cover
3.1 Examination of Appropriate Thickness of a Cover for Compo-Casting Without Preheating of Ceramics

As already mentioned, we can successfully perform the ceramics/cast iron compo-casting if we preheat the ceramics over the temperature determined by the thermal stress analysis based on an transient thermal conduction model and Newman's diagram. However, the preheating temperature can be as high as 1273 K, and it is also not easy to maintain this temperature after preheating until the pouring period; therefore, it does not appear applicable to practical production. We then studied a new compo-casting method which requires no preheating of the ceramic body. We tried to place a cast iron cover over the ceramic body to relieve the thermal shock to the ceramics. In order to prevent the generation of cracks in a ceramic body, the temperature difference between the surface and center should be less than T_{max}. We estimate the appropriate thickness of the cast iron cover which makes the temperature difference between the surface and center of the ceramics less than T_{max}.

Fig.2 shows the analysis results for the relation between the thickness of the cover and the temperature difference between the surface and center of the ceramics. This result indicates that the necessary thickness to make the temperature difference less than T_{max} is 6.7 mm for an alumina ceramic. Therefore, we decided to use the cast iron cover whose thickness is 7 mm for the alumina ceramics in our experiments. Fig.3 shows the relation between the time and temperatures of the surface and center for the 7-mm thick cover for the alumina ceramics, and it also shows the temperature difference between them. This figure indicates that the maximum temperature difference between them is 208 K, which is less than 227 K of its T_{max}. This means that a 7-mm thick cover is considered to be effective to sufficiently relieve the thermal shock to the ceramic body.

3.2 Experimental Procedures

The ceramic materials are alumina ceramics whose chemical composition is Al_2O_3:95mass%, SiO_2:3mass%, and the dimensions of the ceramic body for compo-casting is a cylindrical bar with a 10-mm diameter and 25-mm length. The cast iron material for compo-casting is a flake graphite cast iron equivalent to FC150. The cast iron cover is also made of FC150, and its shape is a hollow cylindrical bar type with one end closed with a cylindrical hole of 10-mm diameter and 15-mm depth, and the wall thickness is 7 mm. The mold for compo-casting is made of CO_2 sand and has a cylindrical hole of 100-mm diameter and 150-mm depth. A ceramic body was placed in the center of the bottom surface with a 10-mm depth and tightly covered with the cast iron cover. The molten metal was heated to 1723 K and kept for 0.6 ks in an induction furnace and then poured into the mold at 1653 K. Based on our calculation of T_{max} as previously mentioned, we assumed that the temperature of the melt is constant and that the heat capacity of the melt was not included. Therefore, in order to clarify the effect of the heat capacity of the melt, we varied the amount of melt from 1.5 kg to 4.0 kg in 0.5-kg steps. The specimen (product of compo-casting) was cut into two pieces along the center line by a diamond cutter, and its cutting plane was precisely examined by an optical microscope.

3.3 Experimental Results and Discussions

Fig.4 shows the cutting planes of the specimens of the alumina ceramic and the 7-mm thick cast iron cover. For the 1.5-kg melt, no crack is observed but a gap (opening) is observed between the ceramic and the metal at the top and the side. The shape of the opening at the top exhibits a trace of the drill hole, which means that the cast iron cover has not yet completely melted. Added to this, the ceramic separated from the metal. For the 2.0-kg, 2.5-kg and 3.0-kg melts, no crack or gap is recognized, which means that excellent compo-casting results are obtained. For the 3.5-kg and 4.0-kg melts, no gap is observed, but cracks are recognized in the ceramics, and the cracks are recognized to be filled with metal, which means that the cast iron cover has completely melted. Though the ceramic has cracks, its bonding is excellent.

Fig.5 shows some examples of the microstructure of the interface area between the cast iron cover and the cast metal. For the 1.5-kg melt, the size of the graphite flakes on the melt side (right) is much smaller than that of the cast iron cover side (left), which means that the melt solidified very quickly near the surface area of the cover. Also, because a clear boundary line is observed between the two sides, the cover has not yet melted. For the 2.5-kg melt, the microstructures of both sides are almost the same and no boundary line is recognized, which means that the cover has completely melted and has combined with the melt.

3.4 Influence of Amount of Cast Metal on Generation of Cracks

The results mentioned above indicate that the generation of cracks is mostly related to the amount of the melt for compo-casting. Our calculation method is useful for estimating T_{max}, but the influence of the melt's heat capacity should be taken into account for a more accurate estimation. Therefore, we tried to measure the actual temperatures at three places in a specimen. These thermocouples (Type K, 0.5-mm diameter) were located using an inorganic adhesive: one is on the surface of the cast iron cover and the other two are on the center and surface of the ceramic body. The temperature of the melt is measured by a thermocouple (Type R, 0.5 mm diameter) which is located on the surface of the mold. The method of compo-casting was the same as the method described in section 3.2.

Fig.6 shows the results of the measurements. For the 1.5-kg melt, the temperature increase in the surface and the center of the ceramic is the lowest among all three diagrams, and the temperature difference between the surface and center of the ceramic is also the lowest of all. The maximum temperature difference is 144 K, which is much smaller than T_{max} (=227K) and means that no crack is generated. For the 2.5-kg melt, the temperature increases in the ceramic become higher than those of the 1.5-kg case, which results in the greatest temperature difference between them. The maximum temperature difference is 259 K, which is near T_{max}, but no crack is generated. For the 4.0-kg melt, the changes and the temperature difference become much higher than in the 2.5-kg case, and the maximum temperature is as high as 499 K, which is much higher than T_{max}, and thus cracks are generated. Though some experimental errors such as the response-delay caused by the thermocouple protection tubes and the heat capacity of the thermocouple itself might have produced some inaccuracies in the diagrams, the tendencies observed in these diagrams are considered to coincide well with the results of our estimations and the experiments. For a smaller amount of melt, the temperature of the melt significantly drops when the melt touches the cast iron cover; therefore, the cover does not sufficiently melt and a gap occurs between the ceramic and the metal. However, for a large amount of melt, the temperature drop in the melt becomes much smaller so that the cover melts, which means that the temperature of the ceramic surface rapidly rises and the temperature difference between the surface and center expands greatly and cracks are generated. The results obtained so far indicate that the amount of the melt affects the temperature drop in the melt. This temperature drop in the melt may be one of the main causes that produced some different results between the experiments and the estimation which assumes that the temperature of the melt is constant. Added to this, Newman's diagram deals only with the thermal conductivity in the radial direction, but the heat transfer from the top surface side of the ceramic body is not included in this experiment. Therefore, we need to consider the most appropriate amount of melt for a given thickness of the cast iron cover.

4.Strength evaluation in compo-casting of ceramics that use a cast iron cover

4.1 Experimental procedure

The compo-casting can be done with no preheating of the ceramics if a cast iron cover is used. However, it is thought that a more quantitative evaluation is necessary when considering about application to a machine material as a compo-casting material. A strength evaluation of the compo-casting material was then done by a sharing examination of the compo-casting specimen. The vicinity of the ceramic member of the compo-casting specimen was cut out. Also, the ceramic compo-casting part was cut into round slices with a diamond-cutter, and a test specimen of 10-mm thickness was made. Fig.7 shows a schematic drawing of the specimen processing. An Amsler-type universal tester was used for the sharing examination. Fig.8 shows the schematic drawing of the piercing examination. The test specimen was placed on a receiving unit. Thereafter, the rated capacity of the ceramic part was determined by punching out a specimen using a 9-mm diameter and then dividing by the compo-casting area of the ceramic and the cast iron which was assumed to be at shear strength.

4.2 Experiment results

The shear strength was calculated from the rated capacity determined by the sharing examination of the test specimen using equation(2).

$$ W \quad A \qquad\qquad 2 $$

Shear strength(MPa) W Rated capacity at the time of punching(N) A Contact area of cast iron and ceramic(mm

Fig.9 shows the sharing examination result. The shear strength rise by the weight of the molten metal increasing. Moreover, it is understood that the shear strength is higher for a specimen that uses the cover compared with a specimen using preheating of the ceramic. Regarding this, the molten metal pressure on the ceramic is added directly to the ceramic because there is no cover. As a result, it is thought that this occurs because big compression is added on the ceramics. Moreover, with 5.0 kg of pouring weight, the maximum shear strength was obtained in the compo-casting with preheating of the ceramics. The compo-casting that used a cover had the same results. Fig.10 shows the bottom of the test specimen after punching of these two specimens for examination. The other side of the test specimen was transformed by preheating the ceramic. When the punched specimen was examined, this transformation was generated. However, the transformation was not seen on the cast iron side under the condition where the cover was used. Fig.11 shows the SEM image on the cast iron side of the contact side of the cast iron and the ceramic. There is a scratch in the vertical direction under the condition of preheating. It is understood that this scratch was occurred at the time of punching the ceramic. However, only a horizontal scratch could be obtained with the drill under the condition of using the cover when the cover was made was seen. It has been understood that the cast iron is joined to the ceramic by the high intensity under the condition of preheating. As for the compo-casting using preheating, application to parts from which strength is required is thought to be possible based on the above-mentioned results. Moreover, it is thought that the compo-casting of ceramics that use the cast iron cover can be applied to parts with no restriction to shape.

5. Conclusions

Using a thermal stress analysis of transient heat transfer, we tried to obtain the critical temperature difference for the generation of cracks during the compo-casting of ceramics with cast iron. From Newman's diagram, we obtained the most appropriate heating conditions to relieve the thermal shock to the ceramics. We also tried to minimize the thermal shock to the ceramics using a cast iron cover whose thickness had been estimated from Newman's diagram, and we could successfully perform the compo-casting of ceramics with cast iron without preheating the ceramics. And, strength evaluation of the compo-casting material was also done. The results obtained are summarized as follows.

(1) The critical temperature difference between the surface and center of a ceramics body for the generation of cracks can be obtained from the thermal stress analysis of transient thermal conduction and Newman's diagram.

(2) We recognized that a cast iron cover reduces the temperature difference and that the compo-casting of ceramics without preheating is possible if a cast iron cover is used. We estimated the appropriate thickness of the cover from the thermal stress analysis of transient thermal conduction and Newman's diagram. Using the cast iron cover and between 2.5-kg and 3.0-kg melts, we successfully performed the compo-casting of ceramics.

(3)The shear strength rise by the weight of the molten metal increasing. Moreover, it is understood that the shear strength is higher for a specimen that uses the cover compared with a specimen using preheating of the ceramic.

References

1. Y Tomita , H Sumimoto , S kiguchi , K Nakamura, *Observation of cast-in inserts of ceramics with molten cast iron and generation condition of cracks*, J.JFS , Vol.74, No 3, Mar 2002, pp143-148.

2. A B Newman , *Heating and cooling rectangular and cylindrical solids* , Industrial and Engineering chemistry, Vol.28, No.5, May 1936, pp545-548.

Figures

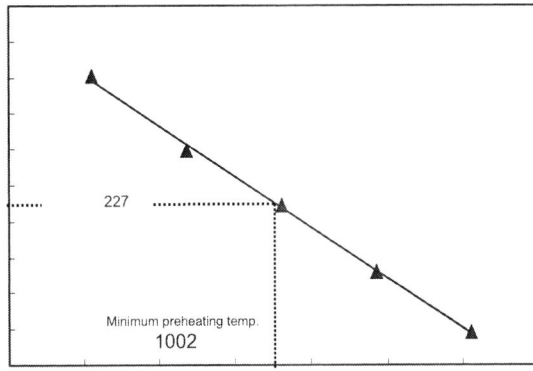

Fig.1 Determination of pre-heat temperature for ceramics by calculated temperature difference and the critical temperature differences for cracking.

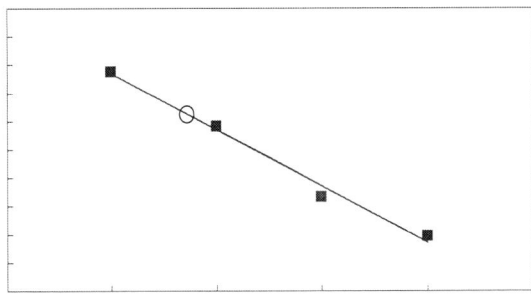

Fig.2 Calculated thickness of cast iron covers prevent cracking in ceramics.

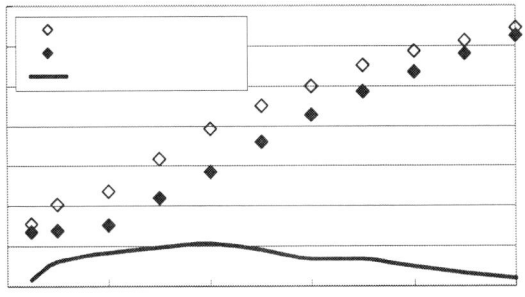

Fig.3 Predicted temperature differences between the surface and the center of the ceramic by Newman's diagrams. (Thickness of cover : 7mm)

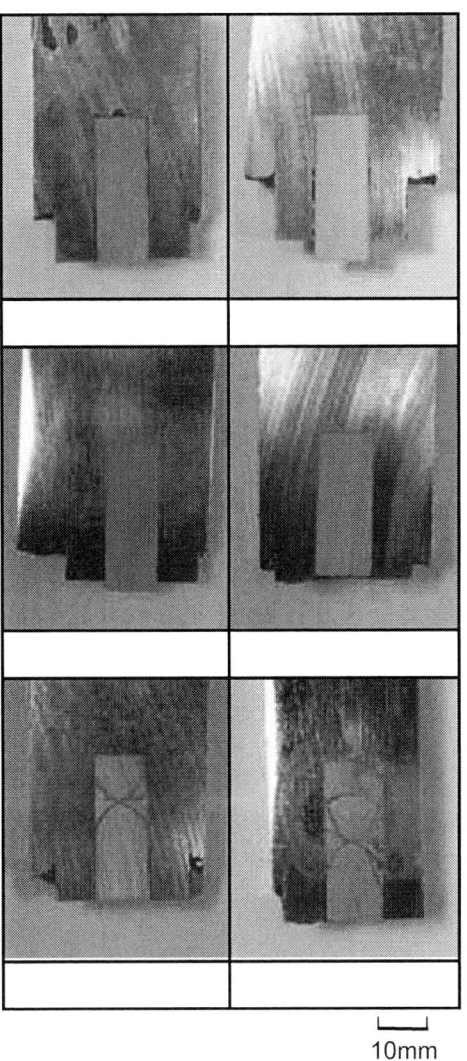

Fig.4 Experimental results of compo-casting with alumina ceramic with cast iron covers (7mm).

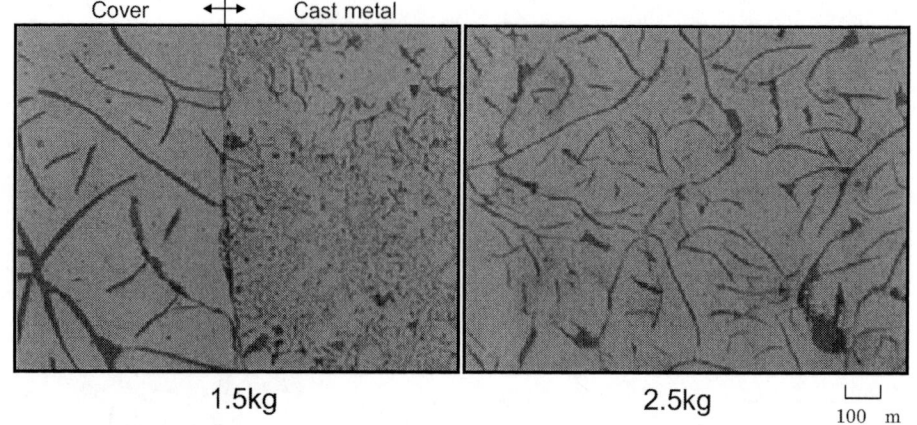

Fig.5 Microstructures near the interface of molten metal and the cast iron cover (7mm).

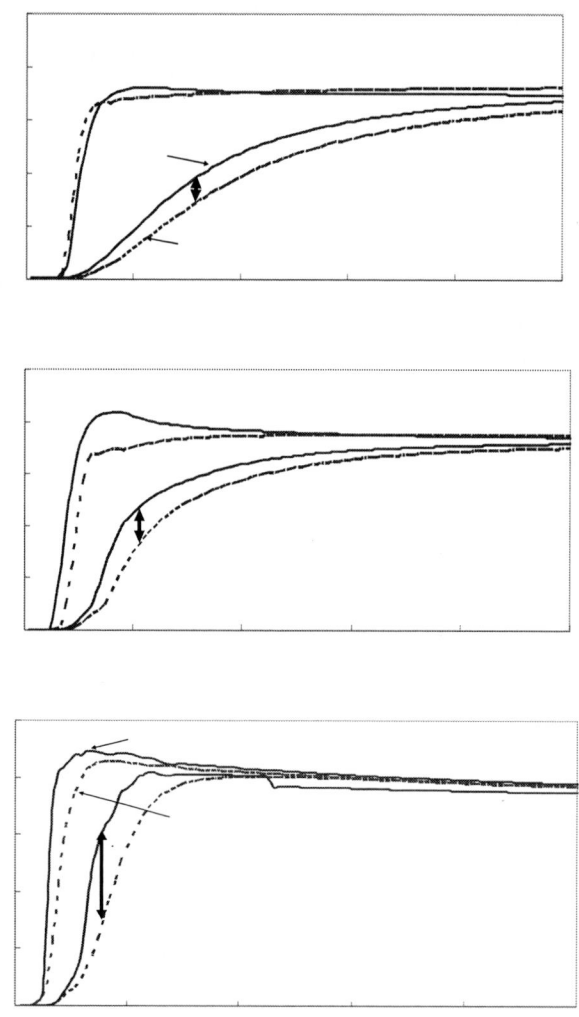

Fig.6 Influence of melt weight on temperature difference between the surface and the center of the ceramic.

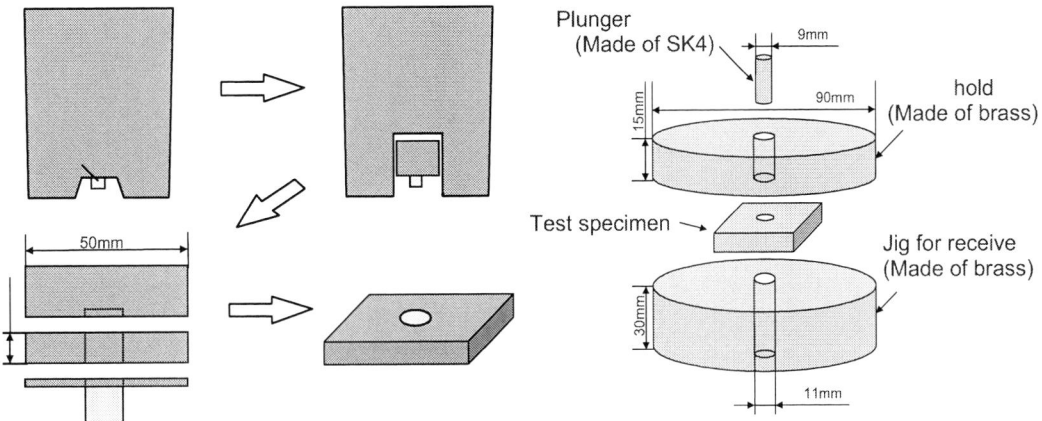

Fig.7 Schematic illustration of processing of test specimen.

Fig.8 Schematic illustration of shear examination.

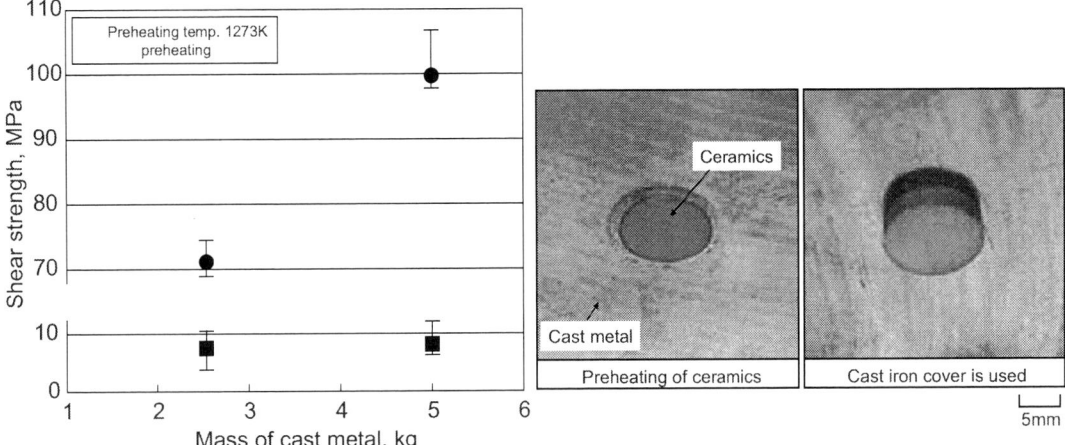

Fig.9 Comparison between preheating result and non preheating result.

Fig.10 Macrostructure of test specimen after shear examination (mass of cast metal : 5kg).

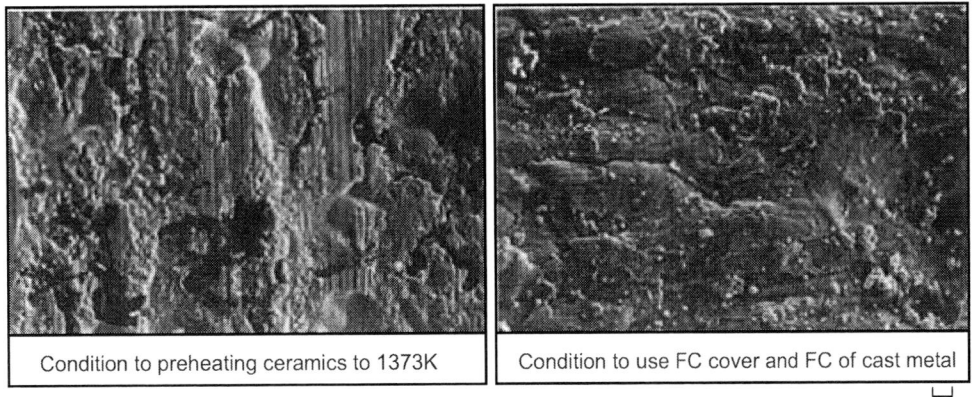

Fig.11 SEM micrograph of cast metal in contact side with ceramics (mass of cast metal : 5kg).

Production of hybrid metal matrix composites and its wear behavior

Mahendra K. V *, and Dr. K. Radhakrishna **.

* Dept. of Mechanical engineering, New Horizon College of engineering, Bangalore, INDIA.
**Dept. of Mechanical engineering, B.M.S. College of engineering, Bangalore, INDIA.

Abstract

Metal matrix composites are engineered materials formed by the combination of two or more dissimilar materials (at least one of which is a metal) to obtain enhanced properties. In the present investigation Al-4.5%Cu alloy was used as the matrix and fly ash & silicon carbide (SiC) as filler material. The composite was produced using conventional foundry techniques. The particulates were added in 5%, 10% and 15% by weight to the molten metal. The composite was tested for fluidity, hardness, density, mechanical properties and dry sliding wear. The microstructure examination was done using scanning electron microscope to know the distribution of particulates in the Al-4.5%Cu alloy matrix. The results show that there is an increase in hardness with increase in particulates content. The density decreases with increase in percentage of particulates. The tensile strength and compression strength increases with increase in percentage of particulates. The resistances to dry wear increases with increase in particulate content. The fracture surfaces were examined under SEM to know their behaviour.

Key words: Hybrid MMCs, Wear, Fly ash, SiC, Mechanical properties.

Introduction

Conventional monolithic materials have limitations in terms of achievable combinations of strength, stiffness, co-efficient of expansion and density. Composites which are combinations of two or more dissimilar materials possess properties that are superior to the properties of the individual materials which constitute them.Composites usually comprise of two parts, the matrix forming the bulk of the material and the reinforcements forming the additions. In recent years, particulate reinforced aluminium based metal matrix composites are finding interest due to ease of their mass production and promising mechanical properties. Commonly used particulates are TiC, SiC, Al_2O_3, Mica, Graphite, SiO_2, etc. Recently fly ash is being used as reinforcement material. Coal fly ash, an industrial solid waste by-product, is produced during the combustion of coal in thermal power plants.

It was reported that mechanical properties increases with addition of particulates. Nair [1] studied Al- 2.5% Si with 20% by volume SiC and reported that the tensile strength increases with increase in percentage SiC. Surappa [2] studied Al-11.8%Si and Al 16% Si, 3% by volume SiC particles of 53-63 μm size. It was reported that hardness and tensile strength of both composites increased. Rohatgi [3] studied Al-Si (A356) with 10% by volume fly ash. The particle size was 63 to 105 μm. The hardness and tensile strength increases with addition of particulates. Guo [4] studied pure aluminum powder with 5% to 20% by weight fly ash composite produced by powder metallurgy process. It was reported that hardness and compacting strength increases with increase in compacting pressure.

Wear is a phenomenon associated with all contact surfaces which are having relative motion between each other. Wear is defined as unintended progressive loss of material under the influence of normal load, relative velocity, nature of contact materials, type of contact surface etc. Wear is characterized by progressive loss of material due to the relative motion between two solid surfaces in contact under load. Pramila Bai et.al [5] in their studies on A356 Al-SiC composites have observed that the composites wear resistance was much better than the base alloy upto pressure of 26 MPa. Alpas et.al [6] reports that wear resistance of Al-Si alloy SiC composite increases with sliding distance. Under conditions where the SiC particles are fractured large strains are transmitted to the matrix material. Bindumadhavan et. al. [7] investigated the wear of Al-Si-Mg alloy with SiC composites and observed that with increase in SiC reinforcement wear resistance increases.

Rohatgi [3,8] has reported that the addition of fly ash particle to the aluminium alloy significantly increases its wear resistance. Improvement in wear resistance was due to the presence of hard aluminosilicates fly ash particles. Guo [9] studied pure aluminum powder with 5% to 20% fly ash particulate composite prepared by powder metallurgy process. It has been reported that during compacting the cenospheres was getting ruptured, where as precipitator fly ash remains unaffected. The wear rate decreases with increase in percentage of fly ash particulates. Accumulation of wear debris leads to two body wear changing to three body wear. Sanmino et.al. [10] reported the beneficial effects of reinforcements in Al matrix during sliding wear.

There is an increasing interest in the development of metal matrix composites having low density, containing low cost reinforcements and having good wear resistance. Conventional MMCs currently contains only one type of reinforcement (monocomposites). Hybrid metal matrix composites (HMMCs) are second generation composites, wherein more than one type, shape and size of reinforcements are used to obtain better properties. Wilson et.al. [11] investigated the wear behaviour of hybrid composite A356 Al-20% volume SiC and 10% volume graphite and reports that there are no severe wear modes at different sliding speeds. Surappa et.al [12] investigated the effect of extrusion on the properties of A356 with fly ash and SiC hybrid composites. It was reported that there was an improvement in the mechanical properties of the hybrid composite. In this investigation hybrid metal matrix have been produced by the addition of fly ash and SiC particulates using stir casting technique and fluidity, mechanical properties and wear properties have been studied.

Experimental

In the present study, Al-4.5%Cu alloy was used as matrix material. The chemical composition of the alloy is given in Table 1. The reinforcements used in this investigation are fly ash particulates and silicon carbide particulates.The morphology of fly ash and SiC were examined using scanning electron microscope. Table2 gives the chemical composition of fly ash in weight percentage. Fly ash consists of cenosphere fly ash and precipitator fly ash. Fig 1 gives morphology of fly ash particulates and Fig 2 gives morphology of SiC particultaes.

Castings were prepared using, cylindrical moulds of different diameters made from S.G. iron. The diameters of moulds are 25mm, 50mm and 75mm and height 200mm. For the production of composites, stir casting method or vortex method was used. The method consists of stirring the molten metal to create a vortex and introducing the particulates into the vortex. Stainless steel blade coated with SiO_2, to prevent direct contact with the metal was used for the stirrer.

The crucible containing the degassed molten metal whose temperature was maintained around 800^0C was placed below the stirrer. Slowly the stirrer was made to rotate and the rpm was maintained at approximately 600 rpm. This results in a proper vortex in the liquid metal. The reinforcement particles fly ash and SiC were preheated to 200^0C to remove the presence of moisture and were added slowly. Along with the addition of particles small pieces of Mg (0.5% by weight of molten metal) were added to the molten metal. This ensures good wettability of particles with the molten metal. The percentage addition of particulates was 5, 10 and 15 by weight. The molten metal was poured into fluidity dies (Spiral & Strip). The length of spiral and strip gives the fluidity of composite. The stirred dispersed molten metal was transferred into pre heated S.G. Iron moulds and was allowed to cool to room temperature. After solidification, the castings were removed from the mould and subjected to heat treatment. The castings were subjected to standard T6 heat treatent.

Density tests were carried out by using Archimedes principle. The hardness of MMCs and base alloy was measured using Brinnel hardness tester with 10mm ball indenter under 500 kg load, as per standard procedure.

The microstructure of specimens were examined using Scanning Electron Microscope (SEM) to know the distribution of particulates. The tensile strength and compression strength were determined using computerized UTM of capacity 200 kN with electronic extensometer. The wear test was carried out using computerized pin on disc wear testing machine as per ASTM G-99 standards. The specimen in the form of pin (5mm dia and 20mm length) was used. A hardened steel disc (60HRC) was used as the counterface. The track velocity was kept constant at 80 m/sec. The pin was loaded with 4.9N load and the disc is rotated with a known rpm. When the pin comes in contact with a rotating disc, there will be reduction in length with time. The same was recorded using LVDT. The test was continued for a total time duration of 2 hours. The data acqusition of the events was recorded on line using a computer. A load cell was used to measure the tangential force to calculate coefficient of friction. Linear wear is measured as a change in length. For every 10 sec interval the change in length was recorded and linear wear Vs time, coefficient of friction Vs time and frictional force Vs time were plotted. The experiment was repeated for loads of 9.8N, 14.7N and19.6N at different track radius of 30, 35 and 40mm respectively. The experiment was carried out for specimens made from different diameters of the casting.

Result & Discussions

Fig 3 gives spiral and strip fluidity length of composites. It was observed that with increase in percentage of particulates the fluidity length decreases. The viscosity of molten metal decreases with increase in particulates and hence fluidity decreases.

Fig 4 shows SEM photomicrograph of hybrid composites. Uniform distribution of particulates can be seen in photomicrograph. Fig 5 gives density of hybrid composites. It was observed that density of composite decreases with increase in percentage of particulates. This may be due to lower density of particulates. Fig 6 gives the average hardness of hybrid composites. The hardness increases with increase in percentage of particulates. This may be due to presence of fly ash & SiC particulates in the matrix. Fig 7 shows the tensile strength of hybrid composites. It was observed that with increase in percentage of particulates the tensile strength increases. This may be due to the presence of fly ash and SiC particulates. It was also observed that smaller diameter castings shows more ensile strength than bigger diameter castings. This may be due to fine grain structure of the smaller diameter castings.

Fig 8 shows the compression strength of hybrid composites. With increase in percentage of particulates the compression strength increases due to the presence of particulates. The smaller diameter casting shows better compression strength than larger diameter castings. Fig 9 shows the fractured tensile surface of hybrid composite.

It was observed that with increase in load the wear increases. Fig 10 gives the wear versus time for different percentage of particulates of 25mm diameter at 4.9N normal load. It was obsereved that with increase in percentage of particulates the wear decreases. This may be due to the presence of particulates in the matrix. Similar trends were observed for other diameter castings. Fig 11 gives the wear of hybrid composite for different diameters. It was observed that smaller diameter castings shows better wear resistance than bigger diameter castings.

Fig 12 and Fig 13 shows the worn surface of base alloy and hybrid composite (25 dia) under load of 4.9N.It was observed that with increase in percentage of particulates the depth of wear scar has decreased.The debris are likely to act as the third body abrasive particles and could be responsible for the higher wear rate of the counterface. The particles trapped between the specimen and the counterface causes a microploughing on the contact surface of the composite. The continuous longitudinal lines parallel to the sliding direction on the worn surfaces of the composites probably result from the ploughing action of the fly ash and SiC particles. At higher loads composites shows delamination due to which, the material loss in the form of plate like debris takes place. Fig 14 shows the wear debris of hybrid composites. When applied load results in stresses higher than the particles, they loose their abilities to support the load. Consequently the aluminium matrix comes in direct contact with the counterface and large plastic strains are imposed on the material adjacent to the contact surfaces. The severe localized deformation give rise to crack formation, a process in which particle/matrix decohesion plays an important role. Subsurface delamination is also the main process of debris formation. Visual observations indicate that the wear debris are dark in colour. It can be seen that the fine powders get agglomerated to particles of 10 to 20μm size. The topography of the wear debris analysis shows that adhesion was a dominant wear mechanism. It was observed that the friction force and co-efficient of friction decreases with increase in percentage of particulates.

Conclusions

1. Spiral and strip fluidity length will decrease with increase in percentage of particulates.
2. With the addition of fly ash and SiC particulates the density of composite decreases.
3. The hardness of composite increases with increase in percentage of particulates.

4. There is uniform distribution of particulates as seen in SEM photomicrographs. More number of particulates were seen with higher percentage of particulates.
5. The tensile strength and compression strength of composite increases with increase in percentage of particulates. The strength values of smaller diameter casting is more as compared to bigger diameter castings.
6. The fractured surface of tensile specimen shows smaller dimples and at some places the particulates have undergone fracture. This indicate very good bonding between matrix and the particulate.
7. Wear resistance of composites are better than base alloy. With increase in particulates wear decreases for all loads. Wear of smaller casting was less compared to bigger diameter castings. The friction force and coefficient of friction decreases with increase in percentage of particulates.
8. With the addition of particulates the depth of ploughing reduces as compared to base alloy. Wear debris analysis indicates adhesion wear mechanism.

References

1. Nair S.V, Tien J.K, Bakes R.C, *SiC reinforced Al-Metal matrix composite,* International Metals review, 30, No 6, 1985, pp 275-288.
2. Surappa M.K Rohagi P.K, *Preparation and properties of cast aluminium ceramic particle composites,* Journal of materials science, 16, 1981, pp 983-993.
3. Pradeep K. Rohatgi, *Low cost, fly ash containing aluminium matrix composites,* Journal of Metals, November 1994, pp 55-59.
4. Guo R.Q, Rohatgi P.K, Nath D,*Preparation of Al fly ash particulate composite by powder metallurgy technique,* Journal of material science, 32, 1997, pp 3971-3974.
5. Pramila Bai B.N, Ramasesh B.S, Surapa M.K, *Dry sliding wear of Al-SiCp composites,* Wear,157, 1992, pp 295-304.
6. Alpas A.T, Zhang J, *Effect of SiC particulate reinforcement on the dry sliding wear of aluminium-silicon alloys (A356),* Wear, 155, 1992 pp 83-104.
7. Bindumadhavan P.N, Heng Keng Wah, Prabhakar O, *Dual particle size (DPS) composites, effect on wear and mechanical properties of particulate metal matrix composites,* Wear,248, 2001, pp112-120.
8. Rohatgi P. K, Guo R. Q, Keshava Ram B. N, Golden D. M, *Cast Aluminum fly ash composites for engineering applications,* AFS Trans, 95-32, 1995, pp 575-579.
9. Rohatgi P.K, Guo R.Q, Huang P, Ray S, *Friction and abrasion resistance of cast Al alloy fly ash composites,* Metallurgical & Material Trans A, 28A, January 1997, pp 245-250.
10. Sanmino A.P, Rack H.J, *Dry Sliding Wear of Discontinuously reinforced aluminium composites review and discussion,* Wear, 189, 1995, pp 1-19.

11. Wilson S, Alpas A.T, *Effect of Temperature on the sliding wear performance of Al alloys and Al matrix Composites,* Wear, 196, 1996, pp 270-278.
12. Sudarshan, Surappa M.K, *Stir casting of hybrid Al-SiC-fly ash composite,* 18th Annual Symposium, Dept. of Metallurgy, IISc, Bangalore, January 2005, pp 11-30.

Tables
Table 1 Chemical composition of base metal in percentage and balance Aluminium

Element	Cu	Mg	Si	Fe	Mn	Ni	Pb	Sn	Ti	Zn
% Wt	4.5	0.06	0.53	0.66	0.13	0.07	0.02	0.02	0.01	0.11

Table 2 Chemical composition of fly ash in weight percentage

Al_2O_3	SiO_2	Fe_2O_3	TiO_2	LOI*
30.40%	58.41%	8.44%	2.75%	1.43

*LOI- Loss on ignition

Figures

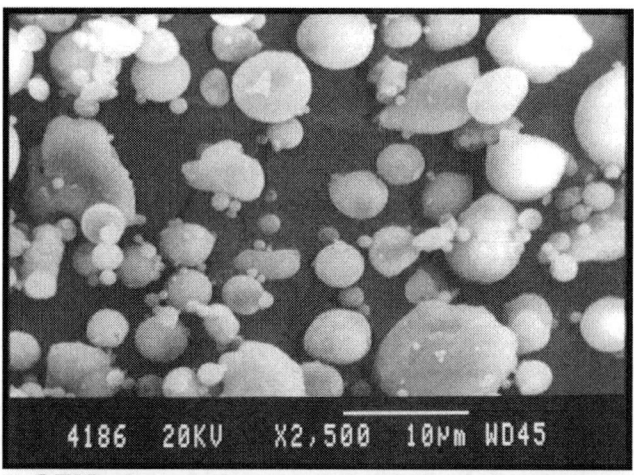

Fig 1 SEM photomicrograph of fly ash particulates

Fig 2 SEM photomicrograph of SiC particulates

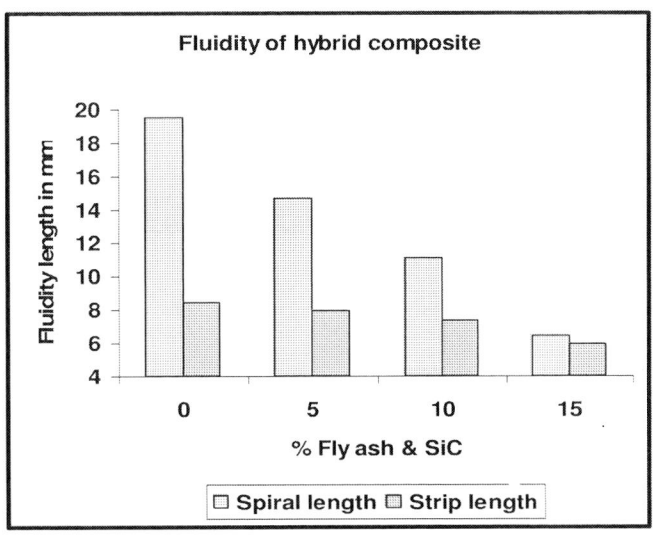

Fig 3 Fluidity of hybrid composites

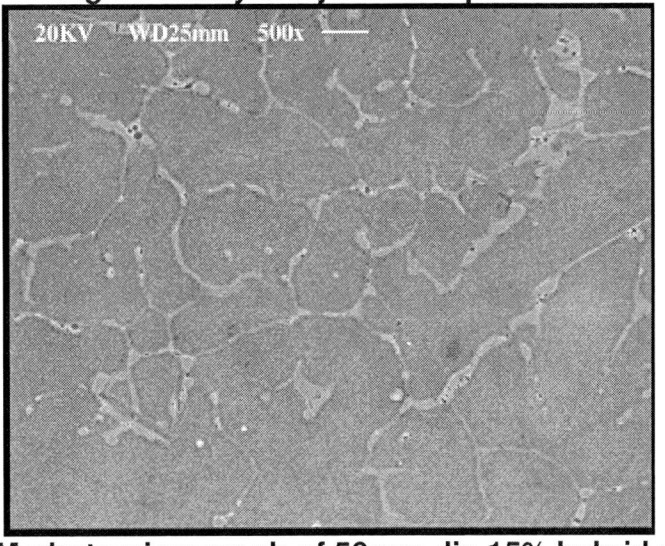

Fig 4 SEM photomicrograph of 50mm dia 15% hybrid composite

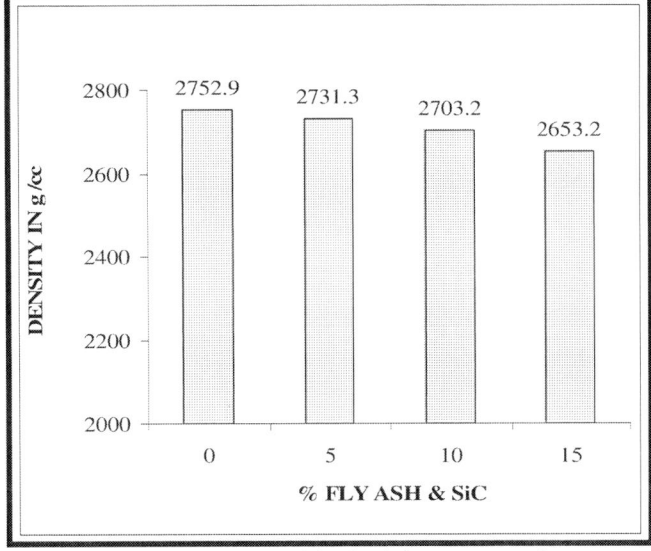

Fig 5 Density of hybrid composites

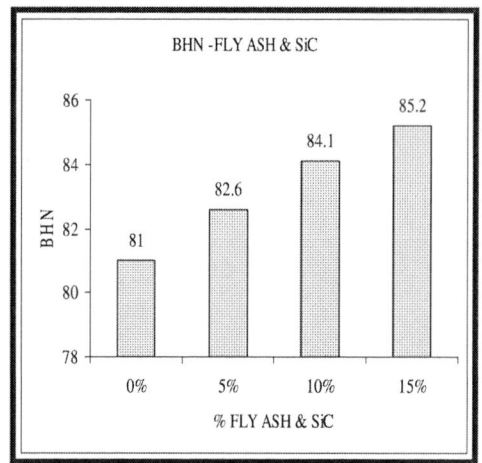

Fig 6 Hardness of of hybrid composites

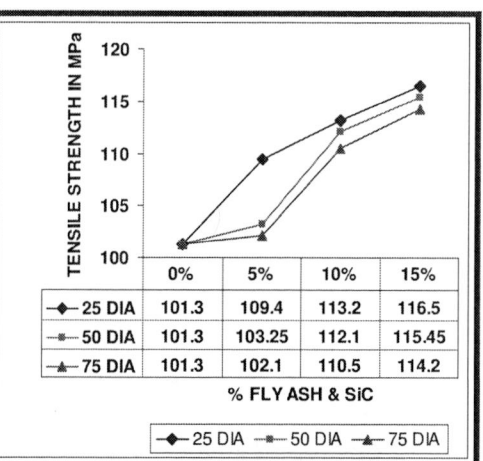

	0%	5%	10%	15%
25 DIA	101.3	109.4	113.2	116.5
50 DIA	101.3	103.25	112.1	115.45
75 DIA	101.3	102.1	110.5	114.2

Fig 7 Tensile strength of hybrid composites

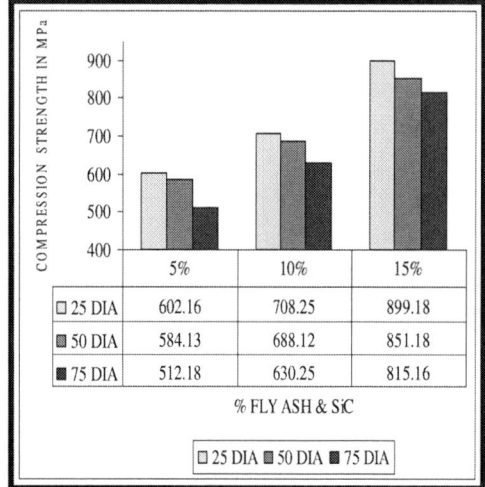

	5%	10%	15%
25 DIA	602.16	708.25	899.18
50 DIA	584.13	688.12	851.18
75 DIA	512.18	630.25	815.16

Fig 8 Compression strength of hybrid composite

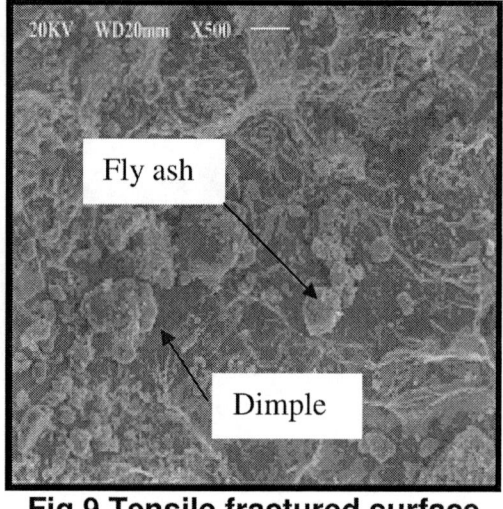

Fig 9 Tensile fractured surface of 10% hybrid composite

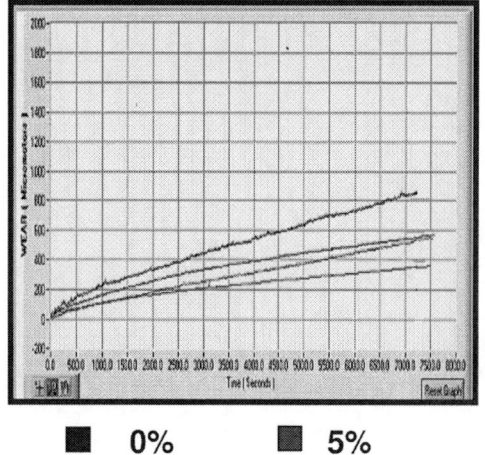

■ 0%   ■ 5%
■ 10%   ■ 15%

Fig 10 Wear Vs Time for different % of particulates

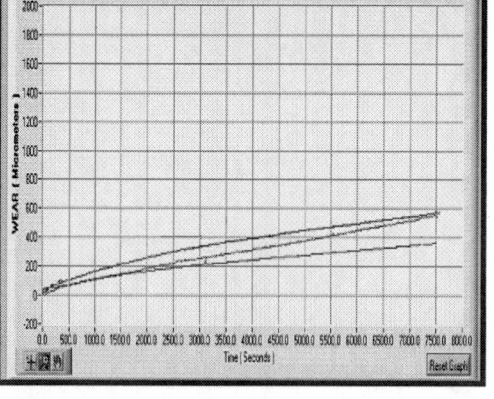

■ 75 dia   ■ 50 dia   ■ 25 dia

Fig 11 Wear Vs Time for different casting diameters

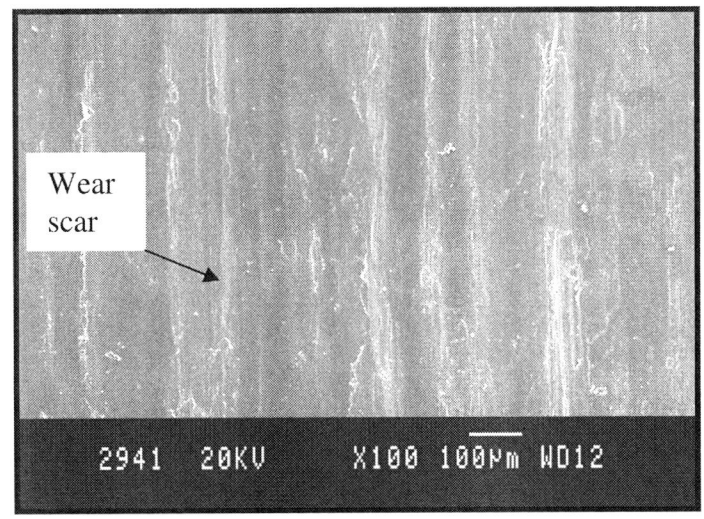

Fig 12 Worn out surface of base alloy at 4.9N (25dia casting)

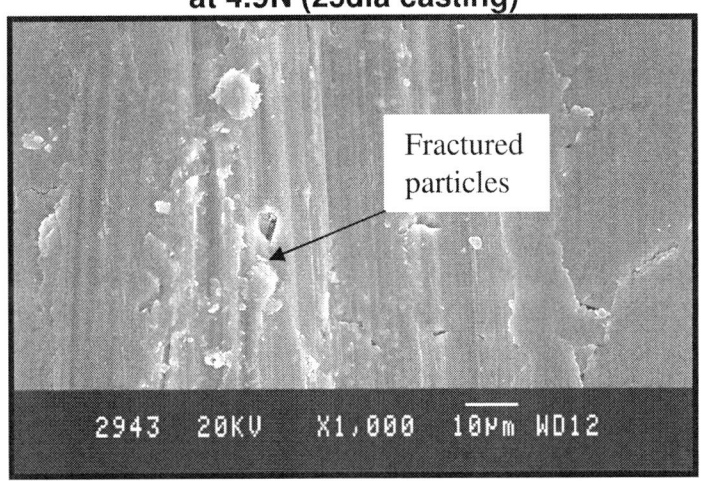

Fig 13 Worn out surface of hybrid at 4.9N load (25dia casting, 10%)

[200X]
Fig 14 Wear debris of 5% hybrid composite at 4.9N load

Mahendra K.V.   email: mahendrakv@rediffmail.com
Dr. K.Radhakrishna   email: drkrk@rediffmail.com

Effect of Volume Fraction and Particle Size of Reinforcement on Thermal Analysis and Heat Transfer Parameters of Gravity Die Cast Hypereutectic Al-22% SI Alloy Matrix Composites.

Subramanya. P. K, Sathyapal Hegde and K. Narayan Prabhu

Department of Metallurgical & Materials Engineering National Institute of Technology Karnataka, Surathkal P.O. Srinivasnagar 575 025, Karnataka State, India

Abstract

The properties of cast metal matrix composites are largely dependent on the solidification behaviour which is dictated by the thermophysical properties of the melt, mould and the interfacial heat transfer from the metal to the mould. In the present investigation the thermal analysis parameters and heat transfer aspects of hypereutectic aluminium alloy matrix composites were studied. As the vol% of SiC_p increases the total solidification time decreases and the cooling rate increases. The morphology of the primary silicon was very much dependent on the presence of SiC_p. The estimated peak heat flux for the composites are lower than matrix alloy melts solidified under similar conditions. The particle size has a negligible influence on the cooling behaviour. However, composite with finer particle size shows slightly higher peak heat flux.

Key words: MMCs, Thermal Analysis, Interfacial Heat Transfer.

Introduction

Al-22Si alloy is a well known casting alloy. Due to the presence of hard primary silicon in its microstructure the alloy exhibits high wear resistance, low thermal expansion coefficient, good corrosion resistance, and improved mechanical properties at a wide range of temperature.. The addition of ceramic particles to these alloys further increases its stiffness, high temperature strength, and wear resistance at ambient temperature[1]. Hence, the SiC_p reinforced Al-22Si alloy composites have considerable potential as an engineering material[2]. Stir casting is the most viable casting technique to produce Al-Si/SiC_p composites[3,4]. The major drawback of stir casting is the non uniform distribution of particles which has a effect on its mechanical strength[5]. In hypereutectic alloys due to the absence of dendrites in the solidified structure there is a fairly uniform distribution of the particles[6,7].

Thermal analysis technique is widely used as a non-destructive tool to assess the quality of molten metal in Aluminium foundries[8]. It can also be used to quantify the amount of SiC_p present in a melt[9], which is a critical aspect of composite production. The technique involves monitoring of the temperature changes in a sample as it cools through phase transformation intervals. During solidification the latent heat released causes changes in the cooling curve, including thermal arrest points, which are characteristics of transformation and reactions, occurring during solidification. The correlation among these cooling curve characteristics and arrest points, considered as thermal analysis parameters and the observed microstructure allows the monitoring of melt quality before pouring[10]. The non equilibrium cooling conditions in the foundries complicate the thermal analysis process.. The cooling rate during solidification affects the thermal analysis parameters[11] . In die casting metal/mould interfacial heat transfer play a dominant role in the removal of the heat from the solidifying melt, with the addition of particles the heat transfer characteristics changes. Also, the addition of particle drastically changes the viscosity of the melt. Particle distribution in the matrix material during the melt stage of the stir casting process depends on viscosity of the melt. Viscosity, which also depends on particle size plays a major role in heat transfer characteristics. In the present study, SiC_p reinforced hypereutectic Al-22%Si alloy matrix composites were processed using stir cast technique. The aim was to investigate the effect of cooling rate, volume fraction and particle size on the thermal analysis parameters and metal/mould interfacial heat transfer characteristics The cooling rate was varied by pouring the molten alloy into moulds having different thermal conductivities. Composites of 10, 15 and 20 vol% SiC_p of two average sizes 80 µm and 40 µm were processed.

Experimental

The metal matrix used in this study is a commercial Al-22% Si alloy. The metal was melted in a clay graphite crucible using a resistance heating furnace. The melt was degassed at 750°C by using hexachloroethane tablets and then stirred using a graphite impeller. SiC_p

preheated at 700°C was added to the vortex created in the melt by the rotating impeller. The addition was completed within 5 minutes. The melt was stirred for 1 minute and poured into metallic moulds instrumented with thermocouples. The moulds used were high conductivity copper (384 $Wm^{-1}K^{-1}$) and low conductivity stainless steel (16 $Wm^{-1}K^{-1}$) with internal diameter of 30mm and height 120mm to obtain different cooling rates Three calibrated thermocouples were used to record temperature data for thermal analysis. A twin bore ceramic beaded thermocouple was placed in the geometric centre of the mould to monitor the thermal history of the solidifying melt. The mould was instrumented with two stainless steel sheathed thermocouples to obtain thermal history of the mould. The thermocouples were connected to a high-speed online data acquisition equipment NI SCXI 1000. Temperature readings were acquired at the rate of 100 samples per second.. Figure 1 shows the schematic sketch of the experimental setup. Experiments were carried out for different volume fractions of SiC_p (10%, 15% & 20%) havimg two sizes. The average sizes of SiC_p were 80 μm and 40 μm. Metallographic test specimens were prepared using a section from the geometric center of the casting. The sections cut from samples were metallographically polished and etched. The samples were then subjected to micro examination. The volume fraction was ascertained by image analysis.

Results and Discussion

Figure 2 shows the effect of volume % of SiC_p of 40 μm on the cooling curve for composites solidified in copper mould. It is observed that, as the volume % of SiC_p increases, total solidification time decreases. Further increase in vol% of SiC_p the eutectic arrest temperature is lowered. It is observed that solidification time decreased sharply on addition of SiC_p upto 15 vol%. The decrease was gradual on further addition of particles upto 20 vol%. Table 1 summarises the eutectic arrest time data for all composites. Figure 3 shows the effect of addition of SiC_p on cooling rate of composite. It is observed that cooling rate increases with increase in the volume% of SiC_p. For example the cooling rate of the alloy solidified in stainless steel mould increased from 12.5 °C/s to 19.23 °C/s when 20 vol% of 40 μm SiC_p was added. The cooling behaviour of the solidifying alloy with varying volume % of SiC_p can be explained on the basis of following phenomena. There is a reduction in the amount of heat to be extracted corresponding to the volume percentage of solid silicon carbide particle present in the liquid metal, which can decrease the solidification time. If Q_m is the total heat content and Q_r is the heat content of the SiC_p reinforcement at the pouring temperature, then the total heat content of the composite, Q_c can be calculated using the rule of mixtures as,

$$Q_c = fQ_r + (1-f)Q_m \qquad (1)$$

where f = volume % of the reinforcement particle.

Thus as the volume % of SiC_p increases there is reduction in the amount of heat to be extracted correspondingly. But there is also reduction in the effective thermal diffusivity of the composite system due to the presence of the low conducting dispersoid which can hinder the heat transfer during solidification. The effective thermal diffusivity is caculated using the equation.

$$\alpha_c = \alpha_m \left(\frac{\alpha_r + 2\alpha_m - 2f(\alpha_m - \alpha_r)}{\alpha_r + 2\alpha_m + f(\alpha_m - \alpha_r)} \right) \qquad (2)$$

where α_c, α_m, and α_r are the thermal diffusivities of composite, matrix and reinforcement respectively.

It is found that with addition of 20 vol% SiC_p the thermal diffusivity reduces from 43×10^{-6} m²/s to 36×10^{-6} m²/s, affecting a reduction of about 16%. In the present study, it is observed that the solidification time decreases even up to 20 Vol% SiC_p. In the low silicon content matrix alloy phenomenon of reduction of amount of heat to be extracted dominates at lower volume fraction and phenomenon of reduction of effective thermal diffusivity dominates at higher volume fraction of SiC_p. But in the present study, for high silicon content matrix, primary silicon initially solidifies and the addition of SiC_p favours the nucleation of primary silicon and hence the time of solidification required will be reduced further because of the absence of undercooling.

From the microstructure shown in Figure 4, it is observed that the primary silicon nucleates on silicon carbide particle, indicating the role of SiC_p as a substrate for nucleation of primary silicon. It is also clear from the microstructure that primary silicon gets refined by the addition of SiC_p which improves heat transfer during solidification. Thus in the present study, there was no increase in solidification time after certain volume% of SiCp instead it continued to decrease even up to 20V% of SiC_p . As seen from Figure 3, cooling rate of the composite increases with the introduction of SiC_p . The mould thermal history was used as an input to the estimation of heat flux transients. The non linear estimation technique utilizes the numerical approach to estimate the mould surface heat flux density from the knowledge of measured temperature inside the heat conducting mould material during solidification[12]. The method analyzes the transient heat transfer at the interface. Estimated values of heat flux for the matrix alloy and composite with different volume % of SiC_p solidified in stainless steel mould is given in Figure 5. Heat flux at the interface showed a maximum, shortly after pouring and dropped rapidly for all moulds and alloy /composite systems studied. The initial increase in the heat flux is due to the good contact of liquid metal with the mould surface. As the casting starts solidifying, the metal/mould interfacial contact region undergoes transformation from an initial conforming to a non conforming state leading to the formation of air gap. Peak heat flux values are summarised in Table 2. The presence of particles in the matrix alloy melt affects its heat transfer behaviour. The estimated q_{max} values for composites are lower than matrix alloy melts solidified under similar conditions. With increase in the volume fraction of the reinforcement, the heat content of the system Q_c decreases as mentioned earlier. This reduction in the amount of heat to be extracted corresponding to the increase in the volume% of SiC_p leads to lower peak heat flux, although the cooling rate increases .

Figure 6 shows the cooling curves of composites with SiCp of size 80µm and 40µm for various volume % SiC$_p$ cooled in two moulds. It is observed that there is no significant change in the cooling behaviour of composites with two different particle sizes of the reinforcement. However composites with finer particles showed a slight decrease in total solidification time. This decrease in solidification time is attributed to the increased refinement of primary silicon due to more nucleation sites offered by the finer SiC$_p$. In the case of solidification of finer particle size composite, more agglomeration of particle was observed. It is also observed that as the particle size decreases, processing becomes difficult due to agglomeration and segregation of the particle and also floating of particle on the slag Thus to obtain a good quality composite particle size also plays an important role.

Figure 7 shows variation in heat flux for 80µm and 40µm average particle size composite solidified in stainless steel mould. Composite with finer particle size shows slightly higher peak heat flux. This is attributed to the lower resistance to heat transfer offered by the solidifying composite. Heat transfer rate is found to be enhanced with finer SiC$_p$ and is due to the easy path for flow of heat through matrix alloy containing finer particles.

Conclusions

1. The increase in the volume % of SiC$_p$ in the matrix alloy reduces the total heat content of the solidifying composite melt. This results in the decrease in total solidification time.

2. Addition of SiC$_p$ favours the nucleation of primary silicon. The refinement of primary silicon would increase the rate of heat transfer during solidification.

3. The estimated peak heat flux (q_{max}) for the composites are lower than matrix alloy melts solidified under similar conditions. The heat flux decreases with vol% of SiC$_p$ owing to the lower heat content of the system.

4. The particle size has a negligible influence on the cooling behaviour. However, composites with finer average particle size showed slight decrease in total solidification time and is attributed to the higher refinement of primary silicon due to more nucleation sites offered by finer silicon carbide particles.

5. Composites with finer particles show slightly higher peak heat flux . This may be attributed to the increase in the thermal conductance of the solidifying composite.

References

1. Y. Sahin, M. Actlar, *Production and Properties of SiC$_p$ Reinforced Aluminium Alloy Composites*, Composites Part A, 34, 2003, pp. 709-718.

2. P. J. Ward, H. V. Atkinson, G. Anderson, G. Elias, B. Garcia, L. Kahlen, *Semi Solid Processing of Novel MMCs Based on Hypereutectic Aluminium Silicon Alloy*, Acta Mater., vol. 33, no. 5, 1996, pp. 1717-1727.

3. A. Mortensen, I. Jim, *Solidification Processing of Metal Matrix Composites*, International Materials Review, vol. 37, no. 3, 1992, pp. 101-128.

4. U. Cocen, K. Onel, *The Production of Al-Si Alloy SiC$_p$ Composites via Compocasting: Some Microstructural Aspects*, Mat. Sci. and Engg. A221, 1996, pp. 187-191.

5. S. J. Hong, H. M. Kim, D. Hub, C. Suryanarayana, B. S. Chun, *Effect of Clustering on the Mechanical Properties of SiC Particulate Reinforced Aluminium Alloy 2024 Metal Matrix Composites*, Mat. Sci. and Engg. A347, 2003, pp. 198-204.

6. N. V. Ravikumar, B. C. Pai, E. S. Dwarakadasa, *Microstructural Evolution in Liquid Metal Processed Al-Alloy/SiC$_p$ Composites*, Int. J. Cast Metals Res., 15, 2003, pp. 573-579,

7. N. Han, G. Polland, R. Stevens, *Microstructural Characterisation of Sand Cast Aluminium Alloy A456-SiC Particle Metal Matrix Composite*, Mat. Sci. and Tech. vol 8. 1992, pp. 52-56.

8. Apelian D., Sigworth G. K. and Whaler K. R., *Assessment of Grain Refinement and Modification of Al-Si Foundry Alloys by Thermal Analysis*, AFS Transactions, vol. 92, 1984, pp. 297-307.

9. C. Gonzalez, J. Baez, R. Chavez, A. Garcia, J. Juarez, *Quantification of the SiCp Content in Molten Al-Si/SiC$_p$ Composites by Computer Aided Thermal Analysis*, J. Mat. Proc. Tech., 143-144, 2003, pp. 860-865.

10. W. U. Shusen, Y. You, A. Ping, T. Kanno, H. Nakar, *Effect of Modification and Ceramic Particles on Solidification Behavior of Aluminium Matrix Composites*, J. Mat. Sci., 37, 2002, pp. 1855-1860.

11. S. Gowri, F. H. Samuel, *Effect of ccoling rate on the solidification Behaviour of Al-7 pct Si-SiCp Metal Matrix Composites*, Met. Trans A, Vol. 23A, 1992, pp. 3369-3376.

12. K. N. Prabhu, W. D. Griffiths, *Metal/mould Interfacial Heat Transfer during Solidification of Cast Iron in Sand Moulds*, Int.J. Cast Met. Res, vol. 14, 2001, pp. 147-15

Acknowledgements

The authors thank Saint Gobain Ltd, Bangalore for providing the SiC$_p$ used in the present work.

Tables
Table 1: Effect of mould material and particle size on eutectic arrest time

Mould	Size	Eutectic arrest time (s)			
		Vol % of SiC$_p$			
		0	10	15	20
SS	240	8.37	5.49	5.3	4.25
	320		5.71	4.47	4.02
Copper	240	7	6.65	5.56	3.9
	320		6.94	4.88	3.85

Table 2: Effect of mould material and particle size on heat Flux Values

Mould	Size	Heat Flux Values kW/m^2			
		Vol % of SiC$_p$			
		0	10	15	20
SS	240	652	487	421	395
	320		446	416	400
Copper	240	953	621	753	527
	320		694	848	279

Figures

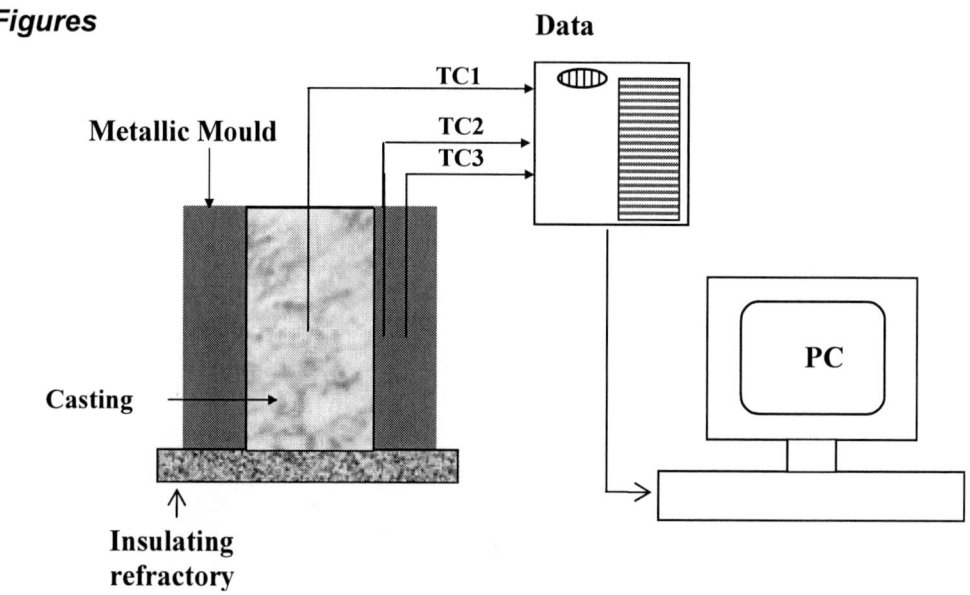

Fig.1 : Schematic sketches of the (a) stir casting and (b) mould set up

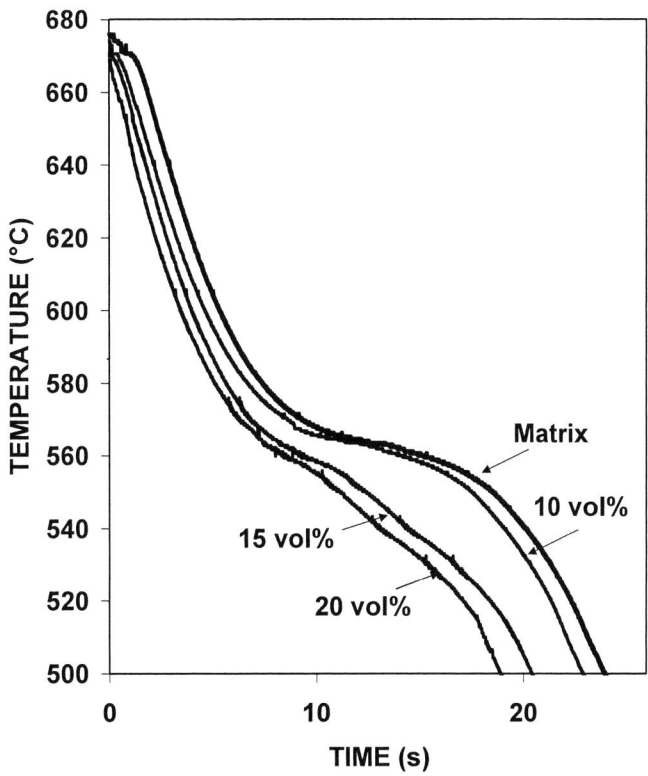

Fig. 2: Cooling curves for composite solidified in copper mould for varying vol % of SiC_p (size= 40μm)

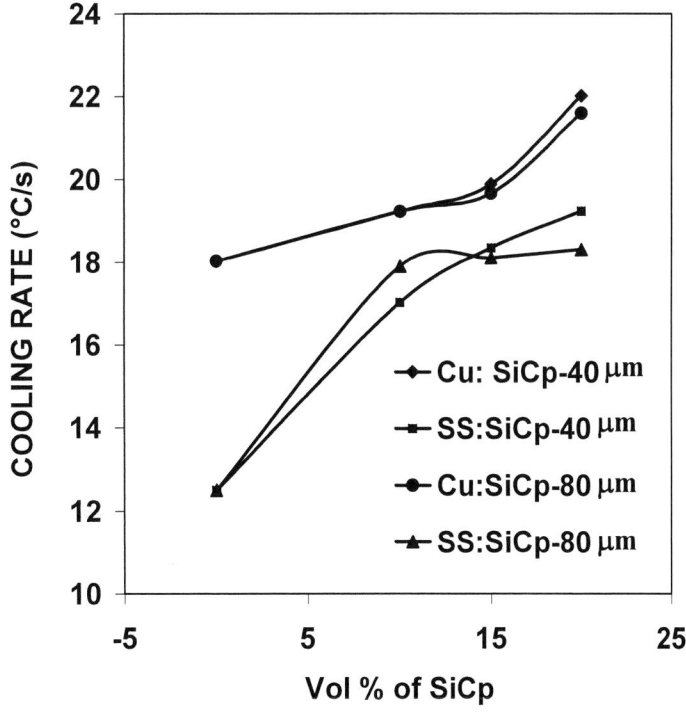

Fig. 4: Variation of cooling rate of the alloy with vol% of SiC_p.

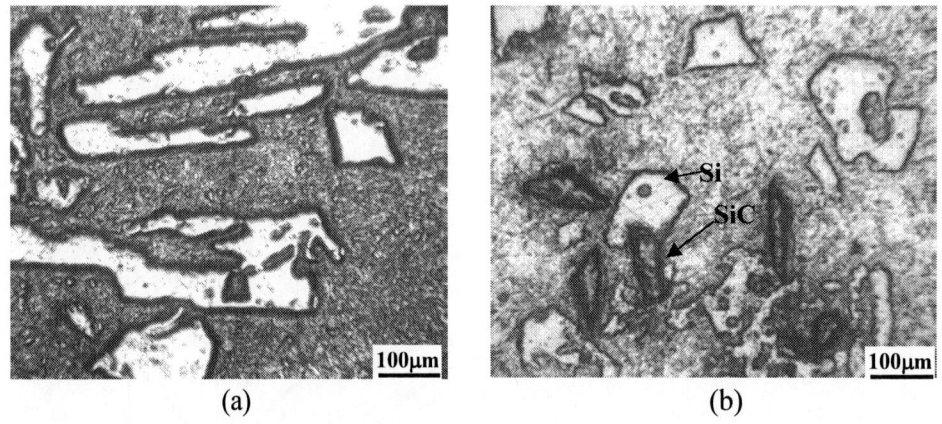

Fig. 5: Microstructure of (a) matrix alloy and (b) composite with 10 vol% SiC_p solidified in stainless steel mould.

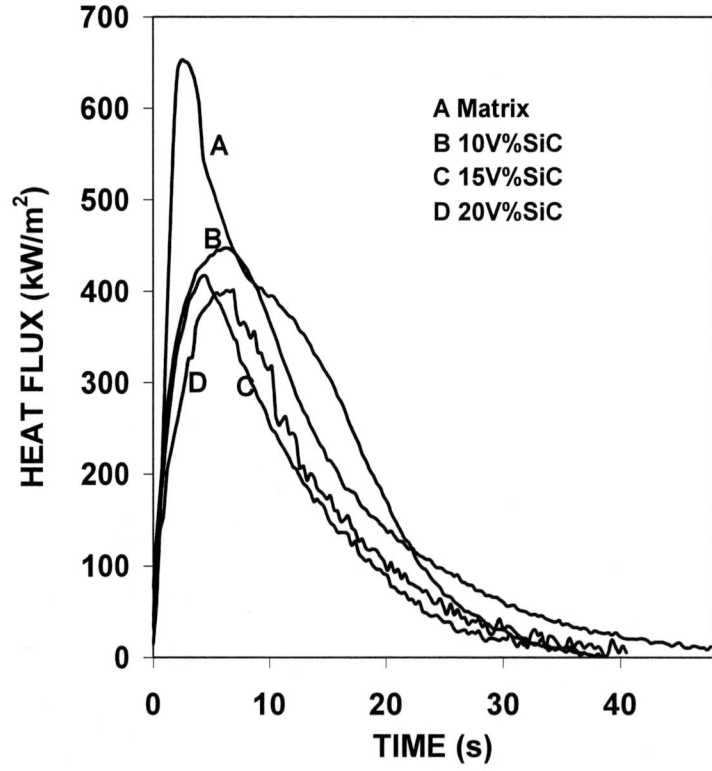

Fig. 6: Variation of interfacial heat flux with time for composite of 80 μm SiCp solidified in stainless steel mould.

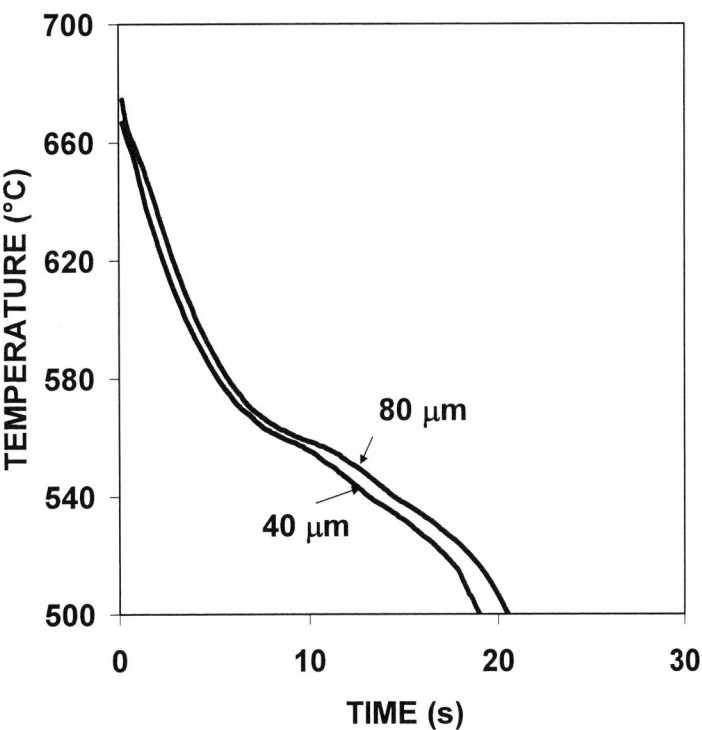

Fig. 7: Cooling curves for composite solidified in stainless steel mould with 20 vol% SiCp.

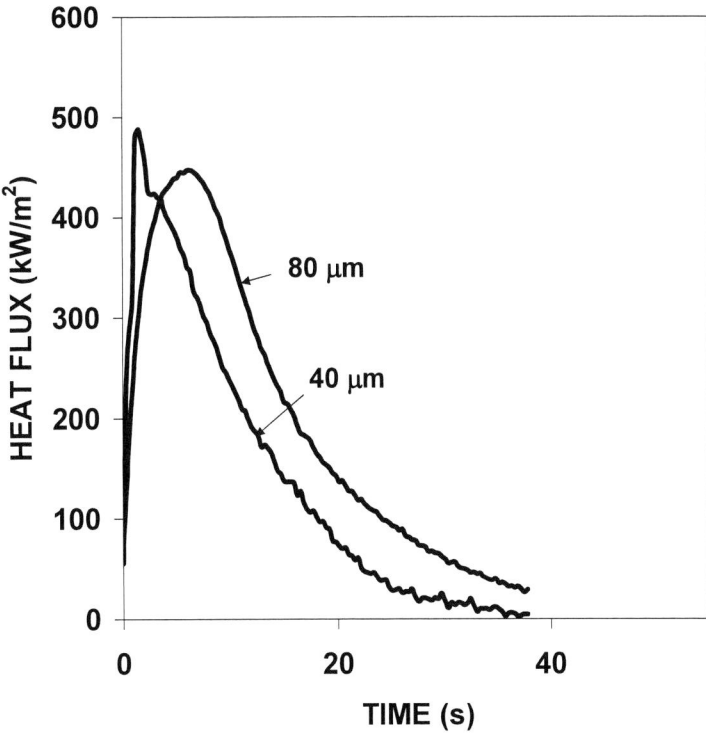

Fig 8: Variation of interfacial heat flux with time for composite of 10 vol% SiCp solidified in stainless steel mould.

Processing of Al-Mg/Al$_2$O$_3$ Interpenetrating Composites by Pressureless Infiltration

H Chang*, J G P Binner*, R L Higginson*, R Sambrook**.

*IPTME, Loughborough University, UK, **Hi-Por Ceramics Ltd, Sheffield, UK.

Abstract

Al-Mg/Al$_2$O$_3$ interpenetrating composites have been produced at atmospheric pressure by infiltrating 2-10 wt.% magnesium content Al-Mg alloys into 20% dense Al$_2$O$_3$ foams with highly interconnected porosity. The processing parameters of temperature, $\geq$900°C, and atmosphere, flowing N$_2$-Ar, were investigated to determine the processing window and infiltration kinetics. *In-situ* observation of the process shows that infiltration is faster at higher temperatures, Mg contents and N$_2$ partial pressures. Both optical and scanning electron microscopy (SEM) have been used to characterize the composites. Analysis using energy dispersive spectroscopy (EDS) has revealed that nitridation occurs in the composites with longer holding times at the processing temperature.

Key words Interpenetrating composites, pressureless infiltration, Al-Mg alloys, Al$_2$O$_3$ foams, microstructure.

Introduction

Metal Matrix Composites (MMCs) with rigid ceramic reinforcements embedded in a ductile metal matrix combine metallic and ceramic characteristics together leading to more attractive physical and mechanical properties, such as higher modulus, strength, hardness, better wear resistance and reduced coefficient of thermal expansion [1]. Amongst the MMCs, interpenetrating composites, which consist of 3-dimensional interpenetrating phases throughout the structure, are emerging as a new class of materials promising to provide superior application-tailored multifunctional properties compared with traditional particulate or fibre reinforced composites due to their unique structures and the capability of varying reinforcement contents.

One of the most widely used methods to fabricate interpenetrating composites is to infiltrate molten metals into ceramic foams or powder beds [2]. Whilst infiltration under pressure, such as squeeze casting, offers a high efficiency, it has difficulty in fabricating complex shaped components and risks damaging the ceramic preform. As a result, pressureless infiltration techniques, which have advantages in overcoming these drawbacks and are potentially cost effective [3], have attracted much attention. In this investigation, Al-Mg/Al_2O_3 interpenetrating composites have been fabricated using a pressureless infiltration technique. Processing parameters have been modified to yield a fast infiltration rate since this is the primary disadvantage of the process.

Experimental

Al-Mg alloys, with magnesium contents ranging from 2-10wt.%, were prepared by metal mould casting of commercially pure aluminum and Mg-Al alloy AZ81. Compositions of the alloys were determined by Scanning Electron Microscopy (SEM) equipped with Energy Dispersive Spectroscopy (EDS) and are listed in Table 1.

Alumina foams, with a density of 20% and completely open porosity in a form of approximately spherical cells interconnected by windows, were prepared using a gel-casting technique [4]. The average cell diameter and window diameter were ~350 µm and ~90 µm respectively.

To infiltrate the molten metals into the ceramic foams, discs of Al-Mg alloy were placed on top of blocks of Al_2O_3 foam and heated together at 30°C/min inside a tube furnace in a controlled atmosphere. A piece of quartz glass was fitted to the furnace tube to enable in-situ observation of the process. Temperatures from 900 to 940°C were used to study the effects of infiltration atmosphere; air, nitrogen and nitrogen/argon mixtures with varying nitrogen contents were used. A holding time of several minutes was typical for total infiltration of a 9 mm height perform when infiltration occurred successfully, otherwise a period of 1 hour at temperature was used before cooling in the flowing gas stream.

After infiltration, the samples were cut longitudinally for the measurement of infiltration height and microstructure observation. Both the sectioned alloys and the composites were carefully ground and polished using diamond paste. For optical examination, the samples were anodized using fluroboric acid and observed using polarized light microscopy. For the SEM study, the samples were given a final polish using colloidal silica before insertion into the microscope. The microscope used was a LEO VP 1530 FEG SEM equipped with TSL EDS data collector.

Results and Discussion

Composite Processing

Average infiltration rates for the alloys at 915°C are shown in Fig.1; at 900°C, only the Al-10Mg alloy infiltrated successfully, no infiltration was observed with the lower magnesium content alloys. At 915°C it can be seen that infiltration rates increased with increasing magnesium content. Given that the only driving forces in the system were from capillary action and gravity (since the infiltration direction was downward) and the latter was the same for all experiments, the results can be explained by the increased wettability of the alloys that contained more magnesium.

As expected, no infiltration occurred in air at any temperature investigated as a result of surface oxidation of the alloy. Hence, work focused on the use of the flowing N_2 and N_2/Ar atmosphere. Infiltration rates as a function of nitrogen content for the Al-10Mg alloy at 915°C are shown in Fig. 2. The highest infiltration rate was achieved in the pure nitrogen atmosphere, whilst no infiltration at all was observed in the pure argon atmosphere. Saravanan *et. al* [5] studied the effect of nitrogen on the surface tension of pure aluminum and concluded that for temperatures higher than 850°C, nitrogen decreases the surface tension well below that obtained in pure argon as a result of the reaction to form aluminium nitride. Hence it is probable that infiltration was facilitated by nitridation.

The influence of temperature on the infiltration process is shown in Fig. 3 for the Al-10Mg alloy in pure flowing N_2. It can be seen that to obtain substantial infiltration a minimum temperature of around 910°C was required. Studies on the wettability of molten aluminium on alumina [6] have shown that the contact angle decreases with increasing temperature and hence less time is needed for wetting at higher temperatures, this is likely to be the explanation of the current results.

Microstructure Observation

The microstructure of the anodized composite is shown in Fig. 4. As the infiltration conditions were not ideal infiltration was not complete and hence excess metal remained on the top of the composite. The infiltration direction is shown in the micrograph. It can be seen that whilst the residual metal is columnar in nature, the infiltrated metal consists of fine grains that nucleated preferentially on the cell walls with grain growth being prevented by contact with other grains and the cell walls.

A typical SEM microstructure of the interpenetrating composite is shown in Fig. 5; it had been processed at 900°C in flowing N_2 for 1 hour. The brighter phase is the aluminium and the darker phase is alumina. Good metal-ceramic interface contacts may be observed and a density of >95% has been obtained. A higher magnification view of a composite produced under the same conditions is shown in Fig. 6, whilst the SEM-EDS analysis in Fig. 7 reveals both a nitrogen and magnesium rich layer at the interfaces.

In contrast, when the temperature was raised to 915°C full infiltration occurred in just 15 mins. Consequently, although the temperature was higher the shorter time period meant that no nitridation was observable in the composite, Fig. 8, a result that is not unexpected [7]. In addition, the magnesium map reveals that this element is preferentially distributed around the alumina cell walls and the grain boundaries between the infiltrated aluminum grains.

Conclusions

From the results, it is apparent that both magnesium and nitrogen are required for the pressurless infiltration of molten aluminium into alumina foams at temperatures in the range 900-915°C. To obtain the highest infiltration rate, a high magnesium-content alloy, pure nitrogen atmosphere and a temperature of at least 910°C are the most favorable conditions; an interpenetrating composite measuring 9 mm in height can be fabricated in just 10 min under these conditions. Whilst premature termination of infiltration has often been reported by other researchers, this was not observed in the current work, which suggests that it is potentially possible to obtain composites several centimeters high. At the same time, since the degree of nitridation will inevitably depend on the holding time, it is clearly necessary to have tight control of the processing parameters.

References

1. Tjong S C and Ma Z Y, *Microstructural and mechanical characteristics of in situ metal matrix composites,* Materials Science and Engineering R: Reports, **29**, [3-4], 2000, pp 49-113.
2. Mortensen A, *Comprehensive composite materials*, in: Clyne T W (Ed.), Metal Matrix Composites, 3, Elsevier, 2000, pp 521-554.
3. Rao B S and Jayaram V, *New technique for pressureless infiltration of Al alloys into Al2O3 preforms*, Journal of Materials Research, **16**, [10], 2001, pp 2906-2913.
4. Sepulveda P and Binner J G P, *Processing of cellular ceramics by foaming and in situ polymerization of organic monomers,* Journal of the European Ceramic Society, **19**, [12], 1999, pp 2059-2066.

5. Saravanan R A, Molina J M, Narciso J, García-Cordovilla C and Louis E, *Effects of nitrogen on the surface tension of pure aluminium at high temperatures,* Scripta Materialia, **44**, [6], 2001, pp 965-970.
6. Ksiazek M, Sobczak N, Mikulowski B, Radziwill W, and Surowiak I, *Wetting and bonding strength in Al/Al2O3 system,* Materials Science and Engineering A, **A324**, [1-2], 2002, pp162–167.
7. Aghjanian M K, Rocazella M A, Burke J T, Keck S D, *The fabrication of metal matric composites by a pressureless infiltration technique,* Journal of Material Science, **26**, [1], 1991, pp 447-454.

Acknowledgements

The authors would like to thank EPSRC and Dytech Co. UK for the financial support of the project and the latter for supply of the ceramic foams. One of the author (H. Chang) would like to thank the ORSAS Committee, UK for financial support.

Tables

Table 1. Compositions of the aluminum alloys as-cast. Other elements present constituted <0.02 wt% for all the alloys, the balance was Al.

Alloy	Actual Mg content / wt.%
Al-2Mg	2.34
Al-6Mg	6.27
Al-8Mg	8.25
Al-10Mg	9.87

Figures

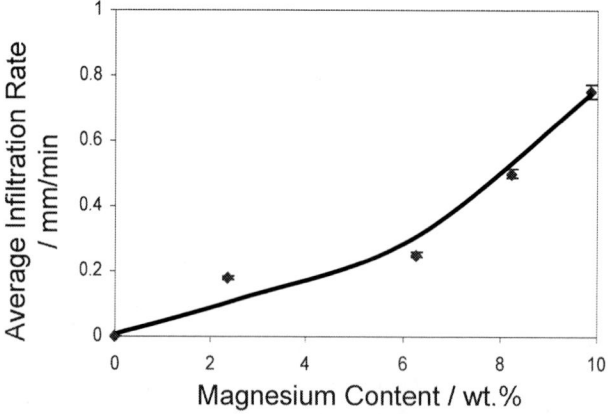

Fig. 1. Average infiltration rates as a function of Mg content in the Al-Mg alloy in pure nitrogen at 915°C.

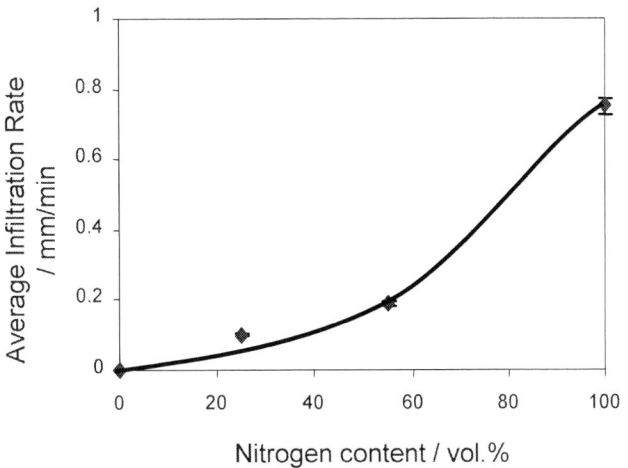

Fig. 2. Average infiltration rates for the Al-10Mg alloy as a function of N_2 content in the N_2-Ar atmosphere at 915°C.

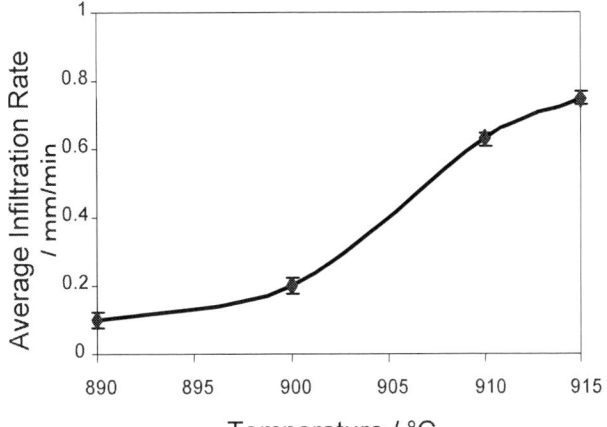

Fig. 3. Average infiltration rates for the Al-10Mg alloy as a function of temperature in a flowing pure N_2 atmosphere.

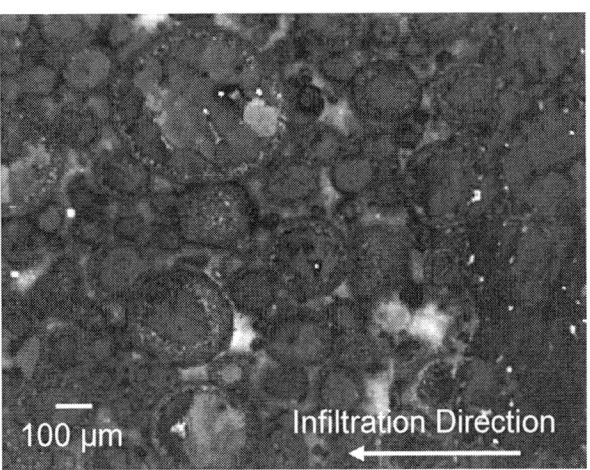

Fig. 4. Optical microstructure of a composite, after anodizing, made with Al-10Mg processed at 900°C in flowing N_2 for 1 hour.

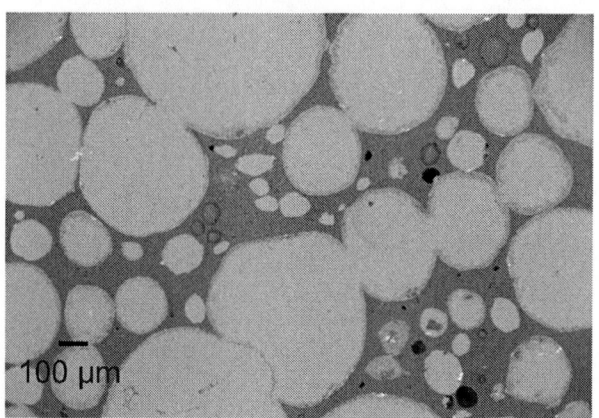

Fig. 5. Typical morphology of a composite. Al-10Mg alloy infiltrated at 900°C in flowing N_2 for 1 hour.

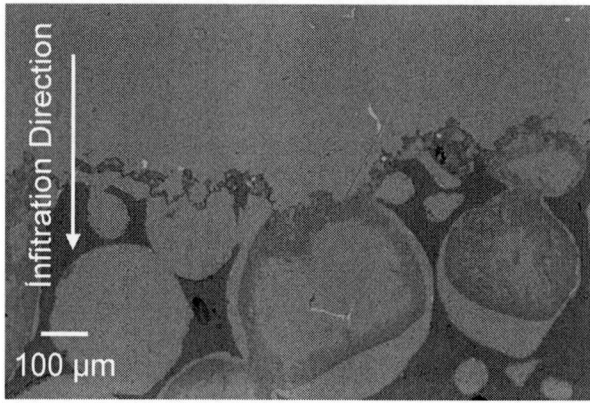

Fig. 6. Metal-ceramic foam interface of an Al-10Mg alloy processed at 900°C in flowing N_2 for 1 hour.

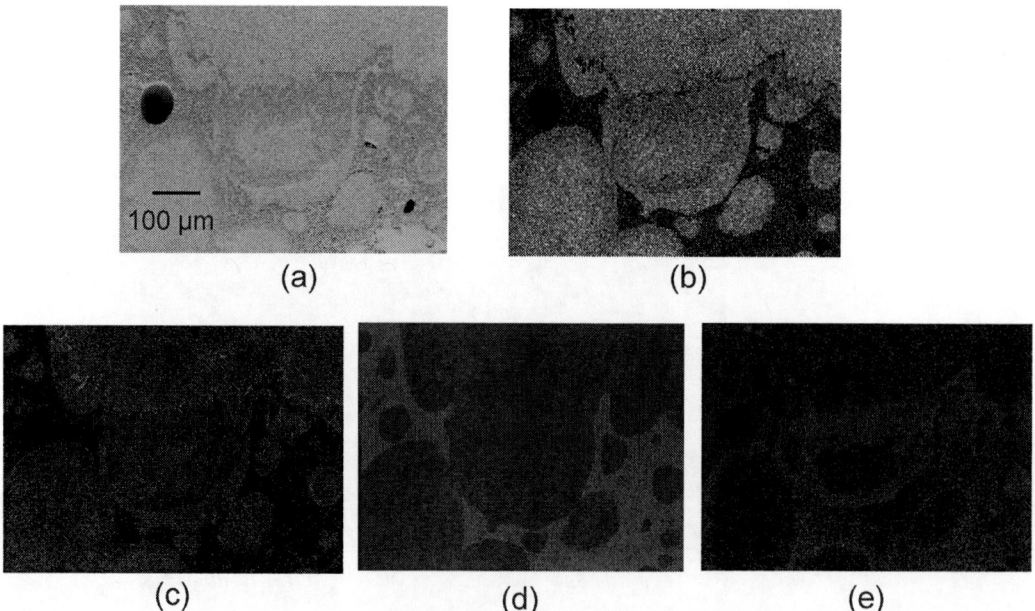

Fig. 7. SEM-EDS map of the composite in Fig. 6 showing nitridation, (a) zero loss image; (b) Al; (c) Mg; (d) O; (e) N.

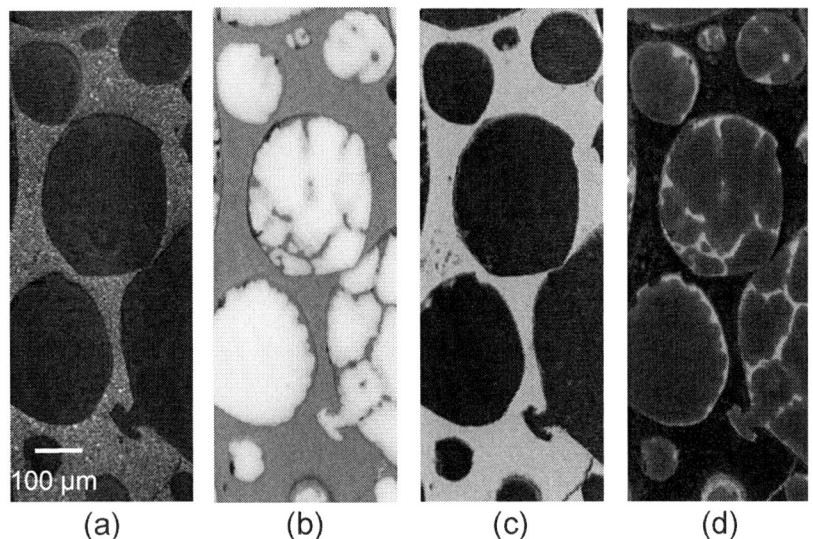

(a)　　　　　　(b)　　　　　　(c)　　　　　　(d)

Fig. 8. SEM-EDS map of a composite processed at 910°C in flowing N_2 for 15 minutes, (a) zero loss image; (b) Al; (c)O; (d) Mg.

H Chang: H.Chang@lboro.ac.uk
J G P Binner: J.Binner@lboro.ac.uk
R L Higginson: R.L.Higginson@lboro.ac.uk
R Sambrook: Rod.Sambrook@dytech.co.uk

Spontaneous Infiltration Mechanism of Al-Si Melt into SiCp Preform

Hideo Nakae[*], Yuuji Araoka[**] and Yuuta Sugiyama[***]
[*]: Dept. Materials Sci. and Eng. Waseda Univ., Tokyo, Japan
[**]: Graduate Student, Waseda Univ., Present Address: NHK Spring Co., LTD.
[***]: Graduate Student, Waseda Univ., Present Address: Nissan Motor Co. Ltd

Abstract

To produce partially composed metal matrix composites, a new spontaneous infiltration process was developed. The infiltration process should be controlled by a wetting phenomenon; therefore, we discussed how to improve the wettability based on the reactive wetting concept. The SiC preforms infiltrated at 1173 K without pressure in air using Al-Si-3Mg melts. The $60mm\phi \times 30mmh$ preforms were fabricated using a water glass binder with Fe_2O_3 powder and for 2h and dried at 393K for 12h before the infiltration. The infiltration rate was measured using a load sensor based on the weight change of the preform with the infiltration. It was confirmed that the infiltration rate was accelerated by the Fe_2O_3 powder addition.

Keywords: spontaneous infiltration, wettability, exothermic reaction, discontinuous MMCs, aluminum composites

Introduction

The further improvement of mechanical properties of metals have been requested. The production process of metal matrix composites (abbreviated MMCs), which are usually superior in their mechanical properties than that of the base metals, has been proposed by many researchers[1-6]. However, their application has not been widely accepted due to their high production costs, especially their poor machinability and the cost of ceramic fibers. Therefore, the production of partially composed near net-shaped composites using SiC particles, SiCp, has been investigated[7,8] using the spontaneous infiltration process.

The famous Lanxide process[1,2] requires a long production time for the production, sometimes up to a dozen hours, in a nitrogen atmosphere. Therefore, the cost is very high for commercial applications except for special cases such as electric vehicles[9]. On the other hand, the infiltration process with a high pressure has been widely adopted for the production of discontinuous MMCs[4,5] for commercial use.

For the production of discontinuous MMCs by the spontaneous infiltration technique, a good wettability between the liquid metals and the solid ceramics should be indispensable, and is one of the most important factors for production[10]. Xi et al.[11] tried to fabricate SiCp/Al-Si composites by a spontaneous process and succeeded at 1673K. One of the authors has investigated the effect of the chemical reaction on the wetting phenomenon, the reactive wetting, for a long time[12-14]. We then applied the reactive wetting concept for the production of discontinuous MMCs consisting of SiCp and a liquid aluminum alloy. We used the Fe_3O_4 powder containing water glass as the binder for the acceleration of the chemical reaction between SiCp and the aluminum melt as described in our previous work[15]. In this research, we discussed the effect of Fe_2O_3 addition due to the high reactivity than that of the Fe_3O_4 during the infiltration[7]. The infiltration was usually completed within 300s at 1173K after dipping, but sometimes, it does not completely occur. Therefore, we have to clarify the main reason why the infiltration sometimes does not take place. We will discuss the infiltration mechanism for the establishment of a reliable production method.

Wettability and infiltration

If the contact angle is less than 90°, the liquid infiltrates into the porous solid without pressure, nevertheless, if the contact angle is greater than 90°, an external force is needed for the infiltration as shown in Fig. 1.

The wettability of SiC by molten aluminum alloys is not very good at temperatures lower than 1223K[16,17]. Also, it is well known that the chemical reaction between a solid and a liquid and the exothermic reaction significantly improve their wettability[12,13,18]. Therefore, we used water glass as the binder, and Fe_2O_3 powder for accelerating the exothermic chemical reaction between the binder and the aluminum melt[7]. Moreover, we added 3mass%Mg to Al-12.6mass%Si alloy for improving the wettability. The chemical reaction and the exothermic calorification of SiO_2, Fe_3O_4 and Fe_2O_3 with pure aluminum are shown by the following equations.

$$4Al + 3\,SiO_2 = 3Si + 2\,Al_2O_3 + 0.62MJ \qquad 0.31MJ/mol\ of\ Si \qquad 1)$$
$$8Al + 3Fe_3O_4 = 9Fe + 4Al_2O_3 + 3.24MJ \qquad 0.36MJ/mol\ of\ Fe \qquad 2)$$
$$2Al + Fe_2O_3 = 2Fe + Al_2O_3 + 0.85MJ \qquad 0.425MJ/mol\ of\ Fe \qquad 3)$$

These equations show that the exothermic calorification of Fe_2O_3 is greater than those of SiO_2 and Fe_3O_4[7]. Moreover, the partial oxygen pressure of Fe_2O_3 is higher than that of Fe_3O_4[19]. Therefore, we selected Fe_2O_3 powder as the additive and studied the effect of the Fe_2O_3 addition on the spontaneous infiltration.

Experimental Procedure

The green SiCps passing through 36, 70 and 120 mesh sieves were used in this research because of their low cost and high hardness. The SiCp was coated with a water glass binder. The binder was prepared by mixing, using a ball mill, $90cm^3$ of water glass with $240cm^3$ of pure water and with or without 50g of Fe_2O_3 or Fe_3O_4 powder. Five cm^3 of the mixed binder was added to the 60g of the SiCp for the production of the preform, having the size of 60mmϕ × 30mmh. The SiCp preform was presintered in a furnace at 1173K for 2h and dried at 393K for 12h before the infiltration, and infiltrated with the aluminum melts in the air from 1073K to 1273K. To prevent the formation of Al_4C_3, we prepared the Al-12.6mass%Si melts, alloyed with or without 3mass%Mg for the improvement of the wettability[15,20], and the Al-Si alloy was made of 99mass%Al and 99mass%Si.

The preform was preheated above the aluminum melt for 5min and dipped into the melt with a weight for canceling the buoyancy. For the confirmation of the infiltration rate and the onset, the infiltration condition was monitored by the weight change during the infiltration using a load sensor, as shown in Fig. 2[15]. The infiltration rate was evaluated by the weight increase in the preform due to the infiltration, i.e., the replacement of the air by the melt. We can confirm not only the infiltration rate, but also the onset of the infiltration at the same time. For the confirmation of the exothermic calorification in the system, we measured the temperature in the preform using a sealed C.A. thermocouple during the experiment.

Most of the infiltrations were carried out perfectly under our experimental conditions, nevertheless, sometimes, the infiltration was imperfect or no infiltration at all occurred. Therefore, we have to clarify the main reason why the infiltration does not take place under identical conditions. At first, we suspected the iron content in the melt due to the exothermic reaction, and the moisture in the preform. To clarify the reason, the infiltration experiments, using an iron contaminated melt and within an inert gas atmosphere, were carried out. We then confirmed the infiltration mechanism, which is based on the wettability and the decrease in the atmospheric pressure in the preform[212,22)]. These details will be discussed later.

Experimental results
Influence of Fe₂O₃ addition on infiltration

Fig. 3 shows the influence of the Fe_2O_3 and Fe_3O_4 powder additions on the infiltration rate at 1173K. The infiltrations finish mostly within 300s. Nevertheless, as can be clearly seen, the effect of the Fe_2O_3 powder addition is more remarkable than that of the Fe_3O_4 addition or no addition. The infiltration times, t_i, of the Fe_2O_3 and Fe_3O_4 powders added systems are 130s and 230s, respectively. On the other hand, without the powder addition, the infiltration time is 300s, nevertheless, their infiltration conditions are different. In the case of no powder addition, the infiltration gradually occurs within 200s, namely the onset time, t_o, is 100s and the finishing time, t_f, is 300s. On the contrary, for infiltration of the Fe_2O_3 system, the onset is nearly identical with that of the non-added one, nevertheless, the t_f is 130s, then the infiltration time, t_i, is less than 30s, t_i = t_f - t_o. If we define the infiltration time, as mentioned above, the Fe_2O_3 system is the shortest. In the Fe_3O_4 system, the infiltration rate is the middle. Moreover, all of these data consist of the incubation time and infiltration time, which shows the infiltration rate. The incubation time is nearly constant, nevertheless, the infiltration time significantly changes with the additives as already mentioned. What does this mean?

The micro-structure of the cross sectional sample in the Fe_2O_3 system infiltrated at 1173K is shown in Fig. 4. We could not find any cavity in the composites. Therefore, it means that the air in the preform must be perfectly thrown out, namely replaced by the infiltrated aluminum or reacted with the Mg or Al in the melt which forms the oxide and the nitride.

The infiltration is usually finished within 300s at 1173K after the dipping under all conditions. Sometimes, it does not completely occur if we repeated these experiments more than ten times or not. Therefore, we have to clarify the main reason why the infiltration does not take place in a few cases. We have to discuss the infiltration mechanism due to this phenomenon for the establishment of the reliable production process.

Influence of infiltration temperature and exothermic reaction

Fig. 5 shows the effect of the temperature at 1073, 1123, 1173 and 1273K on the infiltration for the Fe_2O_3 system. As you see, the infiltration time, ti, decreases and the rate increases with the temperature, nevertheless, the incubation time is nearly constant as mentioned before. The infiltration rate at 1273K can be considered as infinite. What does it means?
For the confirmation of the exothermic calorification in the Fe_2O_3 system, the preform temperature during the infiltration was monitored by the C.A. thermocouple, and the result is shown in Fig. 6. The onset of the temperature increase is consistent with the onset of the infiltration, namely, at the time of 120s, and the rise of 5K. This means that the exothermic

reaction occurred as shown in equation 3.

Reliability of infiltration

Fig. 7 shows the histogram of the infiltration times in our experiment for the Fe_2O_3/system at 1173K. Usually, the infiltration times are less 200s, as already mentioned, nevertheless, in some cases, we could not find any infiltrations for 8 out of 65 samples. An overview photo of one sample is shown in Fig. 8. As you see, the surface of the preform is partially covered with the oxide film of aluminum and the bare SiC preform can be observed in many palaces.

Fig. 9 shows that the influence of the iron contents in the melt on infiltration. We cannot find any significant affect on the infiltration using the 3mass% of Fe in the melt. The difference should be within the error as shown in Fig. 7. Therefore, the main factor is not the iron content. When the infiltration comes into existence, the onset must be the decrease in the atmospheric pressure in the preform due to the chemical reaction between the air and the Mg and Al in the melt to form the oxide and the nitride. We then tried to change the atmosphere from air to Ar gas for the confirmation of our estimation. The infiltration does not perfectly occur in an Ar atmosphere. This is the verification that our estimation is correct. Therefore, we estimated that the influence of moisture in the preform, which was adsorbed during the long storage time in the atmosphere before the dipping, prevent the infiltration due to the evaporation.

We prepared two kinds of preforms, which were dried at 393K for 12h just before the infiltration, and the other ones were left in the atmosphere for 12h after the drying. The seven infiltration experiments were carried out alternately using these two kinds of preforms. The experimental results are shown in Table 1. If the samples were left for 12h in order to adsorb moisture before dipping, the infiltration does not completely occur. Therefore, the main reason for the non-infiltration is the moisture in the preform adsorbed from the air during handling.

Discussion

We first postulated that the spontaneous infiltration mechanism is based on the improvement of the wettability between the SiCp and the Al-Si melt. Therefore, one of the main reasons should be the decrease in the contact angle to less than 90° due to the reaction wetting. Nevertheless, it is not possible to directly measure the contact angle using the sessile drop method, because this chemical reaction does not occur at atmospheric pressure.

We already confirmed[15] the existence of an intermetallic compound in the infiltrated metal matrix by EPMA. There are many needle-like intermetallic compounds, namely, Al_3Fe in the liquid portion. Therefore, we confirmed that the thermit reaction occurs between Fe_2O_3 and the aluminum melt to form Al_3Fe in these systems. By the way, these thermit reactions must accelerate the infiltration.

The chemical reaction of the melt with the trapped air decreases the pressure in the preform as already mentioned. This reduction provides a small infiltration by the suction of the melt into the preform. By this small infiltration, the oxide film of the aluminum melt must be stretched and broken to create a bare aluminum surface, which means an oxide free surface. Therefore, the surface is very reactive for the reaction between the air and the melt. It is well known that the required pressure for the infiltration, P, depends on the radius of the pore, r, and the contact angle, θ. The required infiltration pressure can be calculated as follows:

$$P = 2\gamma_{LV}\cos\theta/r$$

where, γ_{LV} is the surface tension of the melt.

If we look at the surface of the infiltrated SiCp, as shown in Fig. 10, the surface is perfectly wetted[15]. This means that the contact angle between the SiCp and the melt should be nearly 0°. Nevertheless, we think that the slight suction of melt, due to the decrease in the pressure, should accelerate the chemical reaction to make a true contact between the melt and the iron oxide powder. Therefore, the infiltration is accelerated by the reactive wetting with the addition of Fe_2O_3 into the water glass binder accompanied by a decrease in the pressure in the preform. Therefore, the moisture in the preform, which is evaporated during the dipping, makes a thick oxide film at the preform /melt interphase. The oxide film prevents the onset of the chemical reaction, and then the infiltration does not completely occur.

Conclusions

We have developed a new fabrication process for discontinuous metal matrix composites without pressure at 1173K in air using preformed SiC particles formed with the water glass binder. The process is controlled by the thermit reaction between the melt and the additives in the binder, which consisted of a diluted water glass binder mixed with Fe_2O_3 and Fe_3O_4 powders. The onset of the infiltration is controlled by the chemical reaction between the trapped air in the preform and the Mg and or Al in the melt that decreases the pressure in the preform, and this decrease in the pressure induces the infiltration without pressure, namely a spontaneous infiltration.

References

1) M.K.Aghajanian et al.: Proceedings of 34th Int'l. SAMPE Symposium, 1989, pp.817-823
2) J.T.Burke, M.K.Aghajanian, and M.A.Rocazella.: ibid. pp.2440-2454
3) Inorganic Matrix Composites: M.K.Surappaq Ed., TMS, 1996
4) T.Donomoto et al., SAE Paper No.830252, 1983, pp.1-11
5) T.Suenaga, J.J.F.S., Vol.64, 1992, pp.881-886
6) C-H.Andresson and R.Warren: Composites 15, 1984, pp.16-23
7) H.Nakae, K.Ito and Y.Sugiyama: Proc. 8th. AFC, 2003, 383-391

8) H.Nakae, H.Yamaura, T.Miyamoto and T.Yanagihara: 2nd. Int'l Cof. Processinjg Materials for Properties. TMS, 2000, pp.165-168

9) T.Suganuma, J.JFS., 73, 2001, pp.829-833

10) G.P.Martins, D.L.Olson, and R.Edwards, Metallurgical Transaction B 19B, 1988, pp.95-101

11) X.M.Xi, L.M.Xiao, and X.F.Yang: J.Mater.Res. 11, 1996, pp.1037-1044

12) N.Yoshimi, H.Nakae and H.Fuji: Materials Trans.JIM. 31, 1990, pp.141-147

13) H.Fujii, H.Nakae and K.Okada, Metallurgical Transaction A 24A, 1993, pp.1391-1397

14) H.Nakae and A.Goto: Proc. High Temperature Capillarity, 1997, pp.12-17.

15) H.Nakae, H.Yamaura and Y.Sugiyama, J.J.F.S., 75, 2003, pp.29-34

16) V.Laurent, D.Chatain and N.Eustathopoulos, J.Mater. Sci. 22, 1987, pp.244-250

17) W.Köhler: Aluminium 51, 1975, pp.443-447

18) H.Nakae et al.: Proc. High Temperature Capillarity, 1994, pp.306-310

19) L.S.Darken and R.W.Gurry, Physical Chemistry of Metals, McGraw-Hill, 1953, pp.352-

20) T.Iseki, T.Kameda, and T.Maruyama, J.Materials Sci. 19, 1984, pp.1692-1698

21) H.Nakae et al., Proceedings of 2nd. PMP, 2000, pp.165-168

22) T.Kimura, H.Yamaura and H.Nakae, Proceedings of 2nd. PMP, 2000, pp.135-140

Table

Table 1 Influence of preform conditions on infiltration at 1173K

No. of Ex.	1d	2a	3d	4a	5d	6a	7d	8a
ti, s	130	no	140	no	150	no	140	no

d*:dried, a**:moisture absorbed, no***: no infltration

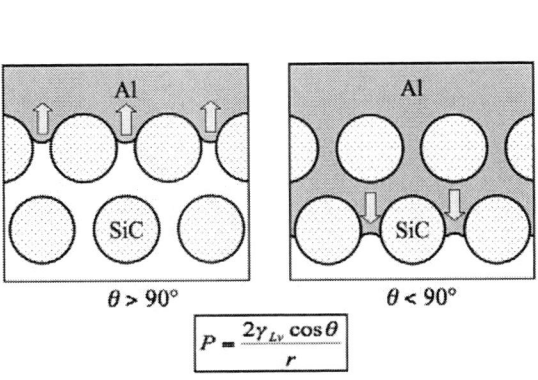

$$P = \frac{2\gamma_{Lv}\cos\theta}{r}$$

Fig. 1 Schematic representation of infiltration conditions with contact angles

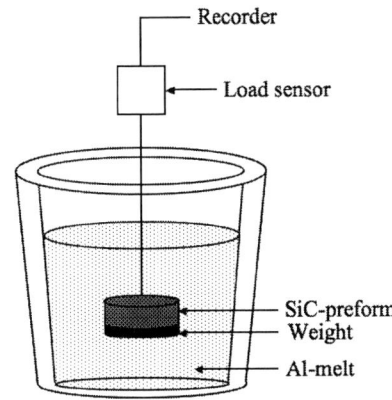

Fig. 2 Measurement method for infiltration rate using a load sensor

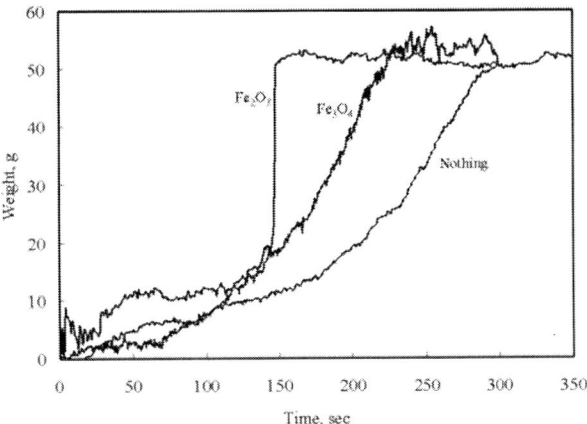

Fig. 3 Influence of Fe_2O_3 and Fe_3O_4 addition on infiltration at 1173K

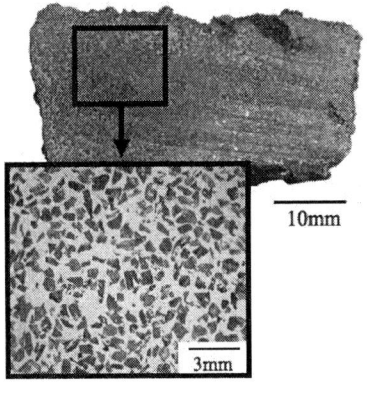

Fig. 4 Cross sectional microstructure of SiCp perform infiltrated with Fe_2O_3 at 1173K

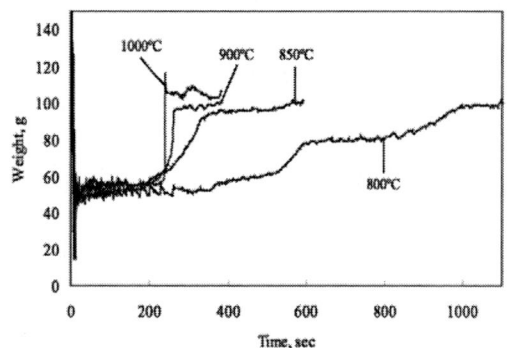

Fig. 5 Effect of temperature on infiltration in Fe_2O_3 systems

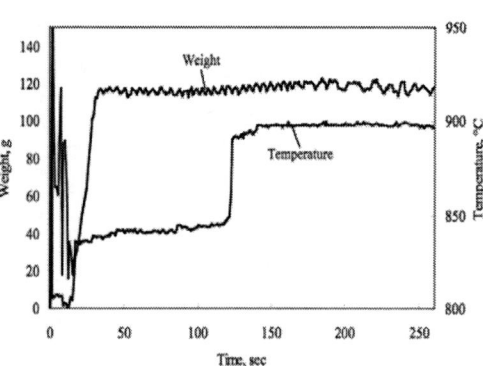

Fig. 6 Change in temperature of preform during infiltration

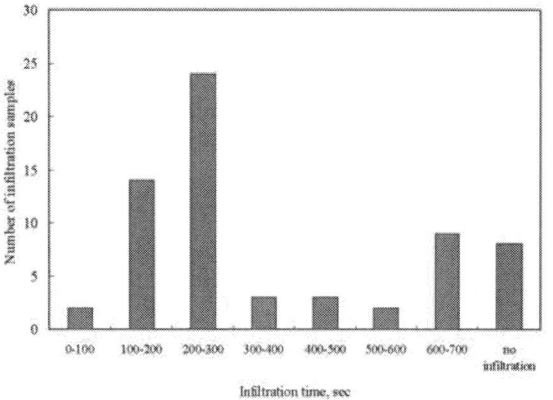

Fig. 7 Histogram of infiltration times at 1173K.

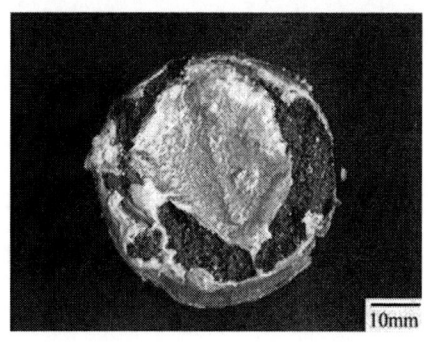

Fig. 8 Over view of non-infiltrated sample at 1173K

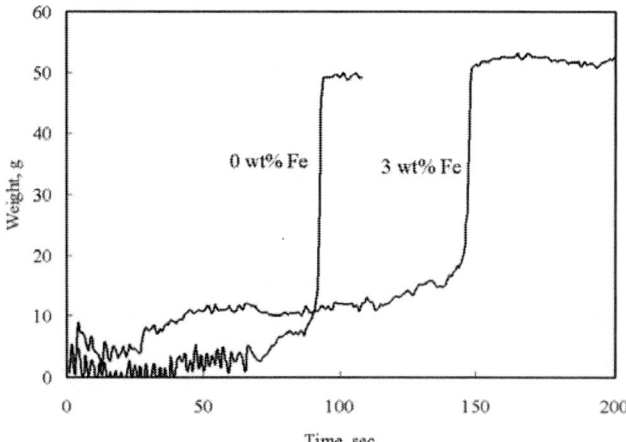

Fig. 9 Influence of iron content in melt on infiltration.

Fig. 10 Over views of preforms before and after infiltration

Fabrication of Short Alumina Fiber and In-situ Mg$_2$Si Particle-Reinforced Magnesium Alloy Hybrid Composite and Its Strength Properties

K. Asano* and H. Yoneda*

* Kinki University, Japan

Abstract

Magnesium alloy matrix composites reinforced with short alumina fibers and *in-situ* formed Mg$_2$Si particles were fabricated in a permanent mold by the infiltration with the molten magnesium alloy into the preforms consisting of fibers having Si particles attached to their surfaces. P or CaF$_2$ particles were used as the refiners of the Mg$_2$Si particles. Fine Mg$_2$Si particles of approximately 5μm grain size were formed due to the rapid solidification in the permanent mold, regardless of the introduction of the refiners. The tensile strengths of the composites that ranged from 293K to 623K were investigated. The strength of the alumina fiber-reinforced composite, in which the Mg$_2$Si particles were homogeneously dispersed in the matrix, was higher than that of the conventional fiber-reinforced composite.

Key words

composite, alumina fiber, magnesium, *in-situ* Mg$_2$Si, high temperature strength

Introduction

To improve the high-temperature strength of magnesium alloys, their reinforcement with heat-resistant ceramics fibers has been presented. We have fabricated an alumina fiber-reinforced AZ91D magnesium alloy composite, and revealed that the strength of the composite at 523K was 160MPa, which was superior to that of the unreinforced alloy [1]. To develop a lightweight composite which has an increased high-temperature strength available for high-performance pistons, we proposed the dispersion of heat-resistant particles in the matrix. The homogeneous dispersion of the particles leads to an improvement of the heat-resistance of the matrix, or the improvement of the stress transmission between fiber and matrix by preventing the fiber-to-fiber contact. We fabricated the magnesium alloy matrix composites reinforced with the short alumina fibers and in-situ formed Mg_2Si, which is hard, and has a low density and high-melting point [2], by the infiltration of the preform consisting of short alumina fibers and pure Si particles with the molten magnesium alloy, and investigated the conditions to finely disperse the Mg_2Si [3]. As a result, it was found that the rapid solidification after the infiltration effectively dispersed the fine Mg_2Si particles, and that the introduction of P or CaF_2 particles as the refiner of Mg_2Si was also effective [4]. We have deduced that the composite with finer Mg_2Si particles could be obtained by squeeze casting, because the infiltration in the permanent mold leads to rapid solidification.

In the present study, magnesium alloy matrix composites reinforced with the short alumina fibers and the Mg_2Si particles were fabricated by squeeze casting. The microstructure and tensile strength of the composites ranging from 293K to 623K were investigated and compared with that of the conventional fiber-reinforced composite. Furthermore, the effect of the P or CaF_2 particles as the refiner was investigated.

Experimental

The AZ91D magnesium alloy (Mg-9.2mass%Al-0.7mass%Zn alloy, hereinafter called AZ91D) was used as the matrix. Short alumina fibers (Saffil, from ICI, diameter: 3.5μm, length: 200μm, tensile strength: 1000MPa, hardness: 700HV) were used as the reinforcement. Pure Si particles (99.9mass%Si) were used as the starting material to form the Mg_2Si. Si particles with an average size of 5μm and 50μm were used to investigate the effect of size on the morphology of Mg_2Si. P or CaF_2 particles with an average size of 5μm was used as the refiner.

The preforms were fabricated as follows. The fibers, Si particles and refiner were dispersed and agitated in an aqueous medium containing an organic binder (polyvinyl alcohol) and an inorganic binder (Al_2O_3 sol), and then the fibers having Si particles and refiner attached to their surfaces were dewatered to fabricate a cylindrical preform of 55mm diameter and 30mm height, followed by drying and sintering. The fiber volume fraction of the composite was set at 18%, because the high-temperature strength of the short alumina fiber-reinforced AZ91D composite became a maximum at 18% [1]. The fiber volume fraction of the preform was set at 15%, because the preform contraction during the infiltration occurs which leads to a 3% increase in the fraction of the composite. The volume fraction of Si particle was set at 3.2%, and that of refiner was set at 0.3%. Fig.1 is the scanning electron micrographs of the preforms, showing that the Si particles and refiner (arrows in the figure) were attached to the fibers and held among the fibers. The composites were fabricated by squeeze casting. The fabrication parameters are as follows: melt and preform temperature, 1003K; mold temperature, 673K; applied pressure, 40MPa. Optical microscopy and a Vickers hardness test (98N,15s) of the composites were performed on the planar section. The tensile test specimens were machined parallel to the planar direction, with a gage length of 10mm, a cross sectional width of 6mm and a thickness of 3mm. The measured tensile strength was ranged from room temperature (293K) to 623K and the fracture surfaces were observed by scanning electron microscopy.

Results and Discussion

Fig.2 shows the macrostructure of the vertical section of a specimen fabricated with the 5μm Si particle-attached preform. In the figure, the dark part is the composite part and the light part is the unreinforced part. It can be seen that the melt infiltration was completely accomplished with no observable defects. The fiber volume fraction estimated from the height of the composite part (approximately 25mm) was 18%, which was the prescribed value. The macrostructures of the composite fabricated with the 50μm Si particle-attached preform (hereinafter called composite 50Si) and the composites with the refiners were almost same as shown in Fig.2.

Fig.3 shows the change in hardness (HV) with the distance from the upper surface of the composite fabricated with the 5μm Si particle-attached preform (hereinafter called composite 5Si). Every region has the same hardness value of 170HV. The hardness of composite 50Si and the composites with refiners was also 170HV in every region of each composite, indicating that the homogeneous composites can be obtained by the present process.

Fig.4 shows the microstructures of the composites. Alumina fibers, which appear black in the microstructures, were oriented as random configurations. Fine granular phases are also observed in the matrices of the composites. Examination of the composite by EPMA and XRD revealed that the granular phase was Mg_2Si. Fine Mg_2Si particles were homogeneously dispersed in the matrix of the composite 5Si (Fig.4(a)). Fine Mg_2Si particles were also formed in the composite 50Si (Fig.4(d)), and this observation suggests that Si dissolves in the melt and the Mg_2Si particles are rapidly formed [2] in the present experiment. However, agglomerated Mg_2Si particles were observed in the composite 50Si (Fig.4(d)). A similar tendency was observed when P or CaF_2 particles were introduced (Fig.4(b)(c)(e)(f)). During the infiltration, the average size of the Mg_2Si particle was approximately 5μm regardless of the Si particle size or the introduction of the refiners. The average cooling rate between the melt temperature and the eutectic point of α-Mg and Mg_2Si in the Mg-Si system (912K) [5] was measured at the center of the preform. It was 36K/s and we think it is high enough to refine the Mg_2Si particles without the introduction of P or CaF_2 particles. Since residual Si was not observed in every composite, it can be stated that the Si particles in the preform changed to the Mg_2Si particles during infiltration and were then dispersed in the matrix.

Tensile tests were performed with the composites without the refiners, because no difference between the microstructure of the composite with and without refiners was observed. The tensile strength of the composites was compared with that of the 18% alumina fiber-reinforced composite without Mg_2Si particles (hereinafter called fiber-reinforced composite) obtained in the previous study [1]. The tensile strength of the matrix (AZ91D) and the composites is plotted versus the temperature in Fig. 5. Since the tensile strength of the fiber-reinforced composite was higher than that of AZ91D at every temperature, reinforcement with the fibers is effective for improving the strength. The tensile strength of the composite 5Si was further higher than that of the fiber-reinforced composite. The strength at 523K was approximately 180MPa, which is higher than that of the heat-resistant aluminum alloy. The strength at 623K was 160MPa, which is higher by 75MPa (approximately 90%) compared to that of AZ91D. On the other hand, the strength of the composite 50Si was lower than that of the fiber-reinforced composite at every temperature, although higher than that of AZ91D.

Subsequently, fractography was used for examining the effect of the dispersion of the Mg_2Si on the strength. SEM micrographs of the fracture surfaces of the composites after tensile testing at 293K are shown in Fig. 6. For the fiber-reinforced composite, many fibers longitudinally oriented in the stress direction fractured without fiber-pullout, and the fiber surfaces transversely oriented in the stress direction were only slightly seen (Fig.6(a)). These results suggest that the interfacial bond between fibers and matrix was strong. However, several bunches of a few fibers were seen (Fig.6(a) circle). On the other hand, there was almost no bunching on the fracture surface of the composite 5Si (Fig.6(b)). Several cleavage planes were observed on the fracture surface of the composite 50Si (Fig.6(c) circle), suggesting that brittle fracture occurred in these areas. It would appear that these areas correspond to the agglomerated areas of the Mg_2Si particles shown in Fig.4(d). Fig.7 shows the fracture surfaces of the composites after tensile testing at 623K, indicating that the variation in the fracture surface morphology due to the Mg_2Si particles was similar to that at 293K.

From these results, the strength mechanism by the dispersion of the Mg_2Si particles can be discussed. The reason for the improvement in the composite strength by the homogeneous dispersion of the fine Mg_2Si particles is considered as follows: (1) Mg_2Si particles reduce the grain size of the matrix (α-Mg) and prevent the grain growth even at high-temperature [6], and (2) a homogeneous dispersion of the fine Mg_2Si particles reduces the fiber-to-fiber contact and the stress concentration at the contact point, and thus the stress transmission between the fiber and the matrix becomes easy. As shown in Fig.6(b) and Fig.7(b), the fiber bunches were not observed on the fracture surface of the composite 5Si, suggesting that the Mg_2Si particles dispersed in the matrix reduced the fiber contact. Previous researches have demonstrated that reinforcement with a large fiber fraction reduced the strengthening effect of the fiber due to the increase in the fiber contact points [7] and increased the smooth fiber surfaces on the fracture surface [8]. However, as we demonstrated in a previous study on the strength properties of the continuous alumina fiber-reinforced aluminum alloy composite, the strength was improved by the particle dispersion because the particles prevented the fiber-to-fiber contact [9]. Also in the present study, the dispersion of the fine Mg_2Si particles would reduce the fiber-to-fiber contact and the stress concentration at the contact point, and thus the stress transmission between the fiber and the matrix becomes easier. The result that the tensile strength of the composite 5Si was higher than that of the fiber-reinforced composite can be explained by the dispersion effect. However, if the agglomerated Mg_2Si particles exist, cracks would first generate in the agglomerated area when the stress was applied because the deformability of this area is poor. A previous study demonstrates that the cracks easily join and thus the strength of the composite decreases when the agglomerated area of the brittle reinforcement is large or these areas extensively exist [10]. A decrease in the strength due to the presence of the coarse brittle phase was shown for the fiber- or whisker-reinforced aluminum alloy composites [11]. In addition, as shown in Fig.4(d), the Mg_2Si particles were heterogeneously dispersed in the composite 50Si. In this situation, a fiber-to-fiber contact might occur in the area without the Mg_2Si particles. The result that the tensile strength of the composite 50Si was lower than that of the fiber-reinforced composite can be explained by these effects.

Conclusions

1. By infiltration in the permanent mold, fine Mg_2Si particles having the average size of 5 µm were formed regardless of the introduction of the refiners, and dispersed in the matrix of the composite.
2. Finer Si particles as the starting material are favorable for homogeneously dispersing the *in-situ* formed Mg_2Si particles in the present process.
3. Ranging from 293K to 623K, the tensile strength of the alumina fiber-reinforced composite, in which the fine Mg_2Si particles were

homogeneously dispersed in the matrix, was higher than that of the conventional fiber-reinforced composite or AZ91D. The tensile strength of the composite with the homogeneously dispersed Mg_2Si particles was approximately 180MPa at 523K, which is higher than that of the heat-resistant aluminum alloy at the same temperature. This high strength is a consequence of the reduction in the fiber-to-fiber contact and the improvement of the stress transmission between the fiber and the matrix.

References

1. K. Asano and H. Yoneda, *Strength of AZ91D Magnesium Alloy Composite Reinforced with Alumina Short Fiber Using Alumina Binder*, J. JFS, 74, No.11, Nov 2002, pp706-713.

2. G. H. Li, H. S. Gill and R. A. Varin, *Magnesium Silicide Intermetallic Alloys*, Metall.Trans. 24A, Nov 2003, pp2383-2391.

3. K. Asano and H. Yoneda, *Refinement of In-Situ Formed Mg_2Si Particles in Fiber-Reinforced Magnesium Alloy*, 66th World Foundry Congress, Istanbul, Turkey, 6-9 Sept, 2004, pp1172-1180, pub Toksad: The Foundrymen's Association of Turkey, 2004.

4. K. Asano and H. Yoneda, *Effect of P and CaF_2 on Refinement of In-Situ Mg_2Si in Fiber-Reinforced Magnesium Alloy*, Report of JFS Meeting, 144, May 2004, p.86.

5. T. B. Massalski, Binary Alloy Phase Diagrams, ASM International, 1990, p.2548.

6. Y. Kojima, Handbook of Advanced Magnesium Technology, Kallos Publishing co., ltd., Japan, 2000, p.71.

7. H. Akbulut, M. Durman and F. Yilmaz, *A Comparison of As-cast and Heat Treated Properties in Short Fiber Reinforced Al-Si Piston Alloys*, Scripta Mater., 36, No.7,1997, pp835-840.

8. T. Shinkawa, H. Kageyama, S. Kamado and Y. Kojima, *Structures and Mechanical Properties of Hybrid Mg-Zn-Ca Alloy Composites Reinforced with δ -Al_2O_3 Short Fiber and $9Al_2O_3 \cdot 2B_2O_3$ Whisker*, J.Jpn.Inst.Light Met., 46, No.12, Dec 1996, pp650-655.

9. K. Asano and H. Yoneda, *Effects of Particle-Dispersion on the Strength of an Alumina Fiber-Reinforced Aluminum Alloy Matrix Composite*, Mater. Trans., 68, No.6, Jun 2003, pp1172-1180.

10. H. Morimoto, H. Iwamura, K. Ohuchi and Y. Ashida, *Fabrication of Thin Wall Tubes of SiC Whisker Reinforced 6061 Aluminum Alloy Composite by Extrusion and Their Mechanical Properties*, J.Jpn.Inst.Light Met.,45, No.2, Feb 1995, pp82-87.

11. C. M. Friend, I. Horsfall and C. L. Burrows, *The Effect of Particulate: Fiber Ratio on the Properties of Short-Fiber/Particulate Hybrid MMC Produced by Preform Infiltration*, J.Mater.Sci., 26,1991, pp225-231.

Figures

Fig.1 SEM micrographs of the preforms.

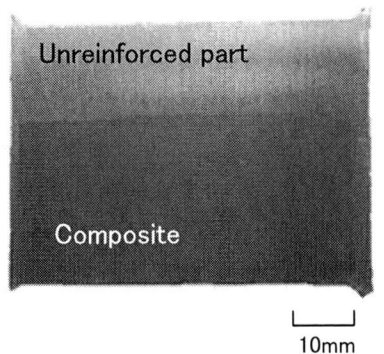

Fig.2 Macrostructure of a vertical section of a specimen (composite 5Si, without refiner).

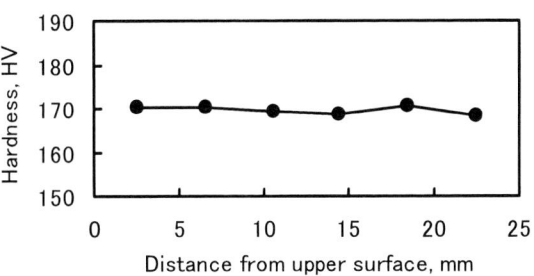

Fig.3 Hardness distribution of the composite 5Si (without refiner).

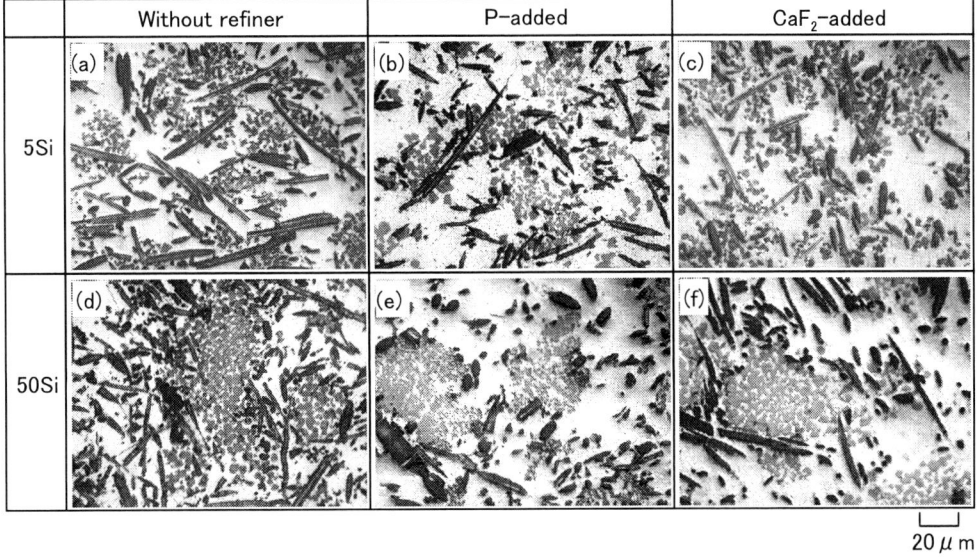

Fig.4 Microstructure of parallel section of the composites.

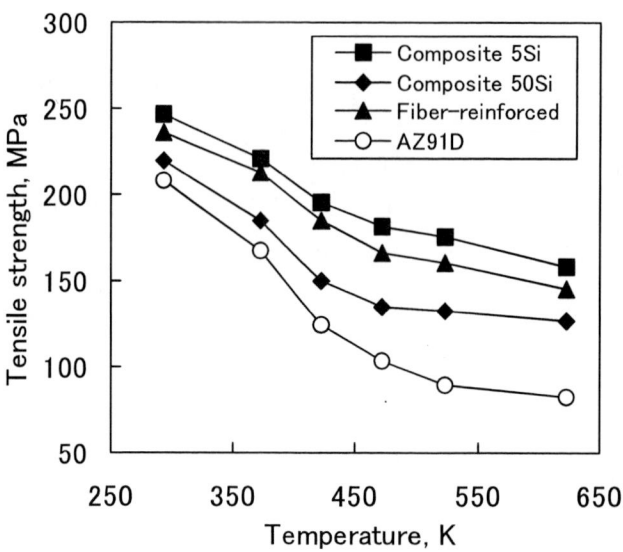

Fig.5 Effect of temperature on the tensile strength of AZ91D and the composites.

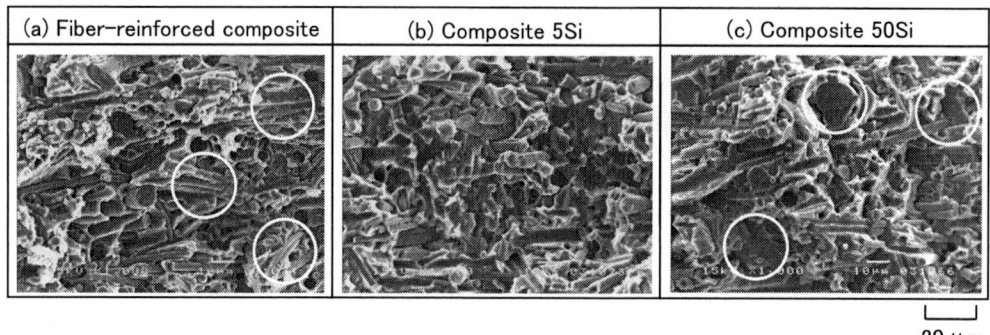

20 μm

Fig.6 SEM micrographs of fracture surfaces of the composites (without refiner) after tensile testing at 293 K .

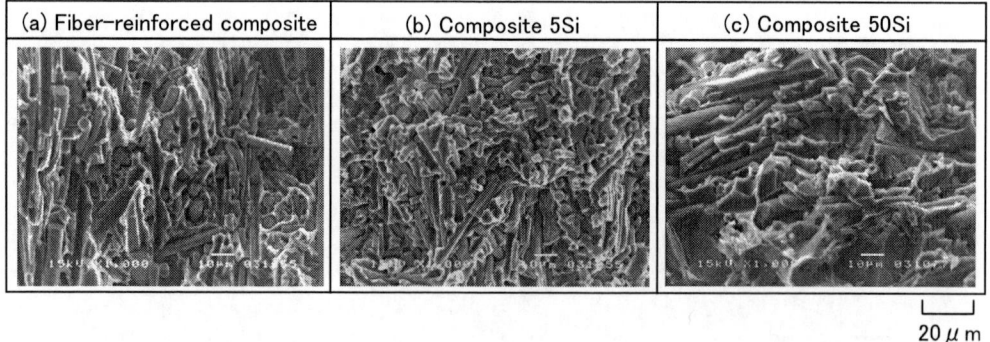

20 μm

Fig.7 SEM micrographs of fracture surfaces of the composites (without refiner) after tensile testing at 623 K.

E-mail : asano@mech.kindai.ac.jp

Production, evaluation and comparison of mechanical, wear, and corrosion properties of Al-Flyash1 and Al-Flyash2 metal matrix composite

M.Ramachandra * and K.Radhakrishna *.

* Dept. of Mechanical Engineering
  BMS College of Engineering
  Bull Temple Road
  Bangalore-560019
  Karnataka-INDIA

Abstract

In the present investigation, Aluminium based particulate composites containing 15 weight percentages of two types of flyash particulates as reinforcement were synthesized using stir-casting method and their properties are compared. The properties like density, sliding wear, hardness, slurry erosive wear and corrosion resistance of the composites were investigated and their properties are compared. The results of wear studies have shown that the resistance to wear increases with the inclusion of flyash as compared to base metal. The addition of flyash particles reduces the density of composite while increasing the hardness. Corrosion resistance decreases with increase in flyash content. The macrostructural and microstructural characteristics of the MMC were investigated with particular emphasis on the distribution of flyash particles in the matrix.

Key words: Aluminium matrix composite (AMC), Flyash1(FA1), Flyash2 (FA2), Wear, Corrosion.

Introduction

The emergence of novel processing techniques coupled with the need for lighter materials with high strength and stiffness has catalyzed considerable scientific and technological interest in the development of numerous high performance composite materials as serious competitors to the traditional engineering alloys. The limitations of conventional monolithic metals and alloys in respect to combination of properties have led the development of hybrid materials in the form of composites [1–4].

The suitability of MMCs as a viable replacement of the conventionally used monolithic materials, however, depends on the acquisition of scientific understanding in order to synthesize them with consistent reproducibility in microstructure and mechanical behavior and their ability to exhibit enhanced performance with cost effectiveness in real time applications [5]. The invention of novel processing techniques coupled with the ability of metal matrix composites to unify the ductility and toughness of metallic materials with the strength and modulus of ceramic materials has been instrumental in the insurgence of global research activities. The reinforced MMCs offer opportunities for sufficient improvement in efficiency, reliability and mechanical performance over traditional base metals. In particular, the particulate reinforced metal matrix composites are attractive because they exhibit near isotropic properties compared to continuously reinforced counterparts and are easier to process using standard metallurgical methods. The primary disadvantage of all MMC's however, is that they suffer from low ductility and inadequate fracture toughness compared to their constituent matrix material [6].

Now a day the main focus is given to Aluminium as matrix material because of its unique combination of good corrosion resistance, low electrical resistance and excellent mechanical properties. For the past few years new particulate composite containing flyash has been developed. Flyash is a light weight byproduct of coal combustion, which is separated from the exhaust gases of power generating plants [7] and used as filler or functional extenders in plastics, paints, resins and additive to cement. Flyash particles are very light materials with a density of 2.1-2.6 g-cm^{-3} for precipitator flyash and density as low as 0.4-0.6 g-cm^{-3} for cenosphere particles [8]. Flyash is being used as filler in aluminium matrices [9] and various components such as pistons, engine cover, and connecting rod castings have been made out of cast aluminium alloy – flyash composite [10]. A Variety of castings such as pistons, connecting rods, and engine covers were made to demonstrate the usage of AMC (aluminium matrix composites). The incorporation of flyash particles in aluminium alloy matrix can lead to the production of low cost aluminium composite which can find widespread applications in automotive small engines, transportation and electromechanical machinery sectors.

In the present investigation aluminium alloy (Al-Si) with two types of flyash as reinforcement material was synthesized and some of the properties like hardness, density and wear resistance and corrosion resistance were measured.

Experimental

Materials:

The matrix material used in the experimental investigation was an Aluminium alloy (**Si-12.2%**) whose chemical composition (in weight percent) is listed in **Table 1**. This alloy has a composition very close to the Al-Si eutectic. It therefore has a low melting point (577°C). Aluminium and silicon have no solid solubility below the eutectic and the microstructure solidifies as silicon particles in an aluminium matrix. Aluminium-silicon alloy in its unmodified state is extensively used in sand casting and die-casting.

The reinforcement material used in the investigation was two types of flyash particulates (Flyash1-FA1 and Flyash-FA2) of assorted size with an average particle size of 10μm. Particle size was estimated by sieve analysis. The spheroidal flyash particles contain both solid spheres (precipitators) and hollow spheres (cenospheres). The cenosphere content was found to be 61 wt % in FA1 and 43% for FA2. Gravity separation method was used to find the cenosphere content. Cenospheres are hard, hollow, free flowing, microspheres found in flyash. **Table 2** and **Table 3** gives composition of FA1 and FA2 as found in electron discharge analysis.

Processing:

The synthesis of the metal matrix composite used in the present study was carried out by using stir casting method. The cleaned aluminium ingots were melted to the desired super heating temperature of 800°C in graphite crucibles under a cover of flux using 3-phase electrical resistance furnace. The molten metal was degassed at a temperature of 780°C. Flyash particulates, preheated to around 500°C were then added to the molten metal and stirred continuously by using mechanical stirrer at 720°C. The stirring time was maintained between 5–8 minutes at an impeller speed of 550 rpm. During stirring, Magnesium was added in small quantities to increase the wettability of flyash particles. The dispersion of the preheated flyash particulates was achieved in accordance with the vortex method [11]. The melt with the reinforced particulates were poured into the dried, coated, cylindrical permanent metallic moulds of size 50mm diameter and 175mm height. The pouring temperature was maintained at 680°C. The melt was allowed to solidify in the moulds. For the purpose of comparison of Al-FA1 and Al-FA2 composites, the base alloy was cast under similar processing conditions as described.

Density:

Density measurements were carried out on the base metal and reinforced samples using Archimedes's principle. This principle is useful for determining the volume and therefore the density of an irregularly shaped object by measuring its mass in air and its effective mass when submerged in water (density = 1gram/cc).

Hardness:

Bulk hardness measurements were carried out on the base metal and composite samples by using standard Brinnel hardness test. Brinell hardness measurements were carried out in order to investigate the influence of particulate weight fraction on the matrix hardness. Load applied was 500 kgs and indenter was a steel ball of 10mm diameter.

Macro and Microstructural characterization:

Macrostructural study was conducted on the as processed and machined composite castings in order to investigate distribution of flyash particles retained in the metal matrix. Castings were plain turned on lathe to remove 5mm of material to reveal the particle distribution on macroscopic scale.

Microstructural characterization studies were conducted on reinforced samples. This is accomplished by using Nikon optical microscope. The composite samples were metallographically polished prior to examination. Characterization is done in etched conditions. Etching was accomplished using Keller's reagent.

Slurry Erosive wear:

Erosive wear is defined as the loss of material from a solid surface due to relative motion in contact with a fluid that contains solid particles [13].

The experimental arrangement for slurry erosive wear consists of stirrer, which can hold 4 specimens at a time, and a water-cooled pot. All 4 specimens were dipped in slurry of distilled water - silica sand and stirred at a speed of 376m/min. The slurry was prepared by mixing 80-micron size silica sand with distilled water with basic, neutral and acidic pH by adding HCl acid and NaOH. The slurry wear test was performed at ambient temperature and testing time was 12Hrs. The specimens for the slurry erosive wear test were cut from composite ingots and plain turned to a diameter of 7mm. Before testing, specimens were weighed to an accuracy of 0.001gms. After testing specimens were dried and re-weighed to determine percentage weight loss.

Fog Corrosion:

A fog of NaCl solution was introduced in to a closed chamber where specimens were exposed at specific locations. The concentration of the NaCl solution used was 3.5%. Corrosive fog was created by bubbling compressed air through hot deionized water - NaCl solution which was maintained at a temperature of 50°c. The specimens for fog corrosion test were prepared by cutting specimens of size 10mmx20mmx5mm from the composite castings. The surface of specimens were abraded by using 600 grit size emery paper and degreased. Before testing, the specimens were weighed to an accuracy of 0.001gms and exposed to corrosive atmosphere for a period of 240 Hrs (10days). The specimens were suspended in corrosive chamber at regular intervals exposing the abraded surface to salt solution fog. After corrosion testing, the specimens were immersed in clark's solution for 5 minutes and gently cleaned with a soft brush to remove adhered particles. After drying thoroughly the specimens were re-weighed to determine the percentage weight loss.

Sliding Wear behavior:

Wear has been defined as the displacement of material caused by hard particles or hard proturberances where these hard particles are forced against and moving along a solid surface [12, 13]. Two body sliding wear tests were carried out on prepared composite specimens. A computerized pin-on-disc wear test machine was used for these tests. The wear testing was carried out at a constant sliding velocity of 95m/min with a normal load of 14.7N. A cylindrical pin of size 5mm diameter and 40 mm length prepared from composite casting was loaded through a vertical specimen holder against horizontal rotating disc. The rotating disc was made of carbon steel of diameter 50mm and hardness of 64 HRC. The principal objective of investigation was to study the coefficient of friction and wear rate.

Results and Discussion:

Density Measurement:

The results of density measurement on the base metal and reinforced materials are shown in Fig.1. The results reveal that reinforcement of flyash particulates in MMC decreases the material density. Lower density results especially because of presence of particles like Cenospheres, which are hollow spheres with very low density of 0.4-0.6gm/cm^3. Earlier studies [13] and the results mentioned above show similar density variations. On comparison, Al-FA1 composite showed lower density.

Hardness Measurements:

The results of bulk hardness measurements conducted on the monolithic and reinforced materials are as shown in Fig.2. The results reveal that an increase in the flyash particulates weight percentage in MMC increases the material hardness. Higher hardness results because of inclusion of flyash particles like cenospheres and presipitators [13] which are very hard particles. On comparison, Al-FA2 composite showed higher hardness.

Macro and Microstructural characterization:

Macrostructural studies revealed reasonable uniform distribution of flyash particles and slight macrosegrigation of particles. The distribution of flyash particles is influenced by the tendency of particles to float due to density differences and interactions with the solidifying metal. It is therefore a strong function of the solidification rate and geometry of castings [9]. Photo macrograph in Fig.3 shows the distribution of flyash particles. Higher concentration of flyash particles was obtained at the top and lower concentration at the bottom of the castings. Central 80% length of castings had near uniform distribution of flyash particles.

Microstructure of matrix material is shown in Fig.4. Microstructural studies of MMC show that there is no void or discontinuities. At some places there was clustering of flyash particulates. The micrograph also shows that there is good interfacial bonding between flyash particles and Al matrix. Good interfacial bonding can be obtained by heating of flyash particulates prior to dispersion and addition of magnesium in small quantities during stirring which improved wettability of flyash particles.

Slurry Erosive wear:

The results of slurry erosive wear test show increase in wear resistance with reinforcement of flyash content. Compared to base metal, composite with 15% flyash showed less weight loss. In all the three mediums: Acidic, Basic and Neutral mediums, Al-FA1 composite showed good wear resistance compared to Al-FA2 composite and base metal. The results are shown in Fig.5, Fig.6 and Fig.7.

Fog corrosion:

The results of salt solution fog corrosion test are shown in Fig.8. The resistance to corrosion is good in base metal compared to flyash reinforced composite specimens. Formation of oxide layer is visible within 6Hrs of commencement of test. The type of corrosion is pitting corrosion. After 24 Hrs it is observed that the formation of pit is more rapid in reinforced samples than unreinforced samples. The presence of flyash particles will act as sites to initiate pits. There will be build up of corroded particle debris in the pits. Pits initiate at flaws within the surface film and at sites where the film is damaged mechanically under conditions in which self repair will not occur. On comparison, Al-FA2 composite showed good corrosion resistance.

Sliding wear behavior:

Fig.9 shows the results of dry sliding wear behavior of base metal and composites with 15 % of FA1 and FA2 in aluminium matrix. From the graph it is evident that the resistance to wear has increased with reinforcement of flyash content compared to base metal. Incorporation of flyash content significantly reduced wear. This is because of the presence of hard flyash particles which will increase the overall bulk hardness of the material. On comparison, it was observed that Al-FA1 showed good wear resistance compared to Al-FA2 composite and base metal.

Conclusions:

1) Two types of Aluminium matrix composites; Al-FA1 and Al-FA2 containing 15 wt. percentage of Flyash1 and Flyash2 was synthesized successfully by using stir casting method.

2) Macro and microstructure revealed near uniform distribution of flyash particles in the center portion of the castings. But there was slight agglomeration of flyash particles on macroscopic scale. The microstructure also revealed good interfacial bond between matrix and flyash particles.

3) The density of composite has decreased with reinforcement of flyash content. Al-FA1 showed lower density compared to base metal and Al-FA2 composite.

4) The bulk hardness of composite increases with reinforcement of flyash content. Al-FA2 showed higher hardness compared to base metal and Al-FA1 composite.

5) The sliding wear resistance of composites has increased with reinforcement of flyash content. Similar results were obtained in slurry erosive wear. Al-FA1 composite showed good slurry erosive wear resistance.

6) The increased wear properties can be made use in the fabrication of bearings, cylinder liners, pistons.

7) Corrosion resistance of reinforced samples has decreased with reinforcement of flyash content.

References

1. S. Suresh, A. Mortensen, Int. Mater. Rev. 42 (3) (1997) 85.
2. S. Schicker, D.E. Garcia, J. Bruhn, R. Janssen, N. Claussen, Acta Met. Mater.46 (7) (1998) 2485.
3. P.C. Maity, P.N. Chakraborty, S.C. Panigrahi, J. Mater. Sci. 31 (1996) 6377.

4. F.M. Hosking, F.P. Folgar, R. Wunderlin, R. Meharbian, J. Mater.Sci. 17(1982) 477.
5. I.A. Ibrahim, F.A. Mohamed and E.J Ievernia, J. Mater.Sci 26 (1991)1137.
6. T.S. Srivatsan, J.Mattingly, J.Met.Sci., 28(1993), 611- 620.
7. Gikunoo, E, Omotoso, O, Oguocha,I.N.A., Mat Sci & Tech, Feb 2005, vol. 21, No. 2, pp. 143-152(10)
8. R.Q. Guo, P.K Rohatgi, D.Nath J. Mat Sci, 32 (1997) 3971-3974.

9. P.K Rohatgi, B.N Keshavaram, P. Huang, R. Guo, D.M Golden, AFS Transactions, 103 (1995) 575.
10. P.K. Rohatgi. J. Metals, Nov, 46 (1994)55.
11. M.K.Surappa and P.K. Rohatgi, Metal Tech., 4, 41(1981).
12. Glossary of Terms and Definitions in the field of Friction, Wear and Lubrication. Research Group on Wear of Engineering Materials, OECD, Paris, 1969.
13. R.L. Deuis,C.Subramanian,J.M.Yellup, Wear 201(1996)132-144

Acknowledgements

The authors would like to thank the authorities of AICTE (All India council for technical education) New Delhi, India for financial support provided during the course of this investigation.

Tables:

Table 1: Chemical composition of Aluminium base alloy

Si	Fe	Cu	Mn	Mg	Zn	Al
12.0	0.2	0.23	0.1	0.4	0.1	Bal

Table 2: Composition of Flyash1 (FA1)

Al_2O_3	SiO_2	Fe_2O_3	TiO_2
30.4%	58.4%	8.44%	2.75%

Table 3: Composition of Flyash 2 (FA2)

Al_2O_3	SiO_2	Fe_2O_3	TiO_2	CaO	Na_2O	K_2O	SO_3
28.06%	6.69%	4.91%	1.37%	3.91%	0.93%	1.51%	0.41%

Figures:

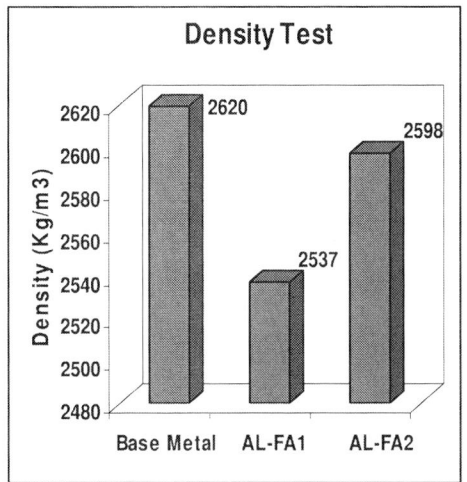
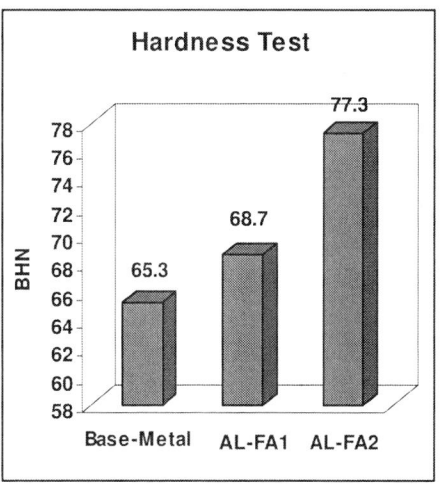

Fig.1: Density variation   Fig.2: Bulk hardness variation

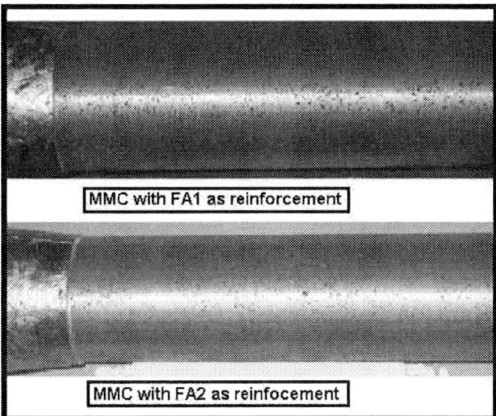

Fig.3: Photomacograph of Al-FA1 and Al-FA2 composites

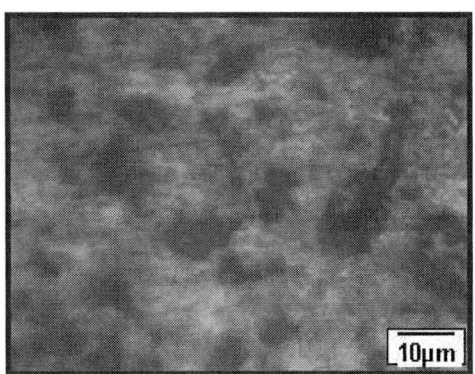
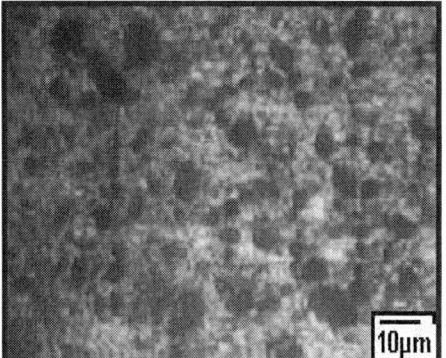

Fig.4: Micrograph of Al-FA1 and Al-FA2 composites

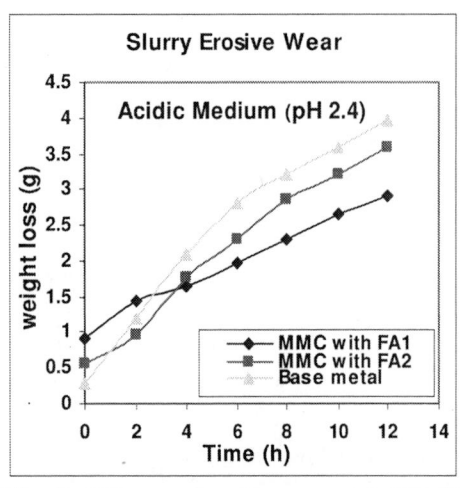

Fig.5: Slurry erosive wear (pH2.4)

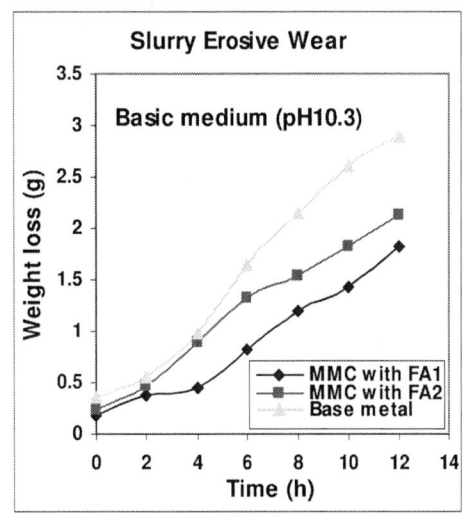

Fig.6: Slurry erosive wear (pH7)

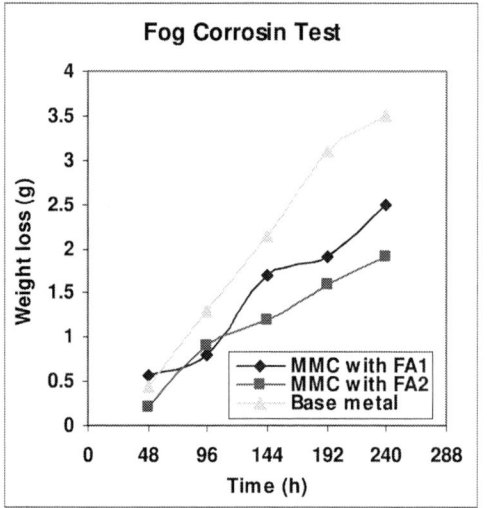

Fig.7: Slurry erosive wear (pH10.3) Fig.8: Fog corrosion test results

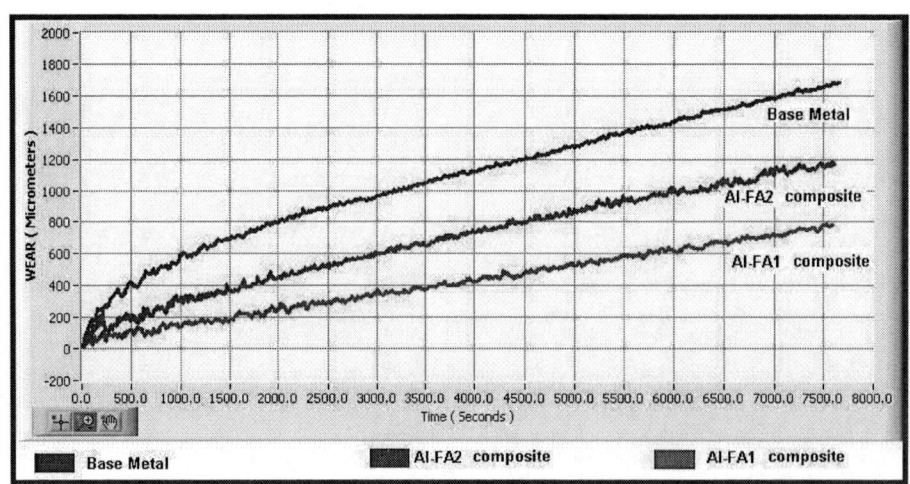

Fig.9: Sliding wear behavior of base metal and composites

Evaluation of mechanical and wear properties of Al-Si(12.2%)-Graphite metal matrix composite synthesized using stir-casting method

M.Ramachandra * and K.Radhakrishna *.

* Dept. of Mechanical Engineering
  BMS College of Engineering
  Bull Temple Road
  Bangalore-560019
  Karnataka-INDIA

Abstract

Aluminium based metal matrix composite containing 15 weight percentage of graphite particulates as reinforcement was synthesized using stir-casting method. The properties like hardness, dry sliding wear resistance and slurry erosive wear resistance of the composites were investigated. Strip fluidity test was conducted to asses the flowability of the melt mixture. The results of wear studies, both sliding wear and slurry erosive wear, have shown that the resistance to wear decreased with the inclusion of graphite as compared to base metal. The addition of graphite particles decreased the hardness. The macrostructural and microstructural studies of the MMC were made with particular emphasis on the distribution of graphite particles in the matrix.

Key words: Aluminium matrix composite (AMC), Graphite particles, sliding Wear, slurry erosive wear.

Introduction

Among the advanced materials that are developed to an increasing extent are metal matrix composites. The space industry was the first sector interested in the usage of these materials and another sector of even greater economic importance for the development of new MMCs is the automotive industry [1]. The flexibility associated with metal matrix composites (MMC'S) in tailoring their physical and mechanical properties as required by the end application have made them suitable candidate for a spectrum of applications related to automobile and aeronautical sectors [2].

MMCs combine metallic properties (ductility and toughness) with ceramic characteristics (high strength and modulus), leading to greater strength in shear and compression and to higher service temperature capabilities. The attractive physical and mechanical properties that can be obtained with MMCs, such as high specific modulus, strength, and thermal stability, have been documented extensively [3-9]. Interest in MMCs for use in the aerospace and automotive industries, and other structural applications, has increased over the past 20 years as a result of the availability of relatively inexpensive reinforcements and the development of various processing routes which result in reproducible microstructure and properties [10].

With the invention of novel processing techniques coupled with the need for lighter materials with high strength and stiffness has catalyzed considerable scientific and technological interest in the development of numerous high performance composite materials as serious competitors to the traditional engineering alloys.The limitations of conventional monolithic metals and alloys in respect to combination of properties have led the development of hybrid materials in the form of composites [11–12].

Now a day the main focus is given to Aluminium as matrix material because of its unique combination of good corrosion resistance, low electrical resistance and excellent mechanical properties. In recent years, attention has been drawn to graphite-particle metal-matrix composites which exhibit superior tribological properties, such as low friction, low wear rate and excellent antiseizure effects. For bearing application, the graphite/aluminum composite saves considerably in cost and weight and has the added benefit of being self-lubricating, compared to the current Cu, Pb-, Sn-, and Cd-containing alloys. These bearings were used for marine diesel engines as they represent one of the most inexpensive and effective bearing materials.

In the present investigation aluminium alloy (Al-Si) as matrix material and graphite as reinforcement material, metal matrix composite (MMC) was synthesized and some of the properties like hardness and wear resistance were investigated.

Experimental
Materials:

The matrix material used in the experimental investigation was an Aluminium alloy (**Si-12.2%**) whose chemical composition (in weight percent) is listed in **Table 1**. This alloy has a composition very close to the Al-Si eutectic. It therefore has a low melting point (577°C). Aluminium and silicon have no solid solubility below the eutectic and the microstructure solidifies as silicon particles in an aluminium matrix. Aluminium-silicon alloy in its unmodified state is extensively used in sand casting and die-casting.

Graphite is a polymorph of the element carbon. Graphite has very different structure and very different properties. Graphite is one of the softest materials and is a good conductor of electricity. It is opaque, good solid lubricant and refractory material. Most graphite is produced through the metamorphism of organic material in rocks.

Processing:

The synthesis of the metal matrix composite (MMC) used in the present study was carried out by using stir casting method. The cleaned aluminium ingots were melted to the desired super heating temperature of 800°C in graphite crucibles under a cover of flux using 3-phase electrical resistance furnace. The molten metal was degassed at a temperature of 780°C. graphite particulates, preheated to around 650°C were then added to the molten metal and stirred continuously. Stirring time was maintained between 5–8 minutes at an impeller speed of 600 rpm. During stirring, Magnesium was added in small quantities to increase the wettability of particles. The dispersion of the preheated graphite particulates was achieved in accordance with the vortex method [11]. The melt with the reinforced particulates were poured into the dried, coated, cylindrical permanent metallic moulds of size 50mm diameter and 175mm height. The pouring temperature was maintained at 700°C. The melt was allowed to solidify in the moulds. For the purpose of comparison, the base alloy was cast under similar processing conditions as described. Simultaneously strip fluidity tests were conducted on the melt mixture. The MMC processing arrangement is shown in Fig.1.

Hardness:

Bulk hardness measurements were carried out on the base metal and composite samples by using standard Brinnel hardness test. Brinell hardness measurements were carried out in order to investigate the influence of particulate weight fraction on the matrix hardness. Load applied was 500 kgs and indenter was a steel ball of 10mm diameter.

Macro and Microstructural characterization:

Macrostructural study was conducted on the as processed and machined composite castings in order to investigate distribution of graphite particles retained in the metal matrix. Castings were plain turned on lathe to remove 5mm of material to reveal the particle distribution on macroscopic scale.

Microstructural characterization studies were conducted on reinforced samples. This is accomplished by using Nikon optical microscope. The composite samples were metallographically polished prior to examination.

Slurry Erosive wear:
Erosive wear is defined as the loss of material from a solid surface due to relative motion in contact with a fluid that contains solid particles [12].

The experimental arrangement for slurry erosive wear consists of stirrer, which can hold 4 specimens at a time, and a water-cooled pot. All 4 specimens were dipped in slurry of distilled water - silica sand and stirred at a speed of 376m/min. The slurry was prepared by mixing 80-micron size silica sand with distilled water. The slurry wear test was performed at ambient temperature and testing time was 12Hrs. The specimens for the slurry erosive wear test were cut from composite ingots and plain turned to a diameter of 7mm. Before testing, specimens were weighed to an accuracy of 0.001gms. After testing specimens were dried and re-weighed to determine percentage weight loss. Fig.2. shows the experimental setup used for slurry erosive wear.

Sliding Wear behavior:
Wear has been defined as the displacement of material caused by hard particles or hard proturberances where these hard particles are forced against and moving along a solid surface [13,14]. Two body sliding wear tests were carried out on prepared composite specimens. A computerized pin-on-disc wear test machine was used for these tests. The wear testing was carried out sliding velocity of 130m/min and a normal load of 14.7N. A cylindrical pin of size 5mm diameter and 40 mm length prepared from composite casting was loaded through a vertical specimen holder against horizontal rotating disc. The rotating disc was made of carbon steel of diameter 50mm and hardness of 64 HRC. The pin on disc wear testing machine is shown in Fig.3.

Results and Discussion:
Fluidity test:
The result of strip fluidity test is shown in Fig.4. Base metal showed good flowability and metal matrix composite with 15% graphite particles showed poor flowability. This may be due to increase in viscosity of the molten metal with the addition of particulates. The addition of graphite particles into the molten metal will impart slurry like characteristic to the molten metal and particles offer resistance to the free flow of liquid metal.

Hardness Measurements:

Fig.5. shows the results of bulk hardness measurements conducted on the monolithic and reinforced samples. The results reveal that the hardness decreases with the addition of graphite particles. This may be due to the lower hardness of graphite particles.

Macro and Microstructural characterization:

Macrostructure revealed near uniform distribution of graphite particles and macrosegrigation. Photo macrograph in Fig.6 shows the distribution of graphite particles. Higher concentration of graphite particles was obtained at the top and lower concentration at the bottom of the castings. Central 70% length of castings had near uniform distribution of graphite particles.

Microstructure of matrix material is shown in Fig.7. Microstructural studies of MMC show that there is no void or discontinuities. The micrograph also shows that there is good interfacial bonding between graphite particles and Al matrix.

Slurry Erosive wear:

The results of slurry erosive wear test show decrease in wear resistance with reinforcement of graphite particles. Compared to base metal, composite with 15% graphite particles showed higher weight loss. The result of slurry erosive wear test is shown in Fig.8. After 8 hours of testing, there is decrease in wear loss. This is due to the formation of protective oxide layer coating on the surface of the specimen. Fig.9. shows the slurry wear specimen with protective layer partially removed.

Sliding wear behavior:

Fig.10 shows the results of dry sliding wear behavior of base metal and composites with 15 % of graphite particles in aluminium matrix. From the graph it is evident that the resistance to wear has decreased with reinforcement of graphite content compared to base metal. This may be due to the reduced hardness of composite with the inclusion of graphite particles.

Optical microscopic examinations of the worn pin surfaces identified different wear mechanisms operating, either singly or in combination, under the various sliding conditions. They are: abrasion, oxidation and thermal softening. Number of grooves, mostly parallel to the sliding direction (Fig.11.), is evident on the worn pins. Such features are characteristics of abrasion, in which hard asperities of the steel counterface, or hard reinforced particles in between the contacting surfaces, plough or cut into the pin, causing wear by the removal of small fragments material. Under the optical microscope, the dark surfaces are found to be covered extensively by a thin layer of fine particles. These characteristics are indicative of oxidative wear, in which frictional heating during sliding causes oxidation of the surface, with wear occurring through the removal of oxide fragments. Under the severe sliding condition of

95m/min and 14.7N, associated with longer contact time of the surfaces, gross plastic deformation of the pin surface occurred and material is extruded from the interface before re-solidifying around the periphery of the pin (Fig.12 (d)).

The worn debris of metal matrix composite with 15% of graphite is shown in Fig.12 (a-c). A substantial quantity of wear debris was generated during the tests. They have three major morphologies, particle-like and flake-like aluminium and black powder of fractured graphite particles. The flakes have irregular shape: some look like sponges; some are like circular lumps; while some have sharp and angular features. Fig.12(c) shows dark powder type mixture containing fractured graphite particles and small Al particles.

Conclusions:

1) Aluminium matrix composite containing 15 wt. percentage of graphite was synthesized successfully by using stir casting method. The fluidity of the molten mixture decreased with the addition of graphite particles.
2) Macro and microstructure revealed near uniform distribution of graphite particles in the center portion of the castings. But there was slight agglomeration of graphite particles on macroscopic scale. The microstructure also revealed good interfacial bond between matrix and graphite particles.
3) The bulk hardness of composite decreased with reinforcement of graphite content.
4) The slurry erosive wear resistance also decreased with the addition of graphite particles. However wear rate has decreased after 8 hours of testing, because of formation of oxide layer.
5) The sliding wear resistance of composite also decreased with reinforcement of graphite content.

References

1. J.Elioasson, R Sandstorm, Key Engineering Materials 104-107(1995)pp3-36
2. P.S Gilman, J.Met, 43(8)(1991)7
3. S.G. Fishman, J. Met. 38 (1986) 26.
4. Y. Flom, R.J. Arsenault, J. Met. 38 (1986) 31.
5. Y. Flom, R.J. Arsenault, Mater. Sci. Eng. 77 (1986) 191.
6. A.H.M. Howes, J. Met. 38 (1986) 28.
7. A. Mortensen, M.N. Gugor, J.A. Cornie, M.C. Flemings, J. Met. 38 (1986) 30.
8. A. Mortensen, J.A. Cornie, M.C. Flemings, J. Met. 40 (1988) 12.
9. V.C. Nardone, K.W. Prewo, Scripta Metall. 20 (1986) 43.
10. T.W. Chow, A. Kelly, A. Okura, Composites 16 (1986) 187.
11. P.K. Rohatgi. J. Metals, Nov, 46 (1994)55.
12. M.K.Surappa and P.K. Rohatgi, Metal Tech., 4, 41(1981).

13. Glossary of Terms and Definitions in the field of Friction, Wear and Lubrication. Research Group on Wear of Engineering Materials, OECD, Paris, 1969.
14. R.L. Deuis,C.Subramanian,J.M.Yellup, Wear 201(1996)132-144

Acknowledgements

The authors would like to thank the authorities of AICTE (All India council for technical education) New Delhi, India for financial support provided during the course of this investigation.

Tables:

Table 1: Chemical composition of Aluminium base alloy

Si	Fe	Cu	Mn	Mg	Zn	Al
12.0	0.2	0.23	0.1	0.4	0.1	Bal

Figures:

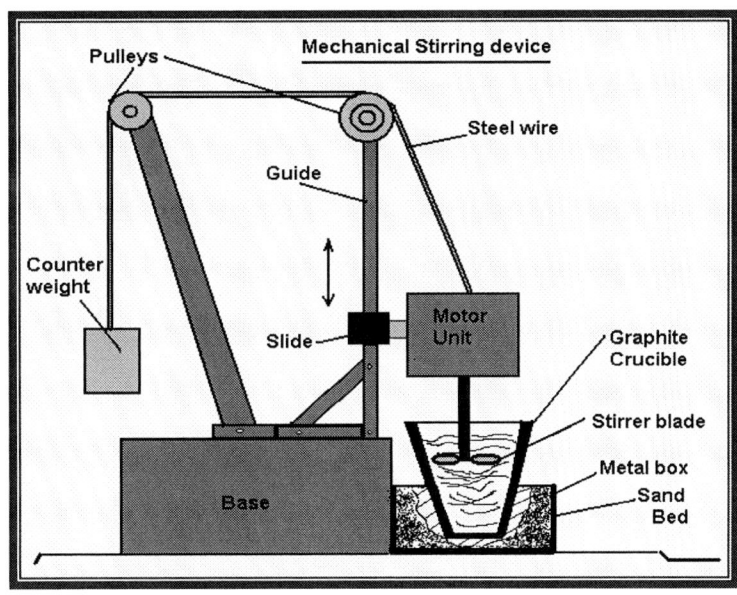

Fig.1.Schematics of Processing of Al-Gr MMC

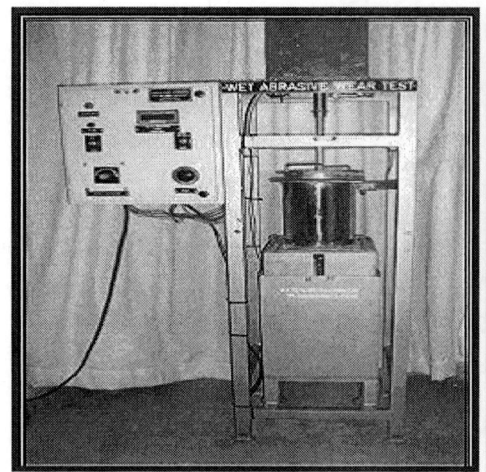 

Fig.2.Slurry wear tesing setup Fig.3.Pin on disc wear testing setup

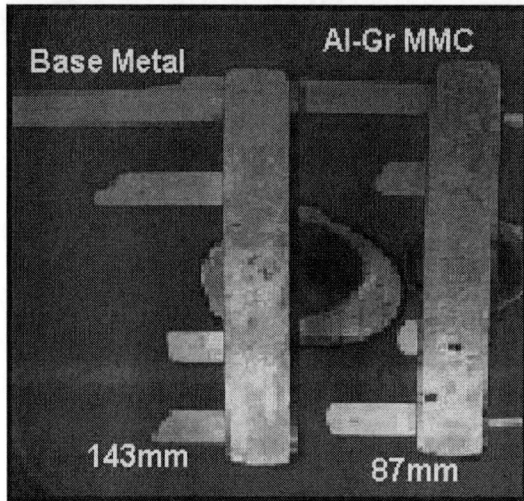

Fig.4. Castings of strip fluidity

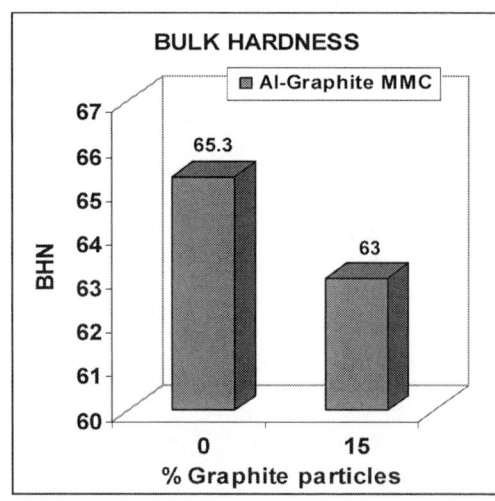 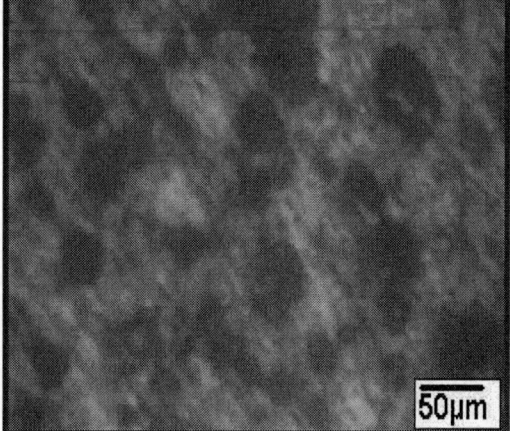

Fig.5: Bulk hardness variation   Fig.7: Micrograph of Al-Gr MMC

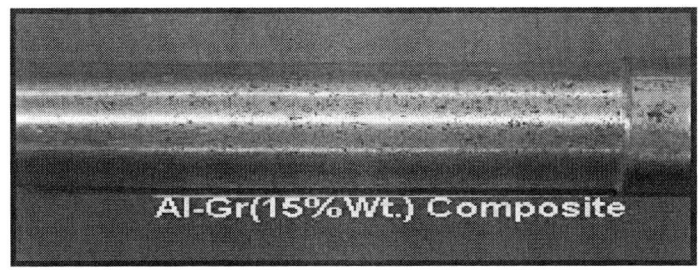

Fig.6. Photomicrograph of Al-Gr (15%Wt.) MMC

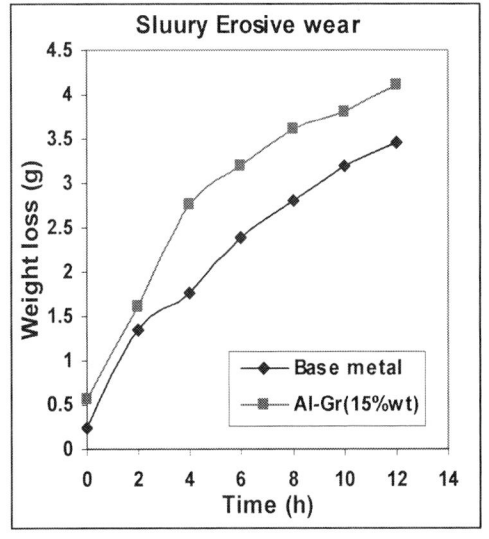

Fig.8. Slurry erosive wear result Fig.9. Slurry erosive wear specimen

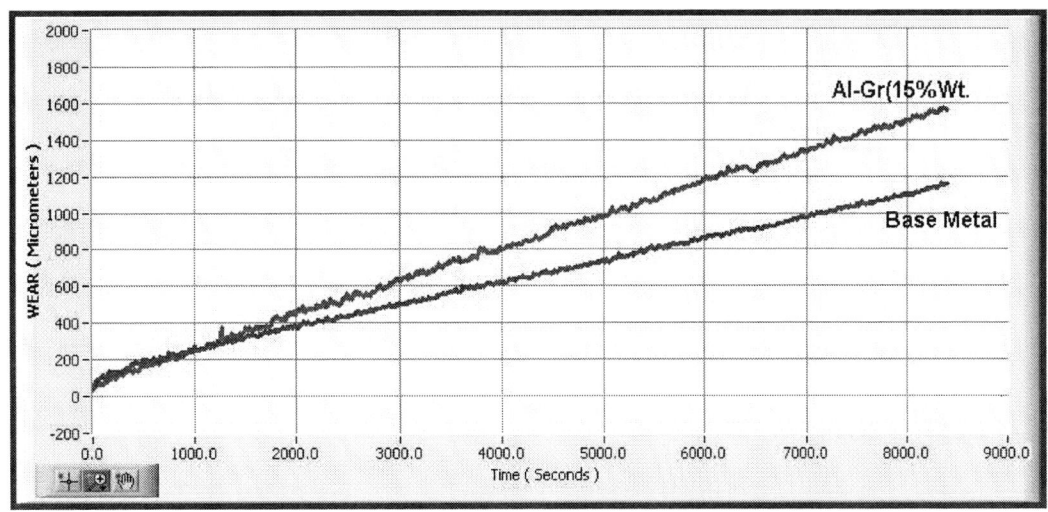

Fig.10: Sliding wear behavior of base metal and composite

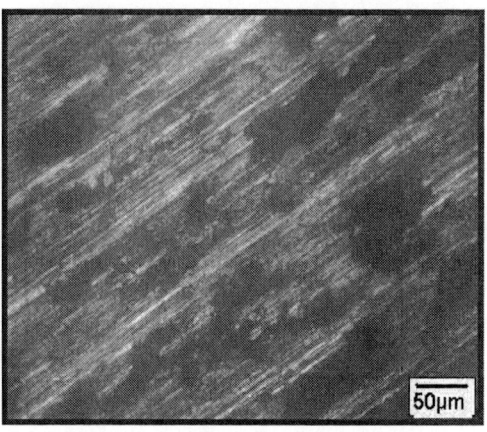

Fig.11: Micrograph of worn specimen

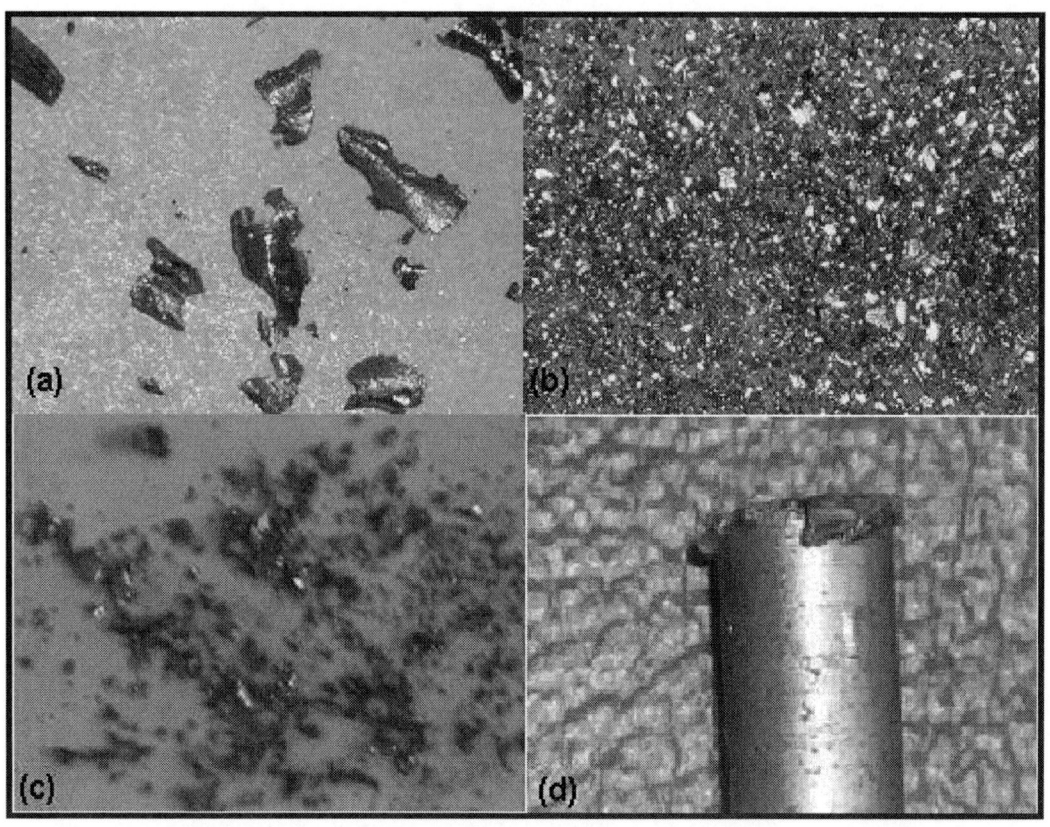

Fig.12.Morphology of wear debris of sliding wear test

AUTHOR INDEX

Ahmed, M.	1185	Changyun, L.	41
Ainsworth, M.J.	341	Chismera, M.	1041
Anglada, E.	126	Cho, G.S.	361, 1157
Araoka, Y.	576	Cho, I.	31
Arrieta, M.	126	Choe, K.H.	361, 1157
Asano, K.	585	Choi, B.J.	412
Ashley, Dr. C.	716	Choi, J.	31
Ashton, Dr. M.C.	760	Choi, J.K.	61, 389, 955
Aufderheide, R.C.	430	Choi, J.P.	939, 946
Averdieck, W.	219	Choi, S.H.	904
Baehr, R.	21	Choi, Y.J.	1157
Bähr, R.	1205	Choi, Y.S.	389
Bako, K.	736	Cilecek, J.	662
Bakrim, M.	405	Costa, J.C.	212
Bale, Eur Ing C.J.C.	716	Creo, R.S.	184
Barnes, C.	824	Cuesta, R.	90
Beeson, A.	126	Czyryca, E.J.	249
Belforte, G.	768	DasGupta, R.	824
Berry, J.T.	100	Datta, G.L.	854
Beziat, A.	1167	Davies, P.J.	331
Biardeau, M.	165	Dayong, L.	914
Binner, J.G.P.	568	De Baerdemaeker, H.	778
Blackburn, S.	351	Debray, C.	165
Blair, M.	155	Delgado, A.	90
Boehm, R.	503	Delorme, H.	396
Bonfield, E.	715	Dequan, S.	914
Branovitsky, A.M.	61	Dicken, P.	808
Brody, H.D.	70, 80	Diószegi, A.	993
Brown, A.	440	Dispinar, D.	874
Buraś, J.	894	Dodd, P.	824
Campbell, J.	874	Donsbach, F.	613, 643
Canon, P.	165	Dussud, M.	165
Carello, M.	768	Eggleston, D.	623
Castro-Román, M.	884	Egner-Walter, A.	1
Chang, H.	568	Elbel, T.	1023
Chang, R.	1195	El-Din, H. Nasr	798

AUTHOR INDEX

Eng, C. 174

Enright, P. 965

Fan, Z. 295, 1109, 1128

Fang, J. 70

Fang, X. 1109

Farnsworth, M. 1091

Froescher, A. 921

Galaz, J. 126

Gnanamurthy, K. 705

Gradowski, A. 108

Greer, A.L. 894

Griffiths, W.D. 331, 341, 864

Güemes, J.A. Goñi 184

Haigh, Dr. P. 778

Haigh, P.M. 11

Hamilton, R.W. 528

Han, Y. 305

Harding, R.A. 239

Harshorne, J. 219

Hashimoto, K. 493

Hayrynen, K.L. 194

Héau, C. 396

Hedge, S. 558

Helber, J.H. 229

Hendley, R.J. 716

Hengzhi, F. 41

Herrera-Trjo, M. 884

Higginson, R.L. 568

Ho, N.U. 955

Hoff, H. 643

Holmen, K. 1176

Holmgren, D. 993

Hong, J.H. 389

Horacek, M. 662

Horvath, L.R. 460

Hu, B. 380

Hwang, H.Y. 61, 389

Hwang, J.H. 136

Ibrahim, M.M. 798

Ikeda, S. 380

Ikengaga, A. 361

Ing, Dr. 1

Jackson, A. 1081

Jakstis, D. 1165

Jin, Q. 324

Jingjie, G. 41

Jo, H.H. 315

Jones, S. 351

Jorge, A. 126

Junjiao, W. 118

Kadoi, K. 371

Kallbom, R. 1071

Kapranos, P. 1091

Karunakar, D. Benny 854

Keena, P. 633

Kelin, Z. 118

Kennedy, M. 1166

Keough, J.R. 194

Khan, R.H. 834

Kiguchi, S. 538

Kiguchi, Shoji 653

Kim, B.H. 816, 975

Kim, I.H. 136

Kim, J.T. 955

Kim, K.H. 136

Kim, K.Y. 1157

Kim, M.H. 904, 1157

Kim, S.K. 315

Kim, S.M. 939

Kim, Y.J. 278, 412, 1148

Kocian, L. 1023

Koss, D.A. 249

AUTHOR INDEX

Krajewski, W.K. 894

Kraly, A. 1119

Krebs, E. 1205

Kurtsiefer, R. 204

Kwak, S.Y. 955

Lauwers, Dr. B. 778

Law, T.D. 517

Lee, D.H. 136

Lee, D.S. 975

Lee, J.K. 315

Lee, K.R. 361

Lee, K.W. 361, 1157

Lee, P.D. 528

Lee, S.M. 816, 975

Lee, S.W. 955

Lekakh, S.N. 1003

Lelito, J. 108

Lengyel, K. 736

Leuven, K.U. 778

Lihua, W. 914

Lim, C. 31

Lim, S.G. 324

Linxe, D. 1167

Liu, G. 295

Liu, Q.M. 681

Löchte, K. 503

Loper Jr., C.R. 1003

Luck, R. 100

Macke-Bart, C. 1167

Macnaughtan, M.P. 174, 623

Maguregi, J.I. 184

Mahendra, K.V. 548

Maroto, J.A. 90

Marukovich, E.I. 61

Matsumoto, Y. 380

Matsuo, Y. 493

Meléndez, A. 126

Metzgar, K. 396

Midson, S.P. 1081

Millan, N. 716

Mnich, F. 1205

Monroe, R.W. 155

Montes-Rodríguez, J.J. 884

Moon, B.M. 816, 975

Morral, J.E. 70, 80

Mozo, D. 90

Mueller, M. 467

Müller, J. 473

Murakami, M. 380

Murata, Hirotoshi 653

Nakae, H. 371, 576

Nakamura, K. 538

Nam, J. 31

Nam, T.W. 939, 946

Narasimhan, V. 510

Neto, E. 396

Ngirabacu, F. 405

Niehoff, T. 633

Noda, Y. 51, 493

Nofal, A.A. 798

Nomura, H. 1099

Nyamannavar, S. 1138

Oh, J.S. 136

Olive, S. 1

Oxley, S. 11

Park, S. 288

Patel, J. 1109

Paterson, T. 527

Pickering, J. 965

Pillai, R.M. 689

Prabhu, K. Narayan 1138

Prabhu, K.N. 558

AUTHOR INDEX

Prat, J. 126

Qi, F.P. 681

Radcliffe, P. 824

Radhakrishna, Dr. K. 548

Radhakrishna, K. 593, 603

Raiszadeh, R. 864

Raji, A. 834

Ramachandra, M. 593, 603

Ramadan, M. 1099

Ransing, M.R. 672

Ransing, R.S. 672

Ravi, M. 1138

Regheere, G. 1167

Reuther, T. 1061

Rhys, Garel 767

Richardson, N. 623

Rimmer, A. 194

Robertson, D.G.C. 1003

Robinson, A.C. 249

Ryou, M. 904

Saewert, H.C. 1205

Sambrook, R. 568

Schmitz, W. 643

Schrey, A. 450

Seifeddine, S. 844

Senberger, J. 1023

Seoane, A. 126

Sharma, A. 983

Shim, S.Y. 324

Shin, J.S. 816, 975

Shiping, W. 41

Showman, R.E. 430

Sillen, R. 145

Song, C.B. 816

Song, Y.L. 681

Sorenson, W.W. 699

Sriram, P. 1033

Staley, J. 929

Stancliffe, M. 483

Steinhäuser, T. 422

Stötzel, R. 473

Stroppe, H. 21

Subramanya, P.K. 558

Suchy, J.S. 108

Sugiyama, Y. 576

Sumimoto, H. 538

Sundarrajan, S. 269

Sung, S.Y. 278, 412

Suzuki, Y. 493

Svensson, I.L. 844, 993

Takita, M. 1099

Tamez, A. 1205

Tayama, M. 371

Terashima, K. 493

Tham, Y.W. 1195

Thompson, P.J. 259

Thomson, Rachel 741

Tiedje, N. 1051

Todte, M. 21

Tomita, Y. 538

Tong, K.K. 1195

Trauzeddel, D. 613

Vandenhaute, C. 405

Vervier, D. 405

Vervier, J. 405

Vicario, T. 126

Viktorov, V. 768

vom Stein, D. 788

Wallis, R. 778

Wang, Y. 295

Wee, K.W. 1195

Wenzhen, L. 118

AUTHOR INDEX

Werrell, S. 219

Wiese, E. 1013

Williams, T.M. 716

Wöhrer, J. 1119

Xiong, S... 380

Yamaguchi, Yasufumi 653

Yeomans, N.P. 430

Yi, F. 80

Yoneda, H. 585

Yong, M.S.. 1195

Yoo, S.. 31

Yoon, E.P. 939, 946

Yoon, Y.C. 136

Yoon, Y.O. 315

Youn, J. 1148

Yousseff, Y.M. 528

Yuan, C.. 351

Yutong, Z. 914

Zadera, A.. 1023

Zakharov, I.L... 61

Zhai, Q.J.. 681

Zhang, S.M.. 1128

Zhang, Y... 681

Zhao, Y... 528

Zhen, Z. 1128

Zych, J.. 894

Institute of Cast Metals Engineers

67th World Foundry Congress

wfc06

"Casting the Future"

June 5 – 7, 2006
Harrogate, UK

Volume 2 of 2

Printed from e-media with permission by:

Curran Associates, Inc.
57 Morehouse Lane
Red Hook, NY 12571
www.proceedings.com

ISBN: 978-1-60423-676-7

Some format issues inherent in the e-media version may also appear in this print version.

Institute of Cast Metals Engineers

The Institute of Cast Metals Engineers, ICME, has been proud to organise the 2006 World Foundry Congress and present to you this CD-Rom of the Congress Proceedings.

ICME is the hub for the casting industry professionals in the UK and worldwide. It offers membership and internationally recognised qualifications as well as a range of technical publications for the castings industry.

ICME, National Metalforming Centre
47 Birmingham Road
West Bromwich
West Midlands
B70 6PY
United Kingdom

Tel: +44 (0) 121 601 6979
Fax: +44 (0) 121 601 6981
Email: info@icme.org.uk

www.icme.org.uk

Institute of Cast Metals Engineers

Membership of ICME

ICME is the professional members institute for individuals in the castings industry. ICME's members are now part of a global industry, involved in a wide variety of sectors within the castings industry. Our members *are* traditional foundrymen, but they are also design engineers, metallurgists, moulders, patternmakers, CAD technicians, methods engineers, researchers and suppliers to the industry.

ICME is able to offer its members a host of tangible and less tangible benefits. This includes guidance with professional development, training and education, technical support (through our network of members and our library; our staff will always aim to help with technical enquiries). Members are able to pursue registration with the Engineering Council UK, as professional engineers at Chartered, Incorporated Engineers or Engineering Technician level. All members also receive the monthly journal, Foundry Trade Journal. The branch network ensures that members are able to meet with like-minded individuals from the industry in both technical and social events.

www.icme.org.uk

Institute of Cast Metals Engineers

ICME Publications

The Institute has several major industry publications in its portfolio, including the **Foundry Trade Journal** of which ten English language issues are published annually and two Chinese language issues, widely distributed in China and the Far East. This journal has a very strong features list and contains industry news, features and technical papers presented in an accessible way. It is also the main voice for a number of other organisations, including PMMMA, the pattern, mould and model manufacturers association and EICF, The European Investment Casters Federation.

Diecasting World, also published by ICME, is received by over 3000 diecasters in all parts of the world.

The Foundry Yearbook and Casting Buyers Directory, produced annually, is a major reference book for all those who work in the castings industry, including, as it does, detailed listings for all the foundries in the UK, contact details for trade associations from around the world as well as international suppliers to the industry categorised in a variety of useful ways.

ICME also publishes the **Castings Buyer**, the only publication whose sole aim is to promote castings and the castings industry to engineers, buyers, component designers and specifiers. The Castings Buyer features case studies showing where fabrications have successfully been replaced by castings and articles on properties and applications of castings alloys.

ICME sees this as an important service to the industry and will be ensuring that our next issue is available as widely as possible.

www.foundrytradejournal.com
www.diecastingworld.com
www.castingsbuyer.com

www.icme.org.uk

67th World Foundry Congress

5-7 June 2006
Harrogate International Centre, Harrogate, UK

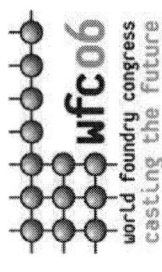

- Welcome
- Organising Committee
- Programme
- Proceedings Sponsor
- About ICME
- Disclaimer

Disclaimer

The opinions expressed in this publication are those of the authors, and do not represent the views of the Institute of Cast Metal Engineers (ICME), its Council or its Officers, the Organising Committee of the World Foundry Congress 2006 (WFC) or the World Foundrymens Organization, except where explicitly identified as such.

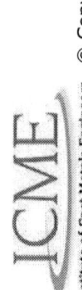

© Copyright Institute of Cast Metals Engineers

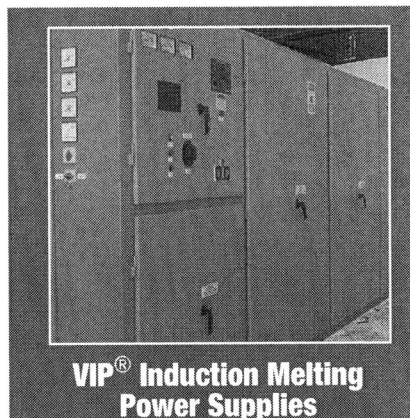
VIP® Induction Melting Power Supplies

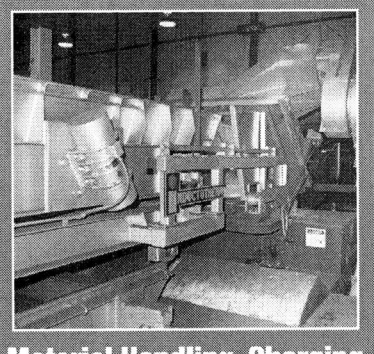
Material Handling, Charging and Preheating Systems

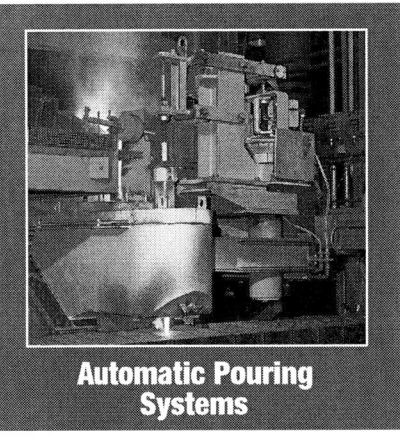
Coreless Melting Furnaces for All Metals

Aluminum Melting Systems

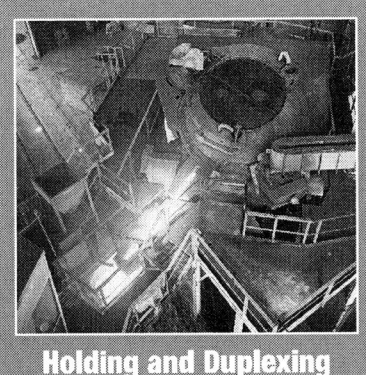
Holding and Duplexing Systems

Automatic Pouring Systems

Computer Control Systems

Cooling Systems

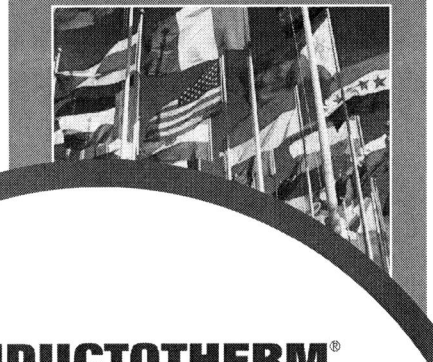

i INDUCTOTHERM® GROUP

Contact Inductotherm Today!

Inductotherm builds induction melting, holding, heating and pouring systems for virtually all metals, including, gray and ductile iron, steel, copper and copper-based alloys, aluminum, zinc, reactive metals and precious metals.

As the world's largest manufacturer of the induction equipment shown here, only Inductotherm can offer **proven** efficient, reliable and effective systems for all your melt shop needs locally with our global presence.

For more information, visit www.Inductothermgroup.com

INDUCTOTHERM® GROUP Leading Manufacturers of Melting, Thermal Processing & Production Systems for the Metals & Materials Industry Worldwide.

Important: Personal Protective Equipment (PPE) must be worn by anyone in proximity to molten metal.

67th World Foundry Congress

5-7 June 2006
Harrogate International Centre, Harrogate, UK

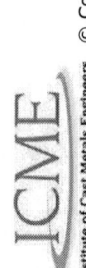

- Welcome
- Organising Committee
- Programme
- Proceedings Sponsor
- About ICME
- Disclaimer

Organising Committee

World Foundry Congress Organising Committees

Chairman	Dr David Rowlands CEng, FICME, MIMMM
Chair of Technical Committee	Dr Pam Murrell FICME
Chair of Funding Group	Mr Peter Nix FICME
Congress Administrator	Mr Matthew Poole

WFC 2006 Steering Committee

Dr David Rowlands CEng, FICME, MIMMM
Mr Bob Brown CEng, FICME, BSc, MIRefE
Mr. Mike Clifford CEng, Hon FICME, BSc, MIM, MCIM
Mr Bob Jordan IEng, FICME, CBIM, MIM, FIDir, FRSA
Dr Pam Murrell FICME
Mr Peter Nix FICME
Dr Tom Paterson FICME, FIMMM, CSci
Dr Philip Ramsell CEng, FICME, FIMechE
Mr Colin Steed IEng, FICME
Mr Barrie Williams CEng, MICME

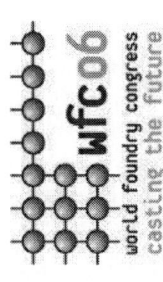

© Copyright Institute of Cast Metals Engineers

67th World Foundry Congress

5-7 June 2006
Harrogate International Centre, Harrogate, UK

Organising Committee

- Welcome
- Organising Committee
- Programme
- Proceedings Sponsor
- About ICME
- Disclaimer

WFC 2006 Technical Committee

Dr Pam Murrell FICME (Secretary General and Chair)
Dr Bill (W D) Griffiths CEng FICME, MIMMM (Vice Chair)
Eur Ing Malcolm Bird OBE, CEng, MIMMM
Mr Roger Davies IEng, AMICME, AIMMM
Mr Martin Fallon CEng, FICME
Mr Roger Kendrick IEng, MICME MIMMM
Mr Malcolm Macnaughtan CEng, FICME, MIMMM
Mr Simon Olive MICME
Dr Philip Ramsell CEng, FICME, FIMechE
Prof Rachel Thomson, MA, PhD Cambridge

Institute of Cast Metals Engineers

President	Mr Barrie Williams CEng, MICME
Senior Vice President	Mr Peter Nix FICME
Junior Vice President	Mr Willie Howson IEng, FICME
Hon Treasurer	Mr David Fletcher FREng, CEng, FICME, FIMMM
Institute Manager	Dr Pam Murrell FICME

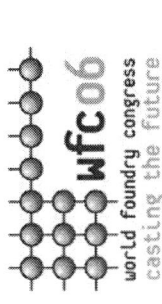

© Copyright Institute of Cast Metals Engineers

67th World Foundry Congress

5-7 June 2006
Harrogate International Centre, Harrogate, UK

INDUCTOTHERM® GROUP

Introduction

The papers can be read on screen or printed using any PDF file reader software. The copyright of the CD contents remains with the Institute of Cast Metals Engineers, from whom permission should be obtained before any paper, or part of a paper is published or reproduced.

I would like to thank all of the authors for their hard work and for the time in preparing and presenting these papers. Special thanks are due to the Technical & Scientific Papers Group of ICME for their long, patient and successful production of such a programme of papers. In particular I wish to acknowledge the support of Inductotherm in the preparation of this CD of Congress Proceedings.

David Rowlands
Organising Chairman
WFC June 2006

wfc06
world foundry congress
casting the future

- Welcome
- Organising Committee
- Programme
- Proceedings Sponsor
- About ICME
- Disclaimer

ICME
Institute of Cast Metals Engineers

© Copyright Institute of Cast Metals Engineers

Institute of Cast Metals Engineers
67th World Foundry Congress
wfc06

TABLE OF CONTENTS

Volume 1

Using Stress Simulation to Tackle Distortion and Cracking in Castings 1
Dr. Ing, A. Egner-Walter, S. Olive

The Use of Different Computer Simulation Software Packages to Predict Casting Filling and Solidification 11
S. Oxley, P.M. Haigh

Prediction of the Infuluence of Microstructure, Porosity and Residual Stresses on Strength Properties of aluminium Casting 21
R. Baehr, M. Todte, H. Stroppe

Intelligent Riser/Chill/Gating Design System Using Simulations and Discrete Optimisation Algorithm 31
C. Lim, J. Nam, I. Cho, S. Yoo, J. Choi

Research on Mould Filling Process of Melt in Vertical Centrifugal Casting 41
G. Jingjie, L. Changyun, W. Shiping, F. Hengzhi

Nonlinear Modeling with Hydrodynamics and Flow Control Using Inverse Pouring Dynamics of Tilting-Ladle-Type Automatic Pouring Process 51
Y. Noda

Three-dimensional Modelling and Simulation of Die-Casting Processes for Al-Si Alloys 61
H.Y. Hwang, J.K. Choi, E.I. Marukovich, A.M. Branovitsky, I.L. Zakharov

Empirical Model for Tensile Property Prediction in Cast and Heat Treated Al-Si-Cu-Mg Alloys 70
J. Fang, H.D. Brody, J.E. Morral

Simulation of Solute Redistribution during Casting and Solutionizing of Multi-phase, Multi-component aluminium Alloys 80
F. Yi, H.D. Brody, J.E. Morral

Computer Simulation Study Upon the Influence of Geometry on the Critical Velocity for Molten Aluminium 90
R. Cuesta, A. Delgado, J.A. Maroto, D. Mozo

Porosity Criteria Functions Revisited 100
J.T. Berry, R. Luck

Simulation of Casting Solidification Using Different Boundary Conditions 108
J.S. Suchy, A. Gradowski, J. Lelito

Mathematical Modelling of Compacting Process of Greensand Molding 118
L. Wenzhen, Z. Kelin, W. Junjiao

Improving Casting Performance Through Customized Insulation Shapes and Advanced Simulation Techniques 126
J. Prat, A. Meléndez, A. Seoane, E. Anglada, A. Beeson, M. Arrieta, J. Galaz, A. Jorge, T. Vicario

Prediction of Shrinkage Defect in Steel Casting for Marine Engine Cylinder Cover by Numerical Analysis 136
K.H. Kim, J.H. Hwang, J.S. Oh, D.H. Lee, I.H. Kim, Y.C. Yoon

Production of Ductile Iron Castings in Green Sand Molds Without Feeders 145
R. Sillen

Developments in the Design of Steel Castings 155
M. Blair, R.W. Monroe

Multiloop Approach for Automatic Mold Filling in Ferrous Foundry 165
C. Debray, M. Dussud, M. Biardeau, P. Canon

Cast Iron Material Standards for a New Millennium 174
M.P. Macnaughtan, C. Eng

Safety Cast Components for the Automotive Industry. the Metallic Charge, the Presence of Micro Elements and Their Most Relevant Effects. 184
R.S. Creo, J.I. Maguregi, J.A. Goñi Güemes

New Engineering and Standards Developments in Austempered Ductile Iron (ADI) 194
K.L. Hayrynen, J.R. Keough, A. Rimmer

How to Link OSHA Requirements to Kyoto's Demands for Protection of Climate 204
R. Kurtsiefer

Occupational Exposure to Chemical Agents in the Portuguese Foundry Industry 212
J.C. Costa

Improving Environmental Performance and Satisfying Regulatory Requirements Through the Continuous Monitoring of Particulate Emissions from Foundry Processes 219
W. Averdieck, J. Harshorne, S. Werrell

Novel Approaches in Reducing Pouring Emissions 229
J.H. Helber

The Use of Tilt Filling to Improve the Quality and Reliability of Castings 239
R.A. Harding

The Influence of Heat Treatment on the Structure and Tensile Properties of Cast Titanium Alloy Ti-511 249
A.C. Robinson, E.J. Czyryca, D.A. Koss

Magnesium Alloy Castings - Past and Present 259
P.J. Thompson

Magnesium Alloy R&D Challenges - Aerospace Spinoff 269
S. Sundarrajan

Alpha-Case Controlled Titanium Casting 278
S.Y. Sung, Y.J. Kim

Surface Tension and Viscosity of Mg Alloys 288
S. Park

Mechanical Properties and Microstructural Evolution of Rheo-Diecast AZ91D Magnesium Alloy During Heat Treatment 295
Y. Wang, G. Liu, Z. Fan

Effect of Grain Refinement on the Mechanical Properties of Magnesium Alloys and Its Alloys 305
Y. Han

Green Manufacturing for Magnesium Alloys ... 315
S.K. Kim, J.K. Lee, Y.O. Yoon, H.H. Jo

Role of Carbon as Solute Element in Carbon Grain Refinement of Mg-Al Alloy ... 324
S.Y. Shim, Q. Jin, S.G. Lim

Wicking of Liquid Polystyrene Degradation Products into the Pattern Coating in the Lost Foam Casting Process ... 331
P.J. Davies, W.D. Griffiths

Mould Filling in the Lost Foam Casting of aluminium Alloys ... 341
M.J. Ainsworth, W.D. Griffiths

Rapid Shell Build for Investment Casting: Revolutionizing an Ancient Process ... 351
S. Jones, C. Yuan, S. Blackburn

Fabrication of Ni-Al Intermetallic Compounds on the Al Casting Alloy by SHS Process ... 361
G.S. Cho, K.R. Lee, K.H. Choe, K.W. Lee, A. Ikengaga

Influence of Viscosity-Increasing Processes on Metal Foam Stability ... 371
K. Kadoi, M. Tayama, H. Nakae

Study on Vacuum Die Casting Process of aluminium Alloys ... 380
B. Hu, S. Xiong, M. Murakami, Y. Matsumoto, S. Ikeda

Investigation on the Flow Pattern in the Shot Sleeve of the Cold Chamber HPDC Process ... 389
J.H. Hong, Y.S. Choi, H.Y. Hwang, J.K. Choi

Definition and Development of an Innovative Coating for Optimised Tooling, Used in Aluminium Die-Casting ... 396
H. Delorme, C. Héau, E. Neto, K. Metzgar

A New Future for Gravity Molding ... 405
M. Bakrim, J. Vervier, C. Vandenhaute, F. Ngirabacu, D. Vervier

Titanium Matrix (TiB + TiC) Composites Shot Sleeve for Al Alloys Die Casting ... 412
S.Y. Sung, Y.J. Kim, B.J. Choi

AWB - An Environment-Friendly Core Production Technology ... 422
T. Steinhäuser

Advances in Thin-Wall Sand Casting ... 430
R.E. Showman, R.C. Aufderheide, N.P. Yeomans

The Significance of Total Carbon in Greensand Systems ... 440
A. Brown

A New Generation of Advanced Polyurethane Coldbox Binders for Aluminium Castings ... 450
A. Schrey

A Technological Advantage for the Environmental Age ... 460
L.R. Horvath

Dispersive Mixing of Natural Molding Sand - An Optimised Preparation Process ... 467
M. Mueller

New Innovative Solutions for Foundries by Inorganic Concepts ... 473
J. Müller, R. Stotzel

Phenolic Urethane Cold-Box Binders - A Study of Global Properties, Variables, Causes and Effects ... 483
M. Stancliffe

Modeling and Identification of Pouring Flow Process with Tilting-Type Ladle for an Innovative Press Casting Method Using Greensand Mold.. 493
Y. Matsuo, Y. Noda, K. Terashima, K. Hashimoto, Y. Suzuki

Inorganic Binders: Properties and Experience.. 503
K. Löchte, R. Boehm

Inclusive, Innovative and Sustainable Environmental Management System.................. 510
V. Narasimhan

Foundry Management by Internet.. 517
T.D. Law

South African Aluminium Foundry Industry: An International Perspective.................... 527
T. Paterson

A Novel Aluminium Matrix Composite Synthesized by Magnetochemical Melt Reactio in the System Al-Zr-O-B... 528
Y. Zhao, Y.M. Yousseff, R.W. Hamilton, P.D. Lee

Compo-Casting Method for Alumina Ceramics Inserted in Cast Iron to Reduce Thermal Stress.. 538
Y. Tomita, H. Sumimoto, K. Nakamura, S. Kiguchi

Production of Hybrid Metal Matrix Composites and its Wear Behavior........................ 548
K.V. Mahendra, Dr. K. Radhakrishna

Effect of Volume Fraction and Particle Size of Reinforcement on Thermal Analysis and Heat Transfer Parameters of Gravity Die Cast Hypereutectic Al-22% SI Alloy Matrix Composites.. 558
P.K. Subramanya, S. Hedge, K.N. Prabhu

Processing of Al-Mg/Al$_2$O$_3$ Interpenetrating Composites by Pressureless Infiltration... 568
H. Chang, J.G.P. Binner, R.L. Higginson, R. Sambrook

Spontaneous Infiltration Mechanism of Al-Si Melts into SiCp Preform...................... 576
H. Nakae, Y. Araoka, Y. Sugiyama

Fabrication of Short Alumina Fibre and In-Situ Mg$_2$Si Particle-Reinforced Magnesium Alloy Hybrid Composite and Its Strength Properties................................ 585
K. Asano, H. Yoneda

Production, Evaluation and Comparison of Mechanical Wear and Corrosion Properties of Al-Flyash1 and Al-Flyash2 Metal Matrix Composite............................... 593
M. Ramachandra, K. Radhakrishna

Evaluation of Mechanical Wear Properties of Al-Si(12.2%)-Graphite Metal Matrix Composite Synthesized Using Stir Casting Method... 603
M. Ramachandra, K. Radhakrishna

Volume 2

Energy Saving Potential of Melting Medium-Frequency Coreless Induction Furnaces... 613
F. Donsbach, D. Trauzeddel

Advances in the Melting of High Quality Grey Iron Automotive Castings at Precision Disc Castings... 623
M.P. Macnaughtan, D. Eggleston, N. Richardson

Oxygen Technologies: Reduce Melting Cost and Emissions 633
T. Niehoff, P. Keena

Development and Use of a New Optical Sensor System for Induction Furnace Crucible Monitoring 643
W. Schmitz, F. Donsbach, H. Hoff

Development of Coke Alternate Material Using Woody Biomass 653
Yasufumi Yamaguchi, Shoji Kiguchi, Hirotoshi Murata

Accurate and Complex NET-SHAPE Castings for Challenging Markets 662
M. Horacek, J. Cilecek

Innovations in Machine Learning and Defect Diagnostics 672
R.S. Ransing, M.R. Ransing

Effect of Ultrasonic Vibration on Structure Refinement of Metals 681
Q.M. Liu, Y. Zhang, Y.L. Song, F.P. Qi, Q.J. Zhai

HIPing - A Potent Post Casting Treatment for High Integrity Aluminium Castings 689
R.M. Pillai

Metal Castings' Secret Ingredient 699
W.W. Sorenson

People and Skills for Today's Industry - An Indian Experience 705
K. Gnanamurthy

How Metals Employers Engage in the UK Skills Agenda 715
E. Bonfield

Latest Trends in Industrial Skills Development Techniques 716
Dr. C. Ashley, Eur Ing C.J.C. Bale, N. Millan, T.M. Williams, R.J. Hendley

OVOTRAIN On-line Virtual Vocational Training System 736
K. Bako, K. Lengyel

Modeling Microstructural Evolution in Cast Alloys 741
Rachel Thomson

Maximising Supply Chain Competitiveness and Market Opportunities by Exploitation of Technology 760
Dr. M.C. Ashton

The Auto Sector in a World of Ultra Competition: Nowhere for the Inefficient to Hide 767
Garel Rhys

Fluidised Bed for Stripping Sand Casting Process 768
G. Belforte, M. Carello, V. Viktorov

Robot Based Oxy-Fuel Cutting and Stub-Grinding for Castings in Low Series 778
Dr. B. Lauwers, H. De Baerdemaeker, Dr. P. Haigh, R. Wallis, K.U. Leuven

Automatic Visual 3-D Inspection of Castings 788
D. vom Stein

Thermomechanical Treatment of Austempered Ductile Iron 798
A.A. Nofal, H. Nasr El-Din, M.M. Ibrahim

New Adaptive Machining Methods for the Foundry Industry 808
P. Dicken

Development of a Low Alloy Cast Steel for Automobile Blanking, Drawing and Trimming Die 816
J.S. Shin, C.B. Song, B.H. Kim, S.M. Lee, B.M. Moon

Squeeze Casting of Aluminium Alloy Safety Critical Components for Automotive Applications 824
R. DasGupta, C. Barnes, P. Radcliffe, P. Dodd

Effects of Pouring Temperature and Squeeze Pressure on the Properties of Al-8%Si Alloy Squeeze Cast Components 834
A. Raji, R.H. Khan

On the Effect of Cooling Conditions and Variation of Alloying Elements on the Microstructural and Mechanical Properties of Al-7%Si Cast Alloys 844
S. Seifeddine, I.L. Svensson

A Study on the Hot Tearing in Al-1% Sn Cast Alloys 854
G.L. Datta, D. Benny Karunakar

A Study of Double Oxide Film Defect Behaviour in a Quiescent Aluninium Melt 864
R. Raiszadeh, W.D. Griffiths

Effect of Melting and Casting Conditions on Aluminium Metal Quality 874
D. Dispinar, J. Campbell

Strontium Effect on the Solidification Path of a 319-Type Aluminium Alloy 884
J.J. Montes-Rodríguez, M. Castro-Román, M. Herrera-Trjo

Effectiveness of Zn-Ti Based Refiner of Al and Zn Foundry Alloys 894
W.K. Krajewski, A.L. Greer, J. Zych, J. Buraś

Effects of Process Parameters on the Morphology of TiAl$_3$ Particle During the Production of Al-Ti-B Master Alloy by Flux Reaction 904
M. Ryou, S.H. Choi, M.H. Kim

The New Method and Device for Fast Evaluating Inoculation Result of Eutectic Al-Si Alloys 914
L. Dayong, S. Dequan, W. Lihua, Z. Yutong

Advances in the Determination of Hydrogen Concentrations in Aluminium Alloys 921
A. Froescher

The Effects of HIP on Bifilms in Aluminium Castings 929
J. Staley

The Effect of Using Bubbling and AlCuP for Refining Primary Silicon at Al-18%Si Alloy 939
S.M. Kim, J.P. Choi, T.W. Nam, E.P. Yoon

A Study of the Structural Controlling of Al-Si Alloy by Using Electomagnetic Vibration 946
J.P. Choi, T.W. Nam, E.P. Yoon

Investigation of Crack Generation Under the Influence of Thermal Stress During Cast Process 955
S.Y. Kwak, N.U. Ho, S.W. Lee, J.T. Kim, J.K. Choi

The Prefil Technique for Molten Metal Cleanliness Measurement 965
J. Pickering, P. Enright

Fabrication of Silicon Ingot by CCCC Method 975
B.M. Moon, D.S. Lee, B.H. Kim, J.S. Shin, S.M. Lee

Microstructure Observations and Refining Performances of Al-Ti-C Master Alloys Prepared by Salt Route ... 983
A. Sharma

Effects of Inoculation and Solidification Rate on the Thermal Conductivity of Grey Cast Iron ... 993
D. Holmgren, A. Diószegi, I.L. Svensson

Thermochemistry and Kinetics of Iron Melt Treatment ... 1003
S.N. Lekakh, D.G.C. Robertson, C.R. Loper Jr.

The Filtration of Large Grey and Ductile Iron Castings ... 1013
E. Wiese

Study of the Occurrence and Suppression of Metal Reoxidation in Ferrous Castings ... 1023
T. Elbel, J. Senberger, A. Zadera, L. Kocian

New Austenitic Flake Cast Iron with Manganese ... 1033
P. Sriram

Carbon Recovery and Inoculation Effect of Carbonic Materials in Cast Iron Processing ... 1041
M. Chismera

Streamlined Gating Systems and Improved Yield - Dimensioning and Experimental Validation ... 1051
N. Tiedje

The Advantages of SiC-Quenching Plates (Chills) When Compared with Grey Iron Quenching Plates (Chills) ... 1061
T. Reuther

Chunky Graphite in Ductile Iron Castings ... 1071
R. Kallbom

A Comparison of Thixocasting and Rheocasting ... 1081
S.P. Midson, A. Jackson

Near Net-Shaping Aerospace Alloys by Thixoforming ... 1091
P. Kapranos, M. Farnsworth

Solidification Structure and Mechanical Properties of Semi-Solid Processed Cast Iron ... 1099
M. Ramadan, H. Nomura, M. Takita

Microstructure and Mechanical Properties of Rheo-Diecast (RDC) Aluminium Alloys ... 1109
X. Fang, J. Patel, Z. Fan

Industrial Application of Thixoforming at SAG ... 1119
J. Wöhrer, A. Kraly

Direct Chill Rheocasting (DCRC) and Extrusion of AZ31 Mg-Alloy ... 1128
S.M. Zhang, Z. Zhen, Z. Fan

Microstructure Evolution in Al-7Si-0.3Mg Alloy During Partial Melting and Solidification from Melt: A Comparison ... 1138
S. Nyamannavar, M. Ravi, K. Narayan Prabhu

Net Shape Forming of Iron and Steel for Clean Production ... 1148
J. Youn, Y.J. Kim

A Study on Semi Solid Squeeze Forging of High Strength Brass 1157
K.H. Choe, G.S. Cho, K.W. Lee, Y.J. Choi, K.Y. Kim, M.H. Kim

The Aerospace Industry and Its Work with Global Suppliers 1165
D. Jakstis

Castings Under Viewpoint 1166
M. Kennedy

New Solutions in Ductile Cast Iron for the Retrievable Storage of Radioactive Waste 1167
C. Macke-Bart, G. Regheere, D. Linxe, A. Beziat

The Intelligent Casting Production: How to Use Automation to Improve the Economics of Near Net Shape Castings 1176
K. Holmen

Double Cavity Casting of Transmission Case: A Techinical solution to a Foundry Capacity Problem 1185
M. Ahmed

Development of Tilt Casting Technology for High Performance Sport Wheels 1195
K.K. Tong, M.S. Yong, Y.W. Tham, R. Chang, K.W. Wee

Process Development for Highly Stressed Aluminium Castings Under Consideration of the Increase in Performance of the Diesel Engines 1205
F. Mnich, H.C. Saewert, A. Tamez, R. Bähr, E. Krebs

Author Index

Energy Saving Potential of Melting in Medium-Frequency Coreless Induction Furnaces

F Donsbach and D Trauzeddel.

Otto Junker GmbH, Simmerath, GERMANY.

Abstract

The available energy saving approaches in induction melting operations are twofold:

- Reduction of the furnace system's electrical losses through improved design
- Use of optimized operating and control regimes

The first part of this paper describes in quantitative terms, based on a review of investigations and test results, how the surplus consumption of power can be reduced markedly through proper feedstock selection, correct charging techniques and an optimum adjustment of furnace parameters.

Furnace design has been improved with the aim of addressing the largest loss factor, i.e., electrical losses in the induction coil, in an effort to cut these losses substantially.

It was demonstrated that the new design yielded energy savings in the range of 5 to 10 %.

Key words

induction melting; cast iron; energy saving; operating regimes; furnace design

Introduction

Optimised medium-frequency technology reduces thermal and electrical losses to a minimum. Exact weighing of the charge materials, a correct calculation and input of the appropriate amount of power based on the use of a melting processor, and a precise computer control of all equipment provide excellent conditions for an energy-saving melting operation.

However, the full benefits of this technology can only be achieved through an appropriate furnace operating and control regime, safe and reliable operating practices, and optimum equipment design.
The overall efficiency of the furnace system can thus be pushed to over 75 %. As a result, the energy input needed to melt cast iron at up to 1,500 °C would be brought down to a mere 490 - 520 kWh/t at an enthalpy of 390 kWh/t.

In practice it has been found that the average power consumption of existing foundries in real-life cast iron melting operations is significantly higher. From UK foundries, for instance, a value of 718 kWh/t is reported [1], whereas French industry statistics show an even higher level of 855 kWh/t [2].

A large energy-saving potential can thus be identified here. Substantial cuts in power consumption are achievable through the use of advanced medium-frequency melting furnaces, but also via modified furnace operating and control regimes. On an existing furnace system, up to 20 % energy can be saved by adopting improved operating and control modes.
This is a cost-cutting potential not to be neglected, specifically in times of ever increasing energy prices.

Influence of the operating and control regime
1. Charge materials and make-up

An accurate calculation of the necessary charge make-up, based on material analyses, and a precise weight determination and metering of charge materials and alloying additives (including correction for set/actual value deviations) are basic prerequisites for minimising melting times and power needs.

The use of clean and dry charge materials will definitely pay off, given that the formation of slag due to sand adhering to uncleaned returns will consume just as much specific energy as is necessary to melt the iron, viz., about 500 kWh/t. With a realistic amount of 25 kg of sand per tonne of iron this adds up to 12.5 kWh/t. Beyond that, of course, the quantity of slag is increased as well.
An even more decisive factor is rusty charge material. Its inferior electromagnetic coupling properties impair the transfer of melting energy and result in much higher melting times. The energy consumption and heat cycles for clean and highly corroded steel scrap, respectively, have been determined in comparative trials [1]. It emerged that rusty steel scrap

took 2 - 3 times as long to melt and required a 40 - 60 % higher power input. Even assuming that these values reflect an extreme case, the negative effect of rusty charge material is quite severe. In addition, there are higher melting losses and greater slag volumes. Hence the use of rusty charge material should be avoided wherever possible.

The level of electromagnetic coupling achieved and hence, the power consumption of the charge, is a function, not least significantly, of the charge packing density. The heat cycle and energy consumption of the charge will thus vary with the packing density.

The nature of this correlation has been examined with charges of different packing density in a high-power medium-frequency furnace operating under production conditions. The system employed for these trials had a capacity of 10 tonnes and a power rating of 8,000 kW at 250 Hz. The empty furnace was filled once with a charge of the specified composition, comprising pig iron, scrap castings, returns, steel scrap and additives. No further charge material was added as the metal was heated to 1,380 °C. The power consumption was measured throughout this period.

Different dimensions of the returns and steel scrap fractions made for packing densities in the 2 - 2.7 t/m³ range. It is evident from the results that a drop in packing density from 2.5 to 2.0 t/m³ caused a 25 kWh increase in power consumption (Fig. 1).

Despite the additional cost and effort, it is therefore advisable to crush all too bulky returns to achieve a higher packing density. This will also facilitate furnace charging and reduce the danger of material bridging in the furnace.

The example of a U.S. foundry demonstrates that this practice can save money despite the costs caused by additional crushing operation [3].

At the same time, a quick and continuous charging workflow is important when it comes to saving operating time and cost. A high filling level should be maintained at all times. Mobile shaker chutes and a bin accommodating the full charge are prerequisite to meeting this requirement. An extractor hood closely covering the chute will minimize radiant heat loss while ensuring that the furnace fumes will be reliably captured.

2.   *Chip melting*

As foundries extend their vertical integration and machine their own castings, they increasingly find themselves with large amounts of chips on their hands – and what would make more sense than to try and use these chips in their own melting operation.

Coreless induction furnaces, unlike other melting processes, are highly suitable for melting down machine tool chips. Since grey cast iron is normally machined without coolants, these chips are dry and clean and can therefore be melted down without any pre-treatment.

However, it should be noted here that the electrical contact between metal chips, despite their good packing density, is notoriously poor as a result of the small contact surface and surface oxidation. This is why the furnace should always be operated with a heel (> 40 %) when chips are melted. If the furnace is operated without heel, the power consumption for melting chips should be anticipated to be 50 kWh/t higher than for lumpy material. An increase in melting time must also be expected.

Chips can be charged on the liquid heel either continuously ('trickle' method) or in one batch up to the top of the active coil. Filling the furnace all the way, without overloading it, will save 2 - 3 % energy and reduce the melting loss. On the other hand, there is the risk of bridge formation in the charge.

Where chips make up only a portion of the charge, the solid or lumpy fraction should be placed in the furnace first, and the chips should be charged into the liquid heel once it has formed.

3. Carburising

Another factor reported [4, 5] to affect power consumption is the method of adding carburising agents. Power consumption will clearly be higher if carburising agents are added into the molten metal bath after melting down rather than along with the solid charge material at the beginning. In-house experience indicates that this practice will consume about 1 to 2 kWh more per kg of carburising agent. This means that with a realistic input of about 2 % of carburising agents, an additional consumption of max. 40 kWh per tonne of iron is to be expected. An average of 70 kWh per tonne of iron for carburisation, as quoted in part of the literature, appears to be unreasonable.

If the carburising agent is introduced into the furnace with the other charge material, this should be done in controlled proportions so that the carbon content of the melt will not rise unnecessarily.

An excessive increase in carbon concentration would cause premature crucible wear. It is also advisable to avoid the use of too fine-grained, low-grade carburising agents which tend to adhere to the crucible wall. Local erosion effects would be the inevitable result.

Furthermore, the input of silicon carriers should not take place until after carburisation is completed because increasing Si content in the melt decreases carbon solubility and also increases silicon losses.

It should be noted in this context that the enthalpy levels of returns and synthetic cast iron differ markedly. When melting synthetic cast iron (steel scrap, carburising agent, silicon carrier), the power consumption should be expected to be 8 - 15 % higher than with a charge of homogeneous returns [6].

4. Melting furnace operating regime

In theory, the most favourable operating regime would be one involving the maximum available electric power and hence, high power densities. The overall efficiency of a melting plant, as can be seen from the following equation

$$\eta_{total} = \eta_{electrical} \left(1 - \frac{\text{holding power}}{\text{rated power}} \right)$$

is also determined, quite decisively, by the ratio between holding power and rated power. For a furnace of the same dimensions, the consumption of electrical energy will decrease with increasing rated power. This has been conclusively confirmed by systematic trials: as the heat cycle is reduced and thermal losses diminish, the electrical power consumption is reduced.

From the calculated power diagram of a 12-tonne-furnace (refer to Fig. 2), it is evident that the electric power consumption increases **exponentially** with decreasing power density since the percentage of energy required to make up for steady-state thermal losses will become disproportionately high when the power density is very low.

A comparison between a 6,000 kW melting operation and one with 3,000 kW reveals (cf. Fig. 3) a substantial power consumption difference of 20 kWh/t. This advantage can be utilized by changing from a mains-frequency to a medium-frequency system, since the maximum power input for a mains-frequency furnace of this size is around 3,000 kW.

The use of medium-frequency technology makes it possible to operate without heel and to melt down small-sized charge material. Thanks to better electromagnetic coupling of the solid charge material (although this applies only to cast iron melting) the energy consumption in batch operation is 8 % less because a much higher coil efficiency is achieved up to the Curie point (Fig. 3). Since mains-frequency systems must be started up with a liquid heel, this energy saving can only be realized by shifting to medium-frequency technology.

The higher power density and superior coil efficiency in batch operation add up to an energy saving of 12 - 15 %, always provided that the switch is made from mains-frequency to medium-frequency melting.

The amount of heat stored in a coreless furnace (i.e., the energy required to heat up the cold furnace to its fully heat stored condition) is normally higher by a factor of 3 - 5 than the holding energy required over a similar period. Thus, the heat stored in an 8-tonne-furnace is 800 kWh, i.e., melting a charge in a still-cold furnace would require 100 kWh/t more energy than is needed to melt the same charge in a unit already storing full heat (Fig. 4).

Since it will take only a quarter of this energy (25 kWh/t) to keep the melt at holding temperature for an hour, it makes sense not to let the unit cool off but to keep it at temperature with a heel of molten metal during interruptions or breaks of less than four hours.

It should also be considered that the service life of the refractory furnace lining can be maximized by keeping the furnace continuously at operating temperatures, or at least by not perpetually switching it off and on. As a general rule, extended holding periods will no longer affect the metallurgical quality today thanks to the advanced treatment and inoculating technologies available.

Energy is wasted also by operating the furnace with its lid open for longer than necessary. The low heat losses achieved by design, e.g., of only 140 kW for an 8-tonne-furnace, would thus rise to 400 kW which means an additional consumption of 4 kW per minute of open lid time. Over 20 minutes this would add up to as much as 80 kWh per charge, equivalent to an additional consumption of 10 kWh/t.

Moreover, energy will also be 'sucked off' the furnace unnecessarily if the exhaust system is run at full capacity even at times when no, or only little, flue gas is produced. Under unfavourable circumstances this may increase the power consumption by as much as 3 %, corresponding to 15 kWh per tonne of iron.

Another issue is superheating of the iron, if one considers that a 50 K temperature rise will consume about 20 kWh per tonne. The melting processor allows the final temperature to be maintained to an accuracy of 5 K, eliminating any unnecessary input of superheating energy.

5. Refractory lining

The wall thickness of the ceramic furnace lining, which in cast iron melting systems will almost invariably be quartzite, always constitutes a compromise between good thermal insulation, adequate mechanical protection of the coil, and good electromagnetic coupling between the coil and the charge.

Reducing the thickness of this lining will increase the coil efficiency and power consumption while causing higher thermal losses through the thinner crucible wall. However, since coil losses exceed the thermal losses across the crucible wall by nearly a factor of 10 in terms of magnitude, coil losses remain the dominant influence here.

Studies have shown [1] a substantial reduction in power consumption with decreasing thickness of the refractory lining, as shown in Table 1.

With increasing furnace operating time and thus increasing refractory erosion the power consumption will decrease by nearly 10 % over the first three weeks. Assuming that a lining having an original thickness of 125 mm loses 30 mm of that thickness within the first three weeks, the coil efficiency would rise by 3 % only according to our calculations (Fig. 5).

It follows that this fact alone cannot explain the above-mentioned energy savings, which are presumably augmented by the increased power input and the resulting shorter melting time.

It might therefore make sense to consider eliminating excessively high "safety margins" on the thickness of the refractory lining with the aid of advanced crucible monitoring equipment such as the OCP optical coil protection system [7].

Further reduction of electrical losses

Continuous progress in the design of inductive melting equipment has yielded a considerable increase in power densities and output, apart from widening the range of technical applications potentials significantly. Much work has also been done to cut thermal and electrical losses through optimized furnace design and improved frequency converter technology. Superior efficiencies of more than 75 % in cast iron melting are the reward of these efforts and now define the state of the art.

Further reductions in power losses have been attained through the new energy saving concept developed over several years of intense R&D and tested for its reliability and practical viability in extensive trials.

The engineers focused their efforts on addressing the largest loss factor, i.e., electrical losses in the induction coil, in order to cut this source of inefficiency by a significant margin. Extensive calculations and numerous model trials were necessary to progress the initial concepts into a valid solution, viz., a special coil design combined with advanced frequency converter technology (Fig. 6).

The new system was integrated into a 1.5-tonne coreless furnace in a stainless steel foundry and subjected to gruelling tests in day-to-day production operations over several months.

The substantial energy savings achieved, in conjunction with high dependability and performance levels, attest to the successful achievement of our design targets and prove the system's industrial viability.

The reduction in energy consumption can be put at 5 - 10 %, depending on specific operating conditions.

Conclusions

The actual average energy consumption in cast iron melting operations, at over 700 kWh/t, holds a significant saving potential if one considers the fact that figures of 490 - 520 kWh/t are now achievable in practical operation. The factors contributing to the excess consumption are exemplified in Table 2.

References

1. *Efficient melting in coreless induction furnaces*, Good practice guide No 50: ETSU, Harwell, Didcot, Oxfordshire, 2000.
2. Jolivot R, Fonderie Fondeur d`aujourd`hui, No 229, Nov 2003, pp36-39.
3. Foundry Management &Technology 131 (2003) No 11, pp14-16.
4. Smith L and Bullard H W, The Foundryman 88 (1995) No 7, pp246-253.
5. Brockmeier K-H, *Induktives Schmelzen*, Brown, Boveri & Cie, Aktengesellschaft Mannheim; Essen: Giradetverlag 1966
6. Duca W J, Trans. Amer. Foundrym. Soc. 81 (1973) pp108/109
7. Donsbach F, Schmitz W and Hoff H, Giesserei 90 (2003) No 8, pp52-54

Tables

Table 1

Power consumption as a function of refractory lining wear

Three-tonne coreless induction furnace; 700 kW/cast iron

	Power input kW	Power consumption kWh/t	Energy comparsion %
Fresh lining	615	656	100
1 week run	650	622	95
3 weeks run	750	598	91

source (1)

Table 2

Energy consumption

Melting of cast iron at up to 1,500°C
Medium frequency furnace 8,000kg/8,000kW/250 Hz

	kWh/t
Melting enthalpy (theoretical)	390
Equipment	
thermal and electrical losses	126
Melting power consumption	516
Excess consumption due to operating and control regime	
sand in charge material, 25 kg/t	12,5
rusty charge material	30
low packing density only 2.0 instead of 2.5 t/m³	25
carburising after melting down 20 kg/t	40
melting with 50 % power density	20
melting with heel	40
unrestricted extraction	15
holding with the lid open for more tahn 20 minutes	10
unnecessary superheating by 50 K	20
Total of excess consumption	212,5
Possible overall consumption	728,5

Figures

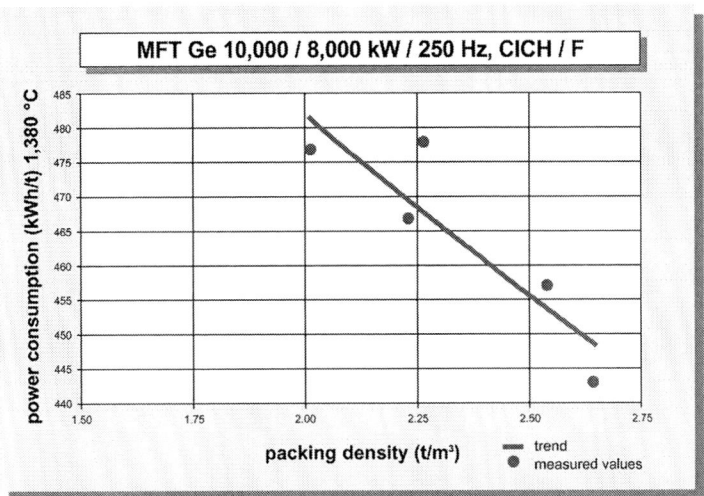

Fig. 1   Influence of charge packing density on electric power consumption

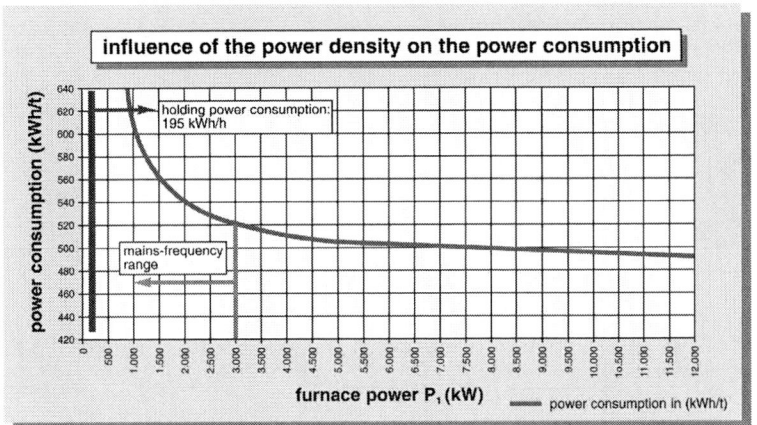

Fig. 2   Power diagram of a 12-tonne-furnace plant

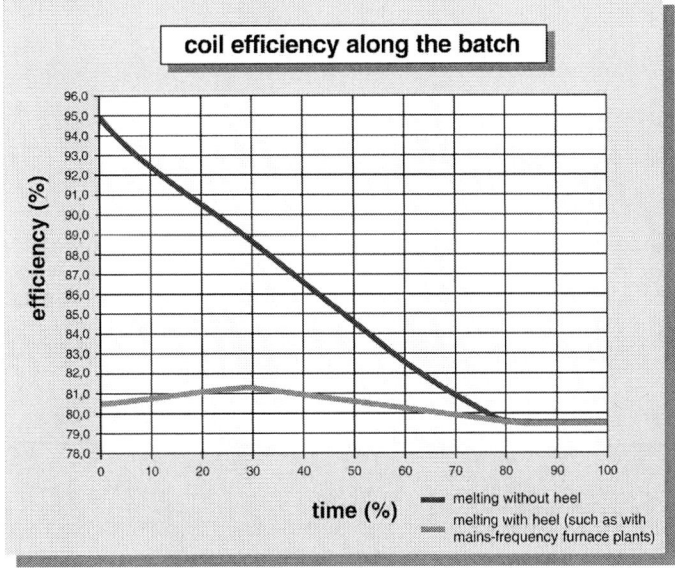

Fig. 3   Influence of operating mode on coil efficiency

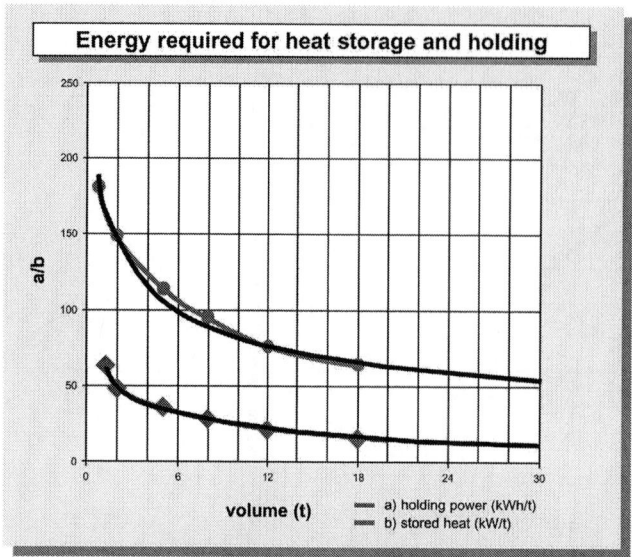

Fig. 4   Stored heat and holding power as a function of furnace size

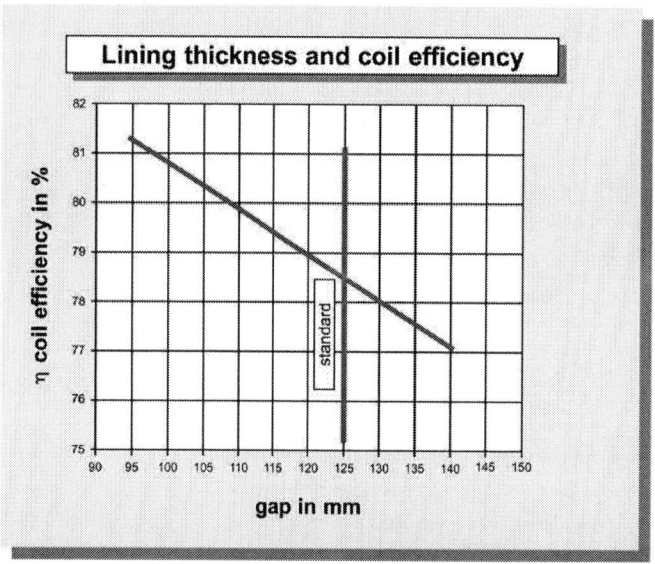

Fig. 5   Influence of lining thickness on coil efficiency

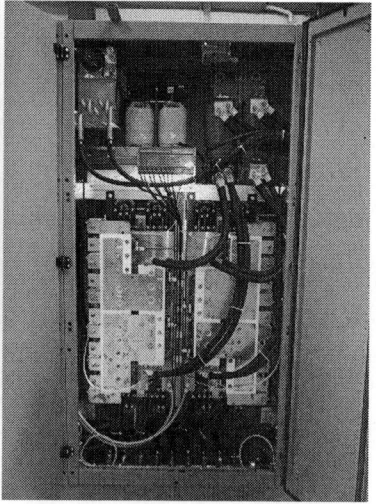

Fig. 6   The new system

Advances in the melting of high quality grey iron automotive castings at Precision Disc Castings

M P Macnaughtan*, D Eggleston and N Richardson**

* Eurac Ltd, UK, ** Precision Disc Castings Ltd, UK.

Abstract

Pressures on the castings industry caused by significant increases in raw material prices during 2004 resulted in Precision Disc Castings performing an evaluation of the melting of grey cast iron in their UK foundry.

Variables inherent in the melting process, melter charging sequence, furnace additions and inoculating materials were evaluated using thermal analysis techniques. Process variables were understood, and procedures introduced to control melt quality. A cost rationalisation process was also undertaken without adversely affecting product quality, the use of pig iron was also reviewed.

The paper describes the work undertaken and details the benefits provided by an in depth study of the fundamentals of cast iron melting and solidification coupled to a clear understanding of the requirements of modern high volume automotive grey iron casting production.

Key words
Grey iron, metallurgy, solidification, inoculation.

Introduction

The EURAC group, established in 1992, is one of Europe's leading independent high volume producers of grey cast iron brake discs for the automotive industry. The group currently have four manufacturing sites in Europe, one foundry Precision Disc Castings and a machining facility High Precision Machining in the United Kingdom, a machining facility Präzisions-Brems-Scheiben in Germany and Brzdové Automobilové Kotouĉe a foundry with machining facilities in the Czech Republic. The two foundries have a combined output of some 10 million components a year.

Precision Disc Castings

Precision Disc Castings (PDC) is recognized as one of Europe's leading ferrous foundries specialising in the production of ventilated brake disc castings. The production of 6 million discs a year, with 80% of these being exported to Europe, has led to international recognition. With a work force of 220, and a capacity of 50,000 tonnes a year, PDC produces alloyed and un-alloyed discs to satisfy customer specifications ranging from grades 150 to 250. The foundry, which is approved to ISO TS 16949, has a catalogue of over 600 references. PDC also supplies brake discs for the racing market, with approximately 90% of the European market share.

Raw materials are melted in four ABB six tonne medium frequency furnaces. Molten metal is poured through Calamari medium frequency automatic pouring furnaces onto two state of the art Disamatic moulding lines. Each autopour furnace is controlled by the latest generation of Selcom laserpour units.

The effect of material structure on the performance of brake discs

The conventional material for automotive brake disc applications is a grey cast iron having a microstructure consisting of lamellar graphite in a pearlitic matrix. Grey cast irons of this type exhibit a great many desirable characteristics, the properties of most relevance to braking applications being:

- Ease of production (castability and machinability)
- Thermal conductivity and vibrational damping characteristics
- Resistance to thermal shock and cycling
- Wear resistance

Grey cast iron is relatively cheap and easy to produce in high volumes to tightly controlled specifications. A further, functional advantage of grey cast iron as a brake disc material is the fact that its specific heat capacity actually increases with increasing temperature, reacting rather like a sponge, becoming increasingly absorbent the wetter it becomes [1]. The microstructue of flake graphite cast iron brake discs is a critical factor in determining their functional performance and consistency under braking, the performance of the disc pad couple depending as much on the grey cast iron used for the brake disc as on the formulation of the friction material used for the brake pad [2] [3].

During braking, the surface of the brake pad attains a higher temperature than that of the brake disc. However the inherently low thermal conductivity of the pad material induces a severe thermal gradient across its thickness resulting in only a very small proportion of the heat generated during braking actually entering the pad, about 98% being absorbed by the disc [4]. In order that this heat is generated and absorbed as uniformly as possible the network of graphite flakes within the disc structure should be as homogeneous as possible. Ideally type A graphite, of a consistent size range with a fine eutectic cell size. An example of such an idealised structure is illustrated in Figure 1.

Failure to produce such a microstructure within the friction ring of the disc can lead to a variety of adverse phenomena that may affect braking performance and, under extreme circumstances, even compromise the integrity of the disc itself. The carbon content of brake disc materials varies from 3.40 – 3.80 mass %, this equates to a free graphite content of 11 – 13 % by volume, the thermal conductivity of this volume fraction being almost double that of the matrix material. In addition to its considerable thermo-mechanical impact, the graphite structure also influences the tribological performance of the friction couple in its role as a solid lubricant acting at the disc/pad interface.

From a practical perspective, brake discs are normally designed to meet specific thermo-machanical performance criteria with a minimum tensile strength requirement of between 150 N/mm² and 250 N/mm². Matrix characteristics such as pearlite refinement contribute overwhelmingly to exceeding the minimum requirements for component strength. Failure to comply with the performance requirements normally indicates the presence of inadequate graphite structures, such as C-type, D-type and E-type graphite forms, all of which facilitate more macroscopic planes of weakness.

The most radical failure mechanism potentially affecting a brake disc is that of catastrophic fracture. Most commonly this will occur around the intersection between the friction ring and the top-hat section or around the outer diameter of the hub face, either failure resulting in a total loss of braking as the constrained section of the disc is effectively separated from the friction ring. This intersection is frequently the thinnest radial cross-section which is exposed to the highest torsional forces under braking, the steepest thermal gradient and the stress raising effects of an inherent profile change. All factors making the area vulnerable to the effects of inadequate graphite structures. Fortunately, such occurrences are rare.

More common are instances of brake judder which can be excited by mechanisms originating from variable microstructures within the disc. Variations around the friction ring in either material hardness or thermal conductivity, each of which can stem from variations in either the graphite form or matrix hardness, can result in uneven wear or heat input around the disc. Non-uniform wear of softer regions around the disc can result in microscopic thickness variations that also cause judder.

Control of metallurgical quality

The control of microstructure is therefore of paramount importance, the final microstructure of a material depending on the thermal history of specific events such as heat transfer, imposed temperature gradients, nucleation and segregation effects that occur during cooling. Such events can be influenced by factors such as:

- Alloy composition
- Furnace charge materials and charging sequence
- Trace element level
- Amount of oxygen in the melt
- Type, amount and addition method of any inoculating materials

Grey iron solidification

The solidification of a cast iron can be illustrated by the cooling curve for a hypoeutectic cast iron shown in Figure 2. Molten iron starts to solidify at the liquidus temperature (TL) at which point dendritic austenite precipitates from the melt. The onset of freezing is accompanied by a thermal arrest caused by the latent heat accompanying the phase change. After this brief arrest, cooling continues accompanied by the further precipitation of austenite until the onset of eutectic freezing at T_{ES}, at which point graphite nucleation commences. The nucleation of graphite again generates latent heat that reduces the rate of cooling until the lowest eutectic temperature TE_{LOW} is reached. At TE_{LOW} the combined heat evolved by the formation of austenite dendrites and the latent heat from the onset of eutectic freezing causes the temperature of the iron to rise. This increase in temperature reaches a maximum at TE_{HIGH} where a second thermal arrest occurs. A balance between the heat evolved and the heat dissipated from the system promotes the steady state condition at TE_{HIGH}, which is accompanied by eutectic cell growth, which occurs initially during recalescence, and is normally accompanied by a degree of undercooling until the steady state temperature is reached.

During this phase of solidification graphite flakes propagate, in preference to phases such as carbides, at nucleation sites provided by inoculants and other impurities. Each time a eutectic cell is formed, a small amount of heat is released to the surrounding liquid. The greater the number of cells formed, the greater the amount of heat that is released. Therefore a lower cooling rate during solidification will tend to indicate a higher eutectic cell count. This reduced cooling rate helps to ensure that the iron freezes at its equilibrium temperature, ensuring that predominantly type A graphite is precipitated in preference to less desirable graphite types such as B, D and E, which are all symptomatic of varying degrees of undercooling. The amount of latent heat generated by the system continues to reduce until eutectic solidification is complete at T_S.

Thermal analysis – practical considerations

In practice, the position of specific temperatures on the cooling curve such as T_L, TE_{LOW}, TE_{HIGH} and the degree of recalescence are influenced by the chemical composition and nucleation potential of the liquid iron. It can be seen therefore that the monitoring and control of these specific parameters plays an important part in the development of new melt additives and inoculants.

The value of T_L is directly related to chemical composition, e.g. C and Si content, whereas TE_{LOW} is influenced by the nucleating potential of the molten iron. The inoculation of liquid iron supplies the large quantity of sub-microscopic nuclei required to stimulate the formation of eutectic cells. Unfortunately however the nucleating efficiency of inoculants fades rapidly with time. Inoculation will also reduce the degree of recalescence, which is a measure of the difference between the TE_{LOW} and TE_{HIGH} eutectic temperatures. A high degree of recalescence indicates that a high degree of undercooling has taken place and therefore nucleation is low, which leads to undesirable D and E type graphite structures. Solidification time, temperature and inoculant chemistry will also determine the graphite shape and distribution in the solidified, cast iron matrix.

The effect of microstructure on the properties of grey cast iron

Graphite flakes of the type found in grey cast iron structures have a weakening effect on the strength of the material. The long thin flakes act as stress raisers and as a consequence reduce tensile strength. In general, the strength of a grey cast iron comes from the condition of the matrix. The harder and stronger the matrix, the harder and stronger is the iron itself. Thermal conductivity is also influenced by graphite content, the higher the carbon level the greater the thermal conductivity. Less desirable graphite types such as B, D and E also reduce the overall strength and mechanical properties of cast iron; these undesirable graphite types will also compromise the thermal performance of the product. It is therefore important to ensure that a grey iron to be used for braking applications solidifies with a uniform distribution, which should be independent of casting section thickness. The good machinability of grey iron is also dependent on the condition of the matrix [5]. Excessive levels of ferrite can, for example, cause CBN cutting tools to overheat the reaction drastically shortening tool life. It can therefore be seen that the ability of the metallurgist to influence and control the final microstructure of grey iron brake discs is extremely important.

Precision Disc Castings have for many years utilised Carbon Equivalent Liquidus (CEL) thermal analysis to control melt quality. However the introduction of computer based predictive thermal analysis software such as NovaCast ATAS has provided the metallurgist with a much more sophisticated and powerful analytical tool to evaluate the production process. Once production variables were understood, a programme to develop the melting process at PDC was undertaken.

The Replacement of Pig Iron in the Basic Charge

The importance of pig iron and the role it plays in the solidification characteristics of cast irons has been well documented. However the rising cost of both steel scrap and pig iron has put pressure on the use of the latter. It was under these circumstances that PDC set out to successfully manufacture high quality grey iron automotive castings without the addition of pig iron to the electric furnace melt charge.

Technical Requirements

To ensure that microstructures of type A graphite are produced consistently there must be a sufficient number of nuclei present in the molten iron to initiate graphite growth. A wide variety of oxides, silicates, sulphides, nitrides and carbides are claimed to serve as nucleation sites, however it is thought that the oxides are the most important for the nucleation of flake graphite. Brake discs are relatively thin section components and as a consequence a high density of eutectic cells are required to promote the desired type A graphite structure and prevent fine undercooled graphite forming during solidification. Gases such as oxygen, nitrogen and hydrogen dissolved in the melt significantly influence final structures obtained during solidification. The role of oxygen is particularly important because of its influence on the nucleation and growth mechanisms that affect the graphite type and distribution during solidification.

In order to ensure that sufficient oxygen is present, it can be added to induction-melted irons in several ways. Melting with a low silicon content base iron will pull in oxygen from the surrounding atmosphere which combines with silicon to form beneficial oxides of SiO_2 that act as nucleation sites according to the equation:

$$Si + 2(CO) = SiO_2 + 2C \qquad (1)$$

The introduction of pig iron with a naturally high oxide content resulting from the smelting and refining processes will also add oxygen to a melt. Therefore to successfully remove the pig iron and still produce the desired microstructures the oxides present in pig iron need to be replaced.

Practical implications - an investigation

For many years the charge make up at PDC consisted 10% pig iron, 30 % cast iron returns and 60% steel scrap (charge1). Typical thermal analysis data generated from this charge 1 melt is listed in table 1. The structure produced from this charge was generally one of predominantly type A graphite with a percentage of type B with isolated pools of D and E type graphite in a matrix of fine lamellar and unresolvable pearlite.

Due to the financial constraints of using pig iron, other materials were introduced into the charge as a direct replacement. With the benifit of thermal analysis it was possible to examine the effect of charging sequence on the solidification characteristics of the iron produced and compare microstructures resulting from differing charge compositions. The first stage in the charge development process was the elimination of pig iron, steel being used as a direct replacement. This revised charge (charge 2), consisted of 30% cast irons returns and 70% steel scrap. To a certain extent this modification to the basic melt worked, however investigations revealed that structures were not consistent. Structures of type A graphite were produced but with areas of fine undercooled graphite type D and an increased amount of interdendritic graphite type E. The rising cost of steel was also making this charge only marginally cheaper than the original charge but at a greater cost to the quality of the product. Typical thermal analysis data from a charge 2 melt is listed in Table 1.

Having reviewed results from the charge 2 trial, the decision was taken to melt cast iron borings, which were cheaper than either steel scrap or pig iron and also gave savings by virtue of their alloy content. A typical charge using this material (charge 3) consisted of 30% cast iron returns, 20% cast iron borings and 50% steel scrap. Typical typical thermal analysis data from a charge 3 melt is listed in Table 1.

Examination of microstructures from charge 3 revealed a marked improvement over those from charge 2, however in general, the microstructures were still not comparable with those from charge 1. The reasons for this improvement were therefore investigated. The obvious answer was that these melts possessed a higher degree of nucleation, however it was suspected that the oxygen potential of the melt was also playing a part in the nucleation process, the borings with their greater specific surface area being coated in a significant layer of oxide. However in order to achieve the desired microstructure consistently, a small percentage of pig iron was still required and a further variant known as charge 4 was developed. Charge 4 consisted of 30% cast iron returns, 20% cast iron borings, 47% steel scrap and 3% pig iron.

The results from charge 4 were acceptable, with microstructures as good as those seen from charge 1 melts. However the main aim of the development program, the removal of pig iron from the charge, had not been completely achieved. At this stage in the investigation, a melt pre-conditioner developed to re-nucleate "dead iron" was investigated. Therefore a final melt, charge 5 was developed which consisted of 30%

cast iron returns, 20% cast iron borings, 50% Steel scrap and a 0.1% addition of melt pre-conditioner.

An immediate improvement in thermal analysis data was noted as a result of using charge 5 (see table 1). Evaluation of the microstructures resulting from this further trial showed that charge 5 gave microstructure results that were favorably comparable with those initially seen in discs produced from charge 1, the initial steel scrap, pig iron and returns charge traditionally employed by PDC. A typical microstructure obtained from the charge 5 melt is illustrated in figure 6.

Oxygen results
In order to evaluate oxygen potential of the various charges, samples were taken from the melts detailed above and analysed. The results of the analysis from melts 3 and 5 are shown in table 2. It can be seen from the results in table 2 that charge 5 had a higher oxygen level that that of charge 3. The melts from charge 5 also yielded the best microstructures. The results also confirmed that charge 5 which consisted of cast iron returns, cast iron borings, steel scrap and a controlled addition of pre-conditioner gave good microstructures, favorably comparable with those of the idealized microstructure (figure 1)

Conclusions
The detailed investigation of melting practice at Precision Disc Castings has shown that pig iron can be successfully removed from an induction-melted iron and substituted with cast iron borings and a melt pre-conditioner without compromising product quality. Oxygen analysis has shown that there is an interaction between the borings and pre-conditioner. It is this interaction that has been the catalyst for work to introduce further improvements in the consistency of product quality at a time of significant change.

The investigation has also resulted in other benefits such as an increase in tensile strength values, and further development in other areas of the EURAC group. Automotive brake discs that satisfy the high standards expected by PDC customers are being produced with the desired graphite morphology and mechanical properties.

The work detailed in this paper has been aimed at reducing the effect of increases in the cost of raw materials on the production process. Without the benefit of thermal analysis technology, the majority of this work would have been performed by trial and error. The benefits of thermal analysis have been significant, especially in the evaluation and prediction of the nucleation potential of melt recipes and the comparative analysis of various melts. These benifits have resulted in greater consistency in the melting process.

References

1. Macnaughtan M, *The history, design and production of cast iron brake discs,* Foundryman, Volume 91, Part 11 Oct 1998, pp321-324.
2. Chapman B J, Hatch D, *Cast iron brake rotor metallurgy,* I.Mech E paper no. C35/76, 1976.
3. Metzler H, *The brake rotor – friction partner of brake linings,* S.A.E. Technical paper no 900847, 1990.
4. Motor Industry Research Association (MIRA), *Unpublished Research programme commissioned by Precision Disc castings, 1998*
5. Marwanga R, Voigt R, Cohen P, *Influence of graphite morphology and matrix structure on chip formation during machining of gray irons,* Paper 99-80 AFS Transactions.

Acknowledgements

The authors would like to thank both Dr J Krosnar, Chairman of EURAC for giving permission to present this work and all those colleagues at PDC who have helped during the investigation.

Tables

Table 1 – Thermal analysis results

Charge No	R	GRF1	GRF2	dTdtTS
1	8.54	60.00	30.60	-3.59
2	8.05	59.28	27.00	-3.74
3	7.17	63.34	28.58	-3.69
4	7.69	68.38	35.23	-3.50
5	7.5	65.52	26.50	-3.82

Where R is the recalescence and is the measure of the temperature increase from TE Low to TE High (Graphic precipitation at the first eutectic with a low value desirable), GRF 1 is the measure from TE High – 10 degrees C (Graphic precipitation during the second eutectic, a high value indicates graphite precipitation over a longer period), GRF 2 and dT/dtTS are a measure of thermal conductivity of the sample (a low value indicating more type A graphite produced).

Table 2 – Comparative Oxygen analysis for charges 3 and 5

Charge No	Oxygen (%)
3	0.008
5	0.025

Figures

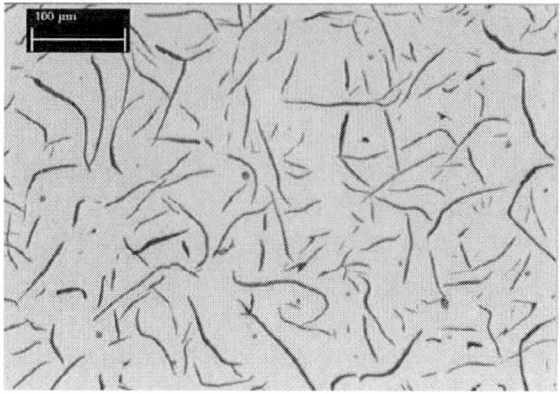

Figure 1 – ideal brake disc microstructure

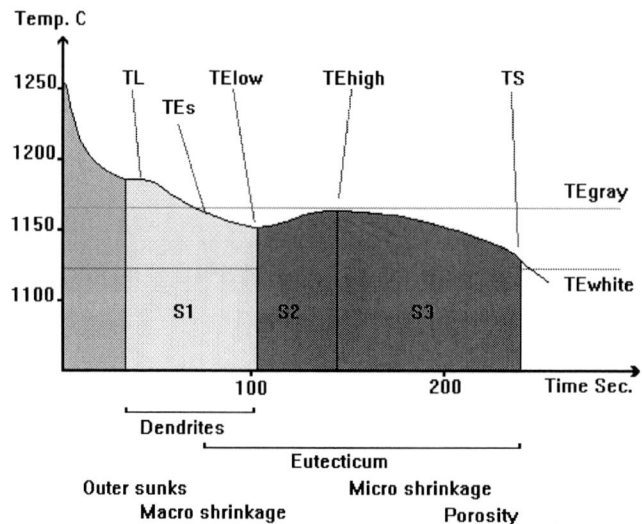

Figure 2 – Thermal cooling curve for hypoeutectic cast iron

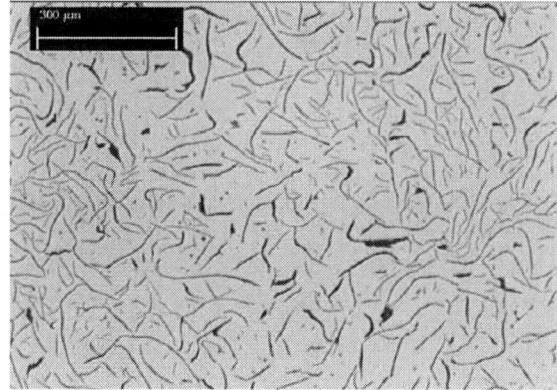

Figure 3 – Microstructure from charge 5

Oxygen Technologies: Reduce Melting Cost and Emissions

T. Niehoff *, P. Keena **

* Air Products GmbH, Germany, ** Air Products plc, UK.

Abstract

This Paper describes the economic and technical case for implementing Air Products' Rapidfire APCOS technology in an iron cupola operation to offset the impact of high and unpredictable costs for coke, steel scrap and alloying agents that all foundries are experiencing.

The Paper includes real examples of this technology in operation today, where it typically delivers net savings of between £1 and £3 per tonne of liquid iron produced. The benefits can include dramatically reduced coke rates, higher silicon yields, reduced particulate and gaseous emissions and higher melt rates where required. The option to directly inject solid material (e.g. silicon fines) into the melt zone without impacting the temperature in front of the tuyere provides an additional opportunity for saving raw material and disposal costs. In short, the technology provides operational flexibility for both cold and hot blast cupola operators and is therefore a valuable tool that foundries can use to remain competitive.

Key words:

Oxygen, process optimization, melting cost, cupola, APCOS.

Introduction

The paper briefly gives the latest experience of applying Air Products' APCOS (Air Products Cupola Oxy-Fuel System) system in cupola furnaces, which is becoming even more attractive as the cost for coke and metallic charge increase. The APCOS technology combines the injection of supersonic oxygen, natural gas and solid particles such as dust or metallurgical powders through the cupola tuyeres. Recent results demonstrate the system can provide increased melt rates, reduce coke rates, improve silicon yields and enable the injection of silicon fines and iron particles without affecting the temperature in front of the tuyere. APCOS can be applied in cold blast and hot blast cupolas.

Coke prices have been high over the last two years and will remain unpredictable in the future. Cost for alloying elements like Silicon (Si) and Manganese (Mn) are high and will remain high for the next years to come. More operational benefits are observed when using APCOS on a cupola these will be described in this paper.

APCOS – Air Products Cupola Oxy-Fuel System

Brief Description

Alternative fuels to foundry coke are currently an interesting subject for foundries to keep their cost optimised and low. Natural gas, oil and coal offer potential to replace foundry coke in a cupola melting process. The APCOS technology does allow to inject natural gas, coal and oxygen into a coke filled shaft furnace. The idea to inject solid particles is not new. An English patent from 1831 describes this invention. Dry and powdery materials such as coal, coke breeze, silicon and cupola ash can be injected.

Combustion of oxygen and natural gas allows flame temperatures of 2900°C which enables to transfer heat into the coke bed which can have temperatures of 2000 to 2300°C. Hence, injection of solids at higher rates is possible without cooling and freezing of the turyere area. The burners can be operated without solids injection to substitute foundry coke. Substitution rates of 30% have been reached on a modern hot blast cupola. Oxy-fuel firing in a coke bed has the benefits of increased stability of runner iron chemistry, reduced coke rates and operating the furnace at optimum gas velocities.

Figure 1 shows the schematic of an APCOS oxy-fuel burner in a cupola tuyere. The burner is typically flanged to the back of the tuyere body and inserted through the water cooled or non water cooled tuyere. At the burner tip oxygen, natural gas and solids are mixed to enhance the effectiveness of the coke bed and the cupola process.

Figure 1:   Schematic of APCOS Burner in Cupola Tuyere.

Oxy-Fuel Firing

Coke is the fundamental fuel for cupola operation that performs the vital task of providing the coke bed structure essential for a cupola to operate. Therefore, if alternative fuels are to be used, they must be used in conjunction with a coke bed or some other structure. The main challenge with using a fuel such as natural gas is that, when burnt in air, the adiabatic flame temperature ($\sim$1900°C) is not high enough to enable sufficient heat transfer to allow iron to be tapped at the temperatures required by most modern foundries (1520°C or higher). The combustion zone of the cupola itself can approach the temperature of the air/gas flame, making heat transfer virtually impossible. The only solution is to burn the fuel in the presence of pure oxygen via an Oxy-Fuel burner, giving a greatly increased adiabatic flame temperature ($\sim$2900°C for Oxygen/Natural Gas). This provides a large enough temperature gradient between the burner and the coke bed/combustion zone to allow heat transfer to take place at a very efficient rate. At higher melt rates only oxygen and oxy-fuel usage allow to operate a given furnace without over blowing. The optimum range of the cupola operation is extended to higher melt rates.

With the reduction of coke in the shaft of the furnace the effects of the Boudouard reaction are lessened as there is less carbon available to reduce the products of combustion.

The addition of Oxy-Fuel burners in selected tuyeres allows the cupola operator to combine the best features of coke, natural gas and oxygen to give a wide range of practical benefits.

A typical APCOS burner installation is shown in the figure 2.

Figure 2: APCOS burner installed in a cupola tuyere

BMBF Kupolopt Project

This project was initiated in 2001 and included several partners:

> Fritz Winter Eisengießerei GmbH & Co. KG,
> E.ON Ruhrgas AG,
> Küttner GmbH,
> IEHK RWTH Aachen,
> FHG Umsicht, and
> Air Products GmbH.

Fritz Winter Eisengießerei GmbH & Co. KG operates two hot blast cupola furnaces in Stadtallendorf (Germany). The cupola used for this project is a modern hot blast cupola furnace (24 t/h) of Küttner design, which mainly produces grey iron. The charge consists mainly of steel scrap (70%) and in house foundry returns (30%). Fritz Winter Eisengiesserei GmbH & Co. KG produces high quality iron castings for the automotive industry.

Six oxy-fuel burners are installed in cupola No. 2 at Fritz Winter Eisengießerei GmbH & Co. KG. Each burner has a firing capacity of 1,000 kW = 33 therms/h (total 6,000 kW = 205 therms/h). The burners have been in use since 2002 and results regarding melt rates, energy and combustion efficiency are now available. In particular, the melt rate has been increased by more than 20%, based not only on short term monitoring and observation but also on long-term results. The increase in melt rate is limited to operating conditions at Fritz Winter Eisengießerei GmbH & Co. KG, as the automated charging system is operated at its capacity limit. Figure 3 shows relative melt rate scenarios for a variation of total oxygen flow rate through the cupola furnace. The total oxygen is the sum of oxygen introduced via hot blast air, oxygen lancing and oxy-fuel burner operation.

Figure 3: Melt Rate as a Function of Total Oxygen Flow.

The melt rate increases due to several factors including the packing of the cupola charge in the furnace shaft and the altered way in which the coke (carbon) is oxidized and gasified. First about 6% less coke and about 10% less silicon (Si) containing alloying briquettes is charged. This gives room for more iron and steel mass per furnace volume. Hence, the melt rate has to increase. Less coke and less silicon briquettes lead to reduced ash and slag of about 8% (≈ 6 kg$_{slag}$/t$_{Fe}$). The energy to melt and vitrify this 8% of slag is available to heat and melt the metal. However, the energy that is provided from the burn up of silicon needs to be replaced by oxy-fuel energy. Overall results are summarised in Table 1.

Table 1: Two Examples of Results of APCOS Oxy-Fuel Burner Operation with different objectives.

The use of hydrocarbon fuels shifts the equilibria of various gas to gas and gas to solid systems (as described in reference [4]). The combination of an increase in temperature and the availability of hydrogen molecules (due to the addition of natural gas) impacts the water gas shift reaction with the presence of coke. This will result in an increase of Carbon Monoxide (CO) and Hydrogen (H_2) in the raw top gas of the cupola operation. For hot blast cupolas the chemical energy contained in the top gas is partially recovered in the post combustion chamber followed by a recuperative heat exchanger to pre heat the blast air. The effect on the cupola energy balance of the cupola is detailed in Table 2, while Figure 4 illustrates the changing composition of the top gas.

Table 2: Results Cupola Energy Balance before and after application of the APCOS system

The modified top gas composition has several benefits for the process. First the Hydrogen (H_2) and (CO) increases the reducing potential in the hot areas of the cupola shaft. This effect reduces the oxidation process of metallic charge components like: Iron (Fe), Si and Mn. Results are reduced losses of metals like iron and alloying agents. To high temperatures and to much oxygen can have negative effects on iron quality and cupola operation. The cooling effect of the dissipating gases controls this in a smart way. Now this allows using higher quantities of oxygen and natural gas to achieve high rates of coke substitution and production rates. Now the blast rate becomes the parameter to decide on the objective which can be coke replacement and melt rate optimisation.

Figure 4: Changing top gas composition before and after application of the APCOS oxy/fuel burners.

Many more positive side effects can be realised when using APCOS oxy/fuel burners in cupola tuyeres. The bed coke which is charged for cupola downtimes can be reduced or even eliminated. When starting up the desired iron temperature and chemistry is reached sooner with the burner. APCOS can enable to run with only half the blast (and top gas) rate at unchanged production rates. This has consequences on dust carryover and dust generation of the cupola process. In addition it leads to underutilised flue gas cleaning system which then allows to conduct maintenance and servicing of heat exchangers and other components which the production continues at 100%. Oxy/fuel assisted iron production leads to more consistent runner iron chemistry particular carbon and silicon. It also reduces the sulphur content of the iron as the coke rate can be reduced.

Solids Injection

Once oxy/fuel firing is established the burners can be used to inject powdery dusts and fines directly into the hot coke bed through the tuyeres, utilising a specialised materials handling machine. In the traditional cupola most materials, if injected in any quantities, would cause cooling of the melt zone, with consequences for metallurgy and temperature and even freezing and blockage of a tuyere. If the material is introduced through the centre of an oxy/fuel flame at about 2600°C it is superheated and therefore has no undesired cooling or metallurgical effects and causes no tuyere blockage. If the material injected is a foundry waste product such as bag house fines, fettling shop waste, etc. then increasingly expensive disposal costs can be saved. Table 3 summarises several types of solids that have tested to date.

Table 3: Composition of several types of solid material injected into the cupola.

In all cases, the injection was performed with no loss of temperature in the combustion zone in front of the tuyeres. Injection of SiC was particularly successful with immediate reaction and Si pickup in the runner iron. This confirms the results previously achieved at Tatra (Tafonco) in the Czech Republic where Si fines have been continuously injected via an APCOS system for 5 years. Similarly, the injection of iron particles (brake disc fettling dust) was also successful. The particles were successful recycled in the cupola melting process with no evidence of the dust in the off-gas system. However, injected coke breeze and sand recycling dust was found in the flue gas due to the combination of low density and high velocities. Further testing is required to enable these materials to provide an economic benefit when they are injected.

Conclusions

The application of APCOS delivers reliable results that allow cupola operators (cold blast and hot blast) to increase melt rate, reduce coke rates, achieved improved Si yields, reduce manganese losses and reduce slag formation. Injection of silicon fines and iron particles has also been demonstrated to deliver positive, repeatable results without impacting the temperature of the melt zone. As a result, the technology offers iron foundries the opportunity to increase their cost competitiveness and flexibility at a time when the costs of crucial input charge materials are rising.

References

1) T. Niehoff, H. Strüning, O. Frielingsdorf, M. Wilczek, T. Wieting, J. Schäfer, M. Lemperle: Oxy-fuel burner technology for cupola melting, 2nd International Cupola Conference, Trier March 18/19, 2004.
2) D. Saha, T. Niehoff, S. P. Smith and O. Frielingsdorf: Oxygen – a versatile tool to enhance cupola operations - 2nd International Cupola Conference, American Foundrymen's Society, Cincinnati, 7 October 1998
3) M. Adamec - Tavení šedé litiny na modernizované kupolové peci Slévárny TATRA, a.s. Kopřivnice, Slévárenství 1999
4) R. H. Nafziger - Alternate Fuels for Cupola Operations, AFS transaction 1989
5) W. J. Peck – Supplementary Cupola Fuels, AFS – CMI Joint Conference, 1980
6) AFS - Cupola handbook , 5th edition
7) F. Neumann, Technologie des Schmelzens für Eisen und Stahlguss – Dortmund 1999

Tables

Objective	Maximise Melt Rate	Coke Replacement
Melt Rate	+ 21%	0%
Coke Reduction	- 6%	- 28%
Silicon Savings	- 10%	0%
Manganese Savings	- 12%	0%

Table 1: Two Examples of Results of APCOS Oxy-Fuel Burner Operation with different objectives.

	Baseline Operation	APCOS
HEAT INPUT	[%]	[%]
Foundry Coke	79.7	74.6
Silicon burn up	10.6	8.9
Hot Blast	7.2	7.1
Natural Gas	0	5.3
Other	2.5	4.1
HEAT OUTPUT	[%]	[%]
Runner Iron	35.9	33
Top Gas	39.5	42.4
Water Cooling	15.7	12.9
Slag	4.1	3.1
Other	4.8	8.6

Table 2: Results Cupola Energy Balance before and after application of the APCOS system

Material	Composition	Injection Rate
SiC	47% SiC, 13% SiO2, 10% CaO Grain Size: 1 – 2.5 mm Density: 1.36 g/cm3	150 kg/hr
Coke Breeze	93% C Grain Size: 0.1 – 3 mm Density: 0.566 g/cm3	300 kg/hr
Sand Recycling Dust	SiO2 – 60%, clay 25%, C 17% Grain Size: 0.1 – 1.5 mm Density: 0.71 g/cm3	400 kg/hr
Iron Particles	Brake disc fettling dust 93% iron content Grain Size: 0 – 0.8 mm Density: 1.43 g/cm3	600 kg/hr

Table 3: Composition of several types of solid material injected into the cupola.

Figures

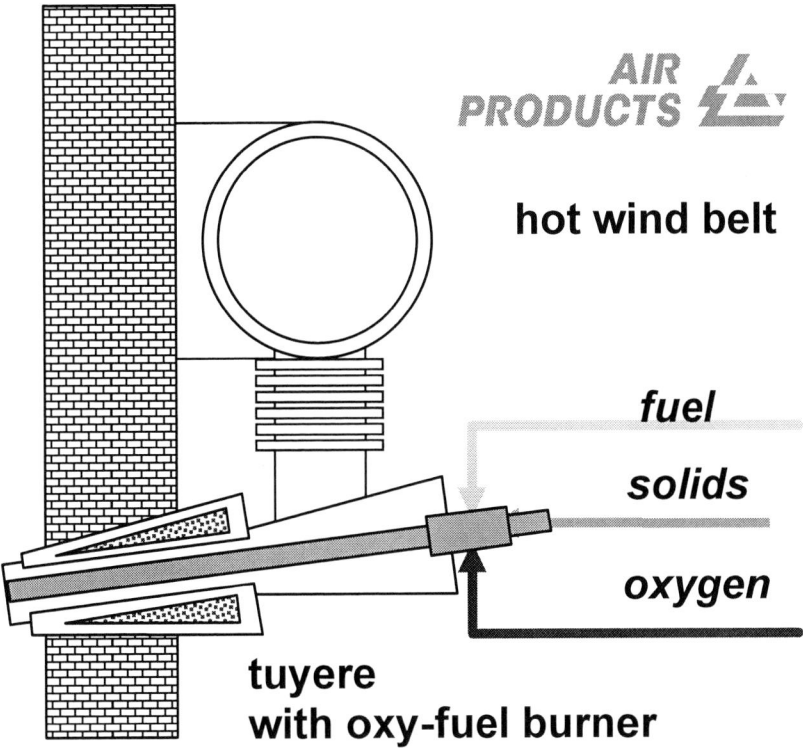

Figure 1: Schematic of APCOS Burner in Cupola Tuyere.

Figure 2: APCOS Burner installed in a Cupola Tuyere.

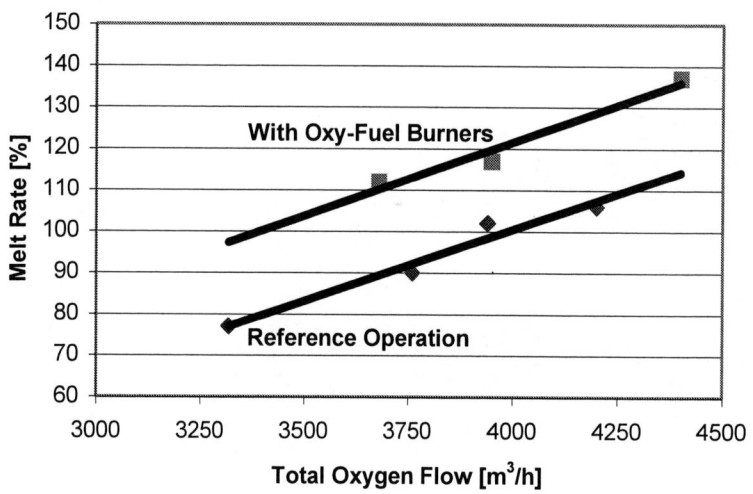

Figure 3: Melt Rate as a Function of Total Oxygen Flow

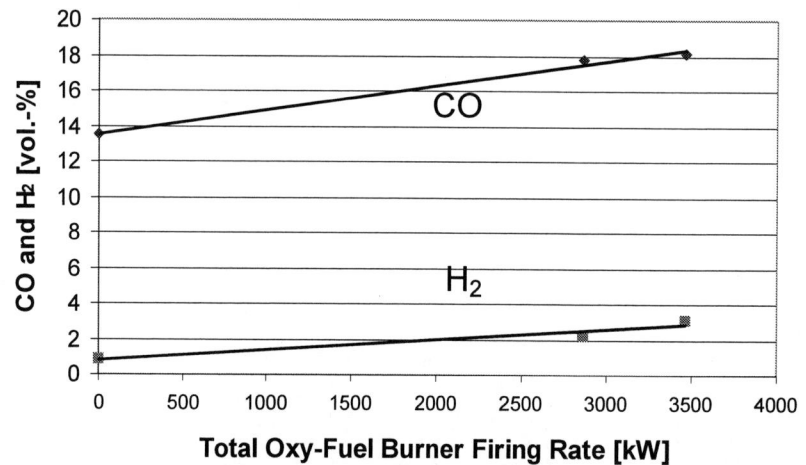

Figure 4: Changing top gas composition before and after application of the APCOS oxy/fuel burners.

E-Mail Contacts: Thomas B. Niehoff   niehoftb@airproducts.com
                    Pete Keena   keenap@airproducts.com

Development and Use of a New Optical Sensor System for Induction Furnace Crucible Monitoring

Wilfried Schmitz *, Frank Donsbach * and Henrik Hoff **.

* Otto Junker GmbH, Simmerath, GERMANY, ** Lios Technology GmbH, Cologne, GERMANY

Abstract

Dependable protection of the induction coil against overheating and, the more so, against contact with molten metal is of vital importance for ensuring safe and reliable operation of an induction furnace.

Addressing this requirement, various technical solutions for monitoring the refractory crucible condition were proposed and implemented in the past but an optimum solution has not been found as yet.

The present paper describes the development and use of a totally new temperature measuring and monitoring system. The new system uses fibre-optical sensors which are particularly suitable for interference-free crucible monitoring in induction melting furnace applications, and provide direct and independent temperature field measurement in the immediate vicinity of the induction coil.

Following months of experiments in the test bay during which all minor problems were duly identified and remedied, long-term tests were conducted in several furnaces under production conditions.

The result can be summarised as follows: Crucible cracks and erosion are detected and localised reliably and precisely and normal refractory wear is under close control.

Key words

induction furnace; crucible monitoring system; fibre-optical sensors

Introduction

One key design feature distinguishing induction furnaces from other heating equipment is the fairly thin ceramic lining between the live water-cooled copper conductor and the molten metal bath (Fig.1). Depending on the furnace size, the thickness of this ceramic lining varies between 10 and 15 cm; it diminishes noticeably as a result of wear or crucible erosion. Inductor insulating materials such as insulating varnish and bandages are heat resistant up to about 150 - 200 °C. If overheating occurs at this point, the insulation may become damaged or even electrically conductive, resulting in interturn short-circuiting of the coil. The coil repair effort required in this case will render the furnace inoperative for several days, even if a spare coil is on hand. In the worst case, which has rarely been documented but is nevertheless a possibility, the melt may penetrate all the way through to the water-cooled coil with all attendant risks of a furnace breakthrough and ultimately, a steam explosion.

These considerations, together with furnace users' economically motivated demands for a maximum service life of the ceramic furnace lining, call for a technology which permits a "visual" inspection of the gap between the ceramic furnace lining and the induction coil. Addressing this requirement, various technical solutions for monitoring the crucible have been proposed and implemented in the past.

Overview of conventional crucible monitoring systems

The most important of these is the classic earth leakage monitoring system. In this technology, a d.c. or a.c. voltage of a defined, fairly low frequency is applied to the induction coil and the system measures the current flow to earth. For this purpose the molten metal bath must be earthed via an earthing rod in the bottom of the crucible. This earth fault monitoring system, although by now a standard feature on virtually all induction furnaces, has a number of disadvantages. For one, it is not selective, i.e., defined tests and disconnection steps must be carried out whenever an earth leakage is detected so as to determine whether the fault has occurred between the coil and the molten metal or in any other part of the equipment, such as in the switchgear or even in the water recooling system. Another disadvantage of this earth leakage monitoring method is that in the event of infiltration or penetration of molten metal to the coil, evidence of this condition will be not be obtained until fairly late. As a result, the furnace must be emptied quickly if a current flow between the melt and the coil is detected. In any event, minor damage to the coil may have occurred already.

The use of thermocouples between the hot-face lining and the coil levelling mix, as well as in the furnace bottom, is another technique employed. However, this method can yield only spot measurements and is therefore not capable of monitoring the entire crucible.

In the past, wire netting in various geometrical configurations has been placed between the coil levelling mix and the hot face lining. The idea is to detect an electrical continuity between an advancing tip of molten metal and the net. One particular disadvantage of this method is that it provides no trend indication, i.e., no advance warning is given.

A further process in industrial use [1, 2] relies on the use of sensor grids comprising an array of metallic electrodes in a comb-type configuration. These electrodes are used to measure the electrical resistance of the ceramic lining. As this resistance is temperature-related, it is possible to infer the temperature in specific crucible segments. Fault locations can thus be identified in relation to the furnace circumference, and an advance warning functionality is obtained. However, spatial resolution is limited (typical system: 8 radial segments, no height information) and the readings are affected by conductivity changes of the refractory not caused by temperature (e.g. moisture, Zn deposits).

OCP Optical Coil Protection System

OCP (Optical Coil Protection System) stands for a latest-generation temperature measurement and monitoring technology which relies on fibre-optic sensors. Given their properties, such sensors are perfectly suited for interference-free monitoring of the crucible on induction melting furnaces. Based on an optical fibre, the system utilizes a quantum-mechanical effect, the so-called RAMAN effect, for temperature measurement [3]. Laser light of a suitable wavelength and modulation frequency is injected into the optical fibre. This laser light scatters on the bonding electrons of the solid state structure over the full fibre length and is detected as a backscatter spectrum. This spectrum contains the RAMAN lines, the intensity of which is a function of vibration levels in the solid state fibre structure, which in turn depend on temperature. A new, patented 'optical radar' technique makes it possible to detect these lines locally and to measure an exact, high-resolution temperature profile through the optical fibre. Thus, OCP is a unique crucible monitoring system which enables us for the first time ever to determine the temperature field in the induction furnace irrespective of refractory type and design. By selecting a radial resolution of 60 measuring points, it is possible to represent the temperature curve in the manner of the familiar analogue clockface (Fig. 2). By adopting an appropriate configuration of the sensor grids, the crucible can be vertically divided into several regions, although only a single optical fibre is used in all cases. Points of particularly high temperature, e.g., due to infiltration, erosion or cracking in the crucible, can thus be accurately localized and checked for potential hazards to the coil insulation.

Design and installation of the OCP sensor cable

The core of the OCP sensor cable, first of all, consists of a commercially available high-temperature glass fibre of the type commonly used in telecommunications. For mechanical protection, this fibre is enclosed in a stainless steel tube measuring 1.2 mm in diameter. The tube, in turn, is coated with a high-temperature insulating compound. The overall diameter of the sensor cable is 5 mm.

In order to provide the fullest possible crucible sensor coverage in the direct vicinity of the coil, it is desirable to have a maximum length of sensor cable in the furnace. This is achieved by placing the sensor cable on the inside of the coil in meandering curves. Fig. 3 illustrates this layout on a 12-tonne cast iron melting furnace. Here we have four meandering cable layers. Once the cable is installed in this manner, the usual former is placed in the coil and a permanent lining made of high temperature resistant corundum concrete is cast, with the sensor cable thus embedded therein. In a next step, the intended measuring end of the sensor cable is connected to an optical fibre transfer cable armoured to foundry standards. This cable is run to the location of the evaluator and connected to one of its ports.

Display of measured temperature data

The main screen of the OCP visualization system is shown in Fig. 2. It displays a schematic top-down view of two furnaces, each comprising two or more meander layers. If more than two furnaces are monitored, the user can freely select which of these should be displayed on the left and right-hand side, respectively. As a general rule, the temperature curves for all individual meander layers are initially rendered in a screen window. For a less cluttered view, individual layers can be suppressed. This has been done in Fig. 2, where the left image shows the current temperature distribution in the upper layer of Furnace 1 while the image on the right gives the current temperature distribution in the bottom layer of Furnace 2. Each of these temperature profiles can be rendered in a polar or linear view. It is also possible to display a "relative" mode, i.e., a profile generated in relation to a given historical (reference) profile. In this case the user will see the current temperature deviation from the selected reference profile (offset view).

By selecting a playback function and entering a date in the respective window, past temperature profiles can be viewed at any time. It is also possible to show temperature profiles in an animated or "video" mode between a user-defined start and end point. Also the temperature history of selected measurement points can be displayed.

Evaluation of temperature measurements

Four types of alarm algorithms can be configured offering several parameters to be adjusted. For every parameter, the user can enter a warning threshold and an alarm trigger threshold. The individual alarm criteria are monitored as follows:

1. Temperature

The temperatures at one or more measuring points are monitored for overruns exceeding these preset thresholds.

2. Deviation from average

The measurements from which the temperature profile is plotted are initially processed into an average value representing the mean temperature in the respective zone. The system then checks whether the temperature at one or more measuring points deviates from this mean value by more than the preset threshold.

3. Temperature change

Here the system determines whether a time-related temperature gradient, defined as a threshold, is exceeded at one or more measuring points. The unit in which this threshold is set in the various input fields is °C/min.

4. Uniformity

The uniformity parameter is largely identical with the "deviation from average" criterion, except that the averaging step and check for threshold overruns is not carried out in a single step over the entire circumference of the furnace. Instead, the system initially examines a radial sector ("pie wedge") whose thickness is defined by the user in angular degrees (°) in the window marked "step". This sector is then analyzed in the same way as for the "deviation from average" criterion. In an interactive process, this evaluation window is then advanced in a clockwise direction one measuring point at a time. The analysis is continued until the evaluation window has covered the entire circumference of the furnace. This is a valuable criterion when it comes to distinguishing local flaws from large erosion areas.

Practical operating examples

OCP systems are now successfully in use in coreless induction furnaces for melting copper alloys, aluminium alloys, cast iron and steel. In the following part of this paper three exemplary cases will be examined in which crucible wear and premature crucible failures were detected in a timely and accurate manner.

2.5-tonne vacuum-type coreless induction furnace for melting copper pre-alloys

This particular furnace is run in three-shifts to produce copper-iron pre-alloys. Such alloys pose exacting demands on the crucible material due to their aggressive chemical characteristics and fluidity. The tapping temperature is in the region of 1500 °C. The furnace is usually operated with a ready-made crucible consisting of refractory concrete. The space between this crucible and the coil is backfilled with a dry ramming compound. A normal crucible campaign lasts about 2 weeks, depending largely on the degree of sintering of the backfill mix. If sintering propagated too far towards the coil the crucible will be very difficult to break out; moreover, there will be an increased likelihood of molten metal penetrating all the way to the coil in the event of a crack formation in the crucible. The degree of backfill mix sintering grows over the crucible campaign, causing the thermal conductivity of the backfilling material to increase progressively. As a result, the temperature in the permanent furnace lining and hence, the temperature local to the OCP sensor cable, will rise steadily. Fig. 4 shows the temperature profile at the start (left) and near the end (right) of a normal crucible campaign in which no apparent local crucible failure occurred. Once a critical maximum temperature had been identified over several crucible campaigns, the OCP system was used as an indicator to identify the need for a scheduled re-lining. Fig. 5 shows the situation for a crucible that had been in use for a week, i.e., half the normal crucible campaign. The graph on the left plots the absolute temperatures; its right-hand side counterpart gives an offset view of the same development. Over a span of a few charges, overtemperatures increasing from one charge to the next were identified in the 5 o'clock position, and the system eventually generated alarms of the "deviation from average" type. The crucible was broken out, and a crack was found at this point which had allowed the melt to infiltrate the ramming compound.

6-tonne coreless induction furnace for melting stainless steel

Fig. 6 shows the temperature profile measured in the lower regions of a 6-tonne induction furnace for steel near the end of a crucible campaign. The "deviation from average" alarm messages (left) indicate general erosion towards the furnace spout. The "uniformity alarm" messages (right) point to the formation of caverns at four points. Dimensional measurements conducted on this crucible prior to break-out confirmed the condition detected by the OCP system (Fig. 7).

Detection of cracks in the crucible

When a crucible cools down, e.g., over a weekend, numerous cooling cracks will form naturally due to volumetric contraction of the crucible material. These cracks will normally close again, due to thermal expansion of the crucible material, the next time the furnace is started up. However, an appropriate heating curve must be used to ensure this. Otherwise, the progressively melting metal may spontaneously penetrate still-open cracks and come dangerously close to the coil. The situation becomes even more critical if the furnace is filled with liquid metal before the cracks have closed.

To simulate this situation, the following test was carried out: A thick-walled steel cylinder of a diameter equivalent to the inside diameter of the crucibles normally used in this application was placed in the middle of an unlined 1-tonne-furnace. This steel cylinder exhibited "artificial cracks" at a level about halfway up the furnace coil, these being in the form of 5 mm thick and 100 mm wide pieces of steel plate welded to the cylinder in the 12, 3 and 9 o'clock positions. The plate in the 3 o'clock position was welded to the surface horizontally and ended about 10 mm short of the furnace's permanent lining. An identical plate was welded on in the 12 o'clock position, but in a vertical direction. In the 9 o'clock position there was another horizontal plate which extended to the permanent lining. Finally, a 100 x 100 x 30 mm steel plate representing a cavern was welded to the cylinder in the 6 o'clock position. Fig. 8 shows a sketch of this arrangement. The space between the steel cylinder and the furnace's permanent lining was filled with a quartzite dry ramming compound. The furnace was then switched on and operated at about 350 kW. This procedure was intended to simulate a cold start with existing metal-filled cracks. After 30 minutes, all "crucible defects" are clearly identifiable and reported by the corresponding alarm messages (Fig. 9). It should be mentioned that the small test furnace allowed to embed only a one-layer sensor cable of limited length, which gives an inferior position resolution. The position resolution will naturally be higher on a larger furnace. However, the unusually good temperature resolution of the measuring method is impressively demonstrated.

Conclusions

The essential advantages of the Optical Coil Protection (OCP) system can thus be summarized in the following key-words:

- Full protection against
  Operational breakdown due to coil damage
  Bodily injury and equipment damage due to molten metal breakthrough
- Recording and visualization of the temperature profile over the entire crucible campaign
  Indication of developments and trends of refractory wear or metal penetration
  Possibility to take action in good time to extend refractory life
- Direct temperature measurement, not resistance-based
  Fully operational with a vast range of refractories and immediately after relining
- Optical (i.e., non-electrical) measuring method
  Eliminates false signals or even sensor grid damage by the magnetic field of the induction furnace
- One single evaluator can monitor up two four furnaces
- Very high resolution, e.g. 60 spots over the circumference of an 8-tonne-furnace crucible, like the second marks on a clockface
- Temperature measurement with a resolution better than 1 K
- This distributed optical-fibre temperature measuring method has evolved into a mature system which has been demonstrating its reliability as a central safety system for years in more than 300 installations worldwide [4].

References

1. Hopf M, Elektrowärme International, Edition B. Industrielle Elektrowärme 50 (1992) No B2, pp229-232
2. Hopf M, Gießerei 80 (1993) No 22, pp746-751
3. Dr.-Ing. Glombitza U, „Verfahren zur Auswertung optisch rückgestreuter Signale zur Bestimmung eines streckenabhängigen Meßprofils eines Rückstreumediums" EP 0692705, 1995
4. Maegerle Rudolf, Siemens Building Technologies Ltd., Cerberus Division, Fire Protection Systems for Traffic Tunnels Under Test, Proceedings 12th International Conference on Automatic Fire Detection

Figures

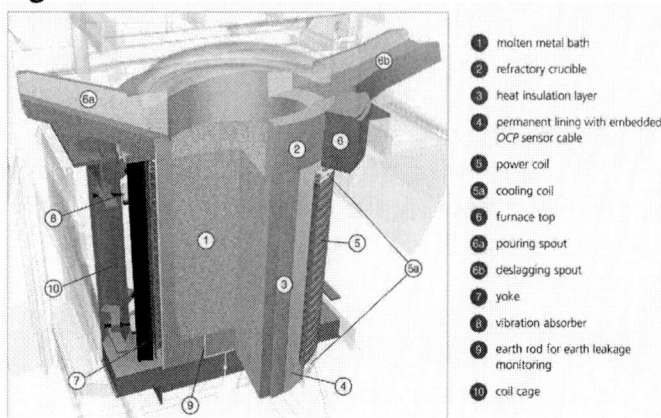

① molten metal bath
② refractory crucible
③ heat insulation layer
④ permanent lining with embedded *OCP* sensor cable
⑤ power coil
⑤a cooling coil
⑥ furnace top
⑥a pouring spout
⑥b deslagging spout
⑦ yoke
⑧ vibration absorber
⑨ earth rod for earth leakage monitoring
⑩ coil cage

Fig. 1   View of a typical crucible structure, showing the permanent furnace lining with the OCP sensor cable embedded ④

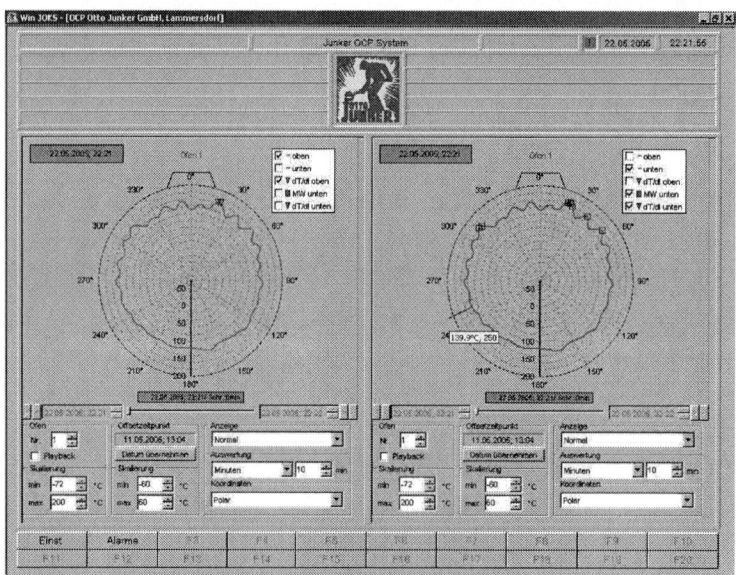

Fig. 2   OCP System monitor screen

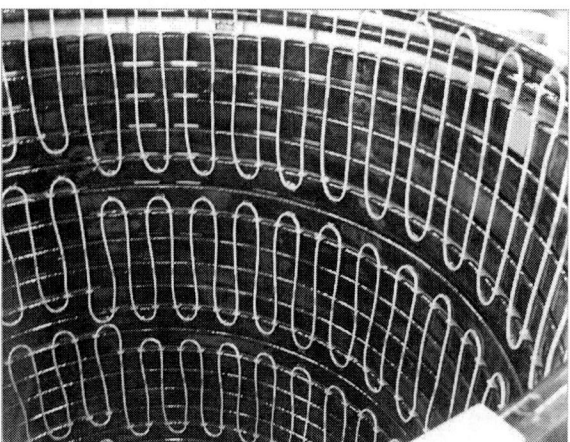

Fig. 3   Arrangement of the OCP sensor cable on the coil of a 6-tonne induction furnace or melting steel

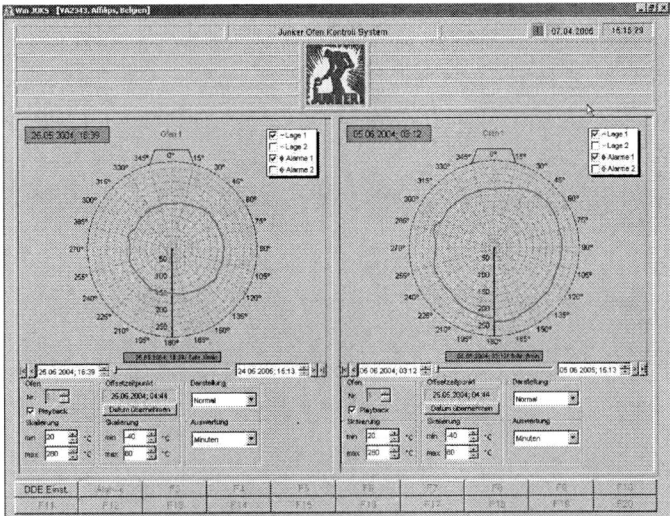

Fig. 4 Temperature profiles on a 2.5-tonne vacuum-type induction furnace at the start (left) and end (right) of a trouble-free crucible campaign

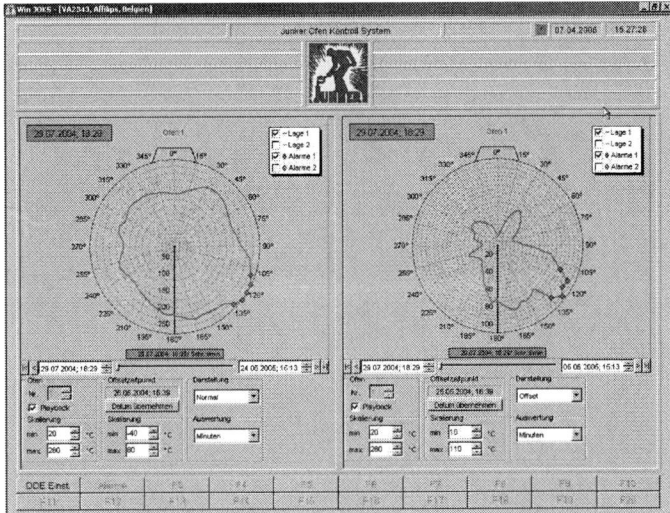

Fig. 5 Temperature profiles obtained after crucible cracking with resultant infiltration; absolute temperature (left) and offset display mode (right)

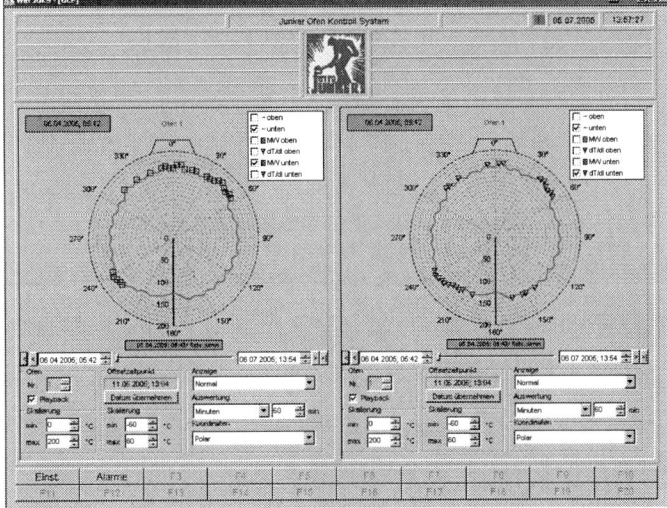

Fig. 6 Temperature profile in the lower regions of a 6-tonne induction furnace for steel near the end of a crucible campaign.

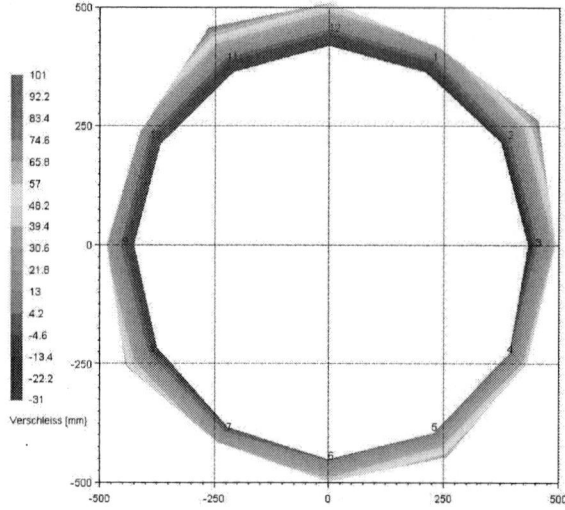

Fig. 7   Result of crucible measurements, showing an integral top-down view of the crucible. Uniform premature wear in the direction of the spout (12 o'clock) and local erosion in the 2, 5, 8 and 11 o'clock positions

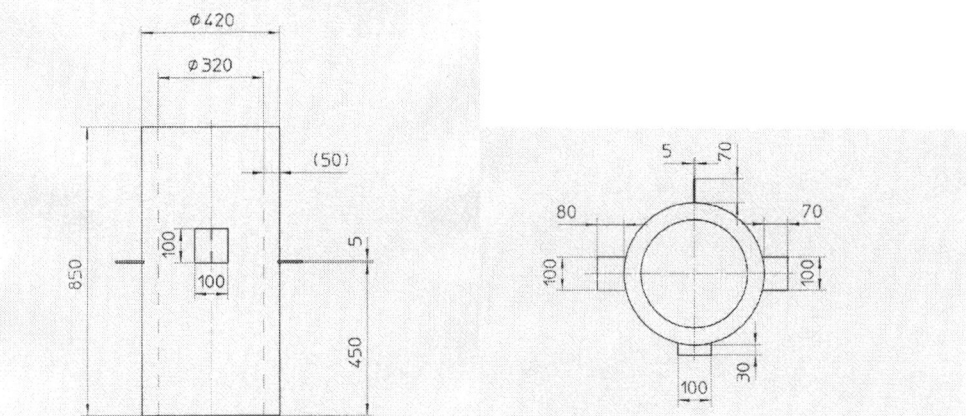

Fig. 8   Thick-walled steel cylinder with welded-on steel plates simulating crucible cracks

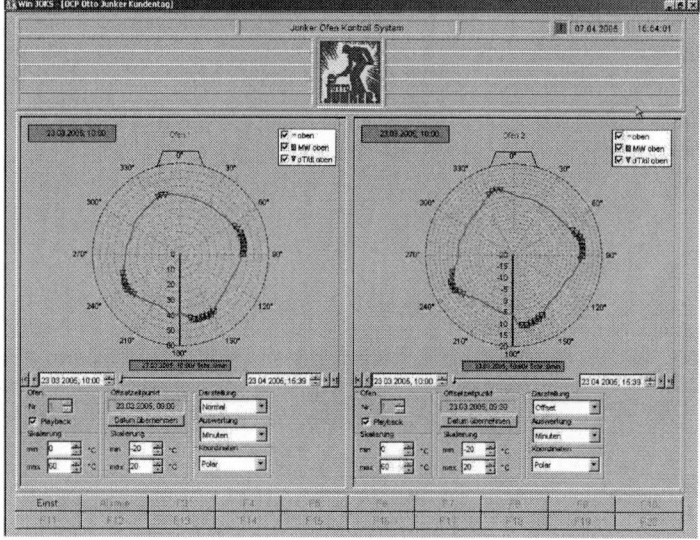

Fig. 9   Temperature profiles after 30 minutes (10:00 a.m.), with all defects duly reported; absolute temperature (left) and offset display mode (right)

652

Development of Coke Alternate Material Using Woody Biomass

Yasufumi YAMAGUCHI * , Shoji KIGUCHI ** and Hirotoshi Murata ***

* Dr. Researcher , Kinki University ,
** Prof. and Dr.
Faculty of Science and Engineering, Kinki University ,
*** Naniwa Roki Co.,Ltd.

Abstract

In the foundry industry, the price of coke rose greatly as well as the inflationary prices of a recent raw material in these 1 or 2 years. As a result, it is greatly influenced that the manufacturing cost increases, the decrease in the competitive edge of the product and the decrease in rate of profit.

Therefore, it is required to use a cheap material to decrease the ratio of coke even a little. So far, it has been much reported that the cheaper smokeless coal or inferior coke is used. However, these inferior materials can cause trouble in casting defect and in cupola operation.

In this study, it paid attention from effective use of the earth resource and the viewpoint of environmental problems to a woody biomass. There are advantages to use the woody biomass that it is easy in the global environment because of the manufacture cost reduction and control of generation of carbon dioxide.

Therefore, it is investigated to make the woody biomass a coke substitution material that manufacturing condition to make the carbonizing woody biomass like briquette, the material characteristic and the combustion process.

As a result, the compression stress of the carbonizing woody biomass is 100 MN/m^2. This value is very hard compared with compression stress 18.7 MN/m^2 of the foundry coke. The calorific density of the carbonizing woody biomass is 21.0 MJ/kg against calorific density 28.9 MJ/kg of coke. This value is an enough calorific value to alternate the coke. In fact, coke and the carbonizing woody biomass were burnt in the carbon crucible. As a result, the heat temperature of the woody biomass is equal to coke in the carbon crucible.

Key words : woody biomass , cupola , cast iron, coke

Introduction

For the foundry industry in Japan, the cost of raw material has been skyrocketing beginning in 2003. Although its cost has been somewhat stabilized now, the cost is still quite high. And this is a global indication, not just Japan. The cause for such indication is the rapid rise in buying power by Chinese economy, accompanying with their exponential economic growth, and expects to progress at higher price tag unless the Chinese economy begins to stall. Also, in regards to coke, its supply is in stringency that purchasing it alone is quite difficult thus not only are there risks of price hike but also the necessity for dealing with stabilized supply was brought out.

In particular, large amount of coke is used in case of cupola operation; therefore the rise in coke price is a problem that is directly related to the pressure of making profit, thus resulting in each company conducting various researches. For example, some places are using lower-priced crude coke or coke for steelmaking purpose and smokeless coal. These materials lack size, intensity and energy density, however, so quality of the product and problems during operation may arise by using them.

This research seen here is done with the purpose of reducing the ratio of coke as much as possible by using biomass originating from the forest resources in place of using coke. In turn, biomass is a natural resource and resulting CO_2 from its combustion can be eliminated from the total amount of CO_2 thus forecasting on the reduction of CO_2 discharge is also possible. In addition, the materials used in biomass are waste matters from sawmill or decomposed, salvage cut branches that are discarded in time, so from the recycling standpoint, it can also be said of having efficient use of earth resources.

In this study, this purpose is to substitute part of coke with biomass during the actual cupola operation. Therefore at first, the type of combustion process by woody biomass is investigated.

Experimental

Ideal size for coke to be used on cupola melting is about one-fifth of the diameter of furnace. However, the size of woody biomass being sold is smaller than 10mm and, according to lack of size, strength and energy density starting from the lowest compressive strength one can easily predict it isn't suitable for cupola melting work[1),2)]. Accordingly, manufacturing of woody biomass suitable for cupola melting is necessary. However, considering the furnace diameter of test use cupola (φ300 mm) and problems from production equipment of woody biomass, manufacturing of φ50 mm woody biomass (to be called as woody biocoke) using high temperature and high pressure was done as the first order of business. Also, investigation on its combustion process was conducted to make decision on whether it is possible to be used in cupola melting process. Here, figure 1 shows the production equipment of woody biocoke.

Woody biomass that was used to create biocoke was made up of chopped up tree barks. Figure 2 shows an example of manufactured woody biocoke. Manufactured woody biocoke shows going half carbonization and turning black by carring out both high temperature and high pressure disposals at once.

The specific gravity of manufactured woody biocoke is from 1.09 to 1.50, as it varies depending on welding force and heating temperature. The caloric value at the specific gravity of 1.09 is 21.0 MJ/kg, which are about two-thirds of coke's value of 28.9 MJ/kg. It can be estimated that as the weight increases greater the caloric value will become, therefore output will rival the coke's caloric value. Also, figure 3 shows a compressive strength of woody biocoke. This is approximately 100 MPa at φ40mm, which has enough strength when compared to coke for cast-iron products. However, this is a strength measured under room temperature setting and it can't be assumed that same degree of strength will be maintained at high temperature.

Therefore, it was investigated how woody biocoke combusts at high temperature. In specific, coke and woody biocoke that were lumped in about 10mm were inserted inside an alumina crucible (φ55mm×180mm) under three different conditions as listed below and each temperature was measured while raising it up to 1173K into the graphite grain.

① coke100%

② woody biocoke100%

Next, to combust the manufactured woody biocoke in the same shape, the firebrick was shaped with inside diameter of 100mm×100mm×600mm, and coke and woody biocoke were placed in layers. Coke (bottom layer) was burned from the bottom portion using the gas burner to recreate the same combustion process as cupola while the process of rising temperature was measured.

Also, the scale was enlarged to φ180mm×600 mm using the fire-proof material to approach the actual conditions of cupola operation, and samples of coke and woody biocoke were placed in layers and burned from the bottom portion like some time ago.

With this result as a foundation, actual cupola melting was conducted with 300mm test furnace with 20% and 50% of coke being substituted with woody biocoke.

Results and Discussion
First, figure 4 and figure 5 show that results of combustion that the coke and woody biocoke cut in 10mm in length and burned inside the alumina crucible.

As figure 4 indicates, the temperature rise in coke was following the temperature rise in graphite morphology. In contrast, woody biocoke in figure 5 showed the rapid rise in temperature between 400K and 800K, and faster temperature rise than coke. Post-exhaustion shape of woody biocoke, however, didn't assume its original shape and showed reduction in volume up to 20% or so. This may be due to its exhausting at once as a whole rather than slowly burning from its exterior, as woody biocoke was broken into small pieces.

Therefore, woody biocoke can burn for rather long period of time depending on whether it's combusting in its original shape. And if it's getting closer to combusting inside the cupola, coke was burned using the gas burner from the bottom portion and combusted woody biocoke with the heat generating from the coke. Figure 6 and figure 7 show these results.

In figure 6, burning temperature of woody biocoke at point B and point C, just as in the case with burning it inside alumina crucible, began rapidly burning around 400K that at point B, the temperature rose until it exceeded the temperature of coke on the bottom. Post-experiment shape is much smaller than still retained its original shape, therefore showing that, unlike the combustion inside the alumina crucible, it is burning slowly from the outside.

In addition, figure 7 shows larger scale experiment, and this shows the combusting tendency of woody biomass maintains its rapid rise about the 400K mark, eventually reaching the temperature beyond 1500K and continuing to keep the same temperature for about two hours. For safety measure, the maximum burning temperature woody biocoke was kept below 1500K.

Accordingly, rather than looking at these results, the melting was done using the actual small cupola for experimental use, concluding that there aren't any operational danger when 20% and 50% of coke are substituted with woody biocoke for cupola melting procedure. The result showed that it was more than enough to melt even with 20% and 50% substitution of coke that was pushed in. Table 1 shows the melting process. The tapping temperature is low experimentally. The reason is that the tapping temperature at the beginning of the first tapped molten metal was low, not the effect that used the woody biocoke. When the woody biocoke increased from 20% to 50%, the tapping temperature rose partly. So, this is difficult to decide that the woody biocoke has an influence on the tapping temperature. At the blast volume and the density of the gas, the woody biocoke doesn't have an influence. Each test piece was picked after the material was changed completely in Cupora. Table 2 shows each chemical composition. The sulfur contents slightly diminishes at 20% and 50% substitution, this being the result of reduction in sulfurization effect from coke. Yet it's still .035% at worst after the reduction, therefore it is still enough not to affect the composition's graphite morphology. Also, even though carbon contents tends to decline a bit from reducing of carburization, its effect on graphite morphology is quite minimal nonetheless.

Accordingly, an assessment can be made that substituting 20% and 50% of coke with woody biocoke in melting of cupola is more than possible as no problems were considered to be found from its melting possibility and chemical composition.

Conclusions
Woody biocoke was manufactured utilizing the high pressure, high heat method to create woody biomass as a form of alternative fuel to coke. The review results on its combustion process are as follow.

(1) Woody biocoke, which cuts into 10 mm size rapidly burned starting around 400 K and achieved the higher temperature peak than the equivalent amount of coke in the end.

(2) Burned woody biocoke in molded form became smaller while attaining the original shape. The temperature at the point of combustion rose until became equal to coke as well.

(3) In case of cupola operation, even if woody biocoke was used as an alternative to coke at 20% and 50% rate, the operation was possible.

References
1. Kunihiko N, Tamio I, Manabu F and Hiroshi S : Pyrolytic and Combustion Characteristics of Woody Bio-Pellets, Journal of Japan Institute of Energy 83, No.10,Oct 2004, pp788-793.
2. Takako H, Tamio I, Manabu F and Hiroshi S : Possibility of New Fuel BCDF, Journal of Japan Institute of Energy 83, No.10,Oct 2004, pp788-793.

Tables

Table 1 Melting process in Cupora

Time	Blast volume (Nm³/min)	Wind blast (mmAq)	Tapping temperature (°C)	CO	CO₂	Remarks	
0:00	Turning on of material						
0:16							5 pass coke
0:18	9	300	1385	1.1	20		fast tapped molten metal
0:45	9	500		21.0	15		7 pass coke
1:01	9			22	15	TP - 1,2	10 pass coke
1:46	8	500		15	18		12pass coke
1:59	8	450	1434	22	15		14pass coke
2:03	8			21.0	15	TP - 3	18pass coke
2:06	8		1437	21	16	TP - 4	19pass coke
2:09	8			21	16	TP - 5	20pass coke
2:24	8	600		30	12		25pass iron
2:40			1453	27	13	TP - 6	29pass coke
2:43				25.0	14	TP - 7	
2:48	8	∞		24	14.0		30pass coke
2:51				21	16.0	TP - 8	Finish
2:54			1412	21	16.0	TP - 9	
2:57				20	16	TP - 10	
2:59			1427	20	16		

20% applies to rows 1:46–2:09; 50% applies to rows 2:24 onward.

Table 2 Chemical composition of specimens. (mass %)

	C	Si	Mn	P	S	
NO.1	3.66	1.99	0.69	0.091	0.058	
NO.2	3.68	1.98	0.73	0.092	0.056	
NO.3	3.58	2.22	0.81	0.092	0.058	
NO.4	3.58	2.37	0.57	0.091	0.042	20%
NO.5	3.60	2.31	0.59	0.092	0.064	
NO.6	3.25	2.89	0.51	0.087	0.048	
NO.7	3.53	2.29	0.50	0.089	0.035	
NO.8	3.52	2.36	0.52	0.091	0.044	50%
NO.9	3.34	2.75	0.60	0.089	0.046	
NO.10	3.55	2.36	0.51	0.091	0.040	

Figures

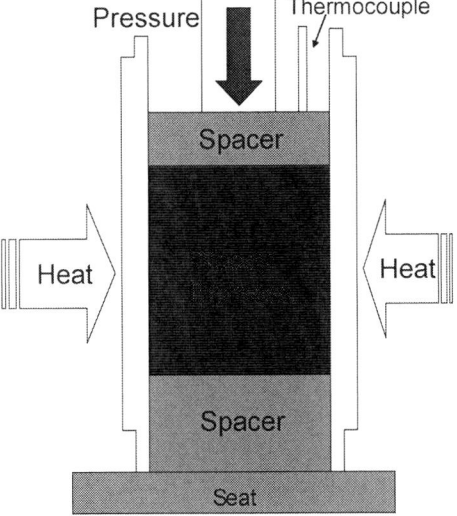

Fig.1 Schematic drawing of electric furnace.

Fig.2 Photograph of woody biocoke.

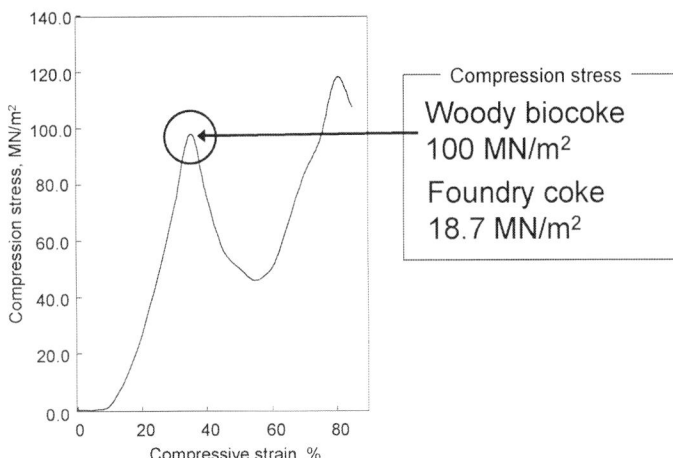

Fig.3 Relationship between compression stress and compression strain. (30ton press machine, 40φ)

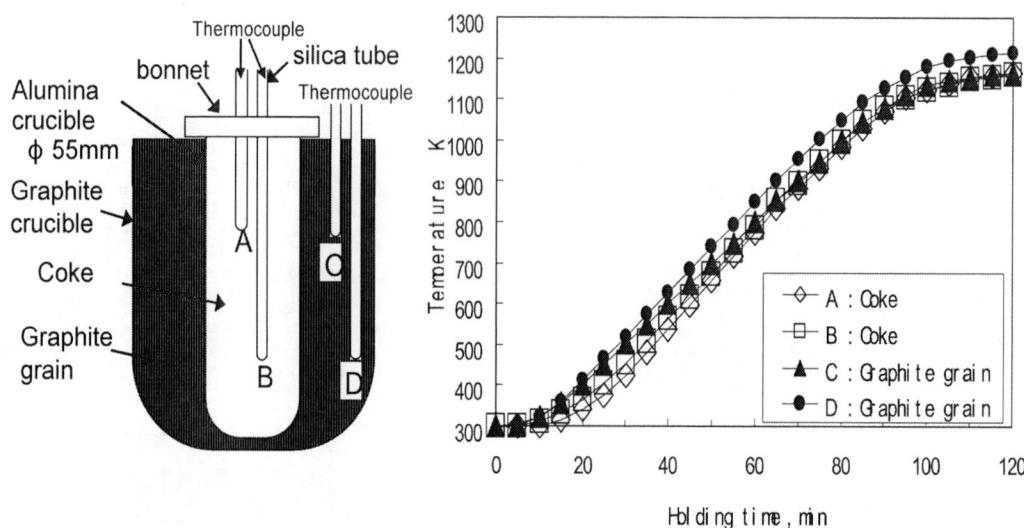

Fig.4 Combustion process in alumina crucible. (coke 100 %)

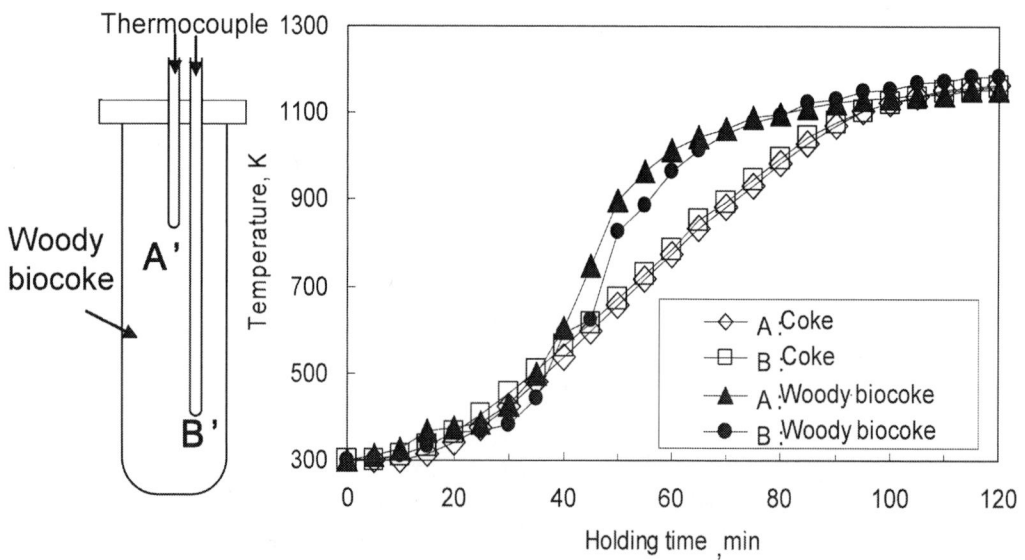

Fig.5 Combustion process in alumina crucible.
(woody biocoke 100 %)

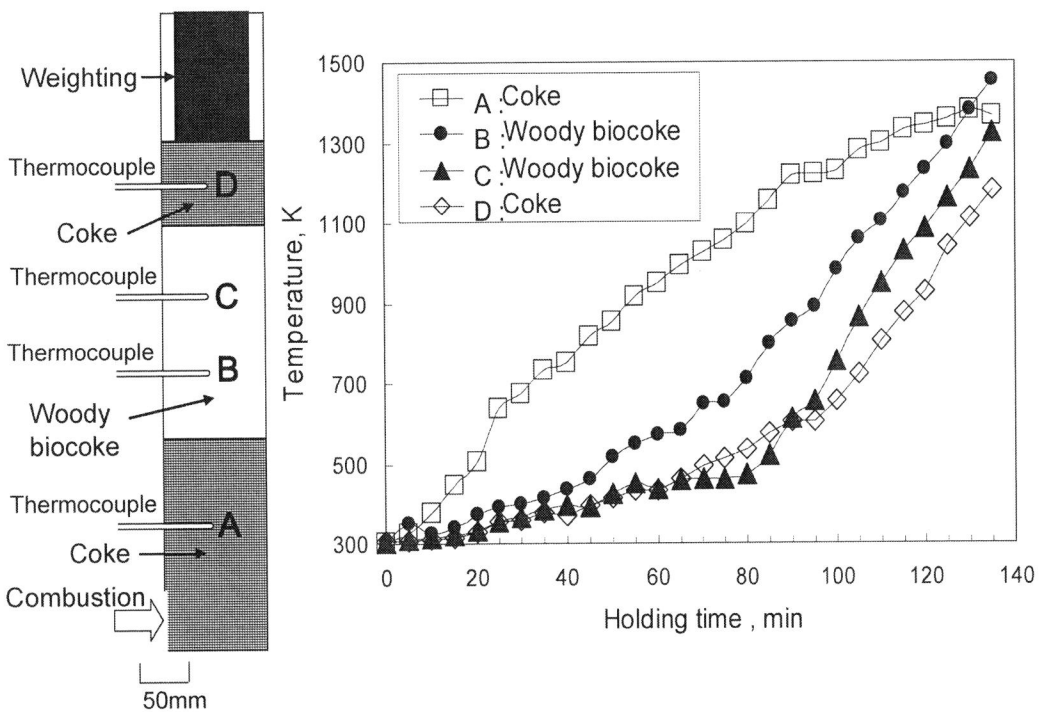

Fig.6 Combustion process in furnance by firebrick.

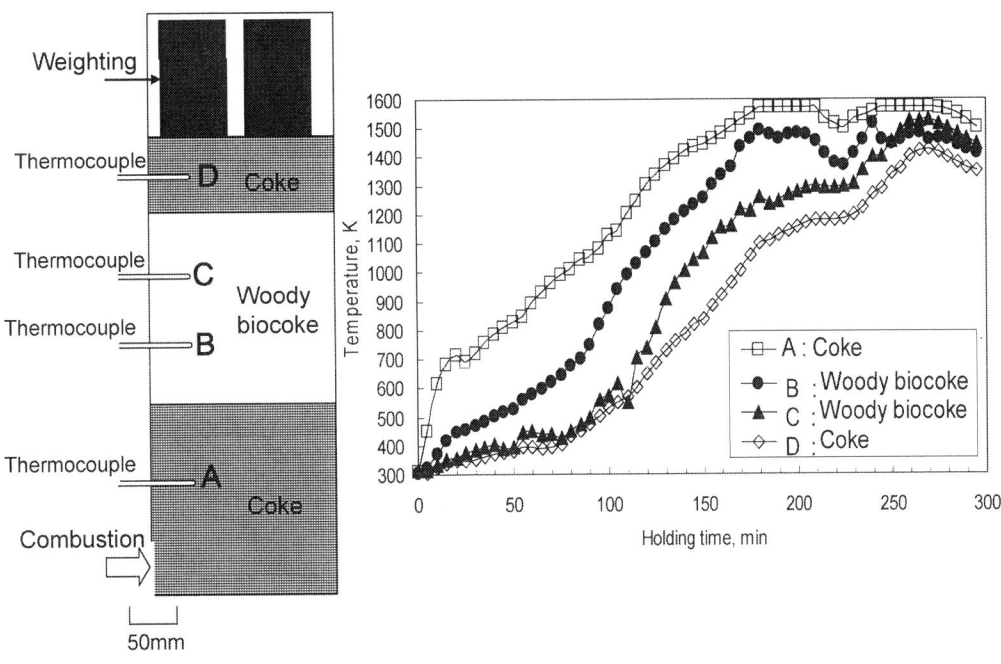

Fig.7 Combustion process in furnace by fireproof material.

"Accurate and Complex NET-SHAPE Castings for Challenging Markets"

M Horacek * and J Cilecek **.

* Brno University of Technology, CZ, ** Alucast Foundry, Tupesy, CZ.

Abstract

In this paper Lost Wax technology will be discussed as one of the „Net-shape" technologies. Achievable final casting dimensional accuracy and how to control/influence this by wax pattern injection parameters and also by other parameters during the Lost Wax process will be discussed. Included are two „case studies/examples"of investment castings, originally manufactured by other technologies. These demonstrate advantages of this very progressive technology. Also the simultaneous engineering approach will be stressed (foundry-customer relationship) during the oral presentation of this paper as a key element for successful business development.

Key words

Investment casting, net-shape technology, casting accuracy.

Introduction

As Beeley and Smart are stating in their book "Investment casting" [1], the process of investment casting or lost wax technology has come to occupy a key position in the range of modern metal casting techniques. Over the half-century dating from 1940, what had been a small and highly specialized sector of casting activity developed into a worldwide and distinctive industry, reflecting the importance of the product in the intensifying search for close accuracy of shape and dimensions in material forming.

The main aim of this paper is to give a general picture about the latest trends in all individual phases of lost wax technology, about achievable casting accuracy and also to demonstrate the capabilities of this technology by presenting case studies of investment castings originally manufactured by other technologies.

DESCRIPTION OF THE LATEST TRENDS IN INDIVIDUAL PHASES OF INVESTMENT CASTING PROCESS

A/ Wax pattern manufacture

Die manufacture – tooling

Tooling is a crucial consideration to investment casters and their customers. The quality of the tool used has a major influence on the price and quality of the casting. There are several tooling methods and types, with various combinations available to be chosen. Many things affect this choice, but the main factor is generally the quantity of components required and the period over which they will need to be produced.
Tooling types range from plaster cast to rubber, resin, metal spray and metal dies, generally in order of increasing casting volume requirement.

Investment casting waxes

Present waxes being used in investment casting technology are complex materials containing following components:

- Natural waxes	*Types of Waxes*
- Synthetic waxes	
- Natural resins	**- Straight (non-filled)**
- Synthetic resins	**- Filled (30% of filler)**
- Organic fillers	**- Emulsified (by water/ air)**

Optimal wax characteristics:

-Melting and congealing point	-Surface quality
-Ash content (‹ 0,05%)	-Expansion/contraction
-Hardness	-Stability to oxidation
-Flexibility	-Solidification rate
-Viscosity	-Possibility to recycle

Wax injection equipment

A wax injector is a machine that takes a preconditioned wax and injects it into a die, creating a wax pattern. Injectors are classified by the state of the wax (namely liquid, paste and solid) that the machine is capable of injecting. The typical design of injection machine is seen on the **Fig.1**

Final dimensions and achievable accuracy of wax patterns depend on injection parameters used (wax temperature and injection cycle parameters, i.e. the time of filling, packing and holding in the die before it is dismantled, and injection pressure). Injection parameter modifications are practically the only possibility for influencing the final dimensions of wax patterns and, consequently, of final castings.

B/ Ceramic shell manufacture

Step-by-step shell building (investing) and drying

I/ Degreasing of Wax Patterns
Removal of remaining separator from the wax pattern surface

II/ Dipping into Ceramic Slurry
Ceramic slurry consists of filler and binder
Filler – heat resistant ceramic flour (fused silica, molochite, zircon,)
Binder – colloidal silica sols based on alcohol (alcosols) or water (hydrosols)

III/ Shell Draining

IV/ Stucco Applied with Ceramic Grit
Fluid or rainfall systems used
Stucco materials – silica, molochite, alumina, zircon, etc.
Grain size according to the coat number:
-First 1-2 „prime coats" fine particles –0,175-0,25 mm (*Casting Surface Finish*)
-Next 3- x „back-up coats" coarser particles – 0,25-0,5 mm (*Mould Gas Permeability*)

V/ Shell Drying
In air-conditioned room 2-4hours – temp. 20 ℃ ±1 ℃, relative humidity 30-60% according to the type of binder used + sufficient air flow

VI/ Repeating (II – V)
Up to the needed number of coats, i.e. 8-12

Shell de-waxing

Key Problem: different wax and shell expansion!
Wax expansion bigger, therefore danger of shell cracking during de-wax process.
Neccesity of „dilatation gap" building on wax pattern surface – through "*Thermal Shock*" in a pressure vessel ,i.e. boilerclave

Ceramic shell firing

Goal: transfer of amorphous type of SiO_2 binder layer into a crystalloid form + removal of volatiles matters (waxes remains)

C/ Metal pouring

On air

a/ Classical pouring
b/„Roll-over" pouring

Under vacuum

a/ Melting and pouring under vacuum
b/ Counter-gravity pouring (CLA, CLV)

D/ Finishing operations

-Casting cut-off from gating system
-Casting heat treatment

-Casting surface cleaning

E/ Final casting quality inspection

-Chemical composition, structure
-Mechanical properties

-Internal casting quality
-Dimensional accuracy

Experimental

ACCURACY of INVESTMENT CASTINGS

The method of manufacturing castings by the „lost wax process„ or „investment casting process„ seems to be one of the best technologies for manufacturing various components, particularly from the point of view of its narrow dimensional tolerances (the so called NET-SHAPE technology).

A/ Dimensional Changes during the Process

A closer look at individual stages of investment casting will, however, show us that such narrow dimensional tolerances are not at all easy to achieve [2].

It follows from **Fig.2 - [3]**, that dimensional changes occur in practically every phase of the technology. From this point of view, the most important ones are the fabrication of the wax pattern, the fabrication of the ceramic shell and the process of solidification and cooling of the cast metal alloy.

Dimensional changes in the last of them, that is to say in the "casting + solidification + cooling" phase, depend to a large extent on the chemical composition and pouring temperature and cannot therefore be "purposefully" controlled (the range of the pouring temperature must be kept as narrow as possible to guarantee perfect metal fluidity of all details in the ceramic shell cavity).

B/ How to Control the Dimensions - Experimental Results

Dimensional changes in the first two technology phases, i.e. during the fabrication of the wax pattern and of the shell mould, have been studied for a very long time and have also been dealt with on a long-term basis at the Dept. of Foundry Engineering of the Brno University of Technology -BUT (in co-operation with foundries in the Czech Republic). Its first results were presented in San Francisco in 1996 [3], Cambridge in 1997 [4], Orlando in 1998 [5], Monte Carlo in 2000 [6], Tokyo in 2001 [7], Chicago in 2002 [8] and Edinburgh in 2004 [9].

A short summary of main results achieved recently at the Brno University of Technology is presented in **Fig.3** and **Fig.4-a, b, c.**

Test Piece - Fig. 3

Detailed results were given in our previous papers [3], [4] and [5]. The **Fig.3** summarises influence of injection time and wax temperature to the pattern contraction. The regression analysis of the results yielded a mathematical model for the calculation of the shrinkage of the upper length of the test piece for individual injection machines used in the experiments.

Fuse Body - **Fig. 4-a**

Based on experiments where the influence of injection parameters **T, p,** τ on dimensional changes was studied, a mathematical formula for the calculation of the D_1, D_2 and **L** dimensions was proposed.

Scraper – **Fig. 4-b**

A scraper body was used for experiments on ceramic shell behaviour. The results of experiments confirmed the well-known fact that the expansion-after-firing of SiO_2 shells is bigger by an order of magnitude than that of molochite shells. The increase in the rate of expansion in relation with the number of coats was observed mainly in SiO_2 shells.

In a similar way, we can see that the tendency to contract during shell drying slightly increases with increasing number of coats (with absolute contraction figures being in the range from 0.15 to 0.40%). From the stucco material point of view, the differences between shell types were negligible due to the mechanism of drying contraction (affected mainly by the binder system type, which was the same in SiO_2 and molochite shells)

Blade - **Fig. 4-c**

Compared to the previous example, injection parameters in this case proved a very important factor influencing final dimensions of the wax pattern. The results clearly demonstrate the relation between a decrease in shrinking and longer injection (filling and packing) times (up to 0.5% in the case of "B" dimension with injection time extension from 5 to 15 minutes). Changes in the injection pressure in the range 800 ÷ 1400 p.s.i., on the other hand, proved insignificant (that is to say changes in shrinking are independent of changes in pressure).

In this experiment, we also studied the influence of the so-called "braced contraction", whereby the wax pattern removed from a die is braced in the longitudinal direction. The length of the brace exactly matches the size of the die. That significantly reduces the extent of the wax pattern shrinking and it is often used in preventing pattern deformations.

INVESTMENT CASTING CASE STUDIES AND APPLICATIONS

Two examples of aluminium investment castings originally manufactured by other technologies are given in **Fig.5** and **Fig.6**; more case studies will be given during the oral presentation.

CONCLUSIONS

Opportunities for castings made by investment casting technology are enormous, nevertheless also competitive technologies to investment casting are not sleeping, therefore only foundries using the latest technology of lost-wax process have got the chance to be successful in producing "high added value" castings featuring in their complex shapes and very high dimensional accuracy.

References

[1] Beeley, P.R. - Smart, R.F.: Investment Casting, Institute of Materials, Cambridge, 1995

[2] Okhuysen,V.F.-Voigt,R.C.: Dimensional prediction and control of investment casting, 51st ICI Conference, 2003

[3] Horáček,M.-Štefan,L.: Influence of injection parameters to the dimensional stability of wax patterns, 9th World Conference on Investment Casting, paper Nr.1, 20 pages, 13/16 October, 1996, San Francisco,USA

[4] Horáček,M.: Accuracy of castings manufactured by lost wax process, 23rd BICTA International Conference, paper Nr.2, Cambridge , 8/10 June, 1997

[5] Horáček,M.-Helán,J.: Dimensional accuracy and stability of investment casting, The ICI 46th annual technical meeting and exhibition, paper Nr.17, 14 pages, Orlando, USA , 11/14 October, 1998

[6] Horáček,M.: Investment Casting Accuracy, 10th World Investment Casting Conference, Monte Carlo, 14/17 May, 2000, paper Nr.13, 16 pages

[7] Horáček, M.: "ICT – The technology for the new Millenium", JACT conference, Tokyo, September 2001

[8] Horáček, M.: "Investment casting technology in the Czech Republic", 50th ICI conference, Chicago, October 2002

[9] Horáček,M.-Michalčík,P.-Wiedermann,J.: "Dimensional changes during investment casting technology", 11th World Conference on Investment Casting, Edinburgh, 23/26 May, 2004, paper Nr.18

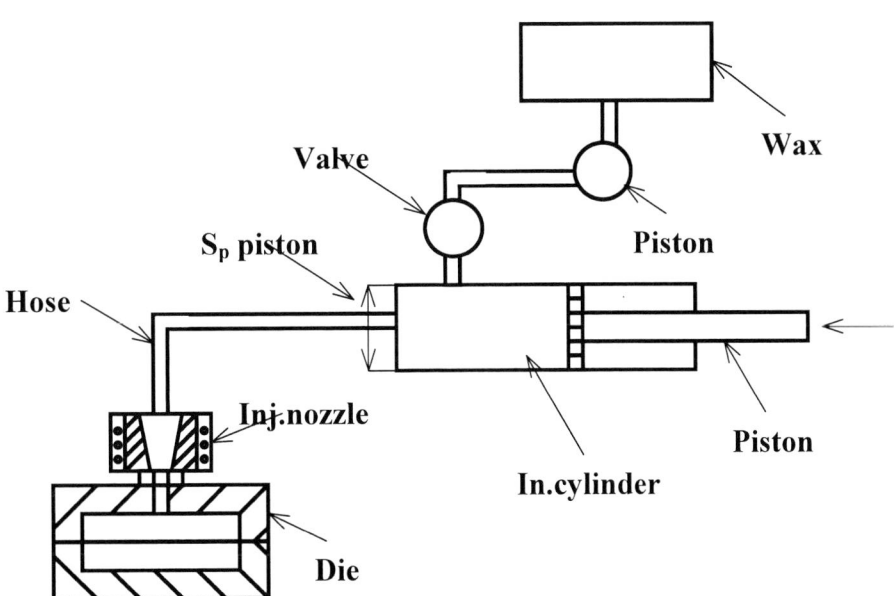

Fig. 1 Basic principle of injection machine

TECHNOLOGY PHASE	ILLUSTRATION OF DIMENSIONAL CHANGES	SHORT PROCESS DESCRIPTION
I. DIE		X_I = **DIMENSION OF DIE** - accuracy depending on die manufacturing method used
II. WAX PATTERN		X_{II} = **WAX PATTERN DIMENSION** - contraction of injected wax $\Delta T : \sim 70\ ^{\circ}C - 20\ ^{\circ}C$ ΔT = temperature range
III. SHELL - including wax		X_{III} = **SHELL DIMENSION - INTERNAL** - practically no dimensional changes $\Delta T : \sim 30\ ^{\circ}C - 20\ ^{\circ}C$
IV. SHELL CAVITY - after dewaxing + 24 hrs drying time		X_{IV} = **SHELL CAVITY DIMENSION** - contraction during process of shell binder system $\Delta T : \sim 25\ ^{\circ}C - 20\ ^{\circ}C$
V. a) SHELL CAVITY - after firing		$X_{V\,a}$ = **SHELL CAVITY AFTER FIREING** - shell expansion during firing process $\Delta T : \sim 20\ ^{\circ}C - 1000\ ^{\circ}C$
V. b) SHELL CAVITY - after cooling **(when applicable – i.e. Al$_2$O$_3$ shell)**		$X_{V\,b}$ = **"COOLED" SHELL CAVITY** - shell contraction during its cooling after firing $\Delta T : \sim 1000\ ^{\circ}C - 20\ ^{\circ}C$
VI. a) METAL POURING - into "hot" mould - metal pouring + + solidification + cooling		$X_{VI\,a}$ = **CASTING DIMENS. – HOT MOULD** - initial slight shell expansion and final metal contraction after metal pouring $\Delta T : \sim 800 - 1500 - 20\ ^{\circ}C$
VI. b) METAL POURING - into "cold" mould - metal pouring + + solidification + cooling		$X_{VI\,b}$ = **CASTING DIMENS. – COLD MOULD** - initial slight shell expansion and final metal contraction after metal pouring $\Delta T : \sim 20 - 1500 - 20\ ^{\circ}C$

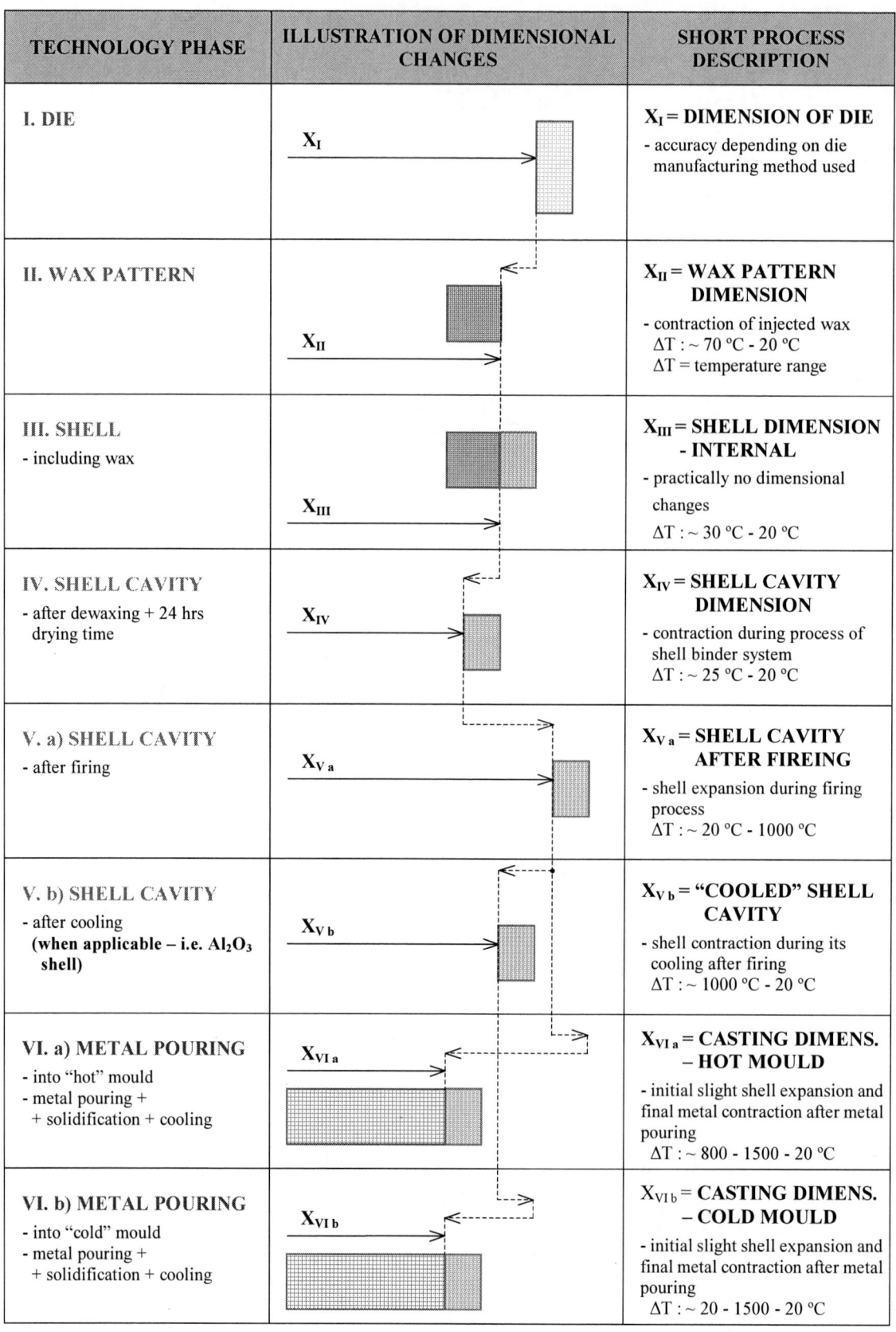

Fig.2 Dimensional changes during investment casting technology [3]

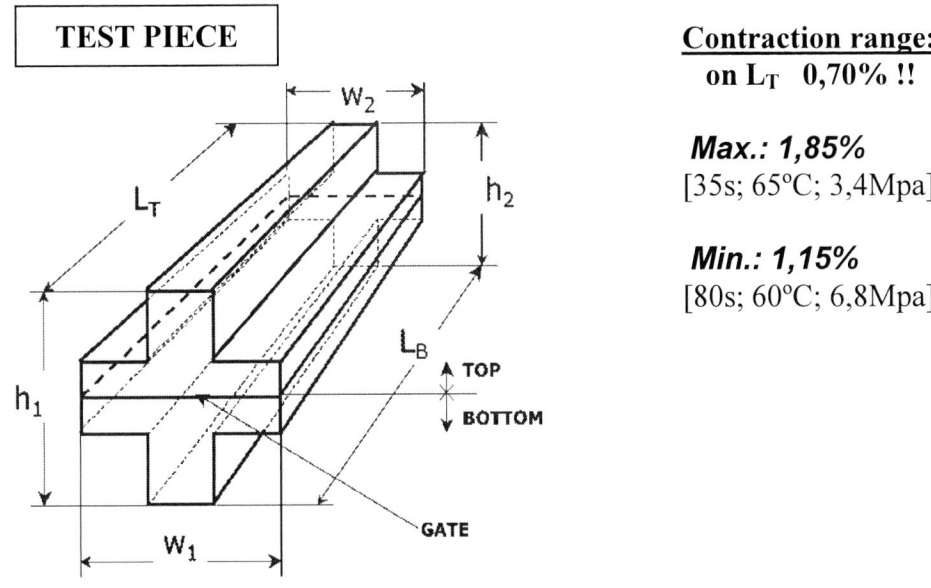

TEST PIECE

Contraction range:
on L_T 0,70% !!

Max.: 1,85%
[35s; 65°C; 3,4Mpa]

Min.: 1,15%
[80s; 60°C; 6,8Mpa]

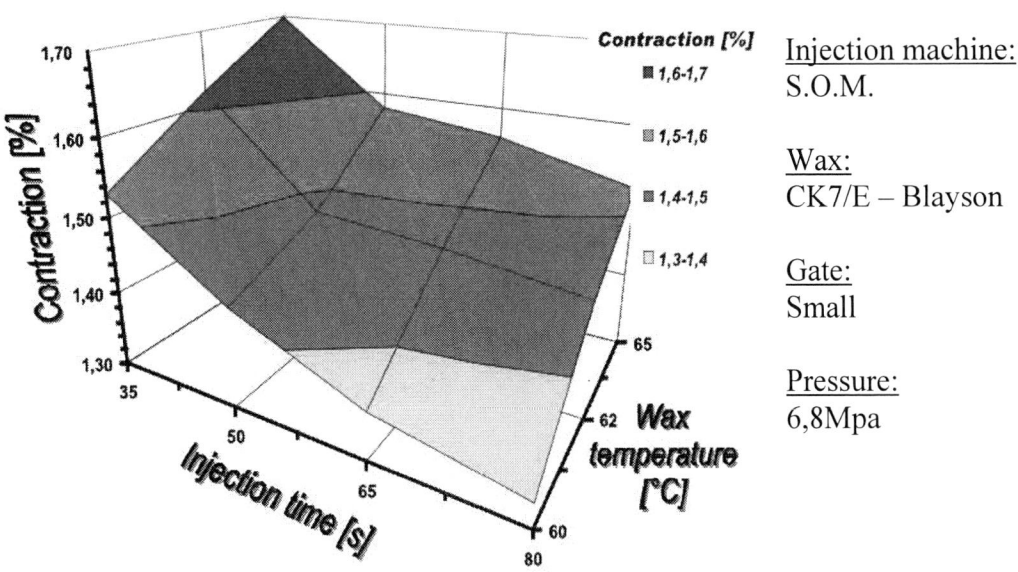

Injection machine:
S.O.M.

Wax:
CK7/E – Blayson

Gate:
Small

Pressure:
6,8Mpa

Fig. 3 Influence of injection parameters to wax pattern L_T contraction

a) **FUSE**

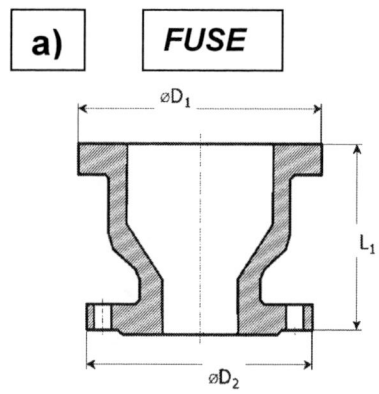

	D_1	D_2	L_1
γ_0	173,204	169,558	115,273
γ_1	-0,045	-0,111	-0,048
γ_2	-0,0066	0,038	0,126
γ_3	0,0014	0,029	0,054
γ_4	0,00013		-0,0731
γ_5			-0,096

$$D_1 = \gamma_0 + \gamma_1 T + \gamma_2 p + \gamma_3 \tau + \gamma_4 Tp$$
$$D_2 = \gamma_0 + \gamma_1 T + \gamma_2 p + \gamma_3 \tau$$
$$D_3 = \gamma_0 + \gamma_1 T + \gamma_2 p + \gamma_3 \tau + \gamma_4 T\tau + \gamma_5 P\tau$$

T..........injection temperature [˚C]
p...........injection pressure [p.s.i.]

b) **SCRAPER**

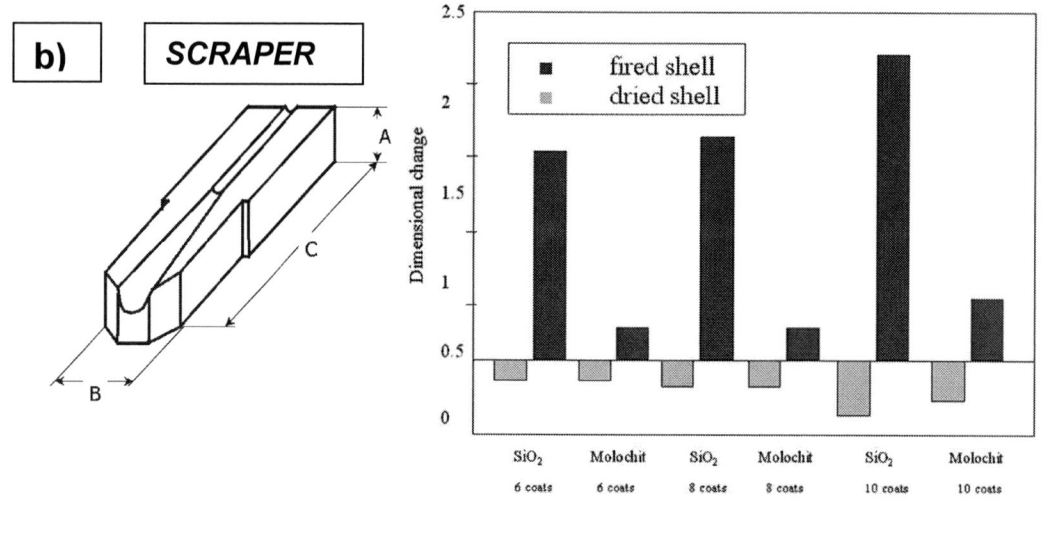

c) **BLADE**

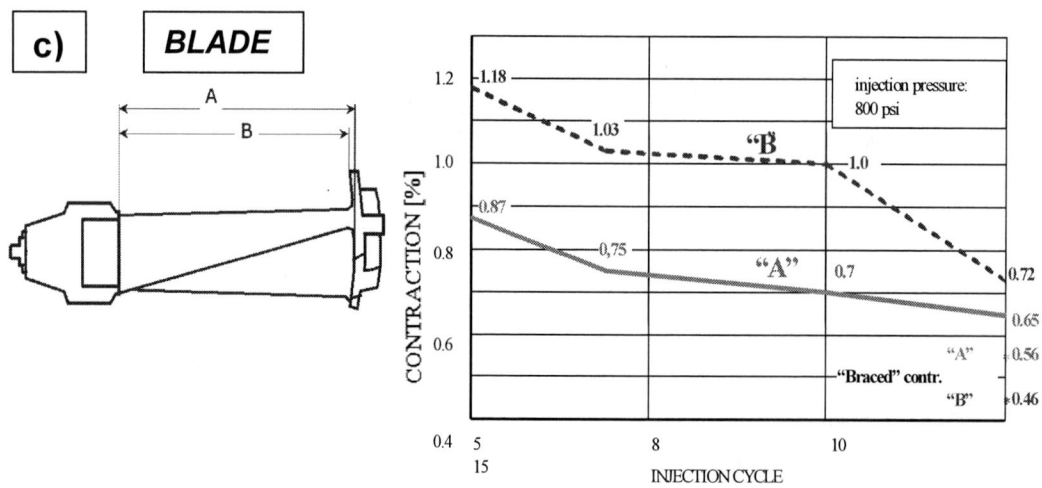

Fig. 4 Summary of results achieved (Fuse body, Scraper, Blade)

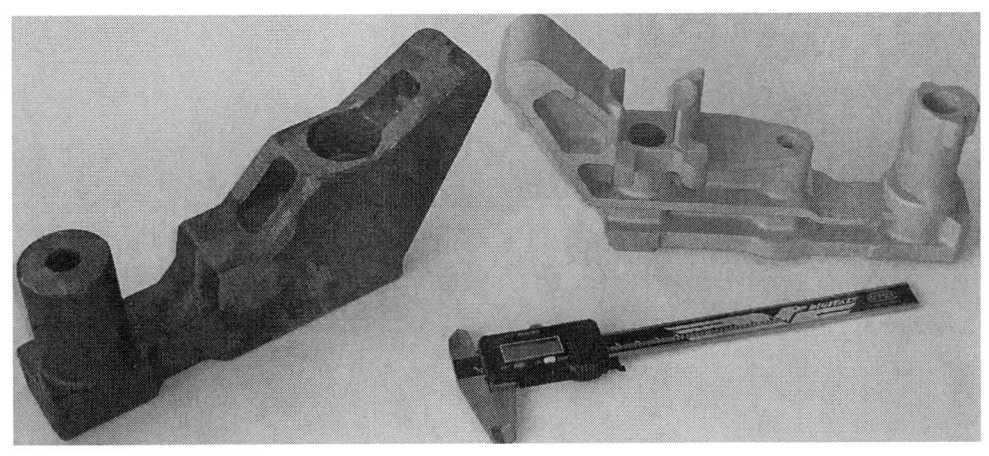

SAND CASTING	INVESTMENT CASTING
Weight - 4,8 kg	Weight – 0,8 kg
Material - grey iron	Material - AlSi7 Mg
Dimensions 240x130x40mm	
Benefit: Labour costs savings	

Fig. 5 ALUCAST example Nr.1

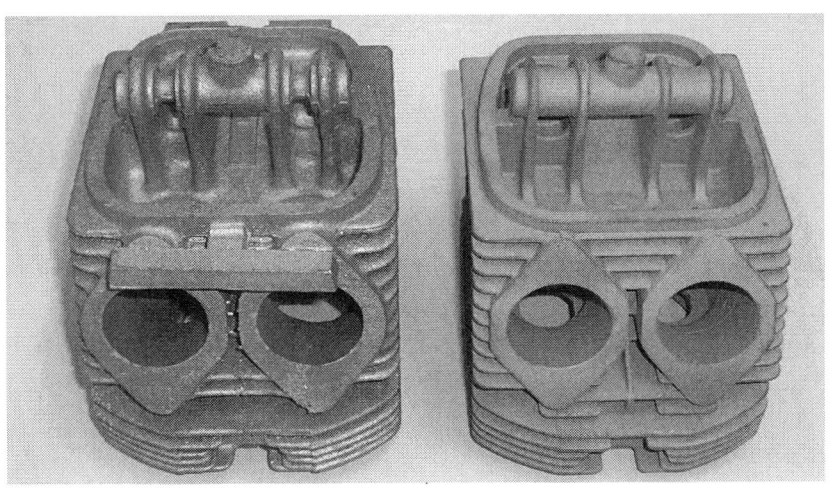

SAND CASTING	INVESTMENT CASTING
Weight 2,4 kg	Weight 1,8 kg
Material RR 350 (AlCu5Ni)	Material RR 350 (AlCu5Ni)
Dimensions: 150x150x110 mm	
Benefits: weight reduction, increase of engine power by app. 15	

Fig. 6 ALUCAST example Nr.2

Innovations in Machine Learning and Defect Diagnostics

R.S. Ransing and M. R. Ransing

Civil and Computational Engineering, School of Engineering, Swansea University, SA2 8PP

Abstract

Analysis of cause and effect relationships for diagnosis has always been a favorite application for the artificial intelligence community. The first paper on expert systems for casting defect analysis has appeared in mid eighties. However, over the last decade, the research has slowed down with very few new publications emerging in this area.

However, the problem is very relevant even today. With globalization and experts retiring from their jobs, the industry is already facing skills shortage. The future casting industry needs to move away from the traditional approach for defect analysis.

This paper will look back twenty years and review the technological advancement. It will also identify why the technology failed to meet foundry-man's expectations.

We have also revealed the patented self-learning knowledge base technology of X1Recall and discussed how it has managed to overcome the limitations of existing expert systems, neural networks or statistical techniques for defect reduction problems. A case study has also been presented.

Key words Defect Reduction, Neural Networks, Expert Systems, Knowledge Representation and Computer Aided Diagnosis.

Introduction

The Quality Control Manager uses a traditional Diagnostic Process, which is the process of identifying one or more causes that are responsible for the occurrence of one or more defects. In a typical set-up situation when rejection analysis is required, the quality control expert will use details about the defective components to diagnose the likely cause. This might include the number of defective components, the defect type that is observed on each defective component, along with information on what happened when the components were manufactured, for example, various process conditions, control chart information, operator data and machine data.

The expert is then required to identify the causes and suggest remedial action to eliminate the future occurrence of such defects. The expert accomplishes this task by using the knowledge he/she has acquired over a period of years, and which has been built up through the results of past trial and error experiments.

In early days, expert systems using rule based architecture were considered by the researchers for diagnosis. These systems relied on question-answer sessions. One of the most widely referred examples is MYCIN [1] – a medical diagnostic expert system developed in mid seventies. Later, researchers such as Creese [2], Roshen [3], Piwonka [4] and Phelps [5] extended this rule based philosophy to foundry defects.

In summary, **the Expert System based technologies** use a set of rules coupled with probabilistic theory to identify causes responsible for the occurrence of one or more defects.

This technology has failed to solve quality control problems because an expert system cannot learn on its own and will always give the same information back to experts which they already know!

The system has other implementation related problems that inhibit successful operation. It requires so much information about causes and defects that it is extremely tedious if not impossible to generate useful feedback.

In mid nineties, the neural network technology received attention [6-12] because unlike the rule-based approach, this technology showed promise of learning and adapting its knowledge from past examples. Smith and her co-investigators [6, 7] did considerable research in this direction. They have presented a case study of back-propagation neural network application for quality control applications. Neural networks were also used for the diagnosis of hydraulic forging presses [8]. Martinez et. al. [9] have investigated its application to relate process conditions to the probable quality rating of casting. This predictive analysis was done for a slip casting process. Spelt et. al. [10] have used neural networks for classifying power plant sensor data and coupled this with an expert system for diagnostic purposes. Zhang and Huang [11] have presented a state-of-the-art survey of neural network applications in manufacturing. These include applications of neural networks for engineering design, process planning, in solving scheduling problems, process modelling and control, in monitoring and diagnosis and quality assurance. Our initial work [12] was also inspired by neural network philosophy. A semantically constrained neural network approach was proposed that modified the three layered 'feed forward' network architecture'. It constrained the connectivity of nodes according to the 'defect-metacause-cause' relationship [13,14].

In summary, **Neural Network based technologies** store defects and causes in a network form with associated rejection data, the network attempts to predict the causes of future rejects from this store of historical data. Neural network represents a family of algorithms that can learn unknown functional relationships from examples or recognise patterns in data. However, they are nothing more than efficient multi-variable and higher-ordered interpolation techniques. One of the major advantages of neural networks is that they do perform better than equivalent regression techniques. However, this technology has following limitations:

- Although, this technology has the potential to learn from past mistakes and it can accept feedback from the user, it too has failed to become commercially successful for solving quality control problems.
- The software requires a large number of past diagnostic examples that are representative of various diagnostic cases to learn correctly. The casting industry does not have such data in the form required by the software.
- For some situations, depending on the quality of the training examples, it could even be completely inappropriate to employ neural network modelling and this is particularly true for cause and effect analysis.
- A skilled person is required to use this technology, which adds to the cost and timescale for meaningful output.

The casting industry is in search of a technology that will:
- Use similar processes to the way a human mind learns
- Give back valuable information about multiple cause and effect relationships, which the human mind finds it difficult to grasp and visualise.
- Not require past diagnostic examples and lots of rules to make it work.
- Start giving plausible results without being required to answer any difficult questions about 'cause and effect' relationships.
- Receive feedback in an intuitive and straightforward way.

X1Recall Methodology

A simple investment casting example is presented to illustrate how the patented software algorithm [15] works in X1Recall. A simple 'defect-metacause-cause' relationship for an investment casting shell drying process is shown in Figure 1. The Metacause or scientific rationale 'uncontrolled expansion of wax in Boiler Clave' influences the occurrence of defects such as 'oversize parts', 'fining defects', 'runouts', 'cracked shell' etc. In a real rejection scenario, not all defects occur at the same time if there is 'uncontrolled expansion of wax'. The defects can also occur at varying strengths and proportions. It is also possible that defects are caused by other Metacauses. E.g. 'cracked shell' can also be caused by the 'low green stength' of primary and secondary coats.

An expert's mind has the knowledge of the degree of influence of each Metacause has on each defect. This can be graphically represented as shown in Figure 2 (b) and (c). The curve shows the belief value that the Metacause 'uncontrolled expansion of wax in Boiler Clave' is responsible for a given proportion of castings rejected due to defects 'runouts' and 'cracked shell' respectively. The expert's main expertise lies in his/her ability to combine these curves such that the belief value in 'uncontrolled expansion of wax' proposition can be predicted for any strength value in all associated defects.

X1Recall has the unique ability of combining these one-dimensional curves (Figures 2 (b) and (c)) to create a belief surface as shown in Figure 2 (a). And like the human mind, the software can also learn to correct its diagnosis, if necessary, automatically. With X1Recall's unique know-how, it generates these one-dimensional curves using the readily available information on histograms or frequency of occurrence tables for defects, metacauses and causes. The software also has the ability of taking a cross-section of this surface as shown in Figure 2(d). Thus, with very little data, a great deal of diagnostic information can be taken from the shape of this belief surface.

Results

X1Recall software has been validated in Rolls Royce Plc who have financially supported this research along with the Engineering and Physical Sciences Research Council in UK. Due to the confidentiality agreements with Rolls Royce Plc, a simple working example has been illustrated.

The interaction with X1Recall is simple. The rejection data can either be entered at various inspection points within a Foundry environment and later picked up by the software for analysis or can be input on-line every time diagnosis is required. The next stage in the rejection analysis process is to input relevant process information. This information can also be automatically loaded in X1Recall via its link to existing on-line process monitoring systems.

X1Recall undertakes the diagnosis based on the knowledge it has stored in surfaces as shown in Figure 2. The remedial actions suggested are shown in Figure 3.

Once the diagnosis is complete, feedback is given to the X1Recall. The software automatically updates its knowledge so that it can provide correct remedial actions with less iterations next time and at the same with will not forget its past knowledge. Mathematically, it achieves it by changing the shape of the surface (as shown in Figure 2) by minimising the error in a least square sense.

Conclusions

X1Recall's mission is to provide an internet based tool for the effective management of 'Transferable Knowledge' with the twin objectives of reducing the total cost of 'knowledge ownership' and the total cost of 'maintaining the desired yield level in a casting process'.

The X1Recall software will be remembered in the foundry world as the first software that has provided a solution for the effective management of 'Transferable Knowledge'. An ever increasing store of corrective production **knowledge** will be **acquired** by the **organization** rather than by individuals. And, with the help of X1Recall the knowledge can be easily transferred from one individual to another within the same foundry.

X1Recall has an ability to rapidly analyse process conditions and rejection data in order to make excellent predictions to eradicate defects. With its self-learning knowledge base, it remembers the corrective actions taken by 'Foundry Experts' to bring the process under control and is capable of expanding its knowledge base to diagnose problems occurring under a variety of circumstances. In other words, it discovers the process knowledge from historical data.

SKYNET may have become self-aware in the TERMINATOR movie but for the authors the inspiration is real. The world's first self-learning knowledge base for Casting Defect Reduction - X1Recall - is on-line at www.x1recall.com and the delegates of this congress are invited to try the technology themselves.

References

1. Shortliffe E. H., *Computer based medical consultations: MYCIN*, New York: American Elsevier, 1976.
2. Creese R. C., *Introduction to Expert Systems to Foundry Application*, AFS Trans, 1988, 96, 443-446.
3. Roshen H.Md., *Expert System for Analysis of Casting Defects: Cause Module*, AFS trans, 1989, 97, 601-606.

4. Piwonka T. S., *Current and Potential Use of Process modeling for Foundry Process Control,* AFS trans, 1989, 97, 465-472.
5. Phelps T. A., *Analysis of Internal Unsoundness Casting Defects using Artificial Intelligence Techniques*, AFS Trans, 1989, 97, 507-512.
6. Smith A. E., *Predicting product quality with backpropagation: a thermoplastic injection moulding case study,* International Journal of Advanced Manufacturing Technology, 1993, 8, 252-257.
7. Smith A. E. & Dagli C. H., *Controlling industrial processes through supervised, feedforward neural networks,* Computers and Industrial Engineering, 1991, 21, 247-251.
8. Lin H., Yih Y. & Salvendy, *Neural network based fault diagnosis of hydraulic forging presses in China,* International Journal of Production Research, 1995, 33(7), 1939-1951.
9. Martinez E. E., Smith A. E. & Idanda B., *Reducing waste in casting with a predictive neural model*, Journal of Intelligent Manufacturing, 1994, 5(4), 277-286.
10. Spelt P. F., Knee H. E. & Glover C. W., *Hybrid artificial intelligence architecture for diagnosis and decision-making in manufacturing*, Journal of Intelligent Manufacturing, 1991, 2, 261-268.
11. Zhang H. C. & Huang S. H., *Aplications of neural networks in manufacturing: a state-of-the-art survey*, International Journal of Production Research, 1995, 33(3), 705-728.
12. Ransing R. S. & Lewis R. W., *A Semantically Constrained Neural Network for Manufacturing Diagnosis,* International Journal of Production Research, 1997, 35 (9), 2639-2660
13. RS Ransing, MN Srinivasan and RW Lewis, *ICADA: Intelligent Computer Aided Defect Analysis for Castings*, Journal of Intelligent Manufacturing, vol. 6, no. 1, 29-40, 1995.
14. RS Ransing and MN Srinivasan, *Computer Aided Analysis for the Identification of Causes of Defects in the Foundry,* 40th Annual Convention of IIF, Madras, pp 285-291, Feb.92, India.
15. RS Ransing, MR Ransing and RW Lewis: '*Diagnostic Apparatus International Patent Application No PCT/GB2002/003805.* (Filing date: August 2002), Patent pending in over 16 countries with single application for European countries.

Acknowledgements

The authors would like to acknowledge the financial support given by three organizations: The Engineering and Physical Science Research Council, Rolls Royce Plc and the Welsh Development Agency.

Authors would particularly like to thank Dr Paul Withey from Rolls Royce Plc and Dr David Ford from European Investment Casting Federation for their invaluable time for technical discussions.

Figures

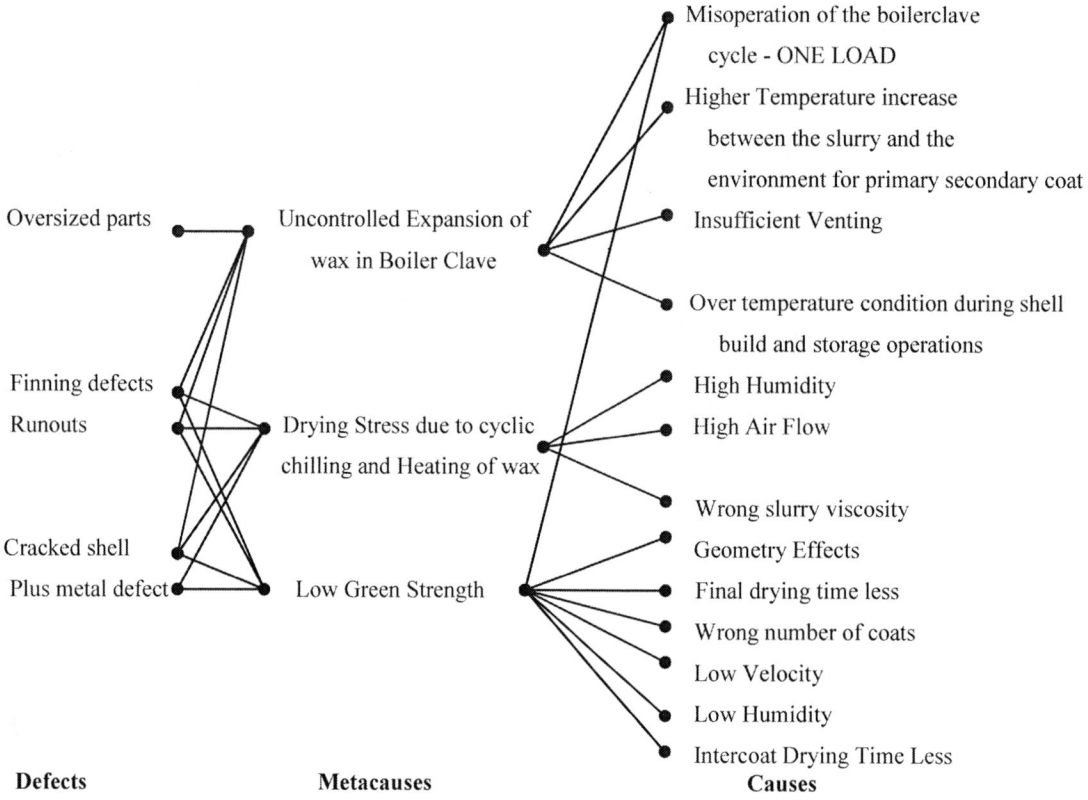

Figure 1: Defect-MetaCause-Cause relationship for investment casting shell drying process

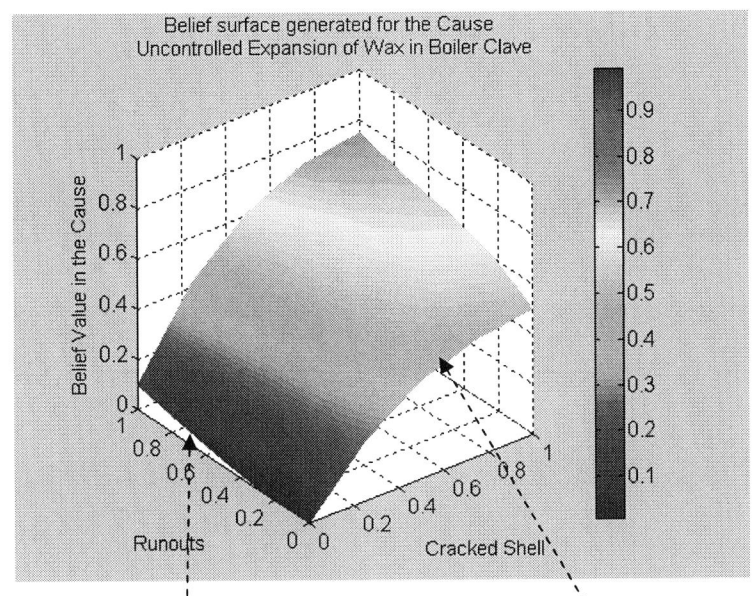

Figure 2(a): Belief Surface created by X1Recall from Belief Curves

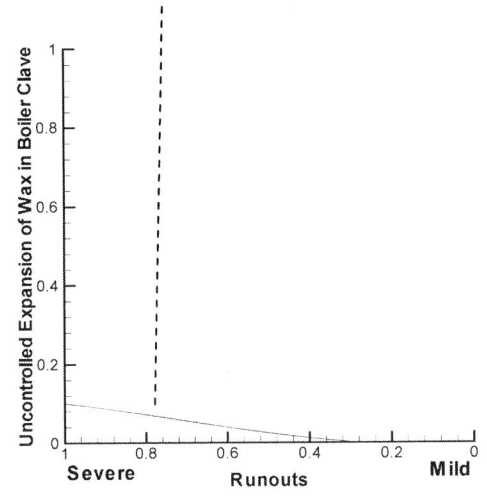

Figure 2 (b) Belief Curve

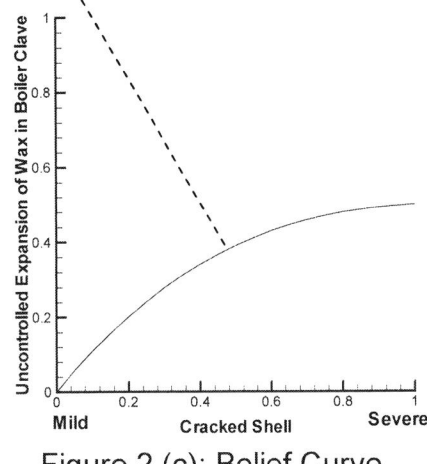

Figure 2 (c): Belief Curve

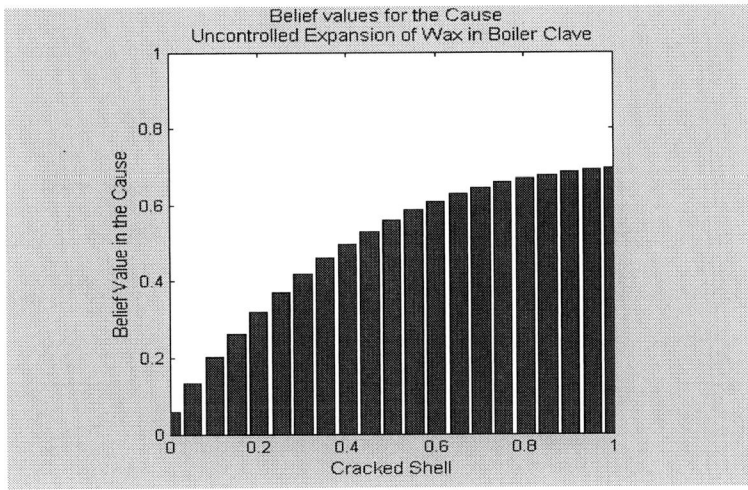

Figure 2(d): Cross section of a belief surface for a high value of defect Runouts.

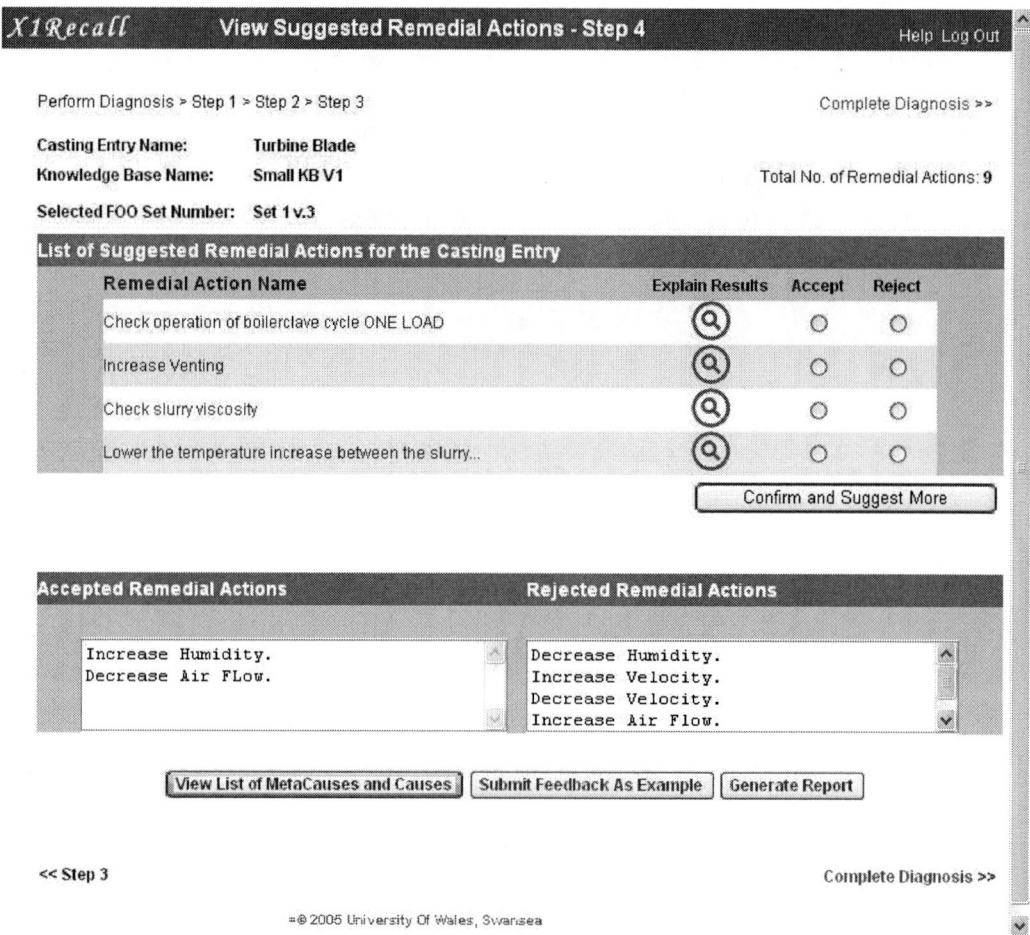

Figure 3: Consultation with X1Recall.

Effect of Ultrasonic Vibration on Structure Refinement of Metals

Q M Liu *, Y Zhang *, Y L Song *, F P Qi * and Q J Zhai*.

* School of Materials Science and Engineering, Shanghai University, P. R. China.

Abstract

In this paper, the recent research progress conducted in the effect of ultrasonic vibration on structure refinement of metals is briefly generalized. As for Sn-Sb alloy, the matrix α phase can be obviously refined by the ultrasonic treatment, with the broken and homogeneously dispersed second-phase β (SnSb). For AZ81 Mg alloy, coarse grained equiaxed structures are refined to spherical ones. As to ZL101 alloy, the introduction of ultrasonic vibration into the melt can eliminate columnar dendrites and refine the equiaxed grains. Employed EBSD and SEM, samples of 1Cr18Ni9Ti steel under various ultrasonic treatments are analyzed. The results show that columnar grains is transmitted into equiaxed ones after ultrasonically treated, with an increase of lattice defects. Furthermore, the mechanism for structure refinement induced by ultrasonic vibration is discussed.

Key words Ultrasonic vibration; Structure refinement;Cavitation effect.

Introduction

As a kind of high-frequency acoustic wave, ultrasonic shows considerable potential for industrial use, such as cleaning and plastic welding [1]. It is demonstrated that ultrasonic creates the cavitation, acoustic streaming and electrical effects when it propagates into the liquid medium. Taking account of these physical natures of ultrasonic, we impose ultrasonic vibration on the melts during solidification and characterize the effect of ultrasonic on metals in virtue of the results of microstructure and other properties. Based on experimental results, fundamental theory gives an insight into the mechanism of ultrasonic vibration on metals.

Experimental

Schematic illustration of experimental setup is shown in Fig.1. The ultrasonic generator with the frequency of 20kHz is attached to the crucible profile in order to introduce ultrasonic vibration into the melt along the vertical direction. Metals are melted into liquid state and poured into the crucible with the dimension of 120mm×60mm×60mm. Castings are treated by different ultrasonic powers during solidification and then cooled at room temperature. In the experiment, platinum-rhodium thermocouples of 0.3mm diameter are positioned into the crucible to test temperature of the melt. The temperature data acquisition system consists of an A/D converter and a computer. Appropriate software is developed for the on-line temperature signal procession.

Results

Effect of ultrasonic vibration on Sn-Sb alloy

The composition of Sn-Sb alloy is 11% Sb and Sn. Fig.2 shows microstructure of Sn-Sb alloy without and with 600W ultrasonic treatments [7]. In absence of ultrasonic vibration, the matrix α phase is coarsened and second-phase β (SnSb) has acute edge angle. In constrast, the application of 600W ultrasonic power makes matrix α phase finer and breaks second-phase β (SnSb) into pieces.

The cooling curves of Sn-Sb alloy without and with ultrasonic treatments are shown in Fig.3. In each experiment, six thermocouples are mounted in the crucible from the crucible wall to the middle of crucible evenly. Therefore, six cooling curves can be obtained at different positions. From Fig.3 (a), it can be observed that in absence of ultrasonic, the solidification time of Sn-Sb is 1218s while with 600W ultrasonic treatment, the solidification time is decreased to 1160s. Furthermore, from Fig. 3, it is observed that ultrasonic vibration can uniform temperature distribution in the melt. That is to say, the ultrasonic processing applied in the melt can decrease temperature gredient of the melt during solidification.

Effect of ultrasonic vibration on AZ81Mg alloy

The composition of AZ81 Mg alloy is Al−8.9%, Zn−0.35%, Mn−0.41%, Fe−0.02%, Si−0.04% and Mg. Fig.4 depicts the microstructure of AZ81 Mg alloy without and with 600W ultrasonic treatments [8]. In Fig.4 (a), coarse grained equiaxed structures show dendrites that are broadly similar to the columnar dendrites. In contrast, grain refinement is so effective that the morphology of α-Mg phase is spherical when treated by 600W ultrasonic power. Fig. 5 is the SEM of Al distribution without and with ultrasonic treatments. From Fig.5 (b), it is found that with 600W ultrasonic treatment, more Al atoms are dissolved into α-Mg phase, decreasing the amount of Al atoms on the grain boundary. Thus, Al atoms play a role in solution strength. Microhardness test shows that microhardness on matrix structure is increased from 86.3HV without ultrasonic treatment to 112HV with 600W ultrasonic treatment.

Effect of ultrasonic vibration on ZL101 Al alloy

The composition of ZL101 Al alloy is Si−7.33%, Mg−0.38%, Ti−0.13%, Fe−0.07% and Al. Microstructure of ZL101 Al alloy without and with 600W ultrasonic treatments is presented in Fig.6. The comparison of Fig. 6(a) and (b) reveals a very large difference in the resultant microstructure. The application of 600W ultrasonic power can transform the microstructure from well-developed columnar structure to equiaxed crystals.

Effect of ultrasonic vibration on 1Cr18Ni9Ti austenitic stainless steel

The composition of 1Cr18Ni9Ti austenitic stainless steel is C − 0.06 wt.% ,Cr−17.45 wt.%, Ni−8.96 wt %, Ti−0.41 wt.% and Fe. Fig.7 is depicted the EBSD analysis of 1Cr18Ni9Ti without and with 600W ultrasonic treatments. The individual grains are clearly revealed in the EBSD map. Fig.7(a) shows coarse columnar and the growth direction is most nearly parallel to the heat-flow direction. In contrast, with 600W ultrasonic treatment, equiaxed crystals is obtained and the growth direction is in almost all directions. Additionally, a mass of subgrains are formed, increasing the density of lattice defects. Tensile experiment is carried out and fracture morphologies of 1Cr18Ni9Ti without and with 600W ultrasonic treatments are shown in Fig.8. By comparison with the SEM observations between Fig.8 (a) and (b), we can see that without ultrasonic treatment, fracture is obviously quasi-cleavage crack and a great number of tearing edges on cleavage planes appear. With the application of 600W ultrasonic power, test sample exhibits the ductile fracture surface covered with deformation dimples.

Discussion

The principal characteristics of ultrasonic waves are particle displacement, particle velocity and sound pressure etc [9]. This is because when ultrasonic wave propagates through the liquid material, a strong nonlinear effect will be generated due to the nonlinear elastic properties of that material [10]. In order to design the ultrasonic processing of melts during solidification, it is very important to know the ultrasonic field established in a liquid metal. In general, the field can be predicted by solving the wave equations described as:

$$\frac{1}{c^2}\frac{\partial^2 p}{\partial t^2} = div \ \mathrm{grad}p \tag{1}$$

Where p the acoustic pressure, c the fluid particle velocity, t the time. With the role of acoustic pressure, cavitation effect is generated in micro-area of the liquid metal. Acoustic cavitation is characterized by the interaction of an acoustic field with bubbles of appropriated size. The transition of cavitation bubble from comparatively mild oscillations to large amplitude oscillations and resulting violent collapse depends upon 1) the initial size of the preexisting bubbles, 2) the closed velocity of cavitation bubbles, and 3) the peak pressure of the applied acoustic field. The acoustic pressure at the transition from stable to transient cavitation is the threshold pressure for transient cavitation. The pressure of the ultrasonic field can be described as:

$$P_B = P_0 - P_v + \frac{2}{3\sqrt{3}}\left[\left(\frac{2\sigma}{R}\right)^3 \bigg/ \left(P_0 - P_v + \frac{2\sigma}{R}\right)\right]^{\frac{1}{2}} \tag{2}$$

Where P_B the acoustic pressure, P_0 the static pressure, P_v steam pressure in the bubble, σ interfacial force and R the initial radius of cavitation bubble. The cavitation bubble is grown up, collapsed and expanded to a new maximum radius instant. When acoustic pressure exceeds the threshold acoustic pressure, shock force induced by cavitation effect can break primary dendritic arm and hard particles. Thus, granulous second-phase β (SnSb) takes the place of coarse blocky second-phase β (SnSb) in the microstructure of Sn-Sb alloy refiner after ultrasonic treatment, as shown in Fig. 2. And results in Fig. 6 show that coarse primary dendritic arm is broken off several parts.

Furthermore, when cavitation bubble generates and expands, it absorbs great energy from the melt and leads to supercooling on the cavity surfaces in local zone. The violent decrease of temperature causes the formation of a mass of extrinsic nucleation. Therefore, coarse grained equiaxed structures are developed to fine spherical equiaxed structures, as shown in Fig.4.

Except the effect of cavitation, acoustic streaming of ultrasonic also plays a dominant role in refining structure of metals. Without ultrasonic treatment, columnar structures are much easier obtained with large grain sizes , as shown in Fig, 7(a). When power ultrasonic is introduced into the melt, a fluid jet streaming is produced at very high speed. Whereas, when introduced ultrasonic vibration into the melt, a fluid jet streaming can bring heat pulses on dendritis arms. These heat pulses accelerate the melting off of dendrites. Hence, it seems that fluid streaming plays a role in the columnar-equiaxed transition, as shown in Fig.7(b). On the other hand, the occurrence of streaming in the melt leads to an accumulated generation of vortices. A high rate of turbulence causes forced convection cooling and dissipate superheat in the melt. Thus, heat transfer is enhanced by the ultrasonic treatment, causing the decrease of solidification time, as shown in Fig.3. As a result, the growing time of grains is decreased by ultrasonic treatment.

Conclusions

- Solidification structure is refined when the melt is subjected to ultrasonic vibration treatment.
- Acoustic streaming effect resulting from ultrasound applied in Sn-Sb alloy accelerates the cooling rate of the melt and shortens solidification time. Furthermore, ultrasonic vibration can uniform temperature distribution of the melt.
- The effect of ultrasonic vibration improves the microhardness of AZ81 Mg alloy.
- EBSD analysis of 1Cr18Ni9Ti shows that ultrasonic treatment on the melt not only plays a role in the columnar-equiaxed transition, but also increases the number of subgrains, causing an increment of density of lattice defects.

References

1. Laborde J L and Gerard A, *Fluid dynamics phenomena induced by power ultrasounds*, Ultrasonics, 38, Issue 1-8, Mar 2000, pp297-300.
2. Laborde J L and Bouyer C, *Acoustic bubble cavitation at low frequencies*, 36, Issue 1-5, Feb 1998, pp589-594.
3. Sehgal and Chandra M, *Non-linear ultrasonic to determine molecular properties of pure liquids*, Ultrasonics, 33, Issue 2, 1995, pp155-161.
4. Nesvijski and Edouard G, *Some aspects of ultrasonic testing of composites*, Composite Structures, 48, Issue 1-3, Jan 2000, pp151-155.
5. Kozhemyakin G N, *Influence of ultrasonic vibrations on the growth of InSb crystals*, Journal of Crystal Growth, 149, Issue 3-4, Apr 1995, pp266-268.
6. Abramov O and Sommer F, *Solidification of aluminium alloys under ultrasonic irradiation using water-cooled resonator*, Materials Letters, 37, Issue 1-2, 1998, pp27-34.
7. Zhai Q J, *Effect of high intensity ultrasonic on solidification structure of Sn-Sb peritectic alloy*, Foundry, 52, No 1, Jan 2003, pp21-23.

8. Zhai Q J, *Effect of side transmission of power ultrasonic on structure of AZ81 magnesium alloy*, Trans. Nonferrous Met. Soc. China, 14, No 2, Apr 2004, pp302-305.
9. Gür C H and Ogel B, *Non-destructive microstructural characterization of aluminium matrix composites by ultrasonic techniques*, Materials Characterization, 47, Issue 3-4, Sep 2001, pp227-237.
10. Jhang Kyung-Young and Kim Kyung-Cho, *Evaluation of material degradation using nonlinear acoustic effect*, Ultrasonics, 37, Issue 1, Jan 1999, pp39-44.

Acknowledgements

The authors would like to thank the National Natural Science Foundation of China (Grant No. 50374046) for financial support.

Figures

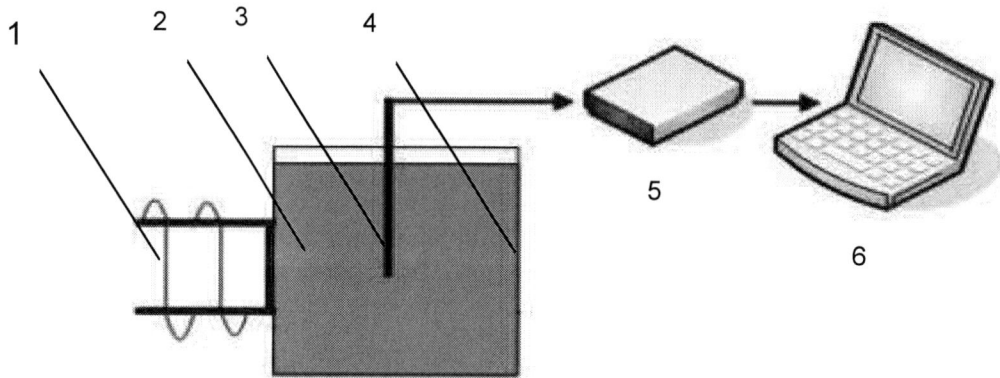

Fig.1 Schematic illustration of experimental setup: 1- amplitude transformer horn, 2- the melt, 3- thermocouple, 4- crucible, 5- A/D converter, 6- computer.

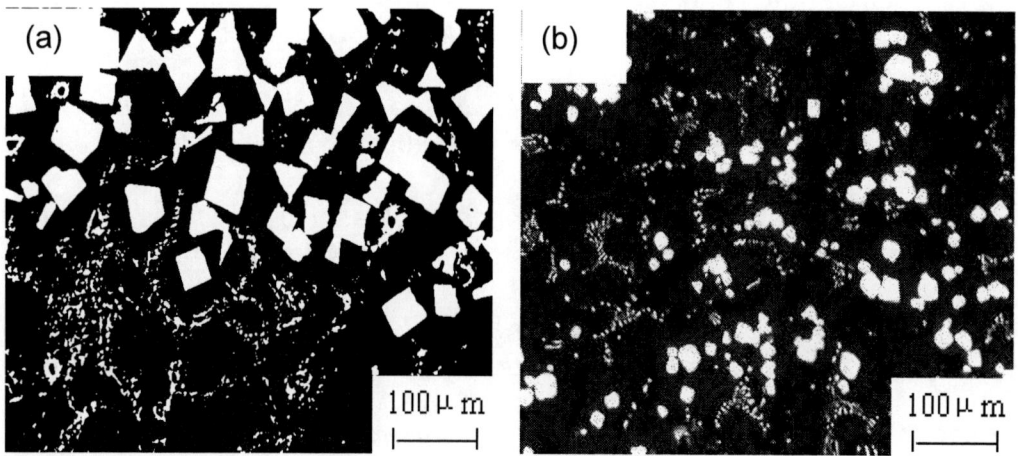

Fig.2 Microstructure of Sn-Sb alloy without and with ultrasonic powers: (a) 0W, (b) 600W.

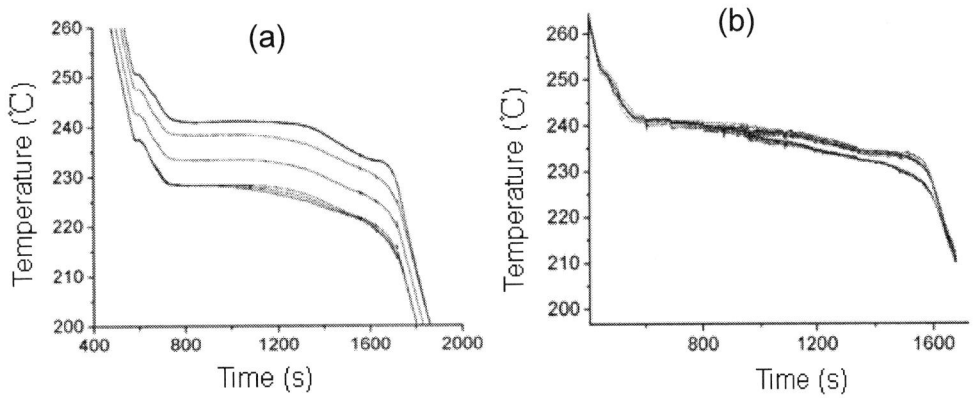

Fig.3 Cooling curves of Sn-Sb alloy without and with ultrasonic treatments: (a) 0W, (b) 600W.

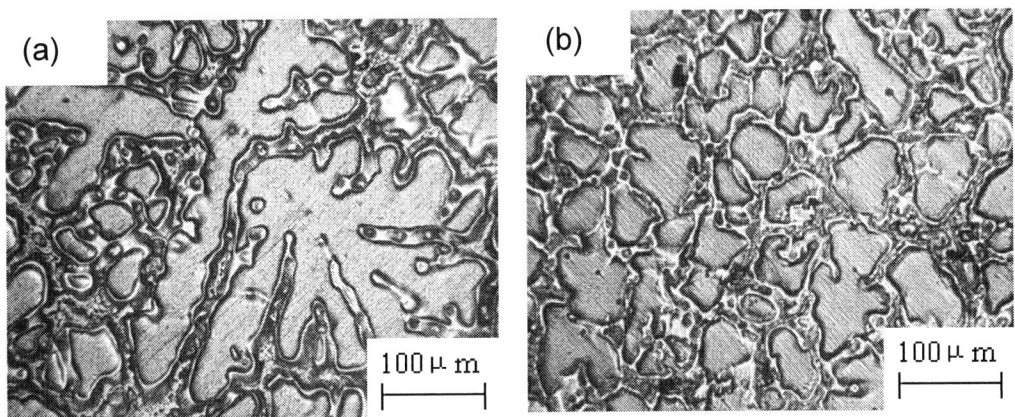

Fig.4 Microstructure of AZ81 Mg alloy without and with ultrasonic treatments: (a) 0W, (b) 600W.

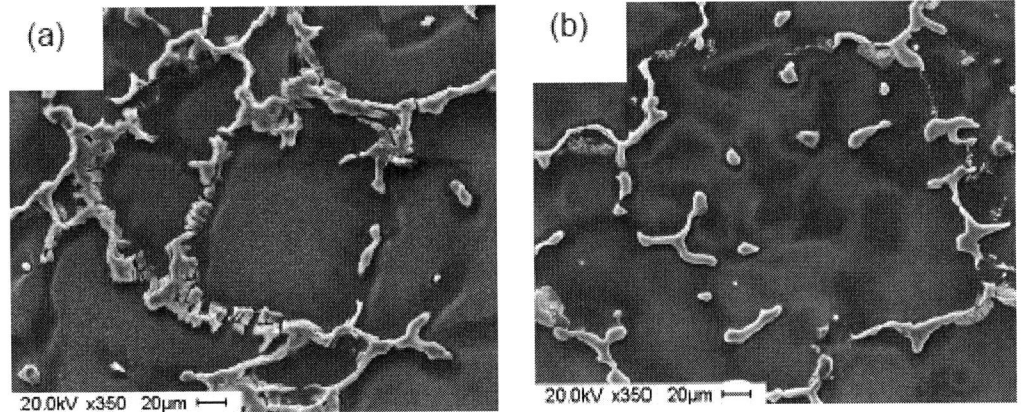

Fig.5 SEM of Al element distribution without and with ultrasonic treatments: (a) 0W, (b) 600W.

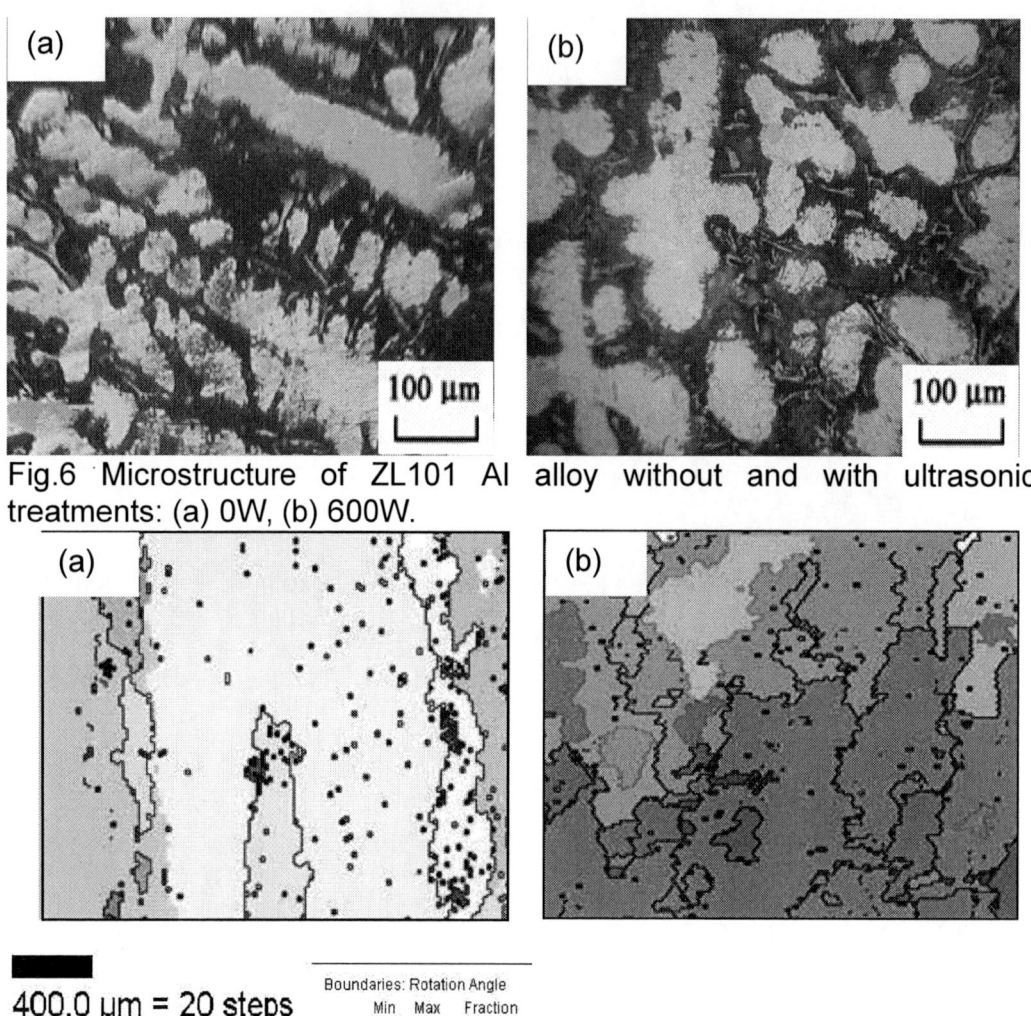

Fig.6 Microstructure of ZL101 Al alloy without and with ultrasonic treatments: (a) 0W, (b) 600W.

400.0 μm = 20 steps

Boundaries: Rotation Angle

	Min	Max	Fraction
—	2°	180°	1.000

Fig.7 EBSD results of 1Cr18Ni9Ti without and with ultrasonic treatments: (a) 0W, (b) 600W.

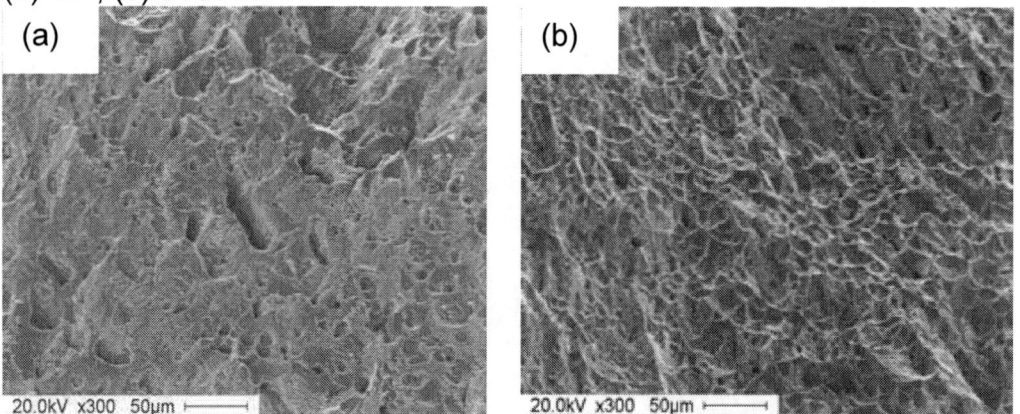

Fig.8 Fracture morphologies of 1Cr18Ni9Ti without and with ultrasonic treatments: (a) 0W, (b) 600W.

HIPing – A Potent Post Casting Treatment For High Integrity Aluminium Castings

R.M.PILLAI
Regional Research Laboratory (CSIR)
Thiruvananthapuram – 695019, India
Email : rmpillai@rediffmail.com

ABSTRACT

The increasing use of high integrity Al castings for repeated loading applications warrants improved fatigue properties as well. The principal cause for the lower fatigue properties and reliability of aluminium castings is the presence of defects and inhomogenities, which are the preferred fatigue initiation sites. Hot Isostatic Pressing (HIPing) is a powerful post casting treatment for eliminating defects like shrinkage porosity and hydrogen pinholes leading to densification and enhanced mechanical properties of the castings meant for high performance applications. This paper brings out the current status of this potent post casting treatment for high integrity castings including basics of HIPing, its variants and their pros and cons and its effects on both microstructure and mechanical properties of aluminium alloy castings.

Keywords : HIPing, LHIPing, Aluminium castings, Fatigue strength,

Introduction

Applications in aerospace, automotive and general engineering, requiring high mechanical properties including resistance to fatigue failure and weight saving, look for aluminium castings. This is because of their excellent casting characteristics, corrosion resistance, near net shape capability and especially high strength to weight ratio. Performance of casting is dictated by many metallurgical features, such as, secondary dendrite arm spacing (SDAS), iron intermetallics, oxides/inclusions and porosity. Castings exhibit defects such as shrinkage and gas porosity, hot tears, inclusion and alloy segregation that generally result in lower and more variable mechanical properties than their wrought counterparts. The fatigue endurance of castings is very sensitive to the size of casting defects. On the other hand, solidification time as reflected by secondary dendrite arm spacing(SDAS) and size of silicon particles also affects it to a lesser extent.These defects can be controlled reasonably by proper mold design and good foundry practice. In many engineering applications, the response to fatigue environment is very critical. Nonetheless, the complete elimination of the shrinkage defects is not always possible without the application of external forces to close voids and porosity. Hot Isostatic Pressing (HIPing) enables this. HIPing helps to reduce porosity in thick-to-thin section transition areas as well. HIPped castings can replace forgings for some high stress applications due to the elimination of porosity and other defects. The elimination of porosity with HIPing increases fatigue life and raises tensile properties. It also significantly reduces scrap, rework and weld repair requirements. Further, mold design can be simplified to save material formerly needed in complex gating and the placement of chills becomes less critical.

Despite the wide application of HIPing for titanium and Ni alloys due to the enormous cost reduction potential, it has not been readily accepted for Al castings meant for aerospace and automotive applications. The reasons include: (i). Low cost of the alloy and high cost of HIPing, (ii).Numerous casting techniques used including investment, die casting, permanent mold, green sand, dry sand, lost foam, thixoforming and squeeze casting, (iii).Very strong tendency of Al alloys to dissolve hydrogen and form oxides, eutectics and intermetallic phases, which have a strong influence on properties and (iv). Unlike Ti and Ni alloys, the predictability of improvement in properties of Al castings is not as exact due to influence of many variables. However, in recent years, the lower cost HIPing processes such as LHIPing, Densal, PIF and quick HIP have emerged with their lower cost enabling them to be considered as a post casting treatment option for aluminium castings for niche applications[1]. Considering the importance of HIPing to high integrity aluminium alloy castings, this paper brings out the current status of this potent post cast treatment tool.

Evolution of HIPing

The very first patent on HIPing dates back to 1913. Table 1[1,2] lists other significant milestones in the evolution of HIPing. In view of the potential improvement realized in castings subjected to HIPing (application of a high inert gas pressure at high temperature), the foundrymen had shown interest in the utilization in the sixties for aerospace and high performance components for racing engines and other niche market. High cost due to high investments and long cycle times (up to 10 hrs) is still the main barrier for extending the use of HIPing for high volume production. In addition, the gas pressurization used in conventional HIPing process is also very dangerous. A variant and recent innovation of Metal Casting Technology Inc and GM, USA termed Liquid HIPing, wherein a molten mixture of salts is used as the fluid for applying isostatic pressure on the components, achieves results similar to those obtained by the much longer gas HIPing cycle time.

Basics of HIPing

Liquid aluminium is prone to both hydrogen absorption and oxidation. Aluminium castings exhibit gas porosity and oxide inclusions inevitably. Further, improper feeding can result in shrinkage porosity. Typically gas and shrinkage pores are spherical and irregular in shape respectively and can associate with aluminium oxides as well. All these significantly deteriorate the fatigue properties of Al castings by shortening both fatigue crack propagation and the initiation period. Figure 1 shows the SEM micrographs revealing the origination of fatque cracks in Sr modified A356 castings from pores and oxide films[3]. The decrease in fatigue life is directly related to the increasing defect size. Porosity is more detrimental to fatigue life than oxide films. A defective casting shows at least an order of magnitude lower fatigue life than sound casting.

In HIPing, the castings placed in a chamber are slowly heated while the pressure of the surrounding inert gas is simultaneously increased. Castings/components are subjected to the simultaneous application of both heat and high pressure in an inert gas atmosphere. This pressure while acting on the casting isostatically enables collapse of any internal porosity left in the castings. Later, the castings are cooled to the room temperature. The simultaneous application of heat and pressure converts the material in to a plastic state leading to the collapse of voids and porosity. It is also to be borne in mind that the collapsed voids do not change either the shape or dimensions of the parts in general. The clean surfaces of the voids enable diffusion bonding together and making a stronger part. HIPing is invaluable in the precision casting, power metallurgical, metal bonding and ceramic industries. It improves the performance and yield of precision castings. The isostatic nature of the applied gas pressure is well suited for defects healing in castings. HIPing of complex shapes parts can be done without complex or expensive tooling.

Relationship between benefits and casting quality

HIPing provides significant improvements in mechanical properties, such as higher strength, enhanced toughness, improved fatigue resistance and longer creep life. Well documented benefits of HIPing include (i) An approximately 50% improvement in ductitity (ii) 3-10 times improvement in fatigue life (iii) Definite improvement in ultimate tensile strength (UTS) (iv). No change in yield strength (YS), (v). Reduction in porosity and thus minimizing the scatter in mechanical properties and (vi). Salvaging the scrapped castings due to porosity and hence improving casting yield especially in castings subjected to radiographic inspection. However, actual benefits to be exhibited in the castings depend on its quality. For example, in test bars cast without serious porosity, HIPing does not affect the tensile properties. If porosity is fine, ductility may be controlled by silicon particles or intermetallics. If the porosity is below a critical value, HIPing will not be effective. However, HIPing results in significant improvement in fatigue properties.

It has been shown[2] that a ten fold enhancement in fatigue life has been achieved in A356 alloys by HIPing for both fine and coarse structured castings (SDAS 30-90μm at 138MPa) due to the effective closure of interdendritic shrinkage porosities even in high quality castings with a very fine structure (30μm SDAS) and corresponding fine porosity. It has also been shown that fatigue properties are influenced dramatically by casting defects other than just porosity, which takes the lead followed by oxide particles. In the absence of these defects, fatigue failure was observed to initiate at slip band leading to significantly longer fatigue life.

Variants of HIPing

Alcoa 359 Process : A process, covered by US Patent No.3.496.624, improves the fatigue strength of aluminium sand and permanent mould castings (up to 300%) over castings without HIPing and enables the castings to meet Class A radiographic specifications[4]. Thus the process makes possible the salvaging of scrapped casting with internal porosity and enhancing he casting yield of a foundry diesel engine pistons, permanent mould cast and treated with Alcoa 359 process, constituted a prime potential market. Any casting component benefited from superior fatigue properties could justify the additional expenditure involved in the Hiping.

Densal HIPing Process : It uses a purpose built unit having a shorter cycle time and much lower acquisition and operating cost than a standard HIPing unit[2,5]. A more cost effective nickel-chromium furnace replaced the costly molybdenum furnace. The lower pressure level makes the pressure vessel less expensive. The less expensive nitrogen gas use as a cover instead of the costly argon used in standard HIPing further brings down the cost. All these brings down the cost to about one third that of a standard HIPing unit. It is also understood that an equipment operated at capacity near to that of high volume automotive foundry can reduce the cost further. Moreover, solution heat treatment is also to be performed in these castings to achieve the required tensile properties. Because of the similarities of the processing temperatures in HIPing and solution heat treatment, attempts have been made to combine these two for greater process efficiency by Boydcote NA, Inc USA, the developer of Densal HIPing process. It is a low cost process specially developed for aluminium alloys in general and the Al-Si alloys in particular. Combining T6 heat treatment with Densal HIPing has been reported to reduce the total processing time and cost even further. Two case studies revealing action of Densal in improving the quality of a high performance air frame casting and an automotive steering knuckle casting are give below:

A large air frame casting, with long, thin interconnected branches of thin and thick sections complicating the feeding of liquid metal is designated as a replacement of a complex sheet metal counterpart. The Densal treatment did improve the ductility of poor quality material by about 60% but not the good quality one. However, the fatigue life of both good and poor quality materials is improved by 3 and 7 folds respectively. Further, the fatigue life of the latter is 2-3 times better than a good quality material in T6 condition without Densal treatment.

An automotive steering knuckle is also a high performance component requiring high quality casting route for its production. Densal treatment of a sand cast steering knuckle has closed internal porosities thus enhancing its quality equivalent to that of a high quality casting route but at a lower cost than the latter.

LHIPing : Instead of inert gas, a molten mixture of salts is used as the liquid / fluid to apply the isostatic pressure on the cast components. A few minutes cycle time is sufficient to achieve the results equivalent to those obtained with the traditional gas HIPing with longer cycle time.

Effect of HIPing on mechanical properties and microstructure
Table 2 summarises some of the researchers findings on the improvement of mechanical properties of aluminium alloys[6-8]. US Naval Air systems command in its attempt to study the effects of HIPing on dynamic properties of castings has shown that fatigue properties of A356 aluminium alloy is significantly enhanced as shown in Figure 2 on HIPing due to closure of the voids[9]. The closure of microshrinkage porosities is clearly reveled in Figure 3 showing the SEM micrographs of fracture A356 alloy with and without HIPing[10].

Discussion
Al castings are finding increasing use in applications warranting fatigue failure resistance. HIPing is paramount in the production of these castings since it eliminates inherent porosity, densifies the castings and thus results in improved fatigue properties. Conventional HIPing using argon gas to apply high pressure for sealing internal porosities is very

effective, but costly (longer cycle time and high investment) and dangerous. On the other hand, liquid HIPing, a recent innovation, where fluid is used to apply the isostatic pressure, achieves similar results of conventional HIPing in a shorter cycle time.

A marked reduction in the statistical spread or scatter usually associated with casting properties is of high significance Minimum observed values are usually increased resulting in improved reliability and efficiency of materials utilization. HIPing can also render castings fit for applications requiring more expensive forged or wrought and machined parts. In addition, alloys prone for hot tears or the formation of deleterious phases during solidification and once considered uncastable can now be redissolved by HIPing.

Poor gating system can introduce surface turbulence and incorporate both oxide films and air in the castings. In the process, the oxide films get folded forming an oxide-to-oxide interface in the liquid and entraining varying quantity of air between them. These double films called bifilms, do not have any significant bonding across the interface and hence act as cracks in the liquid melt as well as in solidified casting. Despite the HIPing of critically stressed Al alloy castings meant for aerospace and automotive applications to achieve closure of shrinkage and gas pores as well as improved mechanical properties and reliability, cracks of bifilms are some what resistant to closure[10].

Effects of HIPing on both fatigue and tensile properties often depend on the as cast quality of the casting – surface connected porosity, alloy chemistry, grain size, dendrite arm spacing and the presence of oxide inclusions. A critical defect size exists for the initiation of fatigue crack, which initiates from other competing initiators like eutectic particles and slip bands below this critical defect size.   The critical defect is in the range of 25-50µm for Sr modified A356 alloy. The fatigue life of sound A356/357 castings depends on the microstructural fineness (SDAS), composition, eutectic modification and heat treatment.   Sr modification results in longer fatigue life in these alloys than unmodified ones. Significant decrease in fatigue life results in both unmodified and Sr modified alloys when Mg increases from 0.4 to 0.7%.   Increasing Fe content too decreases fatigue life particularly in alloys with longer SDAS values. Further, an adequate solution treatment too is beneficial because of dissolution and segmentation of large Fe intermetallic [3] particles.

Friction stir welding[6], an emerging metal joining route for aerospace applications, can also be used to embed wrought microstructure in casting by localized modification leading to dramatic improvement in ductility and strength of A356 alloy (Table II) which can be further improved by HIPing. This approach is aimed at achieving the best combination, low overall cost due to casting and higher performance in localized areas due to the introduction of wrought microstructure by friction stir processing. Further, very significant improvement in Weibull modulus and quality index, the often used parameters for casting quality has also been observed. LHIPing has also been utilized for Al castings made by thixoforming[11,12] and lost foam or evaporative casting process[13].

Conclusions

A defective casting exhibits atleast an order of magnitude lower fatigue life than a sound casting. Now-a-days, foundrymen have been focusing on the production of high integrity aluminium alloy castings meant for critically stressed applications in aerospace and automotive sectors both by improving casting techniques and adopting post casting treatments, namely heat treatment and hot isostatic pressing (HIPing). Variants of conventional or traditional gas HIPing with reduced cycle time are available now making this post casting treatment more economical and safer. Although the development of HIPing is pertinent to the broad range of premium quality Al castings, it is of special relevance to the more difficult to cast Al-Cu series. In a nut shell, a concept of cast-to-fill and HIPing-to-density can be utilized to take full advantages of HIPing. Casting exhibiting superior fatique properties on HIPing could justify the additional expenditure involved. A well designed bottom gating with filters along with a controlled melt quality, wherein oxide films are eliminated or at least greatly minimized in size and number can enable the production of reliable Al alloy castings without hipping. However, foundryman has to go a long way in achieving the above.

Acknowledgements

The author thanks the Director Regional Research Laboratory (CSIR), Thiruvananthapuram, India for his interest and all the authors and publishers of the articles cited.

References

1. Hebeisen JC, Cox BM and Rampulla B, HIP of aluminum castings, Advanced Materials & Processes, April 2004, pp 38-39.
2. Charles Barre, Hot Isostatic pressing, Advanced Materials & Processes, March 1999, pp 47-48.
3. Want QG, Apelian D and Lados DA, Fatigue behaviour of A356/357 aluminium cast alloys. Effect of microstructural constituents – Part II, J. of light Metals, 1(1) 2001, pp85-97.
4. Casting densification process, TMD Report N.5, Alcoa technology Marketing Dn.
5. http//www.machinedesign.com/ASP/strArticle10/55380/strSite/ MDSite/viewSelectedArticle.asp
6. Friction stir casting modification, http://web.umr.edu/~fricstir/cast.htm
7. Premium quality Aluminum castings, Technical update 3D5, Hitchiner Manufacturing Co., Inc.
8. Sergio Gallo and Claudio Mus, Current quality needs for casting in automotive, M.C.Flemings Symposium on solidification and materials processing, 2001, pp 373-378, edrs. Abbaschian R, Brody H and Mortensen A, pub. TMS

9. Dale Moore, Naval aircraft materials and processes, Advanced Materials & Processes, March 1999, pp 27-30.
10. Nyahumwa C, Green NR and Campbell J, Influence of casting technique and hot isostatic pressing on the fatigue of an Al7SiMg alloy, Met. & Matls. Trans. A, 32A, Feb. 2001, pp 349-358.
11. Rosso,M, Mus,C and Chiarmetta,G, Liquid hot isostatic pressing process to improve properties of thixoformed parts, Metallurgical Science and Technology,18,2,Dec 2000,pp16-20
12. Rosso,M, Romano,E and Barone,S, Properties of thixoformed parts by liquid hot isostatic pressing process, Metallurgical Science and Technology,19,1,June 2001,pp28-33
13. Molina,R, Leghissa,M and Mastrogiacomo,L, new developments in high performance cylinder heads : application of LHIP and split cylinder head concept, Metallurgical Science and Technology,22,2,Dec 2004,pp3-8

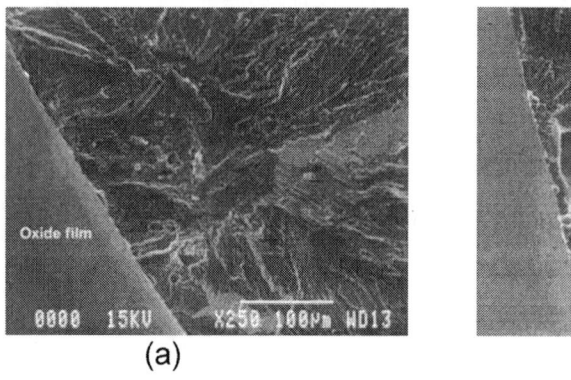
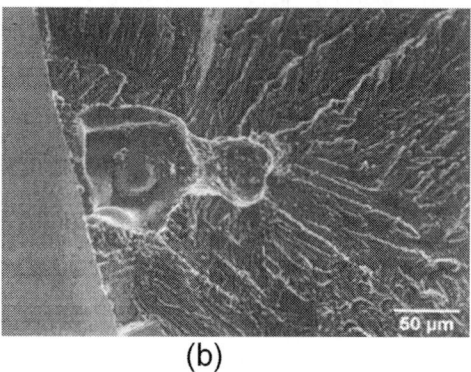

(a)　　　　　　　　　　　　(b)

Figure 1: SEM micrographs showing the origination of fatique cracks from (a). Oxide films and (b). Pore [3]

Table 1 : Evolution of Hot Isostatic Pressing(HIPing)[4,6]

YEAR	EVENT
1913	First patent
1950s	Development by Battelle Columbus Laboratories
1960s	Initially developed for cladding nuclear fuel elements[2]
	Immense interest by Foundryman in utilizing the potential quality improvement of castings by HIPing with high inert gas pressure at high temperature by eliminating defects like shrinkage, porosity and hydrogen pin-holes and using increasing mechanical properties.
	Many Hiping units put in to operation to treat high performance aerospace and racing engines components and other niche markets.
	High cast factor due to huge investments and long cycle time (up to 10hrs) restricted its use for high volume production castings viz. automotive.
	Applied to investment castings for gas turbine engines[1]-Ti and Ni base super alloys
1980s	Hitchiner Manufacturing Corpn. and General Motors, USA patented a low cast version of Hiping coined as Liquid Hot Isostatic Pressing (LHIP).
	Demonstration of innovative liquid HIPing (LHIPing) requiring only few minutes against hours for the same and results by HIPing by Hitchiner Mfg. Co. Inc and General Motors (GM) on a small scale pilot plant level.
1998	Teksid, Italy and Idra Presse was authorized by Hitchner and GM to develop an industrial LHIPing for high volume automotive sector with process cost competitiveness tag.

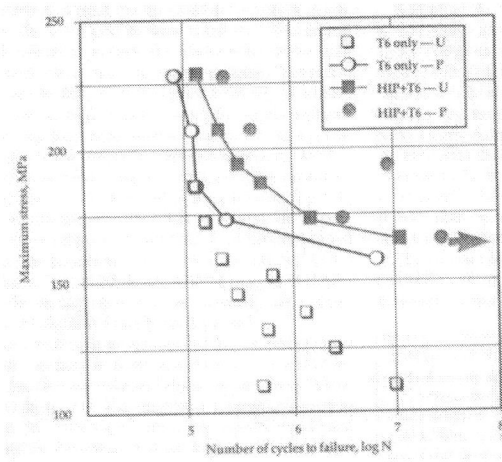

Figure 2 : Effect of HIPing on fatique properties of A356 aluminium
Castings. P- premium quality and U-lower quality[9]

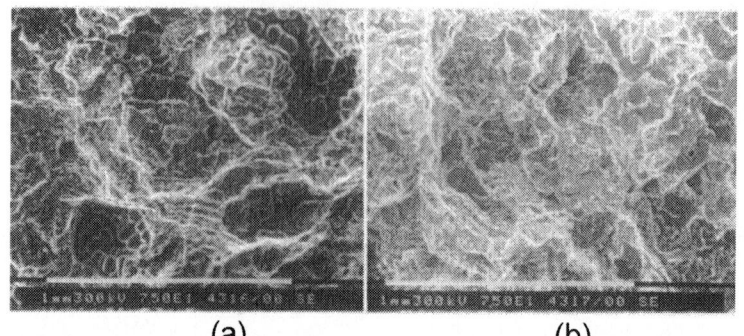

(a)   (b)

Figure 3 : SEM micrographs of fractured surface of A356 alloy (a). without and (b). with LHIPing[12]

Table 2: Mechanical properties of aluminium alloys without and with HIPing treatment

Alloy	Condition	UTS MPa	YS MPa	Elongation, %	Fatigue strength MPa	Remarks
A356 [6]	As Cast	169±10	132±5	3±1	-	Localized property improved by localized modification
	As FSP*	251±5	171±14	31±1	-	
	As FSP+T6	301±7	216±14	28±3	-	
356[7]	Sand cast, unchilled, quenched directly after liquid HIPing and aged	269-275	187-210	6.2-9.1	-	
A356 [8]	Sand cast +T6	230-250	190-210	1-2	80-100	
	sand cast+ LHIP +T6	300-320	230-250	4-6	120-180	

FSP- friction Stir Processed

Metal Casting's Secret Ingredient

William W. Sorensen
Executive Director, Foundry Educational Foundation, Schaumburg, IL USA

Abstract

This paper describes the work of the Foundry Educational Foundation FEF, in the USA to support students working in the metal castings industry.

The paper describes how the FEF works in conjunction with the American Foundrymen's Society, AFS, the castings industry and educational establishments to support the developments of the industry future leaders in the American castings industry.

Introduction

The metal casting industry needs quality people with new ideas and new technologies. That is why the Foundry Education Foundation, FEF, was created. Now, more than ever, the industry ~~we~~ can succeed more quickly and enjoy greater satisfaction by attracting the best college students to work in the metal casting industry.

FEF and the American Foundrymens Society, AFS are two separate organizations in North America that can help students become better acquainted with this industry. I do not know of any other industry that reaches out to college students in America as much as the metal casting industry does with the FEF.

Background and Historical Perspective

For sixty years, this Foundation has brought together ~~consisted of~~ many people—students and professors; trustee boards; company contributors, including foundries and suppliers; AFS chapters; and great leaders.

In 1947, when FEF was established to assure a continuing supply of technical manpower for the metal casting industry, a campaign was organized to provide financial support for three years - 1947,1948 and 1949. Firms were asked to make pledges for the three-year period. Prior to the conclusion of the first three-year cycle, FEF began to demonstrate encouraging results and the Board Members inaugurated a second three-year cycle which was also financed on a pledge basis for 1950,1951 and 1952.

In 1953, the Board Members decided to place FEF on a permanent basis and have it financed through annual contributions similar to the dues assessments of trade associations. A good percentage of the members who participated in the cycle campaigns accepted this proposal and FEF is now on an annual contribution basis.

Its mission statement is:

> *FEF strengthens the metal casting industry by supporting unique partnerships among students, educators and industry, helping today's students become tomorrow's leaders.*

Funding and Support offered to Students

The students are the focus of our efforts. This Industry needs the brightest and the best to enter our workforce. As an Industry, we must be competitive on the college campus. We must protect manufacturing courses designed to teach metal casting because we need the new energy and ideas, as well as workers, to add to our workforce. FEF students have made the single largest impact on metal casting in the United States during the last sixty years.

FEF scholarships are offered currently at 25 colleges and universities in North America. Scholarship recipients are selected by Key Professors at FEF schools. Size (minimum of $500 and maximum of $2,000) and number of scholarships are determined by the professors, based on the amount of money allocated to their school by FEF. FEF "Key" Professors receive financial assistance from the Foundation with discretionary funds and special awards. In addition, many special scholarships are available at FEF website each year.

Many leading foundrymen and many university professors are former FEF scholarship recipients. These former scholarship winners, along with FEF Board Members and other industry people, support the Foundation on an individual basis.

All revenue received from annual contributors goes into FEF's general fund and is used to support the regular scholarship program as well as all other FEF activities. In addition to this revenue, FEF receives funds for special purposes (named scholarships, AFS Chapter scholarships, memorial contributions and year-end contributions. Commissioned to work with full-time college level students, FEF, although independently supported, is an extension of the educational activities of the major trade and technical associations of the cast metals industry.

In this way, FEF does not conflict with AFS, NADCA (the North American Diecasting Association), or other educational programs in the USA. They compliment each other.

The company contributors offer the financial backbone of our operations. Beyond foundries and suppliers, no other large direct funding exists. Their vision for the future is through today's students. They are investing in the lives of young people who will become our future leaders. The provision of summer work, co-op opportunities, and permanent employments are essential components that help make FEF successful. On average, over 75—80% of the FEF students take full-time jobs in metal casting each year.

The students attending independent AFS Chapter meetings get to see our industry in action. The chapter meetings serve as a natural way to promote education and provide a vehicle for technical knowledge to our students. Chapter events often recognize students with scholarships as the Chapter sees the future in the lives of these students. Many times the local Chapter takes an active role in supporting their local FEF School.

Great Leaders guided this Foundation over the years with several volunteer FEF Presidents who were once FEF Students. We have come full circle over the years with today's leadership and strength coming from those who benefited from this Foundation in their years of greatest need.

Endowments through FEF are a special way to honor people. Endowments in the name of individuals will carry on the individual's legacy to future generations. The generosity and tax-deductible gifts and estate planning for FEF is a great way to insure the future. The earnings from these endowments help to support our scholarship efforts.

Students and professors, trustee boards, company contributors, AFS chapters, and great leaders are all a part of FEF.

The Structure and Operation of the Organisation

Through an active Board of Trustees of over fifty-two people, there is 100% support personally and corporately. This FEF dedicated board has pledged their time and their monies to help shape the future of our industry in the lives of students. Great leadership is a key to any great organization. With these dedicated volunteers, we can meet any challenge that comes our way.

The FEF is cost effective. With no large overheads and only a two person staff, FEF is the smallest of any society. We have been protecting contribution dollars long before hard times settled in. Companies and individuals can be confident of their investment in this foundation, which in turn is an investment in the future of our industry.

We have a very active investment committee that meets once a quarter to maintain our portfolio. FEF has an independent audit every year to make sure any possible mistakes are corrected quickly and efficiently.

Companies can be recognised through various levels of accomplished giving. Individuals can also obtain various levels of giving and recognition. Even in Estate planning, this will pass on legacies to countless future generations, and be recognized in the FEF Saugus Society.

Links with Academia through Key Professors

The professors are the key link to the schools. Without a strong student-oriented professor, the metal casting program often fails. This has to be a person who wants to see college students succeed and to succeed in a foundry career. The professors relay FEF information on scholarships and job opportunities. They also provide a role model to these students, guiding and directing the core curriculum to give students the best view of their futures. They are devoted, hard working, and loyal.

FEFs Future

The tendency in good times is to take our organizations for granted and in bad times to forget them. FEF's mission is to strengthen the metal casting industry with unique partnerships, helping today's students become tomorrow's leaders. How awesome! It fits any business cycle.

Do you know of any other industry that supports such a goal? We have no political agenda. Our industry actually has a voice on the college campus through the FEF. We are making sure the future is secure with our greatest asset—people.

Dedicated to People

Companies brag about how important people are to their organization. FEF is a foundation dedicated to people. What other industry has a foundation like FEF? Why would you not support something so good and so noble and so unique?

There is no doubt that even in tough times there is a strong need for engineering students. This need will only increase. Do these students stay in the metal casting industry? Retention over the past ten years at major US companies, which typically can show higher turn-over with other employees, shows FEF students at 90% retention or better!

The FEF student is well prepared. FEF also plays a significant role with curriculum relating to manufacturing, particularly in metal casting. Through an active FEF accreditation process, we keep up with our schools. They have to maintain our standards and demonstrate interest in us to be able to continue.

There must be strong administrative and local support to be an FEF school, along with demonstrated placement of students in our industry. This is why AFS-CMI (Cast Metals Institute) has almost all of their regional labs now placed at FEF schools. Through dedicated Key Professors, we get the word out on the campus. Students also have important role models so essential to the success in their careers.

Through the annual College Industry Conference, (CIC) we bring students into an environment of opportunity. Gathering in Chicago at a great historic hotel, these students have opportunities to meet and network with students from other colleges and universities from across North America. They also have the unique experience to meet face to face with industry leaders who make presentations, offer career opportunities, and provide social interaction. Professors from these institutes of higher learning also have opportunities to network and share ideas in the field of metal casting. What a great way to meet the future, face to face, in the lives of these college students!

Where to find out more

More can be learned about FEF at our website, www.fefoffice.org
The organisation also produces a CD or Video that is called "Casting for a Career." This excellent eleven-minute presentation is great for community groups or classrooms. There is also have a PowerPoint presentation about FEF that can be downloaded from the website.

Please check out our website and encourage metal casting students in North America to register and apply for special scholarships on the web site, www.fefoffice.org.

This is a great industry in which to work and grow, and with FEF you have the secret ingredient for the metal casting industry in North America!

People and skills for today's industry- an Indian Experience

K.Gnanamurthy, Lakshmi Machine Works, Coimbatore

Abstract

The foundry Industry like any other has certain generic but several specific characteristics and hence demands such skills for effective performance. In addition to the normal business practices the foundryman has to deal with many technology issues on a daily routine. This includes metallurgy, refractories, fluid flow, heat and phase transfer.

The dichotomy of increasing demand in skills on one hand and severe shortage of even yesterdays' skills has made this Industry tend to move to geographical locations where employment is available but not the best of skills. Under these circumstances, countries like India, where a very high level of technical education is available, have done reasonably well in consolidating this Industry and providing a workable solution to many casting requirements. Skills in metal casting per se has always been available in the country.

This paper deals with the large number of skills required in the foundry industry, ranging from pure technical ones to skills in management and quality perspectives, and how they are addressed in an Indian Scenario.

Introduction

The foundry Industry has always been considered as one that requires human resources who should be prepared to wear a blue collar most of the times. The need to work with high temperatures of molten metal and sand in a majority of casting processes, though challenging for one who is knee deep in it, does not offer enough incentive for the above average person, who has other options for a career. Not to mention the present day scenario where soft ware and other computer related work can be imbibed quite easily and put in to use for furthering a career, the last twenty to thirty years has shown a declining interest in metal casting even amongst other manufacturing avenues.

As against this not a very encouraging outlook, we have several men and women who love this profession and who would not change places for any thing else in the world. They claim it is an extremely creative field. Like Math, there are a few who love this work, and a vast majority who would not have anything to do with this. And like Math, if not as acutely, you cannot dispense with Foundry just because it does not attract much talent. Incidentally most of those who adore foundry today have either been pushed in to it early in life often without their volition, or those who inherited a foundry from an otherwise brave entrepreneur. The very fact that countries with high unemployment rates have little problem in manning their foundries show the industry is today populated by men who have little options. How do we foresee excellence under such lackluster scenario?

We shall make an attempt to review the various scientific principles that form the mosaic behind this ancient Industry and the faculties and skills required to make a good foundry man at its various hierarchies.

Science of Metal casting

There used to be and there are even today, lively debates whether foundry should be called a science or an art. Many a branch of Engineering has transited through this corridor of arguments and successfully emerged out of the science door. In any case we shall call it science for the simplest of reasons that we are here to understand, evaluate, adapt and replicate good practices, all of these being a process of science instead of calling it an art and getting awestruck, overwhelmed and reduced to a state of helplessness by its complexity.

Casting of a metal involves preparing the liquid metal, making a mould, running the metal in to the mould and ensuring the casting is defect free, cleaning of casting and finally heat treat if necessary to impart certain properties. Said this way, they appear very simple and straightforward. We shall however walk through these processes and understand what it takes to do a good job of it.

The Liquid metal: Melting

Melting is done in cupola, open-hearth furnaces or electric furnaces depending on availability and demands of quality and cost on the liquid metal. Understanding of Cupola or the other fossil fuelled furnaces requires a sound knowledge of the principles of heat and mass transfer besides chemistry of fuels. The various zones of cupola and the calculations to arrive at the liquid metal output and its chemistry are best addressed from first principles rather than remembering thumb rules [1]. Quite unfortunately a large number of such thumb rules exist in literature and offer band-aid solutions to foundry men literally inhibiting their analytical ability. In the present day global sourcing, with material flowing in to foundries from all geographical locations with their own natural characteristics, many working ratios had to be abandoned or at least revisited. There have been instances when foundry men were forced to use very high ash coke, or equally surprising, use very low ash coke. Complete understanding of the Cupola as a contra flow heat exchanger and a knowledge over the effect of each parameter like Cupola geometry, air, its pressure, humidity, coke, its calorific value, the metallics and the non metallics is important. Add to this, high temperature reaction of coke, metal, limestone and refractories. Even with ten variables working at a minimum of four variations each, the result is a mind boggling million. In reality the variables are many times over ten and the variations far too many than four if only they are discrete, and they are not.

The Induction furnaces and Arc furnaces bring churning or its absence along with them and add yet another dimension to the heat and mass transfer during melting.

What line of thinking other than a scientific approach either by rigorous physics and chemistry, with the help of finite methods can produce reliable and repeatable results in such an array of inputs? Can we leave it to thumb rules?

If the answer is no, for it cannot be yes, the next question is how much of a physicist or a chemical engineer or a mathematician you want a foundryman to be? The basic minimum is, good enough of all this to know and appreciate that he cannot do without these skills.

Quality of Liquid Metal

There is the transport of gases like O, H, N which we need to contend with in Al, Cu, Steels and N in iron. Their solubility with reference to temperature of metal and what is available in atmosphere is to be understood. We need, in addition, control S, P, Mn and several other elements. Which metal gets preference over other metals in picking up gases are to be well studied. Foundrymen know by this time we need certain amount of S for effective inoculation which thought has replaced the earlier view of attempting to minimize S. Now we talk about controlling S.

We have not even come in to the mechanical requirements of the metal. We are still in the process of preparing liquid metal in a furnace with a desired chemistry and behold the knowledge and skills requirement for this seemingly simple task.

The Mould

It is generally believed that mould and running system contribute more defects than liquid metal. Mould can be of sand, metal, graphite or other refractory material.

Sand and other refractory material

In a typical Sand casting, one deals with five to ten times weight of casting as sand, plus many times over for material handling. In some special sands like chemically bonded systems or Investment castings, cost of mold can be comparable to that of metal if not more. Apart from the simple mechanical requirements of the mould and core having to take the metallostatic pressure of liquid metal, which itself is not well understood on account of the near total lack of knowledge of high temperature properties of sand systems, we have several other issues like metal mould reaction, gases produced from moulds and cores etc. It is ironical but true that several properties and characteristics we study about sand have little if any relation to the severe demands at the very high temperatures we deal with at the time of mould filling and sustained heating of the sand during solidification. There is also the question of optimizing everywhere instead of the single purpose maximizing or minimizing of goals. Moulds need to be that much strong to take the liquid and solidification pressures and not more, but need to collapse with utmost discipline there after to accommodate shrinking of metal and ensure a crack free casting.

The various organic and inorganic binders we use call for a thorough understanding of their chemical behavior with reference to temperature, humidity and type of metal which is being handled. Every attempt one takes in avoiding certain problem with an alleviator leads in to further frontiers of knowledge which cannot be ignored and by their right require rigorous study. The refractory mould dressing and binders used there off are too important and expensive, for the knowledge base to be totally left with suppliers.

Metal moulds.

Gravity and Pressure die castings form an entirely different branch of metal castings that many a sand casting man would not have estimated the intricacies of die design in his time. Permeability of sand which is taken for granted in sand castings is not available and one needs alternate methods to deal with gases released from metal and air aspired in the metal stream. The die material , its strength, life in dealing with high temperatures, dimensional stability, expansion characteristics, heat transfer are some of the parameters one deals with in Die design.

Metal Flow and solidification.

There is no other aspect of Foundry where the Foundryman has dabbled with more than the Gating and Risering system of castings, often the same casting several times over. Sand as a mould offers so very little resistance to the whims of an imaginative mind that he shapes it to endless variety of gates, exits and riser pools, too many times and too frequently, all to contradict himself and the thumb rules he has read somewhere and swears by.

Every foundryman though less equipped with theory will agree with leading Physicists that fluid flow in open channel is one of the least understood branches in science [2]. Worse is the case with Turbulence[3]. Calculations on Reynolds number and Weber number [4] give insights, but cannot be applied without rigorous understanding. Similarly, it is hard to believe there is no mathematical formula however complex to calculate time it takes for a glass of water to freeze in a refrigerator let alone a metal casting in a mould. We do have a set of non-linear differential equations, initial and boundary conditions and a host of empirical, semi empirical and approximate solutions besides Finite analyses to give somewhat close answers but not 'the' answer as mathematicians may wish to see.

The business of metal filling a mould and then solidifying combines both these difficult to predict aspects of fluid flow with or without turbulence and then solidification. Simulation soft wares address these problems. While the myth entertained by some is that these programs are a panacea for many a foundry's ills, the truth is some where midway. There are benefits, gross benefits, but there are also tall expectations not met with.

Then there are the molecular level dynamics which take place during solidification, like nucleation of an embryo, formation of dendrites, micro and gravity segregation of elements, formation of gas porosity and finally the microstructure.

Cleaning of Castings, Heat treatment etc.

In steel and non ferrous castings where metal has high elongation and toughness, removal of casting appendages call for attention and the understanding of the behavior of metal in response to cutting involving high temperatures like flame, arc or a simple cut off stone. Evaporation of metal, Heat affected zones and the movement of basic elements to and from the cut off grinding stones all lead to dramatic change in material properties.

Then there are stress relieving and heat treatment cycles given to improve material properties. Ductile irons are heat treated often to get ferritic matrix, steels to enhance strength, Stainless steels to improve corrosion resistance and ageing of a variety of non ferrous metals.

Destructive and non destructive testing

Interpretation of Radiography and Ultrasonic testing is generally left to specialists, with the foundry men wondering all the time that there is this superman who classifies his work in to different levels. In order to be effective, the foundry engineer needs to interpret films and data himself so that he gets a feel and X-ray sense of the casting he produces.

Skills required

In the foregoing paragraphs an attempt has been made to bring to focus some of the scientific principles involved in casting a metal. Skills required to cope with the situation are many. The immediate question is whether the foundry man should have all these or distribute specific skills amongst various people in the foundry under different functions. The theme is debatable and depends on the size of the foundry, specialization it focuses on and of course availability of such talent. For example, in a large chemically bonded sand foundry, you may hire some one who has an expert knowledge in polymer Chemistry, or in a small foundry the Foundry Engineer acquires such expertise himself, or alternately as many foundries do, depend on the supplier entirely. It is a matter of choice and the confidence the Foundry has developed over a period of time in one system or the other.

We also have extreme cases where men with little or no experience in Foundry have done very well in Managerial positions and bring about phenomenal improvements. This happens all the time and in every Industry. The basic assumption here is that such skills are available and the leader hires them and puts them to best use. But, how do we deal with a situation when there is a waning interest amongst men to acquire these knowledge and skills? We will have too many managers and few to be managed.

However, what cannot be gainsaid is the fact a top class foundry man has a good deal of knowledge and skills in all the above aspects and many more that we will see later. The bare minimum of course is that he has at least read about them and knows that these are important and cannot be played with without proper skills.

For the present discussions, we will consider middle and fairly senior level Engineers and Managers who are responsible for manufacturing good castings at minimum cost, and skills required for such a position. We do not deal here with senior strategists who conceive Foundries, put up plant, equipment and run it as a business, and the managerial skills they require, which will be a separate study by itself.

Metallurgy

Basics of metallurgy, relevant to the metal that is being dealt with in great detail and about other metals at least as a cursory walk through is mandatory, with particular emphasis on Solidification Dynamics, Phase diagrams, effect of various elements, fracture mechanics and various other aspects which one encounters in Foundry. He should understand and experience melting himself and learn the effects of melting sequences, degassing, deslagging, liquid metal treatments etc. He should be able to identify and segregate scrap for melting and know about alloy and other additives.

He should be conversant with lab procedures particularly special tests like Creep, Corrosion, Impact, Fatigue etc. He should ask analysis for stray and tramp elements and gases. Regular study of microstructure is important as is its relation to chemical composition and thermal analysis.

Heat treatment is a powerful tool in the hands of a foundry man and he needs to use them to full advantage.

Material Science
The foundry man needs to fully understand all the other material he uses often at very high temperatures. The geometric, chemical and refractory properties of sand lead key to the success of many a casting. Organic and Inorganic Binders form an important study material and should be well understood. The relevance of testing sand with its binder at temperatures and conditions close to actual use is important and any other routine test is just ritual.

Furnace refractory is an expensive material and hence its choice, the lining practice and maintenance are important skills. Many a cut through and accident in a foundry and consequent loss and damage to furnace and personnel can be traced to poor furnace lining practice. The behavior of refractory, its sintering, glazing and failure are important study.

Besides these, foundries use a large amount of other material like mould coats, feeding aids, ceramics for running systems, inserts, chaplets, Chills, Resistance elements in heat treatment furnaces, insulating bricks to name a few. Each one of these has a far reaching effect on the quality and cost of castings and cannot be relegated to lesser levels of focus. Foundry men know too well that a change in chemistry of steel used in Chaplets can lead to scrap and still worse time and effort for trouble shooting. Similarly a supplier changing percentage of a particular refractory material used in a mould coat can lead to hours of surface grinding in fettling shop and/or irreparable defects leading to total scrap of casting.

Fluid Dynamics
Starting from flow of sand in to core box, leading up to metal flow in mould, Foundry man addresses fluid flow in its various forms. While the principles of Bernoulli's theorem, Law of continuity etc. need to be well respected, the foundry man should not be naïve to assume that fluid flow is simple and straightforward. Real time situations are too complex and one gains experience and skill only by watching closely innumerable metal flow in mould and take help from finite volume and other techniques of Computational Fluid Dynamics. Though through the eyes of rigorous math, Fluid flow packages appear to take a over simplistic approach towards complex initial and boundary conditions, they are the best we have and Foundry man may well equip himself to learn all about this and get the best out of them.

Solidification

We are discussing here the heat transfer aspect and not the metallurgical aspect of nucleation, cell growth microstructure etc, though some packages attempt to forecast microstructure given the mould and metal conditions.

We will start with the basic math truth that there is no simple formula to predict solidification time which fact should not further bias us from predicting relative solidification time of sections. That's all that matters most of the time and credit goes to Chvorinov and his rule.

Things get complicated however when molding material changes from one to the other. There are chills, inserts, pads etc. When mould filling is complete, the liquid zone is anything but isothermal. Supposing in a complex situation, and by providing proper allowance to each variant in the metal-mold zone, we finally predict the hot spot, the next task is to feed it or coax the system to push the hot spot to a desirable place by further addition and omission of metal here or a chill there, and having succeeded in such an attempt to feed the hot spot conveniently. Present day solidification packages handle very complex shapes and predict solidification accurately as evidenced by actual sectioning of castings. They go on to predict shrinkages and likely places of distributed porosity on the lines of Niyama Criterion [5].

Foundry man needs to confine his experiments to desktop with the help of simulation packages and attempt to produce shrinkage and crack free castings. Talking about cracks, there are packages to predict internal stresses and crack tendency. Foundries have gone to the extent of calculating the buoyancy force on cores and have designed core irons based on the mechanical and thermal load on them.

Drawing and tooling

While there is a wide spread tendency to buy pattern equipment from experts, drawing reading skills and the decisive views of the foundry man on pattern parting and methoding go a long way in the manufacturability of a casting. Drawing reading is such a vital skill that there are foundries where core setting for heavy castings is carried out only by men with such skills. It is a very common experience that many a foundry man will like to tinker with pattern methoding or suggest a totally different parting after the pattern has been brought to the shop floor for use or perhaps still later when the first casting has been produced and found rejected. To reduce such costly and time-consuming afterthoughts, a good foundry man acquires these skills early in life.

It is gratifying to note that several great foundry men have entered this profession through the pattern shop door. With the advent of solid modeling, surface generation and CNC route, pattern making had never been faster and more accurate. Foundry man needs to have hands on experience in all these.

Other skills

We are not listing here, an array of other skills that are required in any case in a manufacturing environment. There is this plethora of writing and emphasis on approaches to quality that foundries need to pick up appropriately. If only by way of caution foundry man like any body else need to protect himself from the overwhelmingly zealous propagandists of one or the other of these generic techniques often over-shadowing the importance of the science we have to deal with which is metal casting. But, given the optimum importance, such agenda like TQM, Six Sigma etc can provide a solid structure to carry out any goal, to produce good castings at low cost in this case.

Availability of Resources

The traditional route to become a foundry man has been through study of Mechanical Engineering or Metallurgy in Undergraduate level and put in experience in Foundry to gain practical experience. One could get courses in Foundry Science and technology as an optional subject where specific Foundry related topics could be learnt. Many world class schools were giving foundry related courses till the sixties. Triggered by the domestic and global requirement for the foundry skills, some top schools in India offered graduate courses and research fellowships in Foundry in sixties and seventies. This was the golden period for the Industry, when brightest young men were attracted to these courses, who later went on to man foundries in India and elsewhere in world. They were in touch with other foundry men across the world and contributed vastly to the theory and practice of Foundry. This was followed by general recession and waning interest in foundry as in many other manufacturing streams. Late nineties and the new millennium witnessed resurgence of Indian foundries. While still some top schools offer tailor made foundry courses, Mechanical Engineering and Metallurgy with Foundry credits are offered in most of the Indian and National Institutes in the country. The growing interest shown by buyers on Indian made castings prompted by favorable experience in the last decade has brought about enormous activity in teaching, training, and dissemination of knowledge by Seminars, Colloquia in regional and national levels. The Institute of Indian Foundry men, has been playing a major role in handling this stupendous task in this vast country.

Conclusion

India has perhaps today one of the largest pools of skilled foundry professionals in the world, supported by technical education in the grass roots, specialist Foundry skills in undergraduate courses and advanced training in post graduate levels. There are also the finest Management Schools to fill in gap in top strategy levels to bring in todays Global Business sense to the Indian Foundry Industry.

References:

1. Heine RW, Loper CR, Rosenthal PC, Principles of Metal Casting, 2nd Ed, McGraw-Hill Book Company, 1967
2. Richard P. Feynman, Six Easy Pieces, 3rd Ed, Addison-Wesley Publishing Company, 1995, ISBN 0-201-40955-0
3. James Gleick, Chaos, Vintage, 1998, ISBN 0 7493 8606 1
4. Campbell J, Castings, Butterworth-Heinemann Ltd, Oxford 1993, ISBN 07506 1696 2
5. Niyama. E, Computer aided design for metal casting, Annual review, Material Science 1990 (20) pp101-105

Email ID of Author:
gnanamurthy@lmw.co.in
drkgnanamurthy@gmail.com

How metals employers engage in the UK skills agenda

E Bonfield

Metskill/SEMTA, UK

Abstract
- not supplied

Please contact the author for more information:
Email: e.bonfield@semta.org
Email: e.bonfield@metskill.co.uk

Latest Trends in Industrial Skills Development Techniques

Dr. C. Ashley*, Eur Ing C.J.C. Bale*, N. Millan*, T.M. Williams * and R.J. Hendley**.

*AutoTrain LLP, UK, ** School of Computer Science, University of Birmingham, UK.

Abstract

This paper describes the application of Internet-based technologies to improve skills within manufacturing industries. Employees need continued training and e-learning offers a more flexible method to complement and replace conventional training. The methodologies and advantages of e-learning are described, with its freedom from time, travel and accessibility constraints, as developed by one of the world leaders in e-learning techniques. Supported by the EC Leonardo da Vinci programme, a multidisciplined engineering and software team has developed innovative ways to deliver vocational education and training to companies and individuals via Managed Learning Environments. Interactive learning material and multilingual video lectures are included in the presentation and the written paper, with references from the international companies and organisations that contribute to, and make use of, this modern approach to knowledge management and transfer.

Key words

Training, e-learning, lean, manufacturing, quality

Introduction

Automotive, engineering and manufacturing industries have had to come to terms with increasing global competition and price pressures that make it possible for only the most skilled companies to remain competitive. Under such circumstances, training time and budgets can be difficult to obtain, especially for small companies. The growth, or even survival, of businesses in the automotive, manufacturing and engineering sectors is threatened by a lack of skills that is perceived as one of the main obstacles for generating a competitive offering against global competition. The AutoTrain and AutoTrain-Europe projects, run jointly through the School of Computer Science and EuroMotor at the University of Birmingham (UK), have developed a friendly system of online training courses, primarily for the benefit of small and medium size companies (SMEs) within the manufacturing, engineering and automotive sectors. The resulting Internet-based technologies are used to generate online learning systems that learners can use to improve skills and to implement company training policies in their required areas in a customised way, at a time to suit the individual, at their own pace and at minimum cost for the business.

Online training

Online training is a type of training system where learners "attend" classes that are delivered through the Internet. EuroMotor (www.euromotor.org) was formed in the nineties to improve the knowledge base of the European motor industry through high-level technical collaborative training, as a network and partnership of European universities, car manufacturers and suppliers, in order to compensate for the increased difficulty in gathering attendees to conference and training events. In the last two years the AutoTrain-Europe programme has acted as a virtual college to bring online training to the European automotive industry and AutoTrain LLP has been established to create Managed Learning Environments (MLE) for individual European institutions, both academic and commercial whilst delivering further value from over one hundred courses available via the Internet, which were created by previous EU-funded projects.

To illustrate the practicalities of online learning, one of the recent projects can be examined in more detail. The AutoTrain-Europe project was set up with its own website at www.autotrain-europe.com (Figure 1) that is used as a portal for the delivery of its online training resources. The website was designed to be both functional and easy to access, even by novice users. The entire system is interactive, user-friendly and easy to navigate even for inexperienced web-users. Furthermore, when it comes to functionality, the design solution chosen by AutoTrain has generated a system that is able to work both effectively and efficiently through a web-browser, going to the project's main page and clicking on the *"Training Centre"* link to gain access to all the resources available.

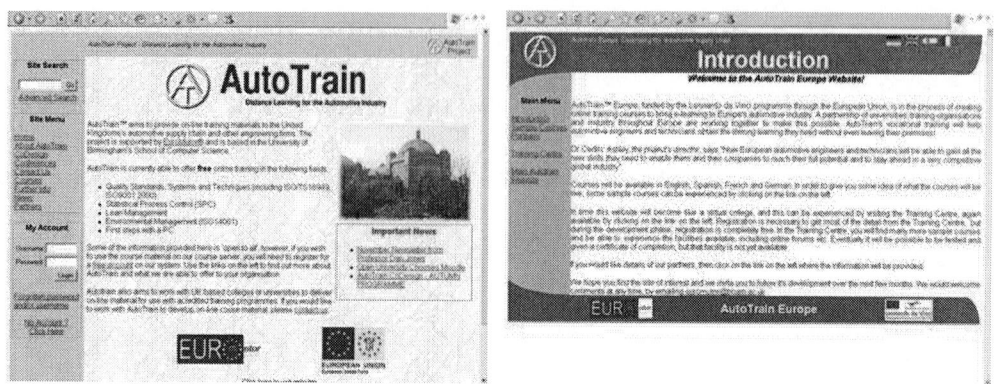

Figure 1. Home pages of the AutoTrain and AutoTrain-Europe Websites

In order to be able to access any of the resources available from the online learning system, users are required to register for an account in the system. Under EU-funded projects, the registration is available at no cost to the end user and it is a quick and simple three-step process that involves the input of some personal and company details in return for a username and a password, which entitle the user to unlimited access to all the online resources available from the site. Outside of EU funding, the access is either provided by the company or institution that hosts the MLE or by subscription payable through the online system.

After initial development work using WebCT, the current training materials are delivered through an "Open Source" MLE named *Moodle*. *Moodle* is a user-friendly virtual learning environment that facilitates the delivery and presentation of courses in an easy-to-browse way and allows educators to create quality online teaching materials. *The Moodle* platform is also used by learners to access courses, to keep track of their progress, to assess their knowledge and to interact with their tutors or fellow learners through the various forums. Key advantages of the *Moodle* MLE are its low cost and the ability of the AutoTrain developers to configure the system to suit different needs.

In its AutoTrain-Europe format, a list with all the materials is available from the *"Training Centre"* link on the home page. Clicking on the titles of any of the training courses gives the user a description of the course contents to enable them to decide whether they want to add a particular course to their account. Once the user has registered for an account, he can add or remove as many resources as he needs, as frequently as he needs, with no time restriction or expiry date. The "My Courses" function enables the learner to access his chosen courses more quickly while leaving all the other available material in the background in a way that does not clutter his learning environment with irrelevant material.

The online training courses produced by AutoTrain (www.autotrain.org) (Figure 1) presently come in two main formats: **e-books** and **video presentations**. E-books are text-based courses that present several levels of interaction. Video presentations are video lectures easily identifiable from the general list of courses by a special icon. Both types of resources are easy to access just by clicking on the title and adding them to each user's account. Nevertheless, in order to be able to watch a video presentation displayed on the computer, the user will need to have additional software such as *RealPlayer* ™ or *QuickTime*™ installed on their machine. For those users who do not have any of these plug-ins installed on their computers, AutoTrain specifies links to the sites where the programs can be legally downloaded at no cost.

e-books

AutoTrain e-books are standard text-based courses that have been digitised (Figures 2 and 3). They are delivered through the Internet so that individuals or companies that require training in a particular area can have the training resources that they need available, on demand, at their convenience. E-books are easy to access and easy to navigate and offer different levels of interaction for the users based on a choice of quizzes and exercises that users can access while they are learning, in order to help them monitor their own progress.

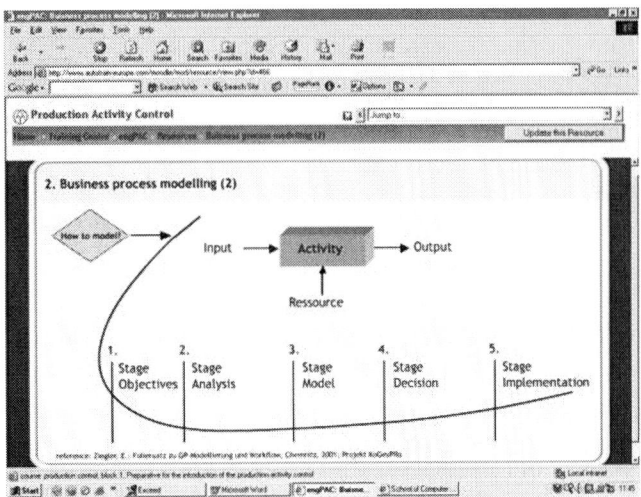

Figure 2. Example of e-book course on Production Activity Control developed by ATB.

Amongst the advantages of the *Moodle* MLE is the way in which the resources are organised, most of which are structured around a single course page. That page can also be used by teachers to add and edit additional activities and features. Learners and teachers can adopt the system very quickly, and, in terms of maintenance and development, there is an international network of developers and users supporting it.

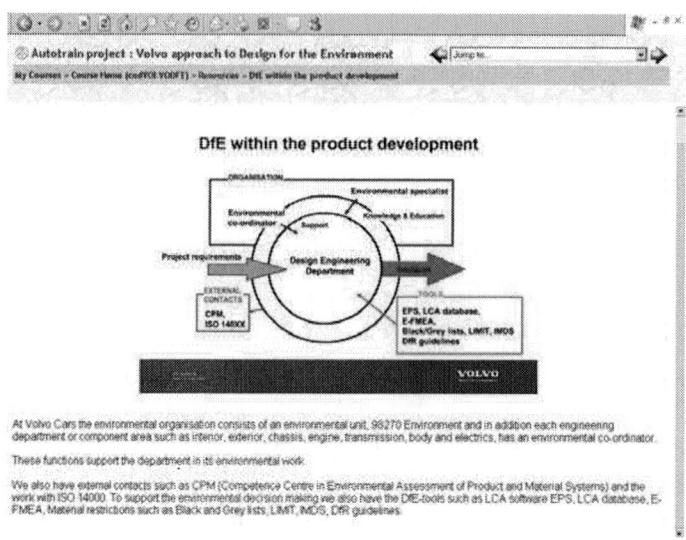

Figure 3. Example of e-book course Design for the Environment by Volvo Cars.

AutoTrain courses delivered through *Moodle* have been provided with additional functionality to encourage learning in a collaborative environment. The main page of each course is organised in sections to give the user access to a range of learning tools. For instance, under the *Activities* section, the user can find access to resources such as glossaries, quizzes or course fora just by clicking on the relevant link. The *Search* function allows running searches on the fora or the course contents, to look for a particular topic or piece of information. The *Administration* link lets the user monitor their grades in the various assessments. The link labelled *My Courses* can be used to help keep track of all the courses that the user is registered for. *Course Tools* includes links to news pages and a "*resume*" function that the learner can use to bookmark the last visited page and start future sessions at the point where he finished the last time he was using a course.

The duration of e-book courses varies depending on the nature of the training. However, a learner may not be required to complete an entire course, since the information that he may need to acquire may be included in just one module. To help customise the information that learners need, courses are divided into modules. Each module contains smaller sections and, ultimately, each of these sections is divided into pages. The purpose of this fragmentation is to help maximise the learner's time and to give them the flexibility and the capacity to choose the specific units of knowledge that are relevant for what they want to learn, helping them in this way to discriminate all that information that may be redundant or unnecessary.   To assist the learner with the navigation through the different pages and modules that make up a course, AutoTrain has incorporated a system of scroll bars, navigation arrows and drop-down boxes that are available to use to move quickly to a particular part of the course.

Most courses are interactive (Figures 4 and 5) and some modules include visual aids and demonstrations in order to illustrate the course content and enable the learners to carry out investigations and exercises to help them reinforce the knowledge that they will acquire throughout the module. A good example of this is the animated "Bead Board" which helps to explain the concept of variation and the factors that influence it, whilst learning the subject of Statistical Process Control and the various types of charts that are used (e.g. histograms and control charts). This is shown in figures 4 and 5. Progress can be monitored via the self-assessment quizzes.

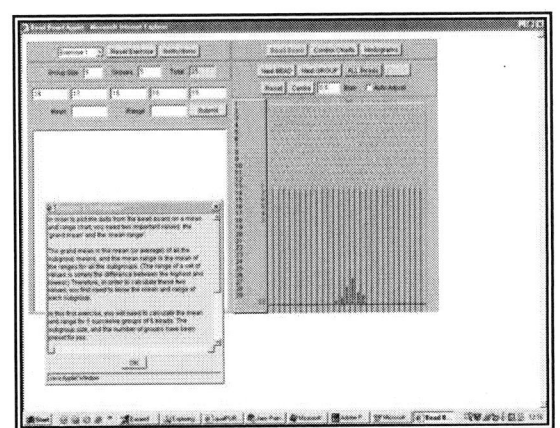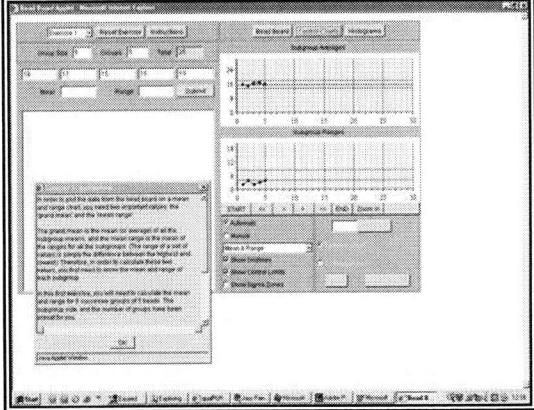

Figures 4 and 5. Examples of interactive applications in 'The Pursuit of Excellence' course.

Video Lectures

The other main type of training material developed by AutoTrain is the video lecture, also known as a video presentation or, informally "talking heads". Video lectures use cutting edge video-streaming technology generated by AutoTrain, to bring the user the expertise of leading figures in the engineering and manufacturing industries in the shape of movies that they can watch from their computers (Figures 6 and 7). From the website, video courses are easily identifiable by a dark icon or by the *RealPlayer*™ logo. The user can choose the video lecture of his interest and add it to his account following the same procedure for the standard e-book courses.

Video lectures are available in different video player formats in terms of the plug-ins that the user needs to have installed in his computer in order to be able to see them. AutoTrain has developed video lectures in two formats. The standard format uses RealPlayer and a combined format that uses other programs like QuickTime or Windows Media Player to display the presentations. Relevant information on how to download and use these tools is available from the AutoTrain website. All the courses are created in a form that will work with different Internet connections. In the worst-case, this is a 56 kB/sec modem but, with systems such as Real Player, it is possible for the learner's computer to recognise the connection speed and use the best available definition for the video format. The software is also capable of pre-loading the slides so that they are readily synchronised with the presentation and are not dependent upon download time during the training session.

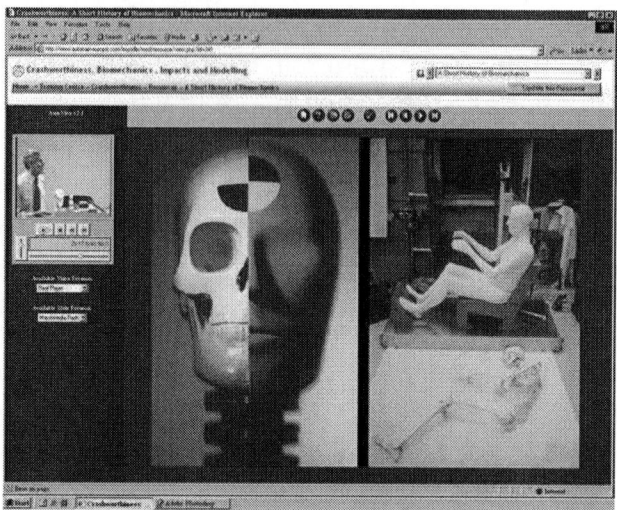

Figure 6. Detail of Video Lecture on 'Crashworthiness'.

The look-and-feel of the video presentations remains uniform regardless of the player format that may be chosen to access them. When the user opens a link to a video presentation, the screen splits into two main frames: the frame on the left side of the screen contains a window to display the movie and a number of control buttons that add functionality to rewind, pause, fast-forward or stop the movie at a particular time. The rest of the screen is dominated by a bigger frame that contains slides to reinforce the contents of the topics that are being explained throughout each of the presentations. Users are given the choice to watch the video presentation and slides in a synchronised or asynchronous way. They can also watch a particular slide or a particular fragment of the presentation. In some cases, some of the presentations include subtitles to highlight the main concepts that are being explained in order to make them easier to follow. Additional functionality includes a *help* function that contains additional information on how to access the video presentations and some troubleshooting information, a *slide menu* that gives an overview of the slides used for a particular presentation, a *print* function that can be used to print a paper copy of all the slides in one presentation, *navigation arrows* to move straight to the beginning or the end of a presentation or to the previous or next slides. There are no log-out buttons when a user is accessing a video lecture. In order to end up a session after watching a video presentation, the user simply has to close the browser, or alternatively, if he wishes to continue using the site, a *home* button is provided that he can use to go back to his course index page.

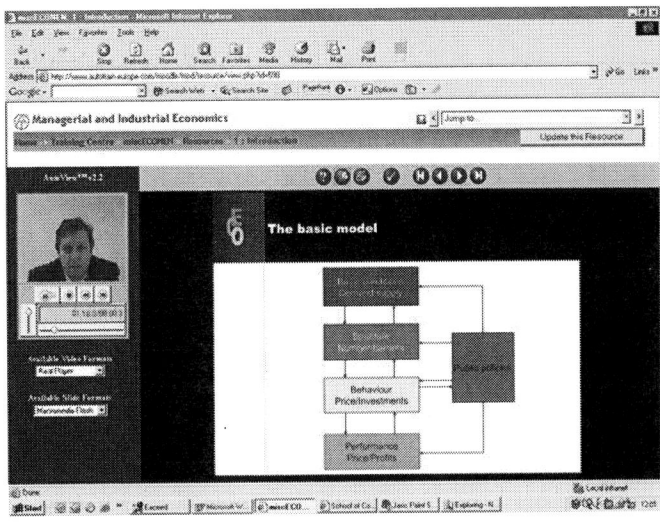

Figure 7. Detail of Video Lecture on' Managerial and Industrial Economics'.

The most recent development to the AutoTrain video lecture is the incorporation of selectable multilingual subtitles (Figure 8). Subtitles were first introduced in order to make training material more available to the aurally disadvantaged with switchable (on-off) subtitles running at the foot of the screen and the functionality to allow printing of the script. However, in the light of the international audience for Internet training, additional functionality was added in 2005 in order to incorporate subtitles in different languages so that lectures delivered by experts who were not of the same nationality as the learner, could be used. In the context of the automotive industry, English is widely spoken but since this is not the native tongue of many, to have subtitles in ones own language can aid effective acquisition of knowledge from the lecture whilst also improving language skills. The subtitles can be switched between languages (presently English, French, German and Spanish) without losing synchronisation and there is no technical reason why many more different languages cannot be available.

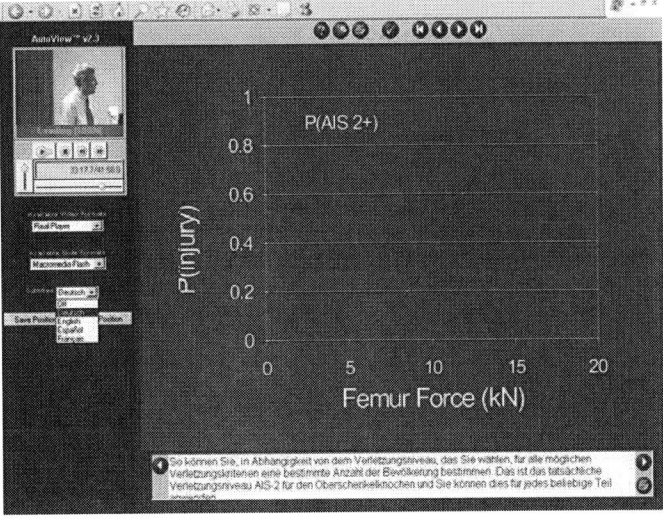

Figure 8. Example of Video Lecture with multilingual subtitles (choice in left screen pane).

Advantages and disadvantages of online learning technologies

Using distance learning technologies for the delivery of training materials has its advantages and disadvantages for both individuals and corporations [1]. The major advantages are that online learning enables the learner to develop a study environment suitable to his needs, online learning allows the user to study at his own pace, at any time and from any location where there is a computer that has access to the Internet.

Organisations can also benefit from the use of online training systems to implement in-house training policies and programmes at very low cost [2]. In its EU programmes AutoTrain delivered online training materials for SMEs completely free of charge, which is extremely beneficial for small companies that otherwise would not be able to afford to send part of their workforce away on training, both in terms of cost and time. Registration fees for conventional training courses tend to be very expensive and often unaffordable for small businesses. More importantly, in most cases as far as time is concerned, training is often beyond what a small company can afford as they may have to close down elements of their operations, even a production line, for the duration of the training course while their key staff are away.

Online training brings the training to the user's desk at low cost. Managing Directors can improve their own skills and those of their of their workforce, as well as the performance of the company, by encouraging online training at times when staff can be available. Staff interested in training may also find it useful that they can start a particular course while they are at work and continue or go back to it from their computers at home, while they are away or while they are travelling to work. Besides, by using online training systems like those developed by AutoTrain, companies can benefit from the flexibility of having customised training programs to meet their specific needs.

A further key advantage to companies and institutions, is the possibility of enabling a low-cost, dedicated learning environment to deliver a structured training programme that is individual to that institution. This can be delivered via the Internet or, more commonly, via the institution's own intranet. Such systems use the company's own training material and can feed user information and assessments directly into the company or academic evaluation area. The online systems can not mitigate the time cost of the employee or individual but it can at least reduce travel time and use the medium to arrange training time around the workplace and home life priorities rather than the other way round.

One of the main disadvantages of online learning is that users may find the discipline of self-learning is difficult. Without a "teacher", answers to their questions may be difficult to obtain. Studying online requires a great deal of discipline and self-control in order to be effective and one of the key courses from AutoTrain is a course to teach some effective techniques to study online.

Other disadvantages of online learning include the lack of human contact, the intimidation that some users may feel when they sit in front of a computer, the lack of skills in IT, and, at a more technical level, some disadvantages which are a result of the computer infrastructure used by the learner. As an example, a user that is accessing video lectures through a modem will find the process less efficient than a user using a Broadband connection, as the amount of time required to download the material will be significantly different. Further disadvantages of e-learning systems [3] are generated by the conformity or non-conformity of the learning resources with appropriate standards and the difficulties of arranging accreditation and testing of the learner against different awards or standards [4]. This latter disadvantage actually has more impact on the developers than on the end users themselves.

In order to alleviate some of these pitfalls some solutions have been developed. To minimise the isolation of the user as a result of a lack of personal contact, human support has been provided through a telephone line and through regular seminars, conferences and workshops. Also, in terms of the compatibility with the system, AutoTrain learning resources are produced in a wide range of formats to match the specifications of the computer system of the end-user and technical help is also available from the development team or the institution's IT department.

Types of courses

Online training uses the Internet as a channel for the delivery of training resources. Courses can be developed in a vast number of areas. AutoTrain-Europe and its sister project AutoTrain provide specialised training in areas that are primarily relevant to small and medium size business, principally within the automotive, engineering and manufacturing sectors. These areas include quality systems, environmental management, lean manufacturing, co-design, crashworthiness and automotive engineering amongst others.

Quality Standards

One of AutoTrain's areas of specialisation is that of quality standards. It offers training courses on the international standards ISO 9001:2000, ISO/TS 16949:2002, the technical specification for the automotive sector and AS/EN9100, the international standard for the aerospace sector. These are standards that have now become accepted around the world as the benchmark for all quality management systems. AutoTrain offers courses on Statistical Process Control, Visual Management, Overall Equipment Effectiveness, Six Sigma and variety of Japanese style production techniques such as Kaizen, Kanban, 5C or 5S and Total Productive Maintenance (Figure 9), with the aim of encouraging small business to implement quality systems that will promote consistency in production and reliability in delivery, as these are essential requirements to keep the business competitive in today's marketplace. The courses also aim to raise awareness of the importance of meeting customers' expectations all the time, every time, to keep them satisfied and loyal, since this is the key to prevent an organisation's customers from going elsewhere.

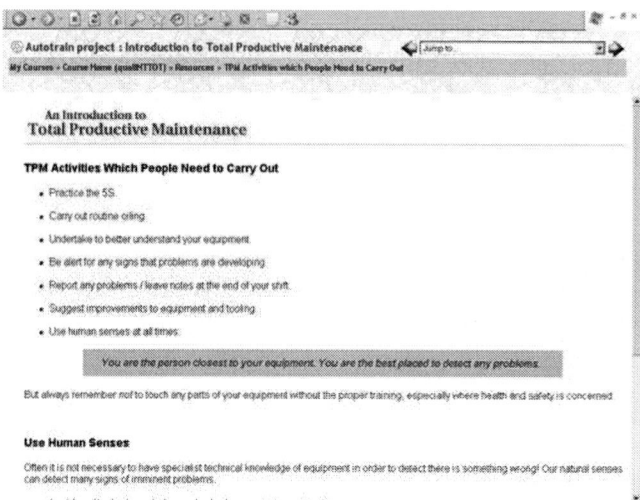

Figure 9. Example of e-book course on Total Productive Maintenance

Some of AutoTrain's training courses in quality standards are focused on the requirements that need to be met by the companies in order to conform to the standards. Other courses concentrate on the auditing process, since the ability to be audited by an independent, third party organisation is the foundation of the worldwide acceptance of these international quality standards. AutoTrain online training resources are an easy and cost effective way to encourage companies to adopt these quality standards that are becoming crucial for the survival of their businesses and to assist companies in the process of becoming compliant. Compliance with the ISO 9001 and ISO/TS 16949 standards can help companies attract more business as customers, both new and old, will have increased confidence in the organisation's ability to meet their expectations.

Environmental Training

AutoTrain also provides online training courses to assist businesses in complying with the environmental standard ISO 14001 (Figure 10). AutoTrain online training courses are essentially focused on environmental management, environmental law and techniques for waste management and pollution control in order to help businesses meet the existing legislation, reduce costs and help them gain a competitive advantage. A course in the AutoTrain CoDesign project, guided learners in designing and assembling products in environmentally favourable ways.

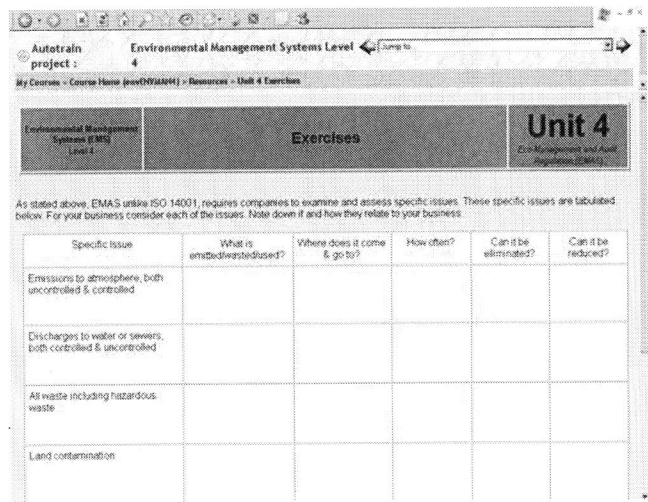

Figure 10. Example of e-book exercise from Level 4 Environmental Management Course

Six Sigma

Six Sigma is a data-driven approach and methodology to eliminate defects in any process, but essentially manufacturing processes. The fundamental objective of the Six Sigma methodology is the implementation of a measurement-based strategy that focuses on process improvement and variation reduction through the application of Six Sigma principles based on defining, measuring, analysing, improving and controlling the process. Both AutoTrain-Europe and AutoTrain have offered several courses on Six Sigma (Figure 11) that include case studies in implementation on the shop floor and bring together the expertise of the leading figures in the field. In recent years these learning materials have been reinforced by face-to-face courses to fast-track learners to recognised standards such as "Green Belt" against supply chain and independent assessor criteria.

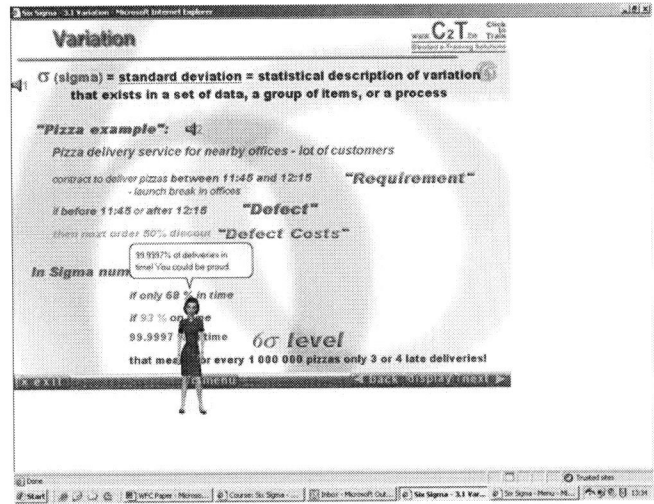

Figure 11. Example of animated e-book course on Six Sigma

Lean Manufacturing

AutoTrain-Europe and AutoTrain have been delivering courses and video presentations for a number of years on Statistical Process Control, Total Quality Management, The Toyota Production System and Six Sigma to highlight the importance of lean manufacturing in specific areas such as organisational infrastructure, analysis tools and methodologies for process improvement, project identification strategy, project management and project review. For those just embarking on the process, concise video case studies in 5C (5S) have been prepared.

CoDesign

AutoTrain also delivered online training aimed at improving the design and manufacturing processes using joint design through the Automotive supply chain with the purpose of sharing the design process between suppliers and customers (Figure 12). All too often "make-to-drawing" operations have very good ideas of how to improve the quality, cost and delivery of the product and are not able to input or are left to resolve manufacturing issues against predetermined tolerances and processes. CoDesign promoted the idea of training driven downwards from Vehicle Manufacturers and Original Equipment Manufacturers (VMs and OEMs) and First Tier Suppliers who want their suppliers to take more responsibility for design in order to improve quality and reduce cost.

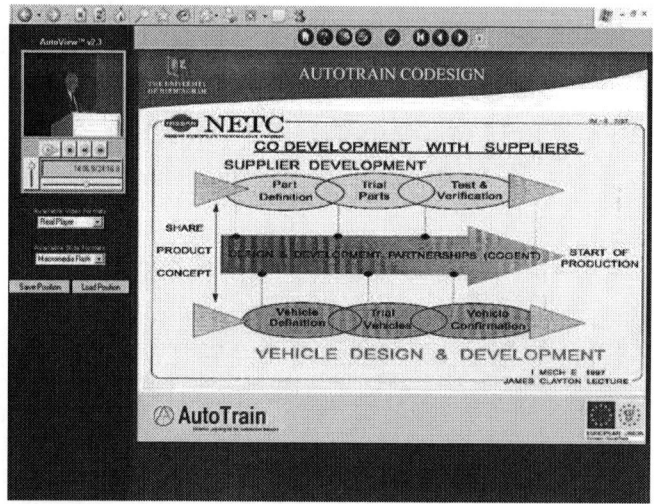

Figure 12. Example of video lecture on CoDesign

Basic IT Skills

There is a dichotomy that online training has to manage. In order to study on-line, you need a degree of IT skill or training. This is becoming less and less of an issue as more and more people from toddlers to pensioners become more proficient with computers and the Internet. AutoTrain has developed Internet training solutions to improve skills amongst the manufacturing and engineering industries. Even if the core of the online training is focused around quality and environmental standards and the implementation of techniques that help improve the manufacturing process, AutoTrain also offers a limited amount of training courses on basic computers skills for those who believe in the potential of online training solutions but cannot make the most of it due to limited knowledge.

Crashworthiness

AutoTrain-Europe has developed a video training programme course that consists of edited recording of a three-day course on crashworthiness held at Aston University (Birmingham, UK) in April 2003. The course offers more than fifteen hours of video material through 28 different video presentations featuring the slide presentations and the recoded delivery of world-class experts in this field.

Automotive Engineering

The Masters degree level course, developed in Germany by IKA Aachen, one of AutoTrain's German partners, provides detailed theoretical knowledge on the areas of power and energy demand, drive-train components and layout, vehicle dynamics and driving performance (Figure 13).

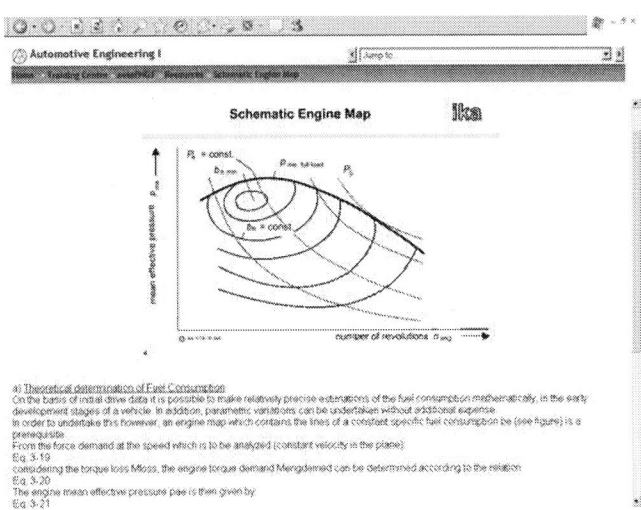

Figure 13. Example of e-book course on 'Powertrain and Vehicle Dynamics'

Training validation

Learning online can be isolating and learners may, at times, feel disoriented and unable to measure the progress that they are making. To help them to get an idea of the nature of their improvements, AutoTrain has developed courses that are interactive and provide the learner with as much feedback as possible. The learner can monitor and validate the progress made through interactive exercises and online quizzes. Other more formal accreditation is also possible by a number of different mechanisms, some of which are being developed further, but often the most effective measure of the success of the learning is demonstrated by production quality, volume, delivery or, most emphatically, by improvements to the company bottom-line earnings.

Some courses are provided with the support of a University or College that can provide workbooks and coursework assessments, which will lead to an accredited qualification. Implementation in the workplace and demonstration of real improvements can lead to the award of National Vocational Qualifications at a number of levels.

An interesting "hybrid" approach from one of AutoTrain's Spanish partners involves remote study of the course (in Mechanics) followed by conventional examination assessment at different places in different countries at the same time.

An issue of online assessment and verification that needs further development, is providing certainty that the registered learner is the same person providing the assessment answers (or that there is not a small committee sitting at the computer during the assessment!). The increased use of biometrics in passport and identification documents, allows scope for the use of these technologies in in-line assessment. This is an area of further development for the AutoTrain team and everyone else involved in the field.

It must be said that one of the main measures of effective training is the "bottom line" performance of a company. Many of the companies that have engaged in AutoTrain programmes in recent years have been less interested in certificates and accreditation than in improved technical and business performance. In fact there are examples of part-modules being studied to deliver a step change in the business.

Interactive exercises

AutoTrain-Europe training resources present various levels of interaction depending on the nature of the course. As an example, each module on a course on Statistical Process Control contains a variety of exercises to reinforce the knowledge that is transmitted through each of the sections in the course (Figure 14). The nature of the exercises varies depending on the topic. In some cases an exercise may require from the learner just the input of a single word to complete a sentence or a definition. In other cases, learners are required to perform a calculation. Some exercises are more sophisticated and may involve printing and plotting a certain type of chart or printing a table a monitor the performance of a certain process within a company. For most of these exercises the system developed by AutoTrain provides immediate feedback and opportunities to go back and revise the knowledge areas where the learner may feel there is a weakness.

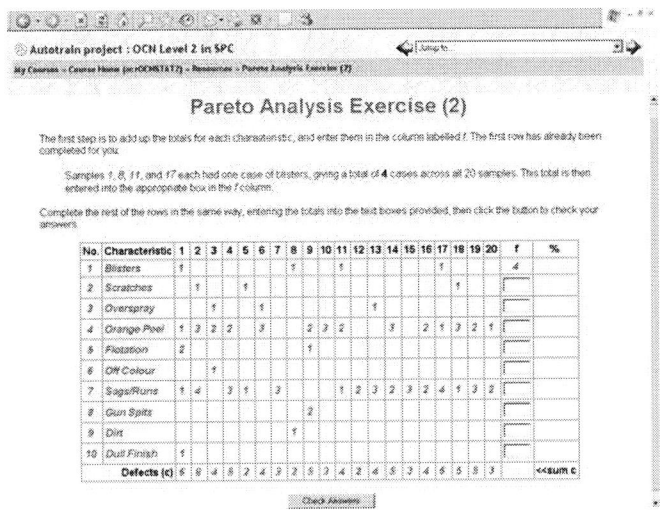

Figure 14. Example of interactive e-book course on SPC

Online Quizzes

Another way to help learners monitor their progress is through the use of online quizzes. Most AutoTrain courses include quizzes at the end of each module that are designed to measure the improvement that the learner has made for each of the sections in the course. The nature of the questions in a quiz varies from multiple choice formats to formats where the learner is required to feed back into the system a short answer or two find matches between related pairs of concepts. The aim of the quiz is to reassure the learner's knowledge. Marks are assigned for every quiz and feedback is provided for every answer. Also the system enables the learner to keep records with the results and time scales from all the courses that he has attempted for a particular module to give him a clearer idea of his performance and achievements (Figure 15).

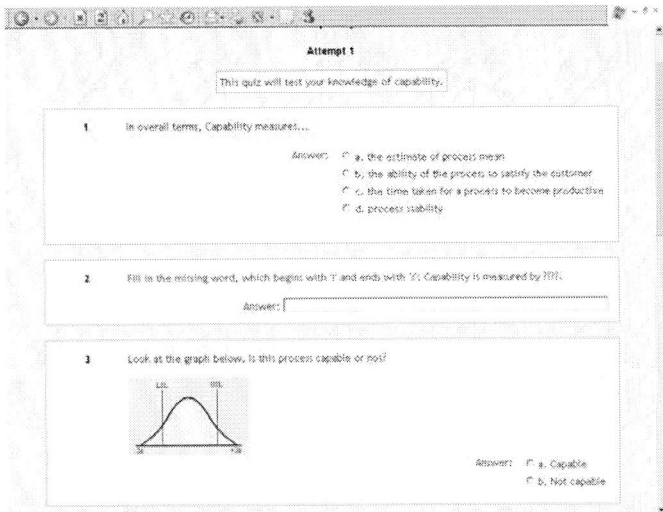

Figure 15. Example of self-assessment quiz from e-book course " Introduction to SPC"

Assessment

AutoTrain has made evaluation forms available via the Internet to assess the status of companies for introduction of quality standards (Figure 16). The completed self-assessment can be interpreted (using standards established with experts) in order that the company gets a judgement of its current status together with a summary of appropriate strengths and weaknesses that can be fed into a management or action plan. Similar assessments against other standard or criteria could be developed.

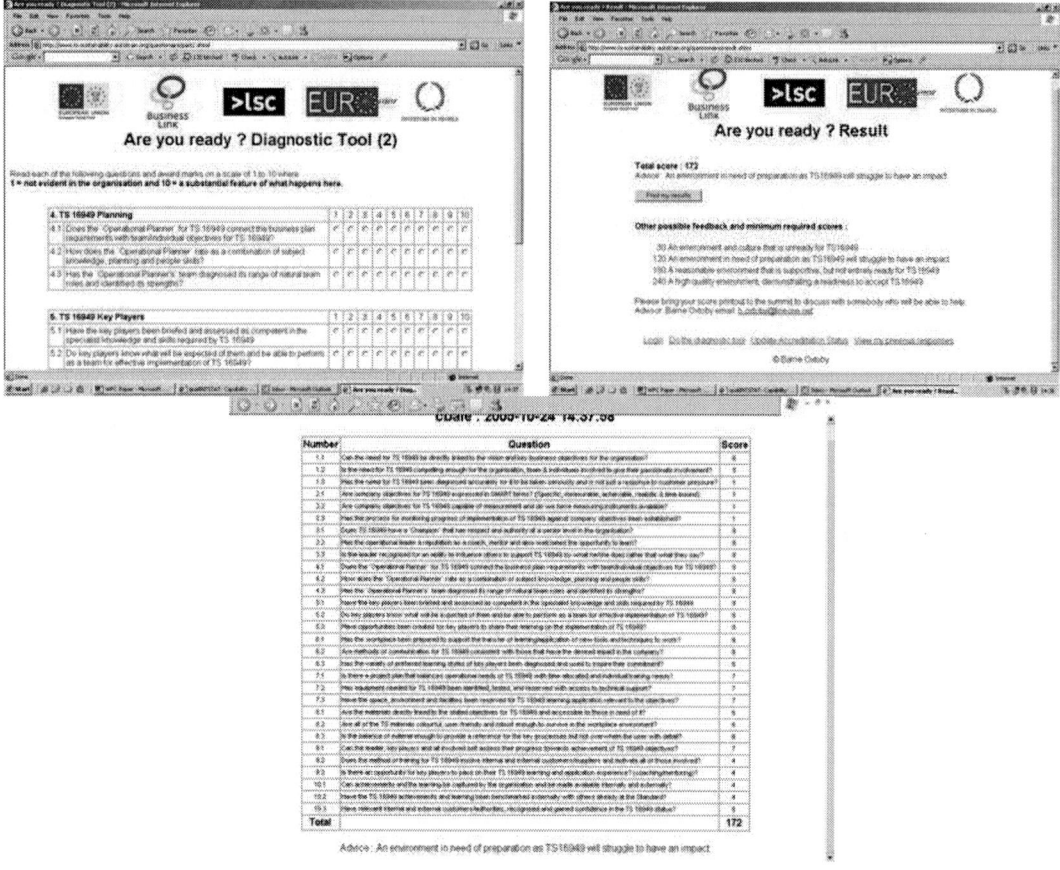

Figure 16. Example of online self-assessment and outputs

Future development

In terms of future developments, the customisation to individual differences in learning and to company-specific training needs will be as critical for AutoTrain as the search for more innovative ways of incorporating interactivity and multimedia capabilities.

The team will continue researching and developing technologies that will provide the user with more responsive systems. One way to achieve this will be by improving the current video and audio technologies that AutoTrain uses for the delivery of training materials. AutoTrain will also concentrate on the development of improved technologies to generate simulations and interactions, especially for self-testing feedback. The use of video conferencing and mobile technologies such as wireless or PDA, leaves an open door for AutoTrain to work on new channels to distribute its courses and to configure learning environments for clients. Adaptability and the search for tools that will identify user requirements and relate them to users' needs is another field where AutoTrain will increase its research activity.

An obvious benefit of an MLE that holds training material and training needs information for a company or group of individuals, is directing the appropriate courses to individuals and groups. Development of the AutoTrain on-line method of assessment of knowledge and training needs can link the outcome evaluation to suitable course material

Conclusions

- AutoTrain has been established for over 13 years as a distance learning resource for the automotive and manufacturing industries. Despite being targeted at specific audiences in Europe and the regional initiative of the UK, the online learning material has been adopted by thousands of users from over 40 countries of the world.

- The implementation of *Moodle*, an open source code for managed learning environments, has facilitated the creation of low-cost, bespoke and generic learning platforms for the public and for specific organisations to enable their in-house training requirements to be met.

- The current resource of more than 100 training courses is being increased through the incorporation of new training material made available to the developers. Technologies are being adopted to make these courses economically available for both the learning community and for AutoTrain to meet the cost on support and ongoing development.

- The leading-edge AutoTrain video presentation technology continues to be developed and to take advantage of the increasing availability of powerful PCs and high-speed data links. The software package and its editing package are being shaped for commercial sale so that others can combine video footage and presentations, such as Microsoft PowerPoint, for their own training purposes. Class-leading presentation and subtitling capability can be increased to incorporate more of the worlds languages and to capture expert lectures from many more nations without loss of clarity or teaching effectiveness.

- The proven expertise of the AutoTrain team is available to the global industrial and academic community to provide MLE implementation and online learning, an important facet of knowledge management and knowledge transfer.

References

1. British Computer Society (2004). *Testing time for eLearning*. The Computer Bulletin. Vol. 46: pp 18-19.

2. Cross, J. (2004). *The future of eLearning*. On The Horizon - The Strategic Planning Resource for Education Professionals. Vol. 12: pp 151-157.

3. Cross, J. (2004). *An informal history of eLearning*. On The Horizon - The Strategic Planning Resource for Education Professionals. Vol. 12: pp 103-110.

4. Devande, O. (2004). *ICTs and the Development of eLearning in Europe: the role of the public and private sectors*. European Journal of Education. Vol. 39: pp 191-208.

Bibliography

Ettinger, A. (2003). *eLearning meets knowledge management under old oak beams: the Ashridge Virtual Learning Resource Centre*. Business Information Review. Vol. 20: pp 51-56.

Fox, S. and MacKeogh. K. (2003). *Can eLearning Promote Higher-order Learning Without Tutor Overload?* Open Learning. Vol. 18: pp 121-134

Fung, C-W. and Leung. E. W-C. et al. (2003). *Efficient Query Execution Techniques in a 4DIS Video Database System for learning*. Multimedia Tools and Applications. Vol. 20: pp 25-49.

Graff, F. (2002). *Providing security for eLearning*. Computers and Graphics. Vol. 26: pp 355-365.

www.autotrain.org (Courses Developed as part of AutoTrain)
- Statistical Process Control
- ISO 9001:2000
- ISO/TS 16494:2002
- Six Sigma
- 5S or 5C
- Lean Management
- Value Analysis and Value Engineering
- Problem Solving Tools and Techniques
- Environmental Management ISO 14001
- Electromagnetic compatibility
- CoDesign (including CAD compatibility, IPR, Project Management, etc.)
- Etc.

www.autotrain-europe.com (Courses Developed as part of AutoTrain-Europe)

- Aerodynamics *(FKFS Stuttgart, Germany)*
- Production Control *(ATB GmbH, Germany)*
- Managerial Economics *(ACES, Grenoble, France)*
- Analysis of Structures *(CIMNE, Spain)*
- Crashworthiness, Biomechanics, Impacts and Modelling *(EuroMotor, UK)*
- Automotive Engineering *(RWTH IKA Aachen)*
- Product Design *(CIDAUT, Spain)*
- Kinematics of Rigid Bodies *(INSA, Lyon, France)*
- Potential Failure Modes and Effects Analysis *(IAA, Belgium)*

Acknowledgements

The authors gratefully acknowledge the financial support of the European Social Fund (ESF) and the Leonardo da Vinci Programme.

OVOTRAIN On-line Virtual Vocational Training System

K Bako, K Lengyel

TP Technoplus Ltd, Hungary

Abstract

OVOTRAIN On-line Virtual Vocational Training System is based on the successfully implemented Leonardo project entitled "Metallurgical Expressions Translation System" (Metaltransys). OVOTRAIN focuses on content and language integrated learning (CLIL) by creating a 7-language-internet based expression dictionary system in metallurgy and mechanical engineering completed with virtual vocational training opportunities. Metaltransys project has developed an illustrated English-German-Hungarian-Swedish language expression dictionary in metallurgy with 11000 terms and their explanations. OVOTRAIN translates dictionary to Czech, Italian and Polish, and develops virtual reality based training systems. Products in virtual reality provide producers and operators of equipment and machines, students and apprentices the opportunity for effective training and short-term adjustment to operating, control and process operations. It can be used on-line on website www.ovotrain.com.

Key words metallurgy, seven language dictionary, virtual vocational training

Introduction

OVOTRAIN (On-line Virtual Vocational Training System) is based on the successfully implemented Leonardo project entitled "Metallurgical Expressions Translation System" (Metaltransys), no. HU/00/B/F/LA-136107. During the test and dissemination of the Metaltransys project, a real demand has been explored to expand the illustrated dictionary toward up-to-date vocational training methods. Therefore the OVOTRAIN, Leonardo project-no. HU/05/B/F/PP-170101, focuses on content and language integrated learning (CLIL) by creating a 7-language-internet based expression dictionary system in metallurgy and mechanical engineering completed with virtual vocational training opportunities. Similar on-line language training tool isn't existing in the market. Metaltransys project has developed an illustrated English-German-Hungarian-Swedish language expression dictionary in metallurgy with 11000 terms and their explanations. It can be used on-line free on website www.metallingua.com with search function from February 2004 without password.

The OVOTRAIN project (www.ovotrain.com) started on 1 October 2005 and will end on 30 September 2007, is based on recognised needs. OVOTRAIN will integrate possibilities like virtual constructing of equipment, machines, production shops with access of the 7 language expression dictionary in metallurgy and mechanical engineering. OVOTRAIN will translate the existing dictionary to Czech, Italian and Polish, and develop virtual reality based training systems. Products in virtual reality provide producers and operators of equipment and machines, students and apprentices the opportunity for effective training and short-term adjustment to operating, control and process operations.

OVOTRAIN: A summary

The on-line dictionary with virtual vocational training tools gives opportunity for learners to put their language skills to immediate use. CLIL will be used by students in the vocational training and workers in lifelong learning. OVOTRAIN develops new ICT tools, which give opportunity to users to apply the system in their professional as well as language learning. Target group: workers at their work in the sector, apprentices, secondary school and university students, translators etc. OVOTRAIN users will exercise assembling through digital mock-ups equipment, running these equipment, production shops supported them by entries from the dictionary, respectively creating new terms with explanations and illustrations in order to enlarge the dictionary escorted by increasing it's value. OVOTRAIN enables them to construct machines and equipment in the framework of virtual reality. Additionally, OVOTRAIN develops an internet based translation-teaching e-learning system to improve the writing skill of users and enhance the application in language learning - teaching.

In OVOTRAIN project the Hungarian University of Miskolc secures professionals in metallurgy and mechanical engineering and language didactics, and will evaluate the translation-teaching e-learning system and secures quality control.. The Fraunhofer Institute for Factory Operation and Automation (Fraunhofer IFF), Germany, will develop the learning software of constructing virtual machines, equipment, production shops with interactive virtual reality-based training, which is far superior to present training systems. The Institut will involve the Hochschule Harz, which has the task to accomplish with it's tutors and apprentices the practical realisation of the continuously developed virtual training methods with feedback to the software. The Italian partner Vemek s.r.l. will translate the dictionary to Italian which gives opportunity to students and workers to apply the internet tool in their professional as well as language learning, creating new, for the time being missing terms in the course of preparations of mission in life. The Czech Foundrymen Society has interest to train their members in the local chapters up to date at European level. The AGH University, Poland, will elaborate the Polish version to be used in vocational training in secondary schools, universities and factories. The TP Technoplus, Hungary, introduces OVOTRAIN to the professional language teaching and virtual equipment, machine and production shop constructing in the in-service, off-school further vocational training for workers in factories, to make the vocational activity with virtual reality possible.

Development of virtual reality based training system

In the OVOTRAIN project new possibilities in supporting the learning process will be developed. The user of the dictionary can use interactive 3D-scenarios to get information about machines and processes, their functionalities and their behaviour. By the help of virtual reality methods the user can interact with the object of learning and is supported in learning in a new way.

Training scenarios as innovations contain the virtual representation of an equipment. Based on CAD-data the virtual models allow not only to ilustrate the geometrical structure but also the functionality and behaviour of a machine. The user can interact with the models like with the real machine.

A virtual reality (VR) platform for the development and the usage of virtual training scenarios created at the Fraunhofer IFF is the basis for the learning modules. In the training environment the scenarios are used for learning purposes.

Learning contents can be rapidly transmitted on user's demand via modern communication systems such as the internet. Here, totally new possibilities are opening with the linkage of common hypertext documents with virtual training scenarios when the contents need to be transmitted via the internet. Besides using the scenarios via the net, they can also be worked on at local PC's.

The on-line virtual vocational training method will help modeling technological processes parallel to teaching the students in foreign languages. Tutors will be trained how to integrate it to lessons. OVOTRAIN will enhance the existing database of terms.

OVOTRAIN will integrate possibilities for virtual reality based training systems. With 3D interactive models of a melting furnace, a portal milling machine, power generator and other examples the user can perform the process operations of the equipment interactively and learn the parts of these in 7 languages and so combine technical and language learning. The models are created in an internet-based form, for example in VRML or X3D and are integrated into the dictionary.

The process of model creation using the VR Platform of Fraunhofer IFF is performed in two steps:

1. Create interactive 3D scenarios of technical processes, machines and systems and their functionality (Figure 1)
2. Transfer important functions in a web-based format (Figure 2)

Conclusions

OVOTRAIN integrates widely and less widely used languages on the base of their interest in metallurgy and mechanical engineering. Parallel to this OVOTRAIN secures innovative teaching methods tailored to specific needs of vocational training and for learners in their lifelong learning efforts. OVOTRAIN enables that personnel with different levels of qualification can be effectively trained in their operation.

The Czech-English-German-Hungarian-Italian-Polish-Swedish language illustrated explanation dictionary with most modern, on virtual reality based vocational training system will be an unique language learning tool in metallurgy and mechanical engineering. Since there is no similar internet-based language-learning possibility, dissemination on this way can create a special network for professional language learning. Target group with medium or advanced level language knowledge can use the system on-line individually in their life long learning or in the frame of vocational training in vocational training schools and at universities. OVOTRAIN gives modern tool for test the knowledge by individuals or institutional level. Metallurgical and mechanical engineering professionals can use the system in both professional and language learning & teaching. New dimension opens by OVOTRAIN because it elaborates a new ICT tool to create equipment and supply with different language terms. The developed internet based translation system promotes the writing skill improvement of trainees.

Figures

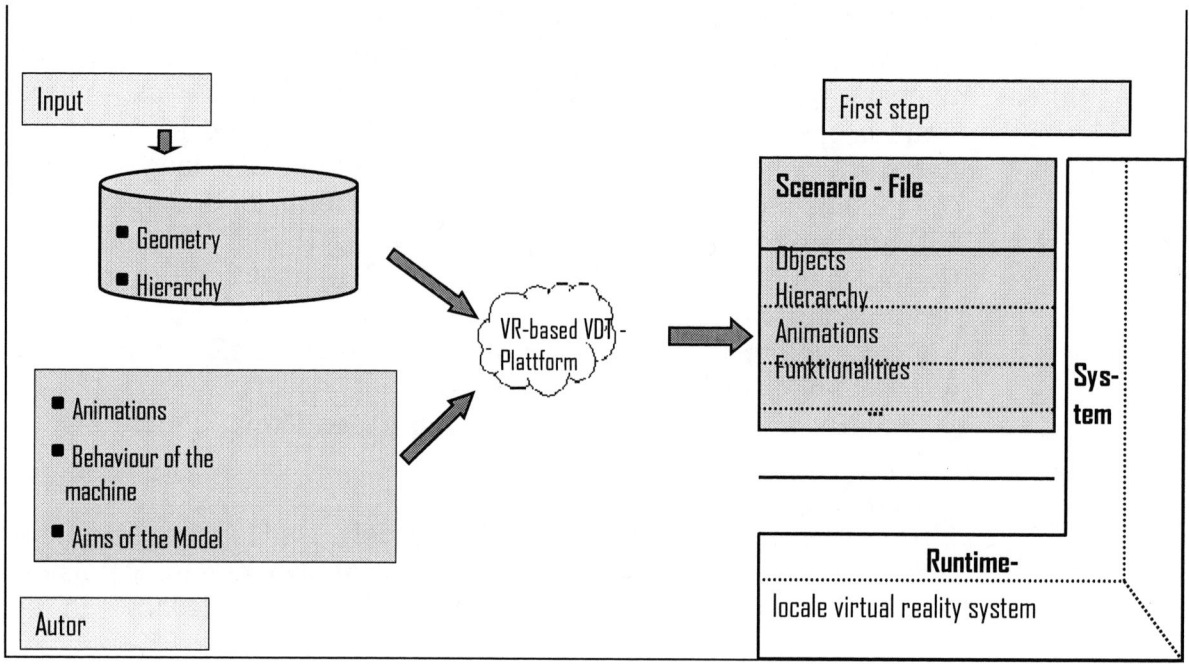

Figure 1 Work of the author of a scenario

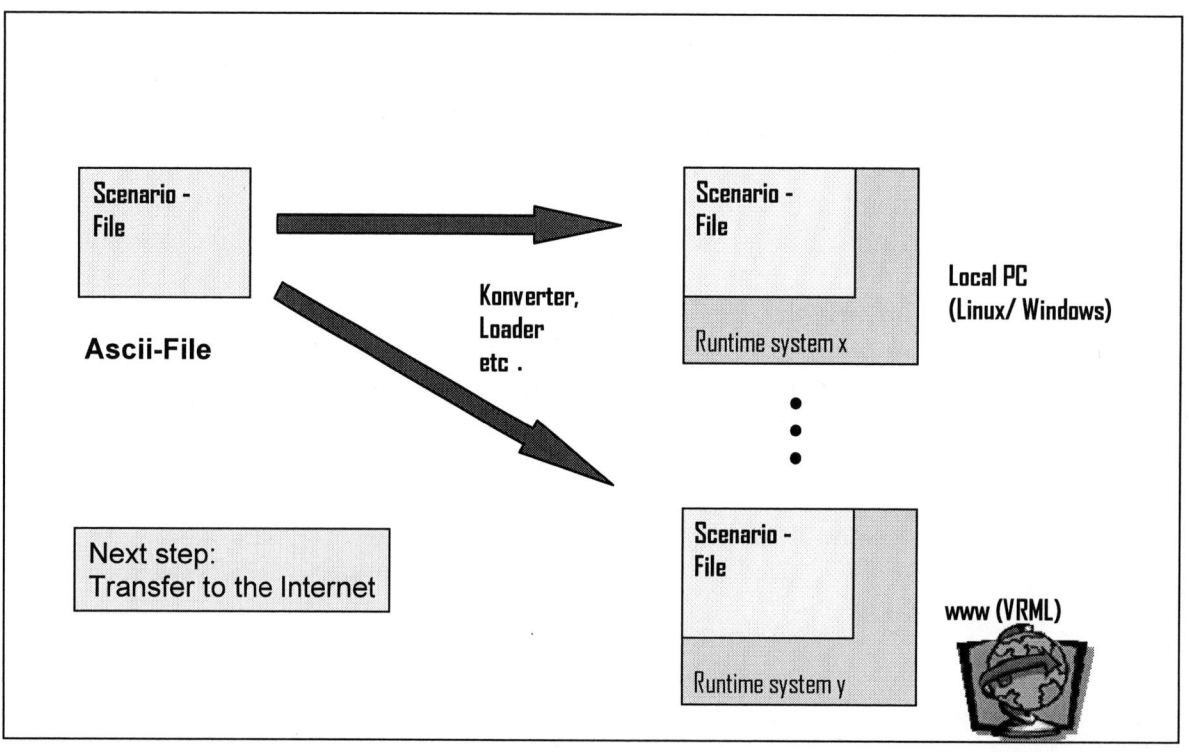

Figure 2 Presentation in the internet

Modelling Microstructural Evolution in Cast Alloys

Prof. Rachel C. Thomson

Institute of Polymer Technology and Materials Engineering,
Loughborough University, UK

Abstract

The most appropriate choice of metal alloy is vital for the successful implementation of engineering components. In many cases, the properties will change as a result both of the initial processing and the subsequent service life of a component, and therefore it is essential to understand how and why such changes occur. In recent years there have been significant developments in the ability to predict both the initial structure of an alloy and its subsequent evolution. This paper highlights the importance of these modelling approaches by considering a number of different engineering components. For example, in the automotive industry, the models have been used to predict and improve the properties of aluminium alloys for car pistons and cast irons for camshafts. The advantage of modelling microstructural evolution is to potentially reduce alloy development lead times by predicting the microstructure and properties of a particular alloy, thereby minimising the number of necessary experimental trials. The ultimate aim is, however, to predict not only the initial microstructure, but also the mechanical properties, and any changes during the service life of a component. This is a significant challenge, particularly for properties such as fatigue behaviour, which are a complex function of a number of variables. In addition to recent advances in modelling capability, there have also been developments in advanced characterisation techniques. These include, for example, electron back scatter diffraction and depth sensing indentation, which allow a more rapid determination of microstructural features, and provide a route to linking microstructure and mechanical property predictions to provide a total product design concept.

Key words

Modelling, Microstructural Evolution, Austempered Ductile Iron, Al-Si

Introduction

The ultimate aim of computational materials science is to model the complete through-process behaviour of materials from primary processing all the way to final component properties and their subsequent service behaviour. This is particularly true for metallic solidification processing in which there are many different phenomena which need to be considered: fluid flow and heat transfer, both liquid/solid and solid state phase transformations, defect evolution, grain structure evolution and recrystallisation, recalesence from latent heat evolution, and the presence of residual stresses. Such models are needed because the properties of industrial components are essentially determined by their microstructure, which results primarily from interactions during and subsequent to solidification at the micro-, meso- and macro-scale. This is a significant challenge, as discussed by Voller [1], and requires the development of new modelling methodologies which can capture the behaviour across micro to macro length and timescales. Nevertheless, there are a number of extremely useful and successful modelling approaches which are used in the production of cast alloys. These have been comprehensively reviewed by Rappaz [2] and Stefanescu [3], and tracked in the proeedings of the MCWASP conference series [4].

There are now commercial software tools which can cope with most of the continuum phenomena involved in solidification processing (e.g. MAGMA, PROCAST, FLOW3D). They take into account heat transfer, fluid flow and can simulate the filling of the mould cavity by molten metal and its subsequent solidification, usually based on finite difference or finite element methodologies. Microstructural features form and evolve during each of the processing steps as a function of the macroscopic heat, mass and momentum transfer. The prediction of grain size has been modelled by deterministic, cellular automata [e.g. 5] and phase field methodologies. Cellular automata methods include effects such as impingement, and can predict not only the average grain size, but also distributions of grain sizes and morphologies. Phase field methods allow the incorporation of thermodynamic data and parameters such as interfacial energies, and are potentially very powerful, although are computationally expensive.

Fine scale features within the microstructure, such as the formation of multiple phases within a particular alloy system have received less attention, but are extremely important in the determination of the properties of the alloy. Thermodynamic calculations can be performed by commercial packages, for example, Thermocalc and MTDATA, which will predict the phases likely to be present in a given alloy. These packages rely upon critically assessed thermodynamic data and use an exceptionally reliable Gibbs free energy minimisation algorithm to predict the phases present at thermodynamic equilibrium. In recent years the quality of thermodynamic databases has improved greatly and there are now a number of databases available which contain data for the large number of elements and phases which are present in cast alloys, including Al, Ni and Fe based alloys of importance in the foundry industry. Predictions generally relate to thermodynamic equilibrium as a function of composition, temperature and pressure, however, it is possible to apply calculations to non-equilibrium phenomena. These include the prediction of precipitation sequences in alloys by making a prediction, then preventing the phase predicted to be stable from being present, and running the calculation again to establish the next most stable phase, or by using a Scheil approach to model segregation during solidification of an alloy.

This paper will demonstrate the application of thermodynamic and kinetic models to the prediction of microstructural evolution in cast alloys using two specific examples. In the first, austempered ductile iron, thermodynamic calculations are used to predict the segregation behaviour in a casting, and combined with a kinetic model for the subsequent solid state phase transformation of the matrix. In the second, multicomponent Al-Si casting alloys, a phase field methodology coupled to thermodynamic calculations is described which can allow for the prediction of the morphology of phases in addition to their formation temperatures and overall mass fraction. There is also brief discussion of advanced characterisation techniques for microstructural assessment and strategies which can be used to determine structure/property relationships.

Microstructural Evolution in Austempered Ductile Cast Iron

Austempered ductile cast iron (ADI) results from the heat treatment of ductile cast iron. After casting, a component undergoes a two step heat treatment which involves an austenitising step, typically in the temperature range 850-1050°C, followed by an austempering step in the temperature range 200-400°C, and is subsequently cooled to room temperature. This results in a microstructure comprising spheroidal graphite, in a matrix of bainitic ferrite together with some retained austenite (often referred to as ausferrite in cast iron). The solid state phase transformations are greatly affected by the composition of the alloy, and in particular the carbon concentration of the austenite prior to the bainite transformation, which in turn is a result of the austenitising temperature and time. ADI originates from a cast microstructure, and therefore there is also chemical segregation present in the alloy as a result of the initial solidification process. In order to model the microstructural evolution, and therefore mechanical properties, it is therefore necessary to consider each step within the process. First, the as-solidified microstructure must be predicted, with respect to both graphite distribution and chemical segregation. The equilibrium between austenite and graphite at the austenitising temperature can then be considered to determine the carbon concentration in the matrix prior to transformation to bainite, and subsequently the kinetics of the bainite transformation itself must be predicted during the austempering process. A final step is to predict whether any martensite will also form within the microstructure on final cooling to room temperature.

Prediction of the As-Cast Microstructure

The prediction of the initial graphite distribution during the solidification of cast iron is difficult, and is a complex function of a number of factors. Nevertheless, it is possible to use thermodynamic calculations [6] to predict the likely chemical segregation using a Scheil methodology in which the solute redistribution is modelled [7]. The Scheil approach assumes no diffusion in the solid and unlimited diffusion in the liquid, and has been shown to agree reasonably with experimental results if the very last solid is ignored [8]. Nastac and Stefanescu developed a model that accounts for diffusion in both the liquid and solid states [9]. Liu and Elliot have also produced a numerical microsegregation model taking into account diffusion in both the solid and liquid, interface movement, non linear growth rates and total solute conservation [10]. Both methods have been shown to be in good agreement with measured segregation profiles.

The starting point for calculation of the likely segregation during casting is the composition of the alloy, which is input into a series of thermodynamic equilibrium calculations performed over a suitable temperature range in order to find the liquidus temperature. Once the liquidus temperature has been found, the system temperature can be set to a small amount below it (e.g. $0.5^{\circ}C$) and a further equilibrium calculation performed. The mass and chemical composition of any solids (austenite and graphite) that are predicted to form in this step are recorded and removed from the system so only liquid remains. The temperature is then reduced by a further $0.5^{\circ}C$, and the process repeated until less than 0.005% of the original liquid mass remains in the system, when solidification is assumed to be complete. It is necessary to 'relocate' the graphite because an artefact of applying Scheil solidification to this system is that graphite forms at each step, the amount of which depends on the solubility of carbon in austenite. The sum of the amount of graphite in the steps is equivalent to the whole of the graphite nodule(s), but must be relocated in accordance with the observed microstructure, therefore the graphite calculated to form in each of the steps was assumed to form in one location only, equivalent to one nodule, and was adjacent to the first solid to form. This process is illustrated schematically in Figure 1.

The first and last steps are assumed to be next to the nodule and halfway between two nodules respectively (i.e. they represent the first and last liquid to solidify). Since the chemical composition of the austenite in each step is recorded, a chemical composition profile can be produced as a function of distance from the nodule. The chemical composition profile can then be divided into an arbitrary number of regions of equal mass starting with material close to the nodule and working progressively outwards. Hence regions of differing composition are obtained which simulate the composition profile found between a nodule and a cell boundary.

It is possible to compare the predictions of the Scheil approach with experimental measurements of the chemical segregation within a casting, typically carried out using energy dispersive X-ray analysis in a scanning electron microscope (SEM). An illustration is given in Figure 2, in which the predictions of the model are compared with experimental measurements for a particular ductile cast iron. Over 1000 composition measurements are included in the graph, which have undergone a statistical analysis following [11], and represent the overall segregation pattern. It can be seen that there is good agreement between the two, with Cu and Si being predicted to segregate to the graphite nodule, and Mn being found in higher concentrations in the last liquid to solidify, half way between graphite nodules. It is acknowledged that the nodule count will have an influence on the segregation behaviour during solidification [12]; if there is a relatively low nodule count, as in the example shown in Figure 2, then this methodology is appropriate, however, for alloys with a higher nodule count, their segregation profile might be expected to be less severe which may result in a reduced accuracy to the model. Additional modifications are necessary to take into account fully the behaviour of the nodules within the ductile iron casting, nevertheless this approach allows subsequent modelling of the austenitising and austempering process across an inhomogeneous material.

Austenitisation

The composition profile of the substitutional elements is assumed not to change during the subsequent austenitisation and austempering heat treatments. In order to model the austenitisation step, each of the regions can then be assigned the alloy carbon content, assuming carbon diffusion is extremely rapid, and an equilibrium calculation performed for each region to determine the carbon content of the austenite and the mass of graphite stable at the austenitising temperature. This simple calculation can also be very important for determining which elements within the specification are most important to control in respect of the subsequent solid state transformations. Figure 3 illustrates a sensitivity study in which the concentration of each of the elements was varied in turn in respect to the amount in the base alloy, and clearly shows that it is primarily the Si, and to a lesser extent the Cu and Mn, concentrations which affect the carbon concentration in austenite the most.

Austempering

The austempering heat treatment is also very important in determining the exact microstructure produced and can itself be considered to occur in a series of stages. In the first stage of the austempering process, the metastable austenite will transform into a mixture of bainitic ferrite and high carbon austenite. The exact temperature employed in the austempering process will affect the structure of the bainitic ferrite – at the higher austempering temperatures within the range, the bainite will be carbide free (c.f. upper bainite in steels) whereas at the lower temperatures the bainite transformation may be accompanied by carbide precipitation (c.f. lower bainite in steels), resulting in a mixture of bainitic ferrite, carbide and high carbon austenite. The high carbon austenite will eventually decompose into a mixture of thermodynamically more stable ferrite and carbide on prolonged heat treatment; this is termed the Stage II reaction. Hence, there is a well-defined processing window during which time a relatively stable structure of bainitic ferrite and high carbon austenite, often termed 'ausferrite', exists between the Stage I and Stage II reactions.

The most significant microstructural changes occur during Stage I of the austempering process, and therefore this part of the reaction has been the primary focus of models to date. It has also been assumed that the high Si content present in ADI will largely prevent the formation of carbide during the Stage I reaction at the lower austempering temperatures, and therefore any carbide formation accompanying the bainite reaction has not been taken into consideration. Predictions are therefore realistic for commercial alloys which are heat treated within a processing window before significant onset of the Stage II reaction.

Supersaturated austenite transforms to supersaturated bainitic ferrite via a displacive mechanism, and following the transformation, carbon diffuses from the bainitic ferrite to the remaining austenite. This leads to an increase in the austenite carbon content and hence to a reduction in the Gibbs energy difference between the two phases, the driving force for the reaction. The diffusionless transformation ceases when the driving force reaches zero, leading to the 'incomplete reaction phenomenon'. The maximum carbon content at which the transformation can occur increases with decreasing austempering temperature. Calculation of this carbon content, x_{T*}, can be carried out by calculating the free energies of austenite and ferrite at the appropriate temperature using thermodynamic data. The maximum volume fraction of bainitic ferrite, V_b, can then be calculated for the alloy composition of interest using the lever rule:

$$V_b = \frac{x_{T*} - x_\gamma}{x_{T*} - x_\alpha} \qquad (1)$$

The matrix carbon content, x_γ, is taken as the value determined from the calculation of the austenite/graphite equilibria at the austenitising temperature, and the carbon content of the (saturated) ferrite, x_α, is calculated at the austempering temperature using a polynomial expression derived from empirical data [13]. In addition to the calculation of the overall amount of bainite formed, it is also possible to calculate its rate of formation by modifying a model developed for low alloy steels [14, 15] and adapted for ADI [16]. A prediction of the transformation kinetics is useful for the determination of production heat treatment times, allowing estimation of the processing window for particular time / temperature / composition combinations. Figure 4 gives an example prediction for the major phases in a particular ADI alloy heat treated under different austenitising and austempering conditions, and shows that the austempering temperature has a greater effect on the proportions of phases within the microstructure than the austenitising temperature, as expected. This combined thermodynamic and kinetic modelling approach, albeit with some simplifying assumptions, allows prediction of microstructure as a result of both solidification and heat treatment in these complex alloys and can be used to reduce alloy development time.

Prediction of Structure/Property Relationships
The ultimate goal of microstructure modelling methodologies is to be able to relate the predicted microstructure to mechanical properties, hence allowing alloy design by computer to produce components with the desired performance.

Properties of particular interest for ADI components include tensile strength, hardness, fatigue [17], ductility, toughness and wear resistance. The prediction of complex mechanical properties, such as fatigue behaviour and toughness is difficult from first principles, and may be approached using methodologies such as neural networks [e.g. 18]. However, 'simpler' properties such as yield strength, can be approached using a law of mixtures approach similar to that for composite materials, i.e. that each phase contributes proportionally to the overall strength weighted by mass fraction. This approach relies on models being available for the strength of each of the individual components within the microstructure [16]. Figure 5 demonstrates the results of such a model, and shows a comparison of the predicted austenite volume fractions and experimental values reported in the literature for a variety of ductile iron compositions, austenitising temperatures and times. There is some degree of overestimation of the yield strength for the lower end of the strength range, probably due to small errors in the prediction of the volume fraction of bainitic ferrite and some underestimation at the higher strength range, possibly due to the neglect of carbide formation within the microstructure. Nevertheless, it can be seen that this simple approach is able to predict the yield strength of different ADI alloys relatively well.

Microstructural Evolution in Al-Si Alloys
Multicomponent Al-Si based casting alloys are used for a variety of engineering applications, including for example, piston alloys. Properties include good castability, high strength, light weight, good wear resistance and low thermal expansion. In order for such alloys to continue operation to increasingly high temperatures, alloy element modifications are continually being made to further enhance the properties. Improved mechanical and physical properties are strongly dependent upon the morphologies, type and distribution of the second phases, which are in turn a function of alloy composition and cooling rate. The presence of additional elements in the Al-Si alloy system allows many complex intermetallic phases to form, which make characterisation non-trivial due to the fact that some of the phases have either similar crystal structures or only subtle changes in their chemistries. These include, for example, $CuAl_2$, Al_3Ni_2, Al_7Cu_4Ni, Al_9FeNi and $Al_5Cu_2Mg_8Si_6$ phases, all of which may have some solubility for additional elements.

Thermodynamic Prediction of the Phases Present

It is possible to carry out simple thermodynamic calculations for multicomponent Al-Si alloys to predict the phases which are likely to be present as a function of variation in chemical composition. Thermodynamic databases are available which contain appropriate parameters for these complex multicomponent alloys [e.g. 19] and can take into account most, if not all, of the elements present. It is recognised that cast microstructures are not necessarily in their equilibrium state, nevertheless such calculations provide a useful insight into phase stability. Figure 6 presents the results of such thermodynamic equilibrium calculations and plots the mass fraction of the minor phases predicted to be present under both equilibrium and Scheil cooling conditions. The majority of the microstructure is predicted to comprise an Al matrix together with approximately 10% Si particles (not shown), with a number of intermetallic phases being predicted to be present in relatively small amounts. It is interesting to note that in general the predictions agree very well with experimental observations, with the two intermetallics (αAlFeSi and Al$_9$FeNi) predicted to be present in the largest quantities being those observed in the highest quantities experimentally. The Scheil calculations provide an insight into the phases which are likely to be found as a result of chemical segregation which occurs on solidification, and indeed one of the phases (βAlFeSi), which was only observed experimentally in small quantities within particular regions of the microstructure, was only predicted to occur using a Scheil methodology. It has been demonstrated, therefore, that thermodynamic calculations are very useful as a tool to guide alloy development. However, their significant limitation is that they can provide no information about the morphology of the phases present, which may have a critical influence on the mechanical behaviour.

Prediction of the Morphology of the Phases: Phase Field Modelling

One possible route to the prediction of not only the amount, but also the morphology, of phases present in a particular alloy system is the use of phase field modelling techniques. Classical solidification models are often termed 'sharp interface' models in which the solid-liquid boundary is described as a two dimensional surface with no internal structure or width. It is then necessary to track the position of the boundary during the entire solidification process in order to apply the relevant equations of motion, and is relatively complex to implement numerically. The phase field method instead relies on an order parameter to describe the physical state of the system (e.g. liquid or solid), and therefore the solid-liquid interface has a finite thickness. Solidification can then be described in terms of the evolution of this parameter, hence removing the need for interface tracking. The phase field method can incorporate the nucleation and growth of matrix and second phase particles, and has the potential to be directly linked to thermodynamic data and phase diagram information, and therefore to the modelling of solidification in complex multi-element, multi-phase systems.

There have been a number of phase-field models published dealing with the solidification. The earliest phase-field models for the solidification of pure substances were developed in the mid 1980s by a number of authors [20-22]. Wheeler et al. developed the first phase-field model for binary alloy solidification [23], and the model has been successfully applied to a number of cases [24, 25]. Steinbach et al. introduced a binary-multiphase field model [26], which has been used to successfully model the solidification of eutectic, peritectic and monotectic alloys [27, 28]. Ode et al. developed a phase-field model for ternary alloys [29]. Miyazaki [30] and Cha et al. [31] have developed phase-field models that are suitable for the study of multicomponent alloys, and recently a phase-field model appropriate for the simulation of a 'real' alloy that contains multiple components and multiple solid phases has been reported [32].

An example of a phase field simulation carried out for a binary Al-Si alloy is presented in Figure 7, in which the development of the eutectic structure can be seen to compare favourably with experimental observations. A more complex phase field simulation is illustrated in Figure 8, for an Al-Si-Cu-Fe alloy. The matrix, Al, forms the majority phase, with Si present at ~10% and additionally two intermetallics are predicted to be stable in this system; the δAlFeSi can form from the liquid, with $CuAl_2$ being predicted to be stable at lower temperatures below 500°C. A number of parameters are required for the phase field simulations, which include interfacial energies, chemical diffusion coefficients, together with a parameter which represents the likely anisotropy of each phase. It is not always possible to find experimental data for each of the parameters needed, and therefore sensitivity studies may be necessary to ensure an appropriate simulation can be carried out for different alloy systems. However, the potential of the phase field method to simulate the morphologies of phases and their relative distributions in multicomponent, multiphase alloys is clear, and there are exciting possibilities to link this to other modelling strategies to provide a complete model of the solidification process.

Advanced Characterisation Techniques

It is essential that models of microstructural evolution are validated to ensure fitness for purpose and to ensure that they accurately represent the observed phenomena within the particular alloy. In addition to 'conventional' techniques, there are a number of characterisation techniques which have developed significantly in recent years such that they are widely available for more routine characterisation. Examples are electron back scatter diffraction (EBSD), an attachment for a scanning electron microscope, and depth sensing indentation (nanoindentation).

Electron Back Scatter Diffraction

Electron backscatter diffraction (EBSD) is used for the crystallographic analysis of fine scale regions within bulk specimens and has primarily been used as tool for mapping the crystallographic orientation of known polycrystalline samples. More recently it has also been used, typically in conjunction with a crystallographic database, for the identification of unknown crystalline phases in bulk specimens. The specimen surface is typically highly polished, and should be strain free. Kikuchi bands are produced from the interaction of the electron beam with the sample. These bands reflect the symmetry of the crystal lattice, and the width and intensity of the bands are directly related to the interplanar spacing [33]. Figure 9 illustrates the conventional use of EBSD in the determination of grain boundaries and grain orientations for a section of a casting in which there is a columnar region at the edge of the casting, and an equiaxed region towards the centre, together with a very fine 'transition' region between the two. This technique is very useful for quantitative studies of grain refinement in cast alloys.

It is also possible to simultaneously collect chemical data by conventional energy dispersive X-ray (EDX) analysis in order to better discriminate between phases using both chemical and crystallographic data. Figure 10 illustrates the benefit of this technique for the rapid identification of phases in a complex multicomponent Al-Si alloys. Figure 10a shows an SEM image of what is apparently a single intermetallic particle, however, the image quality map, a measure of the quality of the electron diffraction patterns, indicates differential contrast across the particle indicative of the presence of more than one phase. The EDX maps in Figures 10b and c for Cu and Ni respectively clearly show a difference in composition across the particle, matched by a difference in orientation in Figure 10e. The phase map in Figure 10f obtained by the combined use of EDX and EBSD to discriminate the different phases present clearly shows that in fact the particle is composed of three different phases which have formed from each other as solidification proceeds. This type of detailed analysis is important in the determination of solidification sequences and provides an insight into the complex nature of phase formation in multicomponent casting alloys.

Depth sensing indentation
Depth sensing indentation (DSI) is commonly referred to as nanoindentation since the technique usually operates in the sub-micron depth range with nanometer resolution [34]. DSI is an important technique for probing the mechanical behavior of materials, particularly hardness and modulus, at small length scales via continuously recording the force applied and the corresponding displacement during an indentation. This technique therefore offers the possibility of the measurement of local mechanical properties for particular phases within an alloy system. Figure 11 presents hardness and modulus data obtained by nanoindentation in the particle analysed in Figure 10 and clearly shows that there is a significant variation in mechanical properties across the complex particle. The Al_2Cu phase is found to be considerably softer than the Cu and Ni containing phases, with the Al_3Ni_2 phase being slightly harder than the Al_7Cu_4Ni phase. It is also possible, through the use of a hot stage, to

obtain the temperature dependence of the mechanical properties of individual phases. These differential mechanical properties may have important implications for the performance of alloys in service and therefore their assessment through the use of advanced techniques are important component parts within an overall modelling strategy.

Conclusions

In the first example of austempered ductile iron, a combination of equilibrium thermodynamics and kinetic theory has been used to successfully predict the amounts of the major phases, austenite, graphite, bainite and martensite, which occur as a function of heat treatment time and temperature. The inherent segregation present in the microstructure has also been considered using a Scheil approach, which enables predictions to be made of the microstructural constituents as a function of position relative to graphite nodules. In the second example, the potential of phase field approaches to predict not only the relative amount, but also the morphology of phases in multicomponent, multiphase systems. The potential contribution that advanced characterisation techniques can make to the validation of through process models has also been highlighted.

The modelling of microstructural evolution during solidification of cast alloys is an exciting field which has developed rapidly in recent years. The next significant challenge is the genuine two-way coupling of models across length and time scales, incorporating links between microstructural evolution and mechanical property prediction, which combined with the continuing availability of increased computer power, is likely to provide new insights into, and potential control of, solidification processing for industrial applications.

References

1. Voller V R, Micro-macro modelling of solidification processes and phenomena, MCWASP Conference Proceedings, TMS, 2001, pp 41
2. Rappaz M, International Materials Reviews, 34(3), 1989, pp 93
3. Stefanescu D M, ISIJ, 35(6), 1995, pp 637
4. MCWASP Conference Proceedings series, TMS (1980 – 2003)
5. Spittle J A and Brown S G R, Acta Met, 37, 1989, pp 1803
6. Davies R H, Dinsdale A T, Chart T G, Barry T I and Rand M, High Temperature Science, 26, 1989, pp 251
7. Scheil E, Zeitschrift Metallkunde, 34, 1942, pp 70
8. Boeri R and Weinberg F, AFS Transactions, 97, 1990, pp 179
9. Nastac L and Stefanescu D M, AFS Transactions, 101, 1993, pp 933
10. Liu J and Elliott R, Materials Science and Technology, 14, 1998, pp 1127
11. Hayrynen K L, Moore D J, Rundman K B, AFS Transactions, 96, 1988, pp 619
12. Liu J and Elliott R, International Journal of Cast Metals Research, 12, 1999, pp 75
13. Bhadeshia H K D B, Metal Science, 16, 1982, pp 167

14. Chester N and Bhadeshia H K D H, Journal de Physique, C5, 1997, pp 41
15. Bhadeshia H K D B, 'Bainite in Steels', 1992, London, Institute of Materials.
16. Thomson R C, James J S and Putman D C, Materials Science and Technology, 16, 2000, pp 1412
17. Reed P A S, Thomson R C, James J S, Putman D C, Lee K K and Gunn S R, Materials Sci. and Engineering A, 346(1-2), 2003, pp 273
18. Yescas M A, Bhadeshia H K D H, MacKay D J, Materials Science And Engineering A, 311 (1-2), 2001, pp 162
19. Saunders N, Materials Science Forum, 217, 1996 pp 667
20. Collins J B and Levine H, Phys. Rev. 31B, 1985, pp 6119
21. Langer J S, in: Grinstein G and Mazenko G (Eds.), Directions in Condensed Matter, World Scientific, Singapore, 1986, p. 164.
22. Caginalp G and Fife P, Phys. Rev. 33B, 1986, pp 7792
23. Wheeler A A, Boettinger W J and McFadden G B, Phys. Rev. 45A, 1992, pp 7424
24. Warren J A and Boettinger W J, Acta Metall. Mater., 43, 1995, pp 689
25. Murray B T, Wheeler A A and Glicksman M E, J. Crys. Growth, 154, 1995, pp 386
26. Steinbach I, Pezzolla F, Nestler B, Rezende J, Seeßelberg M and Schmiz G J, Physica, 4D, 1996, pp 135
27. Nestler B and Wheeler A A, Physica, 138D, 2000, pp 114
28. Nestler B and Wheeler A A, Comp. Phys. Com., 147, 2002, pp 230
29. Ode M, Lee J S, Kim S G, Kim W T and Suzuki T, ISIJ Inter. 9, 2000, pp 870
30. Miyazaki T, CALPHAD, 25, 2001, pp 231
31. Cha P R, Yeon D H and Yoon J K, Acta Mater., 49, 2001, pp 3295
32. Qin R S, Wallach E R and Thomson R C, Journal of Crystal Growth, 279(1-2), 2005, pp 163
33. Schwartz A J, Kumar M, Adams B L, Electron Backscatter Diffraction in Materials Science. Kluwer Acadamic/Plenum Publishers, 2000.
34. Oliver W C, Pharr G M, J. Mater. Res., 7, 1992, pp 1564
35. Putman D C and Thomson R C, International Journal of Cast Metals Research, 16(1), 2003, pp 191
36. Chen C-L and Thomson R C, Private Communication, 2006.

Acknowledgements

The author would like to thank EPSRC for financial support under GR/M38667/01 and past and present members of her research group: in particular Wendy Adams, Chun-Liang Chen, Joss James, Duncan Putman and Rongshan Qin who have all contributed to this research. Fruitful collaboration with Dr Philippa Reed and her group at Southampton University is gratefully acknowledged, together with helpful discussions and support from the MTDATA team, Thermodynamic and Process Modelling Group, NPL, UK, concerning the thermodynamic calculations.

Contact Email: R.C.Thomson@lboro.ac.uk

Figures

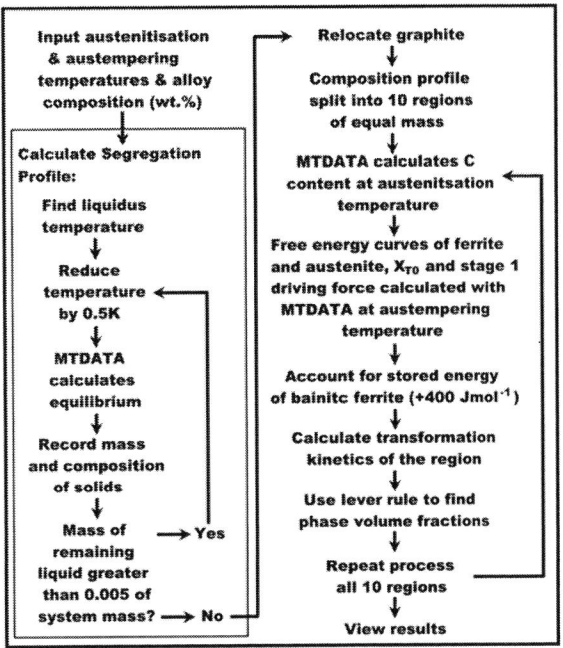

Figure 1: A schematic flow diagram of the Scheil approach to modelling solidification [35].

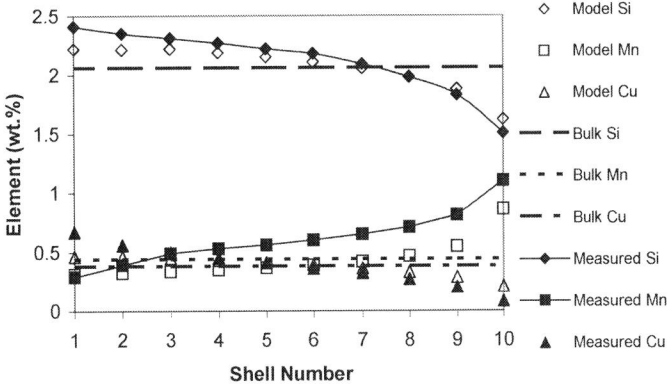

Figure 2: A comparison of segregation profiles predicted using the Scheil methodology and measured using EDX analysis in the SEM [35].

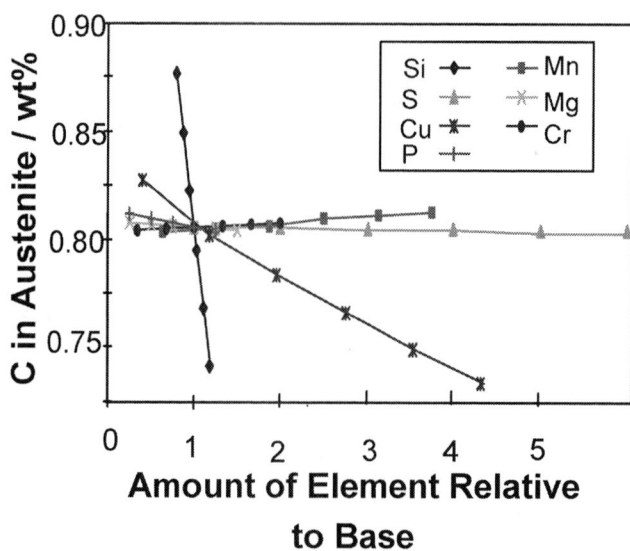

Figure 3: Prediction of the austenite carbon content as a function of alloy composition [16].

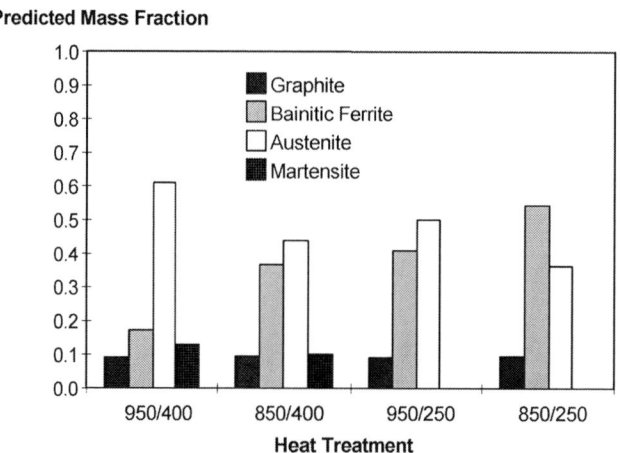

Figure 4: An example prediction for the major phases in a particular ADI alloys. Figures on the bottom axis are the austenitisation temperature and austempering temperature (°C) respectively, with both heat treatments being carried out for 60 minutes [16].

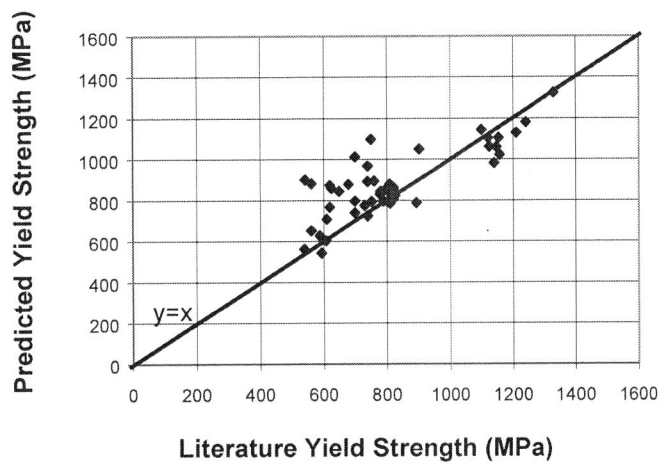

Figure 5: A comparison of the predicted yield strength and experimental values reported in the literature for a variety of austempered ductile iron compositions, austenitising and austempering temperatures and times [16].

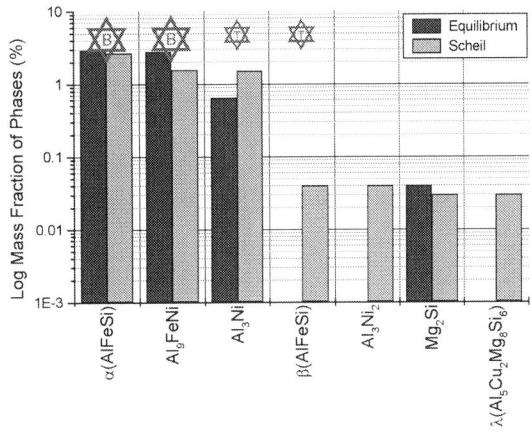

Figure 6: Phases predicted to be present under both equilibrium and Scheil cooling conditions in a multicomponent Al-Si alloys. The phases which were found to be present in the alloy using microscopy techniques in substantial amounts within the bulk of the sample are denoted by 'B' and those observed in trace quantities by 'T'.

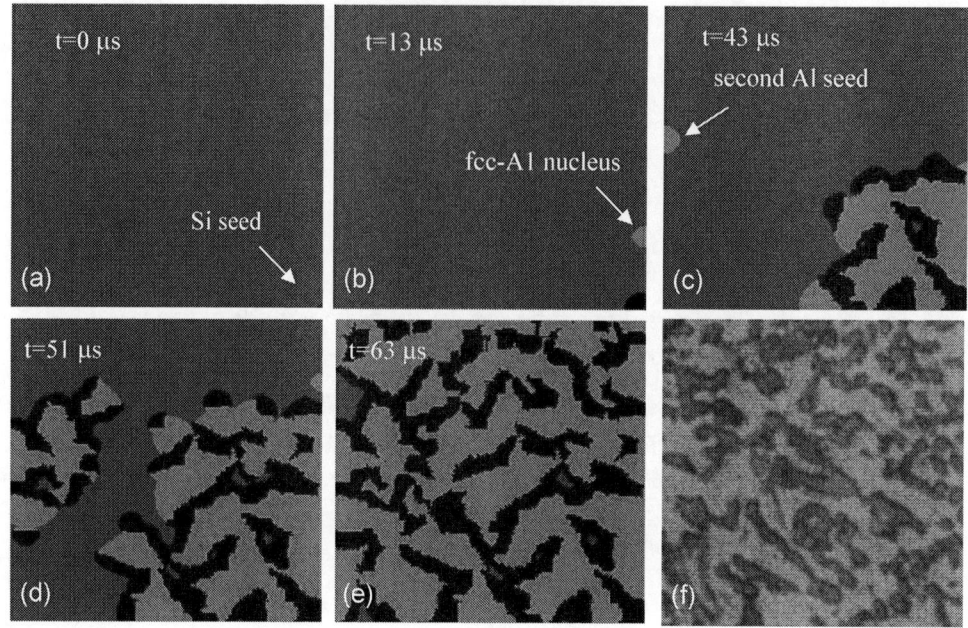

Figure 7: Phase field simulation (images a to e) of solidification in a binary Al-Si eutectic alloy, compared with an optical microscope image (f).

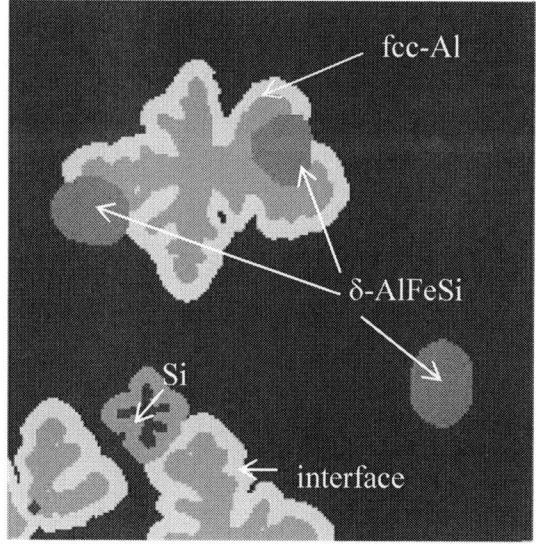

Figure 8: The distributions of phase-field order parameters at t=340 μs showing the morphologies of fcc-Al, Si and δ-AlFeSi particles at 830 K in the simulation of solidification in an AlSiCuFe alloy [32].

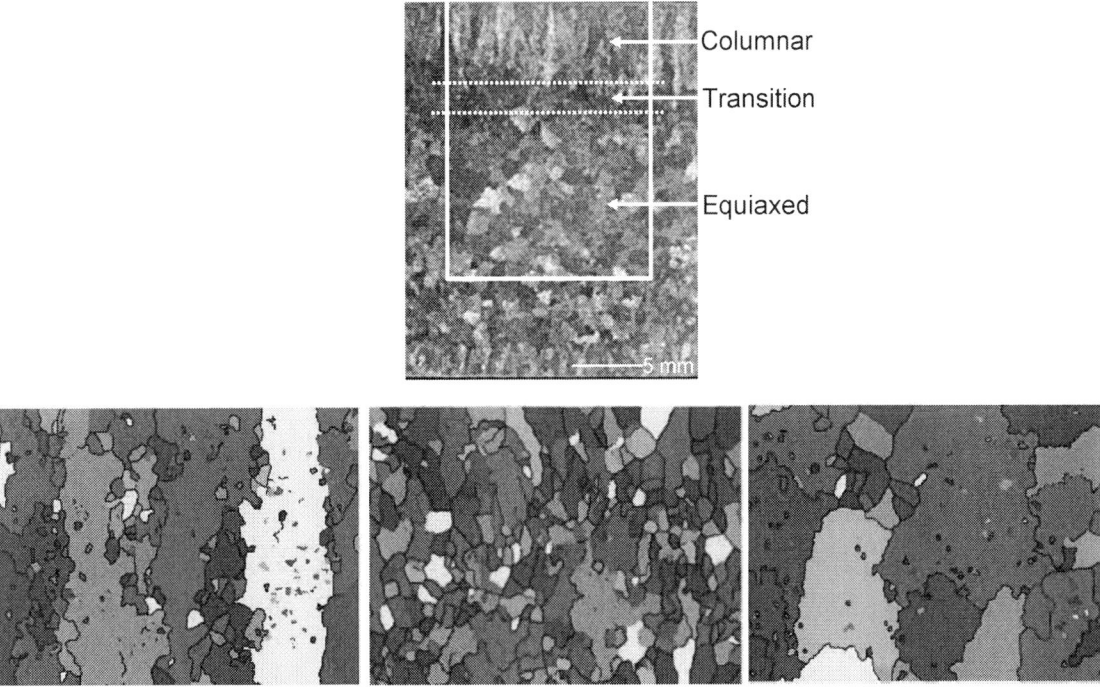

Figure 9: An optical macrograph showing different regions within an Al-Si casting in a step mould (top), and the corresponding grain colour maps obtained using electron back scatter diffraction for the columnar, transition and equiaxed regions respectively (bottom, from left to right).

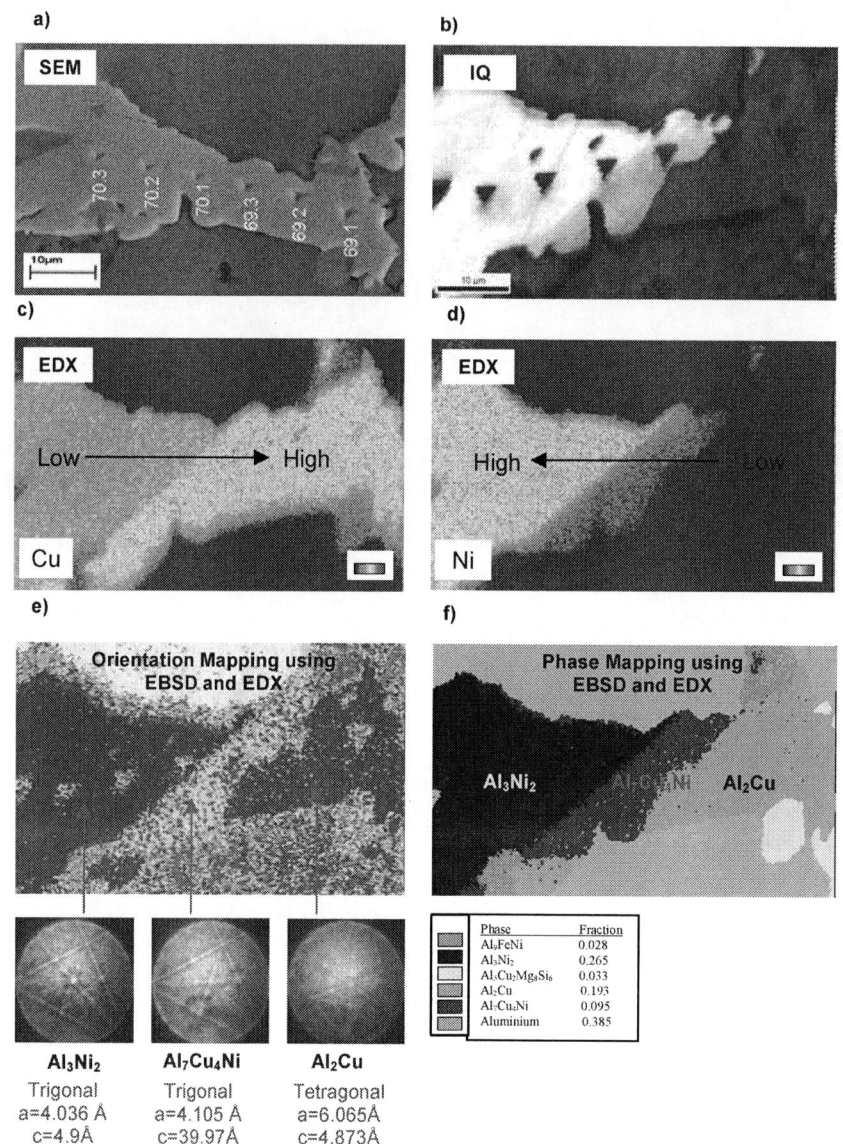

Figure 10: (a) An SEM image of an intermetallic particle in a multicomponent Al-Si alloy, (b) corresponding image quality map, EDX maps of (c) Cu and (d) Ni, orientation mapping (e) and phase mapping (f) using combined EDX and EBSD [36].

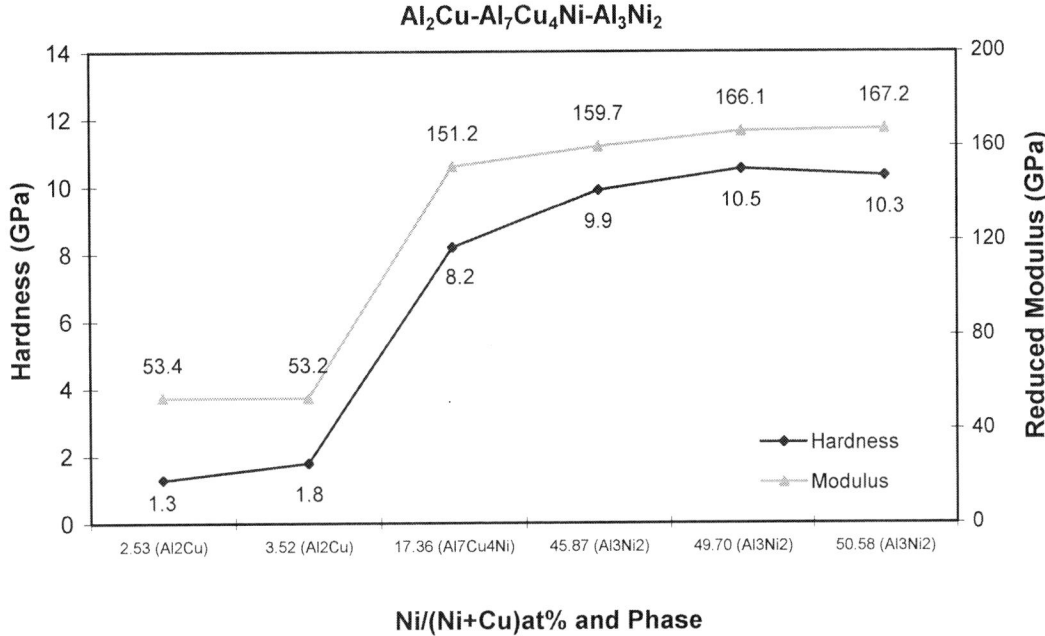

Figure 11: The hardness and reduced modulus of the particle shown in Figure 10 determined using nanoindentation [36].

Maximising Supply Chain Competitiveness and Market Opportunities by the Exploitation of Technology: The Role of a Supply Chain Intermediary.

Dr M C Ashton

Castings Technology International

Abstract

This paper describes how research and technology organisations are in the best position to take measured risks in order to help the foundry industry adapt to the new world order, and to diversify.

The global market for castings is constantly changing and research and development companies, as well as foundries, need to change to embrace new technology, new ways of working and new strategies to compete in this market place. They need to adapt to the rapid changes in their sector to meet the emerging needs of global markets.

New technology can be used to support the industry and provide a lead for the sector, but a comprehensive understanding of technology such as rapid prototyping and fast manufacture methods is necessary if 'right first time' manufacture is to be achieved reliably and consistently. Some of the newer techologies in the area of rapid prototyping are described; in particular stereolithography for ceramic moulded castings and printing for sand moulds and cores as well as direct machining of sand mould and cores. Examples are given of castings where these technologies have been successfully applied.

The paper describes the approach that one international research and technology company has taken to ensure that it can continue to enhance its support for its members and the wider international casting community.

Background

Globalisation continues to impact upon casting production in the Western world. Among others, China, India, central and eastern European nations have all become a significant and growing source of low cost castings. In the UK, this has resulted in a dramatic reduction in the volume of castings produced, and closure of many foundries. The rate at which business is being lost to the industry is accelerated by the decline in manufacturing generally – itself a result of globalisation. In addition the castings industry is always seeking to compete more effectively against the threats from forging and machining.

The UK casting industry, therefore, is confronted by the rapid emergence of immense capacity for low-cost production at the same time as its indigenous customer base is either disappearing to low cost areas of the world or closing. In addition the industry is burdened by meeting the costs of compliance with social, environmental and health and safety legislation. High volume, price-sensitive castings have already migrated to Eastern nations. The remaining low volume, higher margin products will surely follow unless pre-emptive action is taken.

Technology, and services based on the exploitation of technology, can play a vital role in retaining existing business. It can also play an important role in creating new opportunities by providing a more effective response to emerging market needs. Rapid speed to market of low volume, customised castings, exhibiting superior precision, integrity and performance, will increasingly determine the prospect of success in securing future business.

Companies need to look at the newer technologies and consider ways to exploit their potential. For instance rapid prototyping technologies provide the means of accelerating the development of new components. They also provide a platform from which fast manufacturing technologies could be developed for short lead-time, low volume casting production. Indeed, a production-representative 'prototype' casting, fully tested and certified to ensure compliance with a procurement standard, is actually no different to a one-off 'production' casting. Nor should it be different – prototype castings are typically subjected to expensive performance testing, so they need to be faithfully representative of production castings if the test results are to be meaningful.

Prototyping and Fast Manufacturing Technologies

It could be argued, therefore, that rapid prototyping technologies, and the market requirement for fast manufacture of one-off and small batches of castings, are converging. Rigorous assessments have been made of the performance of a wide range of prototyping systems, including tests of their ability to replicate the accuracy, surface finish and integrity of both ceramic and sand moulded production castings. For prototyping ceramic moulded castings, stereolithography has proved to be superior. For sand moulded castings, printing of moulds and cores proved to be the only technique that provided an effective solution. Both of these are complex technologies over which effective control can only be assured through an in-house capability. This is the only way the required precision, surface finish and integrity can be achieved quickly, reliably and predictably, at minimum cost.

Stereolithography

Stereolithography is used to produce a 'QuickCast' replica or model of the desired component. This replica is essentially a honeycomb resin structure produced directly from a 3D CAD file. A ceramic shell mould is then produced in much the same way as in the lost wax investment casting process, the resin being completely removed when the ceramic is fired, leaving an empty mould ready to receive metal.

It is possible to produce large components by assembling several replicas. This technique has been used to manufacture steel housings for submarine optronic masts, with three models used to produce each replica. Its size is evident from the weight of metal, 490kg, required to fill the ceramic mould.

However, the stereolithography technology is capital intensive and material costs are high. For successful operation, it also requires considerable technical resources. Its adoption by typical casting manufacturers is, therefore, difficult to justify, as is borne out by the fact that very few foundries in the world have installed the technology. It is, nonetheless, a powerful prototyping technique and means by which it could be competently and optimally exploited for the benefit of the industry and its supply chains would be of value.

It is also a technique that could be extended so that it can be successfully applied to the manufacture of production, as opposed to prototype, castings. By way of example, a development is underway to enable the short lead-time manufacture of intermediate volumes of lost-wax castings. Techniques are being trialled that produce low-cost 'soft tooling' from stereolithography models in a fraction of the time it takes to produce a wax injection die. The means of filling these 'soft tools' with wax are being developed concurrently.

A further technique developed recently complements those described in that it enables much larger parts to be made from ceramic shell moulds than are possible with either stereolithography or the lost-wax process. This fast, low cost technique uses CNC machined polystyrene replicas to produce ceramic shell moulds into which up to 1 tonne of steel can be poured. It is suitable for prototypes, one-offs or small batch production.

Printing of Sand Moulds and Cores

Sand moulds and cores can be rapidly manufactured by a relatively new printing technology. It is especially suited to complex geometries, and can produce cores in one piece that would otherwise require multiple core-boxes and core assembly.

As with stereolithography, this technology is capital intensive, material costs are high and it is demanding in terms of the technical resources and capabilities required for its operation. For the same reasons, therefore, it too is unlikely to be widely adopted by typical foundries. Yet it is a technology with considerable potential, certainly for the production of complex components in small batches, where the cost of normal patterns and core-boxes would be a significant proportion of the total procurement cost. It is also considerably faster than producing patterns and core-boxes using even the most advanced machining software and 5-axis high-speed milling machines. These no longer offer competitive advantage because they are already being exploited in low cost areas of the world.

Direct Machining of Sand Moulds and Cores
In just the same way that high-speed milling of polystyrene provides a fast, low cost route for the manufacture of ceramic shell moulds, a further development applies the same techniques to the manufacture of sand moulds and cores. The Patternless® Process –in which moulds and cores are machined directly from blocks of bonded sand – is complementary to printing technology in two respects. Firstly, it can produce much larger moulds and cores, for example machines with a working envelope of 3.6 x 3.6 x 1.2 metres are now being used. Secondly, it can be used to produce moulds at the same time as the cores are being manufactured by the printing technology. This reduces overall time and costs, and exploits the respective strengths of each of the techniques to greatest advantage.

This process has recently been used to manufacture 28 fixed-pitch nickel-aluminium-bronze propellers for fast attack vessels for the Royal Greek Navy. Consistent blade-to-blade dimensions and precision of form were more critical in this case because four propellers are used on each vessel. The exercise demonstrated that the process reliably achieved the balance, low vibration and propulsion efficiency required. A reduction in machining stock from the traditional 8mm to 3mm also demonstrated the geometric precision attainable, which clearly impacts upon cost-competitiveness as well as product performance.

These technologies, among others, can address the need for short lead-time manufacture of prototypes, one-off or small batches of castings. However, it is important to recognise that speed alone does not provide the optimum solution. Simultaneous assurance of the highest levels of integrity and performance, through design optimisation and judicious selection of materials, is essential. This is especially so when targeting markets for high duty components, and particularly when challenging alternative solutions based on metal forming techniques such as forging or machining.

Design Optimisation

The widespread availability of 3D CAD models not only underpins the exploitation of fast prototyping and manufacturing technologies, it also facilitates the efficient application of simulation software to predict and address issues associated with the production and performance of a cast component.

Simulation software can improve quality and minimise end-product and life cycle costs by coupling 'design for performance' with 'design for ease of manufacture' *and* with optimisation of the pouring, gating and feeding system.

Clearly, to obtain the greatest benefit from concurrent design, the attributes of the specific manufacturing process selected to produce the part must be taken into consideration. A comprehensive understanding of rapid prototyping and fast manufacturing processes is therefore essential if 'right first time' manufacture is to be achieved reliably and consistently. An 'in-house' capability in these processes provides the level of understanding necessary to underpin rigorous design optimisation of high performance castings.

Material Selection

The performance of a casting is determined primarily by the mechanical and physical properties of the alloy selected. Improvement in the mechanical and physical properties of a casting can provide the differentiation necessary to secure or safeguard business. Vacuum melting, refining and pouring are the most effective technologies by which to achieve significant enhancement in properties. Improved strength, toughness and fatigue, increased corrosion resistance, and enhanced tolerance to thermal cycling and extremes of temperature, for example, can be achieved in a broad range of metals and alloys by vacuum technology.

Higher strength, vacuum melted alloys can, for example, enable wall sections to be reduced to provide weight savings, a primary driver in markets such as transport, aerospace and defence. Pouring under vacuum also allows wall sections to be filled successfully that would otherwise be impossible to fill if poured in air. There are, however, cost penalties and productivity issues associated with vacuum melting and therefore development work is needed to minimise the costs, both capital and operational, and to improve productivity to a level that is sustainable in the marketplace.

A Role for a Technical Intermediary

If it is accepted that a solution to the challenges faced by the casting sector lie, at least in part, in the application of all these technologies, what steps can be taken to ensure their accessibility and exploitation at minimum risk and cost? As has been stated, the significant capital, material and operational costs militate against their exploitation and/or adoption unless and until 'technology demonstrators' justify their commercial application.

Research and technology organisations are in the best position to take measured risks in order to help the industry adapt to the new world order, and to diversify. Indeed, it is essential that research and technology organisations themselves adapt to the rapid changes in their sector and diversify to meet the emerging needs of global markets. They are not immune to change and are just as vulnerable as the companies in the sector to which they belong.

This was recognised by Cti several years ago, and a strategic review was carried out to consider how the organisation could improve support for its Members in the context of the challenges brought about by globalisation. This review took into account the findings of several related studies conducted around the same time, including the USA Metal Casting Industry Technology Roadmap and the UK Casting Industry Foresight Analysis. It also ensured that due consideration was given to studies pertaining to the sector's customer base, such as the Manufacturing Foresight Panel's report – UK Manufacturing 2020: We can make it better.

The review concluded that the organisation should maximise its 'attractiveness to customers' through a strategic re-positioning of the organisation as an impartial 'supply chain intermediary'. The rationale was that the competitiveness of existing supply chains could be improved if Cti was enabled to provide relevant support to customers, as well as to foundries.

The re-positioning was also designed to create new opportunities in 'high technology products' in order to sustain business volumes for indigenous supply chains. By optimising design, material specifications and methods of manufacture, and by supplying certified prototypes and pre-production volumes, it can be compellingly demonstrated that advanced technology castings provide cost-effective solutions to customer needs.

This required the development of new capabilities that build on the traditional services offered in such a way as to reduce the risk, cost and time associated with bringing a new product to market. The new business model also required the company to become proficient in all alloys, both air and vacuum melted, and to be capable of rapidly manufacturing 'production representative' castings in both sand and ceramic moulds.

A five-year plan was developed to create the relevant capacity and capabilities. This called for a total investment of £14.5m ($25m) in R&D, new capital facilities and workshop space. Against the backcloth of a sector in decline, this was clearly a bold initiative for which the risks were high. Yet it was believed that the integration of comprehensive design, materials, rapid prototyping and fast manufacturing technologies, personnel skills and appropriate accreditations, all under one roof, provided the vital key to maximising supply chain competitiveness and to creating new market opportunities.

New headquarters for the company have recently been completed to provide almost 5000m^2 of high-technology workspace in which to integrate all the necessary capital facilities. Its completion coincides with the start of year four of the five-year plan, the first three years having been spent evaluating and investing in all the technologies described, and developing the know-how and skills to operate them competently and competitively in a commercial environment. As the new operations come on stream, there is a high level of confidence that the planned outputs, technical, commercial and financial, will be realised.

Benefits
In summary, the benefits expected from an initiative of this kind are firstly, a practical interface with customers and real products ensures that R&D is relevant to existing markets and to the creation of new markets. Market disciplines maximise the efficiency and effectiveness of R&D investment, and ensure that optimised, market-tested outputs are achieved in the shortest possible time.

Secondly, the risk, cost and time associated with the development of new market opportunities in high performance castings, and their transfer to production operations, is minimised.

Thirdly, the capability of new technology can be demonstrated in commercially relevant terms. Having created a 'market pull' sufficient to justify its application, casting manufacturers can be eased into new technology with all-embracing support available to minimise risk to them.

Fourthly, income is generated which ensures that a critical mass of facilities, technology and expertise are sustained and that they are readily accessible, and also affordable, to Members and their customers. The 'future proofing' of traditional services to Members was a primary objective of the five-year plan.

Last, but not least, this development will help to counter the negative perceptions of both the industry and its products in some quarters, by continually reinforcing the positive attributes of a modern manufacturing environment and the capability of advanced technologies.

It is important that we succeed on all counts. Our collective future is inextricably linked and, at the end of the day, it is not who is right, but who is left.

The auto sector in a world of ultra competition: nowhere for the inefficient to hide

Professor Garel Rhys OBE

Emeritus Professor, Centre for Automotive Industry Research, Cardiff Business School, Cardiff University, Wales

The world motor industry is in a phase where unprecedented competition is facing the vehicle manufacturers and their suppliers. The reduction of protectionism, be it via the fall or elimination in tariff barriers and the easing of non-tariff competition such as quotas and voluntary export restraints on the one hand and the increase in consumer power be it via the information provided to them or the activities of anti-monopoly agencies on the other, has led to the auto sector being subject to the full play of market forces as never before.

The result of all this is not only the consolidation of existing players but the potential for new vehicle firms and component suppliers to enter the industry. Although the industry is over one hundred years old it is on the threshold of unprecedented growth and opportunity for those adroit enough to grasp this.

Abstract only supplied.

Fluidised bed for stripping sand casting process

Guido Belforte, Massimiliana Carello, Vladimir Viktorov

Dipartimento di Meccanica - Politecnico di Torino
C.so Duca degli Abruzzi, 24 – 10129 Torino Italy

Abstract
An innovative application of a non-traditional casting system, in particular a stripping fluidised bed, whose working principle is based on air – sand interaction phenomenon, is presented. A bed's prototype used to analyse the fluidisation phenomena is described.
The purpose of the experimental tests is to measure: pressure, air flow-rate and velocity. Different measurement points inside the prototype have been taken into account. Stripping pieces with different shapes and dimensions have been considered.
Moreover visualisation of air bubbling in the sand is possible and its behaviour is compared to a computational fluid flow simulation.

Key words fluidised bed, casting, air-sand interaction, and fluidisation effect.

Introduction

In the last years the foundry processes have had significant development, connected to the product evolution. This evolution involved aero and earth transports, where faster and larger systems require the realization of structural elements with big dimensions and small manufacturing tolerance. Then the stripping of big pieces becomes very important. In this case it is very difficult to use the traditional techniques based on vibration and shake that require vibration and noise isolation for the safeguard of ambient.

An alternative system for stripping is the fluidised bed with hot air that allows the stripping of piece without stress and damage.

In this paper an experimental bed prototype is described. Experimental tests have been performed to investigate the fluidisation phenomena. The main parameters in terms of air flow-rate and pressure have been identified; also it has been possible to take into account the presence of stripping pieces.

The visualisation of bubbling phenomena is in good agreement with the first results obtained by means of computational fluid flow analysis.

Experimental test bench

When solid particles of sand interact with air flow their behaviour becomes similar to a fluid; if the sand particles are contained in a chamber a relative movement of the particles is possible. [1÷8]

A fluidised bed furnace has been designed and built (scaled down version of a real stripping bed, in particular 1:5) to investigate the fluidisation phenomena.

The prototype is made up of three important parts: a rectangular chamber (riser or bed or furnace); a cover; an air distribution system positioned on the lower part of the chamber. Some glass windows allow the visualisation of the air-sand interaction phenomena during the experimental tests. The dimensions of the chamber are: 800 mm x 470 mm x 370 mm. A frontal view of the prototype is shown in figure 1.

The air distribution system is made by means of 760 nozzles, whose diameter equal is 0.8 mm; the downward realisation of the nozzles on proper distribution pipes avoids their block up. The supply of the nozzles occurs through two flexible pipes positioned in opposed points on the rectangular chamber. Figure 2 shows a lower view of the nozzles.

Figure 3 shows an internal view of the chamber, where it is possible to see the upper part of the air distribution system and the metal crate used to contain the casting piece (internal dimensions 720 mm x 270 mm x 320 mm).

The tests were carried out filling up with foundry siliceous sand the chamber of the bed prototype; in particular sand with particle medium diameter of 0.3 mm (A.F.A. 55 type) was used.

Figure 4 shows the scheme of the experimental test bench, whose the most important components are: the intersection valve 1; the filter 2; the pressure reducer 3; the flowmeter 4 to measure the inlet air flow; the manometer 5 to measure the flow meter upstream pressure, the air heater

6; the manometer 7 to measure the air pressure at the entrance of the furnace; the fluidised bed furnace prototype 8; the differential water manometer 9 to measure the pressure inside the bed; one or more thermocouples 10 to measure the sand temperature.

Two orthogonal slides positioned between the rectangular chamber and the cover allow to position the pressure or the temperature measurement instruments; in this way it is possible to have different measurement points inside the fluidised bed.

The experimental tests were carried out to measure the fluidisation curves in term of pressure vs. air flow-rate for different sand levels H (10; 15, 20 cm). Lower sand levels did not produced interesting results.

The experimental analysis presented in this paper refers to ambient temperature. For this reason the cover of the prototype was not used.

The pressure was measured by means of the manometer 9 of figure 4. Different measurement points, varying the longitudinal, the transversal and the height position inside the chamber, were used.

For air flow-rate before fluidisation Q_1 (Q_1 = 760 dm^3/min (ANR)) and for the minimum air flow fluidisation Q_2 (Q_2 = 840 dm^3/min (ANR)) the piezometric curves in term of pressure vs. distance from the bottom of the chamber are used.

The purpose of other tests presented in this paper is to establish the influence of the presence/absence of a stripping piece on the fluidisation phenomena; in particular rectangular pieces have been taken into account with equal thickness (10 mm) and different dimensions (600 mm x 200 mm; 300 mm x 150 mm; 150 mm x 150 mm). The pieces were fixed in the centre of the furnace chamber, at the height equal to 10 cm from the base of the metal crate.

Experimental results

The outline of the fluidisation curves obtained with the experimental tests reproduces the behaviour shown by other authors [1÷8], in particular varying the flow-rate the pressure increases until it reaches a maximum constant level.

Near the edge of the furnace chamber a marginal bubbling phenomenon may be observed, while in the main internal zone it is very important. In correspondence it is possible to measure the minimum fluidisation flow-rate.

Figure 5 shows a top view of the chamber where it is possible to note the bubbling effect, but also the metal crate and the two orthogonal slides for the pressure system positioning.

The chamber of the furnace was filled with sand until level H = 20 cm was reached. The pressure was measured at the centre of the riser at different distance from the bottom of the chamber h (h = 1, 6, 11, 16, 18 cm). The fluidisation curves in terms of pressure vs. flow-rate are shown in figure 6.

Increasing h the pressure becomes very low, in fact for h = H the pressure is equal to the atmospheric pressure; the fluidisation effect is important for lower level of h, for witch a minimum flow-rate equal to 800 dm^3/min (ANR) is required.

The variation of the sand level H influences the maximum pressure reached inside the furnace, but the minimum flow-rate fluidisation has a negligible variation (800-850 dm^3/min (ANR)).

Similar behaviour were obtained in other measurement points inside the metal crate.

The piezometric curves have been obtained for three different levels of sand H (H = 10, 15, 20 cm) and in each case using two flow-rate Q_1 (760 dm^3/min (ANR)) and Q_2 (850 dm^3/min (ANR)). Figure 7 shows the results in term of pressure vs. distance from the bottom of the chamber h.

It is possible to note a linear trend of the characteristics, because the behaviour of the air-sand mixture becomes similar to that of a liquid.

The influence of the stripping piece on the pressure and flow fluidisation has been investigated. Figure 8 shows the pressure level curve on the surface of the bigger piece considered (600 mm x 200 mm) positioned in the center of the chamber with sand level H = 20 cm and flow-rate Q_2. In this way the passage area between the piece and the crate are constant. The fluidisation of the sand corresponds to the development of big bubbles.

Il is possible to note that on the piece's edge the pressure is bigger than in the central part. It is interesting to consider that in a real application the stripping would be possible in all part of the piece. Then probably, a small pressure corresponds to low or not efficient stripping.

The piece dimensions influence the fluidisation and the bubbling phenomena then it is important to evaluate the better position of the piece to have efficient stripping.

Simulation results

A CFD program was used to model and simulate the behaviour of a fluidised bed using Eulerian - Eulerian model.

Figure 9 shows the countour lines of sand fraction volume with air velocity equal to 0.25 m/s, corresponding to the fluidisation. It is possible to note different sand fraction volume in different zone of the chamber that indicates the bubbling phenomena.

Fluidisation analysis

The study of fluidisation phenomena started with Ergun [9], witch obtained an empirical formula to calculate the pressure drop in a fixed bed, when a gas stream goes through solid particles. Starting from Ergun formula some authors [10÷13] proposed other methods to evaluate the gas pressure drop and the minimum fluidisation velocity or flow-rate.

The minimum fluidisation velocity is reached when the air flow forces are equal to the total mass force of the sand; in this condition the system is balanced and it is possible to consider the equilibrium equation along the vertical direction.

In this paper two different formulations from the literature have been taken into account to calculate the gas mass flow per unit of bed cross section G_{mf} and the theoretical results have been compared to the experimental data G_{exp}.

The first formulation, developed by Levenspiel allows calculating G_{mf1}:

$$G_{mf1} = \frac{a}{2 \cdot b} \cdot \left[-1 + \sqrt{1 + \frac{4b}{a^2} \cdot g \cdot \rho \cdot (1 - \varepsilon_{mf}) \cdot (\rho_p - \rho)} \right]$$

The second formulation, developed by Delebarre [12-13] allows calculating G_{mf2}.

$$G_{mf2} = \frac{a}{2 \cdot b} \cdot \left\{ -1 + \sqrt{1 + \frac{4b}{a^2} \cdot g \cdot \left[\rho_p \cdot (1 - \varepsilon_{mf}) - \rho \cdot H_{mf} \right] \cdot \left[\frac{H_{mf}}{2 \cdot H_0} \cdot \rho_p \cdot (1 - \varepsilon_{mf}) \cdot \rho \cdot H_{mf} \right]} \right\}$$

Where the symbols significance for the two formulations are:
a: first coefficient of Ergun's equation (kg/s·m^3);
b: second coefficient of Ergun's equation (m^{-1}).
g: gravity acceleration (m/s^2);
ρ: gas density (kg/m^3);
ρ_p: solid apparent density (kg/m^3);
ε_{mf}: bed voidage at minimum fluidization (equal to 0.44);
H_{mf}: bed height at minimum fluidisation (equal to 0.22 m);
H_0: atmospheric pressure in gas height at bed surface conditions (m);
In our case has been calculated ε_{mf} equal to 0.44.
Knowing the bed cross-section it is possible to calculate the mass flow-rate and the standard flow volume –rate.
In particular in our case the following fluidisation flow-rate have been obtained: G_{mf1} = 857 dm^3/min; G_{m2f1} = 850 dm^3/min; G_{mf1} = 830 ÷ 850 dm^3/min.
It is important to note a good agreement between the two analytical formulation and the experimental results.

Conclusions

The fluidised bed furnace could be an alternative system for stripping pieces without stress and damage.
In this paper an experimental bed prototype has been presented. The experimental tests have been carried out to investigate the fluidisation phenomena. The main parameters in terms of air flow-rate and pressure have been identified and measured, showing the influence of the presence of stripping pieces.
The glass windows of the prototype allows the visualisation of the air-sand interaction phenomena, in particular the bubbling phenomena, that it is in good agreement with the first results obtained with a computational fluid flow analysis.
A good agreement has been obtained comparing the experimental minimum air-flow fluidisation and the values obtained by analytical formulation of other authors.
Other experimental tests will be carried out to establish the influence of the temperature on the fluidisation curves.

References

1. Arastoopour H, *Numerical simulation and experimental analysis of gas/solid flow system*, Powder Technology, 119, 2001, pp. 59-67.
2. Benyahia S, Arastoopour H, Knowlton T M, Massah H, *Simulation of particles and gas flow behavior in the riser section of a circulating fluidized bed using the kinetic theory approach for the particulate phase*, Powder Technology, 112, 2000, pp. 24-33.
3. Detamore M S, Swanson M A, Freder K R, Hrenya C M, *A kinetic-theory analysis of the scale-up of circulating fluidized bed*, Powder Technology, 116, 2002, pp. 190-203.
4. Gidaspow D, Bezburuah R, Ding J, *Hydrodynamics of Circulating Fluidized Beds, Kinetic Theory Approach*, 7th Engineering Foundation Conference on Fluidization, 1992, pp. 75-82
5. Guenther C, Syamlal M, *The effect of numerical diffusion on simulation of isolated bubbles in a gas-solid fluidized bed*, Powder Technology, 116, 2001, pp. 142-154.
6. Polashenski W, Chen J C, *Normal solid stress fluidized beds*, Powder Technology, 90, 1997, pp. 13-23.
7. Zhang S J, VanderHeyden W B, *High resolution three-dimensional numerical simulation of a circulating fluidized bed*, Powder Technology, 116, 2001, pp. 133-141.
8. Zhang S J, Yu A B, *Computational investigation of slugging behavior in gas-fluidized beds*, Powder Technology, 123, 2002, pp. 147-165.
9. Ergun S, *Fluid flow through packed columns*, Chemical Engineering Prog., 48, 1952, 89.
10. Marthur K B, Epstein N, *Spouted beds*, Academic Press, New york, 1974.
11. Sutherland J P, *The measurement of pressure droop across a gas fluidized bed*, Chemical Engineering Sci., 19, 1964, 839.
12. Delebarre A, *Does the minimum fluidization exist?*, Journal of Fluids Engineering, 124, September 2002, pp. 595-600.
13. Delebarre A, Morales J M, Ramos L, *Influence of the bed mass on its fluidization characteristics*, Chemical Engineering Journal, 98, 2004, pp. 81-88.

Acknowledgements

The authors would tanks you ing. M. Iannuzzi and ing. G. Tagliarini for their cooperation in experimental tests.

The research is part of the Eureka Project "Sand Cast", and it has been carried out in cooperation with IMF (Luino, Italy) and SFU (Ussel, France).

Figures

Figure 1 – Frontal view of the fluidized bed

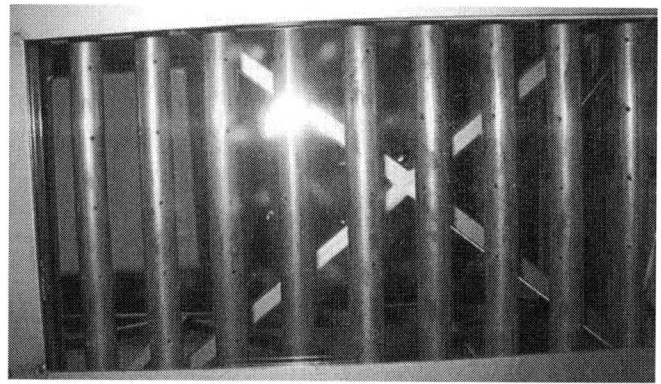

Figure 2 – Air distribution nozzles

Figure 3 – Inside view of the fluidised bed

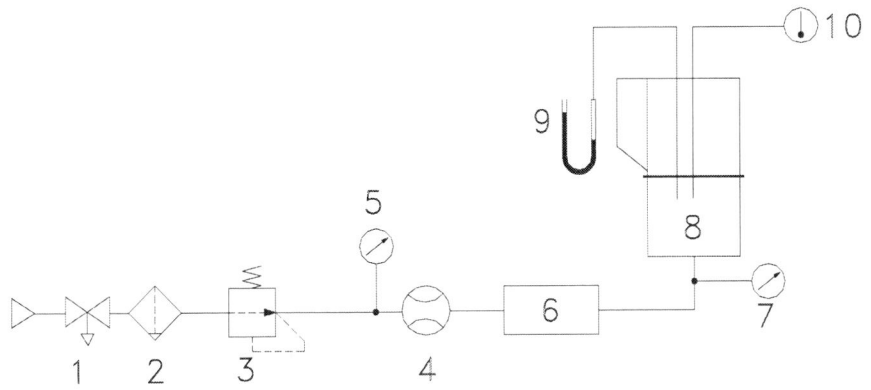

Figure 4 – Scheme of the experimental test bench

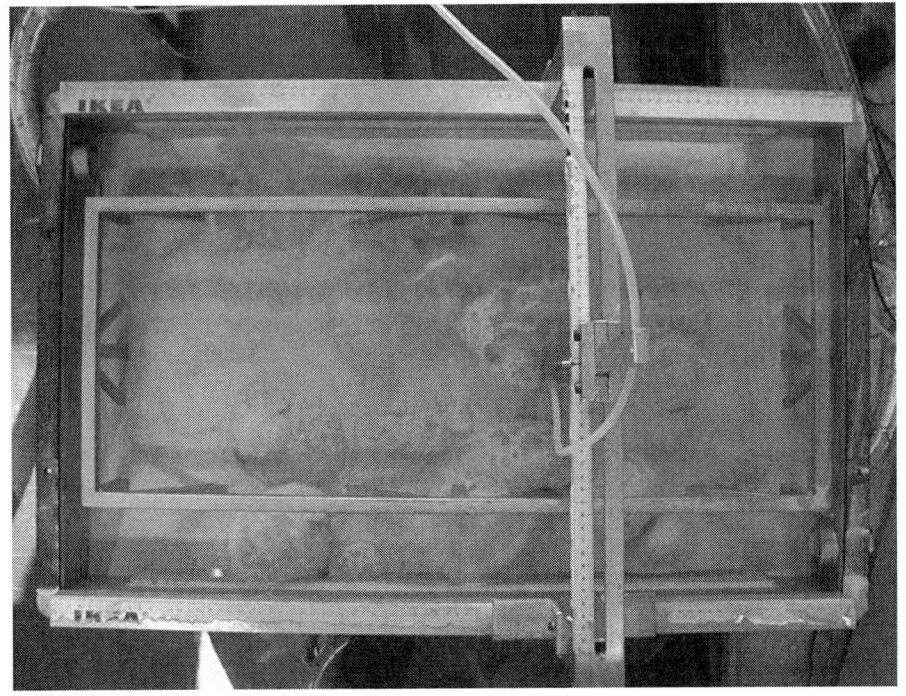

Figure 5 – Example of bubbling phenomenon

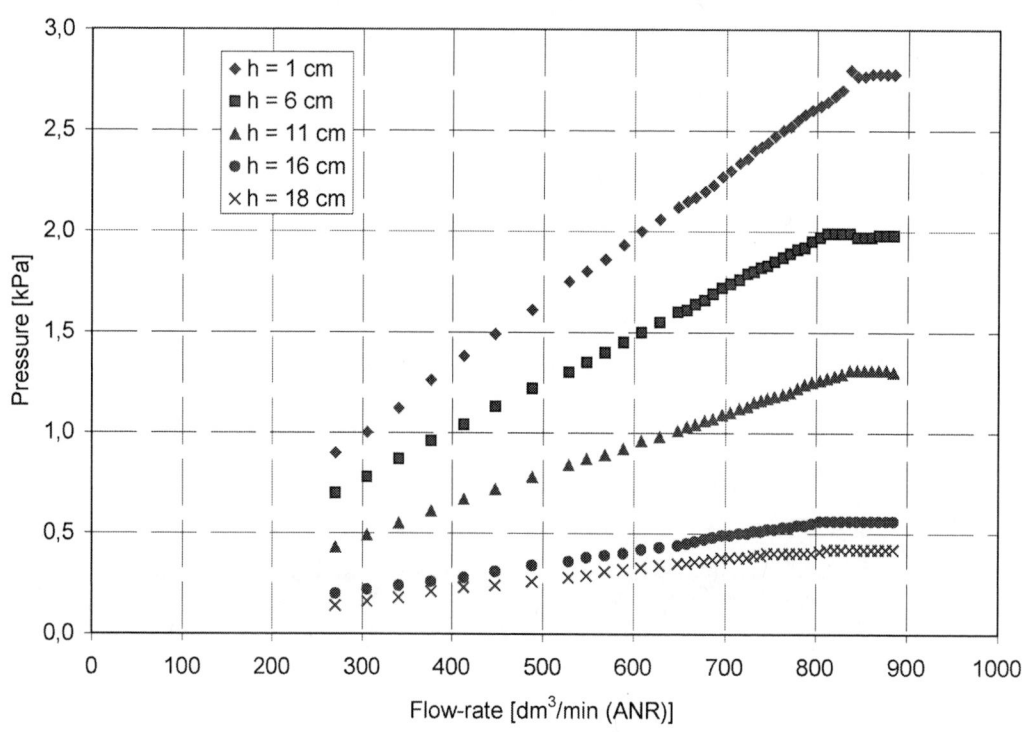

Figure 6- Fluidisation curves pressure vs. flow-rate (sand level H= 20 cm)

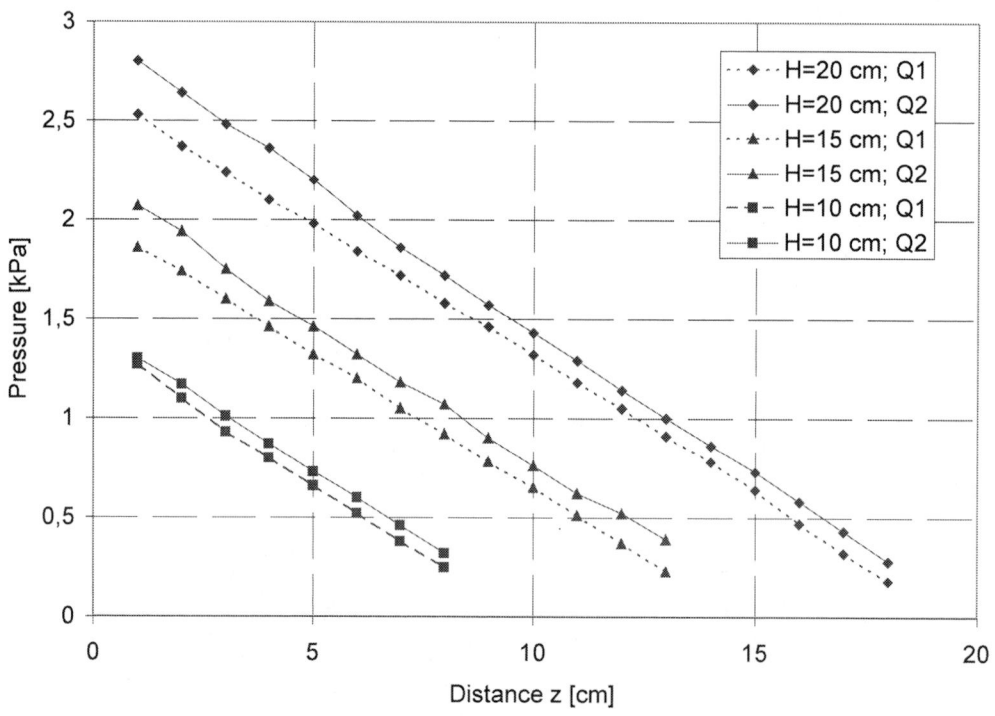

Figure 7 - Pressure vs. distance h (different sand levels H and flow-rate Q_1=760 dm³/min (ANR) and Q_2=850 dm³/min (ANR))

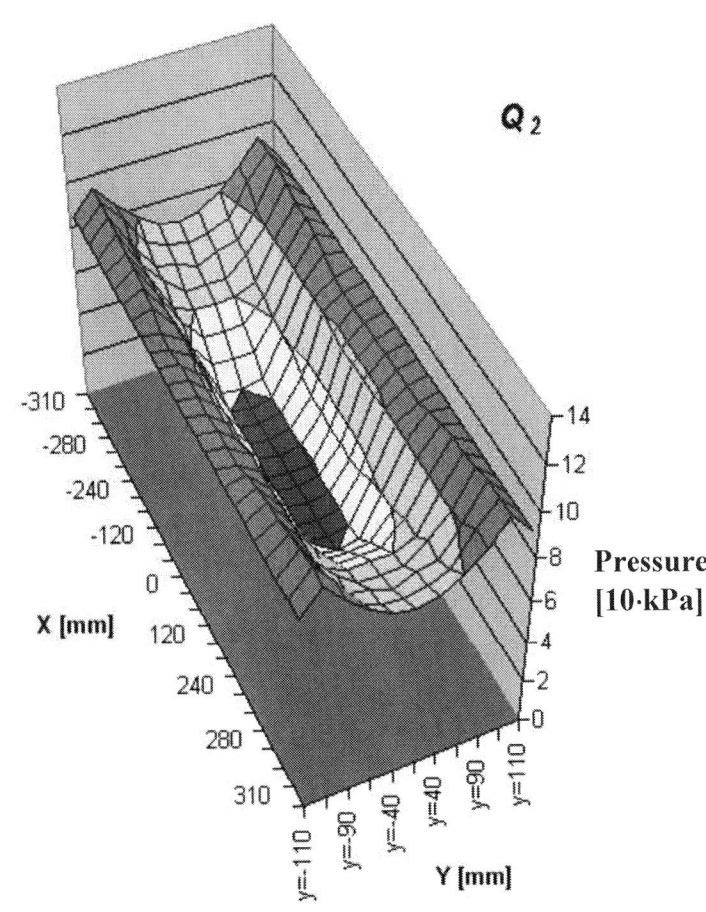

Figure 8 – Pressure level curve on the surface of the stripping piece (sand level H= 20 cm, flow-rate Q_2 = 850 dm^3/min (ANR))

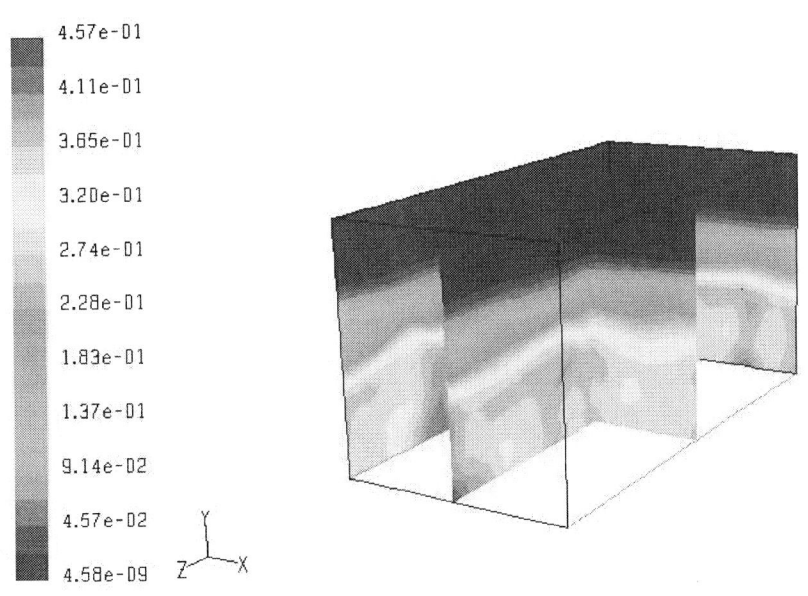

Figure 9 – Contour lines of sand fraction volume in fluidisation conditions

Robot based oxy-fuel cutting and stub-grinding for castings in low series

Prof. Dr. B. Lauwers*, H. De Baerdemaeker*, Dr. P Haigh**, R. Wallis**.

* K.U.Leuven, Department of Mechanical Engineering, Division PMA, Leuven, Belgium, **Castings Technology International, Sheffield, UK.

.

Abstract

This paper describes the development of a robot based fettling cell for medium to heavy section castings in low series. The fettling cell consists of a commercial robot used for foundry applications and a flexible clamping unit, especially developed for this application. The removal of risers and feeders is obtained through an oxy-fuel cutting process, followed by a force controlled grinding operation. Existing off-line robot programming software has been adapted for the off-line generation of a collision free robot path. The cutting path is automatically derived based on the available process knowledge and from two CAD models, one model describing the complete casting geometry including risers and feeders and a second model describing the final product geometry. Process models for force controlled stub-grinding have been developed and will be used for tool path planning.

Key words Robot, Computer aided manufacturing, Fettling, Grinding

1. Introduction

Castings find extensive application in a wide range of industries such as automotive, truck, aerospace, chemical, and many others. Although modern manufacturing process technology enables castings to be made close to net shapes, there is still a need for fettling, finishing and upgrading operations. Today, these operations account for 20 to 40% of the overall casting manufacturing cost. The fettling and finishing operations are often carried out manually, usually by large numbers of people working in an unpleasant and hostile environment. Any automated system for fettling and finishing of castings should accommodate varied shapes and inconsistent profiles [1,2,3]. The castings are usually handled by a robot or counterbalanced flexible arm and presented in different orientations to a variety of fixed tools [2]. Such systems cannot handle medium/heavy castings and systems to program them off-line based on CAD models have never been explored. Most robots used in foundries are programmed by a "teach-in" mode.

This paper describes the development of an affordable and flexible system for the automated fettling and finishing of medium/large castings (up to 1500 kg) made in small numbers. The system has been developed within the EU-project Autofett. At the start of this research project, it was proposed to use a serial robot to perform the different fettling and finishing operations. A robot system has a large work space and is able to reach work piece areas which are often difficult to reach by classical machine tool structures. In addition, the cell to be developed should include a flexible and accurate positioning system for castings and CAD/CAM based generation of collision free robot paths is essential.

Different machining processes for fettling have been considered: arc-air cutting, oxy-fuel burning, slitting, grinding sawing and newer technologies such as plasma, laser, and abrasive jet cutting. Common to all the fettling operations is the requirement to remove the remaining surplus material left after head and riser removal. Today, this operation is usually achieved manually by using a rotating grinding disc. Forces that can be applied during disc grinding are often limited and the occurring wear (decreasing diameter of the disc) makes this process unsuitable for automation. Based on the different applications and requirements of the users within the project, oxy-fuel cutting has been selected to roughly (however better then a manual operation) remove the heads and risers. The remaining part (= the stub) will be removed by a force controlled grinding process using a 'cup' type of disc. This type of disc reduces the effect of wheel wear to essentially one direction (better for automation purposes) and also allows a greater wheel force to be applied during metal removal.

2. Development of the robot based fettling cell

Main components of the developed fettling cell is the flexible clamping unit and a ABB IRB6400® robot system (Figure 1). The robot performs all necessary operations such as part measurement, oxy-fuel cutting and stub-grinding.

Within today practice, the positioning of castings is mostly achieved through jigs and fixtures. Often jigs will not accommodate dimensional variations, they will be costly, and there is no register of part position. The use of low cost fixtures and clamping systems adds to the uncertainty of precise positioning. To cope with this problems, a new clamping system has been developed (Figure 2). This design concept is a result of the following requirements: it should be able to load/unload castings in an efficient way, castings up to 1500 kg and nominal envelope of 1.5x1.5x1.5 m^3 have to be handled and finally the system should have functionality to rotate (index) the casting to different positions. In order to allow flexible clamping of various castings, a V-shaped clamping feature is added on to the casting. The clamping unit itself is a mechanical unit into which the clamping feature (on the casting) can accurately be located. The force to clamp the casting into the unit is delivered by a cable, which lifts the casting into the clamping unit (V-shaped locator). The head of the clamping unit itself can be indexed in different positions. After the fettling operation, the tray is moved upwards to collect the casting.

Automated fettling also needs a proper identification of the part position in relation to the clamping unit. A combination of shape recognition and positional ability is required. Shape recognition is used to see whether the cast does not differ too much from the expected shape. If this would be the case, the cast is rejected and has to be finished manually. Many vision systems for part recognition were tested within this research work for their accuracy and reliability in recognising the casting. The Cognex 4001® camera system has been selected and is able to recognise a part, even with a significant amount of it obscured. This is crucial due to the variability of castings in their "as cast state" due to factors like flash appearing at the split line and solidification distortion of the feed metal. This vision system also proved to be robust for the variability in lighting conditions that are present in a foundry. Developments have been carried out to enable the vision system to obtain images from a safe distance away from the work piece. If the part is accepted, the image data from the vision system is further used to calculate co-ordinate datum points to allow the 3D CAD image to be orientated into the correct plane ready for off-line programming. Final position check and a more exact location of headers and risers is performed by a 3D-touch trigger probe operated by the robot system.

The manual oxy-fuel cutting process is difficult to automate without some customised modification. The first point is that the system has to predict the starting temperature of the preheated surface which is heated by the oxy-fuel torch prior to the cutting process commencing. Secondly is the need to cut the surface as closely as possible, which reduces the subsequent grinding process. Finally, the unit must provide feed back alert when the torch inadvertently collides with the casting. To incorporate all these features, a new torch unit has been developed (the torch system can be recognised on Figure 1). The unit includes a new type of nozzle specially designed to cut close to the casting surface. To achieve, this the torch assembly has built in sensors for collision alert and possesses features which cater for collision situations. The final property is the presence of a sensor which measures the temperature of the preheated casting (e.g.1660 °C) through a lens buried within the torch itself.

For the removal of the remaining stubs (after oxy-fuel cutting), a hydraulic driven grinding system, using a 'cup' type of disc has been developed (Figure 3). These type of discs reduces the effect of wheel wear to essentially one direction and also allows a greater wheel force to be applied during metal removal. Main components of the end-effector are the hydraulic motor including the grinding stone, and the force sensor which is used to feed back force signals towards the robot control system. Because of the required power and the limitations in space and weight, it was decided to use a hydraulic driven end-effector with a maximum power of 18KW and a rotational speed up to 5000 rpm. The force sensor consists of a displacement sensor and a spring. The stiffness of the springs is set to 60 N/mm, but can be increased by a simple replacement of the spring. The stroke of the spring(s) is 20mm, which gives a possible force range from –600N to 600N. The compliance given by the sensor is sufficient to react on irregularities of the casting product and surface roughness, which pass at high frequency. Larger irregularities within the stub or changes of the contour at lower frequency are handled by the force control system.

Force control functionality has been developed within this project by ABB and the University of Lund. The ABB S4CPlus control system has been extended so it can take force signals into account. The force signal is derived from the displacement sensor within the force sensor unit. For the stub-grinding tasks considered in this research, only the motion perpendicular to the surface of the workpiece is required to be force controlled. Therefore, a hybrid force/position control strategy was employed [4].

3. Off-line programming
For the robot based machining of complex parts in low series, such as the envisaged oxy-fuel cutting and stub-grinding process, traditional teach-in programming is not feasible and off-line programming based on CAD-models is needed. Figure 4 shows the information flow starting from a CAD description of the part. The developed CAM system ("Robot Program Generator") is based on an existing system for the collision free programming of welding operations, developed by KPS Rinas. The system has been adapted for automatic generation of robot paths for oxy-fuel cutting and stub-grinding operations. The calculation of robot paths starts from 2 CAD models, which are input as STL-files. The first CAD model is the blank geometry as it is cast (containing all the risers and heads) and

the second CAD model is the final geometry as it will leave the foundry. The input files have the STL-format which makes the use of the software within various industries (using different CAD systems) beneficial.

Prior to the generation of the robot path, information about the position and orientation of the cast within the clamping unit is send to the CAM system (see above). Technological parameters like spindle speed, feed rate,... are described within the process model database. These models describe the influence of controlled process parameters such as force, speed, cut width on resulting parameters such as material removal rate and surface quality. During tool path generation, the complete robot system is checked for collisions and if a collision happens, it is solved by launching a collision avoidance algorithm. The collision avoidance algorithm is totally integrated within the tool path generation algorithm, which is a similar approach as being developed by Lauwers for multi-axis milling operations [5,6]. Dependent on the type of movement, the following collision avoidance algorithms have been implemented. When generating the robot air movements, which means that no tool operation is considered, the system makes use of all available degrees of freedom in the robot system, including external axes, to find a possible collision-free robot movement. In case of planning (generating) process paths, the collision avoidance algorithm will secure that the tool is still following the specified path with the required position and orientation for correct process operation. Still all degrees of freedom, including the external axes and tool angle variations within the given tolerance limits, are used to find collision-free robot positions during that movement. If the optimal tool orientation is not achievable with any robot movement, the "Robot Program Generator" is capable of testing out and finding alternative tool orientations, still within the specified operational tolerances, that can be performed by the robot without collision.

The collision free robot program is send to the cell computer for execution. The syntax of the program is similar to the RAPID language (used for programming ABB robot systems), but includes additional statements related to force control.

4. Machining strategies and process models

A suitable cutting strategy for the oxy-fuel cutting of the risers and feeders has been defined (Figure 5). The aim is to remove to riser/feeder as good as possible, so the size of the remaining stub is limited. First the nozzle is positioned on the riser geometry (not very close to the final geometry), in order to set-up a stable cutting processes (stable flame). Reaching these stable conditions is performed by the cell computer, based on the information received from the sensor build within the nozzle. Once a stable flame is obtained, the nozzle is moved through the riser (via a circular movement) towards the final part, where the real cutting path starts. An initial process model for oxy-fuel cutting, to be used as input for off-line programming, has been set-up based on experiments. Figure 5 shows the relation between the section thickness and the cutting speed.

For stub-grinding, a robust strategy for force control has been developed and implemented. The CAM system generates different grinding tracks and each track contains different sections (paths, Figure 6). Each section can be characterised by different values for speed, change of speed, force and change of force can be defined on different sections of the programmed path. Entry and exit are examples of such sections and make a smooth transition of non-cutting to cutting movements possible. More specifically, a grinding track is broken down into five phases called approach, build-up, process, decline and withdraw. Some paths such as entry and exit, are position controlled, while other segments are position/force controlled. This is also one of the reasons why a combined position/force control algorithm has been implemented (see above).

In order to develop initial data for the process models, grinding experiments have been carried out. A simple geometry (bar) with different thickness values were made (20mm, 30mm and 50mm) in Steel ASTM A216 WCB. Several tests were carried out using a "cup" stone (NCZ 14/16 NOB) with medium grains and a medium to high hardness. The material removal rate has been calculated based on the measurement of the thickness of the removed material measured on a coordinate measuring machine scanning the part. For these experiments, the spindle speed and inclination angle have been kept constant at 5000 rpm and 10°. The results of the tests are shown in Table 1. It is clear from the tests that a higher force results in a higher material removal rate, but a higher force also resulted in a higher surface roughness.

5. Test case and results

The developed fettling cell has been validated based on industrial castings provided by the end-users within the EU-project consortium. Figure 7 shows some important steps of the entire fettling of a pump house. To allow flexible clamping, a clamping feature has been added to the design of the product. A collision free robot path for oxy-fuel cutting and stub-grinding is automatically generated based on the two STL-file descriptions: one for the as cast model and one of the final geometry. The oxy-fuel cutting strategy allows to remove the headers and risers quite accurately, leaving a small, but constant stock to be removed by the stub-grinding process. Due to the accurate cut-off of headers and risers, the stub-grinding process becomes much more easy and controlable.

6. Conclusion

The outcome of this research project has proven that fettling of castings in low series is possible. The different components for obtaining flexibility (flexible clamping unit, measurement system, collision free off-line programming, flexible change of end-effectors, force controlled operations,..) have been developed, but a further and better integration is needed to come up with a commercial system. Further research towards the development of better grinding strategies is needed.

7. References

1   Shelley M. S., *Robotic casting finishing at Hitchcock Industries*, *Modern casting*, 1988, pp. 41-44.

2   Shairi Y., and Pollig O., *Robotic Integrated Cells for finishing operations*, Transactions of the American Foundrymans Society, 1986, pp. 243-248.

3   Spinner D. B., *Advanced automatic fettling*, Transactions of the American Foundrymans society, 1989, pp. 767-768.

4   M. H. Raibert and C. J. J., *Hybrid position/force control of manipulators*, ASME Journal of Dynamic Systems, Measurement and Control, vol. 103, no. 2, 1981, pp. 126-133,

5   B. Lauwers, P. Dejonghe and J.P. Kruth, *Optimal and collision free tool posture in five-axis machining through the tight integration of tool path generation and machine simulation*, Computer-Aided Design, Vol. 35 (5), 2003, pp. 421-432.

6   B. Lauwers, J.P. Kruth, P. Dejonghe and R. Vreys, *Efficient NC-programming of multi-axes milling machines through the integration of tool path generation and NC-simulation*, Annals of the CIRP, Vol. 49/2, 2000, pp. 367-370.

Acknowledgments

The development of this fettling cell, including the off-line programming software, has been developed in the frame of the EU project AUTOFETT (GRD1-2000-25135), including the following partners: Castings Technology International (UK), K.U.Leuven (BE), Wärtsilä (NL), Norton Cast Products Ltd. (UK), Cometna – Compannia Metallurgica Nacional S.A. (P), Kranendonk Industriele Automatisering B.V. (NL), ABB Flexible Automation (S), CRF (I), KPS RINAS (DK).

Tables

Table 1: Grinding test results

	Force (N)	Feed (mm/s)	Thickness (mm)	MMR (mm^3/s)
Pr 1	350	35	25	100
Pr 2	350	45	25	96
Pr 3	350	60	25	60
Pr 4	450	60	25	160
Pr 5	550	60	25	220
Pr 6	350	100	25	80
Pr 7	350	45	30	132
Pr 8	350	45	30	107

Figures

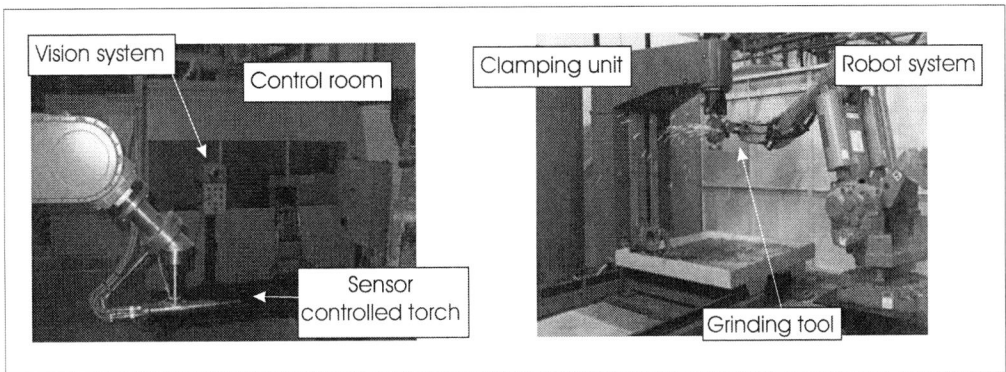

Figure 1: The developed fully automated robot based fettling cell

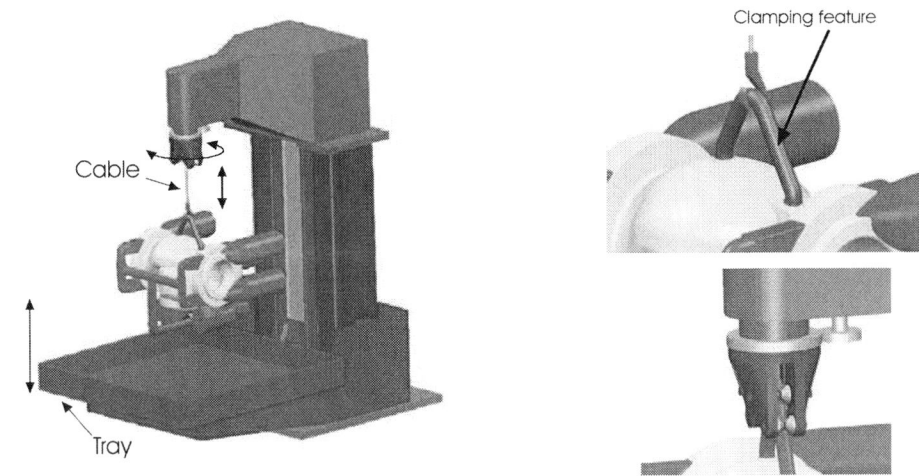

Figure 2: Schematic view of the flexible clamping unit

Figure 3: Grinding end-effector

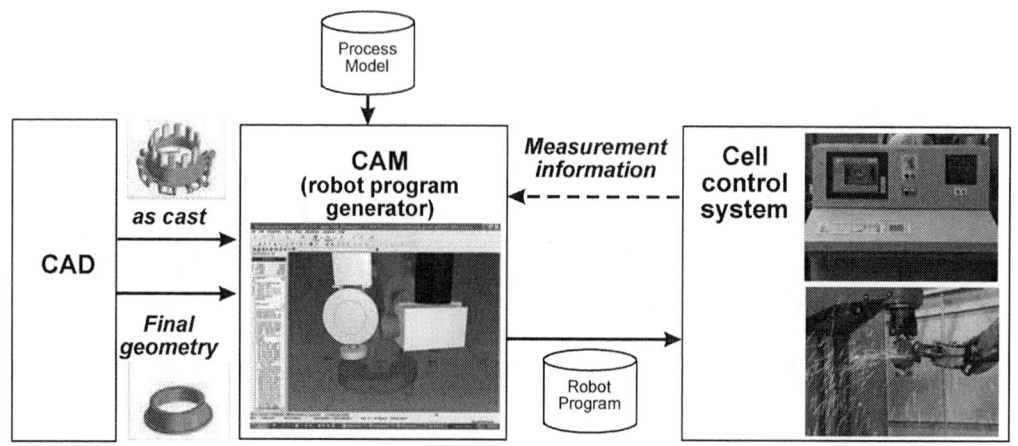

Figure 4: Structure for off-line programming, based on CAD/CAM

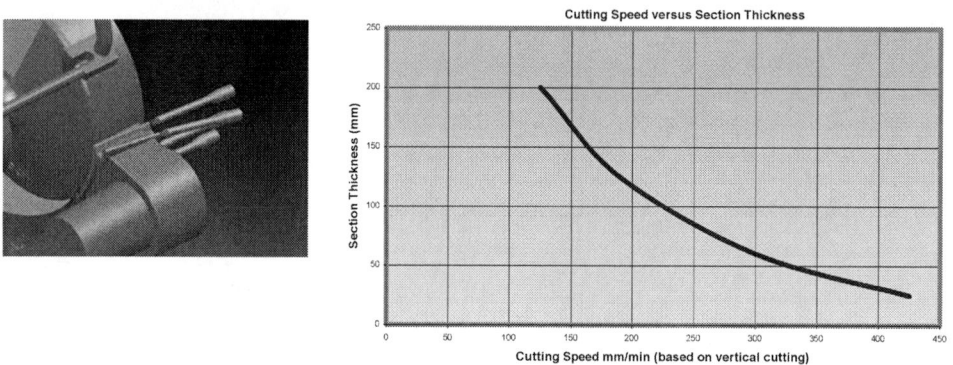

Figure 5: Oxy-fuel cutting strategy – process model

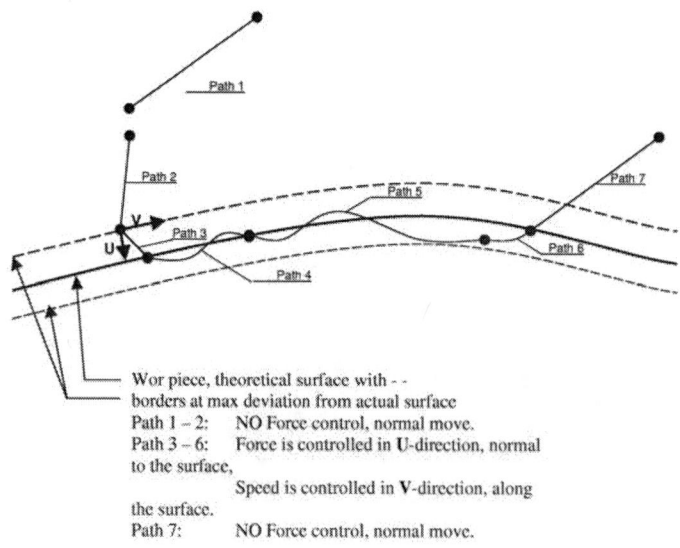

Figure 6: Force controlled stub-grinding strategy

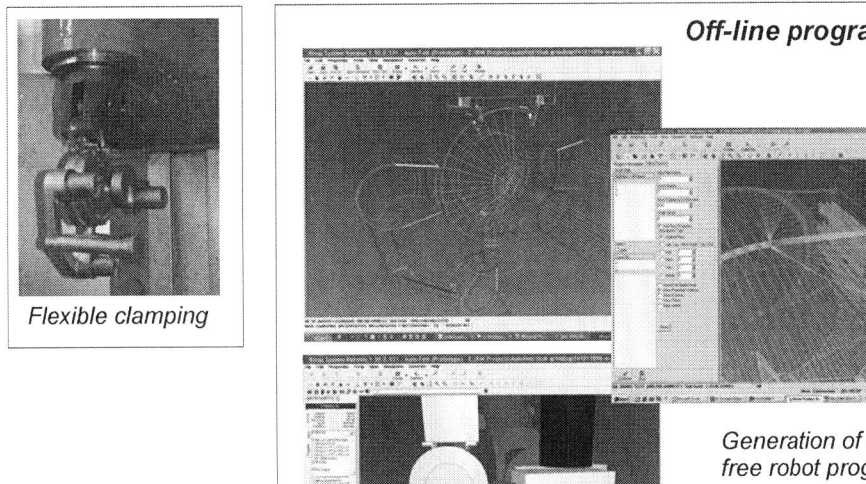

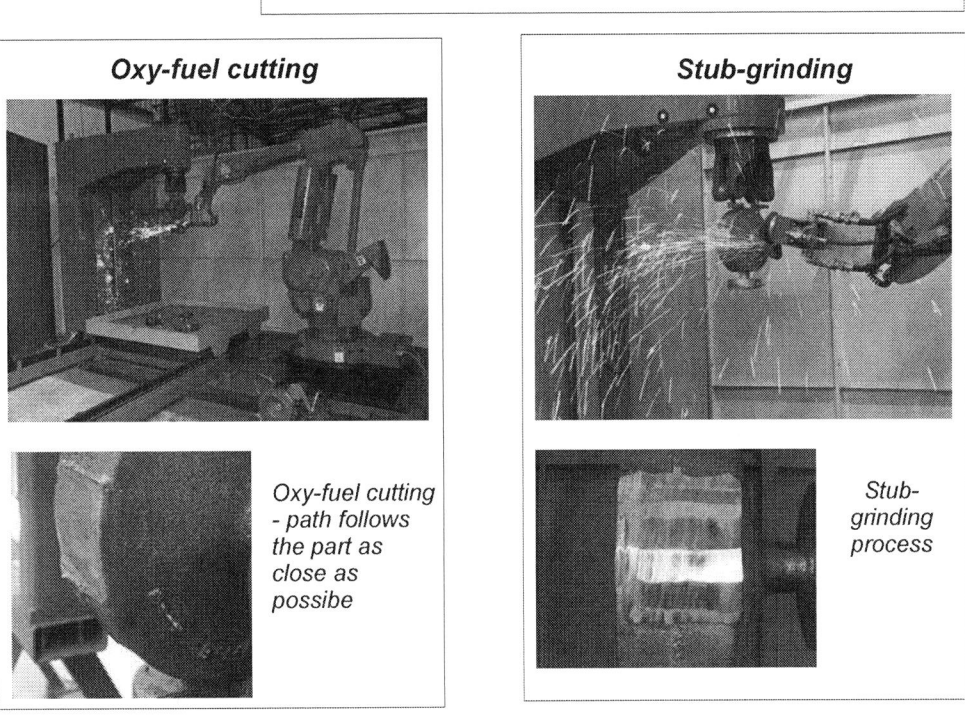

Figure 7: Test case: fettling of a pump housing

Automatic Visual 3-D Inspection of Castings

D vom Stein, inspectomation GmbH, Mannheim, Germany

Abstract

So far, mainly 2-D image acquisition and processing is used for automatic visual quality inspection of castings in foundries. Some tasks however call for 3-D image acquisition and processing, which has become feasible at reasonable costs not until quite recently. The major technique presently in use to acquire so-called range images is based on the triangulation principle that is explained succinctly. Subsequently, two different field-tested applications are presented: Firstly the all side surface inspection of cylinder heads using a very flexible robot based vision system and secondly the recognition of plain text relief information on castings for identification and traceability purposes.

Key words automatic visual inspection, 3-D image acquisition, triangulation principle, surface inspection, recognition of relief information

Introduction

Presently, within the framework of quality control in foundries, automatic visual inspection is predominantly based on acquisition and processing of one or several 2-D images of the casting [1]. Usually, for this purpose monochrome cameras are applied providing intensity images proportional to the reflectivity coefficient of the observed surface. Rather infrequently colour cameras are used acquiring spectrally resolved intensity information.

Actually, in many cases the relevant information consists in the topography of the part and its deviations from a given nominal geometry. Using an appropriate image acquisition setup with special emphasis on lighting and image formation optics, many inspection tasks can be tackled based on 2-D images. Industrially approved examples from our company comprise

- the qualitative inspection of cores and moulds [2] with the aid of a patented shadow modulation technique using several luminous sources turned on and off sequentially resulting in an image series with a certain amount of implicit 3-D information that can be merged into a single result image clearly indicating defective regions or
- locating total or partial blockages in the vents of brake discs [3] using silhouette images of two or more line scan cameras equipped with telecentric lenses [4].

In the aforementioned cases an elaborated setup and a thorough selection of the features separating good parts from bad ones lead to a robust and at the same time competitive solution based on 2-D imaging.

However, various inspection problems can not reasonably be addressed by these means. Particularly for the determination of vertical extents, i.e. distance measures in the line of sight of the camera, 3-D image acquisition techniques are mandatory.

3-D image acquisition

In the following we will only be concerned with the acquisition of so-called range images [5] that are composed of distance information for every camera pixel. This stems from the fact that castings have opaque surfaces, so we can get only surface data (depth or height information) usually correctly designated as 2.5-D data to distinguish it from volumetric data such as the spatial characteristics of the absorption coefficient of transparent materials or the emission or absorption coefficients in computer tomography. Since range image data consists of a point cloud in space, we simply call them 3-D images too.

A common method for 3-D image acquisition is adopted from the human visual system. In stereopsis two cameras are looking at the same part from slightly different angles as outlined in figure 1. A single object point is mapped to different locations in the images of the left and the right camera, see figure 2. From this disparity the spatial location of the object point can be calculated subject to the knowledge of the internal and external parameters of both cameras. In order to compute the disparity, for every image point in the left image the corresponding point in the right image has to be found and vice versa. This correspondence analysis can be carried out easily for points in structured regions, e.g. the corner points of the cubic object in figure 2a. For points located in rather homogeneous regions, it is difficult to find the corresponding pair, confer figure 2b. Thus, this technique only delivers sparse 3-D point clouds.

To impose structure on homogeneous regions, a light source can be added to the setup projecting a grid or another pattern onto the surface as sketched in figure 3. This method is called active stereo and delivers dense 3-D point clouds.

If we accept the necessity of using a structured illumination, we are able to forbear from the second camera, arriving at a solution called triangulation. In the simplest case one point on the surface is illuminated via a finely collimated light ray and observed from a line of sight different from the projection direction. From the location of the light spot observed by a position-sensitive sensor, the height of the associated point on the surface can be computed as illustrated by figure 4a. Provided that sufficient light is scattered onto the sensor, meaning in particular that on the one hand the object is not too specular and on the other hand does not absorb all light, this procedure is pretty independent of the reflectance of the surface.

In order to cover a whole surface, a 2-D scan has to be performed sequentially measuring point after point. Speeding up the measurement can be accomplished by one of the following two generalisations:

1. The so-called light-sectioning method uses a laser line instead of a laser point and a matrix camera instead of a line-scan camera. With every image taken, range information along the whole laser line is computed. Special cameras with on-board processing units are able to locate the laser line and output profile data at rates of up to 20 kHz. Whole surface coverage is met by a 1-D translational or rotary scan movement.

2. If we use several lines at once, we arrive at a stripe pattern. This approach is named coded lighting and projects several fringe patterns with varying spatial wavelengths onto the surface in order to avoid ambiguities at height steps. It covers the whole surface without a relative movement between part and image acquisition unit but needs some time to sequentially project and acquire the image series.

A certain restriction is common to stereopsis and triangulation based methods: Both use two optical paths enclosing an angle of typically 15°–30°. Hence parts of the surface, especially in regions where the normal of the surface is nearly perpendicular to one of the optical axes or even points away from the camera or the projector, can not be gauged as illustrated in figure 4b.

The ultimate 3-D imaging unit would therefore be a monocular system, i.e. one featuring only a single optical axis. This can be accomplished by building an optical radar measuring the delay of an optical impulse (time-of-flight method). In the future such sensors will be available at reasonable prices and resolutions (i.e. PMD device [6]).

3-D image processing and analysis

The acquired 3-D data must be preprocessed in an appropriate way before the actual comparison:

- Outliers should be removed by an adequate filtering operation.
- Different views (partial surfaces) must be combined into a total data record in correct positional arrangement.
- Furthermore, differences in the spatial position between current object and reference data record should be compensated.

The reference data is derived from either one or more good parts or CAD data. The latter is problematic since often the chill-mould or the moulding tool does not conform exactly to the associated CAD data, especially if there are subsequent modifications not updated in the CAD drawings. After the definition of the admissible height deviations from the nominal shape in the form of upper and lower tolerance limits and a minimum lateral extent, the actual comparison is carried out pointwise. Defective regions are marked directly in the 3-D data in order to indicate their location to the operator. Beyond the mere detection of deviations, their sizes can also be accounted for quantitatively. Moreover, the computation of derived values such as lengths, angles, and parameters of geometric primitives, e.g. diameters of circles or cylinders or normals of planes, is feasible. If many measured points contribute to a single result, the stochastic errors level out, and the result will be more accurate.

First application: all-side surface inspection of cylinder heads

The schematic layout of the inspection cell is depicted in figure 5. Whilst the cylinder head is transported from the machining cell into the inspection cell, the camshaft face is scanned. Afterwards, the remaining faces are presented to the visual inspection system by a robot in different views. The acquisition unit in figure 6 contains a light-section sensor moving on a high-dynamic and precise linear actuator. Several lasers are used on opposite sides to minimise the blind areas; different triangulation angles allow to trade off resolution against measuring range. Each scan takes up about 1 s and provides approximately 5 million 3-D points; the fault detection is completed after another second, thus enabling the inline inspection of 100 percent of the production. The attainable height resolution clearly depends on diverse design parameters of the setup. The system at hand achieves a vertical resolution of 0.1 mm on a lateral sampling grid of 0.2 mm × 0.2 mm.

By the use of a robot, this system is extremely flexible: Different products can be examined by the same system, if necessary using different grippers. Beyond that, the test strategy can be chosen very flexibly, selected sides can be scanned several times in different situations using different lasers and thus triangulation angles. If the cycle time is exhausted, several inspections can be alternated in a random sampling way. The check routines are configured by means of a concise graphical user interface shown in figure 7. All inspection results are stored in a database, that can be queried from every networked computer. In particular, the workman at the reworking place only has to scan the data matrix code and immediately gets images of the defective faces with the exact locations of the flaws, see figure 8.

Second application: recognition of relief information

At present, part traceability becomes more important for foundries and their customers. Particularly in order to limit the amount of parts to be called back subsequent to a failure of safety parts, e.g. components of the chassis, traceability up to the cast date or even the pouring ladle is desirable. Also in the context of improvement of process parameters, the part flow should be monitored, at best individually labelling every part [7]. For these purposes they should be marked at the latest during pouring and the label should withstand subsequent processing steps, which could only be achieved by embossed or engraved symbols. They can be generated by mechanical gravure of cores or moulds or contactless via laser inscription.

Since the information does not manifest itself in the reflective properties of the part but rather in the relief, 3-D techniques are the appropriate means providing high recognition rates and robustness even if rust, scratches, or traces of sand are present. By means of the stripe projection technique the relevant shape information can be recorded. The deviation from a reference part without marking presents the sought information (fig. 9a), which has to be projected or unrolled onto a plane (fig. 9b) to obtain a grey scale image with high contrast (almost a binary image, fig. 9c). Afterwards the information of interest can be extracted by standard optical character recognition (OCR) routines (fig. 9d). During the acquisition time of 1 – 2 s the part must not move, during the succeeding calculations taking about 3 s, the conveying can be resumed. Images from one of our installations are reproduced in figures 10 – 12. Existent latitude of design should be used to boost the recognition rate by using fonts optimised for OCR (i.e. OCRB), avoiding similar symbols like 0, O and Q, incorporating validity checks or purposefully introducing redundancy like a check digit. Misreading rates lower than one-tenth of a percent can be achieved.

Conclusion

Techniques for the acquisition and processing of 3-D images certainly will play a major role for the quality inspection of raw as well as machined cast parts since they enable the direct acquisition of the shape that is the relevant quantity for the detection of defects.

References

1. Beyerer J, Bierweiler T, Klawitter T, and vom Stein D: *Automatic inspection of castings*, Casting and Technology International, No 4, 2003, pp24–35.

2. Beyerer J, *A vision of quality*, Cast Metal Times, Vol 3, No 6, Oct/Nov 2001, pp30–33.

3. Beyerer J, vom Stein D, and Klawitter T, *Automated quality control of ventilated brake discs*, Cast Metal Times, Vol 5, No 4, Jun/Jul 2003, pp58–60.

4. Klawitter T, *Automatische Lüfterschlitzkontrolle von Bremsscheiben*, to be published in Giesserei, Vol 92, No 12, Dec 2005.

5. Jarvis R A, *Range sensing for computer vision,* in Jain A K and Flynn P J, *Three-Dimensional Object Recognition Systems,* Elsevier Science Publishers, 1993, pp17–56.

6. Schwarte R, *Breakthrough in Fast 3D-Imaging Using PMD- and OEP-Technology*, IMS Workshop in Duisburg, May 25-26 2004.

7. Bähr R, Ernst W et al., *Teilespezifische Kennzeichnung von Gussstücken*, Giesserei, Vol 91, No 6, Jun 2004, pp40–48.

Figures

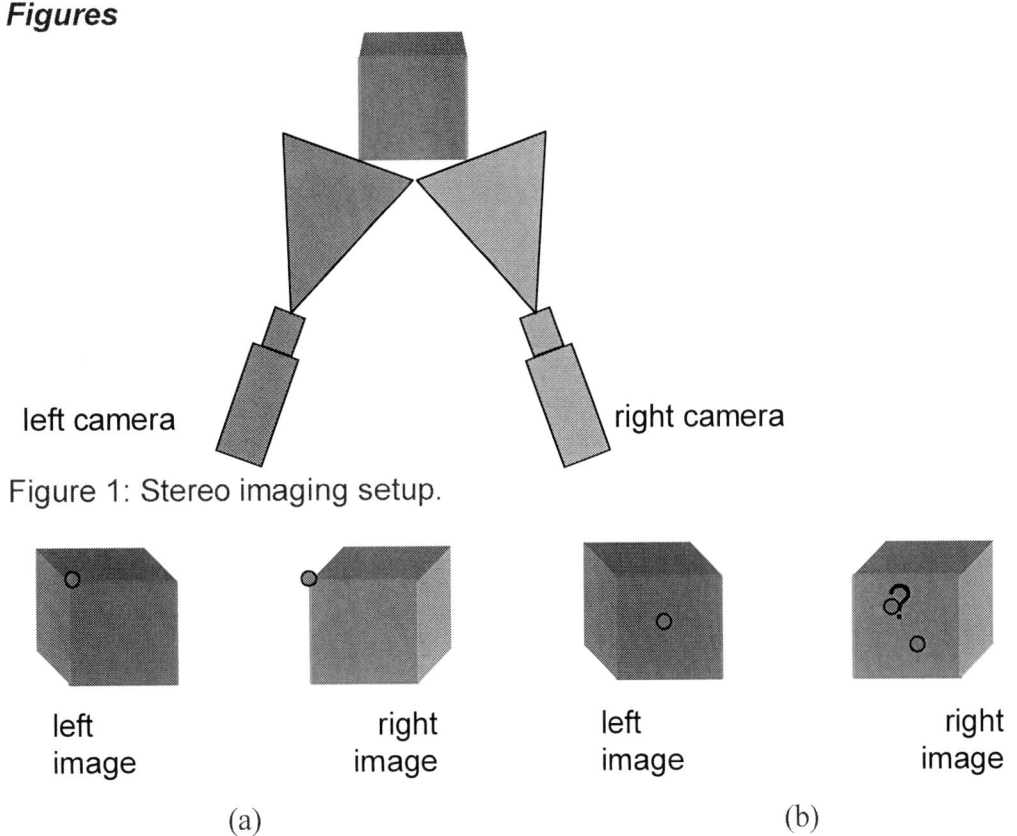

Figure 1: Stereo imaging setup.

Figure 2: Correspondence problem: (a) feature point (i.e. corner); (b) ambiguity in homogeneous region.

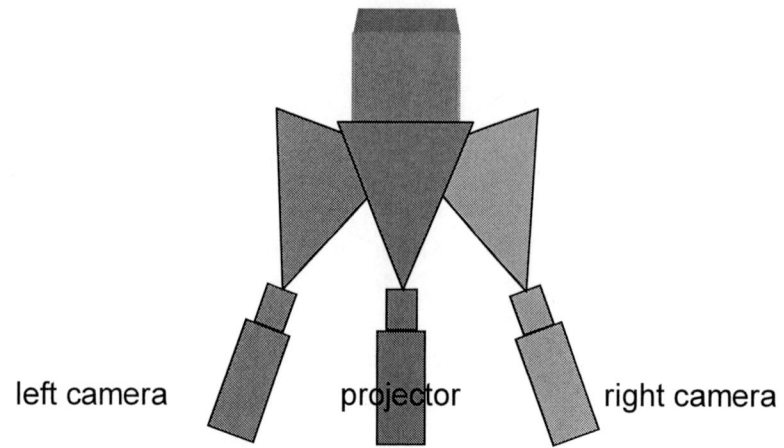

Figure 3: Active stereo imaging setup.

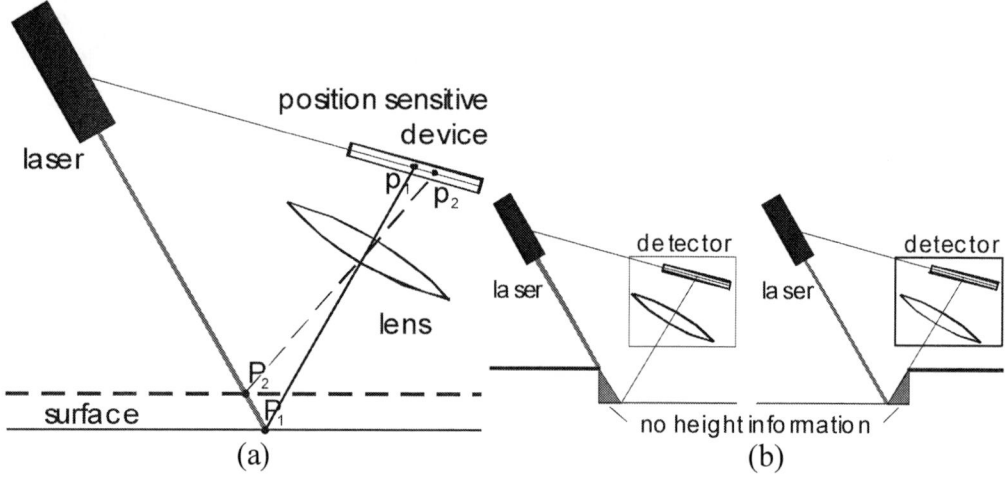

Figure 4: Triangulation principle: (a) setup; (b) shadowing effects at steep edges due to different directions of illumination and observation.

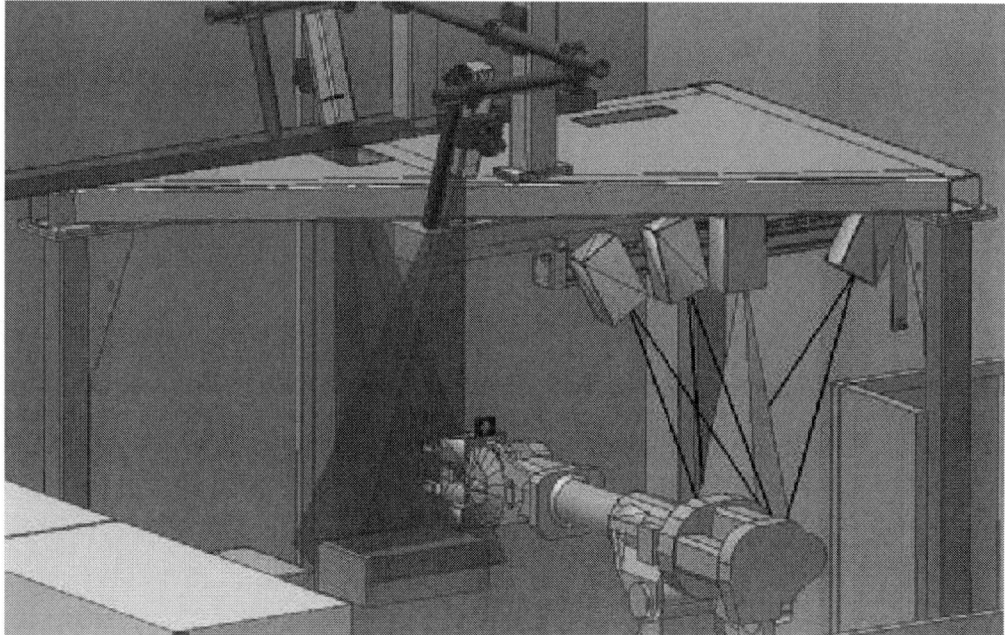

Figure 5: Perspective view of inspection cell.

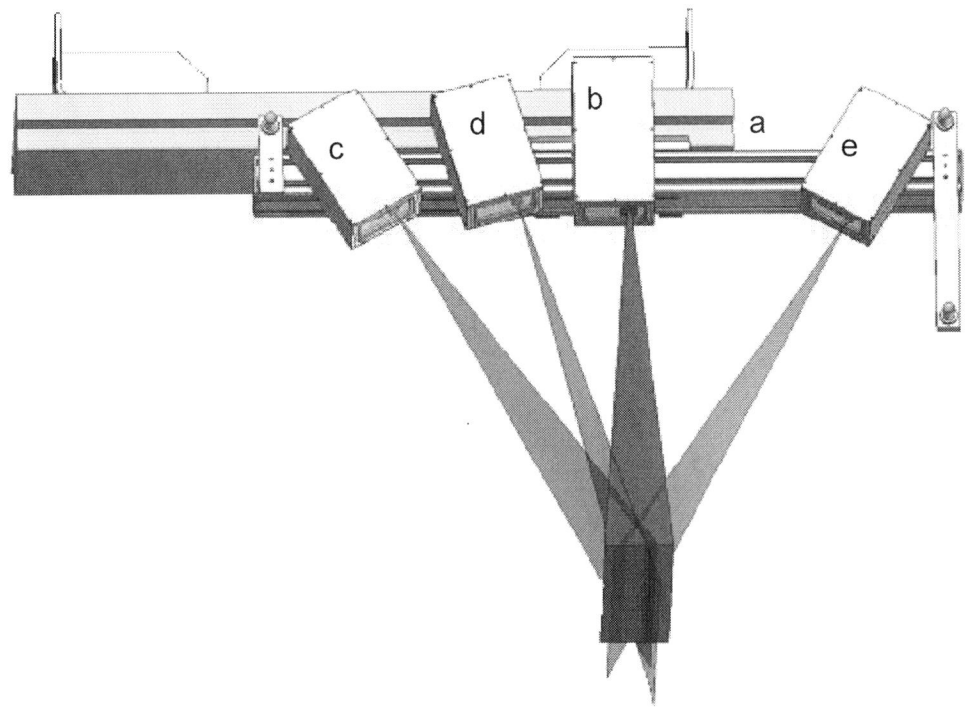

Figure 6: 3-D triangulation scan unit: (a) linear induction motor; (b) 3-D camera; laser line generators inclined at (c) 30°, (d) 15° and (e) –30°.

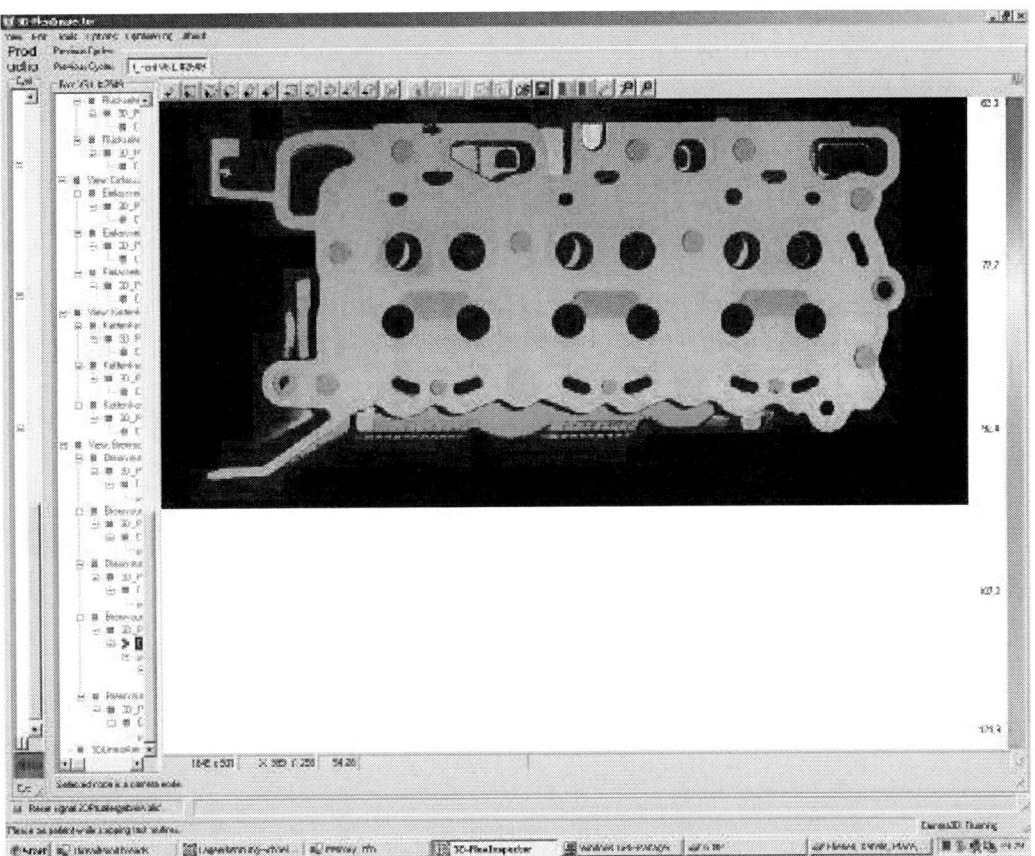

Figure 7: Graphical user interface showing a hierarchic program tree on the left and a colour encoded range image of flat combustion chambers of a cylinder head on the right.

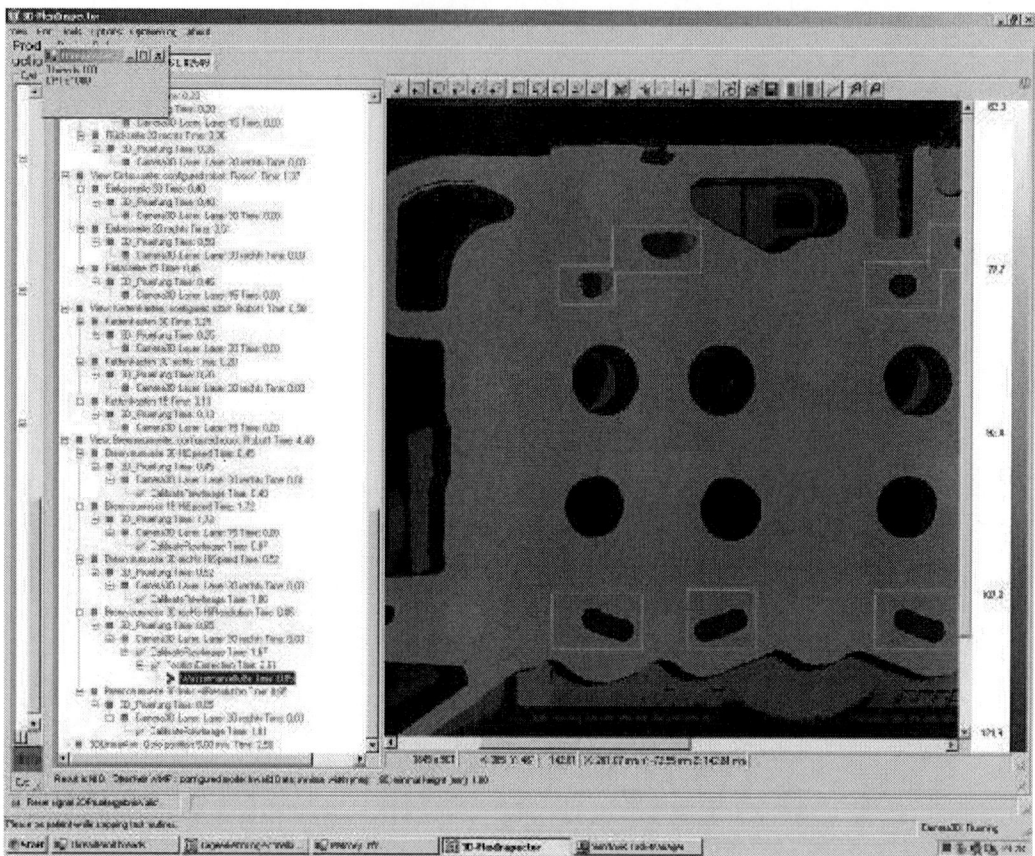

Figure 8: Result image clearly indicating the location and size of partial closures of water jacket in red colour.

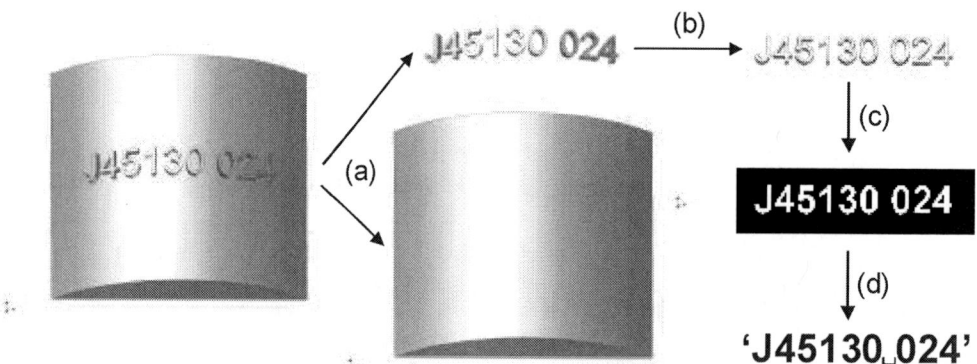

Figure 9: Operating principle of CAST-READER: (a) separation of relief information; (b) unwinding or projection onto reference plane; (c) grey value encoded range image; (d) 2-D optical character recognition.

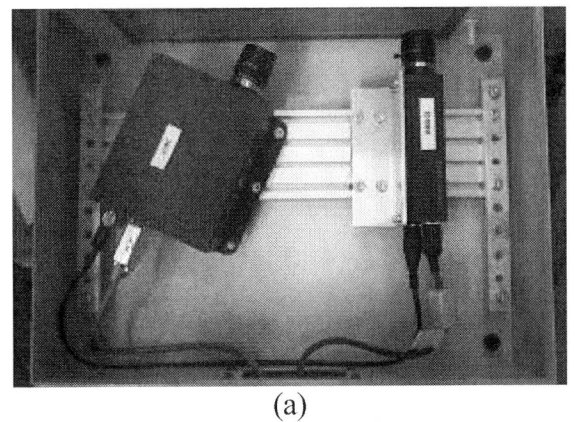 

(a)   (b)

Figure 10: (a) triangulation setup comprising projector (left) and camera (right); (b) read operation applied to crankcase.

 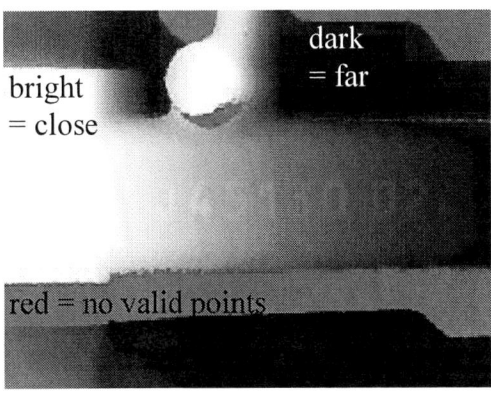

Figure 11: Casting with embossed plain text: (a) grey scale image with visible scratches; (b) raw grey scale encoded range image immune to varying reflectivity (inscription plane tilted with respect to image plane).

(a)   (b)

Figure 12: range image (a) after separation and projection and (b) with OCR result shown as overlay.

Thermomechanical Treatment of Austempered Ductile Iron

A A Nofal, H Nasr El-din and M M Ibrahim

Central Metallurgical Research and Development Institute, (CMRDI), Cairo, Egypt

Abstract

The production of lightweight ferrous castings with increased strength properties became unavoidable to face the serious challenge of lighter aluminum and magnesium castings. The relatively new ferrous casting alloy ADI offers promising strength prospects, whereas the thermo-mechanical treatment of ductile iron may suggest a new route for production of thin-wall products. This work aims at studying the influence of thermomechanical treatment, either by ausforming just after quenching and before the onset of austempering reaction or either by cold rolling after austempering.

In the first part of this work, ausforming of ADI up to 25% reduction in height during a rolling operation was found to add a mechanical processing component to the conventional ADI heat treatment, thus increasing the rate of ausferrite formation and leading to a much finer and more homogeneous ausferrite product. The kinetics of ausferrite formation was studied using both metallographic as well as XRD-techniques. The effect of ausforming on the strength values was quite dramatic (up to 70 and 50% increase in the yield and ultimate strength respectively). A mechanism involving both a refined microstructural scale and an elevated dislocation density was suggested. Nickel is added to ADI to increase hardenability of thick section castings, and ausforming to higher degrees of deformation is necessary to alleviate the deleterious effect of alloy segregation on ductility.

In the second part of this work, the influence of cold rolling (CR) on the mechanical properties and structural characteristics of ADI was investigated. The variation in properties was related to the amount of retained austenite (γ_r) and its mechanically induced transformation. In the course of tensile deformation of ADI, transformation induced plasticity (TRIP) takes place, indicated by the increase of the instantaneous value of strain-hardening exponent with tensile strain. The amount of retained austenite was found to decrease due to partial transformation of γ_r to martensite under the CR strain. Such strain-induced transformation resulted in higher amounts of mechanically generated martensite. The strength and hardness properties were therefore increased, while ductility and impact toughness decreased with increasing CR reduction.

Key words ADI, ausforming, cold-rolling, thermomechanical treatment, transformation induced plasticity.

Introduction

Austempered ductile iron (ADI) is a relatively new engineering material with exceptional combination of mechanical properties and marked potential for numerous applications [1]. The attractive properties of ADI return to its distinct and unique microstructure, which consists of fine acicular ferrite within carbon-enriched stabilized austenite (ausferrite). The relationship between microstructure and mechanical properties of this material has been the subject of extensive studies [2,3]. The morphology of the final two-phase matrix microstructure is determined by the number, shape and size of the initially formed ferrite platelets in the first stage austempering reaction. The control of this stage of transformation will, therefore, ultimately control the final microstructure and mechanical properties.

It has been shown [4] that the rate of ferrite formation during stage I austempering may be controlled by the following processing variables:

- chemical - including alloy content selection, which may be necessary for hardenability, together with the austenitization temperature selection which controls the matrix carbon content
- thermal - including austempering temperature and time
- mechanical - including mechanical deformation before austempering (ausforming) or after austempering (cold rolling).

Naturally, an optimum final microstructure could be produced by including elements of all three processing variables. It has been shown [4-7] that mechanical processing of ADI can act as a control valve for the stage I austempering reaction. In ausformed austempered ductile iron (AADI), mechanical deformation is utilized to affect the microstructure and, consequently, the mechanical properties of ductile iron due to acceleration of ausferrite reaction, refining the microstructure and increase of the structural homogeneity. The first objective of this work is to study the influence of ausforming to different degrees of deformation on the kinetics and mechanical properties of the nickel alloyed ductile iron.

Transformation of austenite to martensite by deformation has been extensively studied in austenite stainless steels, whereas very little has been reported [4] on the martensite transformation induced by cold rolling and its effect on microstructure and hardness of low alloy ADI. In the course of fracture toughness tests of ADI in the upper bainite region, containing high volume fraction of retained austenite, $\gamma_r \rightarrow \alpha$ (martensite) transformation induced plasticity TRIP has been reported to occur [8], leading to superior toughness compared to conventional cast iron. Another objective of the present work is to study the influence of the generated microstructure on the tensile properties of the unalloyed ADI taking into consideration the TRIP effect. The influence of cold rolling on the tensile properties, impact toughness and hardness has been as well investigated.

Experimental

Table 1 shows the chemical analysis of irons under investigation where alloys I and II were subjected to ausforming, whereas cooled rolling was limited to alloy I. Ausforming was performed in the following sequence: austenitizing at 900°C for 60 minutes - quenching into a salt bath to the austempering temperature T_A of 375°C for one minute, rolling to height reductions (RH) of 12.5 and 25% through one or two passes and then austempering in the same salt bath at T_A for different times (0.03-120 min) followed by water quenching to room temperature.

The unalloyed irons were austenitized as mentioned before and then subjected to rapid quenching into the salt bath at 400°C for 60 min, followed by water quenching. Rectangular blanks of 8 x 20 x 220 mm were then cold rolled to different reductions in height of 7, 13, 19 and 25%. Standard flat tensile specimens, DIN 50125 E5 with 16 mm x 50 mm gauge section were prepared along the rolling direction from the unalloyed. The specimens were submitted to uniaxial tensile test on Instron 4112 tensile testing machine at constant cross head speed of 0.5 mm/min.

The percentage of the matrix transformed to ausferrite was determined using the point counting technique. The volume fraction of the retained austenite (X_γ) and its carbon content (C_γ) were determined using X-ray diffraction analysis [5].

Results

Ausforming: Ausforming to 12.5 and 25% thickness reduction was found to have a significant refinement effect on the microstructure of 2.0% Ni alloyed iron austempered for 10 minutes (Fig. 1). Fig. 2 illustrates the effect of ausforming to 25% RH on the refinement of the microstructure. Nickel promotes free austenite and after austempering for very short times, the residual unreacted austenite transforms to martensite on cooling to room temperature (Fig. 3). The rate of stage I transformation was found to be higher in the alloyed ADI compared to the 2% Ni-alloyed irons. For a given short austempering time, the ausferrite transformation was markedly accelerated due to the driving force introduced by deformation (Fig. 4). Ausforming to 25% reduction followed by austempering for short times of 2 seconds up to one minute resulted in extremely high volume fractions of ausferrite of more than 90% and Fig. 4 and 5 show that the transformation in these specimens has almost gone to completion.

Analysis of the XRD results (Fig. 6) of the austempering reaction kinetics of ADI alloyed with 2% Ni indicates that in the conventionally processed irons, the saturated austenite total carbon content $C_\gamma.X_\gamma$ remarkably increases only after 10 minutes austempering. Ausforming resulted in faster progress of the stage I reaction at short austempering times and this effect may be noticed even at as short times as 2 seconds. The $C_\gamma.X_\gamma$ in the ausformed irons undergoes a slight decrease after about 100 minutes austempering which may indicate the onset of stage II austempering transformation. Both the yield and ultimate tensile strength values of AADI were found to be superior to those of the conventionally processed ones. Fig. 7 shows the dramatic increase in strength values of 2% Ni ADI ausformed to 25% reduction and austempered for 10 minutes, where 70% and 50% increase in the yield and ultimate tensile strength values respectively may be noticed.

Fig. 8 shows that the ductility of the conventionally austempered irons alloyed with 2% Ni is rather low at short austempering times, whereas prolonged holding at the austempering temperature improves ductility. Ausforming of this alloy to 12.5% leads to a significant improvement of ductility at austempering times $\leq$ 60 minutes. The increased ausforming to 25% results in a slight decrease in the ductility over the entire range of the austempering times.

Results of tensile tests are given in Fig 9. Each data point represents the average of typically four tensile specimens austempered for a given time. The austempering time in minutes is indicated to each data point. Ultimate tensile strength is plotted against ductility and a curve generated from the minimum specifications for ADI (ASTM standard A897-90) is superimposed on the plot for comparison. Both the UTS and ductilities of ausformed ADIs are superior to those of conventionally process ones, regardless of the austempering times. The improved ductility in ausformed specimens is only shown for the alloyed ADI.

Cold rolling: The austempering conditions of the unalloyed specimens resulted in complete transformation of austenite to fine ausferrite matrix of X_γ = 39% and C_γ = 1.8%. The austempering treatment results in 1065 MPa tensile strength and 7.75% elongation. In Fig. 10 ln true stress versus ln true strain is illustrated. The data are fitted by two intersecting straight lines over the entire range of strain. At plastic strain ε = 0.0094, there is a clear increase in slope of the straight line fitting the data corresponding to the strain hardening exponent "n", which is determined from the true stress-true strain curve.

Increasing cold rolling CR reductions of ADI to 19%, the amount of γ_r decreased due to partial transformation of γ_r mechanically generated martensite (Figs. 11 and 12). The volume fraction of martensite induced by cold deformation was calculated as the difference between the original austenite content and that measured after deformation. The metallographic results were confirmed by X-ray diffraction. Fig. 13 shows the variation of peak intensity for austenite and (ferrite + martensite) phases in the unalloyed ADI. It is evident that the intensity of $(111)_\gamma$ decreases with cold deformation. Figs. 14 and 15 illustrate that the cold deformation results in marked increase of ultimate tensile strength and hardness, whereas elongation and impact toughness decrease.

Discussion

Ausformed ADI: The properties of ADI depend on the ratio of saturated austenite to ferrite as well as the morphology of ferrite in the ausferrite mixture. The remarkable accelerated transformation kinetics of the ausformed austempered irons indicate that the additional energy supplied by deformation adds significantly to the driving forces for nucleation of bainite ferrite. Many more ferrite particles, accordingly, nucleate preferentially on slip bands and grain boundaries. Ausforming results in the elimination of large volumes of austenite, referred to as type II and/or type III, known to reduce the microstructural uniformity of conventionally austempered iron.

Table 2 shows that the austempering time required to develop, for example 50 and 70% of ausferrite increases in the Ni-alloyed ADI. such a delay indicates that the chemical driving force for transformation is reduced by alloying. The effect of ausforming on increasing the rate of ausferrite formation was confirmed by the XRD results. About 30% volume fraction of retained austenite could be obtained after austempering for <1 minute, whereas in the conventionally processed irons, this fraction could be obtained after austempering for about 8 minutes. A slight decrease of both X_γ and C_γ after austempering for time periods longer than 90 minutes may be attributed to the precipitation of some carbide particles at the expense of the carburized austenite (the onset of stage II transformation).

Two different mechanisms may attribute to the increased yield and ultimate strengths of AADI. The microstructural refinement resulting from ausforming can contribute to the higher yield strength, whereas the warm working of austenite would increase the dislocation density in the bainitic ferrite, resulting in elevated yield and ultimate strength values [5].

The rather low ductility of this alloy at short austempering times (Fig. 8) is apparently due to the high amounts of martensite generated during cooling. Prolonged holding at the austempering temperature improves ductility as the austenite carbon content, and, hence austenite stability will decrease leading to lower amounts of martensite. Structure refinement should have little beneficial effect in a material, where cracks initiate at graphite nodules, casting flows or the untempered martensite, which forms from the unreacted or partially stabilized austenite [5]. The acceleration of the stage I transformation by ausforming promotes the formation of more uniform ausferrite throughout the structure and results in an increase of the austenite stability within the ausferrite leading to an improved ductility. Fig. 8 shows that increasing the ausforming reduction from 12.5 to 25% results in a slight decrease in ductility. It is believed that at such rather high degree of warm deformation, the ferrite nucleation in austenite is markedly enhanced. Consequently, a large number of ferrite platelets separated by very thin films of austenite, interspersed with these platelets are formed, such matrix would limit the high plasticity of the retained austenite to be manifested, leading to some decrease in ductility.

Cold rolled ADI: The increase in slope of the straight line fitting the data corresponding to the strain hardening exponent "n" (Fig. 10) was previously observed [8] in the alloyed ADI, which means that the Holloman equation, relating true stress with true strain ($\sigma = k\varepsilon^n$) is not followed. The change in slope of the $\ln \sigma$ - $\ln \varepsilon$ representation can be associated with the transformation induced plasticity (TRIP) effect. As shown from Fig. 10, n^* increases with the tensile strain. The increase is associated with the strain induced martensitic transformation during the tensile test. The generated strain by the volume expansion accompanying the martensitic transformation stimulates new martensitic transformation resulting in the increase in the instantaneous n-value with strain. As a consequence of changes in the structure in the course of tensile deformation, it is believed that the initial segment of ln stress versus ln strain corresponding to the lower plastic strain (Fig. 10) is characterized by the plastic deformation of the retained austenite. At higher strains as previously reported [2] the deformation process is modified by the formation of strain-induced martensite, which takes place when the deformation of austenite has been exhausted.

The increased deformation resulted in a considerable increase in ultimate tensile strength and hardness and decrease in elongation and toughness properties (Figs. 14 and 15). The structural refinement is believed to be the main factor controlling the strength [9]. Hence, the increased strength properties observed after cold rolling can be attributed to the numerous ferrite platelets developed in the matrix. The decreased ductility in the rolled ADI may be related to the morphology of the austenite mixture developed in many zones in the matrix structure, which consists of numerous ferrite platelets separated by very thin films of carbon enriched austenite developed after austempering for longer time (60 minutes). As previously reported [2], these ferrite platelets limit the dislocation movement and do not allow the high plasticity of γ_r to be manifested. It should be mentioned that SEM did not reveal any carbide particles, which may have been associated with the onset of stage II austempering and thus contributing to such decrease in ductility.

The increase in strength and hardness values with the simultaneous decrease in ductility and impact is attributed to the increase of the hardening of the investigated ADI with cold deformation by both the deformation process (deformation bands and twins) and deformation-induced martensite. It must be mentioned that the observed changes in the mechanical properties at light cold deformation (7% reduction) are mainly attributed to the hardening of this alloy by plastic deformation concentrated in γ_r. At this light deformation the amount of mechanically formed martensite is very small (Fig. 12).

Conclusions
1. ADI can be thermomechanically treated either by ausforming or cold rolling, both treatment have significant influence on the microstructural features and mechanical properties.
2. Ausforming refined the ausferritic microstructure which is consistent with an increased ferrite nucleation rate associated with structural defects such as dislocations introduced by ausforming.
3. Both metallographic and XRD techniques indicate an enhanced stage I kinetics of the ausferritic formation and the ausferrite was more uniformly formed throughout the structure.
4. Alloying with 2.0% Ni decreased the rate of ausferrite transformation, particularly in the conventional undeformed ADI. The unreacted retained austenite after short austempering times transformed to martensite on cooling to room temperature.
5. The influence of ausforming to 12.5 and 25.0% thickness reduction on both the ultimate and yield strength was quite dramatic, strengthening mechanism involving both a refined microstructural scale and an elevated dislocation density in both phases of ausferrite was suggested.
6. Ausforming decreased the ductility of the unalloyed ADI, whereas it significantly improved the ductility of Ni-alloyed iron. Higher degrees of deformation are necessary to alleviate the deleterious effect of alloy segregation on ductility.
7. In the course of tensile deformation of the investigated ADI, transformation induced plasticity (TRIP) takes place, indicated by the increase of the instantaneous values of "n" with tensile strain.
8. Cold rolling reduction of 25% of the unalloyed ADI transformed 41% of retained austenite to martensite and a total of 16% martensite volume fraction was mechanically generated. The yield strength, ultimate strength and hardness values were therefore increased whereas elongation and impact toughness decreased.

References
1. Hayrynen K L, Brandenberg K R and Keough J R, Applications of austempered cast irons, Trans. AFS 2002, 110, paper 2002, 084.
2. Dorazil E, High Strength Austempered Ductile Cast Iron, Ellis Horwood Ltd, England, 1991.
3. Hayrynen K L and Keough J R, *Austempered ductile iron - the state of the industry in 2003*, in 2003 Kieth D. Millis Symposium on Ductile Iron, Hilton Head, SC, USA, 20-23 Oct. 2003, pp
4. Moore D J, Rundman K B and Rouns T N, *The effect of thermomechanical processing on boinite formation in several austempered ductile cast irons*, First International Conference on ADI, ASM, 1985, pp13-31.
5. Dela'O J D, Burke C M, Lagather B, Moore D J and Rundman K B, *Thermomechanical processing of austempered ductile iron: an overview*, TMS Symposium Thermomechanical Processing and

Mechanical Properties of Hypereuteutoid Steels and Cast Irons, Sept 14-18, 1997, pp 39-100.

6. Achary J and Venugopalan D, *Microstructural development and austempering kinetics of ductile iron during thermomechanical processing,* Metallurgical and Materials Transactions A, vol 31A, Oct 2000, pp2575-2585.

7. Olson B N, Bruke Ch, Parolini J, Moore D J and Rundman K B, *Ausformed-austempered ductile iron (AADI),* International Conference on ADI, Sept 25-27, 2002, pp 29-60.

8. Aranzabal J, Gutierrez J M, Rodriguez I and Urcola J J, *Influence of heat treatment on microstructure and toughness of austempered ductile iron,* Materials Science and Technology, vol 8, 1992, pp263-273.

9. Naylor J P, *The influence of the lath morphology on the yield stress and transition temperature of martensite bainitic steels,* Metallurgical Transactions A, 10A, 1979, pp861-873.

Table 1: Chemical composition of the investigated alloys

Alloy No.	Chemical Composition, wt %						
	C	Si	Mn	P	S	Mg	Ni
I	3.52	2.80	0.310	0.035	0.0060	0.060	-
II	3.30	2.60	0.270	0.027	0.0053	0.045	2.0

Table 2: Effect of ausforming deformation on the austempering time for developing 50 and 70% ausferrite fractions

Ausforming Reduction, %	Unalloyed		2% Ni-alloyed	
	50%	70%	50%	70%
0	1.5 min	5 min	6 min	8 min
12.5	< 2 sec	< 2 sec	42 sec	1 min
25	< 2 sec	< 2 sec	3 sec	4.2 sec

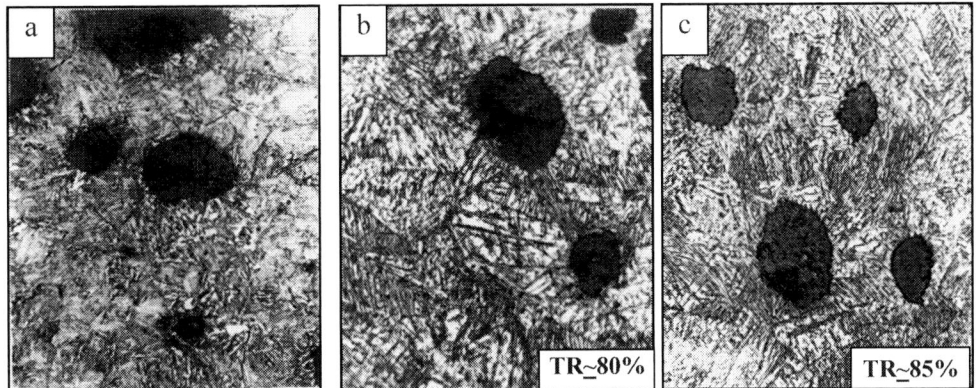

Fig. 1: Photomicrographs of ADI alloyed with 2%Ni austempered at 375°C for 10 minutes after different ausforming reductions (a) conventionally processed (b and c) ausformed to 12.5 and 25% reduction, respectively.

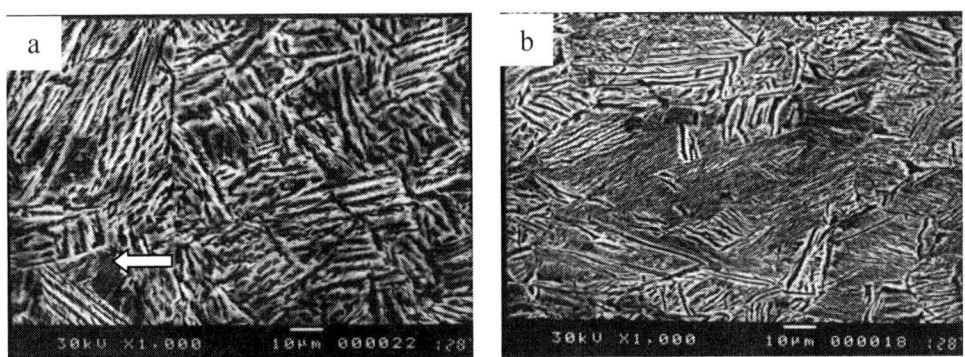

Fig. 2: SEM micrographs of ADI alloyed with 2% Ni austempered at 375°C for 1 minute (a) Conventionally processed; (b) Ausformed to 25% reduction. Arrows indicate the brittle martensite formed in many zones in the conventionally processed ADI.

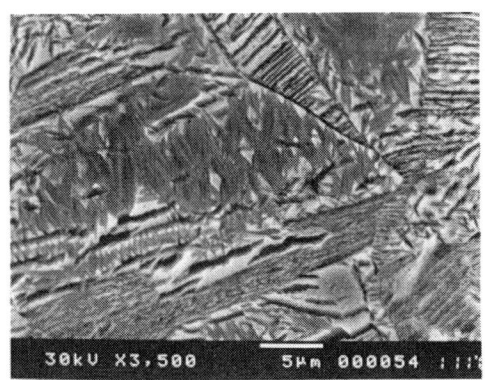

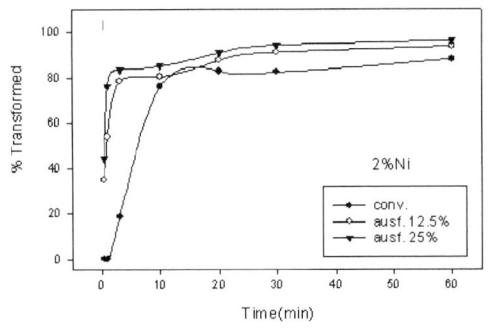

Fig. 3: SEM micrographs illustrating martensite formed in austenite zones between ausferrite platelets in Ni-alloyed ADI, with 3% Ni, ausformed to 25% reductio and austempered at 375°C for 1 minute.

Fig. 4: Volume fraction of transformation versus austempering time for 2% Ni alloyed ADI.

805

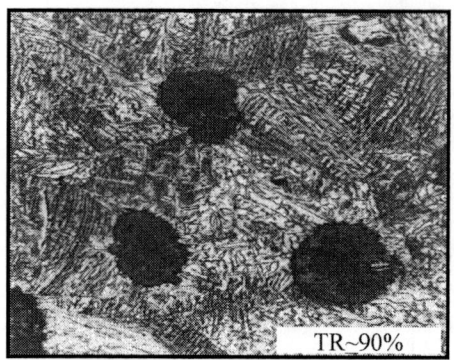

Fig. 5: Alloyed ADI austempered for one minute and ausformed to 25% CR.

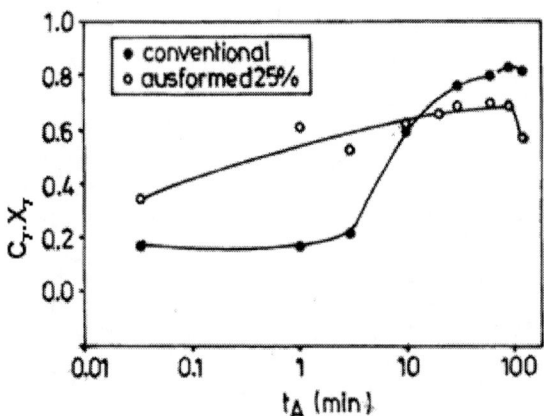

Fig. 6: Variation of total carbon content of saturated austenite C_γ, X_γ with austempering time for conventional and ausformed to 25% ADI alloyed with 2% Ni.

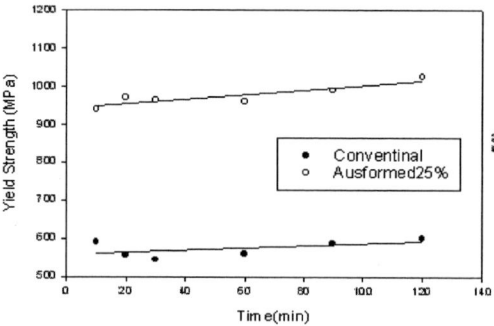

Fig. 7: Yield strength vs austempering time and ausforming reduction for ADIs alloyed with 2% Ni.

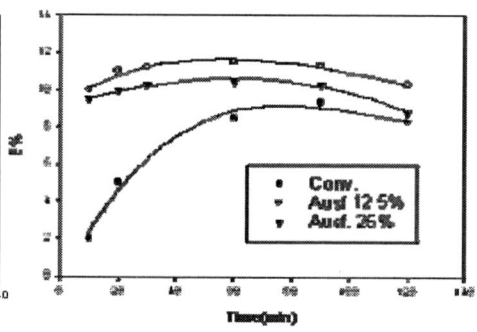

Fig. 8: Total elongation vs austempering time and ausforming reduction for ADI alloyed with 2% Ni.

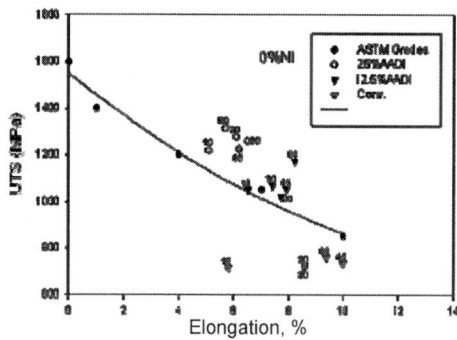

Fig. 9: Ultimate tensile strength versus ductility for 12.5 and 25% ausformed and conventionally austempered specimens along with ASTM minimum requirements for ADI. The austempering time specified (in min.) adjacent of the point.

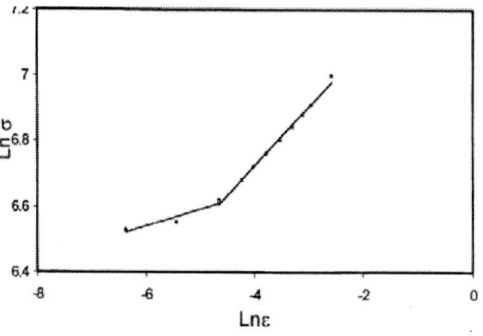

Fig. 10: Ln true stress versus ln true strain

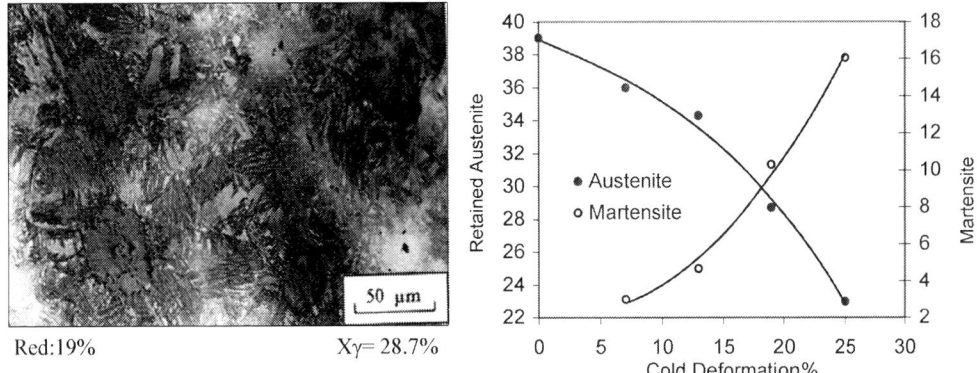

Fig. 11: Microstructures of ADI showing marked decrease of γ_r at 19% reudction.

Fig. 12: Variation of volume fractions of retained austenite and mechanically formed martensite with cold reduction pct.

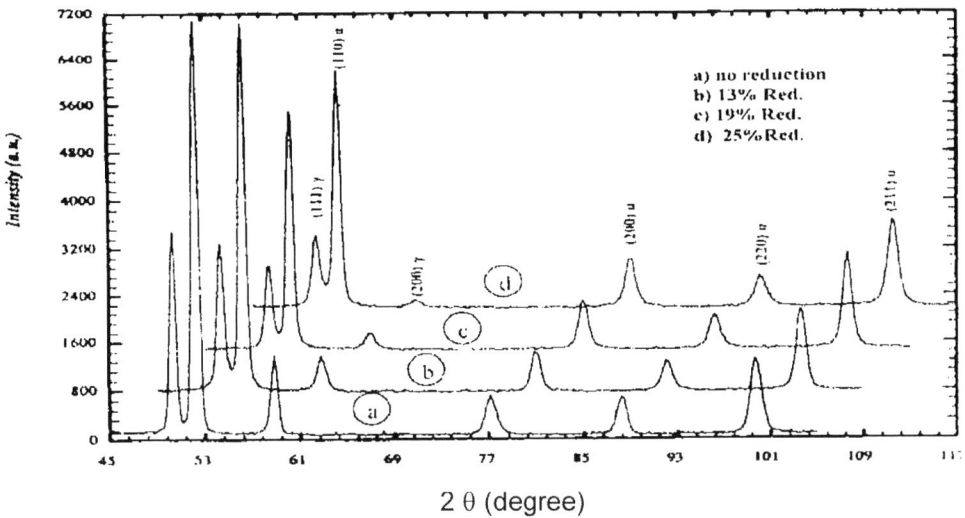

Fig. 13: Variation of peak intensity for γ- and $\alpha+\alpha^{\backslash}$ phases in unalloyed ADI.

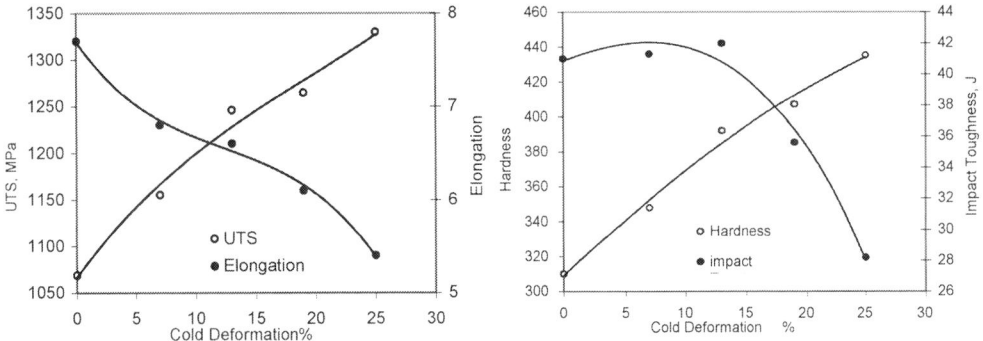

Fig. 14: Variation of elongation and ultimate tensile strength with cold reduction pct.

Fig. 15: Variation of vickers hardness and impact toughness with cold reduction pct.

New adaptive machining methods for the foundry industry

Peter Dickin

Delcam, UK.

Abstract

Computer-based methods for machining and inspection are well established in the foundry industry. More recently, these two technologies have been brought closer together in a new development – adaptive machining. This approach uses a combination of software, such as Delcam's PowerMILL CAM system and PowerINSPECT software, to give new solutions to a range of challenging manufacturing problems.

The programming of most machining operations is based around knowing three things: the position of the workpiece on the machine, the starting shape of the material to be machined, and the final shape that needs to be achieved at the end of the operation. Adaptive machining techniques allow successful machining when at least one of those elements is unknown, by using in-process measurement to close the information gaps in the process chain.

Key words (5 maximum).
Machining; inspection; in-process measurement

Introduction

Computer-based methods for machining and inspection are well established in the foundry industry. While a small number of companies still rely on programming on the machine tool control or still use copy milling, the great majority now use a CAM system to produce CNC code for their machining operations. Similarly, inspection against CAD data is increasingly used for quality control instead of inspection against drawings. With component designs always becoming more and more complex, these computer-based techniques are often the only practical way to undertake their manufacture.

Despite the many benefits these systems bring, there are a range of operations that still remain very challenging, even for companies using the latest technology. The problems usually come from a lack of information at some stage in the process from CAD model to approved component. For successful CAM programming, the user needs to know three things: the position of the workpiece on the machine, the starting shape of the material to be machined, and the final shape that needs to be achieved at the end of the operation. By bringing machining and inspection technology closer together, in-process measurement can be used to close the information gaps in the process chain and allow successful machining when one or more of those elements is unknown. This approach is known as "adaptive machining". It uses a combination of machining and inspection software to give new solutions to a range of challenging manufacturing problems.

Uncertain position

The most common applications of adaptive machining are those where the exact position of the workpiece is unknown. When manufacturing large patterns or when finish machining heavier castings, such as the bodies of large pumps, achieving the correct position and orientation of the component on the machine is a major challenge, taking many hours of checking and adjustment. It is often easier to adjust the datum for the toolpaths to match the position of the workpiece, than it is to align the part in exactly the desired position. This approach has been used for the machining of geometric features for some time. Adaptive machining technology now offers an equivalent solution for the manufacture of complex surfaces that gives the same benefits of shorter set-up times and improved accuracy.

The Delcam process in these cases uses PowerINSPECT, together with a new program, PS-Fixture. First, a probing sequence is created for the inspection software using its off-line programming capabilities. This sequence is used to collect a series of points from the workpiece, which can be used by a range of best-fit routines to determine its exact position. Any miss-match between the nominal position used in the CAM system to generate the toolpaths and the actual position of the workpiece can be calculated in PS-Fixture. The software can then feed the results to the machine tool control as a datum shift or rotation to compensate for the alignment differences.

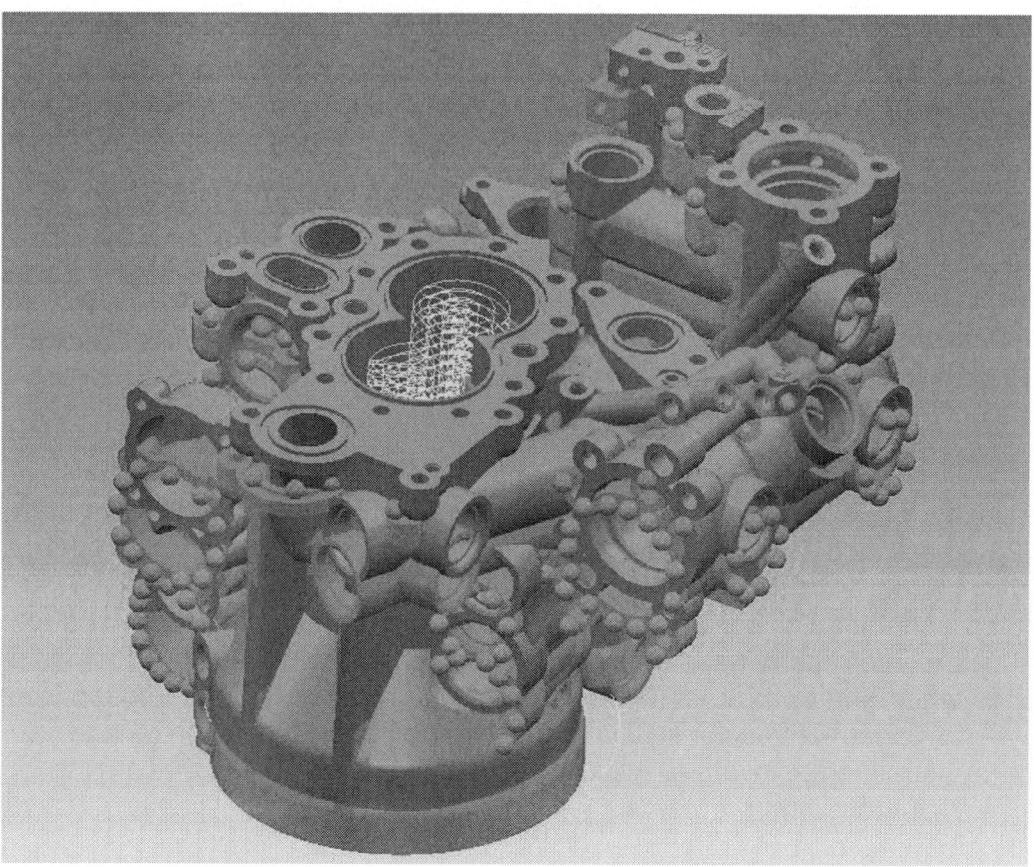

Figure 1. Software can be used to adjust toolpaths to the actual position of the part, rather than aligning the part to a specified position

A similar approach can be used when the relationship between different design elements is more important than the absolute position of each feature based on a single datum for the complete part. For example, it might be necessary to drill a series of holes at a constant spacing around a central bored hole. By basing the drilling operations on the actual position of the central hole, rather than relying on its nominal position within a CAD model, the toolpaths can be adjusted to ensure that the secondary holes are produced to the highest possible accuracy.

As well as giving greater accuracy and reducing set-up times, the use of this approach can significantly cut the cost of fixtures. Since there is no longer the need to locate the component in an exact position, considerably less accuracy is demanded of the fixtures. As long as the component is in approximately the correct position and orientation, the software can compensate for any deviation.

Uncertain starting shape

The second application of adaptive machining comes when there is some uncertainty over the exact shape of the component to be machined, a common problem given the inherent inaccuracy of the majority of casting processes. The main requirement for finish machining is to allow an even distribution of material to be removed around the stock to avoid over-machining in some areas and under-machining in others. This can be achieved by first creating a probing path within the inspection software to determine the form of the casting. The final shape to be reached can then be orientated within this envelope to give an even thickness of material on the surfaces to be machined. Other benefits of knowing where material exists, and where it doesn't, include the ability to give a smooth transition between machined and un-machined areas, a reduction in air cutting and improved control over the feed rate as the cutter enters and leaves the material.

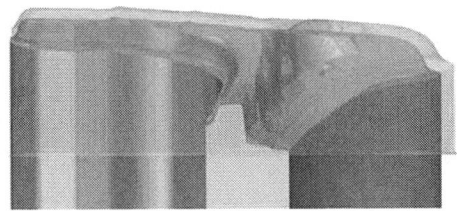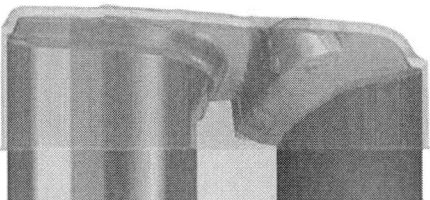

Figure 2. Ensuring a good alignment of the finished shape within the stock can save material and machining time

There is also the potential for material savings. Most castings are designed with worst-case tolerances to ensure that there will always be sufficient material in all critical areas. The ability to tweak the position of the finished shape within the casting means that these tolerances can be reduced, which in turn means that there is less need for excess material. Furthermore, having less material to be removed from the casting reduces the time needed for finish machining.

More comprehensive reverse engineering can be needed for cases where no nominal CAD data exists, for example when repairing older tooling. The surface of the tool must first be scanned to give the data required to create CAD surfaces in the reverse engineering software. Data can also be collected in a similar way to generate a model of the excess weld that needs to be machined away from the repair. The difference between the shape of the weld and the desired surface of the tool can then be used to calculate the necessary machining program. This approach is both faster and more accurate than trying to remove the excess weld by hand grinding.

Figure 3 Adaptive machining can be used for the finishing and repair of cast press tools

This approach is used by a leading automotive manufacturer in Sunderland to replace complete sections in draw tools, especially where laser welding is used to join sheets of different grades of metal together. These areas wear more quickly because of the effects of the laser weld so there is a need to replace them with harder material to extend the overall lifetime of the tool. Surface data from the affected area is collected with a Renishaw probe fitted to the company's machine tool. The replacement section is modelled in the CAD system and then machined on the machine tool.

Unknown final shape

The most challenging adaptive machining operations are those where the final shape of the component is not known precisely. This is most often the case when undertaking repairs to components that have been changed from their nominal CAD shape during service, for example, turbine blades that have been distorted by the high temperatures in aircraft engines.

The initial stage in these cases is to probe the component to determine the extent of its deviation from the nominal CAD data. Then, the CAD system can be used to bring the nominal model into line with the actual geometry. As with the tooling repair examples mentioned above, the position of any excess weld can also be determined. It is then possible to generate a toolpath to remove any surplus material and create a smooth surface on the part.

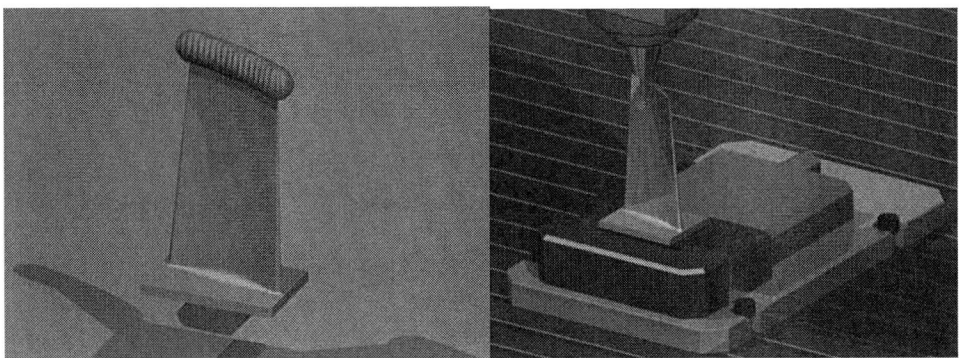

Figure 4 The repair of components that have distorted during service requires machining from an unknown starting shape to an unknown final shape

In all cases, the component can be probed after machining to give a record of the final shape of the part. This data can be used to check the component against a CAD model of the part if one exists, or to create a CAD model for future reference if needed.

Verifying the result
With all of the processes described, and, indeed, with many other machining operations, it is important to check the results against a required specification. That can be undertaken on a co-ordinate measuring machine but an increasing number of companies now carry out an initial inspection on the machine tool using on-machine verification. The most obvious advantages of this technology are for those companies that do not have existing inspection capabilities. However, it is also beneficial for sub-contractors that need to machine bigger components, such as those used in the aerospace or oil and gas industry, and for the production of large patterns.

Most modern machine tools either come with, or can be retrofitted with, probing capabilities to assist in the set-up of the job. With on-machine verification, this same equipment can now be used for initial quality checks at little extra cost. Data for the process can be collected using various spindle probes, such as the established MP 700 or the soon to be released compact OMP 400. Neither of these probes need to be calibrated in all the vectors in which they are to be used, so reducing the number of points required to measure a given part and, therefore, giving shorter verification cycle times.

The inspection software can then use the data both to check surface accuracy or to measure features, such as circles, cylinders, cones, spheres and planes, to a high level of precision. The ability to program complete verification sequences off-line means that there can be minimal interruption of the machining operations.

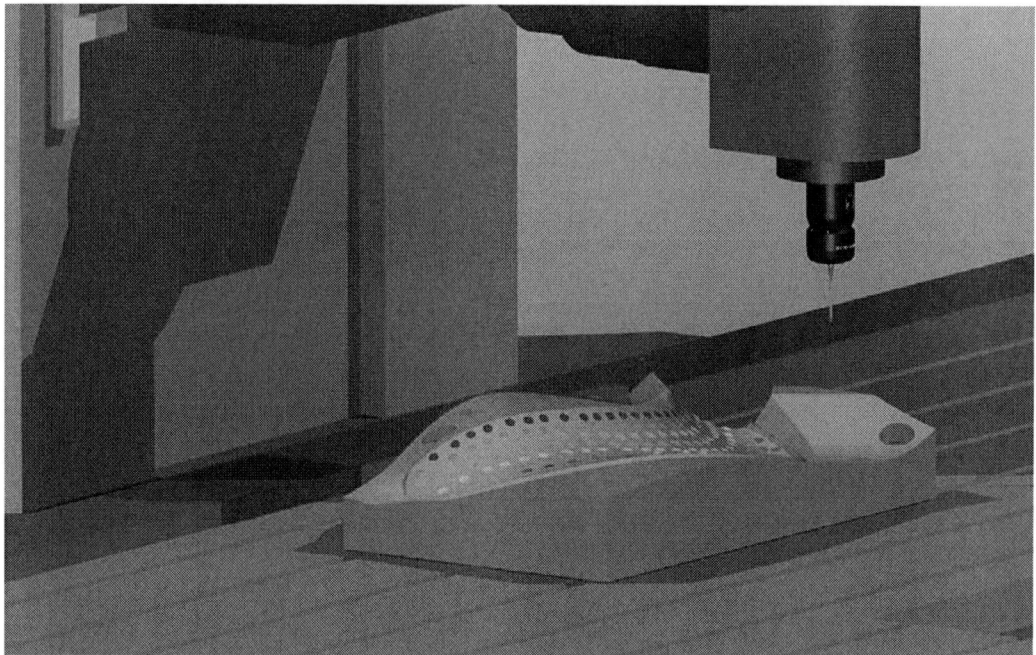

Figure 5 On-machine verification can give significant time savings when manufacturing large components

On-machine verification can give huge timesavings by enabling any errors to be detected earlier, and so corrected more quickly and at lower cost. Similarly, the extent of any damage caused, for example, by a tool breakage, can be assessed accurately and a decision made immediately to determine whether the part can still be completed within tolerance or whether it will have to be scrapped.

On-machine verification will also benefit companies with customers that insist on independent inspection of their work. By carrying out an initial verification on the machine, errors can be detected, and corrected, that might otherwise not be found until after the component had been shipped to the inspector.

Companies already having suitable equipment might think that on-machine inspection is an unnecessary operation that can loose machining time. However, if the whole process is considered, there is considerable potential to reduce delivery times.

If a part has to be transferred to a dedicated CMM and the inspection shows any errors, the component must be returned to the machine tool and re-clamped in position before being machined again. This is time-consuming for any component but can take many hours for any heavy item, such as a large aerostructure or a press tool for an automotive body panel. In addition, any mistakes during the set-up back onto the machine tool could result in a new series of errors in the component, and so lead to a further cycle of inspection and re-machining.

With on-machine verification, the part can be checked before it is moved. Any significant errors can then be detected and corrected before the component is transferred to the CMM for its final inspection. The ability to check that the part is reaching its specifications at the various stages of the manufacturing process will save time, reduce the amount of scrap and increase confidence that time is not being wasted working on components that are already too far out of tolerance.

Conclusions
This paper has described a number of adaptive machining methods that can be used to overcome problems with more challenging manufacturing operations, and the use of on-machine verification to confirm that the desired result has been achieved. Recent software developments allow users to undertake these complex operations more accurately, more reliably and with considerable savings in time and costs.

However, companies wanting to use adaptive machining processes must understand that they tend to be much more complex than conventional CAM programming. They may often be specific to a particular process or even to a single component. There is a core of common functionality at the heart of all the solutions described above, and software developers like Delcam are extending their product ranges to standardise this core functionality. Nevertheless, most adaptive machining projects will still require some specific consultancy and customisation work as part of their implementation.

It now seems inevitable that most standard production operations to produce large numbers of identical parts will move to countries with lower labour costs. There will still remain a large demand for shorter runs of more complex, more individual products and it is this area that the more traditional manufacturing centres can still be competitive, providing that they make the best use of modern manufacturing methods. Of course, the choice of the best approach will be different for each company. However, it is clear that doing nothing is not an option for companies that want to grow their turnover and profitability. Failure to invest in any of these new techniques will see companies losing business to those that do use the latest technology.

For further information, Peter Dickin can be contacted by e-mail to pjd@delcam.com

Development of a low alloy cast steel for automobile blanking, drawing, and trimming die

J. S. Shin, C. B. Song, B. H. Kim, S. M. Lee, and B. M. Moon
Korea Institute of Industrial Technology, Incheon, Korea

Abstract

The optimization of chemical composition with appropriate heat treatments was tried to develop a significantly reduced Mo, V, and Cr containing low alloy cast steel for automobile blanking, drawing, and trimming dies, possessing shorter production time and lower manufacturing cost as well as maintaining working performances compared to the traditional die materials. The effects of Si content on combinations of important properties such as toughness, hardness, hardenability, and resistance to softening on tempering in addition to strength increment were systematically investigated. In order to evaluate the applicability as the insert of cold pressing die, the mechanical properties were measured after annealing treatment, quenching and tempering (Q/T) treatment, and flame hardening, respectively.

Key words low alloy cast steel, cold pressing die, die insert, matrix hardening, flame hardening.

Introduction

With the rapid development of the automobile industry resulting from technological advance and customer's taste, the manufacture of various sheet-type components including the automobile exterior with a short manufacturing time at a low cost becomes one of the key factors to assure the competitiveness of the automobile industry. But, the traditional manufacturing process of the die for the sheet-type components is very complex and thus takes a long time. Especially, SKD11 and flame hardened tool steels, which are widely used as the insert material in a traditional cold pressing die, must accompany casting, forging/rolling, annealing, cutting, roughing, Q/T and flame hardening, finishing, and assembling processes, successively. In addition to that, those high alloy steels contain a large amount of high-priced alloying elements such as Mo, V, Cr, and etc.

Therefore, our study was initiated to develop significantly reduced Mo, V, Cr, and C containing low alloy cast steel materials, which are able to be used as the die insert just by near-net-shape casting and a few following post processes, possessing shorter manufacturing time and lower production cost without appreciable loss of the important mechanical properties such as strength, toughness, and wear resistance. For the achievement of this goal, matrix strengthening according to combined additions of Si, Mn, Cu, Ni, Al, N, and etc. was tried to counteract an impairment of the above mentioned properties caused by the reduction of Mo, V, and Cr. In this paper, a special attention was paid to investigate the influence of increasing Si content on the microstructure and the mechanical properties, since detailed information about the effects of Si content on interesting combinations of other properties such as toughness, hardness, hardenability, and resistance to softening on tempering in addition to strength increment has not been clarified in the low alloy steels. In order to evaluate the adaptability as the insert of cold pressing die, the mechanical properties were measured after annealing, Q/T, and flame hardening treatments, respectively.

Experimental

The chemical compositions of the low alloy steels used in this study are shown in Table 1. The strategy adopted in the design of the chemical composition of the low alloy steels was to minimize the contents of C and expensive alloying elements such as Mo, V, and Cr with varying contents of Si, Mn, Cu, Ni, Al, N, and etc. The C content was kept below 0.5% to improve toughness and to avoid embrittlement problem [1] caused by flame hardening treatment of a mandatory process in manufacturing the die insert and resulting in crack formation. When Mo and V were not alloyed at all, the Si content was systematically varied from 0.8% to 2.3% with a combined addition of 1.0%Mn, 0.45%Cu, 0.1%Ni, 0.036%Al, and 0.017%N to compensate the loss in hardness, strength, and hardenability.

The steels were melted by high frequency induction furnace and the melts were poured at 1,580°C into a Y-block sand mold with the cavity of 35 mm width, considering the maximum thickness of the die insert in automobile industry (Fig. 1). The schedules of annealing and Q/T treatments are given in Fig. 2a and 2b, respectively. Heating process was carried out at a slow rate in two stages to minimize thermal stress, and relatively low austenitizing temperatures (810°C and 920°C in annealing and Q/T treatments, respectively) were utilized to avoid distortion, cracking, and decarburization. In Q/T treatment, tempering at 590~680°C was immediately conducted after air quenching to relieve residual stress and toughen the steels.

Microstructural features were observed by combined analysis of OM, SEM, and XRD. Hardness and tension test were performed to compare the mechanical properties of the low alloy cast steels. In order to evaluate the effects of Si content on surface hardness, hardenability, and resistance to softening on tempering, micro-Vickers hardness was measured after flame hardening treatment as a function of the depth from surface.

Results and Discussion

The most mechanical properties of all alloy steels are determined primarily by their microstructures. In as-cast state, all specimens used in this experiment showed ferritic-pearlitic-banitic structure due to a relatively high cooling rate during casting. After annealing at 810°C for 1 hour, carbides were substantially spheroidized, as shown in Fig. 3. It is interesting that carbides were more effectively spheroidized in the specimen with a higher Si content (especially, specimen A, B, and C), because flame hardening characteristics is significantly dependent on the spherodization and distribution of carbides. This agrees well with other's results [2]: Si was very effective in refining tempered carbides by delaying carbide to cementite, in spite of its deleterious effect on surface decarburization and toughness.

Fig. 4 shows the mechanical properties after the spheroidization annealing treatment. Specimen A, B, and C, in which Si was highly alloyed with properly balanced quantities of the additions of Mn, Cu, Ni, N, and etc., showed the higher hardness and strength values compared to the other specimens. In those three specimens, hardness and strength slightly increased with increasing Si content. It seems be mainly due to solid solution strengthening effect caused by Si alloying. Although the SKD11 specimen showed the highest hardness under the same annealing condition, tensile strength was much low because of its large-sized carbides. This spherodization annealing treatment was able to fulfill delivery requirements (hardness and tensile strength have to be higher than HB230 and 90 kgf/mm^2, respectively) in view of hardness, but not in view of tensile strength. Adequately high tensile strength was necessary to endure repeated pressing process without distortion.

The Q/T treatment effectively increased tensile strength for the all specimens, as shown in Fig 5. Especially, specimen A, B, and C showed remarkably increased hardness and tensile strength, satisfactorily fulfilling the delivery requirements. Similarly to the annealed state, hardness and strength slightly increased with increasing Si content. But, elongation slightly decreased with increasing Si content. Still, the SKD11 specimen showed the lowest strength in spite of the highest hardness due to its large-sized carbides.

In automobile industry, flame hardening treatment must be carried out after roughing process and Q/T treatment to endure repeated blanking, trimming, and pressing processes. In order to evaluate the adaptability as the insert of cold pressing die, flame hardening treatment accompanying air cooling was carried out on the edges of specimen A, B, and C, showing the most excellent mechanical properties and the proper microstructure, as shown in Fig. 6. As a result, surface hardness up to 2 mm depth exceeded HRc60, as shown in Fig. 7. The effectively hardened layer, satisfying an industrial criterion (generally above HRc48), decreased with increasing Si content, while surface hardness also slightly increased similar to the inside bulk hardness.

Fig. 8 shows the representative microstructures as a function of depth from surface after the flame hardening treatment. Almost fully Martensitic microstructure was observed in the edge part, in which flame was directly discharged. Also in XRD analysis of Fig. 9, strong martensitic peaks were observed for all three specimens, except very weak peak, corresponding to M_7C_3 type carbide. These SEM and XRD results reveled that specimen A, B, and C obtained high enough hardenability to permit air cooling during flame hardening treatment.

Conclusions

The development of significantly reduced Mo, V, Cr, and C containing low alloy cast steel materials, enabling the significant cost- and time-savings in manufacturing die insert for automotive components without impairment of the important mechanical properties, was tried by combined additions of Si, Mn, Cu, Ni, N, and etc. After the Q/T and flame hardening treatments, the developed Mo and V free low alloy steels showed excellent matrix strengthening effect and hardenability, satisfactorily fulfilling the industrial criterion of the mechanical properties for the insert of cold pressing die.

References

1. Harold E. McGannon, The Making, Shaping and Treating of Steel, 9th Ed, United States Steel Corporation 1971.
2. Ahn S T, Kim D S and Nam W J, Microstructural evolution and mechanical properties of low alloy steel tempered by induction
3. heating, Journal of Materials Processing Technology, 160, 2005, pp54-58.

Table 1 Chemical compositions of low alloy steels used in the present investigation

	C	Si	Mn	Cu	Ni	Cr	Mo	V	Al	N
Alloy A	0.51	0.8	1.0	0.45	0.10	1.6	-	-	0.036	0.017
Alloy B	0.51	1.6	1.0	0.45	0.10	1.6	-	-	0.036	0.017
Alloy C	0.51	2.3	1.0	0.45	0.10	1.6	-	-	0.036	0.017
Alloy D	0.51	0.5	0.6	0.45	-	1.6	0.40	0.13	0.036	0.010
Alloy E	0.50	0.5	0.6	-	-	1.6	0.20	-	0.036	0.010
Alloy F	0.50	1.2	1.2	-	-	1.6	0.20	-	0.036	0.017
SKD11	1.50	0.25	0.45	0.15	0.15	12.0	0.40	0.40	0.40	0.008

Fig. 1 Photographs showing melting and pouring processes and cast ingots of the low alloy cast steels.

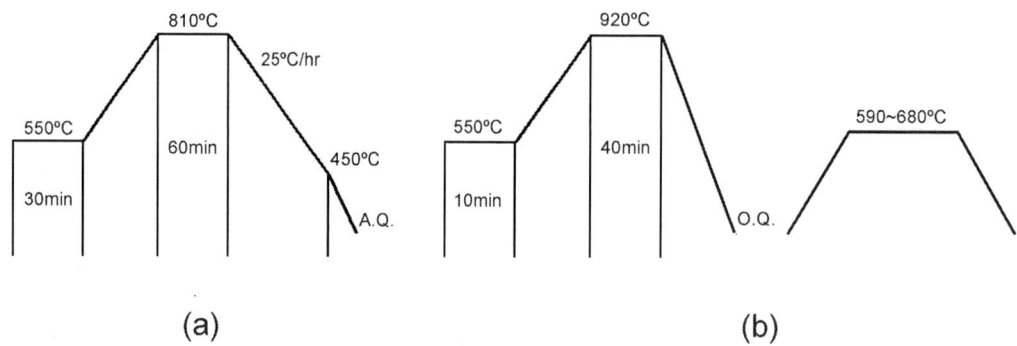

Fig. 2 The schedules of (a) spheroidization annealing and (b) Q/T treatments. (F.C.: Furnace Cooling, A.Q.: Air Quenching, O.Q.: Oil Quenching).

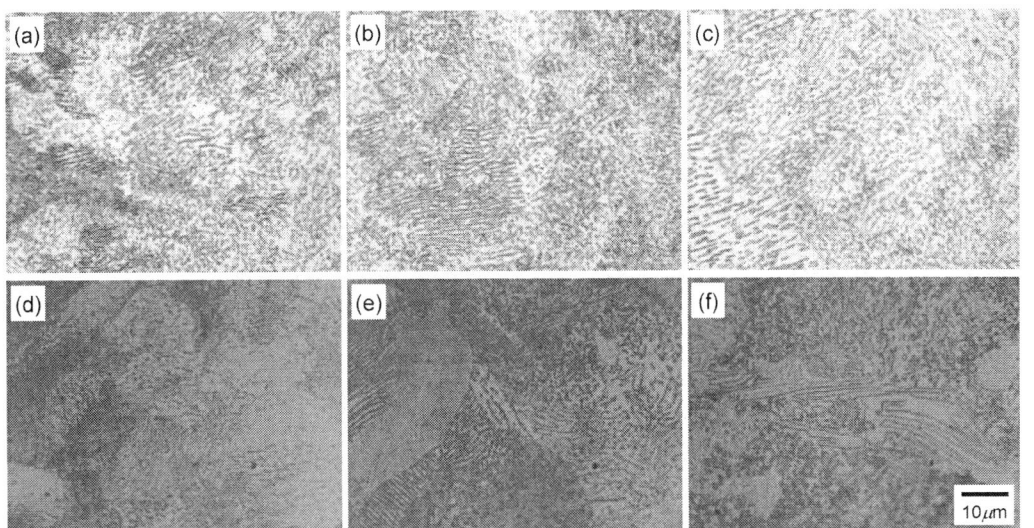

Fig. 3 Optical micrographs of the annealed specimens; (a) alloys A, (b) alloy B, (c) alloy C, (d) alloy D, (e) alloy e, and (f) SKD11.

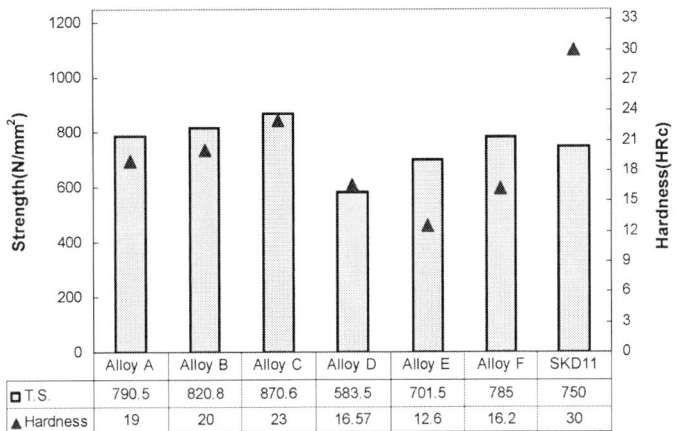

	Alloy A	Alloy B	Alloy C	Alloy D	Alloy E	Alloy F	SKD11
□ T.S.	790.5	820.8	870.6	583.5	701.5	785	750
▲ Hardness	19	20	23	16.57	12.6	16.2	30

Fig. 4 The variation of hardness and tensile strength after the spheroidization annealing.

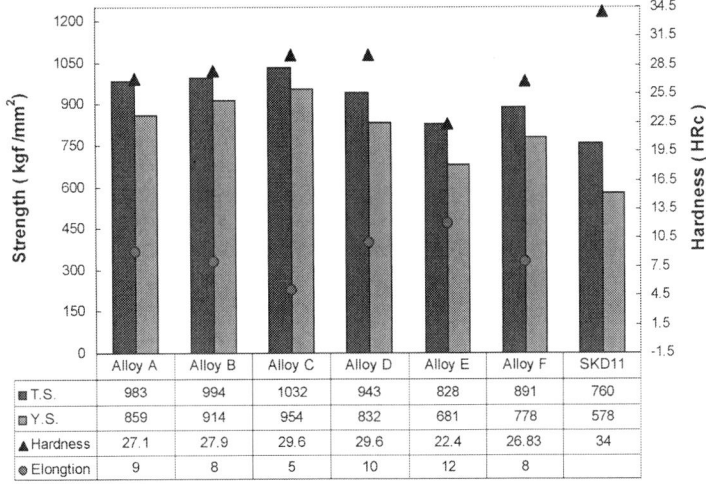

	Alloy A	Alloy B	Alloy C	Alloy D	Alloy E	Alloy F	SKD11
■ T.S.	983	994	1032	943	828	891	760
▣ Y.S.	859	914	954	832	681	778	578
▲ Hardness	27.1	27.9	29.6	29.6	22.4	26.83	34
● Elongtion	9	8	5	10	12	8	

Fig. 5 The variation of hardness, strengths, and elongation after the Q/T treatment.

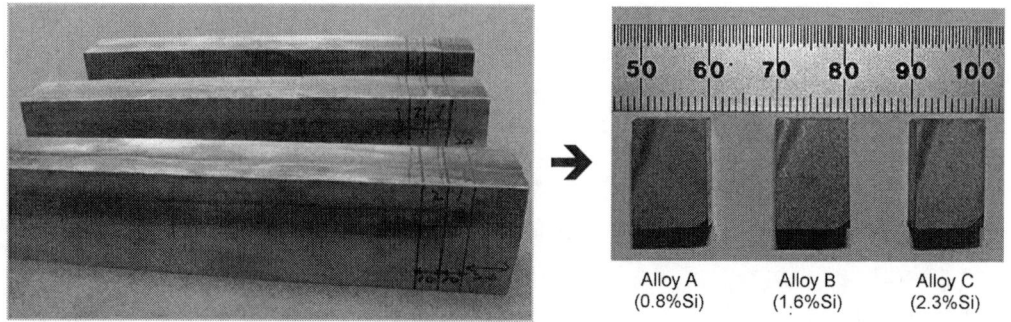

Fig. 6 Flame hardened Mo and V free low alloy cast steel specimens (alloys A, B, and C).

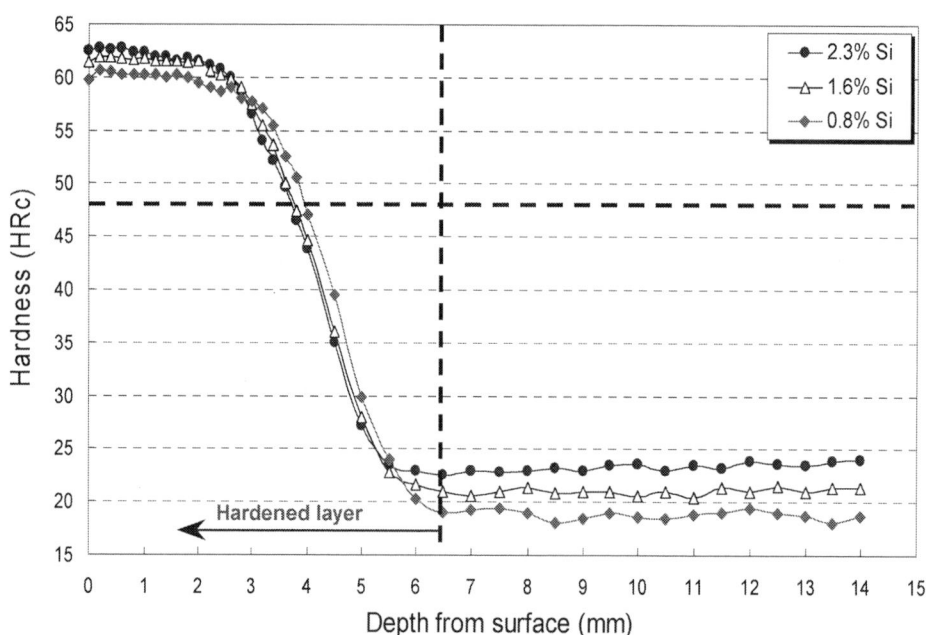

Fig. 7 The depth profile of micro-Vickers hardness of Mo and V free high Si low alloy cast steel specimens (alloys A, B, and C) after the flame hardening treatment accompanying air quenching.

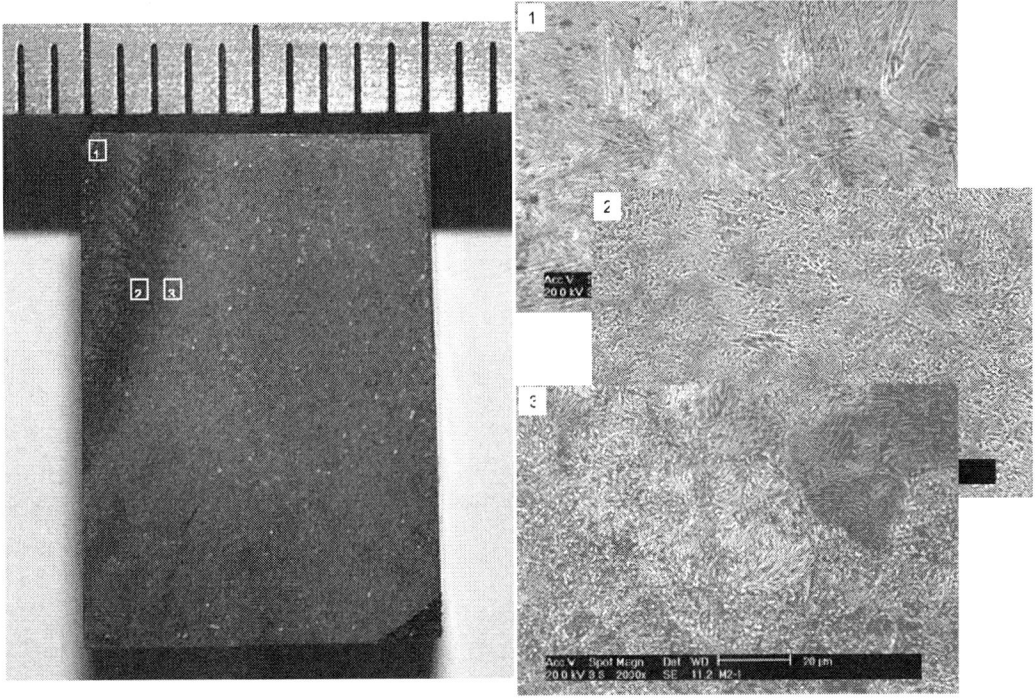

Fig. 8 SEM micrographs showing the phase transformation in a Mo and V free low alloy cast steel specimen (alloy A) caused by the flame hardening treatment accompanying air quenching.

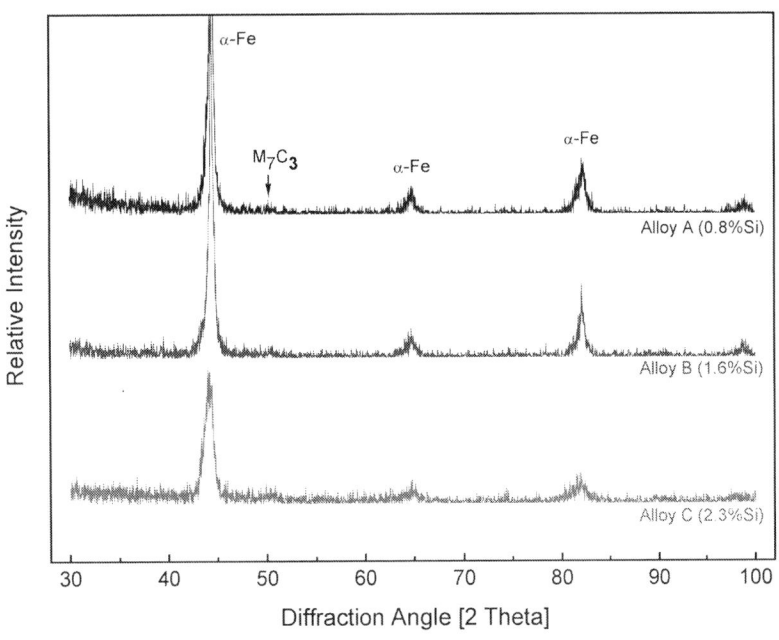

Fig. 9 The results of X-ray diffraction with Cu Kα radiation of alloys A, B, and C after the flame hardening treatment accompanying air quenching.

Squeeze Casting of Aluminum Alloy Safety Critical Components for Automotive Applications

Rathindra DasGupta[*], Chuck Barnes[*], Peter Radcliffe[**] & Paul Dodd[**]
* SPX Contech, USA, ** SPX Contech, UK

Abstract

Squeeze casting is considered a "high integrity" process, and has given materials and design engineers a new alternative to conventional casting techniques: gravity permanent mold (GPM) and conventional (high pressure, high velocity) die-casting. In recent years, the squeeze casting process has been applied to near net shape products requiring high impact strength, high fatigue strength, pressure tightness, or high wear resistance.

This paper briefly discusses the squeeze casting technology (P2000[TM]) at SPX CONTECH and provides examples of the various aluminum alloy safety critical components manufactured. The paper also includes fluid flow simulations and solidification analyses for select components, comparison between predicted and actual microstructures, mechanical properties (tensile and fatigue) based on samples machined from actual castings, and results from strength and fatigue testing of components.

Key words (squeeze casting, safety critical, aluminum).

Squeeze Casting Technology (P2000™)

The unique features of the P2000™ technology include the die design, lubricant type and application mechanism, choice of alloy and heat treatment, melt treatment and finally the casting machine itself. Figure 1 shows the schematic of a typical casting machine used. The casting process consists of pouring degassed and filtered molten metal into a vertical chamber (shot sleeve), slowly forcing the metal into a preheated, lubricated die cavity, and continuous application of pressure during solidification of the melt. The slow injection (speed slower than that used for conventional die-casting) results in reduced turbulence during flow of metal into die cavity, thereby contributing to minimal air entrapment in the matrix. The rapid heat transfer (from the contact of the molten metal with the lubricated die surface) and the continuous application of pressure during solidification of the melt in the cavity also help reduce shrink porosity.

The casting/tooling design process in the P2000™ technology follows much the same rules as for conventional die-casting [1]. However, certain casting/tooling design features do set the P2000™ technology apart from conventional die-casting. For example, the P2000™ casting process requires larger runners and gates to promote non-turbulent metal flow, flat areas for gate since the gate must be sawed off, and thicker walls (>2.5 mm) to allow adequate pressure during solidification [1].

In recent years, the P2000™ technology has been widely used with various aluminum alloys (potential mass reduction of 40 to 60% can be achieved when aluminum replaces cast iron and steel fabrications) to manufacture near-net shape and lightweight safety critical automotive components including control arms, knuckles, wheels, and cross members. Although other casting processes such as gravity permanent mold (GPM) and low-pressure permanent mold (LPPM) are capable of producing these parts, the P2000™ process is now established as a proven, practical and competitive method for producing lightweight structural aluminum castings.

The wide acceptance of the P2000™ technology can be attributed to the following: reduced or absence of shrink porosity in the matrix resulting from high cavity pressure (70 to 100 MPa) and rapid solidification rate, dimensional control similar to that of conventional die castings, heat treatable (can be subjected to solution treatment without the fear of blistering) and improved mechanical properties [1]. Smaller dendrite arm spacing (DAS) and a fibrous silicon morphology resulting from the rapid solidification rate in squeeze casting are factors responsible for the improvement in tensile properties, particularly ductility. For example, Table 1 compares the tensile properties of components made using various casting processes for select alloys [1]. It is evident that for comparable strengths, the P2000™ process yields higher ductility than GPM and conventional die-casting.

The net-shape capabilities, alloy flexibility and ease of automation for high volume production are additional advantages of the P2000™ technology when compared with GPM and low-pressure permanent mold (LPPM) processes.

The benefits achieved from the P2000™ casting process alone, however, cannot guarantee the desired level of part quality. Thus, prior to casting development trials, computer aided process simulation and modeling are used to help optimize tooling design, proper solidification, cooling line configurations, and mold (cavity) fill.

This paper provides examples of the safety critical components, namely, suspension links, front and rear knuckles manufactured at SPX Contech. Also included in this study are process simulation results, mechanical properties based on samples machined from the actual components, and results from product testing.

P2000™ Applications: Suspension Links

Figure 2 shows the various suspension links converted from aluminum forgings to castings. A multi-cavity die for each part number is used to satisfy the customer requirement of 800,000 pieces per year. All parts undergo non-destructive inspection (x-ray and ultrasonic) prior to shipment. Additional information on these castings and the approach to achieving a robust and reliable manufacturing process for the suspension links are outlined below.

Part weight, aluminum alloy type and heat treatment:
a) Tension link = 0.41 kg
b) Compression link = 0.41 kg
c) Camber link = 0.68 kg
d) Aluminum alloy ⇒ A356.2
e) Heat treatment ⇒ T6 temper (solution treated, quenched, and aged)

Material properties attained from actual castings:
The most common material properties measured prior to shipping the production links to customers include the yield strength, tensile strength, percent elongation and hardness. These data are obtained from tensile specimens machined from actual castings. Shown below are typical data obtained from one of the camber links.
- ❑ Yield strength = 241.7 MPa (minimum requirement is 207 MPa)
- ❑ Tensile strength= 305.8 MPa (minimum requirement is 275 MPa)
- ❑ Percent elongation = 10.27% (minimum requirement is 7%)
- ❑ Hardness = 100 BHN (minimum requirement is 85 BHN)

Approach to manufacturing high integrity suspension links:
A critical feature for the proper use of the P2000™ technology is the application of process simulation and modeling (prior to casting

development trials) to help optimize tooling design, cavity fill and solidification. The steps involved in the simulation process typically consist of performing a "natural" solidification (with no external cooling) followed by a thorough investigation into fill profiles, cooling line configurations and cooling sequences to significantly minimize or eliminate the porosity observed during natural solidification. Figure 3 shows the solidification pattern (with appropriate cooling) for a camber link after 80.18% of the melt has solidified. Further optimization of cooling line configurations, cooling sequences and casting process parameters helped produce parts meeting customer porosity specifications.

Component testing during casting development trials was another approach to developing a robust and reliable manufacturing process for the links. Thus, during product testing, attention to consistency in fracture loads (or number of cycles to failure) and failure locations, correlation between actual failure locations and those predicted through finite element analysis (FEA) helped establish the appropriate "process window" required for producing high integrity suspension links. For example, Table 2 shows the fracture loads and failure locations obtained from left-hand (LHD) camber links (made during one of the casting development trials) following strength testing. The fracture loads (ranging from 54.2 to 60.2 kN) are fairly consistent, and failure occurred in desired areas.

Determining the "worst-case scenario" though evaluation of components made <u>intentionally</u> "bad" (specifications for discontinuity size not met) was another avenue for achieving a robust manufacturing process for the links. It was felt that the data from such a test would help in developing preventative measures to combat a potential problem that could occur in production. Table 3 shows the results from left-hand camber links made intentionally bad with inclusions of varying types and sizes. A comparison between Tables 2 and 3 shows that the presence of inclusions in castings has a larger effect on the range of fracture loads than on failure locations.

P2000^{TM} Applications: Front Steering Knuckles

Figure 4 shows a 3.65 kg front steering knuckle, previously made from "direct" squeeze casting. The customer requirement of 60,000 pairs per year is met using a two-cavity die (left-hand and right-hand), strontium-modified A356.2 alloy and a T6 temper. Material property measurement, dimensional checks, x-ray and ultrasonic inspection are required prior to shipping the production steering knuckles. The typical tensile data obtained (three bars per hand from locations shown in Figure 4) are as follows.

- ❑ Yield strength = 244.5 MPa (minimum requirement is 207 MPa)
- ❑ Tensile strength = 302.8 MPa (minimum requirement is 276 MPa)
- ❑ Percent elongation = 10.5% (minimum requirement is 6%)

Although not a customer requirement, it was of interest to determine the low cycle fatigue (cyclic loads are relatively high, significant amounts of

plastic strain are induced during each cycle, and low number of cycles to failure) behavior for squeeze cast A356.2-T6 aluminum alloy using samples machined from these steering knuckles. For a steering knuckle, even if the loads are nominally low, the material at the root of a critical notch may experience local plasticity that is strain-controlled, and the method of low cycle fatigue becomes important in predicting its life. Figure 5 shows the strain-life curve for squeeze cast A356.2-T6 with upper/lower bounds at 95% confidence level whereas Figure 6 shows the cyclic stress-strain curve for the same alloy together with part of the monotonic curves for comparison.

Approach to manufacturing high integrity front steering knuckles:
Various cooling line configurations, cooling sequences and fill profiles were modeled to obtain the desired part quality. In addition to flow and solidification modeling, microstructure simulation of select locations was conducted to predict and compare dendrite arm spacing (DAS) with that obtained from production steering knuckles [2]. Despite an offset between the actual dendrite arm spacing (DAS) and the simulated DAS, a significant correlation (correlation coefficient = 0.95) was observed between the two sets of data [2].

Fatigue and strength testing of the above component were conducted during casting development trials to help achieve a robust manufacturing process for the front steering knuckles. Table 4 reveals the data obtained from one such test [2]. A close examination of the data shows both left-hand (LHD) and right-hand (RHD) knuckles to exhibit very similar fracture loads. Furthermore, failure location was the same for both knuckles during strength testing.

P2000^TM Applications: Rear Steering Knuckles

Figure 7 shows a 2.306 kg rear steering knuckle. The customer requirement of 15,000 pairs per year is met using a two-cavity die (left-hand and right-hand), strontium-modified A356.2 alloy and a T6 temper. Material property measurement, dimensional checks, x-ray and ultrasonic inspection are again required prior to shipping the production knuckles. The typical tensile data obtained (two bars per hand from locations shown in Figure 7) are as follows.
- Yield strength = 241.3 MPa (minimum requirement is 207 MPa)
- Tensile strength = 303.4 MPa (minimum requirement is 276 MPa)
- Percent elongation = 10% (minimum requirement is 6%)

Approach to manufacturing high integrity rear steering knuckles:

Numerous fill and solidification analyses were performed to optimize tooling design, proper solidification and cavity fill. Figure 8 shows the solidification pattern for the rear knuckle with no external cooling (natural

solidification) after 98.56% of the melt has solidified. The "hot spots" (prone to porosity) are clearly evident. Various cooling line configurations,

cooling sequences and fill profiles were, therefore, investigated to eliminate the porosity.

Fatigue and strength testing of the above component were also carried out during the casting development trials to ensure a robust and reliable manufacturing process. For example, Figure 9 shows the typical set-up for fatigue testing (of the upper control arm) and the desired failure location. Results from the various tests conducted are summarized in Table 5. The toe link arms exhibit approximately 50% higher fracture load than the upper control arm when subjected to strength tests. The toe link arms were also observed to have undergone a greater number of cycles prior to failure during fatigue test.

Conclusions

The P2000[TM] technology is considered a "high integrity" process, and has given materials and design engineers a new alternative to conventional casting techniques: GPM and conventional die-casting.

The P2000[TM] technology is currently used for manufacturing a variety of safety critical components including suspension links, front and rear steering knuckles.

The benefits achieved from the P2000[TM] casting process alone, however, cannot guarantee the desired level of part quality. Thus, prior to casting development trials, computer aided process simulation and modeling are necesssary to help optimize tooling design, proper solidification, cooling line configurations, and mold (cavity) fill. In addition, component testing during casting trials is another mechanism to assess integrity (consistency and reliability) of parts.

References

1. Corbit S.A. and DasGupta R., *Squeeze cast automotive applications and squeeze cast aluminum alloy properties*, SAE International Congress and Exposition, paper # 1999-01-0343.
2. DasGupta R., Xia Y. and Szymanowski B., *High volume squeeze cast applications*, Proceedings of the 2nd International Light Metals Technology Conference, St. Wolfgang, Austria, 8-10 June 2005, pp 269-274.

Tables

Table 1: Comparison of tensile propeties [1]

Alloy	Process	Yield strength, MPa	Tensile strength, MPa	Percent elongation
A356.2-T6	P2000™	221-234	296-310	10-14
A356.2-T6	GPM	207-228	283-303	3-5
357-T6	P2000™	241-262	324-338	8-10
357-T6	GPM	248-262	331-345	5-7

Table 2: Left-hand (LHD) camber links fracture loads

Sample identification	Failure location	Fracture load, kN
Camber 1	Top end hoop	57.1
Camber 2	Top end hoop	54.2
Camber 3	Top end hoop	57.5
Camber 4	Top end hoop	59.5
Camber 5	Top end hoop	55.7
Camber 6	Top end hoop	60.2
Camber 7	Midsection	54.7
Camber 8	Midsection	54.8
Camber 9	Midsection	56.2
Camber 10	Top end hoop	58.8

Table 3: Effect of inclusions on fracture loads

Type of inclusion	Range of fracture load, kN	Failure location
Graphite	40.1 - 57.4	Top end hoop and mid-section
Oxide (dross)	23 - 55.4	Top end hoop
Furnace filter	31.2 - 57	Top end hoop and mid-section

Table 4: Strength testing of front knuckles [2]

Sample identification	Left-hand fracture load, kN	Right-hand fracture load, kN	Failure location
1	24.2	24.9	Pad
2	21.4	28.2	Pad
3	23.8	23.6	Pad
4	26.8	24.0	Pad
5	26.7	23.6	Pad
6	25.5	23.4	Pad
7	22.8	24.9	Pad
8	24.8	24.4	Pad
9	23.6	22.6	Pad
10	28.5	20.7	Pad
Average	24.8 +/- 2.1	24.0 +/- 1.9	Pad

Table 5: Rear knuckle testing

Test type	Location tested	Fracture load (kN) or number of cycles to failure	Typical casting failure? (Yes/No)
Strength test	Toe link	47.9 +/- 1.47	Yes
	Control arm	23.9 +/- 3.96	Yes
Fatigue test	Toe link	460,137 to 1,246,014 cycles	Yes
	Control arm	43,033 to 691,659 cycles	Yes

Figures

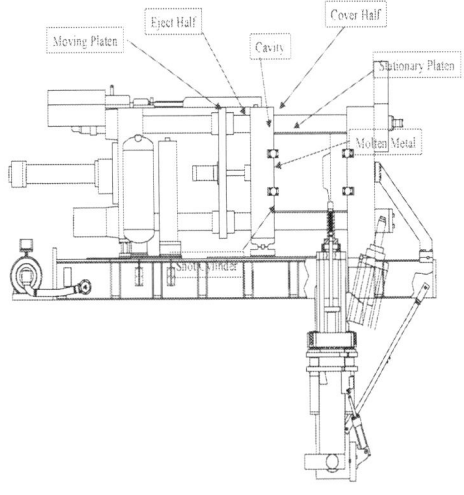

Figure 1: Schematic of P2000™ casting machine [2]

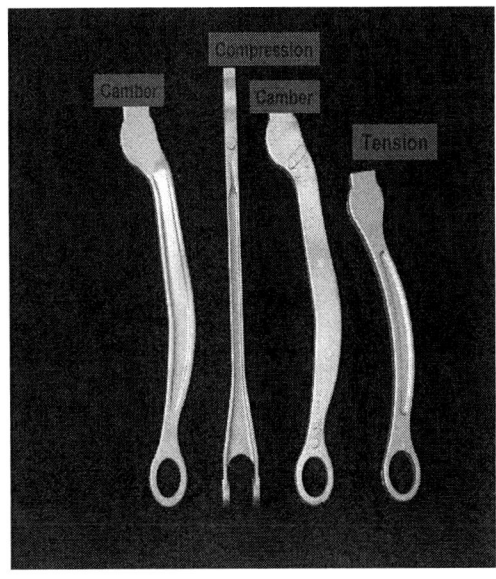

Figure 2: Suspension links [2]

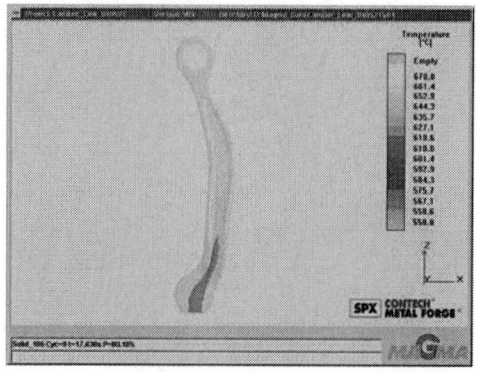

Figure 3: Solidification modeling for
a camber link

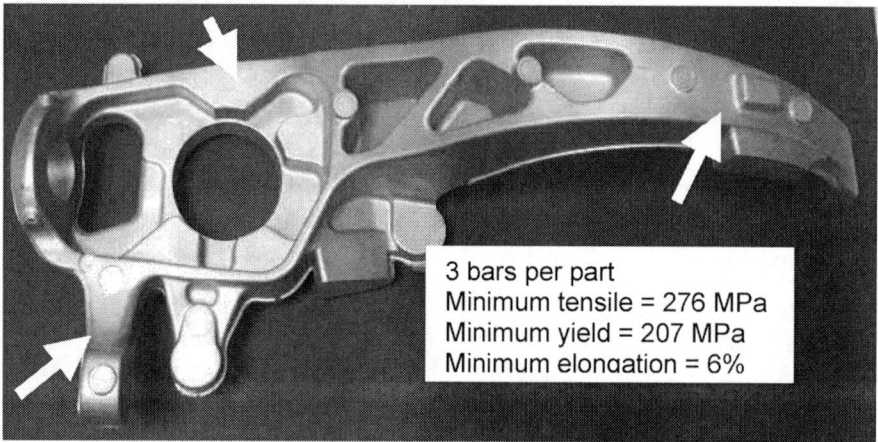

Figure 4: Front steering knuckle

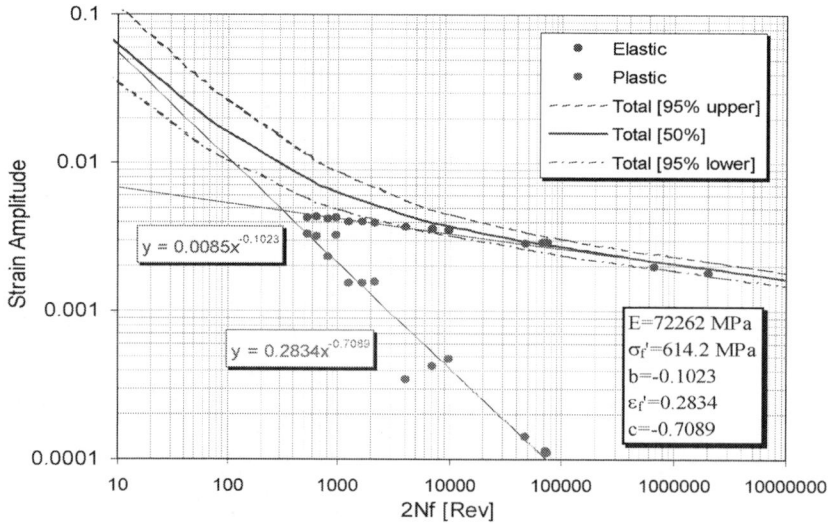

Figure 5: A356.2-T6 strain-life curve

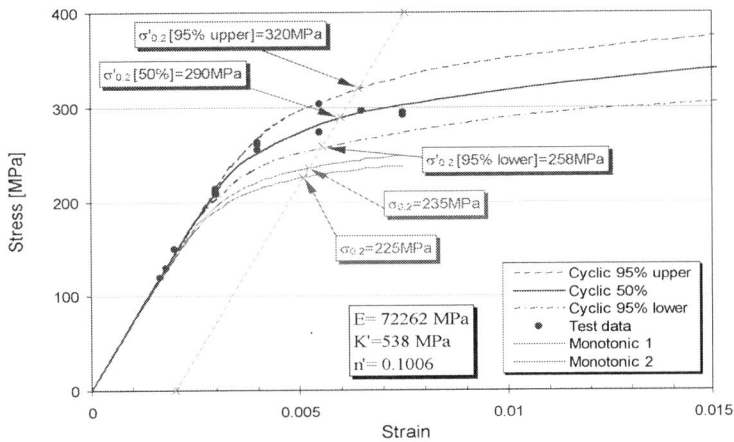

Figure 6: A356.2-T6 cyclic stress-strain curve

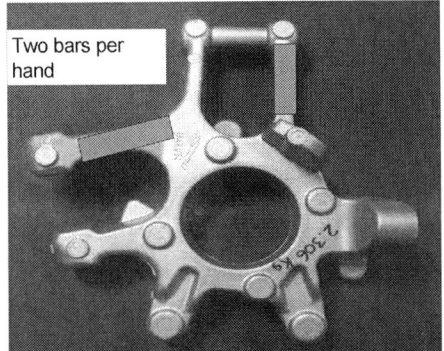

Figure 7: Rear steering knuckle

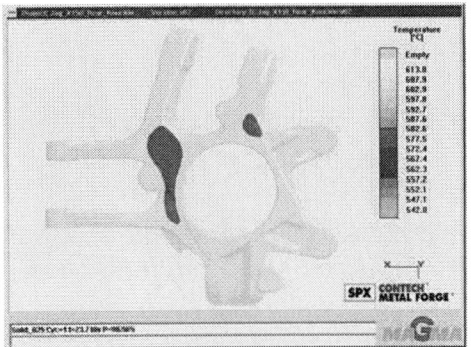

Figure 8: Natural solidification pattern for rear steering knuckle

Figure 9: Set-up for fatigue testing and desired failure locartion

Effects of Pouring Temperature and Squeeze Pressure on the Properties of Al-8%Si Alloy Squeeze Cast Components

A. Raji[*] & R. H. Khan[**]

*Department of Mechanical Engineering, Federal University of Technology, Yola, Adamawa State, Nigeria.
**Department of Mechanical Engineering, Federal University of Technology, P.M.B. 65, Minna, Niger State, Nigeria.

Abstract

This work was conducted to determine the optimum squeeze parameters for producing squeeze castings from the Al alloy and compare the properties of the squeeze castings with those of chill castings. Al alloy Squeeze castings were produced using squeeze pressures of 25-150MPa at steps of 25MPa with the alloy poured at 650°C, 700°C and 750°C for each of the indicated pressure into a die preheated to about 250°C. Squeeze time was maintained for 30 seconds.

It was found that for a specific pouring temperature, the grain size of squeeze cast products became finer; density and the mechanical properties were increased with increase in squeeze pressure to their maximum values while further increase in pressure did not yield any meaningful change in the properties. Squeeze cast sample properties were compared with those of chill cast samples. It was found that optimum pouring temperature of 700°C and squeeze pressure of 125MPa could be used to produce sound squeeze cast Al alloy components with aspect ratio (height-to-section thickness ratio) not greater than 2.5:1.

Keywords:

Squeeze casting, Aluminium alloy, Pouring temperature, Squeeze pressure, Mechanical properties.

Introduction

Squeeze casting, compared with traditional sand casting which dates back to about 2000-3500B.C. It is a relatively new casting technology [1,2]. It is a technology with very bright future, based on its applications and advantages. Yue and Chadwick [3] described squeeze casting as a casting process in which molten metal is solidified under the direct action of a pressure that is sufficient to prevent the appearance of either gas porosity or shrinkage porosity as opposed to all other casting processes in which some residual porosity is left. They further observed that the process is also known, variously as liquid-metal forging, squeeze forming, extrusion casting and pressure crystallisation.

Squeeze casting has a number of advantages which have been discussed by various researchers [4-8]. Some of the advantages include elimination of gas and shrinkage porosities, reduction or elimination of metal wastage due to absence of feeders or risers; ability to cast both cast and wrought alloys; possibility of manipulation of process parameters to achieve the required optimum parameters.

Squeeze casting is a very important manufacturing process, which combine the advantages of forging and casting used for the production of many range of products from monolithic alloys and metal-matrix composites parts. Such parts include vane, ring groove reinforced piston, connecting rod, M6-8 bolt, joint of aerospace structure, rotary compressor vane, shock absorber cylinder, diesel engine piston, cylinder liner bearing materials among many others used in automobile, nuclear, aeronautical components, sports equipment and many industrial equipment [9,10].

Despite its relatively small age, squeeze casting has witnessed a lot of development in the sphere of products and materials cast and quite a number of research studies have been carried out to improve the process particularly in the areas of molten metal metering and metal movement system during pouring into the die, lubrication systems and the use of reinforcement among others. However, in spite of all these researches, it was observed that squeeze casting particularly the relationship between the design, the process parameters and the quality of the squeeze cast components was yet to be fully understood; thus the need for more studies in this area of technology for better understanding of the process [11]. This study was carried out to determine the effect of squeeze pressure and pouring temperature on aluminium-silicon alloy squeeze cast products with aspect ratio (height-to-section thickness ratio) up to 2.5:1.

Materials and Methods

In this study, an Al alloy, the composition of which is given below and lubricant consisting of 10% graphite in lubricating oil of the type 20W/50 were used: Si-8.08%, Cu-1.920%, Fe-.0.686%, Mn-0.173%, Ni-0.086%, Al- Rem.

Melting was carried out in a resistance furnace & a 150t hydraulic press was used for squeezing operation. Series of experiments involving casting of the shape shown in Fig.1 were carried out using squeeze casting, sand casting and chill casting methods. Specimens were then prepared from the castings with the aim of determining the mechanical properties and microstructure of castings by the various techniques and the results were compared with each other.

MELTING OF THE ALLOY
Melting was carried out in an electric resistance furnace. The alloy was charged into a preheated crucible. Covering flux, 2% by weight of charge was used to prevent from oxidation & gas pick up. Degassing was carried out using hexachloroethane tablets (0.5% of melt) before pouring at the desired temperatures of 650, 700 or 750 degree centigrade. Temperature was measured using immersion pyrometer.

CHILL CASTING
A two part permanent mould made from mild steel was employed for making chill castings (at atmospheric pressure and squeeze pressure of 0MPa). The lower half of the permanent mould was mounted on the hydraulic press and its internal surface was preheated to a temperature of 160°C. Simultaneously, the upper part of the mould was placed upside down and the surface was similarly preheated. The surfaces of the mould that were to be in contact with the molten metal were coated with a prepared lubricant (graphite in lubricating oil). The two parts of the mould were preheated to a temperature of 250°C. The two parts were then assembled together. Molten aluminium-silicon alloy of the required temperature was then poured from the crucible into the assembled mould. The metal was allowed to stand for a period of 5mins after which the moulds were separated and the casting ejected out of the mould. Three sets of three chill castings were made with the alloy poured at 650°C, 700°C and 750°C.

SQUEEZE CASTING
Squeeze castings were made using a two- part die, the lower die and the upper die (punch), made from mild steel. The lower die was mounted on a supporting bed of the hydraulic press table. The punch was attached to the ram of the hydraulic press. The assembly was enclosed in a casing to isolate it from the shop atmosphere. With the door of the casing opened, the die heater was placed in between the two halves of the die. Thereafter, the door was closed. The probe for the immersion pyrometer was then placed in the 8mmØ hole located in the lower half of the die through an opening in the door. The die surface heater was switched on to preheat the lower and upper dies. When the temperature of the Squeeze casting die reached 160°C, the door was opened, the punch was raised and the heater was withdrawn from the die. A prepared lubricant made up of 10% of graphite in lubricating oil was applied on the surfaces of the die that were to be in contact with the molten metal. The heater was replaced in its

position, the punch was lowered, the door was closed and the dies were then preheated to the temperature of 240-250°C).

Thereafter, the door of the casing was opened; the punch was withdrawn upward to a position from which it could readily strike. The heater was once more removed away. Measured quantity of the aluminium alloy at the required pouring temperature was poured into the lower die. The punch was then brought down with a velocity of 9.45mm/s onto the lower die and the required pressure was applied for a period of 30s. The punch was, thereafter, withdrawn upward. The solidified casting was ejected from the lower die with the help of ejector pins. Squeeze castings were made using 25, 50, 75, 100, 125 and 150MPa with the alloy poured at 650°C, 700°C and 750°C for each of the indicated pressure. Three sets of squeeze castings were made for each combination of squeeze pressure and pouring temperature. Metallographic examination, density, hardness and tensile properties were evaluated for the samples cast.

Results and Discussion
The results of density, hardness and strength characteristics are shown in Figs.2-6. Properties at squeeze pressure of 0MPa refer to those of chill castings.

DENSITY
The relationship between density of the Al alloy chill castings as well as squeeze castings and squeeze pressure for various pouring temperature is depicted in Fig. 2. The density of the chill castings and squeeze castings varied from 2.712 for chill castings to 2.866 g/cm^3 for squeeze castings that is about 5.68% increase compared to chill castings.

In all pouring temperatures, the density increased with increase in squeeze pressure. There was a very steep increase in the density from 2.718 at squeeze pressure of 0MPa to 2.820 g/cm^3 at squeeze pressure of 75MPa for the pouring temperature of 650°C and thereafter it increased gently (almost horizontally) to 2.830g/cm^3 at 150MPa. Similarly, for the pouring temperature of 700°C, the density increased steeply from 2.720 at 0MPa to 2.842g/cm^3 at 75MPa and thereafter it increased gently to 2.863g/cm^3 at 150MPa. For the pouring temperature of 750°C, the density increased from 2.712 at 0MPa to 2.778g/cm^3 at 50MPa and 2.857g/cm^3 at 75MPa. Thereafter it increased slightly to 2.866g/cm^3 at 150MPa. Hence, the curves for 700°C and 750°C tend towards each other. The trend could be attributed to the fact that squeeze pressure tends to decrease gas porosities and decrease the inter-atomic distances and as these gas porosities and distances decrease the castings become more compact. However, further increase in squeeze pressure leads to more resistance to packing and so the rate of increase in density becomes reduced and density remains almost constant. The initial lower values of density for pouring temperature of 750°C might be due to formation of porosities in the casting at higher pouring temperature and low squeeze pressures.

This agrees with the findings by Clegg [15], which states that there is the possibility of extrusion of molten metal from the die through vents and the formation of porosity in thicker sections, if the pouring temperature is too high. Generally, for the same squeeze pressure, the density increased with increase in pouring temperature as high temperature makes the molten alloy less viscous and hence easier to compress and flow with ease.

HARDNESS

The relationship between hardness of Al-8%Si alloy chill castings as well as squeeze castings and squeeze pressure for various pouring temperatures is shown in Fig. 3. The results showed an increase in hardness of Al-8%Si alloy from Rockwell Hardness, HRF39.5-40.5 for chill castings to a maximum of HRF58.0 for squeeze castings which constitutes about 43 to 47% increase over those of chill castings. The increase in the hardness of squeeze cast products is brought about by the faster cooling rates giving rise to grain refinement and elimination of porosity and hence increased hardness of squeeze cast products. For the pouring temperature of 650°C, the hardness increased from HRF 39.5 for chill casting at 0MPa to HRF 53.5 for squeeze casting at 150MPa. In the case of pouring temperature of 700°C), the hardness increased from HRF 40.0 for chill casting at a squeeze pressure of 0MPa to HRF 58.0 for squeeze casting at squeeze pressure of 125MPa. Further increase in squeeze pressure to 150MPa did not lead to any further change in the hardness of the squeeze castings. The curve for pouring temperature of 750°C is similar to that of pouring temperature of 700°C. The hardness increased from HRF40.5 to HRF57.5 at 125MPa and then to HRF58.0 at 150MPa.

ULTIMATE TENSILE STRENGTH (UTS):

The relationship between UTS of the Al alloy chill castings as well as squeeze castings and squeeze pressure for various pouring temperatures is depicted in Fig.4. The results of UTS showed that squeeze casting enhances the strength of cast materials as can be observed from the graphs. The increase in the strength of squeeze cast products is due to pressure during solidification & higher cooling rates leading to grain refinement. The reduction in the grain size leads to increase in the number of grains and hence increase in the amount of grain boundary. Subsequently, any dislocation moves only a small distance before reaching a grain boundary and the strength of the product is thus increased [12].

The curve for the pouring temperature of 650°C shows an increase in UTS from 115MPa for chill casting at atmospheric pressure to 210MPa for squeeze casting at squeeze pressure of 150MPa. The curves for the pouring temperatures of 700°C and 750°C slightly differ from that of 650°C as they exhibit increase in UTS to maximum values with increase in squeeze pressure up to certain pressure and then remain almost constant with further increase in squeeze pressure. For pouring temperature of

700°C, the UTS increased from 115MPa for chill casting at squeeze pressure of 0MPa to 232MPa for squeeze casting at squeeze pressures of 125MPa and 150MPa. Similarly, for pouring temperature of 750°C the UTS increased from 114MPa at squeeze pressure of 0MPa to 226MPa at squeeze pressure of 125MPa and further increase in squeeze pressure did not yield any meaningful change in the UTS as a value of 225MPa was obtained for squeeze pressure of 150MPa. The increase in UTS to a maximum value of 232MPa obtained at squeeze pressure of 125MPa and pouring temperature of 700°C for Al-8%Si alloy squeeze cast products is similar to that experienced for aluminium casting alloy LM24 (containing, according to Rajan *et al[13]*, 8.5%Si and 3.5%Cu) in which UTS of 233MPa in as-cast condition was achieved [3]. This was also experienced for squeeze cast aluminium casting alloy 356 in which UTS of 212MPa was obtained in as-cast condition [14].

PROOF STRESS
The results of 0.2% proof stresses for the squeeze and chill castings are presented in Fig. 5. The pattern of 0.2% proof stresses is similar to those of UTS of squeeze and chill castings, although with different values. The reasons for the increase in proof stress are the same for those advanced for increase in UTS. The maximum proof stress of 156MPa was obtained for the squeeze casting with UTS of 232MPa made at a squeeze pressure of 125MPa and pouring temperature of 700°C.

ELONGATION
The results of elongation of Al-8%Si alloy chill castings as well as squeeze castings are shown in Figs. 6. The percentages of elongation for the squeeze castings varied between 2.8 to 3.8% as compared to those for chill castings which ranged from 2.4 to 2.7%. The percentage elongation increased for pouring temperature of 650°C from 2.7% at squeeze pressure of 0MPa to 3.4% at 75MPa and 100MPa and finally increased to 3.6% at 125MPa and 150MPa. For the pouring temperature of 700°C, the percentage elongation increased from 2.4 to a maximum of 3.8% at 100MPa and thereafter remains constant. In the case of pouring temperature of 750°C, the percentage elongation increased from 2.5% at 0MPa to 3.8% at 125MPa and 150MPa. The increase in elongation of squeeze cast products is brought about by rapid cooling leading to grain refinement and reduction in secondary dendrite arm spacing so as to speed the evolution of the latent heat. The reduction in secondary dendrite arm spacing is accompanied by increase in strengths and ductility [12]. The obtained trend in elongation is similar to those obtained by Yue and Chadwick [3].

OPTIMUM VALUES OF SQUEEZE CASTING PARAMETERS
Analysis of the properties of the squeeze castings produced showed an improvement in their properties over those of sand and chill castings. Maximum UTS, 0.2% proof stress, hardness and elongation were obtained at squeeze pressure of 125MPa and pouring temperature of 700°C.

Therefore, squeeze pressure of 125MPa and pouring temperature of 700°C were considered as the optimum squeeze pressure and pouring temperature, respectively. The improvement in mechanical properties of squeeze cast Al-8%Si over those of chill cast Al-8%Si is attributable to natural modification (rapid cooling) which causes the silicon phase in eutectic structure to grow as thin interconnected rods between aluminium dendrites [12], and which is achieved by the high cooling rate offered by the die with the assistance of squeeze pressure leading to fine grain structure [3,4] .The results of the study are applicable to squeeze castings of Al-8%Si alloys in as-cast conditions having aspect ratio not more than 2.5:1. This is because castings with higher aspect ratio may lead to entirely different characteristics, particularly for the extruded section of the casting as greater aspect ratio will involve better or additional grain refinement [14].

Conclusions

The following conclusions were made based on the study:

1. The density of squeeze castings increases steeply with increase in pressure initially and then gradually until it becomes almost constant. Generally, the density of squeeze castings is higher than those of chill castings.

2. Squeeze pressure helps to refine the microstructures of castings and hence leads to better mechanical properties of the castings.

3. Squeeze casting significantly increases the density of the Al alloys to about 2.825-2.866g/cm^3 at squeeze pressures of 125 and 150MPa. It also improves the mechanical properties of squeeze castings over those of chill castings.

4. Optimum pouring temperature of 700°C and squeeze pressure of 125MPa have been established for squeeze casting of the Al alloy products having an aspect ratio of up to 2.5:1.

References

1. Amstead, B.H.; Ostwald, P.F. and Begeman, M.L., *Manufacturing Processes*, 7th ed., p739, John Wiley and Sons, New York (1979).

2. Rao, P.N., *Manufacturing Technology: Foundry, Forming and Welding*, p500, Tata Mc Graw-Hill Publishing Co. Ltd., New Delhi, India (1992).

3. Yue, T.M. and Chadwick, G.A., "Squeeze Casting of Light Alloys and their Composites", *Journal of Materials Processing Technology*, vol.58, No.2/3, pp.302-307 (1996).

4. Lynch, R.F.; Olley, R.P. and Gallagher, P.C.J., "Squeeze Casting of Brass and Bronze", Paper No.75-90 *AFS Transactions*, vol. 83 pp.561-568 (1975a).

5. Rajagopal, S., "Squeeze Casting: A Review and Update", *Journal of Metalworking*, vol.1, No. 4, pp.3-14 (1981).

6. Franklin, J.R. and Das, A.A., "Squeeze Casting – A Review of the Status", *The British Foundryman*, vol. 77, No. 3, pp.150-158 (1984).

7. Mortensen, A.; Cornie, J.A. and Flemings, M.C., "Solidification Processing of Metal-Matrix Composites", *Materials and Design*, vol. X, No.2, pp.68-76 (1989).

8. Zhang, D.L.; Brindley, C. and Cantor, B., "The Microstructures of Aluminium Alloy Metal-Matrix Composites Manufactured by Squeeze Casting", *Journal of Material Science*, vol.28, No.8, pp.2267-2272 (1993).

9. Brake, P; Schumans, H &Verhoest,J. Inorganic fibres and composite materials,Pergamon Press, Oxford (1998).

10. Li, Q. F. and McCartney, G. D., "A Review of Reinforcement Distribution and its Measurement in Metal Matrix Composites", *Journal of Materials Processing Technology*, vol. 41, pp.249-262 (1994).

11. Office of Industrial Technologies (OIT), *Metal Casting Project Fact Sheet: Optimisazation of the Squeeze Casting Process for Aluminum Alloy Parts*, p2, OIT, US Department of Energy, Washington, D.C. (2000). www.oit.doe.gov/metalcast/factsheets/cwru_optimize_squeeze.pdf

12. Askeland, D. R. *The Science and Engineering of Materials*, p554, PWS Publishers, Boston, Massachussetts (1985).

13. Rajan, T. V.; Sharma, C. P.; and Sharma, A., *Heat Treatment – Principles and Techniques*, p451, Prentice - Hall of India Private Ltd., New Delhi, India (1988).

14. Lynch, R.F.; Olley, R.P. and Gallagher, P.C.J., "Squeeze Casting of Aluminum", Paper No.75-122, *AFS Transactions*, vol. 83 (1975b).

15. Clegg, A.J. *Precision Casting Processes*, p293, Pergamon Press Plc., Oxford, UK (1991).

16. American Society for Testing and Materials [ASTM]. *1990 Annual Book of ASTM Standards, Section 3 Volume 03.01-Metals- Mechanical Testing; Elevated and Low Temperature Tests; Metallography*, ASTM, Philadelphia PA, pp151-153 (1990).

Figures

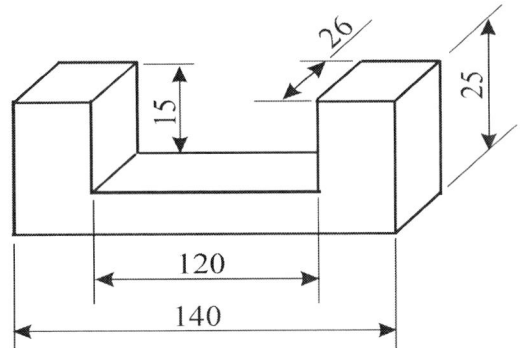

Fig. 1: Details of Experimental Castings
(Dimensions are in mm.)

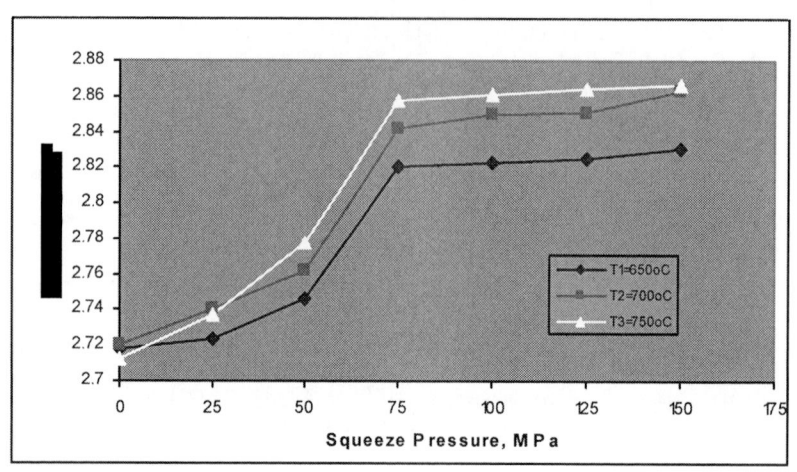

Fig. 2:Squeeze Pressure versus Density of Squeeze Cast Al-8%Si Alloy

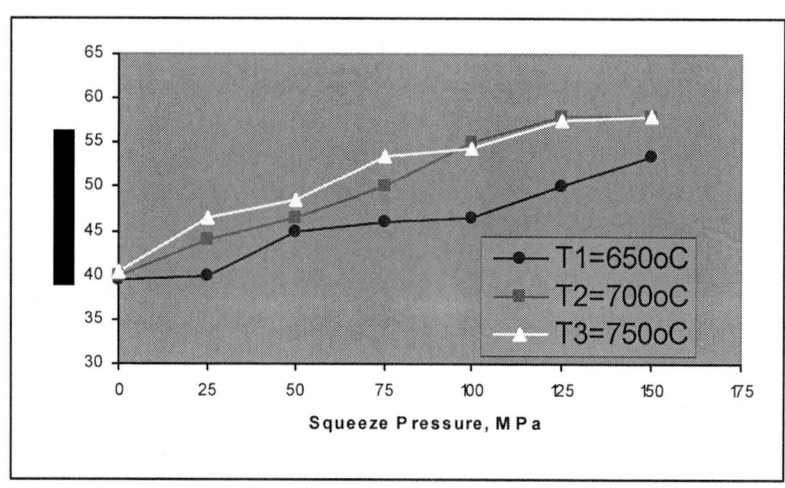

Fig. 3:The Effect of Squeeze Pressure on Hardness of Squeeze Cast Al- 8%Si Alloy

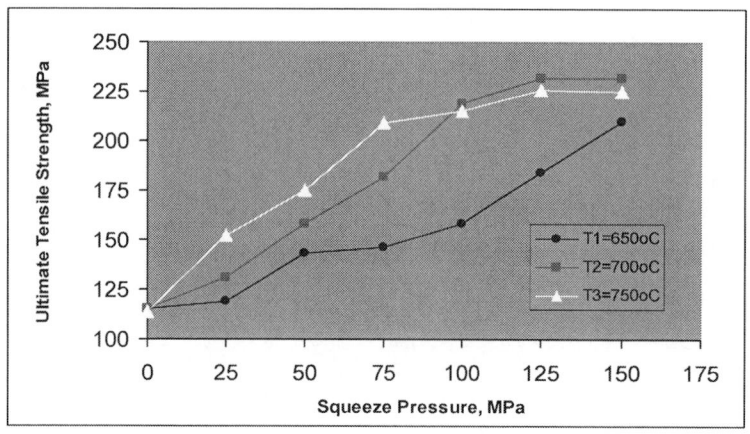

Fig. 4: The Effect of Squeeze Pressure on Ultimate Tensile Strength of Squeeze Cast Al Alloy

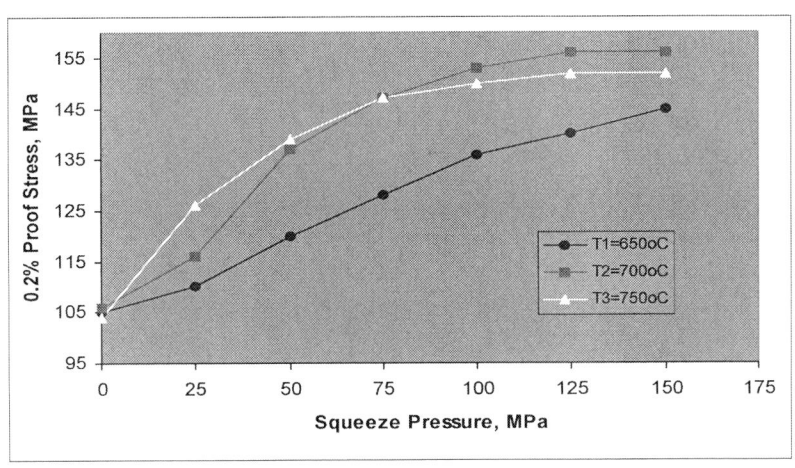

Fig. 5:Relationship between Squeeze Pressure and Proof Stress of Squeeze Cast Al Alloy.

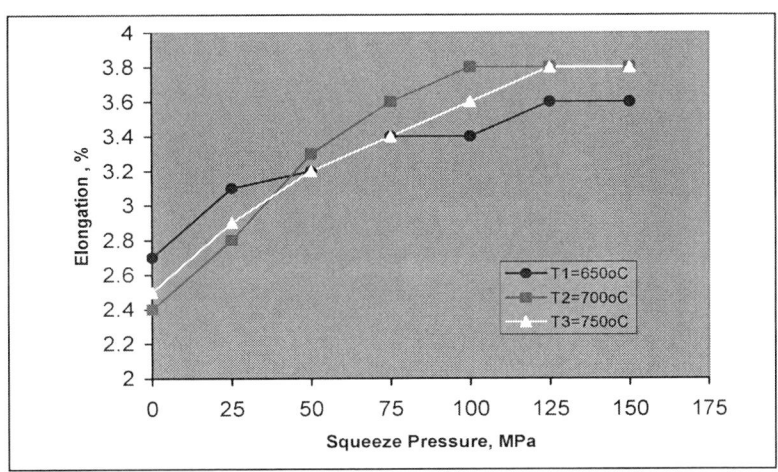

Fig. 6: Variation in Elongation of Squeeze Cast Al Alloy versus Squeeze Pressure and Pouring Temperature

On the Effect of Cooling Conditions and Variation of Alloying Elements on the Microstructural and Mechanical Properties of Al-7%Si Cast Alloys

Salem Seifeddine and Ingvar L Svensson

Jönköping University, School of Engineering, Component technology
S-551 11 Jönköping - Sweden

Abstract

In order to study the isolated influence of microstructure and alloying elemets on the mechanical properties of Al-7%Si based alloys, the level of defects arising from the melt handling and cast process needs to be minimized. Therefore, gradient solidification technology has been applied, providing tensile test samples with a low content of defects and a well-fed homogenous microstructure. In the present work, the influence of cooling rate and Mg, Cu, Fe and Mn on the as-cast microstructure formations and mechanical properties such as ultimate tensile strength, yield strength, elongation to fracture etc. has been elucidated. While Cu and Mg strengthen the material, Fe deteriorates mostly the elongation. The length of Al_5FeSi-needles appears to be a function of the cooling rate and not of the content of iron in the melt.

Key words Al-Si alloys, Tensile properties, Microstructure, β -phase, Gradient solidification.

Introduction

The demands for lighter, more durable, energy efficient and recyclable goods are intensified. In the last few years, significant efforts have been performed to examine and assess the influence of alloying elements on the microstructure and on the mechanical properties of aluminium cast alloys. In order to obtain the required quality and soundness, the mechanical properties of aluminium cast alloys can be tailored by controlling the amount of eutectic in the microstructure, solidification rate as described by the secondary arm spacing (SDAS), adding modifiers or grain refiners or, after manufacturing, applying some kind of thermal treatment etc.

Since aluminium is a very soft and ductile metal, aluminium cast alloys are alloyed with a wide variation of elements, and the major ones are silicon, magnesium and copper. The morphology and distribution of the silicon particles controls to some extent the ductility and strength. Increasing the silicon content hardness, ultimate tensile strength and yield strength will be enhanced. Unfortunately, the ductility will then be lowered [1, 2] .

Adding a slight amount of magnesium, 0,3-0,7 %, the mechanical properties such as yield strength and ultimate tensile strength will gain an appreciable increase. This improvement of strength will be ascribed the compound Mg_2Si that gradually precipitates in the matrix upon aging [2-3]. The main purpose of adding copper to aluminium cast alloys is to achieve higher strength. The mechanical properties, ultimate tensile strength and ductility, depend on whether the copper is present in solid solution as evenly distributed spheroidised particles or as networks at the grain boundaries. Having the copper dissolved in the Al-matrix, the strength will then increase. On the other hand, if the copper is found as continuous network at grain boundaries a loss of ductility will be expected [4].

The presence of iron in aluminium cast alloys is unavoidable. In secondary alloys, the iron content is higher due to the iron containing scrap. Iron can be added intentionally in order to avoid die soldering or unintentionally through for instance, the use of cast equipment made by steel. The intermetallic phases that are formed might be several; but the one that is of vital importance is the iron rich Al_5FeSi. In aluminium cast alloys, iron tends to impair the ductility and ultimate tensile strength due to brittleness of the formed intermetallic compounds. The purpose of this paper is not to study the influence of oxide films on the mechanical properties, which may act as cracks in liquid metals and castings. But it is appropriate to mention that the iron rich β-phase, Al_5FeSi, may nucleate and grow on the surfaces of the oxide films. Sometimes, if the β-phase is precipitated on both sides of the oxide film, the β-phase may exhibit a crack along the centre line or if it is precipitated on only one side, it will lead to an apparent decohesion from the matrix [5]. In order to compensate for the harmful effect of the iron, manganese in Al-Si alloys may act as Fe-corrector by favouring the formation of α- $Al_{15}(Fe,Mn)_3Si_2$ in the shape of Chinese script [6, 7].

Experimental

In order to determine the composition of the alloys that have been produced, see table 1, optical emission spectrometry analysis of one cast coin at the start and another at the end of the casting process was performed. As it is appeared in table1, the alloys are divided into three different groups depending on the elements that are of concern to study.

Each alloy was cast into rods for further solidification studies in the gradient solidification equipment. The gradient solidification techniques give possibilities to achieve a material, with low content of oxide films, shrinkage- and gas porosities and a homogenous and well-fed microstructure all over the length of the sample. In this work a resistance - heated furnace with an electrically driven elevator was used. Three different growth velocities, v, 0.03 mm/s, 0.3 mm/s and 3 mm/s which correspond, generally, to an SDAS of about 60, 20 and 7 µm respectively, were used.

The average tensile properties of three samples for each alloy and condition have been evaluated. Furthermore the samples have been metallographically prepared (mounted, grinded and polished) in order to study the microstructure.

Results

It has been well established in this investigation that the tensile properties are influenced by the amount of the alloyed elements and by the cooling rate. The alloy chemistry plays an important role in determining the quality of the casting due to the intermetallic compounds that may form. The fineness of the microstructure is mostly influenced by the local solidification rate, see figure 1a and b. The microstructure is defined as distance between the secondary dendrite arms, SDAS, the size and shape of the silicon particles and the intermetallics compounds that may precipitate out during the solidification process.

By increasing the cooling rate, the morphology of the β-needles, Al_5FeSi, is refined and a large fraction of β-needles appears to be replaced with Chinese script when manganese is present, figure 2. At higher iron levels and cooling rates, most of the Fe-phases are altered to become more polyhedral and star like. At slow cooling rates, the β-needles are likely to grow and coarsen leading to a remarkable reduction in the overall properties of the casting. Smaller SDAS results in smaller silicon particles, and also in smaller intermetallic phases such as the iron-bearing platelets, see figure 3.

As declared in the bar charts in figure 4, an improvement in ultimate tensile strength and yield strength due to the formation of the precipitated phases such as Mg_2Si and Al_2Cu is obvious. These precipitates block and obstruct the mobility of dislocations; which on the other hand lead to a reduction in ductility of the matrix. The ductility suffers a loss due to the increase of the rate of load shedding onto the Si particles. The high Mg level enhances the potential to form the Mg-containing Fe-rich intermetallics and also increase the occurrence of a peritectic reaction, which alters the previously formed needles with liquid into $Al_9Mg_3FeSi_5$.

The highest tensile and yield strength are obtained when adding both Mg and Cu, alloy 10 and 11 in figure 4a, and alloy 9 in figure 4c. Since the tensile strength and ductility are sensitive to defects, the formation of the iron bearing intermetallics seems to adversely affect these properties. Instead, iron impact positively on the yield strength; see alloys 5-8 and 9-12 in the graph in figure 4. It seems that iron acts as reinforcement in the matrix.

Comparing the influence of Cu and Mg on the cast material, figure 5, the addition of Cu seems to have improved the ultimate tensile strength mostly while Mg exhibits an appreciable effect on the yield strength. As observed, the enhancement in strength has occurred on the expense of ductility. Having Fe in aluminium cast alloys, the overall mechanical properties appear to get severely damaged.

Discussion

Considerable studies have been reported regarding the formation of Fe-bearing phases in relation to the iron content and cooling rate. It is reported that at 0,05%<Fe<0,7% a small needle-like phase is believed to form after the eutectic Si along with Si in a ternary eutectic reaction. At Fe>0,7% most of Al_5FeSi -needles seems to be precipitated prior to the eutectic Si as a binary eutectic reaction. In his article, Lu et al. [8] concluded that higher iron levels not only did enhance the total amount of Al_5FeSi -needles but also shifts to precipitation sequences of this phase to higher temperatures. These phases precipitated at higher temperatures are likely to grow during the solidification process resulting in extremely coarse needle-shapes. At high cooling rates, as in the case of high pressure die casting, the occurrence of primary Al_5FeSi-needles is shifted towards higher iron levels, Fe>1%. The length and thickness of the needles, no matter if they precipitate as primary, binary are ternary reactions, depends to a large extent on the cooling rate and an increase in iron content leads to an increased fraction of iron precipitates which is in accordance to what have been elucidated in this study. From figure 3, it can be revealed that it is the cooling rate that has the largest impact on the growth of the β-needles and that the iron level controls the fraction of β-needles and other Fe-phases that are formed in the casting.

Regarding the copper, an increase in solidification rate may help the formation of elongated Cu bearing phases along the interdendritic regions. Otherwise, as it also has been revealed in this investigation, the Cu will instead form as clusters of lumpy Cu bearing particles being adjacent to the Al_5FeSi- needles and within the Al-Si eutectic [2]. Another phase that Cu may form when Fe is present is the needle like Al_7FeCu_2-phase.

Besides forming Mg_2Si, magnesium promotes also the formation of $Al_8Mg_3FeSi_6$ when iron is present in Al-Si alloys and $Al_5Mg_8Cu_2Si_6$ when Cu is added. These intermetallics can as well be harmful to the mechanical properties since the particle cracking of these phases reduce the ductility and their formations reduce the Mg and Cu available for the age hardening [9].

The primary and secondary dendrites, the eutectic silicon, the iron-bearing intermetallics and copper-bearing phases are all refined with an increasing solidification rate providing improved quality and soundness of the casting. By comparing the performance of the alloys in figure 4 and 5, it is assessed what role the solidification rate plays in determining the tensile behaviour of cast metals.

Alloying elements are also added to change the morphology of the iron rich beta-phase to a more compact, less harmful intermetallic compounds. Among the modifiers that are frequently used for that purpose are Mn, Cr, Co, Sr, Be and Ca. Shabestari et al. [7], has reported that alloying with 0,9% Mn, resulted in apparent fragmentation of the β- Al_5FeSi -needles. In the absence of Mn, the increase in cooling rate and decrease in iron content results in a reduction in the size and volume fraction of Al_5FeSi-needles and a finer dispersion of these particles is appeared. Worth to be mentioned that even if Manganese is present in the alloy, Al_5FeSi-needles and $Al_{15}(Fe,Mn)_3Si_2$ are likely to coexist, see figure 2, and the fraction of each of them, depends on the cooling rate and the ratio Fe:Mn, which is usually recommended to 2:1.

Iron is supposed to facilitates the formation of porosity which seems to be associated with the formation and occurrence of iron intermetallics. The iron intermetallics are expected to cause severe feeding difficulties during solidification. The morphology of the β-phase blocks the interdendritic flow channels, why it is proposed that higher iron contents in the alloy is associated with higher levels of porosity. It has also been suggested that the β- Al_5FeSi -needles are very active sites for pore nucleation and that addition of Mn, which inhibits the formation of these needles, also neutralizers the pore nucleation. Neither of these suggestions has been detected, instead, a relation between pore size and the cooling rate seems to exist. The higher the cooling rate the smaller the pore size. At very slow cooling rates the pores seem to grow some millimetres independent on the iron levels. In this sense, it seems reasonable to assume that the cooling rate and the unfurling process of the oxide films could be responsible for the nucleation and growth process of the porosity [5].

The strain hardening exponent is related to the true flow stress and true plastic strain by equation 1, [2]

$$\sigma = K\varepsilon^n \qquad (1)$$

Rewriting equation (1) as

$$\text{Log } \sigma = \text{logK} + n \text{ log}\varepsilon \qquad (2)$$

the slope n is known as the strain hardening exponent and K is known as the strength coefficient. Generally, n and K are much influenced of the coarseness of the microstructure and less on the alloying element. As observed, the values of n and K are varying from one alloy to another and will also depend on the solidification conditions of the material, see figure 6. Relying on the collected data, Cu seems to influence the n and K mostly. The n and K reaches peak values when alloying with iron together with the copper, alloy 4, in combination with higher solidification rates group 1 in figure 6 a and b. Besides that iron at higher levels weakens the tensile properties such as ultimate tensile strength and ductility of Al-Si cast alloys, its impact on strength coefficient and strain hardening exponent is negligible, see figure 6 group 2 and 3.

Another interesting notice, the fracture path of a fine microstructure is mostly intergranular in contrast to the coarser structure where the fracture path is largely transgranular. In coarser structures, the cell boundaries forms effective obstacles to slip and the cracked particles are located in the cell and the grain boundaries. Concerning the finer structure, the cracked particles are preferentially located in the grain boundaries. So, as the applied strain increases, damage and final fracture are localized to the grain boundaries [10].

Conclusions
Thus, an understanding of the effect of the cooling rate and intermetallic phases on the monotonic tensile strength of aluminium cast alloys is of significance and as an outcome of this investigation:

- With a high solidification velocity a finer silicon eutectic and precipitates are achieved, with small SDAS and finely dispersed β-needles resulting in improved ultimate tensile strength, yield strength and ductility.
- Adding Mg and Cu to Al-Si alloys, the ultimate tensile strength and yield strength are improved at the expense of elongation.
- One of the main reasons for the low values of mechanical properties is proposed to be due to the presence of the β-phase. The higher the solidification rate the shorter the length of the needle.

  - *Higher iron content resulted in higher amounts of iron-rich β-phase, but on an average, the length of the needle seems to be more dependent on the solidification rate.*
  - *At relatively low iron levels, Fe<0,2%, and high solidification rate, SDAS ≈ 8 μm, only few needles could be detected.*

- *The iron content adversely affects properties such as ultimate tensile strength and elongation, since they are sensitive to defects. But the yield strength seems to be positively influenced by the increased level of iron.*
- *Higher iron levels seem neither to damage the strength coefficient nor the strain hardening rate.*

- The SDAS and the shape of the eutectic silicon seem to have a larger impact on the strain hardening exponent and fracture behaviour.
- Regarding the defects as the source of the premature failure might be the inclusions that forms during solidification and the entrainment of oxide films folded over and tangled acting as cracks resulting in poor mechanical properties.

References

1. T. Takaai, M. Koga and Y. Nakayama: *Heat treatment and mechanical properties of A356 aluminium casting alloys*, Light Metals Processing and Applications.
2. C.H. Cáceres, I.L. Svensson and J.A. Taylor: *Strength –Ductility behaviour of Al-Si-Cu-Mg casting alloys in T6 Temper*, International Journal of Cast metals Research, 2002.
3. S. Shivkumar, C. Keller, M. Trazzera and D. Apelian: *Precipitation hardening in A356 alloys*, Production, Fabrication and Recycling of Light Metals.
4. Z. Li, A.M. Samuel, F.H. Samuel, C. Ravindran, S. Valtierra, H. W. Doty: *Parameters controlling the performance of AA319-type alloys, Part I. Tensile properties*, Materials Science and Engineering, September 2003.
5. Campbell J, Castings, 2nd Ed, Butterworth-Heinemann Ltd, Oxford 2003, ISBN 0 7506 4790 6.
6. T.O. Mbuya, B.O. Odera and S.P. Ng´ang´a: *Influence of Iron on castability and properties of aluminium silicon alloys*, Literature study, International Journal of cast metals Research, 2003 Vol. 16, No.5.
7. S.G. Shabestari, M. Mahmudi, M. Emamy, J. Campbell: *Effect of Mn and Sr on intermetallics in Fe-rich eutectic Al-Si alloy*, International Journal of Cast metals Research, 2002, 15, pp17-24.
8. L. Lu and A.K. Dahle: *Iron-rich intermetallic phases and their role in casting defect formation in hypoeutectic Al-Si alloys*, Metallurgical and Materials Transactions A, Volume 36A, March 2005, 819-835.
9. C.H. Cáceres, C.J. Davidson, J.R. Griffiths and Q.G. Wang: *The effect of Mg on the microstructure and mechanical properties of Al-Si-Mg casting alloys*, Metallurgical and Materials Transaction A, October 1999.
10. C.H. Cáceres, C.J. Davidson and J.R. Griffiths: *The deformation and fracture behaviour of an Al-Si-Mg casting alloy*, Materials Science and Engineering A197 (1995).

Acknowledgements

The work has been performed within the IDEAL project supported by the European Commission (5th Framework Programme), which is gratefully acknowledged.

Tables

Table 1. The composition of the investigated alloys is presented.

	Chemical composition (%)							Groups
	Al	Si	Mg	Cu	Fe	Mn	Sr	
1	Bal.	7,04	0,03	0,00	0,06	0,00	0,0171	Group 1
2	Bal.	6,87	0,35	0,00	0,06	0,00	0,0167	
3	Bal.	7,21	0,02	3,49	0,06	0,00	0,0200	
4	Bal.	7,42	0,01	3,07	1,37	0,38	0,0180	
5	Bal.	7,09	0,35	0,21	0,14	0,00	0,0125	Group 2
6	Bal.	7,04	0,36	0,21	0,71	0,08	0,0150	
7	Bal.	7,04	0,32	0,20	1,04	0,12	0,0170	
8	Bal.	7,45	0,32	0,22	1,10	0,41	0,0150	
9	Bal.	7,02	0,40	3,26	0,09	0,00	0,0260	Group 3
10	Bal.	6,96	0,36	3,06	0,63	0,09	0,0240	
11	Bal.	6,77	0,31	2,79	0,88	0,19	0,0100	
12	Bal.	6,90	0,33	3,08	1,27	0,40	0,0200	

Figures

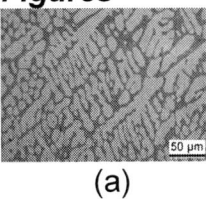 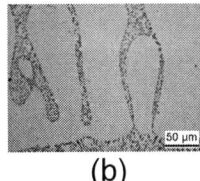

(a)　　　　　　　　　　　(b)

Figure1. Illustration of microstructure of Al-7%Si where a) is exhibiting an SDAS of 8,4 µm and b) having an SDAS of 70,1 µm.

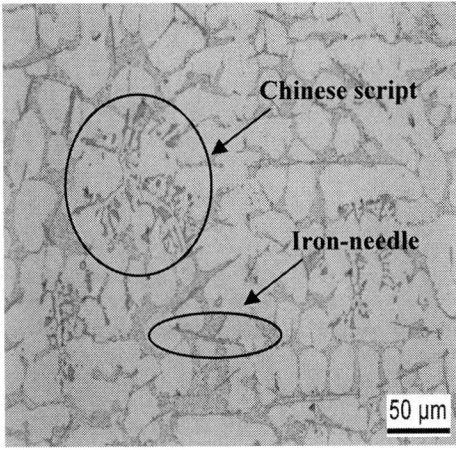

Figure 2. The coexistence of Al_5FeSi- needles and $Al_{15}(Fe,Mn)_3Si_2$ -Chinese script is shown.

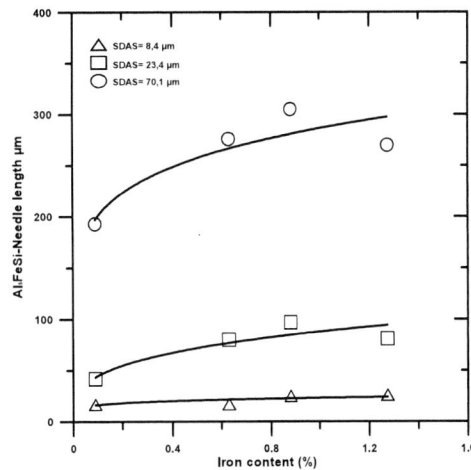

Figure 3. Illustration of the cooling rate impact on the length of the iron bearing Al_5FeSi needles. Alloys in group 3 considered.

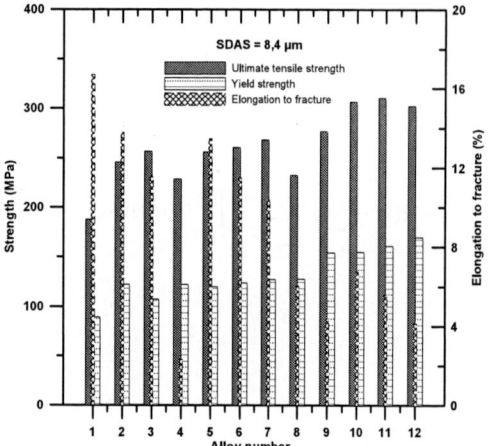

Figure 4a): Tensile strength of the SDAS ≈ 8 µm samples.

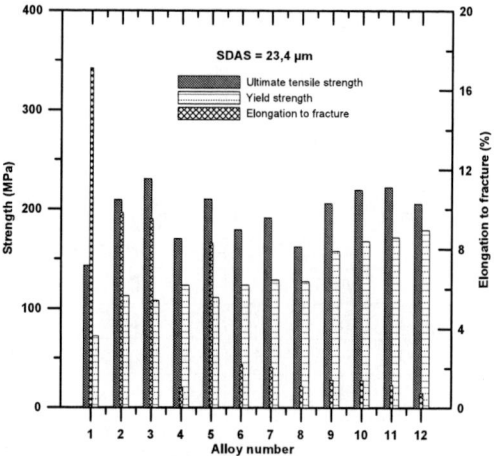

Figure 4b): Tensile strength of the SDAS ≈ 23 µm samples.

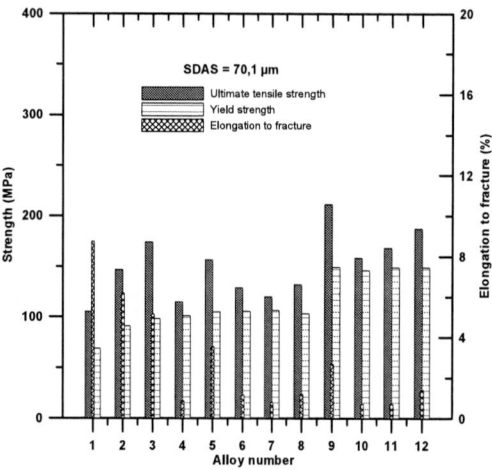

Figure 4c): Tensile strength of the SDAS ≈ 70 µm samples.

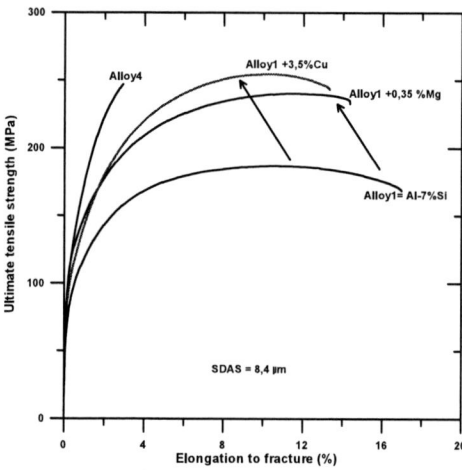

Figure 5a): The effect of Cu, Mg and Fe on SDAS ≈ 8 µm samples.

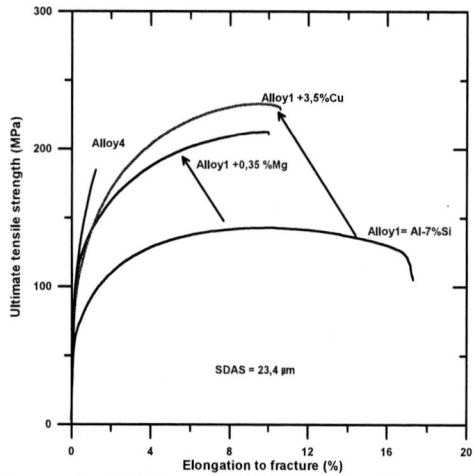

Figure 5b) The effect of Cu, Mg and Fe on SDAS ≈ 23 µm samples.

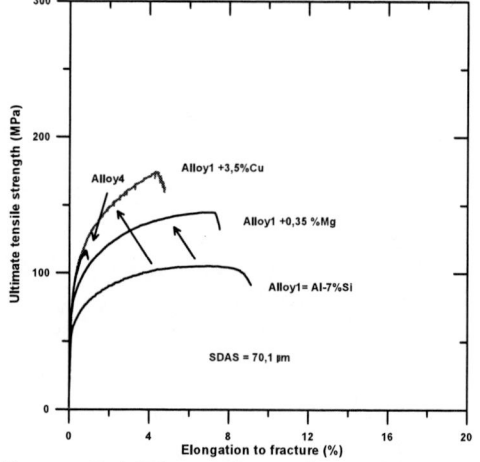

Figure 5c) The effect of Cu, Mg and Fe on SDAS ≈ 70 µm samples.

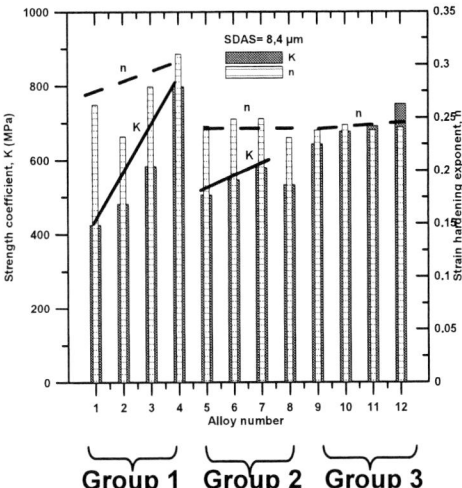

Figure 6a) The influence of alloying elements on K and n for the SDAS ≈ 8 µm samples.

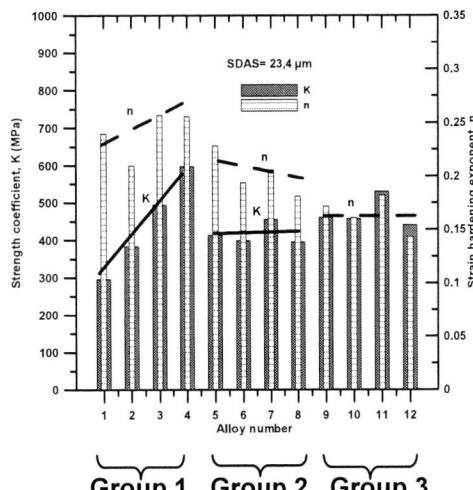

Figure 6b) The influence of alloying elements on K and n for the SDAS ≈ 23 µm samples.

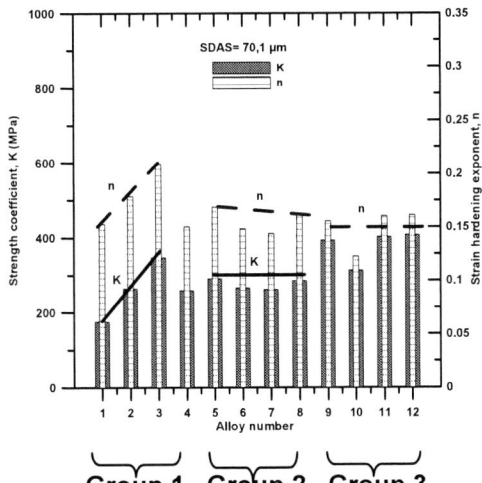

Figure 6c) The influence of alloying elements on K and n for the SDAS ≈ 70 µm samples.

A STUDY ON THE HOT TEARING IN Al-1% Sn CAST ALLOYS

G.L. Datta[+] and D. Benny Karunakar[++]

[+] Professor, Mechanical Engg. Dept., IIT, Kharagpur, India
[++] Research Scholar

ABSTRACT

Hot tearing in certain cast metals continues to bother metal casters. A number of casting parameters affect the susceptibility to hot tearing. Experiments were conducted on Al-1%Sn alloys. Al-1%Sn is a high freezing range alloy known to be extremely prone to hot tear cracking. Effects of grain refinement and presence of iron and nickel have been investigated by making use of a ring type casting with a steel core. Al-5Ti-1B was used as the grain refiner. Optical and scanning electron microscopy was used in the study. The results showed that contrary to some published work, grain refinement alone could not eliminate hot tearing. Average hot tear length decreased with increased grain refinement and the minimum being achieved was at 5% by weight of Al-5Ti-1B addition. Both iron and nickel have positive effects in minimizing hot tear formation. SEM studies have shown that with iron there is an Al-Sn-Fe rich phase in the form of a leaf like structure believed to be providing anchorage to the adjacent grains and thus overcoming hot tear.

Keywords: Hot-tearing, aluminium-tin alloys, grain refinement, effects of iron and nickel.

INTRODUCTION

Hot tearing or cracking is defined as the formation of a macroscopic separation in a casting as a result of distortion due to differential contraction of the casting during solidification. Presence of hot tears may lead to total failure of the cast component. It is a typical casting problem, which is particularly present with the long-freezing range alloys. The phenomenon of hot-tearing has been known to the foundrymen for a long time and has been extensively researched [1-14]. A wealth of published literature is available on hot-tearing on some of the non-ferrous [1-12] and steel castings [13-14]. John Campbell [1] had found that inclusions present in the castings had an important role in the initiation of hot tears. He had further observed that hot-tearing increased with higher freezing range and lower quantity of eutectic at solidification. Davies [3] conducted investigations on different aluminum alloys and concluded that hot-tearing was a singular function of grain size and hence grain refinement, though this contention seems to have not been substantiated by any of the subsequent investigators. Lancaster [4] also investigated the effect of grain size on an Al-2Zn-2Mg alloy with various additions of elements and found a linear relation between grain size and crack length in the restrained cracking test on welded joints. Dodd [6] noticed that the tendency to hot-tearing was most marked when the constitution of the alloy was such that a relatively small proportion of eutectic was present. According to Prokhorov [7] when a metal passes through the brittle temperature range during solidification, it may develop inter-crystalline cracking. Nicholas and Twitty [8] opined that mould wall movement in a green sand mould could cause hot-tearing. Couture and Edwards [9] suggested that the hot-tearing characteristics are largely determined by the freezing range and the amount and the type of liquid present during the last stages of solidification. Sigworth and Guzowski [11] investigated the effects of Al-Ti and Al-B master alloys on grain refinement of hypoeutectic Al-Si alloys and found that Al-B was more powerful than Al-Ti in reducing hot-tearing.

Al-1% Sn alloy has a long freezing range and forms only a small amount of eutectic, making it extremely hot-short. This alloy is an appropriate model for the study of hot-tearing. In the present study an attempt has been made to reinvestigate the effects of grain refinement and presence of iron as a residual element and nickel as an alloying element on hot tearing in aluminum-1% tin alloy castings.

EXPERIMENTAL SETUP

Davies [3] used a ring type metallic mould with a metallic core to investigate the hot-tearing tendency in castings made of aluminum and other non-ferrous alloys having comparatively long freezing ranges. The ring mould with a steel core hinders radial and tangential contraction of the ring causing strain. Heating of the core during casting causes the core to expand, resulting in further tangential strain in the ring casting and consequent cracking in the castings.

In the present work, a similar method has been adopted to evaluate the hot-tearing. A special gated pattern has been designed in the present work, which is similar to the one incorporated by Davies [3]. This gated pattern has been made such that it will produce four rings on four sides to get reasonable and reproducible results. The dimensions of a single pattern are shown in Fig 1. The photograph of the gated wooden pattern designed for the present investigation is shown in Fig 2. The pattern was provided with a core shoe such that the steel cores could be easily placed inside the mould. The mould was made of CO_2 hardened sodium silicate bonded sand. The metallic cores were then placed inside the moulds. The molten metal was poured through a centrally located sprue. The metal during solidification undergoes hindered radial contraction due to the presence of the cores. The average of the sum of the cracks of all the four rings was taken as the length of the hot-tear. Experiments were conducted on Al-1% Sn alloy with and without grain refinement. Experiments were also conducted with addition of iron and nickel, which are the common residual and alloying elements respectively.

EXPERIMENTS WITHOUT ADDITION OF GRAIN REFINER

Commercial aluminum was melted in an induction furnace and 1 % tin by weight was added to the melt. The melt was then poured into the mould. The casting thus obtained, shown in Fig 3, was found to have developed cracks on all the four rings. The cracks were very clear indicating that Al-1Sn alloy is highly prone to hot-tearing.

A section of the fractured portion of the cracked casting (Fig 3) was cut out and examined under SEM. The EDX analysis was carried out to scan several spots for bulk analysis and the average values were obtained to represent the composition. The average crack length along with the actual composition is furnished in Table 1.

The SEM photographs (Fig 4 and Fig 5) show brittle fractured surfaces of the Al-1Sn castings, in which fracture has crossed through interdendritic region. Low melting eutectic had concentrated in the interdendritic region and made it highly prone to cracking. From the Al-Sn phase diagram [5], it is found that the tin rich phase with 67.45 % tin content commences freezing at about 520 °C, though the surrounding α-dendrites of the metal have frozen just below 660 °C.

EXPERIMENTS WITH ADDITION OF GRAIN REFINER

Al-5Ti-1B grain refiner varying from 0.3% to 3.0% by weight was mixed to the melt and stirred thoroughly. The metal was then poured into the sodium silicate mould to make the castings. In each case, the pouring temperature was varied between 750 ^{o}C and 775 ^{o}C. The crack lengths in the four rings of the castings were measured and the average crack length was obtained. A section of the casting in each experiment was metallographically polished and the grain size was measured using Image Analyzer. The average crack lengths along with the actual compositions are furnished in Table 1. A cracked casting has been shown in Fig 3.

The fractured surfaces of the casting obtained with addition of 3 % grain refiner were examined under SEM. Figs 6 and 7 show the presence of a tin-containing phase in Al-1Sn casting with 3 percent grain refiner in which segregations of Si, Ti, Fe and Sn were noted. It was found that titanium from the grain refiner got preferentially segregated in the interdendritic region and formed a complex phase with Al, Si, Fe and Sn. It is worthy to note that Ti got segregated into Sn rich phase even with 3 percent grain refiner addition, which could be due to the formation of iron titanium aluminide. Figs 8 and 9 show that with increasing percentage addition of the grain refiner and consequent reduction in the grain size, the length of crack decreases. The pouring temperature was varied between 750 and 775 ^{0}C and this variation had no significant effect on the crack length.

EXPERIMENTS WITH ADDITION OF GRAIN REFINER AND IRON

From the previous section, it is evident that grain refinement with Al-5Ti-1B could reduce the crack length but failed to prevent hot-tearing completely. In view of these results, further experiments have been carried out with the addition of certain other elements like iron and nickel, which are the most common residual and alloying elements respectively in aluminum castings.

Iron was added to Al-1Sn charge in the form of Al-30Fe master alloy. It was noticed that addition of 6% Al-30Fe master alloy to the charge has resulted in developing a casting, absolutely free from hot-tearing. Thus, iron that occurs as an impurity in the commercial aluminum has a beneficial effect on the reduction of hot-tearing.

A fractured section was obtained from the casting (by bending), in which 6.0% iron master alloy and 3.0 percent grain refiner were added. The fractured specimen was cleaned ultrasonically and was studied using SEM. In Fig 10 a leaf like structure can be seen, in which iron rich phase is anchoring the grains. The iron content of Al-Sn-Fe rich phase is 11.60%. Precipitation of α - phase occurred at below 660 ^{o}C. Thus, the Al-Sn-Fe rich phase that froze much earlier than the α - dendrites actually anchored the α-dendrites and prevented hot-tearing.

In practice, the experimental Al-Fe-Si-Sn system is a multi-component one, in which Ti is also present from the grain refiner. Hence, it may be difficult to pin point the exact composition of the primary α - phase from the binary Al-Fe or Al-Sn phase diagrams. The EDX analysis data, shown in Fig 10, suggests that a relatively low melting iron rich Al-Fe-Sn-Ti phase had precipitated when the melt was cast from a temperature of 750 – 775 $^{\circ}$C.

EXPERIMENTS WITH ADDITION OF GRAIN REFINER AND NICKEL

Commercial ingot aluminum was melted in an induction furnace and 1% tin was added to the melt which was stirred thoroughly. Al-5Ti-1B grain refiner was added to the charge at 2% and stirred uniformly. Next the Al-6Ni master alloy chips were added to the charge at 15%. The entire charge was stirred uniformly; the temperature was then raised and held at 900 $^{\circ}$C for about 10 minutes. Then it was poured into the mould. After the solidification, the casting was taken out from the mould and had shown absolutely no cracking.

A fractured section was obtained from the aforementioned casting by bending for analysis. The fractured specimen was cleaned ultrasonically and examined under SEM. The fractograph of this casting is shown in Fig 11. This figure shows a phenomenon in which Al-Sn-Ni rich phase is interlocking the grains. From the Al-Ni phase diagram [5], it can be inferred that this phase with a nickel content of 16.62% has frozen at a temperature of about 854 $^{\circ}$C. However, the surrounding α-dendrites of the metal have frozen just below 660 $^{\circ}$C. The Al-Sn-Ni phase, that has frozen at much higher than the freezing temperature of α-dendrites, became strong enough to interlock the grains, which ultimately resulted in prevention of cracking.

DISCUSSION

From the Al-Sn phase diagram [5], it can be noticed that the molten aluminum undergoes freezing at 660 $^{\circ}$C and tin at 232 $^{\circ}$C. The freezing range of the binary Al-1Sn alloy is more than 400 $^{\circ}$C. During solidification of this long-freezing range alloy, the growing crystals are at first completely separated by liquid and the alloy has no strength. As the temperature falls, the volume of solid increases relative to that of the liquid, and at some point (the coherent temperature) the growing crystals meet. However, a limited volume of liquid still remains and persists down to the eutectic temperature, causing the dendrite network to be brittle. At the same time, the solid fraction contracts, and is, therefore, subjected to a tensile stress, which may be high enough (depending on the degree of restraint) to cause failure of the weak, probably semi-solid inter-granular regions. The risk of cracking is the greatest when a critically small volume of liquid metal is present below the coherence temperature. Thus, it is to be expected that the alloy will be prone to hot-tearing to the extent that it would be difficult to produce alloys without hot tears. From the phase diagram [5], a low melting eutectic of composition 95.5% Sn forms at 232 °C. This agrees tolerably with the measured composition of 75-80% Sn allowing for the usual effect of matrix dilution.

With increase in grain refiner addition, the grain size gradually decreases. Maximum grain refinement could reduce the crack length in Al-1Sn alloy to some extent but failed to prevent it completely. Thus, it can now be concluded that grain refinement can reduce the hot-tearing but cannot prevent it completely in an exceptionally long freezing range alloy.

From Al-Fe phase diagram it can be seen that α-Al-iron-aluminide eutectic undergoes freezing at a temperature of 652 °C. But tin undergoes freezing at 232 °C. The dendrites undergoing shrinkage at this particular temperature will be held together by the iron-aluminide, which acts as an anchoring agent. This would result in prevention of hot tearing. It can thus be concluded that iron, which is a common impurity in commercial aluminum, enables prevention of hot-tearing.

A similar mechanism is likely to be operative in a nickel bearing Al-1%Sn alloy, although in this case a clear evidence of anchoring could not be obtained. Nickel reacts with aluminum and forms nickel-aluminide. From the Al-Ni phase diagram [6], the α-Al-nickel-aluminide eutectic undergoes freezing at a temperature of 640 °C, which is again much higher than the freezing point of tin. As in the case of iron, nickel-aluminide also anchors the grains and holds them together and thus prevents hot-tearing. It may be noted that nickel is a common alloying element in aluminum castings. Thus, the presence of nickel in the casting not only produces the beneficial effects of alloying but also enables prevention of hot-tearing.

CONCLUSIONS

Based on the present experimental investigation, the following major conclusions can be drawn:

1. Hot-tearing in the Al-1Sn alloy is promoted by the formation of a low melting Al-Sn eutectic at the interdendritic or intergrain boundary and is reduced by the addition of Al-5Ti-1B grain refiner, but grain refinement alone cannot prevent hot tearing.

2. Grain refinement plus addition of iron or nickel can prevent cracking completely. The primary precipitates of iron rich aluminides anchor α-dendrites.

REFERENCES

1. John Campbell, Castings, 2nd Edition, 2003, Butterworth Heinemann, Oxford, UK.

2. H. Chadwick, Hot shortness in Al-4.5 % Cu alloy, Cast Metals, Vol. 4, No. 1, 1991, pp. 367-374.

3. V. de L. Davies, The influence of grain size on hot tearing, British Foundryman, Vol. 43, No. 4, 1970, pp. 93-101

4. J. F. Lancaster, Metallurgy of Welding, Chapman & Hall, London, Fifth edition, 1993, pp. 303-307.

5. T. B. Massalski, Binary alloy phase diagrams, Second edition, Vol. 1, 1990, ASM International, pp 148, 183, 216.

6. R. A. Dodd, Hot tearing of castings – A Review of the literature, Foundry Trade Journal, Vol. 101, Sept 1956, pp. 321-331.

7. N. N. Prokhorov, Resistance to hot tearing of cast metals during solidification, Russian Casting Production, Vol. 4, April 1962, pp. 172-175.

8. K. E. L. Nicholas and M. D. Twitty, The influence of the mould on shrinkage and hot tearing in white irons, British Foundryman, Vol. 57, March 1964, pp. 105-116.

9. A. Couture and J. O. Edwards, The hot-tearing of copper-base casting alloys, AFS Transactions, Vol. 74, 1966, pp. 709-721.

10. S. A. Metz and M. C. Fleming, Hot tearing in cast metals, AFS Transactions, Vol. 77, 1969, pp. 329-334.

11. G. K. Sigworth and M. M. Guzowski, Grain refining of hypoeutectic Al-Si alloys, AFS Transactions, Vol. 93, 1985, pp. 907-912.

12. Mark Easton and David St. John, Grain refinement of aluminium alloys: Part II, Metallurgical and Materials Transactions, Vol. 30A, No. 6, 1999, pp. 1625-1633.

13. J. Van Eeghem and A. De Sy, A contribution to understanding the mechanism of hot tearing of cast steel, AFS Transactions, Vol. 73, 1965, pp. 282-291.

14. A. Chojecki, I. Telejko and T. Bogacz, Influence of chemical composition on the hot tearing formation of cast steel, Theoretical and Applied Fracture Mechanics, Vol. 27, No. 2, 1995, pp. 99-105.

Table 1: Characteristics of Al-Sn Castings

Casting No	Al-5Ti-1B Refiner addition (wt. %)	Pouring Temp (°C)	Actual Composition (wt %)					Grain Size (μ m)	Average crack length (mm)
			Al	Sn	Si	Fe	Ti		
C1	0.0	750	98.01	0.84	0.45	0.65	0	188	56
C2	0.3	775	98.05	0.96	0.55	0.38	0.051	176	57
C3	0.7	750	98.31	0.89	0.15	0.57	0.078	167	52
C4	1.0	775	98.22	0.98	0.28	0.43	0.089	156	49
C5	1.3	750	98.16	0.78	0.56	0.39	0.106	144	45
C6	1.5	750	98.34	0.75	0.12	0.68	0.112	131	41
C7	1.7	775	97.77	0.97	0.48	0.66	0.12	117	43
C8	2.0	750	98.12	0.87	0.61	0.26	0.143	106	39
C9	2.3	775	98.64	0.81	0.16	0.23	0.156	98	37
C10	2.5	775	97.96	0.82	0.35	0.71	0.162	95	36
C11	2.8	750	97.81	0.88	0.67	0.48	0.165	90	33
C12	3.0	750	98.18	0.81	0.16	0.68	0.169	86	31
F0	0.0	750	96.40	1.26	0.72	1.62	0.0	71	55
F1	1.0	750	96.84	0.92	0.58	1.57	0.09	76	42
F2	2.0	750	96.14	1.13	0.61	1.86	0.13	72	20
F3	3.0	750	95.03	2.22	0.38	1.97	0.17	69	0.0

Note: 5% and 6% Al-30Fe master alloy added to F0 & F1 and F2 & F3 respectively.

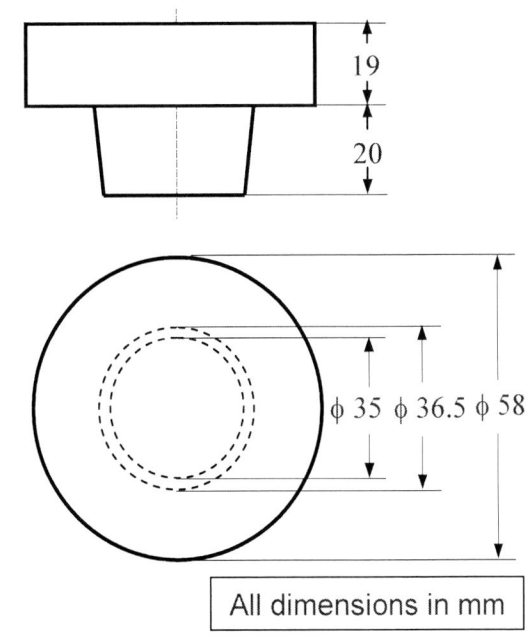

All dimensions in mm

Fig 1: Wooden pattern

Fig 2: Gated wooden pattern with cores

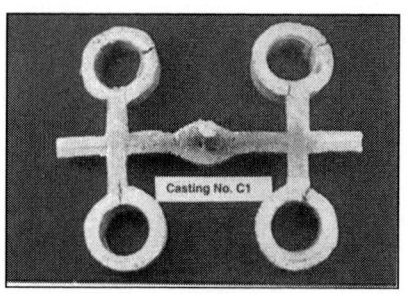

Fig 3: Casting obtained with the addition of 1% tin (average crack length = 56 mm).

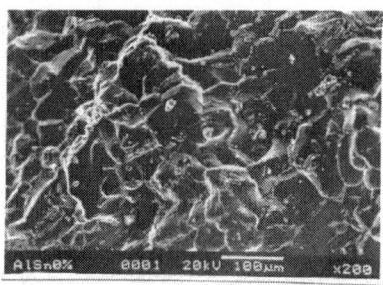

Fig 4: SEM fractograph of Al-1Sn casting (without addition of grain refiner).
EDX spot analysis (wt%):Al-96.38, Sn-1.90, Si-0.66, Fe-1.06.

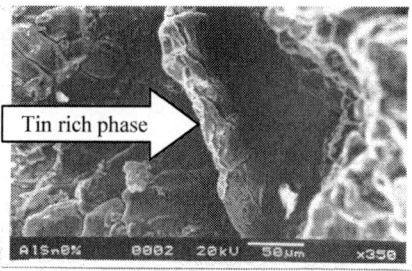

Fig 5: SEM fractograph of Al-Sn rich phase in Al-1Sn casting (without addition of grain refiner).
EDX analysis at tin rich phase (wt%): Al-26.57, Sn-67.45, Si-0.86. Fe-5.12

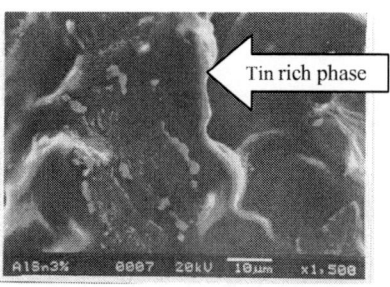

Fig 6: SEM fractograph showing Tin rich phase in Al-1Sn casting.
EDX analysis at tin rich phase (wt%): Al-88.12, Si-1.54, Ti-0.21, Fe - 9.02, Sn– 1.11

Fig 7: SEM fractograph showing Tin rich phase in Al-1Sn casting.
EDX analysis at tin rich phase (wt%):Al -28.30,Si-0.14,Ti-0.19, Fe- 0.75, Sn-70.62

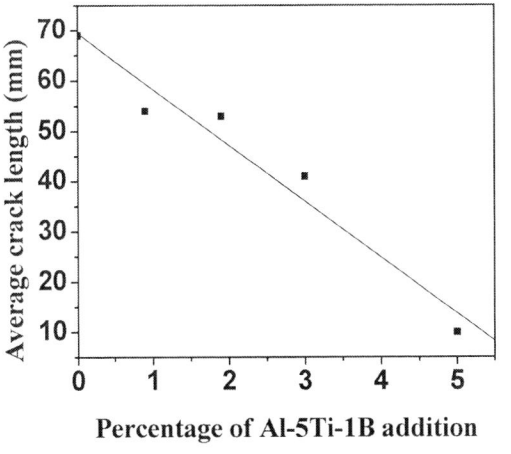

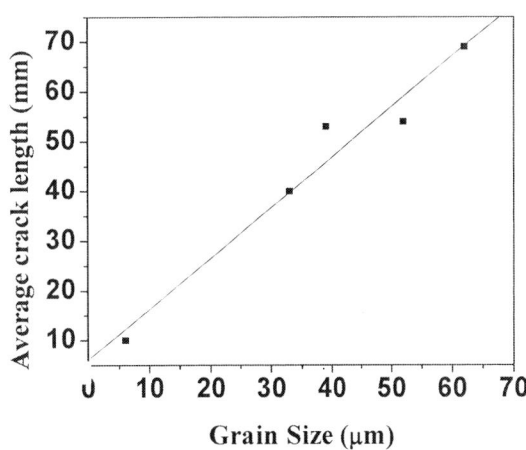

Fig 8: Effect of grain refiner addition on crack length

Fig 9: Effect of grain size on crack length

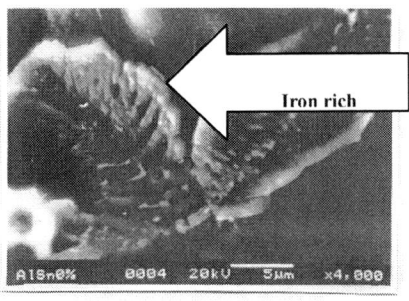

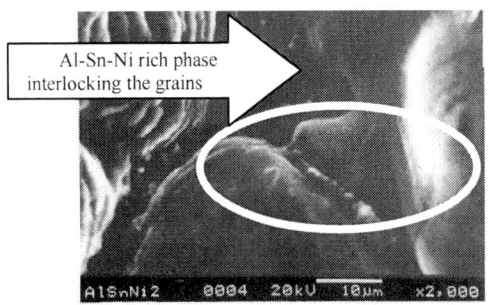

Fig 10: SEM fractograph showing Al-Sn- Fe rich phase anchoring in Al-1Sn casting, at higher magnification (With the addition of 3% grain refiner) EDX analysis at Al-Sn- Fe rich phase (wt%): Al-67.38, Sn-6.28, Ti-0.17, Si-8.96, Fe-17.21

Fig 11: SEM fractograph showing Al-Sn- Ni rich phase interlocking the grains in Al-1Sn–Ni casting with 2.0 % grain refiner. EDX analysis at Al-Sn-Ni rich phase (wt%): Al-76.25, Sn-5.08, Ti- 0.16, Si-0.29, Fe-1.60, Ni-16.62

A Study of Double Oxide Film Defect Behaviour in a Quiescent Aluminium Melt

R. Raiszadeh* and W D Griffiths

Department of Metallurgy and Materials Science, The University of Birmingham, Edgbaston, Birmingham, UK, B15 2TT, (* currently working in the Department of Materials Engineering, Shahid Bahonar University of Kerman, Jomhoori Eslami Blvd, Kerman, IRAN.)

Abstract

The change in the volume of an air bubble held within commercially pure liquid aluminium was recorded by real time x-ray radiography, to estimate the behaviour of double oxide film defects after entrainment in liquid aluminium. This showed that first oxygen and then nitrogen were consumed by the melt to produce Al_2O_3 and AlN respectively. Varying the initial hydrogen contents of the melt also showed that hydrogen could diffuse into the atmosphere within a trapped bubble, and cause it to expand. A mathematical model was developed using the consumption rates obtained from this experiment that included the diffusion of hydrogen, which suggested that the atmosphere within a double oxide film defect should be consumed by the surrounding Al melt in less than two minutes.

In addition, a double oxide film-free liquid aluminium melt was poured into ceramic test bar moulds which then held in the liquid state for 20 minutes. Study of the fracture surfaces suggested that the two non-wetting surfaces of a double oxide film defect might bond to each other, after the internal atmosphere of the film was consumed, and that this could occur in 20 minutes or less.

Key words: Oxide films, Aluminium alloys, Hydrogen porosity

Introduction

Aluminium alloy castings are commonplace in automotive and aerospace applications. It is therefore imperative that internal defects and potential failure mechanisms are fully understood if reliable castings are to be produced consistently. Commonly occuring defects include gas porosity, shrinkage porosity and entrained double oxide film defects. The latter defects occur when the surface oxide film of the melt becomes entrained in the bulk liquid as a doubled-over oxide film in which the internal surfaces are not bonded, but have a layer of air trapped between them [1]. This leads to a crack in the solidified casting, (see Figure 1).

It has been shown [2] that bad running system designs introduced oxide film defects and reduced the reproducibility of the mechanical properties of aluminium alloy castings. Nyahumwa et al. [3] also showed that entrained oxide film defects reduced the fatigue life of Al-Si alloy castings. They observed an improvement in the fatigue life of castings after a HIPping treatment at a temperature close to their eutectic temperature, and suggested that this was due to the collapse of double oxide film defects and diffusion bonding between the two internal non-wetted surfaces.

Nyahumwa [4] also suggested that the air inside a double oxide film defect floating within an aluminium melt might be consumed gradually by reaction with the surrounding liquid aluminium, (after an incubation time associated with the transformation from γ to α-alumina), causing the oxide walls of such a defect to bond together to some extent.

The main objective of the work reported here was to maintain a bubble of air in a mould containing molten aluminium alloy, and study its behaviour with time as the bubble atmosphere reacted with the liquid aluminium. A computer program was also developed to model this behaviour, based on the experimentally obtained reaction rates, to estimate the effect of changes in the ratio of the volume to surface area of a bubble, (i.e., its modulus), and hence to estimate the likely duration of the atmosphere within a double oxide film defect. Tensile tests were also carried out on castings that contained double oxide film defects with known ages, and the fracture surfaces of the tensile samples were studied by SEM and EDS to find the effect of the age of double oxide films on their appearance.

Experimental Procedure

A. A study of the behaviour of an air bubble within an aluminium melt

A 2.5 kW resistance-heated tube furnace was used to heat a cylindrical mould, (225 mm in length, 75 mm external diameter, 45 mm internal diameter), made of a compacted calcium silicate fibre that contained 700 g of liquid commercial purity aluminium at a constant temperature of 700 °C. Figure 2 shows a sketch of the equipment. A fully dense silicon nitride rod, 200 mm in length, 20 mm in diameter, and with a 40 mm deep, 13 mm diameter hole in one end, was used to hold a volume of air in the liquid metal. After the existing oxide film on the liquid Al in the mould was removed, the rod, preheated to 400°C, was gradually plunged into the liquid metal trapping a bubble of air in the melt within the hole in the rod. The apparatus was placed in a real-time x-ray radiography machine to observe and record the volume of the air bubble trapped in the silicon nitride rod and how it changed with time.

During the experiment the volume of the bubble gradually decreased and the liquid metal entered the rod hole. After the experiment, the rod was slowly removed and the liquid metal inside its hole solidified as the rod cooled. SEM and EDX studies were carried out on the surface of this sample. Also, the initial hydrogen content of the melt before pouring it into the mould was measured using a Severn Science Hyscan device.

Experiments with pure aluminium with different levels of initial hydrogen content, (of between 0.1 to 0.3 ml/100g), and different air bubble moduli, (of 5, 7.5, 10 and 40 mm), and different bubble holder shapes, (cylindrical or dome-shaped), were carried out (see Table 1 for a summary of the experiments carried out).

B. The mathematical model
The consumption rates obtained from the experiments were used to create a semi-empirical mathematical model, which included an explicit Finite Difference solution to the diffusion equation in a cylindrical axisymmetric coordinate system. This model was used to extrapolate the results obtained from observing the decrease in the bubble volume in the experiment to the dimensions of a double oxide film defect. Details of this model are reported elsewhere [5].

C. A study of the effect of the age of double oxide film defects
Existing double oxide film defects were removed from about 9 kg of commercially pure aluminium alloy by holding it in a heated chamber under a reduced pressure of 80 mbar at 800°C for one hour, during which the double oxide films present in the melt should inflate and float to the surface to be removed. The liquid metal was then poured, with surface turbulence deliberately introduced, into eight ceramic shell moulds in the shape of tensile test bars, four of which were solidified immediately, (preserving any double oxide film defects created during mould filling), while the other four were transferred to a resistance-heated furnace and held at 800°C for 20 minutes before solidification, (thus any oxide films generated during pouring were about 20 minutes old in these moulds). The UTS and elongation of these test bars was determined using a Zwick 1484 tensile test machine, and SEM and EDS studies were carried out on the fracture surfaces thus obtained.

Results
A. The behaviour of an air bubble held within liquid aluminium
Figure 3 shows the change in trapped air volume with time in different experiments, with different initial hydrogen contents. In Exp. 1, (with a low

hydrogen content of 0.11 ml/100g), the liquid aluminium first began to consume the oxygen in the trapped air atmosphere to form Al_2O_3 (shown in Figure 4(a)), as the free energy of formation of Al_2O_3 is greater than that of AlN. This produced a decrease in the volume of the trapped air and the oxide layer was drawn upwards to compensate for it. After most of the oxygen was consumed, nitrogen began to react, producing the AlN shown in Figure 4(b). At this point a change in the slope of the curve of Exp. 1 was observed, at about 87% of the initial air volume.

An aluminium melt with a high initial hydrogen content of 0.28 ml/100g, (Exp. 3), produced an initial expansion to about 110% of the initial volume of the bubble, (see Figure 3), which was caused by the diffusion of hydrogen into its atmosphere. This decreased the rate of consumption of nitrogen, (and probably oxygen), and hence increased the duration of the bubble in the melt. The expansion of the bubble in Exp. 2, (with an initial hydrogen content of 0.19 ml/100g), was about 104% of the initial volume and no expansion was observed, as was the case in Exp. 1, which had a low initial hydrogen content of 0.11 ml/100g.

Figure 5 shows volume-time curves for experiments with different air bubble moduli, and different bubble holder shapes. This Figure shows that in the experiments with cylindrical bubble holders, (Exps. 1, 4 and 5), as the modulus of the bubble decreased, the duration of the bubble in the liquid aluminium decreased non-linearly. However, Exp. 6, in which the holder of the bubble was dome-shaped and the modulus of the bubble was 7.5 mm, did not obey this trend, instead, the duration of the bubble in this experiment was greater than in Exp. 1, (which had a cylindrical bubble holder and a modulus of 40 mm). In other words, the duration of the atmosphere within the bubble depended not only upon the modulus of the bubble, but also its shape, and how much additional area was created as the bubble volume was reduced. Bubble shapes associated with larger increases in area as the bubble volume was reduced, were associated with shorter durations.

B. The estimation of the duration of the atmosphere within a double oxide film

The mathematical model results were found to be in reasonable agreement with the experimental results [5] for cylindrically-shaped bubble holder, with modulus of 40 mm. The model showed that the duration of the bubble atmosphere in liquid aluminium (t, in hrs) decreased as the modulus of the bubble (M, in mm) decreased. An extrapolation of the results from the model was used to estimate the duration of the atmosphere for air bubbles of different modulus;

$$t = -0.0548M^3 + 0.1155M^2 + 0.972M + 0.002 \text{ (for } H_2 \text{ about } 0.1 \text{ ml/100g)} \quad (1)$$

$$t = -0.01881M^3 + 0.0702M^2 + 1.811M + 0.00369 \text{ (for } H_2 \text{ about } 0.3 \text{ ml/100g)} \quad (2)$$

C. Double oxide film defect age and mechanical properties

A Reduced Pressure Test sample was taken from the melt that had been held under a vacuum for one hour, which showed that the melt was free of previously-introduced double oxide film defects. The hydrogen content of the melt was also measured to be about 0.08 ml/100g.

The Weibull distribution of elongation of the tensile test bars for the quickly-solidified castings and for the castings held for 20 minutes in the liquid state has been shown in Figure 6. Two lines were fit to the elongation data of the specimens from the castings that were held in the liquid state. This distribution suggested that the mechanism of fracture in the four specimens with lowest elongations was probably different from the rest of the specimens. The slope of these two lines, which indicated their Weibull moduli, was 4.4 and 5.0 respectively, both higher than that of the specimens from the castings that were solidified immediately after pouring (which was 3.3).

The average UTS of the specimens from the castings that were solidified immediately was about 10 MPa greater than that of the castings that were held in the liquid state before solidification. The Weibull distribution of UTS for the specimens from the immediately-solidified castings was 27, while for those that were held in the liquid state 21.5, which indicated that holding the liquid metal in the liquid state after pouring seemed to have increased the distribution of UTS properties slightly.

The SEM and EDS study of the fracture surfaces of the tensile test bars from the immediately solidified castings revealed the frequent presence of double oxide films (see Figure 7(a)), or pores with oxide films within them (Figure 7(b)). A symmetrical oxide defect was also found on the same location of the fracture surface of the other half of the tensile test bar shown in Figure 7(a), which indicated that the oxide film was doubled and its crack-like nature produced a weakness in the tensile test bar. 6 out of 13 specimens from the castings that were held in the liquid state before solidification showed no double oxide films or pores in their microstructures, and just one specimen contained an oxide film defect with a symmetrical oxide surface on the other half of the tensile test bar.

Figure 7(c) shows a cavity that was found on the fracture surface of a specimen from one of the test bars that was held in the liquid state before solidification. SEM and EDS study showed that the walls of this cavity were covered with oxide layers, and that the two walls of this cavity were apparently bonded together at several points. One of these connections between the two walls is magnified in Figure 7(d) and an EDS spectrum obtained from point P in this Figure revealed the presence of oxygen indicating it was alumina structure.

Discussion

The experiments with the trapped air bubble suggested that an Al melt should consume the oxygen and nitrogen within a double oxide film, and that this process should be continuous with no incubation time associated with a transition from amorphous to γ or α-Al$_2$O$_3$. The cracks that would be formed during movement of the oxide film, and that allowed contact between fresh aluminium and the trapped atmosphere, eliminated the need for such an incubation time before further reaction takes place.

Experiments with varying hydrogen contents (Figure 3), showed directly, and for the first time, that hydrogen is able to diffuse into the trapped atmosphere within a double oxide film and inflate it into a gas pore.

The experiments that examined the effect of changing bubble modulus and changing bubble shape suggested that the greater the increase in the surface area of the oxide layer as the bubble atmosphere was consumed, (which depended on the shape of the bubble holder), the greater the stress induced in the oxide layer and hence the greater the number of cracks formed in it. This would lead to an increase in the rate of consumption of oxygen and nitrogen, meaning that the deformation experienced by a double oxide defect in the melt may be the main factor affecting the rate of consumption of its atmosphere.

The exact dimensions of double oxide films are not known, but Campbell estimated [1] dimensions of about 5 mm × 5 mm and a trapped atmosphere thickness which would be a maximum of 40 µm. This defect, according to Equations (1) and (2), would have its original atmosphere of oxygen and nitrogen consumed in about 77 and 144 seconds, in a melt with hydrogen contents of about 0.1 and 0.3 ml/100g, respectively.

The decrease in the distribution of the elongation of the specimens from the castings that were held in the liquid state before solidification, compared to those that were solidified immediately after pouring, was probably due to the presence of smaller numbers of double oxide film defects, or perhaps the bonding of the oxide surfaces, which has been shown in Figure 7(c). The tensile test results showed that holding the castings in a liquid state for 20 minutes did not change the distribution of UTS of the specimens considerably, and the difference in the average UTS of the two sets was attributed to a difference in their cooling rates.

Two explanations of the structure seen in Figure 7(c) are possible. The bonding of the walls of this cavity could have occurred by the crystallization associated with the transformation of amorphous alumina to γ-Al$_2$O$_3$, which may happen within a few minutes of the formation of the oxide defect. According to the estimations produced by the mathematical model, the double oxide film defect would have lost its oxygen and nitrogen content by then, and the two oxide surfaces should then be in contact with each other. The diffusion of hydrogen, (rejected into the liquid during solidification), could have then expanded the pockets of argon that remained between the attached parts of the oxide surfaces, and formed the cavities shown in the SEM micrograph of Figure 7(c).

On the other hand, the concentration of oxygen obtained from the EDS spectrum taken from the surface of the connecting structure shown in Figure 7(d) suggested that this connection was not pure Al_2O_3, but was instead, pure aluminium covered by aluminium oxide. This suggested an ulternative explanation, that the penetration of liquid aluminium through cracks at contact points between the two nonwetted oxide surfaces was involved in the bonding of the surfaces. In this explanation, the double oxide film defect was stitched together by liquid aluminium passing through the opposing oxide surfaces. However, neither explanation accounts for the almost regular interval between the connections and the fact that the structure of the pores was tubular.

To summarise, these experiments have shown how the duration of the atmosphere within a double oxide film defect can be estimated, and these initial results suggest that it may be consumed within a few minutes of formation, although this estimate is highly dependent upon the assumptions made about the mechanisms involved. Furthermore, the mechanical property tests have shown, for the first time, that the internal surfaces of double oxide film defects may become bonded together, as has been previously suggested.

Conclusions
1. The oxygen and nitrogen of an air bubble trapped in an aluminium melt were consumed by reaction with the surrounding liquid aluminium to form Al_2O_3 and AlN.
2. If the initial hydrogen of the melt was higher than the equilibrium amount associated with the ambient atmosphere, hydrogen diffused into the trapped air bubble and caused its expansion.
3. The rate of consumption of oxygen and nitrogen within the atmosphere of the trapped bubble depended on the number of cracks induced in the oxide and nitride layers.
4. A mathematical model suggested that the atmosphere within a double oxide film defect in an aluminium melt would be consumed in less than two minutes, depending on its modulus and the amount of deformation it experienced in the melt.
5. Holding of aluminium castings at 800°C for 20 minutes before solidification generally increased the elongation and the Weibull modulus of tensile test specimens. This was suggested to be due to a decrease in the number of double oxide film defects in the liquid metal during its holding period, or due to a partial healing due to the bonding of the surfaces of double oxide film defects at some points.
6. SEM micrographs of fracture surfaces showed that the two non-wetting surfaces of a double oxide film defect might bond to each other, if the casting was held in the liquid state for 20 minutes before solidification.

References

1. Campbell J., Castings, 2nd ed., Butterworth-Heinemann, 2003, ISBN 0 7506 4790 6.
2. Green N. R. and Campbell J., *Influence of Oxide Film Filling Defects on the Strength of Al-7Si-Mg Alloy Castings*, AFS Trans., 114, 1994, pp 341-347.
3. Nyahumwa C., Green N. R., and Campbell J., *Influence of Casting Technique and Hot Isostatic Pressing on the Fatigue of an Al-7Si-Mg Alloy*, Met. and Mat. Trans., 32A, 2001, pp 349-358.
4. Nyahumwa C., Green N. R., and Campbell J., *Effect of Mold-Filling Turbulence on Fatigue Properties of Cast Aluminum Alloys*, AFS Trans., 58, 1998, pp 215-223.
5. Raiszadeh R.and Griffiths W. D., *Estimation of the Duration of the Atmosphere Within a Double Oxide Film Defect in Pure Aluminium*, 12th Int. Metallurgy-Materials Cong., Istanbul, Turkey, 28th Sep.- 2nd Nov., 2005.

Acknowledgements

The authors wish to thank Professor J. Campbell for helpful discussions in connection with the work, Mr. A. Caden for his technical support and the Ministry of Science, Research and Technology of the Islamic Republic of Iran for its financial support.

Table

Table 1. Summary of the experiments carried out

Exp #	H_2 content (ml/100g)	Holder shape	Bubble modulus (mm)	Exp #	H_2 content (ml/100g)	Holder shape	Bubble modulus (mm)
1	0.11	Cylindrical	40	4	0.11	Cylindrical	10
2	0.19	Cylindrical	40	5	0.11	Cylindrical	5
3	0.28	Cylindrical	40	6	0.12	Dome-shp	7.5

Figures

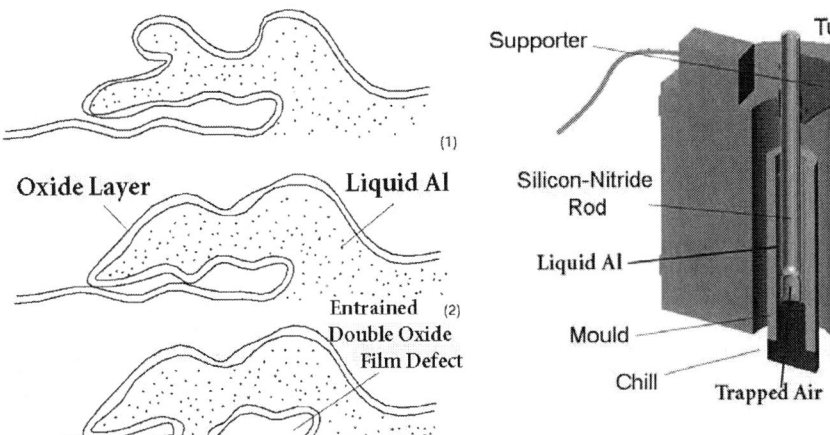

Figure 1. Entrainment of a double oxide film defect.

Figure 2. Schematic of the experimental setup.

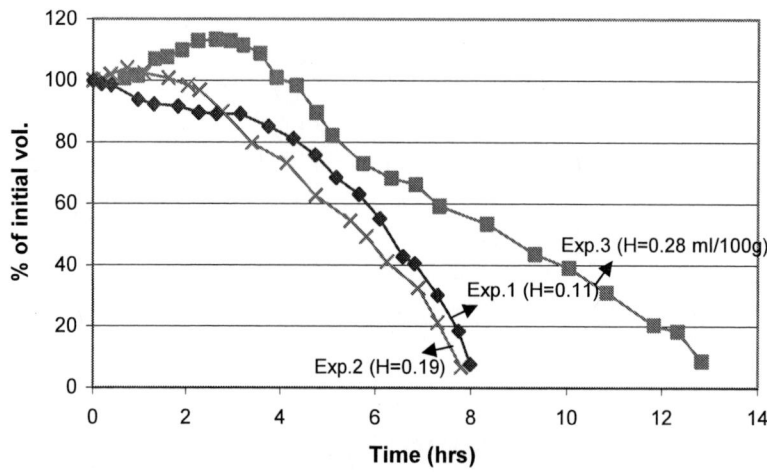

Figure 3. The effect of different initial hydrogen contents on the behaviour of the trapped air bubble in commercially pure aluminium.

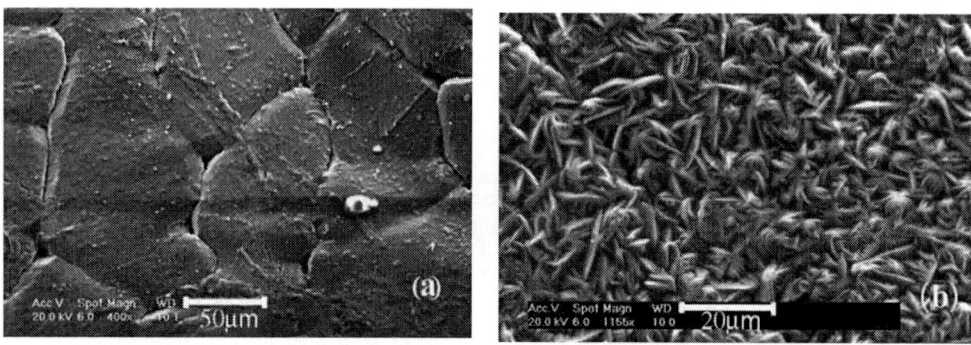

Figure 4. SEM micrograph of the surface of the solidified sample, showing the products of the reactions, (a) an Al_2O_3 layer and (b) an AlN layer.

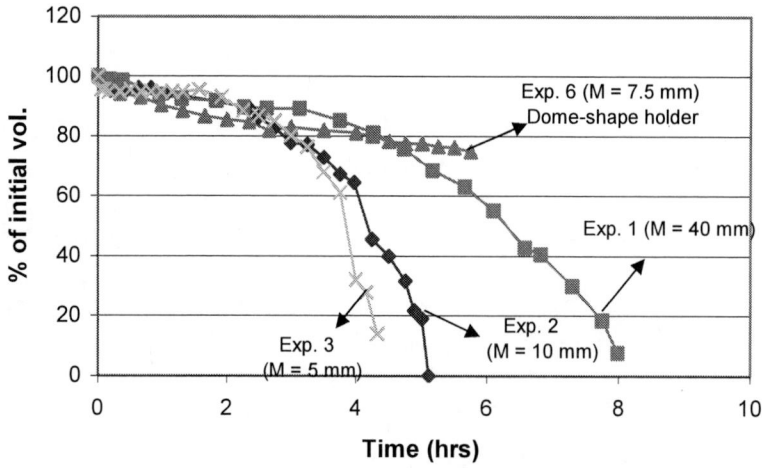

Figure 5. The volume-time curves for the experiments with different moduli

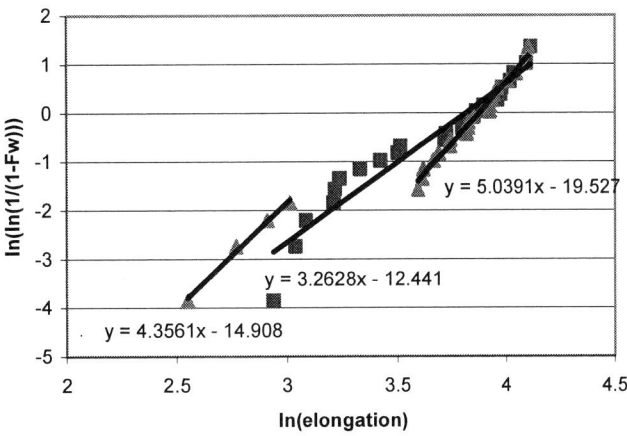

Figure 6. Weibull distribution for the elongation of the immediately solidified castings (squares) and those held in the liquid state (triangles).

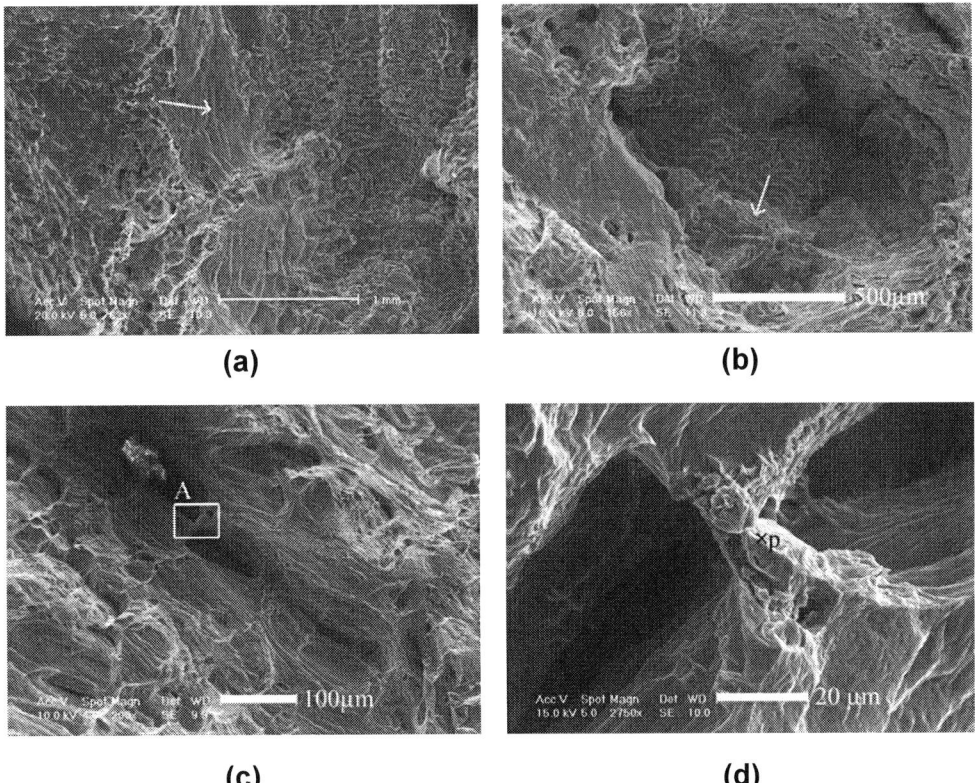

Figure 7. (a) An oxide film (marked with an arrow) and (b) a pore with an oxide film (marked with an arrow) found on the fracture surface of an immediately solidified tensile test specimen, (c) a cavity, the walls of which were bonded to each other at some points, found on the fracture surface of a specimen that was held in the liquid state for 20 minutes and (d) a magnified image of area A.

rraiszadeh@yahoo.com , W.D.Griffiths@bham.ac.uk

Effect of Melting and Casting Conditions on Aluminium Metal Quality

Derya DISPINAR, John CAMPBELL

Department of Metallurgy and Materials, University of Birmingham, UK.

Abstract

A study in a secondary alloy ingot producing plant was targeted to investigate the metal quality change in the holding furnace during secondary remelting of aluminum alloy LM24 (Al-8Si-3Cu-Fe). The investigation was an attempt to understand the effect of (i) the use of diffusers (porous plugs integrated into the body of the furnace) and (ii) casting techniques involving different degrees of turbulence.

Casting density and bifilm index were found to be useful parameters to assess metal quality. It was found that the metal quality was increased significantly, and was maintained high throughout the casting operation when (i) diffusers were used and (ii) turbulence was reduced to a minimum.

Keywords: *secondary remelting, aluminum, metal quality, bifilm index*

Introduction

The cost of recycling of aluminium compared to the cost of primary aluminium is highly attractive as a result of major energy savings [1, 2]. Recycling also has benefits for the environment and for the conservation of natural resources. Today recycled aluminium accounts for one-third of aluminium consumption world-wide [1].

The ultimate goal of the recycling process is to produce clean aluminum to an accurate chemical specification while minimising metal losses [3-6]. For this purpose, in addition to alloying, fluxing and degassing treatments are carried out on the metal in the liquid state in melting furnaces. The melt is then transferred to a holding furnace. Finally, the melt is cast into convenient-sized ingots (pigs) that serves as melt stock for the shaped castings industry. These ingots, formulated according to recognised national or international specifications, go into the manufacture of aluminium cast components. The specifications do not, however, cover the amount of oxides or gas that the ingots may contain. These constituents are suspected of being of central importance to the quality of the final cast products.

In an earlier study [7], the use of different lances and ceramic diffusers in an induction furnace during fluxing and degassing was investigated. In addition, different gas flow rates were also examined. It was found that when high gas flow rates were used, violent surface turbulence was observed on the surface of the melt that introduced bifilms [8]. When low flow rates were used, then the cleaning process was found to be less efficient. Thus the control of the remelting process is not straightforward. In addition, there were still some concerns about the introduction of bifilms during transfer of the melt from melting furnace to holding furnace, mainly due to the heavy turbulence that occurred as a result of the high fall of the liquid.

It is well known that in controlling the concentration of inclusions in aluminium melts, the settling of the melt has been found to be beneficial to metal cleanliness [9]. Therefore, in this study, the aim was targeted to investigate the metal quality change in the holding furnace as a function of time during a pour. In particular, an attempt was made to understand the effect of diffusers and their importance on the metal quality. In addition, the melt was sampled at various points along the production line, from the tapping of the holding furnace until the mould filling station.

Experimental

The process begins with the melting of a charge of selected scrap in a 2500 kg electric furnace (Figure 1) to produce a melt at 750°C. Following fluxing and degassing operations, molten metal is transferred to the holding furnace. Two ceramic diffusers sited in the bottom the holding furnace purge nitrogen gas through the melt for a minimum of 20 minutes. During the tests, different numbers of diffusers were run; 0, 1 and 2. The schematic layout of the holding furnace and position of diffusers is given in Figure 2.

In the second part of this same study, three changes were made in the casting area. These are summarised in Figure 3. The first change was the lowering of the height of the launder so that the fall of the liquid metal into the ingot mould would be minimised (Figure 3a '1'). The second change was made to the casting device that was intended to deliver the melt into the ingot mould at the lowest possible point. The internal geometry of this device (unfortunately not able to be revealed here for commercial reasons) was clearly not altogether satisfactory, involving some degree of turbulent fall. Therefore the design was altered slightly and the rate of production of castings was slowed down. These actions eliminated the worst aspects of the turbulent flow in the device such that the melt was filled from the bottom of the ingot mould with much reduced disturbance (Figure 3a '2'). The third change was made in the tapping procedure of the holding furnace (Figure 3b '3'). When not controlled carefully, the liquid metal tended to jet out from the tap-hole with the result that the melt was violently and turbulently propelled along the launder, clearly creating new bifilms and dross.

The study was performed with alloy LM24 (Al-8Si-3Cu). The composition of the alloy is given in Table 1. Three ingots were taken from each casting trial: one from the start (when operations in the holding furnace are complete, and the furnace is tapped to start the production run), one from middle and one from the end (when the holding furnace is empty) of the casting process. These ingots were then separately remelted in an induction furnace at 750°C and reduced pressure test samples were taken from the melt. (The taking of ingot samples was convenient, allowing the reduced pressure test to be carried out later at a remote location. Furthermore, this approach was not found to affect the RPT result to any detectable extent.)

RPT samples (50 x 35 x 15 mm dimensions) were cast into sand moulds bonded with 1.2% resin and solidified at 100 mbar. Archimedes Principle was used to determine the density of the samples. The samples were then sectioned and polished for image analysis and bifilm indices [10, 11] of each sample were measured.

Results

The density of the RPT samples as a function of the number of diffusers is given in Figure 4. When no diffuser was used, the densities of the reduced pressure test samples fell from 2620 to 2454 kg/m^3 as the casting process progressed (Figure 4a). Once the diffusers were active, the density stayed high between 2700-2550 kg/m^3 and the results were less scattered (Figure 4b, c). This is more clearly illustrated in Figure 4d when averages of the all results are resented in the same graph. The sectioned surface of the RPT samples can be seen in Figure 5a.

The bifilm index results of the same samples are given in Figure 6. At the start of the casting, bifilm index (the total length of defects seen on the polished surface of the RPT sample) lies between 100-150 mm. When no diffusers were operated the bifilm index rose toward 250 mm as the end of casting was approached whereas when diffusers were operational bifilm index fell to around 100 mm. It is interesting to note that there is no clear difference between the results when using one diffuser or two (Figure 6d).

The second study of the action of the diffusers were carried out after the changes in the casting area were made. The experimental conditions are summarised schematically in Figure 3.

The results demonstrate a clear increase in the quality of the castings when these changes were made. The density increases in the reduced pressure test samples are illustrated in Figure 7. In these figures, the term "non-quiescent" indicates the set of results from the original design, and the term "quiescent" indicates the new results after the changes have been made as described in the experimental section. When no diffusers were used (Figure 7a), the increase in the density was approximately +100 kg/m^3 for each section of the casting; from 2620 kg/m^3 to 2704 kg/m^3 at the start, 2538 kg/m^3 to 2673 kg/m^3 at the mid, 2454 kg/m^3 to 2617 kg/m^3 at the end. The average increase in the density when diffusers were applied (Figure 7b, c) was about half of this value, approximately +50 kg/m^3.

The change in the bifilm index results from the improved filling conditions are illustrated in Figure 8. The average bifilm length values for non-quiescent conditions (which is the original fast and turbulent casting) were 188 mm without diffusers and 100 mm with diffusers. In quiescent conditions, these rather high results fell to 39 mm without, and 35 mm with diffusers (Figure 8). These significant improvements can also be seen from the sectioned surface of the RPT samples in Figure 5 a and b.

Discussion

In an earlier investigation of the quality of the melt in the holding furnace during secondary remelting process [7], RPT sample collection was carried out after 30 and 60 minutes during which an increase in the density of RPT samples and a decrease in the hydrogen content were observed [7]. The positive effect of holding time was possibly due to the settling of oxides [9]. However when comparing the RPT results from start to the end the casting process, it was found that the results were not consistent. The results of the density change of RPT samples collected from the holding furnace when different numbers of diffusers were run are shown in Figure 4.

As seen from Figure 4, when diffusers were not used (Figure 4a), the density of the RPT samples decreased from start to the end of the casting. This appears to suggest an increased concentration of inclusions in the lower levels of the melt. It is important to note that the chemical composition of the melt remained unchanged during this period (Table 1). However, once the diffusers were used (Figure 4 b, c), the density results remained practically unchanged from the start to the end. As was expected, the inclusion concentrations were significantly lowered and were, perhaps, uniformly distributed throughout the melt and through the casting process.

The Bifilm Index of these samples are in good agreement with the density results: it increased (Figure 6a) from start to end and once the diffusers were active, the bifilm index decreased dramatically (Figure 6 b, c). For zero, one and two diffusers, the average bifilm index values were 188, 110 and 102 mm, respectively.

In a second phase of these studies, turbulence during the transfer of the melt was reduced at two locations (i) at the outlet of the holding furnace where a stopper was introduced to control jetting, thereby transferring the melt more quiescently; and (ii) the fall of the liquid from the launder into the moulds was reduced to a minimum.

After these changes the tests were repeated, with impressive results. There was a clear increase in the quality of the castings. The density of the RPT samples was increased dramatically (Figure 7). The change in the bifilm index was also remarkable (Figure 8). The average bifilm length values for the original non-quiescent conditions were 188 mm without diffusers and 100 mm with diffusers. These rather high results fell in the improved quiescent conditions to 39 mm without, and 35 mm with diffusers. The clear separation of the results is noteworthy: in non-quiescent conditions the average bifilm length was always above 70 mm whereas for quiescent conditions all results were below 60 mm.

It is interesting to note that when the casting was carried out quiescently (achieved simply by implementing simultaneously all the three changes investigated in this study), the average bifilm index values of quiescent conditions was below 50 mm even when no diffusers were used (Figure 8). Thus the effect of the diffusers was of somewhat less importance than the control of turbulence, confirming the importance of the control of turbulence during the handling and transfer of molten aluminium alloys [8]. One might speculate that given a careful design of plant, theoretical densities and zero bifilm length might be achievable.

Conclusions

1. When no diffusers were used in the holding furnace there was a deterioration in the quality of melt from the beginning to the end of the casting.

2. Metal quality increased significantly when one diffuser was used; however there was marginal additional advantage at the limits of detectability when using two diffusers.

3. The more quiescently the casting was controlled, the higher the quality of the products. Good control includes (i) careful minimisation of turbulence at tapping; (ii) minimised fall of the liquid; and (iii) filling conditions to reduce turbulence in the mould.

References

1. *www.world-aluminium.org*, International Aluminium Institute.
2. *www.eaa.net*, European Aluminium Association.
3. Tenorio, J. A. S., Carboni, M. C. and Espinosa, D. C. R., *Recycling of aluminum - effect of fluoride additions on the salt viscosity and on the alumina dissolution.* Journal of Light Metals, 2001. **1**(3): p. 195-198.
4. Samuel, M., *A new technique for recycling aluminium scrap.* Journal of Materials Processing Technology, 2003. **135**(1): p. 117-124.
5. Khoei, A. R., Masters, I. and Gethin, D. T., *Numerical modelling of the rotary furnace in aluminium recycling processes.* Journal of Materials Processing Technology, 2003. **139**: p. 567-572.
6. Davies, S. B., Masters, I. and Gethin, D. T. *Numerical modelling of a rotary aluminium recycling furnace.* in *Proc. 4th International Symposium on Recycling of Metals and Engineered Materials.* 2000. TMS.
7. Dispinar, D. and Campbell, J., *Metal quality studies in secondary remelting of aluminium.* Journal of Institute of Cast Metals Engineers, 2004. **178**(3612): p. 78-86.
8. Campbell, J., *Castings.* 2nd ed. 2003, Oxford: Butterworth-Heinemann Ltd.
9. Martin, J.-P., Dube, G., Fray, D. and Guthrie, R., *Settling phenomena in casting furnaces: A fundamental and experimental investigation.* Light Metals, 1988: p. 445-455.
10. Dispinar, D. and Campbell, J., *Critical Assessment of Reduced Pressure Test: Part I: Porosity Phenomena.* International Journal of Cast Metals Research, 2004. **17**(5): p. 280-286.
11. Dispinar, D. and Campbell, J., *Critical Assessment of Reduced Pressure Test: Part II: Quantification.* International Journal of Cast Metals Research, 2004. **17**(5): p. 287-294.

Acknowledgements

The authors would like to gratefully acknowledge the financial support of Norton Aluminum, and for their assistance in the use of facilities in the foundry. The diffusers used in the holding furnace were provided by Capital Refractories Ltd.

Tables

Table 1: Chemical analysis of the 2.5 ton melt from start to end

LM24	Cu	Mg	Si	Fe	Mn	Ni	Zn	Pb	Sn	Ti	Al
Start	3.43	0.06	8.17	0.87	0.17	0.05	2.42	0.12	0.11	0.06	rem.
End	3.36	0.06	8.18	0.87	0.16	0.05	2.42	0.12	0.11	0.06	rem.

Figures

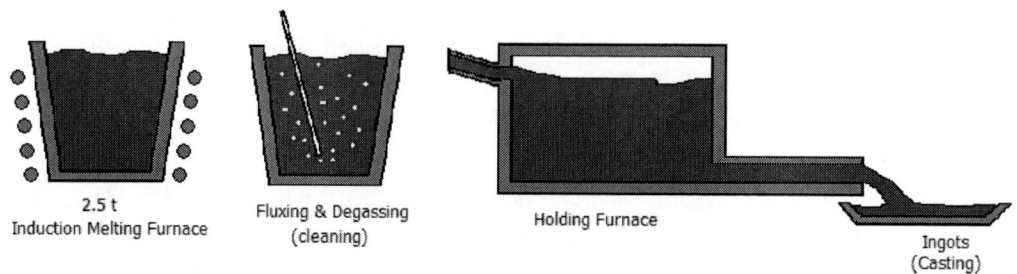

Figure 1: a schematic representation of aluminium ingot production

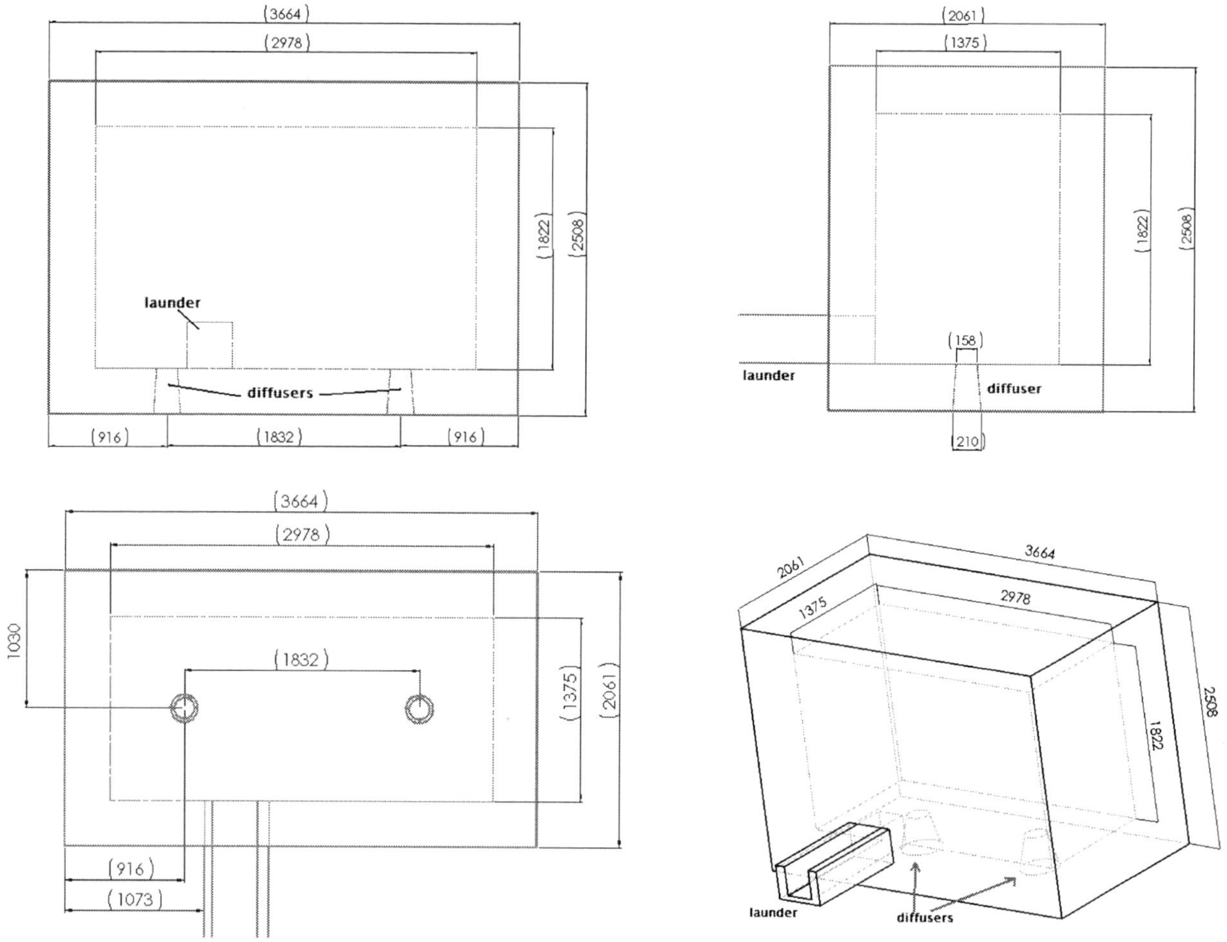

Figure 2: The schematic drawing of the holding furnace and
the position of diffusers

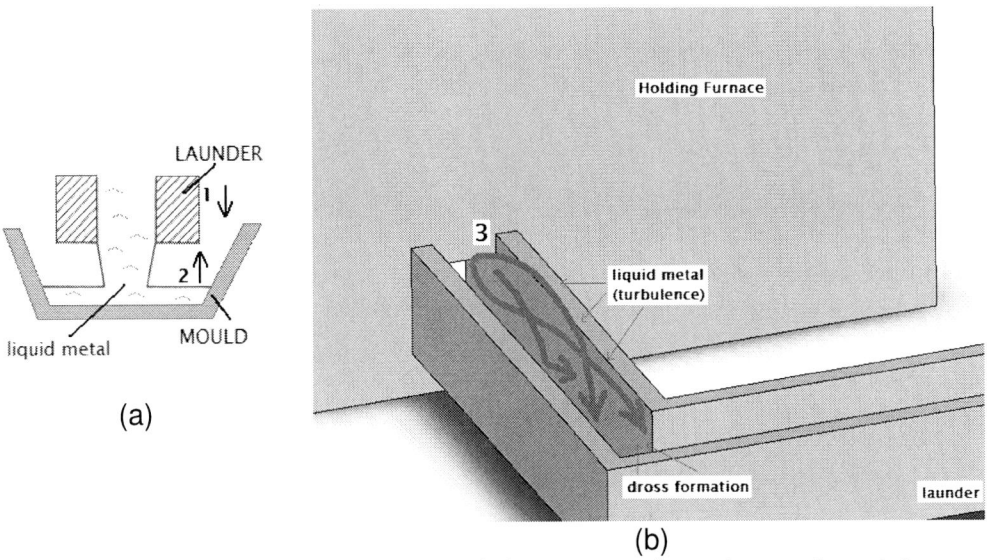

Figure 3: Schematic illustration of changes made at the casting trials
    (a) '1' The launder was lowered to be as close to the casting mould possible,
        '2' The casting and filling speed were decreased
    (b) '3' More care was taken to avoid the severe turbulence on   tapping.

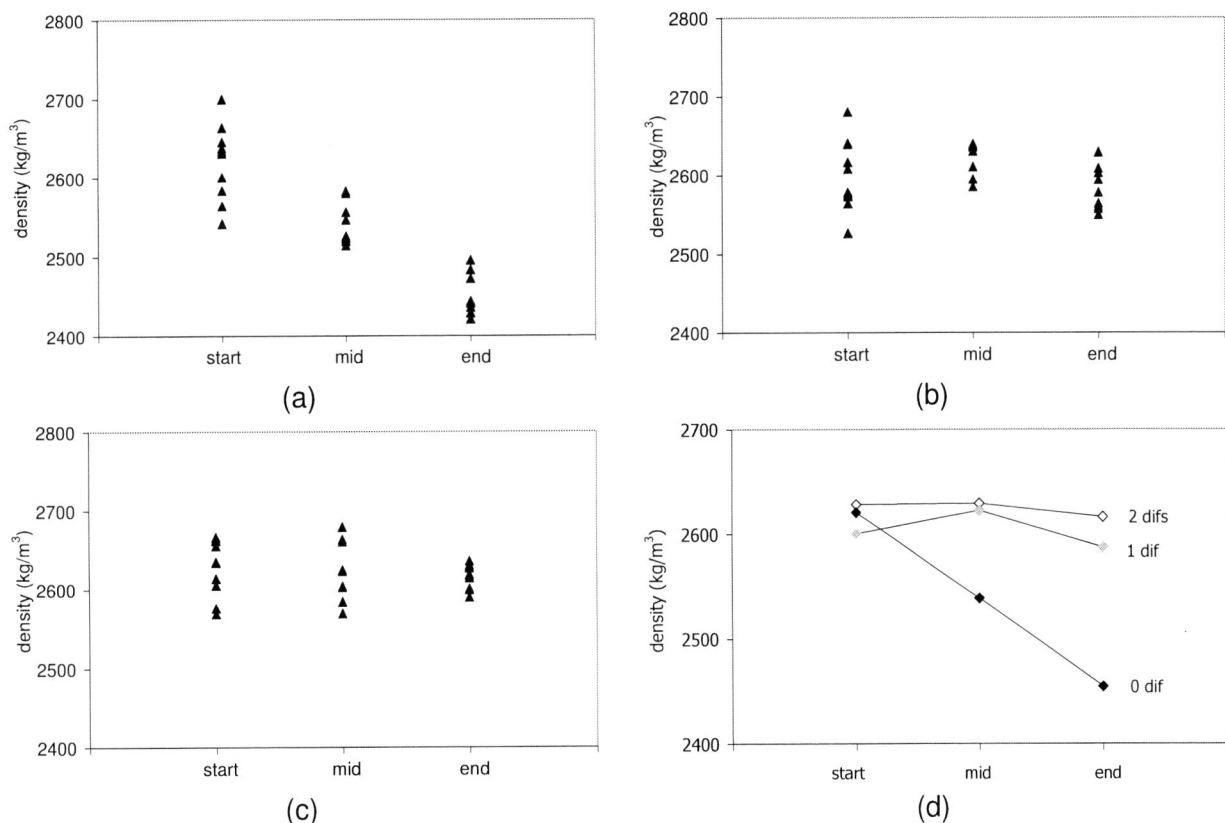

Figure 4: Density change of RPT samples
    (a) no diffusers were run
    (b) one diffuser were on
    (c) two diffusers were on
    (d) average values of density change of RPT samples

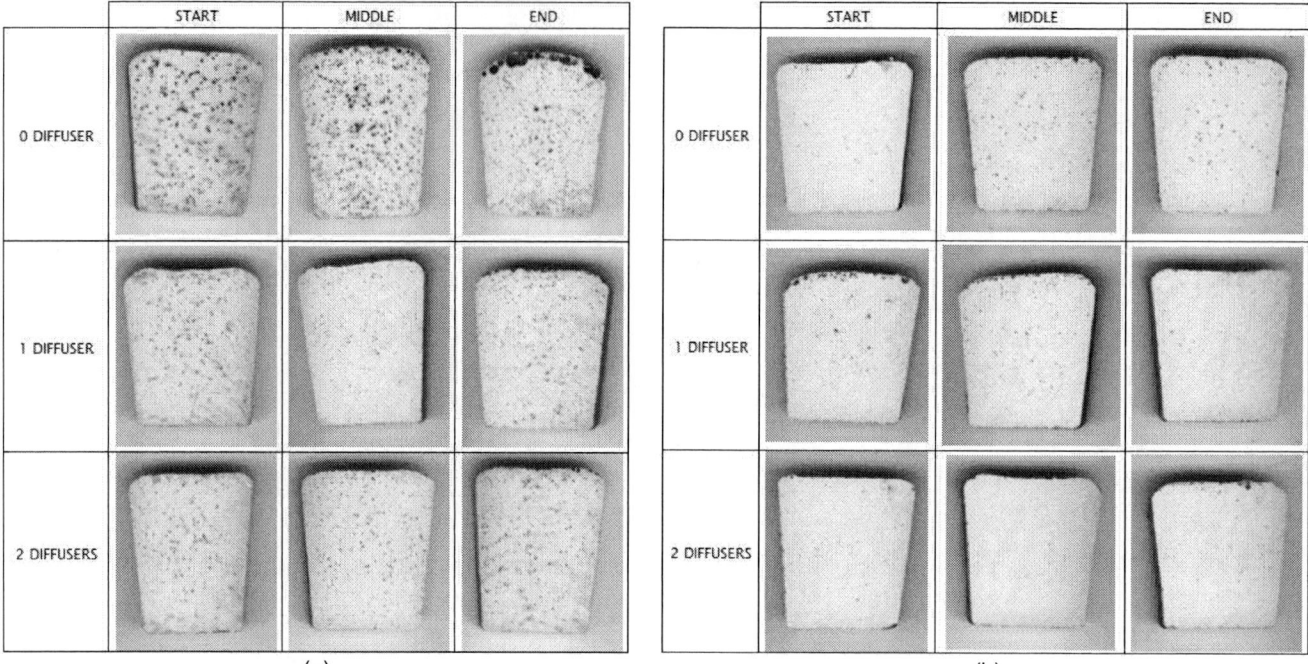

(a)

(b)

Figure 5: Sectioned surface of RPT samples from the trials
(a) non-quiescent castings conditions
(b) quiescent conditions

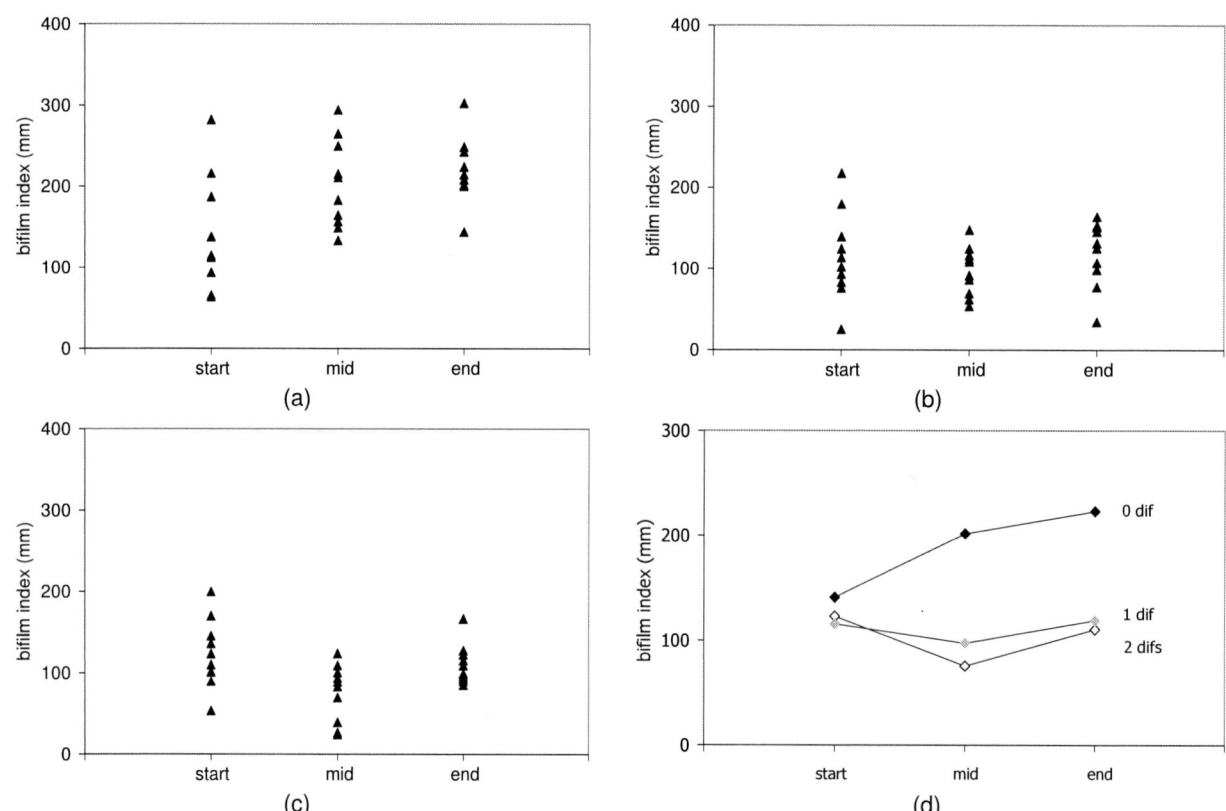

(a)

(b)

(c)

(d)

Figure 6: Bifilm Index comparing different number of diffusers
(a) no diffusers were run, (b) one diffuser were on, (c) two diffusers were on
(d)averages of Bifilm Index comparing different number of diffusers

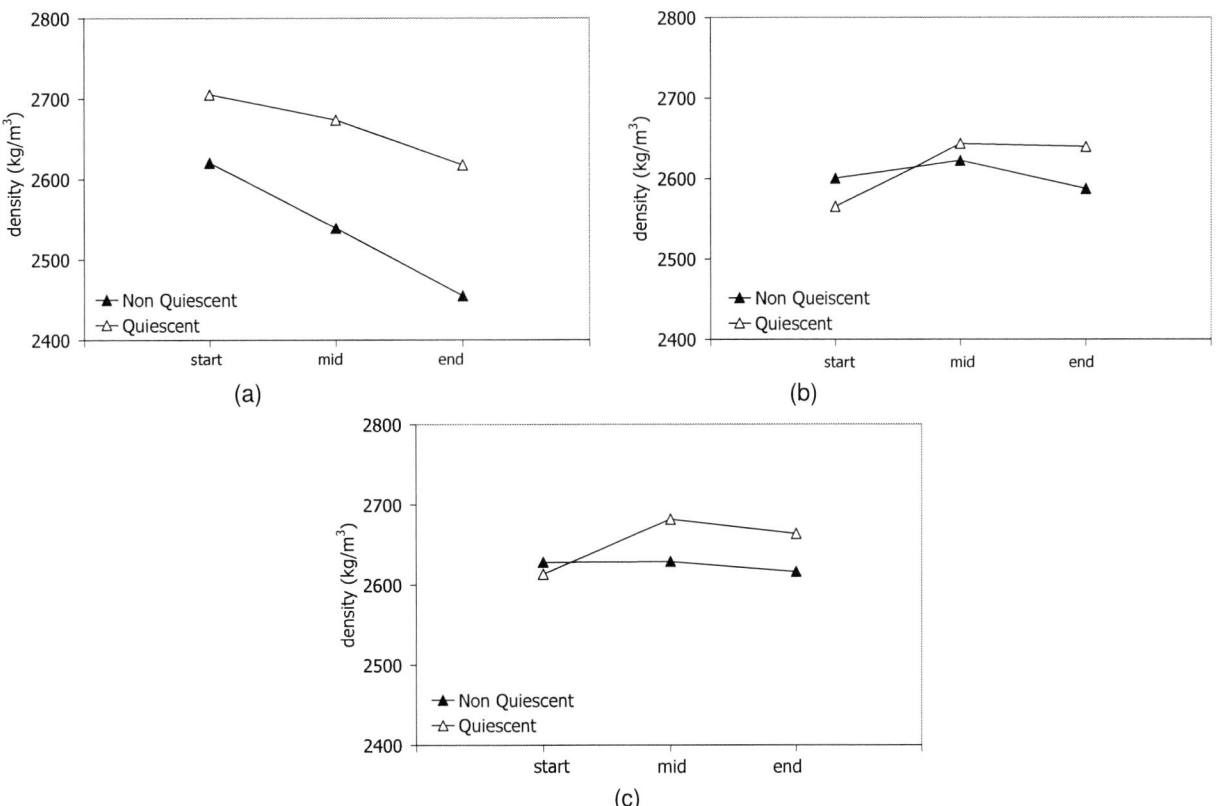

Figure 7: The comparison of the average density values of RPT samples for different techniques
(a) no diffusers were run, (b) one diffuser were on, (c) two diffusers were on

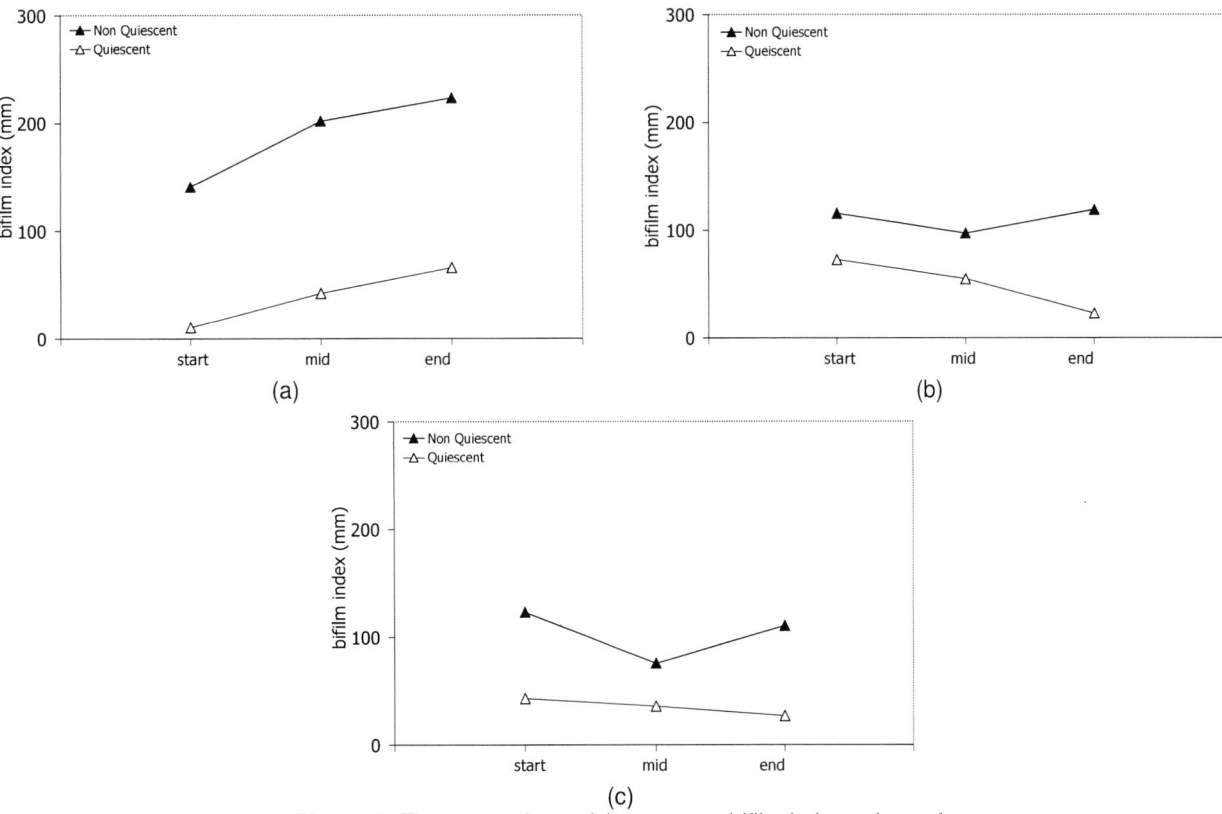

Figure 8: The comparison of the average bifilm index values of
RPT samples for different techniques
(a) no diffusers were run, (b) one diffuser were on, (c) two diffusers were on

Strontium Effect on the Solidification Path of a 319-Type Aluminum Alloy

M. Castro-Román, J. J. Montes-Rodríguez, M. Herrera-Trejo.

CINVESTAV Unidad Saltillo, Carr. Saltillo-Monterrey km 13, 25000 Saltillo, Coahuila, México

Abstract

The sequence of phase precipitation during solidification of a 319-type aluminum alloy was studied by thermal analysis. Melts of the alloy with different Sr contents were prepared in a gas-heated furnace. The liquid metal was poured into graphite cups fitted with two thermocouples. Special attention was paid in the experimental setup in order to obtain cooling curves with low noise. Thermogram data were analyzed using the derivative of cooling curves. Experimental onset temperatures of phase precipitation were compared with simulation results obtained from ThermoCalc and data base Al-DATA V3. In these simulations, the Gulliver-Scheil model was employed to take into account the microsegregation developed during solidification. The simulation results are consistent with experimental data in the sense that the onset reaction temperatures calculated are generally higher than those experimentally measured.

Keywords: Aluminum alloys, Thermal analysis, Thermodynamic calculations, Solidification.

Introduction

The 319 alloy, an important Al-Si alloy, is widely used to produce blocks and heads for internal combustion engines. This alloy finds final application in both the as cast and heat treated conditions. The silicon and copper are the main alloying elements of this alloy, and it can contain smaller quantities of other elements such as iron, magnesium, manganese, strontium, titanium, etc. These elements, in spite of their low contents, affect considerably the microstructure and the mechanical properties of the metal; in this sense, Sr is one of the most important trace elements.

Recent investigations report that strontium, besides of its well-known modifying effect on the eutectic silicon [1], also affects the formation of the intermetallic compounds formed during the solidification of the aluminum-silicon alloys , such as 319 [2,3] and 413 [4] type alloys. Samuel et al. [2,3] have studied the strontium and magnesium effect on the solidification path of 319-type alloys. Their alloys had the following approximate chemical composition in weight: 6%Si, 3.5%Cu, 0.5%Fe, 0.2%Mn, 0.06% to 2%Mg, and 0.0001% to 0.058%Sr, Al balance. These authors pointed out that with strontium addition, dissolution of a large proportion of the needle-like β-AlFeSi intermetallic in the aluminum matrix takes place; no transformation of this phase into any other intermetallic (including α-AlFeSi phase) was observed [3]. However, in the case of the 413-type alloy, Shabestari et al. [4] pointed out that strontium can be considered as a very effective element to change the β-AlFeSi phase into the α-AlFeSi phase.

Regarding thermal analysis facts, it has been observed that the strontium additions diminishes the Al_2Cu formation temperature and causes Al_2Cu to adopt a blocky morphology [1]. Also, since increasing the strontium content decreases the formation temperature of the eutectic silicon, the β-AlFeSi phase formation can be detected more easily by thermal analysis [1]. Finally, Samuel et al. [2] presented data for a 319-type alloy with 0.5% Mg which showed that the eutectic silicon formation temperature slightly increases when strontium is added at contents higher than the one needed for modification purposes.

The present work was carried out to study, by means of thermal analysis techniques, the effect of the strontium in the solidification path of a 319-type alloy. Special attention was paid to develop an experimental technique that allows consistent acquisition of high-quality cooling curves. The experimentally determined phase formation temperatures were compared with temperatures obtained from calculations performed using the Scheil-Gulliver module of ThermoCalc software, with Al-DATA V3 database from Thermotech.

Experimental

A commercial 319 type alloy was used to prepare alloys with 0.2% Mg, and three different strontium contents. For each composition, three melts were prepared from 300 g of base alloy according to the following procedure. The base alloy was first melted in a gas furnace then the required quantities of Al-68%Mg and Al-10%Sr where added and finally the melt was poured into a graphite cup for thermal analysis. Preliminary tests were preformed in order to know the right quantities of the alloying additions necessary to achieve target compositions. Table 1 shows the composition of the base alloy, the target magnesium and strontium contents, and the composition of the alloys obtained.

Thermal analysis tests were performed using an experimental device inspired in the one used by Bäckerud et al. [5]. Experimental setup is shown in Figure 1. The temperature was measured in two different points inside of graphite cup. Two K-type thermocouples were placed on the cup's lid in such a way that a thermocouple junction was located at the center of the crucible's transversal section and the other one next to the crucible's wall. Both thermocouple junctions were located at the middle of the interior cup height.

While the alloy was being melted, the cup assembling used for thermal analysis was preheated at 700°C in a resistance muffle. Once the melt was prepared, the muffle was displaced downwards and then the metal was poured into the cup. Later on, the lid was replaced on the cup and the muffle was lifted to its original position to allow the cup and metal to be held at 700°C for a few minutes in order to stabilize the temperature measurements. Finally, the muffle was displaced down again, allowing the metal to cool down. The cooling of the sample was controlled by means of air injection under controlled conditions.

The acquisition of the analogical signal generated by the thermocouples and conversion of the latter to digital data were carried out by using a 16-bit resolution DAS-TC card, inserted in a computer (plug-in board). This data acquisition system was connected into a ground terminal.

Results

Figure 2 shows a cooling rate curve (CR), accompanied by its respective cooling curve (CC) both corresponding to alloy 1 (0%Sr and 0.2%Mg).

In the cooling rate curve (CR) it can be observed the signals (peaks) corresponding to the formation of aluminum dendrites (A), as well as to the formation of the β-AlFeSi phase (B) and eutectic silicon (C). Two additional peaks were observed toward the end of solidification. It can be seen that even that the signal corresponding to β-AlFeSi formation (B) is very close to that of eutectic silicon formation (C), it is possible to distinguish easily both individual events. Thus, in this case, the temperature difference curve ($\Delta T=Tp-Tc$) proposed by Bäckerud et al. [5] does not provide any additional information with respect to that obtained from the CR curves, as can be observed in Figure 2.

The two peaks observed at the end of solidification can be interpreted according to the results obtained from the Scheil-Gulliver simulation performed. Figure 3 shows a simplified graph for the evolution of the phase fraction as a function of temperature calculated for alloy 1 (0%Sr and 0.2%Mg). In this figure it can be observed that Al_2Cu formation occurs first as a pre-eutectic reaction, and later on Al_2Cu develops simultaneously in a eutectic reaction with the quaternary compound $Al_5Cu_2Mg_8Si_6$ and silicon. That means that the small peak D observed experimentally could correspond to the formation of pre-eutectic Al_2Cu, and that the more intense peak E corresponds to the simultaneous precipitation of quaternary compound $Al_5Cu_2Mg_8Si_6$, Al_2Cu phase and silicon.

Table 2 shows the onset temperatures for the more intense signals observed in the thermal analysis curves, i. e., the peaks corresponding to aluminum dendrites, β-AlFeSi, eutectic silicon, pre-eutectic Al_2Cu, and simultaneous precipitation of Mg quaternary compound ($Al_5Cu_2Mg_8Si_6$) with Al_2Cu and Si. Temperature determination was carried out according to the peak's onset observed in the CR curves.

Discussion

Figure 4 shows the cooling curves (CC) and CR curves for alloys 1 and 3. The Sr content of alloys 1 and 3 is 0% and 0.0325%, respectively. In CR curves of Figure 4, it can be observed that the separation of the signals corresponding to β-AlFeSi formation (B) and to the eutectic silicon (C) is larger in the alloy with higher Sr content, making easier the detection of β-AlFeSi formation, as commented by Tenekedjiev et al. [1].

In Figure 4 it is also observed that when the strontium content is increased the pre-eutectic peak (D), probably Al_2Cu, is shifted to the left, and the signal corresponding to Al_2Cu, $Al_5Cu_2Mg_8Si_6$ and silicon simultaneous precipitation (E) becomes more narrow and higher, i.e., separation among peaks D and E increases when Sr content increase. That means that somehow Sr enhances pre-eutectic Al_2Cu formation, thus more pre-eutectic Al_2Cu could be observed at higher Sr contents. This is in agreement with microstructural observations reported by Samuel et al. [2,3] and Tenekedjiev et al. [1]. They observed that the proportion of pre-eutectic Al_2Cu (with blocky morphology), separated from eutectic Al_2Cu, increases when the strontium content increases. Strontium could also enhance the reaction of eutectic Al_2Cu as suggest by changes in the position of peak D when strontium content increases.

The shape of peak A for alloy 3, with higher Sr content, is slightly different to peak A corresponding to alloy 1. Peak A of alloy 3 shows a small plateau that appears during the formation of aluminum dendrites, which could correspond to the α-AlFeSi phase formation according to calculations carried out with ThermoCalc. These results suggest that α-AlFeSi formation could be enhanced by Sr additions, as pointed out by Shabestari et al. [4] for the case of 413 alloys. However, Samuel et al. [3] conclude that Sr does not have any enhanced effect in α-AlFeSi formation. More work is needed to elucidate this controversial result in the case of 319 alloy.

Figure 5 shows the experimental data regarding the onset reaction temperature as a function of the strontium content. It is also included the corresponding formation temperatures calculated by means of the Scheil-Gulliver module of ThermoCalc considering the actual composition of the alloys.

In this figure it is observed that when Sr content increases, the temperature of aluminum dendrite formation remains practically constant. The agreement of this data with calculations is good considering that the experimental results are affected by a delay in the onset temperatures due to kinetic aspects not considered in simulations.

Regarding the effect of Sr content in the onset temperature of β-AlFeSi formation, in Figure 5 it is observed that this temperature increase slightly when strontium content increases from 0.01% to 0.0325%.

The eutectic silicon formation shows a significant descent in the starting temperature (~9°C) when the strontium content is increased to 0.01%, but later on it shows an increase of about 3°C when the strontium content is increased up to 0.0325%. This observation agrees with previous results reported by Samuel et al. [3].

In Figure 5 it is observed that the onset temperature of pre-eutectic peak (D), probably associated to Al_2Cu precipitation, shows a remarkable increase (~15°C) when the strontium content is increased from 0% to 0.01%, and this temperature increases a little bit more (~3°C) when the strontium content is increased further up to 0.0325%. This behavior could lead to an increase in the proportion of Al_2Cu with blocky morphology (pre-eutectic) when the strontium content increases, as observed by others authors [2, 3].

Eutectic Al_2Cu formation temperature, which precipitates simultaneously with the quaternary compound and with silicon, shows an increase of 5°C, approximately, when the strontium content is increased from 0% to 0.01%, remaining practically constant for a Sr content of 0.0325%.

When the experimental data were compared with the simulation results, it was observed that, in general, the calculated reaction temperatures are superior to the ones obtained experimentally. This was expected because the calculations do not include aspects related to nucleation of phases, i. e., the undercooling required for the formation of the phases was not considered. Also, the experimental detection of the reaction onset depends on the sensibility of the used technique. Considering both aspects, it could be concluded that calculations are in rather good agreement with experience, except for the case of pre-eutectic peak (D), where the experimental onset temperature was higher than the calculated for pre-eutectic Al_2Cu. More work is needed to elucidate this phenomenon.

Conclusions

In a 319 type alloy, the increase of strontium content from 0% to 0.0325% increases the onset temperature for pre-eutectic peak (D), which is probably associated to Al_2Cu precipitation, and enhances the signal for the Al_2Cu eutectic reaction. The effect on the pre-eutectic peak (D) is more evident than that observed on the Al_2Cu eutectic.

The results of simulation performed with Scheil-Gulliver module of ThermoCalc and Al-DATA V3 are in rather good agreement with the experimental data. This kind of simulation is a useful tool to understand the solidification path of complex industrial alloys.

References

1. N. P. Tenekedjiev, M. H. Mulazimoglu, B. Closset and J. E. Gruzleski. *Microstructures and Thermal Analysis of Strontium-Treated Aluminum-Silicon Alloys*. American Foundrymen's Society, Inc. (1995), pp 23-31.
2. A.M. Samuel, P. Ouellet, F. H. Samuel and H. W. Doty. "Microstructural Interpretation of Thermal Analysis of Commercial 319 Al Alloy With Mg and Sr Additions". *AFS Transactions*. Vol. 105, pp 951-962, (1997).
3. F. H. Samuel, P. Ouellet, A. M. Samuel and H. W. Doty. "Effect of Mg and Sr Additions on the Formation of Intermetallics in Al-6 Wt% Si-3.5 Wt% Cu-(0.45 to 0.8) Wt% Fe 319-Type Alloys". *Metallurgical and Materials Transactions A*. Vol. 29A, No. 12, pp 2871-2884, (1998).
4. S. G. Shabestari and J. E. Gruzleski. "Modification of Iron Intermetallics by Strontium in 413 Aluminum Alloys". *AFS Transactions.* Vol. 103, pp 285-293, (1995).
5. L. Bäckerud, E. Król and J. Tamminen. *Solidification Characteristics of Aluminum Alloys*, Skanaluminium, Vol. 1, (1986), pp 63-74.

Acknowledgements

M. Castro wishes to acknowledge the financial support from CONACyT and SEP, project SEP-2003-CO2-44998. Thanks are also due to Mr. E. Córdova, Mr. F. Vázquez and Mr. F. Ortega for their help in the laboratory. J. J. Montes also wishes to acknowledge scholarship from CONACyT for postgraduate studies.

Tables

Table 1. Chemical composition of alloys.

Alloy	Si (%)	Cu (%)	Fe (%)	Mn (%)	Ti (%)	Zn (%)	Mg (%)	Sr (%)	Al (%)	Mg (Target) (%)	Sr (Target) (%)
Base	7.16	3	0.8	0.25	0.09	0.84	0.04	0.0001	Bal.		
1	7.48	3.05	0.868	0.263	0.081	1.08	0.212	0.0005	Bal.	0.2	0
	7.31	3.18	1.04	0.263	0.085	1.11	0.258	0.0001	Bal.	0.2	0
	7.51	3.22	0.839	0.222	0.089	1.02	0.22	0.0009	Bal.	0.2	0
2	7.59	3.1	0.889	0.258	0.094	1.04	0.217	0.0078	Bal.	0.2	0.01
	7.44	3.16	0.861	0.257	0.091	1.02	0.226	0.0139	Bal.	0.2	0.01
	7.32	3.12	0.853	0.259	0.09	1.03	0.224	0.0107	Bal.	0.2	0.01
3	7.4	3.1	0.892	0.267	0.085	1.03	0.257	0.0337	Bal.	0.2	0.0325
	7.26	3.08	0.858	0.264	0.081	0.993	0.225	0.0311	Bal.	0.2	0.0325
	7.34	3.03	0.844	0.279	0.087	0.994	0.204	0.0295	Bal.	0.2	0.0325

Table 2. Experimentally determined onset temperatures for the formation of phases.

Alloy	Reaction														
	Aluminum			β-AlFeSi			Silicon			Al_2Cu (Pre-Eutectic)			Al_2Cu (Eutectic)		
1	597.8	596.6	597.6	576.7	569.1	572.3	565.2	563.4	564.5	525.9	525.8	526.5	502.3	502.5	504.1
2	597.9	595.9	595.8	572	573.7	571.5	556.5	553.8	555.3	540.9	540.3		508.8	506.6	508.2
3	597.4	598.5	598.2	575.3	574.7	575.9	559.1	558.5	556.4	543.9		541.3	508.7	508.7	505.2

Figures

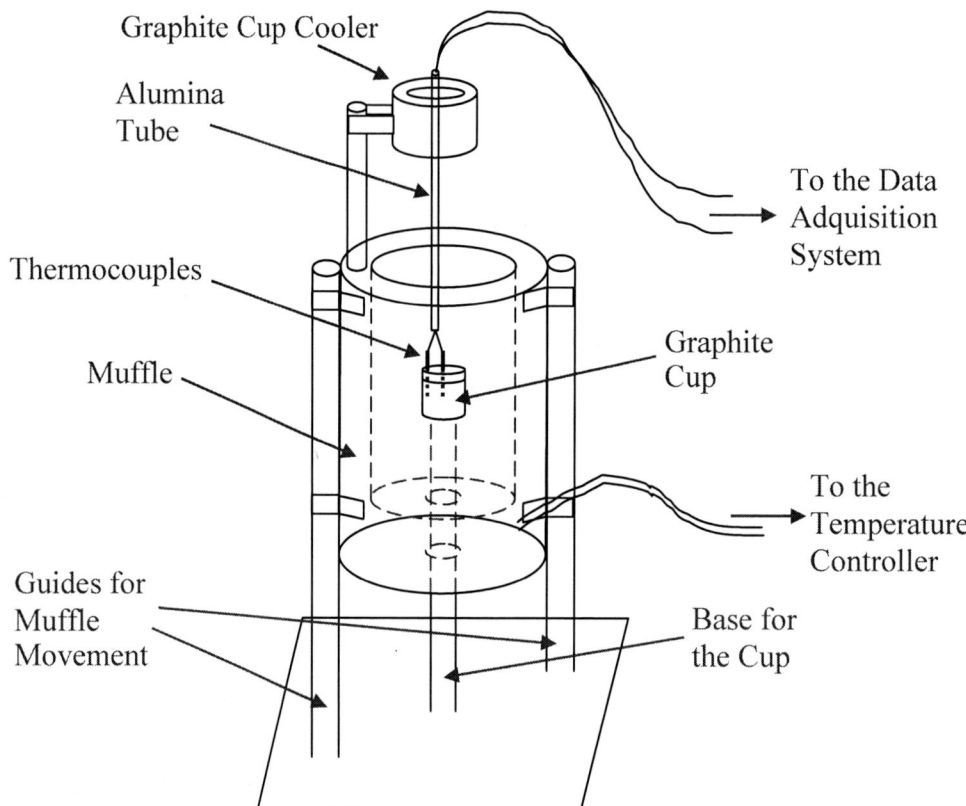

Figure 1. Experimental device.

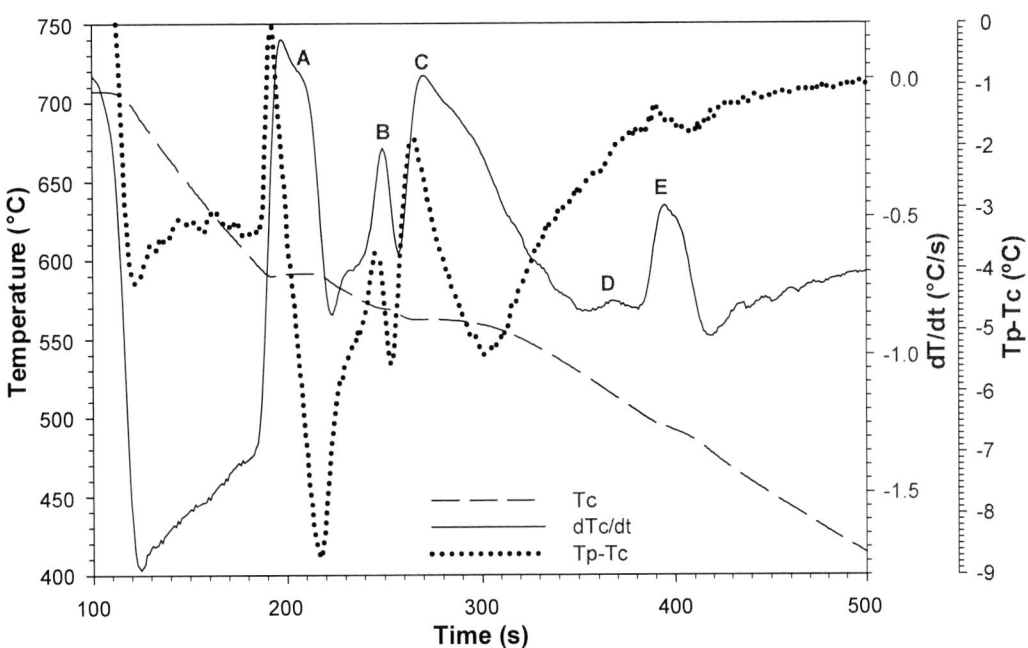

Figure 2. Cooling curve (CC), cooling rate curve (CR) and temperature difference curve (Tp-Tc) for alloy 1 (0.2%Mg and 0%Sr).

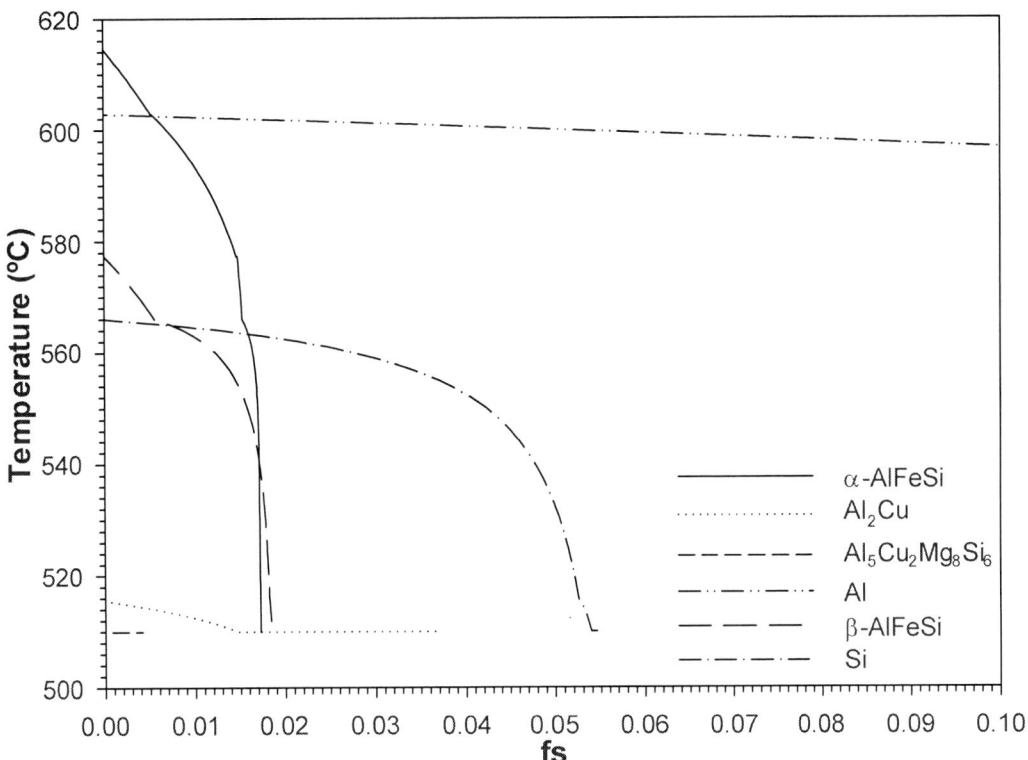

Figure 3. Evolution of phases' fraction as a function of temperature calculated with Scheil-Gulliver module of ThermoCalc for alloy 1 (0.2%Mg and 0%Sr).

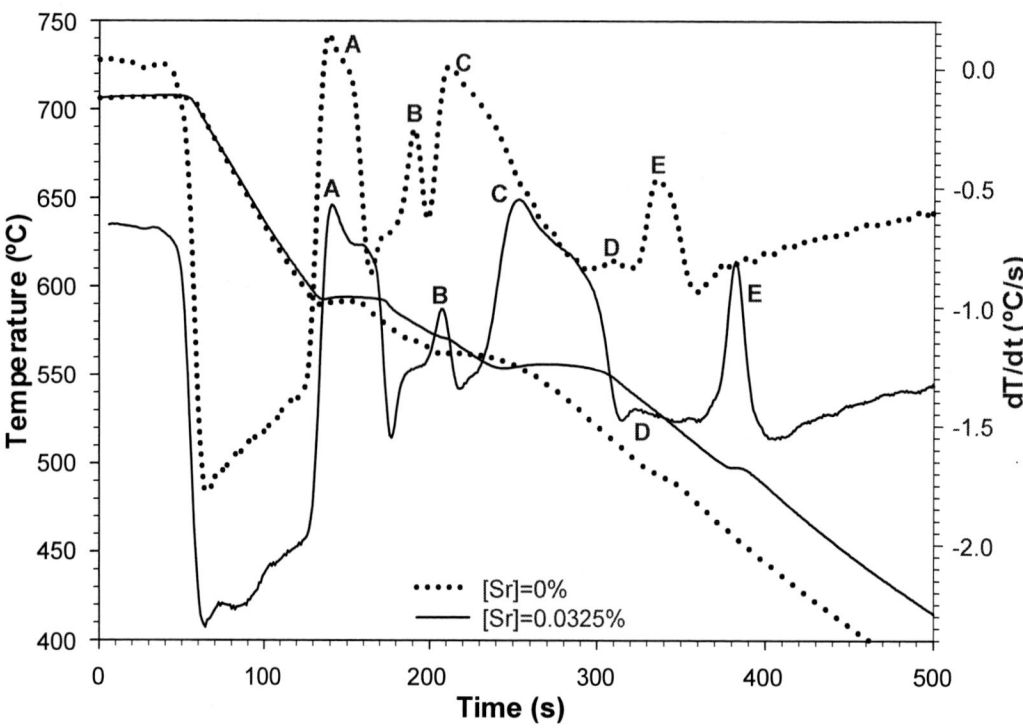

Figure 4. Cooling curves (CC) and cooling rate curves (CR) for alloy 1 (0%Sr) and alloy 3 (0.0325%Sr).

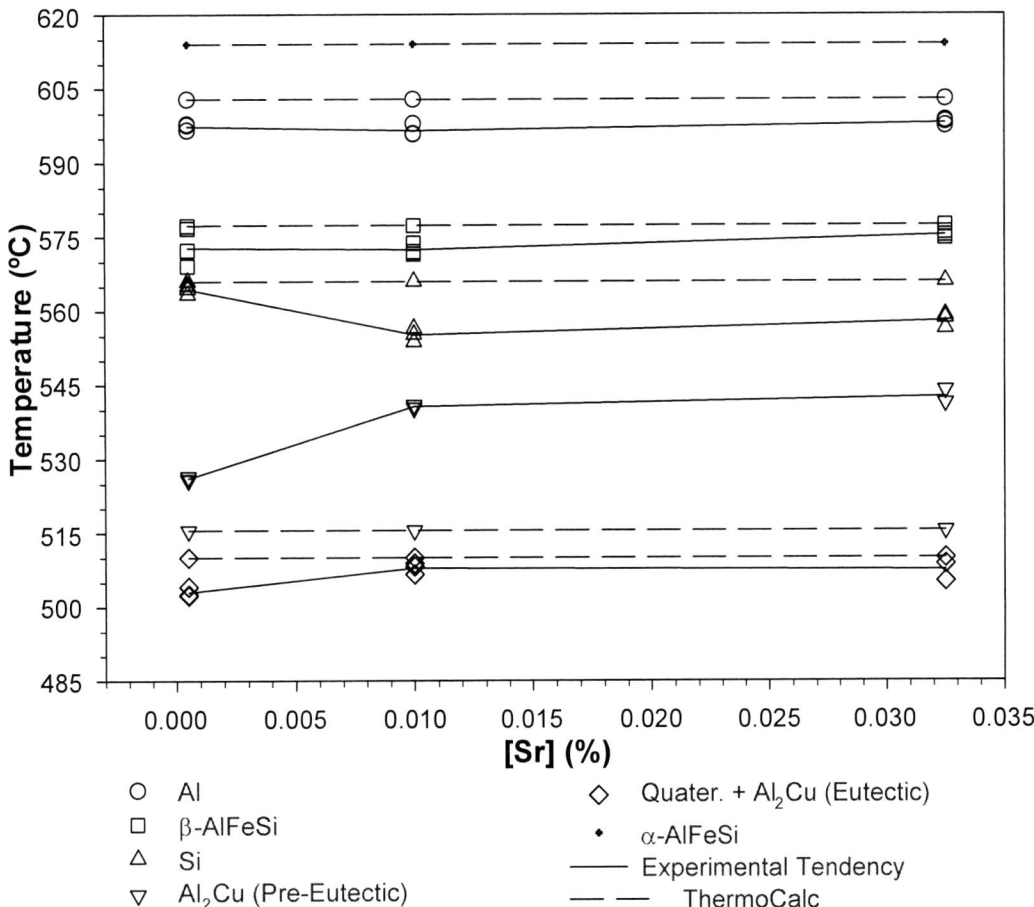

Figure 5. Experimental and calculated onset temperatures of main reactions detected during solidification of 319 type alloy, as a function of the Sr content.

manuel.castro@cinvestav.edu.mx
julian.montes@cinvestav.edu.mx
martin.herrera@cinvestav.edu.mx

Effectiveness of Zn-Ti Based Refiner of Al and Zn Foundry Alloys

W. K. Krajewski[*1], A. L. Greer[**], J. Zych[*] and J. Buraś[*]

* AGH University of Science and Technology, Faculty of Foundry Engineering. Reymonta 23. 30-059 Krakow. Poland.
** University of Cambridge, Department of Materials Science and Metallurgy. Pembroke Street, Cambridge CB2 3QZ, UK.

Abstract

The presented work is devoted to structural characteristics and performance of the new family of grain-refiners, based on the Zn-Ti system. The system studied was Zn-25wt%Al alloy (ZnAl25), Al-20wt%Zn alloy (AlZn20), Zn-4.6wt%Ti master alloy (ZnTi4 MA), ZnAl-4 wt%Ti (ZnAl-Ti4 MA) as well as the traditional refiners Al-5wt%Ti-1wt%B (AlTi5B1 MA) and Al-3wt%Ti-0.15wt%C (AlTi3C0.15 MA). SEM (scanning electron microscopy), LM (light microscopy) and TA (thermal analysis) investigations showed high effectiveness of the ZnTi-based master alloys as refiners of the inoculated with them AlZn20 and ZnAl25 alloys. The obtained structure refinement is comparable to those obtained after addition of the traditional AlTi5B1 MA or AlTi3C0.15 MA. The initial examinations of the damping properties showed, that attenuation coefficient of the refined AlZn20 alloy decreases together with increased structural fineness of the inoculated alloy.

Key words Aluminium-Zinc cast alloys, Grain refinement, Heterogeneous nucleation, Zn-Ti master alloy.

Introduction

Grain refinement of the Al – based cast alloys is a common practice, which allows obtaining fine structure with increased strength properties. On the other hand, many structural materials are required to have improved damping properties. However, damping capacity and strength properties are believed to be contradictory factors [1]. On the other hand, recent literature brings information that grain refinement of A356 alloy increases its damping capacity together with increased mechanical properties [2]. It should be noted that AlZn-based cast alloys are numbered among the structural materials of improved damping capacity. It is well known, that high-aluminium zinc alloys, for example ZA-27 alloy, fall in the category of HiDAlloys (High Damping Alloys) [3]. It was also stated that high-zinc aluminium alloys show increased damping properties [4]. The both groups, i.e. high-aluminium zinc and high-zinc aluminium alloys solidify naturally with a coarse structure and using the refinement process allows obtaining a highly refined structure [5 – 7].

Surprisingly, the literature lacks the information of how the refining process influences damping properties of the AlZn-based cast alloys.

The extensive investigations of the strength properties of sand-cast and chill-cast Al-Zn alloys showed that the latter have higher elongation, most probably due to the finer structure [8], Fig. 1. In practice, there are two main groups of refiners based on the Al-Ti-B and Al-Ti-C systems, commonly used in melting technology of Al-alloys. However, the range of melt temperatures of 700 – 750 °C required for the dissolution of AlTiB and Al-TiC refiners is 150-200 K higher than the 550-600 °C range recommended for the Zn-Al alloys. Also the high-zinc aluminium alloys, for example Al-(20-30) wt%Zn are very prone to surplus overheating, which causes melt oxidation and gases pick-up, detrimentally influencing their properties.

A newly introduced alternative - a master alloy based on the Zn-Ti system - requires a melt temperature of only about 500°C, which avoids detrimental overheating, reducing the costs of energy and material, and improving the mechanical properties of castings [5-7].

This work is aimed at presenting to the foundrymen's community the characteristics of structure and performance of the ZnTi-based refiners. It presents examples of structure and properties changes of the selected Zn-based and Al-based cast alloys inoculated with these refiners. The paper describes also changes of attenuation coefficient of the Al - 20 wt%Zn alloy (AlZn20) after its grain refinement, which was unavailable in literature until recently.

Experimental

The examined alloys ZnAl25, AlZn20 and the master alloys Zn-4.6wt%Ti (ZnTi4) and Zn-20wt%Al-4wt%Ti (ZnAl-Ti4) were prepared from electrolytic aluminium (minimum purity 99.96%), electrolytic zinc (99.995%) and titanium sponge (98-99.8%, from Johnson Matthey Alfa). The AlZn20 and ZnAl25 alloys were melted in an electric resistance furnace, in an alumina crucible of 0.2 litre capacity. The AlZn20 melt was superheated to

~720°C, while the ZnAl25 melt to ~600°C. After introducing a master alloy the melt was held for 2 minutes, then the melt was stirred for 2 minutes with an alumina rod, and the alloy was cast into a dried sand mould (Fig. 2(a)) or into a preheated graphite-chamote crucible (Fig. 2(b)). To monitor the cooling process - two thermocouples NiCr-NiAl0.5 $\varnothing$0.20 mm were mounted in the sand mould cavity or were introduced from the top into the melt in a preheated graphite-chamote crucible. Temperatures (accuracy ± 1°C) were recorded using a multi-channel recorder Agilent 34970A (Agilent Technologies Inc., USA). Microsections for LM examinations were ground on abrasive paper (grit 200-1000) and then were polished using sub-microscopic aluminium oxide in water-alcohol suspension. The AlZn20 samples, used in macrostructure examinations, were etched with Keller's reagent. LM observations of microstructures were performed using Leica-DM IRM microscope. SEM investigations were performed using ESEM Philips XL30 microscope equipped with an EDS system EDAX Gemini 4000. Another microscope used was SEM Jeol JSM 5800 WV equipped with Noran Voyager 3 EDS system.

The examinations of damping properties were performed on the AlZn20 alloy using ultrasonic technique. The used samples were discs $\varnothing$32x7mm cut from the sand-castings (Fig. 2(a)), whose parallel surfaces were ground on abrasive paper of 600 grit. A DI-4P ultrasonic defectoscope (UNIPAN – Poland), operating at constant frequency of 2.5 MHz, was used as a source of the longitudinal ultrasonic wave. The attenuation coefficient α was obtained using an echo-method.

Results

Figs 3(a) and 4(a) show microstructure of the sand-cast, unrefined ZnAl25 and AlZn20 alloys. It can be seen that the both alloys solidify with coarse, branched dendrites of the solid solution of Zn in Al. However, after inoculation the same alloys show highly refined microstructure, as it was shown in Figs 3(b) and 4(b). It should be noted that the refined structures were obtained after using the new ZnTi-based master alloys. Fig 5 and Fig. 6 show microstructures of the master alloys used in inoculation of the examined alloys, i.e. a binary Zn-4.6 wt%Ti (ZnTi4) master alloy – Fig. 5, introduced into the alloy ZnAl25 – Fig. 3(b); and a ternary ZnAl-4 wt%Ti (ZnAl-Ti4) – Fig 6, introduced into the AlZn20 alloy – Fig. 4(b). The chemical compositions of the Ti-based intermetallic particles, existing in microstructures of the ZnTi4 and ZnAl-Ti4 master alloys, are collected in Table 1. In Figs 7 and 8 there are shown results of temperature measurements of the examined alloys slowly solidifying in the graphite-chamote crucible – the ZnAl25 alloy, or in the sand mould – the AlZn20 alloy. The Figures 7-8 contain also first derivatives of the temperature after time.

It can be seen that both the examined alloys solidify with reduced undercooling after inoculation with the ZnTi-based master alloys, which is typical after using a refiner of heterogeneous nucleation. Results of the ultrasonic examination of damping properties of the AlZn20 alloy are shown in Fig. 9. It is evident after comparing results from Fig. 9 with the macrostructures shown in Fig. 10, that attenuation coefficient of the examined AlZn20 alloy decreases together with the decrease of the grain size.

Discussion

As noted in the introduction, the main refiners of Al alloys are those built on the systems Al-Ti-B and Al-Ti-C. For the Al-Ti-B MA the crucial role as a direct substrate of α-Al nucleation is played by Al_3Ti thin layer adsorbed on TiB_2 while TiC is the direct nucleant particle when Al-Ti-C master alloy is used [9]. This is because of the similar crystal structure and lattice parameters of α-Al and Al_3Ti or TiC. The presented here new family of grain refiners based on the Zn-Ti system has $TiZn_3$ phase in its structure – Fig. 5, which has the same crystal symmetry as the α-Al and closely matches its lattice parameter. In a Al-Zn melt the $TiZn_3$ phase transforms into a ternary, more stable, $Ti(Al,Zn)_3$, which has also the same features as $TiZn_3$, and whose particles appear to be active centres of heterogeneous nucleation in the inoculated Zn-Al alloys [10] (the heterogeneous nature of nucleation supports observed reduced undercooling of the inoculated alloys, as it was shown in Figs 7 and 8). This is the most probable reason why the ZnTi-based master alloys are effective as refiners of the examined Al-Zn alloys. It should be pointed out once more, that the ZnTi-based master alloys require relatively low melt temperatures, which is an advantage as compared with the traditional Al-Ti-B or Al-Ti-C master alloys. Macrostructures shown in Fig. 10 indicate that ternary master alloys AlTi5B1, AlTi3C0.15 and ZnAl-Ti4 cause strong refinement of the inoculated AlZn20 alloy, while the binary ZnTi4 master alloy causes rather moderate refinement. However, at the same time the AlZn20 alloy inoculated with the mentioned above ternary master alloys shows also strong decrease of its attenuation coefficient, as it is seen in Fig. 9, which is a disadvantage, when preserved high damping properties are important. As noted in [9], the inoculation is rather aimed at obtaining uniform, equiaxed grain structure, and not very fine grains *per se*. Taking into consideration observed decrease of damping together with the decrease of the grain size, one can conclude that the refinement process should be controlled to obtain a compromise between changes of strength and damping properties.

Conclusions

The elaborated new family of the ZnTi-based grain refiners shows high effectiveness of inoculating of high-aluminium zinc alloys (e.g. ZnAl25 inoculated with binary ZnTi4 MA) and high-zinc aluminium alloys (e.g. AlZn20 inoculated with ternary ZnAl-Ti4 MA). The observed refinement of the examined AlZn20 alloy, inoculated with the ZnAl-Ti4 MA, is of the same order as the observed one after using the traditional AlTi5B1 or AlTi3C0.15 master alloys. However, the ZnTi-based master alloys require lower melt temperatures, which allows avoiding detrimental overheating. The initial investigations of damping properties showed, that attenuation coefficient of the AlZn20 alloy decreases together with the decrease of the grain size. Thus, in case of the damping Al-Zn alloys the refinement should be performed to such extent that would allow improving strength properties with the damping ones preserved. The detail investigations are in progress and will be presented elsewhere in a close future.

References

1. R. Schaller, *Metal matrix composites, a smart choice for high damping materials*. Journal of Alloys and Compounds, 355, 2003, pp131-135.
2. Y. Zhang, N. Ma, Y. Le and H. Wang, *Mechanical properties and damping capacity after grain refinement in A356 alloy*. Materials Letters, 59, 2005, pp2174-2177.
3. I.G. Ritchie, Z-L Pan and F.E. Goodwin, *Characterization of the damping properties of die-cast zinc-aluminium alloys*. Metallurgical Transactions A, 22A, March 1991, pp617-622.
4. S. Rzadkosz, *Effect of chemical composition and phase transformation on the damping and mechanical characteristics of alloys from aluminium-zinc system*. Habilitation thesis (in Polish). University of Mining and Metallurgy, Krakow 1995.
5. W. Krajewski, *Investigation of the high-aluminium zinc alloys grain refinement process due to Ti addition*, Archives of Metallurgy, 44, 1999, pp51-64.
6. W. K. Krajewski, A. L. Greer, T. E. Quested and W. Wolczynski, Solidification and Crystallization, chapter 16, *Identification of the substrate of heterogeneous nucleation in Zn-Al alloy inoculated with ZnTi-based master alloy*, pp137-149, Euromat 2003 papers. Viley-VCH, Ed. D.M. Herlach, Köln 2004, ISBN 3 527 431011 8.
7. W.K. Krajewski, A.L. Greer and J. Buraś, *Grain refinement of Al-Zn alloys by melt inoculation with Zn-Ti based refiner*, European Congress on Advanced Materials and Processes Euromat 2005, Prague, Czech Republic, 5-8 Sept, 2005 (http://www.euromat2005.fems.org).
8. A. E. Wol, Strojenije i swojstwa metalliczeskich sistem. Vol. 1. Fizmatgiz, Moskwa 1959.
9. A.L. Greer, *Grain refinement of aluminum alloys*. In: Solidification of Aluminum Alloys. Proceedings of the TMS Annual Meeting, Charlotte, North Carolina, USA, 14-18 March, 2004, pp131-145.
10. W.K. Krajewski and A.L. Greer, *EBSD Study of ZnAl25 alloy inoculated with ZnTi4 master alloy*, Materials Science Forum. Vol. 508 Solidification and Gravity IV, 2006, pp281-286.

Acknowledgements

The authors acknowledge the KBN - Polish State Committee for Scientific Research for financial support under research grant 4 T08A 040 25. The provision of laboratory facilities in the Department of Materials Science and Metallurgy, University of Cambridge, is also kindly acknowledged.

Tables

Table 1. Chemical composition of the randomly chosen Ti-based particles existing in microstructures of the ZnTi4 and ZnAl-Ti4 master alloys.

	ZnTi4 MA – Fig. 5				ZnAl-Ti4 MA – Fig. 6			
	No. 1	No. 2	No. 3	No. 4	No. 1	No. 2	No. 3	No. 4
Ti, at. %	23.98	22.98	24.63	22.26	24.43	21.90	23.77	23.26
Zn, at. %	75.69	75.23	75.16	76.73	66.03	16.84	67.57	16.27
Al, at. %	0.33	1.79	0.21	1.02	9.54	61.27	8.66	60.46
	NORAN - Voyager 3 EDS				EDAX - Gemini 4000 EDS			

Figures

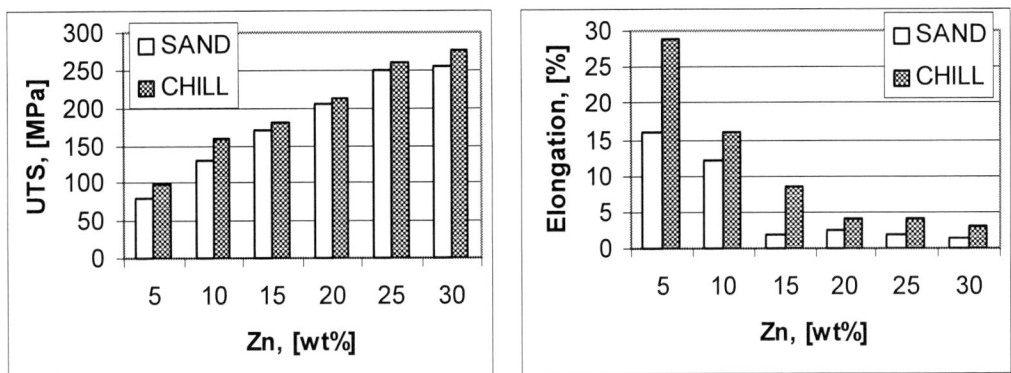

Fig. 1. UTS and elongation of cast Al-Zn alloys as a function of Zn content. (Diagrams based on data published in [8])

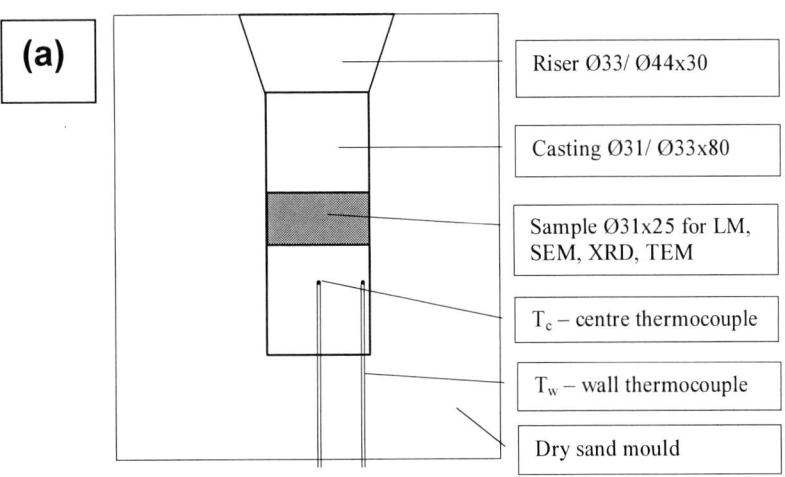

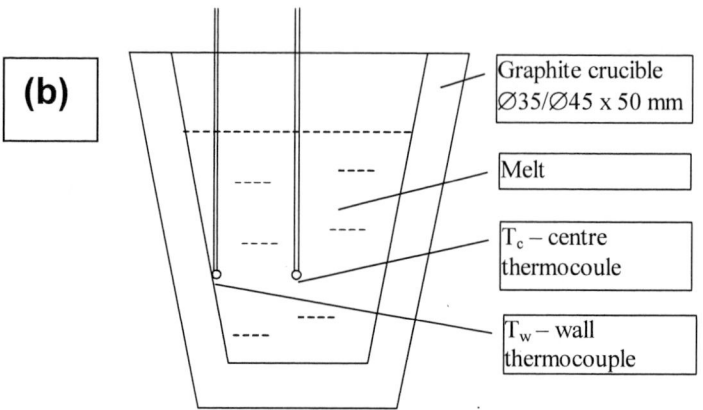

Fig. 2. Schematic sketch of the system: (a) sand mould – casting, with mounted thermocouples. (b) chamote – graphite crucible with thermocouples introduced from the top into a melt.

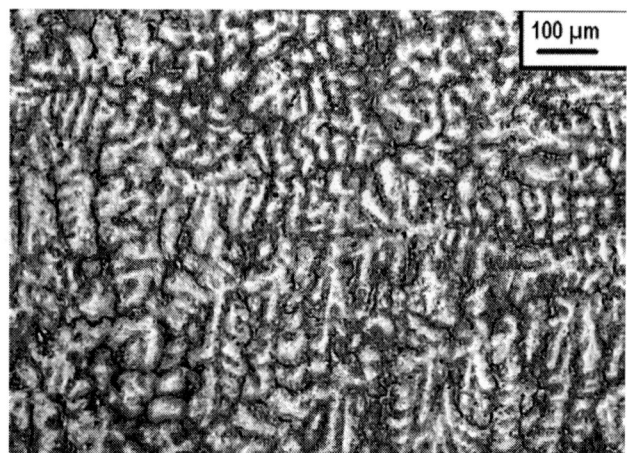

Fig. 3(a). Coarse microstructure of the initial sand-cast ZnAl25 alloy. Leica DM IRM LM.

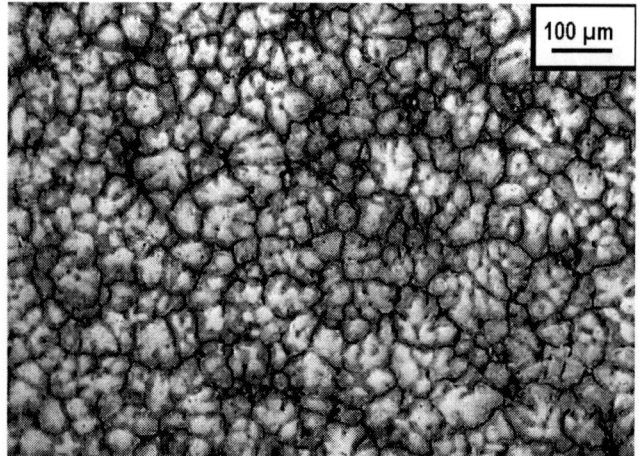

Fig. 3(b). Refined microstructure of the sand-cast ZnAl25 alloy inoculated with ZnTi4 MA (0.05 wt% Ti). Leica DM IRM LM.

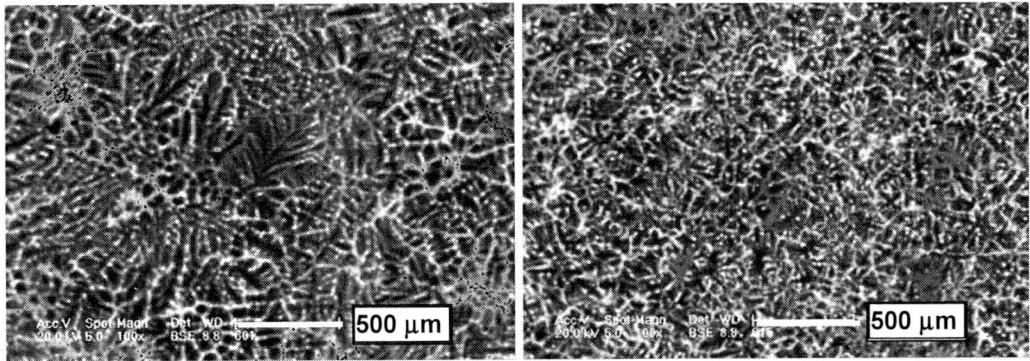

Fig. 4. (a) - Coarse microstructure of the initial sand-cast AlZn20 alloy.
(b) - Refined microstructure of the sand-cast AlZn20 alloy inoculated with ZnTi4 MA (0.04 wt% Ti). Philips XL30 ESEM.

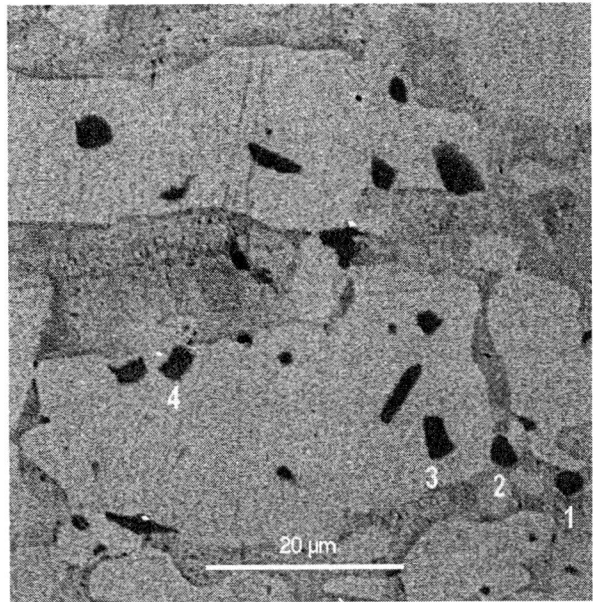

Fig. 5. Microstructure of the binary ZnTi4 MA. JEOL JSM 5800 WV SEM.
Chemical composition of the particles 1 to 4 are collected in Table 1.

Fig. 6. Microstructure of the ternary ZnAl-Ti4 MA. Philips XL30 ESEM.
Chemical composition measured in areas 1 to 4 are collected in Table 1.

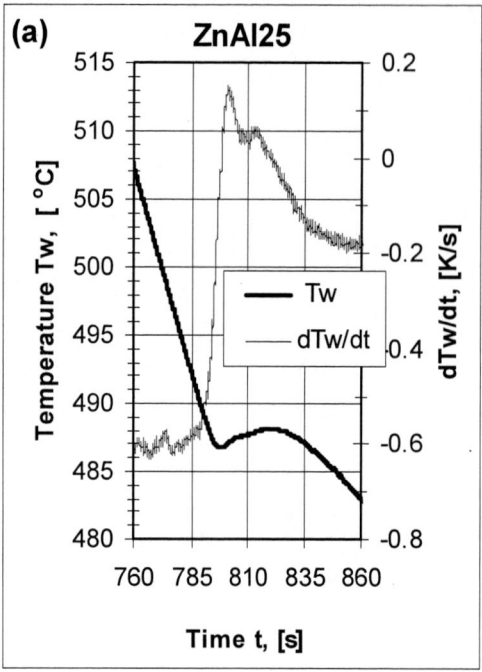
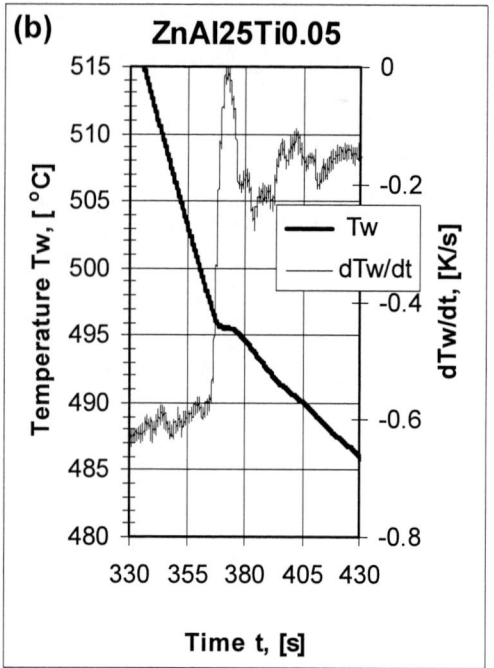

Fig. 7. Cooling curves of the ZnAl25 alloy solidifying in the chamote-graphite crucible shown in Fig. 2(b). (a) ZnAl25 alloy without Ti addition; and (b) ZnAl25 alloy inoculated with the ZnTi4 MA - 0.05wt%Ti addition. Temperatures registered by T_w wall thermocouple, dT_w/dt – first derivative of the T_w temperature after time.

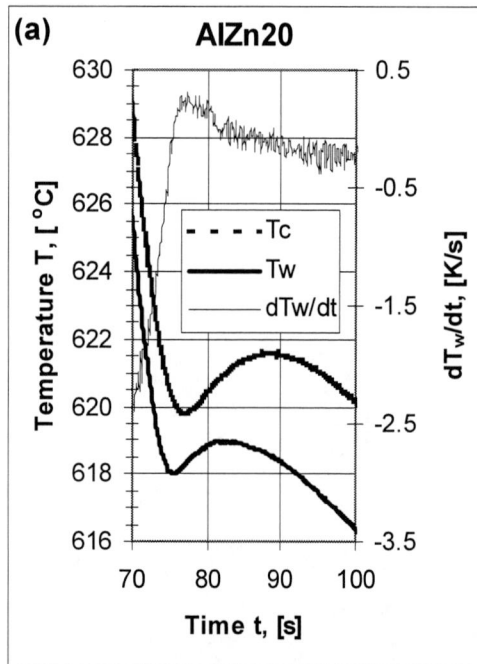
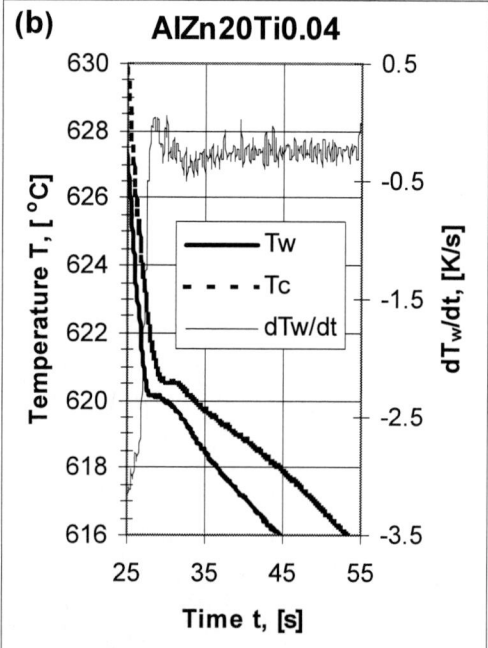

Fig.8. Cooling curves of the AlZn20 alloy solidifying in the sand mould shown in Fig. 2(a). (a) Initial AlZn20 alloy and (b) AlZn20 alloy inoculated with ZnAl-Ti4 MA, 0.04wt%Ti addition, T_w and T_c – accordingly, wall and centre temperatures, dT_w/dt – first derivative of the T_w temperature after time [7].

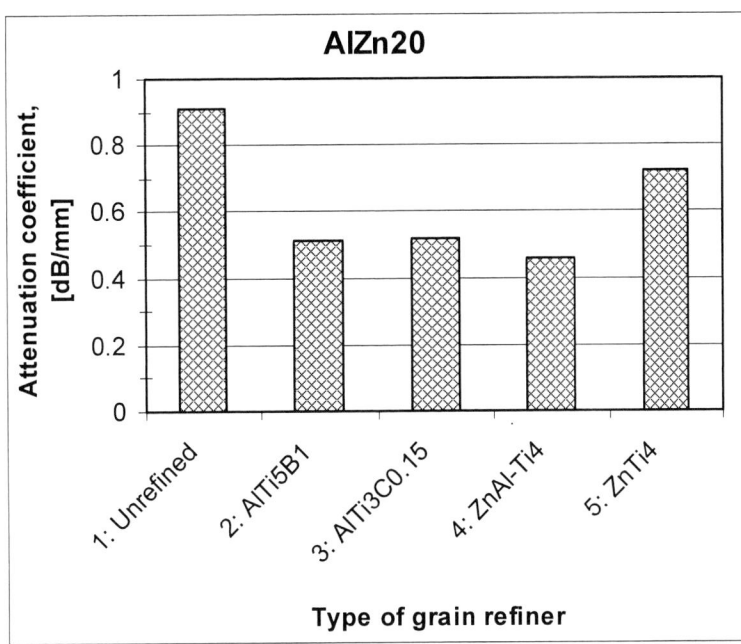

Fig. 9. Attenuation coefficient of the: 1 – initial, unrefined AlZn20 alloy, and 2 to 5 – AlZn20 alloy with addition of 0.04 wt% Ti introduced with the examined refiners.

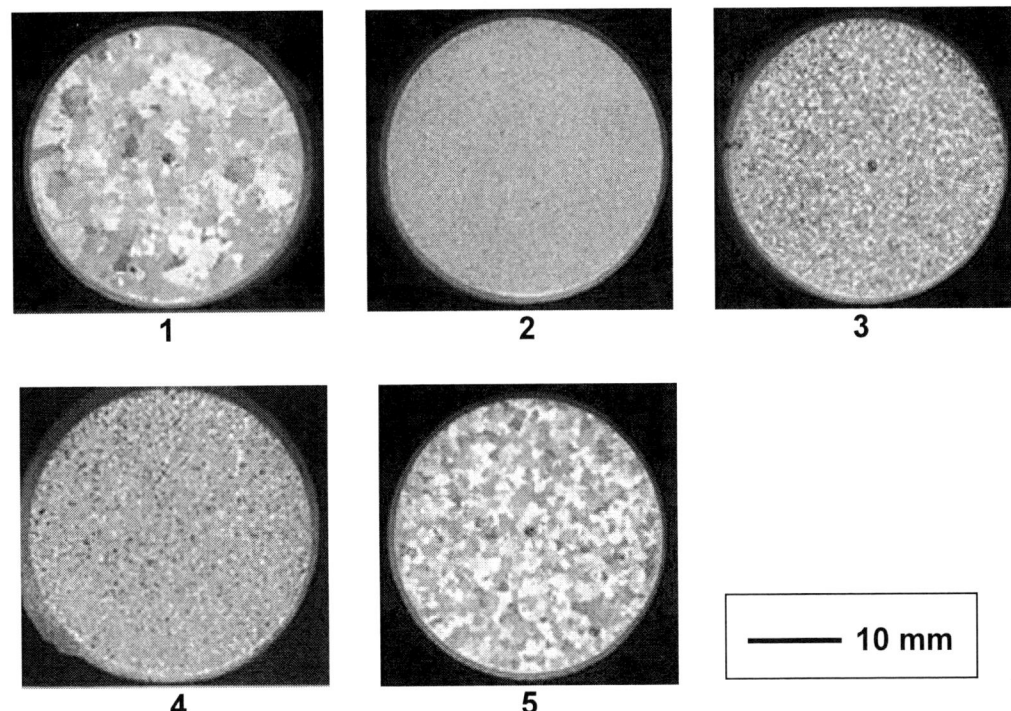

Fig. 10. Macrostructures of the AlZn20 samples which attenuation coefficient is shown in Fig. 9.
1 – unrefined AlZn20 initial alloy; and the same alloy inoculated with addition of 0.04 wt% Ti introduced with the master alloy
2 – AlTi5B1,   3 – AlTi3C0.15;   4 – ZnAl-Ti4,   5 – ZnTi4.

---

[1] To whom all correspondence should be adressed: krajwit@agh.edu.pl (W.K. Krajewski)

Effects of Process Parameters on the morphology of TiAl$_3$ particle during the production of Al-Ti-B Master Alloy by Flux Reaction

Min Ryou*, Sang-Ho Choi** and Myung-Ho Kim *.

* School of Materials Science and Engineering, Inha University
   #253 Yonghyeon-dong, Nam-gu, Incheon, 402-751, South Korea
 ** Dongyang Junior Technical College

Abstract

The effect of process parameters, such as reaction temperature and holding time, on the microstructure of Al-Ti-B master alloy had been studied by using SEM/EDS and EPMA. The flux reactions between K_2TiF_6, KBF_4 and molten aluminum were carried out at various temperatures and holding times. The results indicated that the morphology of TiAl$_3$ particles in master alloy was strongly influenced by reaction temperature and holding time. The TiAl$_3$ particles with blocky type which was known as an effective refiner were observed at lower reaction temperature and in shorter holding time, whereas finer TiB_2 particles were found to be distributed uniformly independent of the reaction temperatures. The size of TiAl$_3$ particles in the master alloy was significantly increased with increasing of the reaction temperature, and the increased temperature resulted in the change of the morphology of TiAl$_3$ particles from blocky to plate-like type. The TiAl$_3$ particles with morphology of blocky type shows longer fading time compare to the plate-like type.

Key words
Grain refiner, Al-5Ti-1B alloy, TiAl$_3$ particle, Flux reaction

Introduction

In aluminum industry, Al-Ti-B Master alloy has been widely used for grain refinement that is an essential process to decrease casting defects, to improve castability, workability, mechanical properties, and so on[1][2]. These effects of grain refinement have been known to be affected by the type and size of the existing phases in the master alloy, such as $TiAl_3$ and TiB_2. The type and size of these particles were mainly affected by the temperature of molten metal and holding time during the production of Al-Ti-B alloy by flux reaction.

Studies on the grain refining ability of Al-5wt.%Ti-1wt.%B master alloy, and the effects of process variables during the flux reaction on the microstructure of the master alloy have been widely studied so far, and especially the effect of $TiAl_3$ phase on grain refinement behavior has been examined by many investigators[3-6]. However, as the reaction mechanism during the manufacture of Al-Ti-B grain refiner by flux reaction is so complex, significant variations in product performance are reported to be existed. And the influence of boron on the formation of $TiAl_3$ intermetallic phases during manufacture of the master alloy, and additional effects of the boride phase on the grain refinement behavior of the $TiAl_3$ intermetallic phases are not clearly understood[7]..

As a first step to investigate the influence of duplex intermetallic phases, $TiAl_3$ intermetallic phases covered with TiB_2 boride, on the grain refinment of aluminum, the effect of process variables during the flux reaction, such as reaction temperature and holding time, on the microstructure of Al-5wt.%Ti-1wt.%B master alloy, particularly for the morphological change of the $TiAl_3$ and TiB_2, and also the grain refining ability of the $TiAl_3$ with different morphology and its effect on the fading time were investigated by using SEM and image analysis.

Experimental Procedure

In this experiment, two kinds of flux, K_2TiF_6 and KBF_4, were selected, and were responded to 700 , 750 , 800 , 850 , and 900 with reference to Fig.1. To produce Al-5wt.%Ti-1wt.%B master alloy, the amount of pure aluminum and added for each runs was K_2TiF_6 132.5g, and KBF_4 56.66g, aluminum 488.95g, corresponding to the stoichiometric ratio. To observe the morphological change of the $TiAl_3$ and TiB_2 particles with holding times, the mixture were induction heated at each temperature and held for 10, 20, 30, 60, 120, 240, and 480 min. respectively.

The master alloy samples were cast in the graphite mold with temperature of 200 , and the microstructure was observed by using Scanning Electron Microscope (SEM), electron probe micro-analyzers (EPMA) after grinding up to #2000 and electropolishing by 5% Percloric Acid etchant. To investigate the effect of $TiAl_3$ particles with different morphology on the grain refinement of commercial purity aluminum, 0.015 wt.% master alloy was added to the molten metal at 800 , and then hold for 10, 20, 30, 60,

120, 240, and 480 min., and then cooled at room atomospher, respectively.

Results

Fig.2 shows the typical microstructure of the master alloy specimen, and Fig.3 shows the SEM/EDS analysis of the microstructure. According to these figures, the particle A is considered to be salt inclusions. In Al-5Ti-1B master alloy, generally two kinds of intermetallics, $TiAl_3$ and TiB_2 particles, were reported to be observed[2,4], However, the size of TiB_2 particles is too small to analyze with SEM/EDS, these boride particles were observed with EPMA, as shown in Fig.4. From Fig.3 and Fig.4, it could be conformed that both the particle B and particles C are $TiAl_3$.

Fig.5 shows the microstructural change of the master alloy specimen with reaction temperatures and holding times, in general. This figure indicates that the higher the reaction temperature, the bigger the size of $TiAl_3$ particles, and the amount of blocky type particles decreased but the amount of plate-like type particles increased. It is also indicated that increase in holding time resulted in increase in $TiAl_3$ particle size. However, the size and type of TiB_2 was not significantly influenced by the reaction temperature and holding time.

According to Arnberg[6], the master alloy with a lots of regular size $TiAl_3$ and TiB_2 particles exhibit very effective grain refining ability. In this experiment, it was found that the blocky type $TiAl_3$ particles with regular size, which is known as a very effective grain refiner could be produced at the reaction temperature of 750 , 800 and holding time of 10 min. 15 min., which was considered to be the optimum conditions for manufacturing of the master alloy.

Fig.6(a) exhibits the variation of the $TiAl_3$ particle size with holding time and reaction temperature. According to this figure, the size of particles tends to be grown with the holding time and the growth rate became accelerated with increase of temperature. Fig.6(b) shows the effect of reaction temperature and holding time on the number of $TiAl_3$ particles. As you see, the number of particles tends to decrease with the holding time and reaction temperatures. These are considered to be due to the development of $TiAl_3$ cluster at higher temperature and longer holding time.

Fig. 7 shows the changes in aspect ratio of $TiAl_3$ particles with holding times and reaction temperatures. At low reaction temperatures, the values of aspect ratio is about 1.5 to 2, and it's doesn't change significantly with holding times. This means that the $TiAl_3$ particles formed at low reaction temperature are blocky type, and the shape doesn't change with holding time. However, in case of the reaction temperature of 900 , the aspect ratio of $TiAl_3$ particles became bigger, and this means the morphology of

TiAl$_3$ particles was converted from blocky type to plate-like type at the higher temperature.

Fig. 8 shows the grain refining ability of the TiAl$_3$ with different morphology of the blocky and plate-like type and its effect on the fading time. From the figure, it could be understood that the TiAl$_3$ with morphology of blocky type exhibits better grain refining ability compare to the particles with plate-like type, as reported before[6]. In addition, interestingly, the TiAl$_3$ particles with morphology of blocky type shows longer fading time compare to the particles with plate-like type. It seems to be due to the change in morphology of TiAl$_3$ particles from blocky to plate-like type with increasing of holding time [7].

Conclusions

From the above study, the following conclusions could be drawn:
1. By reaction of the fluxes K$_2$TiF$_6$ and KBF$_4$ with aluminum Al-5%Ti-1%B master alloy with TiAl$_3$ and TiB$_2$ particles of very effective grain refiner can be produced.
2. The size of TiAl$_3$ particles tends to decrease at low reaction temperature and shorter holding times, and increase in holding time resulted in increase in the TiAl$_3$ particle size. However the size and shape of TiB$_2$ was not significantly influenced by the reaction temperatures and the holding times.
3. The TiAl$_3$ particles formed at low reaction temperature are blocky type, and the type doesn't change with holding times. However, the morphology of TiAl$_3$ particles was converted from blocky type to plate-like type at higher reaction temperatures.
4. The TiAl$_3$ particles with morphology of blocky type exhibits better grain refining ability compare to the plate-like type, and the TiAl$_3$ particles with morphology of blocky type shows longer fading time compare to the particles with plate-like type.

References
1. A. Cibula : J. Inst. Metals. 76 (1949-50) 321.
2. D.G. MCCARTNEY: Int. Mater. Rev. 34 (1989) 247.
3. M.M. Guzowaski, G.K. Sigworth. And D.A. Sentner, Metall. Trans. A.18A (1987) 613.
4. D.G. MCCARTNEY, Metall. Trans. A., 19A (1988) 385.
5. C.D. Mayes, D.G. McCartney, G.J. Tatlock, : Mater. Sci. Technol., 9 (1993) 97.
6. L. Arnberg, L. Backerud, and H. Klang, Metals. Techno. 9 (1982) 14
7. John E. Gruzleski, Bernard M. Closset, The treatment of liquid Al-Si Alloys. (1990) 135

Figures

1. Phase diagram of the binary system K$_2$TiF$_6$ and KBF$_4$.

2. Typical microstructure of Al-5Ti-1B alloy by flux reaction at 850 for 15min..
3. SEM/EDS results of A, B and C phases exhibited in Fig. 2.
4. EPMA results of $TiAl_3$ and TiB_2 phases in the master alloy.
5. Microstructures of Al-5Ti-1B master alloy at the various reaction temperatures and holding times.
6. (a)Size and (b)Number of $TiAl_3$ particles as a function of holding time at the various reaction temperatures.
7. Aspect ratio of $TiAl_3$ particles in the master alloy as a function of holding time at the various reaction temperatures.
8. Grain refining ability of blocky type $TiAl_3$ and plate-like type $TiAl_3$ with time.

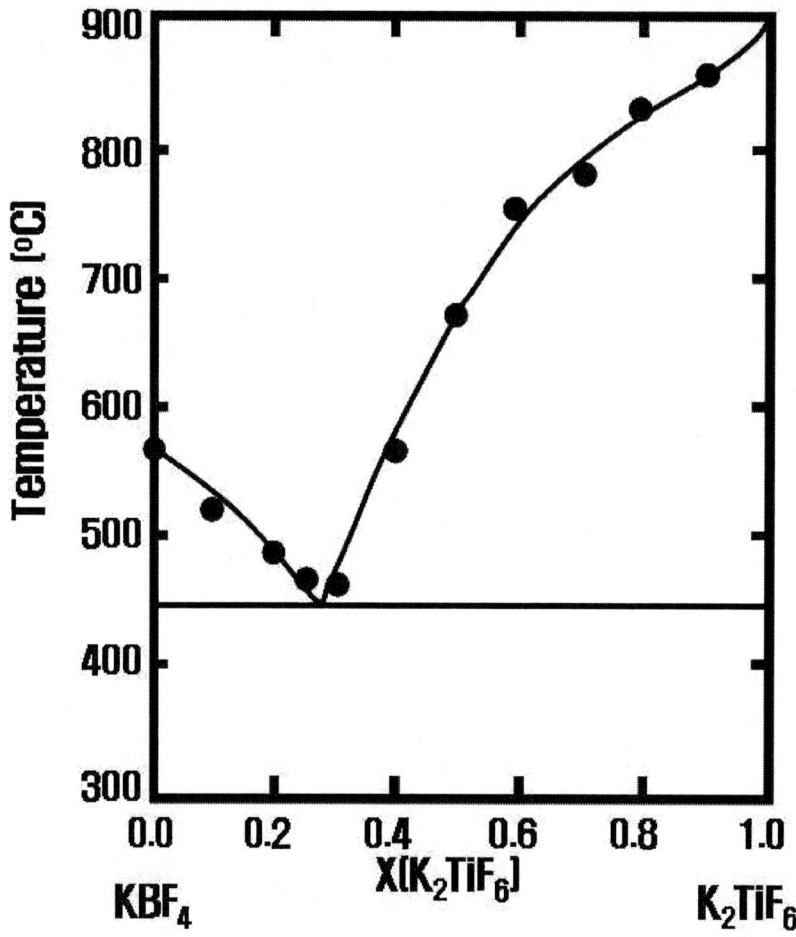

Fig. 1 Phase diagram of the binary system K_2TiF_6 and KBF_4.

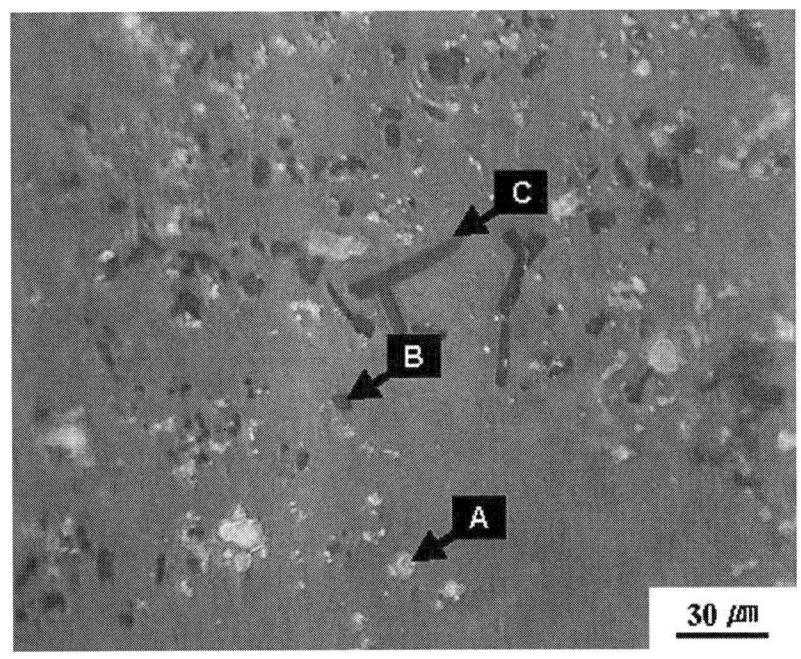

Fig. 2. Typical microstructure of Al-5Ti-1B alloy by flux reaction at 850 deg C for 15min..

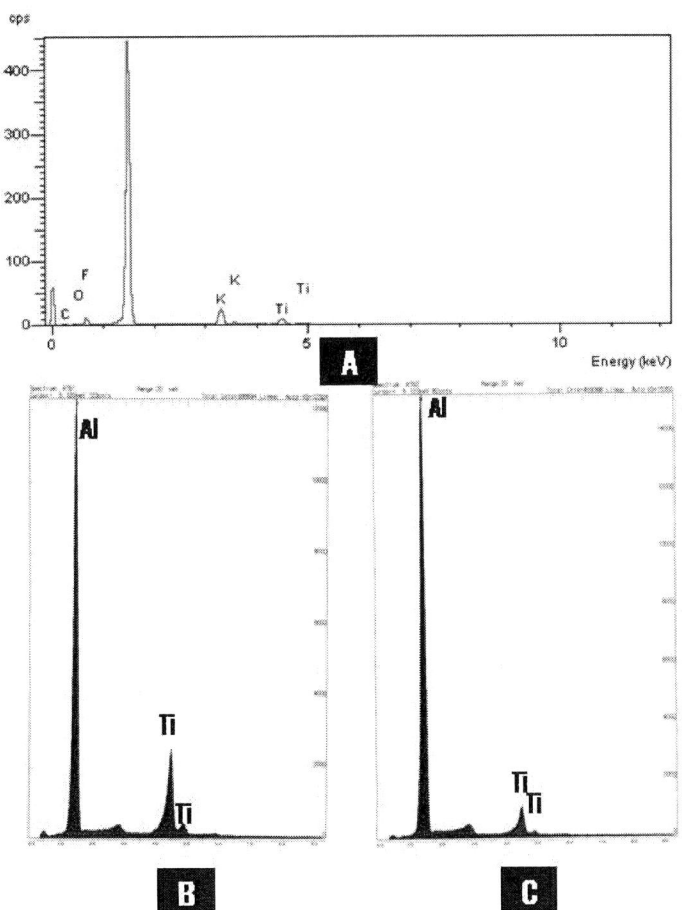

Fig. 3. SEM/EDS results of A, B and C phases exhibited in Fig. 2.

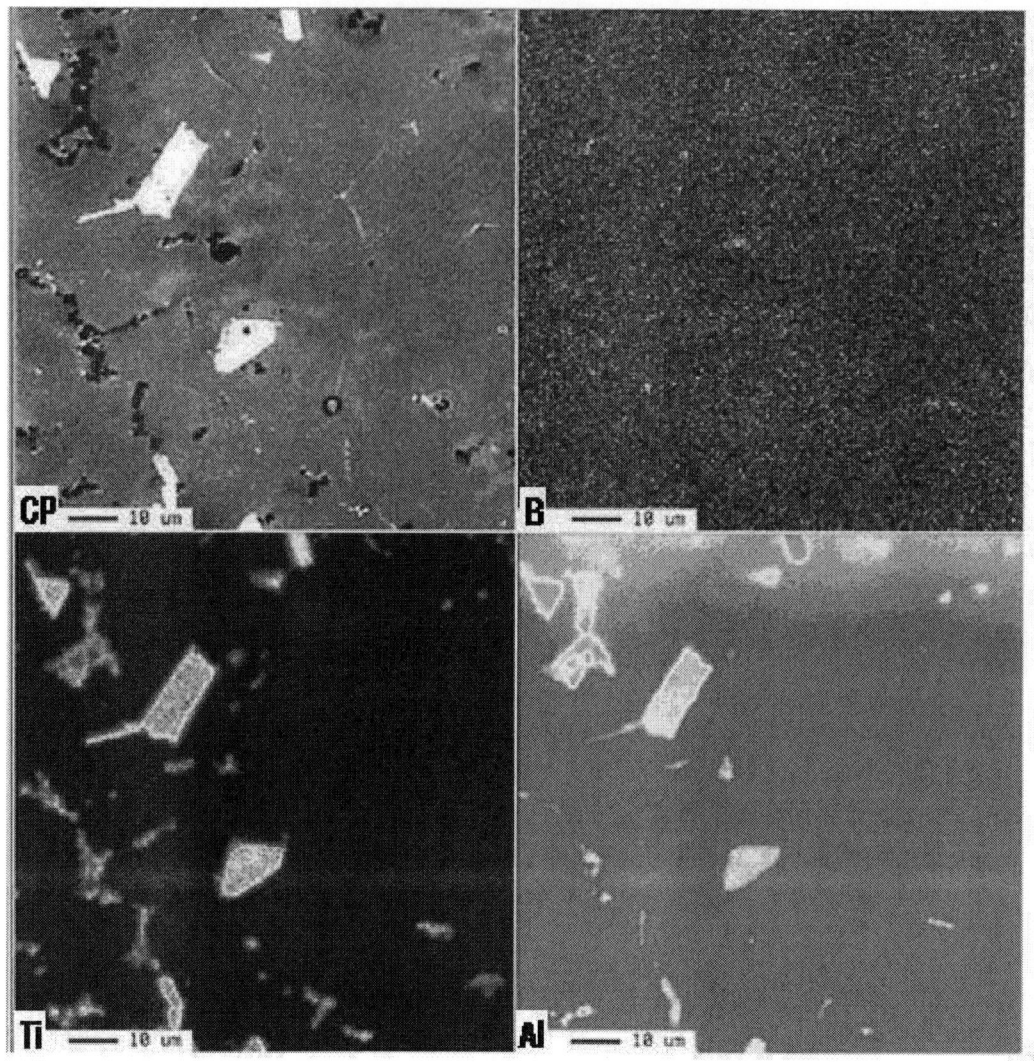

Fig. 4. EPMA results of TiAl$_3$ and TiB$_2$ phases in the master alloy.

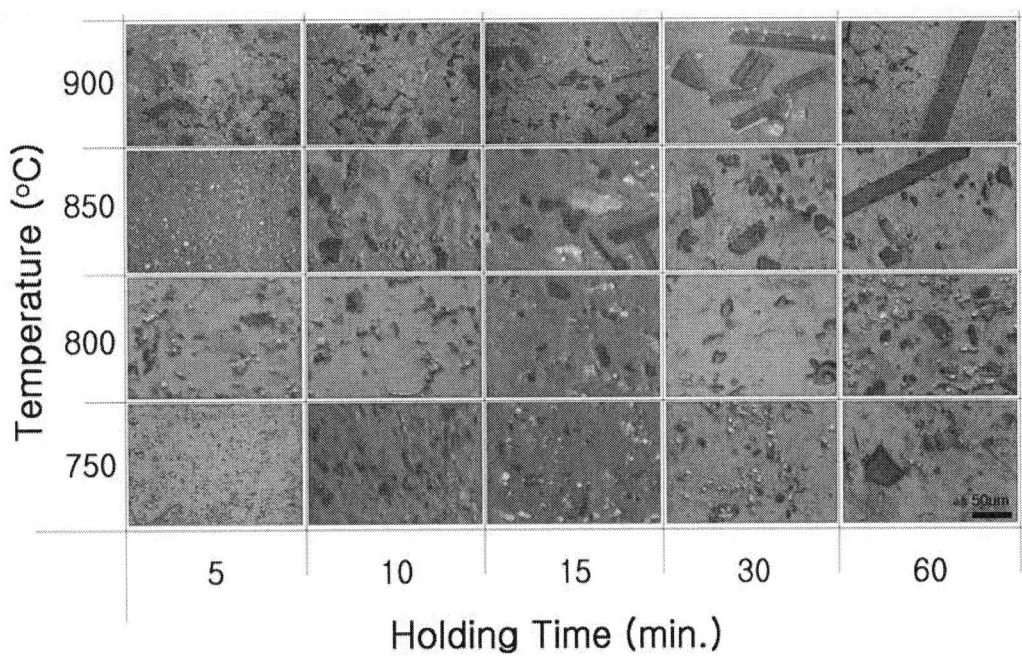

Fig. 5. Microstructures of Al-5Ti-1B master alloy at the various reaction temperatures and holding times.

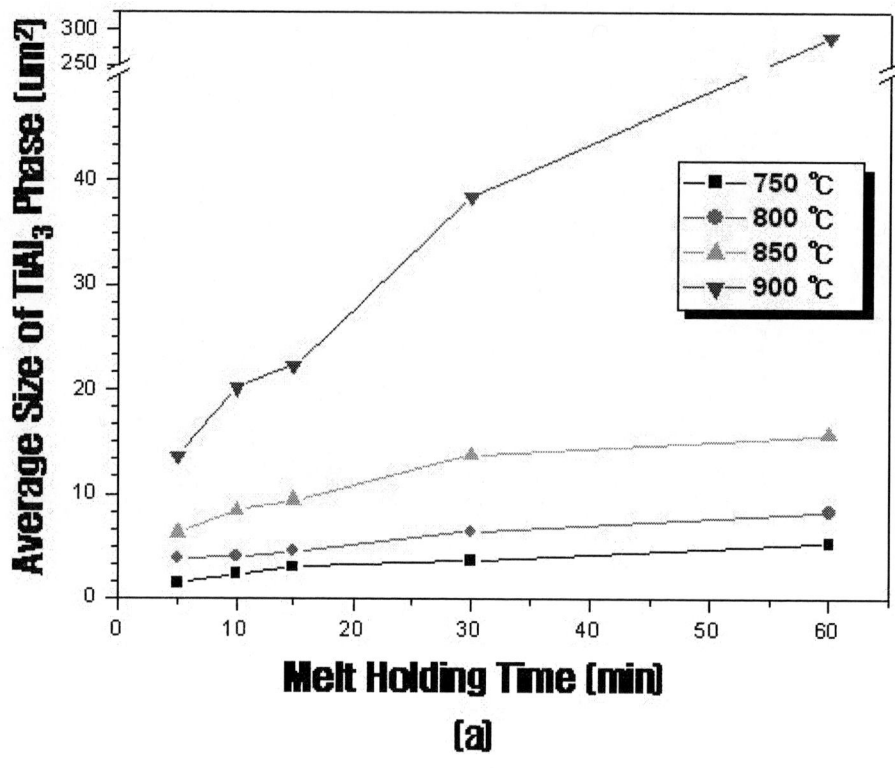

(a)

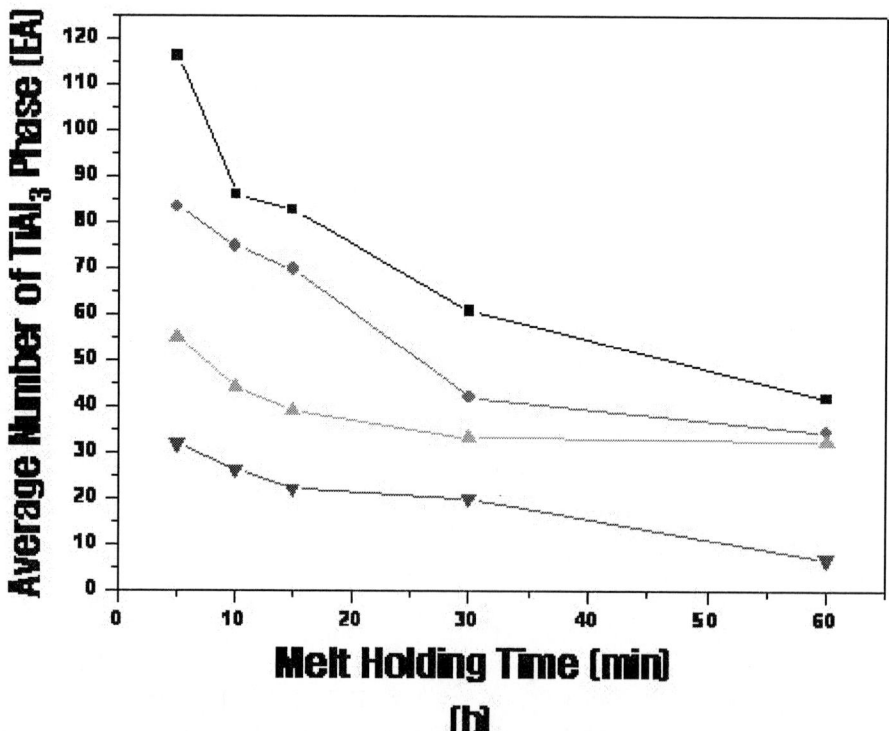

(b)

Fig. 6. (a)Size and (b)Number of TiAl₃ particles as a function of holding time at the various reaction temperatures.

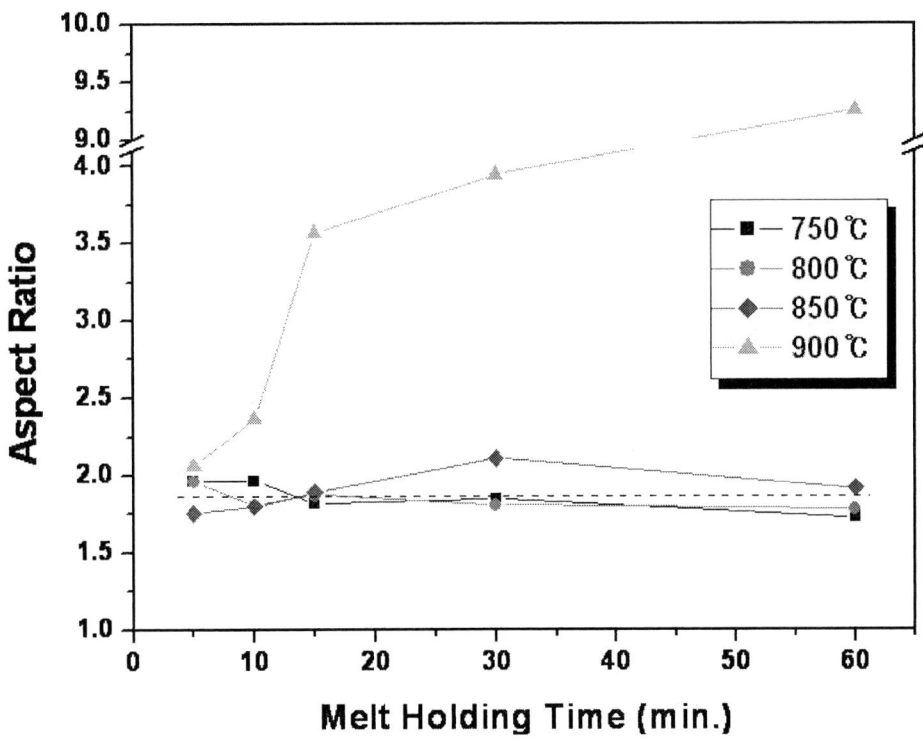

Fig. 7. Aspect ratio of TiAl₃ particles in the master alloy as a function of holding time at the various reaction temperatures.

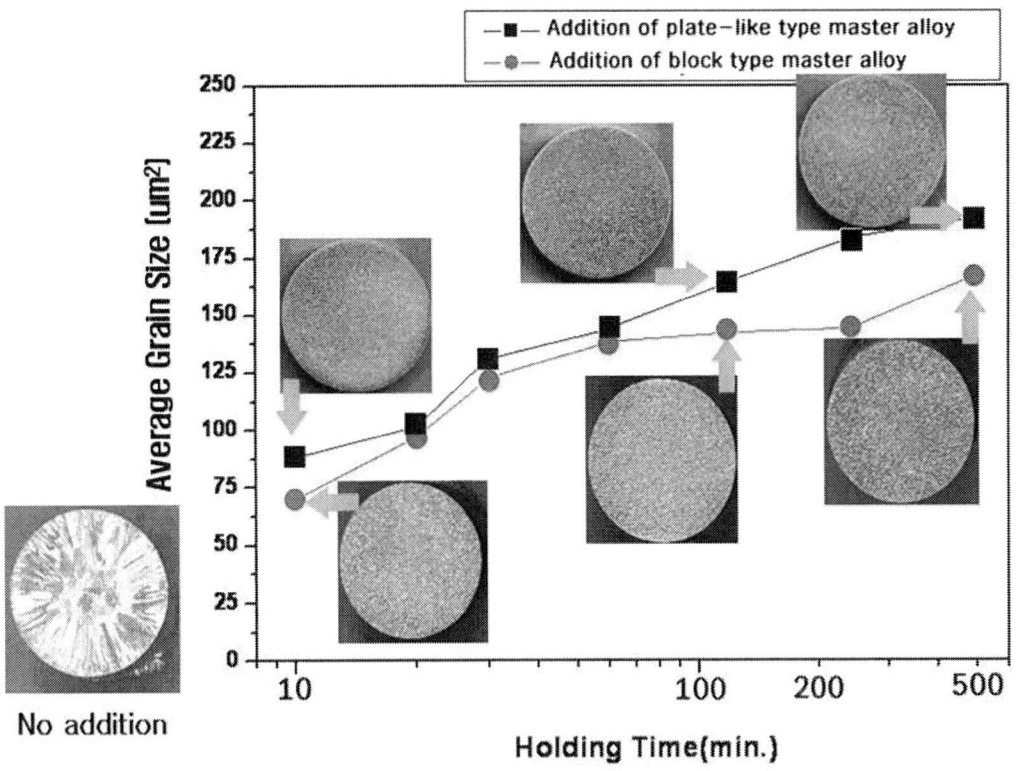

Fig. 8. Grain refining ability of blocky type TiAl₃ and plate-like type TiAl₃ with time.

The New Method and Device for Fast Evaluating Inoculation Result of Eutectic Al-Si Alloy

Li Dayong, Shi Dequan, Wang Lihua and Zhang Yutong

School of Material Science and Engineering, Harbin University of Science and Technology, P.R.China

Abstract

Inoculation is a necessary step to produce high quality eutectic Al-Si alloy castings. Thermal analysis and metallographic observation methods have been being used to evaluate inoculation result of eutectic Al-Si alloy in real production, however they are unable to meet the requirement of real time monitoring in front of furnace because of their relative long measuring period. The authors put forward a new method to evaluate inoculation result of eutectic Al-Si alloy based on melt surface tension quick measurement. To blow the inert gas at appropriate speed into the melt through a capillary and form bubbles in the melt continuously. Then to test the pressure change in the bubbles and the melt temperature in 5 seconds duration with a computerised data acquisition device and to calculate the number and differential pressure of bubbles. The melt surface tension is calculated based on the number and the differential pressure of bubbles and some other parameters. Finally, the inoculation result of eutectic Al-Si alloy can be given by the device according to calculated melt surface tension.

Key words

eutectic Al-Si alloy, surface tension, inoculation result, new evaluation method

Introduction

It is well known that inoculation treatment will increase the mechanical properties of eutectic Al-Si alloy to large extent. It has become indispensable to evaluate the inoculation result in front of furnace for improving eutectic Al-Si alloy casting quality and reducing rejects. At present, the main methods for evaluating inoculation result of eutectic Al-Si alloy are metallographic observation and thermal analysis[1-4]. As we all know, metallographic observation is more accurate, but the detecting period, at least a few minutes or above, is longer. The thermal analysis for evaluating the inoculation result by characteristic values on cooling curve is relatively faster compared with that of metallographic observation, but it is not fast enough to meet with the requirement of production line in speed yet and there is still deficiency in criterion[5,6]. Therefore, it is very significant to develop a new method and device for fast evaluating inocuation result of eutectic Al-Si alloy.

In this paper, a new method for fast evaluating inoculation result based on the surface tension measurement was proposed and an automatic device for fast testing the surface tension of Al-Si alloy in front of furnace was developed. The relationship between surface tension and inoculation result of eutectic Al-Si alloy has been got by experiments. It was proved by comparing with other methods that the new method has merits of short in testing period, simple in operation and cheap in measuring cost and so on.

The Principle of Fast Evaluating Inoculation Result

According to the growth mechanism of Twin Plane Reentrant Edge (TPRE)[7,8], only can silicon phase grow along <211> and <110> orientation of crystal face {111} and grow slowly along the vertical plane of {111}. During the course of growth of silicon phase twin plane will come into being on crystal face {111} and reentrant edge, as a result, will also form in front of twin plane. Therefore, the growth orientation of silicon phase will be changed by reentrant edge, and silicon atom will be absorbed by TPRE which has lower energy level. All of these must accelerate the growth of crystal face {111} and the shape of silicon phase will become massive to the end. In the course of the growth, aluminum phase often lags behind silicon phase.

Sodium, when added in small amounts, has the ability of selective absorption in the molten Al-Si alloy, and it will be absorbed in TPRE firstly and restrain the growth of silicon phase. Consequently, the growth of aluminum phase will exceed that of silicon phase and the number of twin plane defects will increase.

Sodium is a surface active agent to molten Al-Si alloy. The surface tension of the molten Al-Si alloy will reduce remarkably after sodium is added. Experiments denoted that the value of surface tension of the molten Al-Si alloy would reduce from 0.84 N/m to 0.62N/m when 0.1% sodium was added. With the increase of the sodium percentage, the surface tension will reduce further and the number of twin plane defect will increase and the growth of silicon phase will become slow. Appropriate sodium will change the shape of silicon phase from massive to fibrous.

We can conclude from above discussion that melt surface tension reflects inocuation result of Al-Si alloy in great extent. Therefore, the inoculation result of Al-Si alloy can be evaluated by melt surface tension measurement.

New Method and Device for Fast Testing Melt Surface Tension

There exist some methods to test surface tension of molten Al-Si alloy[9-11]. The new method, named Amplitude and Frequency Equivalent of Blowing Bubbles (AFEBB) [12,13] advanced by the authors, will be applied to fast test surface tension of the molten Al-Si alloy in this paper. AFEBB was developed based on Maximum Bubble Pressure method[14], but, in many aspects, it differs from Maximum Bubble Pressure method. The speed of blowing bubbles into the molten Al-Si alloy through capillary is faster than that in Maximum Bubble Pressure method and the depth of capillary being dipped in the melt is alterable, unlike that in Maximum Bubble Pressure method. Therefore, not only the testing period is relative short, but also the operation is very simple by using the new method. It is obvious that the new method is fit for fast testing the surface tension of melt in front of furnace.

In the new method, pressed inert gas is blown into molten Al-Si alloy through a capillary at such a speed that bubbles can form and vanish continuously. The number of bubbles and pressure difference in the capillary during the fixed period are monitored by the pressure sensor and the temperature of the molten Al-Si alloy is also measured by the thermocouple under the control of the computer.

The formula for calculating surface tension of the molten Al-Si alloy based on the new method is created as follows.

$$\sigma_e = aN + b\Delta P + c\left(\phi_x - \phi_0\right) + dT + e \tag{1}$$

Where, σ_e — calculated surface tension of Al-Si alloy at temperature T;

N — the number of bubbles blown into melt during fixed period;

ΔP — average value of pressure difference in bubbles;

ϕ_x — tested capillary diameter;

ϕ_0 — standard capillary diameter;

T — tested molten Al-Si alloy temperature;

$a \sim e$ — coefficients and constants decided by alloy type.

It is very important to test surface tension accurately for fast evaluating inoculation result. The schematic diagram of automatic device for fast testing surface tension of molten Al-Si alloy based on the new method was shown in Figure 1. The device consists of an up-and-down movement mechanism, a heat-resistant detector, a data acquisition and processing unit, a pressure sensor, a thermocouple, a constant pressure inert gas source and a workbench and so on.

The procedure of fast testing surface tension with the automatic device is as follows. To take a proper quantity of the molten eutectic Al-Si alloy with a small graphite crucible and lay them on the workbench under the heat-resistant detector. When the computer starts a reversible motor and turns on the stable pressure inert gas source, the up-and-down movement mechanism is driven towards the molten Al-Si alloy and the pressed inert gas is blown out through the capillary. During this course the pressure sensor will sense the pressure in the capillary and the capillary diameter can be calculated by the data acquisition and processing unit. Once the capillary end meets the melt surface, the pressure in the capillary will increase suddenly, so the depth of the capillary dipping in the molten Al-Si alloy can be monitored by pressure sensor. When the capillary end reaches the setting depth, the up-and-down movement mechanism will stop immediately. In the following fixed period, bubbles will form in the melt one after another and the pressure in the capillary is tested by the pressure sensor, and the data are collected by data acquisition and processing unit. At the same time, the temperature of the molten Al-Si alloy is measured by the thermocouple dipped in the melt. When the period is over, the up-and-down movement mechanism will lift the heat-resistant detector automatically and the computer begins to calculate the pressure difference in the capillary and the number of blown bubbles. Surface tension of the molten eutectic Al-Si alloy is calculated by means of formula (1) and these informational parameters tested by the device, and the inoculation result will be given according to the evaluating criterion saved in the computer. All the testing results including surface tension, melt temperature and inoculation result will be displayed on LED and printed by micro-printer.

Experiments and Evaluation Criterion
Materials for experiments are eutectic Al-Si alloys, and the compositions meet ZL102 criterion of GB1173-86. They were melted in graphite clay crucible via resistance-heating furnace, and the melting and inoculating procedure is as follows. The molten alloy was refined at 700-720℃ by C_2Cl_6 with 0.3% (mass fraction) for about 8-12 minutes, and was inoculated at 740-760℃ by ternary modificator (NaF:NaCl:KCl equals 45:40:15) from 0.0% to 3.0% with 0.25% step for about 12-15 minutes. Metallographic sample was cast at 710-730℃, and at the same time surface tension of the melt was measured by the automatic device.

There is almost no alteration in surface tension when the temperature of the molten alloy changes in fixed range, and the factor of temperature was already included in formula (1), therefor, the unitary relationship between surface tension and microstructure was taken into account in the criterion of evaluating inoculation result. The values of surface tension and microstructures of eutectic Al-Si alloy under different inoculations were shown in Figure 2.

We can see from Figure 2 that sodium has a great effect on microstructure and surface tension of eutectic Al-Si alloy. Surface tension is reduced and silicon grain is refined with the increase of adding sodium. The reason is that sodium with high capability of surface absorption is absorbed in twin plane reentrant edge and changes the growth mode of silicon phase, as a result, the shape of silicon phase is changed from massive or acicular to vermiculate or fibrous.

According to AFS criterion of evaluating inoculation result of eutectic Al-Si alloy by metallographic microstructure, inoculation result is classified 6 levels. In grade 1, eutectic silicon exists in massive or acicular, and it means no inoculation. In grade 6, eutectic silicon exists in vermiculate or fibrous, and it means perfect inoculation.

In this paper, inoculation result evaluated by melt surface tension is classified 3 levels, namely, partial inoculation (including no inoculation), moderate inoculation and perfect inoculation. Three levels correspond to grade 1-2, grade 3-4 and grade 5-6 of AFS criterion respectively. The criterion is shown in Table 1.

Conclusions

The new method for fast evaluating inoculation result of eutectic Al-Si alloy by melt surface tension is brought forward, and an automatic device for fast testing surface tension of molten Al-Si alloy is developed. The principle of fast testing melt surface tension is based on the new method-Amplitude and Frequency Equivalent of Blowing Bubbles.

The relationship between melt surface tension and inoculation result of eutectic Al-Si alloy has been got by experiments, and according to which the criterion of evaluating inoculation result by melt surface tension has been gained as well. If the surface tension exceeds 0.53N/m, it denotes that eutectic Al-Si alloy is not inoculated or inoculated partially and is considered as grade 1-2 according to AFS criterion; If it is below 0.40N/m, eutectic Al-Si alloy is inoculated perfectly and is considered as grade 5-6; If it is between 0.40N/m and 0.53N/m, eutectic Al-Si alloy is inoculated in medium class and is considered as grade 3-4.

References

1. Djurdjevic M, Sokolowski J H, Stockwell T J, *Control of the aluminum silicon alloy solidification process using thermal analysis,* Journal of Metallurgy, 1998, 4(8), pp237-248
2. Djurdjevic M, Sokolowski J H, Jiang H, *On-Line prediction of the Al-Si eutectic modification level using thermal analysis,* Materials Characterization, 2000, 45(2), pp1-8

3. SUN Ye-zan, CHEN Chong-jun, ZHAO Jian-ping et al, *A new progress in technical research for thermal analysis abroad*, Research studies on foundry equipment, 1997(3), pp 28-31

4. Jiang H, Sokolowski J H, Djurdjevic M et al, *Recent advances in automated evaluation and on-Line prediction of the Al-Si eutectic modification level*, AFS Transactions, 2000, 57(3), pp 142-151

5. Fras E, Kapturkiewicz W, Lopez H F, *A new concept in thermal analysis of castings*, AFS Transactions, 1994, 93(5), pp505-511

6. Ho C R, Cantor B, *Modification of hypoeutectic Al-Si alloy*, Journal of Materials Science, 1995, 30(8), pp1912-1920

7. Elliott R, Eutectic Solidification Processing, Butterworths, 1992, pp276-281

8. Gokhshtein M B, Vasileva L S, *The mechanism of the modification of silumin*, Metal Science and Heat Treatment, 1990, 12(7), pp 591-593

9. Sauerland S, Lohofer G, Egry I, *Surface tension measurements on levitated liquid metal drops*, Journal of Non-Crystalline Solids, 1998,158(5), pp833-836

10. FAN Jian-feng, YUAN Zhang-fu, KE Jia-jun, *Development in measuring surface tension of high temperature molten liquid*, Chemistry Online, 2004,(11), pp802-807

11. TIAN Yu-ren, Glynn holt R, Robert E, *A new method for measuring liquid surface tension with acoustic levitation*, Review of Scientific Instruments,1995,66(5), pp3349-3354

12. LI Da-yong, SHI De-quan, LI Feng et al, *Research and application development of the technology for testing melt surface tension*, Foundry, 2004, 53(1), pp12-17

13. LI Da-yong, CHEN Jie, ZENG Xin et al, *Study on new method of fast quality evaluation of nodular cast iron in front of the Furnace*, China Mechanical Engineering, 2002, 13(19), pp1769-1701

14. LI Da-yong, SHI De-quan, ZHANG Yu-tong, *New method and device for fast testing surface tension of liquid aluminum alloy*, The Chinese Journal of Nonferrous Metals, 2005, 15(2), pp179-184

Acknowledgements

The authors would like to acknowledge the support of National Natural Science Foundation of China (50174023) and Heilongjiang Province Key Task Project of Science and Technology (G00A12011).

Tables

Table1 criterion evaluating inoculation result by surface tension

Inoculation result of eutectic Al-Si alloy	Surface tension value / N/m
No inoculation or partial inoculation (AFS grade 1-2)	$\sigma e > 0.53$
Moderate inoculation (AFS grade 3-4)	$0.40 \leqslant \sigma e \leqslant 0.53$
perfect inoculation (AFS grade 5-6)	$\sigma e < 0.40$

Figures

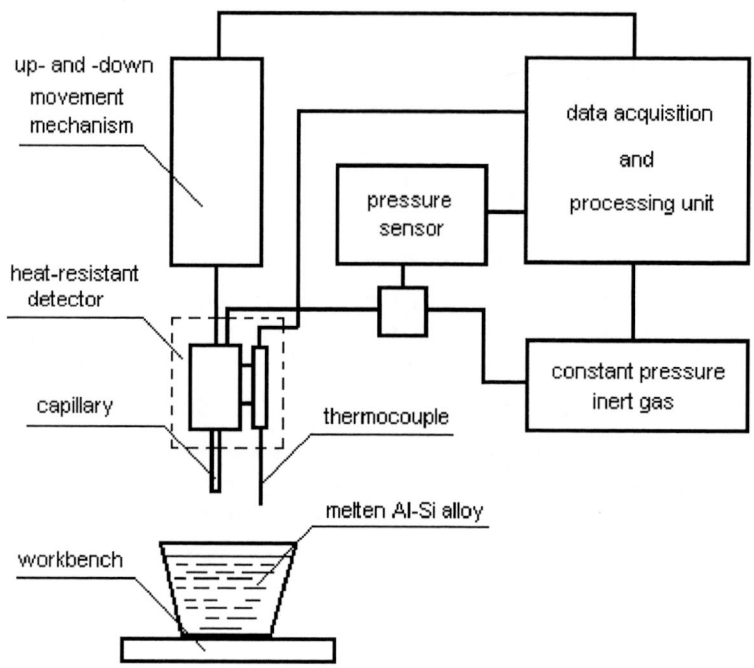

Fig.1 the schematic diagram of automatic device
for testing melt surface tension

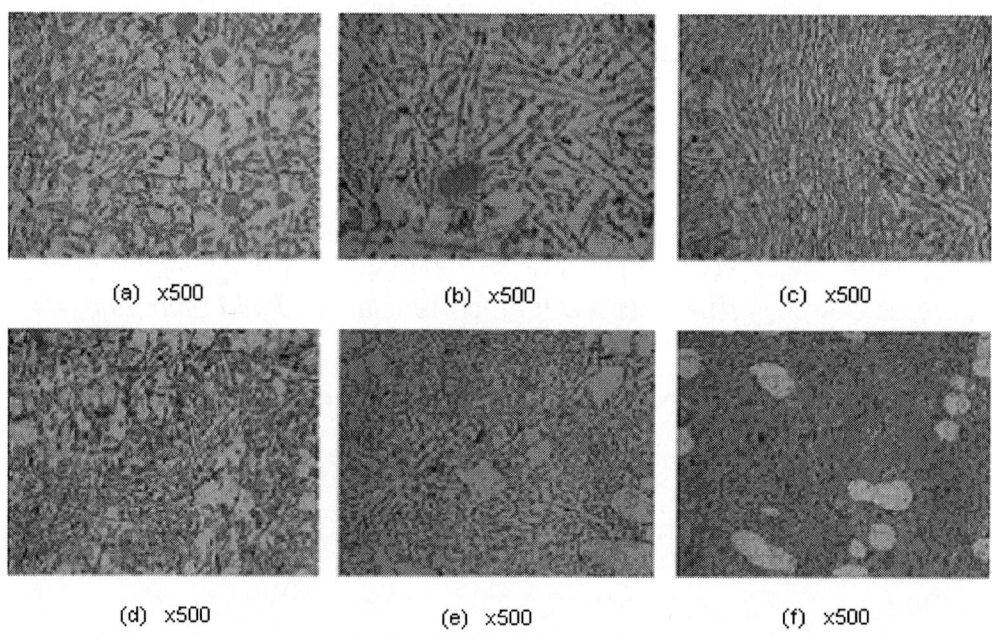

(a) x500   (b) x500   (c) x500

(d) x500   (e) x500   (f) x500

Fig.2 relationship between surface tension and microstructure

(a) modifier 0.0%, σ_e 0.846N/m, no inoculation   (b) modifier 0.5%, σ_e 0.740N/m, partial inoculation
(c) modifier 1.0%, σ_e 0.637N/m, moderate inoculation (d) modifier1.5%,σ_e0.516N/m, moderate inoculation
(e) modifier 1.75%,σ_e0.421N/m, moderate inoculation (f) modifier 2.0%, σ_e 0.354N/m, perfect inoculation

E-mail: dyli@hrbust.edu.cn

Advances in the Determination of Hydrogen Concentrations in Aluminium Alloys

Arndt Froescher
Foseco GmbH, Germany

Andy Moores
Foseco International, UK

Abstract

A new device for determining hydrogen concentrations in aluminium alloys has recently been developed. This new device is based on a calcium zirconate electrochemical sensor and offers the foundryman a quantitative device that is easy to use and gives accurate and reliable results in a relatively short space of time.

This paper will detail the experiences from foundry trials and compare the performance of the new device with other qualitative and quantitative devices commonly used today to assess hydrogen concentration. The potential advantages offered by the new device compared to conventional techniques are discussed.

Introduction

Dissolved hydrogen is a general phenomenon in liquid aluminium and aluminium alloys. The reason for this behaviour is the tendency of liquid aluminium to react with air moisture.

$$2\ Al + 3\ H_2O \rightarrow Al_2O_3 + 3\ H_2$$

As liquid aluminium has a much higher solubility of hydrogen than solid aluminium, the hydrogen releases from the melt during the cooling and solidification process and causes pores and shrink holes.

As the today's end users of castings (e.g. the automotive industry) demand high quality castings for their applications, it becomes more and more important for every foundry to improve their melt quality and to be able to determine the melt quality as accurate as possible.

In practice, most commercial aluminium foundries use some sort of degassing process to reduce the hydrogen level of the melt to an acceptable   level.
Usually there is an optimum level suitable for particular castings and production processes. From there it is extremely important to check the melt quality permanently to maintain the high quality of the casting.

Many different methods of measuring hydrogen have been developed but until today none has been able to give true control over the hydrogen level before, while and after the degassing process. In fact, there are several reliable methods to detect the hydrogen content but either the response time is too long for the daily use in foundries or the handling of the units is too complicated.

The new device

The new device is comprised of three basic components; an electrochemical sensor that can measure hydrogen concentrations in the gaseous phase, a probe that carries this sensor into the molten metal and an analyser that processes the signal from the sensor and calculates the concentration of dissolved hydrogen in the melt.

The Sensor

The function of the sensor has been described in detail elsewhere (1). This paper will concentrate on the potential applications of this new technology in the foundry industry; however, a brief description of the main features of the sensor will be helpful.

The sensor is an electrochemical device based upon a calcium zirconate solid electrolyte. Under certain conditions calcium zirconate becomes a proton conductor allowing it's use as a sensor for hydrogen. To function as a sensor the calcium zirconate needs to encapsulate a reference material with a known partial pressure of hydrogen. When the outer surface of the sensor is then exposed to an unknown hydrogen partial pressure, a voltage is generated that when measured allows the unknown hydrogen partial pressure to be calculated.

A particular feature of this sensor that differentiates it from similar devices is that it includes a solid-state reference meaning that it does not require an external source of hydrogen to provide the reference. This makes the sensor a self contained device and thus an ideal basis for a practical piece of foundry equipment.

The Probe

The sensor cannot operate in direct contact with molten aluminium therefore a probe is required to protect and carry the sensor into the melt. The specially designed probe has a cavity in which the sensor is located and a porous window that allows the diffusion of dissolved hydrogen but not the ingress of aluminium. A schematic of the probe section containing the sensor is given in Figure 1.

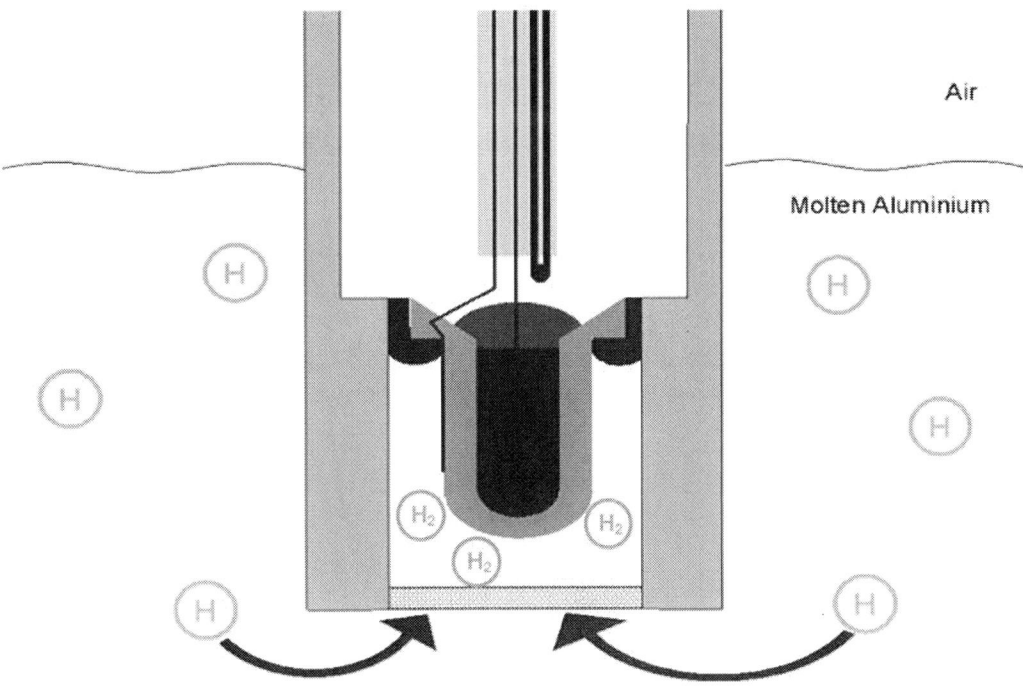

Figure 1 – Schematic of the Probe Section Containing the Sensor

The sensor thus sits in a gaseous environment where it measures the partial pressure of hydrogen in the probe cavity that is in equilibrium with dissolved hydrogen in the melt. The level of hydrogen measured in the cavity then needs to be related to the level of dissolved hydrogen in the melt, which is the value of real interest. This is done by the analyser using an equation based on Sieverts Law.

The Analyser

The third element required to enable the sensor to be used as a practical device is an analyser to convert the electrical output of the sensor to a measure of the hydrogen concentration of the melt. The analyser first processes the voltage output from the sensor and calculates the partial pressure of hydrogen in the cavity. Using data on the hydrogen solubility of the alloy being measured the analyser then calculates the level of dissolved hydrogen in the melt and displays this value in ml/100g. Both of the calculations the analyser performs are temperature dependent therefore a thermocouple is positioned adjacent to the sensor to provide accurate temperature data.

In addition to calculating the dissolved hydrogen concentration, the analyser also has a built in data logger that allows both hydrogen and temperature readings to be recorded and subsequently downloaded onto a PC. These data logs can be plotted to produce real time curves of hydrogen concentration with time.

Figure 2 shows the Home Screen of the Alspek H analyser box.

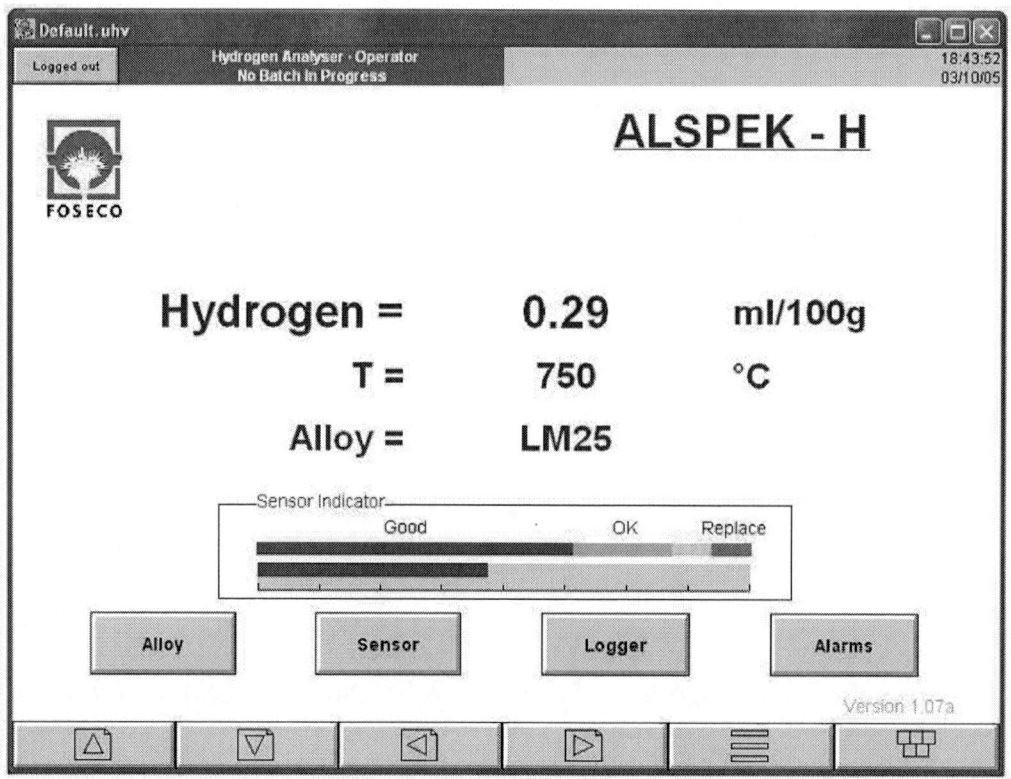

Figure 3 shows the complete ALSPEK H unit consisting of the probe, housing the sensor, and the analyser.

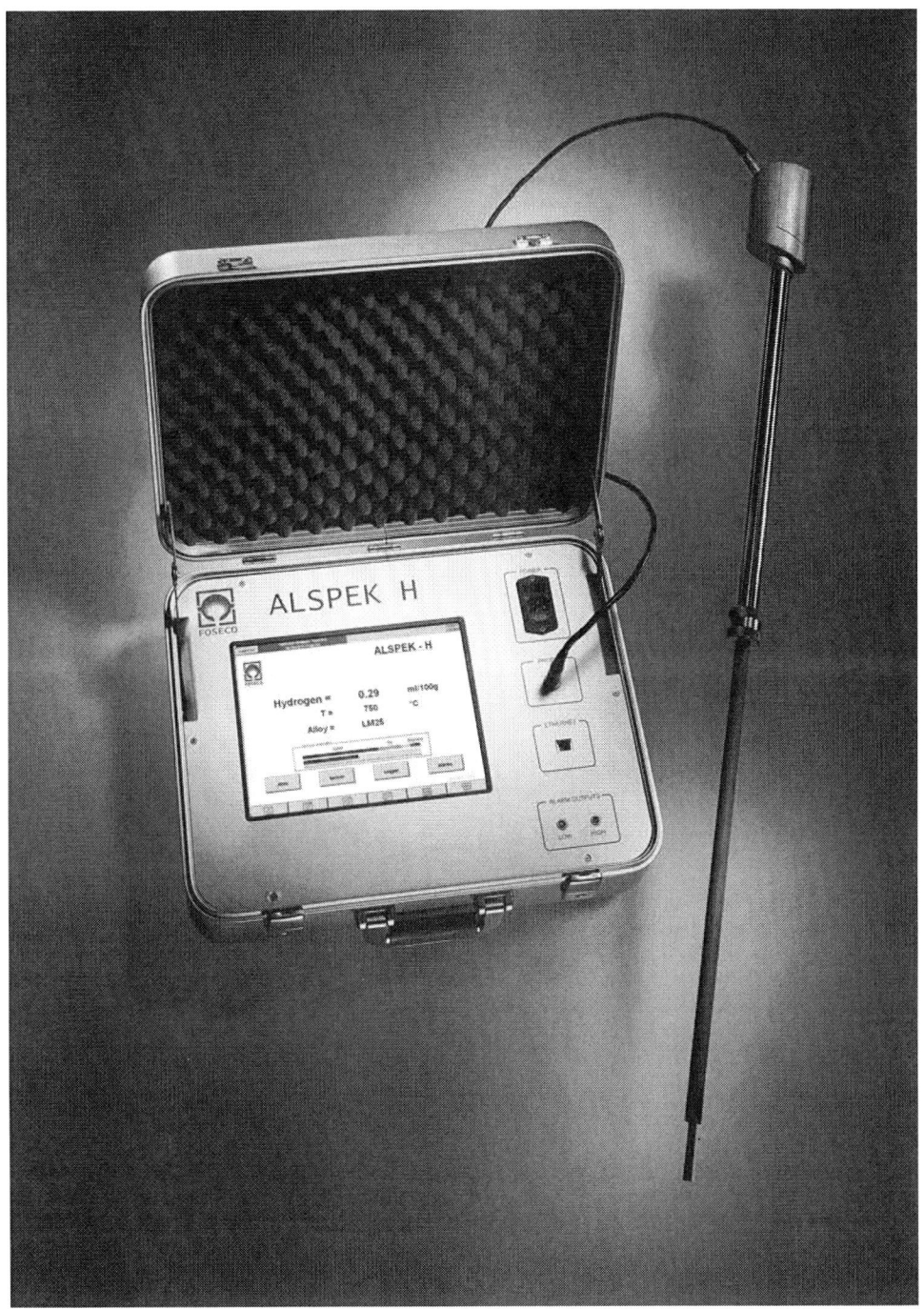

The Practice

To be suitable for foundry applications, a device for measuring the hydrogen content in liquid aluminium has to meet a number of particular requirements:

- short response time
- reliable values
- reproducible results
- longevity in foundry environment
- simple handling

The ALSPEK H hydrogen analyser fulfils all these requirements and makes foundrymen able to control the hydrogen content before, while and after the degassing process.

Probe Accuracy

The quality of each measurement process is determined by it's accuracy. To detect the accuracy of ALSPEK H, a particular melt was brought into equilibrium with a known hydrogen partial pressure by gassing up the melt with Formiergas (30 % H2, 70 % N2). The hydrogen content was calculated from known alloy solubility data.

Figure 4 shows a typical sensor response.

Measurement accuracy: Typical response

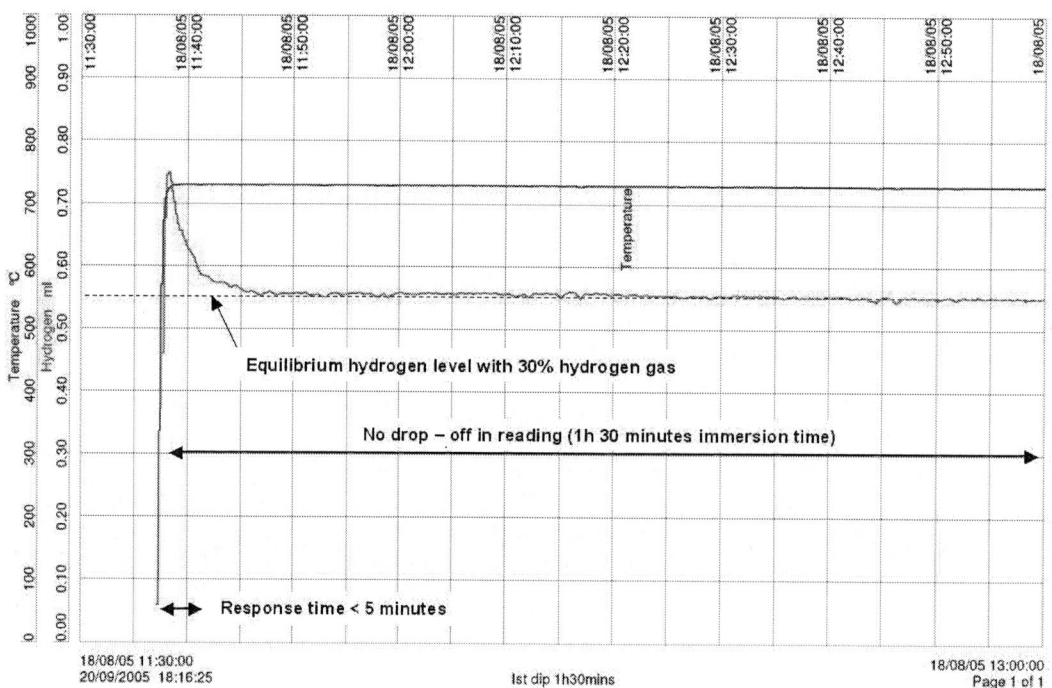

Measurement Reproducibility

In foundries, it is very important to get a reproducible result, e.g. measuring, taking out the probe and re-immerse it. Figure 5 shows a degassing process with removing and re-immersion afterwards.

Degassing and accuracy at low hydrogen levels

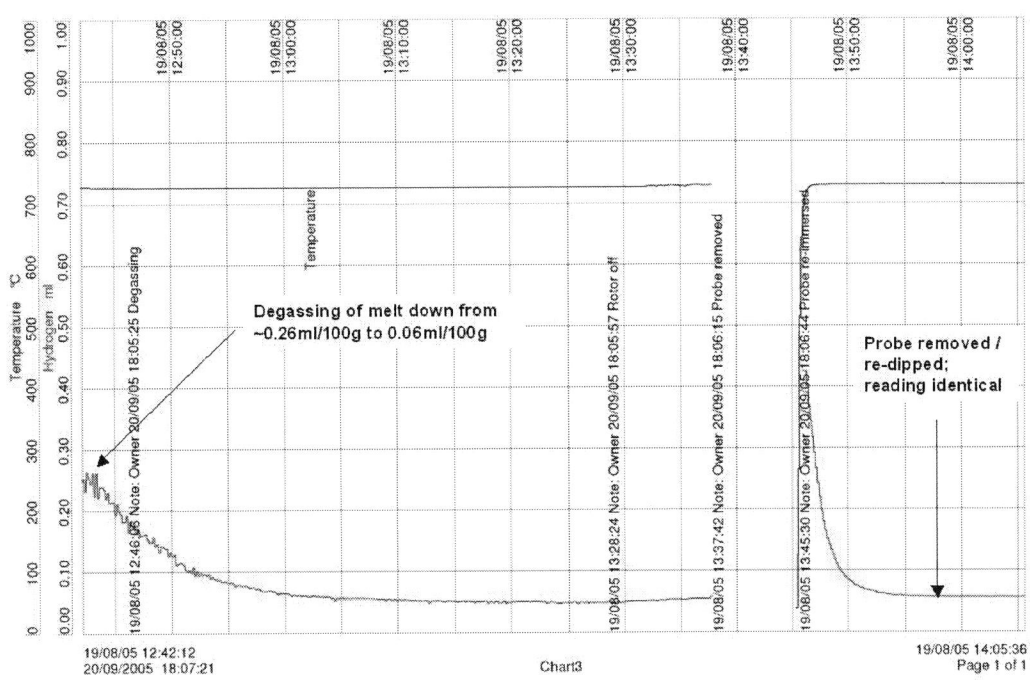

Comparison with other methods

Many different methods of measuring hydrogen have been developed until today. Every new unit, like Alspek H, has to stand the comparison with these references.

The best-known devices to measure hydrogen in aluminium alloys directly are AlScan, Chapel and TYK Notorp. Figure 6 shows a table with the comparison values of all devices:

	Alspek H	Chapel	AlScan	TYK Notorp
Response Time	< 5 min	10 - 15 min	10 - 15 min	10 min
Accuracy	0,01 ml/100g	0,015 ml/100g	0,01 ml/100g	0,01 ml/100g
Lifetime	100x/20h	10-15x	10x/3h	3-5x
Weight	10,2 kg	18,7 kg	17,2 kg	9,8 kg + Gas Bottle
Spot Measurement	yes	yes	yes	yes
Degassing	yes	no	no	no

The table shows, that devices like Chapel, AlScan and TYK are often not suitable for the daily use in foundries. The response time of any hydrogen analyser should be as short as possible to ensure that things run smoothly in a foundry. With a response time between 10 and 15 minutes, Chapel, AlScan and TYK are therefore not suitable.

With Alspek H, foundrymen are able for the very first time to determine the hydrogen content of the melt at every moment of the degassing process and in succession, to adjust a certain hydrogen level that is required for a particular casting.
This is an extremely valuable advantage of Alspek H, as none of the other devices can stay in the melt during the degassing process. The Alspek H hydrogen sensor can only be influenced by hydrogen and not by other gasses like nitrogen or argon, which are mostly used for degassing.

The longevity of Alspek H´s consumables is between 10 and 20 times longer than the competitive units. Again, this is an advantage for Alspek H, as it is often not possible in a foundry to change the consumable more than one time a day to avoid interruptions in the working process.

In addition, Alspek H´s dimensions are small and handy. It has a low weight and can be moved to every place in a foundry. As the probe is connected to the analyser by a cable that could have any length, it is possible to measure at almost every place, while the analyser box can stand in a save distance. In opposite to that, Chapel and AlScan have a limited range.

The article shows clearly, that Alspek H is the most suitable device for detecting hydrogen in liquid aluminium under foundry conditions at the moment.
It has a short response time, a high longevity and a good accuracy. In addition with all the other opportunities the device provides, Alspek H is a universal unit for all foundry applications.

References

1. C. Schwandt et al

The Effects of HIP on Bifilms in Aluminum Castings

J T Staley *, M Tiryakioğlu **, J Campbell ***

* Bodycote Materials Testing, Chicago Laboratory, USA, ** Robert Morris University, USA, *** The University of Birmingham, UK

Abstract

Many aluminum castings contain bifilms, which are cracks that lower the mechanical properties and reliability of castings. The current commercial hot isostatic pressing (HIP) process has proven successful in closing porosity but its success in healing bifilms is questionable. This research investigates the effects of different HIP conditions on the reliability of aluminum castings affected by bifilms. Non-HIPed, typical HIPed, and experimentally HIPed A206 castings with bifilms were mechanically tested and microstructurally examined. As a result of HIP, minimum tensile strength increased by nearly a factor of 3, average elongation improved by up to 8 times, and elongations increased from a maximum of 5 to over 18 percent. However, the elongation reliability essentially did not change as a result of HIP, which suggests that healing of bifilms did not occur.

Keywords Aluminum, Castings, Bifilms, HIP, Tensile

Introduction

The automotive and aerospace industries desire to replace wrought components with aluminum castings to reduce manufacturing and operating costs. Wrought components are well established for use in these industries. They are produced by various processes beyond casting, involving much time and expense. Part consolidation, potential weight savings and reduced tooling costs are the main drivers for use of aluminum castings in the automotive and aerospace industries. Aluminum castings may be able to replace many wrought components without all the additional post cast processing time and expense because of their near net shape. However, there can be problems associated with using aluminum castings.

A thin oxide film forms readily on the surface of molten aluminum in the presence of minimal amounts of oxygen. When the flow velocity exceeds 0.5 m/s in aluminum alloys, these films can fold over on themselves forming bifilms, which get entrained in the casting (Figure 1) [1]. These bifilms are essentially cracks in the structure and are detrimental to mechanical properties; they not only lower the average tensile strength, ductility, fracture toughness and fatigue resistance, but also increase the scatter in those properties [1-4]. Due to the low reliability caused by these oxides, the aerospace and automotive industries cannot be confident enough that a casting will not lead to a catastrophic failure in critical applications. In fact, one of the main reasons castings are not used in critical applications is that their reliability is low due to inherent casting defects such as bifilms and porosity caused by bifilms [5].

Porosity is a common defect in aluminum castings and usually can be formed only if bifilms are present [6,7]. Bifilms are pushed by solidification fronts or dendrites, and may lead to porosity because of solidification shrinkage [1,4,7]. In addition, dissolved hydrogen in the molten metal is released during solidification as hydrogen gas, which causes the bifilms to expand and form porosity. It is well known that hot isostatic pressing (HIP), which uses high temperature and inert gas pressure over a period of time, can close porosity and improve the mechanical properties of castings [8]. To make HIP more economical for aluminum castings, HIP processes with shorter cycle times and lower processing costs were developed [9]. When an aluminum casting is subjected to high isostatic pressure at temperatures near its eutectic temperature (lowest melting temperature) there is enough plastic deformation to close porosity. The typical parameters for HIP of aluminum castings are 2 to 6 hours at 510 to 521°C with an applied pressure of 103 MPa [10]. This high hydrostatic pressure causes non surface connected porosity to be "forged" shut through solid plastic flow of metal. Diffusion across the closed pore is probably necessary to obtain optimal properties. However, bifilms that are attached to the pore surface make it difficult for porosity to be completely healed.

HIP has not been shown to make aluminum castings as reliable as wrought parts mainly due to HIP's inability so far to completely mitigate the effects of bifilms [5,9]. This lack of reliability is a property quantifiable by Weibull statistics. This approach is widely used to model the variability (or reliability) in fracture properties of castings. The probability, P, that a casting will fracture at a given stress or strain, x, or below can be predicted as [11]

$$P = 1 - \exp[-(x/x_0)^m] \qquad (1)$$

Where x_0 is the scale parameter and m is the Weibull modulus or shape parameter. The higher the value of m, the less variable or more reliable are the data. Tiryakioğlu et al. [5] showed that the Weibull modulus for tensile strength of aluminum castings can be increased from a value of around 10 (typical for turbulently filled castings) to a value of 50 (similar to aerospace forgings), by essentially minimizing bifilms by using different gating and filtering techniques. However, Mocarski et al. [12] showed that HIP of A201 aerospace castings did not reduce variability in tensile strength and actually increased variability in elongation.

Thus, aluminum castings with bifilms remain less reliable than wrought parts for critical automotive and aerospace applications. Hence, there is an immediate need in the aerospace, automotive, HIP and aluminum casting industries to mitigate the effects of bifilms in aluminum castings possibly by healing bifilms with HIP [13,14]. This research investigates the effects of different HIP conditions on the reliability of aluminum castings affected by bifilms.

Yeh et al [15] HIPed A206 castings at (460°C, 103 MPa, 2hrs). Here, a drilled hole (sealed from exterior pressure) was not closed by HIP and cracking occurred around the hole. Clearly, the temperature was not high enough to permit sufficient plastic flow of metal to close the porosity. Hence, in this study, temperatures lower than typical (510 to 521°C) were not investigated. We hypothesized that higher temperatures than typical might increase diffusion and flow of material around and possibly even inside bifilms, thus reducing their crack like characteristic.

Experimental
A206 material was chosen as the model alloy to be researched. However, this research approach is expected to be equally valid for the hypo-eutectic alloys such as 319, 354, 356 and 357. Ingots of A206 were procured for testing since it was expected that ingots would exhibit many bifilms and much porosity.

Chemical composition was conducted on one ingot piece using the inductively coupled plasma (ICP) technique in accordance with ASTM E1479. Chemical composition was compared to requirements of Aerospace Material Specification (AMS) 4235A "Aluminum Alloy Castings,

4.6Cu – 0.35Mn – 0.25Mg – 0.22Ti (A206.0-T71) Solution and Precipitation Heat Treated".

One ingot was left non-HIPed, one ingot was HIPed at typical aluminum HIP parameters (516°C, 103 MPa, 4hrs), and the last ingot was HIPed at the material's eutectic melting temperature.

Because of the significant amount of surface connected porosity visible in the ingots, special preparations were made to ensure successful HIP. Ingots were wrapped in thin gauge stainless steel foil and placed into thin gauge welded, stainless steel sheet metal "cans" (one ingot per can). Stainless steel powder was then poured into the cans to take up the remaining space. Cans were welded shut and evacuated. In this way, porosity connected to the ingot surface could be healed as the thin can was crushed under pressure and transmitted pressure through the powder to the skin of foil around the ingot allowing isostatic pressure to close the porosity. All ingots were heat treated to the T71 condition after HIP in accordance with AMS 4235A.

Tensile tests were conducted at room temperature per ASTM E8 on non-HIPed and HIPed castings. Standard specimens of 12.8 mm diameter and 50.8 mm gage length were used. Specimens were strained at a rate of approximately 0.13 mm/minute with an extensometer clipped to the gage length until failure. Percent elongation was determined from the load versus strain curve.

Results and Discussion
Chemical composition of the material meets requirements of AMS 4235A (Table I). Figure 2 compares polished cross-sections of ingots in various HIPed conditions. The non-HIPed ingot contains many bifilms and much porosity. This was confirmed by scanning electron microscopy and energy dispersive spectroscopy of tensile tested fracture surfaces. For the most part, the bifilms are associated with the porosity and are located at dendritic solidification fronts in eutectic areas or the last areas to solidify. The crack-like nature of the bifilms is easily seen with a variety of space existing between the two thin oxide layers that make up the bifilm. The HIPed samples have essentially no measureable porosity suggesting that all porosity was closed during the HIP process in agreement with the observation that there was less visible space between the bifilms in HIPed samples. By closing the pores, some benefit is expected in tensile strength and elongation due to reducing the size of these defects.

Table II and Figure 3 show results of the tensile tests conducted to date. Tests include 35 non HIPed, 20 typical HIPed and 21 eutectic HIPed specimens. For any tensile specimen, the tensile strength and elongation are controlled by the largest effective defect. It is apparent that the average tensile properties increased due to HIP. Average elongation improved by as much as a factor of 8 and elongations of over 18 percent were achieved. Minimum tensile strength increased by nearly a factor of 3. Hence, on the average, HIP reduced the largest effective defect. While the average tensile properties for the non-HIPed samples were well below the AMS 4235A minimum specification requirements, both HIP conditions rendered average tensile properties that met this specification.

Weibull statistics were conducted on the data to determine if the reliability of the castings increased due to HIP. The following probability estimator was used:

$$P = (i – 0.5)/n \qquad (2)$$

Where i is the rank of each data point and n is the number of samples. Figure 4 shows Weibull plots for tensile strength and elongation for castings in various HIPed conditions. The tensile strength reliability modulus m increased slightly from 6 to 16 as a result of HIP mainly due to the two non-HIPed specimens that failed before yielding. Still, these numbers do not approach the reliability of wrought material (m near 50). The elongation reliability essentially did not change as a result of HIP (m = 1 to 2). The Weibull modulus for tensile strength and elongation are similar for typical and eutectic HIP, which suggests that negligible additional healing occured due to increased HIP temperature. Mi et al. [3] calculated Weibull moduli ranging from 13 to 36 for tensile strength and 2 to 8 for elongation for non-HIPed 200 series aluminum casting with various levels of bifilms. The castings with the least number of bifilms gave the highest reliability. HIP of material in this study (with a significant content of bifilms) was unable to render castings as reliable as non HIPed material with significantly fewer bifilms. Hence, these results emphasize the importance of starting with clean metal as stated by rule number 1 of Campbell's 10 rules for quality castings [1].

Future Work
Detailed failure analysis will be conducted on selected samples to see why they failed. In particular, effort will be focused to see what the difference is in HIPed samples that had high and low elongations.

Conclusions
1. Commercial ingots of A206 contained many bifilms and much porosity suggesting that there was significant potential for improvement to be made in the ingot production process.
2. Microstructural examination showed that typical and eutectic HIP parameters closed all porosity.
3. As a result of HIP, minimum tensile strength increased by nearly a factor of 3, average elongation improved by up to 8 times, and elongations increased from a maximum of 5 to over 18 percent.
4. The elongation reliability essentially did not change as a result of HIP, which suggests that healing of bifilms did not occur.

Acknowledgements

The authors acknowledge the helpful comments of Dr. Jim Staley (retired), tensile testing by Bobby Archibald of Bodycote and HIP by Steve Mashl of Bodycote.

References

1. Campbell J, *Castings*, 2nd Ed, Butterworth-Heinemann Ltd, Oxford 2003, ISBN 0 7506 4790 6.
2. Campbell J, *Bifilms-The Most Exciting Discovery Of The Century*, The John Campbell Symposium, Edited by M. Tiryakioglu and P. Crepeau, TMS, 2005, pp 3-12.
3. Mi J, Harding R and Campbell J, *Effects of the Entrained Surface Film on the Reliability of Castings*, Met and Mat Trans A, Volume 35A, September 2004, pp 2893-2902.
4. Campbell J, *The Entrainment Defect: The New Metallurgy*, Advances in the Metallurgy of Aluminum Alloys, Proceedings from the JT Staley Symposium, Materials Solutions Conference, Edited by M. Tiryakioglu, ASM International, Indianapolis, IN, November 2001, pp 35-40.
5. Tiryakioğlu M, Campbell J and Green NR, *Review Of Reliable Processes For Aluminum Aerospace Castings*, Transactions of the American Foundrymen's Society, V96-158, 1996, pp 1069-1078.
6. Fox S and Campbell J, *Visualisation Of Oxide Film Defects During Solidification of Aluminium Alloys*, Scripta Materiala, V43, June 2000, pp 881-886.
7. Campbell J, *The Origin Of Porosity In Castings*, Proceedings of the Fourth Asian Foundry Congress, 1996, pp 33-50.
8. Atkinson H and Davies S, *Fundamental Aspects of Hot Isostatic Pressing: An Overview*, Met Trans A, Volume 31A, December 2000, pp 2981-3000.
9. Hebeisen J, Cox B and Rampulla B, *HIP Of Aluminum Castings*, Advanced Materials & Processes, April 2004, pp 38-40.
10. Hunt W, *HIP Process Makes Aluminum Castings Structurally More Viable*, Aluminum International Today, March 2002.
11. Weibull W, *A Statistical Distribution Function of Wide Applicability*, Journal of Applied Mechanics, V18, 1951, pp 293-297.
12. Mocarski S, Scarich G and Wu K, *Effect of Hot Isostatic Pressure on Cast Aluminum Airframe Components*, AFS Transactions, 1991, pp 77-81.
13. Campbell J, *Entrainment Defects*, Materials Science and Technology, V22 No2, February 2006, pp 127-145.
14. Nyahumwa C, Green NR and Campbell J, *Influence of Casting Technique and Hot Isostatic Pressing on the Fatigue of an Al-7Si-Mg Alloy*, Met and Mat Trans A, Volume 32A, February 2001, pp 349-358.
15. Yeh C, Chen T and Lin Y, *The Effect of HIP on A206 Aluminum Castings*, Transactions of the American Foundrymen's Society, V88-135, 1988, pp 719-724.

Contacts

jstaley@bodycoteusa.com
Tiryakioglu@rmu.edu

Tables

Table I   Results From Chemical Analysis of an Ingot

Elements	Ingot (Wt %)	AMS 4235A (Wt %)
Copper	4.62	4.2 to 5.0
Manganese	0.33	0.20 to 0.50
Magnesium	0.31	0.15 to 0.35
Titanium	0.20	0.15 to 0.30
Silicon	0.02	0.05 max.
Iron	0.07	0.10 max.
Nickel	<0.01	0.05 max.
Zinc	<0.01	0.10 max.
Tin	<0.01	0.05 max.
Others (each)	<0.05	0.05 max.
Others (total)	<0.15	0.15 max.
Aluminum	remainder	Remainder

Table II   Results From Tensile Testing

Tensile Properties		Tensile Strength MPa	0.2% Yield Strength MPa	%El
No HIP	Average	296	266	1.4
No HIP	Minimum	124	227	0.0
No HIP	Maximum	372	307	4.9
Typical HIP	Average	394	291	11.4
Typical HIP	Minimum	355	244	1.8
Typical HIP	Maximum	441	334	18.2
Eutectic HIP	Average	376	292	9.7
Eutectic HIP	Minimum	320	269	2.3
Eutectic HIP	Maximum	417	320	17.0
AMS 4235A	Minimum	345	275	3

Figures

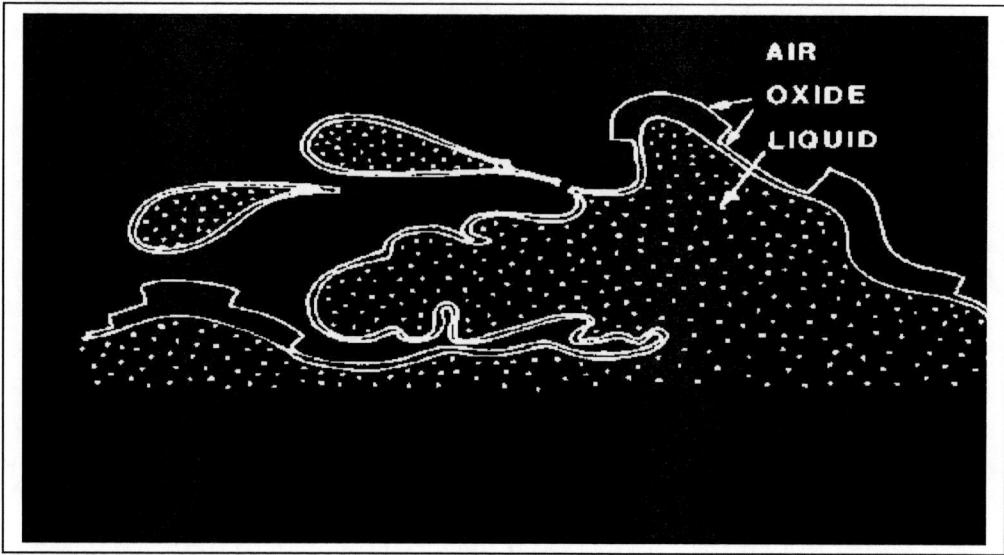

Figure 1: During the mold filling process, oxide films fold over on themselves during turbulent flow forming bifilms [1].

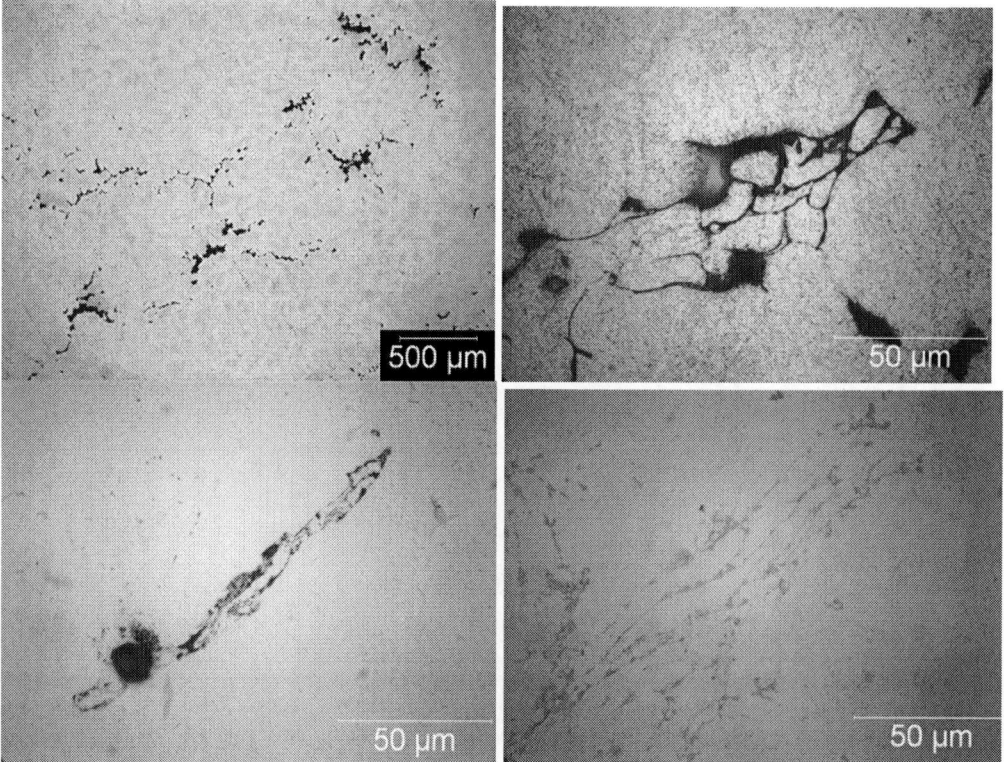

Figure 2: Microstructure of ingots in the polished condition. Upper left and right; as cast, lower left; typical HIP, lower right; eutectic HIP. Note the tangled bifilms and associated porosity in the as cast sample (upper right).

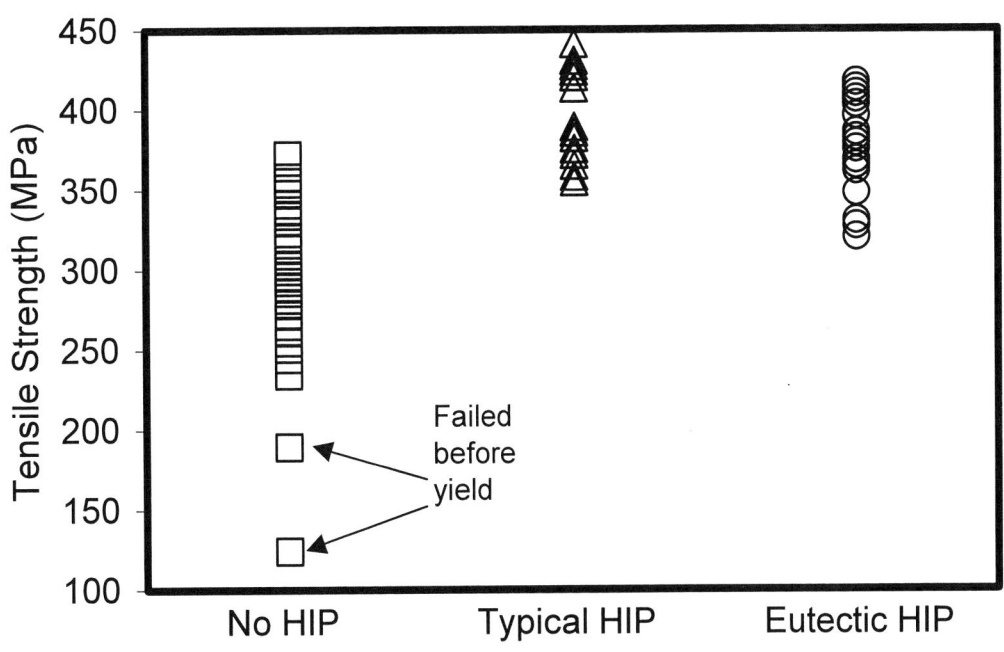

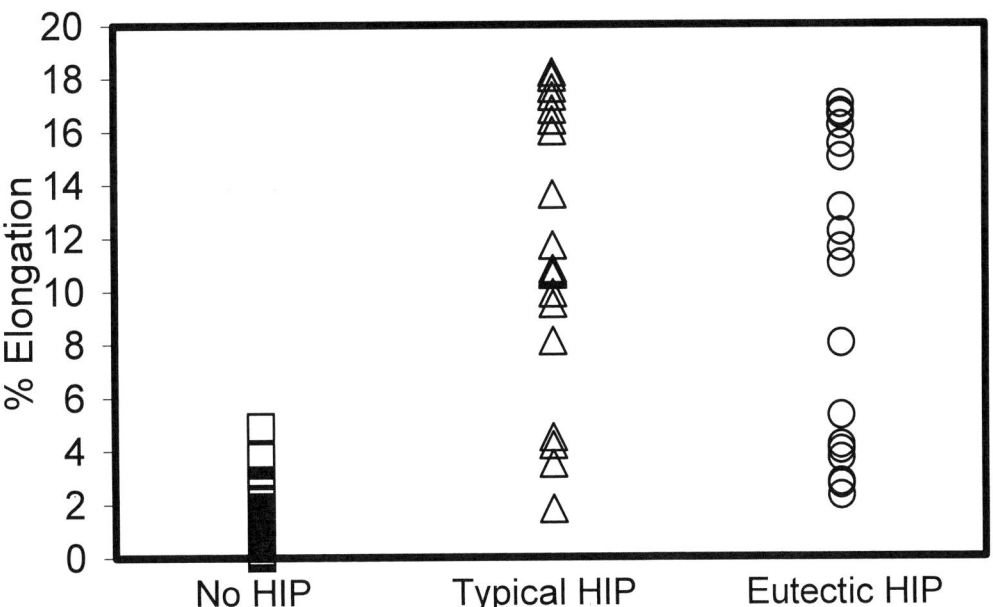

Figure 3: Results of tensile strength and elongation for the various HIP conditions.

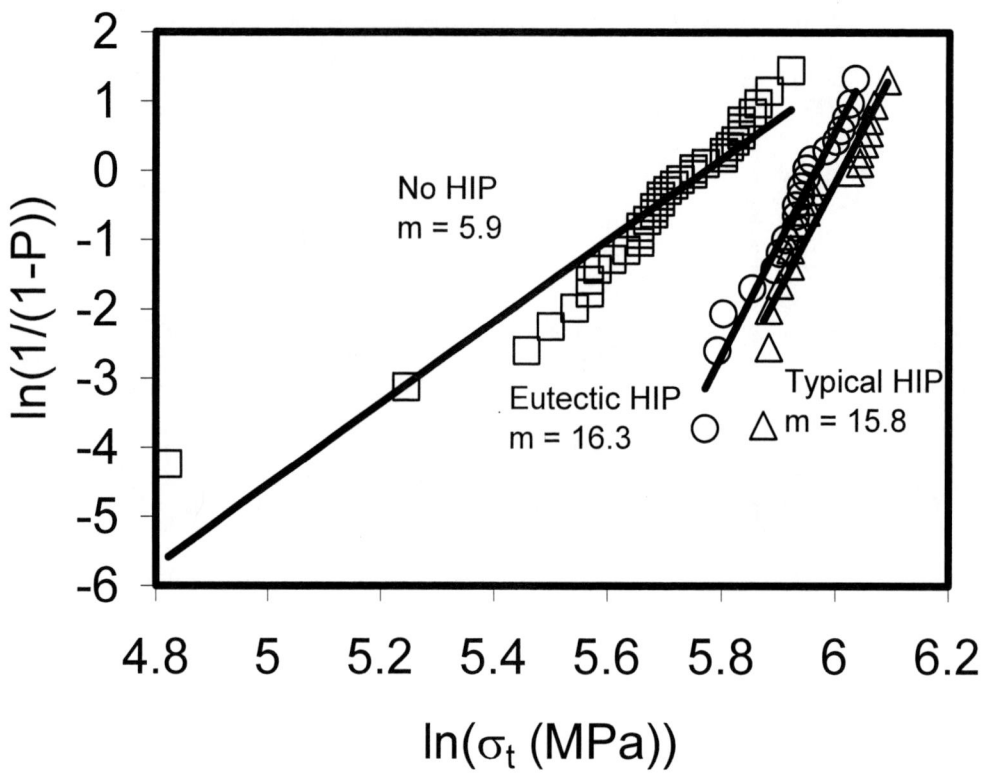

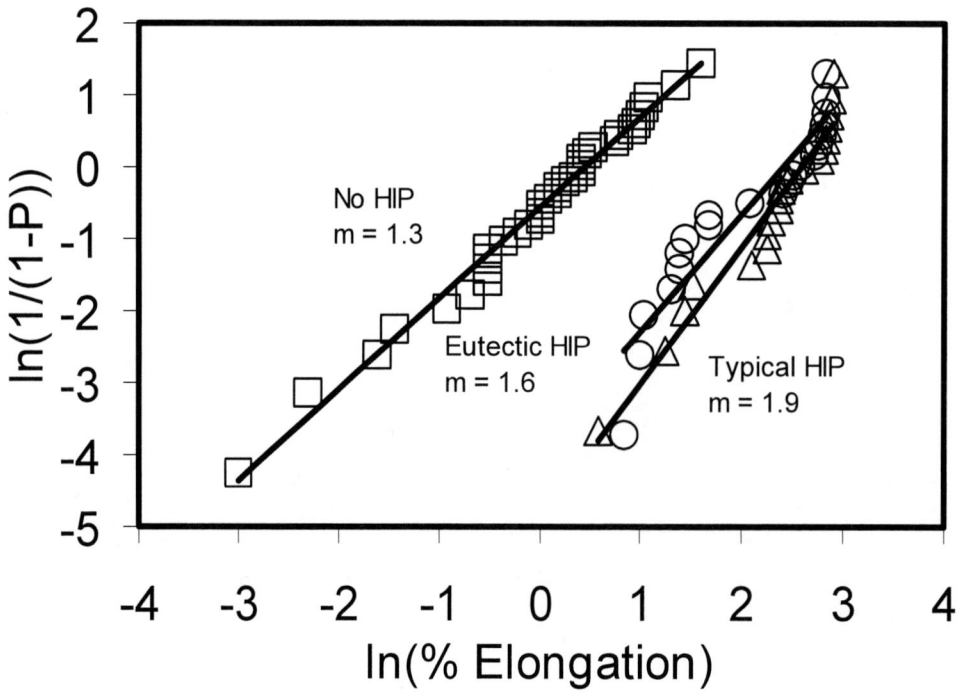

Figure 4: Weibull plots for tensile strength and elongation for castings in various HIPed conditions.

The effect of using Bubbling and AlCuP for refining primary silicon at Al-18%Si alloy

Sang-Man Kim, Jung-Pyung Choi[†], [*]Tae-Woon Nam, and Eui-Pak Yoon

Research Institute of Steel Processing & Application (RISPA) and Division of Materials Science and Engineering, Hanyang Univ., 17 Haengdang-dong, Seongdong-gu, Seoul, 133-791, Korea, [*]RISPA and Department of Metallurgical and Materials Science, Hanyang Univ., Ansan, 426-791, Korea

Abstract

This paper investigates the effect of Bubbling and AlCuP on primary Si size in Al-18%Si alloy. Bubbling process and AlCuP treatment are effective for refining primary Si in hypereutectic Al-Si alloy. Both the bubble and AlCuP can be used as nucleation sites of primary Si. Many nucleation sites can make the size of primary Si decreased. So, this experiment suggests that the mix using of bubbling process and AlCuP treatment are more effective for nucleation site of primary Si. For deciding the optimum treatment order, the experiments are processed 3 kinds of order with varying stirring time and holding temperature; 1)AlCuP treatment and then bubbling process, 2)simultaneously using of bubbling process and AlCuP treatment, 3)bubbling process and then AlCuP treatment.

Key words

aluminum alloy, bubbling, AlCuP, refining, primary silicon, hypereutectic Al-Si

Introduction

In the hypereutectic Al-Si alloy, the primary Si phases have great importance. Because the mechanical properties of alloy are directly related to the size and the shape of primary Si[3]. So, it is essential to control microstructure in order to improve the characteristics of hypereutectic Al-Si alloy.

There exist many refinement methods of primary Si. In the case of primary Si refining, bubbling can be effective for nucleation site[1, 3]. In the case of AlCuP treatment, the AlCuP exist as AlP alloy in Al melt. The AlP has the same lattice parameter of Si[2]. So, when AlCuP is added in Al melt, it helps nucleation of primary Si by forming AlP. It increases nucleation time and reduces growth velocity by dropping the temperature of primary Si crystallization. In this study, the mix using of Bubbling process and AlCuP treatment for refining process in hypereutectic Al-Si alloy is investigated.

Experimental

The base metal was the alloy of Al-18%Si. The alloy was melted in SiC crucible by an electric-resistance furnace. The melt was held at 800 for 30 minutes in the furnace and then cooled at 700 , 730 and 780 , in the furnace. The bubbles made through a stainless tube with a porous plug(2 μm), and for the bubble, Ar gas was used, and it flowed into the melt at the rate of 8ℓ/min. The bubbling process time was equally all 10 minutes in this experiment. The three types of experiment are taken. First, AlCuP treatment and then bubbling process. Second, simultaneously using of bubbling process and AlCuP treatment. Third, bubbling process and then AlCuP treatment. When the AlCuP treatment are taken, the amount of addition AlCuP was 50ppm and stirring time were 5minutes, 10minutes, 15minutes, and 20minutes. And then it cooled at 700 , 730 and 780 in the cast-iron mold. The specimen was inspected with an optical microscope, and the size of primary Si phases measured with an image analysis device.

Results and Discussions

For deciding the optimum treatment order, the experiments are processed 3 kinds of order with varying stirring time and holding temperature; 1)AlCuP treatment and then bubbling process, 2)simultaneously using of bubbling process and AlCuP treatment, 3)bubbling process and then AlCuP treatment.

In all experiments, the best pouring temperature was 730 . So, all figures represents in the case of 730 pouring temperature. In Fig.1, The structures of primary silicon are represented at as-cast, AlCuP treated, and bubbling processed. The AlCuP treated primary silicon has minimum size and then bubbling process one. This result comes from that AlCuP and bubble are acting as nucleation site. In Fig. 2, the microstructure that AlCuP treatment and then bubbling processed (poured at 730) are represented. The silicon size is smallest at 15min stirred and it is smaller than only AlCuP treated or bubble processed silicon.

In Fig. 3, the microstructure that simultaneously using of bubbling process and AlCuP treatment (poured at 730) are represented. The silicon size is smallest at 15min stirred and it is smaller than Fig. 2 (15min) structure.

In Fig. 4, the microstructure that bubbling process and then AlCuP treatment (poured at 730) are represented. The silicon size is smallest at 15min stirred and it is smaller than Fig. 3 (15min) structure.

In Fig. 4, At the melt temperature of 700 , 730 and 780 , the Bubbling was performed for 10minutes, and the adding of AlCuP was 50ppm. The results were shown in Fig. 5. These figures indicate that as the stirring time becomes longer, the size of the primary Si are smaller. And the smallest sizes of primary Si are in 15minutes. As the size of primary Si becomes smaller, the refinement is supposed to not only Bubbling, but also adding AlCuP. This process can be effective. In Fig. 6, It show that Bubble and AlCuP are existent in primary Si. In other words, SEM Micrograph indicate that Bubble and AlCuP are subsistent in primary Si as nucleation sites. This process can be effective.

Conclusions

From an experiment about the Bubbling process and adding of AlCuP in purpose of observing the effect on primary Si in hypereutectic Al-Si alloy, the following results were obtained.

- When the Bubbling process and addition of AlCuP has been done above the liquids temperature, as the stirring time becomes longer, the size of the primary Si are smaller. But over 15 minutes, refinement was showing a slight decline.

- This process temperature becomes high, the size of primary Si becomes smaller. And its optimum temperature is before and after 730 .

References

1.   DUCTILE IRON : The AGNE publisher, 76
2.   Mark. J. D, Phillip. D. J : American Mineralogist vol. 83 (1998), 1008
3.   J. P. Choi : Mat. Sci. Forum Vols. 449-452 (2004), 153

Figures

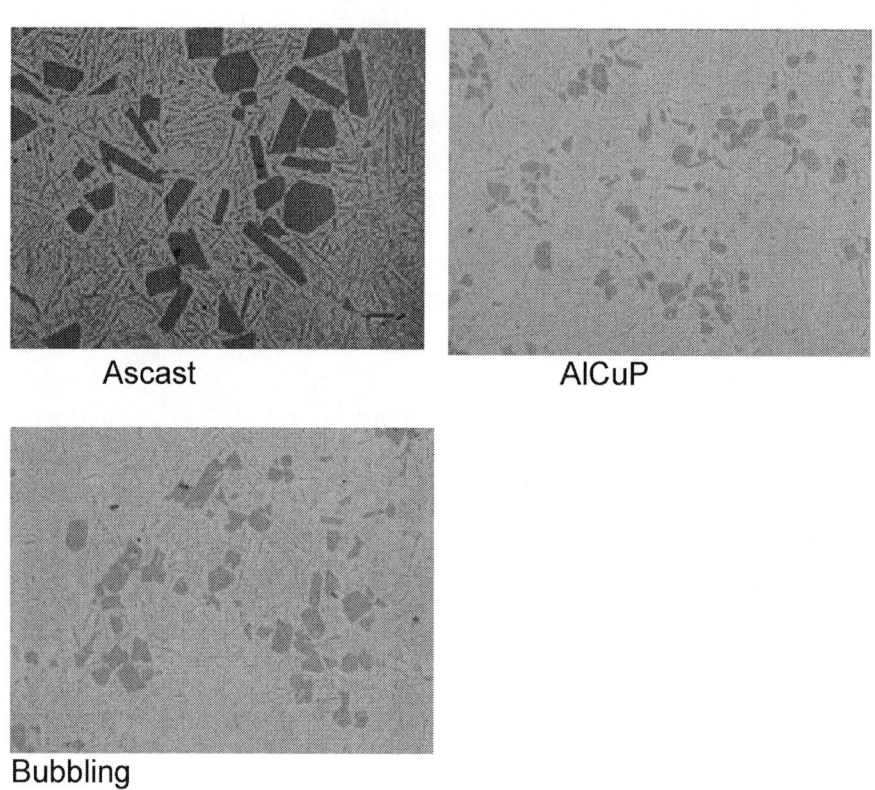

Ascast   AlCuP

Bubbling

Fig. 1 The basic structure of primary silicon

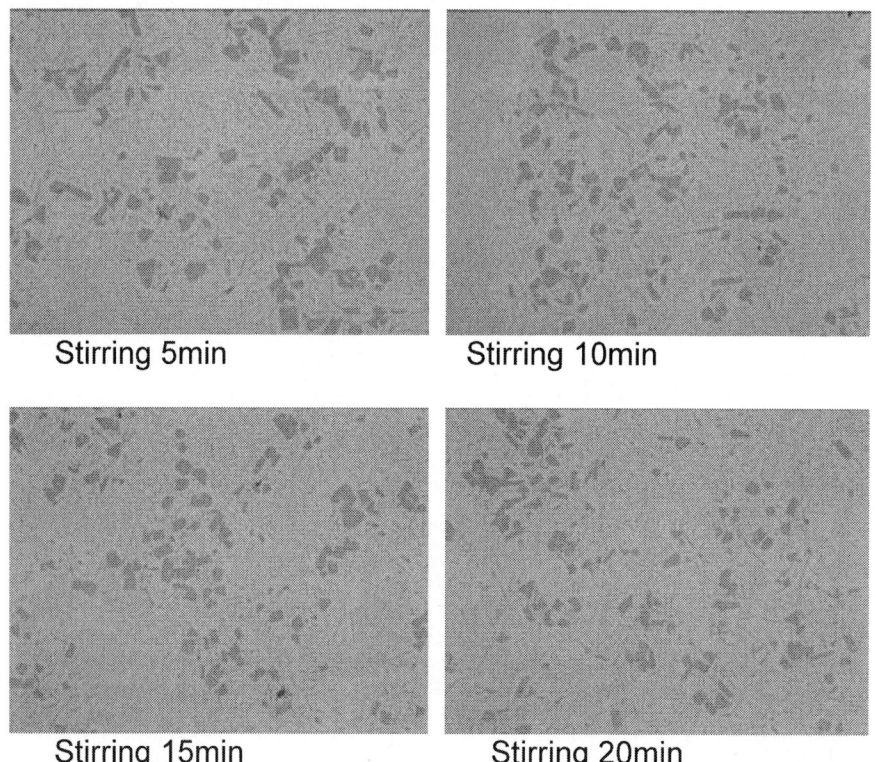

Stirring 5min   Stirring 10min

Stirring 15min   Stirring 20min

Fig. 2 The microstructure of type 1 processed (AlCuP and then bubbling) at 730 (Adding 50ppm AlCuP)

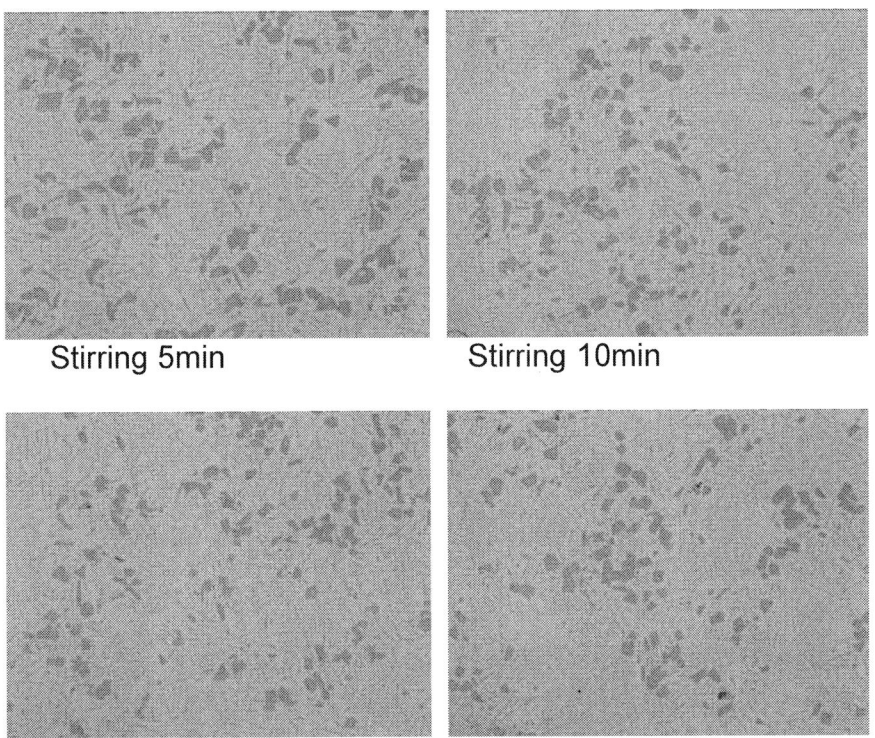

Stirring 5min   Stirring 10min

Stirring 15min   Stirring 20min

Fig. 3 The microstructure of type 2 processed (simultaneously AlCuP and bubbling) at 730 (Adding 50ppm AlCuP)

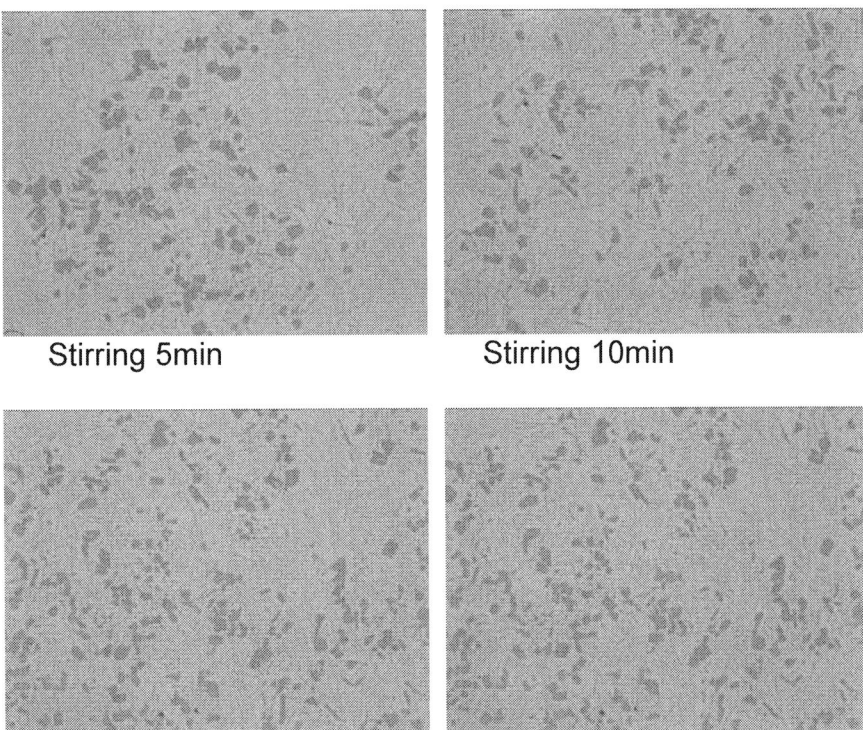

Stirring 5min   Stirring 10min

Stirring 15min   Stirring 20min

Fig. 4 The microstructure of type 3 processed (bubbling and then AlCuP) at 730 (Adding 50ppm AlCuP)

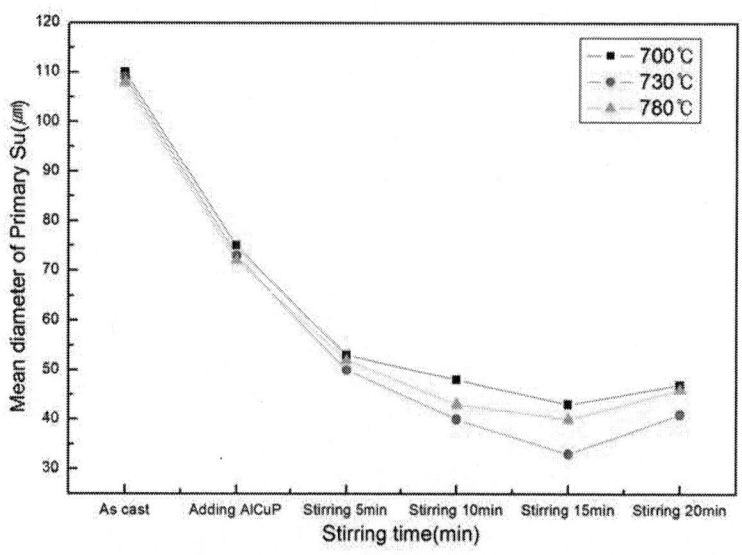

Fig. 5 The change of primary Si size with various stirring time and temperature

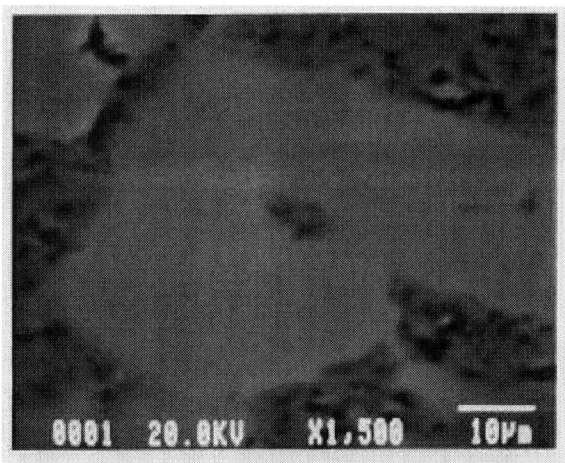

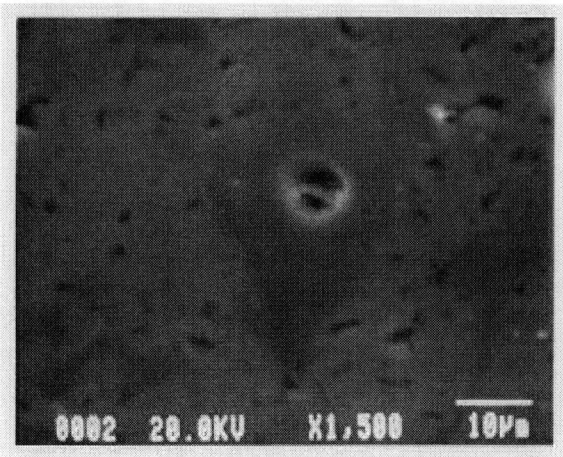

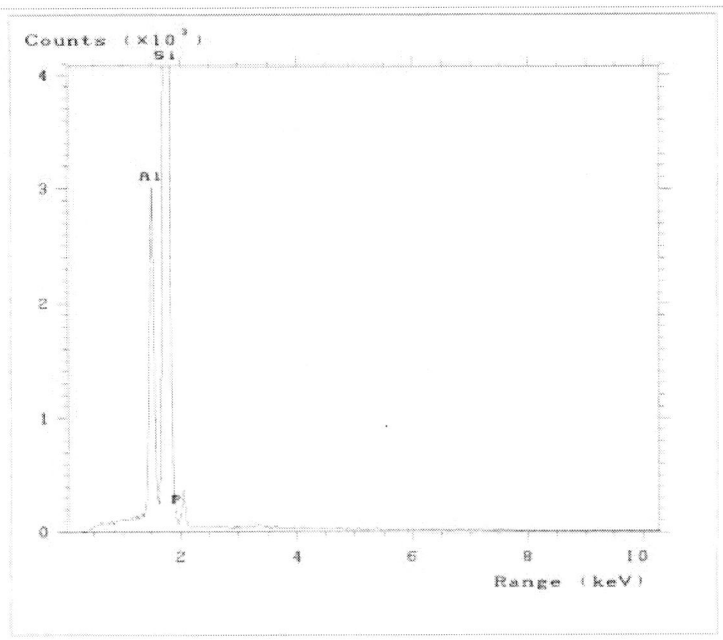

Fig. 6 SEM Micrograph of primary Si at 730 degC

A Study of the Structural Controlling of Al-Si Alloy by Using Electromagnetic Vibration

Jung-Pyung Choi[†], [*]Tae-Woon Nam, and Eui-Pak Yoon

Research Institute of Steel Processing & Application (RISPA) and Division of Materials Science and Engineering, Hanyang Univ., 17 Haengdang-dong, Seongdong-gu, Seoul, 133-791, Korea, [*]RISPA and Department of Metallurgical and Materials Science, Hanyang Univ., Ansan, 426-791, Korea

Abstract

The structural control of Al-Si alloy, which was not studied among various electromagnetic processing of materials, was considered applying the alternating current and direct current magnetic flux density. The main aim of the present study is to investigate the effects of electromagnetic vibration on the macro and microstructure of Al-Si alloy in order to develop a new process of structural control in Al-Si alloy. When the electromagnetic vibration is conducted for changing the shape of primary aluminum, at low frequency (<60Hz), the shape of dendrite is changed speroidal shape. When the electromagnetic vibration is conducted for changing the shape of eutectic silicon, the fact that a morphological change of the eutectic silicon from coarse platelet flakes to fine fiber shape is observed and the improvement of the mechanical properties is achieved with EMV (Electro Magnetic Vibration) process at high frequency(>500Hz).

Key words

aluminum alloy, electromagnetic, eutectic silicon, vibration, primary aluminum

Introduction

Al-Si alloys are of very common use in foundries because they offer good castability and resistance to cracking. Such properties depend on the alloy structures. Especially, such properties depend on the eutectic Si, which may have an acicular or lamellar form at hypoeutectic system, on the primary aluminum, and on the primary Si size at hypereutectic system [1, 2]. For controlling these structures, a number of theories have been proposed, but no theory has attained universal acceptance or really explained the principles of the phenomenon. In these old theories, some element (Na, Sr, Sb, P, Ti, etc) are added in the Al-Si melt. The adding element is different for controlling the structure.

For example, the element of TiB used for refining of primary aluminum, the element of Na, Sr, Sb used for modification of eutectic silicon, and the element of P, AlP, AlCuP used for refining of primary silicon [3, 4].

So, it is difficult to obtain both phenomenons, which is refining and modification. Because of each elements reacting to each other and preventing refining and modification. And this chemical method is difficult to adjust the amount of adding elements and difficult to recycle.

The vibration of the melt has been known to show effects that refine the solidification structure and reduce the internal gas, solidification shrinkage cavities and segregation of the solute [5]. This study considers the grain primary silicon morphology changing process and eutectic modification process with electromagnetic vibration. Because of the electromagnetic process has better merit than the existing other chemical method. The merits of this electromagnetic process are easy control, fast process, continuous process, non-pollutive process. And it has no chemical reaction process[7].

Experimental

Experiments were carried out on commercial casting aluminum alloy, A356. First, a sample was made as shown in Fig. 1 (b). Devices used in the experiment consisted of a D.C. magnetic field generator and an alternating current generator. The stationary D.C. magnetic field generator was fixed to 0.5T. The sample was put into a pre-heated electric furnace, which was located between cores of magnetic field generator as Fig. 1 (a). The sample was first heated to 973K, and then cooled naturally to 923K at which time electric field was induced. Following this, the sample was cooled (about 1.5K/sec) until it fully solidified with changing frequency and current density by 2.1×10^5, 4.2×10^5, 6.3×10^5, 8.4×10^5, 1.05×10^6 A/m^2 coming each 60, 300, 500, 700, 1000Hz at 0.5T.

The center section of the specimen was observed using an optical microscope. Primary aluminum and eutectic Si were observed and to determine the detailed shape of the eutectic Si, the specimen was etched for 10 minutes in NaOH solution of 1 mole and observed its shape by SEM. Then, an image analyzer program (IPP: Image Pro Plus) was used for image analysis and to measure the mean size and the degree of sphericity of the primary Al and the mean size of eutectic Si. A tensile test was carried out to evaluate the mechanical properties of the specimen at room temperature after being set to ASTM E8M regulations (Fig. 2) and samples were measured UTS and elongation with cross head speed by 2.0 mm/min and observed fracture surface with SEM after tensile test. For estimating the mechanism of this phenomenon, an XRD experiment was performed and twin probability was calculated.

Results and Discussions

In this study, the EMV was applied to 2 different structure controlling processes. They are 1) the controlling of primary aluminum phase, and 2) the controlling of eutectic silicon phase.

The controlling of primary aluminum phase

Fig. 3 represent the microstructure of the EMV processed alloy. The intensity of magnetic field , induced current density and cooling rate are each 0.5T, $1.05 \times 10^6 A/m^2$, 1.5K/s. Fig. 3 (a) is present the general microstructure of A356 as-cast. Fig. 3 (d) shows that entire dendrite arms are broken and become spheroidal type of primary aluminum particles. This means that this degree of force ($1.05 \times 10^6 A/m^2$, 0.5T) can entirely change dendrite structure at 60Hz.

Fig. 4 represent the trend curves of mean diameter and degree of sphericity varies with frequency and intensity of EMV in 1.5K/s cooling rate.

Fig. 5 shows the frequency and amplitude map at the liquid Al. If the frequency increases, the amplitude of liquid Al decreases. This result indicates that if experiment is processed in low frequency region, liquid Al lying in mixed or turbulent drag condition. This means that Al dendrite will be bended or fragmented at low frequency region.

The controlling of eutectic silicon phase

Fig. 6 (a) shows microstructures of an as-cast sample, which cooled without the imposition of an electromagnetic vibration. The eutectic Si phase has a traditional morphology growing in a faceted manner. Fig. 6 (b), (c), (d), (e) and (f) shows microstructures of the sample induced at the same current density of $1.05 \times 10^6 A/m^2$ according to a different frequency (60, 200, 500, 700 and 1000Hz). As shown in Fig. 6, there was little difference in the cases between the non-electromagnetic vibration processed sample and EMV processed at a frequency of 60Hz. However, a fine and globular morphology of the eutectic Si phase was observed in the case of a frequency of 1000Hz. It has been noted that the vibration phenomenon due to a low frequency does not affect the modification of the eutectic Si phase. However, according to the increase of frequency, the morphology of the eutectic Si phase was transformed from flakes to a fibrous shape.

Fig. 7 represent the mean diameter of eutectic Si varies with frequency and current density. The mean diameter of eutectic Si decrease with increasing frequency and also decrease with increasing current density from $6.3 \times 10^5 A/m^2$ to $1.05 \times 10^6 A/m^2$.

Fig. 8 represent the mechanical property varies with frequency. The UTS value of the specimen is increased from 126.5MPa to 186.9MPa with increasing frequency. This is the 47% increasing from its basic properties. Especially, the elongation of the specimen at 1000Hz is 8.5%. This shows the rate of significant increase of elongation with 750% from 1% at as-cast to 8.5% at 1000Hz. In this result, the size of eutectic Si effect significantly on mechanical properties of hypoeutectic Al-Si alloy.

It has been well documented that the transformation of unmodified flakes to modified fibrous structures arises from a very different type of growth of the Si phase and this difference appears to lie in the number of twins found in unmodified and modified Si. The electromagnetic vibration was supposed to prevent the Si atom from attaching to its crystallographic site and to cause a drastic increase in the twin probability of eutectic Si. In order to ascertain this assumption, it is necessary to evaluate accurately the twin probability in the eutectic Si. The conventional method for the accurate measurement of twin probability is the X-ray diffraction (XRD) method, which is more convenient to investigate using the bulk of eutectic Si. In order to obtain the precise peak from (220) to (400) reflection, the step scanning is used from $2\theta=45°$ to $72°$ with a time interval of 15sec and angular interval of $0.02°$. The 2θ of each of (220), (311) and (400) reflections, peak separation, α, $1/\alpha$ and the average twin spacing (λ) of the alloy were summarized in Table 1.

XRD Peak displacement caused by twinning in eutectic silicon is expressed as follows by defining α as the probability of finding a twin between any two (111) layers [6].

$$\Delta(2\theta)° = \frac{90\sqrt{3}\alpha \tan 2\theta}{\pi^2 h^2 (u+b)} \sum_b (\pm) L_0 \qquad (1)$$

where $L_0 = h+k+l = 3M \pm 1$, b and u are designated as broadened and not broadened : the component $L_0 = 3M \pm 1$ which are broadened by twinning, and the component $L_0 = 3M$ which are not broadened by twinning. From a measured peak displacement $\Delta(2\theta)°$ (in Fig. 9), we obtain directly the twinning probability α, since all the other quantities in Eq. (1) are readily evaluated. Because of twinning, the reflections of (220), (400) shift toward a larger 2θ.

The peak separation $\Delta 2\theta_{311-220}$ of the electromagnetically vibrated alloy appeared to be -0.06 and $-0.1°$ for 500 and 1000Hz, respectively, while that of strontium modified alloy was as large as $-0.2°$. Since another separation, $\Delta 2\theta_{311-220}$ did not give any consistent results owing to high angle reflection, it was ignored in Table 1.

Measured twin probability of EMVed alloy at a frequency of 1000Hz was approximately six times as high as that of the normal alloy and half of that of Sr modified alloy. Although the twin probability of the electromagnetically vibrated alloy was less than that of sodium or strontium, the electromagnetic vibration during solidification of the Si phase could be believed to increase twin density. Therefore, the mechanism for the increase in twin density due to the electromagnetic vibration may be preventing the Si atom from attaching to the growing interface. It can be supposed from this fact that the preferential growth along <112> in silicon (TPRE growth) was suppressed and twin density was increased by preventing the Si atom from attaching to the growing interface of the Si phase and by changing the solid/liquid interfacial energy of silicon due to the electromagnetic vibration during solidification.

Conclusions

From an experiment about the electromagnetic vibration for observing the effect of electromagnetic vibration on primary aluminum and eutectic Si in hypoeutectic Al-Si alloy, the following results were obtained.

1. At low frequency (<60Hz), the shape of dendrite is changed to speroidal shape. The degree of sphericity decreases with increasing current density and decreasing frequency.

2. If the frequency increases, the amplitude of liquid Al decreases. This result indicates that if experiment is processed in low frequency region, liquid Al lying in mixed or turbulent drag condition. This means that Al dendrite will be bended or fragmented at low frequency region.

3. A morphological transformation of the eutectic Si phase from coarse platelet flakes to fine fibrous shapes was observed when an electromagnetic vibration (EMV) with a different frequency and current density was applied to the solidifying hypoeutectic Al-Si-Mg alloy.

4. The mechanical properties of the electromagnetically vibrated alloy were increased as increasing frequencies due to the fine size and modification of the eutectic Si phase. The tensile strength and elongation of EMVed alloy at a frequency of 1000Hz were measured to 186.9MPa and 8.5%, respectively.

5. Measured twin probability of EMVed alloy at a frequency of 1000Hz was approximately six times as high as that of the normal alloy and half of that of Sr modified alloy.

References

1. L.F. Mondolfo : "Aluminum Alloy: Structure and Properties", Butterworth & Co., London, (1976)
2. J.B. Andrews, M.V.C. Seneviratne : AFS Transactions, Vol. 92, (1984), 209
3. J.E. Gruzleski and B.M. Closset : "The Treatment of Liquid Aluminum-Silicon Alloys", American Foundarymen's Society, Inc. Des Plaines, Illinois, U.S.A., (1990), 19
4. B.M. Closset and J.E. Gruzleski : Metall. Trans. A, Vol. 13A, (1982), 945
5. R. S. Richards and Rostoker : The Thirty-Seventh Annual Convection of the Society, held in Philadelphia, October (1955), 17
6. B.E.Warren and E.P.Warekois ; Acta Metall., Vol.3 (1955), 473
7. J. P. Choi, et al.: Mat. Sci. Forum Vols. 449-452 (2005), 157

Tables

Table 1 Measured diffraction angle (2θ), peak separation, twinning probability(α) and average twin spacing of alloys

	$2\theta_{220}$	$2\theta_{311}$	$2\theta_{400}$
Normal	47.34	56.2	69.36
500Hz	47.44	56.24	69.46
1000Hz	47.7	56.46	69.52
Sr	47.84	56.50	69.50

	Peak separation ($\Delta2\theta_{311-200}$)	α	1/α	Average twin spacing (nm)
Normal	0	-	-	-
500Hz	-0.06	0.0093	107.5	67.4
1000Hz	-0.1	0.015	66.7	41.8
Sr	-0.2	0.031	32.3	20.25

Figures

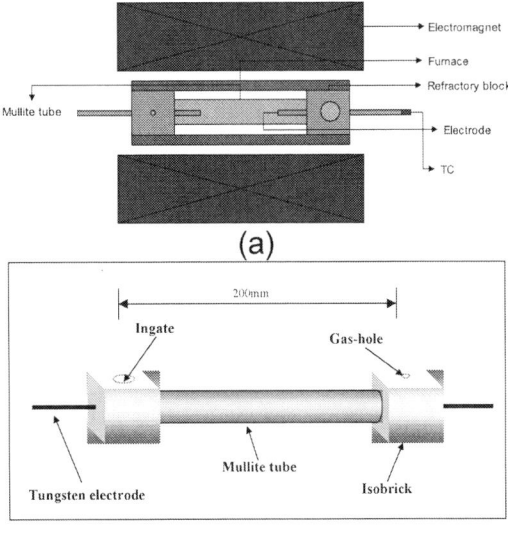

(a)

(b)

Fig. 1 The schematic sketch of the experimental device (a) inside of the yoke (b) mullite tube configuration

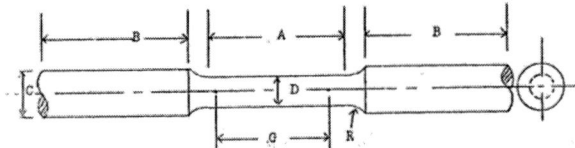

G : gage length   20.0±0.1mm
D : diameter   4.0±0.1mm
R : radius of fillet   4mm
A : length of reduced section   24mm

Fig. 2 The standard design of UTS and Elongation Test (ASTM E8M)
Insert figures here in numerical order.

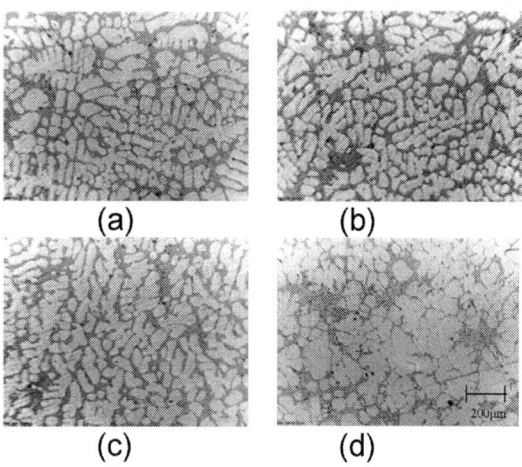

Fig. 3 Microstructures of the A356 alloy with EMV processed to various
frequencies.
(a)As-cast, (b) 1000Hz, (c) 500Hz, (d) 60Hz
(Cooling rate =1.5K/s, $1.05 \times 10^6 A/m^2$, 0.5T)

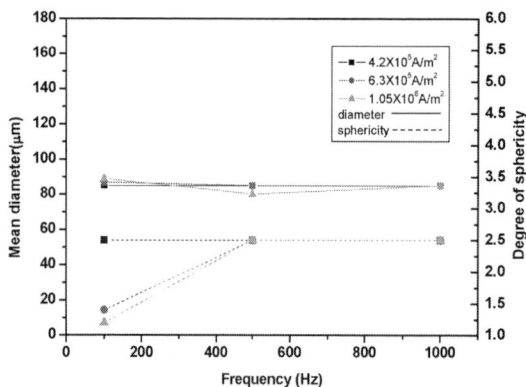

Fig. 4 The trends of mean diameter and degree of sphericity varies with
frequency and intensity of EMV (cooling rate=1.5K/s)

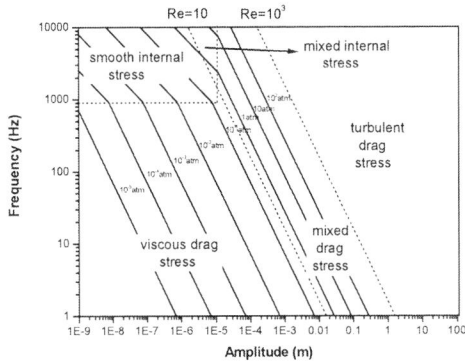

Fig. 5 f-a map showing bending stresses in roots of aluminum alloy

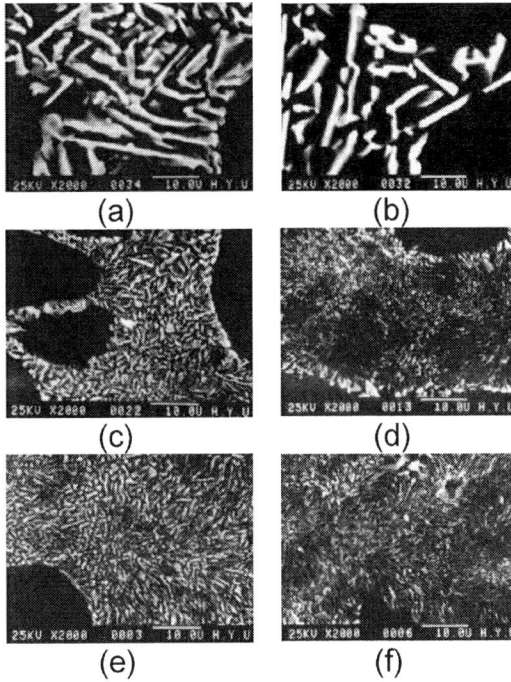

Fig. 6 Microstructures of Al-Si-Mg alloy according to the frequency at $1.05 \times 10^6 A/m^2$

(a) as-cast, (b) 60Hz, (c) 200Hz,(d) 500Hz, (e) 700Hz, (f) 1000Hz

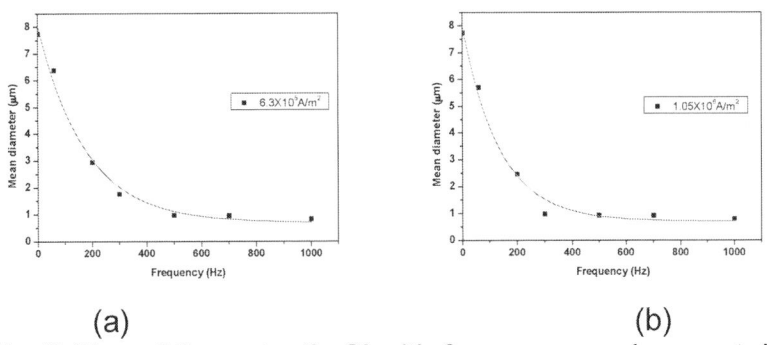

Fig. 7 Size of the eutectic Si with frequency and current density

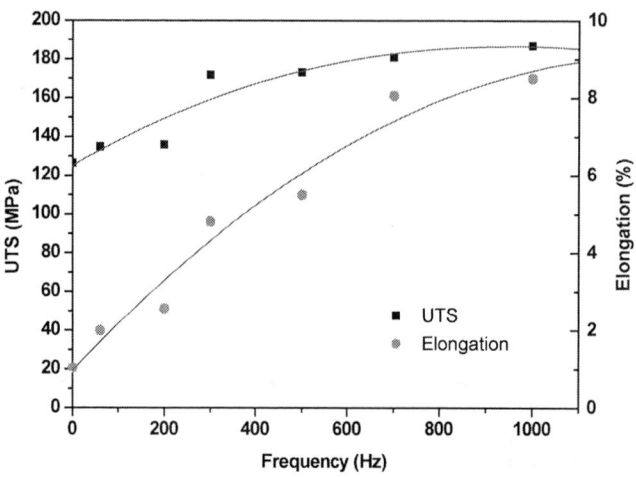

Fig. 8 Variation of the mechanical properties with frequency

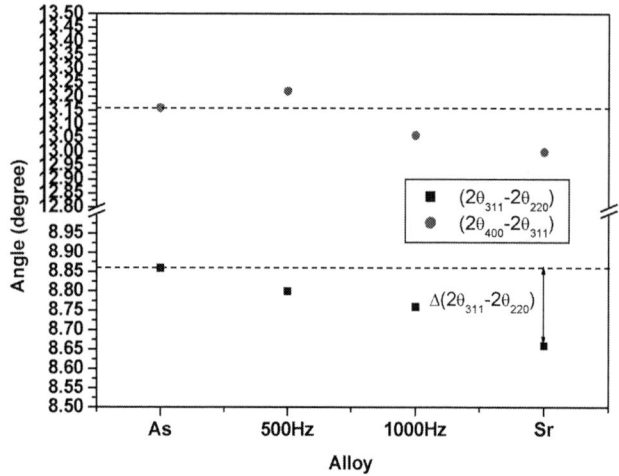

Fig. 9 Peak separations of (311)-(220) and (400)-(311) for normal, EMV modified and strontium modified alloys

Investigation of Crack Generation under the Influence of Thermal Stress during Cast Process

Si-Young Kwak, Nam Jung Ho, Sung-Wook Lee, Jeong-Tae Kim, Jeong-Kil Choi

Center for e-design, KITECH(Korea Institute of Industrial Technology), 994-32, Dongchun-Dong, Yeonsu-Gu, Incheon, 406-800, South Korea.

Abstract

Metallugical properties of cast materials and casting conditions effect on the various defects. Especially the stress has influence on the defect such as hot tear or cold crack. There are lots of needs for the thermal stress analysis in casting process to control the defects concerning crack generation or distortion during casting. In this paper, we studied the relationship between thermal stress and crack generation using numerical methods and experiments. We performed thermal stress analysis by numerical methods and verified the numerical results through residual stress measurement experiment, and investigated the effect of thermal stress on crack generation. It is recognized that a numerical thermal stress analysis is an effective tool for understanding the crack generation mechanism because crack defect is easily generated in the early stage of cooling.

Key words Thermal stress analysis, Casting process, Crack, Hybrid method, FDM, FEM, Residual stress

Introduction

Many engineers want to prevent crack defect because the defect is fatal to casting product and the desire has induced many studies on the crack generation. However, the study is difficult because the casting involves a complicated process; Casting products experience a large thermal variation from melting temperature to room temperature and phase transformation and mechanical contact with mold during the process. Especially, a crack phenomenon like hot tearing involves an interaction of stress and metallugical properties. Despite the difficulties, there have been many studies on the casting process by experimental and numerical methods.

The experimental access to thermal stress is a definite method for the study, but it is difficult to experiment casting phenomenon. It is hard to investigate casting phenomenon during casting process because the casting part is covered by mold. Since only the final state of the casting process can be investigated, it is difficult to understand the thermal stress phenomenon during the whole process. The other reason is that the cast experiences high temperature change and phase transformation from liquid to solid as mentioned above.

Recently, the numerical simulation has been recognized as a proper method in casting field for the analysis. The numerical method is a powerful tool to grasp the whole thermal stress and deformation appearance during casting process. However, the numerical method requires temperature dependent material properties and analysis model considering many factors such as constrains with cast/mold and heat flow considering air gap cast/mold.

The historical review until the middle of 1990's for numerical stress analysis in casting process can be referred by Thomas[1]. Drezet and Rappaz[2] applied the numerical method to predict hot tears of aluminum billets in DC-cast process. Monroe and Beckermann[3] attempted to predict a hot tear based on the physics of solidification and deformation using numerical analysis.

In this study, specimens sensitive to thermal stress were made by sand casting process. In simultaneously, we carried out numerical analysis for the specimens; the numerical analysis procedure calculated the distribution of temperature using FDM and the distribution of thermal stress was used FEM. In order to verify the numerical results, the residual stresses of the specimens are measured by hole drilling and Electron Speckle-Interferometer devices. And we investigate the relation with the thermal-stress state and crack generation.

Experimental method
A. Casting Experiment
Fig. 1 shows the shape and dimension of the specimens for casting experimentation. The shape of the specimens is the most useful in thermal stress generation during casting process. In this experiment, the specimens are made of three different kinds of materials each; specimens are CrMo alloy steel, gray iron and aluminum alloy respectively. But the molds were all sand. The materials and casting process are shown in the table 1. The specimens were shaken out when they were cooled down to room temperature.

B. Residual stress Measurement
The hole drilling method and Electron Speckle-Interferometer device are used for residual stress measurement of casting specimens [4]. Fig. 2 is Electron Speckle-Interferometer device that is developed by E.O. Paton Electric Welding Institute in Ukraine. The measurement sequence is the follows. The material properties for the residual stress measurement are listed table 2.

1) To mark the point on object in which the residual stresses should be determined.
2) To place the drill of selected diameter (1-2mm) into the drilling device and to set the required depth of the blind hole being drilled.
3) To mount a removable speckle-interferometer on the unit base, arranging balls in appropriate grooves.
4) Initial speckle-image of surface of area examined should be recorded into computer.
5) To remove a removable speckle-interferometer from the unit base, then to drill a blind hole of a preset depth.
6) To mount a removable module again on the unit base. To record the speckle - pattern into the computer memory and to obtain information about the change of phase.
7) To calculate of residual stress using obtained interference pattern and material properties.

Numerical method
In this study, we use a hybrid numerical analysis technique: FDM for the heat flow and FEM for the stress. A temperature-field-data-conversion scheme in a three-dimensional space is required in order to employ the FDM/FEM hybrid technique and then the residual stress analysis based on FEM used converted temperature field data. So that, an efficient data conversion procedure based on linear interpolation has been developed. The detail procedure about the FDM/FEM hybrid method and data conversion procedure can be referring to author's papers [5]. The material properties for numerical analysis referred to the material handbooks [6,7] and high temperature tensile test results.

Results and Discussion
A. Casting Experiments
Fig. 3 and Fig. 4 are the pictures of the specimens for thermal stress inspection during casting process. The figures show the deformation shape and crack defect for the specimen 1(CrMo steel) and specimen 2(Gray iron). The specimen shape was designed to be sensitive to the thermal stress, so that the deformation was distinct and cracks occurred. However, there was no defect such as crack in the aluminum alloy specimen.

B. Numerical solution
Fig. 5 is the XX and YY residual stresses of specimen 1(CrMo alloy steel). As for XX stress, the maximum value is about 430 MPa at the upper of center column. And the compressive stress appears at the outside column. After molten metal is poured, the outside column cools at the first, and then the center inside column cools and contracts. But the contraction of center column is districted by the outside column. So the tensile stress appears at the center column, and compressive stress at the outside column.

Fig.6 and Fig. 7 are the XX and YY residual stresses of FC25 (gray iron) and AC4C (aluminum alloy). The appearance of the stresses is same with that of the above specimen 1. Only the values of stresses are different each other.

C. Comparison of Experiment and Numerical solution on Residual Stress

The simulation result was verified by comparison with residual stress measurement result. Fig. 8 shows the location of the measured points. The comparison of experiment and numerical solution was performed at the points.

Fig. 9 is the lateral stress (XX) of the specimen 1(CrMo steel). The stress distribution of simulation is from -300 MPa to 250 MPa and the experimental results show good agreements except outside column (Point no. 16~20).

Fig. 10 is the longitudinal (YY) stress of the specimen 2(Gray iron). The stress distribution of simulation is from -40 MPa to 40 MPa. The experimental results show some differences with the simulation results but the aspect of local-by-local gives the reasonable agreement.

Discussion
A. Stress Distributions as Cooling Time
Fig. 11 is distributions of the longitudinal (YY) stress at specimen 1. Fig. 11(a) is the results at about 300 seconds and Fig. 11(b) is at about 7,000 seconds after pouring. The inside column of specimen experienced compressive stress while the outside column is in tensile state in the early stage of the cooling procedure; Fig. 11(a) was about 1000°C to 800°C. However the inside part was tensile and the outside part was compressive

at the final stage of the cooling. As a result, the stress state was reversed during casting process in these specimens.

B. Prediction of Crack Generation

Fig. 12 shows the location of crack generation of the specimen 1(CrMo alloy) and the specimen 2(FC25). The crack of CrMo steel generated at the outside column and the crack of FC25 generated at the center column. There is no crack at the specimen 3(aluminum alloy). The maximum residual stress (Fig. 7) of specimen 3 is about 85MPa and the value is small considering the tensile strength of about 300MPa.

In case of specimen 2, it is easy to understand or predict why the crack occurred at the center column because the maximum tensile residual stress generated at the same zone (Fig. 6).

But it is difficult to understand or predict why the crack occurred at the outside column in specimen 1 because the compressive residual stress appeared at the same zone (Fig. 5). In generally, the compressive stress of 160 MPa cannot make crack in steel at the room temperature.

This crack generation mechanism can be explained as the stress state at the early stage of cooling. The outside column experiences tensile stress at the early stage of cooling (about 300 seconds). Although the value is small as 27 MPa but the temperature was about 1000°C to 800°C at that time. A material is so weak at such high temperature. We are able to predict the crack generation.

Fig. 13 is the histories of Von-Mises stress and temperature at the location of crack generation in the specimen 2. The maximum residual stress of specimen 2 is about 155MPa (Fig. 6) which is under the tensile strength of 250MPa. Fig. 13 shows that the Von-Mises stress is 30MPa while the temperature is over 1100°C.

However, since the crack generation mechanism like hot tear has to consider the metallugical properties and casting conditions, it is required to consider other factors such as solidification zone.

Conclusions

To investigate the relations between thermal stress state and crack generation, casting experimentation and numerical analysis were performed in the casting process. We made the specimens of CrMo steel, gray iron(FC25) and aluminum alloy(AC4C) by sand casting process and carried out the numerical analysis using FDM/FEM hybrid method. The crack defects occurred in the specimens of CrMo steel and gray iron except aluminum alloy. The results of the numerical analysis were compared with the ones of the residual stress measure test and the comparison between experiment and simulation gives us the satisfactory results.

It is hard to explain the crack generation mechanism by residual stress state only and some crack generation mechanism can be explained by the numerical result of thermal stress analysis at the high temperature.

References

1. B. G. Thomas, 1993, "Stress Modeling of Casting Processes: An Overview," International Conference on Modeling of Casting & Solidification Process –VI, pp. 519~534.
2. J.M. Drezet and M.Rappaz, 2001, "Prediction of Hot Tears in DC-cast Aluminum Billets," Light Metals, Cast Shop Technology, 2001 TMS Annual Meeting, New Orleans, Louisiana, February 2001.
3. Charles Monroe and Christoph Beckermann, 2005, "Development of a hot tear indicator for steel castings," Materials Science and Engineering, Vol. A 413-414, pp.30~36.
4. L.M. Lobanov, V.A. Pivtorak, and N.G. Kuvshinsky, 2000, "Diagnostics of Structures of Metallic and Composite Materials using Holography, Electron Speckle-Interferometry and Sherography", The Parton Welding Journal, 9-10/2000, pp. 72~78.
5. H. M. Si, C. Cho, and S. Y. Kwak, "A Hybrid Method for Casting Process Simulation by Combining FDM and FEM with an Efficient Data Conversion Algorithm," J. Materials Processing Technology, V133, 2003. pp.311-321.
6. High-Temperature Property Data: Ferrous Alloys, ASM International, 1988.
7. ASM handbook, Properties and Selection: Irons, Steels, and High-Performance Alloys, ASM International ,1990, p. 628

Tables

Table 1 Materials, chemical composition and casting condition of the specimens

	Material	Composition	Pouring Temperature
Specimen 1	CrMo steel	0.22%C, 0.42%Si, 0.7%Mn, 0.93%Cr, 0.2%Mo	1600 [°C]
Specimen 2	Grat Iron (FC 25)	3.4%C, 2.0%Si, 0.8%Mn	1350 [°C]
Specimen 3	Al alloy (AC4C)	7.37%Si, 0.34%Mg, 0.17%Fe	700[°C]

Table 2 Material properties for the residual stress measurement of the specimens.

	Material	Elastic Modulus	Poisson's ratio
Specimen 1	CrMo steel	210[GPa]	0.28
Specimen 2	Grat Iron (FC 25)	93[GPa]	0.26
Specimen 3	Al alloy (AC4C)	72[GPa]	0.33

Figures

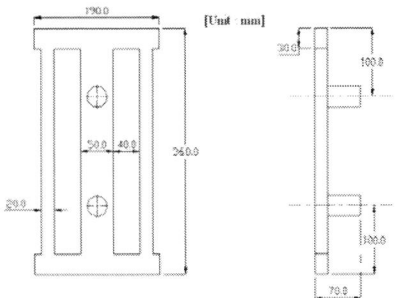

Fig.1 Shape and size of the Specimens for casting experimentation

Fig. 2 Electron Speckle Interfero-meter device

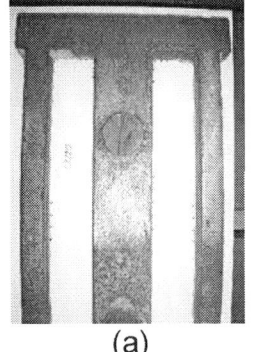
(a)

(b)

Fig. 3 CrMo alloy steel specimen after casting process; (a) the deformation shape, (b) the crack generation

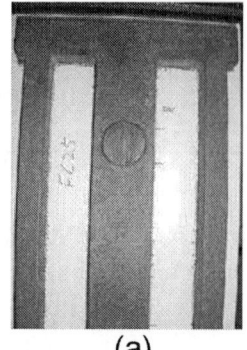

(a)   (b)

Fig. 4 Gray iron specimen after casting process; (a) the deformation shape, (b) the crack generation

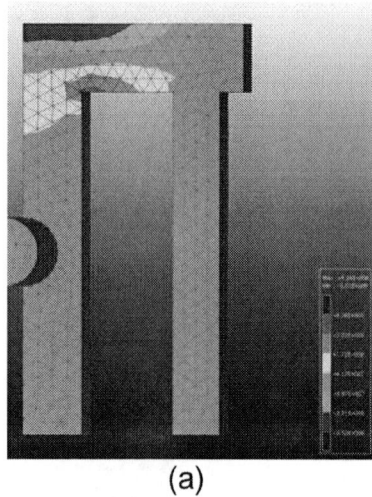
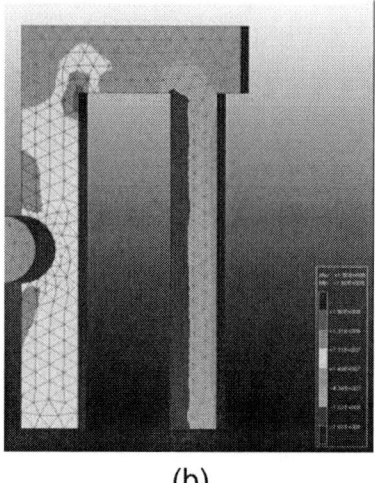

(a)   (b)

Fig. 5 Distributions of the residual stress at specimen 1 at about 7,000 seconds;(a) the result of XX lateral stress(Max 430Mpa), (b) the result of YY longitudinal stress(Max 190Mpa)

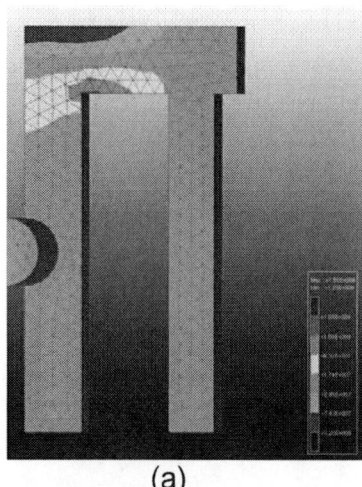
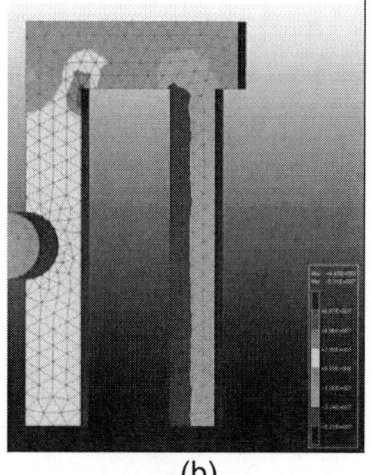

(a)   (b)

Fig. 6 Distributions of the residual stress at specimen 2 at about 7,000 seconds;(a) the result of XX lateral stress(Max 155Mpa), (b) the result of YY longitudinal stress(Max 70Mpa)

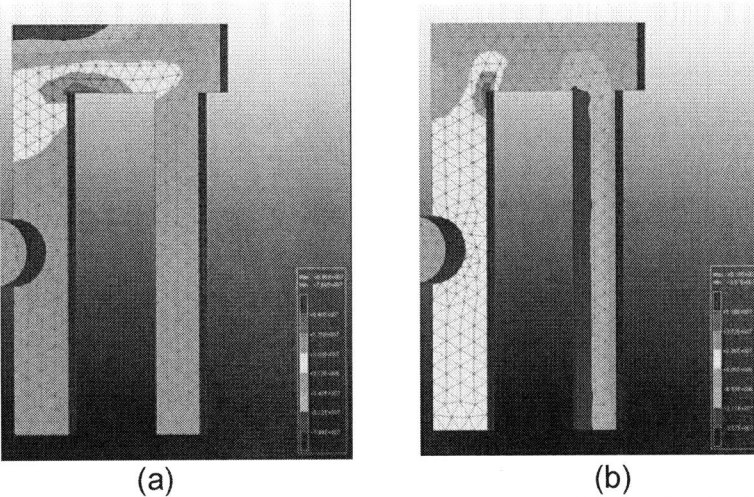

(a)   (b)

Fig. 7 Distributions of the residual stress at specimen 3 at about 7,000 seconds;(a) the result of XX lateral stress(Max 85Mpa), (b) the result of YY longitudinal stress(Max 52Mpa)

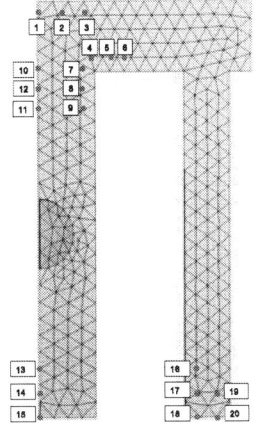

Fig. 8 Locations of points for measurement

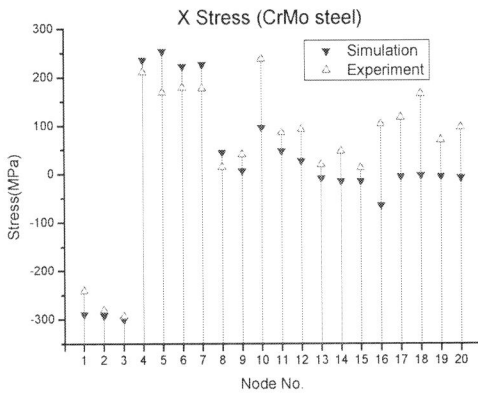

Fig. 9 Lateral (XX) stress of the specimen 1 (CrMo steel).

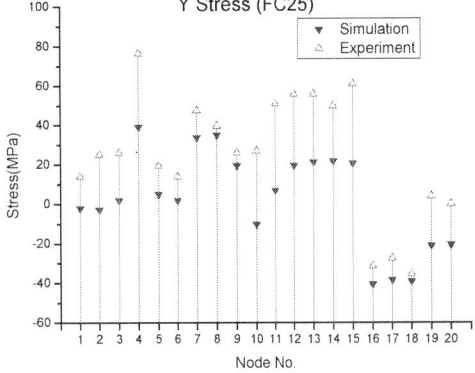

Fig. 10 Longitudinal (YY) stress of the specimen 2 (Gray iron).

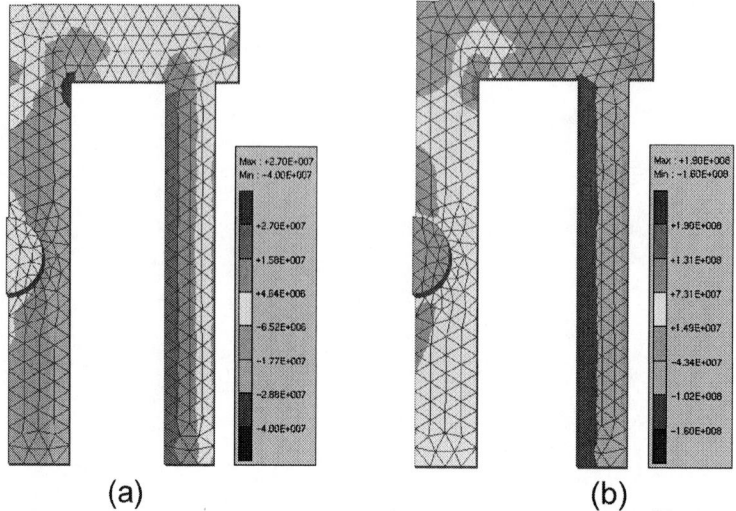

(a)   (b)

Fig. 11 Distributions of the longitudinal (YY) stress at specimen 1;(a) the result at about 300 seconds,(b) the result at about 7,000 seconds

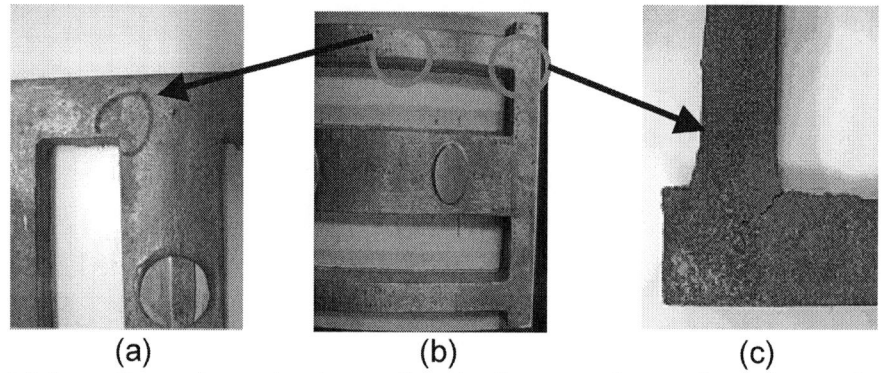

(a)   (b)   (c)

Fig. 12 Location of crack generation in the specimen 1 and specimen 2; (a) Specimen 2(FC25), (b) Shape of specimen, (c) Specimen 1(CrMo steel)

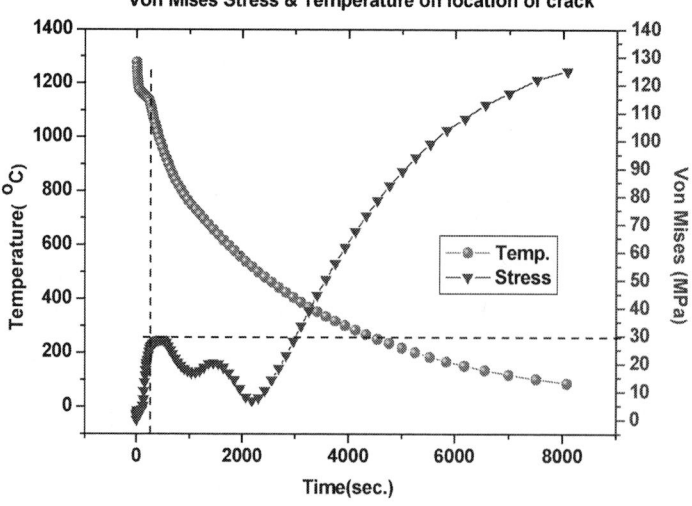

Fig. 13 Histories of Von-Mises stress and temperature at the location of crack generation in the specimen 2.

The Prefil® Technique for Molten Metal Cleanliness Measurement

J Pickering *, P Enright *

* N-Tec Ltd, UK

Abstract

This paper discusses the use of the Prefil pressure filtration technique for the on-line monitoring of molten metal quality. Over the past 10 years Prefil has become established as one of the main metal quality monitoring tools in aluminium. An introduction to the basic technique is followed by a detailed explanation of the results, and their interpretation with reference to specific case studies.

The real power of Prefil is realized when the Prefil characteristic for a particular sample, obtained on-line, is compared with a pre-established window for premium quality. Particular reference is made to the concept of World Class Quality Windows™, which allow the user to immediately determine whether their molten metal quality is at the optimum level for their process and/ or product. Several different types of window are defined, and example uses of each given, with supplementary metallographic evidence.

The use of the Prefil in aluminium is discussed in detail, and recent developments in pressure filtration techniques for both copper and magnesium alloys are introduced.

Key words

Molten metal quality, Prefil, monitoring

Introduction

The Prefil pressure filtration technique for the measurement of molten metal quality was developed in the mid nineteen-nineties by N-Tec Ltd, in the UK. It is now recognized as one of the principal methods for determining and monitoring molten metal quality in aluminium and its alloys, in both foundries and cast houses world-wide.

This paper discusses the technique, and introduces the concept of World Class Windows™, which give the user an immediate quality window with which to compare results, on-line. This means that it is possible to instantaneously see where measured quality lies in comparison to other Prefil users in the same field. Three different windows are identified, and example microstructures and total inclusion contents given in each case.

Recently, the technique has been developed for use in other non-ferrous metals; use in copper and magnesium alloys is discussed here, with examples.

The Basic Technique

The Prefil (Pressure Filtration) technique is used to give an on-line quantitative measurement of oxide films and other inclusions. Figure 1 shows the instrument. Since it can be used to monitor the effect of individual process steps, as well as final quality prior to casting, Prefil gives users the ability to track down inclusion sources, and to 'see' the effect of practice changes in real time. The entire test takes a maximum of 150 seconds, and gives an instantaneous plot used to determine metal quality. The on-screen results can be compared to a World Class Quality Window to give the user immediate feedback on whether their metal quality is fit for purpose.

The flow-rate of molten metal through a micro filter at constant temperature and pressure is monitored and used to plot a graph of weight filtered versus time. Inclusions in the metal, such as oxide films, quickly build up on the filter surface during a test, reducing the flow-rate through the filter. The operating principle of the Prefil can be seen in Figure 2.

In addition to the filtration curve, subsequent metallographic analysis of the residue that is retained on the filter after a Prefil test allows identification and quantification of the types of inclusions present in metal sample to be carried out. Once the inclusion profile of a product line has been characterised it is easier to select the most appropriate monitoring and control procedures for maintaining consistency.

Recent developments to the technique have seen the replacement of the disposable fibrous crucible and filter assembly with a reusable spun steel crucible, and replaceable filter element. Transportation and storage issues are lessened, and filter detachment virtually eliminated.

Off-Line Prefil Testing

It is also possible to carry out Prefil testing on final product, or samples taken during production and solidified prior to testing. This is known as the off-line, or "cold Prefil" technique.

Samples of the appropriate weight and size are rapidly induction melted from bottom to top in a quiescent fashion, usually using a fibrous crucible.

Typically, cold Prefil testing is carried out on 5-6kg notched ingot, or on slugs taken using a specially designed mould. Irregular shapes, such as castings, can also be remelted in this way with good results, although surface area should be minimised in order to avoid surface oxide effects. Experimentation has shown that this type of controlled remelting is consistent, and gives a good correlation with "live" Prefil testing.

Prefil Characteristic

The slope and overall shape of the Prefil flow-rate curve indicates the level of inclusions present in the metal, as illustrated in Figure 3. Clean metal (few inclusions) will result in steep, straight Prefil curve whilst metal containing a large number of inclusions will have a low Prefil characteristic.

Oxide films affect the initial slope of the curve (20-30 seconds). They result in straight lines, with a slope that decreases as the number of oxide films increases. Fine particulate inclusions such as TiB_2, fine Al_2O_3 or carbides cause the curve in the Prefil test to deviate from a straight line. The loading of fine particles can be inferred from the point at which the curve begins to deviate from the initial slope. Example Prefil characteristics can be seen in Figure 4.

As experience is gained with the technique, knowledge of how different inclusions affect the overall shape of Prefil curves (or the Prefil characteristic) is built up, and so it is possible to gain some understanding of the approximate size and number of inclusions present in the metal from the Prefil characteristic alone. As the user becomes more familiar with the technique and their usual Prefil characteristic(s), the need for metallographic analysis is reduced.

Metallographic Analysis and Prefil Curve Reproducibility

It has been established that the repeatability of metallographic analysis is ± 16% for inclusion contents over 1.25 mm^2/kg. The precision of the metallographic analysis decreases with an increase of the metal cleanliness. When the inclusion content is less than 0.07 mm^2/kg, repeatability is +40%. [1]

The Prefil curve has a ±10% error on mass at any time on filtration rate (g/s) curve for fresh A356 at 700°C. [2]

Oxide Films

Oxide films are difficult to measure by metallographic means because of the nature of the metallographic analysis technique. The presence of oxide films in a sample is most probably underestimated. When a metallographer looks at a slice of a Prefil sample, what he sees is often just a slice of the oxide film plane. For instance, he may be looking at several slices of different oxide films or at several parts of the same film. Because of the nature of this type of analysis, discrimination can only be confirmed for changes in the order of magnitude of films present. Also, oxide films have a lower density than aluminium and because of the long solidification time of a Prefil crucible, it is likely that oxide films float out of the cake band area just above the filter disk.

Based on these facts, the Prefil curve has to be trusted as a better mean of measuring oxide film content than metallographic analysis. [3]

Prefil 'World Class Windows'

The real power of Prefil is demonstrated when the Prefil curve for a particular sample is compared with a pre-established window for appropriate quality; this footprint is referred to as the Prefil Footprint or the 'World Class Window'. As soon as new data is compared to the World Class Window it is possible to put the molten metal quality into industrial context. All Prefil data should be referred to an appropriate window. And several different types of Window are commonly used.

The **Industrial Range** represents the full spread of molten metal quality data for a particular alloy, process, or material condition, (grain refined or modified for example). The industrial range database is currently made up of over 10,000 measurements taken from over 200 operating plants worldwide and forms a standard basis for comparing different operations around the world. Often a relatively wide window, the Industrial Range is an important part of the benchmarking process. Not only is the user's position in the data set identified, but also their level of consistency and typical inclusion mix. This type of benchmarking is a first step towards process control, and helps the user to identify what further testing is necessary.

An example Industrial Range for BS 1490:1988 LM27 (Al-Si5-Cu3) gravity die casting alloy is shown in Figure 5. Characteristics are given for two batches of remelted ingot; one batch was filtered during production, and the other was not.

Process Stage Windows™ are used to monitor and optimise intermediate stages of the production process. For example, the effect of in-line treatment systems is cumulative; a flux treatment will remove a certain percentage of inclusions, but if there are higher inclusion levels to begin with then achieving adequate quality will be proportionally more difficult. When a piece of treatment plant such as a degasser or filter is optimised then the Prefil characteristic for the molten metal at the outlet should lie within the relevant Process Stage Window. The ability to monitor production at all stages of the process is useful in identifying and minimising unnecessary operating costs, and in reducing rejections of value added product later.

Figure 6 shows the Process Stage Window for post flux degassing of A356 (Al-Si7-Mg) alloy for general commercial use. The effect of flux degassing and subsequent holding time on metal quality is illustrated.

Target Production Windows are used to monitor or qualify final molten metal quality at the point of casting after all metal treatment processes are completed and additions have been made. Generally, the target production window is a tighter specification than the associated Process Stage Window(s) and directly reflects the ability of the process to deliver consistent material. Target Production Windows are plant and process specific and allow the user to maintain statistical process control on a daily basis. Many studies have shown that if production falls outside the established window on a regular basis there is a real possibility of compromising product quality at a later stage.

Figure 7 shows two Target Production Windows, both for automotive use A356 alloy (Al-Si7-Mg). The upper window is for non-grain refined product, and the lower for appropriately grain refined product. A further data set at the bottom of the plot shows the effect of over grain refining.

Use of the Prefil for Process Control
The use of Prefil to provide Process Control is a well established technique, and is made more effective by reference to relevant World Class Windows. The Prefil characteristic can be used to determine the molten metal quality of a bulk metal sample; for example, if the cleanliness decreases significantly the characteristic will fall outside of the relevant window. It is important to note that care should be taken when making judgements based on one unusual result only.

Use of the Prefil in Copper Alloys
The Prefil technique was first used in copper alloys in 2000, and a number of projects have been carried out in the interim. Figure 8 shows the Prefil results with their corresponding Industrial Range, and accompanying metallographic analyses for work carried out in a 60:40 brass, tested at 950°C. Fibrous crucible and filter assemblies stood up to this temperature with no significant problems. Onboard software in the Prefil machine allows the weight filtered to be altered to provide an appropriate test for metal of different densities.

The introduction of the reusable crucible has brought new challenges to the use of the Prefil technique with copper alloys. The spun steel crucible requires a coating both to protect the crucible from molten metal attack, and to aid release of the solidified residue remaining after the test. Investigations and an experimental program have taken place to source the most appropriate coating material for use with copper alloys.

Gasketing materials on the filter assembly have also had to be replaced with higher temperature alternatives. Currently projects are ongoing with customers, both in the United Kingdom, and Europe.

Use of the Prefil in Magnesium Alloys

Developments to utilise the Prefil technique in magnesium and its alloys have been ongoing since 2004. Several problems have had to be overcome in order to use the system safely and effectively.

It is necessary to pressurise the machine with a protective gas, rather than air, and a protective gas also has to be present in the catch pan area, to protect the filtered metal. To achieve this, engineering work has been undertaken on a Prefil machine dedicated to use with magnesium.

In common with copper applications, the reusable crucible requires a coating to prevent magnesium attack, and allow easy removal of the residue. Again an experimental program has been undertaken and a readily available coating found. A new filter material has been sourced and a new gasketing system designed, since the filters and gaskets for use with aluminium are attacked by molten magnesium. Investigations were made into several different filter materials before the final one was selected.

Work is currently being undertaken with an industrial partner to further develop the knowledge database for Prefil in magnesium alloys. Figure 9 shows early results achieved in magnesium Prefil testing, and associated microstructures. The results are compared to fast neutron activation analysis (FNAA) values. Fast neutron activation analysis (FNAA) is the system commonly used to test magnesium quality at present. This technique detects the presence and determines the concentration of various elements in a given sample using irradiation.

Conclusions

In conclusion, the use of the Prefil technique for the measurement and monitoring of molten metal quality has been greatly enhanced by the introduction of World Class Windows. These windows allow the user to immediately benchmark where their quality lies, and later analysis of the results and more detailed process auditing can identify exact causes for quality issues. Continuous monitoring can, over a period of time, allow the user to tighten their window, as production is brought more under control, and consistency improved.

The Prefil technique has also diversified, with use in copper and magnesium alloys highlighted here. The results have been good, with metal quality issues in these metals distinguished easily and effectively.

Work has also been undertaken in several different alloy systems, with positive results. These include zinc, solders and lead alloys.

References
1. Metallographic Evaluation of PoDFA Inclusions Residue, Alcan, ARDC, Method # 2012-96
2. Prefil specifications
3. ABB Prefil metallographic analysis report

Figures

Figure 1: The Prefil® Footprinter Machine

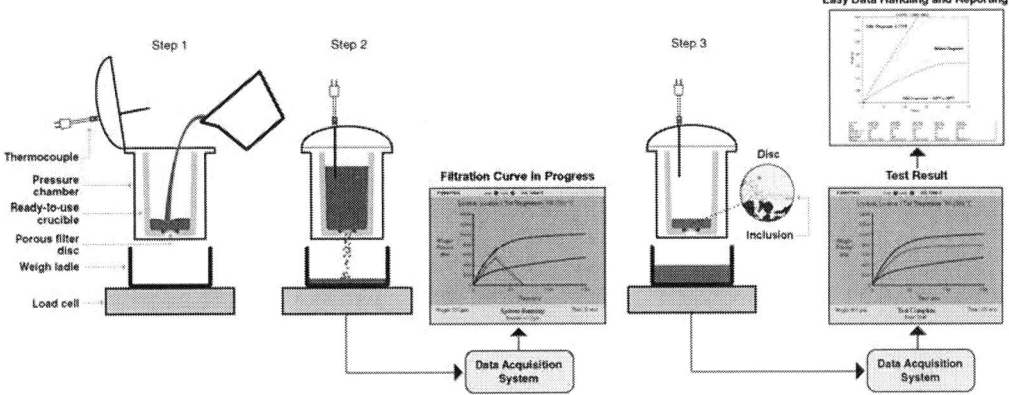

Figure 2: The Principle of Operation of the Prefil® Technique

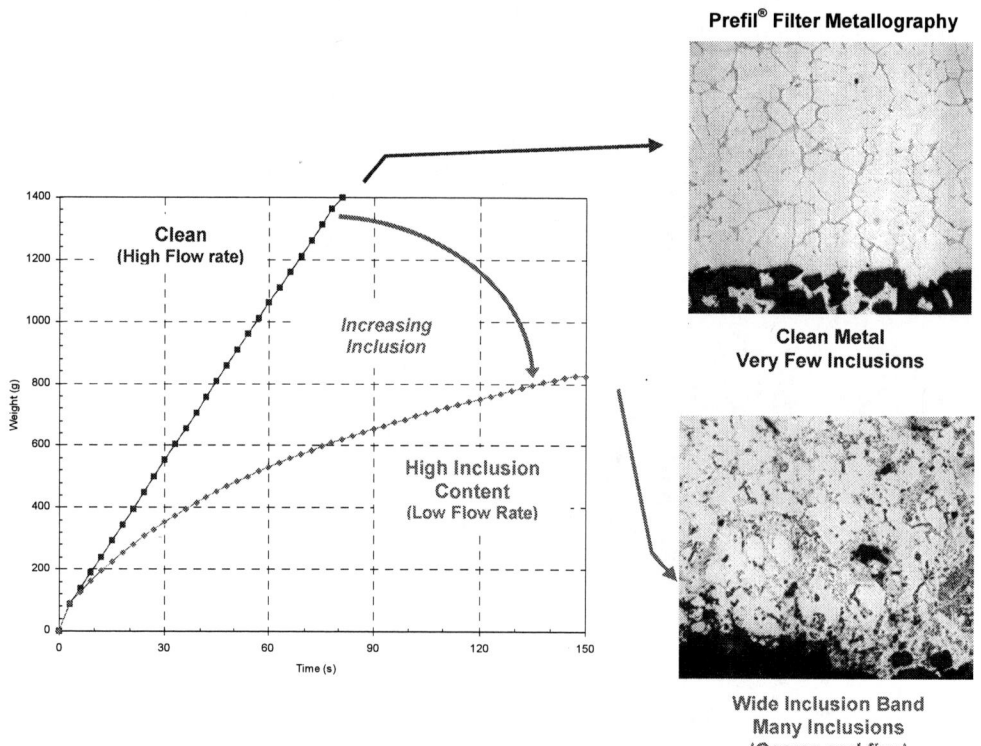

Figure 3: Prefil® Characteristics Observed for Low and High Inclusion Contents

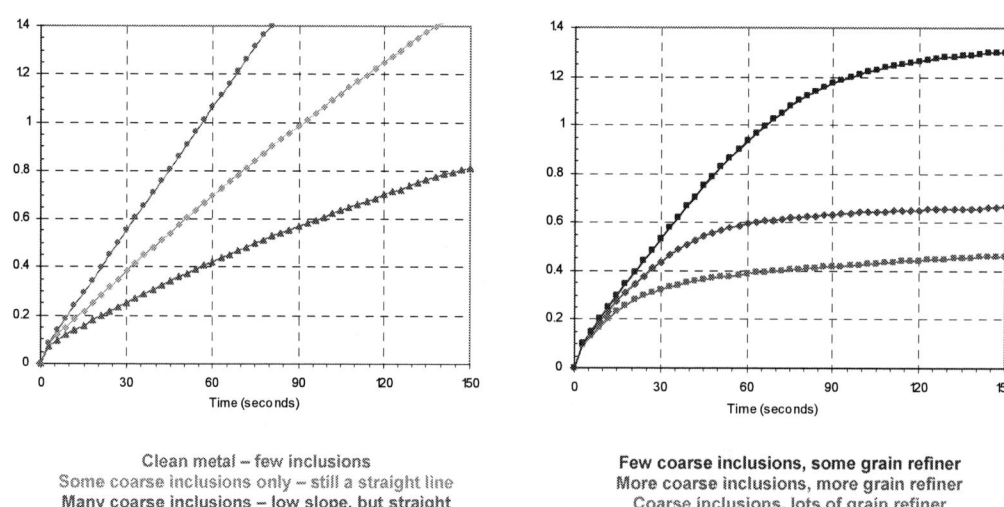

Figure 4: Example Prefil® Characteristics

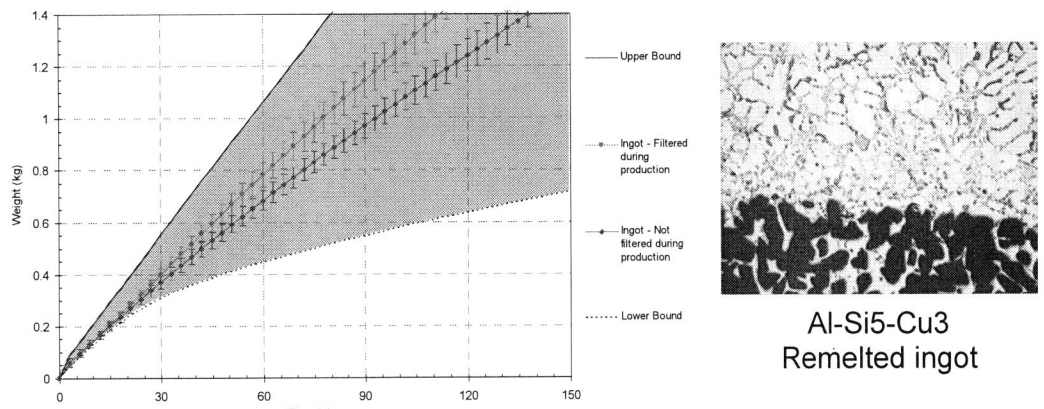

Figure 5: The Industrial Range for BS 1490:1988 LM27 (Al-Si5-Cu3) alloy

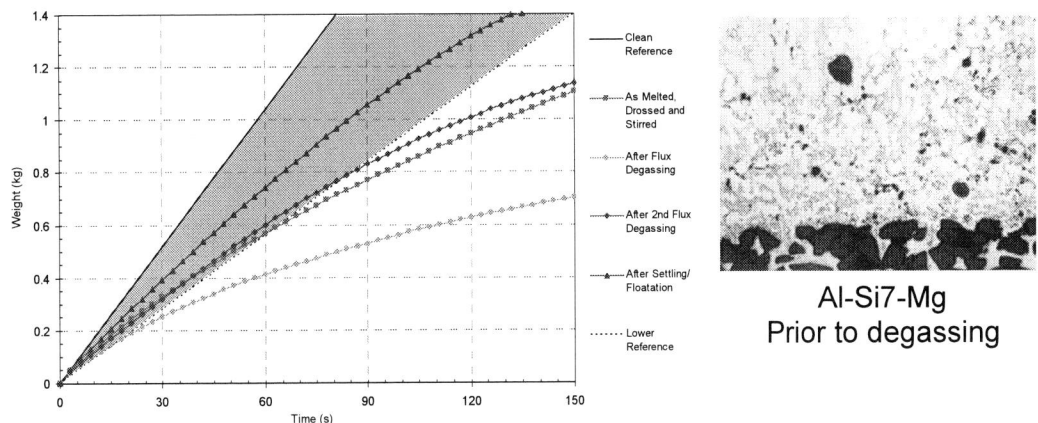

Figure 6: The Prefil® Process Stage Window™ for post flux degasser treatment in A356 (Al-Si7-Mg)

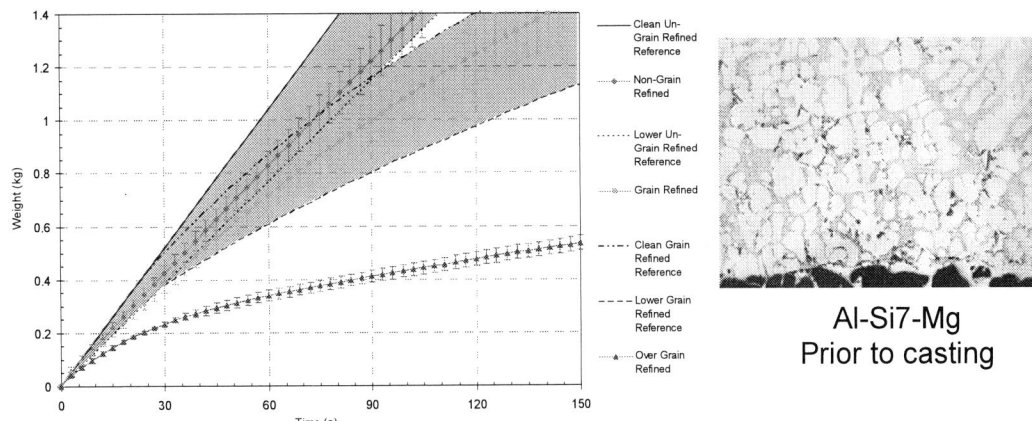

Figure 7: Target Production Windows for A356 (Al-Si7-Mg) automotive components, both grain refined and non-grain refined

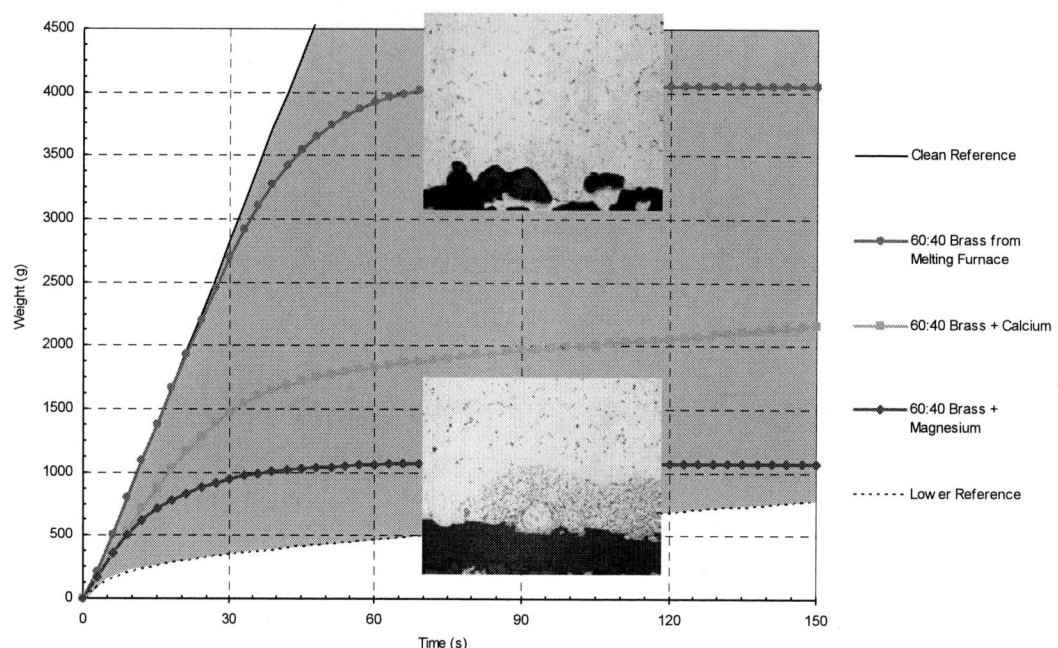

Figure 8: Prefil® results from tests on a 60:40 brass alloy with various additions

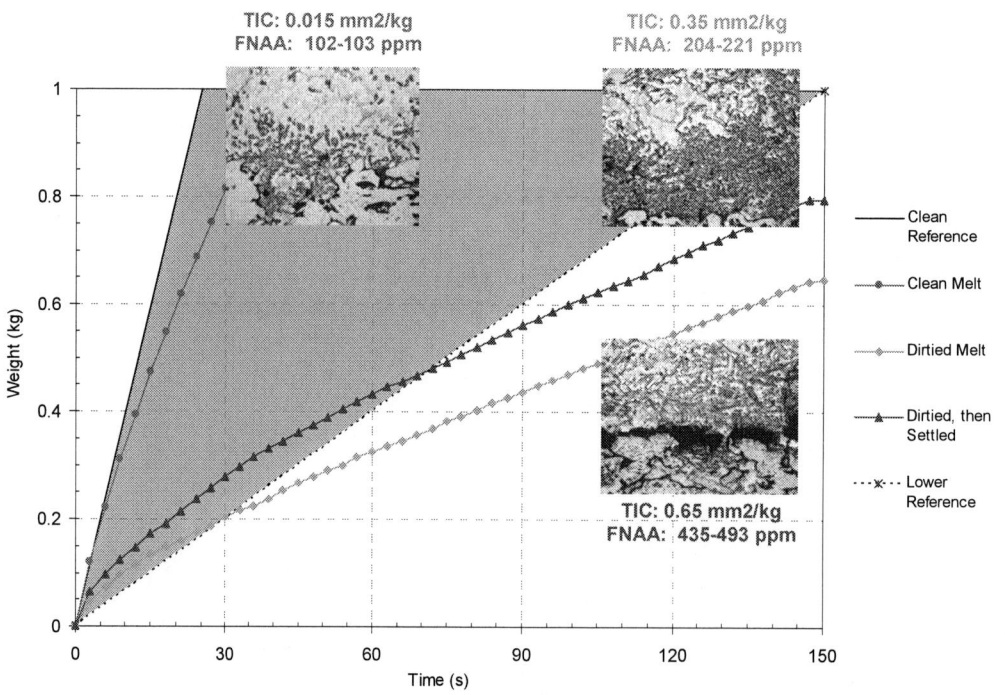

Figure 9: Prefil® results from tests on a magnesium alloy in various conditions

Fabrication of silicon ingot by CCCC method

B. M. Moon, D. S. Lee, B. H. Kim, J. S. Shin, and S. M. Lee
Korea Institute of Industrial Technology, Incheon, Korea

Abstract
In the fabrication process of multi-crystalline Si ingot for solar cell wafer, CCCC (Cold Crucible Continuous Casting) technology was practiced to prevent contamination from crucible wall as well as to overcome the problems of low solidification rate and crucible consumption. The effects of Joule heating and electromagnetic pressure in molten Si were systematically investigated with various processing parameters such as crucible configuration. Throughout the present investigation, multi-crystalline Si ingot was successfully and continuously produced at the casting speed of above 1.5 mm/min under a non-contact condition between molten silicon and the crucible wall.

Key words silicon, cold crucible, continuous casting, Joule heating, pinch effect.

Introduction

The world photovoltaic market entered the age of GW, in a rapid expansion with the average annual growth rate of 35% since the middle 1990s. The strongly growing photovoltaic market is based on crystalline silicon technology. Since 1998, the multi-crystalline silicon has prevailed over mono-crystalline silicon for its cost competitiveness and the popularization of photovoltaic cell generation. More than 70% of the production cost of solar module comes from silicon material. Presently, the larger part of multi-crystalline silicon is produced by HEM (Heat Exchange Method, Fig. 1(b)), one of unidirectional solidification route. In this method, massive ingots are cast in large crucibles and solidified in precisely controlled conditions. However, solidification rate is relatively low and crucible consumption is high. Recently, CCCC (Fig. 1(c)) and EFG (Edge-defined Film-fed Growth, Fig. 1(d)) technologies have been introduced in the production of multi-crystalline silicon for solar cell substrate. Especially, the production by CCCC process is expected to prevent contamination from crucible wall by pinch effect as well as to overcome the problems of low solidification rate and crucible consumption. The characteristics of representative technologies of crystalline silicon ingot are summarized in Table 1 [1-8].

In CCCC process, when an alternating magnetic field is applied in a conductive charge material such as metals, eddy current is induced in the charge, which gives rise to resistive heating (Joule effect) in the charge and as a result it melts. The eddy current also interacts with the applied magnetic field to give Laplace-Lorentz force, which always faces toward the center of the charge. This force repels the liquid from the crucible wall (known as pinch effect).

However, since CMCC process uses a water-cooled cold crucible, it has difficulties in continuously melting and casting silicon into a high-purity ingot. Silicon has a high cooling effect by radiant heat emission due to a high melting temperature and a weak induction heating effect due to a low electric-conductivity. In the present study, numerical and experimental works were systematically carried out to develop CCCC process for the production of multi-crystalline Si with high purity. The effects of Joule heating and electromagnetic pressure in molten silicon were systematically investigated with various processing parameters such as crucible configuration and electric current.

Experimental

The cylindrical cold crucible divided into 12 segments by vertical slits of 0.3 mm width was used in this study. The frequency of 20 kHz was chosen for the stable confinement of a liquid metal dome. At the beginning of melting process, a graphite rod was inserted into the upper part of the cold crucible as a pre-heater. Fig. 2 shows the schematic diagram of the CCCC apparatus used in the present srudy and the actual continuous melting and casting process of Si.

For the optimization of non-contact melting condition, the induction coil current was changed in the range between 840 and 1,230 A. A 30 mm height graphite crucible, which was segmented in the same manner with the cold crucible, was installed at the upper part of the cold crucible to improve Joule heating efficiency.

A commercial software, Opera-3D of Vectorfields Inc., which is based on FEM, was used for electromagnetic analysis. Fig. 3 shows 1 segment 3D modeling for electromagnetic simulation. Magnetic vector potential and electric scalar potential were introduced in the numerical calculation, since Biot-Savart law is not available in conductive media where eddy current is induced. The shape of melt used in electromagnetic analysis was measured by using MgO powder method.

Results

Since the density of Al is similar to that of Si, Al was used to simulate the pinch effect of Si melt. Based on the preliminary experiment using Al system, Si melt was able to be kept apart from the crucible wall at the current of 970 A, while there was a contact between the melt and the crucible wall at the currents below 970 A. However, when a current of 970 A was applied, the surface of Si melt was covered with a solidified scale during adding feedstock after removing the pre-heater, as shown in Fig. 4. Since the hydrostatic pressure of a melt is proportional to the height of the melt, the top surface of a melt, to which a solid raw material is supplied, has a low hydrostatic pressure, so that the melt becomes further apart from the crucible wall. Therefore, near the top surface of the melt, the electromagnetic field is not easily concentrated and eddy current density is significantly low, as shown in Fig. 5(a) and (b) respectively, thus reducing the induction heat value. In a typical CCCC process, generally, 70~80% of the induction heat value is generated in the cold crucible and thus lost through a cooling water medium, without contributing to the heating and melting raw materials [9].

In order to effectively melt silicon, which has a high radiation heat loss due to the high melting point and a low Joule heating effect due to the low electric conductivity, the induction coil current was increased up to 1,230 A, and the segmented graphite crucible with 30 mm height was installed at the upper part of the cold crucible. Since the Joule heat generated from the graphite crucible is not lost by cooling water, it can contribute to heat up the silicon melt. The Joule heat powers calculated from electromagnetic analysis are summarized in Table 2. When only Cu cold crucible was used, the heating efficiency was about 30%; the Joule heat generated in the cold crucible did not contribute to heat silicon melt since it might be diminished by water-cooling. When the graphite crucible was attached, the Joule heat generated from the graphite crucible might not be lost by cooling water, then it contributed to heat up the silicon melt: the heating efficiency for melting silicon was about 50%.

In the above case, although electromagnetic pressure, which was calculated by integrating the radial direction component of Lorentz force, slightly decreased at the upper part of the melt, it was still larger than the hydrostatic pressure over whole melt region under the induction coil current of 1,230 A, resulting in a non-contact melting and casting condition.

Based on these experimental and numerical results, 5 cm circular and square multi-crystalline silicon ingots were successfully produced at the casting speed of above 1.5 mm/min under a non-contact condition by CCCC process. Fig. 6 shows continuously cast multi-crystalline silicon ingots and their representative crystalline growth structure. A strong directionally solidified crystalline structure was clearly observed, as shown in Fig. 6(c).

Conclusions

The optimization of Joule and pinch effects was tried to develop for the production of multiy-crystalline silicon ingot by a CCCC process using a segmented Cu cold crucible under an alternating magnetic field of 20 kHz. A segmented graphite crucible, which was attached at the upper part of the cold crucible, was introduced to enhance significantly the heating efficiency of silicon melt keeping non-contact condition. Throughout the present experiment, multi-crystalline silicon ingot was successfully produced at the casting speed of above 1.5 mm/min under a non-contact condition.

References

1. Scheel H J and Fukuda T, Crystal Growth Technology, Wiley 2003.
2. Perichaud I, Martinuzzi S and Durand F, Solar Energy & Solar Cells, 72, 2002, pp101.
3. Markvart T and Castañer L, Practical Handbook of Photovoltaics: Fundamentals and Applications, Elsevier 2003.
4. Franke D, Rettelbach T, Häßler C, Koch W and Müller A, Solar Energy Materials and Solar Cells, 72, 2002, pp83.
5. Ciszek T F, The capillary action shaping technique and its application, New York 1981. Perichaud I, Martinuzzi S and Durand F, Solar Energy & Solar Cells, 72, 2002, pp101.
6. Schmela M, Photon International March, 2005, pp66.
7. Ehret E, Solar Energy Materials and Solar Cells, 53, 1998, pp313.
8. Boudaden J, Loghmarti M, Ballutaud D, Rivière A, Lüdemann R, Slaoui A and Muller J C, Solar Energy Materials and Solar Cells, 65, 2001, pp517.
9. Colpo P, Zdziobek A A, Driole J, Gagnoud A, Durand F and Garnier M, Proc. of International symposium on electromagnetic processing of materials, Nagoya, Japan, 1994, pp295.

Table 1 Comparison of the major silicon crystal growth techniques used for solar cells [1-3]

| Growing technology | Electric power consumption (kWh/kg) | Production rate (kg/h) | Impurity level | | Typical efficiency (%) |
			O (cm^{-3})	C (cm^{-3})	
Czochralski	18~40	200~400	$< 1 \times 10^{18}$	$< 2 \times 10^{17}$	15
HEM	22~28	600	$< 8 \times 10^{17}$	$< 1 \times 10^{18}$	14
CCCC	12~20	3,000	$< 2 \times 10^{16}$	$< 3 \times 10^{17}$	14

Table 2 Joule heating power with crucible configuration under the induction coil current of 1,230 A

Joule heating power / Crucible configuration	Silicon melt (W)	Cu cold crucible (W)	Graphite crucible (W)
Cu cold crucible	5,386	11,037	-
Cu + graphite crucible	5,478	9,518	4,183

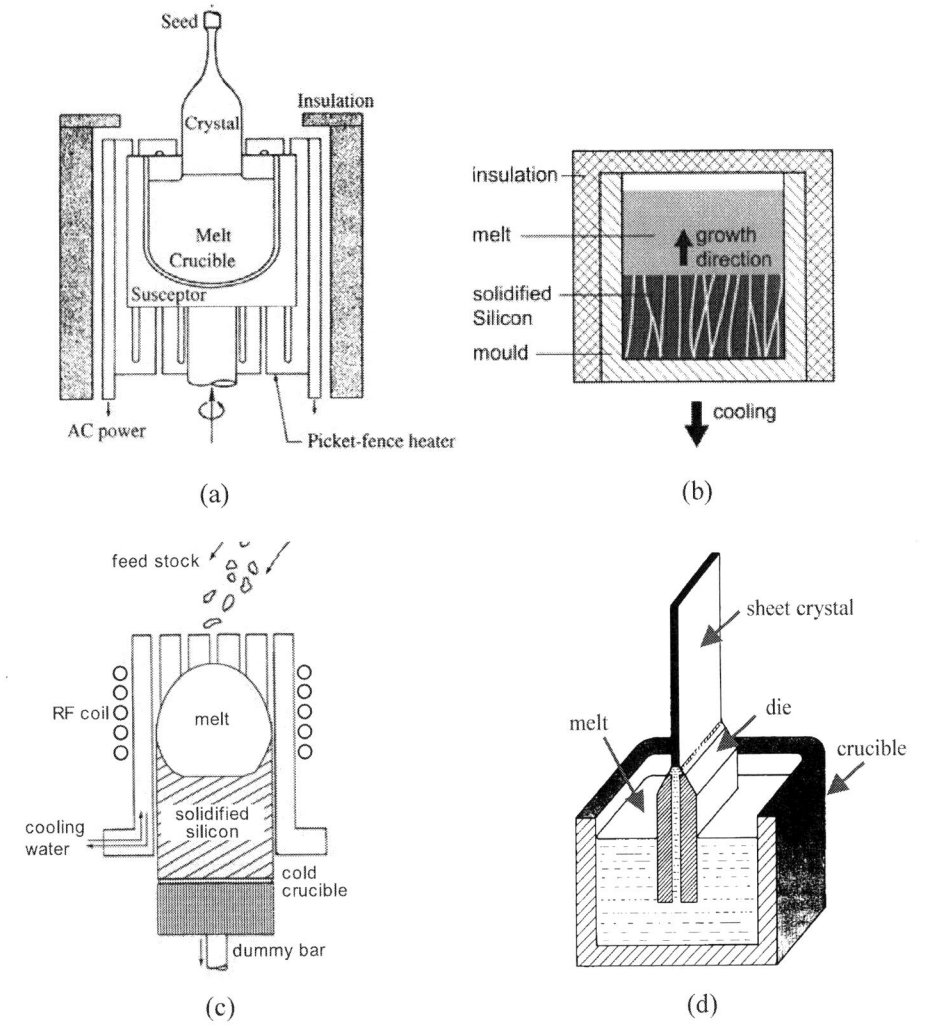

Fig. 1 Representative production technologies of mono- and multi-crystalline silicon ingot for solar cell; (a) Czochralski (b) HEM and (c) CCCC, and (d) EFG [1,4,5].

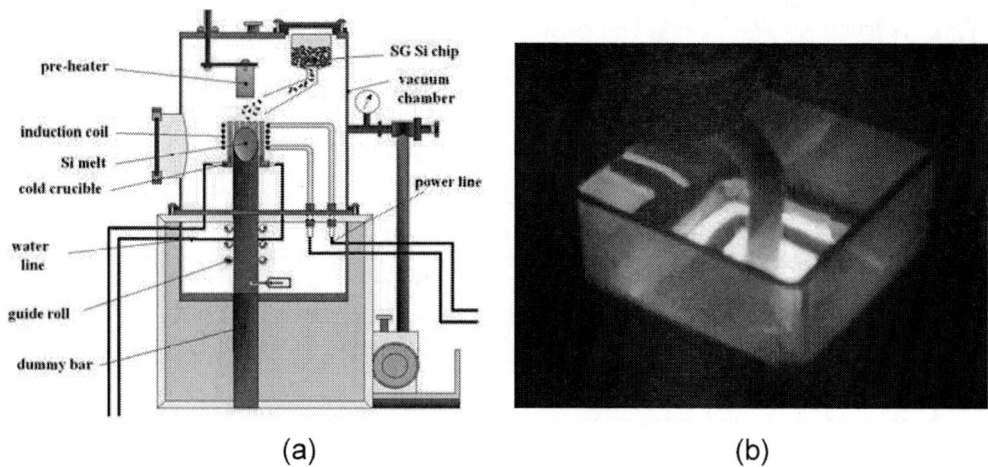

(a) (b)

Fig. 2 (a) Schematic diagram of CCCC method and (b) a scene showing continuous melting and casting process using CCCC equipment developed in this investigation.

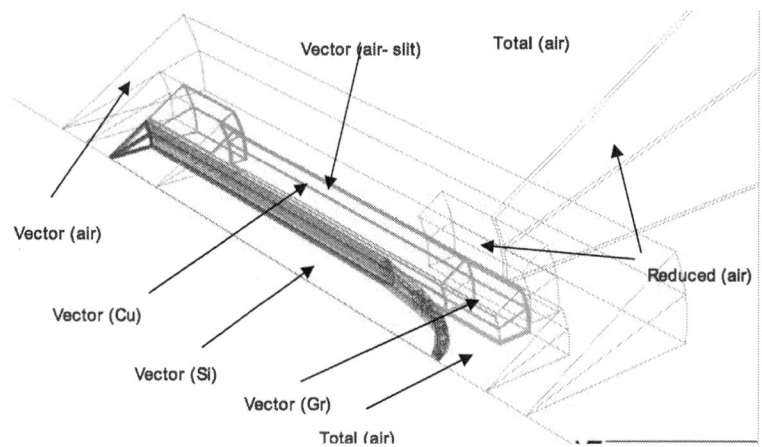

Fig. 3 1-segment model for electromagnetic simulation.

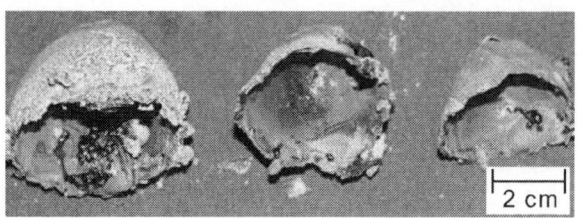

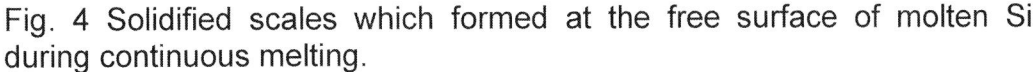

Fig. 4 Solidified scales which formed at the free surface of molten Si during continuous melting.

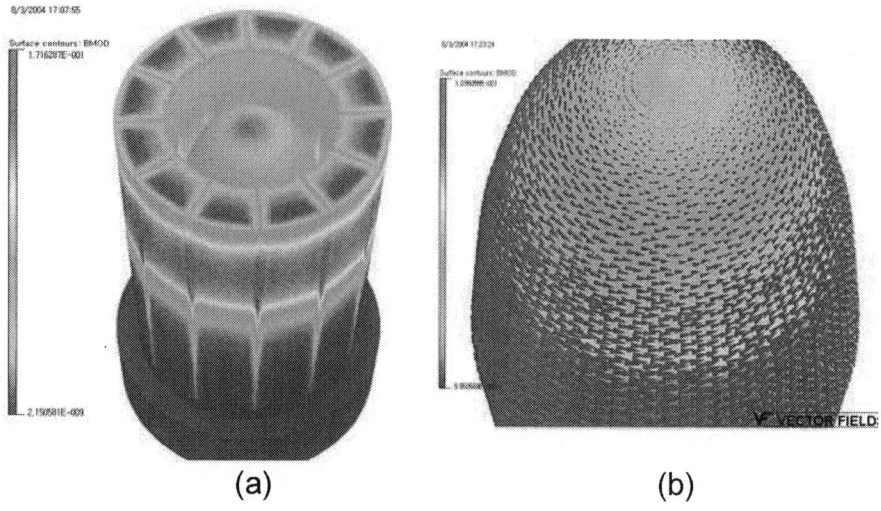

Fig. 5 The distribution of (a) magnetic induction and (b) eddy current density.

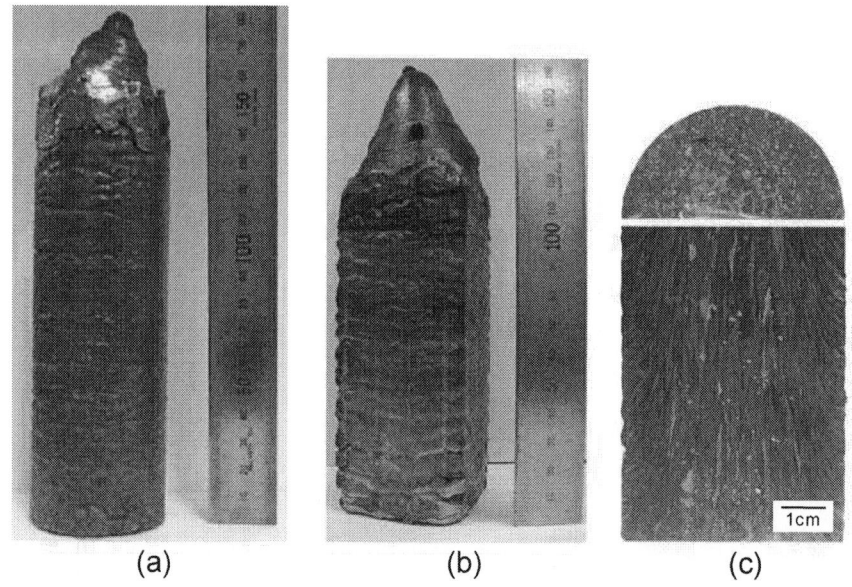

Fig. 6 Multi-crystalline silicon ingots produced by CCCC method; (a) 5 cm circular and (b) square ingots and (c) their representative macrostructure.

Microstructure Observations and Refining Performances of Al-Ti-C Master Alloys Prepared by Salt Route

Dr. Ashok Sharma*

* Professor and Head, Dept. of Metallurgical Engg., Malaviya National Institute of Technology ,Jaipur-302017 (Rajasthan) (India).

Abstract

In the present investigation, influence of temperature on the microstructure of Al-3Ti-0.15C master alloy has been investigated. It is observed that the size, surface characteristics and distribution of TiC particles, as well as the morphology and size of TiAl$_3$ particles are related to melting temperature. The Al-3Ti-0.15C master alloy synthesized at 1200°C and 60 minutes of holding time, shown superior grain refining action as compared when the same has been synthesized at 850°C and 60 minutes of holding time. This is attributed to the favourable surface characteristics, size and distribution of TiC particles synthesized at 1200°C. Conclusions drawn out of this investigation are present in detailed text.

Key words: Al-3Ti-0.15C Master Alloy, Al-7Si Alloy, Grain Refinement, TiC Particles, Microstructures.

Introduction

Grain refinement has been an important technique for improving the soundness of aluminium products for many years. Grain refinement has several benefits in cast alloys like improved mechanical properties that are uniform throughout the casting, distribution of second phase and micro-porosity on a fine scale, better feeding to eliminate shrinkage porosity, improved ability to achieve a uniform anodized surface, better strength and fatigue life. The addition of grain refiners, usually the master alloys containing potent nucleant particles, promotes formation of a fine equiaxed macrostructure by deliberately suppressing the growth of columnar and twin columnar grains [1].

The discovery of carbide – boride theory by Cibula [2 and 3] resulted in the development of Al-Ti-B master alloys as grain refiners for Al and its alloys. These grain refiners contain soluble $TiAl_3$ and insoluble TiB_2 particles. There are some problems with the use of Al-Ti-B grain refiners. These are; agglomeration of borides, blockage of filters, defects during subsequent forming operations and poisoning [4] by certain elements like Zr, V and Cr. These draw backs of Al-Ti-B grain refiners have led to the development of Al-Ti-C master alloys [5 and 6]. Al-Ti-C master alloy, which contain $TiAl_3$ and TiC particles is found to be free from above defects.

In terms of grain refining efficiency, Al-Ti-C is considered more efficient on the basis of the numbers of TiC particles added to the melt, which is only 25% of the number of TiB_2 particles which would be added with number similar addition level of an Al-5Ti-1B grain refiner[7].

Banerji & Reif [5and6] processed Al-Ti-C master alloy for grain refinement through master alloy route. They gave explanations about earlier unsuccessful attempts in respect of poisoning of TiC particles below $1200^{o}C$, and longer holding times as a result of the thin layers of Al_4C_3 on TiC particles. Al-Ti-C master alloys have also been synthesized by salt route. This involves adding graphite powder and K_2TiF_6 in molten Al at elevated temperature [8]. This led to the formation of $TiAl_3$ and TiC particles. There is a effect of temperature on the morphology of $TiAl_3$ particles [9]. If $TiAl_3$ particles are formed at higher temperatures, they have a tendency to become acicular while at lower temperature, they are blocky like.

The present study attempts to develop an efficient Al-3Ti-0.15C master alloy by salt route at $1200^{o}C$ and $850^{o}C$ and to compare the grain refining efficiency of the both. The grain refining efficiency of Al-3Ti-0.15C master alloy so developed has also been compared with commercial Al-5Ti-1B master alloy. The work is assured to satisfying the urgent need of modern aluminium product manufacturing industries.

Experimental Procedure

Al-3-Ti-0.15C grain refiner was synthesized by adding graphite powder of about 50 μm average particle size and potassium titanium fluoride in to aluminium melt at 1200°C and at 850°C followed by gentle stirring for 60 minutes. A low melting-point slag formed contained binary potassium alumino fluorides . The reaction between potassium titanium fluoride and carbon resulted in the synthesis of TiC by this method.

Chemical analysis of Ti was carried out by spectro calorimeter. The carbon content was determined using an automatic combustion apparatus, where the sample is combusted in a stream of oxygen and the carbon of the specimen is converted to CO_2. The nominal compositions of Al-Ti-C master alloys are shown in table 1.

The master alloys so obtained were characterized by X-ray diffraction (XRD), scanning electron microscopy(SEM) and energy dispersive X-ray (EDX) microanalysis.

For grain refinement studies, the Al-7Si alloy (1.00Kg) was melted in a preheated graphite crucible using an electric resistance furnace under a cover flux and held at 720°C. Degassing was carried out with help of hexachloroethane tablets. After fluxing and degassing, 0.2% Al-3Ti-.15C master alloy(sample1) was added to the melt. The melt was gently stirred for 5 minutes, after which no further stirring was carried out. The melt was poured after 60 minutes into an open-ended iron rings (75 mm I.D., 15 mm thick wall and 25 mm high). Similarly, grain refinement was carried out by adding 0.2% of sample-2 in Al-7Si alloy. 0.2% of Al-5Ti-1B was also added in Al-7Si alloy for comparative study. After cooling, the castings were removed from the moulds for further studies. The grain refinement samples were characterized by macroscopy after etching the samples with Poultans reagent and grain size analysis was carried out by the linear intercept method.

Results

The SEM photomicrographs show microstructures of Al-3Ti-0.15C master alloy (sample 1) synthesized at 850°C (Fig. 1a-b). The microstructure (Fig. 1a) reveals TiAl₃ particles which take the form of blocky morphology. This is related to lower melting temperature. TiC particles are closely agglomerated along the α-Al grain boundaries and such regions are shown in Fig.1 (b). It is shown that few TiC particles are coarse and interlaced closely.

Fig. 1 (c-d) show SEM microstructures of Al-3Ti-0.15C master alloy synthesized at 1200°C. The SEM micrographs Fig. 1 (c -d) reveal Al$_3$Ti primary needles and fine TiC particulate phase, which are located at the grain or cell boundaries. This structure is related to higher melting temperature. The TiC particles seem to be agglomerated at lower magnification. However, most of them are disconnected at higher magnifications as shown in Fig. 1(d). Some of the TiC particles are dispersed through out the aluminium matrix, others are inhomogeneous. The cluster of TiC particles having well-developed crystalline morphology and angular to polyhedral shapes. The TiC phase consisted of fine particles usually of submicron size. From these micrographs it is evident that carbon particles of about 50 μm average size have generated a large number of TiC particles which on disintegration from the mother compact disperse as fine particles of submicron size. These observation of morphologies and sizes are quite different from sample 1.

The XRD analysis (Fig. 2) for the aluminium matrix detected two secondary phases, namely, Al$_3$Ti and TiC. Enrichment of Ti in the needle particles and enrichment of both Ti and C in the segregated particulates together with the result from XRD shown in Fig. 2 confirmed the needle particles were TiAl$_3$ and submicron sized particulates are TiC. The EDS microanalysis of these fine particles could be TiC (Fig.3).

Fig.4 (a-d) show photomacrographs of grain refined Al-7Si alloy. Fig. 4 (a) shows macrostructures without any grain refiner. Fig. 4(b) shows macrostructures after grain refinement with sample-1. Fig. 4 (c) shows macrosturctures after grain refinement with sample-2. It is clear from Fig.4 (b-c) that after addition of grain refiners, fine equiaxed grain structure resulted from coarse grain structure Fig.4 (a). Sample-1 showed poor grain refinement in comparison to the grain refinement shown by sample-2. For comparative study, Al-7Si was also grain refined with 0.2% Al-5Ti-1B alloy, Fig. 4 (d). It is observed that sample-2 showed overall superior grain refining results. The average grain sizes of 210 μm were obtained in the grain refined castings which compared well with the 205 μm average grain size obtained with the Al-Ti-B(Table-2).

The influence of grain refinement on mechanical properties is shown in Table-3. The improvement in mechanical properteis is clearly evident from the table. The adition of grain refiners to Al-7Si alloy shows an increase in ultimate tensile strength (UTS) and elongation values.

Discussion

The stoichiometric ratio of (Ti:C in wt%) to from TiC phase is etimated to be 4.Under equilibrium conditions, one part of Ti combines with C to form TiC, while the remaining Ti existing in the form of TiAl$_3$ in the Al-Ti-C master alloys. Fig. 2 shows the XRD pattern of Al-3Ti-0.15C master alloy (Sample-2). The master alloy is composed of three phases; α-Al solid solution, TiAl$_3$ and TiC. The SEM microstructures of four master alloy (Sample-2) revealed the same structures, which are in agreement with the XRD results.

The TiC submicroscopic particles generated from the 50 μm size graphite particles at 1200°C showed well developed crystalline morphology and angular to polyhedral shapes Fig. 1 (d). The inter-particle spacing of TiC particles varied considerably and could not be generalized. However, these particles showed best grain refining efficiency.

Cibula [2and3] postulated the carbide-boride theory which assumed that both TiC and TiB$_2$ particles are virtually insoluble in aluminium melt and can act as heterogeneous nucleants. In the present work, the addition of Al-Ti-C master alloys can markely refine Al-7Si alloy. With the same addition, the refinement performance of Al-Ti-C alloys is similar to that of commercial Al-5Ti-1B alloy. For commercial Al-5Ti-1B master alloys, there seems to be sufficient evidence that TiB$_2$ is the totally dominating phase for the grain refinement [9]. In case of Al-Ti-C master alloys, it has been reported that both the excess Ti (in the form of AlTi$_3$) and TiC phase play important role in grain refinement of aluminium alloys [10]. Two conditions need to be fulfilled to obtain efficient grain refinement : (i) a sufficient number of potential nuclei must be present in the melt; and (ii) a large fraction of potential nuclei must be activated. The Al$_3$Ti and TiC particles act as substrates for heterogeneous nuleation of α-Al dendrites during solidification. The grain refining behaviours of master alloy appears to be sensitive to its microstructure, particularly the morphology and size distribution of TiC and TiAl$_3$ particles, which in turn are influenced by the processing parameters used in preparation of master alloy, such as reaction temperature and reaction time.

Heterogeneous nucleation of α-aluminium dendrites takes place on octahedral shaped TiC particles [11]. There is no evidence that effective nucleation involves any aluminde coating of the particles. The lack of any coating on TiC particles may explain the reduced problems with poisoning and agglomeration as is observed with TiB_2 particulates. Drastic reduction in grain size has been observed after addition of Al-3Ti-0.15C grain refiners to the Al-7Si melt, Fig. 4 (a-c). The alloy shows complete conversion of coarse grain α-Al dendrites to fine grain equiaxed α-Al dendrites due to the presense of Al_3Ti and TiC particles. The results of microscopy supported the DAS analysis and the addition of 0.2% Al-Ti-C grain refiner has completely grain refined the Al-7Si alloy. It is clearly observed that Al-Ti-C are equally efficient grain refiner as Al-5Ti-1B master alloy Fig. 4 (d). Overall, Al-3Ti-0.15C synthesized at 1200°C showed best grain refining efficiency among other master alloys. This is attributed to the well defined crystalline morphology and submicroscopic size of TiC particulates available as heterogeneous nucleants.

Conclusions

In Al-3Ti-0.15C master alloy, the size, surface characteristic and distribution of carbides, as well as the morphology and size of $TiAl_3$, are related to the melting temperature.

The grain refining efficiency of Al-3Ti-0.15C refiners largely depends on the surface characteristic, size and distribution of carbides. TiC has played major role on the refining efficiency. The Al-3Ti-0.15C master alloy, which was manufactured at higher temperature has an intense grain refining action similar to Al-5Ti-1B master alloy.

References

1. McCartney D.G.: International Materials Reviews, 34 (5), 1989, p. 248
2. Cibula: J. Inst. Met., 1951-1952, 80, pp.1-16.
3. Cibula: J. Inst. Met., 1949-1950, 76, pp.321-360.
4. Lee M.S., Terry B.S., and Grievenson P., Metall. Trans. 24B, 1993, p 947.
5. Banerji A and Reif W., Metall. Trans. A, vol 17A, Dec., 1986, pp.2127-37.
6. Banerji A and Reif W.: Metall. Trans. A, vol 16A, 1985, pp.2065-68.
7. Schumacher P. and Greer A.L., TMS Light Metals, 1995, p.869
8. Zhang B.Q., Fang H.S. Lu L., Lai M.O., Ma H.T., and Li J.G.: Met. & Mat. Trans. A, vol 34A., 2003, pp 1727-33.
9. Murty B.S., Kori S.A. and Chakraborty M, Int. Mat. Reviews, 47, 1, 2002, p 1-27.
10. Xiangfa Liu, Zhenquing Wang, Zhang Zuogui and Xiufang Bian, Mat. Sci. & Engg., A 332, 2002, pp 70-74.
11. Small C.M., Prangnell P.B., Hayes F.H. and Hardman A: Proc. 6th Int. Conf. in 'Aluminium Alloys (1 C AA-6) Tokyo Japan Institute for light metals, Vol.1, 1998, pp. 213-18.

Tables

Table 1: Al-Ti-C master alloy samples synthesized in laboratory

Samples No.	Composition	Melting temperature (°C)	Holding time (min)
Sample 1	Al-3Ti-0.15C	850	60
Sample 2	Al-3Ti-0.15C	1200	60

Table 2: Average grain size measured of Micro-samples

Alloy	μm
Without Grain Refiner	2710
Al-3Ti-0.15C (Sample 1)	320
Al-3Ti-0.15C (Sample 2)	210
Al-5Ti-1B (Commercial Grain Refiner)	205

Table 3: Influence of grain refinement on the mechanical properties of Al-7Si after 60 minutes of addition of grain refiner

Alloy	UTS (MPa)	Elongation (%)
Al-7Si (without any grain refiner)	150.4	9.0
Al-7Si (grain refined by sample 1)	170.5	10.5
Al-7Si (grain refined by sample 2)	178.3	11.5
Al-5Ti-1B (Commercial Grain Refiner)	175.2	11.0

Figures

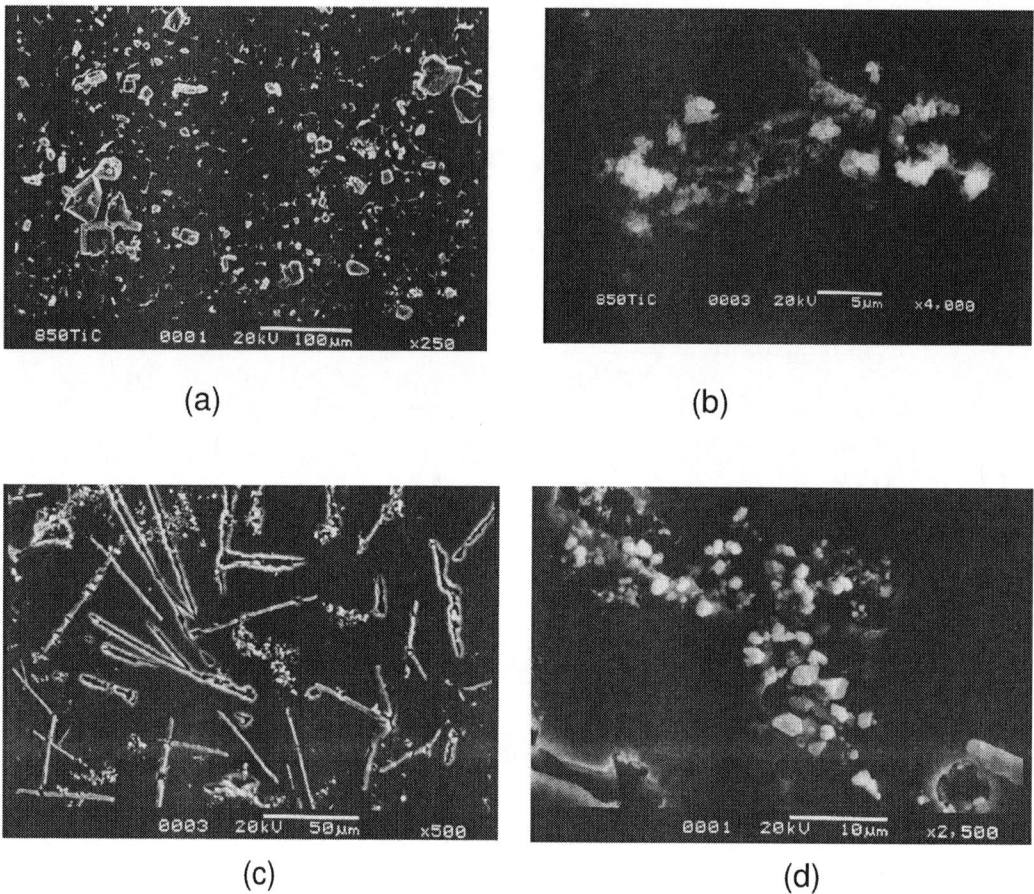

(a)

(b)

(c)

(d)

Fig. 1: SEM examination of Al-3Ti-0.15C master alloy showing typical microstructures and TiC segregated particles synthesized at (a and b) 850°C and (c and d) 1200°C.

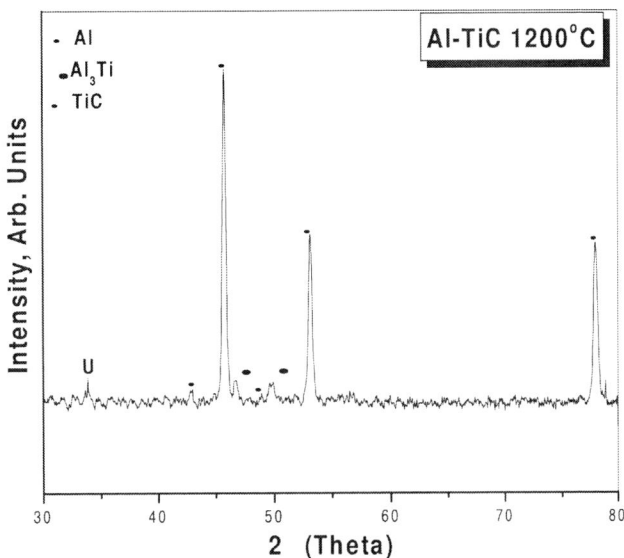

Fig. 2: XRD diffraction pattern of Al-3Ti-0.15C master alloy (Sample-2)

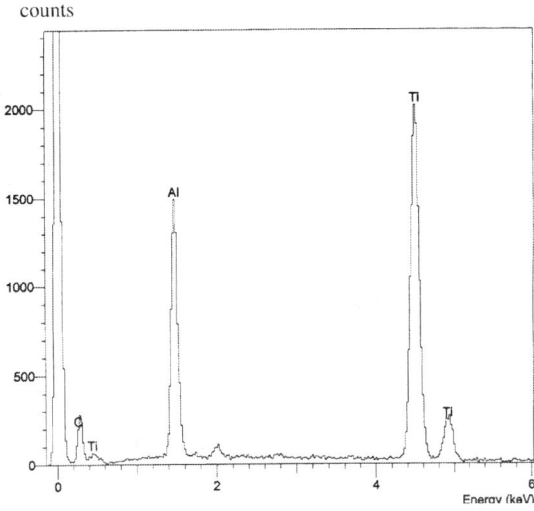

Fig. 3: EDS microanalysis shows Ti and C peaks (Sample-2)

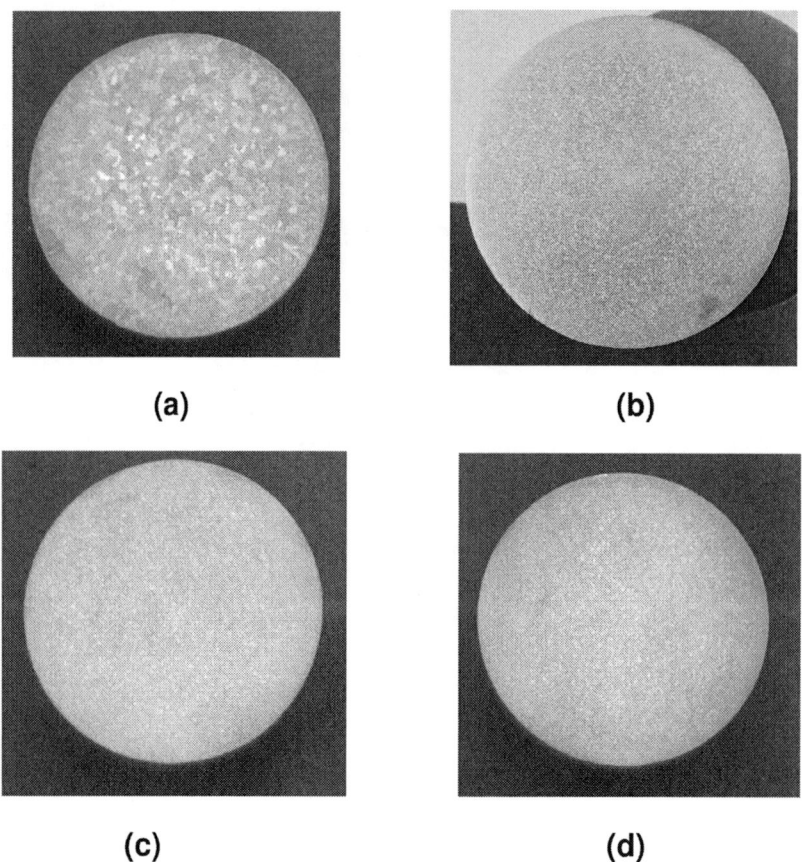

(a)　　　　　　　　　　　　**(b)**

(c)　　　　　　　　　　　　**(d)**

Fig.4:　**Macro structures of Al-7Si alloy**
　　　　(a) without any grain refiner
　　　　(b) grain refined with sample-1
　　　　(c) grain refined with sample-2 and
　　　　(d) grain refined with Al-5Ti-1B master alloy.

e-mail: ashok_mnit12@yahoo.co.in

Effects of Inoculation and Solidification Rate on the Thermal Conductivity of Grey Cast Iron

Daniel Holmgren, Attila Diószegi and Ingvar L. Svensson.

Jönköping University, Dept. of Mechanical Engineering/Component Technology, S-551 11 Jönköping Sweden

Abstract

Laser flash technique has been used to evaluate the thermal conductivity of pearlitic grey cast iron solidified at three distinct cooling rates. Ten different heats were performed with various inoculation treatments. Influences of fraction primary phase and eutectic cell diameter have been revealed by colour etching. SEM studies have been accomplished in order to investigate the graphite morphology of the examined materials.

The thermal conditions during solidification together with the inoculation strongly affect the thermal conductivity of grey iron. In general, a more powerful inoculation increases the thermal conductivity. Increased fractions of primary solidified dendrites lower the ability to transfer heat. Small amounts of titanium increase the thermal conductivity of grey cast iron considerably. The results show that maximum thermal conductivity is achieved at moderate cooling rates.

Key words Grey Iron, Thermal Conductivity, Inoculation, Graphite Morphology, Primary Austenite

Introduction

The thermal conductivity is important in high temperature applications, i.e. permanent moulds, railway-breaks, cylinder heads etc., where the heat has to dissipate quickly. Thermally induced stresses and subsequent distortions can be reduced by having less pronounced thermal gradients within the components, which is obtained by an increased thermal conductivity.

The thermal conductivity and to some extent also the thermal diffusivity of cast iron have been reviewed recently [1]. Only one work dealing with the effects of different inoculation treatments on the thermal conductivity of grey cast iron, accomplished by Holmgren *et al.* [2], has been found in the literature. Some works discuss the beneficial effects of the interconnected graphite skeleton within the eutectic cells of grey cast iron on the thermal transport properties, although no relationships have been established [3,4].

The aim of this work is to correlate the effects owing to inoculation and solidification rate to the thermal conductivity of grey cast iron. In the previous work [2], it was concluded that a high fraction of former primary phase depresses the thermal conductivity of grey cast iron. However, no clear relationship between the eutectic cell size and the thermal conductivity could be established.

In this work, the number of inoculants is extended from two to ten, as compared to the previous work [2]. Three cooling conditions are applied, hence 30 combinations of different solidification rates and inoculation treatments are examined. The mechanical properties as well as the microstructure formation of the materials participating in the study have been investigated elsewhere [5].

Experimental

The thermal conductivity of a series of grey iron samples has been evaluated. The specimens originated from the same hypoeutectic melt, but were treated with ten different inoculants. Three cooling conditions were applied on each combination of inoculant and alloy. The chemical composition can be found in Table 1. The carbon equivalent in Table 1 is calculated as $CE = C(wt\%) + Si(wt\%/3) + P(wt\%/3)$. All inoculants were based on Fe-Si-Ca-Al with additions of different elements; see Table 2 for compositions of the inoculants. The mould, see Figure 1, consisted of three connected cylindrical cavities with different moulding materials surrounding each cylinder; metal-chill, furan sand and insulation. The cylinders cast in metal-chill, sand and insulation had solidification times of approximately 40 s, 300 s and 1100 s respectively. Detailed information concerning the casting procedure can be found in Ref. 5.

The thermal diffusivity, α, was determined by a laser flash apparatus, which is based on the principle presented by Parker et al [6]. The following relation was used in order to calculate the thermal conductivity

$$\lambda = \alpha \rho C_p \qquad (1)$$

where C_p is the specific heat and ρ the bulk density. The specific heat measurements were performed in a Netzsch DSC 404 C Pegasus® differential scanning calorimeter. Only small differences were established between different cast irons and an average value could be applied. The evaluation of the thermophysical properties has been described in detail elsewhere [2].

In order to obtain the density at room temperature, the samples were weight both in air and in distilled water of known temperature. The knowledge of the lifting force acting on the samples enabled the calculation of the individual volumes by means of Archimedes principle. The density at elevated temperatures, ρ_T, was calculated by the following expression

$$\rho_T = \rho_{RT} \Big/ \left(1 + 3\alpha\Delta T\right) \qquad (2)$$

where ΔT is the increase in temperature, ρ_{RT} is the bulk density at room temperature and α is the linear coefficient of thermal expansion. The latter material property was measured by a Netzsch DIL 402 C, dilatometer. The room temperature density and the thermal diffusivity of the samples in Heat 1 and Heat 10 were measured at Netzsch Applications Laboratory.

The primary phase and the eutectic cells were exposed by etching in a reagent based on picric acid. Experimental details have been discussed in Ref. 2. The graphite morphology of some selected deep etched samples was investigated in a scanning electron microscope, model JEOL JSM-5300.

Results
Figure 2 shows the thermal conductivity, calculated with equation (1), of the materials cast in insulation, sand and chill. The diagrams demonstrate the great temperature dependence of the thermal conductivity of grey iron. Generally, the cast irons treated with more powerful inoculants, in general denoted with higher numbers, have a higher thermal conductivity than the less efficiently inoculated materials. Consequently, although having a more extended network of graphite, owing to larger eutectic cells, a lower thermal conductivity is often obtained in the samples having less potent inoculation. Almost exclusively, the sand cast materials have somewhat higher thermal conductivity than the counterparts cast in insulation. Further, independently of solidification rate and inoculation, the thermal conductivity of the various grey irons approaches each other at elevated temperatures.

The thermal conductivity of lamellar graphite cast iron show a pronounced drop if the solidification rate is too high. In the materials cast in chills, the propagation of heat is restricted by a high fraction of primary solidified austenite due to higher undercoolings. Further, all materials cast in chills contain undercooled interdendritic graphite which has been reported as negative for the thermal transport properties [7,8]. The noticeably lower values for specimen Chill_1 can be explained by the presence of cementite, which was detected in the previous work [2].

Discussion

Figure 3 a shows the thermal conductivity at room temperature versus the eutectic cell size of the investigated cast irons. As depicted in the figure, the thermal conductivity does not raise continuously as the eutectic cell size increases. Instead, as mentioned earlier, the thermal conductivity decreases somewhat if the cooling rate alters from moderate to low and at stronger inoculation treatment, which increase the eutectic cell size. The samples inoculated with inoculant No. 7, containing titanium, are treated separately due to its extraordinary thermal conductivity which will be discussed later in the work. The diagram also depicts that the beneficial effects of a decreased cell size owing to more efficient inoculation becomes more obvious as the cooling rate increases.

Figure 3 b shows the relation between the thermal conductivity and the fraction of primary austenite, now transformed into pearlite. Two distinct groups can be distinguished, the first group, consisting of the materials cast in sand and insulation with moderate and low cooling rates, and the second, consisting of the undercooled materials cast in metal-chill producing high cooling rates.

When investigating the mechanical properties of the alloys participating in the study, it was concluded that the ultimate tensile strength and the elongation increases as the rate of heat extraction increases from low to medium [5]. As mentioned above, also the thermal transport properties increase. Hence, the combination of thermal conductivity and mechanical properties can be controlled by means of the cooling conditions. If an optimizing of the strength is desired, high cooling rates are required. On the other hand, in applications obliging good mechanical properties combined with high thermal conductivity, a moderate cooling rate is most favourable. Low cooling rates promote neither high strength nor an optimization of the thermal conductivity.

Figures 4 and 5 show some microstructures, which have been colour-etched. The higher thermal conductivity of the grey irons treated with more powerful inoculants can be explained by different growth modes of graphite which affect the character of the entire eutectic cell. This idea was discussed briefly in the previous work [2].

Consequently, the eutectic cells in these structures contain long graphite lamellae surrounded by thin layers of eutectic austenite, which gives the eutectic cells a scalloped outline, see Figure 5 b. Similar growth modes of the eutectic cells have been observed elsewhere [5]. Hence, the addition of potent inoculants also facilitates long and straight graphite lamellae being efficient in the transfer of heat across the eutectic cells. In the materials with less effective inoculation treatment, having larger eutectic cells, a great fraction of the graphite lamella are shorter, curved and ends up in the central areas of the eutectic cells. Thus, the heat has to pass longer distances in the matrix with lower thermal conductivity until reaching graphite belonging to adjacent eutectic cells. Therefore, the degree of undercooling required for eutectic nucleation and growth is higher if the degree of inoculation is less, which explains the alteration from straight and long graphite flakes to small and branched in the larger cells, which in turn affects the ability of heat propagation throughout the structure.

Similarly, the same mechanism, might explain the observable fact that materials cast in chills, containing highly branched type D graphite in combination with a higher fraction of primary solidified austenite, obtains a considerably reduced thermal conductivity compared to cast irons solidified at moderate- and low cooling rates.

As depicted in Figure 5 b, light metallic areas which probably are solidified late and therefore highly segregated, containing plenty of scattering points, are located between graphite-austenite colonies within the scalloped eutectic cells instead of in the eutectic cell borders. Hence, an improved thermal conductivity is expected if less of the finally solidified melt is positioned between the eutectic cells. In the materials inoculated with inoculant No. 7, containing titanium, exceptionally jagged eutectic cell borders with distinct branches of graphite and former austenite appears in light microscopy, see Figure 5 b. Consequently, these cast irons show evidence of an outstanding thermal conductivity.

The distinctions in graphite morphology become more obvious when investigating deep etched samples by scanning electron microscopy at the same magnification. Branching of the growing graphite flakes and a small lamellar spacing is facilitated by increased undercoolings [9]. Sand_1, with a relatively inefficient inoculation, contains large cells with plenty of curved and branched graphite, see Figure 6 a. In Sand_ 7, which has an extraordinary thermal conductivity, straight extended lamellas and eutectic cells with graphite lamellae sometimes reaching far away from the centre of the eutectic cells can be found, see Figure 6 b. Figure 6 c shows the graphite morphology of Sand_10, which has a thermal conductivity intermediate that of Sand_7 and Sand_1. Figure 6 d shows the graphite morphology of Sand_7 at lower magnification.

However, the superior thermal conductivity of cast irons treated with the inoculant containing titanium, No. 7, can not be explained only by the effects of inoculation since the measured eutectic cells in these samples are relatively large. Hence, the eutectic growth mode seems to be altered by an addition of titanium producing few long and straight graphite lamellae independently of the degree of undercooling. As a consequence, these materials contain both relatively large eutectic cells and extended graphite flakes being effective in the transfer of heat. A decreased number of eutectic cells and presence of type D graphite, owing to increased undercooling due to inefficient inoculation, was observed by Ruff and Wallace in grey irons inoculated with inoculants containing Ti [10]. Except for the samples cast in chills, no evidence of undercooled graphite was identified in this investigation. However, the titanium levels are considerably lower in the present work.

A connection between the nucleation and growth of the eutectic cells and the pre-existing skeleton of primary phase might exist. An increased number of dendrites have been observed in grey irons treated with inoculants containing Ti [10]. Nevertheless, the relation between the formation and growth of the eutectic cells and the existing network of austenite dendrites and the inoculation of the primary phase can not easily be clarified in this work.

Conclusions

(i) A more potent inoculation generally increases the thermal conductivity. This can be explained by a different morphology of the graphite lamellae and the entire eutectic cells containing long and less branched graphite flakes which are favourable for the thermal conductivity.

(ii) An optimal cooling rate exists where a maximum thermal conductivity is obtained. The thermal conductivity increases when decreasing the cooling rate of grey irons until a maximum is obtained at moderate rates of heat extraction. A further reduction of the cooling rate during solidification decreases the thermal conductivity.

(iii) Optimized designs can be realized by altering the cooling surroundings or the dimensions of the component since a coupling between the mechanical properties and the thermal conductivity exists.

(iv) Small amounts of titanium increase the thermal conductivity of grey cast iron considerably, which is a result of long and straight graphite flakes.

(v) The thermal conductivity decreases as the fraction of former primary austenite increases. This can be explained by the lower thermal conductivity of the matrix compared to the graphite.

References

1. D. Holmgren, *Review of thermal conductivity of cast iron*, Int. J. Cast Metal Res., 18, No 6, Dec. 2005, pp 331-345.

2. D. Holmgren and I.L. Svensson, *Thermal conductivity–structure relationships in grey cast iron*, Int. J. Cast Metal Res.,18, No 6, Dec. 2005, pp 321-330.

3. J. Ormerod, R. Taylor and R.J. Edwards, *Thermal diffusivity of cast irons*, Metals Technology, 5, No. 4, April 1978, pp 109-113.

4. K. B. Palmer, *The thermal and electrical conductivities of ductile cast iron and several grey cast irons by J. H. Brophy and M. J. Sinnot*. BCIRA Journals, 8, No 540, 1960, pp 266-272.

5. A. Diószegi, On Microstructure Formation and Mechanical Properties in Grey Cast Iron, Linköping Studies in Science and Technology Dissertation No. 871, 2004, ISBN 91-7373-939-1.

6. W. J. Parker, R. J. Jenkins, C. P. Butler and G. L. Abbot, *Flash Method of Determining Thermal Diffusivity, Heat Capacity and Thermal Conductivity*, J. Appl. Phys., 32, No. 9, 1961, pp 1679-1684.

7. M. C. Rukadikar and G. P. Reddy, *Influence of chemical composition and microstructure on thermal conductivity of alloyed pearlitic flake graphite cast irons*, J. Mater. Sci., 21, No. 12, Dec. 1986, pp 4403-4410.

8. L. Hecht, R. B. Dinwiddie and H. Wang, *The effect of graphite flake morphology on the thermal diffusivity of gray cast irons used for automotive brake discs*, J. Mater. Sci., 34, No. 19, Oct.1999, pp 4775-4781.

9. J.R. Davis (ed.), ASM Speciality Handbook Cast Irons, 1996, ASM International, ISBN 0-87170-564-8.

10. G. F. Ruff and J. F. Wallace, *Effects of solidification structures on the tensile properties of gray iron*, AFS Trans., 56B, 1977, pp 179-202.

Acknowledgements

This investigation is financed by the Swedish Knowledge Foundation, School of Engineering, Jönköping University, Volvo Powertrain AB, Skövde, Volvo Powertrain AB, Gothenburg, and Daros Piston Rings AB, Gothenburg. In addition, the authors would like to thank the following persons: Mr. Leif Andersson, Mr. Lars Johansson, Ph. D student Rickard Källbom and Ph. D student Torsten Sjögren.

E-mail: Daniel.Holmgren@ing.hj.se, Attila.Dioszegi@ing.hj.se
Ingvar.Svensson@ing.hj.se

Tables

Table 1 Chemical composition of alloys (wt%)

Heat No.	C	Si	P	Mn	S	Cr	Mo	Ni	Cu	Mg	Pb	Ti	CE
1	3.380	1.970	0.036	0.680	0.100	0.111	0.062	0.053	0.263	0.003	0.002	0.013	4.05
2	3.350	2.100	0.033	0.670	0.100	0.109	0.060	0.052	0.260	0.003	0.002	0.013	4.06
3	3.330	2.410	0.034	0.660	0.096	0.109	0.060	0.052	0.258	0.003	0.002	0.014	4.14
4	3.370	2.180	0.035	0.670	0.096	0.112	0.058	0.054	0.257	0.003	0.002	0.014	4.11
5	3.350	2.110	0.033	0.670	0.116	0.109	0.057	0.055	0.260	0.003	0.002	0.014	4.06
6	3.430	2.140	0.035	0.680	0.092	0.113	0.059	0.054	0.254	0.003	0.003	0.014	4.16
7	3.410	2.160	0.035	0.670	0.093	0.113	0.058	0.055	0.255	0.003	0.003	0.038	4.14
8	3.450	2.020	0.036	0.680	0.097	0.114	0.058	0.056	0.259	0.003	0.003	0.014	4.14
9	3.480	2.040	0.036	0.670	0.094	0.114	0.058	0.056	0.259	0.003	0.003	0.014	4.17
10	3.500	2.060	0.035	0.670	0.099	0.112	0.057	0.055	0.260	0.003	0.002	0.014	4.20

Table 2 Addition and chemical composition of inoculants (wt%)

Inoculant No.	Si	Ca	Al	Sr	RE	Ba	Zr	Ti	C	Addition (wt%)
1	73–78	0.1 max	0.5 max	0.6 – 1.0	-	-	-	-	-	0.06
2	73–78	0.1 max	0.5 max	0.6 – 1.0	-	-	-	-	-	0.39
3	73–78	0.1 max	0.5 max	0.6 – 1.0	-	-	-	-	-	0.90
4	72 -78	0.5-1.0	0.5-1.3	-	1.5-2.0	-	-	-	-	0.39
5	72-77	1.0-2.0	0.8-1.5	-	-	2.0-3.0	-	-	-	0.39
6	44-50	2.5-3.5	1.0	-	-	-	1.5-2.0	-	-	0.39
7	51-55	1.0	1.0-1.3	-	-	-	-	9.0-11.0	-	0.39
8	32	0.5	0.7	-	-	4.5	-	-	50.0	0.06
9	32	0.5	0.7	-	-	4.5	-	-	50.0	0.19
10	32	0.5	0.7	-	-	4.5	-	-	50.0	0.45

Figures

Figure 1: Casting rig

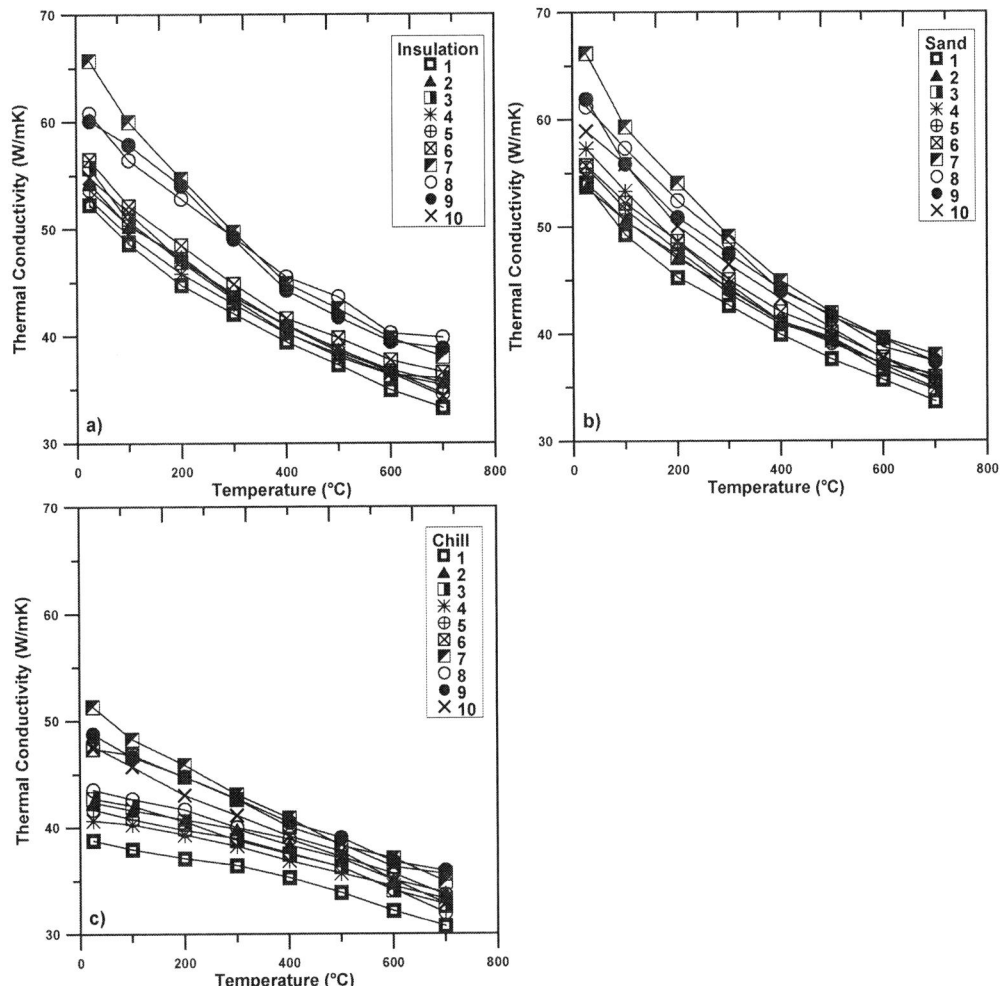

Figure 2: Thermal conductivity of materials cast in a) insulation, b) sand and c) metal-chill.

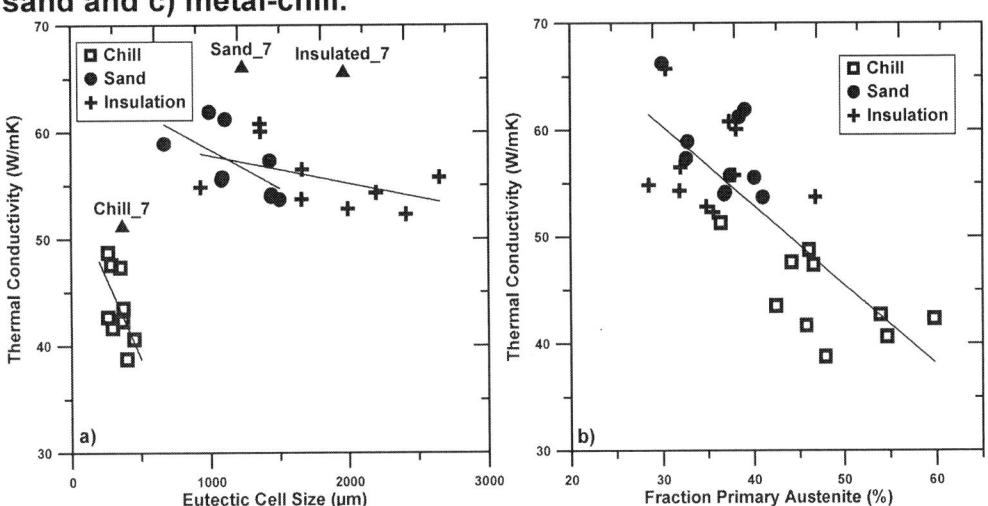

Figure 3: Thermal conductivity at room temperature versus a) eutectic cell size and b) fraction primary austenite.

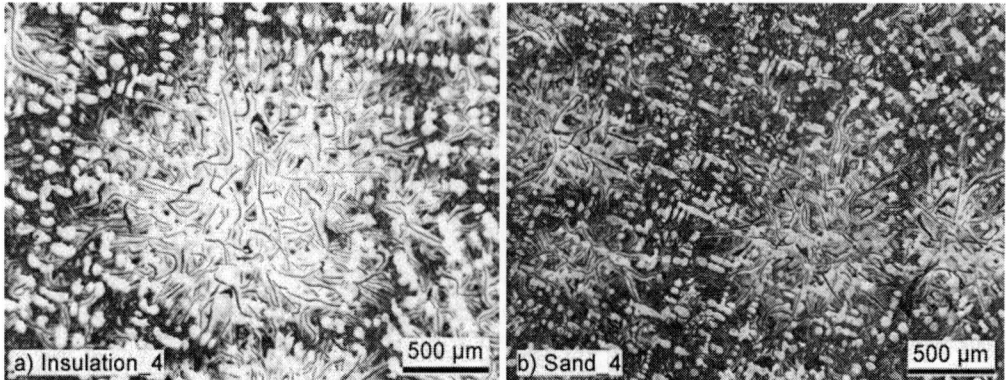

Figure 4: Colour-etched microstructures a) Insulation_ 4 and b) Sand_4.

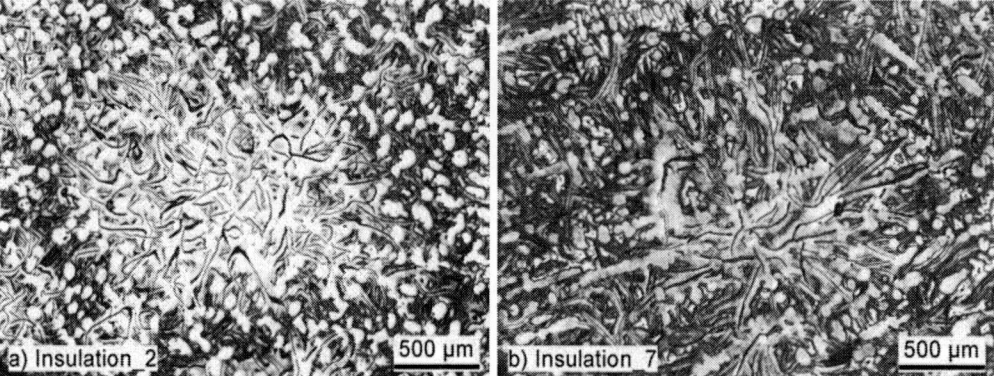

Figure 5: Colour-etched microstructures a) Insulation_ 2 and b) Insulation_7.

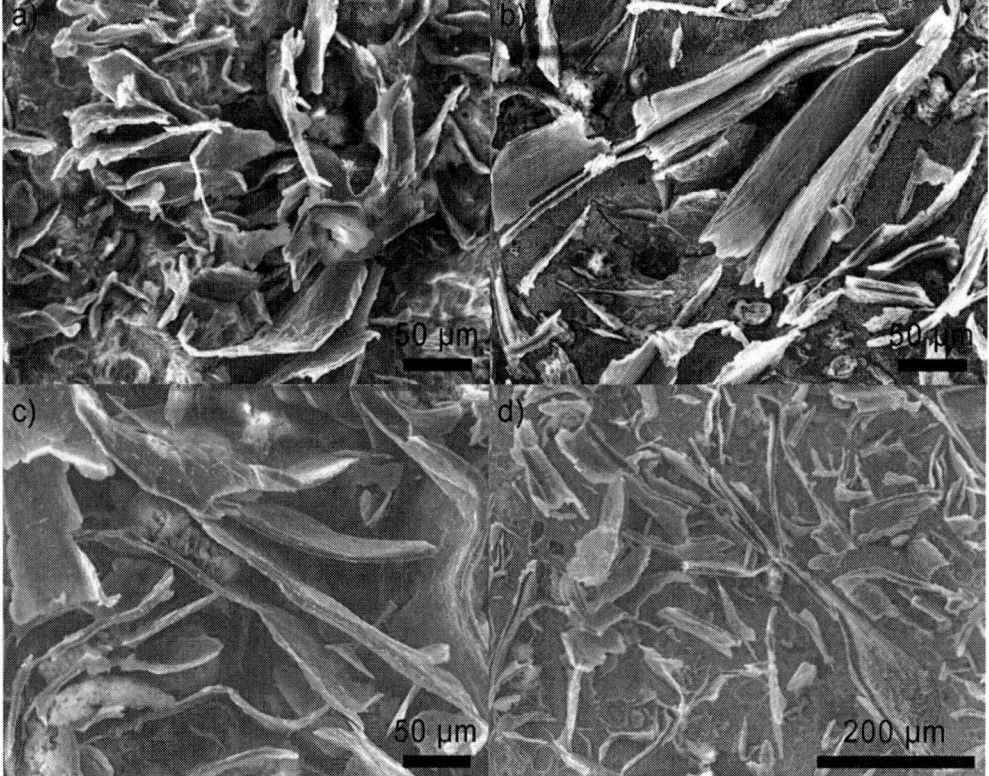

Figure 6: SEM pictures showing the graphite morphology of a) Sand_1, b) Sand_7 c) Sand_10 and d) Sand_7.

Thermochemistry and Kinetics of Iron Melt Treatment

Simon N. Lekakh[*], David G. C. Robertson[*] and Carl R. Loper Jr. [**]

[*]University of Missouri–Rolla, [**]University of Wisconsin-Milwaukee, U.S.A.

Abstract

Nodulization for producing spherical graphite in ductile iron (DI) and inoculation for increasing graphite nodule count are the result of chemical reactions of the additives with the melt. The multi-component equilibria and the kinetics of additive dissolution in the melt were computed and studied experimentally. Measurements of the oxygen activity confirmed the calculated features of the refining reactions. Possible reaction paths were identified, as ferrosilicon-based additives dissolved and reacted with the iron melts. Adding more nuclei to the melt, by C/SiC melt pretreatment and late additions of small amounts of oxides and sulfides, improved DI inoculation. Intensification of inoculant dissolution by changing the additive shape, and by argon stirring, were also investigated for increasing inoculation efficiency.

Key words

Ductile iron, nodulization, inoculation

Introduction

Ductile iron (DI) treatment dramatically changes the structure and the properties of castings. Deep refining of the melt by *Mg* transforms the flake graphite to spheroidal graphite. Because liquid iron treated by magnesium has a tendency to undercool resulting in meta-stable cementite formation, DI treatment also includes inoculation by *FeSiX* additives which create additional graphite nuclei, resulting in a carbide-free structure even in thin wall castings. Therefore, the industrial DI treatment typically consists of two stage – nodulization and inoculation.

Nodulization. Magnesium is the main reagent used in nodulization. Other metals (*Ca, Ba,* and *Ce*) are often used to increase the efficiency of nodulization. These elements have a large affinity for both the sulfur and oxygen impurities in the iron melts and remove these soluble impurities. Also the remaining concentrations of *Mg* and *Ce* promote graphite spheroidization in DI. The processes which take place during DI treatment are often described as analogous to steel refining. However, the high carbon and silicon contents in irons significantly change the sequences of the refining reactions. In this study, the thermodynamic features of the refining reactions of alkali and rare earth metals with impurities of sulfur and oxygen were investigated by using computer simulation and experiments. These methods were also applied to the optimization of the additive compositions for the nodulizing stage of DI treatment.

Inoculation. The nucleation of graphite nodules before the formation of carbides requires the presence of substrates that can initiate solidification [1]. A possible ranking of nucleants is as follows: graphite (highest – least energy required), silicate, oxides, sulfides, carbides, nitrides, and austenite (lowest). The strong inoculation effect of *FeSiX* alloys can be considered to be due to the formation of graphite-containing substrates during dissolution of inoculants in the iron melt. When *FeSiX* dissolves, regions with high *Si* concentration will arise around the inoculant particles [2]. In these regions, favorable conditions for the formation of graphite-containing substrates are created. According to thermodynamic equilibrium [3, 4], graphite will be stable in liquid irons alloyed by *Si* >5-6% at 1300-1500°C and *SiC* will be stable when *Si* >23-28%. A number of kinetic and thermodynamic factors influence these processes, including the rate of inoculant dissolution versus the rate of mixing, and the additional chemical reactions which may create extra non-metallic substrates for subsequent growth of graphite [5,6]. Experimental analysis of graphite nuclei compositions is given in recent research [7].

In this work, the thermodynamics and kinetics of nodulization and inoculation were studied using both computational and experimental methods. Some methods of improving DI treatments were developed on the basis of the additional understanding of the mechanism of the process.

Procedure

Computing. In the actual DI treatments parallel reactions take place between the additives and the impurities already present in the melt. To predict which products will be formed as a function of the composition and the temperature, computer simulation (FACTSage software) based on the minimization of the Gibbs free energy was used. The parallel reactions were simulated step-by step as the additive amount was increased. It was possible to calculate both the products of the reactions as well as the composition of the remaining melt.

The thermodynamic calculations allow us to predict the maximum possible degree of reaction, but in real processes the kinetics also play an important role in determining how the process actually proceeds. Unsteady heat transfer between the iron melt and the initially cold additive particle of FeSiX based nodulizers and inoculants was computed while taking into account the latent heat as well as the heat of possible exothermic reactions. The FLUENT software package was used for simulation of dissolution process with laminar melt flow.

Experiments with high purity iron. High purity iron with 3.7%C, 1.8%Si and controlled concentration of impurities was used in this study. Special alkali and rare earth containing master alloys (with unreactive *Fe, Cu* or *Ni*) were used for the refining experiments, which were carried out under argon. These master alloys had high recovery in the small volume of the treated melts (0.3 kg). An electrochemical method was used to qualitatively measure the oxygen activity in the melts.

Experiments with industrial grade irons. Industrial grade irons were melted in a 50 kg induction furnace. Two types on *FeSiX* additives were used for DI treatment in the ladle. The first type were conventional additives with an equi-axed shape with 8-12 mm particles. The second type were rapidly cooled ribbon-shape additives with a thickness of 0.5-2.0 mm. These were produced directly from the melt by a continuous casting process with the use of a water-cooled copper wheel [8]. The dissolution of the additives was studied with thermocouples placed in the center of the additives submerged into the iron melt. In addition, two special reagents were tested for increasing the efficiency of the DI treatment. Reagent 1 was a treated graphite (75%C, 25%SiC), and was used as a pre-treatment agent before nodulization. Reagent 2 was an additive containing a 1:1 mixture of non-ferrous (*Cu,Fe*) oxides and sulfides, and was used together with the *FeSiX* inoculant after the nodulizing treatment. A step bar casting with thickness from 6 to 50 mm, a chill wedge, and a six-pin core mold with 12mm, 10mm, 8mm, 6mm, 4mm, and 2mm pin diameters, were used for evaluation of the effectiveness of these treatments. The samples were cut, polished, and etched. Structures were quantitatively analyzed using OPTIMAS software.

Results and Discussion

Thermochemistry of nodulization. During nodulization treatment, the additives react with the impurities in the melt, and the sequence of the refining reactions depends on the type and the quantity of additive as well as on the melt composition. The computer simulations and the experiments determined these sequences by taking into account all these parameters at a particular temperature. The calculations showing the possible parallel reactions while increasing the amount of *Mg* added to the iron melts (with different initial sulfur contents) are given in Figure 1a. A small amount of *Mg* produces significant deoxidation of the cast iron melts. But then, with an increase in the amount of magnesium, desulfurization occurs and the two refining reactions take place in parallel. The oxygen potential in equilibrium with the level of free magnesium required for transformation of the flake graphite to spherical graphite (0.02...0.03 wt.%) is shown by a dashed line. This predicted sequence of reactions was experimentally confirmed when the irons with different initial *S* were treated by *Mg* (Figure 1b). A small amount of *Mg* dramatically decreased $a_{[O]}$. In the liquid irons having larger initial sulfur contents the reaction with sulfur forces deoxidation to occur later. In this condition, two parallel reactions take place, and the concentration of sulfur in the melt and $a_{[O]}$ decrease at the same time. When the magnesium addition is sufficient to react with effectively all the sulfur and to decrease the oxygen activity to a value less than 1×10^{-4} wt.%, the flake shape of graphite transforms to the spherical shape during graphite growth.

In contrast to magnesium, calcium reacts first with sulfur in the liquid iron and only after thorough desulfurization will an increase in the amount of the calcium addition allow it to react with oxygen (Figure 2a). Also, the extent of deoxidization will be limited by calcium carbide formation in the liquid iron (dash dotted line). In the high purity *Fe-C-Si* alloy, the measured $a_{[O]}$ decreased with an increase in the amount of calcium additive. When the initial melt contained 0.04% *S*, there was a negligible influence of small amounts of calcium additive on $a_{[O]}$. In this melt, the oxygen activity decreased only after desulfurization with larger amounts of the calcium additive, an $a_{[O]}$ smaller than 1×10^{-4} wt. % was not reached, and the shape of flake graphite did not change.

The experiments qualitatively confirmed the calculated prediction of the sequences of the refining reactions during iron nodulization. In contrast to to liquid steel, where they all first produce deep deoxidation and then further additions produce desulfurization, the behavior is more complex in cast irons, as shown in Figure 3a. The individual features of the reactive species can be exploited in the design of the complex refining additives to be used, for example, for increasing the effectiveness of DI treatment. An example of a 3-dimensional diagram which describes the interaction of $[Mg]^a$ and $[Ca]^a$ additives with impurities in the liquid iron melt is given in Figure 3b. The regions of the reactions of magnesium additives with oxygen and sulfur are indicated as $[Mg]^O$ and $[Mg]^S$ respectively.

Magnesium additions above those required to react with O and S create free magnesium in the melt $[Mg]^f$, the quantity of which depends on the amount of calcium additive $[Ca]^a$. If complex additives are used containing both Mg and Ca then the refining functions are divided between these elements. In general, most of the Ca is consumed in the reaction with sulfur, while the Mg reacts with oxygen. If the amount of calcium is not enough for desulfurization, then any excess of Mg left after deoxidation continues to react with sulfur. It is important to note, that Ca, Ba and Ce can decrease the critical values of the Mg consumption, which are necessary for ductile iron nodulizing treatment.

Kinetics of additive dissolution in iron melts. The transfer of a solid additive to the iron melt can generally be assumed to occur either as dissolution by melting, which occurs when the melting temperature of the additive is below the temperature of the melt, or as dissolution by diffusion, which takes place when the additive has a melting point higher than the melt temperature. The differences between the dissolution mechanisms were studied when the quenched regions around a carbon raiser (dissolution by diffusion) and ferrosilicon (dissolution by melting) were analysed in the iron melts. Primary graphite phases adjacent to the raiser and ferrosilicon show the supersaturated dissolution regions. Also the large variations in the volume of the dissolution regions occur in these cases. During the dissolution of the ferrosilicon based additives, two important features are (a) the strong exothermic reaction (which could decrease the dissolution time) and (b) the formation of "slag" shells around the additive particles. These occur when the alkali earth metals (X) in $FeSiX$ react with with sulfur and oxygen in the melt. These slag shells can have a negative effect on the melt treatment efficiency.

These two effects were studied by computing the melting times of cylindrical and spherical shapes of ferrosilicon additives in the iron melt and by comparison with the experimentally measured melting times. Experimental temperature curves from thermocouples placed in the center of graphite, iron, and $Fe75\%Si$ cylinders (Ø25mm x 50mm) which were simultaneously submerged in the iron melt are given in Figure 4a. These materials were chosen because they exhibit the three principal different melting/dissolution mechanisms. Graphite, with its high thermal conductivity, heats up quickly in the iron melt but the process of diffusion dissolution is slow. The graphite cylinder did not dissolve measurably during the experimental time. Melting of an iron cylinder with the same chemistry as the iron melt is not accompanied by dissolution heat. Finally, the temperature rise of the less conductive $Fe75\%Si$ had a time delay relative to the iron, but then increased quickly when exothermic reaction started at the melting boundary. As a result, the temperatures of the centers of the iron and $Fe75\%Si$ cylinders approached the melt temperature practically simultaneously and then the dissolution heat liberation further increased the temperature around the melted *Fe75%Si* specimen, when compared to the initial melt temperature.

At the same time, the question arises, where is the energy of dissolution released: directly at the additive/melt boundary or partially distributed in the bulk melt? The fraction of the exothermic dissolution heat released at the melting boundary was evaluated by computing the melting time for different values of this fraction and comparing it to the experimental dissolution time. The values of these melting times were equal when approximately half of the dissolution heat was released at the melting boundary and half in the bulk melt. When a *Fe75%Si* additive begins to melt, silicon-rich liquid will be convected away from the boundary as it dissolves in the bulk liquid metal. This effect could be exploited for increasing the inoculation efficiency because the regions of high silicon liquid alloy create conditions for graphite nuclei formation directly in the bulk iron melt.

The effects of argon stirring of the melt, additive compositions and shape of particles on rate of dissolution were experimentally studied. Additional mixing forces had significantly higher influence on the thermal behavior of *Fe75%Si* cylinders compared to carbon cylinders when both were simultaneously submerged in a ladle with a bottom porous plug for producing the active argon agitation of the melt (Figure 4b). Argon stirring intensified dissolution, not only by increasing heat transfer between the melt and the additive surface but also possibly by fragmentation of the mushy zone of the additive.

The effects of the composition and the shape of additives on the dissolution time were experimentally evaluated. The dissolution rate of *FeSiX* complex additive alloyed by calcium was significantly reduced because the *Ca* component reacted with impurities in the melt with the formation of low thermal conductivity slag phases (oxy-sulfide type). As a result, the recovery of this type of additive may be significantly decreased. Changing the traditional equi-axed shape of the additive particles to a ribbon-shape [8] significantly increased the rate of dissolution and the recovery of elements from the additives with the same chemical compositions. Unlike traditional additives, the ribbon-shaped additives heated up quickly due to their high surface area to volume ratio. Also any slag shells that formed did not fully isolate the ribbons from the melt.

Improvements of DI treatments. Improvements in nodulization and inoculation were suggested on the basis of the additional knowledge of thermochemistry and kinetics of DI treatments. Four examples are given bellow.

1. Computing of optimal nodulizer composition and particle shape. The optimal *Fe50%Si5%Mg additionally alloyed with Ca* nodulizers were computed while taking into account the thermochemistry of refining reactions and kinetics of dissolution (Figure 5a). Additional alloying by calcium increased the concentration of free magnesium in the melt after treatment and decreased the consumption on nodulizer for DI treatment. Unfortunately, at the same time, calcium sulfides/oxides were formed around the spherical shape additives, decreasing the dissolution rate and the additive recovery. As a result, a function with a minimum is obtained - the position of which depends on the initial sulfur content in the initial melt. The optimum calcium contents of the nodulizer for the various initial sulfur contents are shown by the dotted line. Because calcium does not have a large negative influence on the dissolution rate of the ribbon-shape nodulizers, they could be used in the optimal range of high-calcium compositions (dashed line).

2. *Pretreatment of DI by C/SiC additives.* These experiments were conducted in a commercial DI foundry, using 0.1% of pretreatment agents (silicon treated graphite with 75%C and 25%SiC) placed on top of the *FeSiMg* nodulizer in the tundish ladle. The effectiveness of this treatment was evaluated by statistical analysis of uniformity of graphite nodule count distribution. It was found [9] that pre-treatment is a technique which enhances the nucleation of graphite and might not exhibit a significant influence on average nodule count but might be expected to result in more uniform graphite nodule distribution.

3. *Improving DI inoculation.* Thermodynamic analysis showed that there are differences in the processes when non-refined and refined-by-magnesium melts are inoculated. In the first case, reactive alkali and rare earth metals create the substrates, as a result of their reactions with the impurities of oxygen and sulfur in the melt. However, melts refined by *Mg* lose this possibility, since they will no longer contain significant *O* and *S* as impurities. As a result, the *FeSiX* inoculants with alkali and/or rare earth metals are not so effective for DI treatment of melts previously deeply refined by *Mg* in order to obtain the graphite nodules. On the other hand *Mg*-treated melts have the important potential possibility of in-situ nucleus formation when small amounts of active impurities *S* and *O* are introduced into the melt. Special additions of a small amount (0.01-0.02%) of a mixture contained copper/iron sulfides/oxides, which can react with Mg, intensifies inoculation of DI by the regular 0.2% addition of *Fe75%Si*. The possibilities of the formation of carbide-free structures in thin-wall castings and of a significant increase of graphite nodule count by using this technique have been experimentally confirmed (Figure 5 b).

4. *Intensification by argon stirring in ladle.* Because the process of inoculation significantly depends on the dissolution kinetics of inoculants in the melt, a new inoculation technique was suggested and tested under lab conditions. This technique combined inoculation in the ladle with argon stirring. The test showed, even in a small 100 lbs. laboratory scale ladle, that active argon stirring increased the inoculation effect and cast structure uniformity.

Conclusions

Computer simulations and measurements of the oxygen activities were used to determine the sequences of the refining reactions during nodulizing treatment of DI. These calculations, together with the experimental data of additive dissolution, were used for optimization of the compositions of the complex additives for DI treatment. The effectiveness of DI treatment may be increased by realizing the advantages that individual active elements have, and by the use of ribbon-shape additions to give faster dissolution and increase recovery.

Iron inoculation was analyzed from the point of view of the non-equilibrium dissolution of *FeSiX* in the melt. Dissolution kinetics were computed and compared with the experimentally measured melting times which confirmed the formation of the regions with the high silicon contents in the bulk melt. As a result, graphite nuclei would form in these regions directly from the melt. Some methods of improving nodulization and inoculation were discussed. Because non-equilibrium regions are responsible for nuclei formation, any methods which can increase the distribution of supersaturated regions in the melt may be used for inoculation improvement. These include in-stream inoculation and forced stirring of the melt.

References

1. Loper C R, Inoculation of Cast Iron – Summary of Current Understanding, AFS Transactions, vol. 107, 1999, pp 523-528.
2. Wang C H and Fredriksson H J, *The Mechanism of Inoculation of Cast Iron Melts*, Proc. 48th Int. Foundry Congress, 1981, pp 16-26.
3. Lekakh S and Bestyzev N, Ladle Metallurgy of High Quality Cast Iron, Nauka&Tekhnika, Minsk, USSR, 1992.
4. Lekakh S and Loper C R Jr, *Improving Inoculation of Ductile Iron*, AFS Transactions, vol. 111, 2003, paper 03-103.
5. Skaland T, Nucleation Mechanism in Dictile iron, Proc. AFS Cast Iron Inoculation Conference, 2005, pp 13-30.
6. Igarashi Y and Senri Okada S, Observation and analysis of the nucleus of spheroidal graphite in magnesium treated ductile iron, Int. J. Cast Metals Res, 11, 1998, pp 83-88.
7. Riposan I, Chisamera M, Stan S and Skaland T, A new Approach to Graphite Nucleation Mechanism in Gray Irons, AFS Cast Iron Inoculation Conference, 2005, pp 31-41.
8. Sverdlin A, Lekakh S, Kalinitchenko A and Sheinert V, *Chips-process for cast iron inoculation*, Foundry Management & Technology, vol. 5, 1994, pp. 31-34.
9. Loper C R, Winardi L and Lekakh S, Experiments in pretreatment of Ductile Iron, AFS Transactions, vol. 110, 2002, paper 02-02.

Figures

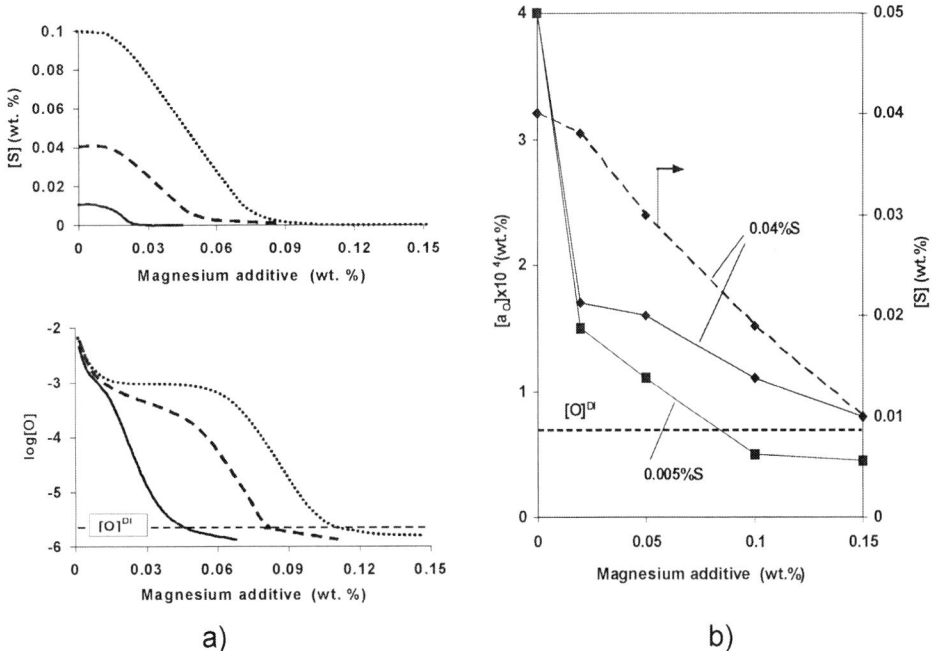

a)   b)

Figure 1. Calculated interactions (a) of *Mg* in liquid iron with initial 0.1%*S* (dotted lines), 0.04%*S* (dashed lines), and 0.01%*S* (solid lines), and experimentally measured $a_{[O]}$ (b)

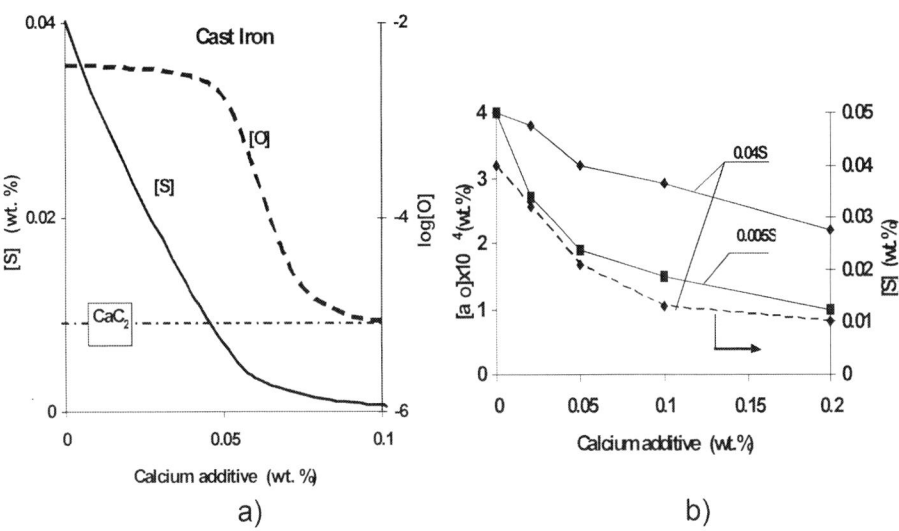

a)   b)

Figure 2. Calculated interactions (a) of *Ca* in liquid iron and experimentally measured $a_{[O]}$ (b)

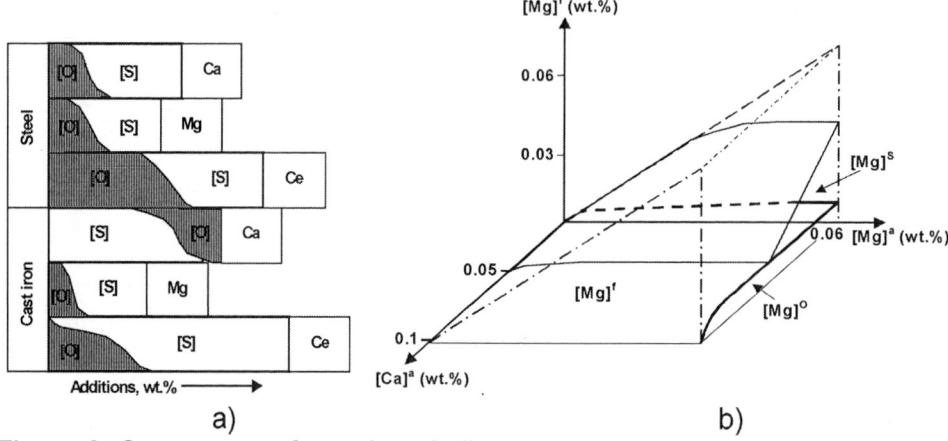

a)   b)

Figure 3. Sequences of reactions in liquid steel and iron (a) and interaction
of *Mg-Ca* additives in iron melt with initial 0.04%*S*, 0.007%*O* (b)

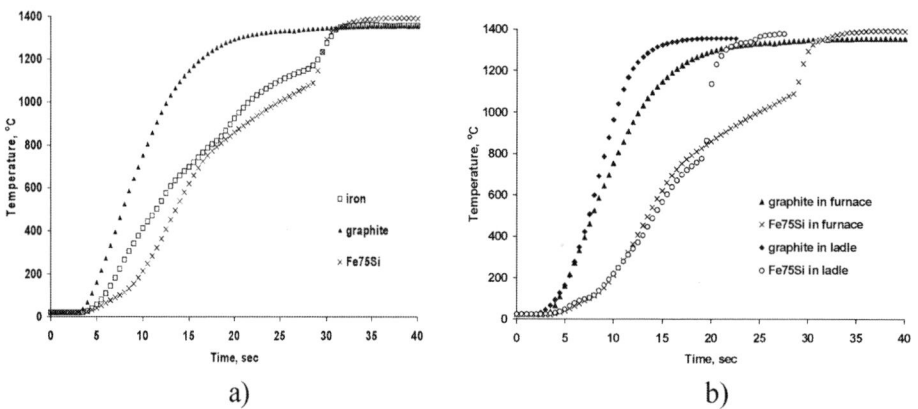

a)   b)

Figure 4. Measured temperature of carbon, iron and *Fe*75%*Si* cylinders
submerged in iron melt in furnace at 1380 °C (a) and influence of argon
stirring in ladle on temperatures of graphite and *Fe*75%*Si* cylinders (b)

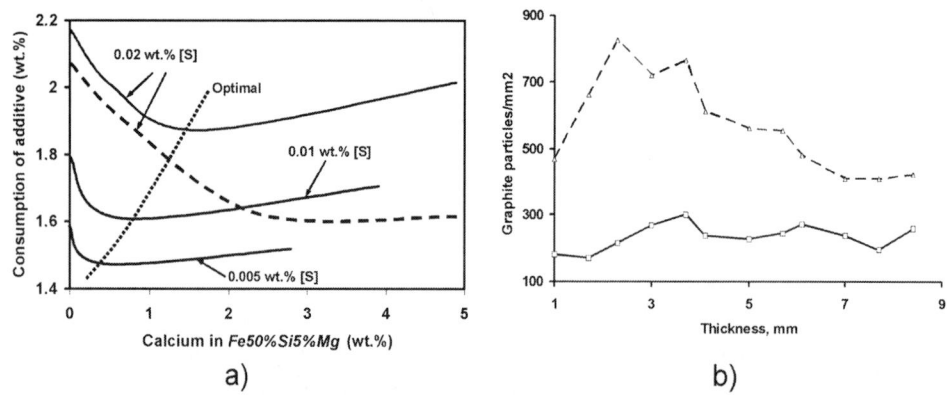

a)   b)

Figure 5. Optimization of *Fe50%Si5%Mg* nodulizer alloyed by *Ca* (a)
and improving of 0.2% *Fe75%Si* inoculation by 0.02% oxy-sulfide
additive (dushed line) treatment (b)

The Filtration of large Grey and Ductile Iron Castings

(author) E. Wiese

Foseco PLC, UK.

Abstract

Filtration of mass produced automobile castings is widely and successful applied, less common and more difficult is the filtration of large castings heavier than 1.000 kg. The paper deals with the special problems of applying filtration to such castings including; filter selection, description of three special filtration gating system concepts, along with ten case studies are given.

Introduction

In order to produce sound inclusion free castings of good quality it has been extensively proven that filtration in the mould is of prime importance.

Dross stringers, especially magnesium-sulphides, oxides and silicates (Figure 1) are the main sources of non-metallic inclusions and a big problem for all ductile iron casters. They can impair mechanical performance and lead to an unacceptable cast surface. This can be avoided by the use of a ceramic foam filter, along with a correctly designed runner system.

However, up until recent times, the filtration of iron castings in excess of 1 tonne has been difficult. Product performance issues, application difficulties and relatively low economic benefits, have all made the use of filters unattractive for large and heavy castings.

Filter Materials

The use of foam filters made of standard SiC-ceramic is usually limited to sizes up to 100x100mm. High ferrostatic pressures, high flow rates and extended pouring times increases the risk of breakage with larger than 100x100 mm filters during the filtration of heavy castings, and there is often not enough space available to place the required number of filters in the gating system/mould.

The need for high-strength filters in large sizes has seen an increasing use of carbon bonded foam filters (Figure 2) or partially stabilized Zirconia filters (Figure 3) over the last two years. The selection of these kind of foam filter ceramic is driven by several factors such as thermal shock resistance, cold compression strength, pouring times, priming performance, cooling rates and its chemical compatibility. The users obtaining a number of technical and economic benefits.

Filter Performance

Size and number of filters used should be adequate to maintain normal flow during mould filling. The quantity of ductile iron that will pass through a given size filter depends on a number of factors which should be carefully considered when designing a gating system:

- Increasing manganese and/or sulphur levels, decreases the quantity of iron that will pass through a filter.
- The cleaner the nodularising alloy used, the higher the degree of iron cleanliness, and consequently a higher quantity of iron passing through the filter.
- Inoculant types and practices influence the capacity of a given filter.

Filter Location

In many cases, little room is available for the gating system on the pattern plate and filter placement is restricted. The correct insertion of the filters into the gating system offers the optimum means of providing effective filtration.

Three different kinds of techniques for positioning the filters are shown below.

A) In-line pouring system – STELEX FP
Horizontal placement (Figure 4) in the drag mould is the preferred orientation for large filters. However, especially for the carbon bonded filters, vertical positioning is a common application. Foseco STELEX filter-position 6 wherein filters are placed vertically (Figure 5), allows large inclusion to flow up before the filter, thus increasing the quantity of metal that will pass.

B) In-line pouring system – (ceramic) hollow ware
For filter placement external of the parting line, ceramic hollow ware prints (Figures 6/7) in different types and different sizes are available. Their entrance and exit diameter are designed to be fitted into ceramic hollow ware tubes forming part of the running system.

C) Direct pouring system – KALPUR ST
Using filter-feeder direct pour systems (Figure 8/9/10) is a most economic method for producing quality filtered castings.

Gating Systems

A gating system designed for ceramic foam foundry filters should take into consideration the technical benefits provided by a filter and the method of application in which the filter will function most effectively. The location in a gating system plays an important part in the level of filtration effectiveness obtained, location and positioning are influenced by casting shape and weight and also by moulding practice and process.

Turbulence and aspiration of air in the gating system after the filter must be minimised to obtain maximum filtration benefits, and filter gating systems should fill as quickly as possible. The best results are obtained when gating systems are designed to completely fill before metal enters the mould cavity.

A correctly designed gating system for ceramic foam filters should provide the following:

- Minimum gating system size
- Consistent mould filling time
- Minimum erosion and turbulence
- Simple filter placement

Practical Case Studies

The following case studies help to illustrate the differences in behavior and performance of the two ceramic compositions. These case studies are also chosen to illustrate the mentioned different methods of positioning of foam filters in gating systems.

The advent of high-value, high technology ductile iron castings e.g. such as wind energy components, has seen an increased interest in filtration because of economic and technical benefits achievable. Ductile iron had been selected by the producer of wind energy for many mechanical components because of its lower density and excellent quality and reliability, even at low temperatures (-20 ℃). Maintenance of this quality level could c onfirm the dominant position of ductile iron in the future in this sector of the market.
A windmill contains about 40 t of different ductile iron castings e.g. rotor hub - blade adapter - gear (box) - machine (bed) frame - shaft - planet carrier - nacelle - beam - front section - front bearing.

1. Cylinder Head

About 70 castings of this cylinder head type are produced continuously per month. Alloy is ductile iron GJS 400-15 with a pouring weight of 980 kg and a casting weight of 930 kg. Figure 11 shows the cope and drag site of the pattern plate.

One 150x200x30 mm STELEX PrO ceramic foam filter is used, vertically located directly after downsprue. Filter capacity 3,27 kg/cm²; pouring temperature 1360-1380 ℃; pouring time 35-40 sec. Downsprue diamete r 50 mm; the 30% enlarged ingate cross section is 2514 mm².

2. Ring-Housing

This casting is produced in GJS with a pouring weight of 9500 kg and a casting weight of 5800 kg. Figures 12 a/b show the cope and drag site of the casting just after shot blasting. Figures 20 a/b show two different views of casting ready for machining.
Filters used are eight pieces STELEX PrO ceramic foam filter size 200 mm diameter, horizontally located on parting line in the inner ring. Filter capacity 3, 87 kg/cm²; pouring temperature 1340 ℃; pouring time 6 8 sec.

Downsprue two times diameter 80 mm; the runner in front of the filters consists of resin sand while the runners beyond the filters are made of ceramic hollow ware. All eight ceramic runners are directed into the six feeders which are placed beneath the inner ring to provide smooth filling of the casting.

3. Windmill Rotor Hub

Figure 13 shows the sketch of a hub and its gating system with filters. This casting is produced in GJS. The pouring weight is 9000 kg. Six zirconia filters of diameter 200 mm are used. More than 90 hubs have been cast in this manner. Pouring temperature 1340 ℃; pouring time approxima tely 50 sec; filter capacity 4,78 kg/cm².

4. Windmill Rotor Hub

Figure 14 a shows the sketch of a hub and it's gating system with filters. Figure 14 b illustrates one of the three hub flanges from where the casting is filled through two STELEX ZR filters which are placed in Hagenburger ceramic prints. This casting is produced in GJS. The pouring weight is 9000 kg. Six zirconia filters of diameter 200 mm are used. More than 90 hubs have been cast in this way. Pouring temperature 1360 ℃; pouring time appr oximately 70 sec; filter capacity 4,78 kg/cm².

5. Windmill Planet Carrier

This casting is produced in GJS with a pouring weight of 3100 kg and a casting weight of 2500 kg. Figure 15 illustrates the casting and its gating system. Clearly seen is the remaining ceramic hollow ware runners and filter prints. Two STELEX ZR filters of diameter 200 mm were used Pouring temperature 1380 ℃ Pouring time approximately 80 sec. Filter capacity 4,94 kg/cm².

6. Housing

This casting is produced in GJS with a pouring weight of 25500 kg. Figure 16 a/b show the pattern and parts of its ceramic gating system. Twenty-two STELEX PrO filters of diameter 200 mm had been used. Pouring temperature 1380 ℃ Pouring time approximately 120 sec; filter capacity 3,58 kg/cm².
A conservative capacity was chosen because of pressure height of almost 3,5 m and because of back pressure during mould filling.

7. Castor

This casting is produced in GJS with a pouring weight of 9000 kg and a casting weight of 7500 kg. Figure 17 illustrates the drawing of casting and its recommend gating system. The application of eight vertical placed 200mm diameter, alumina-carbon filters were recommended. Pouring temperature of first casting 1360 ℃ of the second 1340 ℃; because of s ome feeding difficulties with this large casting, temperature for cast three is planned at 1300℃. Pouring time of approximately 75 sec. was calculated; filter capacity 3,58 kg/cm².

8. Crushing Body

Figure 18 illustrates a ceramic gating system where square shaped rather than round filter prints are incorporated. Four filters with size of 150X150 mm were used.

9. Roller

For this casting the same ceramic square prints and four 150x150x30mm carbon bonded filters were used similar to the previous case history. This casting is produced in an alloy with a carbon content of 1,75 % and 16 % of chrome, with a pouring weight of 3900 kg. Figures 19 a/b show the inner of an unfiltered and a filtered casting. Pouring temperature of casting 1460 °C; pouring time was 110 sec; filter capacity was 4,3 kg/cm².

The last three case histories are focused on direct pour technology. Because there is no need for an extra sprue and runner system, this technique offers an enormous production cost advantage. Filters can be placed in top or side feeders.

10. Clutch Disk

This casting is produced in GJS .... (GGG60) with a pouring weight of 1700 kg; pouring temperature of casting 1330 °C; pouring tim e was 30 sec; Filter used: 1x STELEX ZR Ø200x35mm placed in a filter-feeder ZTAE 23/255; filter capacity was 5,41 kg/cm².
So far 12 castings have been cast successfully. Figure 20 shows a casting with a feeder and the KALPUR direct pour unit remaining.

11. Plaster

Two of this casting type produced in GJS-400-15 were shown on the GIFA exhibition 2003, one as cast and the other one fully machined. Figure 21 shows the casting with remaining feeder. Figure 22 demonstrates the MAGMA simulation model for temperature distribution after mould filling. One ceramic foam filter STELEX PrO diameter 200mm is used, located in a special collar core underneath a cylindrical feeder sleeve KALMINEX X11. Pouring temperature of casting 1400 °C; pouring time is around 20 sec; fil tration capacity 3,39 kg/cm²; yield 86%

12. Valve

This casting is the only sample where the alloy is GJL ..... A direct- pour solution had been introduced and applied in a Czech foundry where many valve castings have been poured this way. The first poured casting with such a pouring system was displayed during a foundry fair in Bruin in the Czech Republic and the foundry received an award called "The Golden Ladle" for this work. Two ceramic foam filter STELEX ZR Ø 150 x 30mm/10 ppi were used, located in two feeders topmost of the casting, a pouring basin like a bridge merged the two KALPUR units together to offer the filter cross section need. Figure 23 shows the start of mould filling simulation. Poured weight 1280 kg.

Conclusions

Large foam structure filters produced of ceramic Al_2O_3/Carbon or ZrO_2 give the foundryman a method for filtering castings higher in weight then 1000 kg, along with the ability to reach the quality standards required. Ceramic foam filters like STELEX PrO and STELEX ZR demonstrate that they are highly efficient in removing inclusions from molten metal and minimising turbulent metal flow.

Different methods are available to incorporate filters in sand as well as in hollow ware gating systems, however direct pour systems are the most efficient related to improvement of yield, however they can lead to higher metal velocities, and other possible problems.

Figures

Mg

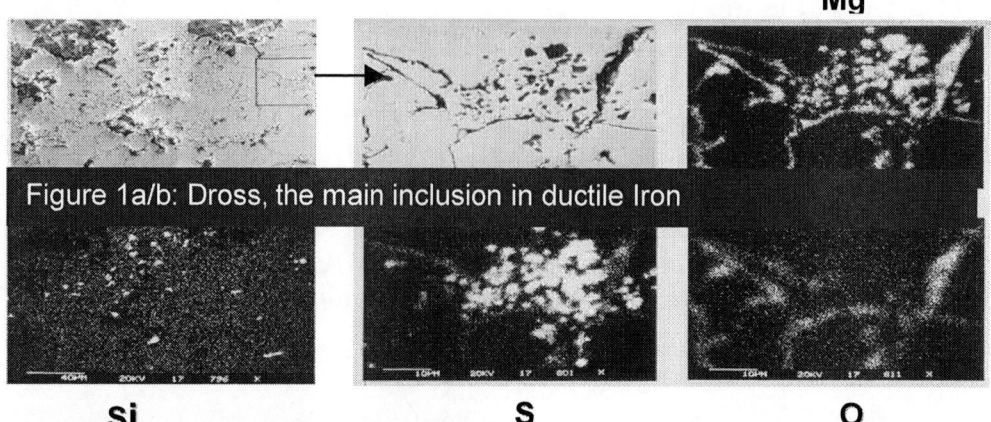

Figure 1a/b: Dross, the main inclusion in ductile Iron

si　　　　　**S**　　　　　**O**

 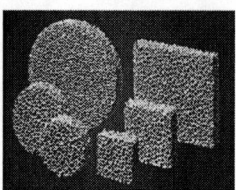

Figure 2: STELEX PrO Carbon bond
ceramic foam filter

Figure 3: STELEX ZR Zirconia
ceramic foam filter

Figure 4: Pattern horizontal placement　　Figure 5: Pattern vertical placement

 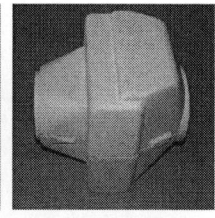

Figure 6:　Ceramic print for
horizontalfilter placement

Figure 7:　Ceramic print for
vertical filter placement

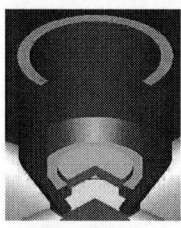  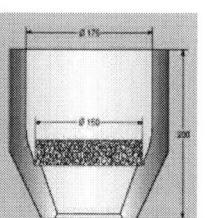

Figure 8: Neck down feeder with
insert filter carrier core

Figure 9: Zylindrik feeder with
filter carrier core

Figure 10: KALPUR filter-feeder

Figure 11 a+b: Drag and cope site
of the pattern plate

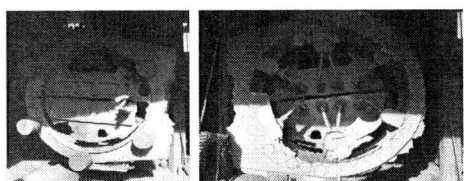

Figure 12 a: Drag and cope site of the ring housing

Figure 12 b: Two different views of the ring-housing

Figure 14 a+b: Filters placed in ceramic prints shortly before the ingates

Figure 13: Filters place at parting line

Figure 15: Ceramic hollow ware gating system with two Hagenburger filter prints

Figures 16 a+b: Two rows of Ø 200 ceramic filter prints are need to cast a 25,5 t GJS casting

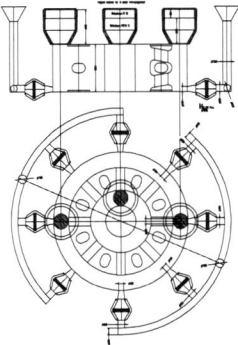

Figure 17: Drawing of recommend gating system for a heavy wheel

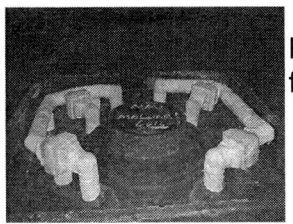

Figure 18: ceramic filter prints for square filters 150²

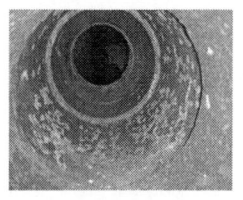 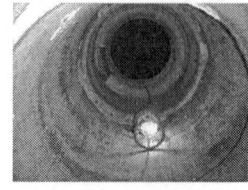

Interior space of roller casting

Figure 19 a: Casting unfiltered

Figure 19 b: Casting filtered

Figure 20: Direct pour through a KALPUR unit

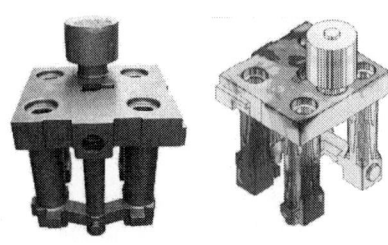

Figure 21: Casting with remaining feeder

Figure 22: Temperature distribution after filling

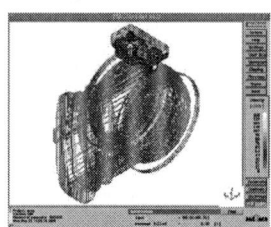

Figure 23: Mould filling simulation of valve

erhard.wiese@foseco.com

Study of the occurrence and suppression of metal reoxidation in ferrous castings

T.Elbel*,J.Senberger**,A.Zadera**,L.Kocian*.

* VSB – Technical University of Ostrava, CZ, ** Brno University of Technology, CZ.

Abstract

Recent advances in understanding the processes of casting metals into green sand moulds have shown that the occurrence of reoxidation plays a key role in the control of the quality of castings. The reoxidation gives rise to solid, liquid and gas products that may further react with the metal and foundry mould, and with various surface and internal defects appearing in the casting. The resultant occurrence of casting defects is affected by the content of deoxidation elements in the metal and by the process of deoxidation reactions in the mould during and after pouring the metal. The present work is concerned with changes in oxygen activity in the metal poured into sand moulds.

Key words Reoxidation, sand moulds, oxygen activity

Introduction

Metal oxidation in the course of melting and metallurgical processing is one of the basic metallurgical reactions in the processing of ferrous metals. In this connection it is referred to as primary oxidation, which usually takes place in the furnace. Oxygen introduced into the molten metal must then be removed. This operation is known as deoxidation and is performed in and out of the furnace. In the production of castings and when casting metal in air there is, from the moment the metal is poured out of the melting furnace, a secondary oxidation taking place in the ladle and in the foundry mould. In addition to the secondary oxidation there is often a tertiary oxidation, which takes place in the cast piece during the solidification between the solidus and the liquidus temperature. The secondary and the tertiary oxidation, which frequently proceed in the mould simultaneously, are referred to as reoxidation.

When ferrous alloys get into contact with air during pouring in air, all elements oxidize that under the given conditions have a greater affinity to oxygen than iron has. These processes take place in both sand and metallic moulds and they also occur in continuous steel casting. They can also be observed when pouring iron with lamellar and spheroidal graphite. Reoxidation processes are particularly strong when casting into green sand moulds with clay binders. A consequence of the reoxidation is the occurrence of solid, liquid and gas products that can react with the foundry mould, accompanied by the appearance of various surface and internal defects. An overview of the defects due to metal reoxidation can be found in Table I. In their practice, the present authors have been confronted with most of these defects in both steel and iron castings, and they had to cope with the problem of suppressing and preventing these defects [1]. This experience has motivated them to study the essence of reoxidation processes.

Experimental

Experimental investigation of reoxidation processes was conducted separately for steel and iron castings.

In steel, the research focused on two groups of unalloyed steel, namely steels with a carbon content of 0.40 to 0.50% and 0.15 to 0.25%.

Two methods of in ladle deoxidation were used: deoxidation using aluminium and deoxidation using aluminium combined with FeSiCa. Steel was cast into moulds of green bentonite mixtures with moisture of 3.2 to 4.0%. The changes taking place in the chemical composition of steel during casting were examined for samples taken from the ladle and from the casings in bentonite moulds. The effect of the capacity of metal flow ratio for a specific casting was also examined.

The progress of reoxidation processes was also followed via measuring the oxygen activity. Real oxygen activity in steels was measured in the furnace and in the moulds, using disposable probes or gauges for long-term measurement (as much as 30 minutes).

In cast irons, the experiments were focused on cast iron with lamellar, compact, and spheroidal graphite. To assess cast iron reoxidation during casting, oxygen activity in the mould was measured up to the solidification temperature. As with steel castings, the oxygen activity was measured using the TSO 6 probes and gauges derived from these probes. The measuring cell of the gauges is in a quartz tube sealed with a refractory mass. The gauge is formed by ZrO_2 electrolyte stabilized by CaO, with a Cr_2O_3+Cr reference mixture. Molybdenum wire was used for the contact with the bath and the lead from the reference mixture. Temperature was measured using a Pt-PtRh thermocouple. It was measured on a cylindrical casting of 15 kg in weight. To extend the measuring time the casting was thermally insulated.

A filter with inoculation agent was placed in the gating system. Two probes with a frequency of 1 Hz measured the electromotive force till the end of eutectic transformation. The time from filling up the mould cavity to the end of solidification was ca. 30 minutes.

Results

Steel castings

In the case of steel castings the difference in the chemical composition was compared for samples taken from the ladle and from the casting after it solidified in the bentonite mould. Statistical significance of selective averages (μ_1, μ_2) was tested on three series at a 95% confidence level [2]. For all three series of heats statistically significant differences in aluminium content were established. In the steel under examination this is (after calcium) the main deoxidation element with the highest activity with respect to oxygen. Actual values are given in Fig. 1, in which individual columns represent the arithmetic average of the set. The relative decrease in Al (expressed in % by weight) after pouring into the green mould was 28.7%, 24.2% and 23.5%. The results of establishing the content of Si and Mn are given in Fig. 2. In the 1st series the relative decrease in Si content, given in wt.%, was 10.23%; in the 2nd series its content increased by roughly 0.09% while in the 3rd series the Si content dropped by 6.48%. The relative decrease in Mn content in the 1st series of heats was 4.31%, in the 2nd series it was 5.38% and in the 3rd series 4.24% by weight. To explain these differences in Si and Mn contents we have to resort to the diagram of so-called silicon isotherms [3] for the Fe-Si-Mn-O system. For silicon, a statistically significant difference between samples from the ladle and from the mould showed only in the 1st series of heats, where its content in the ladle is the highest. This series exhibited the lowest Mn content. This probably resulted in increased oxidation of Si and its statistically significant decrease. In the 2nd series, which by contrast exhibited the highest Mn content and the lowest Si content in the ladle, there was a statistically significant decrease in Mn. For all the three series the Mn/Si ratio of concentrations in samples from the mould was about 1.96. This ratio evidently corresponds to the equilibrium concentration of the two elements under the given conditions, which is in agreement with the silicon isotherm. The behaviour of both elements corresponds with data in the literature, according to which during oxidation the element that will preferably pass into the slag will be that element from among Si and Mn whose content in the melt will be higher.

It was interesting to follow the changes in the chemical composition of these three elements in dependence on the capacity of flow ratio. This indicator, was defined by J. Přibyl [4] as a ratio of the amount of metal that flowed through a certain cross section to the total amount of metal cast into the mould. The investigation was carried out for the same kind of unalloyed steel on a crank shaft casting of 23 kg in weight. Analyses of the chemical composition (Si, Mn and Al) were performed in two places of the casting where the capacity of flow ratio were 0.2 and 0.6. In each place several analyses were performed. Table II summarizes the results of analyses of the test casting. We can see from Table II that in places with a higher degree of capacity of flow ratio (top) the Si and Al contents are lower than in a place with a lower capacity of flow ratio (bottom). In both cases we are concerned with a statistically significant difference, when $\mu_1 \neq \mu_2$. The differences in Mn content are insignificant.

In order to explain these phenomena, oxygen activity was measured in the ladle during pouring the steel and in the moulds subsequent to filling. In the course of casting, the oxygen activity was followed in heats of unalloyed steel containing 0.15 to 0.25% of carbon, produced in a basic arc furnace. The weight of one heat was 7.5 t. Castings weighing 0.5 to 3 t were cast from the heat. Oxygen activity and temperature were measured in the furnace prior to tapping, in the ladle one minute after tapping (in ladle -1), 10 minutes after tapping (in ladle - 2), in the riser of the first mould at the beginning of pouring (in mould – 1) and in the riser of the last cast mould (in mould – 2). The measuring results are given in Table III.

Iron castings

Measuring the activity of carbon in cast irons is used particularly in the production of spheroidal and compact graphite castings to check the modification of metal [5]. The measuring itself is performed in Mg-treatment or pouring ladles prior to pouring the castings.

In contrast to steels, where the dominant deoxidation element is aluminium, oxygen activity in unalloyed cast irons in which inoculation and Mg-treatment were not performed is controlled by the content of silicon.

The present authors measured the progress of oxygen activity in cast irons during the whole manufacturing process, from melting to the end of solidification within a temperature interval from 1160 to 1540°C. The values of oxygen activity measured in an electric induction furnace prior to inoculation and modification were compared with equilibrium oxygen activities calculated for the equilibrium of the reaction:

$$(SiO_2) = [Si] + 2[O] \tag{1}$$

The deviation from equilibrium was evaluated according to the difference in free enthalpy in real and equilibrium state according to the equation:

$$\Delta G = 2RT\left(\ln a_O^r - \ln a_O^n\right) \tag{2}$$

where a_O^r is the oxygen activity in equilibrium state, and a_O^n is the measured oxygen activity. For cast iron in the furnace and in the ladle within a temperature interval of 1350 to 1400°C the calculated values of

ΔG ranged, prior to inoculation and Mg-treatment, from -65000 to -45000 J/mol. If ΔG has a negative value, a reaction will occur between oxygen and silicon that are dissolved in the cast iron. Reaction (1) then proceeds from the right to the left and the volume of silicon oxide increases in the melt or new crystallization nuclei may appear based on SiO_2. With higher temperatures, the deoxidation effect of carbon begins to show and the value of ΔG decreases [3].

The value of equilibrium oxygen activity according to reaction (1) is greatly affected by temperature. For the reaction with silicon, the equilibrium oxygen activity at eutectic reaction temperature acquires values of the order of 10^{-2}, with silicon and carbon content as they are usual in cast irons. The oxygen activities obtained at eutectic temperature of oxygen are lower than oxygen activities in magnesium-modified cast irons at temperatures exceeding 1350°C.

To assess cast iron reoxidation during pouring, the oxygen activity in the mould was measured up to the solidification temperature.

The results of one such measurement on a casting are given in Fig. 3. Plotted in the graph are the oxygen activity values as determined from the electromotive force measured. The upper two curves give the temperature; the lower three give the oxygen activity. The cast irons used for the measurement of oxygen activity in the mould were: cast iron with lamellar graphite, cast iron with magnesium-modified spheroidal graphite, and cast iron compact (vermicular) graphite. For each chemical composition a non-inoculated casting and a casting inoculated on the filter in the gating system were cast. The resultant oxygen activity values measured in dependence on temperature are given in Fig. 4. For the sake of clarity, oxygen activities for selected temperatures are only given in the diagram.

The full line holds for equilibrium oxygen activity, which is in equilibrium with 3.30% Si and 1.8% C. The data given in the diagram are for heats containing 3.25 to 3.49% C and 1.80 to 2.49% Si. Data marked with triangles hold for cast iron with lamellar graphite. Oxygen activity in heats of non-inoculated cast iron with lamellar graphite is above the equilibrium. In castings of the same composition a lower oxygen activity was measured subsequent to inoculation. In two post-inoculation melts a lower than equilibrium oxygen activity was measured, which was on the level of heats with compact graphite. Oxygen activities measured for compact graphite heats are below the equilibrium with silicon and they are marked with squares. As expected, oxygen activities in spheroidal graphite heats are the lowest and they are marked with rhombi. The diagram in Fig. 4 illustrates oxygen activities up to a temperature of 1180°C. In sub-eutectic heats when austenite is being separated the concentration of silicon in the melt decreases and austenite gets enriched. During eutectic transformation, increased oxygen activity was measured.

Low values of oxygen activity were measured in heats of lamellar-graphite cast iron subsequent to inoculation. The cerium content in these heats is high. In spite of the low oxygen activity in the heat up to the solidification temperature, graphite separates in the form of lamellar graphite. Figs 5a to

5c show the morphology of graphite for different levels of oxygen activity at the beginning of solidification. Fig. 5a gives the structure of graphite for lamellar-graphite cast iron. At the beginning of solidification the oxygen activity at a temperature of 1210°C was 0.027 ppm and the cerium content was 0.10%. Fig. 5b gives the structure of compact-graphite cast iron. At a temperature of 1180°C the oxygen activity measured in a non-inoculated casting was 0.055 ppm and in an inoculated casting it was 0.027 ppm. The magnesium content was 0.014%.

Fig. 5c gives the structure of spheroidal graphite cast iron. At a temperature of 1180°C the oxygen activity in a non-inoculated casting was 0.040 ppm while in an inoculated casting it was 0.029, with a magnesium content of 0.046%.

Discussion

The decrease in the aluminium content in the steel being cast, caused by oxidation, is accompanied by increased oxygen activity. If the oxygen activity increases to a value which under the given conditions is higher than the equilibrium activity with silicon, silicon oxidation can occur; this can be described by the equation:

$$[Si] + 2\,(FeO) = 2\,Fe + <SiO_2> \qquad (3)$$

Simultaneous oxidation of aluminium and silicon may give rise to complex oxides. In places where the aluminium content gets "exhausted" and the silicon content drops markedly, the oxidation of manganese sets in. If the total of elements present does not suffice to deoxidize the iron oxides formed, the deoxidation products will contain iron. The composition of deoxidation products can be predicted based on the oxygen activity in steel. Conversely, oxygen activity during metal solidification can be deduced from the composition of reoxidation products.

On the basis of heats measured in the last two years for cast irons [6, 7] a set of statistical data was assembled. Data for selected temperatures are given in Table IV. Based on the set of statistical data the graph in Fig. 6 was plotted. The dependence of oxygen activity on temperature is expressed by a band whose width is ± the standard deviation from the average for cast iron with compact and spheroidal graphite. The higher values in the band hold for non-inoculated cast irons, the lower ones for inoculated cast irons. The bands for spheroidal graphite iron and compact-graphite cast iron touch each other. At temperatures above 1300°C the iron activities in lamellar graphite cast iron are markedly higher. With decreasing temperature, the differences between oxygen activities in lamellar graphite cast iron and in the other types of cast iron get smaller. At a temperature of less than 1200°C the band of oxygen activities in cast irons with lamellar graphite touches the bands of oxygen activities of the other cast irons. For lamellar graphite cast irons the lowest oxygen activity was measured in inoculated heats whose cerium content was ca. 0.10%. The morphology of graphite did not change though.

Conclusions

In the paper, the occurrence of reoxidation phenomena was assessed for steel and iron castings via measuring the oxygen activity, in the case of steels also according to the change in chemical composition of the metal.

Immediately after the deoxidation of unalloyed steels killed by aluminium, oxygen activity in the ladle is usually lower than 10 ppm (4 to 8 ppm). Under these conditions the product of deoxidation is solid aluminium oxide, which can be found after deoxidation in all steels for castings. The morphology of this oxide corresponds to oxygen activity and to the conditions of its appearance. Aluminium oxide crystallizes from the melt as solid phase. At higher oxygen activities in reoxidized steel, fluid oxides can appear in reoxidized steel, which can morphologically be classed by the well-known Sims-Dahle classification in the first type (type I). These oxides can be observed in parts of the casting that are distant from the ingate; indirectly, they point to oxygen activity in the metal in the course of their appearance. If oxygen activity in the steel is higher than equilibrium oxygen activity with respect to the carbon in the steel, inclusions of macroscopic oxides can form in these places of the casting and, after their reaction with carbon, CO bubbles (pinholes) are formed.

In cast irons, in contrast to steels, oxygen activity in the metal at solidification temperatures is controlled by silicon content. In aluminium deoxidized steels the content of the dominant deoxidation element is 0.03 to 0.05% (Al). At some places of the casting during reoxidation processes, all the aluminium that is capable of reacting with oxygen may be consumed. In that case oxygen activity is no longer controlled by aluminium content, and defects can appear in the casting. It is therefore important to observe exactly the deoxidation conditions and, at the same time, to fill the mould uniformly with metal so that there are no places in the castings that would lack deoxidation elements.

Increasing the activity of oxygen in cast iron during reoxidation processes is hindered by the high content of deoxidation element (Si). Also the presence of carbonaceous substances in the moulding mixture limits the appearance of oxidation gases in the mould. Therefore there was no graphite deformation in the heats examined and, in comparison with steel, the oxygen activity measured was very low.

References

1. Elbel,T.;Senberger,J. : *Some aspects of reoxidation occurence in ferrouis castings.* In Proceedings of conference SPOLUPRACA in Zakopane, AGH Krakow 2005, ISBN 83-919232-3-1 p.41-50.
2. Kocian,L.;Elbel,T. : *Analysis of change of chemical composition of steel due to reoxidation of metal.* Acta Metallurgica Slovaka, vol.11, N.3, 2005, p.8-12.
3. Otahal,V. : Slevarenstvi, vol. 32, N. 5/6, 1984, p. 201-207.
4. Pribyl,J. : *Procesy hutniho slevarenstvi.* HUTNICKE AKTUALITY, vol.10, 1969, publ. 8, VUHZ Praha.
5. Hummer, R. : Giesserei-Rundschau 50 (2003), č. 9/10, str. 220 až 226.
6. Elbel, T. Senberger, J. Zadera, A.: *Reoxidation phenomena in ferrous casting.* 45th International Foundry Conference, Portorož 14.-19.9.2005. ISSN 1318-9123.
7. Senberger,J.;Zadera,A.; Elbel,T. : Slevarenstvi, vol. 53, N.7/8, 2005, p. 308-312.

Acknowledgements

The work was accomplished with financial support provided by the Grant Agency of the Czech Republic within project No. 106/05/446.

Tables

Tab. I Casting defects review influenced by the metal reoxidation

INCLUSIONS		
MACROINCLUSIONS	GAS HOLES	
DROSS	SLAG BLOWHOLES	CHEMICAL PENETRATION
CEROXIDES	PINHOLES	
SECONDARY SLAG	ENDOGENOUS BLOWHOLES	STRUCTURAL ANOMALIES
MICROINCLUSIONS		

Tab. II Changes of chemical composition of steel casting in two different places

	Sample	N	E(x) [%]	S(x) [%]	Difference	t-test
Silicon	top	16	0.4056	0.03119	- 12.90 [%]	$\mu_1 \neq \mu_2$
	bottom	14	0.4657	0.01555		
Manganese	top	16	0.7200	0.01592	+1.20 [%]	$\mu_1 = \mu_2$
	bottom	14	0.7114	0.01292		
Aluminium	top	16	0.0706	0.00924	- 9.29 [%]	$\mu_1 \neq \mu_2$
	bottom	14	0.0779	0.00426		

Tab. III Changes in temperature and oxygen activity during tapping, heat transport and steel casting

	Before tapping		In ladle – 1		In ladle – 2		In mould – 1		In mould – 2	
	Temp. [°C]	a/O/ [ppm]	Temp. [°C]	a/O/ [ppm]	Temp. [°C]	a/O/ [ppm]	Temp. [°C]	a/O/ [ppm]	Temp. [°C]	a/O/ [ppm]
x	1657.7	34.6	1625.7	4.6	1608.1	4.3	1517.4	5.1	1550.3	4.6
s	9.1	11.5	4.2	0.8	2.7	0.8	9.6	1.1	4.3	0.7
Xmin	1644.0	19.0	1620.0	3.0	1603.0	3.0	1557.0	4.0	1542.0	4.0
Xmax	1676.0	74.0	1633.0	6.0	1613.0	7.0	1585.0	8.0	1559.0	6.0

Table IV Oxygen activity values in dependence on temperature and graphite shape

Temperature [°C]	Lamellar graphite			Compact graphite			Spheroidal graphite		
	n	X	s	n	X	s	n	X	s
1400	7	1.480	0.181	-	-	-	-	-	-
1350	9	0.966	0.151	7	0.304	0.117	5	0.156	0.037
1300	9	0.621	0.159	8	0.164	0.056	3	0.081	0.009
1250	12	0.254	0.139	10	0.075	0.023	12	0.038	0.024
1200	9	0.090	0.056	10	0.042	0.021	16	0.024	0.023
1180	7	0.096	0.084	5	0.033	0.017	16	0.021	0.020

Figures

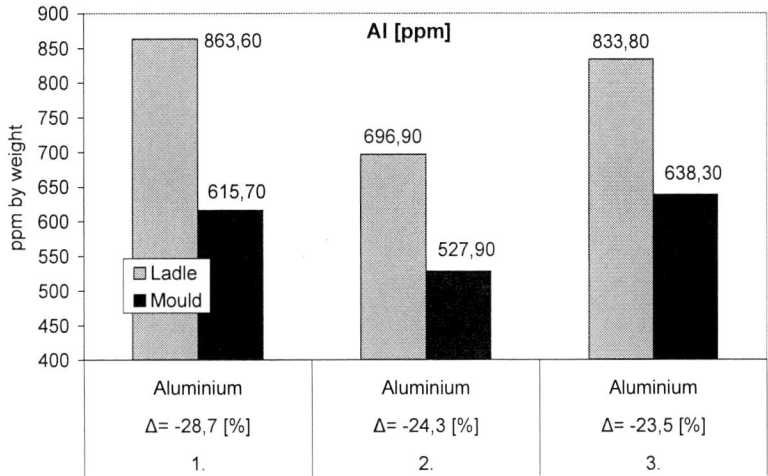

Fig.1 Difference of aluminium content between samples from the ladle and the mould.

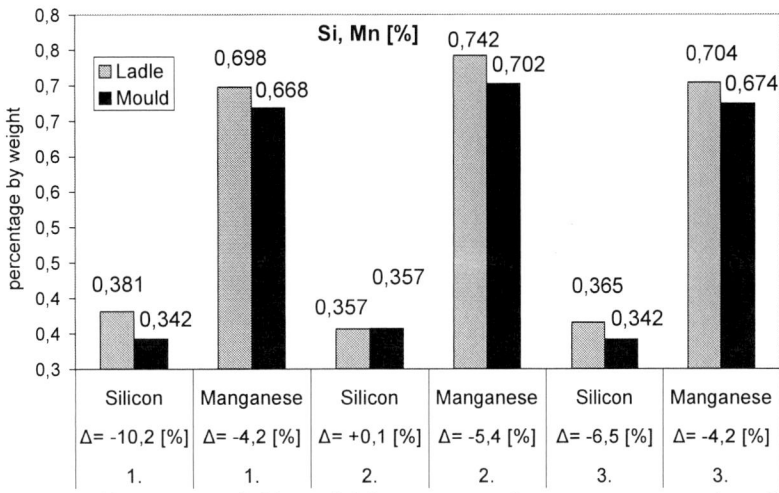

Fig. 2 Difference of Si and Mn content between samples from the ladle and the mould.

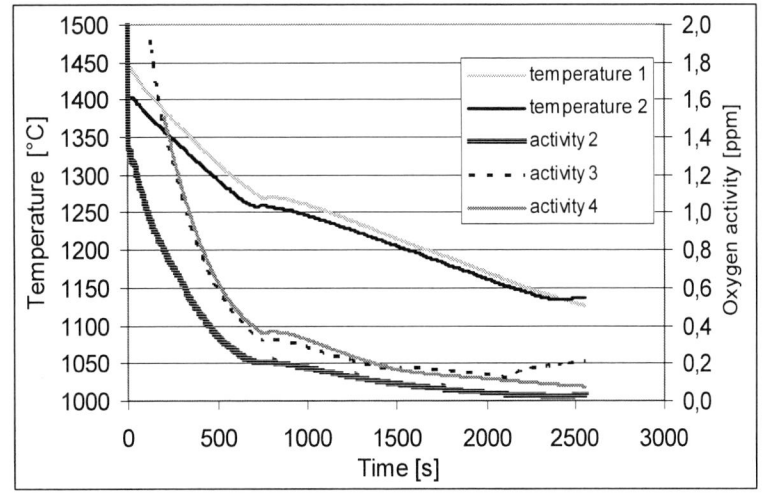

Fig. 3 The reading of oxygen activity measurement up to a temperature of 1150°C.

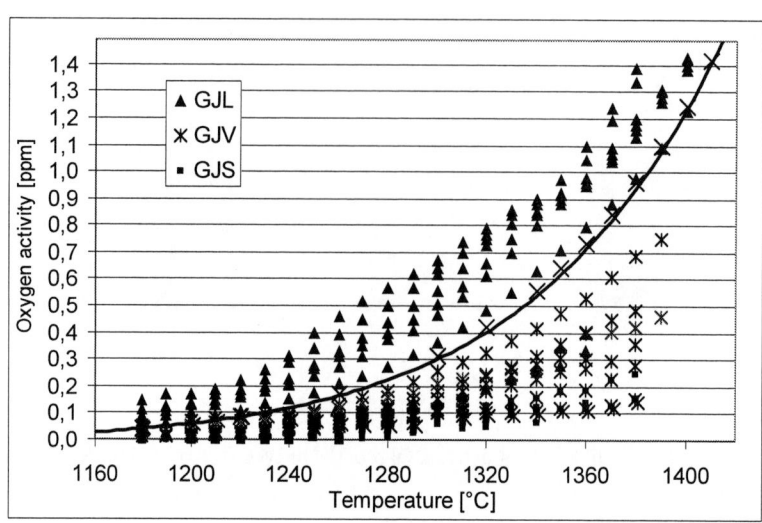

Fig. 4 Oxygen activity vs. temperature, for three types of cast iron.

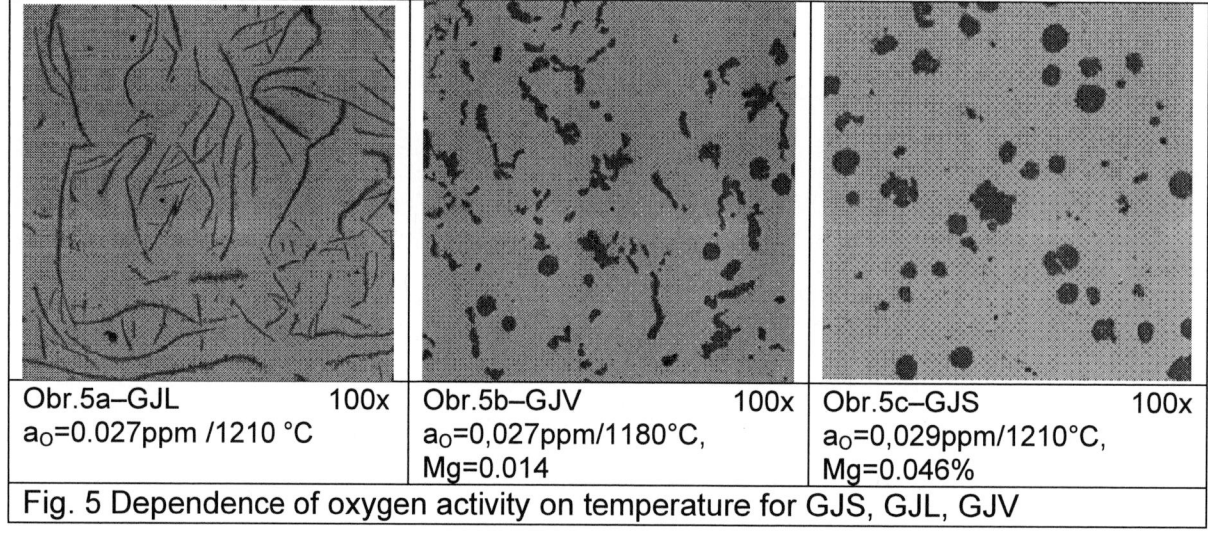

| Obr.5a–GJL 100x | Obr.5b–GJV 100x | Obr.5c–GJS 100x |
| a_O=0.027ppm /1210 °C | a_O=0,027ppm/1180°C, Mg=0.014 | a_O=0,029ppm/1210°C, Mg=0.046% |

Fig. 5 Dependence of oxygen activity on temperature for GJS, GJL, GJV

Fig. 6 Oxygen activity vs. temperature, for three types of cast iron.

New austenitic flake cast iron with manganese

Dr P Sriram
Rapsri, India

Abstract

Austenitic cast irons are used in a variety of applications including automotive, chemical, petrochemical, food handling, marine, power plants, pulp and paper and fluid handling. The alloys are basically of different compositions with varying weight percentages of nickel and other alloying elements. These alloys are a part of national and international specifications and come in the flake and spheroidal graphite varieties. Alloys with supplementary additions of elements for specific application requirements with enhanced properties have been developed as proprietary alloys. Niresist is the name applied to those families of alloy cast irons in which the presence of a substantial amount of nickel along with smaller quantities of other elements like copper, chromium and manganese has rendered the alloy, austenitic in structure. The structure and the alloying elements make the alloy superior to plain cast irons.

This paper describes work carried out on a new austenitic flake cast iron alloyed with manganese for use in top ring inserts for pistons of internal combustion engines.

Introduction

High performance aluminum alloy pistons used in high speed diesel engines require a top ring insert to withstand high temperature, wear and possess a compatiable coefficient of thermal expansion. This requirement is met by an austenitic cast iron, which has the expensive nickel as a major constituent. Extensive alloying with nickel is required for obtaining the austenitic structure but the point at issue is whether a less expensive element in combination with other additions will provide a comparable alloy at lesser cost. The development of a cost effective and yet performance oriented austenitic cast iron with manganese, copper and nickel is described here. A small addition of mischmetal has been made to improve oxidation resistance in the absence of substantial nickel. Consequently, what is needed is an inexpensive alloy having good mechanical properties, high thermal expansion, comparable thermal conductivity and good elevated temperature properties. Further, graphite morphology for correct bonding of aluminum alloy in pistons and also the content and distribution of carbides in the matrix for good machinability are important requirements in the development of the low nickel alloy. Although attempts have been made to develop such an alloy, the author is not aware of any alloy which has been put successfully into commercial practice on an industrial scale.

Literature Review:

Richards in a research report of BCIRA had studied a similar alloy and published the mechanical properties and stability of austenitic flake graphite irons. The report shows that the strain induced martensite forms in the intermediate fracture zone when the alloy is strained to failure at room temperature. Also, austenite becomes unstable at about 50°C. (Ref 1)

SS Dhanjal etc developed a heat resistant cast iron which was non-magnetic and did not contain expensive alloying elements like nickel, chromium molybdenum etc. This alloy showed superior heat resistance properties compared to the conventional Niresists and heat resistant steels. The alloy was machinable, though with some difficulty. (Ref 2)

FK Kies in a report on strengthening of Niresist reported that the yield strength increased by first a refrigeration treatment to form marteniste and followed by a reaustenization treatment at a low temperature. (Ref 3)

AN Volkov studying the influence of chromium on the properties of austenitic manganese cast iron showed that 1.2 to 1.5% chromium considerably improved the mechanical properties like transverse rupture strength, deflection and hardness of the flake graphite irons. (Ref 4)

In a report in Russian Casting Production the authors studied the hot strength and growth resistance of high manganese austenitic cast iron and showed that a 1% nickel with 10.5 – 11.5 manganese and with carbon and silicon at normal levels provides good physical and mechanical properties.

An addition of 0.8 – 1.2% aluminum significantly improves machinability. (Ref 5)

In a related report in Russian Casting Production authors have studied the effect of increasing manganese and silicon on the wear resistance of high manganese austenitic flake cast irons. (Ref 6)

Type of Austenitic Irons Available:
The Ni-resist irons were first developed in 1927 and various grades were developed with nickel contents varying between 14% and 36% along with other elements. Later, the spheroidal graphite austenitic cast irons were developed with better mechanical and physical properties. The various grades of Niresist irons as specified by different countries are tabulated in Table I. The compositions of the various grades are listed in Table II and typical properties are listed in Table III. This is done in order to compare the composition and properties with the low nickel alloy being developed.

Development of alloy and results:
An austenitic cast iron for possible application as top ring insert in pistons of internal combustion engines and containing, by weight 7 to 11% manganese, 1 to 8.0% Nickel, 4 to 5.5% Copper, 0.0 to 1.5% Chromium, 1.5 to 2.75% Silicon, 3% max carbon and balance essentially Iron was selected for study of structure, mechanical and physical properties.

Only few elements like carbon, manganese, copper, cobalt, nickel and nitrogen can extend the austenitic region in the iron-carbon system. However, only nickel in contents around 18% or in combination with copper to obtain a total of 18% can provide the austenitic structure at room temperature. The effect of carbon is to increase carbide precipitation, especially in chromium containing grades. Manganese has an austenitising effect and is half as effective as nickel. Complete substitution of nickel with manganese is possible but the alloy becomes hard and difficult to machine. It is therefore preferable to limit the manganese upto maximum 10% and add copper and nickel to obtain the austenitic structure and yet make the alloy machinable. In order to minimize the massive carbides, the chromium content was limited to 1.5% in the alloy for specific applications. Table IV shows the composition of the alloy selected for experimental purposes.

The optimum composition for new alloy was selected based on the major elements manganese, nickel and copper providing the structure, mechanical and physical properties of austenitic cast irons. After examining the data, it was found that if we retain nickel at low contents for high temperature resistance, use of ample manganese will be required. High manganese content (10-11.5%) results in considerable hardening and poor machinability. A part of the manganese was substituted with copper. Within the range selected and with carbon and silicon at normal levels, the alloy provided a good combination of properties and machinability. The high silicon has a graphitising effect and results in a lower proportion of (Fe Mn)$_3$ C carbides in the structure. Inhibiting carbide formation allows manganese to remain in solid solution in the austenite. The additional manganese in the austenite matrix stabilizes it against transformation at fluctuating temperatures and makes it possible to lower nickel content. The oxidation resistance is maintained by the addition of small quantities of mischmetal (0.05%) and aluminum (0.5% max) which also improves machinability.

Experimental Procedure

The alloy was produced under commercial conditions by melting in a 100 kg, medium frequency electric induction furnace using relatively clean and analysed charge materials like pig iron, scrap, high carbon ferromanganese, ferro-silicon, cathode copper and nickel squares. After meltdown, the melt was processed with mischmetal and inoculated with proprietory strontium bearing innoculant. The melt composition was analysed in a vacuum emmision spectrometer and poured in molds for various test castings. The test castings for the top ring piston inserts were centrifugally cast, under controlled condition, in steel molds.

Production type heats were poured for bulk piston ring insert castings for manufacture of pistons of internal combustion engines and their testing under test conditions and for field testing.

The following tests were made to assess the new alloy and the centrifugally cast insert rings:

Chemistry
Mechanical properties
Microstructure
Oxidation test
Sulphidation test
Lead corrosion test
Salt spray test
Wear tests

The following physical and mechanical properties were evaluated:
Mechanical properties (ultimate tensile, strength, elongation, hardness, compressive strength, transverse strength and deflection and impact)
Density
Coefficient of thermal expansion
Thermal conductivity
Magnetic response
Linear contraction
Section sensitivity
Machinability
Heat resistance

The mechanical properties were conducted on standard tensometer test pieces from centrifugal castings and from standard test bars.

Oxidation resistance was assessed by heating machined 20 mm dia x 25 mm length samples in refractory crucibles in a muffle furnace at 900°C for 24 hours. Specimen weight changes and detached scale weight were determined for each sample. The scales were detached by light brushing and with emery paper and hence the results are subject to some variation.

Sulphidation tests were carried out on machined samples of 10 mm diameter and 20 mm length, by passing hydrogen sulphide at 80°C for a period for 24 hours. More intensive sulphidation test was made on machined samples by immersing in a mixture of Carbon, $BaSo_4$, $CaSO_4$ and Na_2SO_4 mixture at 870°c for 48 hours.

Lead Corrosion test were carried out by immersing machined samples in a mixture of $PbSO_4$ and PbO at 290°C for 1 hour.

Salt spray test was carried out on machined ring inserts in acid condition for 96 hours under standard test conditions.

The section sensitivity of the alloy was analysed by pouring a step bar of 50 mm thick in steps to 10 mm thick and tensometer test specimens machined for tensile testing.

Metallographic specimens were prepared by polishing upto 320 grit paper and finally polished with levigated alumina suspension on polishing pads.

Results
Mechanical and Physical Properties:
The mechanical and physical properties of the new austenitic cast iron are given in tables VI & VII respectively.

The tensile strength is essentially related to the flake graphite present in the matrix and is comparable to that of Ni-resist Austenitic Iron Type 1. All other mechanical properties of the new alloy are comparable to that of the Ni-resist Iron type 1 alloys.

Density:
The density of the new alloy is slightly less than that of Niresist Type1, as given in Table VII. With the comparable properties and an approximate 3% less density than Ni-resist Type1, the new alloy provides an added incentive for reducing weight of castings.

Thermal expansion:
From the Table VII it can be observed that the co-efficient of thermal expansion of 18.49 between 20°C and 200°C and is comparable to the

19.3 value of Type1 Ni-resist. The expansion enables the new alloy to be used in combination with a variety of alloys like aluminum alloys, copper, bronze and austenitic stainless steels. This compatibility prevents the distortion of the joint in the combined metals. This feature allows for the alloy to have potential application as inserts in aluminum alloy pistons.

Thermal Conductivity:
The thermal conductivity as shown in Table VII is comparable to that of Ni-resist Type 1 alloy.

Magnetic Properties:
The new alloy is non magnetic and is similar to Ni-resist Type1 alloy. Its magnetic response can be slightly increased by increasing chromium content within the composition, if required.

Linear Contraction:
The contraction allowance for pattern or die design is about 1.44%. This value of the contraction allowance for the new alloy are also comparable to that of Ni-resist Type 1 alloy, which means existing pattern and die tooling can be used.

Wear Resistance:
The new alloy has excellent wear resistance as the graphite flakes are uniformly distributed in the matrix and makes the alloy highly resistant to wear under metal to metal contact conditions. In the hardness range of 130-180 Brinell, the new alloy has good resistance to metal to metal wear. The adhesive wear test result on a 4.3 diameter pin sample, shows that the wear resistance of the new alloy is superior to that of Ni-resist Type 1 alloy. The new alloy has work hardening characteristic resulting in a work hardened glaze which develops good wear resistance.

Sulphidation Resistance:

In the combustion chamber of diesel engines, the sulphur contained in the fuels forms sulphates with alkali or alkali earth metal compounds added into lubricants. The mixture of sulphate and carbon deposit causes serious sulphidation attack. Two tests have been done for sulphidation:

Weight loss in still hydrogen sulphide gas at 80°C for 24 hours.
Immersion test in a sulphate salt mixture which simulates the sulphidation attack in diesel engines.

The results of the test are tabulated in Table X. The new alloy has comparable resistance to sulphidation when compared to Ni-resist Type 1 alloy.

Lead Oxide Corrosion:
Leaded High Octane Gasolene contains Ethy1 or Methy1 lead in leaded fuels which converts to PbO in the combustion gas mixtures of varying

specifications. The resulting wet deposits of $PbSO_4$ corrodes severely. Tests conducted on machined samples by immersing in a $PbSO_4$ mixture at 290°C indicates that the new alloy and the Ni-resist Type 1 have comparable resistance. The tests were conducted at 290°C as the ring inserts reach 200-220°C in actual practice in engine speed range of 1400 to 2000 rpm. The maximum temperature reached generally is in the range of 240-280°C.

Atmospheric Corrosion:
The new alloy is comparable to Ni-resist Type 1 alloy when exposed to atmosphere. The new alloy is also not rust free but shows very little staining after exposure for 120 days.

Salt Spray Test:
The results show that the new alloy has a weight loss of 2.45% whereas Ni-resist Type I has weight loss of 1.98% which shows that the high Nickel Ni-resist Type 1 alloy is slightly superior to the new alloy in marine atmospheric conditions.

Section Sensitivity:
The section sensitivity tests of the new alloy showed that the alloy was not sensitive to section changes. The results are tabulated in Table XI. The slight differences can be attributed to the relatively low soundness of the thick sections.

Machinability:
The machinability of the new high manganese alloy is comparable to that if Ni-resist Type 1 alloy. The machinability has been assessed by life of carbide cutting tool when machining the high manganese alloy and Ni-resist Type 1 alloy. Studies on high speed machining are in progress and will be reported later.

Microstructure:
The microstructure of the centrifugally cast rings show the following:

Graphite	A	>70%
Structure	B	<10%
	D & E	<15%
	C	NIL
Size	4-5 ASTM	
Matrix	Austenite	
Carbide		<1.0%

The microstructure is comparable to that of Niresist Type 1 alloy. The carbide content is very low and is well dispersed.

Conclusion:

The results of the present work indicates that a cast iron about 1-8% nickel, 10% manganese, 5% copper is essentially austenitic in structure and has properties comparable to the more expensive, nickel rich, conventional austenitic cast irons.

The alloy by virtue of its excellent properties and characteristics is suitable for use for a number of applications and has potential use in the top ring inserts for pistons of internal combustion engines. The major requirement of compatible coefficient of thermal expansion, good wear resistance and elevated temperature properties for the top ring inserts in piston is met with, in the new alloy. Machinability for high speed machining will substantially improve its acceptance for ring insert application in Al alloy pistons. The application can be extended to the manufacture of other automotive components like valve seat inserts, valve guides, manifolds etc. Additional work on the low temperature transformations and strain. Improved hardened martensite formation is in progress and will be provided in the presentation.

References:

1. P.J. Rickards: Tensile properties and stability of austenitic flake graphite irons containing 6 percent manganese, 5 percent nickel and 4 percent copper. BCIRA report No. 1079, July 1972. P 351 – 356.

2. SS Dhanjal, CA Naresh Rao etc., all: Heat Resistant Cast Iron NML Technical Journal Vol 19, May 1977 P 29-32

3. FK Kies and RD Schelleng: Trans AFS Oct 1968 P 83-87

4. AN Volkov and EA Rivnova: Influence of chromium on properties of austenitic manganese cast iron. Russian Castings Production V 1-12, 1976, P-399

5. Hot Strength and growth resistance of austenitic cast irons – Russian Castings Production, V 1-12, 1973, P-171

6. Wear resistance of austenitic manganese cast iron – Russian Castings Production V 1-12, 1972, P-243.

7. Engineering Properties and applications of Ni-resist irons – International Nickel Company.

Carbon Recovery and Inoculation Effect Of Carbonic Materials in Cast Iron Processing

M. Chisamera*, I. Riposan*, S. Stan*, V. Constantin*, C. Diaconu**

*POLITEHNICA University of Bucharest; **ELECTROCARBON Slatina, ROMANIA

Abstract

Metallurgical and petroleum coke materials used as conductive packing medium to surround the carbon products or as resistor medium in electric graphitizing furnaces are compared to natural graphite, scrap electrodes and usually used calcined and graphitized petroleum coke carbon raisers, under laboratory and plant trials conditions. Carbon recovery and dissolution rate, specific sulphur increasing, slag generation, chilling tendency, shrinkage occurrence etc, are considered in cast irons production. The parameters of cooling curves are connected to graphite morphology, metal matrix make-up and chilling sensitiveness of carburized cast irons. Recovered granular carbonic materials, previously used as conductive or resistor medium were found as potential carbon raisers with visible inoculation effects in cast irons, depending on their origin (petroleum or metallurgical coke). Specific characteristics of these products and their application fields are presented (as advantages and limits), compared to conventional calcined and graphitized petroleum coke, especially in grey irons production.

Key words: re-carburizers, by products, recovery parameters, inoculation

Introduction

A large variety of carbonic materials are currently used to correct the carbon content of ferrous melt [1-9]. Generally, it is considered that amorphous phases of carbonic materials are absorbed into the melt as iron carbide. They can supply carbon to the melt, but cannot help the graphite nucleation process, while any crystalline graphite phase is particularly improving graphite nucleation [2-6,8-9]. It was found that using pet. coke the grey cast iron showed no reduction on chill tendency as compared to situations when carbon raisers with different levels of crystallinity were used [5,8]. No difference was found in ductile iron [7].

The main objective of this research is to compare different primary and recovered (by-product) carbonic materials, as far as carbon and sulphur recovery parameters and inoculation activity, under the influence of the crystallinity and purity grade, in lower sulphur (<0.025%) and lower carbon equivalent (3.6-4.0%) un-inoculated grey cast irons.

Experimental Procedure

Two thermally treated groups of carbonic materials were tested: a) new (primary) materials, as calcined (1300-1400°C) or graphitized (>2500°C) petroleum coke and b) packing medium or resistor (insulator) medium materials in electric graphitizing furnaces, as granular petroleum or metallurgical coke (by-products from graphite manufacturing process) (Table 1). Metallurgical coke was usually used as packing medium to surround the carbon products in electric graphitizing furnace; if this material is covered with sand, will act as an insulator, to protect the furnace walls. Calcined petroleum coke was also tested in these positions, especially to define its behavior as recarburizer and inoculation. The purest of the spent by product material is a low ash partially graphitized coke which originates as the packing medium in close proximity to the graphite products. The least pure material is the one which originates as the material at the sand/coke interface (insulator layer) and contains free carbon, silicon carbide (up to 25% SiC) and un-reacted silicon dioxide.

Carbon raisers studied in this work (Table 1) can be characterized more specifically, as follows:

a) Calcined Petroleum Coke (CPC), at relatively low and high sulphur content and high ash level as representative amorphous carbonic material.

b) Graphitized Petroleum Coke (GPC) as crystalline Synthetic Graphite. This list of materials includes two desulphurized petroleum coke products processed in granular form (>2700°C), at very low level of sulphur, ash and volatile matter and scrap electrodes as lower purity product.

c) Recovered Partially Graphitized Packing Coke, as initially granular Petroleum Coke (GPCR) or Metallurgical Coke (GMCR) used to surround electrodes in electric graphitizing furnace. Partially graphitized carbonic materials, they will be different as sulphur, ash and volatile matter content.

d) Resistor Coke, as by-product from graphite manufacture, where calcined petroleum coke was covered with sand to act as an insulator of furnace wall. It was selected a high purity product (MCSX), at more than 95% fixed carbon, less than 0.2%S and 2-3 %SiC.

e) Natural Crystalline Graphite as high purity mined product (96.9% FC).

Equations 1-4 describe the efficiency parameters of carburization treatment, while equations 5 the specific sulphurization parameters:

- Carbon Recovery (CR): $\mathbf{CR} = \dfrac{m \bullet \Delta C}{C_{MC} \bullet q_{MC}} x100 \quad [\%]$ (1)

- Carburizing Efficiency (Ec): $\mathbf{Ec} = \dfrac{m \bullet \Delta C}{q_{MC}} = C_{MC} \bullet \dfrac{CR}{100} \quad [\%]$ (2)

- Carbon Dissolution Rate (CDR): $\mathbf{CDR} = \dfrac{\Delta C}{t} \quad [\%C/\min]$ (3)

- Specific Carburization (Cs): $\mathbf{Cs} = \dfrac{\Delta C}{q^*_{MC}} \quad [\%C/1.0\%MC]$ (4)

- Specific Sulphurization (Ss , S*s):

$$\mathbf{Ss} = \dfrac{\Delta S}{q^*_{MC}} \quad [\%S/1.0\%MC] \qquad \mathbf{S*s} = \dfrac{\Delta S}{\Delta C} \quad [\%S/\%C]$$ (5)

m = treated melt, kg; **ΔC = C_fin - C_in** – dissoluted carbon in iron melt, %; **C_MC** –carbon content, %; **q_MC , q*_MC** – added recarburizer, kg and %; **t** – dissolution time, min; **MC** – recarburizer.

The solubility of these commercial and experimental carbon raisers was evaluated in a series of cast iron heats prepared in coreless induction furnaces under laboratory conditions and plant trials. Two initial 80kg heats were produced from a charge of 60 kg pig iron and 20 kg Steel scrap in a medium-frequency induction furnace (100Kg, 2400Hz, silica lining). The resulted base material from each initial heat was later melted in a small coreless induction furnace (10 kg, 8000Hz, graphite crucible) and two experimental programs were recorded under laboratory conditions (Table 2). The established experimental algorithm consisted in holding the melt at constant temperature for about five minutes followed by de-slagging. The candidate material was then added to the melt while the power was applied to the furnace to maintain the temperature at the same field (1450-1470 and 1480-1500°C, respectively). After a fix holding time (3 or 5 minutes), a mechanical stirring was applied (2 minutes). Finally, FeSi75 was added in the furnace, before tapping. The two programs are differentiated by iron chemistry, carbonic material (type, quantity, grain size), the melt temperature and the quantity of FeSi75 addition (Table 2).

In plant trials low frequency coreless induction furnace was used (6.0t, 50Hz, silica lining). Metallic charge included molten iron heel (17-50%) and steel scrap (50-83%). Each recarburizer was added as 4...6 parts (25-30 Kg) in the steel scrap charge. Finally, the 15min holding time was applied (Table 3).

Results and Discussion

The results of the carbon and sulphur dissolution studies are presented graphically in Figures 1 through 3 for all of the various carbon products studied. In all of cases, the treated iron melts are characterized by low carbon (<3.0%), silicon (<1.5%) and sulphur content (<0.025%).

The first laboratory experiment program was focused on the behavior comparison of amorphous (CPC) and crystalline (GPC, NG) carbonic materials. The carbon dissolution data presented in Figure 1 characterize the manner in which a variety of carbon raisers may be taken into solution in typical iron melt. Essentially, it appears that the carbon recovery (CR), carburizing efficiency (Ec), carbon dissolution rate (CDR) and specific carburization (Cs) depend on the structure (amorphous to crystalline ratio) and the purity of the specific carbon raiser. The laboratory scale experiments at relatively low melt temperature and carburizing time did not allow high level of the tested parameters, but at specific range depending on recarburizer type. The highest level of carbon-recovery parameters is typical for graphitized petroleum coke (crystalline synthetic graphite): higher purity grade, higher the rate of dissolution of these materials (GPC1 vs. GPC3). Despite its crystalline structure and high purity grade, mined natural graphite led to the lowest level of carbon dissolution in molten iron, in the same treatment conditions. Calcined petroleum coke products show an intermediate behavior, but closer to graphitized petroleum coke, in the experimental conditions. Partially graphitized and high purity recovered resistor petroleum coke (special selected) is characterized by a good behavior, as it is on the second place as efficiency in carburizing treatment. It could be comparable to lower purity synthetic graphite. It should be noted that if this material type contains higher amount of SiO_2 and SiC, carbon dissolution rate and recovery are visible lower. Sulphur contribution of tested carbon raisers depends on content in the added materials (see Table 1), but visible level is typical only for calcined petroleum coke products.

The second experimental program developed under laboratory conditions was designed to compare the most representative amorphous and crystalline recarburizers at the one hand and partially graphitized (crystalline) recovered carbonic materials at different origin (packing Pet Coke or Met Coke), on the other hand. It was applied the same experimental procedure such as Program I was, but at higher level of iron melt temperature and carburizing time, for higher addition of recarburizer (see Table 2). The carbon and sulphur recovery parameters are summarized in Figure 2. The rate of dissolution of these materials is higher than in Program I was, due to better conditions as temperature and time: more than 80% C-recovery, compared to 45-64% for the same materials type, in the first program. The crystallinity of these carbonic materials appears to be more important than their purity level to influence the carbon recovery parameters. It is clear that graphitized petroleum coke is better than only calcined petroleum coke. Very good results were also obtained for packing petroleum coke, despite higher ash and lower fixed carbon content (comparable with commercial synthetic graphite). Recovered graphitized metallurgical coke as packing material from high temperature treatment furnace is also better than calcined petroleum coke as C-recovery for the same dissolution rate despite high level of ash content.

As specific sulphurization, the highest contribution is typically for petroleum coke, while the lowest level for graphitized petroleum coke (a little bit higher for recovered packing product). Graphitized packing metallurgical coke has an intermediate position, but closer to graphitized petroleum coke than only calcined material. The visible difference as ash content will differentiate these carbonic materials also as specific slag generation. Good relationship was established between ash content in the recarburizer and slag quantity generated in the melting furnace (CPC3=0.08, GPC2=0.013, GPCR1=0.28 and GMCR1=1.68 kg/%C x ton, respectively).

Prevalent crystalline structure and sulphur removal, as result of high temperature treatment of packing carbonic materials into the electric graphitizing furnaces sustain these materials as good recarburizers, especially for petroleum coke as origin. Also metallurgical coke could be considered in this position, but its specific high ash level will decrease dissolution rate at more slag volume production and higher sulphur contribution. Graphitized packing coke by products could be attractive for cast iron foundries, as possibility to obtain high performance in iron treatment, at lower cost.

Table 3 and Figure 3 summarize the foundry results as carbon and sulphur dissolution and slag volume. The high stirring capacity of the 50Hz coreless induction furnace led to high rate of dissolution of these materials which exhibited recoveries of carbon near 100%, at relatively lower melt temperature (1440-1450°C). Despite its high ash content, graphitized packing metallurgical coke had a good behavior, especially for lower carbon content in the initial melt (see Table 3). No visible difference between calcined petroleum coke and recovered graphitized packing petroleum coke as carbon recovery. Clear difference appears to be as sulphur contribution and slag, but in opposite way: the highest sulphurization and the lowest slag volume are typically for calcined petroleum coke, while the packing by-products are characterized by lower sulphur and higher slag contribution (petroleum coke has intermediate position). The specific production of this foundry by the use of these by-products includes grey and ductile iron castings: GPCR is usually used in both grey and ductile iron field, while GMCR especially for grey iron production (also ductile iron, but desulphurization treatment is usually applied). Plant trials confirmed the laboratory experiment and sustain the possibility to use selected graphitized packing coke by-products as carbon raisers in cast iron foundries (inclusively cheaper metallurgical coke).

LABVIEW6 thermo-analysis system was used, and chill tendency, concentrated shrinkage sensitiveness, mechanical properties and structure characteristics were evaluated during the first laboratory program. Eutectic undercooling degree (ΔTm), recalescence degree (ΔTr), maximum rate of recalescence (TEM) and the minimum value of the first derivative at the end of solidification (FDES) were found to be the representative parameters of the cooling curves capable to distinguish the solidification features of these cast irons. The highest undercooling level (21-22°C) was typically seen on treatments using natural graphite, while synthetic graphite products appear to determine a lower level of undercooling (10-13°C) and higher level of recalescence (9.5-10.5°C). Calcined petroleum coke presents an intermediate position. When used, natural graphite was not effective as carbide avoidance. No obvious difference was found as far as chill tendency between calcined and graphitized petroleum coke materials, but the last group led to lower shrinkage incidence. Resistor petroleum coke as the least partially graphitized material which included silicon carbide as active component appeared to have the highest graphitizing influence, according to the lowest undercooling degree (10.4°C) and FDES level (-3.61°C/s) (Fig.4).

Cast irons microstructure was more influenced by the crystallinity of carbonic materials rather than chill tendency: higher homogeneous grade of both graphite phase and metal matrix make-up for crystalline recarburizers addition in the furnace. Prevalent A-type graphite and pearlitic matrix are typical for synthetic graphite or graphitized resistor coke application. Mechanical properties, as tensile strength (Rm=315-330 N/mm^2) and Brinell Hardness (195-225 HB) are in good relationship with microstructure characteristics and recarburizer type, respectively.

At the same range of initial sulphur and manganese content (0.016-0.020%S, 0.27-0.29%Mn), lower level of equivalent carbon (2.9-3.1%) was tested during the second laboratory experiments program (II-Table 2). Better dissolution conditions (higher temperature and time) and higher recarburizer addition rate led not only to a lower chill tendency level, but also to more evident differentiation between carbonic materials efficiency (Fig. 5). Two types of samples were used, lower cooling rate (W$_{3\,\frac{1}{2}}$ Wedge Test) and higher cooling rate (4C Chill Test) samples (ASTM A367). In the experimented conditions, it is clear that the both graphitized petroleum coke products (granular synthetic graphite-GPC and packing coke as by-product-GPCR) are more effective than calcined petroleum coke for the both cooling rate levels. Graphitized Packing Metallurgical Coke appears to be also more efficient than amorphous calcined petroleum coke.

Conclusions

The results of this research illustrate that the various carbon raisers respond differently when they are applied to ferrous melt, not only from the perspective of carbon recovery parameters, but also as sulphur contribution and inoculation efficiency:

- Medium sulphur and high carbon content amorphous Calcined Petroleum Coke (CPC) is the reference carbon raiser for grey cast iron.
- It was confirmed that Graphitized Petroleum Coke (GPC) products are characterized by a more rapid solubility and higher C-recovery at an evident lower S-contribution and higher inoculation efficiency than CPC in grey cast iron.
- Graphitized Conductive Packing Petroleum Coke (GPCR) is superior to CPC as far as higher C-recovery, lower S-contribution and higher inoculation activity being close to GPC.
- Graphitized Conductive Packing Metallurgical Coke (GMCR) is generally superior to CPC as far as C-recovery performance but at the same level of dissolution rate. CMCR has an intermediary position as S-contribution and inoculation efficiency, usually closer to GPC.
- Graphitized Resistor Petroleum Coke (MCSX) at a high C-level and presence of a low amount of SiC is very efficient from the perspective of C-recovery, inoculation efficiency, and very low S-contribution.

Table 4 illustrates the typical recarburizers, resulted from the present research program and validated by foundry applications.

References

1. Coates R.B., *Types, Selection and Applications of Recarburisers in the U.K. Foundry Industry*, The British Foundrymen, August 1979.
2. Loper C.R.Jr., Liu S.L.,Shrivani S. and Witter, T.H., *The Dissolution of Carbon in Cast Iron Melts as Studied Using Commercial Raisers and Experimental Materials*, AFS Transactions, 1984, pp.323-337.
3. Terrel P.B., *Carbon Raisers and Cast Irons,* Foundry Management and Technology, Aug. 1996 and Sept.1996.
4. Riposan I., *Recarburisers for Cast Iron Foundries*, Romanian Foundry Journal (RO), No.1, 2003, pp. 10-13.
5. Jentsch A., *The Influence of Carburisers on the Microstructure, Quality and Overall Production Cost of Cast Iron Parts*, 66th World Foundry Congress, Istanbul, Turkey, 6-9 Sept.2004, pp.729-741 pub. Toksad: The Foundrymen's Association of Turkey, 2004.
6. Salmon E and Jentsch A., *Analyse Comparative de Produits Recarburants, Effect Inoculants et Trempe des Fontes Spheroidales*, Hommes and Fonderie , April 2003, pp.52-56.
7. Henning W.A, *Comparing Crystalline and Noncrystalline Recarburizers in Ductile Iron Production,* AFS Transactions, 1999, pp. 577-580.
8. Suzuki S, AFS Transactions, 1982, pp 423-434.
9. Riposan I., *High Quality Romanian Recarburisers for Cast Irons,* Romanian Foundry Journal (RO), No. 5-6, 2004, pp. 18-22.

Carbonic Materials Characteristics

Table 1

Type (origin)	Source	Symbol	Characteristics*	Structure
Calcined Petroleum Coke	Calcining Furnace	CPC1	98.7%FC, 2.5%S, 0.5%Ash	Amorphous
		CPC2	95.06%FC, 0.64%S, 1.5%Ash	
		CPC3	98.3%FC, 0.87%S, 0.60%Ash	
		CPC4	97.1%FC, 0.96%S, 1.06%Ash	
Graphitized Petroleum Coke [Synthetic Graphite]	Electric Graphitizing FurnaceProduct	GPC1	99.3%FC, 0.07%S, 0.1%Ash	Prevalent Crystalline
		GPC2	99.7%FC, 0.06%S, 0.10%Ash	
	Turnings	GPC3	99.07%FC, 0.27%S, 0.68%Ash	
Packing Coke (to surround electrodes) [PC – Petroleum Coke; MC – Metallurgical Coke]	Electric Electrodes Graphitizing Furnace	GPCR1	97.4%FC, 0.10%S, 2.4%Ash	Partially Graphitized
		GPCR2	97.86%FC, 0.05%S, 2.0%Ash	
		GMCR1	87%FC, 0.34%S, 11.5%Ash	
		GMCR2	83%FC, 0.19%S, 16.7%Ash	
Resistor Coke (furnace insulator)	Electric Electrodes Graphitizing Furnace	MCSX	96%FC, 3.0%SiC, 0.19%S	Partially Graphitized
Natural Graphite	Mined Graphite	NG	96.9%FC	Crystalline

* Fixed Carbon (FC) = 100 - (%Ash + %Volatile Matter + %H$_2$O)

Table 2

Laboratory Experiment Conditions

Progr.	Chemical Comp. of Iron Melt, wt.% (in/fin)		Equiv. C, CE, (in/fin, %)	Iron Temp. (°C)	Recarburizer addition			Recarb. holding in the furnace, min		Fin. FeSi75 add. in furnace, (%)
	C	Si			Type (Tab.1)	%	mm	Without Stirring	Stirring	
I	2.8-2.9 / 3.0-3.25	1.38-1.40 / 1.78-1.87	3.21-3.24 / 3.55-3.70	1450-1470	CPC1, CPC2, GPC1 GPC3, MCSX, NG	0.6	0.6-1.0	3	2	0.7
II	2.55-2.75 / 3.15-3.53	1.15-1.25 / 1.6-1.85	2.9-3.1 / 3.6-4.1	1480-1500	CPC3, GPC2, GPCR1, GMCR1	0.9	0.4-4.0	5	2	0.8

Table 3

Plant Trials Results

Heat	Metallic Charge, (%)		Recarburizer		Final iron Holding, (°C / min)	C, (%)		S, (%)		C-Recov. Param.		S-Recovery Parameters		Specific Slag Quantity, (kg/%Cxt$_{iron}$)
	Heel	Steel Scrap	Type (Tab.1)	Addition Rate, (%)		Metallic Charge	Final	Metallic Charge	Final	CR, (%)	Ec, (%)	Ss, (%S/1.0%MC)	S*s, (%S/1.0%C)	
1	30	70	CPC4	2.25	1440/15	1.28	3.60	0.026	0.043	100	98	0.0076	0.0073	0.10
2	50	50	GPCR2	1.67	1440/15	1.80	3.62	0.025	0.029	100	96	0.0024	0.0022	0.18
3	17	83	GMCR2	3.00	1450/15	0.61	3.00	0.013	0.021	96	80	0.0027	0.0034	2.10

Table 4

Typical Designed Recarburizers

Recarb. Type	Thermal Treat, (°C)	Origin	Fixed Carbon*, FC, (%)	Sulphur S, (%)	Ash (%)	Volatile Matter (%)	Moisture (%)	Spec. Contribution(1.0wt.% add)	
								C (%)	S (%)
CPC	1300-1400	Selected Calcined Petroleum Coke	98.7-99.3	0.80-0.90	0.40-0.60	0.12-0.40	0.06-0.13	0.75-0.90	0.006...0.008
GPC	>2500	Graphitized Petroleum Coke	99.1-99.8	0.02-0.08	0.15-0.75	0.08-0.12	0.05-0.15	0.90..0.93	<0.0005
GPCR	>2500	Petroleum Coke Conductive Packing Medium, in Electric Graphitizing Furnace	min.96	max.0.20	max.3.0	max.0.30	max.0.30	0.85..0.93	<0.0015
GMCR	>2200	Metallurg. Coke, as Conductive Packing Medium, in Electric Graphitizing Furnace	min.82	max.0.50	max.17	max.1.0	max.0.30	0.70..0.80	0.0025..0.0045
MCSX	>1500	Petroleum Coke as Insulator in Electric Graphitizing Furnace	min.96 FC 2-3 SiC	max.0.2	max.1.0	max.0.2	max.0.30	0.88..0.92	<0.0015

* Fixed Carbon (FC) = 100 - (%Ash + %Volatile Matter + %H$_2$O)

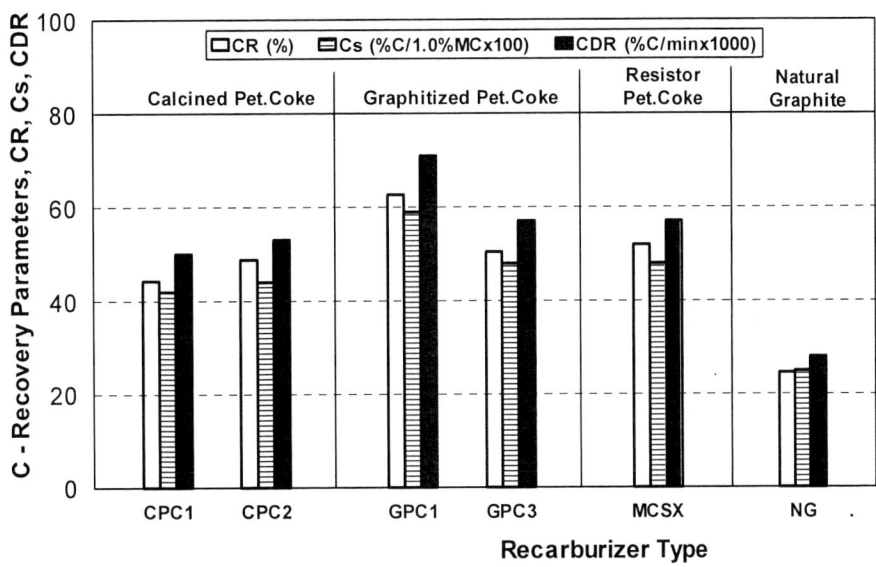

Fig. 1 C-Recovery (CR), Specific Carburization (Cs) and C-Dissolution Rate (CDR) in Iron Melt (Program I)

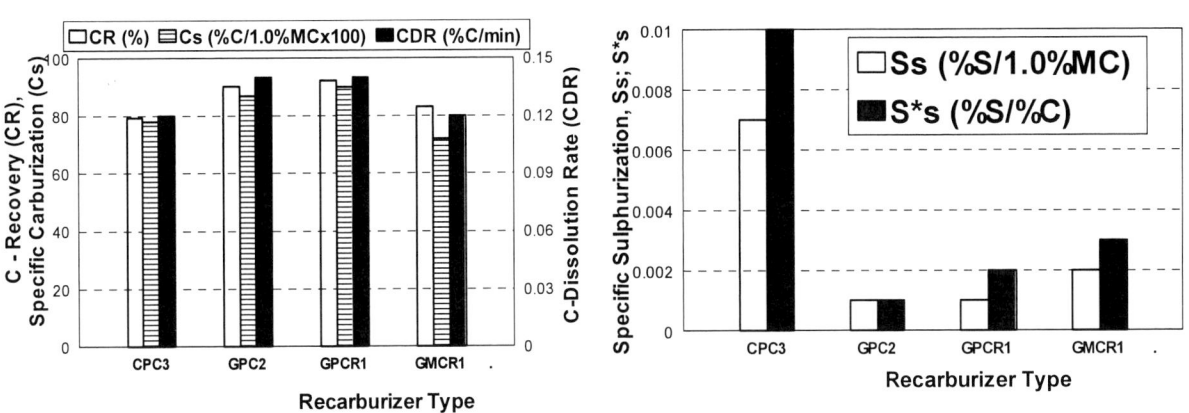

Fig. 2 C-Recovery and S-Recovery Parameters (Program II)

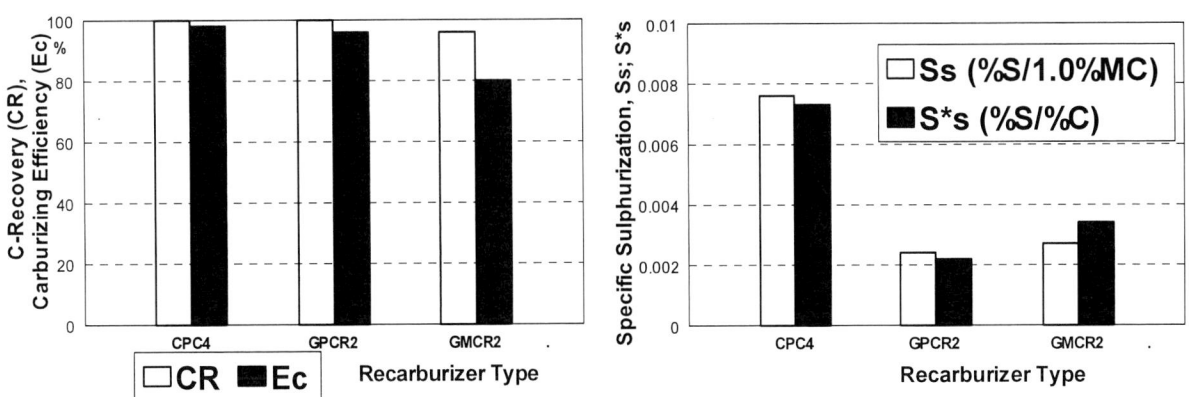

Fig. 3 C-Recovery and S-Recovery parameters in 6.0t/50Hz Coreless Induction Furnace (Plant Trial)

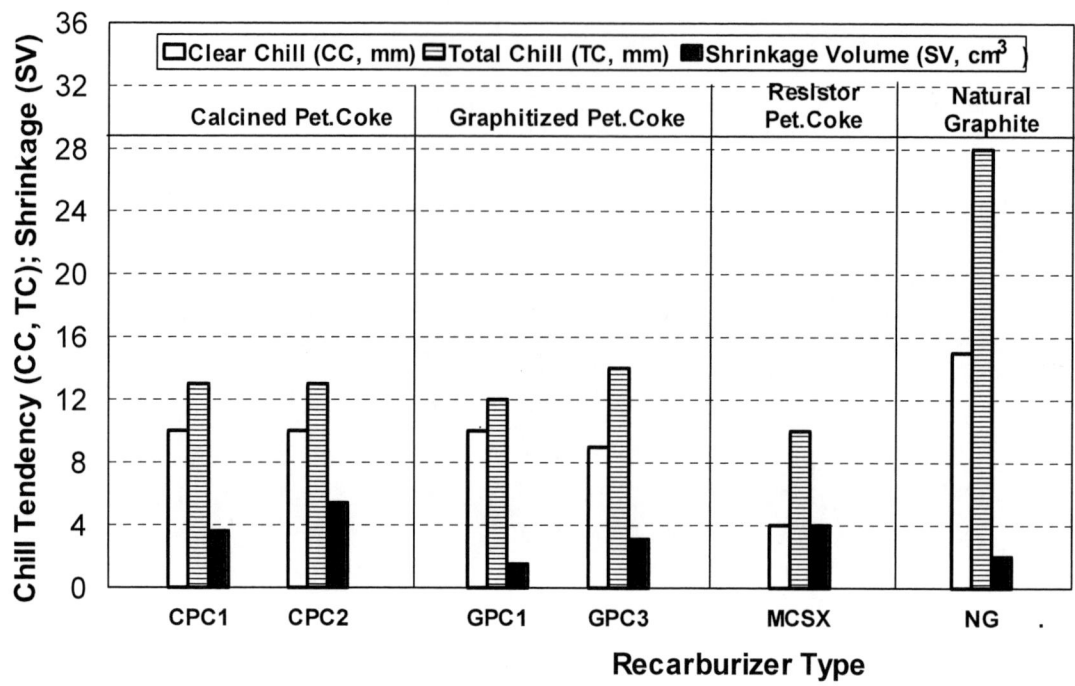

**Fig. 4 Clear Chill (CC), Total Chill (TC) and Shrinkage Volume (SV)
of Un-Inoculated Grey Irons (Program I)**

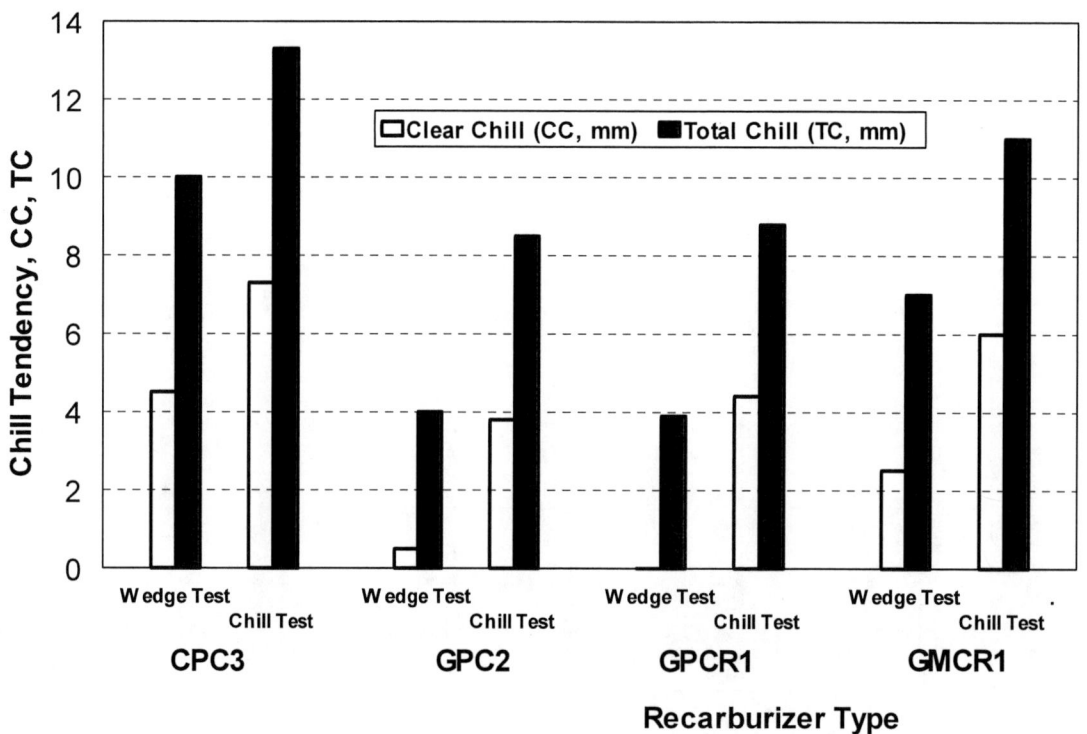

Fig. 5 Chill Tendency of Un-Inoculated Grey Irons (Program II)

Stream-lined Gating Systems with Improved Yield - Dimensioning and Experimental Validation

N Tiedje*, S Skov-Hansen*

* Department of Manufacturing Engineering and Management, Technical University of Denmark, DK.

Abstract

The paper describes how a stream-lined gating system where the melt is confined and controlled during filling can be designed.

Commercial numerical modelling software has been used to compare the stream-lined design with a traditional gating system. These results are confirmed by experiments where the two types of lay-outs are cast in production.

It is shown that flow in the stream-lined lay-out is well controlled and that the quality of the castings is as at least equal to that of castings produced with a traditional lay-out. Further, the yield is improved by 4 % relative to a traditional lay-out.

Key words: Gating systems, Melt flow, Casting quality.

Introduction

The foundry industry is a highly competitive industry where the demand for quality and cost reduction is constantly increasing.

Many foundries strive hard to improve casting quality, reduce scrap rates and to cut costs in all parts in the foundry. In high productivity plants where green sand moulds are made on automatic moulding machines, even small reductions in the amount of metal poured in each mould may lead to large savings in production costs because the number of moulds poured per day is very high.

In recent years work has been done to improve gating systems so that process stability is increased and to reduce the weight of metal that is to be re-melted in the foundry. [1-4]

Some of this work has lead to the development of a new type of stream-lined gating systems. [5] These gating systems have been shown to have very high process stability, and to be able to confine and control the melt well during filling, while at the same time reducing the volume of the gating system compared to conventional systems.

Inspired by this work, a project has been launched that aims at testing, and developing these new stream-lined gating systems for use in high productivity foundries with the aim of improving the yield if the casting lay-outs and improving casting quality.

In this paper design of stream-lined gating systems is described as well as some of the advantages of using such systems. Also a traditional lay-out is compared to a streamlined system, regarding yield, melt flow and casting quality.

Dimensioning the gating system; traditional lay-out

Calculations of gating systems are normally based on two principles: mass conservation and energy conservation.

The mass balance equation is, in theory, very simple. If the gating system is completely filled and there is negligible friction, the flow rate, Q, will be the same in all parts of the gating system, so that: (see figure 1)

$$Q = \frac{G}{t_{fill}} = A_1 V_1 = A_2 V_2 \qquad (1)$$

The average flow rate during filling is equal to the weight of the casting, G, divided by the filling time, t_{fill}.

The assumption that the gating system is filled is an assumption that may be difficult to achieve in practice, particularly in the initial, critical part of the filling sequence. [6,7]

Energy conservation is in its most comprehensive form described by Navier-Stokes equation.[8] For flow along a stream line in completely filled channels this equation may be reduced to Bernoulli's equation:

$$\frac{(V_2^2 - V_1^2)}{2} + g(z_2 - z_1) + \frac{(p_2 - p_1)}{\rho} = \Delta p_{Friction} \qquad (2)$$

Here Δp_{fric} is the loss in pressure head due to friction. The friction can result from internal friction in the melt due to its adherence to the wall of the gating system. In the case of turbulent flow (which is almost always the case in gating systems, [9,10]) the pressure loss can be expressed as: [9]

$$\Delta P_{Friction} = \lambda \; \frac{l}{D_h} \; \frac{\rho \; V^2}{2} \tag{3}$$

The friction factor, λ, can be calculated iteratively from:

$$\frac{1}{\sqrt{\lambda}} = -0.87 \cdot \ln\left(\frac{2.51}{Re \cdot \sqrt{\lambda}} + \frac{0.27 \cdot K_s}{D_h} \right) \tag{4}$$

K_s is the roughness of the surface, typically in the order of the sand grain size for sand moulds. Typical values for λ in sand moulds are: 0.032 to 0.04. [9]

These calculations can be complicated in cases where complex gating systems are dealt with and are therefore not used in practice.

Traditionally, in a gating system where flow is controlled by gravity, the Bernoulli equation boils down to:

$$V = \alpha \sqrt{2 g h} \tag{5}$$

The friction term is substituted by a friction factor, α. [11,12]

And in the case of bottom filled castings, the average velocity of the melt through the controlling cross section can be found to be: [9]

$$V_{avg} = \alpha \sqrt{\frac{g}{2}} \left(\sqrt{h_t} + \sqrt{h_t - h_c} \right) \tag{6}$$

Here h_t and h_c are the total height of the gating system and the height of the casting respectively.

Combining the energy balance and the mass balance gives the following relation between cross sectional area and pressure height in a down sprue:

$$A_2 = A_1 \sqrt{\frac{h_t}{h}} \tag{7}$$

The areas A_1 and A_2, and the pressure height h_t and h are defined in figure 1.

It is customary to use the above equations to calculate key areas in the gating system. Then standard geometries are used to tie the areas together, as shown in the traditional lay-out shown in figure a.

Dimensions of runners and sprues that the melt enters after passing through the area A_1 are normally expanded by multiplying that area by a factor depending on the alloy cast and the experience in the foundry. Typically the factor is between 2 and 6. [11,13].

The traditional lay-out used here contains a filter at the top of the down sprue. The area of the filter was dimensioned to compensate for the pressure loss over it so that the area complies with the area at the top of the down sprue.

The weight of the traditional gating system without feeders is 5.0 kg.

Dimensioning the gating system; streamlined lay-out

It has been shown that traditional gating systems, such as the one shown in figure a, are inherently unstable [5] due to pressure chock waves that form when the melt hits dead ends in the gating system or due to late filling of parts of the gating system.

To avoid such chock waves the gating system in figure a should be redesigned so that it confines and guides the melt into the casting, see figure b. At the same time the friction in the system should be used to reduce velocity at the melt front. [4,5].

The latter is done by restricting the width of sprues and runners so that they do not exceed 10 mm, see figure 2

The down sprue is dimensioned by first calculating A_1 using equations 1 and 6 with a friction factor of 0.5 and choosing the filling time to 6 s. Details of the down sprue are shown in figure 3.

Areas before and after A_1 are calculated using equation 6.

The ingates are dimensioned in the same way the so-called fan gates for high-pressure die-castings. [14] The area at the entrance to the gate is expanded linearly through the gate so that the area where the gate touches the casting is expanded by a factor 2.

In the streamlined gating system it is possible to insert filters at the top of the down sprue. The experiments were conducted using a foam filter and an extruded filter.

The weight of the stream-lined gating system without feeders is 4.2 kg.

Since simulations showed that it was not necessary to feed the casting in the centre, the feeders were reduced and moved to the top flange of the casting as sketched in figure 2.

Experimental – Flow Simulations

Flow in the two types of gating systems were modelled using a commercial software package, Magmasoft, to analyse flow velocities, chock waves and surface turbulence. The simulations were carried out with version 4.4 of the software, where porosity factors are applied to the cartesian grid.[15] This makes it possible to model flow in the true geometry of the gating system rather than using a geometry which is approximated to the real geometry by using a fine rectangular net in places where the geometry does not follow the axes of the coordinate system in the net generator.[16]

Great care was taken to set up realistic boundary conditions so that the filling conditions are as close as possible to the conditions in practice. Flow of melt into the pouring cup was adjusted by controlling the inlet area and the pressure on the melt in the inlet so that the cup is quickly filled to approximately 80 % of its volume. Then the pressure on the melt is reduced gradually as the casting fills at a rate that will ensure that the pouring cup does not fill completely until at the very end of filling simulation. These boundary conditions ensure that simulation of melt flow only depends on gravity rather than on the pressure applied to the liquid in the inlet.

Experimental – Casting Trials

Test castings were made in a commercial foundry where green sand moulds were produced on a Disamatic 2015 mk5. The sand was conditioned according to standards requirements for vertical flask less moulding lines.[11]

Melt was prepared for a ductile iron grade GJS 400-15. It was poured at a temperature of approximately 1410 °C using a laser controlled automatic pouring device.

Both lay-outs were cast with extruded filters and foam filters.

The pouring time was recorded by the pouring device for each casting, and the average was calculated for each series of experiments.

Castings were inspected for surface defects and other visible defects.

Results – Simulations

Figure 4 shows selected pictures of from the filling simulation of the traditional lay-out with filter. Shading shows velocity so that dark areas have the lowest velocity. We see that the down sprue is not filled until the melt flows into the runner and a certain back pressure is created. When the melt enters the ingate the velocity is high because the friction forces acting on the melt are too small in this gating system to reduce the velocity to a significant degree.

The melt flows highly turbulent into the feeders and a fountain of liquid is allowed to form in the feeders.

In figure 5 we see similar pictures of the simulated mould filling of the stream-lined lay-out with filter.

The pictures show that the down sprue is kept filled all through the filling sequence. The melt is confined so that no free surfaces (air pockets) are found behind the melt front. The maximum melt velocity in the gating system is approximately 2.3 m/s, but in the ingate, the velocity is reduced to approximately 1 m/s on entering the casting.

Inside the casting, the melt is allowed to flow more freely due to the geometry of the casting. Because the velocity is reduced before the melt enters the casting, the amount of turbulence is also reduced dramatically compared to what is found in the traditional lay-out.

Figures 4 and 5 show that the temperature distribution after filling is more uniform in the castings made with the stream-lined lay-out than with the traditional lay-out.

Results – Casting Trials

Figures 6 and 7 show the as-cast lay-outs of both types of gating systems. The filling time was measured by the pouring device, and it was found to be on average 6.5 seconds for both lay-outs, with and without filters.

On inspection very few surface defects were found on the other castings. None of these defects were of a nature that would lead to rejection of the castings.

In castings made with the traditional lay-out it was found that on the internal surface opposite the feeder neck there is a high thermal load on the core causing rough surfaces inside the casting though the surface quality is sufficient to meet the requirements of the casting.

In the castings made with the stream-lined lay-out the quality of the internal surfaces are more uniform and smooth, without sign of burn on.

Discussion

The simulations show that in the beginning of the filling the melt velocity is very high. The simulations find maximum velocities around 2.3 m/s. If we calculate the velocity of a free falling melt using equation 5, assuming no friction, the velocity at the bottom of the down sprue will be 2.6 m/s. This indicates that either friction in the initial part of the filling is negligible or that the parameters describing friction in the computer model is not adjusted correctly.

The filling time in the experiments was found to be 6.5 s for all experiments, whereas in the simulations the filling time was found to be 2.6 s and 3.5 s for respectively the traditional and the stream-lined lay-out. This indicates that friction factors in the numerical model should be adjusted, and that the average friction factor as used in equations 5 and 6 should be in the order of 0.5 to 0.7. This is in agreement with results found elsewhere in the literature. [4,9,12] We should therefore expect to find maximum flow velocities in practice in the order of 1.3 to 1.6 m/s, and that the velocity in the ingate will be below 1 m/s.

It is difficult to keep the traditional gating system filled when the first melt flows through it, while the stream-lined gating system is filled right from the beginning of the filling. It is also clear for the simulations that pressure chock waves as described by [5] occurs in the traditional design and not in the stream-lined system.

In practice, the two gating systems have the same filling time. But since the total weight (including feeders etc.) in the stream-lined system is only 15.5 kg compared to the 19.7 kg in the traditional lay-out, the average velocity of the melt will be reduced accordingly. This indicates that the amount of turbulence is reduced too.

The quality of the castings, in the way it is evaluated here, is the same with the two systems, but the yield is improved from 52 % to 66 %. The improvement in yield related only to re-designing the gating system (not taking into account that the feeders are changed too is 4 % of the traditional lay-out.

Conclusions

- The numerical model describes flow in the mould well, but friction in the gating systems should be adjusted (by comparing with experiments) to give realistic filling times.

- The stream-lined lay-out gives good control over the melt flow, so that the gating system is kept filled and shock waves that may cause the melt surface to disintegrate are avoided.

- Comparing filling times in the numerical model to filling times in the foundry shows that friction factors used for dimensioning gating systems are in the range of 0.5 to 0.7, where the stream-lined gating system has the lowest friction factor.

- It was found that the yield was improved from 4 % relative to the traditional lay-out.

References

1. Campbell, J., "Review: Developments in filling system design," CIATF, 1997, pp pp. 1-18.

2. Dai, X., Yang, X., Campbell, J., and Wood, J.. Influence of oxide film defects generated in filling on mechanical strength of aluminium alloy castings. Materials Science and Technology 20[4], pp 505-513. 2004.

3. Jolly, M. R., Cox, M., Harding, R. A., Griffiths, W. D., and Campbell, J.. Quiescent Filling Applied to Investment Casting - The effect of controlled and uncontrolled bottom-filled gating systems on casting reliability are compared with a top-filled system. Modern Casting 92[12], pp 36-38. 2002.

4. Larsen, P. L.. Iron Melt Flow in Thin Walled Sections in Vertically Parted Green Sand Moulds. pp 1-322. 2004. Technical University of Denmark.

5. Larsen, P. L. and Tiedje, N., "Iron Melt Flow in Thin-Walled Sections Using Vertically Parted Moulds," Vol. 1, Tükdöksad, 2004, pp pp. 223-234.
6. Tiedje, N., "Flow through bends in gating systems in vertically parted moulds," Transactions of the American Foundrymens Society, 1999.
7. Isawa, T. and Campbell, J., "Initial Filling Transient of the Running System," Transactions of Japan Foundrymen's Society, Vol. 13, 1994, pp pp. 38-49.
8. White, F. M., Viscous Fluid Flow, 2 ed., McGraw-Hill International Editons 1991.
9. Nielsen, F., Giess- und Anschnittechnik, Giesserei-Verlag G.m.b.H., Düsseldorf, 1979.
10. Tiedje, N., "Flow Through Bends in Gating Systems," Georg Fischer Disa A/S, Herlev, Denmark, Aug. 1998.
11. Disamatic Sand Molding System Application Manual, Dansk Industri Syndikat a/s 1995.
12. Richins, D. S. and Wetmore, W. O., "Fluid Mechanics Applied to Founding," AFS, Buffalo, N.Y., 1951, pp pp. 1-24.
13. Grube, K. and Eastwood, L. W., "A Study of the Principles of Gating," Transactions of the American Foundrymens Society, Vol. 58, 1950, pp pp. 76-107.
14. Herman, E. A., Gating Die Casting Dies, North American Die Casting Association 1996.
15. Bodenburg, M., "MAGMASOFT release 4.4. Improvements and new features.," 2005.
16. Patankar, S. V., Numerical Heat Transfer and Fluid flow, Taylor & Francis 1980, pp. 1-197.

Acknowledgements

The authors whish to thank the Danish transmission system operator, Energinet.dk, who is financing the project as well as the project partners: Dania A/S, Frese Metal- and Steel Foundry A/S and Disa Technologies A/S.

Figures

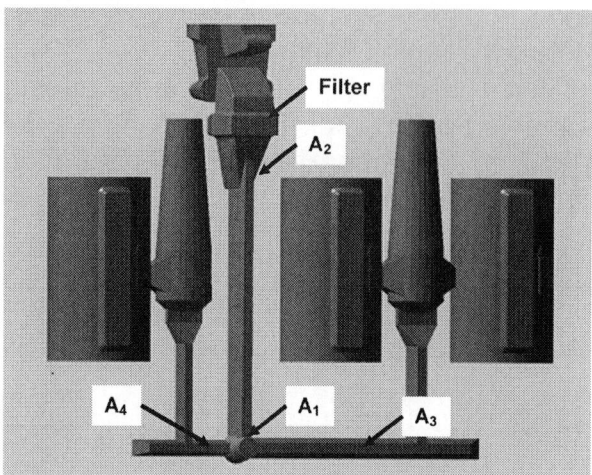

Figure a: The traditional lay-out, used as reference in the experimental work. Key areas which are used in dimensioning the gating system are indicated.

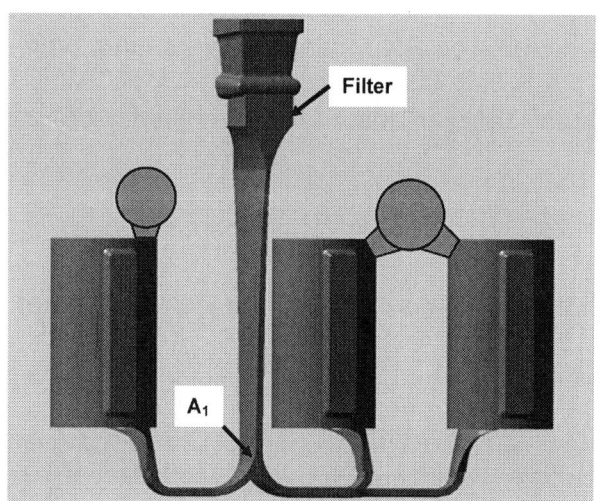

Figure b: Stream lined lay-out designed to confine and control the melt flow.

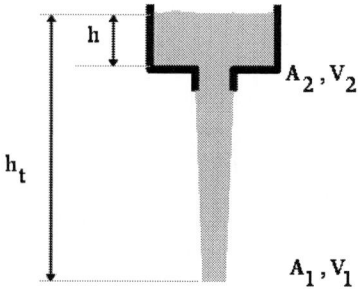

Figure 1: Areas, pressure height and velocities in a liquid flowing under the influence of gravity.

Max. 10 mm

Figure 2: The width of sprues and runners in the stream-lined lay-out is restricted to not exceeding 10 mm.

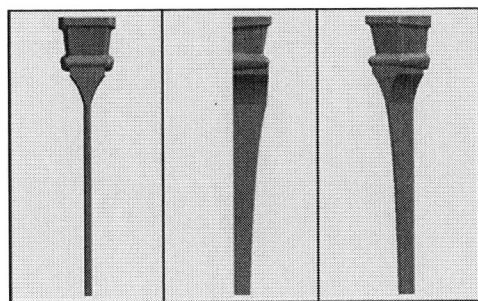

Figure 3: Detail of the down sprue in the streamlined lay-out as it is calculated using equation 6.

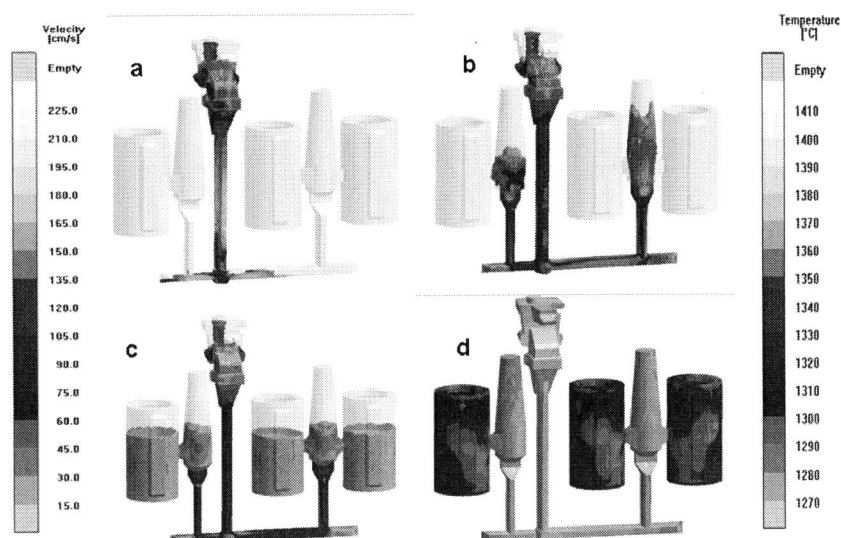

Figure 4: Selected pictures showing velocity of the melt in the filling simulation of the traditional lay-out. Pictures a-c show velocity during filling and d shows temperature after filling.

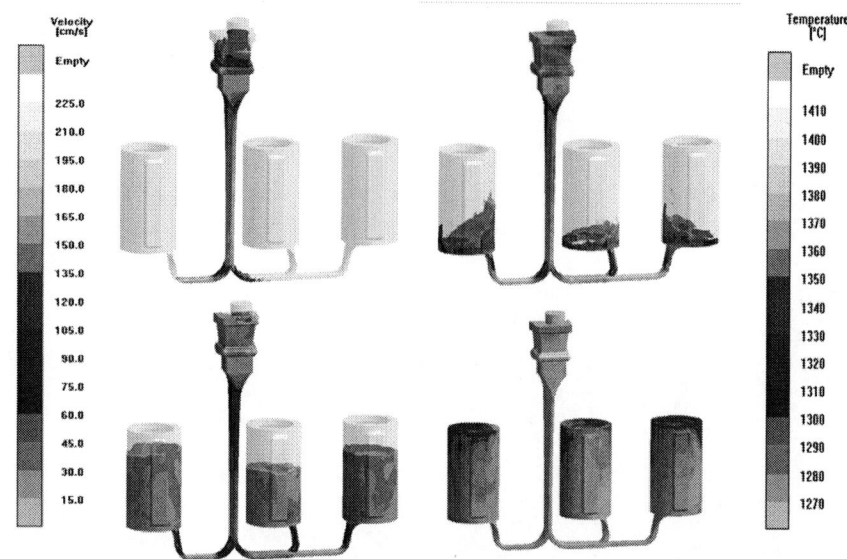

Figure 4a: Selected pictures showing velocity of the melt in the filling simulation of the traditional lay-out. Pictures a-c show velocity during filling and d shows temperature after filling.

The advantages of SiC-quenching plates (chills) when compared with grey iron quenching plates (chills)

T. Reuther *.

* Hofmann CERAMIC GmbH, Germany.

Abstract

This paper shows several influences grey iron plates (chills) and SiC-plates (chills) have on poured rectangular grey iron plate (EN-GJL-250) relating to surface defects, microstructure and hardness. The result of these examinations show that the SiC-plates have a smoother chilling effect and give a better surface finish on the casting because they do not create condensation water.

What this paper demonstrates, with the aid of two practical examples of large hand moulded castings is the application of the two types of chills and the effect on solidification and surface hardness.

It also shows using trials on a rotary furnace, that broken SiC-plates can be used as a Silicon addictive.

This paper demonstrates the advantages of using SiC-plates and that not only the purchase price but the whole cost-effectiveness of a product has to be taken into consideration.

Key words

Controlled solidification, SiC-quenching plates (chills), grey iron quenching plates (chills), chill-effect, microstructure

Introduction

Today, due to casting weight reduction more and more castings are being produced with different wall thicknesses and we find only on rare occasions castings with uniform wall thickness. To eliminate the problem of shrinkage cavities, greater emphasis has been placed on the use of feeder sleeve and chill plates.

This report is the first step of a series of examinations to be made to show how SiC-chills work under different conditions. In this first report we examined the influence of two of the main chill materials on cast iron EN-GJL-250.

Experimental

For all the examination, two plates with the dimensions 300x200x50mm were poured with the same iron quality (EN-GJL-250) each casting having three SiC-plates respectively three grey iron plates in the drag box or in the cope box (fig. 1). After two castings had been poured the surface finish, hardness and microstructure of the plates of EN-GJL-250 were compared.

The casting were cut length-ways down the middle (fig. 2) the areas were the hardness (HB) and the microstructure were compared is show in figure 3.

In a second examination the recycling of SiC- plates was evaluated. One possibility is to sell the broken plates to raw material suppliers for resale to foundries as a Silicon addictive for cupolas, unfortunately this is not always possible due to high transport costs over long distances.

Another possibility is to use the broken plates as a Silicon addictive in other furnaces like a rotary furnace. To examine this possibility a trial was conducted at the foundry of Herborner Pumpenfabrik. The foundry uses a 5 tonne rotary furnace with a oil burner to produce grey iron EN-GJL-250. The basic iron had the analysis shown in table 1 to obtain the required analysis the Silicon content had to be increased to around 1.80%. To achieve this, a charge of 20kg of broken SiC-plates (Si-content approx. 33%) was calculated. Several test using 20kg charges where carried out and the iron analysed to see if the required Si-content had been achieved.

In a third part, two examples of hand moulded castings made with the help of SiC-quenching plates are shown.

Results

The first result of the visual examination of the cast iron plates, shows that the surface of the plates, poured with the help of SiC-plates (chills) have a better finish than the castings pored with grey iron plates (chills), shown in figure 2. This figure shows the surface finish when the chills were in the drag box.

The superior finish can also be seen in the two examples (Fig 4 and Fig 5). In both examples the improved surface produced by the SiC-plates when compared to the grey iron chill is due to the fact that the SiC-plate does not produce water of condensation, which can cause pinholes.

Another advantage when using SiC-plates is the more uniformed matrix and the ability to use them several times without loss of efficiency. This high service life was also noticed during the test with the EN-GJL-250 plates. It was not necessary to change the plates which were used six times, the very good condition of these chills means that they can be used again for the next series of test. The grey iron chills had to be shot blasted to obtain a clean surface for each of the three tests.

The examination of the hardness (HB) was made with a ball diameter of 5mm and a measuring force of 750N. The first measurement was directly onto the surface of the casting. The next measurement was made 5mm under the surface and then in 2,5mm steps .The results of the test series using the three plates (chills) in the cope and drag box are shown in figure 7. The course of hardness stays in a direct relation with the microstructure of the castings.

In figure 7 the microstructure of the cross section of the examined area is shown. On the left, the area influenced by the grey iron and on the right the area influenced by the Sic-plates. In figure 8 the same areas are shown with etched probes to show the matrix. These pictures show the areas influenced by chills placed in the cope box.

It was also found possible to increase the Silicon content of the base iron to 1.80% in the 5t. rotary furnace by adding SiC-plates. This is a very economical way to recycle the broken plates. The plates were put into the charging box without any treatment, i.e. crushing, and they dissolved completely. These results were obtained on numerous trials. Table 2 shows a typical analysis of the melt, charged with 20kg of broken SiC plates to increase the Silicon content from 1.70% to 1.80%.

Discussion

The satisfactory results of good surface finish when SiC-plates (chills) were used are directly connected to their physical properties. The thermal conductivity of SiC-plates is approximately 10W/ (m*K) which is about a fourth of that of grey iron chills. Due to this fact, the air close to the grey iron chills cool faster and the humidity, i.e. steam, in the closed mould can more easily condense on the surface of the grey iron chills, this moisture is responsible for surface defects during the pouring process. In the test case where the grey iron chills were used in the drag box (Fig 2) the first iron which entered the mould solidified too quickly, causing a cold lap defect.

After multiple uses, the SiC-plates can be damaged by mechanical abrasion but not through heat impact. The grey iron chills however will be "thermic subverted" whereby the graphite will be burnt out of the surface and the surface and the chills will have a very rough surface finish. This will result in the chills having to be shot blasted after every use. Also the chill will be attenuated, adversely affecting its chilling properties. For this reason the grey iron chills will have to be replaced after they have been used two or three times

The courses of hardness (fig. 6) is directly connected to the microstructure but one difference is that the surface hardness is always lower were SiC-plates are used. In all cases of use in both the cope and drag, the grey iron chills gave the higher surfaces hardness. Tool abrasion during machining should therefore be reduced when SiC-plates are used.

The hardness was found to be generally higher when the chills were placed in the drag box. This could be due to the chills placed in the cope having more time to heat up before coming into contact with the liquid iron, thereby reducing the chilling effect. The exact thermal mechanisms we want to investigate in further examinations with the help of thermocouples.

After five millimeters under the surface the hardness of both samples is nearly on an equally level. After the hardness of the grey cast iron, were the grey iron chill were used is higher, which is explainable with the finer structure (Fig. 7 and 8).

These pictures show that the grey iron chills have a rough chill effect in contrast with SiC-plates which have a smother chill effect. After 20mm the pictures already show that there is nearly complete A-graphite form with ferritic-perlitic matrix. Comparisons with poured grey iron plates without the use of chills had some defects like contraction craters, these defects were avoided due to an equable and dense microstructure created by using SiC-plates.

Although the SiC-plates will not be destroyed by thermal conditions during the pouring process, they can be destroyed by the mechanical stress during the fettling process and in particular on the vibratory grids were they can be smashed by castings. These broken plates can be reused if the surface finish is smooth enough (fig. 9) or they can be sorted out for other recycling methods.

This investigation shows that using broken SiC-plates as a Silicon addictive is a very economical recycling method and helps to reduce foundry operating costs. When the main costs of both chill materials are compared, SiC-plates are the more cost-effective; table 3. This however, is not the only consideration as the technical advantages of using SiC-plates are also superior.

Conclusions

This paper demonstrates the advantages of using SiC-plates and that not only the purchase price but the whole cost-effectiveness of a product has to be considered

These first examinations generate more interest into the way chills effect the solidification and properties of metals and with the aid of thermocouples more work will be undertaken to look into physical properties of Sic-plates (chills) like thermal capacity and thermal conductivity. The influence of both grey iron chills and SiC-plates on spheroidal graphite cast iron will also be investigated to improve our knowledge of this subject, thereby creating sound guidelines for the foundryman.

Tables

Table 1   Analysis of basis iron in a 5t rotary furnace with a low Silicon content

	Si [%]
Analysis 1	1,66
Analysis 2	1,71
Average:	1,69

Table 2   Analysis of basis iron after a surcharge of 20kg broken SiC-plates

	Si [%]
Analysis 1	1,82
Analysis 2	1,81
Average	1,82

Table 3   Comparison of the total costs of different cooling plate materials

point of interest	costs	
	Grey iron plates	SiC-plates
Purchase	↗	↑
Handling (sort out after using, shot blasting, separation after use of cycles...)	↗	→
Life time	↗	→
Recycling	→	↘

Figures

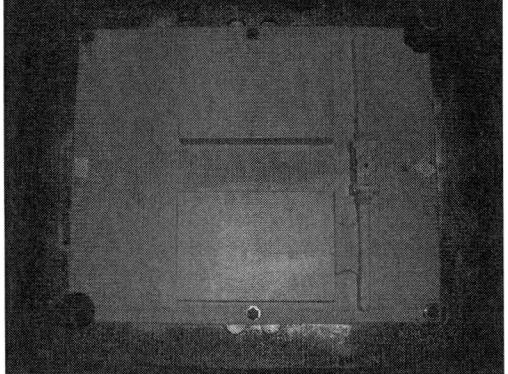 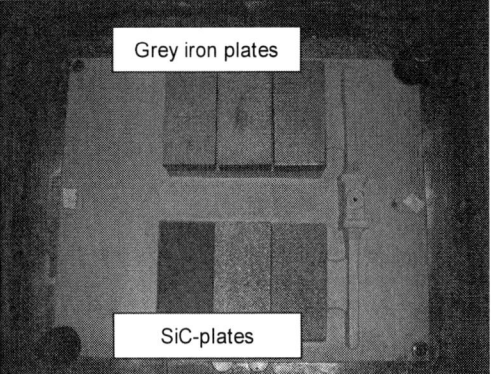

Fig. 1   Pattern without and with chills

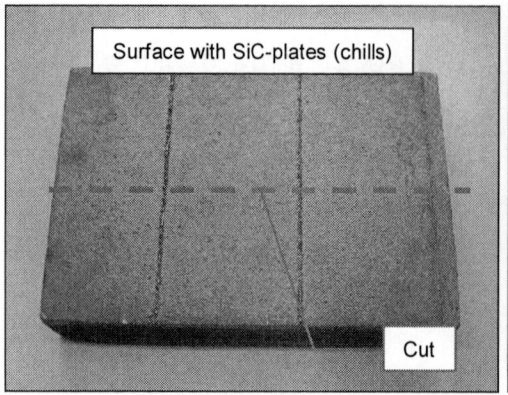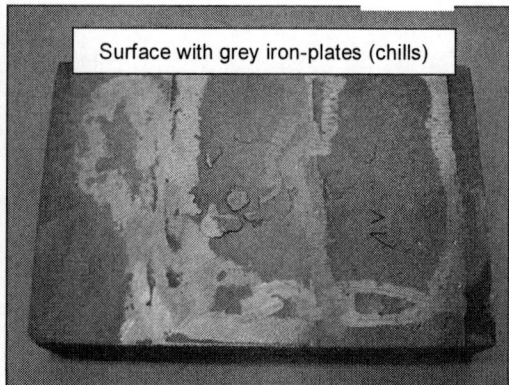

Fig. 2   Surface finishes of castings

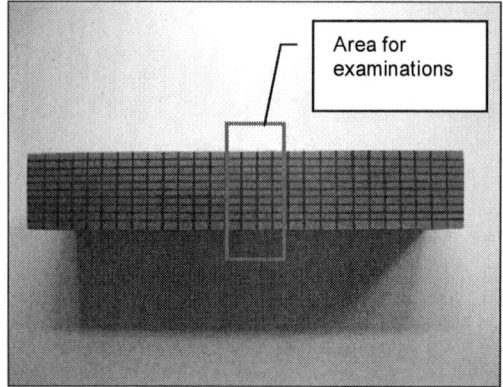

Fig. 3   Areas for comparisons

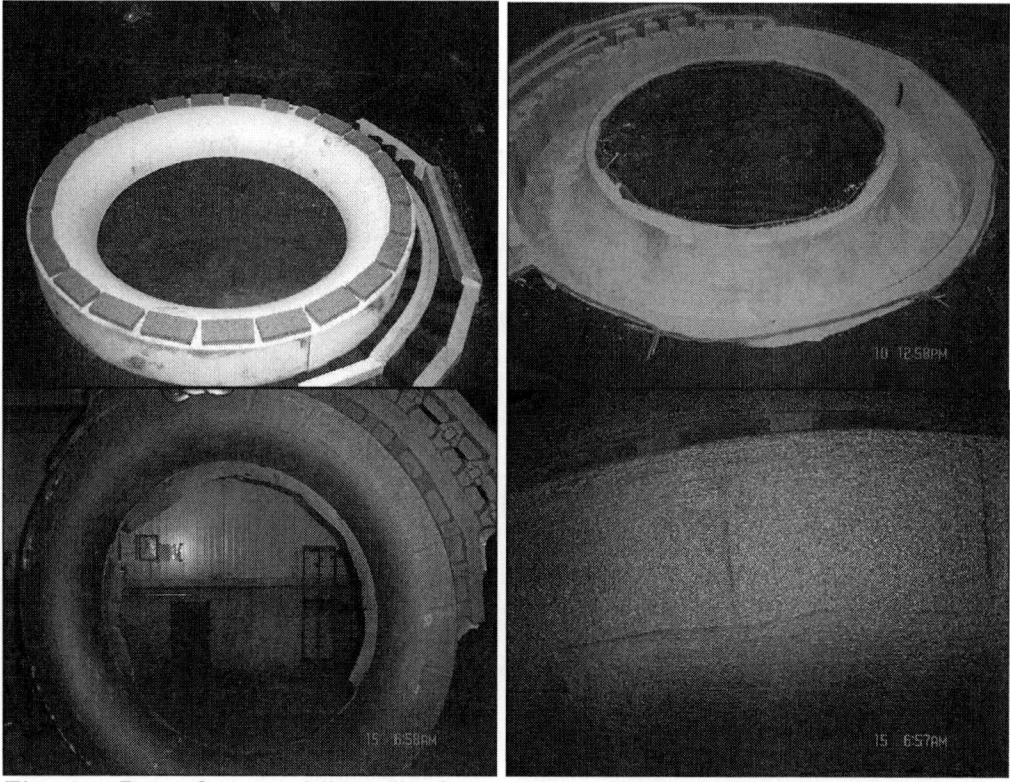

Fig. 4   Boot for moulding die for a deep drawing machine (2,6t weight, EN-GJL-250, 1350-1355°C). Foundry: DOERING, Germany

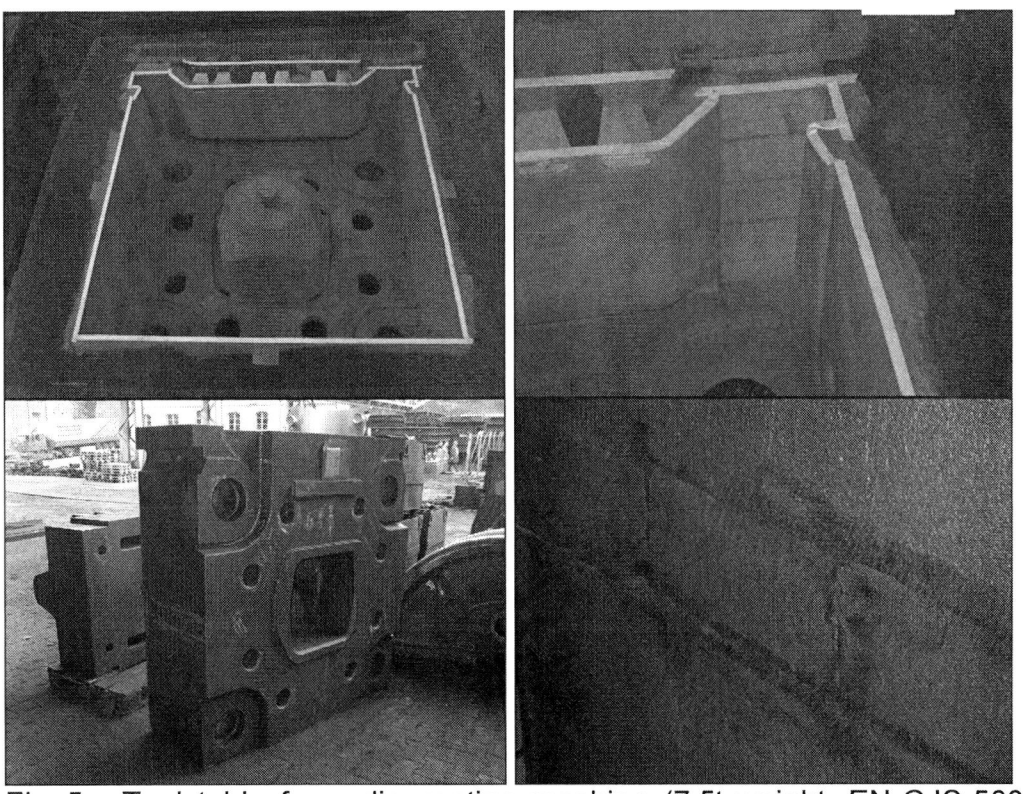

Fig. 5   Tool table for a die casting machine (7,5t weight, EN-GJS-500, ~1330°C). Foundry: DEMAG ERGOTECH, Germany

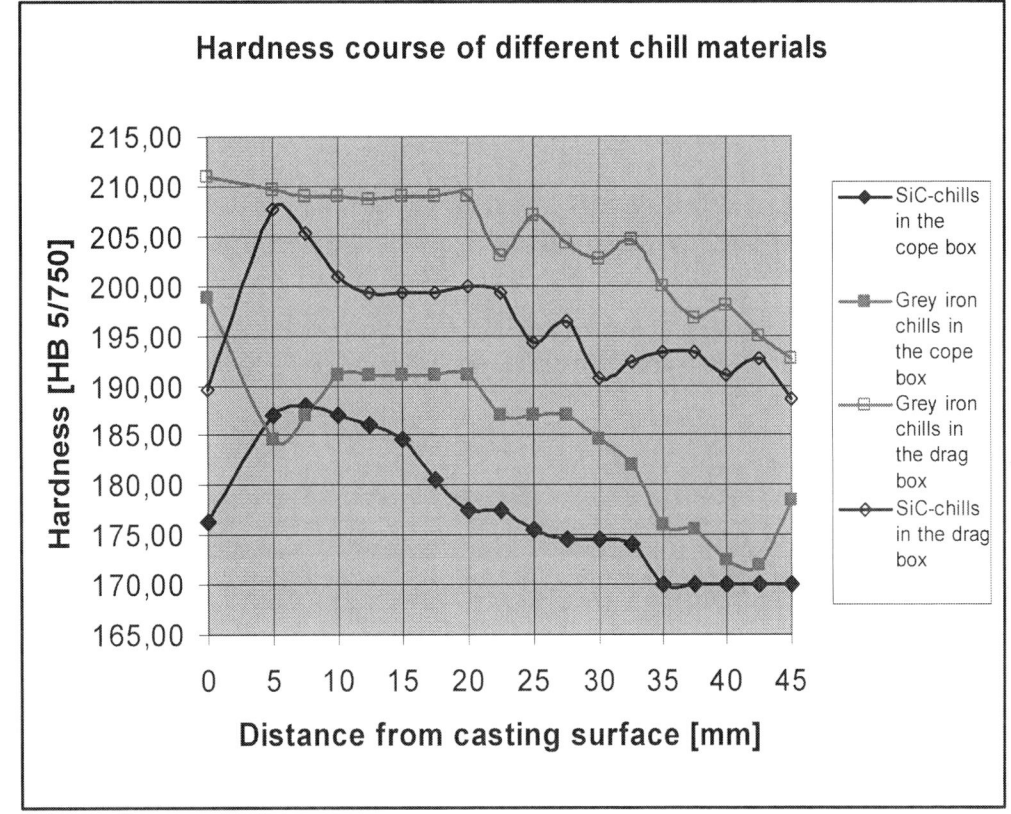

Fig. 6   Course of hardness of SiC- and grey iron plates (chills)

Fig. 7   Micrographs of influenced casting areas (50 times magnification)

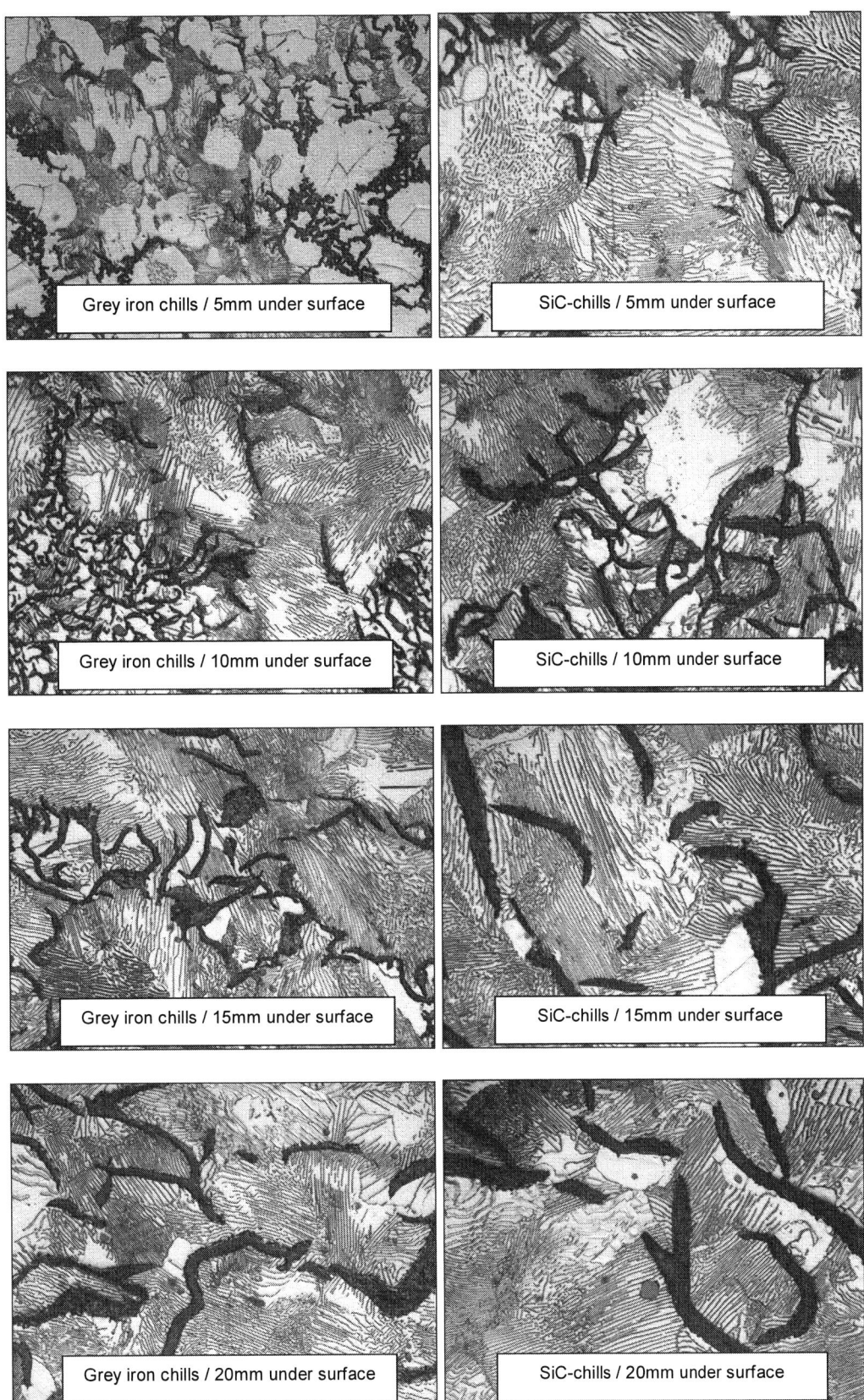

Fig. 8   Etched micrographs of influenced casting areas (500 times magnification)

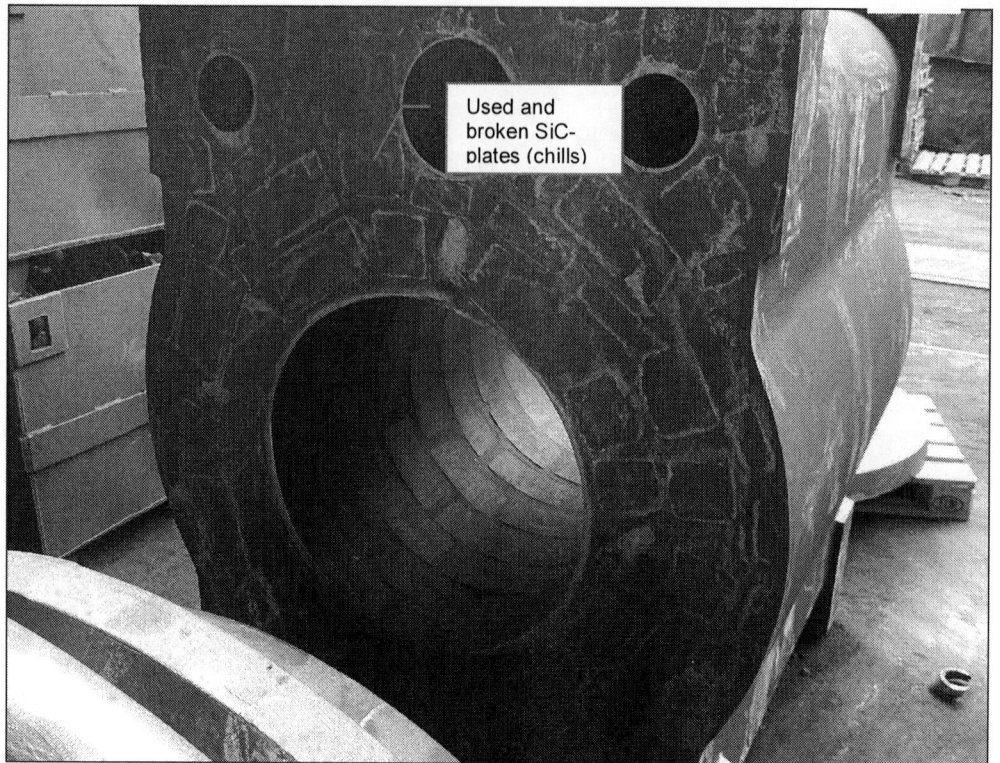

Fig. 9   Component for a hydraulic press (9,5t, EN-GJS-500) where broken SiC-chills were used. Foundry: DEMAG ERGOTECH, Germany

Chunky graphite in ductile iron castings

R Källbom *, K Hamberg ** and L-E Björkegren *.

* Swedish Foundry Association, Sweden, ** Chalmers University of Technology, Sweden.

Abstract

Non-spherical graphite morphology is detrimental on the mechanical properties of ductile iron castings. This includes the branched and interconnected chunky graphite that occasionally occurs in the thermal center of castings. In this work the graphite morphology in ferritic ductile iron that contained chunky graphite was studied. Chunky graphite was shown to be a progressively degenerated morphology of spherical graphite. Attempts to investigate the presence and segregation patterns of elements that might play a role in the still concealed formation and growth mechanism of chunky graphite were made. No macro segregation was detected. The possible role of micro segregation of trace elements was discussed but could not be determined. The graphite nucleation potential seemed to be low in the chunky graphite areas due to the lack of available oxygen and/or sulphur.

Key words ductile iron, graphite morphology, chunky graphite, segregation, nucleation

Introduction

Spherical graphite morphology is an important factor to attain high quality ductile iron castings. Every other type of morphology is detrimental on the mechanical properties. This includes the branched and interconnected chunky graphite that occasionally occurs in the thermal center of ductile iron castings. The presence of chunky graphite decreases the ultimate tensile strength (R_m) and especially the fracture elongation (A_5). The possible decrease in a ferritic ductile iron has been shown to reach 25% and 50% respectively depending on the amount of chunky graphite [1]. However, the hardness (HBW) and the yield strength (R_{p02}) are hardly affected at all by chunky graphite. Regarding the dynamic properties of almost 100% chunky graphite containing material the crack propagation rate is not significantly affected but the fracture toughness of the material will be lower [2].

The risk of chunky graphite formation is increasing with long solidification time. Consequently, the amount of chunky graphite tends to increase with increased wall thicknesses [1]. The call for further research work to determine the cause and growth mechanism of chunky graphite is escalating with increased used of heavy section ductile iron castings in demanding applications within the heavy automotive and the windmill industries among others.

According to Gagné and Argo chunky graphite shows a spiral crystal growth pattern caused by carbon supersaturation and constitutional supercooling as the driving forces [3]. Fast diffusion rate due to the lack of, or partially disrupted, austenite shell around the graphite nodule has also been discussed as one possible cause for chunky graphite formation [3]. According to Itofuji *et al* the chunky graphite forms as a result of the lack of magnesium gas bubbles in the melt and, further on, the growing graphite is in contact with residual liquid iron through thin liquid channels in the austenite [4]. The liquid channels are formed when segregated elements lower the solidus temperature. Liu *et al* regards the chunky graphite to be a deteriorated form of nodular graphite [5]. It has in fact been observed that the transition between the graphite structures type A flake, type B flake, type D undercooled flake, coral, compacted, chunky and spherical graphite is continuous and not intermittent [6]. The different morphologies were stated, by Liu *et al* to occur as a result of change in solidification rate and as a function of alloy addition or segregation [6].

In this work the graphite morphology in ferritic ductile iron that contains chunky graphite has been studied and attempts have been made to investigate the presence and segregation patterns of elements that might play a role of the concealed formation mechanism of chunky graphite.

Experimental

High silicon alloyed ductile iron were prepared in a 250 kg induction furnace. The charge material consisted of pig iron 41 %, returns 17 % and steel scrap 42 %. Furan bonded moulds were used to cast a pattern consisting of five 200 mm x 200 mm blocks with thicknesses ranging from 10 to 200 mm. The temperature change of the melts was logged during solidification using thermocouples placed in the center of each block. The solidification time ranged from 90 seconds to 90 minutes for each block respectively. Besides the blocks a component cast in the same silicon alloyed ductile iron has been investigated. The component is a front axle housing aimed for a dumper. Specimens were cut out from an area with a hot spot with fairly long solidification time (approx. 30 min).

Tensile test bars were machined, perpendicular to gravitational direction, from the center of the blocks and from the components. The graphite morphology was studied in the fracture surfaces of the bars using SEM. The microstructure was studied in different positions within the test bars using conventional optical microscope. The deep etched technique used consisted of 40 minutes etching in a mixture of HCl and HNO_3 (3:1) followed by a cleaning step in Vogel´s etchant and thereafter well rinsed in ethanol. Some specimens were color etched in boiling sodium hydroxide (10g) + picric acid (10g) + potassium pyrosulfite (10g).

Using GD-OES (Glow Discharge Optical Emission Spectroscopy) investigation of macro segregation was made, comparing nodular and chunky graphite areas within the same 200 mm cubic block. Measurements were made in up to six positions located 10, 20, 30, 45, 60 and 90 mm from the cast surface. The analyzed positions are indicated in Figure 1.

SEM with EDS as well as EPMA were used as analyzing tools in orders to investigate micro segregation tendencies.

Results
Graphite morphology
Chunky graphite is mainly located in the thermal center. Nevertheless, the chunky graphite zone can represent a reasonably large volume of the casting. This is exemplified by Figure 1 where the presence of chunky graphite appears as a shaded area in a sawed cross section. Some areas within the dark zone consist of nodules but the main graphite morphology is chunky, Figure 2. From Figure 1 it appears as the transition from nodular to chunky graphite growth happens very sudden in an interrupted manner. However, optical microscopy and SEM studies of the graphite morphology put forward a gradual change.

The graphite morphology in the vicinity of chunky graphite areas in this 200 mm cubic casting is classified to be a mixture of form IV to VI according to the standard EN ISO 945:1994. Large irregular graphite lumps, which cannot be classified by the standard, as well as very small islands of chunky graphite, are also found in those areas, Figure 3.

The examined fracture surfaces of the tensile test bars are located within the chunky graphite zone shown in Figure 1. SEM investigations of graphite in the fracture surfaces indicate gradual degeneration from spherical to chunky morphology. Figure 4 shows well-shaped nodules. Approaching chunky graphite areas different graphite morphologies as in Figure 5 and 6 can be observed. These observations, that chunky graphite is a progressively degenerated morphology of nodules, are in line with the theory of Liu *et al* [5]. The degenerated graphite shape in Figure 7 and the pyramidal growth of chunky graphite branches in Figure 8 confirm the observations of Liu *et al* [5].

All specimens from the front axle housing showed somewhat different graphite morphology compared to the 200 mm thick block. In most locations normal spherical graphite morphology emerged. Roughly 15% degenerated chunky graphite appeared at the most in the hot spots. Figure 9 shows a typical area; here the cell boarders appear in a brown to white color. The blue etching parts in the microstructure contain degenerated graphite and do appear before the brownish cell boarders. Here two variants of degenerated graphite can be seen. One type of graphite that appears as normal chunky graphite (see Figure 10 in deep etched condition) and one more like a stringer of graphite (Figure 11). These graphite stringers lie between the secondary dendrite arms. The classical chunky graphite seems to be placed in the center of the dendrite arms. In all cases the degenerated graphite can co-exist with spherical graphite.

Segregation of elements
Bulk analyzes did not show any significant difference in chemical composition between the different positions indicated in Figure 1. Some variations could be seen in Si content, for example, between different positions but no coupling to chunky graphite could be confirmed. Macro segregation between the nodular areas outside the chunky graphite zone as well as inside the zone was hence not detected, Table 1.

Attempts to investigate micro segregation tendencies of low content elements such as Ce, Ca and S in the blocks by using EPMA turned out to be unsuccessful since the concentrations were below the detection limit of the instrument. This was a fact close to spherical graphite as well as nearby chunky graphite. The average Si content was somewhat higher near the chunky graphite compared to that near a nodule. Nevertheless, the difference was not greater than the Si fluctuation between two nodules.

Closer investigations of the graphite in the front axle housing show some differences. The stringer like graphite had in most cases been nucleated on oxides. Spot analysis of the oxide particles revealed normal oxides containing Si and or Mg. The melts that produced the components had a rather high amount of residual magnesium content, 0.060-0.065 (%), this might explain the amount of particles with high magnesium content. The stringer graphite is not considered to be a chunky graphite variant.

In order to find evidence of micro segregation further investigations were made with SEM – EDS. Mapping, spot and line analysis were tried without much success. The only evidence of segregation was found in the Si and Mn content between graphite particles. No evidence of tramp elements like the elements mentioned in the literature [1], were found. A possible reason is a relatively small concentration of the mentioned elements and an insensitive analyzing method. The method gave the response from a too great material volume that disturbed the analysis.

Further, spot analyses in two types of areas, chunky and nodular, as depicted in Figure 12 were carried out. The focus was on systematic analysis of particles found during the EDS-mapping. A majority of the small particles (<10 μm) were found in the eutectic cell borders between nodular and chunky graphite areas, typical location is shown in Figure 12. However, particles were also found in the nodular area as well as in the chunky area close to the borderline. The amount of particles was greater in the cell border areas than in the chunky graphite areas. The chemical composition of the particles was not the same in all areas. In the cell borders and in areas with nodular graphite, the particles contained Mg, O and Si (see Figure 13). Some particles contained Ti and C. Frequently the particles in those areas also contained P and S. On the other hand, in the chunky graphite areas the number of particles was small. Most of the particles in those areas contained Mg and S but no oxygen. Trace of Al and Ca were found as well.

Discussion
During solidification the condition in the melt is gradually, but rapidly, changing to be more favorable for the chunky graphite growth manner. Different authors have debated the change in melt condition that promotes the chunky graphite growth. Several theories indicate, as mentioned in the introduction, that the cause of chunky graphite is related to the chemical composition of the melt. Heavy section castings with long solidification times are more prone to develop chunky graphite. The graphite precipitation and growth start out to be nodular and then changes towards chunky. Consequently, it can be assumed that variations in concentration due to segregation of certain elements might be a possible reason for the transition of graphite morphology growth. The elements Ca, Si, Al, Ni, Ce and other RE are said to promote chunky graphite, especially in absence of the elements Sn, As, Bi, B, Sb and Pb [1].

However, in this work no macro segregation of elements was found. Further, the role of micro segregation could not be determined since the chemical concentrations were too low to be detected by conventional analyzing methods.

Nevertheless, the evaluation of the results of this work renders a hypothesis that the collaboration between Mg, S and O is important for the chunky graphite formation. Skaland has depicted the nucleation sites for spherical graphite [7]. The substrate contains a MgS core circumscribed by a shell of magnesium silicate, normally $MgO \cdot SiO_2$. Skaland denote this substrate type A. Active elements introduced to the melt by inoculation, such as Ca, Ba, Sr and Al, will react with the magnesium silicate and form a hexagonal substrate that is a favorable site for graphite precipitation. If oxygen (or sulphur) is not present the needed hexagonal nucleus will not form.

The irregular graphite often found in microstructures of castings that contain chunky graphite (as Figure 3) indicates in fact low oxygen content in the melt. The Mg-treatment was experimentally well performed and the Mg content is high enough to produce nodular graphite. No vermicular graphite can be found at all. Therefore the irregular graphite consequently indicates insufficient inoculation. However, since the inoculation procedure was good it can be assumed that the inoculation has not worked properly due to low oxygen content.

In this work, an excess of Mg/O/Si containing particles was found in the areas containing graphite nodules as well as in the borderline between nodular and chunky areas. Therefore, one can assume that the nucleation requirements for spheroidal graphite can be fulfilled in those areas. A consequence, however, is that the areas containing nodules consume most of the oxygen. A strong indication for this is that only MgS particles are found in the chunky graphite areas demonstrating that the oxygen level has been too low to form the hexagonal structure that are needed to favor spherical graphite growth. The graphite nucleation is disturbed and chunky graphite will form between the nodular areas.

Besides Mg, elements as Ca, Al, Si, Ce consume oxygen by forming stable oxides. This further strengthens the hypothesis, that low available oxygen content might be a reason for chunky graphite formation, since these elements also are said to promote chunky graphite.

Conclusions
1. The branched chunky graphite is a progressively degenerated morphology of spherical graphite.
2. Unstable or changing melt condition during the solidification leads to chunky graphite formation, a change that is still not fully defined.
3. In chunky graphite areas MgS particles were found while the amount of magnesium oxides was limited compared to the areas that contained spherical graphite.
4. The lack of available oxygen (or possibly sulphur) to form nuclei for spherical graphite precipitation might be a reason for chunky graphite formation.

5. Macro segregation was not found to be a reason for chunky graphite formation. The role of micro segregation could not be determined.

References

1. Källbom R, Hamberg K, Björkegren L-E, *Chunky graphite – formation and influence on mechanical properties in ductile cast iron*, Gjutdesign 2005 Final seminar, Espoo, Finland, 13-14 June, 2005, VTT Technical Research Centre of Finland, Finland, 2005.
2. Björkblad A, *Conventional vs closure free crack growth in nodular iron,* Gjutdesign 2005 Final seminar, Espoo, Finland, 13-14 June, 2005, VTT Technical Research Centre of Finland, Finland, 2005.
3. Gagné M and Argo D, *Heavy Section Ductile Iron Castings Part I and Part II*, International Conference on Advanced Casting Technology, Kalamazoo, Michigan, USA 12-14 November 1986, pp 231-256, ASM International.
4. Itofuji H and Uchikawa H, *Formation Mechanism of Chunky Graphite in Heavy-section Ductile Cast Irons,* AFS Transactions 90-42, 1990, pp 429-448.
5. Liu P C, Li C L, Wu D H and Loper, Jr, *SEM Study of Chunky Graphite in Heavy Section Ductile Iron*, AFS Transactions 83-51, 1983, pp 119-126.
6. Liu P C, Loper Jr C R, Kimura T and Park H K, *Observations on the graphite morphology in cast iron,* AFS Transactions 80-41, 1980, pp 97-118.
7. Skaland T, *A model for the graphite formation in ductile cast iron,* NTH Trondheim, 1992, ISBN 82-7119-384-8.

Acknowledgements

The Nordic Innovation Centre and VINNOVA partly financially supported this work. The authors thank Volvo Construction Equipment for supplying the front axle housings (prototypes).

Tables

Table 1 GD-OES analyzes of chemical composition inside and outside the chunky graphite zone. The location of analyzed positions is shown in Figure 1.

	MELT 1						MELT 2		
Pos	1	2	3	4	5	6	2	4	6
Si	3.05	3.38	3.27	3.31	3.26	3.29	3.36	3.30	3.55
Mn	0.19	0.16	0.17	0.18	0.19	0.18	0.18	0.21	0.19
P	0.071	0.059	0.055	0.047	0.055	0.067	0.046	0.072	0.062
S	0.016	0.013	0.014	0.016	0.016	0.015	0.017	0.018	0.014
Cu	0.020	0.018	0.019	0.021	0.021	0.021	0.018	0.019	0.019
Al	0.012	0.012	0.013	0.013	0.013	0.013	0.014	0.014	0.015
B	0.0013	0.0006	0.0009	0.0011	0.0011	0.0011	0.0010	0.0016	0.0013
Sn	0.024	0.025	0.024	0.024	0.020	0.022	0.025	0.020	0.020
Ca	0.0001	0.0002	0.0001	0.0002	0.0002	0.0002	0.0001	0.0001	0.0001
Ce	<0.001	<0.001	<0.001	<0.001	<0.001	<0.001	<0.001	<0.001	<0.001
Sb	0.005	0.003	0.003	0.004	0.003	0.003	0.003	0.006	0.006

Figures

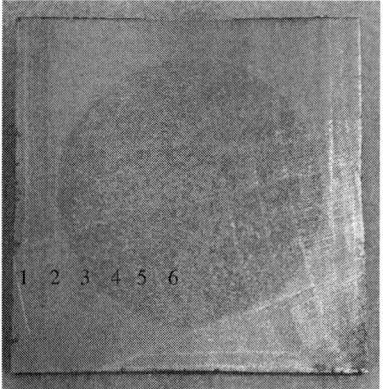

Figure 1 Chunky graphite zone located in the thermal center (200 mm cube). Numbers denote positions of analyzes, see Table 1.

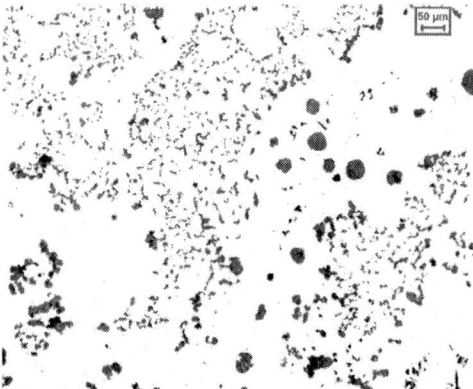

Figure 2 Graphite morphology in the thermal center of a 200 mm cube (105 X)

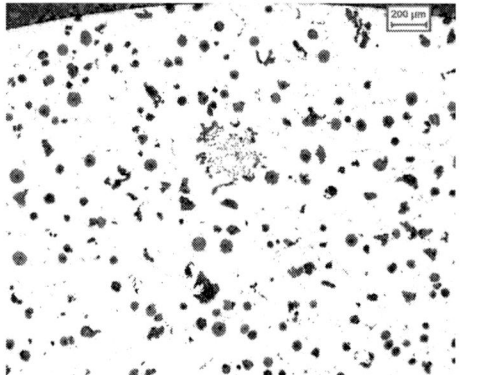

Figure 3 Mixture of graphite shapes just outside the chunky zone (45 X)

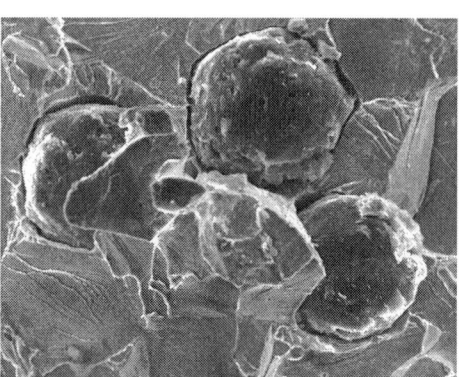

Figure 4 Well-shaped nodules (750 X)

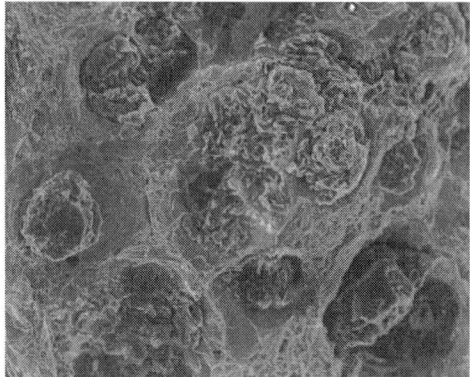

Figure 5 Approaching chunky graphite areas the graphite gradually changes its morphology (350 X)

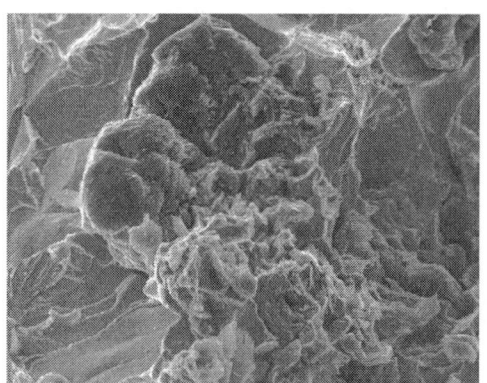

Figure 6 Transition state in the border of a chunky graphite cell (1000 X)

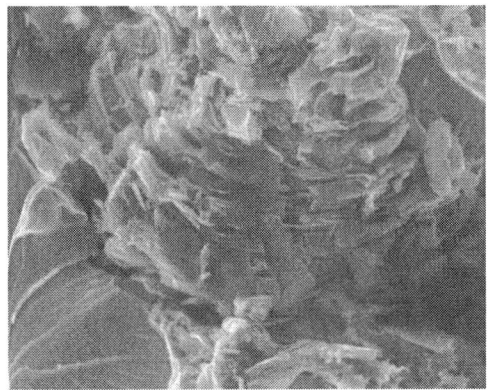

Figure 7 Degenerated graphite (2000 X)

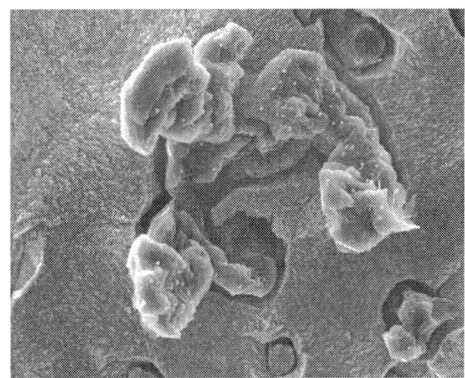

Figure 8 Pyramidal growth of chunky graphite branches (deep etched 1500 X)

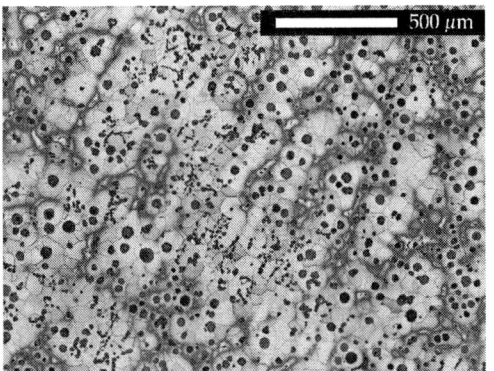

Figure 9 Color etched microstructure containing chunky graphite.

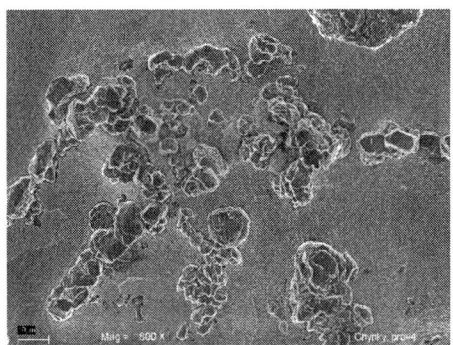

Figure 10 Deep etched graphite that looks like classical chunky graphite.

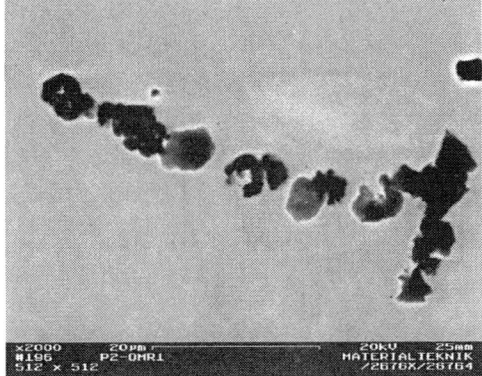

Figure 11 Stringer graphite that is not considered to be chunky graphite.

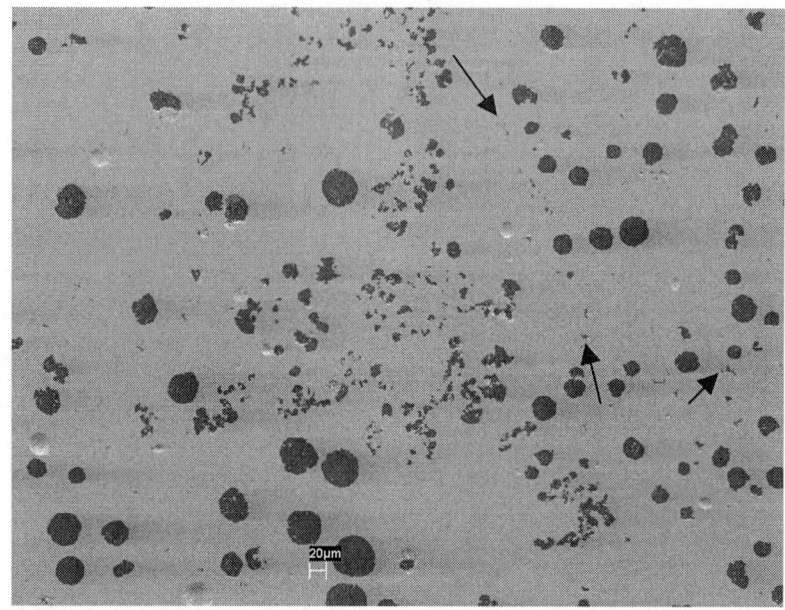

Figure 12 Particles were detected in the borderline between chunky and nodular graphite areas as well as in each area respectively.

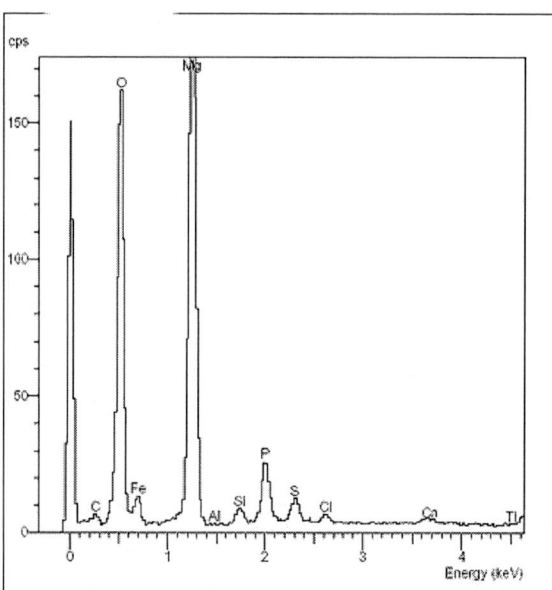

Figure 13 Particles in the borderline and in the nodular areas (see Figure 12) were composed of these elements. The peak to the left is not valid for the analyze.

A Comparison of Thixocasting and Rheocasting

Stephen P. Midson
The Midson Group, Inc.
Denver, Colorado
USA

Andrew Jackson
Arthur Jackson & Co., Ltd.
Brighouse
UK

Abstract

The first semi-solid casting process to be commercialized was thixocasting, where a pre-cast billet is re-heated to the semi-solid solid casting temperature. Advantages of thixocasting include the production of high quality components, while the main disadvantage is the higher cost associated with the production of the pre-cast billets.

Commercial pressures have driven casters to examine a different approach to semi-solid casting, where the semi-solid slurry is generated directly from the liquid adjacent to a die casting machine. These processes are collectively referred to as rheocasting, and there are currently at least 15 rheocasting processes either in commercial production or under development around the world.

This paper will describe technical aspects of both thixocasting and rheocasting, comparing the procedures used to generate the globular, semi-solid slurry. Two rheocasting processes will be examined in detail, one involved in the production of high integrity properties, while the other is focusing on reducing the porosity content of conventional die castings.

Key Words

Semi-solid casting, thixocasting, rheocasting, aluminum alloys

Introduction

Semi-solid casting is a modified die casting process that reduces or eliminates the porosity present in most die castings[1]. Rather than using liquid metal as the feed material, semi-solid processing uses a higher viscosity feed material that is partially solid and partially liquid. The high viscosity of the semi-solid metal, along with the use of controlled die filling conditions, ensures that the semi-solid metal fills the die in a non-turbulent manner so that harmful gas porosity can be essentially eliminated. After the die is filled, high pressure (1,000 bar or more) is maintained on the biscuit to feed micro-porosity.

Semi-solid casting has been used commercially for the past 15 to 20 years for the production of near-net shape aluminum components. Until recently, all semi-solid cast components have been produced by thixocasting, a process which re-heats pre-cast billets to the semi-solid casting temperature.

However, commercial pressures have driven casters to examine alternative approaches to semi-solid casting, where the semi-solid slurry is generated directly from the liquid. These processes are collectively referred to as rheocasting. Instead of re-heating a pre-cast billet, rheocasting cools liquid aluminum into the semi-solid range, while simultaneously generating the globular microstructure necessary for semi-solid forming. Creating the semi-solid slurry directly from the liquid eliminates the need for a special (more expensive) feedstock, as well as permitting biscuits, runners and scrap castings to be recycled in-house.

This paper will provide a brief introduction to semi-solid casting, followed by a description of the technical aspects of both thixocasting and rheocasting. A number of rheocasting processes are either in commercial production or under development around the world, and two of these processes will be examined in detail.

Semi-Solid Casting

Most semi-solid casting processes use metal that is between 25-50% solid and 50-75% liquid, utilizing high pressure, cold chamber die casting machines to inject the semi-solid slurry into re-usable, hardened steel dies[1]. For semi-solid casting to be successful, the slurry must contain the globular primary particles shown in Figure 1a. Conventional, dendritic-type microstructures, such as the one shown in Figure 1b, will not work for semi-solid casting. The main advantage provided by all the different semi-solid processes is that the dispersion of the globular solid particles in the liquid produces a highly viscous semi-solid slurry, and controlling the flow of that viscous liquid without splashing or turbulence is much easier than with fully liquid aluminum.

Conventional die casters accept the turbulence associated with high speed filling of filly liquid aluminum. They inject the liquid aluminum into dies using gates speeds of about 30-60 m/sec, and the resulting turbulence produces high levels of residual porosity in the castings.

Aluminum casting processes such as sand and investment casting attempt to fill the die cavity in a non-turbulent manner by limiting the gate speed to a maximum of about 0.25 m/sec[2]. Such low filling speeds may be acceptable when filling ceramic dies, but attempts to fill thin-walled components using un-coated steel dies at such low speeds would result in rapid solidification of the aluminum and non-fills.

The solid particles dispersed in semi-solid metals increase their viscosities as much as 10,000 times greater than those of fully liquid aluminum. Testing has shown that this higher viscosity allows a semi-solid alloy to be injected into a die using gate speeds of as high as 2.5-5.0 m/sec, while still avoiding turbulence[3]. This allows the production of porosity-free, thin walled castings in re-usable steel dies.

Thixocasting

As noted earlier, the thixocasting process was the first semi-solid process to be commercialized. Thixocasting consists of three separate stages: the production of a pre-cast billet having the special globular microstructure, the re-heating of these billets to the semi-solid casting temperature and the casting of the components (see Figure 2).

Thixocasting is capable of producing extremely high quality components having excellent mechanical and functional properties. The billet feed material is typically produced by aluminum companies in batches as large as 50,000 lbs. These pre-cast bars provide billet-to-billet and lot-to-lot chemistry, cleanliness and microstructural repeatability comparable to forging and rolling stock, and far more consistent than is typically achievable when pouring castings from the liquid in single doses[4]. Thus semi-solid components produced by thixocasting tend to have very consistent properties. The main disadvantage associated with thixocasting is higher manufacturing cost, arising both from the premium attached to special feedstock, as well as the inability to easily recycle biscuits and runners.

Rheocasting

Rather than using pre-cast billet, rheocasting generates the special semi-solid microstructure adjacent to the die casting machine directly from the liquid (see Figure 3). The liquid is cooled into the semi-solid range, while simultaneously generating the globular microstructure. Once the metal has been cooled to the correct temperature, the semi-solid slurry is transferred to the shot sleeve of a die casting machine, and injected into the die using the same type of controlled fill as with thixocasting.

The major advantage of rheocasting is that the semi-solid feed material is produced at the casting machine directly from the liquid. This allows conventional ingot material to be used, eliminating the premium associated with the thixocasting billet. Another advantage is that biscuits and runners can now be recycled directly into the casting stream, again reducing cost (see Figure 3).

Potential disadvantages of rheocasting relate to the consistency of the product and the limited commercial application of the various processes. Questions relating to consistency arise from the fact that rheocasting uses single shot liquid dosing (ie, a single shot of liquid metal is poured to produce each casting), and it is much more difficult to maintain the required levels of metal cleanliness when pouring 5 lbs of metal than when pouring 50,000 lbs[4]. Therefore, it is still unclear whether rheocasting will prove as reliable as thixocasting.

Slurry Generation for Rheocasting

As noted earlier, there are a number of different rheocasting processes in commercial production or under development around the world. These different rheocasting processes generally differ in the manner in which the liquid is cooled and the globular semi-solid microstructure generated. There are four general techniques used to generate the globular, semi-solid microstructure, and most of the different rheocasting processes use some variation of these practices[5]. The techniques are:

❑ Stirring – similar to thixocasting, the liquid aluminum (just enough for one shot) is stirred as it is cooled into the semi-solid temperature range.

❑ Dendrite Fragmentation - a variation to stirring processes is the dendrite fragmentation technique, where the melt is cooled below its liquidus temperature, and the semi-solid alloy is treated in a turbulent manner to break up the dendrites, producing numerous small solid fragments that can be coarsened into globular-shaped aluminum particles.

❑ Pressure Waves – pressure waves generated in the runner system have been shown to generate semi-solid structures.

❑ Numerous solidification nuclei – In this technique, the liquid is poured into a container from a temperature just above its liquidus temperature. The rapid cooling generated during pouring generates a large number of solid nuclei, which prevent the formation of dendrites, instead producing a large number of globular solid particles. Often grain refining techniques are used to assist the generation of the large number of solid nuclei.

A recent publication[5] identified 15 different rheocasting processes, and these are listed in Table 1, showing the organization that developed each process, and the technique used to generate the slurry. The various rheocasting processes are in different stages of commercial development, with some of the processes being used for the commercial production of components, while other processes are in the early stages of development.

Detailed Description of Two Rheocasting Processes

This section of the paper will describe in more detail two of the rheocasting processes listed in Table 1. One of the rheocasting processes is being used for the production of high integrity, safety-critical type castings, while the other is focusing on producing higher quality, porosity-free die castings.

The New Rheocasting Process

The New Rheocasting (NRC) process[6-8] is shown schematically in Figure 4. The process uses a carousel of crucibles on a circular turntable. At the first position, molten aluminum controlled just above its liquidus temperature is poured into the crucible. It is critical for the success of the process that the superheat of the melt has to be low, so that numerous solid nuclei are produced during pouring - for example, Kaufmann et al[6] report that that superheats of only $20^{\circ}C$ are used. It is these nuclei that generate the globular microstructure. As the carousel indexes, air is blown on the walls of the crucible, allowing the liquid to cool into the semi-solid state in a controlled manner. At the penultimate position, the surface of the metal is re-heated using an induction coil. The objective is obtain a semi-solid slug with a relatively consistent solid fraction from edge to center and from top to bottom. The crucible is then removed from the carousel, and the semi-solid slug is transferred into the shot chamber of a vertical cold chamber die casting machine.

The NRC process was the first of the rheocasting process to be commercialized. However, even at the height of its popularity, it never approached thixocasting in tonnage of parts shipped. Typically the NRC process is used for the production of high integrity castings used in safety-critical, structural applications. In Europe, Stampal announced in 2003 that they would convert all their semi-solid cast parts to the NRC process. Giordano and Chiarmetta from Stampal have provided an example of a automotive suspension part (Figure 5) that was scheduled to enter production in early 2005 at a rate of 20,000 sets per month[9].

Semi-Solid Rheocasting

The Semi-Solid Rheocasting (SSR[TM])[10-12] process was originally developed at the Massachusetts Institute of Technology (MIT), and it uses stirring and dendrite fragmentation techniques for generating the slurry. The MIT researchers found that if solid nuclei are present in sufficient quantities in a melt cooled just below its liquidus temperature, further cooling will cause the nuclei to rapidly spheroidze and grow with a spherical morphology, and that vigorous agitation is unnecessary after the formation of only a small fraction of solid.

In Step 1 of the SSR[TM] process (Figure 6), a robot dips a coated ceramic crucible into a holding furnace filled with molten aluminum held several degrees above its liquidus temperature and brings it to the SSR station. In step 2, a rotating, cooled graphite rod is inserted into the liquid metal and rapidly cools the melt for a short time, usually within the range of 5-20 seconds. Yurko et al report that the stirring time is controlled by a PLC utilizing a heat transfer algorithm that can account for variables such as furnace and rod temperature and alloy type. The researchers report that this closed loop feedback system is helpful in the foundry environment where furnace temperature normally fluctuates and die casting cycles are frequently interrupted[10].

Only a small solid fraction is formed during the stirring phase (about 5%), so once the stirring rod is removed (step 3), additional cooling must occur so the semi-solid metal is cooled (without additional stirring) to a solid fraction of about 15-20%. As the melt is cooled, the particles generated in stage two grow to form globular solid particles distributed in the liquid. Once the target solid fraction is reached, the semi-solid alloy is poured from the vessel into the shot chamber of a die casting machine, where it is injected into the die.

One of the commercial focuses of the SSR[TM] process appears to be the production of higher quality die castings. The inventors of the process note that SSR[TM] provides the capability to improve the quality of secondary die casting alloys such as AlSi8.5Cu3 (380), whose high eutectic fraction make them difficult or impossible to process by other semi-solid processes at a solid fraction of 50%. Yurko et al suggest that SSR[TM] can reduce the porosity content, eliminate the need for impregnation, and provide as much as a 25% faster cycle time (as some of the alloy's latent heat is removed at the SSR[TM] station).

Summary and Conclusions

❑ A comparison of thixocasting and rheocasting has been presented, reviewing the advantages and disadvantages of each process.

❑ Due to commercial pressures, casters around the world are currently placing more attention on rheocasting. 15 rheocasting processes have been identified, either in commercial production or under development.

❑ Two of the rheocasting processes (NRC and SSR[TM]) are reviewed in more detail. NRC is typically being used for the production of high integrity, safety-critical type castings, while SSR[TM] is focusing on producing higher quality, porosity-free die castings.

References

1. Fan, Z, Inter. Meter. Rev., Vol. 47, 2002, p 49
2. Campbell, J, Proc. Materials Solutions Conference '98 on Aluminum Casting Technology, 1998, p3
3. Midson S P, Minkler R B & Brucher H G, *Gating of Semi-Solid Castings*, Proc 6th Inter. Conf. on Semi-Solid Processing of Alloys and Composites, Ed. G.L. Chiarmetta & M. Rosso, Turin, Italy, Sept 2000
4. Jorstad, J, *Semi-Solid Metal Processing: A Cost Competitive Approach for High Integrity Aluminum Components*, Proc. 6th Inter. Conference on Semi-Semi Processing of Alloys and Composites, Eds. G. Chiametta & M. Rosso, Sept 2000, Turin, Italy, p 227
5. Midson S P, *Rheocasting Processes for Semi-Solid Casting of Aluminum Alloys*, Die Casting Engineer, January 2006, p48
6. Kaufmann H, Holzl A & Uggowitzer P J, *New Rheocasting of High Strength Aluminum Foundry Alloys*, Proc. 7th Inter. Conference on Semi-Semi Processing of Alloys and Composites, Eds. Y. Tsutsui, M. Kiuchi & K. Ichikawa, Sept 2002, Tsukuba, Japan, p 617
7. Adachi M, Sato S, Harada Y & Kawasaki T, US patent number 6,165,411, *Apparatus for Producing Metal to be Semimolten–Molded*, Dec 26th, 2000
8. European patent number 745,694Al, "Method and Apparatus for Shaping Semisolid", 1999
9. Giordano P and Chiarmetta G, *New Rheocasting: A Valid Alternative to the Traditional Technologies for the Production of Automotive Suspension Parts*, Proc. 8th International Conference on Semi-Semi Processing of Alloys and Composites, Eds. A. Alexandrou & D. Apelian, Sept 2004, Limassol, Cyprus
10. Yurko J A, Martinez R A & Flemings M C, *SSR[TM]: The Spheroidal Growth Route to Semi-Solid Forming*, Proc. 8th Inter. Conference on Semi-Semi Processing of Alloys and Composites, Eds. A. Alexandrou & D. Apelian, Sept 2004, Limassol, Cyprus
11. Yurko J, Flemings M & Martinez A, *Semi-Solid Rheocasting (SSR[TM]) – Increasing the Capabilities of Die Casting*, Die Casting Engineer, January 2004, p 50
12. Flemings M C, Martinez-Ayers R A, de Figueredo A & Yurko J A, *Metal Alloy Compositions and Process*, US Patent number 6,645,323, Nov 11, 2003

Process Name	Organization	Location	Technique Used to Generate Slurry
Gibbs	Gibbs Die Casting	USA	Stirring
Hitachi	Hitachi Metals	Japan	Stirring
Honda	Honda	Japan	Stirring
Induction Heating/Stirring	CSIR	South Africa	Stirring
SEED Process	Alcan	Canada	Stirring
Slurry On Demand	Mercury Marine	USA	Stirring
Rheo-Diecasting	Brunel University	England	Stirring/Dendrite fragmentation
Semi-Solid Rheocasting	IdraPrince	USA	Stirring + numerous nuclei
ATM	CSIRO	Australia	Pressure Waves
Continuous Rheoconversion Process	Worcester Polytechnic Institute	USA	Dendrite fragmentation
Buhler	Buhler	Switzerland	Numerous nuclei
Controlled Diffusion Solidification	Worcester Polytechnic Institute	USA	Numerous nuclei
Direct Thermal Method	University College Dublin	Ireland	Numerous nuclei
New Rheocasting	Ube	Japan	Numerous nuclei
Sub-Liquidus Casting	THT Presses	USA	Numerous nuclei

Table 1: Fifteen different rheocasting processes (after 5)

a) 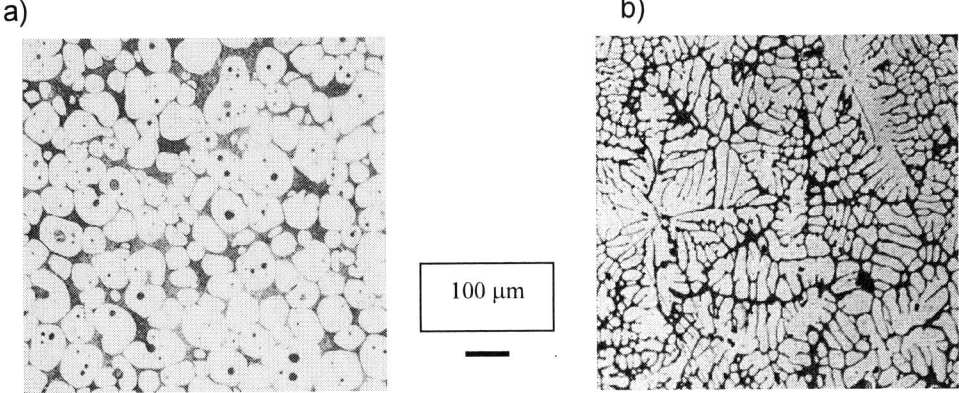 b)

Figure 1: Aluminum alloy 357 microstructures
a) Globular microstructure required for semi-solid processing
b) Conventional cast dendritic microstructure

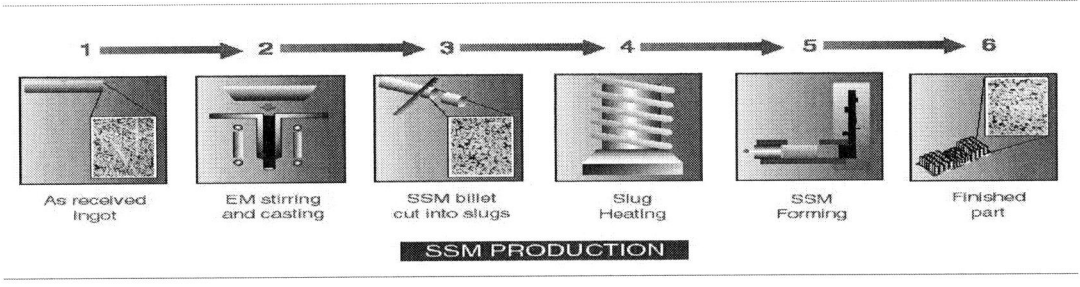

Figure 2: Schematic drawing of the thixocasting process

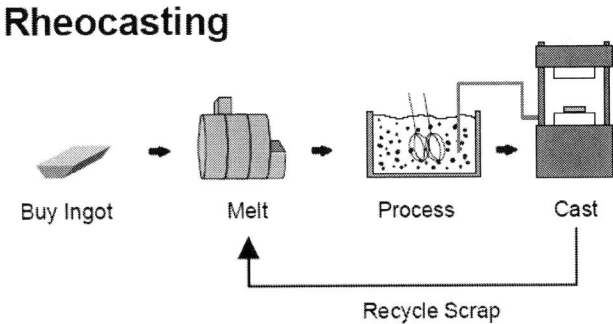

Figure 3: Schematic drawing of the rheocasting process

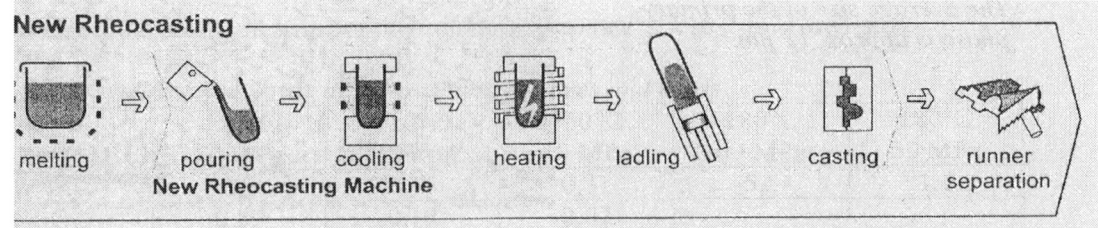

Figure 4 Schematic of Ube's New Rheocasting (NRC) process

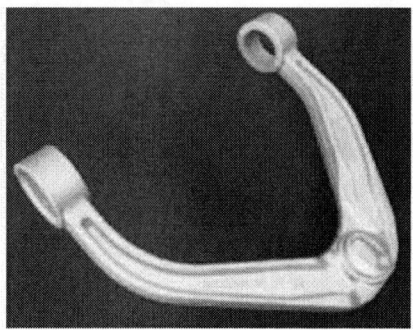

Figure 5: Photograph of an automotive suspension arm scheduled to enter production at Stampal in 2005 using the Ube NRC process (after 9)

a)   b)

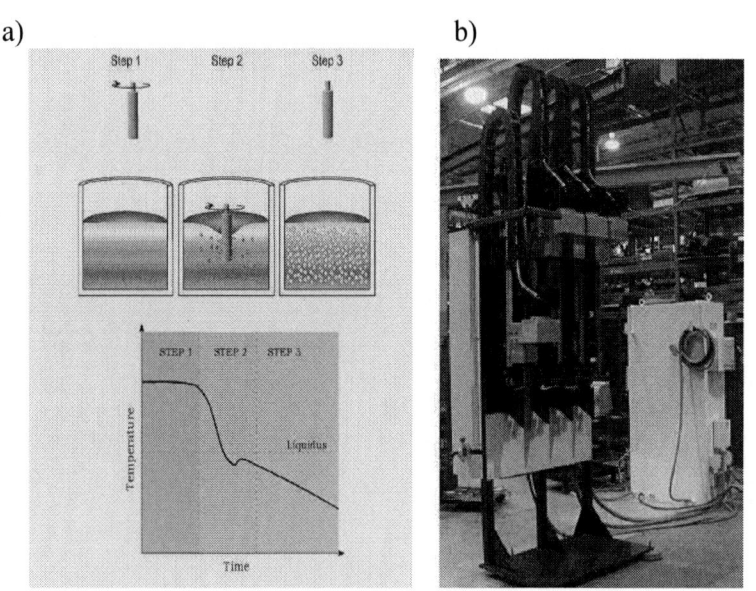

Figure 6: Semi-solid Rheocasting (SSR^{TM}) (after 10)
       a)   Schematic of the Semi-Solid Rheocasting process
       b)   Photograph of IdraPrince's SSR^{TM} Station

Near Net-Shaping Aerospace Alloys by Thixoforming

P. Kapranos & M. Farnsworth

The University of Sheffield, Advanced Manufacturing Research Centre (AMRC/Boeing), Advanced Manufacturing Park, Wallis Way, Catcliffe, Rotherham, S60 5TZ, UK
<p.kapranos@sheffield.ac.uk>

Abstract

Thixoforming, is the method of shaping metal components in a semi-solid state, utilizing non-dendritic alloy feedstock heated to a semi solid state, before being injected into a tooling cavity. The major advantages of this process are the lower energy costs due to lower temperatures involved, the removal of molten metal handling, reduced shrinkage porosity and gas inclusions, fine grain structure and improved mechanical properties, and the ability to cast near-net shape components considerably reduces the need for machining, hence reducing the amount of waste material produced.

A201 is a high strength aluminum-casting alloy but it is also a difficult and an expensive alloy to cast. The application of the thixoforming process to the A201 would allow this alloy to be reliably produced with mechanical and metallurgical properties superior to other cast aluminum alloys, and with substantially reduced fabrication defects.

Key words (thixoforming, non-dendritic microstructures, mechanical properties).

Introduction

Thixoforming is the near net shaping of metals in the semi-solid state, i.e. within the freezing or melting range between the fully solid and fully liquid states. During thixoforming, when the material is in the semi-solid state, it exhibits thixotropic properties, i.e. the unsupported material remains stiff (consistency of butter) and holds its shape so it can be readily handled, but rapidly thins and flows like a liquid when sheared. It is this behaviour that is the key to the thixoforming process where material flows as a semi-solid slurry into a die, as in conventional die-casting. This behaviour is exhibited in all metal alloys that possess a specific microstructure; namely one that consists of metal spheroids e.g. α-Al in aluminium alloys, surrounded by a contiguous layer of eutectic liquid when heated to the semi-solid state. This microstructure is the key to producing successful thixoformed parts.

There are various routes for obtaining the necessary spheroidal microstructures, but until recently, commercially this was done either by continuous casting with magneto-hydrodynamic stirring of the melt during solidification (MHD), or through the New Rheocasting (NRC) route where the melt is cooled down into the semi-solid state prior to thixoforming. The need to be able to re-cycle material in-situ forced the development of the NRC route, based on the original Rheocasting technology developed by MIT back in the 1970's. Currently more companies appear to develop parallel production routes that allow in-situ re-cycling, such as the SSR (Semi-Solid Rheocasting by Idra) and other developments are on the way by the aluminium manufacturer Alcan (Vortex technique) and the die-casting machine producers Buhler.

The thixoforming process has advantages over both the casting and forging processes, such as fine, uniform and virtually free of porosity microstructures, products that may be heat-treated to give mechanical properties superior to those of casting, reduced energy consumption and reduced die thermal shock and therefore longer die-lives, due to the lower heat content of the semi-solid material. In addition, higher melting point alloys, such as hypereutectic aluminium-silicon alloys with very high silicon contents (~25-40%), superalloys or tool steels, which cannot be easily be die cast, may nevertheless, be thixoformed.

As a relatively new process, before proving its value as a commercial success, thixoforming has had to exploit alloys that were already available. However this phase of development is now over and millions of thixoformed automotive parts are now in every day use in the cars we drive. An expanding portfolio of alloys specifically designed to exploit the potential of the thixoforming process is required to strengthen its market potential, but also address the urgent needs in the aerospace industry for near net shape, high strength, aluminium products.

Alloy development work at Sheffield involves developing an understanding of the key scientific principles on which alloy design and development for semi-solid processing must be based, and to produce aluminium alloys specifically tailored to exploit the thixoforming process and with performance approaching that of the wrought specification aluminium alloys.

This work looks at the A201 copper containing casting alloy with additional small quantities of magnesium, silicon and silver. Although this alloy is difficult to cast, it has a particularly high response to age-hardening [1] and therefore offers mechanical properties close to the wrought 2014 alloy.

Experimental

Conventional DC-cast dendritic material was re-cast using a cooling slope (CS) of length 200mm positioned at an angle of 45^0, see Figure 1, in order to obtain feedstock having the necessary non-dendritic, near spheroidal, microstructure for thixoforming [2].

A number of billets were cast using the CS, with the A201 alloy superheated 40°C above the release temperature of 670°C, and typical resulting microstructures are shown in Figure 2. These billets were machined into slugs with length and diameter of 60mm to be used for thixoforming flat products for further mechanical testing using the die shown in Figure 3. The thixoformed fingers were x-rayed and hardness measurements were obtained in the as thixoformed condition as well as the post heat treatment condition. For alloy A201, a conventional T6 heat treatment (two step solution 2 h at 513^0C and 17 h at 527^0C followed by water quenching and then ageing 20 h at 153^0C) has been used and in addition a T7 treatment that differs only in the ageing treatment; 5h at 190^0. From each thixoforging it was possible to obtain two tensile test specimens from the single finger die, see figure 4.

Results

The microstructures of both the as-thixoformed and the T6 heat-treated samples can be seen in figure 5. The as-thixoformed microstructure displays a small liquid fraction however, after the T6 heat-treatment there is clearly a change within the eutectic phase of the alloy. Some liquid entrapment is visible in both samples. When referring to the results of the Vickers Hardness tests (Figure 6), it is interesting to note how the thixoformed and heat-treated A201 recovers to an average value of 150vH, which is comparable to a value expected for as cast A201 from Alcan. Another point to consider is that the Vickers Hardness is relatively consistent along the length of the thixoforging finger, a distance of over 60mm.

The mechanical test results for the thixoformed and heat-treated A201 are very encouraging, see figure 7. The target value for ultimate tensile strength, that of the standard as cast A201, has been matched by the thixoformed alternative, at a value of 490 MPa UTS and 7% E and also the target value of wrought 2014 at 480 MPa UTS and 13%E [3].

More significantly though, the majority of the thixoformed samples show greatly improved percentage elongation as compared to the as-cast target, in the region of 50% in some cases.

Discussion

It is clear from the investigations on the thixoformability of alloy A201 that this alloy appears to behave well under thixotropic conditions, i.e. it can develop the appropriate non-dendritic microstructure through the usual routes, and it holds its shape whilst in the semi-solid state and flows as a viscous liquid when sheared and develops high mechanical properties after appropriate heat treatment. The UTS, YS and E% of the thixoformed A201 products appear to have better properties than the casting versions of this alloy, especially the near doubling of elongation values in the T6 & T7 heat treated conditions. These values are comparable with the values of wrought alloy A2014. In addition, initial fatigue tests of thixoformed A201 specimens have yielded very promising results which we hope to present quite soon. However, these tests have been carried out on polished specimens that have been machined out of thixoformed 'fingers' such as those shown in Figure 8 and the next task for us is to manufacture near net-shape fatigue specimens (Figure 9) that will be tested for fatigue performance in the as thixoformed condition in order to obtain representative fatigue data for aerospace designers.

In addition, the current series of mechanical properties data for thixoformed A201 has been obtained by using non-dendritic feedstock generated by the non-efficient cooling slope process. It is expected that the next series of specimens that will be generated by utilizing feedstock obtained by the Magnetohydrodynamic stirring (MHD) and the New Rheocasting (NRC) processes can only result in even better properties bringing the acceptance of thixoformed A201 alloy parts by the aerospace industry one step closer to reality.

Conclusions

A201 feedstock for thixoforming was cast using the cooling slope method. These billets were then thixoformed into flat products suitable for mechanical testing.

This initial research has shown that the thixoforming process could be an excellent alternative to the standard casting method, with the results of mechanical tests showing comparable ultimate tensile strength to that of as-cast A201, and improvements in percentage elongation of nearly 50%.

It is therefore hoped that the application of the thixoforming process to the A201 alloy will allow this alloy to be reliably produced with mechanical and metallurgical properties superior to other cast aluminum alloys, with substantially reduced fabrication defects. The technology could provide the potential to convert numerous machined parts to A201 castings at a substantially reduced cost.

References

1. D. Liu et al, 'Effect of heat treatment on properties of thixoformed high performance 2014 and 201 aluminium alloys', J. of Mat. Sci., 39, 2004, pp 99-105.
2. T. Haga and P. Kapranos, 'Billetless simple thixoforming process', J. of Mat. Proc. Tech., 130-131 (2002), pp 581-586.

3. P. Kapranos, 'Thixoforming Wrought Alloys', Die Casting & Technology, No 31, Sept. 2004, pp 44-51.

Acknowledgements

The authors would like to acknowledge the UK Engineering and Physical Sciences Research Council (EPSRC) for financial support under Grant GR/M89096- 'Alloy Development for Thixoforming, and also thank Mr. Worawit Jirattitichaorean for his contribution to the cooling slope experiments as well as the die design.

Figures

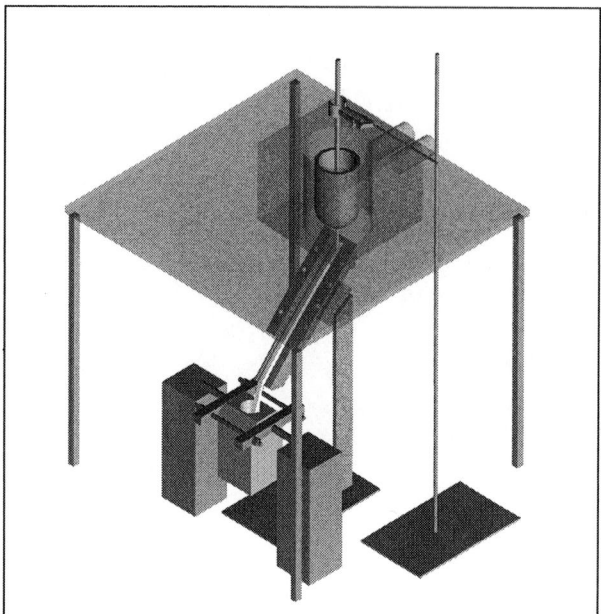

Figure 1. Cooling slope arrangement used for re-casting the A201 alloy.

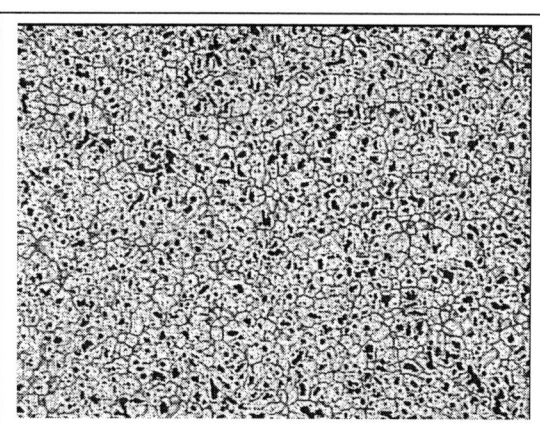

Figure 2. Microstructure of A201 billets cast using the Cooling slope.

Figure 3. Die used for thixoforming, featuring a steel base and replaceable inserts; a single finger cavity was used in this work.

Figure 4. Shows the location of tensile specimens within the forged finger.

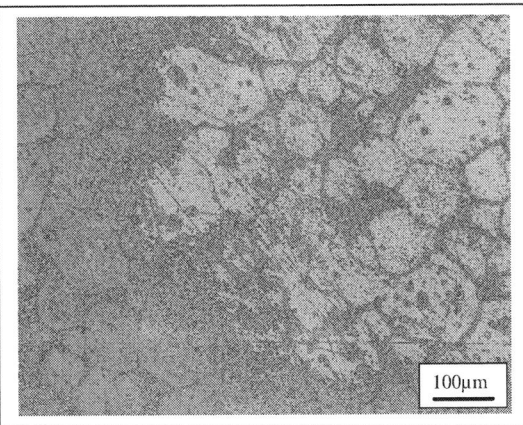 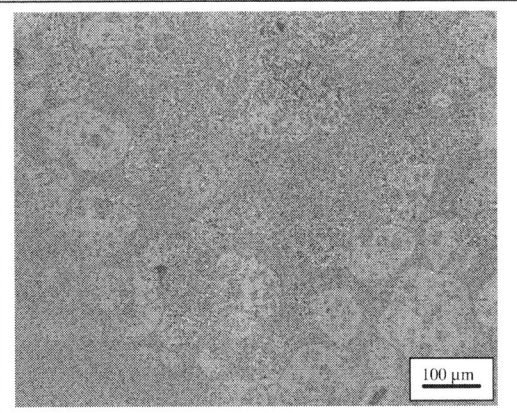

Figure 5. Microstructure of A 201 as-thixoformed and after T6 heat-treatment.

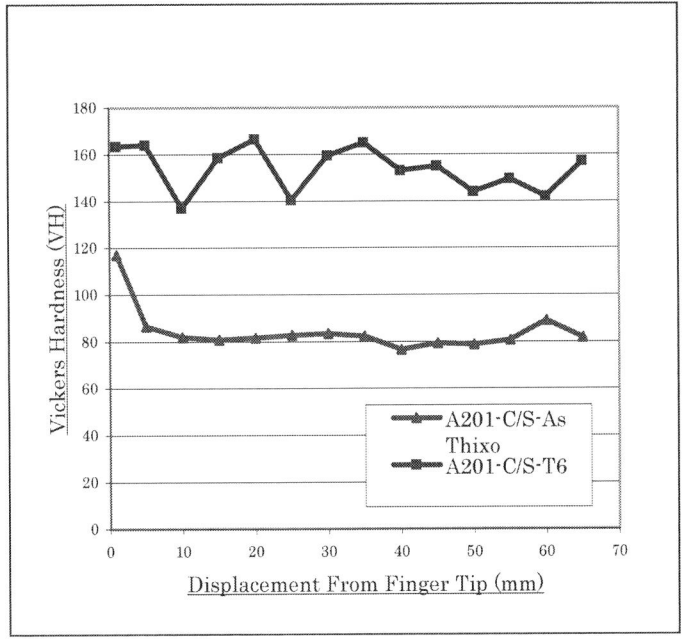

Figure 6. Results of Vickers Hardness tests for as-thixoformed and thixoformed plus T6 heat-treated A201.

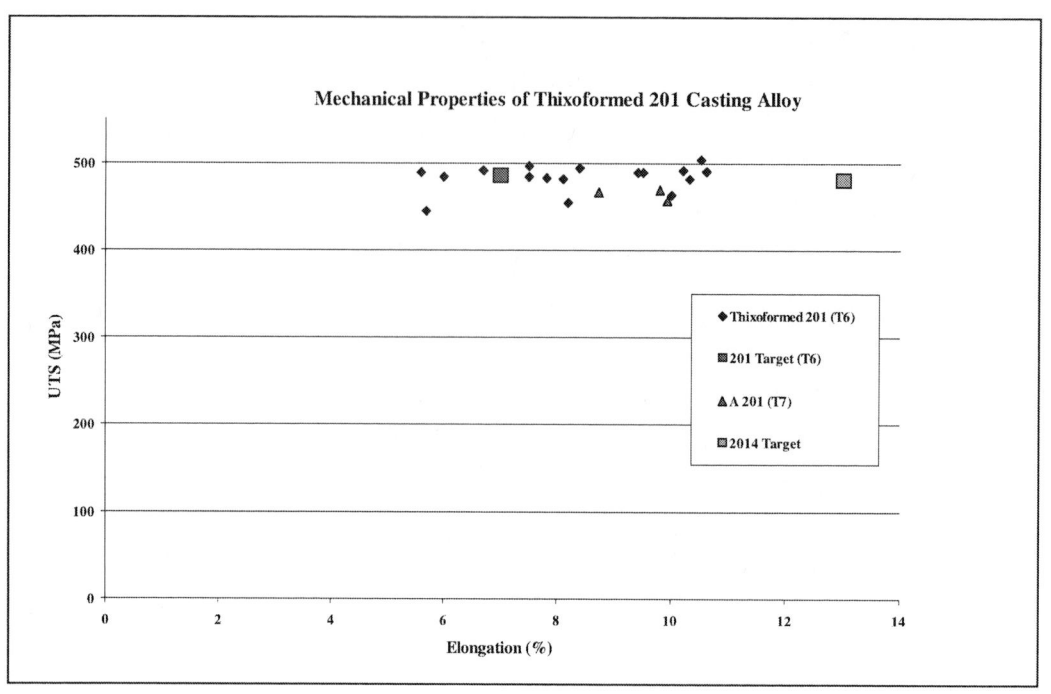

Figure 7. UTS vs Elongation of A201 alloys compared with the as cast and also the wrought A2014 alloy target values.

Figure 8. Thixoformed A 201 specimens, showing the effect of the scrapping ring in the gate area.

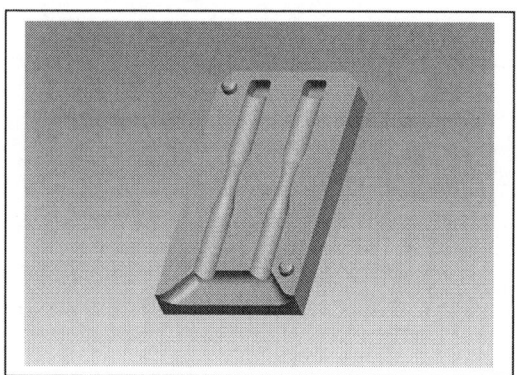

Figure 9. Proposed near net-shape fatigue specimen die.

Solidification Structure and Mechanical Properties of Semi-Solid Processed Cast Iron

Mohamed Ramadan, Hiroyuki Nomura and Mitsuharu Takita

Department of Materials, Physics and Energy Engineering, Materials Processing Engineering, Nagoya University, 464-8603, Japan

Abstract

Semi-solid processing of alloys is becoming one of the key technologies for producing advanced materials. Solidification structure and mechanical properties of semi-solid processed grey cast iron(CE=3.59%) were investigated. Investigation of the resulting microstructure and mechanical properties was carried out, which revealed significant imperovements in primary pariticle size refinement and globularity in components produced through semi-solid casting. Tensile strength of semi-solid processed grey cast iron is relatively high compared with ordinary one. It is found that graphite morphology, primary solid agglomeration and graphite network have a large influence on both the properties in semi-solid processing. The wear resistance of grey cast iron was improved using the semi-solid processing.

Key words semi-solid processing; grey cast iron; microstructure; mechanical properties; cooling plate.

Introduction

Grey iron is a relatively inexpensive and effective material for many engineering applications. All grey cast irons contain flake graphite dispersed in iron matrix including silicon. The properties of the grey cast iron depend on the size, amount and distribution of the graphite flakes and the matrix structure [1-4]. In most of the cases, industrial alloys solidify with dendrites. Mechanical properties of these alloys could be improved by transforming their dendritic structure to the globular one. Application of the shear stress in the semi-solid region leads to the fragmentation of dendrites to small particles. The successive temperature reduction and stress application causes deformation of the particles into globular ones distributed uniformly in the material[5, 6]. There are bright prospects for the use of semi-solid processing materials in many industrial fields because of their attractive properties.

Semi-solid processing of grey cast iron using cooling plate method has been reported in the literature[7-9] where, an improved structure of fine globular primary particles with a high degree of sphericity and phases clearly distinct from adjacent one is obtained. However the previous literature[7-9] did not include any discussion on the distribution of the graphite flakes, the matrix structure, and their effects on mechanical properties for wide range of primary fraction of solid.
In the present study, we investigate the effect of the microstructure changes by the semi-solid processing using cooling plate technique on the mechanical properties of grey cast iron.

Experimental

Strips of the constant width 25 mm and the length 155 mm with the thickness 14 mm as shown in Fig. 1 were investigated in this study. Sand mold was made using silica sand of AFS 69 gfn, 6% sodium bentonite and 1% cereal. After molding, the pattern was resealed and the surface of the sand mold was sprayed with molasses and then dried at 437 K for 2.5 hours[10]. The charge material used for all trials is grey cast iron bar stock. The chemical composition of cast iron samples is shown in Table 1.

Samples weighting approximately 900 g were melted using an electrical resistance furnace. The charge was heated in an argon atmosphere up to temperature of 1773 K, and then soaked for 30 minutes to allow the charge to attain the desired temperature. At 1773 K, the melt charge was removed from the furnace to the controlled pouring system, consisting of isolated box to regulate the melt temperature and speed-changeable motor to adjust the pouring rate, as shown in Fig. 2. At the desired temperature the melt charge was allowed to be poured over a cooling plate inclined at the known angle to the horizontal (10^0) with constant motor speed 600 r.p.m. (constant pouring rate), and to flow into a mold cavity at the end of cooling plate. Before pouring, 0.4 weight percent of Fe-75% Si inoculant was set on the surface of the cooling plate. The temperature of semi-solid slurry was measured by thermocouples inserted within the isolation box and the sand mold flask. The primary fraction of solid corresponding to this temperature is calculated using Scheil's equation and the austenite distribution coefficient k has been determined by the model of Goettsch and Dantzig[11].

Rectangular cast iron samples were cut from the strip casting. Microstructure, SEM observations and hardness measurement for cross section surface at 35 mm distance from pouring base were studied. The primary austenite particles size, primary austenite particle sphericity and graphite length were measured and analyzed with image analysis software. Brinell hardness testes at 750 kg load were also performed. Specimens for tensile test were machined from the strip. Tensile tests were carried out in accordance with the ASTM, A339 standard on round tension test specimens of diameter 6.5 mm and gauge length 25 mm using 50 kN tensile test machine. Two tensile samples were examined for each condition.

Specimens with the size of 2x12x60 mm for wear test were machined from the strip. Sliding wear tests were performed using a block-on-ring wear tester with the test material in contact against a super hard alloy steel ring with 670 Hv hardness number. The tests were conducted at room temperature under dry sliding conditions. The sliding length was 66.6m, while the sliding velocity of ring was changed from 0.51 to 3.62 m s^{-1} under the load of 31.38 N.

Results and Discussion

Microstructure

Solidification of ordinary hypoeutectic grey cast iron begins with the precipitation of austenite dendrites from the melt as temperature falls under the liquidus. Figures 3 and 4 show the effect of fraction of solid on the structure of the semi-solid processed grey cast iron. It is clear that the ordinary grey cast iron has a dendritic structure, on the other hand , semi-solid processed grey cast iron becomes finer and more globular by increasing fraction of solid. The present and previous[7-9] investigations are in good agreement in the point that shape and size of the primary austenite particles is highly affected with the use of cooling plate due to the resultant high cooling rate of the melt. High cooling rate of melt increases the number of the effective nuclei relative to the rate at which latent heat is dissipated. Figure 4 shows the decrease of average primary austenite particles size with increase of fraction of solid, which is considered to be due to increase of solidification rate at higher fraction of solid.

Figures 5 and 6 show the effect of fraction of solid on the size and distribution of graphite in the semi-solid processed gray cast iron. It is clear that increasing fraction of solid decreases the graphite size and changes its distribution. On the other hand, for low fraction of solid(f_s

≤0.12), flaky graphite (graphite type A) was obtained. Further increasing fraction of solid (f_s from 0.14 to 0.18) results in finer graphite morphology (graphite type D). This graphite morphology interferes with the formation of a fully pearlitic matrix by providing short diffusion paths for C, hence aiding ferrite formation [1].

As the metal cools down under the liquidus, austenite is precipitated and the liquid is progressively enriched in carbon until the eutectic composition(4.3%.) is reached. Once this composition is attained, the remaining liquid transforms into two solids, graphite plus austenite in the case of the stable reaction. Once the eutectic solidification is completed, no liquid metal remains, and further reactions take place in the solid state[4]. During the temperature interval between the eutectic and eutectoid transformations, the high-carbon austenite rejects carbon, which diffuses to the graphite flakes. This gives austenite the chance to acquire the composition needed for the eutectoid transformation. This transformation involves the decomposition of austenite into pearlite or pearlite plus ferrite as shown in Fig.5.

Figure 5 shows the effect of fraction of solid on the eutectic structure of the semi-solid processed grey cast iron. It is clear that the graphite morphology changed from coarse flakes of type A to fine flakes of type D by increasing fraction of solid. The eutectic austenite decomposes to pearlite plus ferrite. High ferrite fraction was observed at the fraction of solid 0.18 and graphite flakes became mainly type D.   Figure 7 shows the solidification cooling curves for grey cast iron under the varied fraction of solid. It is shown that by increasing the primary fraction of solid the undercooling is increased and it reaches to the maximum value at the fraction of solid 0.20. The present and previous [4,12] investigations are in good agreement in the point that type A graphite flakes are found in usual cast iron with a minimum undercooling value and type D one forms when the amount of undercooling is high but is not sufficient to cause carbide formation. Increasing of the undercooling may occur by the higher cooling rate due to the higher fraction of solid. Finally, we can say that the cooling plate had a refining effect on primary austenite and graphite, otherwise structure will be similar to that of ordinary casting with dendritic primary austenite and coarse graphite.

Mechanical properties
Tensile strength
Figure 8 shows the effect of fraction of solid on the tensile strength and the elongation of the semi-solid processed gray cast iron. For low fraction of solid (f_s ≤0.12),increasing fraction of solid results in an increment in the tensile strength as well as the elongation, comparing with the ordinary grey cast iron due to the fine graphite formation(fine graphite type A). Further increase of fraction of solid (f_s = 0.14 ~0.18) leads the graphite morphology to become finer graphite (type D), consequently the tensile strength decreases. Increment of fraction of solid (f_s > 0.18) leads again to

increasing of tensile strength due to the agglomerations of primary solid particles. Those agglomerations of the primary solid particles cut the network of the eutectic graphite as shown in Fig. 9. The graphite network concentration ratio, which determined as the ratio of separated particles number to total particles number, rapidly increases over the range of fraction of solid 0.12 to 0.18 and starts to decrease over than fraction of solid 0.18(see Fig. 10). This behavior seems to be due to the udercooling and the agglomerations of primary solid particles consequently.

Previous researches[3,13] also qualitatively described the effects of the graphite morphology and the matrix characteristics on the mechanical properties, where both the small graphite flakes and the maximum primary austenite content is known to increase the strength of cast iron. The crack initiation and propagation in the cast irons are influenced by the degree of absorbed energy by the matrix for plastic deformation. Increase of the energy absorption of the plasticized matrix of fracture is found out in the case of the semi-solid processed grey cast iron($f_s \leq 0.12$), with finer graphite and with matrix bridges between graphite segments wider than those in the ordinary grey cast iron. On the other hand, for semi-solid processed grey cast iron with fraction of solid ranged between 0.14 and 0.18, it was considered that the fine graphite with a very short bridges between the graphite segments and the eutectic ferrite produces decrement in the energy absorption of the plasticized matrix at the fracture.

In this study, semi-solid processing using cooling plate technique takes the advantage of the refinement of the primary grains and the precipitated graphite, through a combination between the rapid solidification and the flow-related fragmentation of the dendrites. Resultant globular structure enhances the tensile strength of the cast iron products. Types of graphite and matrix structures can be controlled by adjusting fraction of solid. Although, type D graphite flakes which can be produced at fraction of solid ranged between 0.12 and 0.18, give a low tensile strength, small type D flakes are reported to yield a fine machined surface finish by minimizing the surface pitting[14]. Due to this good surface finishing, the grey cast iron molds for glass processing are considered to be one of its most important applications[15,16].

Hardness

Hardness values are shown in Fig. 11 as a function of fraction of solid. Hardness generally increases by increasing fraction of solid. Hardness starts to decrease for fraction of solid more than 0.12 due to the ferrite formation with type D graphite morphology. However this high hardness values compared with that of ordinary cast iron are attributed to a characteristic microstructure with a globular primary phase and fine graphite flakes.

Wear behavior

In Fig. 12, the wear curves for ordinary and semi-solid processed grey cast irons show that the wear mechanism differs with the change of sliding speed under a constant low load (31.38 N). It can be seen that for any sliding speed, the wear rate of ordinary grey iron is higher than those measured for all tested semi-solid processed grey iron. In ordinary casting the wear rate increases as sliding speed increases until 1.63 ms^{-1} and further increasing of sliding speed (2.38 ms^{-1}) decreases the wear significantly. As the sliding speed increases, surface damage in specimen increases causing more feeding of graphite to the surface and resulting in a better lubrication and lower wear rate. On the other hand, the feeding of graphite to the surface in case of the semi-solid processed grey cast iron is much less due to its fine graphite, globular matrix structure and high hardness.

Conclusions

Solidification structure and mechanical properties are investigated for grey cast iron by semi-solid processing using sand mold. The process has several merits are follows:

1- Increasing fraction of solid makes the microstructure become finer and more globular.
2- Increasing fraction of solid increases the undercooling and changes the graphite morphology.
3- Tensile strength of the semi-solid processed grey cast iron is relatively high compared with ordinary one. Both the tensile strength and elongation closely connect with the type and shape of graphite changed with fraction of solid.
4- Increasing fraction of solid increases the hardness values, which is explained by the same reason for tensile strength.
5- Wear resistance of the semi-solid processed grey cast iron is relatively high compared with ordinary one. Wear resistance closely connects with the microstructures and the graphite morphology changed with fraction of solid.

References

1. Elliott R, Cast Iron Technology, Butterworths,London, Boston, 1988.
2. Bates C E, effects of alloy elements on the stength and microstructure of grey cast iron, AFS Transaction, 92,1984, pp923-945.
3. leube B and Arnberg L ,Influence of different microstructure properties on mechanical properites of gery iron, Int. J.Cast Metals Res.,11,1999,pp 507-514.
4. Metals Handbook, Casting, vol.15, 9th ed., ASM International, Ohio, USA, 1988.

5. Pahlevani F, Salarfar S and Nili-Ahmadabadi M, Proc. 8th S2P international conference on semi-solid processing of alloys and composites, Limassol, Cyprus, Sept., 2004.

6. Flemings M C, Behavior of metal alloys in the semi-solid state, Metallurgical transactions A, 22A,1991, pp957-981.

7. Muumbo A, Nomura H and Takita M, Mechanical properties and microstructure of semi-solid processed cast iron, Proc. 7th International Symposium on Science and Processing of Cast Iron, Barcelona, 2002.

8. Muumbo A, Takita M and Nomura H, Processing of semi-solid grey cast iron using the cooling plate technique, Materials Transactions, 44, 2003, pp 893-900.

9. Ramadan M, Nomura H and Takita M, Semi-solid processing of thin section grey cast iron in sand mold, Proc. 8th S2P international conference on semi-solid processing of alloys and composites, Limassol, Cyprus, Sept., 2004.

10. Heine R and Rosenthal P, Principles of Metal Casting ,MC Grow-Hill Book Co., Japan, 1955.

11. leube B and Arnberg L, Int. J. Cast Metals Res.,11,1999, pp 505-514.

12. Glover D, Bates C E and Monroe R,The relationships among carbon equivalent, microstructure and characteristics and their effects on strength and chill in grey cast iron, AFS Transaction, 90,1982 pp 745-757.

13. Sjogren T, Vomacka P and Svensson I L, Int. J. Cast Metals Res., 17, 2004, pp65-71.

14 L. R. Jenkins, R. D. Forrest, Ductile Iron Metal Handbook, 10th ed., 1(1990)33-55.

15 Cingi M, Arisoy F, Basman G, Sesen K,The effects of metallurgical structures of different alloyed glass mold cast iron on the mold performance, Materials letters, 55, 2002, pp360-363.

16 Barton R,Cast iron for glass molds, J. Res. Dev.,7, 1985, pp146–156.

Tables

Table 1 Chemical composition of cast iron samples (wt.%)

C	Si	Mn	P	S	C.E.[*]	T_L (liquidus)	T_S (solidus)
2.92	1.98	0.59	0.022	0.085	3.59	1543 K	1418 K

[*]C.E.: carbon equivalent (= %C+1/3 (%Si + %P))

Figures

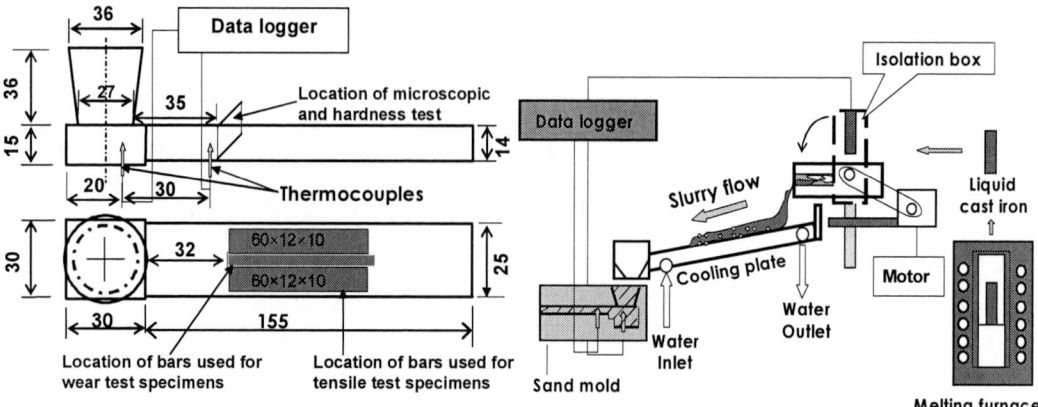

Fig. 1 Design used for casting specimens (unit in mm).

Fig. 2 Pouring system.

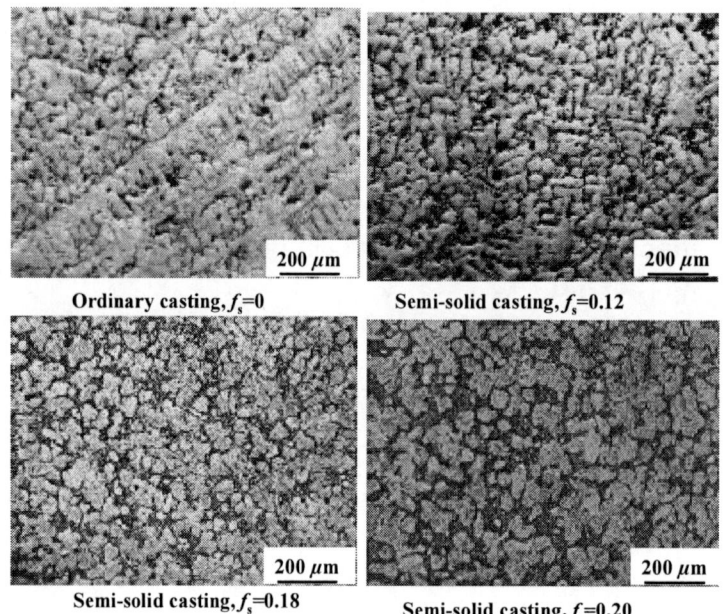

Fig. 3 Effect of fraction of solid on the structure of semi-solid processed grey cast iron.

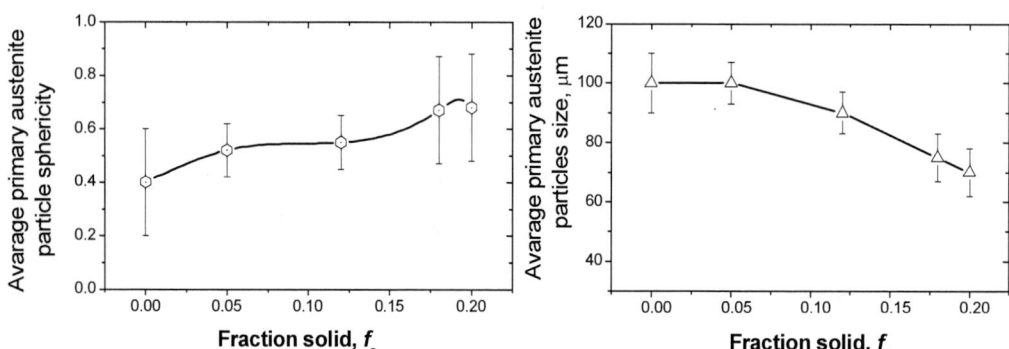

Fig. 4 Effect of fraction of solid on average primary austenite particle size and primary austenite particle sphericity.

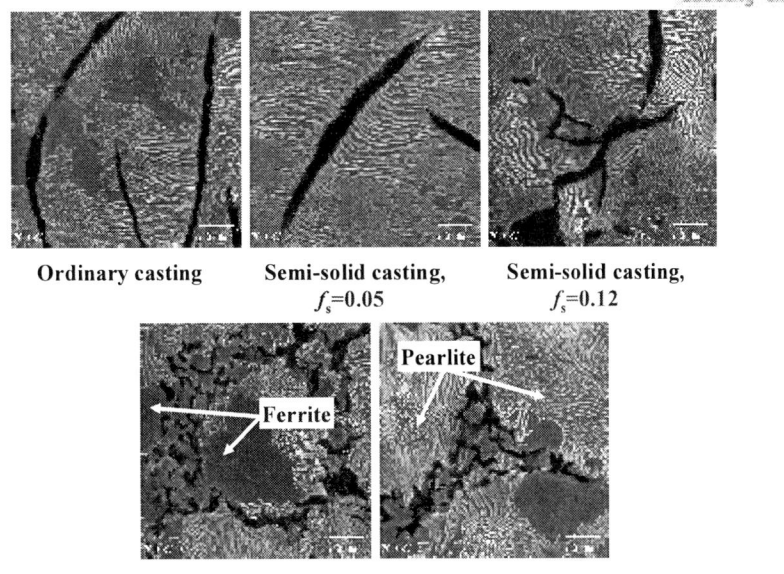

Ordinary casting   Semi-solid casting, $f_s=0.05$ Semi-solid casting, $f_s=0.12$

Semi-solid casting, $f_s=0.18$ Semi-solid casting, $f_s=0.20$

Fig. 5 SEM photographs of eutectic zone of cast irons of different fraction of solid.

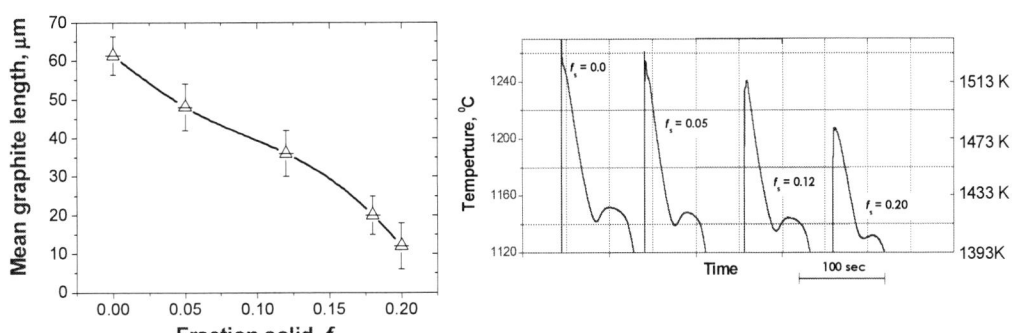

Fig. 6 Mean graphite length as a function of fraction solid.

Fig. 7 Representative solidification cooling curves for grey cast irons for varied primary fraction of solid.

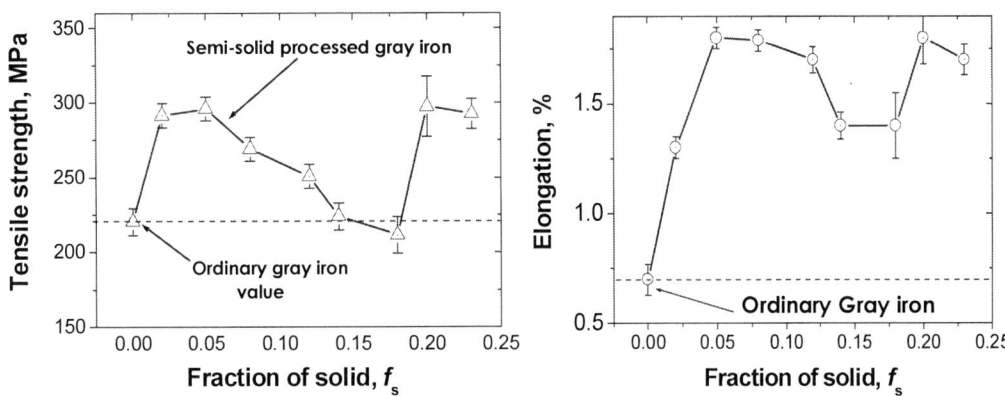

Fig. 8 Effect of fraction of solid on the strength of semi-solid processed grey cast iron.

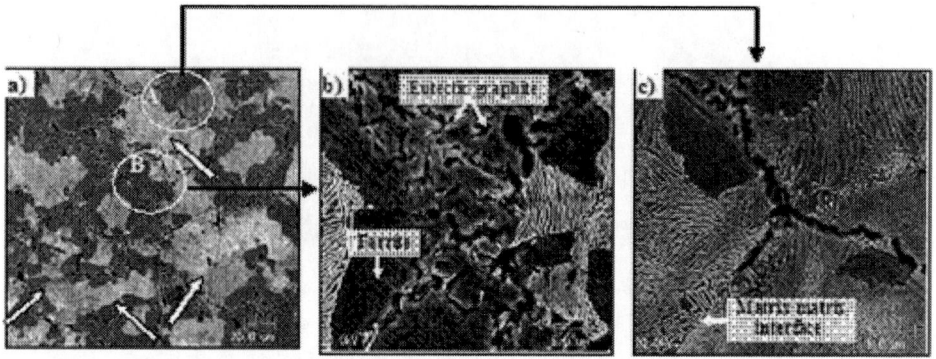

a) Microstructure showing particles agglomerations; b) wide eutectic zone of spot B
shown in a); c) narrow eutectic zone of spot A shown in a).

Fig. 9   SEM photographs showing particles agglomeration for fraction of
solid 0.23 (the arrows indicate the positions of the particles
agglomerations).

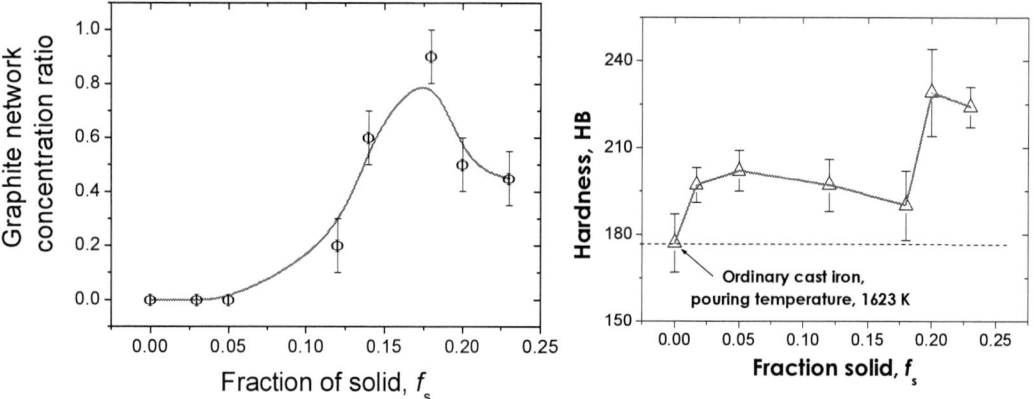

Fig.10 Graphite network concentration as
a function of fraction of solid.

Fig.   11 Brinell hardness as
a function of fraction of solid.

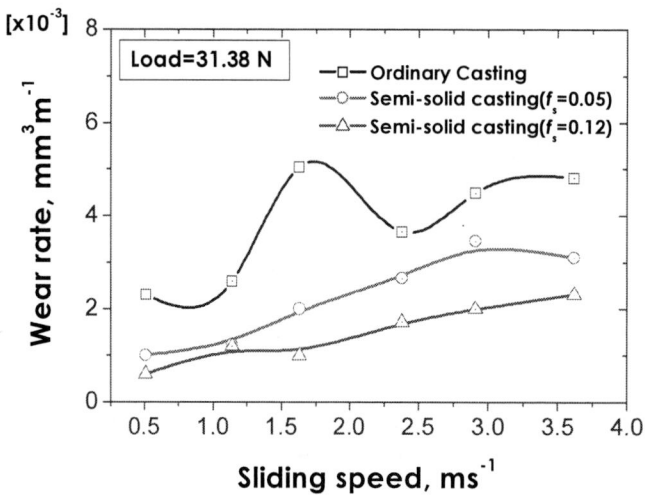

Fig. 12 The change of wear rate as a function of sliding speed.

Microstructure and mechanical properties of rheo-diecast (RDC) aluminium alloys

X. Fang, J. Patel and Z. Fan

BCAST (Brunel Centre for Advanced Solidification Technology) Brunel University, UK

Abstract

A new semisolid metal (SSM) processing technology, the rheo-diecasting (RDC) process, has been developed for manufacturing near-net shape components of high integrity directly from liquid Al-alloys. The RDC process innovatively combines the well-established twin-screw mechanism with the conventional cold chamber die casting process. In this paper we present the microstructure and mechanical properties of both the as-cast and heat-treated RDC Al-alloys. The results indicate that the in the as-cast condition, RDC samples have a fine and uniform microstructure throughout the entire sample and close-to-zero porosity. The RDC samples have improved tensile strength and ductility, and further heat treatment of the RDC Al-alloys, can substantially improve strength, with a slight sacrifice of ductility. Discussions will be made on the effects of intensive forced convection and the effects of the resulting microstructure.

Key words Al-alloys, Rheocasting, Microstructure, Mechanical Properties

Introduction

Al-alloys, as lightweight structural materials, play an important role in achieving vehicle weight reduction and improving fuel economy in the automotive industry. Since 1990, the use of Al has been doubled in cars and tripled in the light truck market [1]. Currently, 85% of all Al castings are used by the automotive and mass transport industry, a large proportion of which are produced by the high-pressure diecasting (HPDC) process. However, the quality of components manufactured by the HPDC process is limited by the presence of a substantial amount of porosity, which not only excludes the application of HPDC components in the high-safety and airtight systems, but also denies the opportunity for further property enhancement by heat treatment. It is clear that a further increase in Al applications in the transport industry will require a major advance in processing technologies. The new processes need to be capable of producing components of high integrity and improved performance, while being comparable with the HPDC process in terms of production cost and efficiency.

Porosity due to turbulent mould filling could be reduced or even eliminated if the viscosity of the melt could be increased to reduce the Reynolds number sufficiently minimise the trapped air [2]. This is the concept of semisolid metal (SSM) processing. In addition to entrapped gas porosity, SSM processing can also reduce the shrinkage porosity due to the much reduced processing temperature and the formation of a given volume fraction of the primary phase prior to mould filling.

In this paper we present a new rheo-processing technique, the rheo-diecasting (RDC) process, and the experimental results on microstructures and mechanical properties of the RDC Al-alloys. The discussions will focus on the solidification behaviour of Al-alloys under intensive forced convection, and its consequences on microstructural evolution and mechanical properties.

Experimental

The RDC process innovatively adapts the well-established high shear dispersive mixing action of the twin-screw mechanism to the task of in situ creation of SSM slurry with fine and spherical solid particles, which is followed by direct shaping of the slurry into a near-net shape component using the existing cold chamber diecasting process.

Fig. 1 schematically illustrates the RDC equipment for manufacturing Al-alloy components. It consists of two basic functional units, a twin-screw slurry maker and a standard cold chamber HPDC machine, with no modification. The twin-screw slurry maker has a pair of screws rotating inside a barrel. The specially designed screw profiles are co-rotating, fully intermeshing and self-wiping. The screws and barrel are made from special materials in order to prevent reaction with molten Al. During the slurry making process, there is an enormous amount of ever-changing interfacial area between the solidifying alloy, the screws and the barrel. This makes the slurry making process extremely efficient for heat extraction.

The alloys used in this work include commercially available cast Al-alloys, A357 and A380, and a commercially available wrought alloy, 2014. The chemical compositions were analysed using an optical spectrometer system and DSC analysis was used to measure the solidus and liquidus of the alloys. The compositions and freezing range of the Al-alloys used in this work are listed in Table 1.

The alloys were melted in graphite crucibles using a top-loading resistance furnace and held at temperature for approximately two hours for homogenisation of temperature and composition. During the RDC process, a predetermined dose of liquid alloy from the melting furnace is fed into the slurry maker. The liquid alloy is rapidly cooled down to the SSM processing temperature while being mechanically sheared by a pair of closely intermeshing screws converting the liquid into semisolid slurry. The barrel temperature determines the solid fraction in the slurry. The semisolid slurry is then transferred to the shot chamber of the HPDC machine, for component shaping. In order to prevent Al-alloy from oxidation, nitrogen gas may be used as the protective atmosphere during the slurry-making process. The die used for casting had six cavities, of which four were tensile test samples and two were Charpy test samples. The dimensions of the tensile test samples are 6mm in gauge diameter, 60mm in gauge length and 150 mm in total length. The temperature of the die was kept at 220ºC during processing

Once the SSM slurry is transferred to the shot chamber of the HPDC machine, the slurry supply system starts to prepare slurry for the next shot. The slurry making process is more efficient than the component shaping process; the component shaping process therefore, dictates the cycle time of the RDC process. Since SSM slurries have less heat to be removed than a fully liquid feed, the RDC process has a shorter cycle time than the conventional HPDC process.

Solution treatment (T4) was carried out for the RDC A357 alloy in a resistance furnace at 540ºC for 4 hours and then water quenched to room temperature. In order to prevent oxidation during heat treatment, samples were covered with graphite powder. The aging for both T5 and T6 heat treatments was done at 165ºC for 6 hours.

Samples for the metallographic examination were prepared from the cross section of tensile test samples, and were observed using an optical microscope (OM) and a scanning electron microscope (SEM). The volume fraction and particle size of the primary α-Al particles were determined using an image analysis system. Mechanical properties were tested using a universal materials testing machine at a crosshead speed of 2mm/min.

Results

Fig. 2 shows the microstructure from the cross-section of the RDC A357 alloy bar (6mm in diameter), processed at 595ºC and sheared at a shear rate of 890s^{-1} for 30 seconds. At 595ºC, the measured solid fraction was 0.25, but with further decrease of processing temperature, the solid fraction was increased to 0.41 at 575ºC. It is interesting to note that the average particle size only varied between 39-41 μm and the particle shape factor 0.89-0.87. Further microstructural examination indicates that there was no entrapped gas in all the RDC samples, only very fine shrinkage pores were occasionally observed. Generally, the total porosity volume fraction was less than 0.2%. The primary particles were uniformly distributed throughout the entire cross section of the 6mm bar.

Fig. 3 shows the detailed microstructure produced by the solidification of the remaining liquid inside the die, which is referred to as "secondary solidification". Secondary solidification contributed a further volume fraction of fine and spherical primary particles, which are uniformly distributed in a eutectic matrix. The average size of the primary particles produced by the secondary solidification was less than 10μm.

The A357 alloy used contains 0.46 wt% Fe. SEM examination shows that the Fe-containing intermetallic compound existed as fine particles (most of them are less than 1μm) with a compact morphology, as shown by the backscattered SEM image in Fig. 4. Fine and spherical Fe-containing intermetallic particles are less harmful to mechanical properties, particularly on ductility, than large plates or needles formed during conventional HPDC.

The RDC process was also used to cast A380 alloy with 1.57%Fe. Although the Fe-containing intermetallic compound particles were larger, 40μm, they still had a compact morphology and a uniform distribution. The tensile test results indicate that both strength and elongation only had a slight decrease when Fe contents increased from 0.5 wt% to 1.5 wt%. This means that the RDC process is much more tolerant to Fe contents, and more scrap can be recycled in an industrial environment.

The suitability of the RDC process for wrought Al-alloys was also investigated. Fig. 5 shows the microstructure of 2014 wrought Al-alloy processed at 636ºC for 30 seconds. Similar to the cast alloys presented previously, the primary solid particles were fine in size, 50μm, spherical and a uniformly distributed. The secondary solidification process seems to be more pronounced in wrought alloys than in the cast alloys. Primary particles produced by the secondary solidification had a size less than 10μm, and an extremely uniform distribution. The implication from Fig. 5 is that wrought alloys should be processed near their liquidus temperatures to promote microstructural refinement through secondary solidification. By doing so, the inherent confliction between temperature sensitivity of solid fraction and hot tearing can be addressed at the same time.

The obtained mechanical properties of the RDC A357 alloy under different processing conditions are tabulated in Table 2. Compared with the conventional HPDC process, the RDC process provides samples with slightly lower yield strength, similar ultimate tensile strength and much higher tensile elongation. The elongation of the RDC samples is approximately 30% higher than that of the HPDC samples. This is a very important feature of the RDC process.

The low porosity of the RDC samples made it possible to enhance mechanical properties through heat treatment. Both T5 and T6 heat treatments were carried out on RDC A357 samples. The tensile test results of the heat-treated samples are summarised in Table 2. Generally, in comparison with the mechanical properties of the as-cast samples, the heat-treated RDC samples have substantially higher yield strength and tensile strength, but a lower elongation.

Component production trials have been conducted to confirm the reliability of the slurry maker and consistency of the RDC process for continuous component production. The component die (the identity of the component is omitted here) used for this work had two cavities and four sliding cores. The thickness of the component wall varied between 2-6mm. The runner had a thickness of 10mm and the biscuit had a diameter of 60mm. The results indicate that similar to the tensile test sample presented previously, RDC components, produced under the optimised conditions, had a very good surface finish, close-to-zero porosity and a very fine and uniform microstructure throughout the entire casting (Fig. 6).

Discussion
The fluid flow inside a twin-screw slurry maker is unique. The liquid or semisolid metal inside the slurry maker is divided into a larger number of C-shaped pockets. Due to the rotation of the twin-screw, an existing pocket near one screw will reduce its volume from a maximum value to a minimum value, and at the same time, a new pocket is created near the other screw. During this volume reduction process, the fluid inside the pocket is squeezed to escape at a high speed from the narrow gaps confined by the screws and the barrel. This process repeats periodically with the rotation of the screws. The consequence of this periodic elimination and creation of the fluid pockets can be summarised as follows:

(1) The fluid flow inside the slurry maker is of high shear rate and high intensity of turbulence. The intensity of turbulence is unique to the twin-screw slurry maker, and the shear rate here can be an order of magnitude higher than that offered by other mechanisms used in semisolid processing so far.

(2) This flow pattern makes the twin-screw mechanism extremely powerful for dispersive mixing at a very fine level. The exact scale of

the turbulent eddies is difficult to quantify. However, our experimental results on shearing liquid metal indicate that effect of the turbulence could reach nano scale.

(3) There is an extremely large, ever-renewing interfacial area between the fluid and the solid surface of the screws and the barrel. This makes the twin-screw slurry maker a very efficient heat exchanger. The change from the pouring temperature to the semisolid processing temperature can be achieved in a few seconds, somewhat similar to quenching.

(4) The temperature and chemical composition of the alloy inside the slurry maker are extremely uniform.

The extremely uniform temperature and composition fields, created by the intensive mixing action in the twin-screw slurry maker, eliminates the typical 3-zone structure produced by conventional casting processes, by promoting not only heterogeneous nucleation throughout the whole volume of the undercooled liquid, but also the survival of all the nuclei. Compared with conventional solidification, the actual nucleation rate may not be increased but all the nuclei will survive resulting in an increased effective nucleation rate [3].

It has long been believed that the nondendritic particles are developed from the initial dendritic morphology under dynamic agitating conditions [4]. However, the recent theoretical analysis of the morphological evolution under forced convection reveals that this might be only applicable to the case of a simple shear flow with a low shear rate. With increasing shear rate and the intensity of turbulence, the growth morphology changes from dendrites to spheres via rosettes due to the change in the diffusion geometry in the liquid around the growing solid phase [5]. This has been verified by experimental observations in the ref [6] and the experimental results presented in the previous section.

The remaining liquid in the SSM slurry will solidify in the die cavity without shearing. However, this remaining liquid has been intensively sheared in the twin-screw slurry maker. It has a uniform temperature and composition throughout the liquid. According to the previous analysis, nucleation would occur throughout the entire remaining liquid, and every single nucleus would survive and contribute to the final microstructure. However, different from the nucleation in the twin-screw slurry maker, nucleation in the die cavity will occur with a much higher nucleation rate due to the high cooling rate provided by the metallic die (in the order of 10^3 K/s) of the HPDC machine. Under such conditions, each nucleus would not have much chance to grow before the remaining liquid is completely consumed, giving rise to a very fine structure (Figs. 3, 5). Solidification of the remaining liquid in the semisolid slurry has been called secondary solidification [7]. A detailed account of the solidification behaviour under intensive forced convection in the RDC process has been presented elsewhere [3].

Microstructural non-uniformity, large intermetallic compounds, large pores caused by entrapped gas and shrinkage pores affect the mechanical properties of die cast samples [8]. Minimising these factors will enhance the mechanical properties of cast samples. The extremely low porosity, fine size and equiaxed morphology of the Fe-containing intermetallic compounds, and perhaps more importantly, the fine and uniform microstructure throughout the entire sample of the RDC samples produces a good combination of strength and elongation.

A357 is a heat treatable alloy. Strength can be enhanced by precipitation of Mg_2Si in the supersaturated Al solid solution. In Table 2, compared to the as-cast condition, the strength of the RDC samples is substantially improved by both T5 and T6 treatments, though the elongation is reduced. Artificial aging enhances strength but reduces ductility of the primary Al particles. At the solution treatment temperature, the lamellar eutectic Si undergoes a spheroidising process eliminating the continuous eutectic network in the as-cast condition. A uniform dispersion of fine and spherical Si particles is more ductile than a eutectic network. However, solution treatment under the T6 condition will lead to substantial grain growth, which in turn results in a decrease in ductility. Therefore, both T5 and T6 treatment provide similar ductility.

Conclusions

A new semisolid metal processing technology, rheo-diecasting (RDC), has been developed for the production of Al-alloy components with high integrity. The RDC equipment can be easily achieved by adding a twin-screw slurry maker to an existing HPDC machine. The fluid flow inside the twin-screw slurry maker is characterised by high shear rate and high intensity of turbulence, which in turn promotes effective nucleation and spherical growth during solidification, resulting in a fine and uniform microstructure. Under optimised conditions the RDC samples have close-to-zero porosity, fine, spherical and uniform distribution of intermetallic compounds and are free from other casting defects. Compared with HPDC process, RDC offers components with improved strength and ductility in the as-cast condition. The close to zero porosity of the RDC components mean that they are weldable and that heat treatment can be used to enhance the mechanical properties of RDC samples. It was found that both T5 and T6 treatments substantially improve the strength but slightly decrease the ductility. This technology is also capable of processing wrought alloys and alloys that are difficult to cast, i.e. with large or narrow freezing ranges.

The changes in morphology of Fe-containing compounds caused by the RDC process mean that up to 1.5 wt.% Fe can be tolerated as impurity without any severe damage to the mechanical properties. Therefore more scrap can be recycled in the melting furnace. This is in addition to the inherently lower scrap rate and higher material yield of the process itself. Additionally longer die life can be expected, due to the much-reduced processing temperature. All of these factors contributing to the lower overall component production cost.

References

1. J. Hirsch, in: J.F. Nie, A.J. Morton, B.C. Muddle (Eds.), Aluminium Alloys - their physical and mechanical properties, Proc. ICAA9, Institute of Materials Engineering Australia, Brisbane, 2004, pp15-23.
2. S.A. Metz, M.C. Flemings, AFS Trans. 78 (1970) 453.
3. Z Fan, G.J. Liu, Acta Materialia, 53 (2005) 4345-4357
4. R.D. Doherty, H.I. Lee, E.A. Feest, Mat. Sci. Eng. A65 (1984) 181-189.
5. A. Das, S. Ji, Z. Fan, Acta Materialia 50 (2002) 4571-4585.
6. S. Ji, Z. Fan, Met. Mater. Trans. 33A (2002) 3511-3520.
7. S. Ji, A. Das, Z. Fan, Scripta Mater. 46 (2002) 205-210.
8. A.L. Bowles, J.R. Griffiths, C.J. Davidson, in: J. Hryn (Ed.), Magnesium Technology 2001, TMS, 2001, pp161-168.

Acknowledgements

The authors acknowledge contributions to this work from Mr. M. Hitchcock, Mr. G. Liu and Dr. Y. Q. Liu, and the financial support from EPSRC (UK).

Tables

Table 1. Chemical compositions and freezing ranges of the Al-alloys used.

Alloy	Si	Cu	Mg	Mn	Fe	Zn	Cr	Ti	ΔT_{L-S}
A357	8.18	0.296	0.300	0.164	0.463	0.246	0.013	0.048	65
A380	9.30	3.25	0.255	0.295	0.70	1.584	0.044	0.048	89.7
2014	0.7	4.9	0.8	-	0.3	0.14	0.02	0.03	128.1
Al-6Si-2Mg	6.11	-	2.19	0.001	0.1	<0.003	-	-	48.8

Table 2. Mechanical properties of the RDC A357 alloy under as-cast and different heat treatment conditions.

Processing conditions	Yield strength (MPa)	UTS (MPa)	Elongation (%)
HPDC, F	143-150	273-298	6.2-9.6
RDC, F	120-125	280-298	9.1-13.1
RDC, T5	188-202	306-327	5.4-9.5
RDC, T6	250-279	319-349	5.9-8.2

Figures

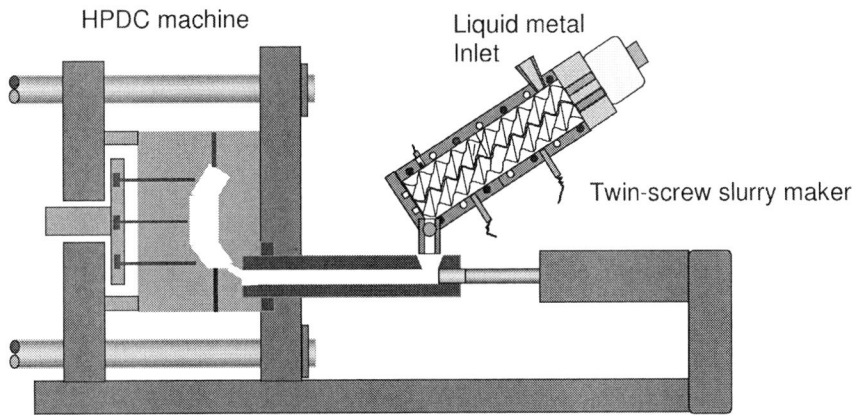

Fig. 1. Schematic illustration of the RDC process Al-alloy.

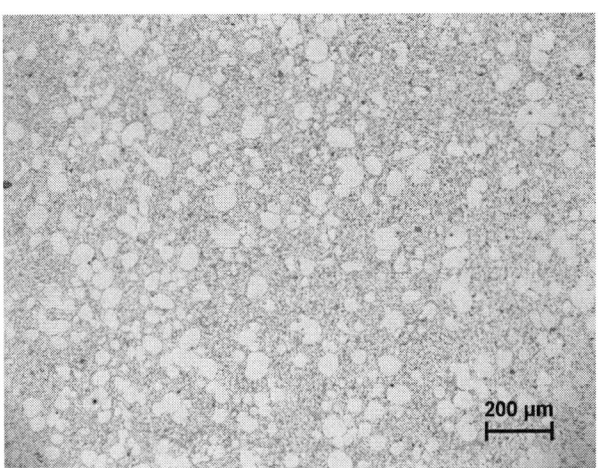

Fig. 2. Unetched microstructure of the RDC A357 alloy processed at 595°C, sheared at a shear rate of 890s^{-1} for 30 seconds.

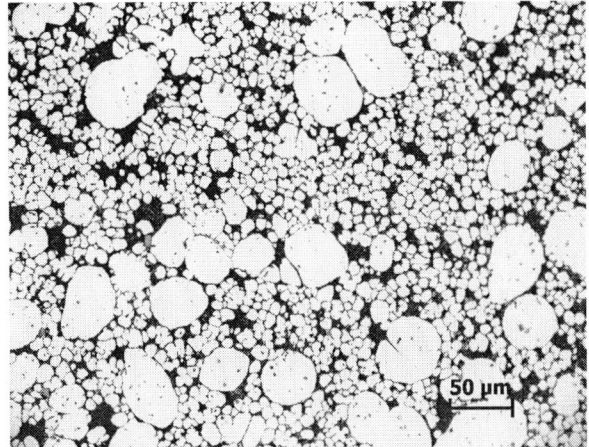

Fig. 3. Microstructure of the RDC A357 alloy, showing the morphology and distribution of the primary particles produced by secondary solidification inside the die without shearing.

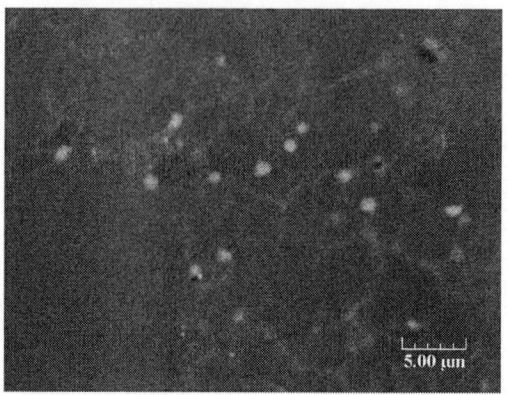

Fig. 4. Backscattered SEM micrograph showing the compacted morphology of Fe-containing Al compound (the bright phase) in the RDC A357 alloy.

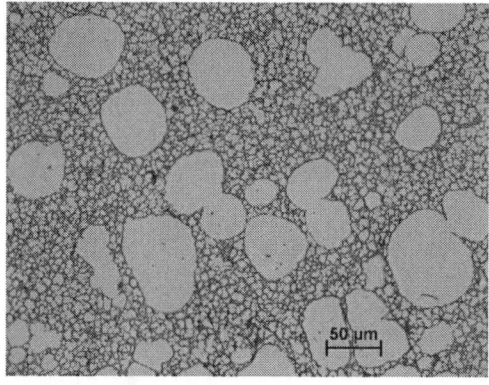

Fig. 5. Microstructure of the RDC 2014 wrought Al-alloy processed at 636°C for 60 seconds.

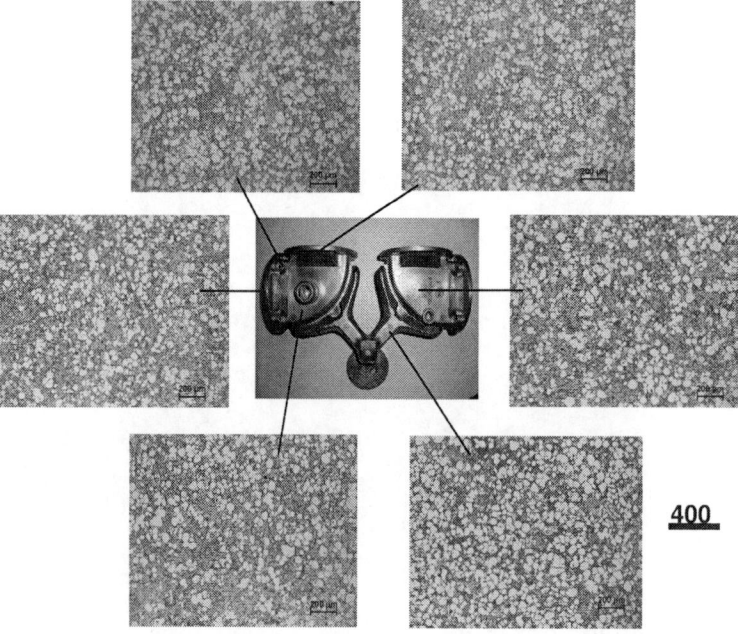

Fig. 6. Optical micrographs showing the microstructures at different locations of a RDC A380 casting

Industrial Application of Thixoforming at SAG

Josef Wöhrer and Andreas Kraly

Salzburger Aluminium AG in Lend (Austria)

Abstract

SAG is a producer of thixoforming raw material for worldwide use at different producers of Semi Solid Metalforming serial parts.

The SAG subsidiary Thixalloy® Components Company produces by itself complex designed Thixoforming parts in Alloy A356/A357 for different applications. The production of weldable structure parts and weldable pressure tight parts plays an important role as well as the production of surface treated parts for industrial use.

The customer benefits are tight dimensional tolerances, superior weldability with high process ability, good mechanical properties in the thixoformed condition with and without heat treatment, pressure tightness up to 80 bar after welding, good surface quality and good behaviour in plating.

Yield strengths of 200 MPa can be reached with a very cost efficient T5 heat treatment on TX 540 AlMg5Si2Mn or TX630 AlSi7Mg0.8. The so called Silicon Spheroidization Treatment (SST) is described which enables real parts to fulfil the high demands and to reach the values which are needed on suspension parts and wheels.

Additionally in this paper are shown parts which are produced with alloy type MAXXALLOY TX540 Al5Mg2Mn in Thixoforming. So Thixoforming works out as a more and more cost efficient process for Aluminum parts with high quality requirements.

Key words:

Thixalloy, Thixoforming, SST, Serial parts, Structure parts, Maxxalloy

Introduction

The increasing need of lighter and safer cars is the driving force behind research in the area of high strength light metal safety part applications. This also influences dramatically the technological changes to the automobil and has consequences for manufacturers, component suppliers and equipment manufacturers (Figure 1).

Since 1898 the Salzburger Aluminium AG (SAG) in Lend, Austria has been a competent partner in aluminum processes. SAG has been on the forefront when it comes to introducing new technologies in smelting, casting and treating aluminum. SAG prides itself in continuously building upon experience and innovation in both technological advances as well as human resources. The close cooperation with universities, research centers and partners constantly keeps the company abreast with the latest findings in development and know-how resulting in state-of-the-art applied technologies mirrored in the high standard and range of the products.

The plant has an annual capacity of 45,000 tons whereas 25,000 tons are solely based on recycling activities. Furthermore, SAG has a casting line, independend from regular production, equipped with own furnace and horizontal casting machine for training as well as research and development purposes. Process-ability and quality control are of highest importance to the company leading to the following process steps:

- Scrap material control by melting tests;
- Alloy determination;
- Melting in any one of the three hearth furnaces or the closed well furnace;
- Metal treatment and stirring;
- Metal quality control;
- Casting process on the horizontal casting machines with continuous metal treatment (degasser, filter box and rod feeder);
- Product quality control by analysis of material density, conductivity,
- Stacking and packaging including marking ready for transport.

The process leads to the high quality of the products such as foundry alloys, extrusion billets, thixotropic and cast forging billets/slugs, cast plates and bus bars. SAG has a recognized experience in casting special alloys, specifically using billets the standard diameters varying from 67 up to 203 mm, with developed technology to cast 50 or 32 mm diameter. The reasons for the excellent quality and variety of the products are based on the long-time experience of SAG and the advanced constructions and designs of casting moulds. Based on comparing investigations with other casting processes (e.g. vertical die casting) the horizontal casting technology gives SAG further advantages in their production. E.g. to a length of 18 m, enabling cutting billets to length during the casting process according to customer requirements. Other benefits include the well known economical aspects such as low energy consumption, low investment costs, lower necessity of consumables. The duration of the cast can last up to 120 hours casting time and more. The mould technology with high cooling rate and optimized lubrication assures a high productivity and quality. The mould design furthermore allows a higher number of strands on smaller equipment sizes. Process control automation and documentation, automatic starting programs are standards.

Starting Material – Billets – Slugs

The future development of lightweight construction (automotive and transport industry) depend especially on material properties of the used components. The claim of better material properties with simultaneous weight reduction gives rise to the development of technologically advanced casting and forming methods. Horizontal continuous casting (HCC) and thixoforming meet these requirements. SAG has 30 years of experience in HCC and more than 10 years with HCC materials. (One by SAG registered THIXALLOY® material). Starting material for the thixoforming process billets with a uniform and finely globulitic structure are produced by magnetically stirred HCC process according to the MHD technique (Magneto-Hydro-Dynamic Stirring), in diameters from 2.5 up to 6 inches.

Although alternative processes have recently appeared, they havenot yet made any decisive impact in the industrial context because the quality of the starting material is decisive for that of the components produced. In SAG's case the process is mainly used for weldable structure parts, surface treated parts, suspension parts and wheels in the automotive industries. In-line melt treatment and the continuous control of all the casting parameters ensure that the structure is the producible. Since no melting is involved in the thixoforming process, no metal purification and structure adjustment measures are needed there either.

The main characteristics of billets made by the company's MHD-process are:
- Extremely fine grained billets: max. 130 μm grain size;
- Good billet surface: max. class 2 (according to SAG's internal standard);
- High process ability and high metal purity during billet production;
- Non-metallic inclusions: max. 40 μm;
- Content of hydrogen: DI max. 5 %;
- Porosity: max. 100 μm;
- Peripherical zone: max. 7 mm.

A list of available diameters of the billets for industrial processing depends on the used thixoforming process and mainly on the locking forces of their machines. (See Table 1)

Table 1: *A list of available diameters of the billets for industrial processing depends on the used thixoforming process and mainly on the locking forces of their machines.*

Diameter	2.5 inch	3 inch	3.5 inch	4 inch	5 inch	6 inch
Metric in mm	63.5	76.2	88.9	101.6	28	152.4
Locking Force	300-700 tons	500-800 tons	700-100 tons	800-1200 tons	100-1600 tons	1600-2700 tons
Shot Weight	1.0-2.0 kg	1.5-3.0 kg	2.0-4.0 kg	3.5-7.0 kg	6.0-12.0 kg	10.0-20.0 kg

A list of available alloys mainly used for industrial thixo-processing with respecting reached mechanical properties in raw-parts can be found in Table 2.

Table 2: *A list of available alloys mainly used for industrial thixo-processing with respecting reached mechanical properties.*

Alloy	Alloy Type	Condition F:		
Thixalloy ® 615	AlSi7Mg0.15	Rp0.2: 100 Mpa	Rm: 180 Mpa	A: 15%
Thixalloy ® 630	AlSi7Mg0.30	Rp0.2: 120 Mpa	Rm: 230 Mpa	A: 12%
Thixalloy ® 640	AlSi7Mg0.40	Rp0.2: 130 Mpa	Rm: 240 Mpa	A: 10%
Thixalloy ® 650	AlSi7Mg0.55	Rp0.2: 140 Mpa	Rm: 250 Mpa	A: 8%
Thixalloy ® 680	AlSi7Mg0.80	Rp0.2: 150 Mpa	Rm: 260 Mpa	A: 5%

The continuously cast billets are sawn into slugs within close weight tolerances. The slugs are dispatched to the further processors in containers optimally adapted for their machine loading logistics.
The customer benefits with MHD stirred billets are mainly given by:

- A homogenous distribution of silicon and accompanied alloyed elements in the cross section of the billet;
- A low free liquid phase at the end of the heating cycle;
- Saving of induction heating time during the preheating of slugs in the followed thixoforming process;
- Reduced investment for induction furnaces;
- A low scrap rate; and at least above all;
- Better mechanical properties of parts;
- High and constant serial quality -> 1 batch = 10,000 slugs and constant properties through the MHD process.

Thixoforming Process

Between molten and solid state, alloys show thixo-tropic behavior. The homological basis of this behavior is due to the fact that with increasing shear forces, the solid-phase skeleton of the metallic structure can be completely dismantled (2) This is the most important precondition to achieve low-viscous flow, which finally allows the production of extremely complex shaped parts with very small forming forces. Under low shear forces, the material shows a solid-state behavior and therefore can be handled like a solid part.

SAG heats slugs while positioned horizontally in an inductive heating unit for individual billets. This avoids having to tilt them (as would be necessary in a carousel unit) once they have been softened. The power input is determined and controlled by a computer system. This enables the required homogeneous molten fraction to be produced in the slug with pinpoint accuracy, even without temperature measurements. The final adjustment of the structure is carried out during this heating process.

The α-phase is of globular shape, and the metal will only behave thixotropically if this is so. The benefits of induction billet heating and characteristics of the company's thixoforming process compared to other processes (e.g. die casting units) lies mainly in:

- No liquid metal process (therefore less hydrogen);
- No melt house, dross, fumes and losses in process environment;
- Horizontal billet heating and sure handling;
- Exact heat content billet by billet;
- Semi solid forming under low shear stress.

Why is Thixoforming a Practical and Economic Solution for the Industry?

Thixoforming is a metal forming process which utilises the thixotropic properties of certain alloys when they are heated into a semi-solid state. In this semi solid-solid state the alloys consist of a fluid phase on a globolitic solid phase. Modern casting processes take advantage of these special characteristics, they are called "Thixoforming" processes. Thixoforming is a new process for producing safety parts for the automotive industry. It permits the near net shape production of complexe shaped components combined with a high degree of productivity and reliability. Apart from these economical advantages, components produced by Thixoforming exhibit outstanding mechanical properties.

Thixoforming of aluminium alloys has come to play an increasing role as manufacturing route for complexe components of high quality.

Having regard to the properties described above, SAG's process experience shows that thixoforming can achieve optimum cost/utility characteristics and can considerably improve customer utility.

The essential advantages and customer benefits of Thixoforming are:

- Weight savings can be achieved through optimised component design.
- Good internal quality gives the best possible static and dynamic strength.
- Production-line components have excellent dimensional accuracy.
- Articles of complex shape with large wall thickness differences can be produced.
- Small wall thicknesses, in part down to 1.7 mm and averaging 3 mm can be produced, as well as combined in parts with very large wall thicknesses are also possible.
- Pressure-tight components, thanks to low porosity up to 80 bars after welding.
- Reliable weldability (laser and MIG welding) thanks to low gas content and a lowporosity structure which is based on inline degassing during the HCC process.
- Good surface quality for decorative or functional surface finishing & plating.
- Fine uniform microstructures give enhanced component properties.
- It is an energy efficient process which can be easily automated to achieve consistency.
- Production rates are similar to or better than die casting.

The following table shows examples of dimensional properties of thixoforming serial parts with tolerances for customer needs on length and diameter:

Table 3: *Dimension tolerances of thixoforming parts*
Radius down to R1, some areas with jump, in wall thickness up to higher radius.
Draft angle: 1° to 3° depending from die depth and location.
Gate system: wall thickness in gate zone: at least 3 mm thick.

Nominal Dimension (in mm)	30	60	120	240	480
Thixoforming tolerances for space-diagonal of 500 mm					
No Parting Line Included	± 0.07	± 0.12	± 0.18	± 0.25	± 0.30
Over Parting Line	± 0.12	± 0.15	± 0.25	± 0.33	± 0.45
Comparison to: HPDC-tolerances DIN 1688 GTA 13/5 (over 50 up to 500 mm)					
No Parting Line Included	± 0.2	± 0.3	± 0.35	± 0.45	± 0.60
Over Parting Line	35	± 0.45	± 0.50	± 0.60	± 0.75

SAG is a producer of thixoforming raw material for worldwide use at different producers of semi solid metal forming serial parts.

The SAG subsidiary THIXALLOY® Components Company produces weldable structure parts and weldable pressure tight parts. This plays an important role as well as the production of surface treated parts for Industrial use.

Thixoforming Parts in the New Audi A8 and the New AUDI A6 Car

The use of thixoforming technology and its advantages for lightweight structures parts is a main reason to use it for structural elements of doors (where they replace parts in sheet and castings) or other functional parts in cars.

The SAG group proves the competence of their processes by delivering more than 34 different parts for the new Audi A 8 car (Figure 1). Of all those parts, 13 are produced in the thixoforming process. SAG uses TX 630 alloy (AlSi7Mg0,3) which fulfils the high demand of the customer needs mainly in the field of mechanical properties, good weldability, low porosity and narrow dimensional tolerances. The need of lighter and safer cars is the driving force behind research in the area of high strength light metal safety part application. The new AUDI A6 will have two new thixoforming parts with high ductile alloy type MAXXALLOY TX540 for the air pressure vessel of the air suspension system (Figure 2).

The prerequisite for the above mentioned developments was the big change and challenge in the abilities of SAG's entire workforce and management and the installation of a highly automated production process.

Additionally in this Paper are shown development parts which are produced with alloy type MAXXALLOY TX540 Al5Mg2Mn in thixoforming. The parts fulfill the requirements with up to 15% elongation in wallthickness of 3 mm together with higher yield strength of 170 MPa which can be used for structure parts and additionally for suspension parts or motorcycle wheel production. The complete heat treatment can be saved by using alloy, test for mold-war are in progress.

Benefits with Thixoforming Parts by Billet

What we have described is a significant example of redefining and optimizing a production process at SAG according to new market needs and customer demands. The evolutionary development, from the production of the primary feed stock to supply producers, weight of light components and systems for the automotive industry was SAG's mission to increase the quality of parts.

Table 5: *Alloy: TX630 corresponds with A356; Alloy: TX650 corresponds with A357.*

Serial Part / Wall Thickness in mm	Alloy	State	Rp 0.2 (Mpa)	Rm (Mpa)	A5 %
Side door part	Thixalloy 630®	F	120	230	11-14
Side door part	Thixalloy 630®	T6	260	310	10-15
Side door part	Thixalloy 630®	T6x3	229	317	16-20
Rear door hinge	Thixalloy 650®	T6	270	330	8-12
Rear door hinge	Thixalloy 650®	T6x3	245	325	13-18
Steering wheel	Thixalloy 630®	T4x3	118	247	20-26

Conclusion

So thixoforming works out as a more and more cost efficient process for aluminum parts with high quality requirements.

Our understanding of quality for thixoforming parts is the summary of:

- High static and dynamic mechanical properties.
- Very low dimensional tolerances -> tailor made net shape quality.
- High complexity design and variability of wall thicknesses of parts is practicable.
- Wall thicknesses partial of 1.7 mm to a medium range of 3 mm is possible.
- Pore free quality in thick wall areas.
- Superior Laser, MIG, WIG e.g. weldability secured through low gas content in the raw material and low gas porosity in parts.
- Very long tool life in comparison to (Vacuum) - high pressure die casting.
- Competitive price through low process costs for complex thixoforming parts.

The additional use of aluminium in automobiles can be rapidly introduced to reduce energy consumption and pollution. In order to achive cost-effective solutions a close partnership between the car industry and the aluminium industry is necessary. The aluminium and automotive industry have a history of collaboration on developing the use of the most common light weight metal in cars. The aluminium content of cars is predicted to double within the next ten yours. Aluminium is the key to light weighting; aluminium is the preferred material of the 21st century.

Therefore thixoforming is ideal for automotive components; those products that demand a high level of complexity, weight savings and thight dimensional tolerances. This is SAG's chance!

References

[1] MERCER Management Consulting
„Automobile technology 2010, Technological changes to the automobile and their consequences for manufacturers, component suppliers and equipment manufacturers"

[2] Dipl. Ing. H. Luechinger, Dipl. Ing. B. Wendinger, SAG
„Thixoforming – der rationelle Weg zur Herstellung von Premium-Bauteilen aus Aluminiumwerkstoffen"
Giesserei-Rundschau 49 (2002)

Acknowledgements

The work was based thanks to a gradual change in the use of ability of partners within the internal and external group, the adoption of automated production processes, groups dedicated to innovation and to activities aimed at guaranteeing quality of SAG's products.

Josef Wöhrer
Chairman & CEO
Salzburger Aluminium AG (Lend / Austria)
josef.woehrer@sag.at

Andreas Kraly
Product and Technology, Development in Continuous Casting
SAG Aluminium Lend GmbH & Co KG (Lend / Austria)
andreas.kraly@sag.at

Figures

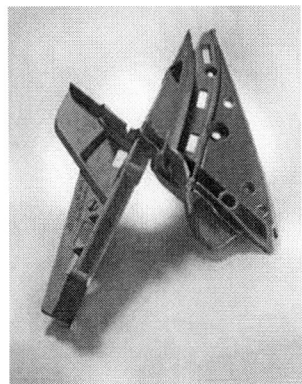

Figure 1:
A-pillar for Audi A8 front door

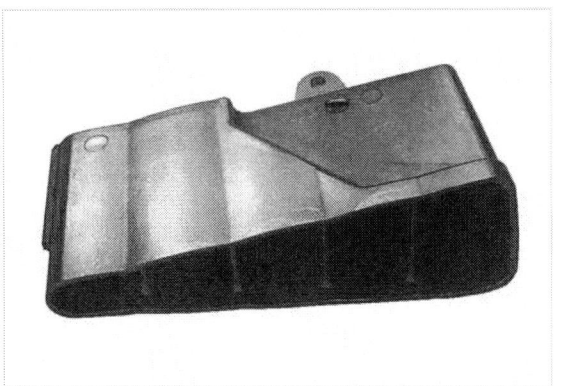

Figure 2:
Air Suspension for Audi C6 – Top of the air tank

Direct Chill Rheocasting (DCRC) and extrusion of AZ31 Mg-Alloy

S. M. Zhang*, Z. Zhen and Z. Fan

BCAST (Brunel Centre for Advanced Solidification Technology) Brunel University, UK,
*On leave from General Research Institute for Non-Ferrous Metals, Beijing, P.R.China

Email: zhongyun.fan@brunel.ac.uk

Abstract

The Direct Chill Rheocasting process (DCRC) and experimental results on microstructure and mechanical properties of DCRC billets are presented. The DCRC process uses a high quality semisolid slurry supply system, continuously feeding a conventional direct chill (DC) caster, to produce billets. Experimental results show that the DCRC billet microstructure is fine and uniform, the average grain size being around 50µm. Direct extrusion of the DCRC billets was conducted to assess and compare the deformability and mechanical properties after extrusion to conventional DC billets. As extruded, the average grain size is <3µm, and strength and ductility is much improved. The increased deformability has been attributed to the fine and uniform microstructure of the DCRC billets, while the improved mechanical properties are credited to the fine grain size of the extruded product.

Key words Magnesium, Casting, Extrusion, deformation, Microstructure

Introduction

As the lightest of all the structural metallic materials, magnesium (Mg) alloys offer excellent combinations of mechanical and physical properties, which make them attractive for applications in many industrial sectors. For instance, the application of Mg sheet products in a vehicle body can achieve 65% weight saving over steel and 20% over aluminium. There is a great expectation for future applications of wrought Mg products in vehicles [1]. However, the main barriers to the penetration of wrought Mg alloys into motor vehicles are poor deformability, low productivity and high cost [2]. Technologies for processing wrought Mg alloys are "copied" directly from the Al industry with little modification and have proven unsuitable [3]. This is due to the fundamental difference between the FCC Al crystal structure and the HCP Mg crystal structure. Therefore, it is essential to develop new processing technologies, which not only offer higher quality, but also provide competitive price in comparison with other engineering materials.

BCAST at Brunel University has recently developed a new semisolid processing technique, the rheo-diecasting (RDC) process [4,5]. The twin-screw slurry maker, the key technology of the RDC process, has been extended to process wrought Mg-alloys. The new development includes direct chill rheocasting (DCRC) process, rheoextrusion (RE) process and twin-roll rheocasting (TRRC) process, which have been collectively named rheoforming processes [6]. In this paper we present the DCRC process and the experimental results of DCRC AZ31 Mg-alloy. The DCRC billets were also extruded to assess their deformability.

Experimental

The DCRC equipment consists of three basic functional units: twin-screw slurry makers, a slurry accumulator and a standard DC caster (Fig 1). The twin-screw slurry maker has a pair of co-rotating, fully intermeshing and self-wiping screws. The screws have specially designed profiles to achieve high shear rate and high intensity turbulence [7]. The basic function of the slurry maker is to convert the molten Mg alloy into high quality semi-solid slurry, where fine and spherical solid particles with a predetermined volume fraction are uniformly dispersed in a liquid matrix. However, such slurry makers work in a batch manner. In order to convert the batch process into a continuous process, a slurry supply system has been designed, in which one or more slurry makers (2 shown in Fig. 1) are used to supply slurry to a slurry accumulator in a cyclic manner. A stirrer is used in the accumulator to (1) prevent the solid particles from agglomeration, (2) clean the inner surface and (3) force the SSM slurry to flow downwards. The slurry makers and the slurry accumulator form the slurry supply system, which provides semisolid slurry continuously to a DC caster (Fig. 1). To prevent oxidation, protective gas should be used in both the slurry makers and the accumulator.

A batch of 8.5kg commercial grade AZ31 ingots, supplied by MEL (Manchester, UK) was melted in a top-loading resistance furnace in a steel crucible, under the protection of N_2-0.5%SF_6. The furnace temperature was controlled at 700°C. The alloy compositions are given in Table 1.

To cast conventional DC billets and DCRC billets in this experimental work, a water-cooled copper mould was used as a DC casting simulator. A water pump was used to regulate the water pressure and to keep the cooling power consistent. During the DCRC experiment, molten AZ31 alloy was fed into the twin-screw slurry maker and processed at 500rpm screw rotation speed for 35 seconds. The semisolid slurry was then transferred to the water-cooled copper mould for the final solidification.

Extrusion of the ingot samples was conducted in an 800ton direct extrusion machine. The DCRC billets were heated at 350°C for an hour before extrusion. The container temperature was set at 350°C as well. Unfortunately, the extrusion machine used did not have a speed control, so extrusion speed could not be assessed in this work. The extrusion ratio used was 16:1 with only one pass for the DCRC billets. Extrusion of the gravity cast billets was also carried out under the same conditions to establish the advantages of DCRC billets. This includes the 2-pass extrusion of the gravity cast billets with the extrusion ratio for the second pass being 16:1.

For evaluation of mechanical properties, tensile test samples were machined from the extruded bars. The dimensions of the tensile test samples were 5mm in gauge diameter, 60mm in gauge length, and 75mm in total length. The tensile tests were conducted at room temperature on an Instron tensile test machine with a strain rate of 10^{-3}/s. For each extrusion condition, at least 5 samples were tested, and the mechanical properties were average values of all the samples tested.

Samples were cut from the cast billets and extruded bars, to prepare for microstructural examination. The microstructure of the samples was examined by optical microscopy (OM) with quantitative metallography, scanning electron microscopy (SEM) and X-ray diffractometry (XRD).

Results
The DCRC billet showed a finer and more uniform cross section compared with that of the DC billet (Fig 2). The DCRC billet was not only structurally uniform, but also chemically uniform, as shown by the variation of the chemical compositions of the major alloying elements along the radial direction in Fig 3. Within the experimental uncertainty, the concentration of Al, Zn and Mn had no change after the DCRC process (see Table 1). This is in contrast to the usual chemical segregation of the alloying elements in the centre of the DC billets.

The conventional DC AZ31 alloy microstructure showed an extremely non-uniform cast dendritic structure. In contrast, the DCRC process offered a much finer and more uniform globular microstructure through the cross-section (Fig. 4). Fig. 5 compares the solidification cell size in the conventional DC billets with the particle size of the primary globules in the DCRC billets. In the DC billet, the solidification cell size increased from 120µm near the surface to 230 µm at the centre of the billet, while in the DCRC billet, the globule size was fairly constant, being around 50 µm. Fig. 5 indicates that in comparison with the DC process, DCRC process could achieve grain refinement with a factor of 5.

In order to check the deformability of the DCRC billets, both DCRC and gravity die cast billets were extruded using a direct extrusion process. The extrusion conditions are listed in Table 2. A course-grained structure was observed in the extruded microstructure of the gravity die cast billet. Deformation twins were present in nearly all the grains. In contrast, the microstructure of the extruded DCRC alloy was much finer and more uniform. The average grain size on the transverse section was measured to be 3µm (Fig 6). Large grains corresponding to the primary particles in the DCRC samples were substantially elongated along the extrusion direction. Deformation twins were only observed in such elongated grains.

After Heat Treatment at 350°C for 3 hours some of the elongated grains in the extruded DCRC samples were not completely recrystallised. However, after 3 hours at 400°C the recrystallisation process was completed. The fine equiaxed grains had grown from 3 to 5µm after heat treatment, and the recrystallised grains were finer than 1µm.

In the as-extruded conditions, the basal plane (0002) was absent from the X-ray trace along the transverse direction, while the basal plane diffraction is very strong along the longitudinal direction, indicating a strong texture in the as extruded sample. In addition, XRD revealed that this strong texture remained unchanged after heat treatment at 350°C for 3 hours, only a small peak from the basal plane diffraction appeared after heat treatment at 400°C for 3 hours. The return of the small basal plane peak may be attributed to the recrystallisation of the large and heavily twinned grains in the as extruded condition.

Tensile properties of AZ31 alloy extruded under different conditions are given in Table 2. Compared with the samples extruded from the gravity diecast billets, extruded DCRC samples had much improved mechanical strength and ductility, with the ultimate tensile strength (UTS) increase being 24% and the ductility increase being 44%.

To investigate the origin of the improved mechanical properties of the extruded DCRC samples, gravity cast billet was extruded from 93mm diameter to 50mm diameter in the first pass, followed by extrusion from 48mm diameter to 12mm diameter in the second pass. Table 2 indicates that it was impossible to achieve the mechanical properties of extruded DCRC samples by 2-pass extrusion of the gravity cast billet. Therefore, it can be concluded that the DCRC billets have superior deformability over the conventional billets produced by either DC casting or gravity casting.

The results indicate that the yield strength follows the well-known Hall-Petch Relation, so elongation increases with the decrease in grain size. Consequently, it can be concluded that the improved strength and ductility of the extruded DCRC samples can be attributed to the refined grain size.

Discussion

In the conventional casting processes, heterogeneous nucleation takes place in the undercooled liquid, close to the mould wall. The majority of the nuclei are transferred to the overheated liquid region and re-melted; only a small proportion (as low as 0.3% [8]) of the nuclei can survive and contribute to the final microstructure. In order to achieve finer grain size, conventional wisdom is to add a grain refiner to enhance heterogeneous nucleation [9], (a chemical approach to grain refinement).

However, the solidification process in the DCRC process is completely different. The fluid flow inside the twin-screw slurry maker is highly turbulent, even if the shear rate is low (low screw rotation speed). Consequently, solidification in the DCRC process can be divided into two distinctive stages; primary solidification, inside the slurry maker under high shear rate and high intensity of turbulence to create semisolid slurry, and secondary solidification, in the DC caster with high cooling rate but without forced convection.

Primary solidification occurs in a melt with extremely uniform temperature, uniform chemical composition, and well-dispersed nucleation agent, due to the dispersive mixing power of the twin-screw mechanism [7]. Nucleation occurs continuously throughout the entire volume of the liquid during the continuous cooling to the semisolid temperature, all the nuclei can survive and grow very rapidly in a spherical manner. During the subsequent isothermal shearing stage, particles coarsen through Ostwald ripening, decreasing the particle density with time. However, the coarsening rate is extremely slow due to the spherical particle morphology and narrow particle size distribution [10]. Such growth phenomena are in good agreement with the theoretical results from our analytical modelling [11] and Monte Carlo simulation [12]. This effective nucleation and spherical growth, produces semisolid slurry, in which fine and spherical particles are uniformly dispersed in a liquid matrix.

During secondary solidification, the semisolid slurry has previously been sheared in the twin-screw slurry maker, having a uniform temperature and uniform chemistry. Once the copper mould wall triggers nucleation, it will take place throughout the entire volume of the remaining liquid, and all the nuclei will survive resulting in an extremely high effective nucleation rate.

Our recent experimental results showed that increasing the intensity of forced convection increases the nucleation rate during the secondary solidification [10]. However, without forced convection these nuclei grow dendritically. The observed globular particles from secondary solidification can be attributed to the fact that solidification completes before or shortly after the occurrence of interfacial instability. Particles simply do not have time to develop a fully dendritic morphology. Consequently, secondary solidification produces a fine and fairly uniform structure across the entire cross section of the DCRC billets.

As a result, the DCRC process produces a fine and uniform microstructure, the grain size being as fine as 40-50μm. The grain size is 1 or 2 orders of magnitude finer than the conventional DC billet; such an effect is denoted as a physical grain refinement. Different from the chemical approach, physical grain refinement offers no chemical contamination, consistent results and more uniform final microstructure.

In large-grained Mg alloys, the dominant deformation mechanism at room temperature is $\{0002\}<11\overline{2}0>$ basal slip, and will be assisted to some extent by secondary slip (prismatic and pyramidal) and at elevated temperature $\{10\overline{1}2\}$ tension twinning and $\{10\overline{1}1\}$ compression twinning [13]. This deformation mechanism has been confirmed by the presence of a great amount of twins in the large grains after extrusion. However, recent research has revealed that grain refinement improves substantially the deformability of Mg-alloys by introducing new deformation mechanisms [14]. Enhanced deformability through grain refinement can be achieved through the DCRC process. The average grain size in the DCRC billet is around 50μm (Fig. 4). In addition to the fine grain size, uniform microstructure and spherical particle morphology would promote dislocation cross slip to non-basal planes and enhance grain boundary sliding, consequently improving deformability of the DCRC billets.

Conclusions
In this paper we have presented a new processing technology, the direct chill rheocasting (DCRC) process. Basically, the DCRC process is a combination of a semisolid slurry supply system and a conventional DC caster. The slurry supply system produces high quality semisolid slurry under intensive forced convection, provided by the twin-screw mechanism, and feeds the DC caster continuously to produce billets or slabs. Experimental results show that the DCRC billets have a fine and uniform microstructure through the cross section, with the average grain size being around 50μm, which is much finer than that offered by either gravity casting or conventional DC cast billets. Direct extrusion of both the DCRC and DC billets was conducted at 350ºC to assess the deformability and mechanical properties after extrusion. The average grain size is less than 3μm in the as extruded state. Compared to conventional DC billets, the DCRC billets have much improved strength and ductility after extrusion. The increased deformability of the DCRC billets has been attributed to the

fine grain size in the DCRC billet, while the improved mechanical properties of the extruded product is attributed to the fine grain size in the extruded materials.

References

1. H. Friedrich and S. Schumann: in Proc. IMA 2001 Magnesium Conf., Brussels, Belgium, 2001, p8.
2. S. Schumann and H. Friedrich: Mater. Sci. Forum, 419-422 (2003), p51
3. S. R. Agnew: J. of Metals, May 2004, p20.
4. Z. Fan, M. J. Bevis and S. Ji: PCT Patent, WO 01/21343 A1, 1999.
5. Z. Fan, S. Ji and G. Liu: Mater. Sci. Forum, 488-489 (2005), 405-412.
6. Z. Fan, S. Ji and M. J. Bevis: PCT Patent, WO 02/13993 A1, 2000.
7. Z. Fan, G. Liu and Y. Wang: J. Mater. Sci., in press, 2005.
8. P. D. Lee, Private communication, Imperial College, London, 2003.
9. M. Easton and D. StJohn: Met. Mater. Trans. 30A (1999), p1613.
10. Z. Fan and G. Liu: Acta Mater., 53 (2005), p4345.
11. R. S. Qin and Z. Fan: 6th Inter Conf on semi-solid processing of alloys and composites, Turin, Italy, 27-29 Sept 2000, pp819-824.
12. A.Das, S. Ji and Z. Fan: Acta Mater., 50 (2002), p4571.
13. S. R. Agnew, M. H. Yoo and C. N. Tome, Acta Mater., 49 (2001), p4277.
14. T. Mukai: Mater. Trans., 42 (2001), p1177.

Acknowledgements

The authors would like to thank Dr S. Ji and Dr Y. Wang in BCAST at Brunel University for their help with the experiments, Dr Ma Qian for useful discussions. The financial support from both EPSRC and MEL is also gratefully acknowledged.

Tables

Table 1. Chemical composition of the AZ31 alloy used in this work

Element	Al	Zn	Mn	Be	Cu	Fe	Ni	Si	Mg
Wt %	3.411	0.698	0.318	.0007	.033	.004	.002	.053	Bal

Table 2. Mechanical properties of the AZ31 alloy produced by direct extrusion, extruded at 350°C from Φ48→Φ12 and heat treated as listed.

Processing conditions	Yield strength (MPa)	UTS (MPa)	Elongation (%)
DCRC Billet	264	309	19.5
DCRC Billet, 350°C for 3hrs.	231(12)	295($\pm$7)	20.6($\pm$2.8)
DCRC Billet 400°C for 3hrs.	197($\pm$5)	275($\pm$3)	18.6($\pm$1.5)
Gravity cast Billet	152	250	13.5
Gravity cast Billet, two-pass Φ93→ Φ50, Φ48→Φ12.	172	291	15.5

Figures

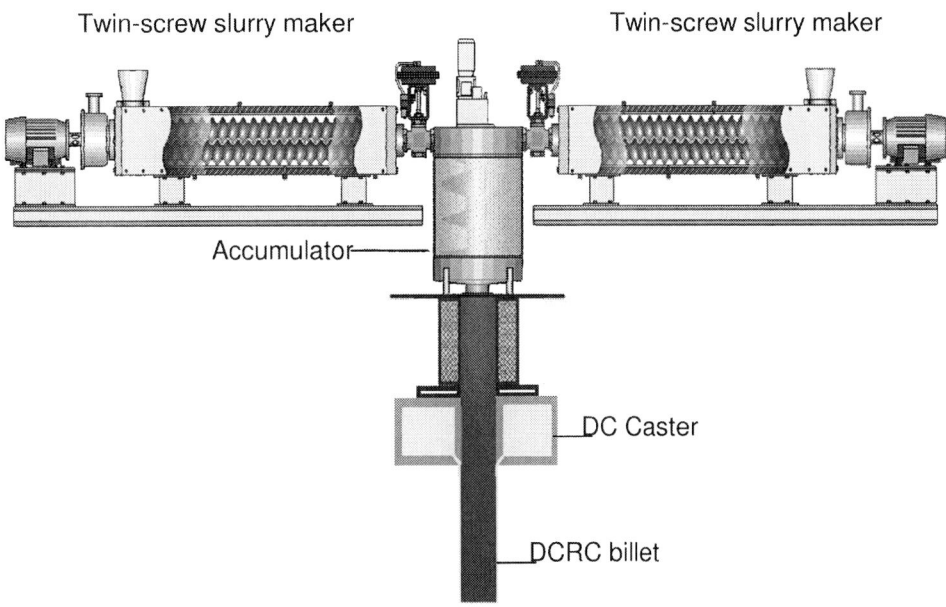

Fig. 1. Schematic illustration of the DC rheocasting (DCRC) process

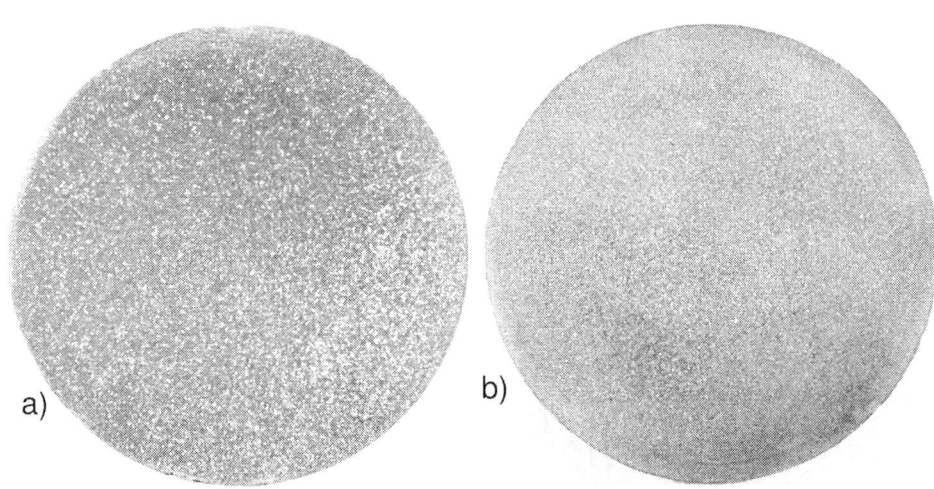

Fig. 2. Macrostructures from the cross section of 80mm billets produced by (a) DC process and (b) DCRC process

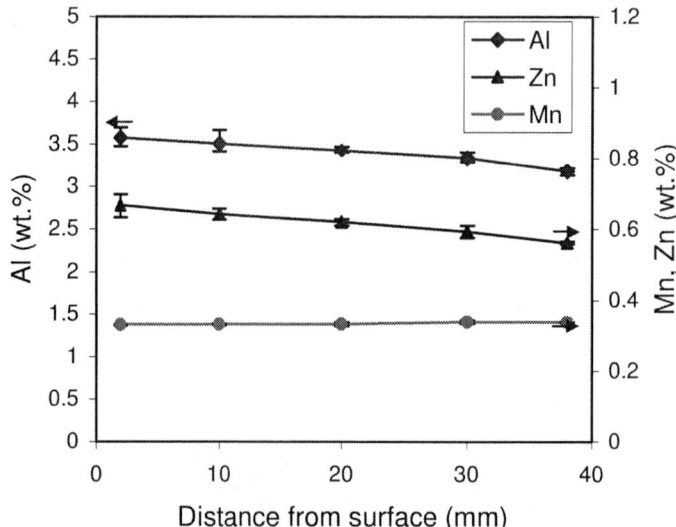

Fig.3. Chemical compositions through the cross section of 80mm billets of AZ31 alloy produced by the DCRC process.

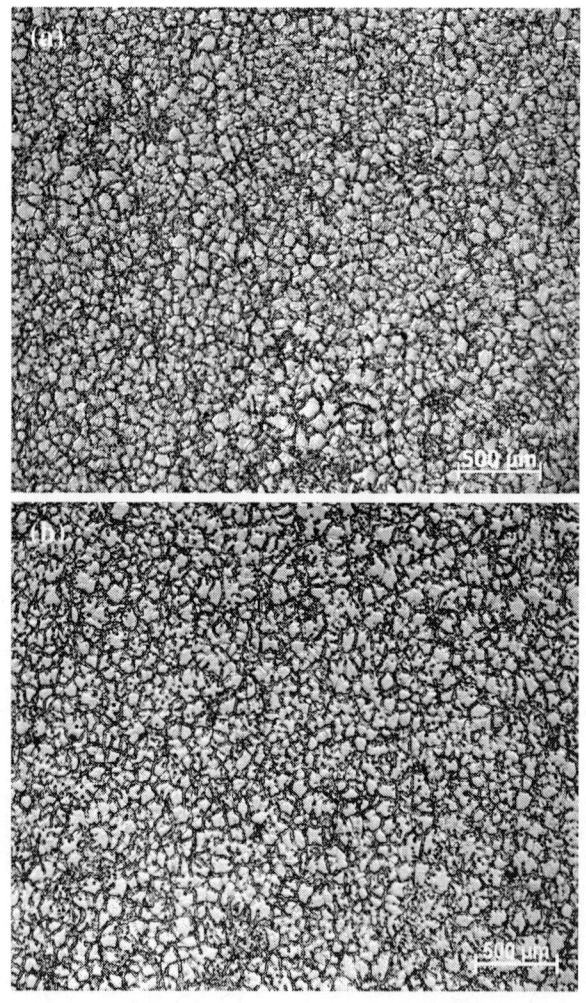

Fig. 4. Microstructure of DCRC AZ31 alloy billet (Ø80mm) (a) Near the surface (b) At the centre processed at 631°C.

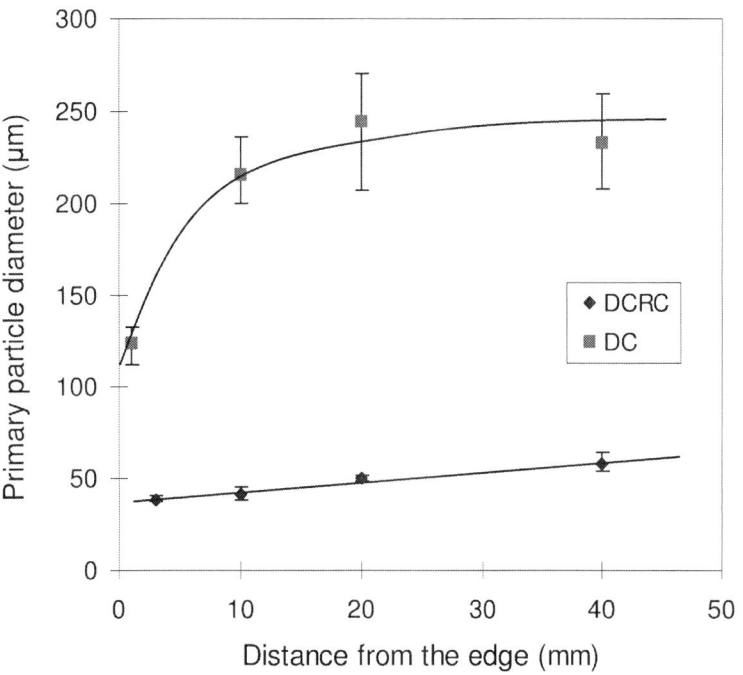

Fig. 5. Solidification cell size and primary particle size of AZ31 alloy billet (Ø80mm) produced by DC casting and DCRC process as a function of distance from the surface along the radial direction.

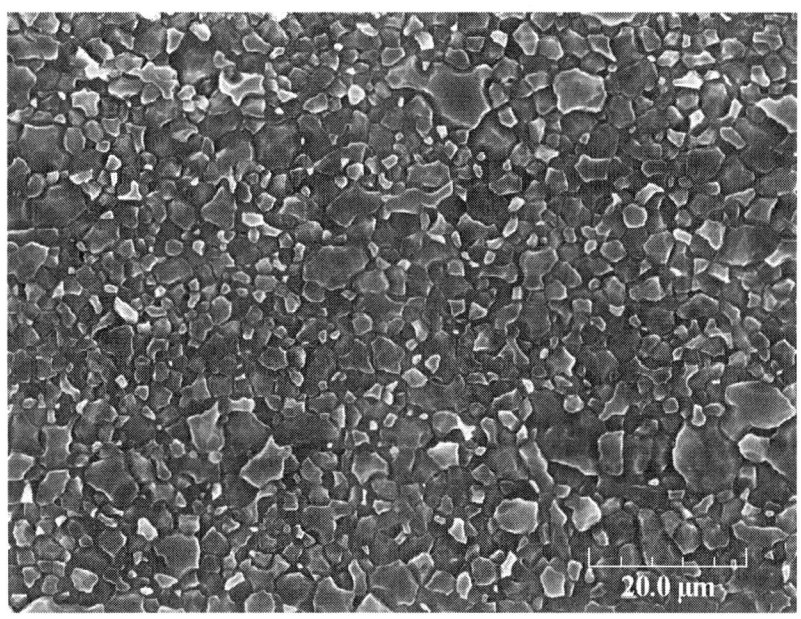

Fig. 6. SEM micrograph showing the microstructure of transverse section AZ31 alloy extruded from the DCRC billet. The extrusion ration was 16:1.

Microstructure Evolution in Al-7Si-0.3Mg Alloy During Partial Melting and Solidification from Melt: A Comparison

Shankargoud Nyamannavar*, M. Ravi and K. Narayan Prabhu*

Regional Research Laboratory, Trivandrum-695019, India, *National Institute of Technology Karnataka-Surathkal, P. O. Srinivasnagar-575025 India.

Abstract

In the present work, a comparative study of microstructure evolution in Sr modified Al-7Si-0.3Mg alloy isothermally held at semi-solid state under conditions of (i) cooling from liquid state and (ii) partial melting from solid state to the semi-solid temperature was carried out. The effect of cooling rate (0.01 to100 K/s) on the microstructure during solidification of semi-solid alloy is studied. Partial melting of alloy results in the fine and more spherical solid phase compared to cooling of the same alloy from liquid state. Chemical modification of the eutectic silicon by Sr addition was found to remain same for both cooling the melt from liquid as well as partial melting from solid state, contrary to the reported results. The morphology of eutectic silicon corresponding to the liquid entrapped in solid phase is finer compared to that in interconnected liquid channel.

Key words SSM, partial melting, modification, a-particles, and instability.

Introduction

In semi solid metal (SSM) processing the alloy is processed in a temperature range where it is partly liquid and partly solid. The process makes use of semi-solid slurry containing spherical solid phase particles (a-particles) dispersed in the liquid matrix. The desired semi-solid slurry is obtained either by controlled solidification of the melt as in the case of Rheocasting/Rheoforming or by partial melting and isothermal holding of preprocessed solid as in the case of Thixocasting/Thixoforming. Semi-solid metal processing of aluminium-silicon alloys has process advantages of casting and product advantages of wrought forming [1-5].

In Thixocasting/Thixoforming processes, preprocessed alloy billet is heated to semi-solid temperature and held isothermally, to get desired solid fraction and microstructure, and then die-cast/forged. Hence partial melting prior to forming is an important step in SSM processing. During isothermal holding the system reduces its surface energy by changing the morphology of solid phase from dendrite to spherical one. A schematic representation of various mechanisms of morphological change, in the solid phase during isothermal holding of semi-solid alloy, is shown in the Figure 1 [6].

Al-Si alloys A356 (Al-7Si-0.3Mg) & A357 (Al-7Si-0.6Mg) are being used widely in SSM processing for general as well as automotive applications [7-10].The coarse acicular needle shaped eutectic silicon of Al-Si alloys promotes crack initiation and propagation thereby degrading the mechanical properties. The properties of these hypoeutectic Al-Si alloys can be improved by inducing structural modification in the eutectic silicon. Addition of trace elements like, sodium, strontium, antimony, rare earths etc. leads to fine silicon morphology in the eutectic phase. Strontium is widely used as modifier for hypoeutectic Al-Si alloys due to its semi-permanent modification effect, ease of handling and non-toxic nature. Melt holding of the modified alloy results in fading i.e level of Sr present in melt decreases with time. Time rquired to fade to one half of the orginal value is normally of the order of few (10-14) hours [11]. Stucky et al. [12] have reported that partial melting and air cooling to room temperature of Sr modified A356 alloy nullified the effect of chemical modification. However a fully modified eutectic microstructure was retained when the same alloy was completely melted and solidified by air cooling. The reason for the absence of modification in the partially melted alloy was not stated. However modification of the eutectic silicon by Sr is a chemical phenomenon resulting from impurity induced twinning [13]. Hence the morphology of Si should be same for both solidification from the melt and partially remelted conditions.

In the present work, a comparative study of microstructure evolution in Sr modified Al-7Si-0.3Mg alloy isothermally held at semi-solid state under conditions of (i) partial melting from solid state and (ii) cooling from liquid state to the semi-solid temperature was carried out. The effect of cooling rate on the microstructure during cooling from semi-solid state was studied.

Experimental

Sr (0.02%) modified Al-7Si0.3Mg alloy was prepared from the commercial A356 alloy by the melt addition of Al-10Sr master alloy. The chemical composition of the alloy is given in Table 1. Modification effect was ascertained by microstructure examination (Figure 2).

The gravity die cast cylindrical samples with diameter of 10mm and height 10 mm were used in the experiments. These samples were coated with ceramic slurry (Aron Ceramic from Togosei Co., Ltd) which forms a shell after drying. The shell acts as a container for the liquid metal during experiment. After inserting the Chromel-Alumel thermocouple in the sample it is connected to a Keithley data acquisition system interfaced to a computer. The sample was then placed in a vertical gradient furnace and heated to semisolid temperature of 590°C, held isothermally at this temperature and then quenched or cooled at different cooling rates. Cooling of the samples at different cooling rates was achieved by quenching in water, blowing air, air cooling and cooling in furnace.

Samples were held at 590°C (37% solid fraction) for 8000 s and cooled at different cooling rates varying from 0.01 K/s to 100 K/s. Samples were brought to the semi-solid temperature by two different methods:
1. Heating of sample to 680°C (complete melting) and then cooling to 590°C.
2. Partially melting of the sample by controlled heating to 590°C from room temperature i.e. from solid state.

Microstructures of samples obtained from both the conditions of partial melting and solidification from melt were then compared. The specimens were etched with 0.5% HF solution. A Leica DMRX microscope and a Clemax image analysis system was used for microstructure examination and analysis.

Results

Microstructure of unmodified and 0.02% Sr modified Al-7Si-0.3Mg alloy cast in metal mold is shown in Figure 2. Microstructure of the modified alloy with fine eutectic silicon compared to unmodified alloy reveals the modification effect by Sr addition. Microstructure of the samples held at 590°C for 8000 s and quenched in water, for both conditions of cooling from melt and partial melting from solid state, are shown in Figure 3. The white and the dark regions correspond to the solid and liquid phases in the alloy before quenching. Isothermal holding at the semisolid temperature resulted in the formation non-dendritic microstructure. The dendritic solid phase reduces its surface energy, during isothermal holding at semi-solid temperature, by changing to a spherical shape by coarsening and ripening process. When such an alloy is quenched to room temperature, to freeze the high temperature structure, instabilities form on the a-particles and fine primary dendrites nucleate in the liquid region between a-particles [14, 15]. Figure 4 shows microstructure of the semi-solid alloy cooled at various cooling rates for both the conditions of partial melting and solidification from the melt. The variation of eutectic silicon particle size with cooling rate for the eutectic phase of the interconnected liquid region is shown in Figure 5.

Discussion

Primary aluminum phase, in partially melted sample, shows more spherical and finer morphology compared to the sample solidified from melt. The difference in the morphology of the solid phase in two conditions can be attributed to the difference in the initial microstructure i.e one existing prior to the isothermal holding. In the case of solidification from liquid state, as the temperature falls bellow the liquidus temperature of 613°C, solid phase starts forming in a dendritic morphology. Due to slow cooling rate adopted during cooling in furnace, coarse dendritic solid phase is obtained. During isothermal holding this coarse dendrite solid phase undergoes morphological change to form non-dendritic particle. Thus microstructure shows coarse a-particles dispersed in liquid phase. In the case of partial melting process, samples originally having fine microstructure (from die-casting) are partially melted from solid state to semi-solid state. During partial melting, as temperature increases above the solidus temperature, melting of eutectic phase takes place. With further increase in temperature, part of the primary aluminium phase melts to attain the equilibrium composition of solid and liquid phases as dictated by the phase diagram. Thus after partial melting to semi-solid temperature i.e. 590°C, the alloy will contain a mixture of fine dendritic solid phase dispersed in the liquid phase. During isothermal holding, this fine dendritic solid phase changes into globuletic solid. Thus fine dendritic structure of samples obtained in gravity die casting is retained after partial melting and therefore the partial melted samples show finer a-particles (Figure 3(b)) compared to the sample solidified from melt (Figure 3(a)).

To investigate the effect of cooling rate on the microstructure (particularly the eutectic silicon) these samples were cooled at different cooling rates for both partial melted and the solidified from melt conditions. Eutectic silicon morphology at different cooling rates for partial melting and the solidification from melt conditions is observed to be the same (Figure 4) for a given cooling rate in the range of cooling rates investigated. The average size of eutectic silicon particles, in the region of interconnected liquid, decreases with increase in the cooling rate (Figure 5). There is no significant difference in the eutectic silicon particle size for condition of partial melting and solidification from melt at all the cooling rates studied. Thus the degree of modification of eutectic silicon by Sr remains same, whether the alloy is solidified from liquid state or from partial melted state. In both the cases the modification of eutectic Si is retained in contrast to the reported observation by Stucky et al. [12] that the modification effect is lost during partial melting. Even though there is significant difference in the morphology of a-particle for both the conditions of partial melting and solidification from the melt, the eutectic volume is in liquid state signifying the same initial condition hence the result about the modification effect are conclusive.

In the microstructure of partially melted sample large amount of liquid entrapped in the solid phase was observed. It is observed that at lower cooling rates the eutectic exhibits very different morphologies. Eutectic phase corresponding to entrapped liquid shows fine silicon and the eutectic corresponding to interconnected bulk liquid shows coarse lamellar silicon. However this difference was not observed in water quenched (100 K/s) semi-solid alloy. This difference indicates different solidification conditions during nucleation and growth of eutectic in two locations. As reported by Valer *et al.* [16] for Al-21.8Ge alloy eutectic corresponding to the entrapped liquid probably forms under large undercooling by homogeneous nucleation. However homogeneous nucleation should result in the coarser eutectic morphology. Probable reason for the fine structure is the quench modification of the entrapped eutectic liquid by the surrounding solid. The surrounding solid acts like a heat sink. The interface between an entrapped liquid and the surrounding solid is convex and the heat flow from the entrapped liquid to the solid is divergent. However in the case of interconnected liquid channel the solid phase is surrounded by the solidifying eutectic liquid and the situation is similar to a liquid metal solidifying around a core. The interface between the interconnected liquid and the solid phase is concave and the heat flow is convergent. Thus the entrapped liquid experiences higher cooling rates compared to the interconnected liquid channel even when the semi-solid alloy is subjected to furnace cooling (0.01 K/s).

Conclusions

1. Partial melting of the alloy results in fine and more spherical solid phase compared to cooling of the same alloy from liquid state and this is influenced by the initial microstructure of the alloy.
2. Chemical modification of the eutectic silicon by Sr remains the same for both cooling the melt from liquid as well as partial melting from solid state.
3. The morphology of eutectic silicon corresponding to the liquid entrapped in solid phase is finer compared to that in the interconnected liquid channel. The effect is significant at lower cooling rates during solidification of the semi-solid alloy.

References

1.   M. C. Flemings, R. G. Riek and K. P. Young, *Rheocasting*, Material Science and Engineering, 25, 1976, pp103-107.
2.   M. C. Flemings, *Behavior of metal alloys in the semi solid state*, Met. Trans. 22A, 1991, pp957-981.
3.   D. H. Kirkwood, *Semisolid metal processing*, Int. Mat. Rev. 39, 1994, pp173-189.

4. Z. Fan, *Semisolid metal processing*, Int. Mat. Rev. 47, 2002, pp49-85.

5. W. R. Loue, M. Suery, *Microstructural evolution during partial remelting of Al-7Si-Mg alloys,* Material Science and Engg. A203, 1995, pp1-13

6. Andreas Mortensen, *On the influence of coarsening on micro-segregation*, Met. Trans. 20A, 1989, 247-253.

7. P. R. G. Anderson, J. C Summerill and A.R.A. McLelland, *The view of potential users*, Proceedings of 4th International Conference on Semi Solid Processing of Alloys and Composites, The University of Sheffield, England, 9-21 June 1996, pp.208-214.

8. G. Chiarmetta, *Thixoforming of automobile components*, Proceedings of 4th International Conference on Semi Solid Processing of Alloys and Composites, The University of Sheffield, England, 9-21 June 1996, pp204-207.

9. R. Moschini, *Mass production of fuel rails by pressure die casting in the semi-solid state,* Proceedings of 4th International Conference on Semi Solid Processing of Alloys and Composites, The University of Sheffield, England, 9-21 June 1996, pp248-250.

10. T. Basner, *Rheocasting of semi-solid A357 Aluminum,* SAE 2000 World Congress Detroit, Michigan, 6-9 March, 2000

11. John E. Gruzleski, Bernard M. Closset, *The treatment of liquid aluminum-silicon alloys*, American Foundrymen's Society, Inc. des Plaines, Illinois. USA 1990 ISBN 0-87433-121-8.

12. M. Stucky, M. Richard, L. Salvo and M. Suery, *Influence of electromagnetic stirring, partial remelting and thixoforming on Mechanical properties of A356 alloys*, Proceedings of 5th Int. conference on semi solid processing of alloys and composites. Colorado, 23-25 June 1998, pp.513-520.

13. Lu, Shu-Zu and Hallawell, *The mechanism of silicon modification in aluminium-silicon alloy: Impurity Induced Twinning*, Met. Trans. 18A. 1987, pp.1721-23.

14. M. Rettenmayr, O. Pompe *Interface instabilities on solidifying globuletic particles*, Journal of Crystal growth, 173, 1997, pp182-188.

15. O. Pompe, M. Rettenmayr, *Microstructural changes during quenching,* Journal of Crystal growth, 192, 1998, pp300-306.

16. J. Valer, F. Sant-Antoinin, P. Meneses and Suery, *Influence of processing on microstructure and semi-solid Behavior of Al-Ge alloys*, Proceedings of 5th International Conference on Semi Solid Processing of Alloys and Composites. Colorado, 23-25 June 1998, pp.513-520.

Acknowledgements

Authors acknowledge the technical assistance provided by Mr. S. G. K. Pillai, Mr. Antony, Mr. S. Ramakrishnan, Mr. Sreekantan, Mr. Sisupalan and Miss. Sindhu of the Regional Research Laboratory-Trivandrum, India towards this study.

Tables

Table 1: Chemical composition of the alloy (by weight pct.).

%Si	%Mg	%Cu	%Fe	%Ti	%Al
6.61	0.391	0.0102	0.1790	0.0023	Rest

Figures

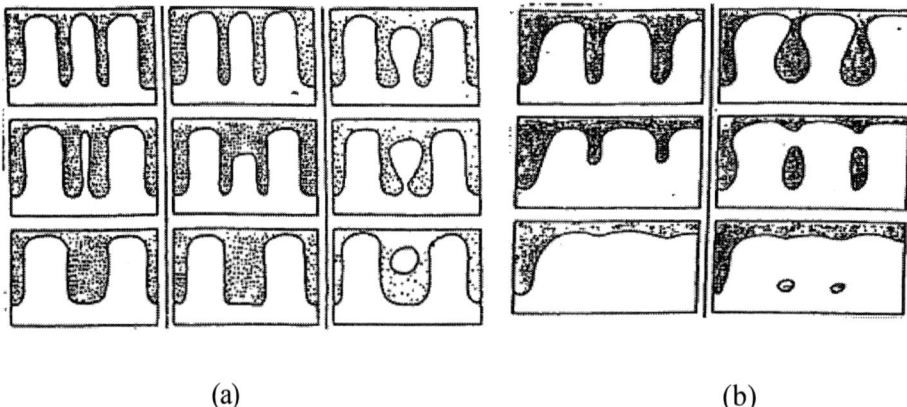

        (a)   (b)

Figure 1: Mechanisms of morphology change in the solid phase during isothermal holding of semi-solid alloy. [6]
a) Coarsening mechanisms (b) Coalescence mechanisms

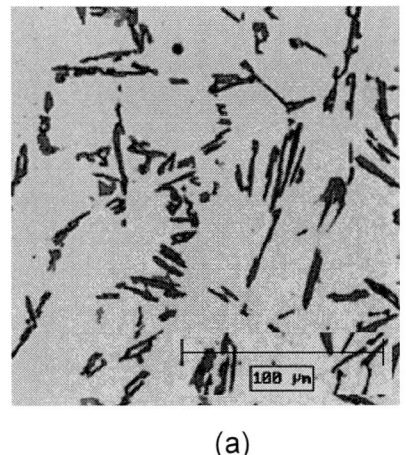 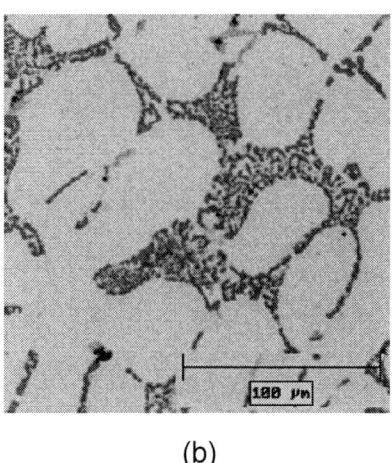

(a)   (b)

Figure 2: Microstructure of Al-7Si-0.3Mg alloy
a) Unmodified (b) 0.02% Sr modified

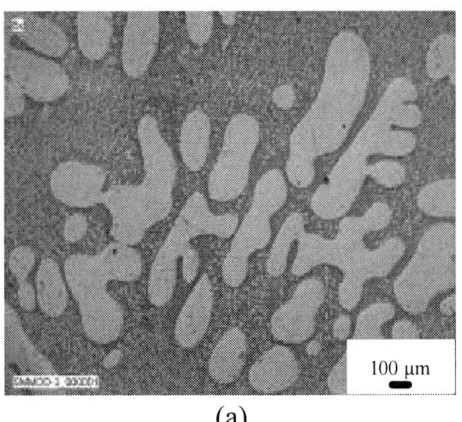 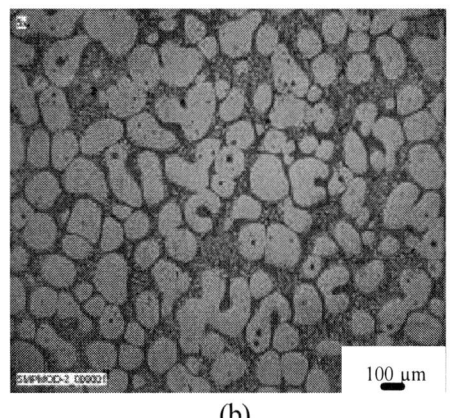

(a)   (b)

Figure 3: Microstructures of 0.02% Sr modified Al-7Si-0.3Mg alloy
isothermally held at 590°C for 8000 s and quenched in water
a) Cooled from liquid b) Partial melted

Cooled from liquid　　　　　Partial melted

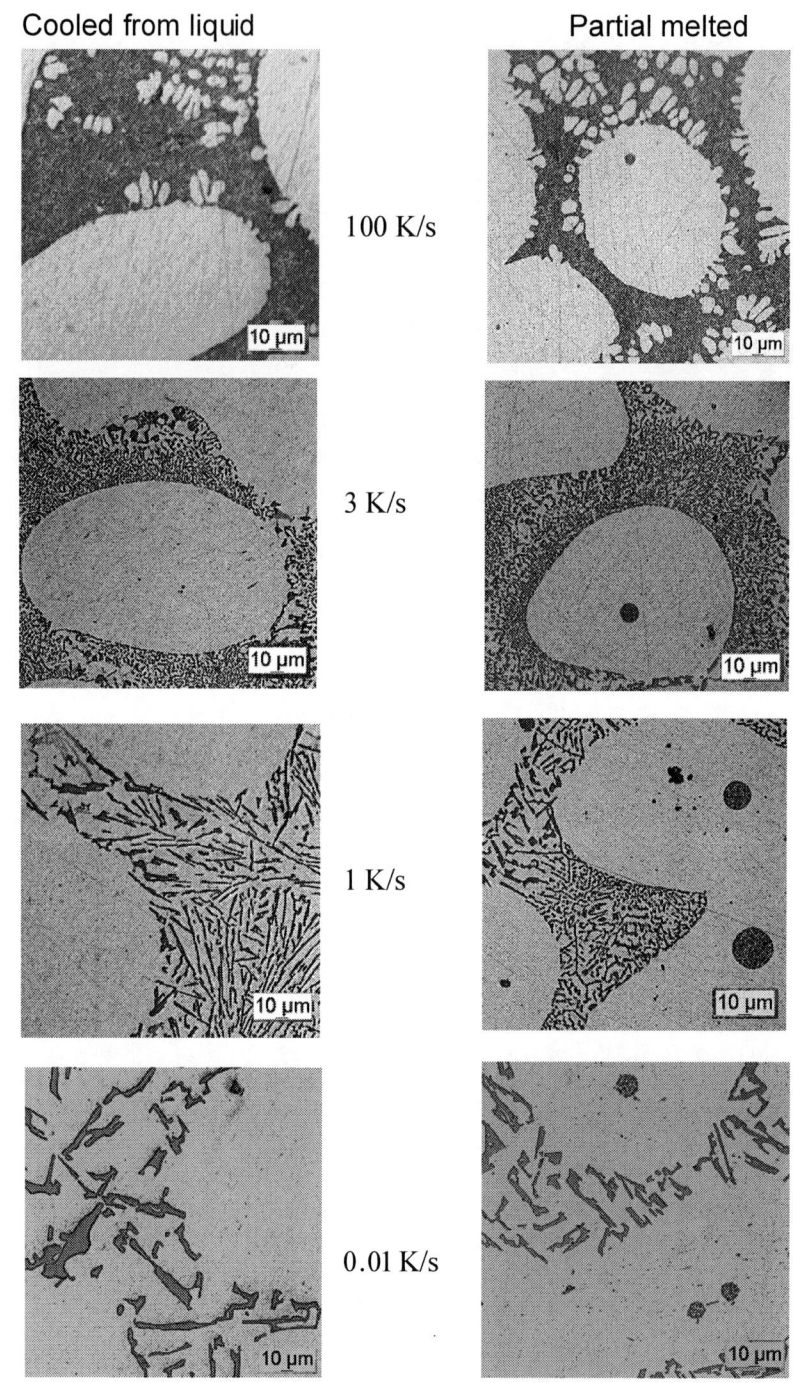

100 K/s

3 K/s

1 K/s

0.01 K/s

Figure 4: Microstructures of 0.02%Sr modified Al-7Si-0.3Mg alloy isothermally held at 590°C for 8000 sec. and cooled at different cooling rates.

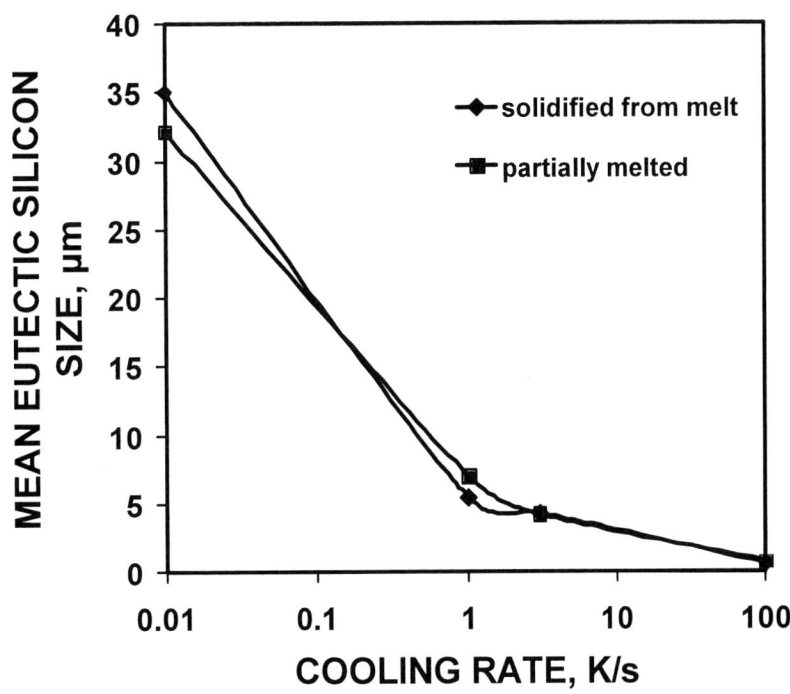

Figure 5: Variation of size of eutectic silicon particle in the interconnected liquid region with cooling rate.

Net Shape Forming of Iron and Steel for Clean Production

Jeong-Il Youn* and Young-Jig Kim*

* Department of Advanced Materials Engineering,
Sungkyunkwan University, Korea

Abstract

The net shape forming of iron and steel by semisolid processing (especially thixoforming) would be an optimum process. The process is able to develop parts having advance qualities that combine the superior material properties of ferrous alloys with the dimensional accuracy and without harmful by-product. In this paper, the microstructural evolution of steel and iron was investigated for making a technical map to get a proper thixoforming condition, such as solid and liquid fraction, size of globular grains, holding temperature and time. Based on the map, D2 tool steel could be thixoformed by the machine which was developed to be controlled an atmosphere. The demonstrator component shows that thixoforming is an attractive manufacturing method to produce mechanical parts which have a complicated shape and curved surface.

Key words Iron, Steel, Semisolid, Microstructure, Thixoforming.

Introduction

Iron and steel have always played a significant role in industrial productions for a long time. Nowadays, the usage of non-ferrous metals - especially aluminum and magnesium alloys- has been increasing to substitute the ferrous alloys with the point of view of weight reduction of parts [1, 2]. However, iron and steel are in a better position in terms of the manufacturing cost and mechanical properties until now.

Iron and steel have been cast by sand casting mainly, so a lot of waste and dust were derived from the mold sand, as like the furan, during casting for parts manufacturing. Increasing health concerns about the dusty working environment, as well as the waste of mold sand, have stimulated development efforts to recycle of the sand and produce new eco-sand. However, it is hard to remove the fundamental problems of the mold sand. The clean production of iron and steel can be achieved through the change of process to produce parts. Among the new processes, the net shape forming by semisolid processing (thixoforming) is proper process. Because, the process is able to develop parts having advance qualities that combine the superior material properties of ferrous alloy with the dimensional accuracy, and there are no harmful by-product [3-7].

In the present study, the microstructural evolution during partial remelting and net shape forming conditions of ferrous alloys, especially, D2 and M2 tools steels and cast iron, were investigated. In a viewpoint of low energy consumption, special attention was paid to the microstructural control of materials and net shape forming using the as-received and as-cast specimen without additional processes.

Experimental

Materials

D2 and M2 tool steels in the as-received hot rolled and spheroidizing annealed condition were provided by Changwon Steel Ltd, Korea, and the cast iron has chemical compositions of hypo-eutectic and it is shown in Table 1. The cast iron has a flake type graphite, which are randomly distributed and oriented throughout the iron matrix and classified type A by ASTM A247.

Microstructural evolution

The as-received tool steels and as-cast iron slugs were cut approximately and a cylinder type (Ø10×20 mm) samples were prepared as a specimen. The partial remelting experiments were carried out in a SiC electric resistance tube furnace. The specimen was held in a quartz tube in order to avoid oxidation. The conditions of isothermal holding temperature and time are shown in table 2. The heating rate was controlled to be 1.4 °C/s in the solid state and 0.25 °C/s in the semisolid state. The change of temperature in the specimen was measured and monitored by R-type thermocouple which be inserted into the specimen. At the end of the isothermal holding, the specimen was rapidly quenched in a water bath. The specimen was metallographically prepared and etched using a dilute nitric acid solution (distilled water 50 ml + nitric acid 5 ml) for 10-30 s, and then examined using optical microscopy.

Thixoforming of D2 tool steel

The thixoforming machine, as shown in Fig. 1, has a vertical configuration consist of two chambers. A slug is reheated in the lower chamber and the die to thixoforming is set up in the upper chamber. The vertical configuration allows in-situ heating, which means that the billet does not require transfer from external equipment to the machine while heating, and thus the heating and forming cycle can be performed while keeping the controlled atmosphere. The machine has a ram of maximum load 40 t, maximum velocity of 1.5 m/s, and the maximum die clamping force 100 t. The machine is capable of control the ram speed and casting pressure. Heating of alloy slugs in the semisolid state is provided in-situ, using a 20 kW medium frequency induction heater. The D2 tool steel slug (Ø80×50 mm) was placed on an insulator on top of the ram. Based on the experimental results of the microstructural evolution, the time-temperature profile was built with two thermocouples; one at the center and another near the periphery of the slug. For practical thixoforming, the slug was reheated by a heating program using the standardized heating procedure of the time-power profile built by the time-temperature profile. In order to alleviate the oxidation of the slug, the machine was evacuated to 10^{-3} torr and refilled with argon to 1150 torr. When the correct condition was attained, the slug was thixoformed into a die at a ram speed of 0.08 m/s with a casting pressure of 5 MPa.

Results and Discussion

Microstructural evolution of D2/M2 tool steels

Figures 2 and 3 show the as-quenched microstructures of D2 and M2 after isothermal holding for 0 s at the given temperatures, respectively. The microstructure evolutions of D2 and M2 with isothermal holding at 1300 °C for the given times are given in Fig. 3. It is clearly seen that the thixotrophic structure, the fine solid globules with less intragranular liquid, can be obtained by using the as-received D2 and M2 tool steels, even through such a simple reheating procedure within a length of time suitable for commercial production. The driving force for the thixotrophic structure evolution in the semisolid state is the tendency to reduce the overall free energy of the system by dropping the total area of the internal interface [8]. The microstructures shown reveal the perfect thixotrophic structure only after 0 s of isothermal holding even at 1300 °C. The white areas in the figures correspond to the austenite solid phase in the semisolid state while darker regions, which have resulted from the solidification of the liquid phase, are complex phase rich on Cr, V, Mo and W. The average diameter of solid globules increases gradually with increasing isothermal holding time, that is, the size of the globules is less dependent of isothermal holding temperature. The average size of globules of D2 increase form 60 to 100 μm while that of M2 from 40 to 70 μm with isothermal holding time.

Microstructural evolution of cast iron

Compared with as-received cast iron, Fig. 5 (b) shows that the partial remelting starts at the regions where dissolution of eutectic graphite took place at the eutectic temperature just above.

Partial remelting of a dendritic structure should lead to globularization of the primary phase, coarsening or ripening processes leading to the reduction in surface area will be operating whereby regions of high curvature are eliminated or reduced by diffusion of solute in the liquid [8-10]. This discussion is demonstrated in Fig. 1, which shows that the initial microstructure of partially remelted cast iron transforms to globular microstructure with increasing of isothermal holding temperature by diffused and redistributed carbon. In case of Fig. 5 (c) held at 1150 $^{\circ}$C for 300 s, the globules appear in patches, and it is thought that the transformation, however, depends on the size of the primary dendrites and on the crystallographic related surface energy [11]. Almost techniques to produce pre-materials for semisolid processing applied the shear force at the semisolid slurry in order to cut off the dendrites and to reshape them to a globular microstructure. The size and distribution of globules in these techniques are uniform, however, in case of this research; the size and distribution are not uniform. This result was also considered as the effect of the size of the dendrites. As a result of XRD analysis, as-received cast iron of ferrite phase is transformed to austenite and eutectic phase (liquid in mushy zone), however, iron carbide is not detected when the partial melting arises at very small region. Fig. 7 shows the EPMA elemental profile of carbon and silicon. The concentration of carbon – distribution coefficient, k, small than 1- around globule is higher than it, to the contrary, the concentration of silicon is lower than globule. These results imply that globule of austenite phase is surrounded by the liquid of high concentration of carbon, and silicon is redistributed during partial remelting.

Thixoforming of D2 tool steel

The die to thixoforming of D2 tool steel was made from alumina-based ceramic and no lubricant was used. Fig. 8 shows the thixoformed automobile valve of one-side shape, which is to evaluate the design flexibility and dimensional accuracy of thixoforming of D2 tool steel. It is not a real part, however, Fig. 8 demonstrates that thixoforming is an attractive manufacturing method to produce mechanical parts, which have a complicated shape and curved surface.

Conclusions

1. The examination of the microstructure evolution and thixoforming has proved enough evidence for the reheating system, utilizing the standardized reheating procedure of time-power profile that is deduced from the optimum time-temperature profile obtained experimentally.

2. The thixotrophic structures of D2 and M2 tool steels can be easily obtained by simple partial remelting of original microstructure and their maintenance in the semisolid state. The average size of globules of D2 increase form 60 to 100 µm while that of M2 from 40 to 70 µm with isothermal holding time.

3. When the cast iron was held in semisolid state, partial remelting starts at the eutectic phase with dissolution of graphite by the difference of melting temperature of primary austenite and eutectic phase. Also, carbon and silicon were redistributed and the globular grain was appeared. The globularization of the primary phase, coarsening or ripening processes leading to the reduction in surface area will be operating.

4. The demonstrator component shows that thixoforming is an attractive manufacturing method to produce mechanical parts, which have a complicated shape and curved surface.

References

1.   A. Jambor and M. Beyer, *New cars-new materials*, Materials & Design, 18, 1997, pp203-209.
2.   D. Brungs, *Light weight design with light metal castings*, Materials & Design, 18, 1997, pp285-291.
3.   D. M. Stefanescu et al., eds., Metal Handbook, vol. 15, Materials Park, OH: American Society for Metals, 1998, ISBN 0 87170 007 7.
4.   D. H. Kirkwood, *Semisolid Metal Processing*, International Materials Reviews, 39, 1994, pp173-189.
5.   G. Hirt, R. Cremer, T. Witulski and H. C. Tinius, *Lightweight near net shape components produced by thixoforming*, Materials & Design, 18, 1997, pp315-321.
6.   M. C. Flemings, *Behavior of Metal Alloys in the Semisolid State*, Metallurgical Transactions B, 22B, 1991, pp269-293.
7.   D.H. Kirkwood, *Semisolid processing of High melting Point Alloys*, 4th International Conference on Semi solid processing of alloys and composites, Sheffield, United Kingdom, 19-21 June 1996, pp 320-325.
8.   M. Ferrante and E. de Freitas, *Rheology and microstructural development of a Al-4wt%Cu alloy in the semisolid state*, Materials Science & Engineering, A271, 1999, pp172-280.
9.   W. Lapkowski, *Some studies regarding thixoforming of metal alloys*, Journal of Materials Processing Technology, 80-81, 1998, pp463-468.
10.   D. R. Poirier, S. Ganesan, M. Andrews and P. Ocansey, *Isothermal coarsening of dendritic equiaxial grains in Al-15.6wt.%Cu alloy*, Materials Science & Engineering, A148, 1991, pp289-297.
11.   M. Tsuchiya, H. Ueno and I. Takagi, *Research of semisolid casting of iron*, JSAE Review, 24, 2003, pp205-214.

Tables
Table 1. Chemical compositions of D2, M2 tool steels and cast iron (wt.%)

	C	Si	Mn	P	S	Ni	Cr	Mo	V	Cu
D2	1.52	0.21	0.30	0.027	0.001	0.15	12.34	0.87	0.26	0.05
M2	0.85	0.38	0.25	0.026	0.001	0.09	4.05	4.87	1.85	0.09
Iron	3.21	2.15	0.64	0.025	0.021	-	-	-	-	-

Table 2. Isothermal holding temperature and time

	Holding temperature (°C)	Holding time (s)
D2 / M2 tool steels	1300 1320 1340	0 300 600 1800 3600
Cast iron	1142 1150 1160 1170 1180	300

Figures

Fig.1. Photograph of the thixoforming machine.

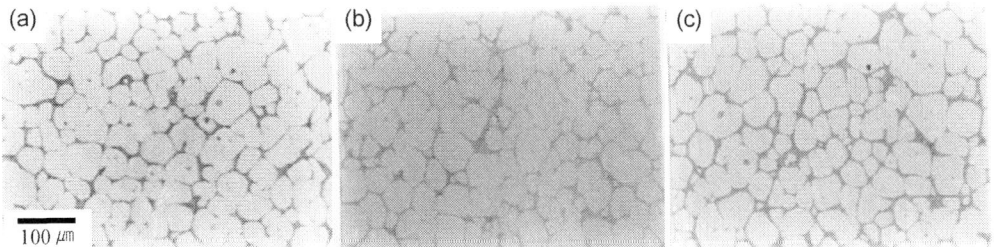

Fig. 2. As-quenched microstructure of D2 after isothermal holding for 0 s at the given temperature; (a) 1300 °C, (b) 1320 °C and (c) 1340 °C .

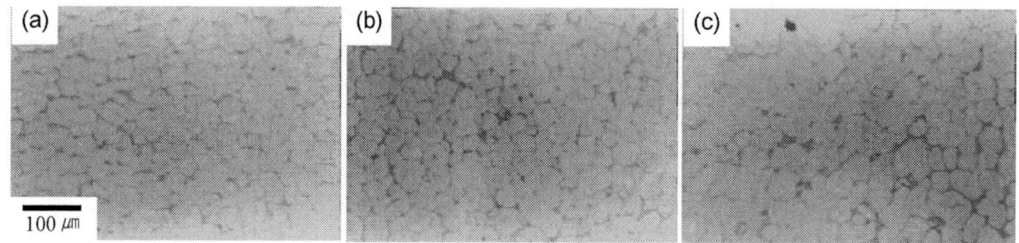

Fig. 3. As-quenched microstructure of M2 after isothermal holding for 0 s at the given temperature; (a) 1300 °C, (b) 1320 °C and (c) 1340 °C .

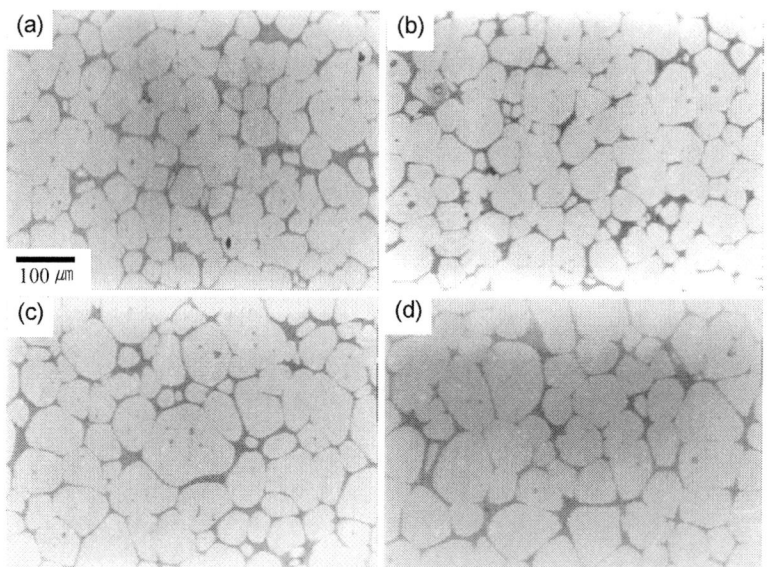

Fig. 4. As-quenched microstructure of D2 after isothermal holding at 1320 °C for (a) 300 s (b) 600 s (c) 1800 s and (d) 3600 s.

Fig. 5. Microstructure of (a) as-received cast iron and as-quenched microstructure after isothermal holding at (b) 1142 °C, (c) 1150 °C, (d) 1160 °C, (e) 1170 °C and (f) 1180 °C for 300 s.

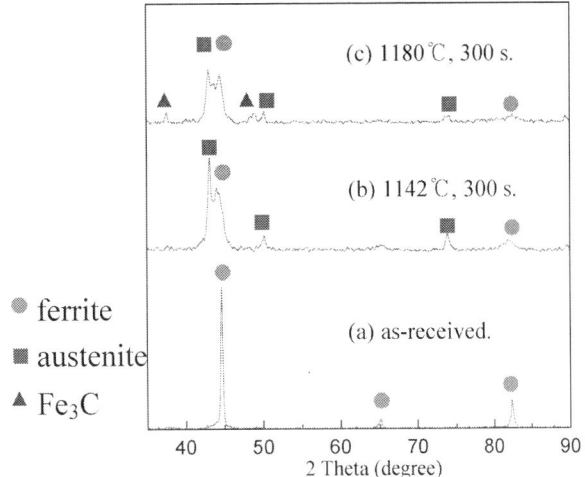

Fig. 6. XRD results of (a) as-received cast iron and isothermal holding at (b) 1142 °C and (c) 1180 °C for 300 s.

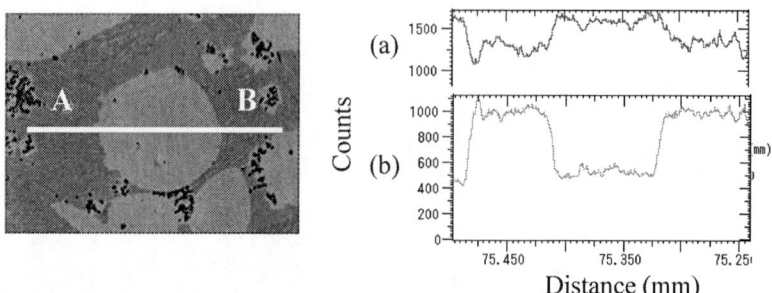

Fig. 7. EPMA elemental profile across AB; (a) Si and (b) C after isothermal holding at 1180 °C for 300 s.

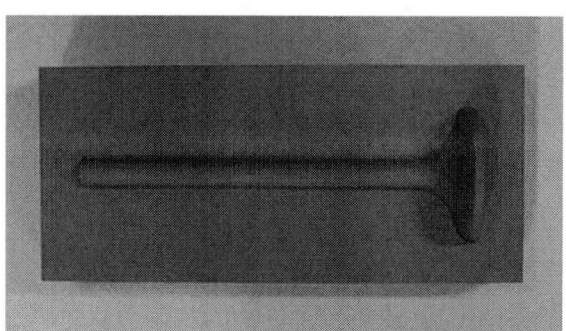

Fig. 8. Photograph of thixoformed D2 tool steel.

A Study on Semi Solid Squeeze Forging of High Strength Brass

K.H. Choe*, G.S.Cho*, K.W. Lee*, Y.J. Choi**, K.Y. Kim*** and M.H. Kim****

*Advanced Material R/D Center, KITECH, 994-32 Dongchun-dong, Yeonsu-gu, Incheon, 406-130, Korea
**Hanbat National University, San16-1, DuckMyoung-dong, Yuseong-gu, Daejeon, 305-719, Korea
*** Hankuk Aviation University, 200-1, Hwajeon-dong, Deogyang-gu, Goyang-city, Gyeonggi-do, 412-791, Korea
****Inha University, 253, Yonghyun-dong, Nam-gu, Incheon, 402-751, Korea

Abstract

The microstructures and mechanical properties of high strength brass made by semi solid squeeze forging were investigated and compared with those of conventionally hot forged product and gravity die castings. No shrinkage or gas hole was found in squeeze forgings. Fine equiaxed crystals developed at the center of squeeze forgings, while grains in the corner of squeeze forgings were elongated perpendicular to the pressure direction. The grains of squeeze forgings were smaller than those of hot forgings and gravity die castings. It is suggested that a rapid heat transfer condition due to applied pressure is responsible for grain refinement. Tensile and yield strengths of squeeze forgings were as high as those of hot forgings but elongation was positioned between that of hot forgings and gravity die castings.

Key words: Semi solid squeeze forging, High strength brass

Introduction

The name "high strength brass" is given to the wrought and cast alloys indicating their particular virtue of high strength, which can be achieved by additions of Al, Fe, Mn and Sn. These alloys were formerly known, and are still sometimes referred to, as "manganese bronzes." For many applications, their extra strength is needed for parts to be used in corrosive environment[1]. Forgings made from copper base alloys offer a number of advantages over products made by other processes. Dimensional accuracy is greater than by casting, working the alloys develops improved strength, and overall cost is modest. However, because for forging more heat must be applied to the ingot which was solidified once, there are some disadvanteges in the economy of energy and time.

Semi Solid Squeeze Forging is a hybrid process which the casting and the forging are done consecutively. This process is operated as follows. Firstly, the molten material is poured into the metal mold and starts to be solidified. When the thickness of solidified shell reaches at a certain level, semi-solidified material is removed from the die and transported to a hydaulic press machine. Then, the pressure was applied on the semi-solid alloys to solidify and shape final products. Becuase of the consecutive operation of casting and forging, this process has some advantages such as simple process and no need of additive heat for forging. In addition, the grain refinement is expected due to rapid cooling by forging during solidification.

In this study, we investigated the microstructures and mechanical properties of high strength brass made by semi solid squeeze forging and compared them with those of conventionally hot forged product and gravity die casting.

Experimental

In this experiment, KS CACIn304 copper alloy(similiar with JIS HBsC4, UNS C86300) was used and its chemical composition was shown in Table 1. The alloy was molten using induction furnace. The molten material was poured into the die which was preheated at 573K. Pouring temperature was 1273K. When the ratio of solid reaches at 0.5, semi-solid material was removed from the die and transferred to a hydraulic press. In a hydraulic press, one-dimensional pressure is applied to shape final products. The applied pressure was 600MPa and the pressing time was 10 seconds.

The microstructures and mechanical properties of this forgings (squeeze forging) were investigated and compared to those of hot forged product (hot forgings), and as-cast ingot (gravity die casting). Samples were etched by nitric acid(50%) and their macrostructures were observed. Microstructures were also observed. For the observation of mechanical properties, mechanical tests were done and their fracture surfaces were observed by SEM.

Results and Discussion

Macrostructure

The macrostructures of semi solid squeeze forgings, hot forgings and gravity castings are shown in Fig. 3. In squeeze forgings, the area labeled as "A" is the portion that remained at liquid phase before being forged and fine equiaxed crystals were found. The area labeled as "C" was previously solidified in the die, so the grains of these areas were deformed by applied pressure. No shrinkage or gas hole was found. Among area "A", area "B" and area "C", there were the differences in the charactersitic of solidification, so it was expected that there was discontinuity in the macrostructure, However, no distinct difference of macrostructures was found. It is considered that because the ingot was not fully solidified, grains are not deformed severely by applied pressure and previously solidified shell was annealed by the latent heat from unsolidified areas. In hot forgings, the all areas were severely elongated during forging process. In die castings, shrinkage was found at the top of castings and numbers of gas holes were also found.

Microstructure

Fig. 4 shows the microstructures of semi solid squeeze forgings and hot forgings. A, B, and C in Fig. 4 correspond to the areas labeled as "A", "B" and "C" in Fig. 3. It was found that very little α solid solution was present in a matrix of β phase. In copper alloy, β phase is stable in binary alloys containing more than 39.5% Zn, but strong β stabilizers such as Al promote its presence at lower zinc contents[2]. Guillert[3] suggested Zn equivalent of Cu alloying elements. From this relation, Zn equivalent of the alloy used in this study is 48.4%Zn. Therefore, the matrix consists of β phase.

In squeeze forging, though the area "A" was pressurized directly, equiaxed crystals developed. At the area B, equiaxed grains were found as well. However, the grains in the area B had facets, but in the area A grain hadn't. This means that in the area B solid state transformation and diffusion reaction occurred during cooling but grains of the area A had insufficient time for this reaction. On the other hand, grains in the area C, which were solidified in the die, were deformed perpendicular to pressure direction. Compared with the grains in same position of hot forgings, the grains in squeeze forgings have different shape. The grains in area B of hot forgings were deformed by forging pressure. On the other hand, forging pressure didn't affect the shape of grains in same position of squeeze forgings. It is considered that the area A remain as liquid phase until forging, so it acted as buffer to forging pressure.

In hot forgings, all grains were deformed perpendicular to forging pressure direction and grain size was increased by applied heat during hot forging process.

Compared with gravity die castings, particles of α solid solution and grains of β phase matrix were finer. In the case of squeeze castings, the applied pressure and the instant contact of the molten metal with the mold surface produce a rapid heat transfer condition that yields a fine-grain casting.[4] Similarly, in the case of squeeze forgings, finer grains was found at the area whose heat transfer rate is high.

Mechanical properties
Mechanical properties of high strength brass made by various processes are shown in Fig. 5. In the case of squeeze forging, test results showed that its mechanical properties are equivalent to those of sand casting. Except for block, all specimens showed similar values of tensile and yield strength. The elongation of squeeze casting fell between that of forging and gravity die casting. Fig. 6 shows the fracture surface of high strength brass made by various processes. In semi solid squeeze forgings, typical ductile fracture surface was found dissimilar to in gravity casting. Similiar structure was also found in hot forgings.

Conclusions
The microstructures and mechanical properties of high strength brass made by semi solid squeeze forging were investigated and the following conclusions were drawn from this study.
1. No shrinkage or gas hole was found in squeeze forgings.
2. Fine equiaxed crystals developed at the center of squeeze forgings, while grains in the corner were elongated perpendicular to pressure direction.
3. Grain size of squeeze casting was smaller than that of hot forgings and gravity die castings. It is due to a rapid heat transfer condition by applied pressure.
4. Tensile and yield strengths of squeeze forgings were as high as those of hot forgings but elongation fell between that of forgings and gravity die casting

References
1. http://www.copper.org/innovations/2000/01/brasses.html
2. ASM International, Copper and Copper alloy, (2001), 91 Campbell J, Castings, 2nd Ed, Butterworth-Heinemann Ltd, Oxford 2003, ISBN 0 7506 4790 6.
3. S.Oya, Nonferrous Metals Casting, Nikan Industry News, (1968), 74
4. ASM, Metals Handbook vol.15 Casting, (1998), 323

Acknowledgements
Authors would like to acknowledge Gabsan Metal Co. for technical support and help with squeeze forging.

Tables

Table 1 Chemical composition of copper alloy

Cu	Al	Sn	Zn	Pb	Fe	Ni	Mn	Si
63.6	5.50	0.10	24.3	0.19	3.10	0.21	2.95	0.07

Figures

Fig. 1 Semi solid state before squeeze forging

Fig. 2 Final product after squeeze forging

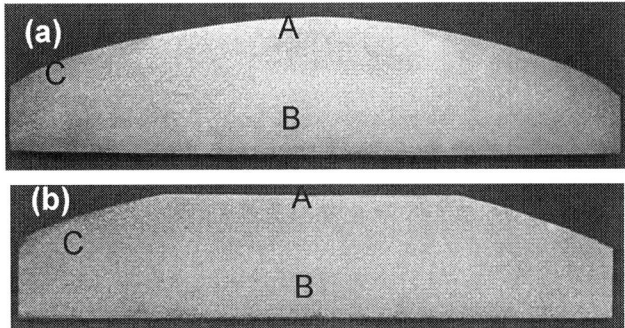

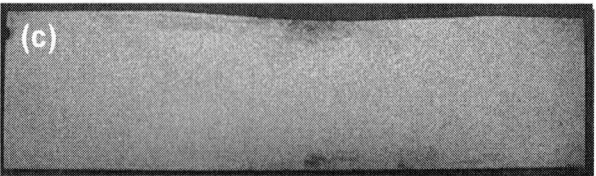

Fig. 3 The macrostructures of semi solid squeeze forgings(a), hot forgings(b) and gravity casting(c).

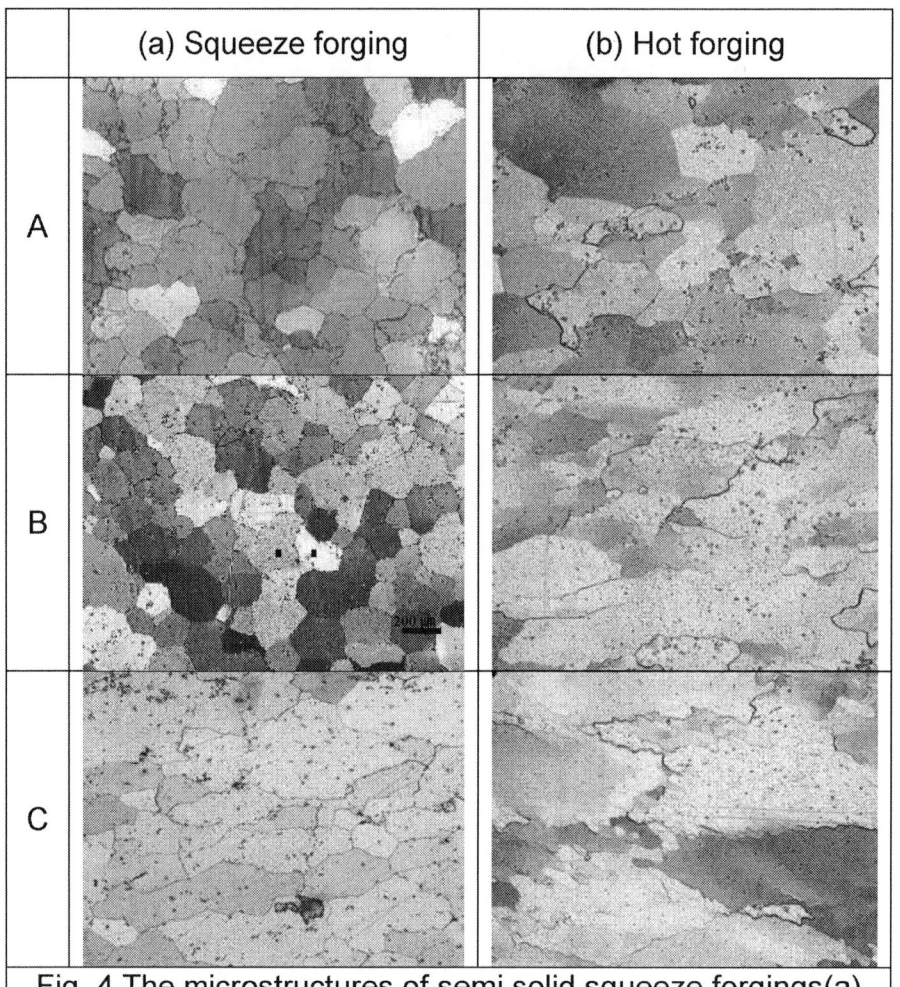

Fig. 4 The microstructures of semi solid squeeze forgings(a) and hot forgings(b)

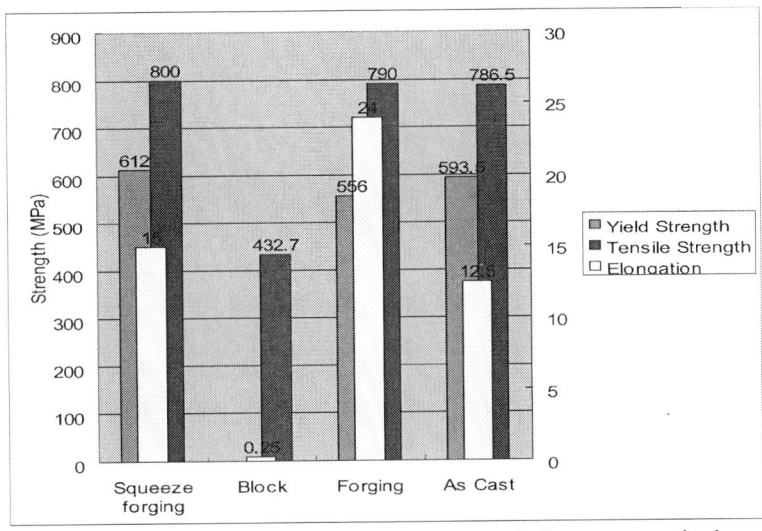

Fig. 5 Mechanical properties of high strength brass made by various processes.

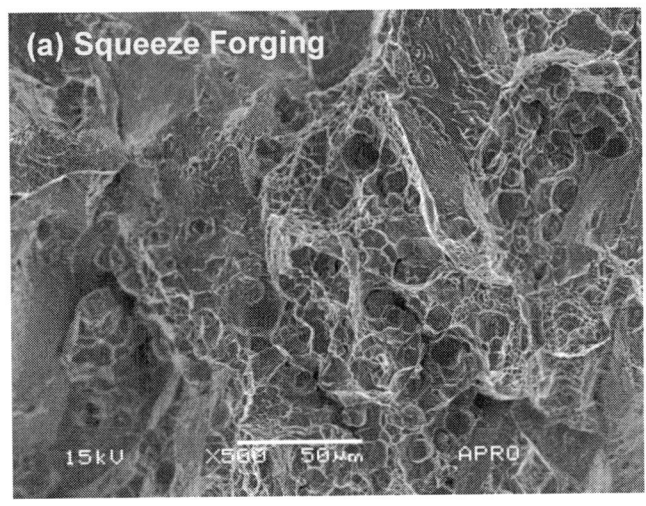

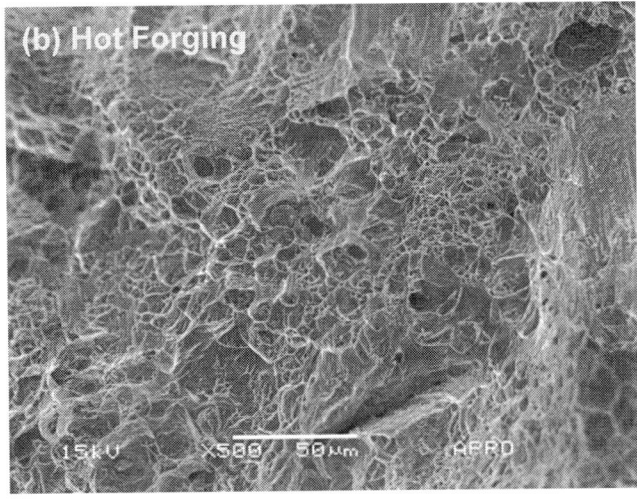

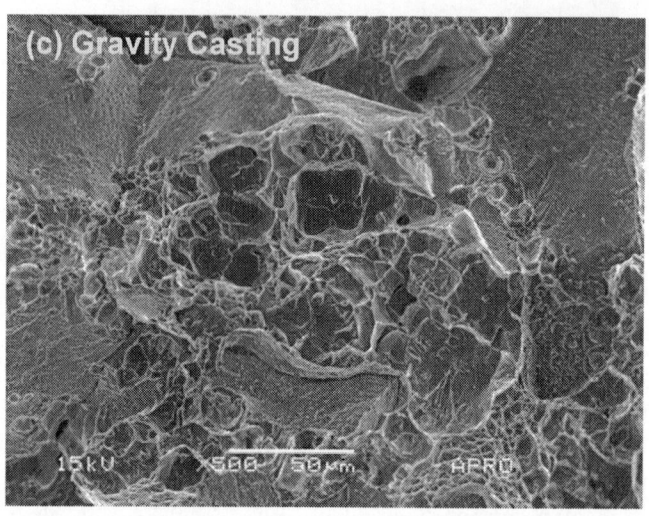

Fig. 6 The fracture surfaces of semi solid squeeze forgings(a), hot forgings(b) and gravity casting(c)

Aerospace Castings – The Future Challenge

David E. Jakstis

Spirit Aerosystems Wichita, Kansas USA

Abstract

Castings have been employed on the earliest airplanes produced and continue to contribute to the structural systems of aircraft today. With the competitive drive to reduce aircraft costs, castings have been able to provide production solutions that reduce manufacturing costs, reduce part requirements, reduce assembly cycle time, and even reduce the weight. Despite these contributions, there has been a reluctance to design aircraft with castings, especially in critical structural applications. This paper will address the issues regarding the use of premium quality castings for aerospace and the challenges that the aerospace and the casting industry face together to achieve confidence in critical cast structures.

Key words: Casting Factors, Aerospace, Premium Quality.

A Global Castings Users Viewpoint

Mandy Kennedy

Cummins Turbo Technology Ltd.

Abstract

Cummins Turbo Technology Ltd are major users of castings. Mandy Kennedy is the Global Quality Executive for the company and will present her views on how the industry is changing and how suppliers to the company will need to adapt. This will include changes to alloys and specifications, surface finish requirements and the trend towards purchasing sub-assemblies and finished machined sub-assemblies. The presentation will also focus on how the company will work with suppliers to develop tighter controls on quality and to share product liabilities in the future.

New Solutions in Ductile Cast Iron for the Retrievable Storage of Radioactive Waste

C Macke-Bart *, G Regheere *, D Linxe * and A Beziat **.

* CTIF (Development Centre for Materials Forming Industries), France,
** CEA (French Energy Atomic Commission), France.

Abstract

Within the framework of the French radioactive waste management act, the CEA has carried out a major R&D program on long-term interim storage (temporary deposit) of high-level nuclear waste. The program concerns container development, design of storage facilities, durability and functional performance demonstrations.

This article focuses on the high level waste container development.

In addition to the geometry, the essential container design aspects are the material selection and the package closure method. Ductile cast iron was selected as the reference material and a full-thickness welding process was developed to seal the container.
The development program has been performed to demonstrate the technical feasibility of the container including specific metallurgical and process improvements.
CTIF and SAFEM (cast iron foundry) participated technically in this program in which several full scale prototypes were realized.

Key words
> Cast iron
> Steel
> Insert
> Storage

Introduction

For safe management, nuclear wastes are conditioned in containers. The CEA developed a specific container for long term interim storage of high level waste. The key criteria for the container are:

- Maximum size: height 5m, diameter 1.2m,
- Possibility of creating seven compartments to accommodate seven primary waste packages,
- Lifespan: 300 years,
- Uniform corrosion metallic material,
- Good bending, tensile and impact properties,
- Resistance to irradiation,
- Competitive cost.

The design studies addressed the materials, the mechanical and thermal properties, corrosion resistance of the package. The container will be made of carbon steel or cast iron, sealed by full-thickness welding. Cast iron is the most suitable metal: it ensures a homogeneous container with an integral bottom plate, and the manufacturing cost is much lower than for rolled/welded carbon steel. The only drawback is that it is impossible to weld cast iron thicknesses of a few centimeters. The CEA has overcome this difficulty by developing a process in which a carbon steel rim insert is metallurgical bonded to the top of the container. A carbon steel cover is welded directly to the rim.

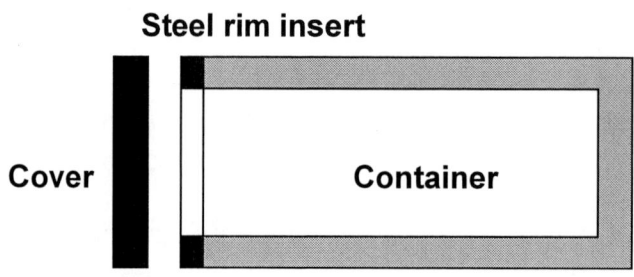

The target of the casting process was to obtain, in a repeatable way, a perfect bonding between the cast iron and the steel rim, at the inteface of the two metals, and to control, on samples first, and then on prototypes and finally on full scale demonstrators, parameters and conditions to guaranty the reliability of the integrity of this bonding on the complete surface of the iron-steel interface.

Beyond other trials made to test different solutions for the closure of the container – not detailed here -, the program was developped in 4 main stages :

> A first stage was dedicated to the **adjusting of metallurgical parameters**. It was driven by CTIF in its experimental casting facilities, to validate the thermal and metallurgical conditions to get a perfect bonding between a steel insert and a cast iron cast on it. Solutions developed on samples had to be transferable on industrial means at full scale.

> A second stage was dedicated to **casting design,** to transfer the results of the first stage at full scale. The target of this phase was to determine the specific improvements to apply on the casting process to get in the area of the insert (steel rim), the thermal and metallurgical conditions recommended in the previous stage. The so defined casting design was then validated by **thermal simulations**, at full scale (1.20 meter diameter, and representative hight of 1 meter).

> A third stage was dedicated to the **validation on prototypes** of the process parameters and recommendations defined in the previous stages. This third stage was developed in partnership with the French foundry SAFEM, specialized in the casting of large cylinders. Prototypes at full scale –apart the hight reduced to 1 m– were produced in that step, and submitted to destructive and non destructive testings, to confirm the ability, under industrial conditions, to get the requested quality at the interface between the iron and the inserted steel rim.

> A fourth stage was dedicated to the manufacturing, under industrial conditions, of **full scale complete demonstrators**. Casting parameters were defined from the conclusions and recommendations issued from the previous steps.

I – Adjusting of metallurgical parameters

In this first stage, experimental casting trials were driven by CTIF to validate, on samples, the thermal and metallurgical conditions to obtain a reliable bonding between a steel insert and a EN-GJS-400-15 iron cast on it.

Testings were made on a steel sample representative of the steel rim dimensions which is to be inserted on the container : 100 mm large, 80 mm thick, and 400 mm long. The insert in E24 steel was not submitted to any specific surface treatment before casting. Tooling was designed to ensure a turbulent flow of the cast iron on the rim surface corresponding to 1/4 of what will be obtained on the full scale casting.

Since the carbon migration duration depends on the involved masses (2 hours on the sample against over 48 hours expected on the full scale part), the insert was heated before casting.

Controls made on the bonding after solidification, showed a at least 2 mm thick dissolution of the steel insert, and the surface which originally was flat got a "U-profile" (Tab.1).

Metallographic inspection (Fig.1) revealed a reliable bonding between the 2 metals, without oxide inclusions or cementite precipitation.

Tensile test pieces were then machined from the cast iron, the steel, and the bonding area (Fig.2) : it appeared that the bonding area is not weakening compared with the 2 metals (cast iron and steel) but offers a compromise (Tab.2) : strength is kept at the level of the steel's one, and elongation is upper than the cast iron's one. The interface between steel and cast iron is continuous and so guarantees a real tightness.

II – Casting design & thermal simulations

In this stage, specific improvements have been designed on the casting process at full scale, to get in the area of the insert (steel rim), the thermal and metallurgical conditions obtained in the first trials. **Thermal simulations** were driven on PAM-CAST/SIMULOR software, at full scale (1.20 meter diameter, and representative hight of 1 meter), and allowed to display the filling and solidification conditions.

Observations were essentially made on :

- the flow of the molten cast iron at the surface of the insert, with different configurations to modify this flow (Fig.3), and optimize the heat exchanges with the insert,
- the profile of heating and cooling of the insert, in core and at its surface covered by the cast iron.

In the configuration which was finally chosen as the most efficient, 400°C are reached in core of the insert after 5 minutes, and 820/860°C after 1h20' (Fig.4). Cooling of the insert begins only after 2h30', with a surface temperature of the insert which is still above 580°C after 24 hours.

The right conditions are reached to get a reliable bonding at the interface between the cast iron and the steel rim.

Recommendations could be issued from these simulations for the next step : casting on prototypes.

III – Validation on prototypes

This third stage was dedicated to the casting on industrial means of first prototypes at full scale –apart the hight reduced to 1 m-.

Casting conditions were defined from the recommendations issued from the previous stage.

The cast iron grade was EN-GJS-500-7.

Temperatures recorded at the surface of the part were lower than the temperature reached in the previous simulations (460°C after 24 hours instead of 580°C in simulation) for a comparable pouring temperature of cast iron (1350°C). So cooling of the insert and the part is faster than expected in simulation.

Destructive and non destructive testings were carried out on the prototypes in order to validate under industrials conditions the quality of the bonding obtained between cast iron and steel rim.

Testings of the bonding show like in the first stage, a complete dissolution of the steel insert at its surface (Fig.5).

Metallographic inspection (Fig.6) reveals a reliable bonding between the 2 metals, without visible defects. Tensile test pieces were machined from the cast iron and the bonding area (Tab.3). Once again it appears that the bonding is not weakening : the test piece broke in its cast iron part. The interface between steel and cast iron is continuous, and so guaranties a real tightness.

Ultrasonic control on the complete surface of the rim confirmed that tightness, except on one point of the 3700 mm long insert. This means that the thermal and metal flow conditions on the surface of the insert are reached, but that quality of the cast iron's flow and cleanliness of the steel surface must be controled on the whole surface of the insert.

IV – Full scale complete demonstrators

This fourth stage was dedicated to the manufacturing, under industrial conditions, of **full scale complete demonstrators**. Casting parameters were defined from the conclusions and recommendations issued from the previous stages.

Manufacturing and control operations can be summarised as follows :

- casting on E24 steel inserts (vertical inserts for supporting the internal structure, and rim on the top of the container)
- machining and non destructive testings
- manufacturing of the internal structure
- assembling and inspections

The cast iron grade was EN-GJS-400-15.

Another main change compared with the tests on prototypes, is the integrated bottom. Only cylinders were cast in the prototype phase.

Temperatures recorded at the surface of the part were higher than the temperature reached on the prototypes for a comparable pouring temperature of cast iron (1365°C). The area of the rim was still above 400°C after 40 hours, 10 hours more than on the prototypes. These conditions are in favour of the bonding between the cast iron and the steel rim.

Non destructive testings were carried out on the demonstrator in order to validate under industrials conditions, and in full scale, the quality of the bonding obtained between the cast iron and the steel rim :

- visual examination of the bonding, before and after machining showed a continuous bonding, without visible defect (Fig.7),
- the integrity of the bonding was confirmed by ultrasonic control on the complete surface of the rim.

The internal structure was then assembled into the container.

After this first full scale demonstrator, another full scale part was manufactured in the sames conditions, but with changes on the internal structure assembling.

Both demonstrators complied with the part's specification.

V- Conclusion

A four years development program was conducted to validate a new concept of nuclear waste container. This container is made of ductile cast iron and a steel rim inserted at its top in order to allow the welding of a steel cover.

The validation was driven :

- on samples first, to identify the thermal and metallurgical conditions for a reliable bonding between the cast iron and the steel rim,
- by thermal simulation, to transfer the experimental results on samples, to the geometry of the container,
- on prototypes, to confirm the results of the simulation, and check on part by destructive controls, the integrity of the bonding,
- and finally on two full scale demonstrators.

The feasibility is demonstrated and thermal aging test on test coupons and representative container mockups have been conducted since 2002 and are still in progress.

A patent has been already taken out by the CEA on the part and its manufacturing process (French Patent- INPI 03 50182, 27-05-2003)

Acknowledgements

This demonstration supported by a development program including the fabrication of full-scale prototypes, qualification and durability testing, is the results of a partnership between the CEA, CTIF and SAFEM, independently of the contractual positioning of each one.

Tables

Tab. 1. Hight of the insert after casting (stage 1 : metallurgical specification)

Position on length (mm)	80	160	240	320
Hight of the insert (mm)	74	77	78	78

Tab. 2. Iron, steel & bonding characterization (stage 1 : metallurgical specification)

	Iron	bonding	Steel
UTS (Mpa)	440	276	278
YS (Mpa)	314	146	189
A% (%)	22.2	33.1	52.5

Tab. 3. Iron & bonding characterization after heat treatment (stage 3 : prototypes)

	Iron	bonding
UTS (Mpa)	427	427
YS (Mpa)	279	294
A% (%)	24.1	10.4

Figures

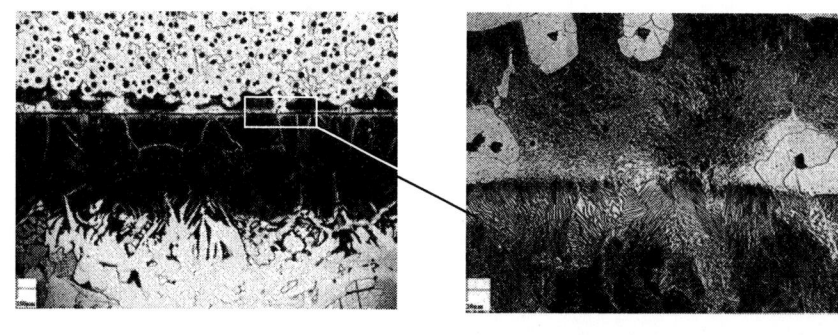

Fig.1. Micrography on the bonding (stage 1 : metallurgical specification)

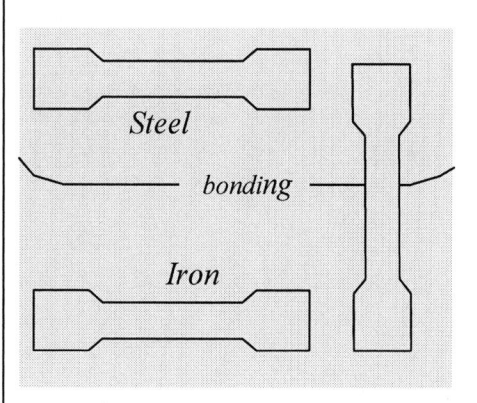

Fig. 2. Testing samples & Results (stage 1 : metallurgical specification)

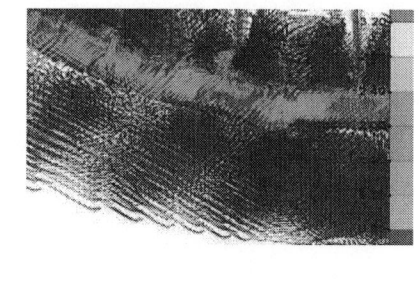

Fig. 3. Turbulent flow of the cast iron on the insert (stage 2 : simulation)

Fig. 4. Insert temperature at t ~ 5 min & t ~ 1h15 (stage 2 : simulation)

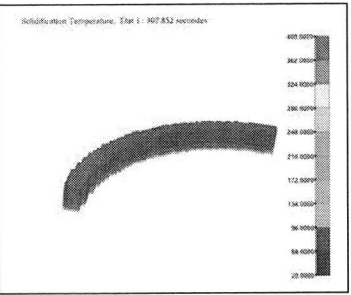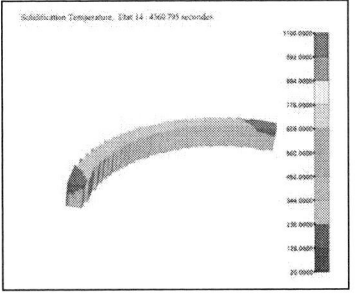

Fig. 5. bonding between steel insert and cast iron (stage 3 : prototypes)

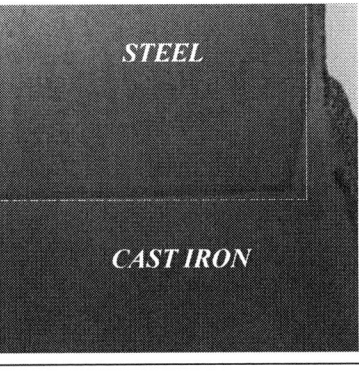

Fig.6. Micrography of the bonding (stage 3 :prototypes)

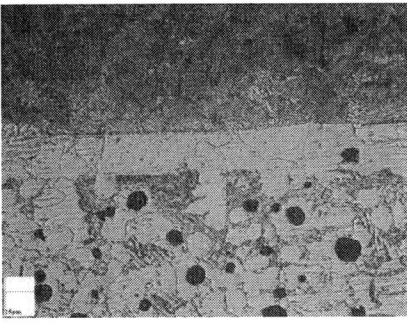

Fig.7. Bonding after machining (stage 4 :demonstrator)

Title:

"The Intelligent Casting Production: How to use automation to improve the economics of near net shape castings"

Author: **Product Manager, Kjell Holmen, DISA Industries, Denmark.**

Abstract:

In the foundry and metal working industry, globalisation has meant new and larger markets, accelerated production cycles, increased demands on product quality and standardisation, and, of course, keen competition among foundries. In order to stay competitive in this scenario, this paper suggests automation as an intelligent and cost-effective solution to time-consuming and arduous manual routines in the foundry. Following the introduction of the central automatic functions in the core production area, along the moulding line and in after-treatment, the paper verifies the level of regularity and process control achieved by using an automatic data collection and exchange system. The paper documents how automation affords an efficient production cycle, consistent product quality and significant improvements in labour functions and the working environment within the foundry.

Key words:

Automation, cost savings, process control, data collection, working environment

Introduction

Competition is keen and challenges are many in today's foundry and metal working industries. The demands on quality and efficiency are constantly on the increase – whereas the opposite is the case for prices. Most managers in the foundry industry are only too familiar with the discrepancy between product prices and production costs. In addition to this, the health and safety at work legislation has been tightened in recent years, making the compliance with these regulations a necessary, but costly, investment for most foundries.

At DISA with our market knowledge, we believe the *intelligent* solution to these challenges is to automate central procedures in the casting production. Automation ensures a consistent product quality, a higher degree of process control within the entire foundry, and last but not least a much healthier working environment for foundry workers. Consequently, automation becomes not only an intelligent but also the most cost-effective approach to casting production in the 21st century.

Within the foundry industry, the primary consequences of globalisation are growth of markets, accelerated production cycles and keen competition among producers and suppliers. Globalisation has also increased demands on product standardisation to the extent that manual manufacturing processes become difficult to maintain. The end-user will expect the same product quality and uniformity whether the production of the given parts is situated in China or Germany. Although the craftsmanship of foundry workers is indisputable, the level of accuracy and regularity demanded is difficult to achieve by human hand.

In order to meet these challenges and stay competitive a new generation of foundry equipment is needed, where automated processes replace central manual functions in the casting production. Tasks that are time-consuming - and no longer permissible for factory workers to perform according to current working environment regulations - are replaced by robotic functions. These automated functions transform the foundry from a workplace with many monotonous and laborious functions to a workplace in which the production is govern by mechanical regularity, monitored processes and greater safety. Let us examine these processes in greater detail.

Experimental

In a modernized foundry, the core production, transport and moulding line processes are performed by sophisticated robotic machinery, thus reducing the degree of labour and the amount of time involved in manual processes.

Automatic core handling: from core shooter to moulding line

We begin our tour in the core production area. Traditionally situated adjacent to, or in the vicinity of the moulding line, the core production area is a separate production unit with relatively complex logistics.

In a non-automated foundry, cores are extracted, inspected and coated manually; they are moved into and taken out of the drying oven, moved onto transport vehicles, placed in storage or drawn over to the moulding line. These activities are not only many and arduous for the workers; they are also time-consuming and involve the risk of creating bottlenecks, given that the cycle time for the core shooter is longer than that of the moulding line. In a fully automated core production area, the cores will be extracted from the core box, transported from the core shooter cell, inspected and treated, passed through the drying oven, transported to a storage position, and finally from storage directly to the core setter robot at the moulding line - without any manual labour involved! Automation is thus the most effective way of dealing with a production area with many separate functions. It also ensures a smooth transition from one production area to another.

Figure 1.

The automatic pattern plate exchanger

The first automated function at the moulding line is the automatic pattern plate exchanger, which is mounted onto the moulding machine. The pattern plate exchanger replaces one set of pattern plates with another. Consequently, the lift onto the moulding line and the manual fixing of the pattern plates are no longer necessary. These manoeuvres are not only time-consuming; they are also hazardous for the workers and involve the real risk of damaging or even dropping the pattern plate. An automated pattern plate exchanger eliminates such risks.

The robotic core setter

The second automated function on the moulding line is the automatic core setter. Usually, core setting involves the manual fixing of cores into the core setter, and the subsequent action of activating the deliverance of the cores into the mould. Once again, the time cycle of the new generation of moulding line makes manual core setting more demanding. Workers have to perform the core setting manoeuvre within 4-5 seconds. Given that this manoeuvre might involve many small cores – or larger heavy ones, as the case may be - the manual process is both stressful and monotonous for the worker. By allowing a robotic arm to grip the cores and place them in the core mask, the foundry ensures that the core setting process stays in synch with the cycle time of the moulding machine. In this way, the foundry derives optimal benefit from the increased efficiency of the new generation of moulding lines, and spares their workers an unnecessarily stressful job function.

The robotic casting extractor

The final automated process on the moulding line is the robotic casting extractor. In the shake out area, the foundry needs to find rational solutions to the execution of the following processes: the separation of castings from gating systems, a mechanical handling of clusters that does not damage the castings, and a positioning of castings that enables further automatic treatment. The automatic casting extractor offers solutions to these processes.

The system consists of a moveable table that controls the movement of the last mould in the sequence, a detection system that identifies the position of the cluster on the moulding line, and finally a robot that grips the cluster from the moulding line, and moves it to a new position.

When casting in aluminium, a gentle mechanical handling of the cluster is of great importance so that the movement from extraction to conveyor does not damage the relatively fragile casting. To ensure this, the robot grips the cluster on its pouring cup on the one side of the mould, and places it gently on the transport conveyor.

As regards the separation of castings from gating systems, the automatic casting extractor may be programmed to handle the clusters in different ways. When casting in SG-iron, the robot grips the pouring cup and brings the cluster to the separation unit; when casting in grey iron, the robot grips both the casting and the gating system and brings the entire cluster to a separation unit. To meet the cycle time of moulding machine, a two-robot system is used.

The use of robotic assistance not only effects a rationalisation and an increase in the level of process control in the shake out area. When using a robotic casting extractor, this area no longer needs to be manned but merely to be supervised, typically by the same person who oversees the other automated functions along the moulding line.

Automated after-treatment

Due to the labour intensity, time and cost involved for manual trimming and fettling, the modernisation of casting finishing is of paramount importance. The after-treatment area is an extremely unhealthy and hazardous area to work in. It involves many strenuous and monotonous functions, which do not comply with the health and safety regulations for industrial workplaces. As regards the castings, the automated fettling, grinding and shot blasting procedures ensure a level of standardization that the human hand cannot achieve.

In a fully automated after-treatment area, the trimming and fettling processes are executed by robots and interconnected. All activities are combined in groups and linked by various automatic transport and handling systems.

When planning the depth of automation in the after-treatment area, there are various significant factors that need to be taken into account. Among these are the size of the serial production, the complexity of the operations on the component, the diversity of products produced by the foundry, and finally, the structural and financial parameters of the foundry itself. Having carefully considered these factors, a fully automatic production sequence may be organised that is both technically and economically manageable for the foundry.

Data collection and data exchanging for various processes

To secure an efficient running of the casting production, it is necessary to have all information about the production process available at any time. The availability and exchange of online data between various units allows the monitoring of the entire production process, thus making it easier and faster to make adjustments in the respective production areas.

In order that online data may be used systematically, DISA has developed a wide range of computer integrated manufacturing modules – CIMs – that are installed on a PC, which is connected to the control system of the moulding machine. The PC - or the "cell controller" – collects, stores and distributes information between the moulding system, the other process equipment (e.g. the pouring unit, the cooling drum and the sand preparation plant) and the foundry computer system. In this way, a communication circuit is established which performs the collection, control, validation and storing of data from the various production sections.

The electronic interface between the different units in the foundry improves the productivity of individual units and the casting production as a whole. Primarily, this is achieved through five different data collection modules: the process control module, the management module, the preventive maintenance module, the mould and database module, and the external pattern plate module. Let us look at the advantages of these CIM modules in greater detail:

- The *process control modules* allow the foundry to coordinate and trim the operations of the various units (i.e. the pouring unit, the sand preparation and the cooling drum). [1]
- The *management module* produces up-time reports and production reports based on continuously collected data. This module locates bottlenecks or faults in the production line, which allows the operator to take immediate action to remedy the problem. Delays less than 0.1 sec. can be detected. By using this information, the

---

[1] The pouring unit interface will specify the next pouring position in advance, so that no time will be lost searching for the pouring cup. A pre-pouring signal allows the pouring sequence to start *before* the mould transport is completed. DISA figures show that this may cut 5% or more off the cycle time, which effectively results in a 2% increase in productivity.

foundry will effectively be able to increase its production rate.[2] Apart from identifying problems, the management module also enables the foundry to provide its clients with immediate and accurate data on the status of a given order.

- *Preventive Maintenance Module.* This module produces maintenance lists that identify and schedule the necessary maintenance to be performed on the various units (i.e. cleaning, checking, lubricating, replacements, etc.). This is done on the basis of the actual time or number of moulds produced, or any other criteria the foundry wishes to use.
- *Mould and casting database* is a statistic and quality assurance module. It stores the production conditions for every mould produced on the moulding system. This allows the foundry to relate the quality of individual castings to the conditions under which the given item was produced. In the case of irregularities in casting quality, these faults can be traced immediately, and the necessary corrections made.[3]
- *The external pattern plate module* stores data from the set-up of each pattern plate. This module optimises the set-up time for each individual pattern plate, ensuring identical settings every time the set is used. This minimises the risk of wrong settings and significantly reduces the level of scrap rates caused by human errors.

Figure 2.

Results

As is evident from the process descriptions above, automation is not only an intelligent, but also a cost-effective solution to many of the technical challenges and environmental demands that affect modern casting productions. Automation ensures a reliable production process, where all processes are synchronized; this optimises the lead-time for the entire foundry. The availability and exchange of online data from the various production units makes the systematic monitoring and adjustment of the casting process far easier. It allows for effective troubleshooting procedures and the presentation of reliable production data at any time. The customer derives direct benefit from this, as it shortens the foundry's response-time on process enquiries.

As for the products themselves, automation guarantees a more consistent product quality, meeting the customer's demand for accuracy and uniformity in the castings.

---

[2] By cutting 0.5 s. off the cycle time, the production rate will be increased by 3-5%.

[3] For every mould, this module records the pattern plate number, compressibility of the mould, whether the mould has been poured or not, exact time for pouring, and the specific machine set-up during pouring.

Complementing the automation of the production process is the improvement of the workers' labour content and working environment. By performing many of the monotonous, strenuous and hazardous manual functions, automation affects an upgrading of job functions from manual labour to technical supervision and maintenance functions. Judging by the feedback DISA has received from many foundry managers, such upgrading has had an immediate effect on the self-esteem of the workers. When combining their experience and skills from the manual processes with their newly acquired technical knowledge, workers are able to contribute constructively to the implementation of technical adjustments in their respective areas. Prior to automation, workers have not had the incentive or the energy to provide the management with such feedback.

Automation is indeed an intelligent solution – technically, professionally and environmentally. Most importantly, however, automation makes economic sense. Manual labour processes are time-consuming and, for this reason, costly for the foundry. Bottlenecks in the production, inaccurate process reports or irregularities in the finished products jeopardize the foundry's production flow and, in consequence, its competitiveness in the market. By automating the central core handling and casting functions, these processes will match the time cycle of the moulding line, thus optimising the production flow in the foundry. Shorter production times and the possibility of quick changeovers among different products allow the foundry a greater level of flexibility in the planning of its production.

Discussion

Pertinent to this presentation of the merits of automation is the question of which industrial circumstances warrant the investment in automated equipment?
Several internal and external conditions make a certain level of automation almost inevitable for the modern foundry. The first concerns the size of the casting production. In the case of large series of similar castings, automation would be the only rational and cost-effective choice. However, even with smaller, more complex casting series, automation would still be desirable, since it would significantly reduce the manual handling processes and ensure greater precision in the finished product.
Automation is not only an investment in a smooth production process; it is also a way of meeting the demands for standardisation in a globalised market. For this reason, automation is not only the way to survive in competition with low salary countries; it is a relevant solution for any foundry that wishes to compete internationally.

The replacement of human labour functions with electronic processes is obviously not without challenges. This conversion involves a significant restructuring of the foundry itself, as well as qualified training of the foundry staff. Added to this is the degree of vulnerability that the addition of automated equipment involves. However, it is a mistake to assume that the risk of breakdown is proportionate to the electronic complexity of the equipment. In the fully automated foundry, where processes are linked, there is obviously a high level of interdependency between the various automated functions. It goes without saying, however, that well-trained staff and methodical maintenance routines are just as important in a fully automated foundry as they are in a traditional production. Skills and maintenance are the only reliable measures a foundry can take against production disturbances.

Conclusion

The primary objective for any foundry is to stay competitive. The formula to follow to achieve this is well known among foundry managers: providing the best product and the most competitive price! Automation ensures that this familiar formula also makes financial sense of the foundry.

An investment in automation involves careful planning on the part of the foundry management. It is necessary to accurately assess the foundry's financial resources, its technical capacities and the structural and physical changes automation will involve at all levels. However, the long-term benefits derived from an automated production line are clear: product and process regularity, a systematic data collection, and a substantial upgrading of labour functions and working environment.

Email address: kjell.holmen@disagroup.com

Figure 1 showing a core production cell (incl. defining and coating)

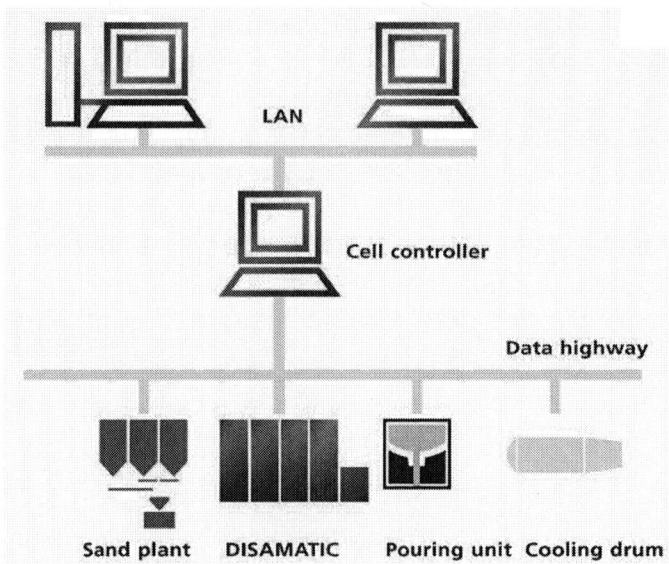

Figure 2 illustrating different equipment linked to a computer network

Double Cavity Casting of Transmission Case: A Technical Solution to a Foundry Capacity Problem

Muneer AHMED

HONDA of the UK manufacturing Ltd., Swindon, SN3 4TZ United Kingdom

Abstract

Honda of the UK Manufacturing (HUM) has a medium sized High Pressure Die Casting facility at its Swindon plant. Capacity utilization was over 90% in 2002 when the "Engineering Theme" was set up to challenge the facility's productivity and profitability. The aim was to create spare capacity in existing equipment and to fill it by producing castings which were being outsourced. Consequently this led to finding and solving problems, which were both conventional and innovative in terms of management control and technical complexity. The lessons learned have brought the company closer to realizing its 'power of dreams', successfully creating a near ideal business cell. The experience has been shared worldwide and set new benchmarks for future planning within the global organization.

Keywords: Diecasting, up-time, cycle time, capacity utilization, computer aided engineering

Introduction

Creating capacity
It is generally accepted that there will be down-time related to casting machines, casting dies and the casting process. It is usual to allow for a total uptime of 85% [1] when calculating capacities; it is also usual that in practice, only 80% up-time is achieved. The challenge at HUM was to exceed this widely accepted standard. A short study highlighted that this 5% allowable time was being interpreted rather generously and also revealed other hidden losses.

Many are familiar with the many recurring problems that cause disruption to production plans and HUM is no exception. A simple strategy was applied to identify production time losses and provide countermeasures. The strategy included a visual, live down-time monitoring system which was called *'Round by Round'*, whereby production equipment is checked for its output on two hourly rounds against its 100% capability, (see Figure 1). The objective was to identify gaps in performance and to install immediate remedies. Over a period of one month, repeated and avoidable problems were identified where a permanent countermeasure could be planned using the *'5 why method'*, *(see* Table 1).

Experimental
A number of experiments were planned and conducted to test the best possible performance target. From this, a realistic capacity plan was created for the various options that might be considered, (see Table 2). Some of the changes implemented are listed below.

Process cycle time
- 10% improvement in cycle time.
- Improved cooling capacity.
- Regular thermal imaging.
- Advanced die lubricant application.

Dies and Tooling
- Installed real time coolant flow monitoring for critical core pins and core faces.
- Ability to detect core pin and critical core failures.

Casting machine
- Fully flexible part extraction and sleeve insertion mechanisms.
- Integrated cycle time monitoring.
- Automatic monitoring of critical parameters.

Management

- Continuous overview of daily achievement including *'Round by Round'*.
- Strong and focused leadership to resolve the top three problems.

Having become confident of achieving 35% spare capacity, new products were considered which could meet the following criteria:

1. Must acquire technical know how for the company's business world wide.
2. Use advanced design and simulation techniques, actual design must be validated to rectify any gaps.
3. Component for in-house-built power train.
4. Minimum investment (except tooling, and research and development (R&D).
5. Only one die set allowed.
6. No increase in operating costs.
7. Clear profitability.
8. Must utilise 2250 tonnne high pressure diecasting (HPDC) machine efficiently (energy & environment).
9. 100% volume switch (vendor non-reliance).
10. Guaranteed minimum risk on switching.

Major powertrain components, including cylinder blocks and cylinder heads for were already manufactured in the engine plant. Cost analysis showed that clutch housings or transmission cases would be the most profitable to manufacture. This offered a new challenge. There was a requirement to make 470 units per day to meet the demand of the company's Civic range, but simulations had shown that there was capacity for only 278 units. This constraint led to the *'double cavity'* idea.

Conventional multiple cavity diecasting of large components raised major concerns due to high reject rates and un-reliable tool performance. Experience showed that overall losses sometimes reach 40% of total production and it is not uncommon to see that only one good casting is extracted out of every two. In the author's opinion, multi cavity diecasting had the potential to provide a cost benefit of up to 30% if common problems could be solved. Solutions to these problems formed the basis for our tooling and process design.

Initially, a single cavity clutch housing die was introduced for in-house production as this enabled identification and validation of capacity issues. However, the clutch housing was found to be too wide for a double-cavity die on the 2250-tonnne HPDC machine. The clutch housing was later sub-contracted to a British foundry when the double cavity transmission case design was ready for testing.

Performance requirements for double cavity diecasting

- Over all die performance *must* match or exceed single cavity dies already in use.
- Both cavities must behave in the same manner under mass production conditions.
- Cycle time: 90 seconds for two castings.
- Die up-time: 97%.
- Total Pass Ratio (TPR) 95% including preheat parts, and maximum leak rate 3%.

To achieve these performance requirements the design team had to implement the technical features and process criteria outlined below.

Up-time & Cycle time (process)

Generally, process up-time is severely affected by 'die flash' from die parting lines. Sealing of parting lines is extremely difficult due to mechanical and thermal distortion, particularly in larger dies where metal force is near to machine locking force. The problem was analysed and appropriate countermeasures were determined as follows:

Downtime $\Rightarrow$	Flash $\Rightarrow$	Parting distortion $\Rightarrow$	Cavity cooling Bolster cooling

Therefore, the new tooling design incorporated cooling for the cavities to remove energy so that the target cycle time could be accommodated. Cooling channels, (see Figure 2) were added to the die body to maintain even temperatures and limit distortion below 0.3mm.

For the purpose of computer aided engineering, (CAE) design calculations, distortion limits were set at 0.26 mm. All cooling channels were designed as individual circuits so that they could be adjusted and monitored during mass production conditions. Also, the problems of cooling deficiencies which often occur due to accumulations of water born elements in a high temperature environment, could be monitored and controlled.

Up-time (Die)

Since only one die was to be available, up-time was of major concern. Aspects of die design had to be robust and were based on the analysis below.

	Cavity distortion $\Rightarrow$	Insert type design $\Rightarrow$	Distortion < 0.5mm
Downtime $\Rightarrow$	Bolster strength $\Rightarrow$	Added strength pillars reduced stress $\Rightarrow$	Max. stress < 120Mpa
	Pin breaks $\Rightarrow$	Cooled, break detect $\Rightarrow$	Easy to replace

From the analysis, the following aspects of tooling design were closely adhered to:

- Construction: Both cavities were designed identical to each other as inserts installed into a single bolster to meet future maintenance issues.
- Rigidity: Designed distortion limited to below 0.5mm; split insert design was applied to achieve the required rigidity level.
- Strength: 500% of requirement. This stipulation was evaluated by CAE simulations that led to the addition of a bolster centre support pillar as illustrated in Figure 3.
- Core breaks: die and cores are monitored constantly for coolant circulation, any coolant failure is alarmed to indicate breakage and cycle stoppage. Most cores and their hydraulic cylinders were designed for quick replacement. Some critical pins could be replaced without removing the die from the machine.

Quality (metal flow)

The design incorporated identical placement and orientation of each cavity. This created the problem of unequal runner routes to the cavities. Equal metal flows into the two die cavities had to be identical to avoid misfills. The following countermeasures were determined to prevent problems from this cause:

Misfills $\Rightarrow$	Flow route $\Rightarrow$	Identical and balanced flow $\Rightarrow$	Multistage runner design

The multi-stage runner design allowed balanced gating so that aluminum could reach all gates simultaneously, (see figures 4 & 5).

Quality (gating & venting)

Gating is one of the most important aspects to consider when designing for the quality of a casting. Gate design has a critical impact on cavity filling and uniform solidification control as well as porosity due to gas entrapment.

Porosity misfills $\Rightarrow$	Fill time $\Rightarrow$	< 0.13s $\Rightarrow$	Optimum gate speed 42 - 45m/s increased venting

The analysis above led to the application of the company's patented gate design to counter quality problems. A simulation of the result is shown in Figure 4. An increased gas vent area was included to aid the gate velocity and gas removal from both cavities. It was decided to use surface vents for design simplicity.

Quality (dimensional)

With an average wall thickness of just 3.0mm, it is very easy to create die distortions unless influences such as metal flow (gating), cooling, filling and solidification are considered methodically as below.

Distortion <0.3mm $\Rightarrow$	Die cooling $\Rightarrow$	Consistent and uniform $\Rightarrow$	Fully monitored individual cooling channels
	Cure time. $\Rightarrow$	Biscuit cooling $\Rightarrow$	Shot ring cooling
	Mechanical $\Rightarrow$	Ejector design $\Rightarrow$	Part & runner ejector mechanism

As a result, all cooling circuits were carefully designed to guarantee consistent and uniform performance. A solidification time of 14 seconds was achieved due to the stamp cooling circuit.

Furthermore, ejector positions were selected to minimize retrieval distortion. Among the most important locations selected for ejectors, were the runner areas which effectively reduced excess pressure on the gates during die opening and casting ejection.

Results
In November 2002, eight months after concept launch, Honda R & D and HUM were able to conduct initial trials. The double cavity die was put into mass production during April 2003. Further adjustments with cooling and metal flow were conducted during the next three months with the aim of improving the original performance targets.

The double cavity casting die continued to produce until model run-out in September 2005, achieving nearly 100,000 cycles (200,000 castings). It was then de-commissioned and remained in complete operating condition for a further 80,000 cycles.

The results against the planned performance after six months of mass production are shown in Table 3. In addition to performance improvements, further significant cost savings were achieved through:
- Increased productivity;
- Overall effect on unit cost (depreciation);
- Improved manpower efficiency;
- Efficient usage of energy;
- Direct saving in transport and importation costs;
- Direct saving on supplier's profit.

Conclusions
Taking this idea from concept to implementation proved to everyone involved in the project, that which sometimes seems impossible can be challenged. This spirit keeps HUM competitive through innovation.

References
1.   Metal Castings Ltd.

Acknowledgements

The author is grateful to the many people at HUM who contributed to the success of this project. In particular K.Ariyoshi, the R & D group, H.Kadota and Ian Hughes and the autor's team of production engineers – especially Mark Malham.

The author also thanks his family for their support during the many long days and weekends spent at the plant whilst working on the project.

Tables

Table 1. Problem analysis by '5 why method'.

Area	Problem	Effect	Why 1	Why 2	Why 3
Tool/die	Pin breaks	Down time	Stress	Overheat	Cooling failed
		Scrap parts	Galling	Overheat	
		Failed planned events			
Machine	Part extractor failed	Down time	Stamp clamp non detect	Clamp LS set for one size	Multi model machine. Not flexible
		Damaged castings scrapped			
Process	Low P3	Leak rate High			
	Damaged cast hole	Leak rate high	Galling	Overheat	Cooling failed
		Scrap parts	Hole oversize	Galling	

Table 2. Production capacity simulations.

Test conditions	Work time (s)	Cycle time (s)	Up-time	Pass rate	Capacity (per shift)
1. Conventional plan	77700	108	85%	95%	581
2. Failed conventional plan	77700	108	79%	95%	540
3. @100% capacity	77700	108	100%	95%	683
4. Improved cycle time (CT)	77700	100	100%	95%	738
5. Improved CT + Up-time	77700	100	90%	95%	664

Table 3. Performance comparisons.

Task	Target	Actual
Uptime	97%	99%*
Cycle time	90s	86s $\Rightarrow$ 83s
Total pass ratio	95%	97.5%
Leak rate	3%	2.2%
Good castings/hour	73	83
Good castings/day	1591	1807
* Main down time: ejector seizure in runner area $\Rightarrow$ increased clearance		

Figures

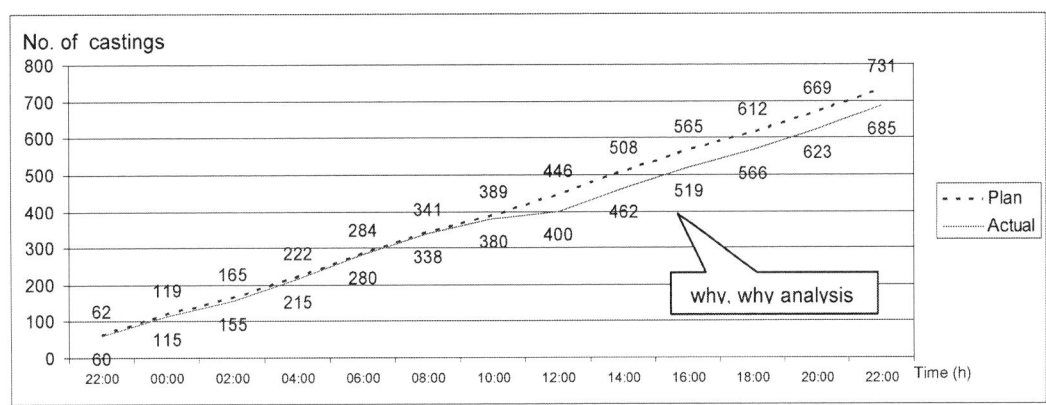

Figure 1. Graph showing actual output versus planned output.

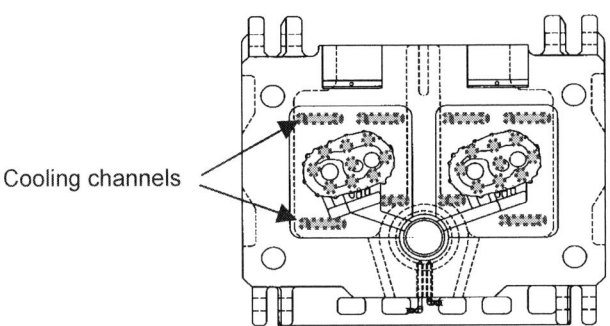

Figure 2. Arrangement of additional cooling channels.

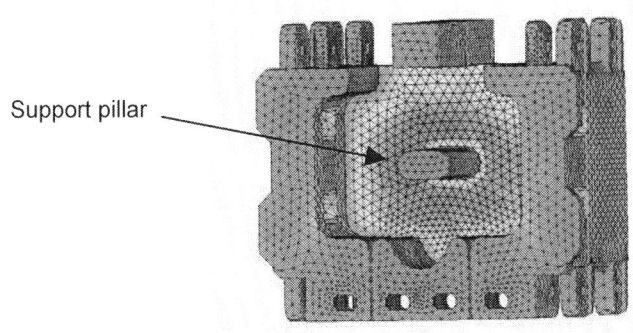

Figure 3. Location of bolster centre support pillar.

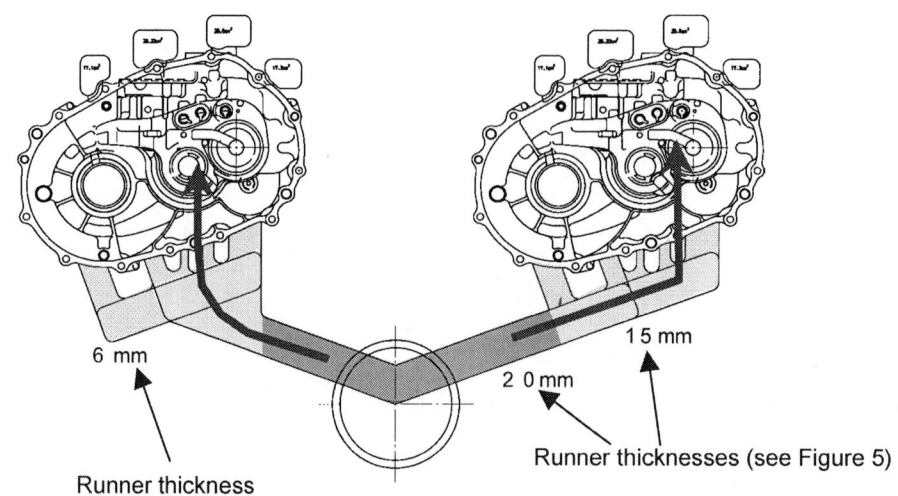

6 mm

Runner thickness

15 mm

2 0 mm

Runner thicknesses (see Figure 5)

Figure 4. Multi-stage runner for balanced flow & gating.

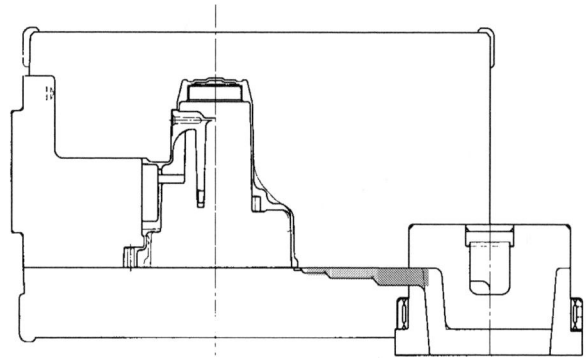

Figure 5. Cross section of multi-stage runner method.

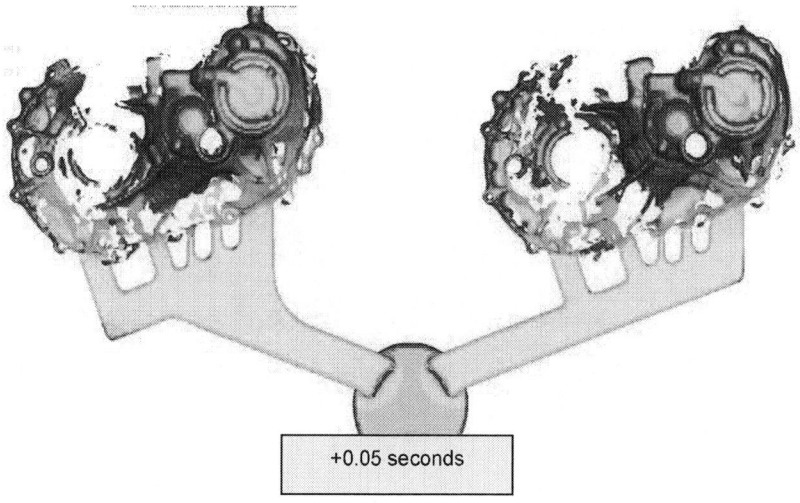

+0.05 seconds

Figure 6. Gating and filling simulation using patented gate design.

Development of Tilt Casting Technology For High Performance Sport Wheels

K K Tong*, M S Yong*, Y W Tham*, R Chang** and K W Wee**

* Singapore Institute of Manufacturing Technology, Agency for Science, Technology and Research, Singapore, ** Stamford Tyres Corporation Ltd, Singapore.

Abstract

Stamford Tyres, a Singapore-based regional distributor for a complete range of tyres, wheels and supporting services has collaborated with SIMTech's metal forming technology group to design and instal its new casting plant in Thailand. The three-year partnership has enabled Stamford Tyres to build new capabilities for manufacturing its proprietary brand of lightweight aluminium alloy wheels. SIMTech has supported Stamford Tyres with: a joint survey and evaluation of manufacturing facilities; technical training; plant installation and commissioning; computer aided design; melt treatment; experimentation; and defect and process analysis through advanced CAE and diagnostic and processing facilities. The successful collaboration has provided Stamford Tyres with the confidence to accept the demanding challenge to design, develop and manufacture high performance sport wheels from its Singapore base.

The paper describes how the lightweight, high performance sport wheels are produced using the latest tilt casting technology and identifies the practices necessary to ensure that the highest product quality is achieved.

Key words Aluminium wheels, tilt casting, light alloy wheels, quality control

1. Introduction

Aluminium alloy wheels produced using A356 (Al-7.5-0.45Mg) are one of the most difficult automotive castings to produce because of the stringent cast surface and mechanical property requirements. Tilt casting technology utilizes the latest state of the art programmable logic controls, hydraulic systems and precision metal parts to ensure rigidity, stiffness and smooth transition of tilting with a significant reduction in vibration. The technology optimizes wheel quality with respect to time to market of new designs, productivity, and efficiency and cost benefits.

The successful manufacture of high quality aluminium alloy wheels requires an understanding and application of casting technology knowledge. In order to deliver the required mechanical properties and fatigue performance, the alloy is typically modified and heat-treated. Alloy modification is used to alter the acicular morphology of the Al-Si eutectic into a finer, more interconnected fibrous phase that improves toughness and ductility [1]. Although both sodium and strontium modification have been used historically, the trend in industry is toward the use of strontium, as it has a tendency to fade less, is less toxic, and less corrosive to refractories [2].

This paper describes the various processing steps that are taken to manufacture high quality aluminium alloy wheels that are light and have a good surface appearance. The tilt casting technology produces wheels with diameters of 13" (330mm) to 26" (660mm). The paper also reports a case study of a competitor's wheel that had failed in service due to poor microstructure control and casting defects.

2. *Manufacturing Steps for High Quality Wheels*

The process flow diagram for wheel manufacture is shown in Figure 1. The quality of a wheel is affected by each step of the production process, namely the melt preparation, casting process, heat treatment, machining and painting (surface finishing). Every production step in wheel manufacture is closely monitored and controlled to deliver the best possible quality that can be achieved in the final product. The following briefly describe the processes required to deliver the optimum combination of strength and ductility while maintaining an excellent surface finish.

The alloy used in the plant is a strontium pre-modified primary A356 ingot with the chemical composition shown in Table 1. Careful preparation of the aluminium melt is the basis of high quality wheel production. Tight control is exercised over melting procedures to ensure the use of clean scrap, good quality raw aluminium ingots and chemical treatments or fluxing that ensure the cleanliness of every melt of aluminium poured into the wheel moulds. The main factors that affect melt quality are: a) charge material, b) melt holding, c) melt transfer operations, d) gas content and e) modification and grain refinement [3]. The reduction of hydrogen gas pick-up was achieved with a continous N_2 degassing by rotary impeller.

The typical conventional gravity casting process produces turbulence in the metal stream during pouring. This tilt casting technology provides quiescent mould filling to minimize inclusion generation and entrapment. Tilt pouring is analogous to pouring beer into a tilted mug, where the formation of bubbles and turbulence is greatly reduced. The benefit can also be demonstrated by computer simulation, as shown in Figure 3.

The tilt casting line includes a complete 6-axis industrial robotic arm able to ladle the melt into the mould with consistent melt volume control and little temperature loss (Figure 4), precision hydraulic tilt control system and mould opening (Figure 5), automated pick-up of castings and a linear control quenching mechanism. This automation ensures product consistency and quality.

The heat treatment process is used to obtain the optimum combination of strength and ductility in the wheels. Either a T4 or T6 heat treatment procedure can be applied to achieve the required mechanical property specifications. Upon completion of the heat treatment, the wheels were machined to its correct dimensions. All wheels are finally painted or coated to specified paint colours including customized colours with high quality finish and visual aesthetics.

The wheels that leave the plant are tested and certified to very strict international specifications such as VIA, JIS and JWL. Testings such as salt spray for corrosion resistance, tensile testing for mechanical strength, radial endurance testing, cornering bending endurance testing, leak testing and impact testing were conducted to ensure wheels are of a highest standards.

3. *Experiments to characterize the effects of cooling rate on structure and mechanical properties*

It is essential to understand the relationship between cooling rate, microstructure and mechanical properties if an effective process and quality control system is to be implemented. This requirement was highlighted by the failure study conducted on a competitor's product as part of a bench-marking exercise. An experimental programme was initiated and comprised of the following activities:

- **Microstructural Examination**
  Metallographic techniques were employed to determine the general microstructure of the wheel casting. Microstructure examinations were taken from the sample areas indicated in Figure 6.

- **Dendrite Arm Spacings (DAS) measurement**
  DAS was measured on the micrographs using the random linear intercept technique [4] and image analyzer software.

- **Failure Investigation of broken wheel in service**
  Metallurgical analysis comprising of visual examination, optical microscopy and scanning electron microscopy was conducted.

- **Impact testing (Drop test)**
  The wheels were oriented at 13^0 and 30^0 to the horizontal. A static load of up to 1 tonne was dropped and the wheels must meet the required international standards against permanent strain in a repeated test.

4.   Results and Discussion

Microstructure and DAS Measurements
The microstructure at points 1 and 6 (as indicated in Figure 6) are shown in Figures 7 and 8 respectively. In general, the microstructures of all the points indicated in Figure 6 revealed no visible porosity. DAS is controlled by solidification rate and the solidification time is given by equation 1 [5]:

$$t_s = (DAS / 7.5)^{2.56}$$ Equation (1)

Where: DAS = Dendrite Arm Spacing: and t_s = solidification time.

The average DAS and solidification time for each point on the sample was calculated and plotted as shown in Figure 9. In a wheel casting, the solidification is complicated by the spoke/rim intersection, which has a larger mass (volume) than the adjacent rim and spoke areas. The DAS of a casting is directly affected by solidification rate and this in turn will affect mechanical properties. With reference to Figure 6, liquid metal at zone A, the end of the spoke, (points 1, 2 and 3) freezes first. Zone C (points 5-10), which is located at the intersection of the rim and spoke and is connected to the riser, is a much thicker section in comparison to either zone A or B. Thus the solidification time for zone C is much longer than either zone A and B and consequently the DAS is much greater.

The shorter solidification time/higher solidification rate in zones A and B will result in a finer structure (measured by the DAS) and better mechanical properties. Figure 10 illustrates how a finer DAS (faster solidification rate) will improve tensile strength and elongation. The mechanical properties of the casting at zone C (Figure 6) are inferior when compared with those at zones A & B. A faster solidification rate at zone C can be achieved by extracting heat using a cooling mechanism to provide a smaller DAS, thereby resulting in better mechanical properties.

CASE STUDIES

A.   Failure of wheels in service
A cracked A356 aluminum alloy sport wheel from a competitor was subjected to a metallurgical failure investigation to determine the possible cause of the premature failure (shown in Figure 11). The cracked wheel was reported to have failed after about 1.5 years of use under normal road service conditions.

Figure 12 shows the presence of beach marks at the fracture surface, which are typical characteristics of fatigue failure. SEM was employed to study the fracture surface and evidence of fatigue striations confirmed the cause of failure (illustrated in Figure 13). Microstructural examination of the cracked spoke showed significant porosity in the structure as shown in Figure 14. Porosity can arise from either shrinkage or the presence of gas or, indeed, a combination of the two causes. Hydrogen gas is easily soluble in liquid aluminum particularly at high temperatures above 750^0C. Therefore, upon solidification any dissolved hydrogen will come out of solution forming pores in the casting. The presence of porosity reduces mechanical properties and will especially reduce fatigue performance if the pore is located at the surface. Hydrogen levels less than 0.20cc/100g are typically required to minimize the detrimental effects of the porosity [3]. The microstructure also exhibited acicular eutectic silicon suggesting the absence or ineffectiveness of modification and the primary dendrite arm spacing was observed to be large.

These microstructural features lower the ability of the wheel to carry a higher load during its normal service condition. The results indicated that the fractured sport wheel cracked as a result of fatigue failure under cyclic loading. The crack originated from the inner side of the spoke at the sharp corner in that region. The casting defects and poor quality microstructures inherent in the wheel are not acceptable and were the likely causes of the premature failure.

B.   Failure during Impact / Drop Test

Wheels were subjected to stringent reliability tests and impact testing is one such test. The shrinkage porosity is the weakest point where the fracture or cracking occurs during the drop load test. One of the wheels that failed the impact test was sent for failure investigation. Figure 15 shows the fracture surface of a wheel at the spoke area after impact testing. Microstructural examination of the fracture surface shown in Figure 15 consisted of shrinkage porosity (as shown in Figure 16) and a coarse DAS (Figure 17). As was observed in the previous section, these inherent casting defects combined with a coarse structure will lead to poor mechanical properties.

The shrinkage porosity is due to poor directional solidification and insufficient provision for feed metal. As the liquid aluminium alloy begins to solidify, the volumetric shrinkage occurring due to the change of physical state is not compensated by sufficient liquid metal. This problem can be overcome by the provision of a feeder head or riser [6]. The feeder must be designed to solidify after the casting proper so that whilst the latter is solidifying liquid metal runs from it to compensate for the shrinkage. When the casting and riser are cool, and have been removed from the mould, the feeder that will contain all the shrinkage can be cut off from the casting.

5. Recommendations

The following recommendations should be followed to ensure the production of high quality wheel castings.

- The melting practice should minimise the presence of oxide particles or films because melt cleanliness is very important for high quality wheel castings.
- Effective degassing must be practiced since hydrogen gas is readily soluble in liquid aluminum and any dissolved hydrogen will come out of solution to form porosity in the castings.
- Modification, preferably with strontium, should be practiced to avoid acicular eutectic silicon, reduce the presence of stress concentrators and thus improve ductility.
- The molten alloy must be handled and transferred in a non-turbulent manner to prevent the formation and entrapment of inclusions in the castings.
- Directional solidification, heat transfer management and effective feeding must be implemented to reduce shrinkage porosity and promote the preferred DAS.

6. Summary and Conclusion

The following summary and conclusions based on the reported work can be drawn.

- The fractured wheel failed as a result of fatique failure under cyclic loading. The failure was initiated from porosities in the casting and at sharp corners, which acted as stress concentrations points for cracking to start and propagate along a weak microstructure.
- The microstructure examined in the broken wheels consisted of larger DAS, resulting from poor solidification patterns and insufficient grain refinement, which lowered the ability of the wheels to sustain higher loads in service.
- The shrinkage porosity seen in the castings were caused by incorrect solidification patterns and poor thermal gradient along the rims which has restricted feeding to the spoke areas, which are generally thicker in sections.

References

1. Thompson S, Cockcroft S L and Wells M A, *Advanced light metals casting development,* Materials Science and Technology, Vol 20, Feb 2004, p194-200.
2. Colmaco Aluminium Ltd, *Modification of foundry Al-Si alloys,* Technical Report No.4, Dec.1997.
3. Colmaco Aluminium Ltd, *Factors affecting the quality of Al wheels,* Australian Die Casting Association Report, 2004, p27-33.
4. D.G. Eskin, J. Zuidema Jr., V.I. Savran, L. Katgerman., Structure formation and macrosegregation under different process

conditions during DC casting, Mat. Sci. Eng., A384: 232-244, (2004).

5. ASM Handbook, Volume 9, Metallography and Microstructures, Ninth Edition, American Society of Metals, pp. 629, (1998).

6. Webster P D, *Fundamentals of foundry technology*, Portcullis Press, Redhill, pp376-381, (1980).

Acknowledgements

The author wishes to acknowledge the great support given by Stamford Tyres, particularly Mr Wee Kok Wah (President) and Mr Roger Chang (Senior Vice President) for this collaborative partnership.

Table 1: Chemical composition of A356 ingots

Si	Fe	Mg	Sr	Zn	Ti	Mn	Cu	Na	Al
6.5-7.5	0.12 max	0.30-0.45	0.015-0.030	0.05 max	0.20 max	0.05 max	0.10 max	0.05 max	Rem

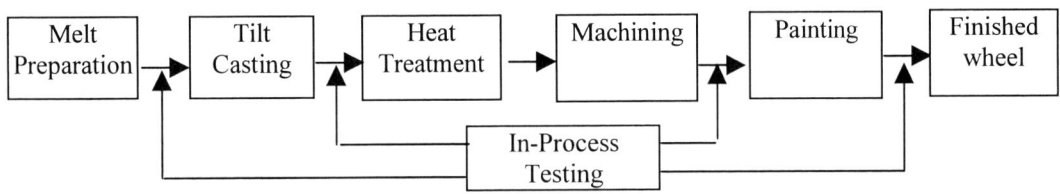

Figure 1: Brief schematic of the process flow of wheel manufacturing.

Figure 2: Tilt casting machine will ensure smooth liquid metal filling into mould.

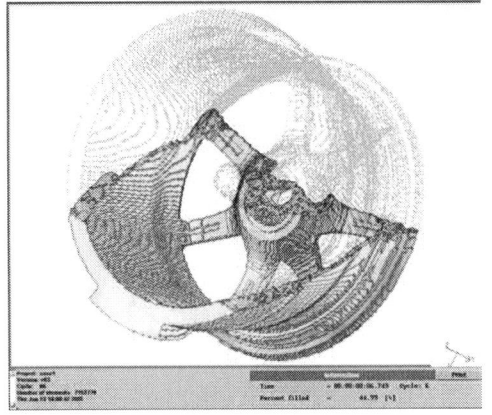

Figure 3: CAE simulation of tilt casting showing smooth flow of melt filling.

Figure 4: Melting furnace with robotic arm ladling for efficient melt pouring.

Figure 5: Mould opening with wheel casting ready to be extracted by pick-up mechanism.

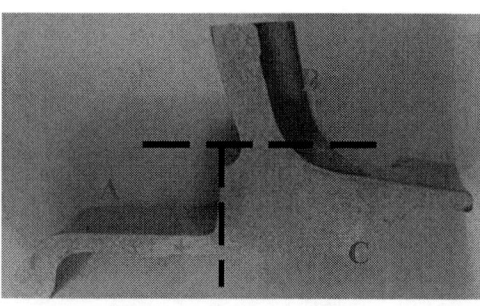

Figure 6: A wheel section having different rates of solidification zones A, B and C resulting in different DAS microstructure.

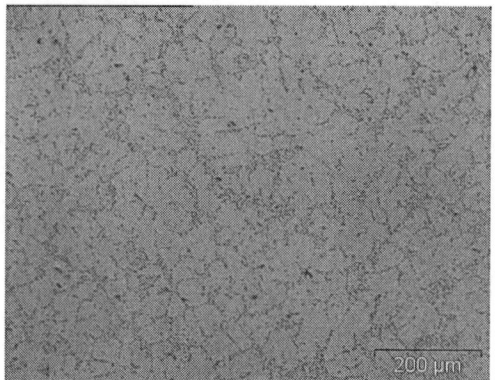

Figure 7: Microstructure at zone A (Point 1) showing a finer DAS (100X).

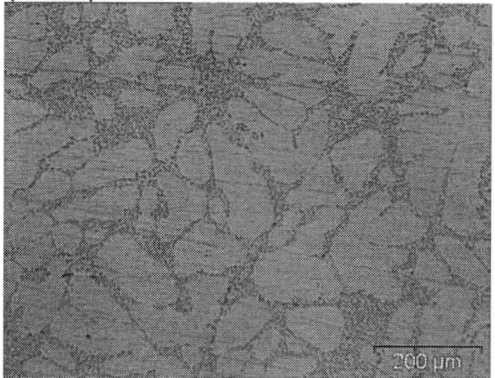

Figure 8: Microstructure at zone C (Point 6) with a coarser DAS. (100X)

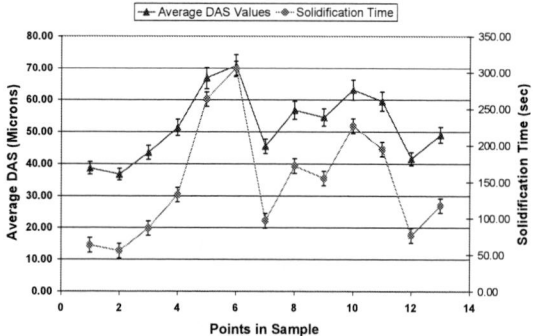

Figure 9: Graph illustrating DAS and solidification time vs points on wheel section in Figure 6.

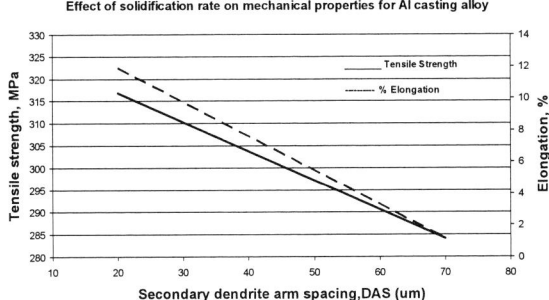

Figure 10: Graph illustrating the effect of DAS on mechanical properties for Al casting alloy [3].

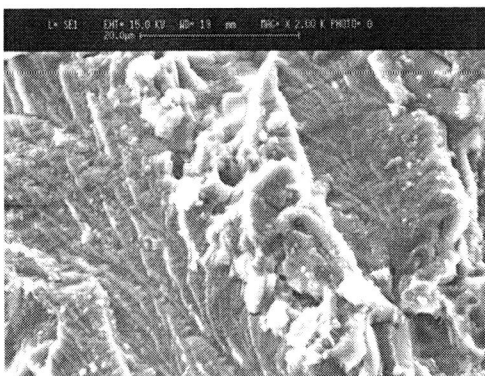

Figure 13: SEM image showing fatigue striations on fracture surface as shown in Figure 12.

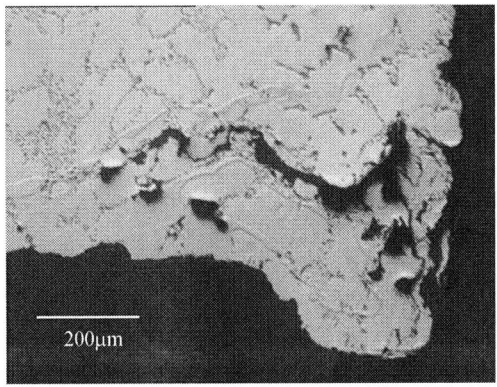

Figure 14: Microstructure taken at the crack region with the presence of porosities (MX100).

Figure 11: General and close-up views of crack on the spoke of a sports rim.

Figure 12: Evidence of 'beach marks' characterizing typical failure by fatique can be seen at the crack surface of the spoke shown in Figure 11.

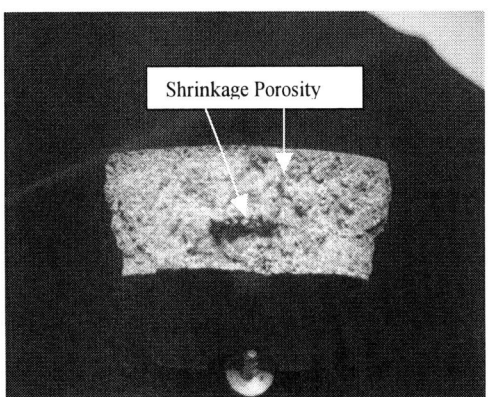

Figure 15: Fracture surface taken at spoke areas after impact tests showing shrinkage porosity that weaken the wheel strength.

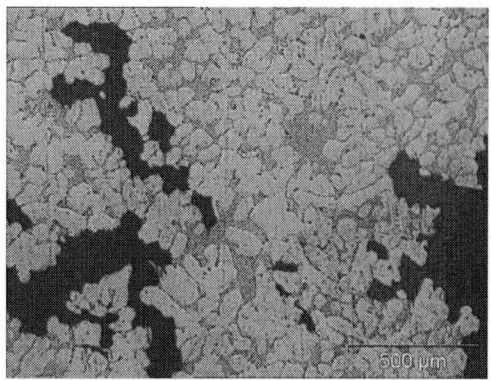

Figure 16: Cross-sectional microstructure of fracture surface taken after drop test showing shrinkage porosity.

Figure 17: Coarser DAS and script-liked eutectic silicon observed which gave rise to lower impact properties during the drop test.

Process development for highly stressed aluminium castings under consideration of the increase in performance of the diesel engines

F. Mnich*, H.C. Saewert*, A. Tamez**, R. Bähr***, E. Krebs****

*NEMAK-Rautenbach Europe; **NEMAK S. A. de C. V. Mexico, *** University of Magdeburg, Casting Technologies, Germany, ****Rautenbach-Guss Wernigerode GmbH, Germany

Abstract

This article will focus on the development of engine technology and its effect on innovation of a foundry. With the example of a diesel engine, which has experienced highest growth rates throughout the past years, the development of the process of the Nemak-Dynamic-Casting-Systems will be introduced. Its components, the cylinder head and the engine block, are subject to high stresses. Due to economic and scientific advantages that have been achieved so far, this principle will quickly be introduced into series production.

Key words

cylinder head, diesel engine, stress, mechanical properties, new casting process, NDCS

Introduction

Cylinder heads for passenger cars are nowadays exclusively made of aluminium alloys. Automotive producers are impelled to offer customers products with growing comfort, growing functionality and safety; at the same time, high reliability has to be provided. Cast parts therefore constantly become more and more complicated, while still having to be up to the standards in terms of mechanical properties. The automotive supply industry therefore has to face a strong cost pressure, which it copes with by realizing innovation. The predominant part of cast engine components made of aluminium is nowadays produced in metallic permanent moulds. For this, a broad variety of manufacturing methods can be applied. Dynamic processes have so far only rarely been used, though. The reasons for this are the high facility and maintenance costs. In order to meet the ever growing expectations when it comes to cast cylinder heads, it is necessary to break new grounds in manufacturing processes
(/1/ to /3/).

Since these are very complex, complicated and variform tasks, an intensive cooperation and alliance of practical works of foundry enterprises with universitary research have been proved to be of great advantage.

The Nemak Company is the world's biggest independent cylinder head founder and operates three development centers in Monterrey/Mexico, Windsor/Canada and Wernigerode/Germany (**figure 1**). Based on the long lasting experience with high-performance cylinder heads, a new dynamic casting method could be developed in Wernigerode, in close collaboration with the other two development centers. This method is especially apt for highly stressed diesel cylinder head developments.

Nemak Europe is continuing its close collaboration with the Otto-von-Guericke-University in Magdeburg. This collaboration has already been established with the "Rautenbach-Guss Wernigerode GmbH" (Rautenbach Casting Wernigerode Ltd) more than ten years ago. Apart from being situated nearby, one of the main reasons for continuing the successful work was the connection of scientific work with practical research and realization in production.

Current market condition in the automotive industry, role of the suppliers, especially development competence and production know-how

The demands on components for the automotive industry are significantly growing due to two facts: There is first a considerable pressure on increased efficiency and lightweight construction; and secondly, on cost

reduction. In **figure 2**, the development of diesel engines is illustrated in this context. This situation is most likely to become even more intense.

High performance cylinder heads have already been produced in lost moulds (sand castings) (**figure 3**) in 1934, yet this casting method is of no importance for modern high-power cylinder heads anymore - even though it is still used for Formula 1 cylinder heads (**figure 4 and 5**).

Growing key performance indicators and decreasing consumption values have led to a fast distribution of diesel engines in Europe (market share: over 40%); the market entry for these engines for Light Truck Vehicles in the USA is generally supposed to happen soon.

Further performance increase (partly in connection with downsizing concepts) and ever stricter exhaust-emission laws ask for a different diesel cylinder head design.

The cylinder head is one of the most complex and most stressed components of an engine. During the camshaft drive system of valves, not only the areas of the combustion chamber, but also the column for the camshaft drive system, are mechanically stressed. These stresses develop in higher temperatures where inhomogeneities within the material structure are of a great importance.

High Cycle-, Low Cycle- and Thermomechanical fatigues (HCF, LCF, TMF) demand a consistent high quality structure with high ductility and minimal inhomogeneities. Cavities and inclusions – carried-along oxide films – develop germs for crack initiation.

Furthermore, germs that cause blowholes can develop in the melt due to impurities. Thus, the more unclean the melt, the more porous are the castings. In the glory hole, the melt is metallurgically provided in the desired quality. Every further transport of the melt can only degrade its mechanical properties.

In order to realize acceptable component temperatures with the presently available aluminium alloys, cosiderably constructive efforts are necessary. Thus, there is a two-part water jacket which, over the channels, is drawn near the outer wall.

In classic gravity die casting processes, these design solutions can only be realized with a limited operational stability or reduced structural demands.

The Nemak-Dynamic-Casting-Process (NDCS) is based on further developments of the classic tilt casting method (see /4/ to /8/; **figures 6 and 7**). A coordinated ingate system, together with a special spoon and downsprue geometry, make it possible to track the spoon at the beginning of the turn-tilt movement of the die-casting in a way that the filling speed of the mould with the melt is optimal and adjustable.

That way, a mould filling without any turbulences is repeatably provided. **Figure 8** shows that, by selective control of the filling speed, the SDAS can optimally be adjusted.

Due to the advantages of the NDCS method and the emerging trends in terms of design in the field of diesel development, a quick introduction into mass production is planned.

References

1. Bähr R; Mnich F; et all: Optimization of the mechanical properties of highly stressed castings through direct control of the casting process, Proceedings of the 65th World Foundry Congress, Gyeongju, Korea 2002.

2. Nyahumwa, C., Green, N.R., Campbell, J.: Effect of Mold-Filling Turbulence on Fatigue Properties of Cast Aluminium Alloys. Transactions of the American Foundrymen's Society (1998)

3. Rezvani, M.; Yang, X.; Campbell, J.: The Effect of Ingate Design on the Strength & reliability of Al Castings. American Foundrymen's Society, 103. AFS-Casting Congress, St. Louis 1999

4. Gosch, R., Stika, P.. Innovatives Gießverfahren für hochwertige Gussstücke. Gießerei-Rundschau 48 (2001) Nr. 5/6

5. Gruneberg, N.; Escherle, E.; Sturm, C.: Simulation as an intergral tool for process design and optimization for Aluminium cylinder head manufacturing. American Foundrymen's Society, 103. AFS-Casting Congress, St. Louis 1999

6. Hasse, S.: Giesserei Lexikon, 18. Auflage. Schiele & Schön Verlag Berlin 2001

7. Ingerslev, P., Andersen, S.T.: Flow Analysis of Mould Filling High Speed Motion Pictures of the Metal, Watermodelling and Numerical Calculations. 54th International Foundry Congress, New Delhi, India 1987

8. Kim, S.B., Yeom, K.D., Hong, C.P.: Simulation of mould filling sequences in gravity tilt-pour casting. Int. J. Cast Metals Res. Nr. 10 (1997)

figures

Overview:

> 13 Foundries with 200.000 Tonnes sold casting
> workforce: 6.715

Mexico	Canada	Germany	Czech Rep.	Slovakia
6 Plants	2 Plants	3 Plants	1 Plant	1 Plant

Figure 1) the Nemak-Company

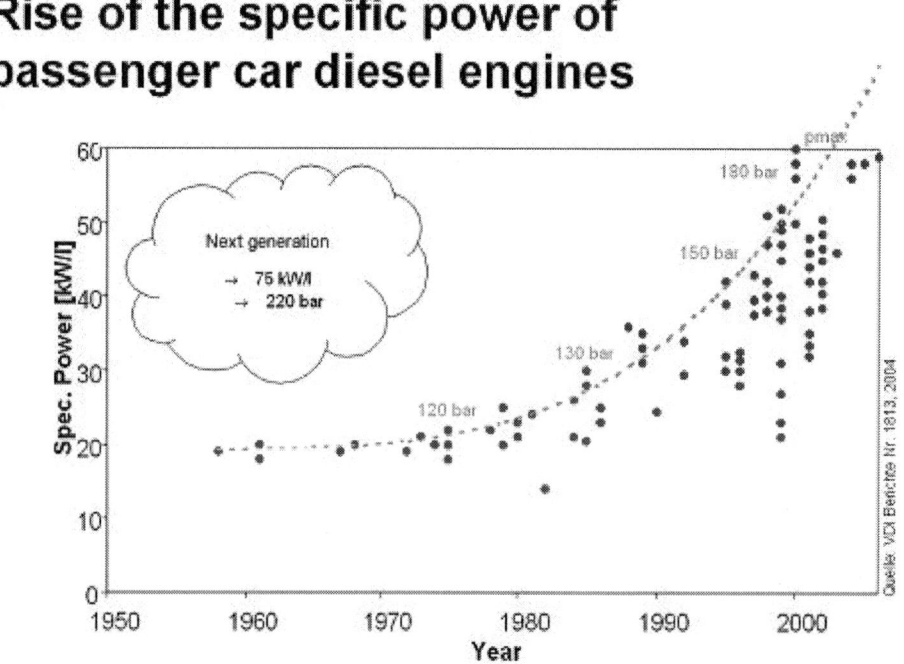

Figure 2) strong increase of diesel engine efficiency

**Cylinder head by Rautenbach:
1934 for Auto Union Type P**

Figure 3) increase of the specific power output from 67.7 to 87 HP/Liter
from 1934 to 1938, to 520 HP

Share of materials and casting processes for passenger car cylinder heads

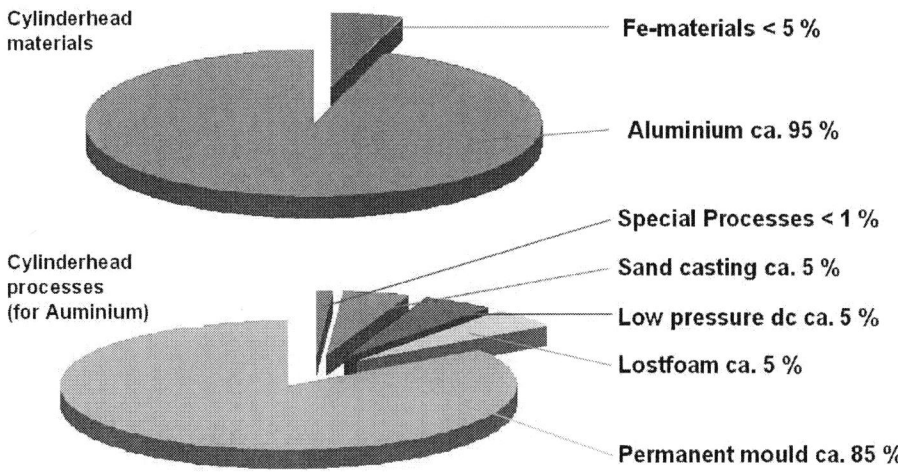

Figure 4) distribution of material and casting processes for the production
of passenger car cylinder heads

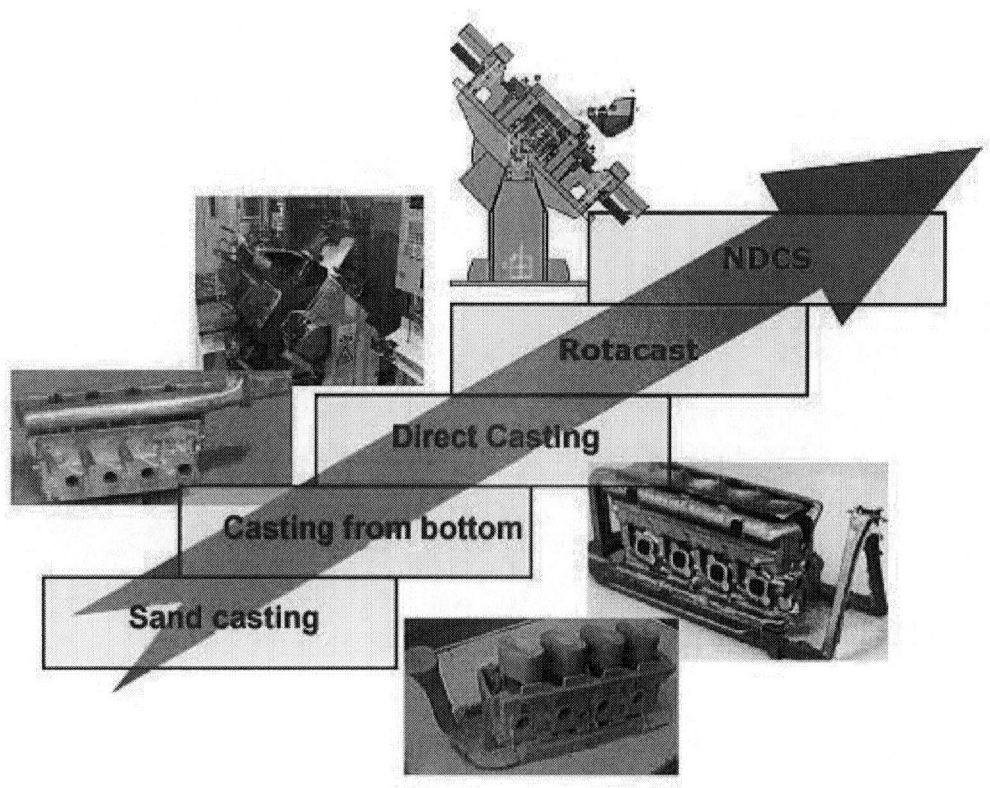

Figure 5) development of casting methods for cylinder heads

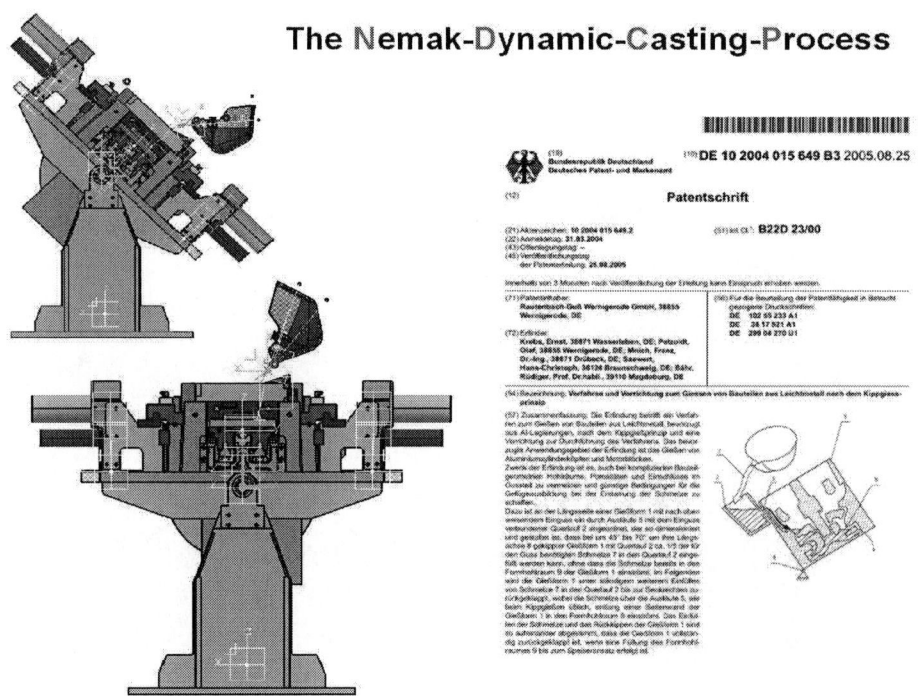

Figure 6) schematic diagram of the NDCP installation engineering during the beginning and end

Comparison of the gravity die casting processes				
Casting process	Bottom casting	Rotacast	Direct casting	NDCS
Mechanical properties	↘	→	→	↗
Filling / Oxide	→	↗	↘	↑
Riser	→	↑	↗	↑
Patent	no	yes	no	yes

Figure 7) tendential comparison of different casting methods for cylinder heads

Casting-Properties

SDAS [µm]

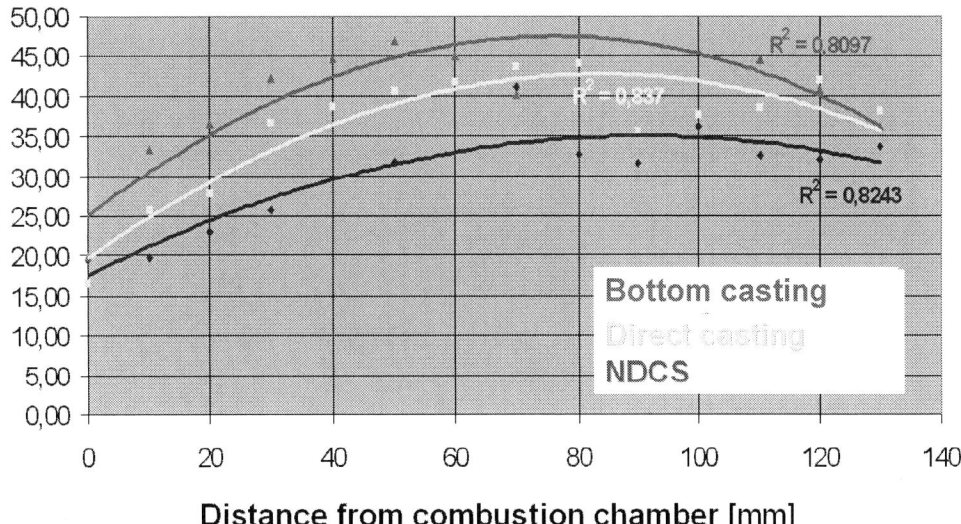

figure 8) obtained mechanical characteristics in dependence of casting methods

AUTHOR INDEX

Ahmed, M.	1185	Changyun, L.	41
Ainsworth, M.J.	341	Chismera, M.	1041
Anglada, E.	126	Cho, G.S.	361, 1157
Araoka, Y.	576	Cho, I.	31
Arrieta, M.	126	Choe, K.H.	361, 1157
Asano, K.	585	Choi, B.J.	412
Ashley, Dr. C.	716	Choi, J.	31
Ashton, Dr. M.C.	760	Choi, J.K.	61, 389, 955
Aufderheide, R.C.	430	Choi, J.P.	939, 946
Averdieck, W.	219	Choi, S.H.	904
Baehr, R.	21	Choi, Y.J.	1157
Bähr, R.	1205	Choi, Y.S.	389
Bako, K.	736	Cilecek, J.	662
Bakrim, M.	405	Costa, J.C.	212
Bale, Eur Ing C.J.C.	716	Creo, R.S.	184
Barnes, C.	824	Cuesta, R.	90
Beeson, A.	126	Czyryca, E.J.	249
Belforte, G.	768	DasGupta, R.	824
Berry, J.T.	100	Datta, G.L.	854
Beziat, A.	1167	Davies, P.J.	331
Biardeau, M.	165	Dayong, L.	914
Binner, J.G.P.	568	De Baerdemaeker, H.	778
Blackburn, S.	351	Debray, C.	165
Blair, M.	155	Delgado, A.	90
Boehm, R.	503	Delorme, H.	396
Bonfield, E.	715	Dequan, S.	914
Branovitsky, A.M.	61	Dicken, P.	808
Brody, H.D.	70, 80	Diószegi, A.	993
Brown, A.	440	Dispinar, D.	874
Buraś, J.	894	Dodd, P.	824
Campbell, J.	874	Donsbach, F.	613, 643
Canon, P.	165	Dussud, M.	165
Carello, M.	768	Eggleston, D.	623
Castro-Román, M.	884	Egner-Walter, A.	1
Chang, H.	568	Elbel, T.	1023
Chang, R.	1195	El-Din, H. Nasr	798

AUTHOR INDEX

Eng, C.	174	Hwang, H.Y.	61, 389
Enright, P.	965	Hwang, J.H.	136
Fan, Z.	295, 1109, 1128	Ibrahim, M.M.	798
Fang, J.	70	Ikeda, S.	380
Fang, X.	1109	Ikengaga, A.	361
Farnsworth, M.	1091	Ing, Dr.	1
Froescher, A.	921	Jackson, A.	1081
Galaz, J.	126	Jakstis, D.	1165
Gnanamurthy, K.	705	Jin, Q.	324
Gradowski, A.	108	Jingjie, G.	41
Greer, A.L.	894	Jo, H.H.	315
Griffiths, W.D.	331, 341, 864	Jones, S.	351
Güemes, J.A. Goñi	184	Jorge, A.	126
Haigh, Dr. P.	778	Junjiao, W.	118
Haigh, P.M.	11	Kadoi, K.	371
Hamilton, R.W.	528	Kallbom, R.	1071
Han, Y.	305	Kapranos, P.	1091
Harding, R.A.	239	Karunakar, D. Benny	854
Harshorne, J.	219	Keena, P.	633
Hashimoto, K.	493	Kelin, Z.	118
Hayrynen, K.L.	194	Kennedy, M.	1166
Héau, C.	396	Keough, J.R.	194
Hedge, S.	558	Khan, R.H.	834
Helber, J.H.	229	Kiguchi, S.	538
Hendley, R.J.	716	Kiguchi, Shoji	653
Hengzhi, F.	41	Kim, B.H.	816, 975
Herrera-Trjo, M.	884	Kim, I.H.	136
Higginson, R.L.	568	Kim, J.T.	955
Ho, N.U.	955	Kim, K.H.	136
Hoff, H.	643	Kim, K.Y.	1157
Holmen, K.	1176	Kim, M.H.	904, 1157
Holmgren, D.	993	Kim, S.K.	315
Hong, J.H.	389	Kim, S.M.	939
Horacek, M.	662	Kim, Y.J.	278, 412, 1148
Horvath, L.R.	460	Kocian, L.	1023
Hu, B.	380	Koss, D.A.	249

AUTHOR INDEX

Krajewski, W.K.	894
Kraly, A.	1119
Krebs, E.	1205
Kurtsiefer, R.	204
Kwak, S.Y.	955
Lauwers, Dr. B.	778
Law, T.D.	517
Lee, D.H.	136
Lee, D.S.	975
Lee, J.K.	315
Lee, K.R.	361
Lee, K.W.	361, 1157
Lee, P.D.	528
Lee, S.M.	816, 975
Lee, S.W.	955
Lekakh, S.N.	1003
Lelito, J.	108
Lengyel, K.	736
Leuven, K.U.	778
Lihua, W.	914
Lim, C.	31
Lim, S.G.	324
Linxe, D.	1167
Liu, G.	295
Liu, Q.M.	681
Löchte, K.	503
Loper Jr., C.R.	1003
Luck, R.	100
Macke-Bart, C.	1167
Macnaughtan, M.P.	174, 623
Maguregi, J.I.	184
Mahendra, K.V.	548
Maroto, J.A.	90
Marukovich, E.I.	61
Matsumoto, Y.	380
Matsuo, Y.	493

Meléndez, A.	126
Metzgar, K.	396
Midson, S.P.	1081
Millan, N.	716
Mnich, F.	1205
Monroe, R.W.	155
Montes-Rodríguez, J.J.	884
Moon, B.M.	816, 975
Morral, J.E.	70, 80
Mozo, D.	90
Mueller, M.	467
Müller, J.	473
Murakami, M.	380
Murata, Hirotoshi	653
Nakae, H.	371, 576
Nakamura, K.	538
Nam, J.	31
Nam, T.W.	939, 946
Narasimhan, V.	510
Neto, E.	396
Ngirabacu, F.	405
Niehoff, T.	633
Noda, Y.	51, 493
Nofal, A.A.	798
Nomura, H.	1099
Nyamannavar, S.	1138
Oh, J.S.	136
Olive, S.	1
Oxley, S.	11
Park, S.	288
Patel, J.	1109
Paterson, T.	527
Pickering, J.	965
Pillai, R.M.	689
Prabhu, K. Narayan	1138
Prabhu, K.N.	558

AUTHOR INDEX

Prat, J. 126
Qi, F.P. 681
Radcliffe, P. 824
Radhakrishna, Dr. K. 548
Radhakrishna, K. 593, 603
Raiszadeh, R. 864
Raji, A. 834
Ramachandra, M. 593, 603
Ramadan, M. 1099
Ransing, M.R. 672
Ransing, R.S. 672
Ravi, M. 1138
Regheere, G. 1167
Reuther, T. 1061
Rhys, Garel 767
Richardson, N. 623
Rimmer, A. 194
Robertson, D.G.C. 1003
Robinson, A.C. 249
Ryou, M. 904
Saewert, H.C. 1205
Sambrook, R. 568
Schmitz, W. 643
Schrey, A. 450
Seifeddine, S. 844
Senberger, J. 1023
Seoane, A. 126
Sharma, A. 983
Shim, S.Y. 324
Shin, J.S. 816, 975
Shiping, W. 41
Showman, R.E. 430
Sillen, R. 145
Song, C.B. 816
Song, Y.L. 681
Sorenson, W.W. 699

Sriram, P. 1033
Staley, J. 929
Stancliffe, M. 483
Steinhäuser, T. 422
Stötzel, R. 473
Stroppe, H. 21
Subramanya, P.K. 558
Suchy, J.S. 108
Sugiyama, Y. 576
Sumimoto, H. 538
Sundarrajan, S. 269
Sung, S.Y. 278, 412
Suzuki, Y. 493
Svensson, I.L. 844, 993
Takita, M. 1099
Tamez, A. 1205
Tayama, M. 371
Terashima, K. 493
Tham, Y.W. 1195
Thompson, P.J. 259
Thomson, Rachel 741
Tiedje, N. 1051
Todte, M. 21
Tomita, Y. 538
Tong, K.K. 1195
Trauzeddel, D. 613
Vandenhaute, C. 405
Vervier, D. 405
Vervier, J. 405
Vicario, T. 126
Viktorov, V. 768
vom Stein, D. 788
Wallis, R. 778
Wang, Y. 295
Wee, K.W. 1195
Wenzhen, L. 118

AUTHOR INDEX

Werrell, S. 219

Wiese, E. 1013

Williams, T.M. 716

Wöhrer, J. 1119

Xiong, S. 380

Yamaguchi, Yasufumi 653

Yeomans, N.P. 430

Yi, F. 80

Yoneda, H. 585

Yong, M.S. 1195

Yoo, S. 31

Yoon, E.P. 939, 946

Yoon, Y.C. 136

Yoon, Y.O. 315

Youn, J. 1148

Yousseff, Y.M. 528

Yuan, C. 351

Yutong, Z. 914

Zadera, A. 1023

Zakharov, I.L. 61

Zhai, Q.J. 681

Zhang, S.M. 1128

Zhang, Y. 681

Zhao, Y. 528

Zhen, Z. 1128

Zych, J. 894